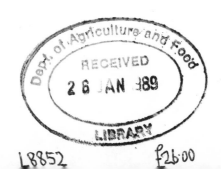

The Agricultural Notebook

Originally compiled by Primrose McConnell in 1883 this standard work has been completely re-written and enlarged by the contributors listed on p. vii, under the editorship of R. J. Halley and R. J. Soffe.

Primrose McConnell's

The Agricultural Notebook

18th edition

Edited by
R. J. Halley
R. J. Soffe
Seale-Hayne College

Butterworths

London Boston
Singapore Sydney Toronto Wellington

First published, 1883
17th edition, 1982
Reprinted, 1983
Reprinted, 1984
Revised reprint, 1986
18th edition, 1988

© **Butterworths & Co. (Publishers) Ltd, 1988, excluding chapter 27.**
Chapter 27 Crown copyright 1988, published by permission of the Controller of Her Majesty's Stationery Office.

British Library Cataloguing in Publication Data
McConnell, Primrose
Primrose McConnell's the agricultural notebook.—18th ed.
1. Great Britain. Agriculture – Manuals
I. Title II. Halley, R.J. (Richard John), *1928*– III. Soffe, R.J.
630′.941

ISBN 0-408-03060-7

Library of Congress Cataloging-in-Publication Data

McConnell, Primrose.
    [Agricultural notebook]
    Primrose McConnell's the agricultural notebook.—18th ed./
edited by R.J. Halley and R.J. Soffe.
    p.   cm.
    Includes bibliographies and index.
    ISBN 0-408-03060-7:
    1. Agriculture—Handbooks, manuals, etc. 2. Agriculture—Great Britain—Handbooks, manuals, etc. I. Halley, R. J. (Richard John) II. Soffe, R.J. III. Title. IV. Title: Agricultural notebook.
S513.2.M33 1988
630′.941—dc19   88-9503
                                                                             CIP

Typeset by the Alden Press, Oxford, London and Northampton
Printed and bound by Hartnolls Ltd, Bodmin, Cornwall

Preface to 18th edition

The 17th edition introduced a completely new approach and format and this has been popular with readers, particularly students. This new edition builds on that experience by continuing the format and introducing new material to cover developments in the industry over the last six years. It is more difficult in these uncertain times to anticipate in detail needs over the next decade. The Editors believe the current main UK agricultural enterprises will remain the sheet anchor of the industry, supplemented by new developments of the type covered in the additional chapter on alternative enterprises.

A number of new authors have been recruited and the Editors are grateful to all contributors for their efforts and enthusiasm.

Preface to 17th edition

The first edition of *The Agricultural Notebook* appeared in 1883. It was compiled by Primrose McConnell, a tenant farmer of Ongar Park Hall, Ongar, Essex, who found, when he was an agricultural student under Professor Wilson at Edinburgh University, the great want of a book containing all the data associated with the business of farming. This is the seventeenth edition and all but coincides with the centenary of the publication. The Agricultural Industry has changed so much since the last edition it was decided to abandon the 'pocket book' size and use the new, easy-to-read, format.

Clearly no one person is capable of writing on all the aspects of agriculture necessary for achieving the objectives of *The Agricultural Notebook*. The Editor is most grateful to the team that have contributed to the new edition. Contributors have not been constrained to a common editorial mould but rather, following discussion with the Editor, have been encouraged to let their own informed aspirations guide the content and layout of their particular sections. Contributors have collaborated and, where appropriate, adequate cross-referencing is given. Compared with the previous edition, some new topic areas have been added and some eliminated to reflect development within the industry and to contain the total volume within an acceptable limit.

The book is aimed at meeting the needs of students of agriculture, be they full-time students in colleges and universities, or practising farmers and advisers. It is hoped the reader will perceive that the essentials of modern science, technology, management and business studies have been carefully blended to give the information which experience suggests is necessary for successful farming today.

Contributors

Peter Beale, BSc (Hons), PhD, CBiol, MIBiol, MBIM.

Present appointment: Principal Lecturer in Landscape Planning and Applied Ecology and Head of Environmental Studies Section, Seale-Hayne College, Newton Abbot, Devon.

Held appointments as a river authority fisheries officer in Cornwall, then Devon between 1964 and 1969. This was followed by research at Exeter University, leading to a PhD. Appointed assistant county conservation officer and ecologist to the planning department of the Devon County Council. A lectureship at Seale-Hayne College followed. Now head of the environmental studies section and director of the HND course in Rural Resources and their Management. Keenly interested in landscape and wildlife conservation in relation to the effects of land uses. Past Chairman of the Environment Division of the Institute of Biology and Chairman of the Devon branch of the Council for the Protection of Rural England.

P.H. Bomford, BSc (Hons), MSc, CEng, MIAgrE.

Present appointment: Senior Lecturer, Engineering Department, Seale-Hayne College, Newton Abbot, Devon.

Educated at Reading University and the University of Newcastle upon Tyne. Previous appointments have included assistant lecturer in Farm Mechanisation, Wye College (University of London), lecturer in farm mechanisation, Essex Institute of Agriculture and head of Engineering Department, Ridgetown College of Agricultural Technology (Canada). Directed and carried out research into field drainage, environmental control, vegetable crops mechanisation, farm machine performance, reversible tractors, maize harvesting, mulch laying. Current professional interests are cereal and forage harvesting and storage, tractors and power units and alternative sources of energy.

Carl Boswell, TD, CEng, MIStructE, FIAgrE, MWeldI

Present appointment: HM Chief Agricultural Inspector, Health and Safety Executive

Worked as a Director of an engineering company and then as Chief Structural Engineer in the CWS Ltd before starting a career in health and safety in 1986. Since then he has mainly specialised in the construction and agricultural industries. A member of the Governing Body of the AFRC Institute of Engineering Research, he is Chairman of the Agricultural Industry Advisory Committee and has served as Chairman of EEC, ISO and BSI Committees.

R. A. Bourne, BSc, PhD.

Present appointment: Senior Lecturer in Animal Production Science, Seale-Hayne College, Newton Abbot, Devon.

Graduated in zoology from Sheffield University and then completed a PhD at the School of Agriculture, Nottingham University. Spent the next three years at Michigan State University engaged in research in lactation and reproductive physiology in dairy cattle. After a short period lecturing in Brazil, returned to the UK to teach biology in secondary schools. Appointed to post of lecturer in animal production at Seale-Hayne College in 1980. Current research interests include animal behaviour and welfare.

Paul Brassley, BSc (Hons), BLitt.

Present appointment: Senior Lecturer in Agricultural Economics, Seale-Hayne College, Newton Abbot, Devon.

After reading agriculture and agricultural economics at the University of Newcastle upon Tyne went on to research at Oxford. Following this he farmed in Devon until appointed to his present post at Seale-Hayne College. Current research interests are in agricultural policy and agricultural history.

J. S. Brockman, BSc, PhD.

Present appointment: Principal Lecturer in Agriculture, Seale-Hayne College, Newton Abbot, Devon.

After obtaining a PhD for a residual study of the classic Treefield experiment at Cockle Park, he helped to found Fisons North Wyke Grassland Research Centre in Devon. Has published many papers on the nutrition and utilisation of grassland. Since 1978 has been at Seale-Hayne. President of the British Grassland Society, 1982–83. Deputy Editor of *Grass and Forage Science*; Editorial Board of *Plant Varieties and Seeds*; Chairman of British Grassland Society Publications Committee.

J. L. Carpenter, BSc, MSc, CEng, NDAgrE, FIAgrE.

Present appointment: Senior Lecturer in Agricultural Structures, Seale-Hayne College, Newton Abbot, Devon.

Educated at Leeds and Newcastle upon Tyne Universities. Taught horticultural mechanisation and farm buildings at Writtle Agricultural College. Now teaches courses on farm structures and environment. Past holder of the Top Fruit Cultivation Scholarship from Worshipful Company of Fruiterers. Research interests include dairy energy use, water utilisation on farms and the economics of farm utility vehicles.

Alan Cooper, CDA (Hons), NDA, MSc, PhD.

Present appointment: Senior Lecturer in Animal Production, Seale-Hayne College, Newton Abbot, Devon.

Previous appointments have included lectureships at Shropshire Farm Institute and the University of Malawi. Research interests cover growth, meat production from goats and animal behaviour.

J. C. Eddison, BSc, PhD, CBiol, MIBiol.

Present appointment: Senior Lecturer in the Computer Unit, Seale-Hayne College, Newton Abbot, Devon.

Having graduated in zoology from Leeds University, conducted PhD research at Aberdeen University into the effects of environmental variations on ecological communities. Post-doctoral research followed at the Edinburgh School of Agriculture investigating the grazing behaviour and ecology of hill sheep which involved extensive computer analysis of time-lapse photography. Took up current position in 1983. Research interests lie in the areas of quantitative behaviour and ecology.

Michael P. Fuller, BSc, PhD.

Present appointment: Senior Lecturer in Crop Physiology, Seale-Hayne College, Newton Abbot, Devon.

Graduated from Leicester University and then completed a PhD at the Welsh Plant Breeding Station on frost resistance in grasses. Employed as a research demonstrator at Leeds University and as a Research Trials Officer at the Sports Turf Research Institute before taking up his present position in 1980. He currently has research programmes on the physiology of cereals, sunflowers and cauliflowers and manages a tissue culture suite.

R. John Halley, BSc, MSc, MA.

Present appointment: Vice-Principal, Seale-Hayne College, Newton Abbot, Devon.

After gaining first degree was awarded a Ministry of Agriculture postgraduate scholarship to the University of Reading and studied at the National Institute for Research in Dairying. First appointment was at Seale-Hayne College where teaching duties were combined with research on grassland utilisation. Appointed as university demonstrator, School of Agriculture, Cambridge and also acted as assistant to farm director. Managed a large farming estate in Hertfordshire before returning to Seale-Hayne as senior lecturer in agriculture and farm manager. Later appointed vice-principal at Seale-Hayne. Mr Halley co-edited the sixteenth edition of *The Agricultural Notebook* with the late Dr Ian Moore, edited the seventeenth edition and is joint editor of this new edition.

Richard P. Heath, BA, NDA, NDAgrE, TEng, MIAgrE, Cert Ed.

Present appointment: Senior Lecturer, Seale-Hayne College, Newton Abbot, Devon.

Trained at Shuttleworth and Writtle Colleges. Joined the Engineering Department staff in 1970, teaching TEC diploma, higher diploma and degree levels. Interests are mechanisation aspects of crop production and dairying. Research, investigation and materials/structures study extends his teaching interests. A member of the Institution of Agricultural Engineers he has held branch and national officer positions.

A. D. Hughes, BSc (Hons), PhD.

Present appointment: ADAS R&D Programme Manager and Senior Professional Officer for Soil Science, ADAS/WOAD Trawsgoed, Aberystwyth, Dyfed.

Trained in soil science at The University of Newcastle upon Tyne and subsequently researched aspects of phosphate nutrition at Leeds University. From 1970 to 1984 he has worked as an ADAS Soil Scientist at the Starcross Sub-Centre, at the Reading Regional Centre and also in the ADAS Midlands and Western Region. He has recently held the post of ADAS Regional Agricultural Scientist for Wales and his present post includes overall responsibility for professional aspects of Soil Science throughout ADAS.

David Iley, BSc (Hons).

Present appointment: Senior Lecturer in Zoology, Seale-Hayne College, Newton Abbot, Devon.

Trained at King's College, Newcastle upon Tyne and Imperial College Field Station, Sunninghill. Professional interests include development of integrated systems for control of crop pests, apiculture and landscape and wildlife conservation.

Anita J. Jellings, BSc (Hons), PhD.

Present appointment: Senior Lecturer, Department of Agriculture, Seale-Hayne College, Newton Abbot, Devon.

After reading Applied Biology at the University of Bath, completed a PhD at the Plant Breeding Institute with a cytogenetic study of cereal ovules. This was followed by a research fellowship at the University of York investigating constraints on photosynthetic capacity. Current interests include cellular development of *in vitro* cultures, intracellular variation associated with frost tolerance in cauliflowers and sunflower agronomy.

J. A. Kirk, BSc, PhD.

Present appointment: Senior Lecturer in Animal Production, Seale-Hayne College, Newton Abbot, Devon.

Spent nine years farming before reading agriculture with agricultural economics at the University College of North Wales, Bangor. Postgraduate research on the growth and development of Welsh Mountain lambs led to a PhD. Current research interests are the growth and development of lambs and bull beef.

G. Moule, BSc (Hons), MSc.

Present appointment: Senior Lecturer in Botany and Crop Pathology, Seale-Hayne College, Newton Abbot, Devon.

Has held present appointment since 1971. Main interests include crop pathology and weed biology, with particular interest in the epidemiology of crop fungal and viral pathogens.

R. M. Orr, BSc, PhD, MIBiol.

Present appointment: Senior Lecturer, Seale-Hayne College, Newton Abbot, Devon.

Read agriculture with animal science at the University of Aberdeen. Postgraduate research at Edinburgh University on appetite regulation led to a PhD. Teaching and research interests include food science and nutrition.

Robert Parkinson, BSc (Hons), MSc, PhD.

Present appointment: Senior Lecturer in Soil Science, Seale-Hayne College, Newton Abbot, Devon.

Studied at Leeds and Reading Universities before taking up post as Research Assistant at Birkbeck College, University of London working on soil water dynamics and agricultural drainage. Obtained PhD in 1984. Took up present appointment on leaving Birkbeck College. Current research interests include efficiency of drainage systems, methods of slurry and sludge disposal and environmental impact of agricultural practices.

John I. Portsmouth, NCP, NDP, DipPoult, NDR, FPH.

Present appointment: Executive Director, Peter Hand (GB) Ltd, Stanmore, Middlesex.

Had been employed extensively in the poultry feed industry before present appointment. Is author of several books and many articles on poultry and rabbit feeding and management. Presented original research on calcium metabolism in pre-lay hens at World Poultry Science Association meetings. Spent last fifteen years studying the micronutrient requirements of farm livestock, particularly poultry.

David Sainsbury, MA, PhD, BSc, MRCVS.

Present appointment: Lecturer in Animal Health, University of Cambridge, Madingley Road, Cambridge.

After a period of research and lecturing at the Department of Veterinary Hygiene at the Royal Veterinary College, London, has worked at the Department of Clinical Veterinary Medicine at the University of Cambridge in the Division of Animal Health. Main interests have been concerned with the health and well-being of farm livestock especially under intensively managed conditions. Author of five textbooks and some 120 scientific papers.

R. W. Slee, MA, PhD, Dip Land Economy

Present appointment: Senior Lecturer in Land Use, Seale-Hayne College, Newton Abbot, Devon.

Teaching and research interests in rural policy and farm diversification. Publications include chapters in *The Changing Countryside*, (eds. J. Blunden and N. Curry, Croom Helm, 1985); *A Future for the Country-side*, (eds J. Blunden and N. Curry, Blackwells, 1988). Author of *Alternative Farm Enterprises* (1987) Farming Press.

John Sobey, MIPM, MBIM, MRIPHH.

Present appointment: Academic Registrar and Senior Lecturer in Management, Seale-Hayne College, Newton Abbot, Devon.

Had been involved in staff training schemes in large companies before joining Seale-Hayne College. Main professional studies are now in interpersonal skills in agriculture and related industries.

Richard J. Soffe, HND, MPhil, MBIM, M Inst M.

Present appointment: Senior Lecturer in Farm Management, Seale-Hayne College, Newton Abbot, Devon.

Previous appointments have included a lectureship at Hampshire College of Agriculture, Winchester and assistant farm manager of a large estate in Hampshire. Research interests include financial management and the marketing of financial products to farmers. Mr Soffe is co-editor of this new edition of *The Agricultural Notebook*.

G. H. T. Spring, MA.

Present appointment: Senior Lecturer in Law, Department of Business Studies, Plymouth Polytechnic, Drake Circus, Plymouth.

After reading law at Cambridge, spent several years in East Africa in government and business employment. Has taught for the last twelve years in the Business Studies Department of Plymouth Polytechnic and also at Seale-Hayne College. Is an active member of the Agricultural Law Association.

R. D. Toosey, BSc, MSc, NDA.

Present appointment: Principal Lecturer in Crop Production, Seale-Hayne College, Newton Abbot, Devon.

Graduated from Reading University. Did post-graduate research on sprouting of potatoes and subsequent work on brassica fodder crops. Has spent nine years in practical agriculture and managed Seale-Hayne College Farm for six years. Formerly lectured at Rodbaston and Writtle Colleges of Agriculture before taking up present appointment. Is author of several publications.

M. A. Varley, BSc, PhD, C Biol.

Present appointment: Lecturer, Department of Animal Physiology and Nutrition, University of Leeds.

Before his present position, spent three years as a lecturer in animal science at Seale-Hayne College and then moved to the Rowett Research Institute where he was a Senior Scientific Officer for six years. Research interests include reproductive physiology of the sow, neonatal immunology, piglets.

A. T. Vranch, BSc, PhD.

Present appointment: Head of the Computer Unit, Seale-Hayne College, Newton Abbot, Devon.

After graduating in chemical engineering from the University of Birmingham, moved to the University of Exeter to complete PhD research work in two-phase fluid mechanics related to coolant flows in nuclear reactors. Initially appointed as lecturer in physical science/engineering, later set up the Computer Unit and is currently responsible for computer developments at Seale-Hayne College. Main interests include development of teaching software, techniques in computer communications/networking and sensor interfaces for a range of applications in agriculture and food technology.

Martyn Warren, BSc (Hons), MSc.

Present appointment: Senior Lecturer in Farm Management, Seale-Hayne College, Newton Abbot, Devon.

Started teaching farm management at Harper Adams Agricultural College in 1974, before taking up his present post at Seale-Hayne College where he is Course Director of the Diploma in Farm Management. A member of the Editorial Board of *Farm Management*, a visiting assessor for C&G Farm Business Management courses in the South West, and an experienced tutor of courses for farmers and other business owners. His book *Financial Management for Farmers*, first published in 1982, is now a standard student textbook.

Contents

Part 1 Crop production 1

1 Soils 3
What is a soil? 3
The mineral components of soil 4
Clay minerals 5
Soil organic matter and the carbon cycle 10
Chemically active surfaces and the chemistry of soil colloids 13
Soil organisms 17
The major nutrient elements 20
Calcium and liming 25
Trace elements and micronutrients 27
Soil sampling, soil analysis and soil nutrient indices 28
Physical properties of the soil 32
Soil survey and land classification 40

2 Field drainage 45
The need for drainage 45
Drainage benefits 46
Soil water 47
Drainage systems 48
Problem diagnosis and system layout 55
Drainage system maintenance 56
Economics of drainage 56

3 Crop physiology 58
Fundamental physiological processes 58
Growth 60
Crop development cycle 60
Environmental influences on growth and development 63
Competition 67
Modification of crops by growth regulators and breeding 69

4 Crop nutrition 70
Summary of soil-supplied elements 70

Organic fertilisers and manures 74
Inorganic fertilisers 78
Forms of fertiliser available 81
Compound fertilisers 82
Fertiliser placement 83
Fertilisers and the environment 83
Optimum fertiliser rates 84
Fertiliser recommendations 85
Liming 87

5 Arable crops 90
Crop rotation 90
Choice of crop and sequence 90
Crop duration and importance 91
Methods of crop establishment 91

Brassica crops 99
General 99
Brassicas grown primarily as vegetables 100
Brassicas grown primarily for fodder 110
Brassicas grown for seed processors 115

Graminae – cereals, maize, cereal and herbage seed production 118
Cereals grown primarily for grain 118
Wheat 125
Barley 128
Oats 129
Rye 130
Triticale 131
Maize 131
Sweet corn 133
Seed production 135
Cereals 136
Herbage seed production 136

Miscellaneous non-leguminous combinable cereal break crops 140
Borage 140
Evening primrose 141
Linseed 141
Sunflowers 142

Leguminous crops   142
Leguminous seedcrops for human and/or
  animal consumption   142
Legumes grown for fodder only   151
Root and bulb crops   153
Beetroot (red beet)   153
Carrots   155
Leeks   158
Dry bulb onions   161
Parsnips   164
Potatoes   165
Sugar beet, mangels and fodder beet   172

6   **Grassland   177**

Distribution and purpose of grassland
  in the UK   178
Grassland improvement   179
Characteristics of agricultural grasses   183
Herbage legumes   184
Basis for seeds mixtures   185
Fertilisers for grassland   188
Patterns of grassland production   191
Expression of grassland output on the
  farm   193
Output from grazing animals   194
Grazing systems   196
Conservation of grass   198
Grassland farming and the environment   205
Glossary of grassland terms   206

7   **Trees on the farm   209**

Choice of site and species   209
Measurement of timber   209
Length of rotation   224
Timber quality   225
Systems of woodland management   225
Methods of establishment for broadleaves
  228
Preparation of site   229
Planting   230
Fertiliser application   232
Plantation aftercare   232
Thinning   233
Felling and marketing   236
Shelterbelts and windbreaks   237
Timber preservation   239
Protecting woodlands   240
Insect pests and fungi   240
Agencies for assisting woodland owners   241
Grants for woodland owners and farmers
  241

8   **Weed control   243**

Occurrence of weed problems   243
General control measures – applicable to all
  systems   243
Principles of chemical (herbicidal) weed control
  244

Choice of herbicide   245
Arable crops and seedling leys   247
Weed control in cereals   248
Root crops, row crops and other broadleaved
  crops   252
Weeds from shed crop seed or groundkeepers
  252
Weed control in established grassland   258
Weed control in forest and woodlands   258
Treatment of individual large trees   260
Weed control (unselective) in non-crop
  situations   263

9   **Diseases of crops   265**

Air-borne diseases   265
Fungicide resistance   267
Soil-borne diseases   268
Seed-borne and inflorescence diseases   268
Vector-borne diseases   268
Diseases of wheat   269
Diseases of barley   272
Diseases of oat   274
Diseases of rye/triticale   275
Diseases of forage maize and sweet corn   276
Diseases of potato   276
Diseases of brassicas   280
Diseases of sugar beet, fodder beet, mangold
  282
Diseases of peas and beans (field, broad/dwarf,
  navy)   283
Diseases of carrots   284
Diseases of onions and leeks   285
Diseases of grasses and herbage legumes   286

10   **Pests of crops   288**

Cereal pests   288
Potato pests   292
Sugar beet pests   295
Oilseed rape pests   298
Kale pests   301
Pea and bean pests   301
Field vegetable pests   303
Schemes, regulations and Acts relating to the use
  of insecticides   303

11   **Grain preservation and storage   304**

Conditions for the safe storage of living grain
  304
The time factor in grain storage   305
Grain moisture content and its measurement
  305
Drying grain   306
Fans for grain stores   309
Grain cleaners   310
Management of grain in store   310
Handling grain   311
Storage of grain in sealed containers   312
Use of chemical preservatives to store grain
  313

Part 2   Animal production   315

12   Animal physiology and nutrition   317
Regulation of body function   317
Chemical composition of the animal and
  its food   319
Digestion   325
Metabolism   329
Voluntary food intake   340
Reproduction   342
Lactation   354
Growth   359
Environmental physiology   363

13   Livestock feeds and feeding   365
Nutrient evaluation of feeds   365
Raw materials for diet and ration formulation
  371
Nutrient requirements   381

14   Cattle   388
Definitions of common cattle terminology
  390
Calf rearing   391
Management of breeding stock replacements
  394
Feeding dairy and beef cattle   396
Beef production   399
Beef production systems   401
Dairying   404
Cattle breeding   410

15   Sheep and goats   411
Sheep   411
Goats   426
Appendix 1 Body condition scores   428
Appendix 2 Calendar for frequent-lambing
  flock   429

16   Pig production   430
Introduction   430
The structure of the UK pig industry   430
Pig housing and animal welfare   432
Genetics and pig improvement   434
Sow and gilt reproduction   437
Weaner piglets   443
Finishing pigs   445
Conclusions   449

17   Poultry   450
Meat production   450
Commercial egg production   452
Broiler breeding industry   453
Environment of layers and broilers   456
Turkeys and waterfowl   457
Conclusion   460

18   Game conservation   461
Introduction   461
Habitat requirements   462
Breeding, release and holding strategies   464
Predator control   464
Keepering   464
Options for shooting   465
Economic benefits   465
Conclusion   465

19   Animal health   466
Introduction   466
Disease and immunity   470
The health of cattle   474
The health of pigs   485
The health of sheep   489

Part 3   Farm equipment   493

20   Services   495
Drainage installations   495
Energy   496
Fire   498
Heating   498
Lighting   499
Lubricating oil and grease   500
Water supplies   500

21   Farm machinery   503
The agricultural tractor   503
Cultivation machines   507
Fertiliser application machines   510
Planting machines   511
Crop sprayers   514
Harvesting forage crops   517
Grain harvesting   521
Root harvesting   523
Irrigation   524
Feed preparation machinery   530
Milking equipment   532
Manure handling and spreading   533

22   Farm buildings   537
Design criteria   537
Stages in constructing a building   541
Building materials and techniques   545
Environmental control   554
Specific purpose buildings   559
Drawings and site measurement (surveying)
  581
Surveying   581
Mandatory and advisory design requirements
  (excluding finance)   584
Advisory organisations   586

Part 4   Farm management   589

23   The Common Agricultural Policy of the European Economic Community   591

Background, institutions and the legislative process   591
CAP price mechanisms   593
Monetary arrangements in the CAP   597
Structural policy   599
The problems of the CAP   600

24   Farm business management   602

What is business management?   602
Setting primary objectives   602
Planning to achieve primary objectives   603
Information for financial management   604

25   Farm staff management   624

The need   624
Farm requirements   624
Workforce planning   625
Obtaining the right labour   629
Reduction of labour   632
Motivation theory   632
Working conditions   636

26   Agricultural law   637

Introduction   637
The English legal system   637
Legal aspects of the ownership, possession and occupation of agricultural land   638
Employer's liabilities   642
Conclusion   643

27   Health and safety in agriculture   644

Introduction   644
The Health and Safety at Work, etc. Act 1974   645
Some of the statutory requirements of the HSW Act   646
Regulations applying to agricultural activities   647
Food and Environment Protection Act 1985   648
Conclusions   649

28   Computers in agriculture   650

What is a computer and in what applications can it be used?   650
How are computers used in agriculture?   655
What are the implications of installing an on-farm computer system and what are the criteria for its selection?   659
What developments might be seen in the future for computer applications in agriculture?   660

29   Alternative enterprises   661

Defining alternatives   661
The policy climate   662
Types of alternative enterprise   663
Management of alternative enterprises   666
Summary   667

Glossary of units   668

Index   671

Part 1

Crop production

1

Soils

A. D. Hughes

WHAT IS A SOIL?

A soil is an ordered combination of minerals, organic matter, air, water and living organisms.

Simple chemical analysis tells us the gross composition of soil but takes no account of the ordered nature of its constituents. Such analysis therefore obscures more useful knowledge than it reveals.

The relative proportions of mineral material, organic matter and 'pore spaces' (which may contain either air or water) vary within very wide limits. For example, peat soils are almost devoid of mineral material, whilst some sands have only very low levels of organic matter. Subsoils usually have lower organic matter levels and fewer pore spaces than topsoils. A useful figure to remember is that in fertile agricultural soils, pore spaces account for about 50% of the total volume. The subdivision of the other 50% between mineral and organic fractions is complicated by the difficulty of knowing the volume occupied by organic matter in soil. By weight (dry soil basis), organic matter levels are typically in the range 2–10% for mineral soils. *Figure 1.1* shows the proportions of the various soil constituents on a percentage volume basis.

Formation of soils

Soils originate from the rocks of the earth's crust. Rocks are classified as follows:

(1) *igneous* formed by cooling of 'molten magma', e.g. granite, diorite and gabbro;
(2) *sedimentary* formed by consolidation of sediments from some previous weathering cycle, e.g. sandstone, shale, dolomite and limestone;
(3) *metamorphic* formed by action of heat and pressure on sedimentary rock, e.g. slate, marble and quartzite

Note that igneous rocks are formed *de novo* from the earth's molten core whereas material in the other (secondary) types must at some time have existed as igneous rock.

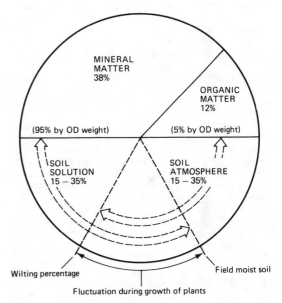

Figure 1.1 Percentage composition of soil on a volume basis (after Jackson, 1964) (from *Chemistry of the Soil*, p. 72, ASC Monograph No. 160, 2nd Edn. Ed. by F. E. Bear. New York: Reinhold, reproduced by courtesy of the author and publishers)

Weathering

Weathering of exposed rock surfaces is due to both physical and chemical agencies.

Physical weathering

Forces of expansion and contraction induced by diurnal temperature variation cause rock shattering, whilst freezing and thawing of any entrapped water greatly accelerates the process. In glaciated areas the major physical agency is crushing and movement by ice and water.

Chemical weathering

Rock minerals are slightly soluble; dissolution of cations and their removal over thousands of years leads to a gradual weakening of the rock's fabric. Chemical weathering is intensified by the acidity resulting from dissolution of atmospheric carbon dioxide, and from organic acids produced during the decay of vegetation. Lichens (which are a symbiotic association of an alga and a fungus) have the ability to fix atmospheric nitrogen as well as being able to fix carbon by photosynthesis. During their growth they accelerate weathering by acidifying the 'soil solution', and rapidly removing the products of weathering as inorganic nutrients. Following their death, carbon is returned to the embryonic soil and provides its first addition of organic matter.

The processes described above can be observed at any time by careful study of the stonework of many ancient buildings and monuments.

Peat soils

A special kind of soil development occurs when very wet conditions prevent the normal breakdown of organic matter by aerobic decomposition. Instead, anaerobic decomposition occurs and the end product is peat. There are two distinct peat formation mechanisms.

Blanket bogs occur in upland areas of the UK where temperatures are low and rainfall exceeds evapotranspiration for at least a part of the year. Heathers and sphagnum moss decay slowly and the organic residues are acid. Any bases (nutrient cations) which may be present are continually leached away by the high rainfall.

Basin peats form in areas where drainage water collects. Native plants are typically sedges and grasses. The incoming drainage water provides ample nutrients so the peat is fertile, having a high base status. Such fen peats are ideal for agricultural cropping when drained, but the act of drainage initiates 'peat wastage', since normal oxidative decay becomes possible. Raised bogs occur when the level of a basin peat surface exceeds the water table height and rainfall is sufficiently great for the blanket bog type vegetation to take over, leading eventually to a blanket bog over the top of fen (sedge) peat.

The soil profile

The vertical section of soil seen in the sides of a pit is the soil profile. Individual layers composing it are called horizons.

Two set of processes are involved in soil formation:

(1) physical and chemical weathering giving rise to 'parent material';
(2) profile development from parent material.

Usually the former precedes the latter but they may proceed simultaneously. Soils where development has proceeded without disturbance exhibit distinctive profiles, their characters being utilised for soil classification and survey.

The upper layer called the A horizon (or 'plough' layer) generally contains appreciable amounts of organic matter which produces a dark colour. This horizon merges into a layer markedly weathered, but comparatively free from organic matter, known as the B horizon. At its base the B horizon, or subsoil, merges into the C horizon or parent material. The upper part of the C horizon is often considerably weathered and merges gradually into unaltered rock.

THE MINERAL COMPONENTS OF SOIL

The mineral matter in soil is derived by weathering of pre-existing rock – the soil parent material. The mineral fraction of the soil can be subdivided in two principal ways:

(1) according to particle size; into sand, silt or clay;
(2) according to mineral composition.

The two are closely related, e.g. fine particles (clay) are composed predominantly of clay minerals, whilst the coarse particles (sand) are normally composed of primary minerals, derived from rock without chemical modification. For practical agricultural purposes in the UK, particle size is the more important attribute.

Sand is largely inert and a very sandy soil is usually well drained, drought prone and lacking in nutrient reserves. Sand grains are usually composed of unaltered primary minerals inherited from the soil parent material; quartz (SiO_2) and feldspars ($MAlSi_3O_8$ – aluminium silicates of basic cations M^+) are the most abundant of these. The so-called accessory minerals also occur in smaller amounts in the sand fraction. They have a high density and are therefore easily separated from the rest of the sand fraction, and can be used as 'fingerprints' to help identify soil parent material. Weathering of some accessory minerals releases nutrients of importance to agricultural crops. Pyroxenes and amphiboles provide a number of minor nutrients (trace elements) whilst phosphate is released by the weathering of apatite.

Clay is composed of very fine particles with an enormous surface area and great chemical activity. Clay soils hold large amounts of water, much of which may not, however, be available to plants. They are frequently poorly drained, but have good reserves of nutrients.

Silt is insufficiently fine to have significant surface chemical activity, but also lacks the agriculturally attractive properties of sand. Hence silty soils have many of the undesirable properties of both sands and clays and the desirable ones of neither.

Real soils are always a mixture of sand, silt and clay, with the relative proportions governing soil properties. An ideal soil would have balanced proportions of sand, silt and clay, giving adequate water holding capacity, free drainage, stable structure and a plentiful supply of nutrients.

Soil texture and particle size distribution

The relative proportions of sand, silt and clay determine the texture of the soil. Texture may be derived from particle size distribution obtained by 'mechanical analysis' in the laboratory, or assessed directly in the field by hand.

Hand texturing

For normal agricultural purposes hand texturing is rapid, cheap and reliable. Simply take about a dessertspoonful of soil. If dry, wet up gradually kneading thoroughly between finger and thumb until aggregates are broken down. Enough moisture is needed to hold the soil together and for the soil to exhibit its maximum cohesion:

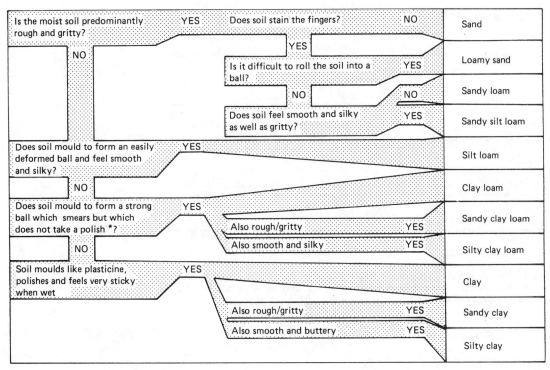

* Polish = Smooth shiny surfaces to soil when rubbed

Figure 1.2 The assessment of soil texture based on the feel of moist soil

By reference to *Figure 1.2* a texture is assigned to the soil on the basis of its feel.

Particle size distribution (psd)

The process of determining the psd of a soil is referred to as 'mechanical analysis'. The principle is that soils are dispersed in an aqueous solution and separated into size fractions by a combination of sieving (down to about 0.05 mm) and timed sedimentation.

Stokes' equation states

$$V = \frac{g(\sigma - \rho)d^2}{18\eta}$$

where V is the settling velocity in cm/s,

g is the acceleration due to gravity (about $980\,\mathrm{cm/s^2}$)

σ is the density of the particle,

ρ is the density of the water

η is the viscosity of water (about 0.010 at 20°C),

d is the diameter of the particle (cm).

If an aqueous suspension of soil is thoroughly agitated, then left to stand, the various components will settle at different velocities in accord with Stokes' equation. By carefully sampling from a given depth in the suspension at successive intervals of time, it is possible to calculate the proportion of particles with a given 'equivalent spherical diameter' (ESD). (To make this possible particle density is assumed, by convention, to be $2.6\,\mathrm{g/cm^3}$, so that the term $(\sigma-\rho)$ is numerically 1.60.)

Various standard procedures differ in detail, for example:

(1) pre-treatment – drying and sieving techniques may differ,

(2) dispersion – use of ultrasonic equipment, treatment with deflocculating chemicals or exchange resins, destruction of organic matter or dissolution of carbonates,

(3) sedimentation and sampling techniques are fairly standard but an alternative procedure (the Bouyoucos method) uses a hydrometer to measure the density of the suspension at a given depth.

Procedures in common use are described by Russel (1974) and the method used for routine purposes by The Soil Survey of England and Wales and by ADAS is described in MAFF (1986) and Avery and Bascomb (1974). There is no universal agreement on the boundaries of the particle size classes. Systems in common use are illustrated in *Figure 1.3*.

Since numerical percentage composition data are not easily appreciated in speech, it is customary to use a triangular diagram to relate psd data to named textural classes. Again, worldwide standardisation is lacking and a number of systems have been used. In England and Wales both soil surveyors and ADAS advisers use the triangular diagram shown in *Figure 1.4*

CLAY MINERALS

The minerals of the clay fraction may be primary, inherited from the parent material or secondary, formed as a result of chemical processes in the soil. The term 'clay mineral' is conventionally reserved for clay formed as a result of chemical changes in the soil.

The study of clay minerals is a very complex subject and much of the information contained in specialist publications

6

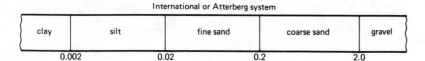

International or Atterberg system

clay	silt	fine sand	coarse sand	gravel

0.002   0.02   0.2   2.0

USDA system

clay	silt	fine sand	medium sand	coarse sand	very coarse sand	gravel

0.002   0.05 0.1   0.5 1.0 2.0

ADAS/Soil Survey of England and Wales, British Standards and Mass. Institute of Technology

clay	silt	fine sand	medium sand	coarse sand	stones

0.002   0.06   0.2   0.6   2.0

Diameter (mm) (log scale)

Figure 1.3 Particle size class systems in common use (after White, R. E. (1979)
Introduction to the Principles and Practice of Soil Science. **Oxford: Blackwell Scientific Publications, reproduced by courtesy of the author and publishers)**

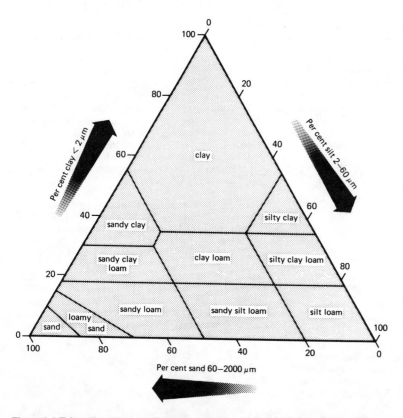

Figure 1.4 Triangular diagram of soil textural classes (after Hodgson, J. M. (1974)
Soil Survey Field Handbook, **p. 23. Soil Survey of England and Wales, Rothamsed Experimental Station, Harpenden, reproduced by courtesy of the author and publishers)**

on the subject is of limited practical use to agriculturalists. Agriculturally important properties of soil clays are:

(1) cation exchange capacity and ability to absorb anions;
(2) surface area available for chemical reactions;
(3) ability to shrink and swell in response to changes in moisture content.

Structure of clay minerals

The commonly occurring clay minerals are layer silicates and may be thought of as being built up from planes of oxygen or hydroxyl units. There are two types of oxygen/hydroxyl planes.

(1) *Complete planes* in which each oxygen touches six others (*see Figure 1.5*) and some or all the oxygen position may be occupied by hydroxyls.
(2) *Hexagonal planes* which differ from the 'complete' arrangement (*see Figure 1.6*) in that (a) there are no hydroxyls, (b) every fourth oxygen position is vacant, (c) the inter-oxygen separation is reduced in the ratio $\sqrt{3}:2$, so that the two types of plane contain the same number of oxygens:hydroxyls per unit area.

Tetrahedral sheets

If a single oxygen is placed on top of a complete plane it will fit comfortably in a position above a space in the plane and beneath it there will be what is called a tetrahedral space (so-called because it is between four spheres whose centres may be thought of as the apices of a tetrahedron). If, instead of one single oxygen, a hexagonal plane is placed above a complete one, a network of tetrahedral holes will occur. In clay minerals these holes contain either silicon (usually) or aluminium cations (occasionally). The two planes and their associated cations are said to form a tetrahedral sheet.

Octahedral sheets

If two complete planes are brought together, a network of octahedral spaces will be formed between three oxygens (or hydroxyls) of one plane and three of the other. A proportion (ranging from 66 to 100%) of these octahedral spaces are occupied by either aluminium or magnesium cations. The two planes and associated cations are said to form an octahedral sheet. If magnesium is the predominant cation all the octahedral spaces are filled and the sheet is said to be tri-octahedral (sometimes called a brucite sheet – brucite is $Mg(OH)_2$).

If the predominant cation is aluminium, two-thirds of the spaces are filled and the sheet is said to be di-octahedral (sometimes called a gibbsite sheet – gibbsite is $Al(OH)_3$).

Isomorphous replacement

In the sheet structure just described the arrangement of oxygens, hydroxyls and cations is such that electrical neutrality is maintained. In 'real' clay minerals some cations are replaced by others of similar size but different charge. This replacement does not alter the fundamental structure (form) of the layer but does leave the layer with a net charge.

Thus, aluminium can replace a proportion (up to about

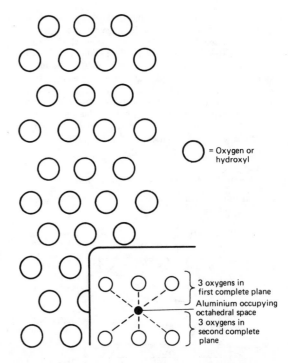

= Oxygen or hydroxyl

3 oxygens in first complete plane
Aluminium occupying octahedral space
3 oxygens in second complete plane

Figure 1.5 Complete plane of oxygens or hydroxyls. Inset – aluminium occupying the octahedral space between oxygens of two complete planes

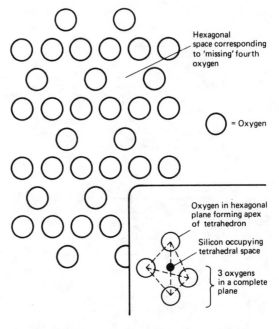

Hexagonal space corresponding to 'missing' fourth oxygen

= Oxygen

Oxygen in hexagonal plane forming apex of tetrahedron

Silicon occupying tetrahedral space

3 oxygens in a complete plane

Figure 1.6 Hexagonal plane of oxygens. Inset – silicon occupying the tetrahedral space between oxygens of a complete plane and a hexagonal plane

5%) of silicon in tetrahedral sheets. The larger octahedral spaces can be occupied by a wide range of cations. Isomorphous replacement normally leaves the sheet with a negative charge since every silicon replaced by an aluminium is conceptually the removal of one unit of positive charge, equivalent to a gain of one unit of negative charge. In the same way every octahedral aluminium replaced by a divalent cation contributes one unit of negative charge. Note that isomorphous replacement is a feature of a sheet incorporated at the time of its formation. 'Replacement' implies a departure from an idealised model *not* a dynamic soil process.

Combinations of sheets – layer lattice minerals

The clay minerals may now be visualised as combinations of tetrahedral and octahedral sheets, some of which may be negatively charged due to isomorphous replacement.

Interlayer bonding

The bonding between adjacent layers may be only oxygen – hydroxyl linkages between uncharged layers which are close together (*see* Kaolinite group *below*), or by means of cations in the interlayer space neutralising negative charges on the sheets. This type of bonding is readily subdivided into three categories on the basis of the magnitude of the negative charge.

High charge – mica type bonding

The outer sheets of adjacent layers are close together and arranged so that potassium ions can be located where

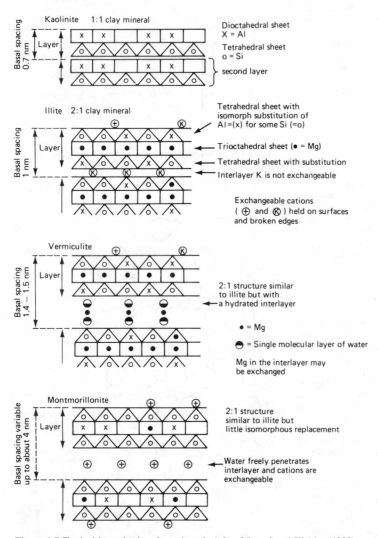

Figure 1.7 Typical layer lattice clay minerals (after Mengel and Kirkby (1982)
Principles of Plant Nutrition. **Bern: International Potash Institute, reproduced with permission of the authors and publishers)**

'surface depressions' in adjacent sheets are opposite one another. This gives a strong bond since inter-ionic distances are small and the basal spacing is 1 nm.

Medium charge – vermiculite type bonding

The adjacent layers are further apart than in the mica type and the interlayer space is occupied by two molecular layers of water containing magnesium or aluminium ions. The basal spacing is 1.4–1.5 nm and the larger inter-ionic distance gives a weaker bond, although the interlayer water contributes some hydrogen bonding.

Low charge – montmorillonite or smectite type

The structure is similar to vermiculite, but the bond is weaker and the interlayer cation can be exchanged with solution cations.

Typical layer lattice clay minerals (Figure 1.7)

Kaolinite is known as a 1:1 clay mineral composed of one tetrahedral and one octahedral sheet. There is no interlayer and successive layers are held by oxygen–hydroxyl linkages. Water is not able to move between the units and only exposed surfaces and broken edges contribute to the cation exchange capacity.

Illite is a 2:1 mineral (two tetrahedral sheets sandwiching one octahedral sheet). Adjacent layers are bonded by potassium which forms an integral part of the structure and is not exchangeable.

Vermiculite is similar to illite, but interlayer bonding is due to magnesium (Mg) in association with two monolayers of water. The Mg^{2+} is exchangeable, so intergrades occur and leaching with KCl solution leads to replacement of Mg^{2+} by K^+ and adoption of the illite structure (i.e. 1 nm basal spacing)

Montmorillonite is the commonest of the 'smectite' or 2:1 swelling clays. Lattice structure is again a variation of that of illite; the charge on layers is low (little isomorphous replacement) and bonding is sufficiently weak that interlayer cations are readily exchangeable, and water may enter the interlayer so that there is no unique basal spacing.

Hydrous oxides

Iron (Fe) and aluminium (Al) form hydrous oxides with no recognisable structure (amorphous) and a gross composition represented by $Fe_2O_3 \cdot H_2O$ and $Al_2O_3 \cdot H_2O$. These hydrous oxides occur principally as surface films on clay minerals in soils of temperate lands, and contribute to the cation exchange capacity (CEC) and phosphate fixing capacity. In tropical soils the hydrous oxides may dominate the colloid fraction; phosphate fixation capacity is then extremely large.

Clays in soils

The clay particles present in soils differ from the pure examples already described. In particular, soil clay minerals are formed *in situ* during natural weathering processes, so in all but the oldest and most mature soils intermediate stages of clay mineral formation, as well as end products, are found. In general, soil clays have a less well ordered structure than the pure varieties (usually of geological origin) and both the layers and interlayer ions may differ within a single clay particle. Similarly, adjacent clay particles are not expected to have the same composition, and the picture is further complicated by the ubiquitous occurrence of hydrous oxide films.

Because of this large range of variability, soil clays should be thought of as smectites (montmorillonite type), vermiculite type, illite type or kaolinite type. When successive layers (and interlayers) have different forms and compositions the mineral is said to be interstratified.

Figure 1.8 shows the important features of a soil clay complex formed by weathering of mica.

Clays and soil shrinkage

Laboratory observations show that minerals of the smectite group expand as they are wetted and contract on drying. Casual examination of field soil surfaces or soil profile pits during dry weather reveals the presence of cracks which are entirely absent under wet conditions. Since the cracking

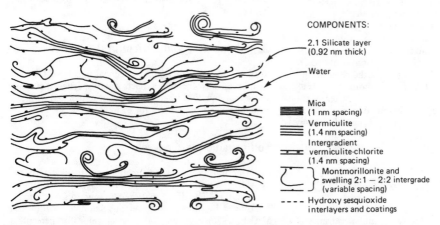

COMPONENTS:

2.1 Silicate layer (0.92 nm thick)

Water

Mica (1 nm spacing)
Vermiculite (1.4 nm spacing)
Intergradient vermiculite-chlorite (1.4 nm spacing)
Montmorillonite and swelling 2:1 – 2:2 intergrade (variable spacing)
Hydroxy sesquioxide interlayers and coatings

Figure 1.8 Important features of a soil clay complex formed by weathering of mica (from Jackson, M.L. (1964) *Chemistry of the soil,* **p. 93. ACS Monograph No. 160. Ed. by F. E. Bear. New York: Reinhold, reproduced with permission of the author and publishers)**

observed in the field is most extensive on clay soils, it is tempting to assume that it is due to the same mechanism as the expansion and contraction seen under laboratory conditions. This may be the case in hot climates with a long 'dry season', but under UK conditions field cracking behaviour is not governed by the proportion of smectite in the clay fraction (Reeve *et al.*, 1980). The explanation for this is that the drying experienced in the field is not sufficiently severe to remove the interlayer water removed by laboratory drying. The observed cracking in the field is largely due to 'bulk shrinkage' caused by reduction in the size of soil pores as the soil dries out.

SOIL ORGANIC MATTER AND THE CARBON CYCLE

Soil organic matter (OM) is an extremely emotive topic, associated in many people's minds with 'the goodness of the soil' and man's presumed ill treatment of it. It is also a difficult subject for study. This is because OM is not a single entity – it is a complex system whose individual components are not all capable of precise characterisation. Furthermore, the soil's organic fraction is never static – transformation from one form to another is one of its fundamental characteristics. At any one time, some 2% of the organic matter in an arable soil and up to 4% under grass or woodland are accounted for by living organisms.

The formation of organic matter

Living organisms (plant and animal, including micro-organisms) are the sole precursors of the soil organic fraction. Visible plant remains are not usually considered as 'soil organic matter' but micro-organisms are included, if only for want of any means of isolating them from the soil matrix.

Chemical analysis of organic matter (OM)

In a previous section it was noted that simple chemical analysis of soil obscures more useful knowledge than it reveals. This is particularly true for OM.

Classic techniques for fractionation of OM involve successive treatments with alkali and acid or extraction with ether and other organic solvents. The 'fractions' isolated are not homogeneous and they do not exhibit the recognisable properties of soil organic matter. Furthermore, many fractionations leave the majority of the organic matter in the residue which is resistant to the methods of the analytical organic chemist, and therefore only readily characterised by its total elemental analysis.

Properties and functions of organic matter in the soil

'Chemical' classifications of OM are of limited relevance to agricultural problems. The serious agriculturist should beware of being misled by the great volume of specialist literature relating to them.

Organic matter and soil structure stability

The bonding together of fundamental soil particles to form the structural units which are such an important feature of productive soils is a complex process involving clay mineral particles bonded to each other and to organic matter by chemical forces to form so-called 'microaggregates'. Similar mechanisms together with the mechanical actions of plant roots and fungal hyphae bind the microaggregates together to form larger units.

Polysaccharide and polyuronide gums (mostly of bacterial origin) may be adsorbed onto soil particles. Such gums typically have a molecular weight between 100000 and 400000 so one molecule can be adsorbed onto more than one mineral particle, forming a very strong link.

Additionally, charged groups on organic molecules (e.g. amino groups) can form bonds by electrostatic linking with opposite charged sites on clay minerals and again one large molecule may interact with more than one mineral particle. The so-called 'cation bridge' which involves linking of a negative charged clay site to a negative charged organic anion through a polyvalent cation can be viewed as a variant of this theme.

Exchange capacity

Many of the chemical properties of soils are dependent upon cation exchange capacity (CEC – sometimes called base exchange capacity). Organic substances contain carboxyl groups which can dissociate to provide negatively charged sites. The charge is typically 2–4 mEq/g compared with 0–2 mEq/g for clay minerals. Thus even though the OM content of most soils accounts for a small proportion of the total soil mass, it may make a significant contribution to the CEC.

Mineral nutrition of crops

Since nitrogen, phosphorus and sulphur are important constituents of OM, eventual breakdown by micro-organisms releases quantities of these elements to the soil in a plant available state. (Nitrogen, phosphorus and sulphur cycles are discussed more fully on pp. 20–25.) Some organic substances form complexes with copper, and hence its availability to crops is also influenced by OM formation and breakdown.

Organic matter and agricultural practice

Notwithstanding the complex chemical nature of OM and the mechanisms whereby it affects soil properties, there is no doubt that high OM levels are generally associated with ease of management of soil (*see* for example Low, 1972).

Soil organic matter level

In the laboratory, soil organic matter is normally determined by wet oxidation with a dichromate/sulphuric–phosphoric acid solution. The dichromate remaining after oxidation is determined by titration with ferrous sulphate solution (*see* MAFF, 1986). The method gives a value for percentage organic carbon: by convention this is multiplied by 1.72 to give 'organic matter percentage'.

The oxidation reaction is invariably incomplete since a

Table 1.1 Quantities of organic material returned by roots of various crops (after Davies, Eagle and Finney, 1972)

Crop	kg/ha of dry roots in top 20 cm	% increase in total organic matter in 20 cm of soil before decomposition
1-year grass ley	4500–5500	0.2–0.3
3-year grass ley	6750–9500	0.3–0.5
Winter cereals	2500	0.1
Spring cereals	1450	less than 0.1
Sugar beet	550	less than 0.1
Potatoes	300	less than 0.1
Red clover	2200	0.1
25 tonnes farmyard manure	4500	0.2

From *Soil Management*, Table 27, p. 194 by courtesy of authors and the publishers. Farming Press: Ipswich.

proportion of the OM is protected by intimate association with clay, but the figures obtained are well established in the literature and serve adequately for most comparative purposes.

The OM level in a soil at any given time is the result of a balance between additions and losses over its recent history. The factors affecting the balance are as follows.

Climate

In higher rainfall areas or where drainage is poor, the rate of decomposition is lower and OM levels are higher. Thus for any given soil texture and cropping sequence the OM level will be higher in the west of the UK, and at higher altitudes.

Soil type

Because of the intimate association between humus and clay particles, the OM in clay soils is better protected against attack and degradation than in sandy soils. Also clay soils are in general more likely to remain moist for longer periods of the year. Both these factors lead to clay soils having higher OM levels than lighter soils under otherwise similar conditions.

Cultivation

Frequent cultivation is thought to increase the rate of OM loss. This is not now believed to be a simple effect of better aeration leading to faster oxidative breakdown of organic materials. Powlson (1980) has suggested that the disruptive forces of cultivation result in exposure to micro-organisms, of organic matter which had previously been inaccessible to them. Also some micro-organisms are killed when soil is disturbed and their premature decay contributes to the loss of OM.

Less soil disruption and consequent slower OM breakdown in the surface horizon may be a beneficial effect attributable to 'non-plough' cultural methods.

Cropping

There has been interest in the effects of crop remains returned to the soil since chemists first started to look at agriculture. Typical quantities involved are shown in *Table 1.1*.

The main point illustrated by this table is that the amounts of organic material returned to the soil from crop roots or manure dresssings are too small to affect soil OM levels appreciably, even if no adjustment is made in recognition of the fact that the majority of organic residues returned to the soil are lost by respiration of micro-organisms, and only a small proportion is eventually incorporated into humus.

Note also that contrary to common belief, the root remains from a good winter cereal crop makes a sizeable contribution to the soil's OM economy (equivalent to about 14 t/ha of farmyard manure).

Effects of leys on soil OM levels have been studied by a series of experiments – at the ADAS (formerly NAAS) experimental husbandry farms. Results are summarised in *Table 1.2* and show that the increases due to a three-year ley are small and those following nine years of ley are only modest; note also that the gains of nine years' grass were virtually wiped out by a mere three years of arable cropping.

These results are consistent with the view that the structure stabilising powers of leys are due to gums produced by actively growing soil bacteria, the clay binding powers of bacteria themselves, and the mechanical effects of both grass roots and fungal hyphae. All of these agents are ephemeral and their decline following return to arable cropping would be reflected by a fall in total OM level.

Permanent pasture usually has higher OM levels than leys under similar conditions, and at least a proportion of it is of

Table 1.2 Effect of leys on level of organic matter in soils from ley fertility experiments (from Davies, Eagle and Finney, 1972)

Experimental husbandry farm	Location	Soil	% organic matter in topsoils			
			at start of trial	after 3 year ley	after 9 year ley	after 3 year arable following 9 year ley
Boxworth	Cambs	Clay loam	3.1	+0.1	+0.5	+0.0
Bridgets	Hants	Chalk loam	4.4	+0.1	+0.6	+0.1
Gleadthorpe	Notts	Sand	1.6	+0.1	+0.36	+0.3
Rosemaund	Hereford	Silt loam	3.5	+0.2	+1.0	+0.0
High Mowthorpe	Yorks	Chalk loam	3.8	+0.4	+1.0	+0.5

From *Soil Management* Table 28, p. 196 by courtesy of the authors and the publisher Farming Press: Ipswich.

a more durable nature, so that it is lost only slowly following a change to arable cropping.

The carbon nitrogen (C:N) ratio of organic matter of cultivated soils in temperate regions is fairly constant, having a value between 10:1 and 12:1. When manures with wider ratios than this are added to soil, either a disproportionate amount of carbon must be lost by microbial respiration, or additional N taken from some other source to compensate. (This is dealt with in more detail on p. 21.)

The carbon cycle

The transformations of carbon from gaseous carbon dioxide in the atmosphere to plant, animal, and thence soil OM and finally back to the atmosphere, form a cycle depicted in *Figure 1.9*.

Carbon dioxide (CO$_2$) in the atmosphere is the source of carbon for plant growth, and the 0.03% CO$_2$ in the world's atmosphere represents a total quantity of about 7×10^{11} tonnes of carbon. This is a relatively small amount by comparison with the estimated 9.6×10^{13} tonnes of carbon in fossil fuels (coal and oil) and 12×12^{15} tonnes in sedimentary rock (Ehrlich *et al.*, 1977).

The rate of fixation by photosynthesis depends on climatic conditions, soil and crop. Maximum rates (worldwide) are about 50 g of dry weight/(m^2 d) (200 kg/(ha d) of carbon). In the UK winter wheat makes the majority of its growth in the period April–July (120 d) and the average rate of increase in

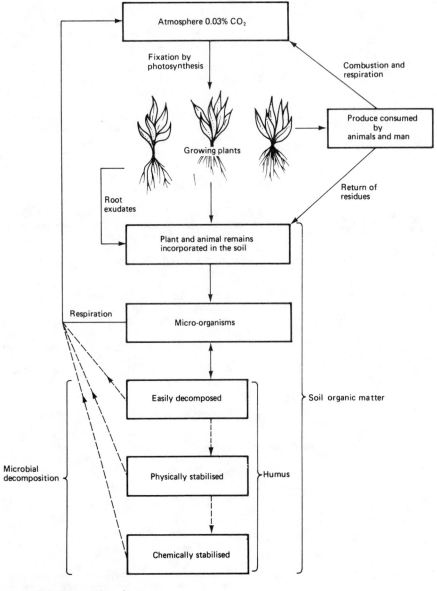

Figure 1.9 The carbon cycle

grain dry weight over this period is 0.1 tonnes/(ha d) or 10 g/(m^2 d). Making allowance for leaf and root growth, 20 g/(m^2 d) is a reasonable figure for this crop's total dry matter (DM) production. This corresponds to a carbon fixation rate of 80 kg/(ha d).

Plant and animal remains (including excreta) are the source of carbon from which soil organic matter is formed and quantities returned by roots of various crops have already been quoted (*Table 1.1*). Additionally, root systems contribute to soil organic matter by production of root exudates. These comprise both soluble (sugars, amino acids, etc.) and insoluble compounds (polysaccharides, sloughed-off cells, etc.). They make a direct contribution to the nutrition of micro-organisms in the rhizosphere and are thought to be responsible for the very high level of microbial activity in that zone.

Because of the close association of soil, micro-organisms, roots and their exudates, measuring the quantity exuded has proved extremely difficult. Estimates vary, but 10% of the total above ground dry weight over a season seems a likely order of magnitude.

This input of organic matter to the soil may be visualised as a flux of carbon being 'processed' through a volume of soil according to the equation

$$\frac{dC}{dt} = A - kC$$

This assumes that decomposition is what is known as a 'first order reaction' and k is its 'rate constant'; A is the annual addition of residues and dC/dt is the rate of change of the soil's organic carbon level (by convention organic matter (% OM) is derived from a laboratory determination of %C by multiplication by 1.72). For an equilibrium state $dC/dt = 0$ and the 'turnover time' is given by

$$\frac{C}{A} = \frac{1}{k} \text{ (years)}$$

Annual inputs of carbon for old arable soils at Rothamsted are typically 5% of soil organic carbon, so turnover time is calculated to be 20 years.

Radioactive carbon-14 dating techniques have allowed the mean age of carbon in existing soil OM to be measured, and results are in the region of 2000 years. Closer study has shown that different components of soil OM have differing turnover times.

Easily decomposed humus has a turnover time of about three years.

For physically stabilised humus – that which is protected from easy microbial attack by its location in fine pore spaces – the average figure is about 100 years.

Chemically stabilised 'ligno humus complexes', which may be protected by adsorption onto clay mineral surfaces, have mean turnover times as large as 4000 years.

The population of micro-organisms at any one time accounts for 2–3% of the soil's total organic carbon and this fraction has a turnover time of two to three years.

The 'priming action' of fresh organic matter additions

At one time it was thought that the enhanced microbial activity occurring when fresh OM was added to the soil could lead to accelerated breakdown of the old and more stable humic substances. By use of carbon-14 tracer techniques, Jenkinson (1966) was able to show that this 'priming effect' generally had a negligible effect on the overall rate of decomposition, any losses being compensated for by the biomass of the enlarged population of micro-organisms.

CHEMICALLY ACTIVE SURFACES AND THE CHEMISTRY OF SOIL COLLOIDS

The seat of chemical activity in the soil is the colloid fraction. This comprises the clay minerals and organic matter.

The soil's power to hold (adsorb) cations

Cations, represented as M^+ (monovalent, e.g. Na^+, K^+) or M^{2+} (divalent, Ca^{2+}, Mg^{2+}) are held by electrostatic attraction to negatively charged sites in the soil colloid fraction.

The negative charge on soil colloids is due to:

(1) *isomorphous replacement* of trivalent by divalent cations within clay mineral lattices, leading to a deficit of charge;
(2) *unsatisfied charges* at broken edges of lattices;
.(3) *dissociation* of protons
    (a) from organic acid groups

$$- COOH \rightarrow COO^- + H^+$$

    (b) from hydrous oxide surfaces

$$M\!-\!OH \rightarrow M\!-\!O^- + H^+$$

(M is Si (silicon), Mn (manganese), Fe (iron) or Al (aluminium).

Note that such dissociations are pH dependent, so that the negative charge (i.e. the CEC) is greater at high pH, and some protonation may occur at low pH to give sites of positive charge capable of interacting with anions (*see* the section 'The soil's power to hold anions', *below*).

Exchange of cations

Ions held by the electrostatic forces at charged surfaces are also subject to the kinetic forces of thermal motion, and the strength of their retention is the result of the balance between the two. At equilibrium a characteristic distribution pattern described by the Gouy–Chapman model is observed.

The Gouy–Chapman model says, in effect, that distribution of charged species within a solution is distorted near to a negatively charged colloid surface, there being a disproportionate concentration of cations near to the surface. An inner conceptual layer, next to the charged surface, is composed solely of cations (Stern layer) with a gradual transition (the Gouy or diffuse layer), to free solution. As shown in *Figure 1.10*, these two layers have an excess positive charge, balanced by the negative charge on the surface and cations in the 'double layer' are in equilibrium with those in the solution. Increasing the concentration of the external solution compresses the Gouy layer, so that the change from the all-cation Stern layer to free solution occurs over a smaller distance. Changing the composition of the solution leads to corresponding changes in the double layer system. Thus, if a clay is saturated with Ca^{2+} and in equilibrium with a $CaSO_4$ solution addition of a K^+ salt will lead to replacement of some Ca^{2+} by K^+ ($2K^+$ replacing each Ca^{2+}), i.e. cation exchange. Some cations exchange (or form Stern

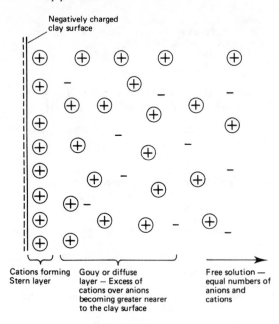

Negatively charged
clay surface

Cations forming	Gouy or diffuse	Free solution —
Stern layer	layer — Excess of	equal numbers of
	cations over anions	anions and
	becoming greater nearer	cations
	to the clay surface	

Figure 1.10 The double layer system associated with a negatively charged clay surface

layers) more readily than others. Small, highly charged ions are held preferentially to large ones with low charge, but small, strongly hydrated ions behave similarly to larger ones. Exchange reactions are reversible and the thermodynamically less favourable ones are able to occur, providing the concentration in the external solution is great enough. Thus strongly bound Ca^{2+} will be replaced by less strongly bound K^+, provided that the K^+ concentration in the external solution is high enough (i.e. a mass action effect).

The Gapon Equation is a physical relationship that governs the equilibrium between exchangeable and free cations of two species. For a monovalent cation C^+ and a divalent cation C^{2+} the equation states:

$$\frac{C^+ \text{ ads}}{C^{2+} \text{ ads}} = k\,\frac{a_{c+}}{\sqrt{a_{c2+}}}$$

C^+ ads = adsorbed monovalent cation

C^{2+} ads = adsorbed divalent cation

a_{c+} = activity of monovalent cation in solution

a_{c2+} = activity of divalent cation in solution

$\dfrac{a_{c+}}{\sqrt{a_{c2+}}}$ is called the activity ratio, AR.

Hence

$$\frac{C^+ \text{ ads}}{C^{2+} \text{ ads}} = k \cdot AR$$

k is primarily dependent upon the strengths of adsorption of the two species.

Adsorption of water

Water is a dipolar molecule with an asymmetrical distribution of charge:

$$\delta^+ \;\; \overset{H}{\underset{H}{>}}\!O\delta^-$$

Thus a layer of water molecules may be attracted to a negatively charged colloid surface and they will all be oriented with their δ^+ ends towards the surface. They will thus expose a new surface of δ^- charges which may in turn attract a second layer, and so on. The proximity to the negative colloid surface will increase the effective charge on the innermost layers of water by induction, but this effect decreases with distance from the colloid and so the strength of bonding also decreases towards an outer surface at which electrostatic attraction is balanced by the forces of thermal motion.

The soil's power to hold anions

Positively charged colloid surfaces

Although soil colloids are predominantly negatively charged, localised positive charged sites also occur, particularly on clay mineral edges and on hydrous oxide surfaces.

Of the nutritionally important anions in the soil NO_3^- is not adsorbed to any appreciable extent and sulphate is adsorbed only slightly. Although phosphate is fixed and tightly held by soils, it is incorrect and misleading to think of phosphate being exchanged by negative surfaces (*see below*, 'Ligand exchange'), so overall the positive charge sites on soil colloids are relatively unimportant.

Ligand exchange

Phosphate, silicate, fluoride and molybdate are bound by ligand exchange. This reaction is very important as the mechanism for 'phosphate fixation'. On iron or aluminium hydroxide surfaces, some oxygen atoms are less strongly bound to iron or aluminium than those in the body of the material. They may be replaced by negatively charged species such as phosphate anions. This reaction is characterised by an increase in the negative charge on the surface. It is not reversible and the exchange species can only be displaced by another anion capable of further increasing the negative charge.

Quantity intensity relationships

Although roots take up nutrients from the soil solution, as already explained, the solution is in equilibrium with a solid phase capable of replenishing losses. Thus nutrient availability to plants depends not only upon concentration in the solution, but also on the ability of the solid phase to sustain that concentration against depletion. The term Quantity (Q) refers to the amount of a nutrient, and Intensity (I) reflects the strength of the retention whereby it is held. Concentration in the equilibrium solution is a convenient measure of intensity, since if a species is weakly adsorbed it will support a high concentration in the equilibrium solution. *Figure 1.11* shows two contrasting Q/I relationships, or isotherms.

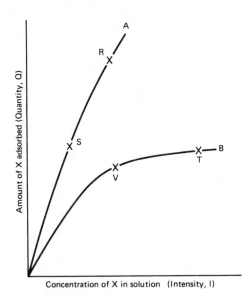

Figure 1.11 **Quantity/Intensity relationships for adsorption of 'X' by two contrasting soils**

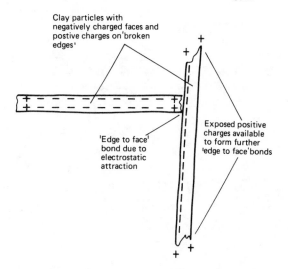

Figure 1.12 **'Edge to face' bonding in flocculated clay**

Going from R to S along A shows that the soil is able to sustain the intensity of X as uptake proceeds and the quantity (Q) in the reserve is reduced. From T to V along B a small uptake (reduction in Quantity) leads to a large fall in intensity and the soil is said to be 'weakly buffered with respect to X'.

Flocculation and deflocculation of clays

Clays have properties that enable their particles to either repel one another, or to attract one another and form stable groupings called 'flocs'. Flocculation of clay is vitally important in agricultural soils and fortunately deflocculation rarely occurs in the clay of UK soils.

The mechanism of flocculation

A clay mineral may be visualised as a lattice structure carrying negative charges. It has already been pointed out that positive charges may occur on broken edges. If two such particles are brought together edge to face they will attract one another and many such would flocculate into larger units (*see Figure 1.12*). This mechanism is responsible for the structural stability of clay soils.

If anything happens to suppress the positive charges, flocculation is prevented. Consider, for example, a clay saturated with Na^+ ions in a strong NaCl solution (e.g. flooded with sea water). A Gouy–Chapman double layer will form where there are negative surface charges, and it will be compact due to the concentrated solution (*see Figure 1.13a*).

If the external solution becomes less concentrated (e.g. rainfall, or irrigation after flooding) the double layer expands and the clay's positive charge sites virtually become part of it and are quite definitely 'blocked' by it as shown in *Figure 1.13b*. In this condition individual clay particles repel one another and the clay's contribution to soil structural stability is lost.

Calcium on an exchange complex is tightly bound, Ca^{2+} ions are to be found in the Stern layer rather than in the double layer. Consequently Ca^{2+} does not prevent flocculation in the way that sodium does (Ca^{2+} forms a compact double layer system, even with a dilute external solution). Thus the clay in a well limed soil can be relied upon to remain flocculated and therefore stable.

Prevention of deflocculation after salt water flooding

Deflocculation in salt water flooded soils is prevented by applying gypsum ($CaSO_4$) to ensure that the Na^+ is replaced by Ca^{2+} *before* the electrolyte concentration in the soil solution is allowed to fall.

Soil acidity and pH

The soil's acidity, measured on the pH scale, is of fundamental importance to farmers. When pH is too high, the availability of most trace elements to crops is reduced, whilst at low pH (acid conditions) crop yields are reduced drastically, due to excessive availability of aluminium and manganese.

In formal terms, the pH of a solution is defined as the negative logarithm of the activity of H^+ ions. Soil pH values range from about 4.0 (very acid) to around 8.0. The chemist's concept of pH 7.0 being neutrality is not appropriate to soils, and pH 6.5 is the value at which a soil normally exhibits neither acidic nor alkaline properties.

Measurement of soil pH

A sample of soil is shaken with 2.5 times its own volume of demineralised water and the pH of the suspension determined with an electronic meter which relies on the effect of H^+ ions on a thin glass membrane. Alternatively a coloured dye solution which changes colour according to solution pH may be used. There are a number of difficulties relating to measurement and interpretation of soil pH.

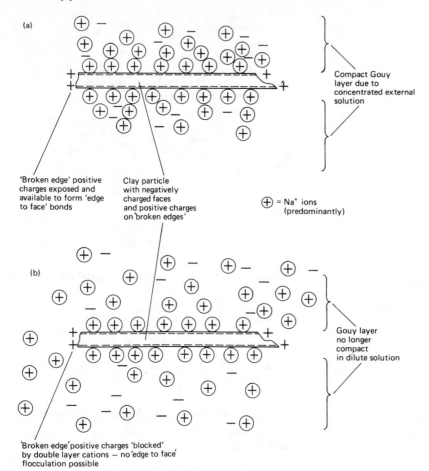

(a)

Compact Gouy
layer due to
concentrated external
solution

'Broken edge' positive
charges exposed and
available to form 'edge
to face' bonds

Clay particle
with negatively
charged faces
and positive charges
on 'broken edges'

⊕ = Na$^+$ ions
(predominantly)

(b)

Gouy layer
no longer
compact
in dilute solution

'Broken edge' positive charges 'blocked'
by double layer cations — no 'edge to face'
flocculation possible

Figure 1.13(a) 'Broken edge' charges available for bond formation when the external solution has a high Na$^+$ concentration. (b) 'Broken edge' charges 'blocked' by the cations of the double layer with a weak Na$^+$ external solution

pH is particularly variable within fields and it is important to have a representative sample (*see* p. 28).

Duplicate pH measurements at the same point may vary over a time scale of days or weeks due to changes in moisture content, soil solution concentration or carbon dioxide level in the soil atmosphere.

The presence of a Gouy–Chapman double layer means that H$^+$ ion concentration is not uniform throughout the whole suspension.

Thus interpretation of a pH reading is imprecise — no one can know whether a particular root environment corresponds to the external solution or something near to the composition of the Stern layer.

Large variations in the soil:water ratio affect the result obtained. On average, the pH sensitive surface of a glass electrode is nearer to the inner (Stern) layer in a more concentrated suspension and so records a lower value (more H$^+$ ions) than in a more dilute suspension.

An electrical junction effect due to clay particles may occur in the reference electrode system of an electronic pH meter, giving a slightly erroneous reading. This effect is most pronounced when measurements are made in thick suspensions or pastes.

Hydrogen ions are cations and therefore subject to the Gapon equation (*see* p. 14) so that their activity is controlled in part by the activity of other ions in the system. If, for example, K$^+$ ions are added to the solution, H$^+$ ions will transfer from the exchange sites into the solution, increasing its acidity (i.e. lowering the pH). When pH is measured by the water suspension method described, the cations in the external solution are derived entirely from the sample itself. The cation concentration will be governed by the previous history of the soil and the soil:water ratio. This uncertainty and its consequent effect on pH may be reduced by measuring pH in a salt extract having a concentration similar to that of a typical soil solution; 0.01 mol CaCl$_2$ is frequently used.

pH measurement using 0.01 mmol CaCl$_2$

Results obtained by using 0.01 mol CaCl$_2$ in place of water for making the suspension are generally about 0.5–1 pH unit lower at the acid end of the range. Differences are smaller at high pHs. The Soil Survey of England and Wales routinely uses both methods and an idea of the relationship between them can be obtained from the analytical data on typical soils (*see Table 1.17*). Despite the theoretical advantage of CaCl$_2$ as an extractant for pH measurement, it is not used for agricultural advisory purposes in the UK — most people working with crops are familiar with the numerical values given by the 'water method', and the advantages of CaCl$_2$ are not thought sufficient to justify any change. Because of all the sources of possible error in measurement or interpretation, there are few practical circumstances where it would be advantageous to know a pH value to an accuracy better than ±0.2 pH units.

The effects of soil properties on activity and degradation of pesticides

Adsorption

The concentration of a pesticide in soil solution is determined by the extent and strength with which it is adsorbed by soil colloids. This in turn determines its effectiveness or toxicity to weeds or insects, its rate of degradation, leachability and uptake.

Organic matter (OM) content is the soil property with the greatest influence on pesticide adsorption, so %OM must be taken into account when deciding the dose necessary to give the required activity for pest or weed control without causing crop damage. At one extreme, on soils of high OM level, the dose necessary to give the required effect may be prohibitively expensive or undesirable. For many pesticides adsorption by peat soils may be virtually complete, so the pesticide is entirely without effect. By contrast, on some sandy soils with low OM levels, certain soil-acting herbicides cannot be used because of the risk of damage to the crop.

The bipyridyl herbicides, paraquat and diquat, are strongly and irreversibly adsorbed by clay minerals and are thereby rendered inactive on contact with the soil. However adsorption onto organic surfaces is reversible, so that bipyridyl herbicides adsorbed by peats or humose layers can subsequently be taken up, and may damage crops.

Soil texture is closely related to both the soil's colloid surface activity and its ability to retain moisture. Consequently, texture is important in determining suitability and application rates for residual herbicides. *Figure 1.14* illustrates this schematically.

Pesticide degradation

Most pesticides in common use are degraded by microbial action. Bacteria, actinomycetes and fungi are able to utilise even chemically stable pesticides as carbon sources. Typically, such microbial degradation is preceded by a 'lag phase', the time taken for a population capable of performing the particular degrdation to build up in the soil. Thereafter, the soil is 'enriched' with adapted organisms which enjoy a competitive advantage over their non-adapted counterparts whilst the substrate (pesticide) is present. Subsequent additions of the same pesticide to enriched soil are degraded without any lag phase so long as the adapted organisms remain viable.

Some pesticides may be protected from microbial attack by adsorption onto soil organic colloids. Thus residue levels are frequently much greater in peaty soil than in mineral soils. Because these residues are adsorbed, they do not

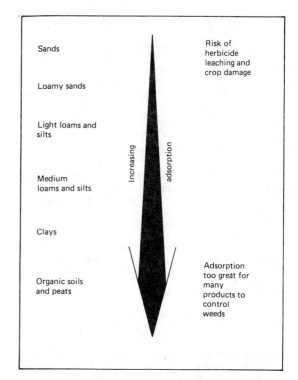

Figure 1.14 Effect of soil texture on adsorption of residual herbicides

normally support high concentrations in the soil solutions, so are not taken up by roots.

SOIL ORGANISMS

Chemical processes in the soil can be imagined as a number of cycles. Carbon is cycled between atmospheric carbon dioxide and plant tissue, and nitrogen is fixed from its gaseous form, incorporated in living tissue then released again to the atmosphere. Similarly, other elements are taken up by plants, become part of their structure and are eventually released to be utilised over and over again. All such cyclic processes in soil are dependent on the processes of decay and degradation brought about by living organisms.

Recently dead plant (or animal) remains are either consumed by animals, worms, slugs, etc. (the fauna) or decomposed *in situ* by smaller micro-organisms. Fungi generally mount the initial attack on resistant organic remains (which include the excreta of the faunal population), whilst bacteria and actinomycetes require chemically less complex nutrients such as simple sugars. Algae, at least when found on or near the soil surface, carry out photosynthesis in the same way as green plants, and some are able to fix nitrogen. The smallest of the fauna, the protozoa, feed largely by capture of other micro-organisms, particularly bacteria, whilst the larger macro-fauna (e.g. earthworms) feed on plant and animal remains. They also have a mechanical effect on the soil, improving aeration and drainage as well as mixing soil constituents.

Note that the various soil organisms are not to be regarded in isolation. All compete for food; some are predatory on others. Whilst some exist together in symbiotic relationships, others exude antibiotics which may confer some competitive advantage. The guts of saprophytic invertebrates commonly support a protozoal population to facilitate cellulose decomposition, and there are numerous examples of symbiosis between micro-organisms and the plant population.

Nutrition of soil micro-organisms

Historically micro-organisms have been thought of as either autotrophs, needing no organic substances, or heterotrophs needing organic compounds of various degrees of complexity. This classification leads to confusion, for example, when dealing with bacteria *capable* of functioning without any organic substances, but only showing maximum growth when obtaining energy from oxidation of carbon compounds. Such difficulties are avoided when it is understood that nutrients have three separate functions:

(1) provide the building blocks from which the organism is formed;
(2) provide a source of energy;
(3) accept electrons released during the oxidation reaction which provides energy.

Autotrophic organisms are either chemoautotrophs, obtaining energy from inorganic oxidation reactions, or photoautotrophs carrying out a photosynthetic reaction similar to that of higher plants. As well as the complexity of their carbon source requirements, heterotrophs have differing electron acceptor requirements. In aerobes the electron acceptor is oxygen, whilst anaerobes utilise some organic products of their own metabolism or an inorganic substance.

Classification of soil micro-organisms

Classification may be according to taxonomy, morphology (form as revealed by observation) or physiology (function). In the case of soil bacteria the activity of the various species is well documented. Hence, from the viewpoint of soil studies, it is profitable to adopt a classification based on physiological criteria. When dealing with actinomycetes, fungi, algae and protozoa it is not always possible to ascribe clearly defined functions to the various species, and only taxonomic or morphological classifications are used.

Bacteria

Due to difficulties in reliably distinguishing bacteria from the rest of the soil, estimates of their population vary widely. The commonly accepted range is in the region of 10^7–10^9 organisms/g of soil, equivalent to a live weight of about 3 t/ha (i.e. less than 1% of the soil organic matter).

Morphological classification distinguishes between cocci (diameter about $0.5\,\mu m \times 3\,\mu m$ length) and rods of diverse shape, but is of little use since many bacteria are able to change from one form to another and there is no correlation between morphology and function. A more useful type of classification for heterotrophs is that proposed by Lockhead

Table 1.3 Lockhead and Chase's classification of heterotrophic bacteria according to complexity of nutrient requirement

Group	Types included
1	Those growing in a simple glucose–nitrate medium containing mineral salts.
2	Those that can only grow when ten amino acids are added to this medium.
3	Those that can only grow when cystein and seven growth factors (aneurin, biotin, etc.) are added.
4	Those needing both the amino acids and the growth factors.
5	Those needing yeast extract.
6	Those needing soil extract.
7	Those needing both yeast and soil extract.

and Chase (1943), based on nutrition requirements and summarised in *Table 1.3*.

It is observed that fertile soils contain a greater proportion of groups 5, 6 and 7 (having complex requirements), whereas the immediate root zone of plants is characterised by the presence of groups 2, 3 and 4. Large dressings of farmyard manure have been shown to increase populations of bacteria having the more complex requirements, whereas inorganic fertilisers are without any detectable effect.

The chemoautotrophic bacteria may be similarly classified on the basis of the oxidation reaction used for energy supply.

(1) Nitrogen compounds oxidised
   (a) Ammonium oxidised to nitrite   *Nitrosomonas*
   (b) Nitrite oxidised to nitrate   *Nitrobacter*
(2) Inorganic sulphur compounds
   converted to sulphate   *Thiobacillus*
(3) Ferrous iron converted to ferric   *Thiobacillus ferroxidans*
(4) H_2 oxidised   Several genera

The microbial oxidation of nitrogen compounds, in particular, is vital for crop nutrition see p. 21.

Bacterial fixation of nitrogen

Free living bacteria notably *Azotobacter* and *Beijerinckia* are capable of fixing atmospheric nitrogen, but quantities fixed are small. The process is inefficient in terms of nitrogen fixed per unit of carbohydrate oxidised and the rate of fixation appears to be limited by the carbohydrate supply. It is not therefore surprising that when such bacteria establish themselves in close association with the roots of plants which exude carbohydrate, they perform rather better. On maize, rice and some tropical crops, such associations can give fixation rates in excess 1 kg/(ha d) of nitrogen (N). Quantities of N fixed in this way under UK conditions are minimal.

The notion, once popular in Russian literature, that N fixation could be increased by inoculation with *Azobacter* has been shown to be fallacious — it is almost universally true that if the rate of a process is limited *only* by the size of a microbial population, that population will increase until some other factor, such as substrate availability, becomes limiting.

Symbiotic nitrogen fixation is a characteristic of legumes and a small number of non-leguminous plants. Agriculturally, interest is confined to the legumes and in the UK the crops concerned are clovers, lucerne, peas and beans. The fixation occurs in root nodules formed from plant tissue, in response to invasion by *Rhizobium* bacteria. Up to 200 kg/ha of N may be made available to a UK clover crop in this way and rather lesser amounts to peas and beans. In grass–clover swards, grass is able to benefit from the 'clover nitrogen', although the mechanism involved has not been clearly established. In such mixed swards the *Rhizobium* can only start to fix N in the spring time once clover growth has become sufficiently vigorous to provide a carbohydrate supply.

Although attractive as a means of providing herbage without using high inputs of nitrogen, grass/clover swards have substantial management problems. The N supply from rhizobium does not become available until well into the growing season. Whilst use of fertiliser nitrogen can 'fill the gap' this may prove undesirable since N inhibits fixation and may also have a detrimental effect on clover survival. Clover is also very sensitive to grazing management and considerable skill is needed to maintain the clover content of a grazed sward. Current work at the Welsh Plant Breeding Station is addressing the joint problems of early growth and survival and leading to the development of early growing matched grass and clover varieties. Such developments should introduce new opportunities for exploitation of biological N fixation.

Bacteria and soil structure

Polysaccharide gums are thought to be of great importance in stabilising soil structure (*see* p. 10). Most of these are thought to be formed as products of bacterial metabolism.

Actinomycetes

In any strict sense actinomycetes are classified within the bacteria, but on the basis of morphology and physiology they represent a transition between true bacteria and the fungi. A large number of genera of actinomycetes have been recognised in the soil. Many of these are of interest because of their ability to synthesise antibiotics and the soil has been the primary source of organisms for commercial production of these agents in the pharmaceutical industry.

In agricultural soils, abundance is in the order 10^5–10^8/g but only *Streptomyces* (70–90% of the population) and *Nocardia* (10–30%) are present in sufficient numbers to be of importance in soil studies. The nocardias are sufficiently similar to bacteria that they are frequently (and perhaps inadvertently) included in estimates of bacterial populations. Some of their number have the ability to modify their metabolism quite rapidly, to decompose synthetic chemicals.

Streptomyces are responsible for the characteristic smell of freshly turned earth. The role of actinomycetes in soil processes is still not clearly defined, but it appears that they have a lesser biochemical importance than bacteria and fungi. They have been observed to attack resistant components of plant and animal tissues some time after addition of fresh remains, presumably when competition from organisms with more fastidious nutritional requirements has declined. Their metabolites are of interest since they are very similar to the complex molecules described as 'humus' when occurring in the soil. Actinomycetes are favoured by abundance of organic materials, relatively dry soil conditions and high soil pH. Under waterlogged conditions or at pHs less than 5.0 they are virtually absent and some authorities claim that it is possible to estimate the pH of arable soil on the basis of the smell associated with *Streptomyces*.

Thermoactinomyces and some species of *Streptomyces* have an important degradative function at elevated temperatures (50–65°C) in compost heaps.

Fungi

Fungi occur in all soils and they are commonly associated with the initial breakdown of organic debris. Many are saprophytes, tolerant of acidic conditions and capable of flourishing in the litter layer of forest soils. They are highly efficient at converting organic carbon into fungal tissue, even to the extent of taking nitrogen from the soil solution in order to metabolise organic matter with a large C:N ratio. Since fungal tissue is quite resistant to breakdown, N locked up in this manner may be unavailable for some time. Like the actinomycetes, fungi can incorporate N into humic-like substances which normally have a very long half-life in the soil.

Fungi are heterotrophs and require oxidisable carbon compounds as energy sources. The requirements of various species range from simple carbohydrates, sugars, organic acids and starch through cellulose and lignin (a substance resistant to attack by most other organisms) to those requiring complex chemical growth factors, or only capable of growing as parasites on living plant tissue.

The soil is a highly competitive environment and there are many organisms capable of metabolising simple sugars. Those of the fungi which have sugar as substrate maintain a competitive advantage by growing very rapidly, so consuming the sugar before it is lost to other organisms. The cellulose decomposers, Ascomycetes, Fungi Imperfecti and Basidiomycetes face less competition and have slower growth rates. Since there is virtually no competition for lignin as a substrate, the higher Basidiomycetes which function as lignin decomposers are able to grow very slowly and appear to build up a food reserve prior to attacking any new lignin source. Apparently the initiation of decomposition of lignified material is an energy-consuming process.

Common parasitic fungi cause several well known diseases of crop plants (*see* p. 264). Some other fungi are normally saprophytic, but may occasionally invade living tissue and produce disease symptoms. The mechanisms controlling this 'occasional parasitism' are not understood, but adequately fertilised, vigorously growing crops enjoying good cultural conditions rarely seem to be affected.

Symbiotic fungi — mycorrhiza

For many years it has been appreciated that a symbiotic relationship exists between some fungi and forest trees. The commercially important relationships such as occur between Scots pine and fungi such as *Boletus*, *Amanita* and *Lactarius* have been carefully studied. Generally the fungus is one requiring sugar as substrate and the most abundant supplies of sugars are found in trees whose protein synthesis is limited by nitrogen or phosphate deficiency. Thus the fungus invades the root tissue of the deficient trees and then 'repays' the tree by increasing the efficiency of uptake of N and P.

The relationship is in delicate balance: if the fungus is too successful supplying N + P to the tree its sugar supply will be reduced, whilst if the growth of the tree is checked for any reason, the fungus may become parasitic.

A number of crop plants can sustain populations of vesicular–arbuscular mycorrhiza in their root systems. Although these mycorrhiza are able to contribute to phosphate uptake at very low soil phosphate levels they do not appear to have any beneficial function under normal agricultural conditions.

Fungi and soil structure

Although fungi lack any significant capacity to produce 'gummy' substances, some have considerable power to stabilise soil structure by the mechanical effect of their mycelia binding soil particles together.

Algae

Algae are photoautotrophs using light to provide energy and absorbing atmospheric carbon dioxide. Thus their preferred habitat is the soil surface. Strictly, many soil algae are facultative photoautotrophs having the ability to metabolise some carbohydrates, and significant numbers are found in sub-surface habitats. Since algae only have any recognisable competitive advantage when exposed to light at the soil surface, the significance of these sub-surface populations is not understood. Algae form a relatively small part of the soil biomass, typically 10–300 kg/ha (but up to 1500 kg/ha where a visible surface bloom occurs). When growing at the surface, multiplication of algae is only limited by lack of light, moisture or inorganic nutrients and competition for other factors does not affect them. Thus an algal bloom on soil is indicative primarily of moist surface conditions and a lack of crop competition for light. Algae have no significance in UK agriculture, but the blue-green algae in particular, are important as nitrogen fixers in rice paddies. Algae and lichens (symbiotic alga + fungus combinations) are important as primary colonisers of bare rock and they provide the sole source of organic matter in embryonic soils.

Protozoa

The Protozoa are the smallest of the soil fauna, distinguishable from all the organisms previously described by having means of locomotion. They are very small, 10–80 μm in length, to enable them to move in soil water films. Nutrition may be photoautotrophic (e.g. *Euglena*) and some may exist as saprophytes. However, the major mode of nutrition is by predation on bacteria, small algae, yeasts, etc. The estimation of protozoal populations is extremely difficult, but estimates of low precision are acceptable since their sole claim to agricultural significance is their role in reducing the numbers of organisms that constitute their prey. 'Order of magnitude' calculations indicate that the weight of the protozoal population is about 1% of the weight of bacteria in the soil.

Other animal life

The fauna is capable of near infinite sub-division. For the present purpose it merely needs to be pointed out that anything that eats, be it elephant or earthworm, is participating in the chain of events whereby vegetation is transformed to yield soil organic matter and plant nutrients. Those animals which have their being within the soil mass, rather than above it, fulfil the additional function of physically mixing the soil constituents. In this context, earthworms are worthy of particular mention. They ingest organic remains and excrete an intimate mixture of soil and fresh 'humus', whilst leaving behind them channels which provide easy access for plant root growth, and improve both aeration and drainage. Where earthworms thrive, they dominate the faunal population, exceeding the sum of all other fauna by up to five-fold, on a weight basis. The three million or so worms which may be found under one hectare of deciduous forest weigh about 2 tonnes. Where earthworms do not dominate the fauna, total populations are normally much smaller and the soil appears 'lacking in life' when examined in a profile section.

Manipulation of micro-organisms

Since micro-organisms occupy such a fundamental place in the cyclic processes involving nutrient elements in the soil, the idea of improving crop nutrition by influencing microbial activity is an attractive one. Although much has been written, particularly in Russian literature, about inoculation of soil with organisms such as the nitrogen fixing Azotobacter, research in UK (and elsewhere) has failed to demonstrate any benefits. This is thought to be due to inability of added organisms to compete with 'native' populations in what is a very competitive environment.

A notable area of success is the use of inoculated seed for legumes and the general rule seems to be that this technique is beneficial in environments lacking suitable indigenous rhizobia.

At least a part of the success of traditional crop rotations is attributable to the way in which rotations influence micro-organisms and a current research goal is to introduce a more direct element of 'biological control' into the control of root diseases. The introduction of organisms which produce growth promoting substances in the soil is another possibility, but such organisms would need to be able to compete with the indigenous population.

THE MAJOR NUTRIENT ELEMENTS

These are nitrogen (N), phosphorus (P), potassium (K), magnesium (Mg) and sulphur (S). They are present in soils and are taken up by plants in relatively large quantities. Calcium (Ca) is also a major element, but is dealt with separately (p. 26) because of its importance in relation to liming.

The major elements have a role in building the structure of plants, whereas the micronutrients (*see* p. 27) are important in enzyme systems and contribute to the plant's function rather than its structure.

Nutrient offtake and nutrient cycles

The quantities of phosphate and potash removed by normal yields of crops are given in *Table 1.4*.

Table 1.4 Phosphate and potash removal by normal yields of common crops

	Fresh yield (t/ha)	P_2O_5 (kg/ha)	K_2O (kg/ha)
Cereal – straw burnt or ploughed in	7 (grain)	55	39
– straw removed	7 (grain)	63	79
Sugar beet – tops ploughed in	40 (roots)	32	84
– tops removed	40 (roots)	63	246
Potato – haulm not removed	40	40	232
Oilseed rape	3	48	33
Grass – silage	50	70	240
– hay	5	30	90
Kale	60	72	300
Swedes	40	28	96

ADAS data.

According to the type of farming, a part, or nearly all of the crop's major element uptake may eventually be returned to the soil in plant remains or animal excrement.

Nitrogen

Nitrogen exists in the soil either in the inorganic form, as ammonium (NH_4^+) and nitrate (NO_3^-) ions, or as organic compounds.

Inorganic nitrogen (also called mineral nitrogen)

This is the form of nitrogen used by crops. Nitrate (NO_3^-) is the main form taken up by plants, and ammonium (NH_4^+) is quickly converted in the soil to nitrate by the action of autotrophic bacteria:

$$NH_4^+ \xrightarrow{\textit{Nitrosomonas}} NO_2^- \text{ (Nitrite)}$$

$$NO_2^- \xrightarrow{\textit{Nitrobacter}} NO_3^- \text{ (Nitrate)}$$

These reactions can only occur under aerobic conditions. Since the second is potentially able to proceed faster than the first, NO_2^- does not normally accumulate in soils.

Nitrate ions are not held by soil colloids, so if not taken up by roots, nitrate is 'leached' by rainfall and lost in drainage water. Ammonium ions, however, are adsorbed by negatively charged exchange sites and so are held against leaching. Adsorption does not provide protection against oxidation to nitrate, so adsorbed ammonium is readily lost from the soil when conditions (temperature and aeration) are suitable for nitrification. Thus, with the exception of the driest parts of eastern England, it may be reliably assumed that between one autumn and the following spring, virtually all the organic nitrogen in a soil will be leached from the rooting zone unless taken up by a crop.

Levels of total mineral N ($NH_4^+ + NO_3^-$) in soils fluctuate widely over short time periods, so although it is a relatively simple matter to determine total mineral N in the laboratory, the result is of no use for determining fertiliser application rates. Levels normally fall in the range 5–50 mg/litre (mg of N/litre of soil), corresponding to about 10–100 kg/ha of N. The highest levels are frequently found under heavily fertilised grassland at the end of the growing season, before the onset of winter rains.

Organic nitrogen

Soil reserves of organic nitrogen are large. A soil with an OM level of 5% would have an organic N content of about 0.3%, or 6 tonnes/ha of N. This is some 60–600 times larger than the total mineral N.

The N in organic compounds is not available to plants until it has been released by the decomposition of organic matter. The fraction of the soil OM most actively involved in decomposition reactions contains most of its N in the form of proteins, and many micro-organisms are able to break these down to amino acids and then to ammonium ions, a process known as ammonification. In more general terms the conversion of nitrogen from organic to inorganic forms is referred to as mineralisation.

Since the small amount of mineral N is derived from the large heterogeneous reserves of organic N by biological degradative processes, it is not surprising that mineral N levels are highly variable. In general, soils with a high total N content are likely to need less N fertiliser to achieve maximum yields than those with low N levels, but the relationship is not good enough for predictive purposes. A 'total N' determination measures the reservoir of nitrogen potentially available to be mineralised, but tells nothing of the rate at which the micro-organisms are likely to carry out the mineralisation.

Nitrogen in crop residues and organic manures is not available to crops until it has been converted to ammonium. Subsequent conversion to nitrate is normally rapid and build-up of ammonium N only occurs under anaerobic (waterlogged) conditions. The ratio of C:N in the organic matter of agricultural soils in temperate lands is remarkably constant at between 10:1 and 12:1. If a manure or crop residue has a ratio less than this (i.e. more N relative to C) microbial action will release N quite rapidly. Materials with a large C:N ratio, such as cereal straw, require an additional source of N for their decomposition. In the absence of any other source, micro-organisms decomposing high C:N ratio materials will remove mineral N from the soil. Under present day conditions of high nitrogen fertiliser use this is not necessarily a bad thing. Post harvest mineral nitrogen levels in soil are usually high and this N is readily lost by leaching. However, when straw is incorporated this potentially wasted N may instead go to satisfy the micro-organisms' demands. This N is eventually released again when the biomass decays, so that as well as preventing pollution it can make a contribution to subsequent nutritional needs.

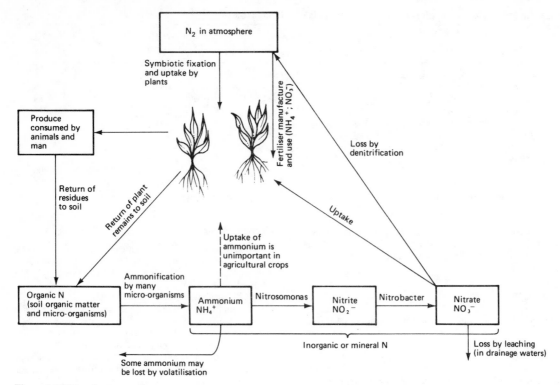

Figure 1.15 The nitrogen cycle

The nitrogen cycle

A diagrammatic representation of the N cycle appears in *Figure 1.15.*

The total nitrogen in the earth's atmosphere is estimated at about 4×10^{15} tonnes. It exists as molecular N_2 (di-nitrogen — the gas that makes up four-fifths of the atmosphere). This is only available to plants after *fixation*, a process which involves large inputs of energy.

The Haber–Bosch process is used industrially. N_2 and H_2 are reacted together at high temperature and pressure to form ammonia, which is the precursor of modern nitrogen fertilisers.

Microbial N fixation involving the symbiosis between a legume and *Rhizobium* bacteria can make a useful contribution to agricultural production.

Peas and most types of beans are grown without any fertiliser N and would not respond if it were given. A grass/clover ley without N yields about as much dry matter as an all grass sward given 120 kg/ha of N. This nitrogen is not 'free' to the farmer since the *Rhizobium* uses carbohydrate that would otherwise contribute to yield.

Soil OM breakdown typically provides about 100 kg/(ha annum) of N to agricultural crops. The quantity is greater on soils with higher OM levels and where the OM is of recent origin. On cultivated fen peat soils the N supply from the soil organic matter is great enough that little or no fertiliser N may be needed for cereal crops.

Nitrogen is lost from the soil by leaching and denitrification. Denitrification occurs when soils are waterlogged whilst biological activity is high. Most of the soil micro-organisms are aerobes utilising oxygen as final electron acceptor. Under anaerobic conditions many are able to utilise NO_3^- reducing it to nitrous oxide (N_2O) or di-nitrogen (N_2). Both of these are gases and are therefore immediately lost.

Waterlogging in winter does not cause anaerobic conditions to develop rapidly, since biological activity is low, so winter waterlogging does not lead to denitrification.

The denitrification process is difficult to study experimentally but it is usually assumed that N losses are large during short periods of waterlogging in the growing season. Leaching losses may be determined experimentally by collecting drainage water from large confined cylinders of soil called lysimeters. Over winter most N (including any that is mineralised during the winter) is lost by leaching. In the growing season, leaching is greatest in areas of high rainfall. Only small quantities of N are lost from grass crops but shallower and less vigorously rooting arable crops allow more N to be leached. Losses are greatest under fallow and are particularly large when land used for heavily manured early crops is fallowed through the summer, as is common practice in the Isles of Scilly and Channel Islands. Leaching losses represent more than just a waste of nitrogen. It is becoming more difficult for water authorities, particularly in eastern England, to meet their obligation to provide supplies having a nitrate level within the range regarded as acceptable.

Nitrate and water quality

In recent years the fate of leached nitrate has attracted much attention and in July 1985 an EC Directive (The Drinking

Water directive) set a maximum admissible concentration (MAC) of 50 mg/litre, which is half the earlier limit based on the 1970 World Health Organisation 'European Standards for Drinking Water'. The whole issue of 'nitrate in water' is reviewed in the Report of the Nitrate Co-ordination Group (1986) and the recent history of the subject is to be found in The Royal Commission on Environmental Pollution's Seventh Report 'Agriculture and Pollution' (1979), the Government's response (Pollution Paper No. 21 — DoE, 1983), the fourth Biennial Report of the Standing Technical Advisory Committee on Water Quality (1984) and the Royal Society's study group report 'The Nitrogen Cycle of the United Kingdom' (1983).

From the viewpoint of UK agriculture two features should be noted:

(1) Whilst it is clear that agriculture contributes to the nitrate load of potential water supplies in the UK, the extent of the agricultural contribution in comparison to other sources, in particular sewage, is uncertain. There can be no doubt that excessive and wasteful use of nitrogen can have a serious effect on nitrate in water, but even with small applications of N or no use of fertiliser at all, some leaching of nitrate into water will occur whenever land is farmed. There is strong evidence to suggest that a substantial amount of nitrate in supplies taken from some aquifers at the present time has its origins in the major postwar ploughing up of old established grassland.

(2) Notwithstanding recent publicity and legislative activity there is little evidence to suggest that nitrate in the water supply is potentially harmful under Western European conditions. However, Water Authorities are now subject to legal controls and so in turn must look to agriculture to avoid unnecessary contamination of potential water supplies. Statutory powers in this respect are provided by the Control of Pollution Act 1974.

The Control of Pollution Act 1974

This Act makes it an offence for a person to cause or knowingly permit:

(1) any poisonous, noxious or polluting matter to enter any stream or controlled waters or any specified underground water; or

(2) any matter to enter a stream so as to tend (either directly or in combination with other matter which he or another person causes or permits to enter the stream) to impede the proper flow of the water of the stream in a manner leading or likely to lead to a substantial aggravation of pollution due to other causes or of the consequences of such pollution;

(3) any solid waste matter to enter a stream or restricted waters.

The provisions of the Act are directed at all sections of the community. In general, any discharge of polluting material may only occur under the terms of a consent from the Water Authority. The special position of agriculture is recognised in that a person is not considered to be guilty of an offence if pollution is due to an act or omission which is in accordance with good agricultural practice. This provision covers the leaching of nitrate from manures and fertilisers and for the purpose of this legislation 'Good Agricultural Practice' (GAP) is documented in the form of a Code (the Code of Good Agriculgural Practice, MAFF and WOAD, 1985). It should be noted that the GAP provision does not extend to anything discharged directly through a ditch or drain. Such a discharge would require Water Authority consent. The Control of Pollution Act makes provision for additional restrictions which may be applied if a Water Authority considers that pollution may occur as a consequence of practices which come within the definition of GAP.

Current ADAS fertiliser recommendations (*see* p. 85–87) are a part of GAP and they are designed to ensure optimum yields without wasteful over-application of fertiliser nitrogen which could lead to high leaching losses. Thus is it always unwise to exceed the recommended amount of nitrogen fertiliser for any crop. In addition, use of nitrogen in excess of GAP may lead to legal action if pollution ensues.

Phosphorus (P)

Phosphorus is unusual amongst the major nutrient elements in that concentrations in the soil solution are frequently extremely low, in the order of 1 μg/ml, but it is taken up by crops in large amounts. This is explained by the replenishment of soil solution phosphate from solid phase reserves.

The soil's reserves of phosphate

In addition to phosphate in solution it is possible to recognise by experiment two fairly distinct pools of phosphate.

Labile phosphate

This is held on colloid surfaces but remains in rapid equilibrium with soil solution phosphate. ^{32}P labelled phosphate added to the soil solution is found to exchange readily with the non-labelled P in this fraction.

Insoluble phosphate

This part of the soil's reserve of P is not readily available to crops, but it can release P slowly into the 'labile pool' and conversely labile pool P may be 'fixed'.

Phosphate fixation and fertiliser placement

Phosphate fixation is the process whereby soil solution/labile pool P is rendered insoluble by reactions occurring at colloid surfaces. Under suitable conditions phosphate reacts with calcium surfaces, to form layers having properties similar to the almost insoluble calcium phosphate minerals. Under more acidic conditions, iron and aluminium surfaces predominate, and ligand exchange reactions occur (*see* p. 14). It appears that phosphate films on iron surfaces are stable, but phosphate adsorbed into aluminium surfaces is able to move into the solid phase and become part of a crystal structure, leaving a fresh aluminium surface for further adsorption of phosphate.

Older texts on soils and fertilisers place great emphasis on fixation of applied phosphate fertiliser. Thus, it was said that phosphate in cereal seedbeds should be placed (i.e. combined drilled) near to the seed, to prevent fixation throughout the whole bulk of the soil. Similarly, a granule, which gave rise to a discrete volume of high P availability, into which roots could grow, was preferable to a powder. Further, it was argued that since water soluble P quickly became part of the insoluble phosphate reserves if not taken up immediately by plants, the same effect could be achieved at a lesser cost by use of insoluble phosphate fertilisers.

In advanced agricultural countries such as the UK, there is now a sufficient history of phosphate fertiliser use and phosphate import in animal feeds (much of which eventually finds its way into the soil in animal manures) that the most powerful sites of phosphate fixation have already been satisfied, and newly applied P is fixed less strongly than was the case in years gone by. Placement of phosphate thus assumes a lesser importance and should only be regarded as desirable where soil phosphate reserves are known to be low.

Quantity/intensity isotherms for phosphate (*see* p. 14) have a characteristic shape such that in a low solution concentration it is difficult to increase the concentration because any P added is rapidly adsorbed. At higher concentrations this adsorption is less pronounced and thus less phosphate needs to be added to the system to bring about a further increase. This effect is illustrated by curve B, *Figure 1.11*.

Phosphate in soil organic matter

Organic matter contains phosphate and so its decay eventually liberates phosphate into the soil. This phosphate contributes to the labile pool and is, of course, liable to fixation. Where soil phosphate levels are very low, release of P from OM assumes great importance; the growing plant may be regarded as competing with unsatisfied fixation capacity for any available phosphate, and organic matter decomposition favours the plant by releasing P during

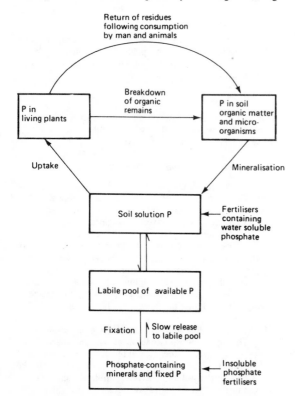

Figure 1.16 Schematic representation of soil phosphate relationships

periods of high biological activity, when the plant's needs are likely to be greatest. The organic fraction thus protects phosphate from fixation in these circumstances.

A schematic representation of soil phosphate relationships is given in *Figure 1.16*. No crops grown in the UK normally remove more than 70 kg/ha of P_2O_5 per annum and most take considerably less. Thus fertilisers used at the rates recommended by ADAS (*see* pp. 85–87) are more than sufficient to replace offtake.

Phosphate in the soil solution, even in fertile soils, amounts to only 1–5 kg/ha of P_2O_5. Since a rapidly growing crop can take up about 3 kg/(ha d), the solution phosphate must be capable of rapid replenishment, more than once per day in some cases. The size of the labile pool is largely governed by previous fertiliser history, but for most UK soils it is in the order of several hundreds of kg/ha.

Soil OM has a P:C ratio of 1:100–1:200, so that the organic P in a soil of 5% OM represents a reserve of up to 1700 kg/ha of P_2O_5 and a small proportion of this is made available each year as OM is mineralised.

Potassium (K)

In the absence of applied fertilisers, K in the soil is derived from weathering of minerals in which potassium is either an integral part of the mineral structure, as in the feldspars, or forms a tightly held interlayer as in illite (*see* p. 9). Potassium thus released becomes part of the soil's reserve of exchangeable K.

Exchangeable K

The K^+ cation may in theory occupy any exchange site, satisfying the negative charge on a soil colloid and maintaining an equilibrium with other soil solution cations as required by the Gapon equation (*see* p. 14). Experiments have shown that two different Gapon coefficients must be used to account for the behaviour of exchangeable K.

Weakly held K

Gapon coefficient 2.2 $(mmol/litre)^{-1/2}$ relative to Mg^{2+}. This is the K adsorbed onto external layer silicate surfaces and other sites which are not specific for K.

Strongly held K

Gapon coefficient 102 $(mmol/litre)^{-1/2}$. This is the K adsorbed at the edges of exposed interlayers in illite-type structures and these sites are specific for K^+, since larger ions could only be accommodated by a separation of the layer structure.

Potash fixation

To a degree, the weathering processes may be considered as slowly reversible, and K^+ in the soil solution may diffuse into the interlayer of vermiculite-type 2:1 minerals. Eventually, if sufficient K^+ enters the interlayer (or more correctly a part of an interlayer) it may revert to an illite-type structure with 1 nm basal spacing, and the K is then 'fixed'. This is sometimes represented as involving a third type of exchange site (with infinite Gapon coefficient), but there is little justification for use of the term exchange since, in the absence of any weathering, the reaction is reversible.

Schematic representation of soil K relationships

Potassium is peculiar amongst crop nutrients in that although it is retained in the soil in an available form (*see Figure 1.17*), annual crop uptake may be quite large in relation to soil reserves. Hence the soil 'K index' may change rapidly and this is particularly so under intensive grassland. For example, grass DM may contain up to 3% K (3.6% K_2O) so that a conserved crop of 10 t/ha dry matter could remove 360 kg/ha of K. Two or three years of removal at this rate would easily take a sandy or silty soil from ADAS K index 2, down to index 0. Clay soils are able to release K from clay minerals so their reserves are not so easily reduced, and for crops which are not responsive to high levels of K it is good practice to keep 'available K' at only a moderate level. Maximum benefit can then be obtained from clay mineral derived K.

Sodium

Sodium may contribute to the nutrition of some crop plants, and is occasionally used as a fertiliser. In the quantities used for this purpose the behaviour of Na^+ in the soil may be considered analogous to K^+. At the higher concentrations which may be achieved in laboratory experiments, or occur in soils which have been flooded by sea water, sodium in the exchange complex brings about very profound changes (*see* p. 15).

If a fertiliser dressing of salt (sodium chloride) remains in the surface few cm of an unstable soil it can produce surface capping and reduced germination. Autumn application of salt, and subsequent cultivation, both reduce this risk.

Magnesium (Mg)

Like potassium, magnesium is a constituent of the minerals of the clay fraction and is released by weathering. Water soluble and exchangeable forms are available to plants, but Mg^{2+} is relatively easily lost from the soil by leaching. In clayey soils, weathering can usually be relied on to keep pace with magnesium, but sandy soils are frequently deficient. Where a potential deficiency is recognised, it is normal practice to apply Mg to the soil by use of liming materials containing magnesium, and rates of addition can be quite large. Mg^{2+} typically accounts for 10–20% of the soil's cation exchange capacity, and contributes to the control of soil pH in the same way as calcium.

Sulphur (S)

Although crop offtakes of sulphur are of a similar order of magnitude to phosphate offtakes, there has been little interest in soil sulphur until quite recently. Sulphur has been described as the forgotten element. For many years it has been applied unwittingly as an impurity in commercial fertiliser, particularly superphosphate (12%S) and ammonium sulphate (24%S). These two materials are not constituents of modern compound fertilisers and no appreciable quantities of S now come from inorganic fertilisers.

It is thought that a large proportion of present day crop S need is being provided by S in rainfall (i.e. atmospheric pollution from industrial processes and combustion of fossil fuel). Typically, inputs from this source in industrialised

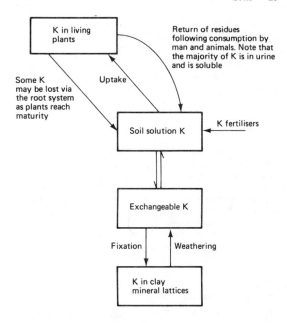

Figure 1.17 Schematic representations of soil potassium relationships

countries are in the range 10–20 kg/ha, greater near to industrial centres, and possibly considerably less in more remote areas. Sea spray carried by wind is another source of S, and in the UK at present either proximity to industry or to exposed coasts ensures adequate supply of S in many areas.

Grass in particular may respond to added sulphur in areas receiving little sea spray or industrial pollution. Responses to S in such localities in Southern Ireland are well known (Hanley and Murphy, 1970; Murphy, 1975).

Sulphur in the soil occurs in both organic and inorganic forms, but organic S provides the major reserve. Soil C to S ratios are about 100:1.

Approximate calculations assuming a soil OM level of 5% and a mean turnover time for OM of 20 years, show that annual mineralisation from OM provides about 30 kg/ha of S.

Hydrogen sulphide (H_2S) is produced during microbial breakdown of organic matter. Under aerobic conditions this is oxidised (autoxidation) to SO_4^{2-}, but in anaerobic soils it is oxidised to elemental S by the bacteria *Beggiatoa* and *Thiothrix*. These bacteria and *Thiobacillus*, in aerobic soil, can also oxidise S to H_2SO_4, the mechanism whereby additions of sulphur can be utilised to lower soil pH.

CALCIUM AND LIMING

On the basis of its role in plant nutrition, calcium (Ca) should be included amongst the list of major nutrient elements, but for practical agricultural purposes it is hardly ever considered as a nutrient at all. This is because calcium deficiency is unknown to agriculture (it occurs occasionally in horticultural crops grown in small containers of peat compost). Calcium has two distinct functions:

(1) an essential nutrient for plants;
(2) the dominant ion in determining soil pH.

The second of these is the critical one, and depletion of soil calcium is always evidenced by acidity damage to the crop, long before the level is low enough for Ca deficiency to occur.

Calcium in the soil

Calcium exists in water soluble form in soils, in equilibrium with exchangeable Ca^{2+} held on soil colloids. It also occurs in solid form as $CaCO_3$ and as a constituent of clay minerals.

Levels of calcium in the soil solution are governed by chemical equilibria between solution, exchangeable and solid forms. Weathering of primary Ca containing minerals and dissolution of added lime ($CaCO_3$), is greatly assisted by 'carbonic acid' – CO_2 (produced by respiration of organisms), dissolved in water.

$$CO_2 + H_2O \rightarrow H_2CO_3$$

$CaCO_3$ is only very slightly soluble (about 0.3 mmol) but in the presence of carbonic acid:

$$CaCO_3 + H_2CO_3 \rightarrow Ca(HCO_3)(calcium\ bicarbonate)$$

This bicarbonate is much more soluble than $CaCO_3$.

Exchangeable calcium is the dominant ion on soil colloid surfaces and its function of replacing H^+ ions from exchange sites governs the H^+ concentration in the soil solution. It therefore determines soil pH.

Particles of calcium carbonate (CaCO₃) in soil may be of natural origin, derived from weathering of limestone, chalk or chalky boulder clay, or they may arise from use of agricultural lime.

Feldspars, amphiboles and calcium phosphates of the clay fraction release Ca on weathering, and constitute the main source of Ca in soils other than those formed from chalk and limestone parent materials.

Lime in the soil

Wherever rainfall exceeds evapotranspiration, a steady downward movement of water through the soil removes (leaches) cations from the soil exchange complex. In its simplest form this reaction may be represented as follows.

$$Ca^{2+} + 2H_2CO_3 \rightarrow 2CA(HCO_3)_2 + H^+H^+$$
$$-\quad- \qquad\qquad\qquad -\quad-$$
clay surface   clay surface

The calcium remains in solution as Ca^{2+} ions and is lost in drainage water. Eventually Ca (and other cations) are lost, and H^+ takes their place on exchange sites, i.e. the soil becomes acid.

This is a natural tendency and, under UK weather conditions, all soils whose parent materials are incapable of releasing cations from the clay fraction at an adequate rate, eventually become acid unless

(1) lime is applied,
(2) the soil is subject to flushing by water of a high base status,
(3) bases are re-cycled by deep rooting trees (as leaf fall) or added in the excrement of animals.

pH and liming

The acidity of soils is expressed on the pH scale. *Table 4.6* (p. 87) shows the pH requirements of various crops grown on mineral soils (slightly lower values are acceptable on peats).

Determination of pH may be by use of either an electronic meter (*see* p. 16) or a coloured dye solution called indicator. Whichever method is chosen, taking the sample from the field is more likely to lead to significant error than is any shortcoming in 'technique'. Careful attention should be paid to points mentioned on p. 31. When using indicator on field moist soil, mixing and sub-sampling is not generally feasible, the correct procedure is to examine *each core* with indicator and take an average value as the pH of the field.

The lime requirement is the amount of lime (expressed as $CaCO_3$) needed to maintain a pH of 6.5 for arable crops, or 6.0 for grassland. Although the lime requirement (LR) is expressed in tonnes/ha of ground limestone ($CaCO_3$), equivalent quantities of any other lime may be used and the magnesium is magnesian limestones is also able to replace H^+ ions. It therefore has a liming effect similar to that of Ca^{2+}. In general, for any given pH, clay soils and soils with a high organic matter will have higher lime requirements than sandy or low OM soils. Lime requirement can be determined accurately in the laboratory using the method described by MAFF (1986). However, for practical purposes adequate precision can be achieved by use of known relationships between pH, texture and lime requirement. These relationships are summarised in *Table 1.5* and it is apparent that at any given pH the lime requirement of a clay soil is substantially greater than that of a sandy soil.

Table 1.5 Lime requirement (t/ha ground chalk or ground limestone) for soils of different texture

pH	Sands A[1] G[1]	Light A G	Med + clay A G	Organic A G	Pty + peats A G
6.5	0 0	0 0	0		
6.0	4 0	5 0	6 0	4	
5.5	7 3	8 4	10 4	9 3	8
5.0	10 5	12 6	14 7	14 7	16 6
4.5	13 7	15 7	18 7	19 7	24 7
4.0	16 7	19 7	22 7	24 7	32 7

Notes:
1. A = Arable LR (based on 20 cm soil);
   G = Grassland LR (based on 15 cm soil)
2. Maximum surface application for grass 7 t/ha. Larger dressings may be applied to arable land, and should be split with a part applied prior to cultivation and 'ploughed down', and the remainder spread over the ploughing.
3. Grass reseeds should normally be limed as an arable crop.

Overliming can lead to serious financial losses. At high pH values trace element availability is drastically reduced. Deficiency of boron in brassicas is frequently induced by over-zealous use of lime. For the dairy farmer, too much lime can lead to low herbage manganese levels and this may contribute to problems of infertility in the herd. Most livestock are susceptible to deficiency of copper and this condition also may be aggravated by overliming.

In cereal crops, deficiencies of manganese and copper which remain untreated may lead to serious yield loss. Overliming is the most frequent cause of these deficiencies. Iron deficiency is also induced by overliming, but no major agricultural crops are affected. Note that lime-induced deficiencies are common where the pH has been increased by liming at high application rates; problems are less widespread on soils where pH is naturally high.

Loss of lime from the soil

Calcium is removed from the soil whenever crops are harvested, since it is a constituent of all plant tissue. Amounts removed by individual crops are given in *Table 1.5*. *Table 1.6* shows lime removals associated with four different cropping systems in terms of kg/ha $CaCO_3$ removed. Note that these amounts are small by comparison to normal rates of liming and crop removal does not make a major contribution to lime loss.

Loss in drainage water occurs all the time due to the effects of slightly acidic rainwater (containing small quantities of dissolved CO_2, and SO_2). The amount lost depends on rainfall, soil texture and the amount of calcium in the soil. Losses are estimated to range from 100 to 1000 kg/ha of calcium carbonate annually.

Since lime moves downwards in drainage water, it is quite possible for the surface soil to become acid, irrespective of the nature of the subsoil. The movement of drainage water tends to be uneven and some areas of soil lose lime more rapidly than others; this may lead to variations in acidity and hence the first observable effects of acidity in a crop are often patchy.

Use of fertilisers affects loss of lime from the soil. Nitrogen fertilisers have an acidifying effect whenever ammonium nitrogen is converted in the soil to nitrate nitrogen. Urea breaks down to ammonium salt in the soil and therefore behaves similarly. Aqueous ammonia and anhydrous ammonia increase acidity in the same way. Experimentally determined values for the amount of lime ($CaCO_3$) needed to counter the effect of 1 kg of nitrogen range from 1.4 to 2.9 kg.

Some ammonium nitrate fertiliser is sold mixed with lime. Research in the Netherlands has shown that 100 kg of an ammonium nitrate–lime compound containing 20.5% N has a liming effect equivalent to 8 kg $CaCO_3$. A product of this type containing 23% N has only a slight effect on soil pH whilst the 26% N product commonly sold in the UK may be expected to have a slight acidifiying effect.

Potassium fertilisers have an acidifying effect on the soil, but it is much less than that due to nitrogen, and in practice can be ignored.

Phosphate fertilisers do not cause any lime loss from the soil. Basic slag contains lime, and any phosphate fertiliser containing basic slag will act as a liming material.

Normal dressings of farmyard manure and slurries have very little, if any, effect on soil acidity.

Soil texture and organic matter level influence the retention of lime. Light, sandy soils have the lowest capacity for holding calcium, and, as they are usually freely drained, it is rapidly lost. On these soils troubles from acidity are most common and most acute, but easily remedied. Since comparatively small amounts of lime will raise the exchangeable calcium content towards saturation point, small dressings generally suffice to restore very acid sandy soils to neutrality. By contrast, clay soils and those with high organic matter levels hold very much larger quantities of calcium. This calcium is tightly held and many years may be required for its removal. When liming becomes necessary, heavier dressings are needed.

Lime and soil structure

It is widely held by farmers that liming improves the structure of heavy soils, reduces stickiness, lightens cultivations and makes it easier to break down clods and obtain a satisfactory tilth. In the days of horse cultivations, it was often said that after very heavy applications of burnt lime, 'four-horse land' became 'three-horse land' or even better. In experiments it has been observed that soil structure can be improved on clay soils by heavy applications of lime, to give a pH of over 7. It is also well known that naturally calcareous clays are generally much less intractable and better drained than naturally acid clays. However, it is usually inadvisable to purposely overlime, because nutrient deficiencies may be induced.

Other beneficial effects of maintaining the optimum pH are indirect. The addition of lime to an acid soil increases the activity of the carbon cycle bacteria and enhanced decomposition of organic matter assists soil crumb formation and increases soil structural stability.

TRACE ELEMENTS AND MICRONUTRIENTS

Elements that occur in very small amounts in soils are called trace elements. A sub-set of these, the micronutrients, are essential for plant growth.

Table 1.6 Lime removals associated with four different farming systems

Farming system		Calcium carbonate removal/annum (kg/ha)
Arable	70 ha cereals, 20 ha sugar beet, 10 ha potatoes $\{$all straw retained	14
	$\{$all straw sold	24
Mixed	30 ha cereals, all straw retained, 10 ha cash roots, 60 ha temporary grass, milk, calves and culled stock sold from 110 milking cows, plus followers	44
Dairy	All grass; milk, calves and culled stock from herd of 160 cows plus followers	56
Grass drying	All grass, very high nitrogen fertiliser usage; all herbage sold	220

From MAFF Reference book No. 35 'Lime and Liming' with permission.

Toxicity

Most trace elements (including micronutrients) are toxic to plants at high concentrations; they are also toxic to animals, but poisoning of farm livestock is usually attributable to causes other than the consumption of home-grown feeding-stuffs having high tissue levels of trace elements.

Occurrence

Table 1.7 gives median trace element contents of soils in England and Wales (*see* p. 32 for explanation of the term 'extractable').

The 'National Soil Inventory', produced by the Soil Survey of England and Wales (now the Soil Survey and Land Resource Centre of Cranfield Institute of Technology) in conjunction with ADAS and Rothamsted Experimental Station, contains trace element data for sites on a 1 km grid covering England and Wales.

Since trace elements are largely bound in mineral lattices and released by weathering, the parent material very largely determines the abundance of individual elements in soil. Copper, manganese, cobalt, nickel and lead are associated with the ferromagnesian minerals of basic rocks. Boron is derived from the weathering of the mineral tourmaline. Soils derived from unmineralised granite and coarse grained sandstones tend to have low levels of most trace elements; levels are high on serpentine derived soils such as those of the Cornish Lizard Peninsula.

Table 1.7 Median trace element content of soils in England and Wales (total values expressed as mg/kg; extractable as mg/litre air dry soil)

Element	Median	Range	Number farms	Number samples
TOTAL				
Boron	33	7–71	72	227
Cadmium	< 1.0	0.08–10	204	689
Cobalt	8	< 1.0–40	125	421
Copper	17	1.8–195	226	751
Nickel	26	4.4–228	226	752
Lead	42	5–1200	226	752
Zinc	77	5–816	225	748
Selenium	0.6	0.2–1.8	34	114
Mercury	0.04	0.008–0.19	17	53
ACETIC ACID EXTRACTABLE				
Nickel	1.0	0.12–22.7	198	647
Zinc	6.6	0.4–97.6	204	664
EDTA EXTRACTABLE				
Copper	4.4	0.5–74.0	201	662
Selenium	0.04	0.01–0.59	32	112
HOT WATER EXTRACTABLE				
Boron	1.0	0.1–4.7	153	493

From Archer, F. C. In *Inorganic Pollution and Agriculture* Table 1, p. 184–190 MAFF Reference Book No 326. (1980), London: HMSO with permission.

Trace elements not essential for plant growth

Iodine and selenium (which are essential to animals), arsenic (the chemistry of which has similarities to phosphorus in the soil) and a host of 'heavy metals' such as cadmium, chromium, nickel, tin and lead are not required by plants. Interest in the heavy metals stems largely from their occurrence in sewage sludges originating from urban areas. Because of the possibility of heavy metal contamination, professional advice should be sought before applying sewage sludge to farm land (*see* ADAS, 1987).

The essential micronutrients

Copper, zinc, manganese, boron, molybdenum and iron are essential for plants. Deficiencies in agricultural crops may be due to:

(1) an absolute lack of the element (e.g. soils derived from granite). This type of deficiency is common on peats where the current soils surface is too far removed from weathering minerals for replenishment by weathering to be effective;
(2) inadequate availability, frequently due to overliming or adverse soil physical conditions.

Information on inidividual elements is summarised in *Table 1.8*.

SOIL SAMPLING, SOIL ANALYSIS AND SOIL NUTRIENT INDICES

The methods of calculating fertiliser requirements for common UK crops are described on p. 85. Knowledge of the soils 'index' of P, K and Mg is basic to accurate and reliable calculations. Soil pH and lime requirements must also be known if liming policy is to have a rational basis.

Soil sampling

Where any chemical analysis is employed as a basis for reaching a decision, the decision can only be as good as the sample submitted to the analyst.

The essence of a good soil sample is that it is representative of the whole situation. In order to achieve this, account must be taken of soil variability.

Variability and sampling error

Variability is a fundamental property of biological materials and thus constitutes a potential major source of error when sampling.

Variability in soils has been studied extensively. If a large number of soil cores were taken from a 'uniform' field and analysed individually, a spread of results would be obtained. More useful information may be obtained by mixing the soil cores (sub-samples) and then doing a single analysis. Such a 'composite sample' may be described as representative if there is a high mathematical probability that another similar sample obtained in the same way would give similar results. Experiments have shown that 25 cores or sub-samples will constitute a representative sample provided that the following instructions are followed carefully.

Table 1.8 Summary of information on micronutrients

Element	Forms in the soil	Remarks	Threshold values[1] for susceptible crops (mg/litre)
Copper (Cu)	Clay mineral lattices Organic compounds Exchangeable Cation Cu^{2+} In soil solution	Soil solution concentrations are low. Cu occurs in clay mineral lattices but this form is not available. Cu^{2+} is tightly held by inorganic colloids and this form is not easily available. Complexes with soil organic matter largely control availability to plants. This is why the chelating agent EDTA is frequently used to determine 'available' copper. Availability is reduced by excessive liming.	0–1.6 deficiency likely 1.6–2.4 slight deficiency likely 2.5–4.0 deficiency unlikely > 4.0 adequate For chalk soils deficiency levels are related to organic matter content OM > 6% deficiency possible when Cu < 2.5 OM 0–5% deficiency possible when Cu < 1.0
Zinc (Zn)	Substituted for Fe^{2+} and Mg^{2+} (isomorphous replacement) in ferromagnesian minerals. Exchangeable Zn^{2+} Organic complexes	Zink deficiency is virtually unknown in the UK	
Manganese (Mn)	Clay mineral lattices Nodules and manganese containing iron pans Exchangeable Mn^{2+} and oxides of trivalent and tetravalent Mn	Mn^{2+} is taken up by plants and this is the form stable under reducing conditions. Mn^{2+} stability decreases 100-fold for each one unit increase of pH, and formation of Mn-organic complexes unavailable to plants is greatest at high pH. Thus Mn availability is governed by soil reaction (pH) and aeration, more than by the presence of 'adequate quantities of Mn'. Deficiency is frequently related to weather conditions, being common during warm periods following cold spells. Lack of consolidation of the seedbed predisposes a crop to Mn deficiency. Chemical determination of manganese is useless as a predictor for deficiency.	
Boron (B)	Contained in tourmaline and may substitute for Al or Si in clay minerals Adsorbed onto soil colloid surfaces	Adsorption of B by soil colloids decreases as pH is lowered Availability to plants is minimal at high pH values, hence B deficiency can be induced by liming. The range between deficiency and toxicity is very small.	< 0.5 severe deficiency likely 0.6–1.0 severe deficiency possible 1.1–2.0 satisfactory 2.1–4.0 high > 4.1 toxicity possible
Molybdenum (Mo)	Adsorbed on clay mineral surfaces. As calcium molybdate. May also occur in organic forms and as hydrated Mo oxides	Occurring anionic form Mo behaves differently from other micronutrients. Availability is high at high pHs and in soils derived from high Mo sediments (Teart soils). Toxicity in livestock (Mo induced copper deficiency) is consequent upon use of lime. Deficiency does not normally occur unless pH is below 6.0, and is rectified by liming	In relation to possible deficiency in crops Low 0.1 Satisfactory 0.2 In relation to possible Mo induced copper deficiency in stock Normal 4 Induced deficiency possible 4–8 Induced deficiency likely > 8
Iron (Fe)	Ubiquitous in soils — an essential constituent of ferromagnesian minerals	Soluble Fe levels are very low compared with the total Fe present. Except where microbiological activity under anaerobic conditions leads to reduction to Fe^{2+}, iron availability in the soils is governed by the solubilities of Fe^{3+} hydrous oxides. Oxide precipitation is strongly pH dependent and Fe^{3+} activity in solution decreases 1000-fold for every unit increase in pH. Thus deficiency occurs on soils of high pH and where lime use has been excessive. Iron deficiency does not occur in agricultural crops in UK but some fruit trees are seriously affected.	

[1] These values relate to analysis by standard 'available' methods.

Method of taking a soil sample

The area to be sampled should be examined carefully and, if necessary, sub-divisions made to accommodate areas which are not uniform in crop growth, soil drainage, soil texture or colour, or areas which have been cropped, limed or fertilised differently, and areas of marked variations in acidity and alkalinity (which can be delineated by use of soil indicator).

Soil should be taken from at least 25 spots in the field. To ensure even distribution of the sample spots, follow a 'W' shaped path as illustrated in *Figure 1.18*.

The sampling technique must be appropriate to the particular crops being grown.

Arable soils and short-term leys (including permanent grass about to be re-seeded) should be sampled to a depth of 15 cm using a screw auger (approx. 25 mm diameter), tubular corer, or a cheese-type corer for very dry powdery soils. The 25 cores obtained may be put straight into the sample bag to give an acceptable size of sample.

Permanent grass and long leys (unless they are soon to be ploughed up) should be sampled to a depth of 7.5 cm, preferably with a tubular corer, a cheese-type corer or pot corer. The core should be of uniform diameter and must include the surface mat or turf. This requirement cannot be met using a screw auger. The 25 cores obtained may be put straight into the sample bag to give an acceptable size of sample.

Direct drilled crops require special attention. When sampling either grassland or the stubble of a previously direct drilled crop, prior to direct drilling, it is a wise precaution to make a separate examination of the surface 2–3 cm of soil. This avoids the possibility of undetected acidity in the immediate surface, adversely affecting crop establishment. It may be best done using soil indicator, but where doubt still remains a separate sample to 2–3 cm should be taken. Be especially careful to detect possible patches of acidity in such cases.

Avoidance of contamination

Soil samples for chemical analysis should not be put into old fertiliser bags. Even bags which have been washed will usually contain sufficient phosphate to ensure that any analytical result is completely erroneous.

Samples for trace element analysis must be carefully protected from possibility of contamination by sealing in new polythene bags. Such samples should not come into contact with copper, brass or other metals which might interfere with the analysis.

Samples for special purposes

The foregoing sections refer specifically to samples to be used as a basis for making fertiliser recommendations. For other purposes (e.g. diagnosis of a crop disorder or determination of a herbicide residue) requirements may be different. Instructions for taking samples in most circumstances likely to be met by farmers have been given by Hughes (1979).

Soil analysis

On receipt at the laboratory, it is normal practice to air-dry soil samples, crush them mechanically to separate soil aggregates without breaking stones, and then retain as 'fine earth' all the soil which will pass a 2 mm sieve. This may be sub-sampled and analysed to ascertain pH and a variety of nutrient levels. Elements may be determined either as 'total' or 'available'

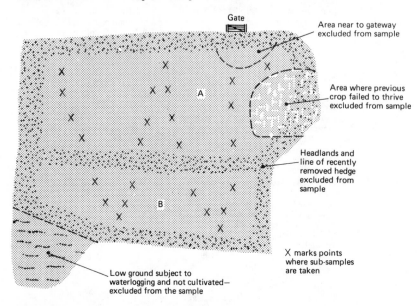

Note — If the soils of parts A and B are known to be different, or the sampler believes them to have been farmed differently in the recent past, two separate samples should be obtained.

Figure 1.18 Path to be followed when soil sampling a field

Table 1.9 Summary of standard ADAS laboratory methods for analysis of soils

Determinant	Extraction procedure	Analytical finish
Available phosphorus	Shake for 30 min with 0.05 mol sodium bicarbonate solution at pH 8.5, containing polyacrylamide.	Spectrophotometric (880nm). Phosphomolybdate formed when acid ammonium molybdate reacts with phosphate, is reduced by ascorbic acid to a blue coloured complex.
Available potassium	Shake for 30 min with 1 mol ammonium nitrate.	Flame photometry on filtrate.
Available magnesium	As above	Atomic absorption spectrophotometry at 285 nm using a strontium chloride releasing agent.
Available copper	Shake for 1 h with 0.05 mol EDTA diammonium salt at pH 4.0	Atomic absorption spectrophotometry at 325 nm using calcium nitrate compensation in standard solutions.
Water soluble boron	Boil for 5 min with water (detailed instructions must be followed).	Spectrophotometric, as a methylene blue complex with fluroborate (in 1, 2-dichloroethane extract) after removal of nitrates and nitrites.
Available molybdenum	Shake for 16 h with ammonium oxalate/oxalic acid solution	Spectrophotometric, as a complex formed by Mo, iron and thiocyanate in the presence of a reducing agent.
pH	pH measurement is described on p. 18.	
Lime requirement (LR)	Uses the same suspension as pH determination. LR is calculated from the new equilibrium pH attained when *para*-nitrophenol buffer is added to the extract.	

Determination of 'total' element levels

Because of the occurrence of the major nutrient elements as constituents of mineral lattices, their total levels in soil have no bearing on potential for crop growth. Total levels of major nutrients only need to be known by those studying soil formation processes.

In the case of trace elements, knowledge of total levels can be of use in predicting possible deficiency situations, and where use of different techniques for 'available' analysis would otherwise invalidate comparisons.

Determination of 'available' element levels

For most elements, the amount available to plants comprises that which is water soluble, together with some exchangeable and probably also some insoluble form (which may come into solution or enter the exchange complex during the crop's growth). To date it has not proved possible to draw any clear distinction between such 'available nutrients' and the non-available remainder. Because of this, developments in soil analysis have relied on the fact that even if chemists don't know what is available to plants, the plants themselves do . Over the years, a number of empirical procedures for determining available nutrients have been tested by comparison with crop performance in field experiments.

Those which correlate best with crop performance have been retained and are considered to be standard techniques within the UK. Different techniques are frequently used in other countries or for special purposes in UK. Whilst various techniques have their protagonists, it should always be remembered that, owing to the empirical nature of 'available nutrient' analysis, the procedure for which the largest and most reliable range of interpretive information is available is likely to be of greatest use. Brief details of standard ADAS procedures are given in *Table 1.9*.

The methods are published in full by MAFF (1986).

Determination of nitrogen in soil

There is no satisfactory chemical means of determining the nitrogen likely to become available to a crop during its growth and so chemical analysis is not normally of use for determining rates of application of nitrogen.

The nutrient index system

The nitrogen index

ADAS advice on nitrogen fertiliser use is based on the Nitrogen Index. This is calculated from knowledge of previous cropping and automatically makes allowances for the likely contributions of crop residues to the supply of available nitrogen under normal circumstances.

How to determine the nitrogen index of a soil

(1) If lucerne, long leys (three or more years) or permanent pasture have not featured in the last five years' cropping, use *Table 1.10* only.
(2) If the last crop grown was lucerne, long ley or permanent pasture, use *Table 1.11* only (use the 'Years since ploughing out' column 0 only).
(3) If one of the above crops has been grown in the last five years, but was not the last crop, look up the values from both *Tables 1.10* and *1.11*. The higher of the two values found is the Nitrogen Index.

Table 1.10 Nitrogen Index — based on last crop grown

Nitrogen Index 0	Nitrogen Index 1	Nitrogen Index 2
Cereals	Peas or beans	Any crop in field receiving
Sugar beet	Potatoes	large frequent dressings
Maize	Oilseed rape	of farmyard manure or slurry
Vegetables receiving	Vegatables receiving	Lucerne
less than 200 kg/ha N	more than 200 kg/ha N	
Forage crops removed	Forage crops grazed	
		Long leys, grazed or
		cut and grazed,
		high N^2
Leys (1–2 year),	Leys (1–2 year),	Permanent pasture, cut
grazed or cut and	grazed or cut and	only, grazed or cut
grazed, low N^1	grazed, high N^2	and grazed
Leys (1–2 year),	Long leys, cut only	
cut only		
Permanent pasture,	Long leys, grazed	
poor quality, matted	or cut and grazed,	
	low N^1	

From HMSO Reference Book No. 209 (1988)
[1]Low N = less than 250 kg/ha N per year or low clover content.
[2]High N = more than 250 kg/ha N per year or hbigh clover content.

Table 1.11 Nitrogen Index — following lucerne, long leys and permanent pasture

	1st crop	2nd crop	3rd crop	4th crop	5th crop
Lucerne	2	2	1	0	0
Long leys, cut only	1	1	0	0	0
Long leys, grazed or cut and grazed, low N^1	1	1	0	0	0
Long leys, grazed or cut and grazed, high N^2	2	2	1	0	0
Permanent pasture poor quality, matted	0	0	0	0	0
Permanent pasture, cut only, grazed only or cut and grazed, low N^1	2	2	1	1	0
Permanent pasture, grazed or cut and grazed, high N^2	2	2	1	1	1

[1]Low N = less than 250 kg/ha N per year and low clover content.
[2]High N = more than 250 kg/ha N per year or high clover content.

Table 1.12 Soil indices for phosphorus, potassium and magnesium

Index	Phosphorus (mg/litre)	Potassium (mg/litre)	Magnesium (mg/litre)
0	0– 9	0– 60	0– 25
1	10– 15	61– 120	51– 100
2	16– 25	121– 240	51– 100
3	26– 45	245– 400	101– 175
4	46– 70	405– 600	176– 250
5	71–100	605– 900	255– 350
6	101–140	905–1500	355– 600
7	141–200	1510–2400	610–1000
8	205–280	2410–3600	1010–1500
9	over 280	over 3600	over 1500

From *Fertiliser Recommendations* publication RB 209 MAFF 5th Edition (1988). London: HMSO with permission.

Phosphorus, potassium and magnesium indices

Indices for these elements are derived from analytical data as shown in *Table 1.12*.

Interpretation of results of soil micronutrient analyses

Soil analysis may be used to predict likely deficiency of boron, molybdenum and copper in crops. The data in *Table 1.8* is helpful in interpreting these results. Soil analysis may also be used to assist in diagnosis of deficiencies of copper and cobalt in grazing livestock, but the interpretation of the results obtained is not straightforward.

Levels of manganese (Mn) found by analysis of soil do not correlate with Mn uptake by crops. Soil analysis for Mn is of no use to farmers, since results obtained are incapable of interpretation.

PHYSICAL PROPERTIES OF THE SOIL

Soil physical properties affect crops as follows:

(1) directly – soil structure, porosity, aeration, moisture holding capacity are part of the root enviornment;
(2) indirectly – since soil physical properties determine suitability and timing of cultivations.

Soil structure, air and water

Soil structure

The term 'soil structure' relates to the arrangement of 'primary particles' of sand, silt and clay into ordered units. Some examples of types of structural units (peds) are illustrated in *Figure 1.19*.

Two types of structure may be regarded as end members of a continuum.

Single grain structure (of sandy soils) is exemplified by dune sand and the bulk of the soil is not organised in any perceptible manner. Roots may grow through the intergranular spaces and these spaces afford adequate aeration. Water holding capacity is very low, since the only zones where water can be retained against gravity are adjacent to points of contact of the individual grains.

Massive structure is the opposite extreme. In this case the soil constituents are tightly packed together to form a continuous mass with no visible planes of weakness. Soils of widely differing textural composition are capable of assuming a massive structure, usually in response to ill-timed trafficking or cultivation.

Soils with massive structure do not allow development of root systems, and although they may contain considerable water within their small pore spaces it is very tightly held by capillary forces and not available to plants (*see* p. 37).

Granular, subangular blocky, angular blocky and prismatic structures represent the middle range of the continuum and have features conducive to biological activity and crop growth. These structures are further subdivided on the basis of size (*see Table 1.13*).

The stability of structure is an important agricultural property. The ideal soil for crop production has a highly stable structure not easily destroyed by cultivation, water movement or treading by livestock. Structural stability is favoured by high organic matter levels, loamy or clayey soil texture, and high levels of calcium. Silty soils with low levels of organic matter and particularly unstable.

Adequate aeration with rapid drainage through the large spaces between the 'peds', (the name for the individual structural unit) and water retention in the smaller voids, is provided by a fine granular structure. Since roots are readily able to explore the spaces between peds they extract both water and nutrients from the whole volume of the soil.

Aeration, drainage and ability to provide water and nutrients to plants generally decrease as structural units become larger, and the regular-faced angular blocky and prismatic units are less favourable to root activity than the irregular surfaces of granular peds.

Structure of subsoils

Most crops in the UK remove water from the soil to a depth of at least 1 m, and the deeper-rooted ones noted for drought resistance (e.g. lucerne), to 2 m or more. It is thus important that the subsoil provides a satisfactory rooting environment. Adequate cracks and fissures must be present for roots to grow through, and to allow free movement of water. Many subsoils in their natural states have prismatic structures which are perfectly adequate, provided that the spaces between the prisms are large enough.

Usually, in subsoils, the faces of one ped more or less fit those of its neighbours. When the fit is good, changes in the width of fissures between peds, as a result of shrinking and swelling, become important.

Narrow fissures may close completely on swelling causing impeded drainage. This is evidenced by polished ped surfaces indicating contact between adjacent peds, and the mottled colours associated with drainage impedence.

Subangular and columnar structures are frequently of significance, since the absence of the more pronounced angular features of prismatic and blocky peds may indicate a degree of instability, and the possibility of displaced material flowing between the peds and causing serious drainage impedence. A transition from angular blocky or prismatic to this more rounded type of ped is usually indicative of

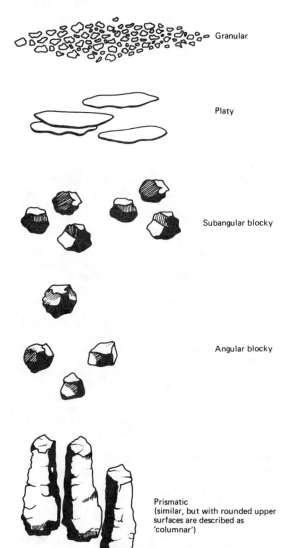

Granular

Platy

Subangular blocky

Angular blocky

Prismatic
(similar, but with rounded upper surfaces are described as 'columnar')

Figure 1.19 Some examples of common types of structural units (*after Modern Farming and the Soil* (1970) Agricultural Advisory Council. London: MAFF, reproduced with permission)

Table 1.13 Size classification of soil structural units (after SSEW, 1974)

	Platy[1]	Prismatic[2]	Angular blocky[3]	Subangular blocky[3]	Granular[3]
Fine	Fine platy; < 2 mm	Fine prismatic; < 20 mm	Fine angular blocky; < 10 mm	Fine subangular blocky; < 10 mm	Fine granular; < 2 mm
Medium	Medium platy; 2 to 5 mm	Medium prismatic; 20 to 50 mm	Medium angular blocky; 10 to 20 mm	Medium subangular blocky; 10 to 20 mm	Medium granular; 2 to 5 mm
Coarse	Coarse platy; 5 to 10 mm	Coarse prismatic; 50 to 100 mm	Coarse angular blocky; 20 to 50 mm	Coarse subangular blocky; 20 to 50 mm	Coarse granular; 5 to 10 mm
Very coarse	Very coarse platy; > 10 mm	Very coarse prismatic; > 100 mm	Very coarse angular blocky; > 50 mm	Very coarse subangular blocky; > 50 mm	Very coarse granular; 10 mm

From Soil Survey Technical Monograph No 5 *Soil Survey Field Handbook* (by Hodgson J.M.) Harpenden 1974 (Table 3, p. 31) by courtesy of the Soil Survey of England and Wales, Rothamsted Experimental Station, Harpenden
[1]Thickness of plate.
[2]Diameter in the horizontal plane.
[3]Diameter.

transient waterlogging and the cause should be sought and rectified, since the effect is a self-perpetuating one.

Platy structures are particularly detrimental, since roots can only pass through them by very tortuous paths, and they restrict water movement. In arable soils a platy layer can quickly develop into a serious pan, since soil above the platy layer will be subject to transient waterlogging and is therefore likely to 'slake' into its constituent fine particles. These move downwards with drainage water and progressively build up a massive impermeable layer.

Platy structures are rare in unfarmed soils as they are formed by loading with machinery, or the smearing action of cultivation equipment.

Timely and correctly executed cultivations should always lead to an overall improvement of the structural state of a soil profile.

Recognition of poor soil structure

Soil conditions can only be studied by examination of a soil profile. A drainage trench or a pit purposely excavated by machine provide ideal opportunities for soil profile inspection. Much additional information is often obtained by using a spade to dig a rectangular pit to a depth of 45 cm, or deeper if interest centres on possible 'natural' subsoil conditions, as well as those produced by agricultural practices. In the writer's experience this is best done during the months of April and May when it is possible to follow the distribution of roots of actively growing crops.

The following points should be observed.

Poor or weedy patches in the crop, often accompanied by a hard and rutted or very cloddy surface layer, indicate structural problems.

Overgrown, blocked or infilled ditches indicate that the problem may be primarily due to lack of maintenance of an artificial drainage system.

The plough layer should be friable and of uniform colour. On arable land, rusty staining on root channels indicates waterlogging (some rusty staining is normal under grass on heavy soils) and mottling (colour variation) on structure faces indicates transient waterlogging. Man- (or machine-) made clods are very dense, not able to be explored by roots

and they do not subdivide along natural planes of weakness. Their occurrence is clear evidence of inappropriate or ill-timed cultivation.

The base of the plough layer should merge gradually into subsoil. Any abrupt change is easily identified with the spade and a platy structured layer, or plough pan, is recognised as 'difficult digging'. Such layers are best examined by gently disturbing the soil in an exposed profile, using a penknife. The stronger platy layer will then stand out from the rest, and be clearly visible.

The subsoil should have adequate large pores and fissures. In heavy textured soils, ped faces should be examined for evidence of polishing due to adjacent units rubbing together. Mottled colours on structural surfaces indicate transient waterlogging. When a ped is broken open by hand, the exposed internal surface should be of a similar colour to the structure face and either living roots, or the channels of dead roots and worms should be visible. A uniform grey internal colour and absence of visible channels is indicative of water-logged conditions. Beware of relying on colour criteria when examining soils developed on strongly coloured parent materials such as Kauper marl.

Symptoms of waterlogging may be due to either a soil structural problem or the need for artificial drainage, and it is often difficult to identify the primary cause. If symptoms are found within 50 cm of the surface, it is always worth examining the subsoil to a greater depth; if no symptoms are found in any deeper horizon the problem is almost certainly one of soil structure.

Soil air

That part of the total soil porosity which is not occupied by water contains the soil atmosphere. This differs from 'air' in two main ways.

(1) Its carbon dioxide level is many times higher.
(2) Its composition varies.

The magnitude of the variation of composition of soil air is shown in *Figure 1.20* which gives data obtained from a

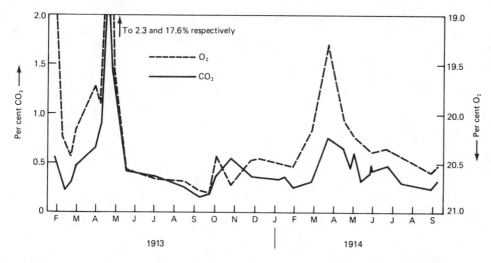

Figure 1.20 The oxygen and carbon dioxide content of the soil air in the dunged plot of Broadbalk under wheat (after Russell, E. W. (1973) *Soil Conditions and Plant Growth*, p. 411. London: Longmans, reproduced with permission of the author and publishers)

dunged wheat plot on the classic Broadbalk field at Rothamsted in 1913 and 1914.

The importance of oxygen in the soil

Most of the soil micro-organisms which are beneficial to agricultural crops (e.g. those involved in organic matter production, and the bacteria of the nitrogen cycle) require aerobic conditions. When conditions are anaerobic (lacking oxygen) the following occur:

(1) normal oxidative decomposition of carbohydrate gives way to fermentation and putrefaction. Organic acids and ethylene may accumulate and seriously damage crop roots,
(2) denitrification occurs. Nitrate (NO_3) is reduced microbially to N_2O or N_2 which are lost to the atmosphere (*see* p. 22),
(3) other inorganic species are reduced (*see* p. 18).

Aeration and soil structure

The main mechanisms whereby oxygen is supplied to sites of biological activity in the soil, and carbon dioxide removed, is diffusion through soil pores. Diffusion is restricted when:

(1) pores are few in number,
(2) pores are small,
(3) pores are very tortuous or not continuous,
(4) pores are filled with water.

In arable soils particularly, such limitations on diffusion often ensure that the internal zones of soil peds remain anaerobic. Since such zones are not normally explored by roots this may be of little consequence, apart from being a waste of rooting volume. However, any nitrate in such anaerobic zones is likely to be lost by denitrification.

Cultivations are often undertaken 'to improve soil aeration'. Whilst any movement of soil clearly increases the flow of air to the surface of structural units that are disturbed, cultivations frequently do nothing to improve air

movement *within* structural units, and problems may be exacerbated by the production of dense impermeable clods.

The major means that a farmer has at his disposal for improving soil aeration is the improvement of drainage.

The use of water by crops, soil water reserves and irrigation

Water and crop growth

In order to 'fix' atmospheric carbon by photosynthesis, plants need to maintain their stomata in the open position so that carbon dioxide gas can be taken up. As a consequence, water vapour is lost by diffusion in the opposite direction. Since water is vital to a plant's well-being, in times of stress it protects itself from further water loss by closing stomata. As well as preventing further (potentially catastrophic) loss of water, this stops photosynthesis. Inhibition of photosynthesis through drought normally occurs during bright sunny weather when the potential for photosynthesis is at its greatest, so crop loss may be considerable.

The efficiency with which crops utilise water depends on many factors and for grass it ranges from 1 kg of dry matter produced per 1000 kg of water transpired (low N input) to 1 kg per 400 kg of water with high nitrogen use. Efficiencies quoted for other UK crops generally fall within this range.

Water in the soil

Unless a soil is so wet that water will drain freely from it by gravity, water may only be removed from the soil by the expenditure of energy. Work must be done to remove water, and the drier the soil the more work must be done to remove a given quantity of water. Conversely, when water is added to dry soil it is distributed within the soil so as to maximise its attraction to the soil; the system achieves a maximum reduction of free energy, and this may be measured as the soil water potential, ψ.

Table 1.14 Water potential, hydraulic head and pF of a soil of varying wetness (after White, 1979)

Soil moisture condition	Water potential (ψ) (bars)	Head (m)	pF	Equivalent radius of largest pores that would just hold water (μm)
Moisture held after free drainage (field capacity)	−0.05	0.51	1.7	30
Approximately the moisture content at which plants wilt	−15	153	4.2	0.1
Soil at equilibrium with a relative vapour pressure of 0.85 (approaching air-dryness)	−220	2244	5.4	0.007

From White, R.E. *Introduction to the Principles and Practice of Soil Science* (1969) p. 68, by courtesy of Blackwell Scientific Publications, Oxford

Table 1.15 Average available water (Av) per cent for mineral soils in relation to horizon, particle-size class and packing density

Particle-size class[3]	Horizon	Av(%)[1] for different packing density classes[2]		
		Low (< 1.40 g/cm^3)	Medium ($1.40–1.75$ g/cm^3)	High (> 1.75 g/cm^3)
Clay	A	22	18	(19)
	E, B, C	(19)	15	13
Sandy clay	A	—	—	—
	E, B, C	—	—	14
Silty clay	A	23	17	—
	E, B, C	—	16	12
Sandy clay loam	A	(25)	17	—
	E, B, C	—	17	16
Clay loam	A	25	20	(17)
	E, B, C	19	15	12
Silty clay loam	A	27	20	—
	E, B, C	21	17	12
Silt loam	A	—	(25)	—
	E, B, C	—	(22)	—
Sandy silt loam	A	23	21	—
	E, B, C	20	18[4]	—
Sandy loam	A	20	16[4]	(19)
	E, B, C	(20)	15[4]	11[4]
Loamy sand	A	13	14	—
	E, B, C	(16)	12[4]	(8)
Sand	A	—	—	—
	E, B, C	—	9[4]	(4)

From Soil Survey Technical monograph No 9 *Water retention, porosity and density of field soils*, by D. G. M. Hall, M .J. Reeve, A. J. Thomasson and V. F. Wright, by courtesy of The Soil Survey of England and Wales, Rothamsted Experimental Station, Harpenden.

[1] 10 mm/m = 1%.

[2] Packing density is derived from bulk density as follows: Packing density = Bulk density + 0.009 (% clay). Packing density is considered easier to estimate in the field than bulk density.

[3] Particle-size classes are determined by use of the triangular diagram of *Figure 1.4*.

[4] If mainly fine sand add 2%: if the majority of the fine sand is of 60–100 μm grade Av will be 20% except for sandy silt loam (about 30%).

Brackets ( ) indicate limited data.

A dash indicates insufficient information.

The soil water potential (ψ)

At any point in the soil this may be defined as the energy required to transfer to that point, 1 mol of water from a free liquid state at the soil surface, and free from dissolved solutes. Since such a transfer would release energy, ψ is always negative.

Units of measurement

The most appropriate units for ψ, on grounds of theoretical chemistry, would be joules per mole of water (J/mol). In practice the normal unit is J/m^3 since values expressed in this unit are numerically equal to pressures expressed in Newtons per square metre (N/m^2). Any pressure or suction may be expressed in terms of the height of a water column, usually measured in cm. For soil suctions it is traditional to avoid large numbers by use of the term pF when $pF = \log_{10}h$ (h measured in cm). A convenient unit for measurement is the bar $(10^5 N/m^2)$ which approximates to the pressure exerted by a column of water 10 m high. The equivalence of the various units is indicated in *Table 1.14*.

When water is progressively added to dry soil, initial increments are adsorbed onto soil colloid surfaces (*see* p. 14). Subsequent additions are held in the soil pores by forces of attraction. The finest pores are filled first and coarser ones at later stages. The equivalent radius of largest pores that would just hold water are given in *Table 1.14*.

Field capacity is the moisture state at which free drainage just ceases following thorough wetting of the soil. The concept is strictly applicable only to freely draining soils.

Plant available water has a potential between -0.05 bars and -15 bars (field capacity – permanent wilting point).

Available water capacity (AWC) is a soil's capacity to store plant available water.

It may be expressed as a percentage of soil volume (Av%) or, more usefully when it is intended to perform computations involving both available water and rainfall – mm/m depth of soil (mm/30 cm is also used since 30 cm represents a reasonable depth of plough layer). To determine the AWC of a soil profile it is necessary to sum the individual values for identified soil horizons to the maximum depth of rooting. *Table 1.15* provides all the necessary information. Typical values for common soil series are given in *Table 1.16*.

Water uptake by crops

Water is not removed from soils in a uniform manner. During spring growth uptake is normally from the surface horizon, and appreciable amounts of water start to be removed by roots at greater depth as the work done in extracting the water from surface soil increases. Where a water table exists within the depth of a profile, root systems may use this water, provided that they can get near enough to it. A water table is of no use to crops if its depth varies

Table 1.16 Mean values of available water for some common soil series

Soil series	Depth of integration (to 100 cm or to rock if shallower) (cm)	Total available water (0.05–15 bar) (mm)	Easily available water (0.05–2 bar) (mm)
Aberford	60	110	70
Ardington	100	150	100
Bardsey	100	130	80
Bridgnorth	80	85–105[1]	65–90[1]
Bromsgrove	80	105	75
Clifton	100	150	100
Crewe	100	140	75
Denchworth	100	160	82
Eardiston	60	105	70
Elmton	30	65	40
Flint	100	120	70
Gresham	100	140	80
Marcham	35	45	30
Newport	100	75–150[1]	65–130[1]
Oak	100	130	65
Ragdale	100	130	75
Ross	100	155	110
Salop	100	135	80
Sherborne	35	55	30
Spetchley	100	120	65
Tickenham	100	165	120
Urchfont	100	195	160
Wick-Arrow-Quorndon	100	100–200[1]	70–160[1]
Wilcocks	100	215	135
Worcester	100	115	60

From Soil Survey Technical Monograph No 9 *Water retention, porosity and density of field soils:* Appendix II (p. 75) (1977) by D. G. M. Hall, M. J. Reeve, A. J. Thomasson and V. F. Wright by courtesy of The Soil Survey of England and Wales, Rothamsted Experimental Station, Harpenden
[1]Varies greatly according to content of 60–100 μm sand and surface organic matter levels.

during the growing season in such a way as to cause transient waterlogging within the rooting zone.

The ease of uptake of water is governed by the water potential, ψ. Water in coarse pores in the topsoil is more readily available than water held tightly by fine pores deep in the subsoil.

Potential transpiration (PT), is the amount of water lost from a complete canopy of a uniform green crop adequately supplied with water. It is not easily measured but may be calculated from readily available meterological data. The method for doing this is described in MAFF (1967). Monthly values for England and Wales are tabulated in MAFF (1976) and range from 0 (January in NE England) to 100 mm + (June and July in SE England).

Current values are calculated daily by the UK Met. Office and published as weekly summaries.

Actual transpiration is less easily determined, since a real crop may transpire either at the potential rate, or at some lesser rate due to:

(1) incomplete crop cover,
(2) crop water use reduced by water stress.

Calculation of irrigation need

Planning an irrigation programme involves calculating how much water to apply and when to apply it. A number of different systems, mostly computerised, are available for doing this. All rely on the construction of a water balance but they differ in the detailed treatment of individual factors in the balance.

The important factors are as follows.

The amount of water stored in the soil at the outset. For simplicity this is recorded as a soil moisture deficit, SMD (i.e.

the amount already lost). Initially SMD may be taken as zero (i.e. soil at field capacity), or some other value calculated from historic data. Note that undisturbed bare soils are unable to transpire more than 25–30 mm of water, irrespective of weather conditions.

A record of rainfall and irrigation inputs of water. Rainfall *must* be measured 'on site' – figures obtained as near as 1 km away can lead to major errors. Similarly, adequate metering of irrigation water is essential, since errors are cumulative throughout the season.

A figure for potential evaporation is needed in order to know how soon an SMD will reach some critical value if irrigation is not applied. In practice two figures are needed:

(1) An average daily rate of potential transpiration. This is taken from MAFF (1976). Since potential transpiration values change only gradually from place to place and do not vary excessively between successive dry days, this figure may be used for daily calculations.
(2) At the end of a 7-d period a weekly potential transpiration figure is provided by ADAS and the UK Meteorological Office, based on daily climatological measurements. This figure is used to apply a correction to the previous 7-d calculations.

Crop requirement

Detailed recommendation for various crops will be found in ADAS (1979). Differing programmes are given for soils of low, medium and high AWC (less than 60 mm, 670–100 mm and greater than 100 mm within 500 mm depth of soil, respectively).

When all the above information is to hand a water balance may be constructed, without the aid of a computer, by daily updating of a simple graph and the procedures involved

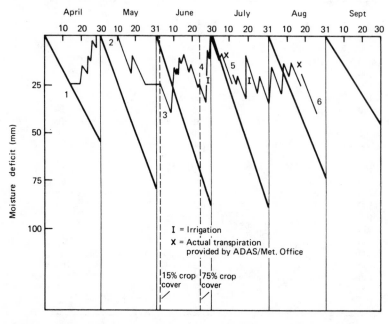

Figure 1.21 Soil water balance for maincrop potatoes grown in a class B soil in South Devon (Agroclimatic area 43S)

provide a useful insight into the working of the more sophisticated computerised techniques now available.

A sheet of squared paper is prepared as in *Figure 1.21* by
ruling lines to represent the long-term average PT for each
month. Each dry day the line is extended parallel to the
pre-ruled line, and on wet days rainfall is entered as a
decrease in SMD. Until the crop cover reaches 15% water
loss from bare ground sets an upper limit of 25 mm SMD.
Between 15 and 75% crop cover record keeping is exactly as
described above (irrigation is counted as rainfall). After 75%
ground cover is attained the long-term average PT figures
cease to be sufficiently reliable and each week the calculations should, if necessary, be corrected to allow for any
discrepancy between the *previous week's* ADAS/Met. Office
PT figure and the long-term average. (Note that this
correction must be done retrospectively since a very large
SMD could build up before the Met. office figure becomes
available.) Note that in these calculations it is assumed that
irrigation will be applied when needed. If a large SMD is
allowed to build up transpiration rates need to be adjusted
to allow for lower transpiration by crops under stress.

Figure 1.21 gives an example of a water balance record
using fictitious data designed to illustrate the various steps
involved. The crop was potatoes (maincrop) grown on a
Class B (medium AWC) soil in South Devon (Agroclimatic
Area 43S). A 15% crop cover was attained on 2 June and
75% on 24 June. The irrigation plan was 25 mm applied at
a 35 mm SMD. The following notes are numbered to
correspond with *Figure 1.21*:

(1) Although the first 20 d in April were dry the SMD did
    not exceed 25 mm since this figure represents the
    maximum loss under bare ground conditions.
(2) SMD returned to zero on 28 April. When this occurs it
    can only depart from zero following a day in which PT
    exceeds rainfall.
(3) An irrigation requirement arose in the period 3–9 June,
    but no water was applied since the crop was not thought
    to have reached a responsive stage.
(4) Irrigation applied 27 June. Note that 10 mm SMD
    remained unsatisfied. If the entire irrigation need has
    been applied, the rainfall on 29 June would have been
    wasted.
(5) On 13 July information was received that the actual PT
    for the previous week was 5 mm less than average. An
    adjustment was made on 13 July to compensate. A
    similar situation occurred on 16 August.
(6) The planned deficit was reached during the last week in
    August, but irrigation was delayed since the crop was
    near to harvest and rainfall following irrigation could
    have led to difficult harvest conditions.

An example of a computerised irrigation program is the
ADAS 'Irriguide' operated by ADAS in conjunction with
the Meteorological Office. This system has a number of
useful refinements:

(1) It uses three day weather forecast information to modify
    recommendations in the light of expected rainfall/trans-
    piration.
(2) It can provide individual recommendations for
    sequential sowings/plantings on the same land.
(3) Allowance is made for additional water loss consequent
    upon any 'weed control' or other cultivations.

Soil and cultivations

Traditionally there have been many reasons for cultivating
soils, with weed control and burial of trash amongst the most
important. Now that herbicides can take care of much of the
weed problem, the reasons for continuing to cultivate are to
do with soil structure, straw disposal and rooting conditions.
Cultivation should normally:

(1) reduce the bulk density (mass per unit volume) of soil,
    thereby increasing the proportion of pore spaces (par-
    ticularly large pores) able to contribute to water
    movement and aeration;
(2) break up any soil pans allowing greater root penetration;
(3) break up clods to form a fine tilth. This increases the
    proportion of the soil volume which can be explored by
    roots.

The success or otherwise of cultivation operations is
strongly influenced by soil moisture content.

The lower plastic limit is the minimum moisture content at
which puddling is possible and the maximum moisture at
which the soil is friable. It is defined as the moisture content
at which the soil can be rolled into a 'worm' about 3 mm
diameter without breaking. The optimal working range for
most implements is below the lower plastic limit. Smearing
and soil compaction (loss of structure) will occur if cultivations are carried out when the soil is too wet.

It is important to estimate the working properties of soils
at the depth at which an implement will work. Surface soils
dry very rapidly in windy weather, but below plough depth
the subsoil is likely to remain too moist to support the weight
of a tractor without compacting.

Primary cultivations

Ploughing inverts the soil and buries trash. In so doing the
surface structure is disrupted and under good conditions the
ploughed soil may need little additional disturbance to form
a seedbed.

When soil is too wet, the furrow slice is smeared by the
mouldboard and clods are formed. The tractor wheel
running in the furrow bottom causes smearing and
compaction. On heavy soils the use of crawler tractors
running on unploughed ground avoids this problem.

Ploughing year after year at the same depth will eventually
cause soil compaction in most soils.

Use of special shallow ploughs working to only 12–20 cm
depth can achieve burial of trash without such a great risk of
causing soil structural damage under poor conditions.

Powered digging machines are able to incorporate straw
under moist conditions where ploughing would not be
desirable.

Chisel ploughs and tine cultivators disturb the upper layer
of soil without inversion and burial of trash. The action of
tines has been studied by Stafford (1979). As with other
cultivation equipment, the action in any particular soil
depends on moisture content. Working at excessive depths in
soils which are too wet produces clods which are difficult to
break down by subsequent secondary cultivations.

Secondary cultivations

Following the primary cultivation, some secondary
operation is usually needed to produce a seedbed. Such
secondary cultivations should be kept to the bare minimum
since even when no compactive forces are produced by the

implement, frequent tracking by a tractor quickly reduces the beneficial effects of primary cultivations. On soils other than clays a furrow press will give initial consolidation after the plough and reduce or eliminate the need for further secondary cultivations.

Harrows and tine cultivators used at shallow depths break down the coarse structural units left by primary cultivations.

Disc harrows and some rollers are used for the same purpose but involve more drastic disruptive forces. These implements have a compactive effect on soil and should not be used unnecessarily. However, the extra compaction produced by these implements may be desirable where puffy seed beds are a problem.

Direct drilling

Where the soil structural conditions following harvest of an above ground crop (i.e. not potatoes, swedes, sugar beet, etc.) remain adequate to support a further crop, no cultivation is necessary and the seed may be sown by 'direct drilling'. The technique has several advantages.

Work rates are high, so there is no temptation to work under adverse conditions and crops can be established rapidly.

The risks of causing soil structural damage due to cultivation when soils are too wet are removed.

Fuel use is reduced.

Winter cereal crops are particularly suited to establishment by direct drilling. Not all soils give consistently reliable results when direct drilled. Cannell *et al.* (1978) have classified soils according to their suitability for sequential direct drilling.

Direct drilling is not possible where cereal straw remains in the field. Whilst it is possible to direct drill successfully into cereal stubble after the straw has been carted off, this usually involves undesirable delay and the heavy vehicles used frequently damage soil structure so reducing the success of the direct drilling operation. From the agronomic viewpoint burning straw in the field is the best preparation for direct drilling, but burning is now considered undesirable in many situations. When straw is burnt, the provisions of the Straw Burning Code must be observed fully.

Disturbance of the subsoil

Subsoiling is undertaken to disrupt the subsoil to a greater depth than achieved during normal cultivations. Compacted layers or pans may be shattered and water movement and root penetration greatly improved. Spoor and Godwin (1978) have studied the working of subsoilers and shown that in a particular soil any implement has a critical working depth. Working at a shallower depth, subsoiling is achieved, whereas below the critical depth upward movement of soil is prevented and compaction occurs along with the formation of a channel, sometimes called a 'square mole'. The correct subsoiling action can be achieved when working at depth, by loosening the surface soil beforehand (this may be by means of tines immediately preceding the subsoiler). Such loosening has the effect of increasing the critical depth. In general terms subsoilers work most satisfactorily in dry soils (critical depth is influenced by soil plasticity).

Mole drainage is really a drainage treatment but is frequently confused with subsoiling. When mole draining, the mole plough should be operated at a depth exceeding the critical depth for subsoiling, so that a channel is formed without appreciable upward heave.

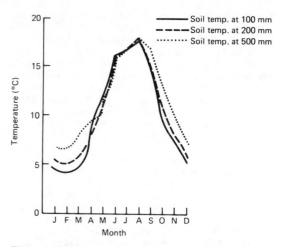

Figure 1.22 Average underground temperatures at 100, 200 and 500 mm (after Dougall, B.M.)

Soil temperature

With the important exception of the effect of drainage, soil temperature is beyond the control of the farmer for practical purposes.

The effect of land drainage on soil temperature

There are two reasons why wet land warms up slowly in spring:

(1) heat that would raise the temperature of a dry soil is used (as latent heat) in evaporating water from the surface of a wet soil;
(2) the specific heat of water is high and additional heat is required to raise the temperature of the water, over and above that which would be needed to warm a similar dry soil.

Heavy land which has been artificially drained is frequently up to two weeks 'earlier' in spring than its undrained counterpart.

Variations in soil temperature

Through the year mean temperatures of surface soils follow a cyclical pattern roughly in phase with increasing solar radiation. At depth the magnitude of the cycle is reduced and lag occurs in the timing of maxima and minima.

The nature of the surface affects the reflection of incoming radiation and hence temperatures of the surface soil. Temperatures are lower under vegetation than bare soils. In experiments, darkened soil surfaces (spread with soot) warm up more rapidly than where lime is used to whiten the surface. This effect is illustrated in *Figure 1.22*.

SOIL SURVEY AND LAND CLASSIFICATION

Soil survey

It is common experience that soils differ from one another.

Table 1.17 SSEW* published data on a selection of topsoils

Soil series	Wye Downs	Malling	Batcombe	Denchworth	Highweek
Topsoil texture	SL to SCL	L	ZL to ZyCL	ZyCL to C	CL
Classification	Brown calcareous soil	Brown earth	Gleyed brown earth	Surface-water, gley soil	Brown earth
Analyses (%)					
Sand ⎧ 500 μm–2 mm	1.2	2.5	3.3	⎫ 6.4	9
⎪ 200–500 μm	8.3	16.2	8.7	⎭	8
⎨ 100–200 μm	51.3	28.5	4.8	⎫ 5.2	10
⎩ 50–100 μm	9.3	4.4	5.9	⎭	
Silt 2–50 μm	12.3	28.5	52.8	51.6	46
Clay > 2 μm	17.7	20.5	24.5	36.2	27
Loss on ignition	8.0	4.3	9.1	10.2	—
$CaCO_3$ equivalent %	0.6	Nil	Nil	—	—
Organic carbon	—	1.3	—	3.1	3.1
pH in water	7.2	6.1	6.1	5.3	7.0
pH in $CaCl_2$	7.0	5.6	6.1	4.6	6.9
CEC	—	15.3	—	22.6	31.6
Depth of 'A' horizon (cm)	7	23	20	19	—
Available water in 'A' horizon (cm)	1.3	4.1	4.6	4.6	—

— = Data not available.

It is thus to be expected that our understanding of them will be enhanced by systematic classification.

Soil series

All classifications of soil are founded on the soil surveyor's perception of the 'unit of soil', referred to as the 'pedon' in the United States Department of Agriculture (USDA) nomenclature. This is the soil as observed in a single pit, usually 1 m square and up to 1.5 m depth (shallower if parent material is encountered at a lesser depth). A second 'pedon' would certainly be different from the first, but it is possible to define a 'soil series' such that the pedons comprising it have sufficient in common to be thought of as a homogeneous entity.

Producing a soil map

Depending on the local complexity of soils in the landscape and the scale of mapping, the soil surveyor may use the series as 'mapping unit'; alternatively he may be unable to distinguish individual series at the particular scale of mapping, and so use a 'complex mapping unit'. In either case, the boundaries of the mapping units must be located – these are what actually form the map. Two methods may be used.

(1) *Free survey* involves the soil surveyor using his knowledge of the relationship of soil with landscape features to guide him to areas where boundaries are expected. He then employs his efforts in making auger borings around such areas to obtain clearly defined boundaries. Areas that the surveyor considers to be uniform are given relatively less attention. Where work is undertaken by skilled surveyors, this approach gives quite accurate results and work rates are high. There is always the danger of a change in soil type not related to visible features passing undetected, and very little information is obtained about the degree of purity of the mapping units. (High purity implies a uniform mapping unit with few variations or inclusions of different soils).

(2) *Grid survey* as the name implies, involves collecting data (usually soil descriptions based on auger borings) at systematically chosen points (grid line intersections) throughout the area to be mapped. Mapping units are subsequently identified, and boundaries inserted on the map so as to group like with like and delineate differences. This type of survey does not demand the employment of highly trained staff on field work, but speed of mapping (per person employed) is less than using free survey. It has the considerable advantage that the field records may be retained and consulted at a later date, or entered into a computer store to facilitate use of numerical classification techniques (see for example Webster, 1978).

Soil survey publications

The usefulness of any soil survey is greatly enhanced if it is accompanied by text giving background information. The Soil Survey of England and Wales (SSEW)* publish 'Memoirs' and 'Records'. These provide a description of the methods used by the surveyor, major features of the landscape and much else relevant to the particular locality. All contain a number of specimen soil profile descriptions and analytical data. SSEW data relating to topsoils of a variety of soil series is reproduced in *Table 1.17*.

Soil classification

As an aid to understanding differences between soils, or to facilitate mapping at small scales (e.g. 1 : 1 000 000 soil map of England and Wales) a system of soil classification is needed. This may be based on:

(1) inferred processes of soil formation (genetic classification),
(2) observable or easily detectable attributes (morphological classification),

*Soil Survey of England and Wales: As of 1 July 1987 is the Soil Survey and Land Resource Centre of Cranfield Institute of Technology.

(3) suitability for one or more specific purposes (suitability classification).

The soil series is normally the lowest unit of classification and the higher units, groups of series, have different names in the various classification systems. In England and Wales the classification is into subgroups, groups, and major groups.

Genetic classification was used almost exclusively in Europe and the UK until the 1960s. Such systems have the considerable advantage of focusing attention on the natural processes of soil formation.

Morphological systems have become more popular in recent years, because the need for rapid surveying of soils throughout the world has demanded systems which are precise, and do not rely on the surveyor's judgement to interpret his observations in terms of soil formation processes. There has been a change of emphasis to use 'observable or easily detectable attributes' – the philosophy basic to the USDA 'Comprehensive System' (Soil Survey Staff, 1975). There is general agreement amongst soil scientists that morphological systems are most likely to be of use if the diagnostic criteria are chosen so as to be related to processes of soil formation. Such choices of criteria will of course provide some links with genetic systems of classification. In the UK, the change to a morphological type of system has been made in a way that retains a large measure of consistency with previous surveys and at the same time facilitates easy reconciliation with similar systems in use elsewhere (Avery, 1973).

Suitability classification has frequently been a major part of soil survey work. Some of the earliest soil surveys in England were concerned with suitability for fruit cropping.

The United States Department of Agriculture (USDA) soil taxonomy (Soil Survey Staff, 1975)

This is probably the most completely developed system in the world, with some 10500 known series grouped into familes, subgroups, great groups, suborders and orders. It was developed by a process of publication, distribution to working soil scientists and modification in the light of comments received, so giving rise to successive 'approximations'. It was the first major morphological classification with soils assigned to the various series, and series assigned to the higher units of classification, on the basis of the occurrence of diagnostic horizons which may be recognised by the surveyor, using only techniques likely to be readily available to him in the field. This system has the virtue that, properly applied by any surveyor, it will provide the same unique classification of any soil.

The world soil classification

This classification was produced as an adjunct to the FAO–Unesco Soil Map of the World, published in 1974. Classification is on the basis of 'diagnostic horizons' as in the USDA 'Comprehensive System'. Higher order names are traditional ones with unambiguous, internationally accepted meanings, or newly coined ones which can be relied upon to translate without change of meaning, so the system is highly suitable for communicating soil information across national boundaries.

Soil classification in England and Wales

The classification used in England and Wales (by The Soil Survey of England and Wales) has been evolved over many years through successive refinements, based on the concepts of Robinson (1943). Since the early 1960s, further modifications have been made to change the basis of classification from genetic to morphological, without loss of continuity. The system used at present is described by Avery (1973).

In Scotland, soil survey work is undertaken by the Scottish Soil Survey and soil information in Northern Ireland is the responsiblity of the province's Department of Agriculture.

The concept of good land

It is a common misconception that the quality of land can be assessed by a chemist working on a soil sample at his laboratory bench. Nothing could be further from the truth. Most (but not all) chemical properties of soil can be modified by agricultural practice, usually at minimal expense. Physical properties are of prime importance to the farmer and cannot be changed readily.

The following are the attributes of good land:

A deep soil to facilitate adequate depth of rooting and storage of moisture. The best soils have 1 m or more of depth favourable to roots.

A satisfactory soil texture so that moisture retention is adequate, but cultivations are possible throughout much of the year.

Adequate drainage so that the soil may dry quickly and root exploration will not be restricted by anaerobic conditions.

Artificial drainage may be installed to alleviate problems but

(1) it is expensive,
(2) maintenance and eventual replacement is necessary,
(3) even the best artificial drainage is seldom an equal to natural good drainage.

A stable soil structure to facilitate arable cropping.
Freedom from risk of erosion.
A climate suitable for crop growth.
Level or with uniform gentle slope and relatively stone free, to enable machinery to be operated safely and efficiently.
Freedom from any concentration of noxious chemical.

Land classification

The aim of land classification is to provide an interpretation of the results of soil survey work, combined with knowledge of physical factors that affect land use. The aim is to locate 'good land' within the area, usually using some numerical scale that relates to 'degree of goodness'. Different classifications have different objectives. They are either general purpose or special purpose.

General purpose classifications use some generalised concept of 'good land'. The MAFF Agricultural Land Classification of England and Wales provides an example of a general purpose classification.

Special purpose classifications relate to specific soil

properties (frequently electrical conductivity in arid lands) or some proposed land use.

Land classification in England and Wales

The Agricultural Land Classification of England and Wales divides agricultural land into five main grades on the basis of the type of physical limitations to land use already discussed. Existing provisional maps at a scale of 1:63360 (1 inch to 1 mile) cover all agricultural land.

The *ADAS/Soil Survey of England and Wales Land Capability System* has been used in conjunction with soil surveys in some parts of the country. It has seven capability classes. These classes are further subdivided into subclasses on the basis of the limitations which exclude the land from higher classes.

For example, much of inland Devon and Cornwall (excluding the high moors and areas affected by poor drainage) is classified as 3c since only the climate (high rainfall and exposure, in this case) prevents classification in class 2. A further subdivision into capability units is made, such that all land in any one capability unit responds in a similar way to management and improved agricultural practices. It should be noted that this classification deals with *potential* land use, so relies on skilled judgement of likely responses to different levels of input. For example, an overgrown waterlogged area of alluvial soil may be quite worthless under present land use; its potential would be assessed assuming it had been drained if this were judged practicable, and class 2 could well be appropriate. In judging the level of input to assume (e.g. drainage in the above example) the appropriate yardstick is 'under a moderately high level of agricultural management'. The system has been described in detail by Bibby and Mackney (1969).

References

ADAS (1979). *Irrigation Guide*, Booklet 2067. Pinner, Middlesex: MAFF

ADAS (1987). *The Use of Sewage Sludge on Agricultural Land*, Booklet 2409. Alnwick, Northumberland: MAFF

AVERY, B. W. (1973). *Journal of Soil Science* **24**, 324–338

AVERY, B. W. and BASCOMBE, C. L. (1974). Soil Survey Laboratory Methods. *Soil Survey Technical Monograph No. 6*. Harpenden

BIBBY, J. S. and MACKNEY, D. (1969). Land Use Capability Classification. *Soil Survey Technical Monograph No 1*. Harpenden

CANNELL, R. Q., DAVIES, D. B., MACKNEY, D. and PIDGEON, J. D. (1978). *Outlook on Agriculture* **9**, 306–316

DOUGALL, B.M. (1976). *The Agricultural Notebook*, 16th Edn. London: Butterworths.

EHRLICH, P. R., ERHLICH, A.H. and HOLDEN, J.P. (1977). *Ecoscience*, San Francisco: W. H. Freeman & Co.

HANLEY, P. K. and MURPHY, M. D. (1970). Crop Response to Sulphur in Ireland. In *Sulphur in Agriculture*. Dublin: An Foras Talúntais

HUGHES, A. D. (1979). *Sampling of Soils, Soilless Growing Media, Crop Plants and Miscellaneous Substances for Chemical Analysis*. Booklet 2082. Pinner, Middlesex: MAFF

JENKINSON, D. S. (1966). *Journal of Soil Science* **17**, 280–302

LOCKHEAD, A. G. and CHASE, F. E. (1943). *Soil Science* **55**, 185

LOW, A. J. (1972). *Journal of Soil Science* **23**, 363–380

MAFF (1976). *The Agricultural Climate of England and Wales*. Technical Bulletin No 35. London: HMSO

MAFF (1986). *The Analysis of Agricultural Materials*. RB 427. London: HMSO

MAFF (1967). *Potential Transpiration*. Technical Bulletin No 16. London: HMSO

MURPHY, M. D. (1975). Sulphur on Grassland. *Soil Research Report*, 26–28. Dublin: An Foras Talúntais.

The Nitrogen Cycle of the United Kingdom (1983). London Royal Society

Pollution Paper Number 21 (1983). *Agriculture and Pollution*. The Government response to the Seventh Report. London: HMSO

POWLSON, D. S. (1980). *Journal of Soil Science* **31**, 77–85

REEVE, M. J., HALL, D. G. M. and BULLOCK, P. (1980). *Journal of Soil Science* **31**, 429–442

ROBINSON, G. W. (1943). *Discovery* **4**, 118–121

The Royal Commission on Environmental Pollution Seventh Report (1979). *Agriculture and Pollution*, Cmnd 7644. London: HMSO

RUSSELL, E. W. (1974). *Soil Conditions and Plant Growth*, 10th Edition. London: Longmans

SOIL SURVEY STAFF (USDA) (1975). *Soil Taxonomy Agricultural Handbook*. Washington: US Department of Agriculture No 436

SPOOR, G. and GODWIN, R. J. (1978). *Journal of Agricultural Engineering Research* **23**, 243–258

STAFFORD, J.V. (1979). *Journal of Agricultural Engineering Research* **24**, 41–56

Standing Technical Advisory Committee on Water Quality (1984). Fourth Biennial Report 1981–1983. DOE/National Water Council. London: HMSO

WEBSTER, R. (19978). *Journal of Soil Science* **29**, 389–402

Further reading and sources of information

There are a number of modern texts which have achieved the status of 'standard works' in soil science. Probably the most comprehensive is *Russell's Soil Conditions and Plant Growth* (11th Edn, edited by A. Wild, Longmans, London, 1988).

Chemical aspects are covered by D. J. Greenland and M.H.B. Hayes in *The Chemistry of Soil Constituents* (Wiley–Interscience, Chichester 1978), K. Mengel and E. A. Kirkby in *Principles of Plant Nutrition* (International Potash Institute, Berne 1982) and F. E. Bear in *Chemistry of the Soil* (American Chemical Society Monograph No 160, Reinhold Publishing Corp, New York 1964). The latter publication includes a description of methods for chemical analysis of soils. C. A. Black in *Methods of Soil Analysis, Agronomy No 9* (American Society of Agronomy, Madison USA 1965) deals with physical chemical and microbiological analysis of soil. The 'standard' chemical analyses used for soil science advisory work in UK are described in *The Analysis of Agricultural Materials* (MAFF RB 427, HMSO, London 1986).

A wide range of Government publications on agricultural subjects including many concerning soils are listed in the catalogues of MAFF publications issued periodically and available from MAFF (Lion House, Alnwick, Northumberland NE66 2PF).

Much of the interest to those concerned with soil structure and the general well-being of soil is to be found in *Modern Farming and the Soil – Report of the Agricultural Advisory Council on Soil Structure and Soil Fertility* (HMSO, London 1970). *Soil Management* by D. B. Davies, D. J. Eagle and J. B. Finney (Farming Press Ltd, Ipswich 1972) is a guide to the maintenance of overall soil fertility by farming practices.

Microbiological aspects of soils are dealt with by M. Alexander in *Introduction to Soil Microbiology* (John Wiley, London 1977).

A wide range of information on the properties and distribution of individual soils is provided by Soil Survey and Land Research Centre publications.

A large number of student texts are available, such as R. E. White's *Introduction to the Principles and Practice of Soil Science* (Blackwell Scientific, London 1987).

To various degrees, all the above have been consulted during the preparation of this section of 'The Notebook' and acknowledgement is due to the authors concerned.

2

Field drainage

R.J. Parkinson

THE NEED FOR DRAINAGE

The removal of excess water from the soil by an artificial drainage system is essential for many soils, particularly those with heavy textures. Rapid water disposal will minimise crop damage that could arise from high water tables. The most widely employed method of water table control in the UK is underdrainage, whereby a system of pipes are laid in the soil at such depths and spacings that the water table can be controlled for all but the most severe rainstorms. In developing countries, particularly where subsistence agriculture is practised, open ditches provide a cheaper and less sophisticated alternative to pipe drainage, but one that does not use the land available in the most efficient manner, and is not suitable for more mechanised agriculture. This chapter describes the methods of field drainage used in the UK, although the principles outlined are applicable to any heavy textured soil with a drainage problem.

Why drain?

Heavy textured soils, because of their ability to retain nutrients and water to satisfy the needs of a growing crop, are capable of producing high yields, particularly of cereals and grass. For example, the majority of 10 t/ha winter wheat crops have been obtained from clay loam or clay textured soils, but the agricultural productivity of these slowly permeable soils depends upon the rapid disposal of winter rainfall so that waterlogging does not occur.

Annual patterns of water loss and gain are very variable across the UK. As a result of higher rainfall and less sunshine, the soils in the west tend to pose more drainage problems than those in the east of the country. An indication of the difference in soil water regime can be obtained by comparing the duration of field capacity (*see* p. 37) across the UK (Thomasson, 1981). In soils which do not shrink and swell in response to changes in water content, drainage starts when autumn rainfall has recharged the soil such that it will store no more water. The soil is then said to be at field capacity or the winter mean water content. Drainflow may occur from clay soils before the winter mean water content

is reached because of the presence of cracks. The duration of the winter mean period, during which drainage will occur in response to rainfall, varies across the UK, as is shown in *Figure 2.1*. Although this map is generalised and takes no account of different soil types, it can be seen that many soils

Figure 2.1 Duration of winter mean water content period, in days (after Thomasson, 1981)

in south-west England will be at winter mean water contents for > 200 d, almost twice as many days as soils in East Anglia (< 125 d). In consequence, cropping patterns will be dictated by the weather, and cultivation will be difficult within the winter period unless soil water contents can be controlled.

Historical perspective

The use of ditches and drains in the UK to reduce soil water content probably dates from Roman times, but it was only after the large-scale enclosure of common lands occurred in the eighteenth century that drainage became widespread (Trafford, 1970). Techniques used varied widely, with most systems relying on stones, straw or other organic material to form and stabilise a channel to remove excess water. The mechanical production of clay tiles in the mid-nineteenth century led to dramatic improvements in the quality of drain pipes, and 100-year-old systems can still be found that are working well today. Estimates of areas drained in the nineteenth and early twentieth centuries are inevitably speculative, but significant areas of land, up to 50% of the total agricultural land in many parts of southern England, were drained, all by hand. Mechanical drain installation became the norm in the middle part of the twentieth century, with the annual area drained varying from 50000–100000 ha per year (Trafford, 1977).

DRAINAGE BENEFITS

Lowering soil water contents by drainage results in several changes in the rooting environment, all of which are beneficial to the crop and soil management.

Duration of waterlogging

Autumn sown crops rely on efficient water table control to allow development of a root system that is ready to take advantage of the warmer spring and summer weather. It is therefore important that waterlogging is kept to a minimum during the winter. Failure to do so may result in a stunted root system that will be unable to exploit deeper seated water and nutrient reserves during the following summer. Crops benefit from keeping the water table below the depth of active root growth during the autumn, winter and spring.

Soil workability and trafficability

A reduction in water content will result in an increase in soil strength. Successful cultivation usually depends upon the soil water content being below the lower plastic limit (*see* p. 39). Drainage lowers the water content of the topsoil, resulting in increased opportunity for landwork during the critical autumn and spring months. In grazing systems, the benefits from drainage will be the increased number of days that livestock can graze a sward without risk of damage to the soil structure. It is very difficult to quantify the benefit thus derived, as climate and hence soil water status will vary widely from year to year. Perhaps as important is the increased flexibility in terms of management options that arise from the longer grazing or cultivation periods in the autumn and spring.

Soil temperature

It is often quoted that draining a soil reduces the specific heat capacity and therefore results in higher soil temperatures, particularly in the spring, when compared with similar but undrained soils. For example, this effect has been shown on clay soils at Drayton Experimental Husbandry Farm, where combined tile and mole drainage produced elevated temperatures, particularly in the spring (*Table 2.1*). However, it is important to record that other researchers have found no significant differences in soil temperatures as a result of drainage. If drainage does increase soil temperature, particularly during the spring months, then the main benefit will be the more rapid emergence and growth of spring sown crops. For example, spring barley may emerge a week or more sooner on drained soils that warm up more quickly in March and April.

Table 2.1 Effect of drainage on soil temperatures at Drayton EHF, 9 April 1974

Soil depth (mm)	Soil temperature (°C)	
	Drained	Undrained
10	11.0	9.2
40	10.2	9.0
160	9.4	8.8
320	8.6	8.4
640	7.7	7.7

(After Castle, McCunnal and Tring, 1984)

Efficiency of fertiliser usage

Increased soil temperature and aerobic conditions will lead to more efficient use of applied fertilisers, particularly nitrogen top dressings, and stimulate the release of nitrogen from organic matter by bacteria. Under these conditions, actively growing roots will absorb nutrients and proportionately less will be leached from the soil by rainwater to pollute groundwater and watercourses.

Arable crop yield benefits

The benefit obtained as a result of installing a drainage scheme can be most easily measured in terms of crop yield, but the variability of the British climate often produces a wide range in yield benefits from year to year. Thus, it is necessary to express benefits in terms of average responses, although in a wet year efficient drainage can make all the difference between crop failure and success. The yield advantage for most crops when comparing drained with undrained treatments is typically 10–25%. For example, average winter wheat yields can be increased by 1.0 t/ha (Armstrong, 1978) although year to year variations can be large.

Grassland/livestock benefits

In the wetter regions of the UK drainage is essential in order to maximise grass utilisation. The benefits can be expressed in terms of dry matter yield or liveweight gain, but it is also important to recognise other advantages. In particular, drainage and careful management can alter the composition of the sward (*Table 2.2*) as well as increasing the response of the sward to nitrogen. *Table 2.3* displays the benefits of

Table 2.2 Effect of drainage on the botanical composition of a grass sward three years after reseeding, on a clay textured soil

	Perennial ryegrass	Weed grasses (% composition)	Other species	Bare ground
Undrained	65	11	14	10
Drained	72	7	17	4

(After Castle, McCunnal and Tring, 1984)

Table 2.3 Livestock grazing days and liveweight gain as controlled by drainage of a clay soil in Devon

	Bullock grazing days (d/ha/year)	Bullock LWG (kg/ha/year)
Undrained	239	240
Undrained, fertilised	380	300
Drained, fertilised	451	433
Drained, reseeded, fertilised	433	428

(After Trafford, 1971).

drainage, with laterals spaced at approximately 30 m in a clay textured soil, in terms of bullock grazing days and liveweight gain. Note that in this example reseeding gave no advantage over the original sward. In addition to weight gain, other benefits in a livestock production system can include a reduction in the incidence of both liverfluke and foot problems.

SOIL WATER

Water retention by soils

Soil water content changes continually in response to additions by rainfall and from upslope, and losses by drainage, evapotranspiration and downslope water movement. The change in storage of soil water, assuming additions from upslope equal downslope losses, can be expressed as:

$$\Delta S = R - D - E$$

where ΔS change in soil water content (mm)

> R rainfall (mm)
> D drainage (mm)
> E evapotranspiration (mm)

In the UK the climatic pattern results in an excess of rainfall over evapotranspiration in the winter months, with

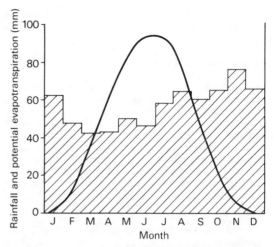

Figure 2.2 Average monthly rainfall (shaded area) and evapotranspiration (smooth curve) for South Oxfordshire

the reverse normally being true for the summer months. The result is that the storage of water by soils, normally assumed to be in pores $< 60\,\mu m$ in diameter (*see* p. 36) varies with the time of year. An example, based upon standard meterological data for the south Oxfordshire area, is given in *Figure 2.2*. Once the soil water deficit has been reduced to zero in the autumn, and the soil is at field capacity, then drainage will commence, and carry on until evapotranspiration creates a soil water deficit the following summer.

As stated previously, the main reason for drainage is because heavy textured soils retain water during the winter, leading to a reduction in the air filled porosity, restricted crop growth and low soil bearing capacity. In addition to soil texture, structure development is also important; two soils with similar clay contents but different degrees of structure development can have very different water retention characteristics. Several examples are given in *Table 2.4*. Air-filled porosity is an expression of the percentage of total pore space that is $> 60\,\mu m$ in diameter and hence available to transmit water through the soil to the drains under the influence of gravity (*see* p. 38).

Table 2.4 Examples of the effect of texture and structure on winter mean water content and air-filled porosity (i.e. % pores $> 60\,\mu m$)

Textural class	Organic matter (%)	Structure development	Winter mean water content (%)	Air-filled porosity (%)
Clay loam	10	Good	50	10
Clay loam	5	Poor	55	5
Sandy loam	10	Good	20	15
Sandy loam	5	Poor	25	10

The consequence of a low air-filled porosity ($< 10\%$) is that oxygen exchange with the atmosphere becomes restricted and the soil can become anaerobic. This leads commonly to the development of mottles of contrasting orange and grey colours, due to the process termed gleying. The presence of rusty staining in the topsoil, particularly around roots, is an indication that the soil is waterlogged for at least part of the year.

Water movement in soils

During the winter months, water in soil flows mainly in response to gravitational gradients; that is water will move down through the soil profile and laterally downslope to the lowest point in the field, which ideally coincides with the outfall of a drainage system. A well maintained system provides a rapid direct route to transfer water from the soil surface to the drain. In a uniformly permeable material, such as a sandy textured soil that is underdrained because of a high regional water table, as for example would be found on a river flood plain near sea level, water movement to the drains takes place through the whole soil (*see Figure 2.3a*). Rain falling on the soil surface moves down to the water table by unsaturated flow. Below the water table flow processes are saturated, with no air present. As a result of rain falling on the soil, the water table rises and the drains discharge water. When the rain ceases, the water table falls until there is no hydraulic potential pushing water out of the soil. Analysis of the quantity and timing of drain discharge can provide useful information concerning the efficiency of a particular drainage system, and may indicate whether a

system is deteriorating with time (for example, Reid and Parkinson, 1984; Parkinson and Reid, 1986)

In heavy textured soils with low rates of water movement, particularly in the subsoil, water tends to move in the topsoil and then into the trench created by the drain laying operation (*see Figure 2.3b*). These trenches are often backfilled with gravel to increase the rate of water movement. The bulk of the subsoil plays little part in the disposal of excess water.

The rate of water movement through soil is termed the hydraulic conductivity. High values for hydraulic conductivity indicate rapid water movement, such as are found in coarser textured soils, for example sandy loams and gravels. Finer textured soils usually possess low hydraulic conductivities, unless they are well structured, in which case flow rates can be larger than would be expected. Examples are given in *Table 2.5*. These classes of saturated hydraulic conductivity are used in drainage design even though a substantial proportion of sub-surface water movement is by unsaturated flow.

DRAINAGE SYSTEMS

Land can be permanently drained by a system of ditches or pipes laid in the soil, or a combination of both. In the case

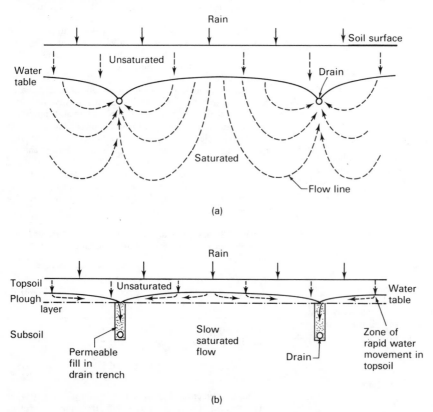

(a)

(b)

Figure 2.3 (a) Water flow routes in a soil of uniform high hydraulic conductivity. (b) Water flow routes in a soil with low subsoil hydraulic conductivity

Table 2.5 Saturated hydraulic conductivity classification and soil type

Saturated hydraulic conductivity (m/d)	Class	Example soil types
< 0.01	Very slow	Clay, silty clay[1]
0.01–0.1	Slow	Clay loam[1]
0.1–0.3	Moderately slow	Silty clay[2]
0.3–1.0	Moderately rapid	Sandy clay loam, clay loam[2]
1.0–10	Rapid	Loamy sand
> 10	Very rapid	Gravel

[1] Poor structure
[2] Good structure

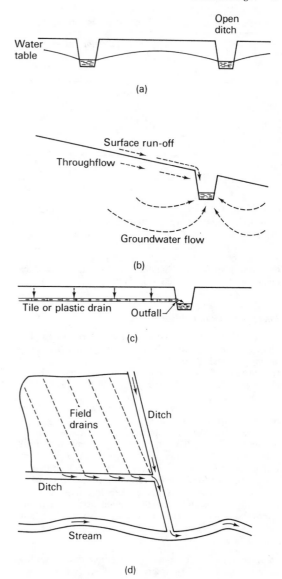

Figure 2.4 Ditches as water collectors and carriers. (a) Water table. (b) Surface or groundwater interception control. (c) Water removal from a pipe drainage system. (d) Water movement to a stream or river

of pipe drainage, additional short-lived measures, such as mole draining or subsoiling can be carried out to improve the efficiency of the drainage system. Both permanent and temporary systems serve specific purposes and must be installed following recommended guidelines (*see* Castle, McCunnall and Tring, 1984). The principles and some of the practical points are outlined here.

Open ditch drainage

Most drainage systems find outlets into an open ditch which will usually lead to a larger watercourse. These ditches are a vital component of a drainage system and may in some cases be the sole method of water removal. Careful design, construction and maintenance is therefore very important.

Ditches can have a number of functions, as are illustrated in *Figure 2.4*. In some cases a ditch may be the only method of draining land (*Figure 2.4a*), and relics of such systems can be seen in the ridge and furrow landscapes of the English Midlands. More commonly today, ditches are receptors of water from an underdrained system (*Figure 2.4c*). Ditches have several advantages and disadvantages:

Advantages:	Direct access for water
	Large capacity to carry storm flows
	Easily maintained
Disadvantages:	Hinder cultivations and other machinery operations
	Need fencing to exclude livestock
	Obstruction due to wall collapse and vegetation growth common

Ditch specifications

Ditch width and depth

Design standards require that ditches must be of sufficient capacity for the catchment area drained. Within any catchment area weather conditions, soil type and ground slope will be the main factors controlling the volume of water carried. There are guidelines for ditch dimensions in particular situations that will ensure that a ditch functions effectively for a long time. Recommended channel side slope depends upon soil type. The more stable the soil, the steeper the permissible slope. Some examples are given in *Table 2.6*, where it can be seen that ditches in sandy materials, with low inherent stability must have lower slopes than ditches in more stable finer textured soil.

Ditch floor gradient

This will, in practice, depend upon the local topography, although there are theoretical specifications to allow water to move quickly to a larger watercourse. The gradient should be uniform and not too steep (leading to channel erosion) or too shallow (leading to silting). A gradient of 0.5–1.0% is generally considered to be adequate.

Ditch sidewall protection

Bank collapse can be minimised by guarding against water

Table 2.6 Ditch channel side slope ratios and soil type

	Channel side slope ratio[1]	
	Channel < 1.3 m deep	Channel > 1.3 m deep
Fen peat	Vertical	0.5:1
Heavy clay	0.5:1	1:1
Clay loam or silt loam	1:1	1.5:1
Sandy loam	1.5:1	2:1
Sand	2:1	3:1

(After Castle, McCunnal and Tring, 1984)
[1] Horizontal:vertical ratio (h/v)

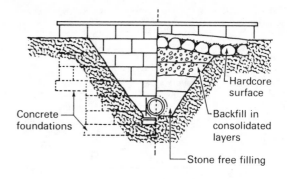

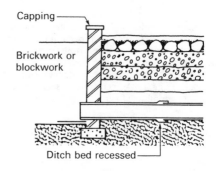

erosion and livestock damage. Pipe drain outfalls must be correctly installed with splash plates. Ditches must be fenced to prevent animals damaging vulnerable sidewalls. Protection of sidewalls against 'natural' erosion can be achieved by several methods, the simplest of which is to grass them down. In zones of high erosion risk, such as at ditch junctions, stoning the ditch walls will provide more effective protection.

Inlets, outlets and culverts

Ditches must be piped under roads and farm tracks. The pipes used, normally concrete, must be large enough to carry anticipated peak flows. The design flow can be calculated (*see* MAFF, 1982), but the minimum recommended pipe diameter is 225 mm. This size pipe will serve a catchment area of up to approximately 12 ha. Larger catchments will generate higher peak flows, and the carrying capacity of the pipe must increase accordingly.

The culvert inlets and outlets must be constructed to maximise the useful life of the system, so attention to detail is important. The pipe must be strong enough to carry farm traffic and be firmly bedded in stone-free soil. Examples of culvert and inlet design are given in *Figure 2.5*. If correctly installed, such a culvert will be unlikely to collapse or cause downstream erosion, although maintenance, in particular removal of vegetation causing blockages, is very important.

Hill drains

In upland areas of low potential economic return, shallow open ditches at 15–20 m intervals provide a low cost alternative to more intensive drainage systems. Such ditches are normally 450–500 mm deep, 500–750 mm top width with a 150–250 mm wide base. Using a hill-draining plough such ditches can be constructed rapidly, although high rainfall and uneven soil depth with stones make them liable to deteriorate much more rapidly than is the case for conventional ditches.

Pipe drainage

Pipe or underdrainage removes excess soil water without reducing the area of land cropped or interfering with field operations. Installing a permanent system of clay tiles or plastic pipes to carry water below plough depth can solve drainage problems efficiently and be a worthwhile investment. It is particularly important that the principles of operation, design, installation and maintenance are all

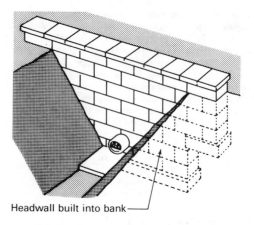

Figure 2.5 Example culvert and inlet design (by courtesy of MAFF, 1983)

understood and followed, as a buried drainage system is difficult to inspect and maintain once installed.

Materials

Water movement into and in pipe drains

An effective underdrainage system maintains the water table adjacent to the drains at drain depth. Between drains, the water table will be higher due to the restricted movement of water through soil (*see Figure 2.3*). Water enters a pipe system either through the cracks between clay tiles or

Clay tiles

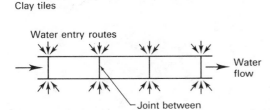

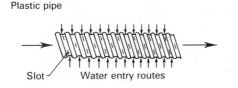

Plastic pipe

Figure 2.6 Water entry routes into clay and plastic pipes

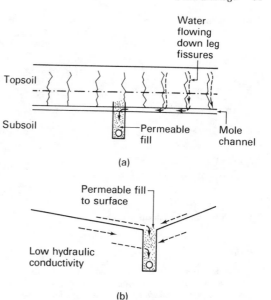

Figure 2.7 Permeable fill over a pipe drain, showing water flow routes. (a) Permeable fill to within 300–350mm of the soil surface. (b) Permeable fill to the surface

through the slits in plastic pipes (*Figure 2.6*). Once in the drain, water flows downslope into a larger main drain or out into a ditch. A drain will only conduct water if the water table is at or above drain depth.

Pipe hydraulic capacity

The quantity of water which a pipe can carry depends upon its diameter, gradient and the resistance to flow or roughness of the pipe. The internal diameter of pipes commonly used for laterals is 65–75 mm. Design calculations will determine the size appropriate for a particular area to be drained by a lateral. At diameters in the range 65–75 mm, the greater internal roughness of corrugated plastic pipe results in the clay tile being able to carry up to 50% more water.

Clay or plastic?

Clay tiles are normally cylindrical and 250 mm long. The outer surface may be smooth, corrugated or octagonal. Tiles are commonly supplied in a palletised form, which allows easy mechanical handling. The most common sizes are 75, 100 and 150 mm internal diameter. Good quality tiles are essential, as the failure of one tile can cause the failure of a complete lateral. The tiles are laid in the trench to butt up to one another, but with a gap of 0.5–2.0 mm resulting from the uneven ends of successive tiles.

Clay tiles have, to a large extent, been superseded by slotted plastic drains. The latter have several advantages, being easier and lighter to handle than clay tiles, and are more suitable for mechanical laying. In addition disjointed drain runs are unlikely. However, as pointed out earlier, hydraulic capacity and also strength characteristics are advantages in favour of clay tiles. Plastic drains are most commonly supplied in 200 m lengths, either of smooth or corrugated sides. Slots run the length of the pipe so water entry is relatively unrestricted. Standard joints and junctions are available to fix lengths together, or to lead laterals into mains. Plastic pipes are available prewrapped with a filter, which may be necessary in silty soils to prevent blockage of the slots by sediment.

Permeable fill

Some form of permeable backfill is placed over pipes in about 60% of field drainage systems installed in England and Wales. In the past a wide variety of materials, such as stone, brushwood or straw were used. Nowadays gravel is the dominant fill material. As permeable fill may account for 50% of the cost of a drainage system, it is important to justify its use. Permeable fill acts as a connector to allow water movement from the topsoil via mole drains, or subsoiler fissures, to the pipe drain. In addition, backfill forms a permeable surround, to improve water entry rates to the pipe, and acts as a filter to prevent soil particles reaching the drain.

Washed gravel with a mean particle diameter of 20–50 mm is the most suitable permeable fill material. The function of permeable fill is illustrated in *Figure 2.7*. The use of permeable fill is not recommended for medium and coarse textured soils.

Outfalls

Where a pipe drain discharges directly into a ditch, precautions must be taken to ensure the longevity of the system, as it is the outfall that is most likely to be the cause of failure. Pipes should be able to discharge freely — therefore the invert of the pipe should be at least 150 mm above the normal ditch water level. At least the first 1.5 m of pipe should be sealed, rigid and frost resistant. Soil should be packed along this length of pipe to prevent water travelling outside the pipe into the ditch, causing erosion.

The outfall should ideally be supported by a headwall to stabilise the bank and prevent erosion by the drain outflow. These are expensive, and good quality turfing is an alternative, provided a splash plate is installed (*see Figure 2.8*).

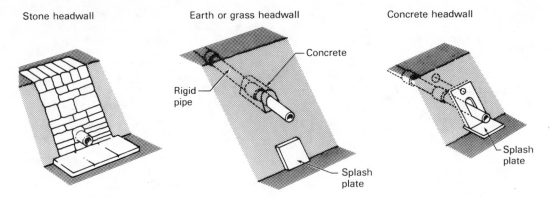

Stone headwall   Earth or grass headwall   Concrete headwall

Concrete

Rigid
pipe

Splash
plate

Splash
plate

Figure 2.8 Examples of drain outfalls

Installation

The success of an underdrainage scheme depends upon careful installation under optimum soil conditions. Drains can be laid either by excavating a trench, laying a pipe and backfilling (trenched) or by creating a slot into which a pipe is placed immediately before the soil falls back in place (trenchless). There are guidelines that must be adhered to for both methods of installation, as well as points to look out for on each individually.

Soil and site conditions

Drainage work should be carried out when the soil is dry and hence strength is high, thus resisting the compacting and smearing of the soil surface and trench sides that would occur under wetter conditions. In some cases, drainage through a crop can be justified if disruption to the farming operations can be kept to a minimum. If, for example, drainage can be carried out through a cereal crop before the flag leaf stage, loss of yield will be relatively low.

Drain laying

Pipes must be laid on a firm smooth bed, shaped to support the pipe. Permanent pipes must be placed at least 600 mm deep to avoid damage by moling or subsoiling operations. Loose soil must not be allowed to fall beneath the pipe. Drains must be plugged at the upper end to avoid the ingress of soil or animals. In the case of clay tiles, it is important that they are firmly butted together, and located at one side of the drain trench, against the side wall to ensure continuity of flow from one pipe to the next. Care must be taken with plastic pipes when installing them in air temperatures below 5°C, as they become brittle.

It is essential to level accurately during the drainage operation so that a continuity of flow occurs. Laser levelling techniques allow grades of 0.1% to be attained, although for practical purposes 0.2% is the limit to which most machines work. Lateral slope should not exceed 2.0% if possible, with 4.0% being the maximum permissible before erosion problems are likely to occur.

Trenched drainage

Machines which excavate trenches come in many forms — tractor attachments, self-propelled trenchers or winched

trenchers — which are described in more detail in Castle, McCunnal and Tring (1984). Several important features favour the trenched system of drain laying.

Advantages: Pipes laid in the trench can be inspected before backfilling

Old drainage systems can be found and tied in to the new system

Permeable fill can be placed in the ditch from a hopper to a uniform depth

Grade of trench can be checked easily

Disadvantages: In unstable soils, trench wall collapse may restrict the use of trenched methods

Open trench requires more permeable fill than trenchless

Slower workrate than trenchless

More expensive than trenchless

A trench, normally 150–300 mm wide, is excavated using a vertical endless chain with blades attached to its links. Spoil is brought up by the chain and pushed out to either side of the trench by augers. Work rates of up to 30 m/h are possible, although in weakly structured or stony soils progress may be considerably slower.

Trenchless drainage

Trenchless drain laying machines feature a large plough which cuts a slit, to a predetermined depth, and drops a plastic pipe down a narrow box behind the plough. If required, permeable fill is supplied from a hopper mounted above the pipe shute. *Figure 2.9* illustrates some of the major points relating to trenched and trenchless systems. Currently, 75% of drain installation is by the trenched method. The trenchless method is most commonly used in closely spaced systems where small plastic pipes (< 75 mm) are laid. The degree of surface disruption varies with conditions and machinery, but typically a ridge of up to 0.3 m high will be left by the machine, and will need to be flattened, for example, by the track of a crawler tractor, before farming operations can continue.

There are several other items of machinery which are less commonly used for drainage operations. Tractor mounted backacter trenchers can be used for short runs of drain provided that maintaining grade is not vital. Large 'V' trenchers can be used to lay drains by the trenchless method with very little surface disturbance. The pipe drain is passed down one hollow leg of the V, while gravel is passed down

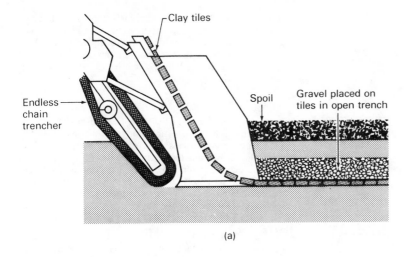

(a)

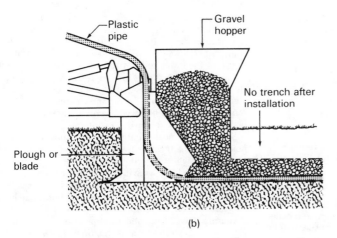

(b)

Figure 2.9 Trenched (a) and trenchless (b) drain installation methods

the other. This form of trench is most suited to close spaced systems (< 10 m) where small (< 60 mm diameter) pipes are needed.

Secondary drainage

Heavy textured soils often require, in theory, a drain spacing of 5 m or less in order to dispose of excess rainfall in the winter months. In addition, near surface compaction by vehicles or animals may reduce the porosity of the surface soil and hence cause ponding of water. Under these conditions, it is possible to enhance the efficiency of an existing drainage scheme by either moling or subsoiling, depending upon the particular soil and site conditions. Artificial creation of temporary cracks and channels allows wider spacing of drains and hence reduces the costs of installation of a drainage system, although the recurrent costs of re-moling or subsoiling every four to five years must be borne in mind.

Moling

Principles and soil suitability

A system of permanent drains spaced 30–40 m apart in a dense clay soil can work efficiently if a system of mole drains at 2–3 m apart is pulled across the permanent system, such that water flows from the soil surface through the cracks into the mole channel and hence via the permeable fill into the drain (*see Figure 2.10a*). A mole plough is essentially a circular bullet followed by an expander which leaves a channel connected to the soil surface by a series of cracks. The soil conditions at moling are vital to the success of the operation and hence the expected effective life of the moles.

Moling can only be successfully carried out in soils with a substantial clay content; 30% clay is commonly quoted as being the minimum allowable, although it is important to note that clay type and organic matter content are also important (Spoor, Leeds-Harrison and Godwin, 1982). Clays which expand and contract in response to changes in water content are more likely to provide stable long-lasting

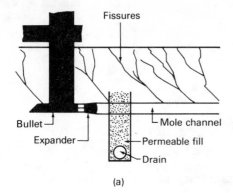

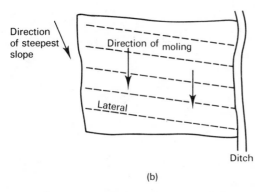

Figure 2.10 Drawing a mole channel. (a) Section through soil illustrating connection with permeable fill. (b) Field plan of moling direction relative to laterals

channels than non-expanding clays. Similarly, heavy textured soils with a high organic matter content and hence a high aggregate stability are more likely to be successfully moled. Generally speaking, clay loam and clay textured soils with a low stone content are likely to hold a mole channel. Silty soils, and those with a high stone content (causing premature channel collapse) are unlikely to hold a mole, in which case drainage by a closer spaced system of permanent drains is more appropriate.

Moling should be carried out at a depth of 0.5–0.6 m, at an angle approaching 90 degrees to the field drain (*Figure 2.10b*). The minimum gradient should be 0.5%. Unlined mole channels are prone to collapse if the channel slopes at < 0.5% because of standing water, and also at > 5% because of erosion.

When to mole

Figure 2.11 demonstrates the effect of moling under ideal conditions, which will normally be when the topsoil is drying out with the subsoil remaining moist. Such conditions are likely to occur during a drying cycle, such as in May or June, although this varies across the country. As *Figure 2.11* shows, increasing overburden and water content at depth results in plastic failure and the formation of a channel, while nearer the surface brittle failure results in extensive cracks forming. It can be seen then that in order for a mole drain to function efficiently there must be formation of a stable channel at depth linked to the surface by a continuous system of cracks. A mole channel can be expected to last four to five years if drawn in good conditions, and they have been known to last more than ten years, although with a much reduced cross-sectional area, and hence water carrying capacity, as time goes on.

Gravel tunnel moling

By placing pea gravel in the mole channel at the time of formation, it is possible to mole stony or less stable soils with a greater chance of success. This technique, using a modified mole plough without an expander but with a hopper to supply the gravel behind the bullet, has been used successfully in Northern Ireland but the extra cost of installation has resulted in few drainage contractors employing this system in other parts of the UK.

Subsoiling

In contrast to moling, subsoiling is not specifically a drainage operation. Subsoiling can be any form of soil cultivation that is intended to shatter the soil beneath the normal cultivation depth. When carried out as a drainage operation, the intention is to work the soil when at its driest, so that extensive shattering occurs, thus creating a network of artificial cracks that will conduct water to the drain or at least into the lower soil horizons. In contrast to moling, a channel is not necessarily formed at depth. Subsoiling may also be carried out in soils that do not need draining but have a compaction problem; for example a soil with a high fine sand content which has been heavily trafficked may need to be subsoiled prior to sowing a root crop.

Subsoiling is most suitable for those soils which will not hold a mole channel; generally soils with a high silt content or with a high stone content. The winged subsoil shoe 'heaves' the soil, allowing it to fail under tension. The result is extensive cracking, providing the soil is sufficiently dry. Normally this operation is carried out immediately after harvest. If soil conditions are not suitable, restricted shattering may occur, and the operation is less effective. *Figure 2.12* illustrates various situations. In *Figure 2.12a* the soil is dry, hence the critical depth is deep enough to allow extensive shattering. The addition of wings to the tine greatly improves the shatter and allows wider leg spacing. Critical depth depends not only on soil water content, but also on overburden pressure and subsoiler configuration. In *Figure 2.12b* restricted shattering and plastic failure are illustrated as a consequence of working below the critical depth. A set of shallow leading tines can lower the critical depth and therefore allow deeper loosening (*see* p. 507 for more details of subsoiler design and operation).

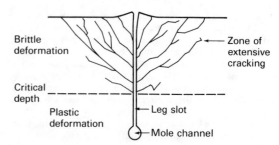

Figure 2.11 Ideal mole channel configuration

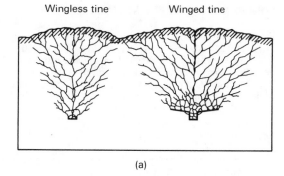

Wingless tine   Winged tine

(a)

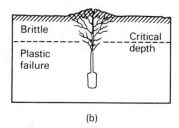

(b)

Figure 2.12 Subsoiling (a) above, and (b) below critical depth, with and without wings on the tine (after Farr and Henderson, 1986)

Ideally the subsoiler should work just below the depth of the pan to be shattered. Tine spacing will depend upon conditions but shattering should be extensive enough to allow efficient water disposal. It may be necessary to repeat the operation every two or three years, depending upon the rate of recompaction.

PROBLEM DIAGNOSIS AND SYSTEM LAYOUT

Site investigation

Drainage problems can arise from a number of causes, some of which have already been mentioned. Before embarking upon the installation of a drainage system it is important to carry out a thorough appraisal of the causes of the drainage problem and the likely benefit to be derived from the work.

Examination of the soil, both in known wet patches and also in other areas in the same field, can yield much useful information. The presence of compacted pans, ochreous mottles and greyer colours can all indicate that there is a problem, even during the summer months. Soil examination is most easily carried out by excavating a large pit with a backacter. Following a detailed soil investigation it should be possible to describe soil texture, structure and the nature of the drainage problem. The information is useful when a drainage adviser or contractor presents a solution to the problem. The drainage adviser will prepare a detailed plan showing the location of the poorly drained areas, cause of the poor drainage and suggested remedy. It is important to check this plan to make sure the recommendations suit the particular farming methods and are financially sound in terms of the likely benefits.

Drainage systems

Regularly spaced system

Soils with low hydraulic conductivity on level or uniformly sloping land are usually drained by a regularly spaced system, with drains spaced according to the rate of water movement through the soil; closer spacing being used for a more slowly permeable soil. In heavy textured soils such as clay loams or clays, which can hold a mole, permanent drains may be installed at 30–40 m spacing, with a secondary treatment superimposed upon that system. A regularly spaced system can be designed to use the slope of the land to collect water, in a herring-bone design.

Irregular/random systems

Complicated geology or undulating relief may necessitate a change in design from a regular to a random system. Such a system will be designed to tap spring lines, using interceptor drains, or to drain wet valley bottoms. The resulting design will leave much of an area undrained, only removing water from natural water collecting zones. Examples are shown in *Figure 2.13*.

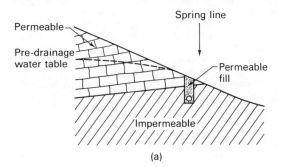

(a)

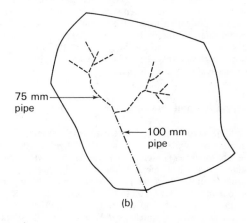

(b)

Figure 2.13 Examples of drain layout for irregular/random systems. (a) Spring-line interception, in section, (b) random layout, in plan

System design

Recommendations for the design of field drainage pipe systems are given in detail in MAFF (1982). Only the factors taken into account in the design are discussed here. Design criteria are based upon determining the size of pipe needed to conduct away the water so that damage to crops is minimised. Pipe sizes for laterals, main drains, interceptors and ditches can be calculated knowing the characteristics of the site. For a lateral, the following information is needed:

Land use: Design rates assume that horticultural crops are more valuable and more susceptible to damage than arable crops, with grass being most tolerant of waterlogging.

Soil type: Soil texture and structure control relative rate of water movement such that, for example a well structured clay loam will drain more rapidly than a poorly structured silty clay loam.

Rainfall probability: The probability of a certain amount of rain falling in one day varies from region to region across the country. Drainage design is based on keeping the water table below a certain depth in the soil. A given system will therefore be able to dispose of a given quantity of rainfall in one day.

Ground and pipe gradient: On steeper slopes, water disposal is more rapid. However, pipes are often laid across the slope so it is important to know both these gradients when designing a system.

Given this information, and assuming a certain pipe spacing and drainage system to be installed, i.e. 40 m spacing with permeable fill and moles, the pipe size needed can be calculated. This calculation will allow the system to deal with most (but not all) of the rainfall which can, on balance of probability, be expected in any year. There will be occasions when very heavy rainstorms may occur, say every two or three years, which will cause the system to surcharge (drains run full) but economics dictate that a limit must be placed on pipe size based on smaller, more frequent rainstorms.

Arterial drainage systems

In low-lying areas, such as the Fens, it may be necessary to pump drainage water over a sea defence system to maintain water levels in ditches low enough to allow field drain outfall. Pumped drainage is usually the responsibility of either Water Authorities or Internal Drainage Boards. Pumping is a capital intensive process and beyond the scope of this chapter, but must be borne in mind when designing a field drainage system in such areas.

Drainage and conservation

Wildlife habitats may be affected adversely by land drainage operations. While the principal aim of field drainage is to reduce soil water contents to enhance crop growth, there are many established methods of reducing the impact of drainage on the flora and fauna of an area. Most important is to plan for conservation so that, for example, an old pond in the corner of a field can remain undisturbed during a drainage operation. Ditches can be designed with wildlife in mind, for example with a wide shallow section to carry stormflow. (*See* Castle, McCunnal and Tring (1984) for further details.)

DRAINAGE SYSTEM MAINTENANCE

Drain maintenance is simply a method of protecting an investment and it is absolutely vital to ensure that the system functions efficiently by carrying out an annual maintenance plan. Particular attention should be paid to outfalls, which should be checked both in the summer, when dry, and in the winter when the drains are running.

Ditches

Several points should be borne in mind when carrying out inspections of ditches. Excessive weed growth reduces the rate of water flow, resulting in siltation and a rise in the water level, which may lead to pipe outfalls becoming submerged. Vegetation should be controlled by using the appropriate herbicides or by flailing. If the bank sides are too steep or unconsolidated bank slips may have occurred and re-excavation may be needed. It is particularly important to fence ditches in fields that are stocked in order to avoid damage by animals.

Pipes

The effective functioning of a pipe drain depends upon a clear outfall. It may therefore be necessary to clear obstructions and fit vermin traps to prevent further blockage. In unstable soils siltation within the pipe may be a problem, particularly if the pipe is laid to a low grade. Blockages can be cleared using an extendable system of rods (rodding) or a system of employing water sprayed at high pressure up the drain (jetting). The latter is particularly useful for clearing deposits of iron ochre which may form in peats and poorly drained marine clays. A well-laid pipe system in a clay soil with good structure might be expected to function effectively for 50 years or more, although in poorly structured silty soils an effective life of 20–30 years is more realistic. It is possible to delay the effects of siltation in unstable soils by wrapping pipes in a fibrous material, such as coconut matting, which will prevent fine soil particles reaching the drain itself. These filters have been used to good effect in the Netherlands during polder reclamation, but are only occasionally used in the UK.

Secondary treatment

Moling and subsoiling are both temporary methods of drainage, but the working of a permanent system may depend upon good mole channel formation or subsoiling. It is recommended that moles are redrawn every four to five years, while subsoiling is carried out as conditions and time allow and require.

ECONOMICS OF DRAINAGE

In recent years, there has been a steady reduction in the level of government grants available for drainage work, from 65% in the mid-1970s to 15% in 1985 (this does not include schemes in Less Favoured Areas, which are eligible for 30%

grants). In view of the increasing cost of drainage to the farmer, both in absolute terms and relative to the price of land, the economics of the drainage operation must be taken into account. The increased cost of drainage to the farmer has partially been to blame for a reduction in the total area of agricultural land which is being underdrained, whether for the first time or more commonly to replace old defective systems. From a peak of approximately 100000 ha/year in the 1970s, it is estimated that only 50000–60000 ha/year is being drained in the mid 1980s in England and Wales. Approximately 40000 ha/year must be replaced just to keep up with the rate of failure of old systems, some of which may be 50–100 years old.

In the current economic climate, and in view of recent reductions in the level of grant available for drainage work, it is necessary to assess the financial benefits from drainage. It is most likely that at present an economic return following the installation of an intensive drainage network will only be obtained from a change of cropping, such as from a grass to an arable system.

Costs

It is very difficult to estimate the cost of a particular system, as it will vary widely according to field size, amount of work to be done and other factors. Typical installation costs per metre of excavating a trench, supplying and laying an 80mm plastic pipe and backfilling with soil vary from £0.90–1.00/m. Installation of 75mm clay tiles cost £1.05–1.20/m. Supplying and laying permeable fill over a pipe to within 360mm of the ground surface adds £0.85–1.0/m (Nix, 1987). An intensive regular spaced system, say 30–40m spaced laterals with permeable fill, may cost between £500–800/ha. Such a system could be expected to have a useful life of at least 25 years, and this can then be considered to be the write-off period. Mole draining may be necessary with a wide spaced system, adding a cost of £40–50/ha. It must be remembered that this operation, which is eligible for grant-aid, will need to be repeated every four to five years. Digging a new ditch (1.8m top width, 0.9m depth) costs £1.50–1.75/m.

In general, the closer spaced systems are more expensive to install. In addition, it must be remembered that permeable fill can constitute up to 50% of the total cost of a system, although this varies according to the local geology and hence the supply of aggregate. Trenchless systems are much cheaper to install but quality control on the operation is not as good, so the system may not have such a long life expectancy as a drain installed in an open trench.

References and further reading

ARMSTRONG, A. C. (1978). *Journal of Agricultural Science, Cambridge* **91**, 229–235

CASTLE, D. A., McCUNNAL, J. and TRING, I. M. (1984). *Field Drainage, Principles and Practices*. London: Batsford

FARR, E. and HENDERSON, W. C. (1986). *Land Drainage*. London: Longmans

MAFF (1982). *The Design of Field Drainage Pipe Systems*. Reference Book No. 345. London: HMSO.

MAFF (1983). *Technical Notes on Workmanship and Materials — Field Drainage Schemes*. London: HMSO

NIX, J. (1987). *Farm Management Pocketbook*. 18th Edn. London: Wye College, University of London

PARKINSON, R. J. and REID, I. (1986). *Journal of Agricultural Engineering Research* **34**, 123–132

REID, I. and PARKINSON, R. J. (1984). *Journal of Hydrology* **72**, 289–305

SMEDEMA, L. K. and RYCROFT, D. W. (1983). *Land Drainage: planning and design of agricultural drainage systems*. London: Batsford

SPOOR, G., LEEDS-HARRISON, P. and GODWIN, R. J. (1982). *Journal of Soil Science* **33**, 411–426

THOMASSON, A. J. (1982). The distribution and properties of British soils in relation to land drainage. In *Land Drainage*. Ed. Gardiner, M. J. Rotterdam: Balkema, pp. 3–10

TRAFFORD, B. D. (1970). *Journal of the Royal Agricultural Society of England* **131**, 129–152

TRAFFORD, B. D. (1971). *Agriculture* **78**, 305–311

TRAFFORD, B. D. (1977). *Journal of the Royal Agricultural Society of England* **138**, 27–42

3

Crop physiology

M. P. Fuller and A. J. Jellings

FUNDAMENTAL PHYSIOLOGICAL PROCESSES

Analysis of green plant material shows that about 60–90% of fresh weight is made up of water. Of the remainder, the dry weight, about 45% is carbon, 45% oxygen, 5% hydrogen, 2–3% nitrogen, and 2–3% other elements such as potassium, phosphorus, calcium, magnesium and sulphur.

The crop obtains the carbon, oxygen and hydrogen from photosynthesis, and water, nitrogen, potassium and other mineral elements by uptake from the soil. These two processes are fundamental to crop physiology, and thus to efficient crop production.

Water and nutrient uptake

Water enters the roots near the root tips by diffusion across the cell wall. It must reach the xylem vessels embedded in the root by passing through or around the cells of the root to be delivered to the pores in the leaf surface (stomata) via the leaf veins (vascular bundles). It is the evaporation of the water from the stomata (transpiration) which creates the main driving force for water uptake from the soil. The moving column of water from the soil to the atmosphere through the plant is the transpiration stream. The creation of the transpiration stream is largely a physical, not a biological process, in which the plant exists as a tube for the flow of water.

The pathway of water from soil to atmosphere thus involves four key steps: (i) through the soil to the roots, (ii) in and out of cells across membranes, (iii) along the xylem, and (iv) out of the leaves into the air. Water moves along the pathway down gradients of water potential, (water potential may be defined as the concentration of water, and is thus always a negative value since pure water is said to have a water potential of 0), encountering resistance to flow in the soil, at membranes and cell walls, and at the stomata. Water uptake will be at a maximum when the stomata are fully open in conditions of warm moving air, in full sunlight, with high soil water availability. Water uptake will be reduced by low soil water availability, frozen soil conditions, low soil temperatures, low light levels, high humidity, and low air

temperatures. A reduction in water uptake will eventually result in wilting, and death.

Of the water taken up from the soil 95% is not used directly by the plant, but moves through and out of the plant (a plant may transpire its own weight of water in 1 h). Approximately 5% of the water taken up is used by the plant to give turgidity to the cells, to provide a medium for metabolism and to participate in the reactions of the cell.

Water uptake is the means by which plants gather mineral ions from the soil (or other root medium). Mineral ions are taken up by the roots in solution and thus the ions travel by the same pathway in the plant to be delivered to the cells of those leaves which are actively transpiring. Mineral ions in excess of the requirement of transpiring tissues are passed into the phloem to be retranslocated to non-transpiring plant tissues.

Root cells can be selective in their uptake of mineral ions, i.e. the relative concentrations of ions in different parts of the plant will be different from each other, and different from that of the soil solution. In this way plants can absorb elements in relative proportions different from that of the soil mineral composition.

Selective uptake operates through selective permeability of cell membranes. Active uptake up the electrochemical gradient requires energy from respiration (some passive uptake occurs by diffusion down the electrochemical gradient), so that a plant under stress, or whose roots are in anaerobic conditions will be unable to acquire an adequate mineral supply. In addition mineral uptake is reduced by the same factors which reduce water uptake, e.g. crops often exhibit symptoms of mineral deficiency during a cold winter as soil water has been frozen and mineral uptake prevented.

Photosynthesis and respiration

Ninety-five per cent of the dry matter (DM) of plants is created from the gaseous environment, by the process of photosynthesis.

The molecules of the green pigment, chlorophyll, of plants absorb radiation in the visible light range 400–700 nm (photosynthetically active radiation or PAR), and by means of the electron transport chain is used to make ATP (adenosine triphosphate) and NADPH (nicotinamide adenine dinucleotide phosphate). The energy thus conserved in these compounds can be used to drive the reactions of the carbon fixation cycle which fixes atmospheric carbon dioxide by converting it to sugars. Photosynthesis is the only natural process to use energy from outside the earth and convert it into a chemical form.

The summary equation for both parts of photosynthesis is:

$$6CO_2 + 6H_2O \xrightarrow{\text{light}} C_6H_{12}O_6 + 6O_2$$

All the reactions of photosynthesis occur in the chloroplasts. The key enzyme to the fixation of carbon dioxide is ribulose bisphosphate carboxylase/oxygenase or rubisco. This is the most fundamentally important enzyme to life on earth, and it makes up a large part of the world's soluble protein fraction; in crop production it is one of the main beneficiaries of added fertiliser nitrogen.

The rate of photosynthesis depends on both internal and external factors. Internally the amount and activity of the participating enzymes, the amount of chlorophyll, and the anatomy of the leaf all dictate the photosynthetic capacity. The rate achieved at any one time will be affected by several environmental factors, of which the most important are light intensity, temperature and carbon dioxide concentration, the rate always being determined by the factor in shortest supply. *Figure 3.1* demonstrates the general relationship between these three main factors.

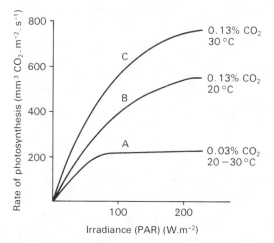

Figure 3.1 The general relationship of phytosynthetic rate with light intensity, carbon dioxide concentration and temperature in leaves of temperate crops. Curve A demonstrates that light saturation occurs at very moderate light levels under atmospheric carbon dioxide concentrations. Curve B shows that if carbon dioxide concentration can be raised above atmospheric levels photosynthetic rate can be stimulated, thus carbon dioxide concentration is the limiting factor to photosynthetic rate under normal conditions (curve A). In curve B temperature becomes limiting, as at higher temperature (curve C) there is a further increase in photosynthetic rate; this latter case is only applicable in controlled environments where carbon dioxide concentration may be artificially raised, e.g. glasshouse lettuce production

The sugars produced by photosynthesis are used by the plant to provide energy through the process of respiration which occurs in the mitochondria and is essentially the reverse of photosynthesis. Photosynthesis and respiration are thus the means by which the plant utilises light energy (which cannot itself be stored) to provide a continuous and translocatable energy source. This means that not all of the carbon fixed is translated into DM increase as a portion will be respired to provide energy and carbon dioxide, which must be refixed or lost.

Thus, growth analysis which measures the DM accumulation estimates net photosynthesis, and to measure gross photosynthesis more sophisticated techniques must be employed, such as gas exchange analysis under varying conditions, or the incorporation of radioactive carbon dioxide. In addition the process of photorespiration, which occurs under conditions of high light and high temperature, also takes up oxygen and releases carbon dioxide which may further disturb the interpretation of gas exchange measurements.

In most common temperate crops at average and above average levels of light intensity, carbon dioxide concentration is the commonest limiting factor to photosynthesis as light saturation is achieved at less than the light intensity experienced on most UK summer days, and photorespiration can be severe under good summer conditions (above 30°C) making carbon dioxide compensation points high.

Many tropical crops, e.g. sugar cane, sorghum, which use the modified (C4) photosynthetic pathway with little or no accompanying photorespiration, are not limited in their photosynthetic rate by carbon dioxide concentration until much higher light intensities. The only UK crop with this tropically adapted metabolism is maize, which, on hot summer days, may gain the advantage of higher photosynthetic rates and faster DM accumulation than its neighbouring C3 crops.

Atmospheric carbon dioxide concentration changes very little with time of day or of year, but the internal leaf carbon dioxide concentration can be controlled to some extent by the opening and closing of the stomata. The stomata may open passively in response to changes in the gradient of water potential, but are mostly controlled by light levels, leaf temperature and internal carbon dioxide levels, which cause active, energy-requiring, stomatal opening.

The carbon fixed by photosynthesis is moved around the plant as sugars in the phloem. The extent and direction of translocation is very important in crop production since it determines the partitioning of assimilates within the plant. A tissue actively importing sucrose is termed a sink, and one which is exporting sucrose is termed a source. Rate of growth and metabolic demand affect the strength of a sink, and positive feedback between sources and sinks allows increased demand by the sinks to stimulate increased rate of assimilation by the source tissues. Thus, as tubers form within a potato crop the photosynthetic rate of the leaves tends to increase (under standard conditions), but falls again if tubers are removed. The rate of translocation, determined by the size of the phloem vessels, also has some control over assimilate partition.

For temperate crop plants photosynthesis is only about 1% energy efficient. This low efficiency occurs for a number of intrinsic reasons: the conversion of light energy into chemical energy is about 25% efficient; about half of the total radiation is unusable by the plant; about half of the PAR is lost by reflection and transmission, and through respiration and photorespiration. This gives a possible maximum photosynthetic efficiency of about 5% which is further reduced to 1% by imperfect crop production practice, e.g. stress from pathogens, climatic factors, etc.

GROWTH

Growth is strictly the accumulation of dry matter, though the term is often used loosely to describe any increase in size.

Accumulation of dry matter (DM) occurs fundamentally at the cellular level. DM fixed by photosynthesis is laid down as new compounds within the cell so that individual cells tend to increase their mass with time. At the same time cells may divide to form new cells thus redistributing DM. Thus, when an individual leaf increases in DM this is through a combination of both an increase in cell size and an increase in cell number. The way in which the cells are organised will determine whether the increase in DM is visible as an increase in leaf area, in leaf thickness or in leaf density.

An overall increase in size will include a certain amount of water accumulation – this is not true growth. In order to analyse crop growth it is usual to determine dry weight and leaf area at known time intervals. These data can be used to obtain the mean crop growth rate (*Figure 3.2*) between successive sampling dates (t_x and t_y) and mean leaf area index:

Mean Crop Growth Rate $(\overline{CGR})$ = $(W_y - W_x) \times 1/P$
(between t_x abd t_y)

Mean Leaf Area Index $(\overline{LAI})$ = $(A_x + A_y)/2P$
(between t_x and t_y)

where W = dry weight,
A = leaf area,
P = cropping area.

From the measurements of dry weight and leaf area an

estimation of crop photosynthetic efficiency (Net Assimilation Rate) can be made:

Mean Net
Assimilation Rate $(\overline{NAR})$ =
$$\frac{(\log A_y - \log A_x)(W_y - W_x)}{(A_y - A_x)}$$

The relationship between these three parameters of growth is:

$$\overline{CGR} = \overline{LAI} \times \overline{NAR}$$

There are a number of other growth analysis formulae which are used to analyse growth in more depth and these have been the foundation of mathematical modelling of crop growth.

Potential yield

The total DM of a plant is termed its biomass. Potential biomass is determined largely by photosynthetic capacity. The proportion of the potential biomass which is economic yield is determined by the extent to which assimilate is partitioned into useful plant parts; this gives rise to the expression Harvest Index.

$$Harvest\ Index = \frac{Economic\ yield}{Total\ biomass}$$

Both photosynthetic capacity and Harvest Index are very largely genetically determined so that potential yield for any crop can be theoretically calculated for any given climate.

Many factors may limit photosynthesis to less than capacity, and may influence the partitioning of assimilates to give a less favourable Harvest Index, so that actual yields rarely reach potential yields, this shortfall is termed the Yield Gap. The practice of efficient crop production tries to close this yield gap by understanding the relationships outlined in this chapter, and acting upon them as described in other chapters within the Crop Production section of this book.

CROP DEVELOPMENT CYCLE

From germination to maturation crops go through a developmental cycle which involves an increase in complexity usually accompanied by an increase in biomass. The cycle is a result of the interaction of the genotype (*see* p. 69) of the plant and the environment it is grown in. The environment provides not only the basic needs of the crop for growth, i.e. light, nutrients, water and heat, but also provides stimuli to trigger developmental sequences.

During its development the crop exists in several 'states' and passes from one state to another in an organised manner via a number of 'processes' (*Figure 3.3*).

Germination

The germination of a non-dormant seed will be initiated given the presence of water, an aerobic medium and sufficient warmth. (For most temperate crops the temperature must be above 0–5°C for germination, with an optimum of about 30°C. A few cold-sensitive crops will not germinate below a higher temperature, e.g. maize requires a temperature of 15°C for germination.) Water is imbibed

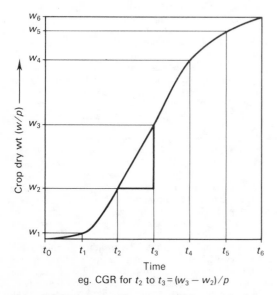

eg. CGR for t_2 to $t_3 = (w_3 - w_2)/p$

Figure 3.2 Typical crop dry matter growth curve showing calculation of mean crop growth rate (CGR)

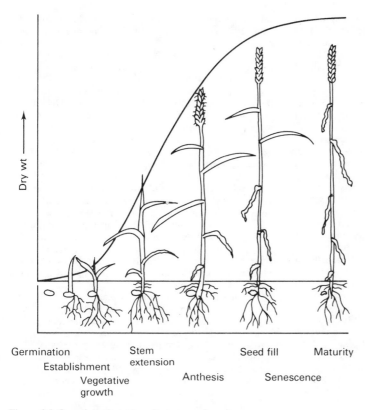

Germination

Establishment

Vegetative
growth

Stem
extension

Anthesis

Seed fill

Senescence

Maturity

Figure 3.3 Growth cycle pattern in an annual seed crop

through the testa or through the micropyle and returns the existing cells to full turgor. When a seed enters germination it passes from a quiescent state with extremely low metabolism and a high resistance to drought and frost stress into a state of high metabolic activity and extreme susceptibility to stress. Initially water content of the plant rises from 5–10% to 30–40% as water is taken up by imbibition (the colloidal attraction of water by large molecules in the seed). At this point many plants progress directly into active metabolism but in those which show dormancy mechanisms germination may cease until dormancy has been broken. Thus, many weed seeds exist in the soil in an imbibed state but ungerminated. In the absence of dormancy or when dormancy has been broken, germination can continue and metabolic rate rises dramatically, with an increase in protein and nucleic acid synthesis. Food reserves are mobilised, often as a result of hormone stimulation of enzyme synthesis, e.g. barley, in which gibberelic acid stimulates the mobilisation of protease and synthesis of α-amylase, which catalyse the breakdown of starch in the endosperm for translocation as sugars.

The expansion of the embryo and its development into seedling roots and leaves ruptures the testa, from which time the seedling ceases total reliance on its stored food reserves and begins to harvest the resources of its environment. Initially, a greater proportion of the reserves are directed to root growth in most species to ensure an adequate supply of water and minerals before the photosynthetic machinery is fully synthesised and operational.

The size of the seed, which tends to be related to the

volume of food reserves, dictates the length of time for which the developing embryo can delay becoming fully independent, e.g. germinating field beans can survive on seed reserves for much longer than germinating oilseed rape. This factor has considerable influence on the optimum depth of sowing for each crop species. In order to achieve early photosynthesis to supplement the seed reserves the cotyledons are frequently expanded above ground in small seeded crops.

The germination percentage of a seed lot is an important concept in terms of both the calculation of correct sowing rates, and for complying with the regulations governing the sale of seed.

Vegetative development

During vegetative development, the shoot apex develops primordia which develop into leaves. A leaf first appears as a bump (primordium) on the side of the apex and then undergoes cell division and differentiation. A mid-rib consisting of protoxylem and protophloem develops and establishes contact with the vascular system of the stem so that it can obtain assimilates and water for its continued development. As the leaf emerges into the light, cell components differentiate, proplastids develop into chloroplasts and the leaf appears green. Assimilation can proceed as soon as the chloroplasts and stomata are functional.

The point of attachment of the leaf on to the stem is important as it is here that the vascular connections are made allowing water to pass to the leaf for transpiration and for assimilates to pass from the leaf for relocation to other growing parts or storage organs. This connection of leaf and stem is called a node and the stem between two nodes an internode. In many plants particularly cereals and oilseed rape, there is a stage of development when the internodes are not expanded and the plant takes on a prostrate or rosette appearance. This adaptation keeps the apex of the plant close to the ground where it is more protected from frost and is a useful feature of over-wintered crops. In the spring, such plants demonstrate internode expansion and the apex is pushed up into the air as the crop approaches flowering. Internode expansion is primarily expansion growth of stem cells and is followed by the strengthening of the cell walls to give the stem stiffness. At any one time usually only one internode is expanding rapidly although expansion phases of successive internodes frequently overlap. In most plants, this internode expansion is under the control of gibberellin in the plant and chemicals which interfere with its synthesis or action are commercially available as stem shorteners.

The final number of leaves that a plant bears (assuming it does not branch) is dependent on the duration of the vegetative phase of the stem apex. At some point the stem apex will receive a stimulus which will trigger a change from producing leaf primordia to producing floral primordia. This change in function of the apex is invisible to the naked eye and usually does not alter the outside appearance of the plant for some considerable time, e.g. in winter barley, floral induction often occurs in November or December but the ear does not appear until the following May. Microscopical examinations of stem apices have allowed the development of growth keys which can have direct bearing on agricultural practices, e.g. nitrogen timing to winter wheat. The stimuli which trigger such changes in the development of the stem apex include photoperiod and low temperature (vernalisation) and in some plants, growth for a set number of days of thermal time.

At the same time as above ground growth and development is occurring, root growth and development is also progressing. Root apices branch to provide a network of structures specialised for the uptake of water and soil nutrients. Apices are located behind a protective root cap and produce cells which go through a cycle of development. Early in development new external cells have root hairs which increase root surface area and therefore absorptive area for water and nutrients.

Normally in a crop the root system develops in a progressively deeper and more extensive manner maintaining a balance with the shoot size (root:shoot ratio); however, often as rapid spring growth occurs root growth declines. It becomes important therefore to establish a well-rooted crop prior to rapid spring growth.

Reproductive development

Reproductive development starts with the induction of flowering. Externally the plant will continue to expand leaves which have been initiated up to the point of floral induction. By the time the floral structures are visible to the naked eye they are already fully developed with stamens bearing immature pollen and ovaries with developed ovules. The actual process of flowering involves the opening of the flowers, rupture of the stamens and release of pollen (called anthesis), pollination of the stigma, growth of the pollen tube and, finally, fertilisation of the ovules. In relation to the whole period of floral development, flowering is only a brief phase.

Some crop plants show remarkable synchrony of flowering, e.g. cereals, with all florets flowering within a few days (determinate flowering). Others, however, have a protracted flowering period lasting weeks (indeterminate flowering), e.g. field beans and oilseed rape. Determinate flowering is a desirable feature in a crop plant since it leads to an evenness in ripening.

Seed development

Following fertilisation cell proliferation of embryo and storage cells occurs. The embryo tissues develop into a polarised structure which defines the sites of the stem apex and the root apex. Often the shoot apex undergoes further development and will produce leaf primordia, e.g. in wheat the first three to four leaves are already initiated in the embryo before the seed is shed from the plant. The mass of storage tissue cells become filled with storage compounds – starch, lipids, protein.

The duration of seed filling is temperature dependent, lasting longer at cooler temperatures. If the supply of assimilates during this time is limited the seed may not fill to its maximum capacity and will appear shrivelled or may lead to seed abortion. In a crop plant therefore, greater seed filling capacity will occur in a cool environment where the length of seed-fill will be protracted, provided the crop is not stressed by drought and provided photosynthesis is not impaired. This contributes to the higher yield potential of winter wheat in Scotland compared with southern England. As seed filling progresses, the moisture content of the seed falls. At the end of the seed filling, the process of maturation begins with continued loss of moisture from the seed, the cessation of development of the embryo and frequently the induction of dormancy. Vascular connections between the seed and the mother tissues are discontinued and the seed becomes an independent body. Some further seed development can occur after seed shed but is usually completed when moisture content declines to 5–10%.

Agriculturally, seeds are of major importance as the main food source for animals and man (e.g. wheat, barley, rice maize, beans), as well as providing many beverages (e.g. coffee), medicines (e.g. evening primrose), spices (e.g. mustard), and industrial oils (e.g. linseed).

Senescence

Senescence is a two phase process the first of which is an organised breakdown of the tissues where many of the products of the breakdown can be re-mobilised to other parts of the plant. In this way, certain nutrients can be re-used for new growth. The initial stages of senescence are not visible to the naked eye but involve a depression of photosynthetic activity followed by a dismantling of the photosynthetic apparatus leading to a breakdown of chlorophyll. The second phase of senescence is the rapid self-destruction of cellular integrity by powerful hydrolytic enzymes.

Following senescence, leaves are often shed from the plant after the formation of an abscission zone in the stem at the point of attachment of the leaf. The cells in this zone become filled with inert substances which effectively act as a barrier to pathogens at the scar left when the leaf is shed. Abscission layer formation is a feature of dicotyledonous plants; monocotyledonous plants such as cereals do not form such layers.

Senescence is occurring throughout the life of a crop plant as old leaves are being replaced by young leaves. However, 'crop senescence' is a term which is usually applied to a crop where the rate of senescence is greater than the rate of growth and the net effect is that the crop is dying. Crop senescence is usually noticeable from about mid-seedfill onwards.

ENVIRONMENTAL INFLUENCES ON GROWTH AND DEVELOPMENT

Temperature

Temperature and growth rate

The growth rate of most plants responds to changes in temperature (*Figure 3.4*). The maximum temperature for growth (T_{max}), the optimum (T_{opt}) and the base temperature (T_{base}) vary between species but for the UK crops typically all fall between 0 and 35°C. For crops native or well adapted to the UK, e.g. perennial ryegrass and winter wheat, T_{base} is around 0–1°C. However, for crops originating from warmer climates, e.g. maize and sugar beet, T_{base} is around 10°C. As a consequence, winter wheat and ryegrass are able to grow early in the spring (March and April) while maize and sugar beet do not commence rapid growth until much later (June/July). In a temperate climate this has important consequences with respect to canopy development (*see* p. 65), radiation capture and yield potential.

Temperature and development

The progression of a plant through a sequence of development is closely related to the temperature it experiences over a period of time. Temperature varies both diurnally (within a day) and on successive days and plants integrate their response to temperature over time. Thus, the amount of heat a plant experiences in a day can be called its daily heat unit and a set number of heat units is required between developmental stages (*Table 3.1*).

Some heat unit systems are simply the sum of the average daily air temperatures while others use the daily maximum and minimum air temperatures in a more complex formula,

Table 3.1 Examples of heat units in relation to various crop development stages

Crop	Developmental stage	Heat unit requirement	T_{base} (°C)
Maize	Sowing to silking	1400 OU	10
Maize	Sowing to 30% DM	2500 OU	10
Winter wheat	Sowing to 50% crop emergence	110–150°Cd	1
Winter wheat	Successive leaves	120–150°Cd	0
Winter barley	Grain-fill	650°Cd	0
Lupins	Sowing to first flower	260–390°Cd	4

Notes: OU = Ontario Unit
OU = X + Y where
Y = 3.33 (T_{max}–10)–0.084 (T_{max}–10)
T_{max} = Daily max. temp.
X = 1.8 (T_{min} = Daily min. temp.
°Cd = degree centrigrade days

e.g. Ontario Units for maize (*Table 3.1*). With very short crops soil temperature may give a better indication of the temperature experienced by the crop.

Heat units may be used in crop production in at least two important ways: firstly, with reference to past meterological data, areas can be mapped for the suitability for growing particular crops. For example, forage maize can only be grown successfully in the UK in areas which experience 2500 Ontario Units (OU) or more during the months of May to

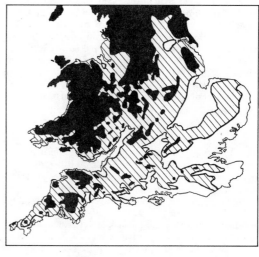

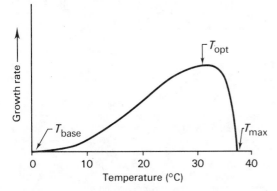

◼ < 2300 OU

▨ 2300 – 2500 OU

☐ > 2500 OU

Figure 3.5 Areas of England and Wales mapped for suitability for Ontario Unit accumulation nine years in ten (1941–1970). Successful forage maize requires in excess of 2500 OU (After Hough, M.N. (1978). *ADAS Quarterly Review*, 31)

Figure 3.4 Growth rate response of plants to temperature

October inclusive. This restricts potential production areas to southern lowland England (*Figure 3.5*). Secondly, by using up-to-date meterological data the stage of development of a crop in the field can be pinpointed and its subsequent development over the forthcoming week or two can be predicted. Where husbandry inputs are very time responsive such prediction can be of great benefit in helping to achieve optimum use of inputs. The use of such prediction schemes are available for nitrogen and growth regulator timing to cereals.

Heat units may be modified by photoperiod which can effectively delay or hasten the thermal response. Temperature can also affect development by acting as a trigger for a change in the development sequence. In particular, low temperature can trigger flowering (vernalisation) and dormancy.

Vernalisation

Many biennials (e.g. sugar beet), winter annuals (e.g. winter cereals) and perennials (e.g. perennial ryegrass) require a period of vernalisation over winter before flowering is possible the next spring. In the UK, this is normally achieved quite adequately during the winter but if winter cereals are sown very late (after February–March) then flowering and yield may be seriously impaired. Within winter wheat there exists varietal differences in the vernalisation requirement and care should be taken not to sow varieties with low vernalisation requirement early in the autumn and conversely, not to sow varieties with a high requirement too late.

Maincrop sugar beet production exploits only the first year of the biennial cycle of this plant and flowering in the first year (bolting) reduces crop yield, interferes with machinery and can introduce a weed beet problem. The vernalisation requirement of the crop usually prevents flowering in the first year but if the crop is sown early in the spring or if unusually low temperatures are experienced, flowering of a proportion of the crop may occur. Plant breeders have attempted to produce varieties of high vernalisation requirement so that early sowing can be practised.

Dormancy

Low-temperature induced dormancy can be regarded as a survival strategy in certain plants. True dormancy is common in woody perennials and bulbs but is rare in UK agricultural crops. Perennial ryegrass exhibits a winter quiescence which appears similar to dormancy but upon close study it is found that new leaves are continually being produced at the same time as the death of old leaves. Low temperatures in winter and the lack of net growth are often mistakenly referred to as dormant periods. Close inspection would reveal that development has progressed whereas in true dormancy development ceases.

Dormancy is also apparent in seeds and helps prevent germination in unfavourable environments. Thus weeds which have little or no frost resistance in the vegetative state require a cold shock to the seed (obtained over winter) to break their dormancy before germinating in the following spring. In this way the weed avoids frost which may kill off the species.

Other dormancy breaking stimuli include extremes of heat, exposure to light, exposure to high concentrations of specific ions, degradation of seed coverings, washing away of inhibitors and the passage of time to allow seeds to fully mature or ripen. Seed dormancy is common in weed species and this, coupled with the prolificacy of seed set of many weeds, makes it virtually impossible to eliminate completely a weed species from a field. In contrast, crop species have had dormancy mechanisms bred out of them to ensure rapid and even germination after sowing. Even so, cereals retain some dormancy and it helps prevent sprouting in the ear before harvest which is extremely important both for bread-making wheat and malting barley.

Temperature induced stress

Temperature stress to plants can be caused by extreme heat or cold although heat stress is not normally encountered in UK crop production. Cold stress however, particularly frost damage, commonly occurs and can lead to complete plant death or to a degree of damage. Some crop plants have little or no inherent frost tolerance, e.g. maize and potato, and for this reason are restricted in their cropping to the frost-free months (May–October) in lowland areas. Other crop species have varying degrees of frost tolerance and there is frequently varietal variation within a species.

The obvious advantage of frost resistance in a crop species is that it allows the crop to be planted in the autumn which generally raises the yield potential. Species of plants which are capable of withstanding frosts in the winter are not equally frost-hardy throughout their life cycle. During the declining temperatures and daylengths of autumn they undergo a conditioning process known as hardening. This raises the frost-hardiness in 'anticipation' of the ensuing winter. In the following spring, as temperatures and daylength increase this hardiness declines and the crop may become susceptible to damage by late frosts. Hardening temperatures are typically in the range 0–8°C.

Some plants escape damage by frost by frost-avoidance mechanism. The extreme of this is for the plant to exist in a very resistant form, e.g. a seed or a bulb. Another avoidance mechanism is undercooling (or supercooling) where the tissues attain a temperature below their freezing point without the formation of damaging ice crystals. This may only be effective for mild frosts (−1 to −2°C) of short duration and can be dependent upon the absence of ice nucleators (e.g. certain bacteria). Other plants protect their sensitive tissues with layers of hardy tissue which delay and therefore prevent the penetration of damaging frosts, e.g. the wrapper leaves on winter cauliflowers and the bud scales on woody deciduous perennials. With some high value crops it can be worthwhile investing in frost protection measures by the use of polythene sheeting, e.g. early potatoes, or by fogging or misting techniques, e.g. blossom protection in orchards.

Light

Photoperiod

Plants have adapted to use photoperiod in many ways because it is the only true predictable parameter of the environment detectable by the plant. Notable examples occur with flowering and germination which help a species to synchronise critical phases of development. All photoperiodic responses involve the compound phytochrome as the detector compound at the cellular level. Phytochrome has a light absorbing pigment (chromophore) and a protein fraction comprising of many reactive amino acids. It has two

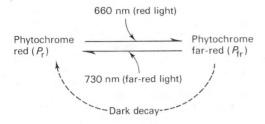

Figure 3.6 Relationship of phytochrome with light

forms, P_r (which absorbs red light) and P_{fr} (which absorbs far-red light), which are interchangeable on the absorption of the correct wavelength of light (*Figure 3.6*). Furthermore, P_{fr} decays back to P_r in the dark. The exact mechanism of action of phytochrome is still not clear but there is considerable evidence to suggest that P_{fr} is the active form and is involved in gene regulation and hormone activity.

In respect of flowering, plants can either be photoperiodic responsive or day-neutral. Day-neutral plants flower after a set time from germination or dormancy breaking, e.g. annual meadow grass. Photoperiodic responsive plants are either 'long-day' or 'short-day' types requiring exposure to the correct daylength to either start flowering or to accelerate flowering. Temperate plants are frequently long-day types which helps ensure flowering in the summer months.

Light capture by crop canopies

The crop canopy is defined as the structure of leaves, stems and flowers found above ground. At full development, the depth of the canopy can vary from 30 cm (grass) to more than 30 m (agroforestry).

It is important that for a crop to achieve high growth rates and high yield potential it must (i) fully intercept the incoming solar radiation, (ii) make efficient use of that radiation within the canopy, and (iii) produce its canopy at a time when incoming radiation is at its maximum.

It is found that the growth rate and the dry matter production of many crops is linearly related to the quantity of radiation intercepted during the life of the crop. Incoming radiation and therefore cropping potential varies considerably throughout the year in temperate regions with average daily radiation totals of 16 MJ m^{-2} d^{-1} in June/July falling to only 2 MJ m^{-2} d^{-1} in December/January.

Within a day, actual radiation receipts (the daily insolation) depend on the time of year, which dictates the daylength and maximum radiation intensity, and the degree of cloud cover. Cloud cover may reduce the actual radiation reaching the crop by as much as 75% compared with a clear sunny day. In this way excessive cloud cover can reduce cropping potential whilst in years of high radiation receipt (little cloud) cropping potential is raised. This was certainly a major contributory factor in the record-breaking yields obtained from cereals in 1984.

Measurements of crop canopies have revealed two important expressions which are useful: Leaf Area Index (LAI) and Leaf Area Duration (LAD). LAI is a measure of the surface area of the photosynthetic area of a crop per area of ground occupied whilst LAD is the integral of LAI with time (LAI × time). Thus, on a graph of LAI *versus* time, LAD is the area under the LAI curve. The value of LAI that has to be achieved by a crop before efficient interception of radiation occurs varies with the crop and the plant spacing

but commonly needs to exceed 3 to 4. Some crops that are well adapted to improving radiation conditions in the spring, e.g. cereals, are able to adapt their canopy structure and raise their LAI to more than 10 and achieve high growth rates and high physiological efficiency.

Factors affecting light interception by crops

Early sowing Advancing the sowing date of most crops leads to a better chance of early ground cover and a better timing of the peak LAI closer to mid-summer when radiation levels are higher. This effect is maximised if the crop is frost resistant and sowing can be made the previous autumn, e.g. cereals, oilseed rape. Such early sowing usually also prolongs the growing period allowing more radiation to be intercepted by the improved LAD. This can be illustrated with reference to wheat (*Figure 3.7*) where autumn sowing leads to increased yields through improved canopy timing, higher LAI and increased growing period leading to greater light interception over the life of the crop. Many crops are, however, limited in the degree to which the sowing date can be advanced by temperature limitations, e.g. sugar beet, potatoes, maize, peas. Here, the advantages of early canopy development are offset by the risks of frost damage or poor slow germination. In some high value cash crops temperature protection can be supplied by plastic film which act as a 'mini-greenhouse', e.g. first early potatoes, early vegetables and sweetcorn.

Seedrate Increasing seedrate can have the advantage of producing an early rise in LAI with ground cover occurring early. This may be of particular importance in crops with short growing periods, e.g. spring and summer sown crops. Indeed, in spring cereals seedrate becomes more and more critical with later and later sowing. For crops with long growing periods compensatory growth (side-branching, tillering) frequently diminishes the advantages of high seedrate. Using high seedrates can produce disadvantages to the crop by leading to excessively high competition between plants. This has the effect of lowering the size of important

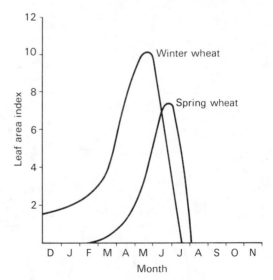

Figure 3.7 Leaf Area Index development in winter and spring wheat

structures, e.g. grain size, root size, and can lead to lodging (canopy collapse).

Nitrogen fertilisers   The application of nitrogen fertiliser to most arable crops raises yield potential. This is achieved through a stimulation of cell growth, particularly in the developing leaves, which increases leaf size and thus leaf area index. Also, leaf longevity is improved, i.e. senescence is delayed, which helps to improve LAD. Nitrogen also improves photosynthetic capacity (*see* p. 59).

If nitrogen is available in very high amounts then over-stimulation of growth can occur which can be counter-productive. Leaf Area Index may be increased to such an extent that lower leaves in the canopy are severely shaded and fall below the compensation point (respiration greater than photosynthesis) and such leaves effectively become 'parasitic' on the plant. Also, excess nitrogen may cause a weakening of stems leading to complete canopy collapse (lodging). Furthermore, stimulation of vegetative growth by nitrogen can be at the expense of the growth of important structures, e.g. tubers in potatoes, roots in sugar beet.

Canopy structure   Photosynthesis in the leaves of most crop plants is saturated at radiation intensities of about

$300 \, Wm^{-1}$ (or $150 \, Wm^{-1}$ of photosynthetically active radiation). This means that on many days in spring and summer leaves at the top of a canopy are saturated in radiation for maximum photosynthesis. Unfortunately, such leaves will continue to absorb radiation but will be unable to utilise it and the radiation is therefore wasted because it is not available to the rest of the canopy. There is benefit to crop photosynthesis if the extra radiation can be allowed to penetrate the canopy and be intercepted and used by lower leaves. Canopy structures that allow such penetration have been shown to improve crop growth rate and yield (*Figure 3.8*). It is perhaps not surprising to find that crops such as cereals which are well adapted to the UK climate show the best type of canopy structure.

Common problems depressing canopy effectiveness

Lodging   The collapse of the canopy caused by wet and/or windy weather conditions and/or the weakening of the stem and/or roots will drastically reduce crop growth rate. Lodging is most serious in deep well-structured canopies with high yield potential, e.g. cereals. As well as depressing yield, lodging may also lead to a spoiling of the economic product by soil contamination, rotting diseases and can lead to a surge of weed growth up through the crop. Lodging may

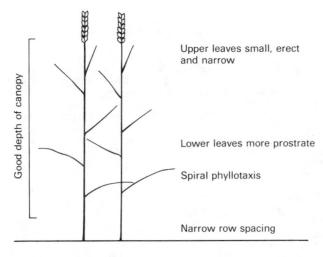

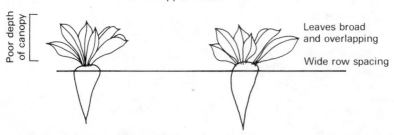

Figure 3.8 Crop canopy structures

not be serious if the crop is nearly ripe but will frequently slow down and delay harvest.

Lodging commonly results from weakened stems through high nitrogen use, high seedrates, stem based diseases or weak stemmed varieties and from poor supportive root systems occurring in unconsolidated seedbeds.

Foliar diseases Foliar diseases depress crop yield by depressing the effective leaf area and thus lowering the LAI. Many diseases are actually affecting a greater leaf area than is apparent from their visual symptoms and this means that they have to be controlled at a relatively early stage in their infection. Infection on lower leaves in the canopy is generally insignificant since these leaves are contributing least to the gross production of the crop. Similarly, in autumn sown crops autumn disease infection is frequently not worth controlling since crop growth rate is slow and protection of leaf area is not as necessary as during the main growth period in spring.

Unfortunately, well structured crop canopies are frequently ideal environments for disease multiplication and spread and in attempting to achieve high yields from good canopies, disease control will normally be necessary.

Drought Most UK crops can withstand a good deal of water stress without injury. However, even mild stress can provoke a drought response which lowers productivity of the canopy. Stomata close to conserve moisture under drought stress and at the same time restrict carbon dioxide uptake into the leaf and depress photosynthesis. If drought stress is severe or prolonged, it will have an adverse effect on canopy structure since older leaves will senesce and die whilst younger leaves may wilt or roll-up. New leaves produced during drought-stress will be smaller and less photosynthetically responsive.

Since the risk of drought-stress is frequently towards the end of a growth cycle it is likely to affect the crucial partitioning of assimilates to the economic portions of the crop and can drastically affect economic yield.

COMPETITION

Competition occurs when two or more plants are growing in an environment and the combined demands of the plants exceed the supply of one or more of the limiting factors for growth and development. These factors include water, soil nutrients, soil oxygen, carbon dioxide and light. Space is frequently referred to as a limiting factor but in reality space embraces two or more of the factors already listed.

Within a crop, competition occurs between plants of the same species and is termed intraspecific competition, and between plants of different species is termed interspecific competition.

Intraspecific competition (plant populations)

In the extreme case of a crop plant growing in complete isolation, i.e. in the absence of competition, its individual yield will give an indication of the maximum yield possible per plant. As the population is increased, competition between plants will increase and individual yield per plant will decrease. However, yield per unit area (crop yield) will increase (*Figure 3.9*). This is because, despite each plant not

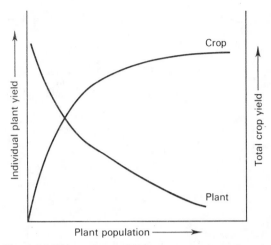

Figure 3.9 Plant and crop yield in response to increasing plant population

fulfilling its full potential production, a more efficient use of the limiting factors for growth is being made by the community of plants (the crop). This is particularly the case with respect to the efficiency of utilisation of light and the ability of the crop canopy to capture more light at higher populations. As population increases the yield response starts to diminish until a plateau is reached and no further response to plant population can be achieved (asymptotic response curve – *Figure 3.10*). In biological terms the optimum plant population in such a case is at the point where the plateau starts whereas in practice the optimum is lower than this when seed costs are taken into account. A dense community of plants also has the advantage of synchronising and accelerating ripening.

Plant density affects the partitioning of assimilates within the plant with high densities tending to lead to a greater vegetative component and a lower reproductive or storage

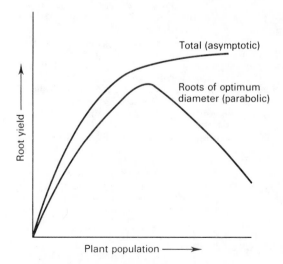

Figure 3.10 Response of root crops to increasing plant population

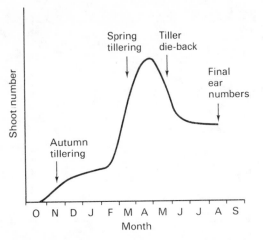

Figure 3.11 Variation with time of the unit of density in a winter cereal crop

component per plant. If the adverse effect of partitioning away from an economic portion is not outweighed by the advantage of increased plants per unit area, then economic yield will decline beyond a certain point (parabolic response curve – *Figure 3.10*). Often an asymptotic yield response is associated with total biomass production and crops grown for complete utilisation, e.g. grass and other fodder crops. In contrast, the parabolic yield response is associated with crops grown for economic yield, e.g. cereals, oilseed rape, potatoes, sugar beet, peas.

When considering crops showing a parabolic yield response obtaining the optimum plant population is very important. However, if a suboptimal population is established then compensatory growth may occur as individual plants can have a greater share of the limiting resources and attain a larger size. In crops able to show a high degree of compensatory growth, e.g. winter cereals (by way of tillering), the parabolic curve becomes a wide flat-topped response curve.

In many crops, especially those which show a high degree of compensatory growth, the unit of population may not always be simple to define. For example, in potatoes it is the stem population and in cereals the tiller population that is of the greatest value. Competitive demands in the crop are not constant but change with time and may lead to alterations in the density of the unit of population (*Figure 3.11*). In the case of cereals, the mortality of stems in the spring is a result of competitive demands within the plant for assimilates and can be greatly offset by applying nitrogen to improve leaf area and thus assimilate supply.

Increasing plant populations can also lead to higher mortality rates within a crop (thinning response). Usually, in a highly competitive situation 'weaker' genotypes or individuals located in an unfavourable position will be disadvantaged and may well die. This could be considered to be a waste of seed but may be a necessary loss to achieve the desired plant population.

Interspecific competition

Competition for resources occurs between species in crop mixtures, e.g. grass/clover swards, cereal/legume arable

silage crops, where two or more crops are harvested (or grazed) together, and between a crop and its weeds. (Intercropping, where two or more crops are grown together but harvested separately, e.g. maize and beans, is very little used in the UK but is common in other countries. The physiological principles of intercropping are identical to those of mixtures.)

The total biomass produced per unit area of ground when more than one species is present will frequently be greater than that produced by one of the species grown alone at the same density. This is because species of different habit are able to utilise the available resources more efficiently, chiefly by better spatial distribution within both the aerial and soil environments, e.g. by forming a more efficient leaf canopy, or by tapping a greater depth of soil for water and nutrients. In general the greater the difference in habit of the component species the greater will be the yield advantage of the mixture, and conversely the less will be the degree of competitive pressure between the component species. This yield advantage is exploited by crop mixtures, but may be disadvantageous in crop/weed situations, depending on the degree of weed competition.

Quantitative analysis of the degree of competition, and its effects, is important to determine (i) the optimum mixture proportion, and (ii) the economic threshold for weed control. Many forms of analysis have been devised appropriate to each type of interspecific competition situation. One of the simplest and most widely useful is the replacement series analysis for two-species competition. Yield data are collected from the monoculture of each species, and from a range of mixture proportions (it is usual to employ the same plant density throughout the experiment though this may lead to bias in the case of two species with widely differing single plant sizes), and analysed using a replacement series diagram (*Figure 3.12*). This form of analysis can

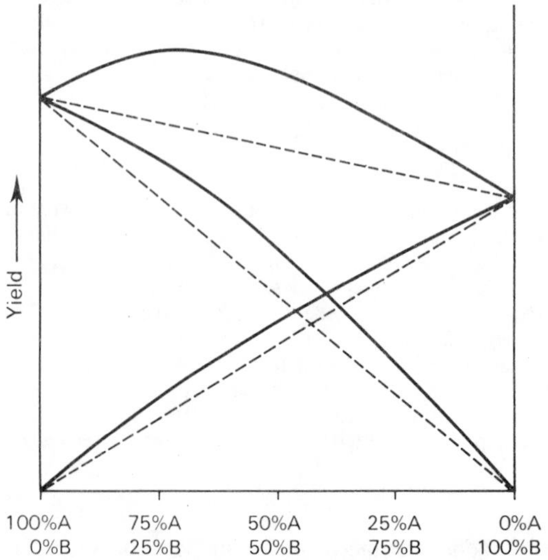

Figure 3.12 Theoretical replacement series diagram to analyse the performance of a mixture of species A and B. Solid lines indicate observed yields, broken lines indicate expected yields (extrapolated from monoculture yields) if no competition were to occur. Diagram demonstrates mutual cooperation with a maximum mixture yield benefit from 70%A and 30%B

indicate the optimum mixture proportion in yield terms, the extent of the yield advantage, and the behaviour of the component species within the mixture. Three cases are possible:

(1) mutual cooperation, in which the yield of both components are enhanced giving an overall yield advantage;
(2) mutual inhibition, in which the yield of both components is depressed giving an overall yield reduction;
(3) compensation, in which one component is enhanced and the other depressed; this case may lead to a yield advantage, a yield depression or a balanced result, depending on the relative effects on each component.

Analysis of this type often demonstrates that one of the component species has competitive superiority. Competitive superiority is conferred by general spatial superiority, e.g. greater height, a broader leaf canopy, better root distribution, all enabling the species to gain a disproportionately high share of the light, water and nutrients, by shading its neighbours, achieving a higher photosynthetic rate, increasing its transpiration rate and thus its rate of mineral uptake, resulting in higher growth rates and even greater competitive superiority. Competitive advantage may also be achieved through temporal means, e.g. weeds emerging prior to crop emergence will tend to dominate, and vice versa, though this relationship will be mediated by intrinsic, and temperature influenced, growth rate and habit differences. Yield advantage from competition may be enhanced by symbiotic relationships, e.g. legumes donate atmospherically fixed nitrogen from their decaying nodules to their non-legume partners. Competitive advantage is gained by some species through allelopathy, the exudation of toxic substances from their roots.

Spatial and temporal distribution interact, and are profoundly influenced both by the environment and the overlying plant density, making physiological interpretation, and general prediction of competitive situations extremely difficult, particularly when more than two species are involved.

MODIFICATION OF CROPS BY GROWTH REGULATORS AND BREEDING

Plant growth regulators

Regulatory compounds known as plant hormones (*Table 3.2*) influence many of the processes of plant growth and development. These compounds act both independently and interactively at several levels of concentration to achieve control. The internal (endogenous) levels of some of these hormones can be manipulated either by applying further hormone to the plant (exogenous application) or by applying synthetic regulators which depress the endogenous hormone levels.

Plant growth regulators may be regarded as a short-term answer to a plant breeding problem though they may in effect be a more cost-effective solution.

Discovery and characterisation of plant growth regulators can be extremely difficult since effects on plants are frequently transitory and compensated by subsequent growth phases.

Table 3.2 The plant hormones and their major functions

Auxins	Cell elongation, cell division, prevents abscission, apical dominance, directs translocation, RNA and protein synthesis
Gibberellins	Cell elongation (particularly stem extension), inhibits organ formation, accelerates germination, enzyme induction
Cytokinins	Releases apical dominance, prevents senescence, mobilises nutrients, stimulates organ formation (in association with auxin)
Abscisic acid	Stomatal closure, gibberellin antagonist induces dormancy, control of embryo development (with cytokinins and gibberellins)
Ethylene	Ripening of fruit, senescence, epinasty, abscission, geotropism
Florigen	Flowering stimulus following photoperiod induction (hypothetical hormone)
Vernalin	Flowering stimulus following vernalisation (hypothetical hormone)
Anthesin	Ripeness to flower stimulus, needed before vernalin and/or florigen are effective (hypothetical hormone)

Plant breeding

The form of any organism (phenotype) is fundamentally determined by its genes (genotype). The degree to which any character is expressed is modified by the environment.

Crossing of individuals or within populations, followed by a lengthy programme of selection enable the plant breeder to change the genotype of an organism and so to modify its phenotype, thus creating a new variety.

The objectives of any breeding programme will be legion but will usually include those leading to an improvement in yield, e.g. improved climatic adaptation, pest and disease resistance, partition of biomass, and those leading to an improvement in quality of product. A considerable proportion of the post-war increase in UK crop production is attributable to the plant breeders. For example, it has been estimated that more than 50% of the yield increase for winter wheat has been due to improved varieties.

Further reading

EVANS, L. T. (1975). *Crop Physiology: Some Case Histories*. London: Cambridge University Press

HARPER, F. (1983). *The Principles of Arable Crop Production*. London: Granada Publishing

HURD, R. G., BISCOE, P. V. and DENNIS, C. (1980). *Opportunities for Increasing Crop Yields*. London: Pitman Advanced Publishing Programme

MILTHORPE, F. L. and MOORBY, J. (1979). *An Introduction to Crop Physiology*. London: Cambridge University Press

WAREING, P. F. and COOPER, J. P. (1971). *Potential Crop Production: A Case Study*. London: Heinemann Educational Books

4

Crop nutrition

J. S. Brockman

All the essential nutrients required by green plants are taken up by the plant in inorganic form, unlike animals where some organic compounds are also needed. Indeed, organic materials have to be broken down to inorganic before they can become of value to the plant. There are 16–19 elements known to be essential for plant growth and these are listed in *Table 4.1*, together with a summary of their main functions in the plant and the form in which they are taken up by the plant. Much further information on the uptake and role of plant nutrients is contained in Mengel and Kirkby (1982).

An essential nutrient is defined as one which is required for the normal growth of the plant, is directly involved in the nutrition of the plant and cannot be substituted by another nutrient. Not all plants have the same essential requirements, e.g. sodium (Na), silicon (Si) and cobalt (Co), are needed by some plant species only. Sometimes essential plant elements are divided into macronutrients (the top 11 elements in *Table 4.1*) and micronutrients (the remaining eight elements in *Table 4.1*). This division is based on the probable content of each element in the plant, but plants differ in their nutrient composition so such a distinction is not always meaningful. Also the importance of a nutrient is not related to the quantity contained in the plant tissues, and an insufficient quantity of any nutrient can lead to a restriction in crop growth and/or quality.

SUMMARY OF SOIL-SUPPLIED ELEMENTS

Nitrogen (N)

Nitrogen is an essential constituent of all proteins, so is vital to any living plant.

Broadly, there are three stages in the utilisation of N by the plant:

(1) *stage 1* uptake of NO_3^- and NH_4^+ ions with N present in the plant in inorganic form;
(2) *stage 2* N present in low molecular weight organic compounds such as amino acids, amides and amines;

(3) *stage 3* N in high molecular weight organic compounds such as proteins and nucleic acids.

If there is plenty of N available in the soil, the rate of N uptake (stage 1) can exceed the rate of protein synthesis (stage 2–stage 3). Whilst this is not harmful over a short period because protein production can catch up with a short-term flush of N uptake, if the imbalance continues for a long period there can be:

(1) excessive elongation of vegetative cells of high moisture content, leading to soft, weak growth, e.g. lodging in cereals;
(2) decrease in carbohydrate production; this can lead to unpalatability in grazed grass, fermentation problems in grass cut for silage and poor quality in plants that store carbohydrate in roots or tubers, such as sugar beet and potatoes.

Deficiency symptoms

These first show in older leaves, as young tissues get priority for any limited N supply available to the plant: older leaves become senile' prematurely, show a general yellow/brown colour and probably become detached from the stem. Nitrogen deficient plants beome spindly, lack vigour and are dwarfed, often featuring less vegetative development, including tillering, and are lower yielding; also they have a pale colour and a low leaf:stem ratio. Often N deficient plants ripen earlier than those receiving adequate N nutrition. Nitrogen deficiency in young plants can be corrected rapidly by application of nitrate-containing fertilisers, but over-application of these fertilisers should be avoided as this will lead to the imbalance problems listed above.

Phosphorus (P)

Phosphorus is associated with the transfer and storage of energy in the plant, as such it is vital for living processes within the cell. A supply of P is essential to the plant during

Table 4.1 Essential nutrients, form of uptake and functions in plant

Element	Form of uptake by plant	Main functions in plant
C	CO_2 from atmosphere (HCO_3 from soil solution)	Major constituent of organic material; can account for 40% of dry weight of plant
H	H_2O from soil water (and from atmosphere if wet)	Linked with C in all organic compounds
O	CO_2 and H_2O; O_2 during respiration	With C essential in carboxylic groups; with H in oxidation-reduction processes
N	NO_3^- and NH_4^+ from soil solution (N_2 from air if fixed by micro-organisms)	Essential constituent of all proteins
S	SO_4^{2-} from soil solution; SO_2 absorbed from atmosphere	Constituent of some essential amino acids, e.g. cysteine and methionine; also forms S–S bridge in enzyme proteins
P	HPO_4^{2-} and $H_2PO_4^-$ from soil solution	Vital constituent of living cells, associated with storage and transfer of energy within the plant
K	K^+ from soil solution	Essential for efficient water control within the plant; formation and translocation of carbohydrate; found in some enzymes
Na[1]	Na^+ from soil solution	Essential in some plants only (e.g. of marine origin); partially interchangeable with K^+ in most plants
Ca	Ca^{2+} from soil solution	Essential role in cell walls and biological membranes
Mg	Mg^{2+} from soil solution	Essential constituent of chlorophyll, some enzymes and some organic acids
Si[1]	Probably $Si(OH)_4$ from soil solution	Used in cellulose framework; interacts with P in plant (mechanism uncertain)
B	H_3BO_3 or BO_3^{3-} from soil solution	Assists in carbohydrate synthesis, uptake of Ca^{2+} and absorption of NO_3^-. Big variation in need for B; risk of toxicity
Mn	Mn^{2+} from soil solution (availability to plant affected by soil pH)	Associated with chlorophyll formation and some enzyme systems
Cu	Cu^{2+} or copper chelates in soil solution	Small quantities needed in chloroplasts and for enzyme systems coverting NO_3^- to protein
Zn	Zn^{2+} from soil solution	Assists with starch formation and some enzyme systems
Mo	MoO_4^{2-} from soil solution	Small quantity essential for enzymes controlling N nutrition (also for N-fixing bacteria)
Fe	Fe^{2+} of Fe^{3+} or Fe chelates	Essential in chlorophyll and enzyme activities; associated with enzymes in photosynthesis
Cl	Cl^- from soil solution	Involved in evolution of O_2 during photosynthesis (excess Cl more common than a deficiency)
CO[1]	Co^{2+} or Co chelates[2] from soil solution	Not essential for most species; essential for N fixation by bacteria and could be used in N nutrition in plant

[1] These three elements are not essential in all plant species.
[2] Chelates are organic-type compounds that can maintain the availability of some metal ions over a wide soil pH range

the early stages of growth and also during seed formation — the latter because P is found in phytin seed reserves that can supply P to the seedling in the next generation. Most P is taken up from the soil in inorganic form but can be converted to organic forms in a matter of minutes: hence most of the P found in young tissue is in organic compounds. Phosphorus in older tissue is mainly in inorganic form.

Deficiency symptoms

These are not easily detected from appearance. Early stages of deficiency cause a reduced growth rate and limitation in root development, often with leaves looking a rather healthy dark green-bluish green colour as the plant is less efficient in responding to other nutrients such as N. Also there will be poor seed formation and poor vigour from any seed that is subsequently used. The plant will appear stunted only when the deficiency is severe, and at this stage the stem can be reddish in colour. Phosphorus deficiency is hard to correct in existing plants once visual symptoms are found, because

such symptoms appear at a late stage in the deficiency. Where plants are to be sown in P-deficient soil, it is essential some water-soluble P, such as superphosphate or ammonium phosphate, is placed near to the seed.

Potassium (K) and sodium (Na)

Potassium is essential for efficient water relationships in the plant, both in controlling water content of cells and movement of water through tissues and in transpiration (by influencing the stomatal guard cells). K is also associated with the formation and translocation of carbohydrate and activation of some enzymes. Plant requirements for K vary from high to very high and many plant species contain more K than any other soil-supplied nutrient. As size of plant increases, so does the need for K; N × K interactions can occur and lack of K will limit full yield increases from N application. The only known form of K uptake by the plant is K$^+$ and this is true whether the K is applied in inorganic or organic materials.

Some plants, particularly those of marine origin, need Na$^+$ as an essential nutrient (e.g. sugar beet, fodder beet, mangolds and spinach). Other plants (e.g. maize, rye, soya and perhaps most cereal species) do not require any Na. For other species there is some acceptable replacement of Na$^+$ for K$^+$ and vice versa. However, in *all* species K$^+$ is needed to fulfil the enzyme activities associated with K.

Deficiency symptoms

For K these are seen as a loss of plant turgidity because the water control mechanisms become impaired, leading to reduced growth rate and lower plant metabolic efficiency. Older leaves show yellow/brown spots on leaves and/or whole leaves beome brown at the edge.

In potatoes and brassicas initial K deficiency is shown by a blue-green colour, leading to bronzing and marginal necrosis of the leaves. In barley the necrosis of leaves can be spotted as well as on the leaf margins and can be whitish-brown or purplish-brown. In legumes the first symptoms are often whitish spots on leaves followed by marginal browning or bronzing. Generally, application of fertilisers containing muriate or sulphate of potash will correct the deficiency provided the deficiency has not reached an advanced stage.

In Na-demanding crops, deficiency of Na shows in reduced plant water movement and an apparent moisture stress. Correction is by the application of an Na-containing material such as Kainit.

Sulphur (S)

Most of the sulphur acquired by plants is taken from the soil as SO_4^{2+} but plants can utilise atmospheric SO_2 absorbed through stomata. S is used in essential amino acids such as cysteine and methionine, and also the presence of disulphide (S–S) bonds is an essential feature of many protein enzymes.

Deficiency symptoms

Sulphur nutrition is closely linked to N and in most crops receiving fertiliser N, a deficiency will show as a failure to respond fully to N. Also a shortage of S will lead to S-demanding proteins and enzymes being in short supply and this can affect crop quality (e.g. insufficient S can lower baking quality of flour).

Crops particularly sensitive to S are brassicas (high content of S-dependent proteins), grass receiving high rates of N (need for S to support action of N) and legumes (high requirement by rhizobial bacteria). Where S is deficient, S-containing fertilisers (superphosphate, ammonium sulphate and potassium sulphate), gypsum or S-coated fertilisers of any type should be used. Normal application rate is 10–50 kg/ha/year of S.

Calcium (Ca)

Calcium is essential for plant growth, being used in cell walls and biological membranes. As soils differ widely in their natural Ca content (and resultant pH) some plants have evolved either a very low need and tolerance for Ca (*calcifuges*) or a high need and tolerance for Ca (*calcioles*).

Apart from this, problems from soil Ca content as such are rare as Ca is the major element controlling soil pH: soil pH is far more important in the overall nutrition of plants than the soil content of Ca. Whilst vegetative growth of plants is restricted rarely by lack of Ca, in some plants poor translocation of Ca within the plant can cause *calcium deficiency* in fruit, e.g. bitter pit in apples and blossom-end rot in tomatoes. Prevention of such Ca deficiency is by spraying calcium nitrate on the leaves when the fruit is setting.

Strontium (Sr)

Strontium is chemically related to Ca and is taken up and deposited in cell walls in much the same way as Ca. If excessive Sr uptake becomes a problem, it can be avoided by heaving liming.

Magnesium (Mg)

Some 20% of the magnesium in the plant is found in chlorophyll and much of the rest is in enzyme systems or associated with the movement of anions (mainly phosphates). Mg uptake is rapid in the early stages of growth. Where the plant has access to abundant levels of K$^+$ and NH$_4^+$ the uptake of Mg can be reduced, leading to induced Mg deficiency. Sensitive crops include sugar beet, potatoes, tomatoes, grass and many fruit and glasshouse crops. In the past many fertilisers contained Mg as an impurity; now this is not the case and Mg deficiencies are most likely on highly leached acid soils and heavily limed sandy soils.

Deficiency symptoms

These are always seen first in older parts of the plant, as young tissues get priority. Typical deficiency symptoms are intervenal chlorosis, with the leaf veins standing out as dark green. In sugar beet the chlorosis first shows on the margins of older leaves and can be followed by black necrotic areas. In potatoes, mottling on the leaf can be followed by reddish/purple tinting and loss of older leaves.

Low levels of Mg can cause foliar symptoms without reduction in yield, although yield loss has been found in severe deficiency in sugar beet and potatoes. In grass, Mg deficiency does not affect the yield of grass, but it can lead to Mg deficiency in stock eating the grass (hypomagnesaemia).

Where Mg deficiency is caused by a genuine lack of available Mg in the soil, the application of 250 kg/ha of magnesite (45% MgO) or 400 kg/ha of kieserite (27% MgO) will help maintain Mg supplies. For rapid action, foliar application of 25–35 kg/ha of Epsom salts (16% MgO) in 40 litres water is recommended. If Mg deficiency is *induced* by K$^+$ or NH$_4^+$, then foliar application is essential for plant deficiency to be corrected, although with grass it is better and more certain to feed an Mg supplement to the stock at risk.

Boron (B)

Boron acts in a similar way to P in hydrolysis reactions and it facilitates sugar translocation. It is an important minor element and its availability is much reduced at high soil pH (e.g. as a result of overliming). Adequate B is essential for good storage quality in root crops.

Deficiency symptoms

They have not been found in UK cereals or grass, but they occur commonly in root crops (particularly swedes and sugar beet), legumes and leafy brassicas. Boron deficiency is characterised by the death of the apical growing point of the main stem and failure of lateral buds to develop shoots. Leaves may become thickened and sometimes they curl. In sugar beet, fodder beet and mangolds mis-shaped, young leaves are the first sign, followed by death of primary leaves and a 'rosette' of secondary leaves covering a scarred root crown. Hollowing may extend from the crown into the root and fungal decomposition can occur. In swedes and turnips plant growth may appear normal, but at harvest the roots may have brown hollow areas known as 'crown rot'. In lucerne terminal leaves become yellow or red, plants are stunted and adopt a rosette form with death of terminal buds and in fruit, formation can be impaired.

Treatment is by use of a boronated fertiliser or borax at a rate equivalent to 20 kg/ha of borax. Alternatively, foliar application can be made using 'solubor' at 5–10 kg/ha in 250–500 litres/ha of water using a wetting agent.

Toxicity symptoms

Boron is toxic to some plants at a level only slightly above that needed for normal growth. Leaf tips become brown and rapid necrosis of the whole leaf can follow. The problem is worse in dry areas and is exacerbated where irrigation water high in B is used. Sensitive crops include runner beans and grapes; semisensitive crops are barley, potatoes, tomatoes and legumes.

Manganese (Mn)

Manganese is involved in chlorophyll formation and enzymatic control of oxidation–reduction processes and in some ways is similar to Mg. Most UK soils contain adequate available Mn, but some organic soils of high pH are deficient and so are some heavily leached podzols. Manganese deficiency can also be found in badly drained soils and can be induced by overliming or deep ploughing of calcareous soils or by the presence of high levels of Mg.

Deficiency symptoms

They are chlorosis of younger leaves (*cf*. Mg, where effects are on older leaves). In oats this deficiency is called 'grey speck'; in barley the leaves have brown spots and streaks; in wheat there are intervenal white streaks. In all three of these cereals maturity is delayed and ear emergence reduced with a high incidence of blind ears. In sugar beet the chlorosis is called 'speckled yellows' and is most severe in early stages of growth, accompanied by an upright leaf habit with curling edges. Brassicas show chlorotic marbling; potatoes have stunted leaves with small terminal leaves rolled forward; peas and beans often show brown lesions on the inner surface of cotyledons, the term 'marsh spot' referring both to the condition on the plant and its occurrence on badly drained organic soils. Although Mn deficiency may be controlled by soil a̶ 125–250 kg/ha of manganese sulphate ng effect

as the Mn is oxidised rapidly. Recommended treatment is foliar application of 6–10 kg/ha of manganese sulphate in 225–1000 litres/ha of water using a wetting agent.

Toxicity symptoms

These are brown spots and uneven chlorophyll on older leaves. Silicon (Si) can minimise the harmful effects of excess Mn.

Copper (Cu)

Only small quantities are needed and the function of Cu in the plant is uncertain, but it is associated with enzymes that convert N to protein and also the redox processes in cells. Cu is a constituent of chloroplasts and aids the stability of chlorophyll.

Deficiency symptoms

They can be found in fruit and some cereal crops, particularly wheat growing on some peats in well-defined areas, on leached acid sandy soils such as heaths and on some shallow, puffy chalks with high organic matter content. In cereals, deficiency symptoms are not seen until the plants are well-developed, at which time the symptoms can change rapidly from yellowing of the tips of youngest leaves to spiralling of leaves, giving a stunted, bushy appearance to the plant, ears have difficulty emerging, have white tips and are devoid of grain. Wheat grown on copper-deficient chalk soils specifically can show blackening of ears and straw.

Treatment can be by application of 60 kg/ha of copper sulphate to the soil — a treatment that can last two to three years on peaty and sandy soils but not on the deficient chalks: otherwise foliar application of 2–3 kg/ha of copper oxychloride, cuprous oxide or chelated Cu can be effective when applied at a fairly late stage of growth (and this is when symptoms may first show).

Toxicity symptoms

When present in excess, Cu^{2+} replaces other metal ions, particularly Fe^{2+}; root growth is restricted and chlorosis occurs.

Iron (Fe)

Iron is essential for the proper functioning of chlorophyll and related photosynthetic activity. Iron uptake is strongly related to soil pH and some species, such as sugar beet, brassicas and beans, can show lime induced Fe deficiency, particularly where soil aeration is poor (e.g. in a badly structured soil).

Deficiency symptoms

These are chlorotic markings on younger leaves (in contrast to Mn deficiency which can affect all leaves irrespective of age). Iron deficiency is found more widely in horticultural crops than agricultural ones, being found particularly in fruit and calcifuge plants such as azalea and rhododendron. The only effective treatment for deficiency is foliar application of an iron chelate (*see Table 4.1*).

Molybdenum (Mo)

Molybdenum is essential for the enzymes nitrogenase and nitrate reductase. It is also important for N fixation by rhizobial bacteria. Mo deficiency can occur on some acid soils and on soils of high pH. Excessive Mo availability in the soil can induce low Cu uptake, where in grassland it gives rise to 'teart' pastures with Cu deficiency in grazing stock. Excess Mo in herbage can cause toxicity in animals.

Deficiency symptoms

They are found in cauliflowers where hearts fail to form in plants nearing maturity, leading to 'whiptail'. Sometimes liming will alleviate whiptail, otherwise plants should be sprayed early in their life with 0.25–0.5 kg/ha of sodium molybdate solution.

Zinc (Zn)

Plants need Zn in very small quantities only and there are signs that an increasing number of crops may respond to Zn application. Apart from fruit trees, other sensitive crops are maize, flax and field beans (and perhaps other legumes). Excess Zn can be toxic to plants and induce Fe deficiency. Some sewage sludges contain large amounts of Zn and regular use of such materials on the same area of land can lead to an undesirable build up of Zn. Although grass is not itself affected by excess Zn, it will raise the concentration in the herbage and this can lead to high levels of Zn in milk. Zn deficiency can be corrected by using 4–10 kg/ha of $ZnSO_4$ as a fertiliser or by using a Zn chelate.

Silicon (Si)

Silicon is the second most abundant element in the lithosphere after O, and so it is not likely to limit plant growth. Si is used by plants in the cellulose framework of cells and plants deprived of Si under experimental growth conditions have a very limp growth habit. Application of up to 450 kg/ha of sodium silicate has been shown to increase P availability on some soils.

Cobalt (Co)

Cobalt is essential for micro-organisms which fix N, but is probably not essential for higher plants. Excess Co can induce deficiency of both Fe and Mn.

Chlorine (Cl)

Chlorine is essential to plants in small quantities only, where it is used in processes connected with the evolution of O_2 during photosynthesis. As large amounts of Cl are applied with K in muriate of potash, the effects of excess Cl are common, particularly on soils that have been affected by salt. Cl toxicity is seen as burning of leaf tips, bronzing and premature yellowing of leaves. Sugar beet, barley, maize and tomatoes are tolerant, but potatoes, lettuce and many legumes are sensitive to excess Cl. Where there is a risk of a sensitive crop being affected, sulphate of potash should be applied in place of muriate.

Potentially toxic elements

Some elements such as iodine, fluorine, aluminium, nickel, chromium, selenium, lead and cadmium are not essential for plant growth and their presence can lead to toxicity. For more information *see* Mengel and Kirkby (1982).

ORGANIC FERTILISERS AND MANURES

Organic fertilisers are derived from either plant or animal materials. Not all the nutrients contained in such materials are in organic form and those that are in organic form are not readily or completely available to plants. Complex organic compounds will become part of the soil organic cycle and could perhaps have an eventual nutrient value, depending on the activity of the soil biomass.

Compared with inorganic sources of nutrients, organic sources have the following features:

(1) not immediately soluble in water and so not readily leached;
(2) because they have to break down to become partially soluble, they can act as a slow-release source of plant nutrient;
(3) they can be applied at heavy rates without risk of injury to roots or germinating seeds as they have little ionic activity;
(4) they can stimulate microbial activity;
(5) they are much more costly per unit of plant food (unless by-products);
(6) the recovery of nutrients contained in the materials is low.

Because of the above features, organic fertilisers have little general use in agriculture, but they do have a place in market gardening and horticulture where the slow-release characteristics have application for some of the high-value crops

Table 4.2 Fertilisers accepted by British Organics Standards Committee (1983)

(1)   *Manures and composted manures* from animals kept on non-intensive systems. *Slurry* should be aerated before application, particularly when used on green crops for human consumption. Some *spent mushroom compost* may be approved and *sewage sludge* can be used on grass and forage crops at a maximum of one year in three.
(2)   Of *animal by-products* only *dried blood* is approved.
(3)   *Vegetable by-products* can be used only if they do not contain residues of agrochemical treatment: this applies also to wood products.
(4)   The following *mineral fertilisers* are approved: all other mineral fertilisers, including *urea* are prohibited:
      basic slag, gro___d phosphate rock, dolomite (magnes___ ___ ___ ___ ___ ___ ___ ___ dularian shale (ro___ ___ ___ ___ ___ ___ ___ ___resh *seaweed*

that are grown. In recent years there has been increasing interest in 'organically grown' food: at present there is no generally accepted standard code of practice. *Table 4.2* sets out the approved fertilisers recommended by the Soil Association and from this it can be seen that not all organic materials are approved.

Organic nitrogen fertilisers

Hoof and horn meals, hoof meal, horn meal

These contain 12–14% N and can be obtained either coarsely or finely ground: fine materials release N more quickly. They should be worked into the soil before planting or sowing.

Dried blood

This contains 12–13% N, is very expensive but of great value in glasshouse crops where it is quick-acting.

Shoddy

Analysis varies from 3 to 12% N, depending on the proportion of wool contained amongst other wastes. Shoddy is a very slow-release material, should be worked into the soil before planting and should be analysed before purchase.

Organic phosphate fertilisers

Bone meals

These contain 20–24% P_2O_5 (insoluble in water) together with 3–4% N. The phosphate acts very slowly and is of most value on acid soils.

Steamed bone meals and flours

These contain 26–29% P_2O_5 (insoluble in water) together with about 1% N. These materials are made from bones that are steamed to obtain glue-making substances and a good deal of the N is removed in the process. The bones are generally ground after extraction and the phosphate acts more quickly than in ordinary bone meal.

Meat and bone meal (also meat guano or tankage)

They contain 9–16% P_2O_5 (insoluble in water) together with 3–7% N. These are made from meat and bone wastes and analysis varies; the phosphate has slow availability.

Fish and meals and manures (also fish guano)

These contain 9–16% P_2O_5 (insoluble in water) together with 7–14% N. They are waste products from fish processing.

Organic potash fertilisers

Potassium does not occur in chemically organic form but it does occur in organic materials (such as livestock manures and bird guano). Also potassium occurs in some 'natural' materials (such as wood ash, mica and adularian shale).

Livestock manures

On stock farms large quantities of faeces and excreta are accumulated from housed animals. Quite apart from the environmental and health aspects of the storage and disposal of these products, all animal manures have some value as plant nutrients. The application of these manures to farmland can represent both a relatively safe and a money-saving method of disposal. Before outlining the probable nutrient value of animal manures (including slurry) it must be stressed that all such materials can be extremely variable in composition for two main reasons:

Variation in nutrient content

The main factors affecting the nutrient content of the manure as collected for field distribution are:

Table 4.3 Guide to composition and nutrient value of manures

Type of stock	Body weight (kg)	Faeces and urine voided (l/d)	Composition					Available nutrients[1]		
			Approx. DM%	% of fresh weight			N	P_2O_5	K_2O
				N	P_2O_5	K_2O			
Slurry (undiluted)							(kg/m³ or kg/1000 litres)		
Dairy cow	550	42							
Two-year bullock	400	27	10	0.5	0.2	0.5	2.5	1.0	4.5
One-year bullock	220	15							
Pig (dry meal feed)	50	4	10	0.6	0.4	0.3	4.0	2.0	2.7
Pig (pipeline fed)	50	7	6–10	0.5	0.2	0.2	2.0–3.5	1.0–2.0	1.5–2.7
Pig (whey fed)	50	14	2–4	0.3	0.2	0.2	0.8–1.6	0.4–0.8	0.8–1.5
1000 laying hens	2000	114	25	1.4	1.1	0.6	9.1	5.5	5.4
Farmyard manure (FYM)							(kg/t of material)		
Cows and cattle			25	0.6	0.3	0.7	1.5	2.0	4.0
Pig			25	0.6	0.6	0.4	1.5	4.0	2.5
Poultry (deep litter)			70	1.7	1.8	1.3	10.0	11.0	10.0
Poultry (broiler litter)			70	2.4	2.2	1.4	14.5	13.0	10.5
Poultry (dried house droppings)			70	4.2	2.8	1.9	25.0	17.0	14.0

[1] Availability from spring application; for availability of N from autumn and winter application see text. Table based on MAFF (1979)

(1) the quantity of excreta produced, based on the size and type of animal involved (*see Table 4.3*);
(2) the composition of the excreta is influenced by the animals' diet;
(3) the method of collection and storage of excreta, involving the degree of dilution by rainwater and/or washing-down water or straw.

Variations in losses during storage

The main factors are:

(1) gaseous loss of N as ammonia; the quantity lost will vary with the conditions of storage and temperature. Loss of 10% N is average but loose-stacked farmyard manure (FYM) that is turned prior to spreading can lose up to 40%. Where slurry is stored for long periods, 10–20% of N can be lost and agitation before removal will aggravate the loss;
(2) leaching losses from FYM stored in the open are in the range 10–20% of N, 5–8% of P_2O_5 and 25–35% of K_2O: the flatter the heap the greater the likely loss;
(3) seepage losses from slurry can result in 20–25% loss of N, a little loss of P_2O_5 and 20–30% loss of K_2O.

The above sources of loss can be additive, so that material that has been stored outdoors for a long period will have a greatly reduced manurial value due to gaseous, leaching and seepage losses.

Table 4.3 sets out the standard ADAS guidelines for estimating the composition and likely nutrient value of livestock manures. Points to note are the relatively high K value associated with cattle manures and the high N value of poultry manure.

Sometimes a simplified approach to the estimation of the available nutrients will be of value, based on the number of days over which the material is collected. For example, if dairy cows are kept indoors over a 200-day winter, then each cow will produce 200 × 42 = 8400 litres of undiluted slurry. From *Table 4.3*, it can be seen that each 1000 litres of cow slurry should be worth 2.5 kg N, 1.0 kg P_2O_5 and 4.5 kg K_2O. So the following estimates can be made:

	Per cow	
	Over 200 d	*Daily rate*
N (kg)	21.0	0.105
P_2O_5 (kg)	8.4	0.042
K_2O (kg)	37.8	0.189

If this slurry is spread over a known area, then a quick estimate can be made. For example, if there are 80 cows kept indoors for 200 days and all their slurry is spread over 25 ha of grass, then the potash made available is (80 × 37.8)/ 25 = 121 kg/ha of K_2O.

The nutrient values in *Table 4.3* are based on assumed spring application, as this is normally the most effective time for nutrients contained in manures — particularly those liable to leaching over winter. If manures are applied at other times in the year, then the N value should be reduced to the following comparative effectiveness:

> if spring application = 100 then
> autumn = 0–20
> early winter = 30–50
> late winter = 60–90

Phosphate and potash values are not affected by leaching and should not be reduced.

Whenever livestock manures are applied to a crop, their nutrient value and cost saving potential should be recognised and adjustment made in the quantity of inorganic fertilisers used. Standard fertiliser recommendations normally allow for this, for example:

Fertiliser recommendations for maincrop potatoes
N index = 1,   P and K index = 2

	kg/ha to apply			
	N	P_2O_5	K_2O	
Total required	160	250	250	(A)
From 50 t/ha of FYM	75	100	200	
From fertiliser	85	150	50	(B)

It should be noted that not only is much more fertiliser needed where no muck is applied (situation A), but also the N:P:K ratio is different when the nutrients in muck have been deducted (situation B).

Risks from application of livestock manures

Hypomagnesaemia in grazing stock

Cattle farmyard manure and slurry are rich in potash and where high rates of such manures are applied in late winter and spring to grassland, there will be an increase in the potash content of the grass. Spring herbage is naturally low in Mg, and as a high plant content of K depresses Mg content, then there is an enhanced risk of hypomagnesaemia in animals grazing grass that has received cattle manures. Where such manures have to be applied to grass that is allocated to spring grazing, then it should be applied well before Christmas.

Animal disease

The main hazard that exists is from the infection of pasture with bacteria of the *Salmonella* group — normally as a result of applying infected cattle or poultry manures and slurries. The following advice is offered to minimise the risk:

(1) as most problems arise where fresh material is applied to grass, store the manure for two to three weeks before application;
(2) allow rain to wash the manure from the herbage before grazing;
(3) ensure water courses are kept free from contamination.

Toxic elements

The main risk is from manure obtained from fattening pigs, as this can contain high levels of Cu and Zn, both derived from feed additives. Cu and Zn build up slowly in the soil and frequent application of pig manure to the same field can lead to potential toxicity problems. Particular risk arises on grazed grassland, where evidence suggests that stock suffer more from slurry contamination on the herbage than high Cu and Zn levels in the herbage itself: so physical contamination should be allowed to disappear prior to grazing.

Specifically, sheep should *never* be allowed to graze grass contaminated with Cu as extensive liver damage can occur.

Pollution

All animal manures have a high moisture content and under some storage treatments this can lead to loss of effluent from the store. To minimise the risk of effluent it is vital that additional water (rain water, washing-down water) is kept out of manure and slurry stores.

There are strict legal requirements concerning the discharge of effluent into water courses, quite apart from the loss of potential plant nutrients that results. The legal aspects of effluent also apply to fields where heavy dressings of manure lead to effluent being lost into field drains or from surface run-off. Consqently it is unwise to:

(1) apply very heavy dressings of manures during autumn and winter;
(2) apply any manure at all in winter to sloping fields near ditches and water courses.

(*See also* later section in this chapter, 'Fertilisers and the environment'.)

Management aspects of use of livestock manures

Mixed grass/arable farms

Where suitable arable crops are grown, the most beneficial crops for much application are potatoes, maize and cereals.

All grass farms

These produce more manure per total farm hectare than mixed grass/arable farms and have fewer suitable crop situations for its use. Suggested areas during the year are:

(1) *November/December* apply to next year's spring grazing area;
(2) *January–March* apply to areas for spring conservation cuts;
(3) *May/June* apply thinly to areas cut for conservation in spring.

Other points to note are:

(1) High rates applied to grass can cause physical shading of the grass and loss of stand, particularly if the grass is not growing rapidly at the time.
(2) Cattle manures are high in potash and its use in spring can predispose hypomagnesaemia.
(3) Most manures cause taint and subsequent refusal problems by grazing stock: some farmers find sheep graze behind cattle slurry better than cattle.
(4) There is a *Salmonella* risk where fresh cattle slurry is applied to grass regrazed by cattle.
(5) Pig slurry is often high in Cu and Zn: grass should be free from slurry contamination before grazing, particularly with sheep.

Pig and poultry units

These are always 'exporters' of manure and this manure requires disposal at regular intervals throughout the year as the stock are nearly always housed. A mixed farm can more easily accommodate and efficiently utilise pig and poultry manure than either an all-grass or all-arable farm. Poultry manure has a higher N content than other manures and is valuable for application to grassland in the March–September period.

Straw contamination

Where manure contains much fresh, unrotted straw, it is posible that the soil bacteria breaking down the straw need more N than is contained in the manure: thus the material can *deplete* N status in the short term. To correct this, additional fertiliser N may be used at the rate of 10 kg of N for every 1 tonne of fresh straw used: note that this N should be applied to the next crop and *not* to the manure or the soil at the time of manure application.

Other organic manures (*see Table 4.4*)

Straw can be incorporated into the soil, preferably after it has been chopped, and this will provide some phosphate and potash but as noted above, it may cause a deficiency of nitrogen that will need correction in the first year or so. However, in the long term, regular annual incorporation of straw will stimulate the N cycle in the soil so that 20–30 kg/ha of *extra* available N per year can be anticipated.

Where straw is burnt a small quantity of phosphate and potash residues are left but no N.

Composts can be made from straw alone, straw mixed with another crop waste, or from other wastes alone. As a general rule all such materials suffer from lack of sufficient nitrogen for proper bacterial decomposition and it is recommended that ammonium nitrate is added at a rate of 30 kg of fertiliser/tonne of composting material. Subsequent rotting down may well release N as ammonia unless the mass is kept under anaerobic conditions, and so the resulting compost will still be deficient in N. For this reason good compost is made from very short-cropped material that is well-consolidated and kept in a compact heap. Even so, composts are not highly rated for their nutrient value and it is often easier to apply waste materials to suitable fields at the earliest opportunity as a disposal exercise, plough them in and let decomposition take place within the soil: in this way any available nutrients find their way into the soil/plant complex at minimum cost.

Table 4.4 Analysis of some organic materials

	DM%[1]	% composition as spread[2]		
		N	P_2O_5	K_2O
Straw compost	25	0.40	0.20	0.30
Fresh seaweed	20	0.20	0.10	1.20
Liquid raw sludge	4	0.17	0.15	trace
Liquid digested sludge	4	0.21	0.20	trace
Dewatered digested sludge	50	1.50	1.10	trace-0.50
Dewatered sludge with lime	40	0.90	0.90	trace-0.50

[1] Sometimes expressed as the synonymous 'per cent total solids'
[2] As with all organic manures, composition can vary widely and not all the nutrients are either immediately or eventually available

Sewage sludge is often available, either from cesspits on the farm or from urban treatment works. Depending on the treatment process different types are produced: *liquid raw sludge* is material taken straight from sedimentation tanks: this material can be allowed to digest anaerobically to remove oils and reduce pathogens, giving *liquid digested sludge*. The above materials may be *dewatered* and may also be *conditioned with lime* or other chemicals.

Domestic sewage sludge often contains high levels of potentially toxic elements, notably Zn. Sludges arising from industrial plants should be analysed as one or more of some 12 high-risk elements may be present.

Fresh seaweed should not be confused with the liming material calcified seaweed. Seaweed is rich in potash (and sodium) as shown in *Table 4.4*, and a dressing of 25 t/ha would provide some 300 kg/ha of K_2O. Sometimes seaweed is used to make composts, but like so many organic manures, it is bulky material and expensive to transport. Thus seaweed is used only in close proximity to the coast and is valuable for some horticultural crops and potatoes. Also it is used to provide humus on some light soils.

INORGANIC FERTILISERS

Inorganic fertilisers form an important part of the basis of modern crop production for very few soils are able to supply sufficient quantities of all the nutrients necessary for high yields and quality in crops. However, profitable use of inorganic fertilisers depends on careful assessment of the total level of nutrients required by the crop on the one hand and the extent of supply from the soil and organic sources on the other.

This section lists the commonly available inorganic fertilisers and concludes by summarising guidelines for their profitable and environmentally safe use.

The three nutrients needed from inorganic fertilisers in greatest quantity are nitrogen, phosphorus and potassium. By law every bag of inorganic fertiliser that contains one or more of these three nutrients must state the content of that nutrient on the bag — in terms of percentage composition of N, P_2O_5 or K_2O. There are strict EEC regulations that specify the narrow permitted tolerance between the stated composition on the bag and the actual analysis of the contents, thus it can be assumed that there is great consistency in the composition of different batches of a similar fertiliser.

It is important to realise that fertilisers do not contain N, P_2O_5 or K_2O: these are merely a convenient way of giving a common basis for expressing the composition of all inorganic fertilisers. Nitrogen is usually in the form of nitrates or ammonium compounds, P_2O_5 in the form of phosphates and K_2O in the form of potassium chloride or sulphate. In the case of phosphates, some are water-soluble (and rapidly available in the soil, at least for a short period) and some are insoluble in water. If these latter are to have some agronomic value, they must become available in the soil and sometimes the solubility of these materials in 2% citric acid or neutral or alkaline ammonium citrate is given as a guide to probable value as a fertiliser.

In some specific crop and soil situations other nutrients are needed, such as Mg, B or S. Other nutrients are either applied on their own or mixed with one of the normal fertilisers, in either case the composition must be stated.

Nitrogen fertilisers

The major N fertilisers are as follows

	Formula	% N
Ammonium nitrate	NH_4NO_3	34.5
Urea	$CO(NH_2)_2$	46
Ammonium sulphate	$(NH_4)_2SO_4$	21
Calcium ammonium nitrate	$NH_4NO_3 + CaCO_3$	21–26
Anhydrous ammonia	NH_3	82
Aqueous ammonia	$NH_3 + H_2O$	25–40
Potassium nitrate	KNO_3	14
Sodium nitrate	$NaNO_3$	16
Calcium cyanamide	$CaCN_2$	21

Ammonium nitrate (AN)

This is an important fertiliser material and is sold as a straight fertiliser or as a component in compound fertilisers. Half the N is in ammonium form and half in nitrate form; in this way it has only half the acidifying effect of ammonium sulphate for any given amount of N applied. Practically pure ammonium nitrate is sold as a straight fertiliser containing 33.5–34.5% N: this material is very soluble in water and also hygroscopic, so it must be stored in a dry place in sealed bags. It is a very powerful oxidising agent and if subject to heat or flame it can explode: thus straight ammonium nitrate and compounds containing a high proportion of ammonium nitrate should not be stored in barns with hay or straw and *never* stored in bulk, as the risk of explosion is thus enhanced.

Being soluble in water, ammonium nitrate is often used as a major N source in liquid fertilisers.

Urea

Urea is the most concentrated solid source of N that is currently available on any wide scale. It is very soluble in water and hygroscopic. Sometimes its granules are softer than those of ammonium nitrate and do not spread so well. When urea is applied to most soils it breaks down rapidly to NH_4^+. However, ammonia (NH_3) can be formed if the soil contains free Ca, or is very dry.

(1) If the ammonia is lost to the atmosphere then the fertiliser is less effective as a source of N for crop nutrition; this loss of ammonia is most likely when the soil is alkaline and if the urea is applied to the soil surface in dry weather.
(2) In seedbeds the high concentration of ammonia near germinating seeds can cause toxicity and serious loss of plant stand.
(3) On poorly structured soils liable to surface capping the released ammonia can be trapped in the soil and not only kill germinating seedlings but also severely check the growth of roots of established plants.

It follows that urea is best used on acid-to-neutral soils, surface-applied to established crops during periods of frequent rainfall. Thus, in the British Isles, urea has a place on grassland and for top dressing cereals in spring. Compared with AN, the cost per unit of N in urea varies widely. Each tonne of urea contains 33% more N than AN. It follows that even if it is assumed that N in urea is 15% less effective than N in AN, when urea is less than 20% dearer per tonne than AN, urea is a better buy.

Urea is widely used as a major N source in liquid fertilisers, as liquid application overcomes some of the physical problems of urea noted above and when mixed with phosphoric acid to make a compound fertiliser the acid nature of the material further reduces the risk of ammonia loss.

Ammonium sulphate

Ammonium sulphate was once the most important source of inorganic N; it is soluble in water and although all the N is in NH_4^+ form it is quick-acting under field conditions. However, because of its high ammonium ion content it has an acidifying action in the soil and also its N concentration is limited to about 21%. Ammonium sulphate has been replaced by ammonium nitrate in UK and some other temperate countries and by urea in tropical and subtropical countries and some high-rainfall temperate areas.

Calcium ammonium nitrate (CAN)

This is a mixture of ammonium nitrate with calcium carbonate and is sometimes called nitrochalk. It has a higher N content than ammonium sulphate and has little acidifying action in the soil. Also the hygroscopic and explosive properties of ammonium nitrate are nullified by the presence of chalk, making it easy and safe to store, even in bulk. However, CAN tends to be more expensive per unit of N purchased than ammonium nitrate.

Anhydrous ammonia

Anhydrous ammonia is the most concentrated source of N available, being pure ammonia. At normal temperature and pressure this material is a gas, but when stored at high pressure it is a liquid containing the equivalent of 82% N. At this concentration it is very economical to transport on a weight-to-value basis, but the whole process of storage, transport and application needs special equipment, both to maintain up to the point of application the high pressure needed and to prevent hazards from uncontrolled loss of ammonia. Anhydrous ammonis is applied below the surface of the soil using special injection equipment and because of the high cost of this operation, it is economical to apply only fairly high rates of N (at least 100 kg/ha of N) at each application. Also it is essential that loss of ammonia to the atmosphere is kept at a minimum during application by careful sealing of the slits. For these reasons the use of anhydrous ammonia is best restricted to row crops that need high individual N dressings. Although widely used in some countries where suitable crops are grown in stone-free, friable soils, anhydrous ammonia has not established itself in the UK; mainly because it was not found suitable for the potentially major market on grassland due to sward damage caused by injection equipment and loss of ammonia from the slits in the soil.

The very high concentration of ammonia released in the soil in the vicinity of the slits causes partial sterilisation of the soil and has the effect of retarding the action of the bacteria which convert NH_4^+ to NO_3^- so that anhydrous ammonia can act as a slow release fertiliser.

Aqueous ammonia

Aqueous ammonia is a solution of ammonia in water, a solution of normal pressure containing about 25% N. Aqueous ammonia has most of the advantages of the anhydrous form except high N concentration, but to offset that it does not need such specialised equipment. In crop situations where a liquid straight N fertiliser is required, then aqueous ammonia has an important place. Its concentration can be increased by:

(1) partial pressurisation, where storage and application under only a modest pressure can enable concentration to increase to 40% N;
(2) mixture with urea and/or ammonium nitrate to give 'no pressure' solutions of about 30%N.

To prevent gaseous loss of ammonia all these materials should be either injected into the soil or worked in immediately after application: they should *not* be used on calcareous soils.

Potassium nitrate and sodium nitrate

These are both very quick-acting and highly soluble forms of N which are expensive but valuable in some horticultural crops and in particular for use in foliar feeds.

Calcium cyanamide

This is not important in the UK but is a fertiliser containing N in both amide and cyanide forms. It is soluble in water and in the soil it is converted to urea: during this conversion some toxic products are released which can kill germinating weeds and slow the rate of nitrification. Breakdown requires water and so the product is not effective in dry conditions.

Slow release fertilisers

Most of the commonly used and cheaper forms of N are rapidly available to the growing plant and yet many crops would gain greatest benefit from applied N if it were available over a period covering most of the vegetative growth of the plant. This is probably the most convincing reason for considering organic N sources. When inorganic sources are used there are three ways in which this objective can be met:

(1) Apply frequent small dressings of conventional N fertilisers during vegetative growth, for example as with spring N applications on some winter wheat crops and on grassland.
(2) Use a conventional fertiliser material which has received a coating to reduce its rate of breakdown into plant-available forms, for example sulphur-coated urea.
(3) Application of complex compounds of N which require considerable chemical change in the soil before they are available to the plant.

There is much research to find suitable materials in the last of the above categories. To be successful such a compound must be economic in price and supply available N at a rate required by the crop. The materials on the market tend to be expensive, make available to the crop a relatively low proportion of the total N they contain and have rates of release controlled by soil temperature and moisture so that during humid weather most of the N is released too rapidly for full crop benefit.

Among synthetic slow release fertilisers are:

Urea formaldehyde (urea form) 40% N: this is made by reacting urea with formaldehyde and its rate of breakdown to release available N is controlled by soil bacteria.

Isobutylidene di-urea (IBDU) 32% N: rate of release of available N depends on differential particle size and soil moisture status.

Sulphur coated urea (36% N) and *resin coated AN* (26% N): both depend on differential times for the coatings to break down before releasing available N.

Urea and IBDU are used in commercial glasshouse production, because not only is the slow rate of N release of particular value to some of the crops grown, but also the rate of breakdown can be controlled to a large extent by adjustment to the management of the house.

Phosphate fertilisers

The major P fertilisers are:

	Formula	% P_2O_5
Water soluble		
superphosphate	$Ca(H_2PO_4)_2$ $+ CaSO_4$	18–22
triple superphosphate	$Ca(H_2PO_4)_2$	45–47
mono-ammonium phosphate	$NH_4H_2PO_4$	48–50
di-ammonium phosphate	$(NH_4)_2HPO_4$	54
Water insoluble		
basic (or Thomas) slag	$Ca_3P_2O_8 \cdot CaO$ $+ CaO \cdot SiO_2$	10–20
Rhenania (sinter) phosphate	$CaNaPO_4 \cdot$ Ca_2SiO_4	25–30
ground rock phosphate	$CaPO_4$ (apatite)	30
Senegal rock phosphate		30

In addition to the above, which can all be used in solid forms, there are a number of liquid products based on *polyphosphates*. The basic component of these solutions is superphosphoric acid, made from orthophosphoric acid and one of a number of polyphosphoric acids. An ammonium compound is used to neutralise the appropriate acid and NP solutions of ratios in the order 11:37:0 can be obtained. These products are expensive but do provide a highly-concentrated source of available P in liquid fertilisers. There can be problems when potassium salts are mixed with polyphosphates, as they can cause crystallisation and precipitation.

Superphosphate

Superphosphate is made by treating ground rock phosphate with sulphuric acid and producing water-soluble monocalcium phosphate and calcium sulphate. Usually some rock phosphate remains in the product as the quantity of sulphuric acid is restricted to prevent the final product containing free acid. Superphosphate was once the most widely used form of water-soluble phosphate and formed the basis of most compound fertilisers. Because of its relatively low P_2O_5 content it limited the concentration of phosphate in compounds, but because of the $CaSO_4$ present, it did apply useful quantities of sulphur.

Triple superphosphate

This is made by treating rock phosphate with phosphoric acid. This produces mainly water-soluble monocalcium phosphate with no calcium sulphate. Thus, triplesupers contains nearly two-and-a-half times as much phosphate as supers, but no sulphur. Because of its high concentration, triplesupers is used widely in the production of high-analysis compound fertilisers.

Ammonium phosphate

This is made by adding ammonia to phosphoric acid, producing both mono- and di-ammonium phosphates. These compounds are very soluble in water, quick-acting in the soil and supply both N and P in a highly available chemical combination. They are not used as straights to any extent because of the few crop situations that require a low N:high P ratio, but they do form an important base for many compound fertilisers.

Basic slags

Basic slags are by-products of the steel industry and contain phosphates, lime and trace elements. In the smelting process, P contained in iron ore is held in the furnace bound to CaO, becoming a solid 'slag' as the material cools. The phosphate is in complex chemical combination with calcium and has to be ground before it can be used as a fertiliser. Although the phosphate in slag is not water soluble, it has been found that it can release phosphate slowly over a period of years on soils of pH range 4–7, although the rate of release is greater on acid soils. The trace element content of basic slag is also important, particularly Mg, Fe, Zn, Si and Cu.

Recent changes in the steel-making process have resulted in a drastic reduction in the quantity of slag available with a reasonable phosphate content. None of the 'slag substitutes' on the market seem able to emulate all four of the main attributes of traditional slag, namely:

(1) steady release of available P over several years on both acid, neutral and alkaline soils;
(2) supply of valuable trace elements;
(3) some liming value;
(4) cost per unit of P about half the cost of water-soluble P.

Rock phosphates

These are now being offered to farmers as a source of phosphate where rapid release of P is not essential but soil P status needs to be maintained. Rock phosphates will only release P under acid conditions, although some experiments have shown that soil pH is not necessarily a true indication of pH that might exist around root hairs, so that at times rock phosphate has shown some value on neutral soils. Rock phosphates vary considerably in their composition and potential agronomic value, and in all cases they must be ground finely to have any value at all, current legislation being that 90% of the material should pass a 100-mesh sieve.

Some of the hard crystalline apatites have very little fertiliser value even when crushed finely and used under acid conditions. On the other hand, soft rock phosphates such as *Gafsa* do have value and are regarded as the best of untreated rock phosphates.

Rhenania phosphate is produced by disintegrating rock phosphate with sodium carbonate and silica in a rotary kiln:

this 'sintered' product contains much of the P in the form of calcium sodium phosphate and as such the P is rendered a little more available than in the original rock.

Senegal rock phosphate contains aluminium phosphate and when this is heated in a kiln and then allowed to cool in a humid atmosphere, an expanded type of rough granule is produced that greatly enhances the surface area of the material in contact with the soil after application. This calcined Senegal rock is claimed to be an effective P source, even on alkaline soils.

No rock phosphate has a rapid release of available P, so that these materials should not be used in situations where the need of the crop for P is rapid, for example at crop establishment or on crops that have a big demand for P, such as potatoes. Also, the crop recovery of P from rock phosphates is less than the recovery from water-soluble sources, so that the amount of *useful* P in rock phosphate is less than in the water-soluble types: this fact should be remembered when the costs of rival products are considered.

As a guide, the cost of each kg P_2O_5 in a water-insoluble phosphate should be less than half that in a water-soluble product before it becomes a better buy for most uses.

Potash fertilisers

The major K fertilisers are:

	Formula	% K_2O
Muriate of potash	KCl	60
Sulphate of potash	K_2SO_4	50

Other potassium-containing fertilisers sometimes used are:

	Formula	% K_2O
Potassium nitrate	KNO_3 (13% N)	44
Potassium metaphosphate	KPO_3 (27% P_2O_5)	40
Potassium magnesium sulphate	K_2SO_4, $MgSO_4$ (18% MgO)	22
Magnesium kainite	$MgSO_4$ + KCl + NaCl (6% MgO; 18% Na)	12

Muriate of potash (potassium chloride)

This is mined and is sold in either a powdered form or a fragmented (granular) form. It is the source of K in nearly all compound fertilisers and its use as a straight fertiliser in the UK is very limited. Where crops require substantial amounts of K it should be remembered that equal quantities of Cl are also applied if muriate of potash is used so that in a crop like potatoes it is sometimes advisable to use potassium sulphate instead of muriate.

Sulphate of potash

This is made by treating muriate with sulphuric acid and so is a more expensive source of K. However for Cl-sensitive crops the extra cost of the sulphate salt is recommended. Some manufacturers make a series of compound fertilisers containing potassium sulphate and these are widely used by horticulturists, not only because of some sensitive crops but because of the risk of a build-up of Cl$^-$ ions in the soil where regular heavy manuring is carried out.

Potassium nitrate (saltpetre)

This is expensive and its main fertiliser use is in foliar applications on fruit trees and some horticultural crops.

Potassium metaphosphate

This compound is a water-insoluble source of K (unlike all the other sources mentioned which are soluble in water), and it can be used where it may be necessary to keep ionic concentrations at a low level in the root vicinity.

Potassium magnesium sulphate and magnesium kainite

These are sometimes used where some Mg is required in addition to K; also kainite is often used for sugar beet because of the Na content.

Agricultural salt (mainly NaCl)

This can replace KCl in some situations, for example where sugar beet or mangolds are grown. Sodium chloride should not be applied to poorly structured heavy soils as it may predispose deflocculation, and it should not be applied less than three weeks before sowing lest it affect germination: in fact it is often applied in autumn and ploughed in. Application of Na reduces the need for K.

FORMS OF FERTILISER AVAILABLE

Solids

This is the most common form in which fertilisers are available and normally manufacturers go to great lengths to ensure that the material stores without becoming compacted and spreads evenly and freely at the time of application. Water-insoluble materials, such as some types of phosphate, have to be in solid form, but most other materials could be sold in liquid form if necessary. The main advantages of solid types are:

(1) high concentration of nutrients in the material, reducing transport and storage costs and leading to high rate of work at spreading;
(2) when sold in plastic bags, very cheap storage is possible and the material will keep in good storage condition for many months;
(3) when sold in bulk, relatively simple modifications to existing buildings will enable successful storage and make possible spreading systems which have low labour requirements;
(4) can buy large quantities at times when market conditions are favourable;
(5) the farmer can have a range of fertiliser types available, can use only that quantity which is needed at a given time, and can judge the application rate easily by counting the number of bags used per hectare.

Where water-soluble ingredients are used in solid fertilisers, it normally takes very little soil moisture to render these ingredients available. Similarly, where the same ingredients are applied to a dry soil, the quantity of water applied with the fertiliser is so minute that it has no irrigation effect. So it can be assumed that '*once brought into contact with the soil, liquids behave in the same way as comparative solid fertilisers, and generally no differences are observed in relation to growth and crop yields'*. (Mengel and Kirkby, 1982)

Solid materials are sold as powders, granules, prills or fragments.

Powders

These can vary in degree of fineness and for some materials (e.g. water-insoluble phosphates) powders are essential to ensure activity of the material in the soil; powders can be difficult to spread evenly and this is critical if relatively low rates/ha are needed; under these conditions full-width spreaders are advised.

Granules

They are made during manufacture by passing the fertiliser powder through a rotary drum, often with an inert granulating and coating agent; sometimes several fertiliser materials are mixed at the same time, so that each granule contains a mixture of materials. All granules made in this way are designed to break down quickly in the soil and release their nutrients, but sometimes the inert carrier remains visible in the soil for some days. The size and texture of the granules will have a marked effect on the sowing rate and spreading width of fertiliser distributors, and both these granule characteristics are specific to the type of material being granulated and the conditions under which the granulator is working: it follows that granules will vary in spreader performance and each consignment should be checked when being used.

Prills

Prills are made by the rapid cooling of a hot fertiliser liquid, are homogeneous in character and usually have a very smooth surface. Like granules, they can vary in size and great care is necessary at spreading time as prills usually flow more freely than granules and spreader-setting is critical to get correct rates and even spread. Prills normally consist of one fertiliser material only but this may contain more than one nutrient, for example a prilled ammonium phosphate would contain both N and P.

Fragments

Fragments are made by very coarse grinding of a solid material, usually followed by a screening process to produce material of a certain size range, for example some samples of muriate of potash are made in this way. Fragments behave in a similar way to very uneven granules (but note that granules are always spherical in shape because of the way they are made), but fragments and granules should not be mixed and spread together as marked segregation will occur both in the spreader and during the spread itself. Some fragmented material has a very abrasive action on working parts in the spreader and can cause rapid wear.

Most manufacturers take great trouble to ensure compatability within each fertiliser type, but mixing of different fertiliser types on the farm prior to or during application can lead to poor spreading performance.

Liquids

Some materials must be used in liquid form, either because they do not exist as solids at normal temperature and pressure (such as polyphosphates) or because their intended use precludes solid form (e.g. foliar nutrient application).

The main features of liquid fertilisers are:

(1) easy handling on the farm, as they can be pumped in and out of store and spread by sprayer fitted with non-corrosive working parts and nozzles;
(2) relatively low concentration of nutrients, as some chemicals will crystallise out during storage and at low temperatures if concentration is raised;
(3) storage tanks required, which can be expensive; this can mean that storage capacity is restricted so that an annual requirement for fertilisers cannot be purchased at the lowest price;
(4) the possibility that herbicides can be mixed with fertiliser to further reduce spreading costs.

In the UK liquid fertilisers hold a fairly small share of the total fertiliser market, unlike the situation in some other countries. There is no simple guide as to whether a farmer should use liquids or solids as the decision rests on several aspects of general farm management, such as fertiliser purchasing policy, availability of existing stores for solid fertiliser storage, contacts for contra-trading with merchants selling either solid or liquid fertilisers, types of crop grown and labour and/or machine availability for spreading.

It is likely that the use of liquid fertilisers will continue to increase, both as a result in changing farm management practices, and as a consequence of developments in the production of liquid fertilisers. For example, there has been much research in the USA on the use of fertiliser *suspensions*: these are not true liquids as not all the material is completely dissolved, but they do offer the possibility of much more concentrated fluid fertilisers than at present. However, these materials will need very carefully controlled transport, storage and spreading, so that they will be based on a contractor service rather than a farm operation.

Gases

As mentioned earlier, anhydrous ammonia can be used as a source of N (but because of the high pressure needed for storage and the special injection equipment needed, this is a contractor-based service) and is valuable on relatively few crops and soils in the UK.

COMPOUND FERTILISERS

When fertilisers are sold with a single main chemical ingredient they are known as *straights*: examples are ammonium nitrate, superphosphate and muriate of potash. Sometimes straights can supply more than one nutrient, for example ammonium phosphate supplies N and P and potassium nitrate supplies N and K. Occasionally straights may be comprised of a mixture of ingredients, but if they still supply only one main nutrient they are still known as straights, for example some phosphatic fertilisers contain a mixture of water-soluble and water-insoluble phosphate, but still supply only P.

Many crops need more than one of the three major nutrients, N, P and K, and in most of these situations it is convenient to apply all the nutrients at the same time. For this purpose, manufacturers have produced a wide range of *compound* fertilisers, that is fertilisers which contain more than one ingredient and supply more than one nutrient. Compound fertilisers are more expensive than straights, but many farmers find the extra cost is justified because if several straights were used the cost of application would be increased and it is difficult to mix and effectively spread straights on the farm.

As a guide, the extra cost of nutrients in a compound compared with straights should reach 15% before the compound looks expensive.

Comparing fertiliser costs

The easiest way to compare the relative costs of two compounds is to relate each of them to the cost of straights, using the following steps:

(1) establish the cost of 1 kg of each nutrient in straights;
(2) calculate the nutrient value of the compound, costing all nutrients in the compound as if they were straights;
(3) express the extra price of the nutrients in the compound as a percentage of the cost as straights.

Example

Assume ammonium nitrate (34.5% N) costs £125 per tonne. Then 345 kg N cost 12500 p = 36.2 p/kg.

Triplesuperphosphate (45% P_2O_5) costs £160 per tonne. Then 450 kg P_2O_5 cost 16000 p = 35.5 p/kg.

Muriate of potash (60% K_2O) costs £130 per tonne. Then 600 kg K_2O cost 1300 p = 21.7 p/kg.

A 26:5:5 fertiliser costs £140 per tonne;
B 22:4:4 fertiliser costs £125 per tonne;
Which is the better buy?

A				B		
N	260 × 0.362 =	94.12		220 × 0.362 =	79.64	
P_2O_5	50 × 0.355 =	17.75		40 × 0.355 =	14.20	
K_2O	50 × 0.217 =	10.85		40 × 0.217 =	8.68	
Value	=	122.72		Value	=	102.52
Cost	=	140.00		Cost	=	125.00
Difference	=	17.28		Difference	=	22.48

$$\% \quad \frac{17.28}{122.72} = 14.1\% \qquad \frac{22.48}{102.52} = 21.9\%$$

so fertiliser A is the better buy.

FERTILISER PLACEMENT

Band placement

Most fertilisers are spread evenly over the surface of the soil or crop, and are used at the period when the crop is most in need of the nutrients contained. For example, PK fertilisers are worked into a seedbed prior to sowing and N fertilisers are top-dressed on to some growing crops, particularly winter cereals in the spring and to grassland.

In some row-crops, placing fertiliser in a band near to the zone of the soil in which the majority of the plant roots will feed can both reduce the total quantity of fertiliser needed and check weed growth outside the crop rows. Fertiliser placement in this way tends to reduce the speed of fertiliser application and if carried out at sowing time by *combine-drilling* will also slow down the rate at which drilling can be carried out. Also, if the fertiliser rate is too high or too close to the seed or plant root, the material can cause scorch and this can reduce plant survival and check growth in established plants. Band placement of fertiliser is most advantageous where:

(1) soil nutrient status is very low, for here responses to placed fertiliser can be greater than to very high rates evenly spread over the field;
(2) crops need very high rates of nutrient application and are grown in discreet rows set well apart. Here, placement not only reduces the overall amount of fertiliser needed but also ensures the roots come into contact with sufficient material. A good example of this is the main-crop potato grown on normal mineral soils;
(3) the applied nutrient has limited mobility in the soil: this applies particularly to P which scarcely moves in the soil even when applied in water-soluble form.

Rotational placement

Some soils are reasonably well supplied with nutrients but further fertiliser application is recommended to maintain the adequate level: this situation applies particularly to P and K. It is thus suggested that it may not be necessary to use P and K fertilisers for *every* crop but adopt a *rotational* fertiliser programme where PK fertilisers are used periodically and used immediately before a PK sensitive crop is grown. This approach can save money but needs careful monitoring to ensure that soil nutrient levels do not decline: also it should be noted that PK fertilisers will be needed on some fields on the farm each year as the rotation dictates.

Deep placement

In dry weather the growth of crops slows down, mainly due to lack of water for growth and transpiration but partly because the roots cannot absorb the soil nutrients which are located near the surface. Experiments have shown that *deep-placement* of fertilisers can maintain a faster growth rate in crops in dry weather than where the same nutrients are placed on the surface in a conventional way. It should be noted that for this deep-placement to work, the nutrients should be at least 150 mm below the surface and this is deeper than normal injection of anhydrous or aqueous ammonia.

FERTILISERS AND THE ENVIRONMENT

When fertilisers are used, consideration must be given to the environmental consequences. Phosphate is remarkably immobile in the soil and scarcely any is leached. Some potash is present in the soil in very soluble forms but the K^+ ion is held from leaching by both clay particles and organic matter, so that significant leaching of potash will only occur on sandy soils low in organic matter. Nitrogen is the most likely nutrient to leach, a specific problem being caused by NO_3^- being leached into water used for drinking. Excess nitrate in water can lead to *methaemoglobinaemia* (blue baby disease) in which the action of the red blood corpuscles is impaired. A considerable amount of nitrate is lost from agricultural land in the natural course of events (e.g. when grassland is ploughed or when there is heavy application of farmyard manure or slurry). Use of high rates of fertiliser N, particularly when used at periods in the growth cycle of the plant when N uptake is low, will exacerbate the problem. To minimise loss of fertiliser N by leaching:

For grassland apply no N from mid-September to the expected start of spring growth in February or March;

For arable crops apply no autumn N to seedbeds; make spring applications at the appropriate recommended growth stages in March and April (sometimes in February) and not earlier.

OPTIMUM FERTILISER RATES

Fertiliser recommendations are based on consideration of both physical response and financial return.

Physical response

This is obtained from field experiments, where the shape of the crop response curve to increasing levels of nutrient input is studied. *Figure 4.1* gives a typical crop response curve:

although the exact shape of this curve will vary with crop, soil type and climate, the following principles hold good:

(1) Under field conditions, some yield is obtained where no fertiliser nutrient is used, the yield A depending on the level of available nutrient in the soil. Fertiliser recommendations take this into account by allowing for soil index and making specific recommendations for each crop.
(2) There is a maximum yield (B) which is obtained from a certain level of nutrient input (M): using more nutrient than M will not give a greater yield — it may give the same yield as M or less depending on the nutrient and the crop (shown by the shaded area in *Figures 4.1* and *4.2*).
(3) The shape of the response curve is such that the yield return per each extra kg of nutrient declines: for example it can be seen from *Figure 4.1* that the first half of the nutrient dressing M gives a much greater yield increase than the second half of the rate concerned. Thus maximum yield is reached at a point where yield return per unit of extra nutrient is so low that it would not pay for the extra nutrient used.

Financial return

Financial return is considered in establishing fertiliser recommendations by giving monetary value to the crop yield and nutrient costs. *Figure 4.2* is based on the data shown in *Figure 4.1* and shows a curve for 'profit from fertiliser' which is taken as value of crop minus nutrient cost. The point O is very important as it is the highest rate of nutrient application that will show an increasing financial return. It is called the

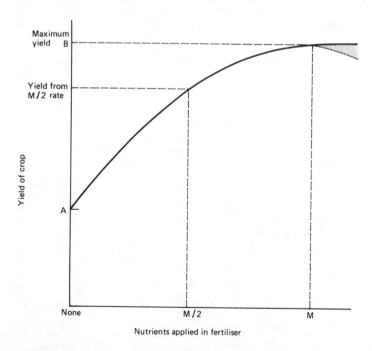

Figure 4.1 Response of crop yield to increasing rate of fertiliser nutrient

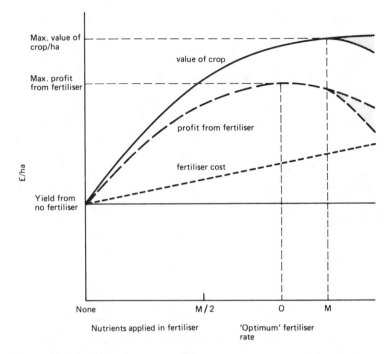

Figure 4.2 Calculation of optimum fertiliser rate

optimum rate and is below the maximum rate (M). Optimum rate depends on the ratio of crop value: nutrient cost, as well as on the shape of the agronomic response curve.

Optimum rates are not very sensitive to changes in nutrient costs or value of crop output. For most crops in the UK there is a large bank of experimental data that gives reliable information on nutrient responses over a wide range of soil and climatic conditions. Thus, ADAS and manufacturers' recommendations are sound guides to probable optimum rates of application. The periodic update of recommendations takes steady trends into account.

Where a new crop situation develops, either by the extension of an existing crop into a new soil type, climate area or by the introduction of a new crop into the country, then soil scientists and agronomists soon conduct experiments to establish agronomic criteria for economic optima to be calculated. Good examples in recent years have been the work on forage maize and oilseed rape, where recommendations were updated regularly for several years and now the fertiliser recommendations for these crops in the UK are different from those suggested in other countries where soil, climate and cropping systems are not the same as in the UK.

FERTILISER RECOMMENDATIONS

Figure 4.3 illustrates the principles used to link knowledge of soil nutrient availability with understanding of crop nutrient need, to reach a fertiliser recommendation.

For all nutrients other than N, soil analysis will give a good indication of the quantity of readily available nutrients in the soil. Assuming a crop needs the same total quantity of nutrient for optimum growth, *Figure 4.3* shows how the need for inorganic fertiliser supplementation will vary with soil index and also how organic manures may or may not give high yields. The same principle is used in deciding N recommendations, except that the soil index is determined by the previous cropping history (*see* Chapter 1).

Sometimes the concentration of nutrients within the plant is assessed by *tissue analysis*, as there is a good general relationship between the concentration of available nutrients in the soil and the quantity found in the growing plants. However, results of tissue analysis need careful evaluation, as the quantity of any nutrient found in a plant will be affected greatly by:

(1) physiological age of plant when tested;
(2) part(s) of plant taken for analysis;
(3) species (and variety) of plant;
(4) growing conditions prior to testing;
(5) interactions with other nutrients.

Soils differ in their ability to release available nutrients and crops vary in their nutrient requirements. ADAS and major fertiliser manufacturers publish crop fertiliser recommendations that take into account not only soil differences but also differences in the management of the crop: for example whether grass is cut or grazed, whether cereal yields are consistently above or below average, whether straw is burnt, removed or ploughed in. It should be remembered that fertiliser recommendations are not based only on chemical aspects of plant nutrition but also on the results of many field trials and farmer experience: thus recommendations are not

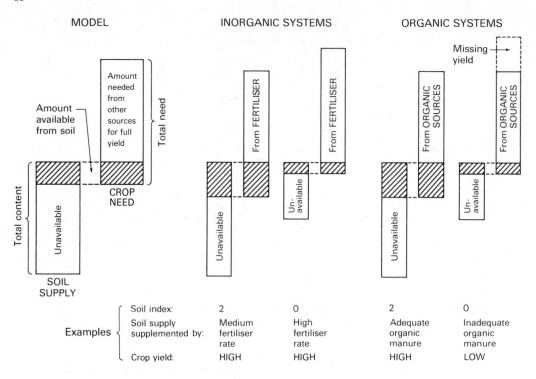

Figure 4.3 Basis for need to supplement available nutrients in soil

Table 4.5 ADAS fertiliser recommendations for wheat

Yield level (t/ha)	Soil P and K index	kg/ha of P_2O_5 or K_2O to apply			
		Straw incorporated or burnt		Straw removed	
		P_2O_5	K_2O	P_2O_5	K_2O
5	0	90	80	90	110
	1 & 2	40	30	40	60
	3	40	nil	40	nil
7.5	0	110	95	110	140
	1 & 2	60	45	60	90
	3	60	nil	60	nil
10	0	130	110	130	170
	1 & 2	80	60	80	120
	3	80	nil	80	nil
At all levels	over 3	nil	nil	nil	nil

N rates for spring sown wheat			kg/ha of N to apply		
N index	Sandy soils, shallow soils over chalk or limestone	Deep silty and clay soils	Other mineral soils	Organic soils	Humose soils
0	150	125	125	40	70
1	100	50	75	nil	35
2	50	nil	30	nil	nil

clinically correct but best approximations which are updated from time to time as new information is obtained.

For example, the ADAS recommendations for wheat, given in *Table 4.5*, illustrate the way in which recommendations should be used.

Calculating product use from recommendations

By law every bag or consignment of fertiliser must contain on it a statement of the plant food value of its contents, and under EEC regulations the phosphate value must be as P_2O_5 and the potash value as K_2O. Thus a compound fertiliser might be called 20:10:10 and state on the bags that its contents were:

20% N   10% P_2O_5 10% K_2O

If the bag weighs 50 kg, then dividing the percentage value on the bag by two gives the number of kg of the nutrients in that bag. So the bag mentioned above contains the equivalent of 10 kg of N, 5 kg of P_2O_5 and 5 kg of K_2O: if a farmer were to apply 5 bags/ha of this compound then the nutrient application would be 50 kg/ha of N, 25 kg/ha of P_2O_5 and 25 kg of K_2O.

The following example is based on *Table 4.5* and is designed to help those who wish to check their ability to convert recommendations into actual product use.

Example

Spring wheat is to be grown on a farm where yields are 4.5 tonnes/ha and the field in question has N index = O, P index = 2, K index = 2. Straw is removed and it is a mineral soil. All the P and K are put in the seedbed, with some N, using a 15:10:15 compound and the remainder of the N is applied later as 34.5% N.

From recommendations, N usage should be 125 kg/ha, P_2O_5 should be 40 kg/ha and K_2O 60 kg/ha.

Seedbed

P_2O_5 = 40 kg/ha and each bag of 15:10:15 applies 5 kg, so 40/5 = 8 bags/ha needed. K_2O = 60 kg/ha and each bag applies 15/2 = 7.5 kg so 60/7.5 = 8 bags/ha needed (400 kg/ha of product). So 8 bags/ha 15:10:15 applies to all the P and K needed, plus (8 × 7.5) = 60 kg/ha of N.

Top-dressing

Top-dressing needed is (125 − 60) = 65 kg/ha. Each bag of 34.5% N applies 34.5/2 = 17.25 kg, so rate needed = 65/17.25 = 3.8 bags/ha (190 kg/ha of product).

Note that in the example given, the fertilisers applied have met the recommendation exactly. In practice this will not always happen: the essential nutrient to get nearly right is N, followed by either phosphate *or* potash depending on which has a lower soil index and to which the crop grown may have a greater sensitivity.

Some textbooks give phosphate and potash recommendations in terms of P and K and not P_2O_5 and K_2O. Great care should be taken when reading literature to ensure confusion does not occur and the following conversions can be used:

(1) to convert P to P_2O_5 multiply by 2.29 (P_2O_5 to P multiply by 0.44);

(2) to convert K to K_2O multiply by 1.20 (K_2O to K multiply by 0.83).

LIMING

Adequate lime status is essential for the following reasons:

(1) it reduces soil acidity, and most plants will not thrive under acid conditions (*Table 4.6* gives pH guidelines for major crops);
(2) it increases the availability of certain plant nutrients in the soil;
(3) adequate soil pH encourages soil biological activity, thus enhancing the organic matter cycle in the soil, often releasing available N;
(4) it improves the physical condition of the soil by increasing the stability of crumb structures.

As explained in Chapter 1, there is a constant loss of lime from the soil, partly by crop removal and partly through leaching. Thus it is necessary to have a regular liming policy on all soils other than those with a natural Ca content. The frequency of liming will depend on soil type, rate of leaching (soil texture, rainfall, fertiliser N use), and sensitivity of crops grown. Soil analysis gives a good indication of soil pH and each field should be sampled on a regular basis — say every four years.

On *grassland farms* soil pH should not be allowed to fall below 5.5 and it should be limed to bring the pH up to 6.0.

On *arable farms* soil pH must be kept above the level

Table 4.6 Soil pH and crop tolerance

pH	Agricultural crops	pH	Horticultural crops
6.2	lucerne	6.6	mint
	sanfoin	6.3	celery
6.1	trefoil	6.1	lettuce
6.0	beans	5.9	asparagus
5.9	barley		beetroot
	peas		peas
	sugar beet	5.8	spinach
	vetches	5.7	brussels sprouts
5.8	mangolds		carrot
5.7	alsike clover		onion
5.6	rape	5.6	cauliflower
	white clover	5.5	cucumber
	maize		sweet corn
	wheat	5.4	cabbage
5.4	kale		mustard
	linseed		parsnip
	mustard		rhubarb
	swede		swede
	turnip		turnip
5.3	cocksfoot	5.1	chicory
	oats		parsley
	timothy		tomato
4.9	potato	4.1	hydrangea
	rye		
4.7	wild white clover		
	fescue grasses		
	ryegrass		

At a pH below the indicated level the growth of the crop may be restricted required by the most sensitive crop grown and general advice is to apply lime when pH falls to 6.0 and bring the pH up to 6.5.

Where grass is grown in rotation, it is essential that soil pH is kept up in preparation for succeeding arable crops, as it is difficult to restore rapidly a low pH. If acid grassland is ploughed for arable crops, not only must some lime be worked into the soil, but sensitive crop such as barley should not be grown for several years.

Liming policy can affect the two diseases — 'club root' of brassicas (*Plasmodiophera brassicae*) and 'common scab' of potato (*Streptomyces scabies*). Club root is less severe on well-limed soils whereas common scab is more prevalent where lime has been applied recently. Thus, in a rotation where brassicas and potatoes are grown, lime that may be required should be applied before the brassica crop, leaving at least two years before potatoes are grown.

Purchasing lime

Under the Agriculture Act 1970 the purchaser of liming materials must be given a warranty by the vendor. Under present regulations there are three main considerations:

(1) fineness of grinding;
(2) neutralising value;
(3) permissible variation (normally $\pm$ 5%).

Fineness of grinding affects the speed of action of the lime and the total quantity of the lime applied that may become useful: the finer the material the more rapid its action and some very coarse materials may never break down sufficiently to be of value. Not only is fineness important, but also the hardness of the material; it is more essential for hard material such as limestone to be ground fine than a softer form such as chalk.

Neutralising value (NV) is the standard basis for comparing liming materials. NV can be determined by laboratory analysis and is expressed as a percentage of the effect that would be obtained if pure calcium oxide (CaO) had been used. For example, if a sample of ground limestone has NV = 55, then 100 kg of this material would have the same neutralising value as 55 kg of CaO.

Lime recommendations, based on soil pH and the buffering capacity of the soil, are given in terms of the weight of CaO needed.

Example

A field has a lime requirement of 1.0 tonne/ha of CaO and two materials are available, ground limestone (NV 55) at £20/tonne delivered and hydrated lime (NV 70) at £30/tonne delivered. The two materials can be compared by calculating the cost of NV:

$$\text{ground limestone} = \frac{2000}{55}$$

$$= 36.4 \, \text{p/kg CaO equivalent}$$

$$\text{hydrated lime} = \frac{3000}{70}$$

$$= 42.8 \, \text{p/kg CaO equivalent}$$

Thus the ground limestone is the cheaper buy, except that the hydrated lime may be easier to apply (if not being applied by a contractor) and is quicker-acting, factors that may override the price difference in some circumstances.

Some liming materials contain magnesium as well as calcium; where the addition of Mg is considered beneficial, then the use of such material may be justified, even though the cost of each unit of NV is greater than a straight Ca lime.

Liming materials

Liming materials occur naturally over a wide area of the UK and they are mostly sedimentary rocks laid down as a result of:

(1) accumulation of shells and skeletal remains of animals and plants;
(2) deposition by marine calcareous algae;
(3) precipitation of Ca (or Mg) carbonates from water;
(4) aggregation of fragments from pre-existing limestones.

Ground limestone

This is a sedimentary rock consisting largely of calcium carbonate and containing not more than 15% of MgO equivalent. One hundred per cent of material must pass through a 5 mm sieve, 95% must pass a 3.35 mm sieve and 40% through a 150 μm sieve. NV is about 50 (maximum for pure $CaCO_3$ is 56). Legal declaration is NV and % through 150 μm sieve.

Screened limestone (limestone dust)

This is a by-product of grading limestone for non-agricultural purposes such as road metal and railway ballast. One hundred per cent must pass 5 mm sieve, 95% must pass 3.35 mm sieve and 30% through a 150 μm sieve. NV is about 48 and legal declaration is NV and % through 150 μm sieve.

Coarse screened limestone (coarse limestone dust)

This is a by-product of grading some road stones and is similar to the above but coarser in that 100% must pass 5 mm sieve, 90% pass 3.35 mm sieve and 15% through a 150 μm sieve. Legal declaration is as above and expected NV is about 48. However, not all the lime in this material is likely to be available: it should be applied at 120% of normal rate and not used where a rapid liming effect is needed.

Ground magnesian limestone

This is similar to the above materials, and subject to the same legal requirements, except that it must contain not less than 15% magnesium expressed as MgO. These materials are useful where there is a long-term need to maintain soil Mg level, however they do not provide rapid treatment for an existing Mg deficiency.

Ground chalk

This is natural chalk (cretaceous limestone) ground so that at least 98% will pass through a 6.3 mm sieve. The legal declaration is for NV only, which is usually about 50. Some chalks are quite wet when quarried and have to be dried before grinding: the NV value should refer to the material as supplied (i.e. partially dried) and not to a completely dried sample.

Screened chalk

This is produced from chalk outcrops by scarifying the surface with tractor-drawn implements and passing the collected loose material through a screen. This material is produced widely in south and east England. Ninety-eight per cent must pass a 45 mm sieve and the legal declaration is for NV only (usual value is about 45). However, producers often guarantee fineness to pass 25.4, 12.7 or 6.3 mm sieves.

Calcareous sand (shell sand)

This is found on some beaches, particularly in Cornwall and Scotland, and arises because of the large proportion of shell fragments in the sand. In such localities, this material is cheap and effective for arable and grass crops. All the material must pass 6.3 mm sieve and the legal declaration is for NV only (variable in the range 25–40).

Calcified seaweed (Lithothanmion calcareum)

These exist in some coastal waters and this material, of coral-like consistency, is dredged, screened to remove shells and either sold as graded material or dried and sold as milled powder. A number of properties have been ascribed to these products from time to time, but they are very effective liming materials, with an NV of 40–50. They tend to be much more expensive than other liming materials.

By-product liming materials

A number of industrial by-products can be used as liming materials; their availability will depend on production in the locality as transport costs can soon cancel out any price/ tonne advantage of these materials.

Sugar beet factory sludge

This contains a large quantity of calcium carbonate, but is very wet with about half its weight as water. Nevertheless, this wet material has an NV of 16–22 and contains some N, P and Mg as well: it can be spread by normal farm muck spreaders.

Sometimes this sludge is dried; it is then more expensive and has an NV of 50.

Water works waste

This waste occurs in some areas and consists of a very wet material that contains both calcium carbonate precipitated from water and some calcium hydroxide. Some works partially dry the waste to a sludge containing 25–30% water.

Whiting sand

Whiting sand is a by-product of the manufacture of whitening. It is a pure form of calcium carbonate arising from the process where chalk is ground under water, but normally has a high water content.

Other materials

They include waste from soap works, paper works and bleach works. Many of these materials are wet, but can be spread by contractor and they will dry out to form a fine powder. Some have a slight caustic quality and should be applied to the soil rather than to a crop.

Historically other liming materials have been used. For example when fuel was cheap, it was common practice to burn limestone or chalk in kilns to produce burnt lime (calcium oxide). With an NV of 100, this material was the cheapest form of lime to transport, but it had to be kept free from contact with water. When water is added to burnt lime it forms hydrated (or slaked) lime (calcium hydroxide) and the chemical reaction can evolve considerable heat if allowed to proceed without control. Hydrated lime has an NV of about 70, and although more expensive than the liming materials mentioned earlier, it can be used in cases of emergency as it is extremely fine and quick acting.

Blast furnace slag and basic slag

These materials can have an NV of 35–45 and were widely available until changes in the steel-making processes rendered scarce slags with good NV.

Overliming

Overliming can cause problems and arises when either a regular liming policy is carried on without regard to soil pH or when the soil is so acid that substantial quantities of lime are needed. In the latter case, it is essential that some lime is worked into the soil, so that extreme acidity can be alleviated and no part of the soil is overlimed. A good rule is never to apply more than 3 tonnes/ha of CaO equivalent at one dressing.

Overliming can cause the following deficiencies: Mn — causing grey speck in oats, speckled yellows in sugar beet and marsh spot in peas; B — causing heart-rot in sugar beet and crown rot in swedes; Cu — causing severe stunting, particularly in cereals; Fe — causing chlorosis in many crops, particularly fruit-bearing species. Overliming also encourages scab in potatoes.

References

MAFF (1979). *Profitable utilisation of livestock manures.* ADAS Booklet 2081. London: HMSO

MAFF (1985). *Fertiliser recommendations.* ADAS Reference Book 209. London: HMSO

MENGEL, K. and KIRKBY, E. A. (1982). *Principles of Plant Nutrition,* 3rd Edition. Berne: International Potash Institute

SOIL ASSOCIATION (1983). Paper from British Organic Standards Committee. Soil Association

5

Arable crops

R. D. Toosey

CROP ROTATION

Rotational practice is valuable for

(1) weed control;
(2) control of certain soil- and residue-borne pests and diseases;
(3) maintenance of soil structure and organic matter and recycling of plant nutrients.

Many soil- and residue-borne diseases are specific to a crop or group of crops and related weeds, which must be present for pest multiplication. Growing one crop too frequently allows pests to build up to infestation levels, so that subsequent crops of the same type fail completely or produce uneconomic yields. The pests must then be starved out by growing crops not susceptible to the pest or disease in question. Unfortunately some pests and diseases survive in the ground without a host plant for many years and starving out becomes a long-term project of dubious practicability; growing field-resistant or immune varieties (if available) is of value in certain cases. Hence it is cheaper and safer to adopt a sound rotation, with other measures to prevent disease building up to serious proportions. Other measures include:

(1) sound general sanitation, using clean seed, avoiding infected tubers or transplants and carrying infected material from one field to another, e.g. feeding clubroot (*Plasmodiophora brassicae*) infected swedes on leys;
(2) providing suitable conditions for vigorous crop growth and where possible making the environment unsuitable for development of pests, e.g. by good drainage, enough lime, ample nitrogen;
(3) destroying all alternative sources of carry over, e.g. trash, self-sown progeny or ground keepers and suscept-ible weeds.

Rotating crops around the whole farm ensures all fields benefit from restorative crops like leys, clover, lucerne, ap-plications of farmyard manure and deeper cultivations. The three to four year ley improves structure and though it scarcely lives up to the results originally claimed, farmers who have leys in rotations rarely suffer soil structure problems. It is possible to manure many crops exclusively with purchased artificial fertilisers but cost is so high that maximal conservation and re-cycling of nutrients is of prime economic importance. Residues of leguminous crops, heavily stocked leys, folded roots and green crops provide substantial quantities of nitrogen and cut fertiliser bills con-siderably. Slurry and farmyard manure, properly conserved and applied, provide substantial quantities of all nutrients.

CHOICE OF CROP AND SEQUENCE

When deciding whether to introduce a new crop check:

(1) climatic requirements – temperature and water needs, tolerance to exposure and wind;
(2) suitable soil types, depth and subsoil requirements; in-teraction of soil and rainfall on particular site; crop tolerance of acidity;
(3) total and seasonal labour requirements, availability of suitable casual labour, e.g. vegetable production; likelihood of labour peaks coinciding with other crops. Feasibility of mechanising production and harvesting, storage and capital requirements. Availability of contractors;
(4) market demands and transport, marketing costs. It is pointless to produce crops which cannot be sold at a profit possibly because of excessive transport and marketing costs. Some crops are subject to area quotas, e.g. potatoes and sugar beet; others can only be grown safely on contract, e.g. herbage seeds, oilseed rape, vegetables for freezing;
(5) suitability of fodder crops to type of livestock, farming system, method of utilisation, length of growing period and time of year feed is provided;
(6) soil- or residue-borne pests and diseases already on farm, e.g. clubroot (*Plasmodiophora brassicae*) may preclude production of brassica vegetables – brussels sprouts, cabbage, cauliflower. Some pests may attack a range of unrelated crops, e.g. stem and bulb nematode (*Ditylen-chus* spp.). Various races may attack onions, peas, beans,

vetches and oats; sugar beet cyst nematode (*Heterodera schachtii*) attacks *all* members of beet family, e.g. mangels, fodder beet and red beet and is harboured by *all* brassica crops.

CROP DURATION AND IMPORTANCE

Maincrops

These occupy land for the majority if not the whole of the growing season, are the most important financially and take precedence over all others. Maincrops include all major crops, e.g. cereals for grain, sugar beet, maincrop potatoes, brussels sprouts. Always stick to maincrops when it is difficult to establish a second crop, especially in dry areas (e.g. East Anglia and on heavy soils).

Doublecrops

Double cropping refers to growing two crops in a year when both are equally important. The system applies to intensive situations (e.g. quick growing vegetable crops, Italian ryegrass for silage followed by kale). Second crops are highly dependent on adequate moisture and good tilth for satisfactory germination or establishment and the system is only suitable where there is ample rainfall or irrigation and a free working soil. Timely removal of first crop is essential.

Catch crops

These are of secondary importance and are snatched between two maincrops when the land would be otherwise unoccupied. They are usually grown to provide short-term fodder: e.g. *Year 1* winter barley harvested in July, then sow continental turnips to graze in autumn. *Year 2* spring barley. Only fast growing crops are suitable as catch crops.

Summer catch crops

These are sown in summer, make all their growth during summer and autumn and are usually consumed by Christmas. Use forage brassicas, selected according to date of clearance of first crop, e.g. after early potatoes there is a wide choice; after winter barley in southern England the choice is restricted to continental or grazing turnips or forage rape; rape kale for early spring keep, sown early in August is also suitable.

Winter catch crops

These are sown in late summer or early autumn, must be frost hardy to stand the winter and produce a large bulk the following spring. Mainly forage rye and Italian ryegrass.

Effect on maincrops

Catch crops must not interfere with maincrops, particulary:

(1) they must not harbour pests or diseases of any maincrop in rotation: e.g. rape is highly susceptible to clubroot (*Plasmodiophora brassicae*) which affects all maincrop brassicas; rye or Italian ryegrass should be preferred as catch crops in rotations including swedes, cauliflowers or oilseed rape. Avoid brassica catch crops in rotation containing sugar beet;

(2) catch crops must not delay sowing of subsequent maincrop by occupying land too long; or

(3) dry out soil.

Resources to grow catch crops

Plan well ahead for early harvested preceding maincrop to ensure timely sowing of catch crop, give adequate fertiliser to catch crop and only catch crop clean land. Catch cropping is unsuitable for dry areas where there is no irrigation, on heavy soils or where the growing season is very restricted, e.g. much of the North and cold uplands. If the crop is to be grazed or folded, use only well drained land with good access.

METHODS OF CROP ESTABLISHMENT

The seed and its treatment

As the cost of seed is a very small proportion of the total cost of producing a crop, it is false economy to sow any but certified seed of high germination and vigour; seed must also be suitable for the purpose for which it is sown, e.g. seed production, when basic grades of seed must be sown or with cellular tray systems when special, high germination lots are advisable.

Under EEC rules it is legal only for licensed seed merchants to sell seed. Always check the germination with the vendor; knowledge of this is required when calculating seedrates for vegetables and many other crops. Germination of merchants' cereal seed is high and no allowance is made in the calculation. When using home grown seed *always* have the germination tested and, if suitable, the seed should then be cleaned and dressed with an appropriate seed dressing.

Vigour

The statutory germination test is carried out under ideal growing conditions in a laboratory, when all seed capable of germinating will do so. A parcel of seed may have a high laboratory germination but subsequently fails to establish adequately in the field; such seed is said to be of 'low vigour'. Seed vigour reacts strongly with soil conditions. All but the lowest vigour seed will usually establish well under good conditions but with cold wet soil, as with early spring sowings, all but high vigour seed is unreliable and may fail completely. It is generally wisest to buy high vigour seed which should always be used for sowing in adverse conditions. There is as yet no statutory vigour test and merchants use their own versions.

There is no single factor determining seed vigour. It is believed to be a composite effect, jointly attributable to factors such as ageing and deterioration of the seed, environment and nutrition of the mother plant, mechanical integrity, seed-borne pathogens, seed size, weight and specific gravity and genetic constitution of the seed.

Seedrate and plant population

The concept of 'standard' seedrates is largely obsolete, as the same weight or volume of seed per unit area can result in vastly different plant populations. With the need to control populations to obtain optimal results, a calculated seedrate is much more effective. The variables include the following:

Required plant population/m²

This may vary greatly with the same crop according to size or grade of produce required, e.g. small canning carrots or large dicers; pickling or ware onions.

Weight or size of individual seeds, expressed as 1000 grain weight or number of seeds per unit of weight: small seeds are stated per gram; large seeds per kilogram. Variations between parcels, even of the same variety, can be very large, e.g. carrots 500–1700 seeds/g; maize 2600–4000 seeds/kg.

Percentage germination is usually high in cereals but can be extremely variable in some vegetables.

Field factor

Crop establishment is greatly affected by field conditions, i.e. soil, temperature and moisture and condition of the tilth. Field losses under adverse conditions can be large and it is necessary to allow for them. The field factor is an assessment of the likely reduction in germination from that obtained in a laboratory germination test, stated on a 0.1 to 1.0 scale or as a percentage scale, according to type of crop. In practice, the figure varies from 0.5 or 50% establishment for poor conditions to 0.9–90% in ideal conditions. This allowance is only a 'best estimate' and cannot cater for changes in seedbed conditions *after* sowing.

The following formula is widely used for all small seeds:

$$\text{Seed requirement (kg/ha)} = \frac{1000 \times \text{no. of plants required/m}^2}{\begin{array}{c}(\text{no. of seeds/g}) \times \\ (\text{laboratory germination}) \times \\ \times \text{ field factor}\end{array}}$$

For detailed field factors and formulae for cereals and legumes *see* appropriate sections.

Seed preparation

Natural seed of any species varies considerably in size, both between and within different parcels; the seed of some species is also irregular in shape. With the exception of some pneumatically operated drills, most precision drills and block or cellular tray seeding mechanisms require seed of even size and regular shape to operate effectively; there are two main approaches.

Graded seed

Graded seed is especially suitable for round seeded crops, e.g. all brassicas (brussels sprout, cabbage, cauliflower, swedes, turnips) and some others; each size grade is allocated a letter which must match that of belt or wheel on the precision drill.

Pelleted seed

Pelleting is especially suitable for irregularly shaped seeds, e.g. monogerm sugar beet. Seed is enclosed in a pellet of regular size and shape, composition being based on kaolin or similar substances designed to disintegrate on contact with moisture. It is important to drill the seed deeply enough to place it in the soil moisture and take steps to conserve moisture during seedbed preparation. Fungicides and insecticides for crop protection may be included in the pellet, which may also be treated with fluorescent dye to highlight any pellets left uncovered; if these are found by field mice they will hunt along the row, so ensure all seed is covered.

Split pellets

These are preferred when an extremely uniform product and hence uniform germination is required, especially in some salad and vegetable crops. These pellets split open on contact with moisture, allowing rapid access of water and oxygen to seed and offering no resistance to the emergence of the young root and shoot. *Pregerminated seed* is also offered in pellets (*see* Pricking out p. 94).

Seed dressing

It is a routine precaution to dress seed before planting with a suitable fungicide and if necessary an insecticide, to protect the young plant against certain soil- or seed-borne diseases (e.g. fungi causing damping off *Pythium* spp.) and pests (e.g. flea beetle, *Phyllotreta* spp.). For choice of suitable dressings *see* Chapter 9, Diseases of Crops, and Chapter 10, Pests of Crops.

Seed dressing is best carried out by the merchant, who has the necessary equipment to ensure an accurate and even coating. Dressings can be in powder or liquid form; the latter is often preferable as all seed is more evenly coated and does not become dusty. Dressed seed should *never* be fed to livestock, even if diluted. Surplus should be safely disposed of or sold for manufacture of rat poison. Do not sow seed treated the previous year without a further germination test. After handling treated seed wash hands thoroughly before eating or smoking.

Methods of plant raising and crop establishment

Choice of system (Figure 5.1)

Sowing ungerminated seed directly into the open field without transplanting is the quickest and cheapest method of plant establishment. It is standard practice with most agricultural and many vegetable crops, but is unsuitable for many early and high cash output crops and some intensive field vegetables. The use of transplants ensures a well developed crop and, during the raising period, releases the land for other crops.

Some early vegetable crops are sown directly into the field in late summer for overwintering, e.g. Japanese type dry bulb onions, salad onions and direct drilled spring greens or cabbage, but varieties selected must be winter hardy and be resistant to bolting. A free-draining soil is also essential.

Problems with early spring sowing include: difficulty of getting on land due to wet soil; damage to soil structure and unreliability of the season, often causing late drilling and delayed crop harvest or even no crop and sheer failure to grow owing to low soil temperatures. Some crops are very slow starters, e.g. celery, celeriac which succumb to weed competition; others are susceptible to bolting if sown very early, e.g. sugar beet, red beet, round summer cabbage.

Transplants may be either 'peg' or 'bare rooted' (i.e. lifted from transplant bed with little or no soil attached) or raised in individual peat blocks, cells or containers where the roots are contained in the growing medium. Peg transplants are cheaper but there is serious root disturbance and losses in some crops can be excessive.

Open field plant bed

This is the traditional method of raising peg transplants for maincrop brassicas, spring greens or cabbage and maincrop leeks. Unsuitable for early summer harvested brassica crops

METHODS OF CROP ESTABLISHMENT

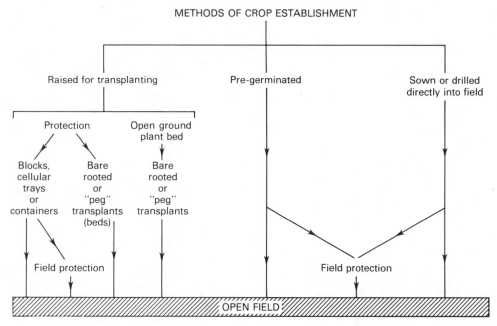

Figure 5.1 Methods of crop establishment

and early leeks, when a *protected plant bed* is preferred. Autumn sowings of early summer cauliflowers have been made in frames. Early spring sowings before early April under polythene tunnels or glasshouses require initial heat.

Peat blocks, cellular trays and other modules

Cuboid peat blocks are made by a block making machine using specially blended blocking compost; a wide range of block sizes can be turned out. Blocks are fairly highly compressed and there is a depression in the centre to take the seed, which is placed by a special seeder unit mounted on the machine. Plants are then raised in a glasshouse and are planted out by planter with special attachments. Irrigation after planting out is required.

Cellular trays are multi-celled units made from expanded polystyrene or rigid polyethylene usually tapering towards the drainage hole at the base. Cell size is variable. Each cell is an individual 'mini-container' which prevents roots or water from entering adjacent cells. Rigid polyethylene trays are much less readily damaged than expanded polystyrene and easier to sterilise.

Cellular trays have considerable advantages over peat blocks: compost is not compressed, allowing free root growth; growth of seedlings is totally under plant raiser's control, as strength of liquid feed can be adjusted at will. Absence of intercell root penetration allows a much longer holding period than blocks if required and avoids root damage at planting, giving better crop establishment and uniformity. Trays are easily handled and allow faster transplanting (ADAS, 1985).

There are also systems using 'linked' modules, e.g. linked paper pots (Japanese paper pot system used for transplanting sugar beet in Japan; cylindrical pots are 20 mm diameter and 130 mm long filled with uncompressed compost); also bandoliers of cylindrical peat blocks (ADAS, 1981).

These systems are for high output automatic transplanters, especially suited for large scale operations on reasonably level ground.

Management methods for systems for raising transplants (Table 5.1)

Plant beds

Aim to produce thick stemmed sturdy well rooted plants 175–225 mm long, which must not be soft, blind, 'leggy', or infected with disease, especially clubroot (*Plasmodiophora brassicae*).

Open field plant beds

Choose clean clubroot-free site, which has not grown brassicas (or leeks and onions if applicable) for previous five to six years. For large areas an ADAS clubroot test in previous summer is advisable; plants may follow cereal or ley. Control any couch (*see* Chapter 8) or other grasses and subsoil if required. After cereal give 40 t/ha FYM and plough before Christmas, in dry autumn soil conditions. In spring broadcast fertiliser:

| | *Nutrient* (kg/ha) | | |
	N	*P₂O₅*	*K₂O*
After cereal	30	30–60	30–60
After ley	0		

Excess N gives soft sappy plants. Quality of plants is improved by giving slow release N, as in organic compounds. Lime should also be applied, e.g. 5 t/ha hydrated lime or equivalent as ground limestone to bring pH

Table 5.1 Choice of suitable plant raising system for various crops

Bare rooted or peg transplants
Open field
  Maincrop brussels sprouts; spring, late summer, autumn and winter cabbage; winter, late summer and autumn cauliflower; maincrop leeks.
Protected beds or frames
  Early brussels sprouts; early summer cauliflowers; early summer cabbage and early leeks.
Peat blocks and cellular trays
  All brassicas if required and early crops of brussels sprouts, cabbage, calabrese and early summer cauliflowers; early leeks; Chinese cabbage; earliest crops of multiseeded red beet, multiseed cell or block onions, sweet corn, lettuce, celery and celeriac, Florence fennel.

up to 7.5–8.0, to discourage clubroot. Prepare fine seedbed with minimal working, avoiding compaction. Seed must be treated with fungicidal and insecticidal seed dressing, especially against flea beetle (*Phyllotreta* spp.) in cotyledon stage and protect against cabbage root fly (*Erioischia brassicae*). When ready plants are pulled and counted into bundles. Protect against drying out and plant firmly as soon as possible into fine moist tilth. Always ensure land is waiting for the plants, never the plants waiting for the land. Treat transplants against cabbage root fly (*Erioischia brassicae*) and irrigate in dry weather if possible, especially summer and autumn cauliflowers.

Protected plant beds

If bed has previously been used for plant raising, soil sterilisation is essential before sowing. Prefer drilling to broadcasting. Harden off plants thoroughly before transplanting; plant out as for open field reared transplants, but handle with care as they are more brittle.

Peat blocks, cellular trays and paper pots

Principles of management are similar but there are differences in detail. The components of each system are: (i) size of block, container, module or cell; (ii) the compost and (iii) the seed. Raising of plants by these methods is a highly specialised job and all but the largest growers, and often some of them, arrange for a specialist nursery to raise plants for them.

Choice of cell or block size

This is a compromise between the needs of the crop and cost of production, cost increasing with individual size.

Peat blocks Field crops will tolerate much smaller blocks than glasshouse crops, where larger plants are required. Main sizes of block in production are 27, 35 and 45 mm cubes; the latter are largely used for glasshouse crops, early marrows or courgettes and cauliflowers; large cauliflower plants raised in 60 mm blocks or 75 mm pots will crop earlier than those in cells or bare-rooted transplants. Early crops of field vegetables may justify block sizes up to 38 mm but most growers prefer the smallest size.

Cellular trays for field vegetables Volume of these is smaller than blocks and more economical of compost, varying from 9 to 15 ml/cell, but larger sizes are available.

Appropriate cell volumes (ml/cell) are: *brassicas*, 9–15 ml; larger sizes do not increase earliness; *celery*, early crops 16 ml or more; for maincrops 13 ml is adequate; *multi-seeded dry bulb onions*, 9–15 ml; *multi-seeded early leeks*, 14 ml, single-seeded smaller cells; *early bunching red beet*, 14 ml; *sweet corn*, 14 ml.

Compost, physical requirements

Composts may be peat or soil based, the former being essential for blocking, where special composts are required and preferred for cellular trays. Peat composts usually contain a proportion of inert substances, e.g. sharp sand, perlite or washed vermiculite. Proportions and fineness vary according to purpose of compost. For seeded cells finely milled peat with 5–10% sand or perlite is used to ensure free running but not so fine that it loses structure when watered.

Compost, lime and nutrient status Moss peat is acid and requires neutralising. It is also deficient in major nutrients and trace elements. With the exception of lime, magnesium, phosphate and trace elements, it is only possible to supply a small part of the total NK requirement, as high concentrations lead to excessive salinity and damage or death of the germinating seed. Growers may either buy ready made composts or make their own. Only buy well known reputable brands; cheap composts give poor results. For delicate seeds, especially flowers, seed and cutting compost (weak in NK) is preferred; for most work dual purpose seed/potting composts are satisfactory; for potting on, buy (stronger NK) potting compost. Packs are also available for converting peat bales, or growers can make their own from the following recipe: *Thorough mixing is essential.* To fine peat 95% + 5% fine sand by volume, add:

Ingredient	kg/m³
Potassium nitrate[1]	1.0
Superphosphate (single)	2.25
Ground chalk/limestone	2.25
Ground magnesian limestone	2.25
Fritted trace element (WM 255)	2.25
(ADAS, 1985)	

[1] For brussels sprouts and calabrese use instead 0.5 kg/m³ potassium nitrate + 0.5 kg/m³ sulphate of potash.

Seed

Use only seed of the highest germination and vigour; some seedsmen offer special grades for plant raising in cellular trays and this is well worth the extra outlay. Seed which merely complies with minimum statutory requirements is unlikely to be good enough.

Pricking out Traditionally small seeds are germinated in seed trays and as soon as large enough to handle are lifted with a small knife, plant label or similar and replanted individually in cells or blocks. Method is slow and laborious and only used for crops which germinate erratically, e.g. celery, celeriac and many amenity flower species where other facilities are not available.

For celery, pregerminated seed is available from at least one firm in split pellet form, given minimum notice of six weeks. Seed is delivered and stored at 5°C, for maximum of 14 d. Sow directly into blocks or cells. Growth at four weeks is more even and well ahead of pricked out plants.

Direct seeding Wherever possible crops are direct seeded into blocks or cells, either multi-seeded e.g. dry bulb onions, earliest crops of red beet or single-seeded, e.g. brassicas, lettuce. Hand seeding is laborious and often gives poor results. Choose type of seeder suited to type of seed and seeding (i.e. single or multiple).

Operating cellular tray system

The following facilities are required:

(1) covered area for storing sterile trays and composts;
(2) working area for filling and seeding trays;
(3) facilities for applying water or fungicidal drenches to trays;
(4) temperature controlled germination room;
(5) glasshouse or tunnel equipped with facilities for heating and for watering and feeding cells uniformly;
(6) system to keep bottoms of trays off floor and 'air prune' the roots to prevent growth through drainage holes.

A very high standard of expertise and hygiene is required; before starting to study the subject in depth. Hand filling cells with compost is only suitable for small scale operations; otherwise use a filling line. After seeding cover with perlite, vermiculite or other inert material; any excess above the cell division is scraped off to prevent cross rooting. Trays are then removed to unlit insulated germination room or cabinet with temperature controlled around 21°C, the optimum for most crops. With the exception of leeks and onions, which may be left in the chamber for up to one week after emergence, provided the trays are not stacked on top of one another, all crops must be removed to the light *before* emergence; otherwise etiolated seedings result. Precise time in chamber depends on individual crop, e.g. remove all brassicas after 36–48 h. For new crops try a few trays some time before the main sowing.

Trays are transported by rack or stillage to the glasshouse; place batches to allow for quickest removal, provide inspection alleys and isolate trays from floor to prevent rooting.

Watering As there is no lateral movement between cells, even watering is essential. Overhead static nozzles with overlapping cones are useless, as some cells are overwatered and some receive no water at all. A travelling overhead gantry is the most efficient. Hand watering is appropriate for small scale work. Outside trays tend to dry out first. *Do not water at predetermined times* but when cell surfaces are starting to dry. Frequency of watering need increases with plant size and in warm sunny weather. Small plants in overcast conditions need much less.

Liquid feed This is applied through the irrigation system. Once the plants have exhausted the small amount of NK in the cell compost they are totally dependent on liquid feed. The operator thus has absolute control over growth of plants in cells and can manipulate the nutrient supply to obtain any rate between fast growth to holding the plants in good condition with minimal growth. Do *not* starve plants during a 'holding period'. Examples of suitable liquid feeds are shown in *Table 5.2*. Individual crops vary in their requirements and high K feed may be desirable every second or third watering in some cases. Growers may find it more convenient to use suitable proprietary feeds; these may include a marker dye to assist in monitoring application.

Hardening off Plants raised under protection, especially where heat has been applied, are soft and will be heavily damaged by frost or desiccation if suddenly moved outdoors. They require gradual 'acclimatisation' (i.e. hardening off). The heat is lowered gradually, then removed and plants held at ambient temperature. Open all doors and windows by day, then at night and move plants out in mild weather to site sheltered from cold or drying winds. Frames are ideal, as cover can be removed in daytime. Continuous low (plastic) tunnels can be raised gradually. *Never* transfer soft plants to adverse conditions.

Table 5.2 Liquid feeds

Stock solution		Dilution	Feed strength		Uses
Potassium nitrate (kg/100 litre)	Ammonium nitrate (kg/100 litre)		N (mg/litre)	K_2O (mg/litre)	
9	8	1:200	200	200	To produce fast growth or to re-activate plants held with low N
14		1:200	100	300	Given two/three times a week will give sturdy growth
9	2	1:200	100	200	Given two/three times a week will give sturdy growth
4.5	1	1:200	50	100	Given every watering will produce slow, sturdy growth. Given once a week will 'hold' plants

(ADAS, 1985)

Transport plants in stillages or trays. Protect against drying out. Once hardened off plant as soon as possible, with short holding period on headland. If not planted within 48 h unload and give liquid feed. Plants are utterly dependent on nutrient and water and are easily ruined by lack of care at this stage.

Transplanting Soak trays immediately before transplanting, preferably in high N feed. Treat brassicas against cabbage root fly (*Erioischia brassicae*) with chlorpyrifos or other suitable insecticide. Plant *firmly* into moist crumbly tilth and plants will establish rapidly and evenly.

Planting pregerminated seed

The major commercial application of this method is the presprouting or chitting of seed potatoes (*see* Seed treatment, p. 169). Method also includes 'fluid drilling' (Gray and Salter, 1980) used for drilling pregerminated seeds of some horticultural crops, e.g. lettuce.

Sowing ungerminated seed directly into open field

Seedbed quality

This is all important, especially for small seeds and highly sensitive species e.g. maize and french beans, which should be surrounded by fine moist soil of adequate temperature (maize and french bean min. 10°C). With early sowing an excessively fine surface can cause capping on some soils, so leave some *small* clod on the surface. Rough seedbeds cause uneven germination, with a thin gappy stand and the resultant plants lack vigour and develop only slowly; severe cases end in crop failure. Soil applied (residual) herbicides do not work satisfactorily. Precision drills sow inaccurately and at an uneven depth. Seedbeds should be neither compacted nor loose but reasonably firm.

Depth of sowing

This depends on (i) species and seedsize and (ii) depth of soil moisture. In general, sow seed no deeper than necessary to ensure good coverage and, during spring and summer, cut into the soil moisture. Small seed has small food reserves and *must* be sown shallowly (e.g. onions, timothy grass 12–15 mm), depth increasing with size of seed, brassicas up to 25 mm, field beans 75 mm or more. Shallow drilling ensures quick emergence and a vigorous plant.

Broadcasting

Broadcasting seed is quick and cheap, saving the cost of a drill and can be done with broadcast machines or suitable fertiliser distributors. Frequently used for grass seed, sometimes for cereals and forage brassicas. Broadcasting on a cambridge rolled surface gives a good cover for grass seeds mixtures, but is most unreliable in dry conditions. *Drilling* is then much safer, as the seed is cut into the soil moisture, and is normal practice for arable crops.

Drilling to a stand

Precision drills are used where accurate seed placing and precise plant populations are required. Singling and chopping out have now largely been superseded by drilling to a stand, where graded or pelleted seed is spaced individually in the row to give a final crop stand, allowance being made for non-germinating seeds. Success depends on a high quality seedbed, precise drilling, effective protection against seedling pests and diseases and a satisfactory herbicidal programme.

Direct drilling refers to two distinct situations:

(1) where surface vegetation is killed off by herbicide (paraquat or glyphosate) and the seed is sown with special drills and minimal or zero cultivation. Applies mainly to cereals, oilseed rape, forage brassicas and sometimes grass. Only cereals and oilseed rape are suitable for sequential direct drilling and then only on well drained and structured soils; winter crops are safest.

(2) Certain field vegetable crops, mainly brassicas and leeks, which are traditionally raised in a seedbed and transplanted, are instead drilled directly into their final cropping position, with minimal subsequent chopping out. The technique is normally used for calabrese, which is not suited to bare rooted transplants; may also be used for autumn sown spring greens and cabbage, autumn and winter cabbage and savoys, high density brussels sprouts for freezing and maincrop leeks. Labour costs are lower than transplanting and special machinery is not . required, but a very high standard of seedbed preparation is necessary. The very light seeding makes the crop especially vulnerable to adverse seedbed conditions, i.e. cold rain and capping of soil, when total loss of plant can occur. The crop also occupies the land longer than transplants and seed requirements are higher.

Direct drilled crops are drilled about one month later than plant beds. With wider spaced crops, i.e. hearted cabbage and brussels sprouts, sow groups of three seeds 25 mm apart with specially cut belt longitudinally with or laterally across rows as required within row spacing. Chop out to leave a single plant when crop is well developed, about the four leaf stage.

Bed systems (Figure 5.2a)

These involve growing crops in strips or beds which the tractor straddles. All wheelings are restricted to the same area between beds, thus eliminating compaction in the bed where the crop is growing. After ploughing, the beds are marked out and tractors always follow the same wheelings between beds. It is essential to match all equipment, i.e. harrows, drill, sprayer and fertiliser distributor to bed width. Matching harrows and drills in multiples of three beds is commonplace; wider machinery covers a larger number of beds. With root crops, width of bed and arrangement and number of rows is determined by the number of rows lifted by the harvester or the width of the share where applicable, e.g. a four-row bed of sugar beet requires a four-row harvester. It is essential that everything fits the system.

Advantages of the system include: elimination of fanging in root crops, better germinating conditions and crop establishment, quicker ground cover with higher yields and some reduction in reliance on ground conditions for some jobs, e.g. spraying. Narrow row systems, e.g. carrots, are totally reliant on effective deployment of herbicides.

Beds are widely used for field vegetable crops: carrots, parsnips, onions, brassica seedbeds, early brassicas and

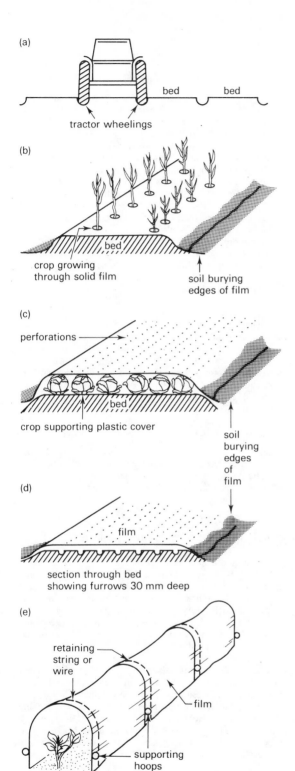

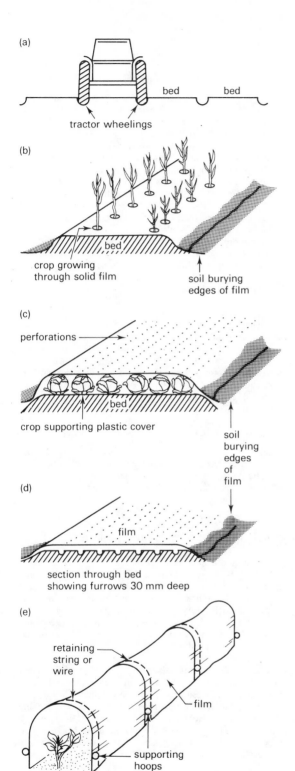

Figure 5.2 (a) Bed system, (b) surface mulch, (c) floating plastic film, (d) seed furrows, (e) continuous low tunnel (b–e, after ADAS, 1984c)

spring greens, plastic film protected crops and others. Suitable farm crops include sugar beet and potatoes.

Plastic covers for field protection (*Figure 5.2*)

Covers are made of either (i) transparent material, which may be solid or perforated to allow ventilation and top drainage, or (ii) opaque material, normally black, used only for weed control on small areas or winter storage in the field (ADAS, 1984).

Transparent plastic film covers

Their prime purpose is to produce early crops in spring; occasionally used in autumn to aid ripening and maturation before frost, e.g. Chinese cabbage. Covers give protection against cold desiccating winds and raise soil temperatures in early May by as much as 3–4°C, giving earlier emergence and growth and advancing crop maturity by one to three weeks. Technique can widen potential growing area of marginal crops, e.g. sweet corn; it also gives larger better filled cobs. Cover protected crops are ready after glasshouse but before unprotected field crops. There is also a quicker turnround between crops, which aids double cropping. Financial returns can be spectacular but only for wellgrown quality crops which achieve their target date.

Covers are expensive. Once this cost is incurred, the grower is locked into a high input–high output system. Everything must be done really well.

Site quality

Covers are only suitable for making an early site even earlier, not for improving late sites. Choose a sheltered site with flat or southerly aspect and low frost risk. Covers are susceptible to wind damage; winds also cool site. For long-term use plant shelter belts.(*see* Chapter 7, p. 235). Light sandy soils which warm up early are ideal.

General management

Use high vigour seed or uniform high quality plants, adequate fertiliser and ensure a high standard of weed, pest and disease control. Match bed system of culture with width of film. Apply fertiliser and granular pesticides before covering.

Eliminate perennial weeds and aim to reduce weed seed populations. Annuals can be controlled by 'stale seedbed' techniques. Residual herbicides can behave differently under covers and manufacturer's advice should be sought before use. Allow at least a week after removal of covers for crop to harden off and form a waxy protective layer before applying a contact herbicide. Also apply irrigation if necessary to make good any moisture deficit after cover removal.

Types of film or cover

Surface mulches (*Figure 5.2b*)

Simplest and cheapest method, using solid transparent film laid in direct contact with soil surface in bed widths and held down by soil at sides. Sunlight passes directly through film and warms soil; coloured or opaque film is less effective. Photodegradable film gradually becomes brittle, cracks and disintegrates in sunlight; life period is variable, e.g. 30, 60, 80 or 100 d according to manufacturer's specification. Crop grows *through* film, *not* underneath it. Machines are available which lay solid film on top of bed after drilling seed or transplanting plants into furrows; film must then be pierced to allow growing points to emerge. Other machines can lay cover and drill or transplant through it. Film covered by soil does not disintegrate and needs collecting.

Suitable crops Sweet corn, runner beans, marrows and courgettes.

Transparent floating film covers (*Figure 5.2c*)

Unlike low tunnels, these have no rigid support. As the crop grows it lifts the film off the ground, so that the film is supported by the crop, giving a 'cloche' effect. This type is used mainly in spring, when the soil is moist but low temperatures and cold winds delay or prevent germination and crop growth.

Except when used for short-term protection of seedlings, all floating covers contain perforations or slits, which permit gaseous exchange and drainage of rainwater. They also prevent overheating of the crop. To prevent wind drag, which may tear the cover off the crop, it must be laid tightly and the edges well secured by a covering of soil.

Choice of film Low ventilation films (< 2% perforation) retain more heat, accelerating germination and crop growth in early spring. Removal date is critical, as temperatures under cover can become excessive. There is little gas exchange or water drainage through holes. Effects of wind drag are accentuated.

High ventilation films (> 5% perforation) give smaller temperature increases and initial growth is slower than under low ventilation films, but excessive temperatures are less likely. Growth is not so advanced and plants are hardier and suffer less check on removal of cover; timing of removal is not so critical, permitting some delay in frosty or windy weather. There is reduced liability to wind drag and more uniform penetration of irrigation and rain.

Wide films (e.g. 10 m wide) are available for leafy crops, e.g. lettuce, giving fewer outside rows, thus reducing crushing. Must be laid by hand. Using 'flip over' technique, three adjacent crops can be covered in succession.

Management, sowing Drill small seeds into furrows 30 mm deep to give 60 mm gap under cover (*Figure 5.2d*) to prevent seedlings adhering to wet or frosted film and being uprooted when film moves in wind. Parsnips and carrots, with tiny taproots, are most vulnerable, small brassicas get chafed. With furrows young plants are adequately established when they reach the film. The greater airspace also gives better ventilation and insulation.

Removal of cover cannot be done in stages. Timing is a difficult decision, as protected crops tend to be soft and subject to desiccation and are readily damaged. Try to remove on a still moist evening, when frost is not expected. Most crops have an optimum removal time; with some it is critical. Although it may involve a yield loss, it is better to err on the early side as late removal causes crop deterioration and increased yield loss; timing is less critical with highly ventilated films. Removal normally occurs some time before harvest.

Suitable crops Early potatoes, carrots, lettuce, marrows and courgettes; also early summer cabbage, parsnips, radish, runner beans, early brassica seedbeds and others.

Continuous low tunnels (*Figure 5.2e*)

These were developed before floating film to replace glass cloches. They are normally made of solid clear plastic (as opposed to perforated) and are supported on rigid wire hoops and held down by twine. The system is very expensive, due to high costs of steel hoops and plastic and heavy labour demands for erection and removal. Tunnels are very susceptible to wind damage unless securely anchored.

Low tunnels are especially suitable for the earliest spring crops as they produce higher temperatures underneath than floating film. Hoops are adjustable, allowing adequate headroom to delay removal for taller growing crops or allowing harvesting while crop is still protected. Wire supports prevent crop abrasion, while twine anchorage allows hardening off either by gradually raising bottoms of plastic or cutting a slit along the top and allowing crop to grow through, e.g. bush tomatoes. Beds allow ready access for spraying and harvesting.

Suitable crops Owing to high cost, system is best suited to very early crop production to follow imports, glasshouses and larger tunnels. Suitable crops include: very early lettuce, especially overwintered crisp lettuce; earliest field crops requiring more headroom, e.g. runner and french beans, marrows and courgettes, outdoor bush tomatoes, strawberries, rhubarb and seedbeds for peg rooted transplants of early harvested brassicas. Low tunnels frequently replace cold frames.

Opaque plastic covers for winter field storage

Crop is covered with straw and solid black plastic film secured over it, e.g. celery. Owing to laying difficulties with carrots, plastic is laid first and straw on top. System is relatively expensive. Allow 25 t/ha straw for carrots, 15 t/ha for other crops where film covers straw.

'Cool chain' for horticultural produce

Vegetable and fruit crops deteriorate rapidly after harvest if field heat is not removed quickly to prolong shelf life, especially green produce. Once cooled, fresh produce must be kept cool and transported in refrigerated vans or containers to the buyer's cool store. The most advanced retailers have refrigerated display shelves to keep the produce cool to the last possible moment, which is essential for the freshest produce; others argue that turnover is so rapid that the extra cost of refrigerated shelves is not justified, at least for non-salad crops. This sequence is known as the 'cool chain' through which most multiples and supermarkets require fresh produce to be handled. They do not pay extra but merely refuse to buy produce not passing through the 'cool chain'. Cooling equipment is expensive and involves the grower or the packhouse in heavy capital expenditure. The following rapid cooling systems for newly harvested crops are availabe (ADAS, 1984d).

Hydrocooling

Produce is cooled by immersion or drenching in chilled water. Process is continuous and may be followed by a clean rinse. Cooling time 20–40 min. Process makes heavy demands on electric power but is suitable for use with an ice bank to reduce peak electricity demand. No weight loss, but produce must be removed from cooling plant to a conventional cool store for holding. Spreading disease within treated produce is a major risk of this treatment.

Suitable crops   Only those that can be washed or handled in water, e.g. carrots, leeks, salad onions, parsnips, radish.

Wet air cooling with forced or positive ventilation

Produce is placed in bins through which air, cooled by passing over an ice bank or refrigerated tubes, is blown. Cooling time: salads 2–8 h, brassicas and other vegetables 10–17 h. Minimal weight loss, produce cannot freeze and can be stored in plant. Ideal for a wide range of crops, especially brassicas, peas, beans, leeks, new potatoes, rhubarb and sweet corn; never for dry bulb onions. System is widely used.

Vacuum cooling

Vacuum cooling is a batch system requiring specially built cylindrical chamber and pump unit to achieve vacuum; relies on evaporating some water from produce for cooling effect, time: 20–40 min. Some weight loss, produce can freeze.

Especially suitable for fast cooling of leafy crops, e.g. lettuce, butterhead and crisp ('iceberg') lettuce, celery and leafy cabbage. Requires conventional cool store to hold produce once cooled.

Air-cooling: conventional cool store

Apart from dry bulb onions, for which it is the only suitable method, this system is not ideal for the rapid cooling of fresh produce. It is most suitable as a holding store for produce already cooled and an essential adjunct to hydrocooling and vacuum cooling systems for maintenance of cool chain. Produce can freeze. Performance of store is improved with forced draught ventilation, as cool air then comes into close contact with produce. For extended period of exposure, protect produce with plastic film to prevent excessive evaporation.

Cooling packed produce

Always ensure package is suitable to the system. Most types of package are adequate for vacuum coolers and conventional cool stores, but hydrocooling requires water resistant packs while high grade materials are essential for wet air systems.

Brassica crops

The group includes:

(1) brassicas grown primarily for market as vegetables: brussels sprouts, cabbages, calabrese, cauliflowers, kohl rabi;
(2) brassicas grown primarily for fodder: kales, rapes, fodder radish, white mustard, turnips and swedes (including 'market' swedes);
(3) brassicas grown for producing seeds for manufacture or processing: oilseed rape and mustard for condiment manufacture.

GENERAL

Rotation

Do not grow brassicas more frequently than once in five years at most, preferably longer, to avoid build up of persistent diseases in soil, notably clubroot and dry rot or canker (*Phoma lingam*). Infected land must be rested much longer.

Pests and diseases

Clubroot (*Plasmodiophora brassicae*)

This is worst in wet and/or acid soils, i.e. high rainfall areas and poor drainage.

Prevention   Good sanitation; feed infected root crops only *in situ*. Ensure clean transplant beds. Improve drainage and ensure soil pH of at least 6.5 for all brassica crops by liming if needed.

Susceptible varieties   All horticultural brassicas, oilseed and most forage rape varieties and many varieties of turnip and swede are highly susceptible and should not be grown on infected land.

Degree of resistance   This depends on races of clubroot present and level of infestation and is not absolute. Marrow stem and thousand head kales, fodder radish and some varieties of continental and hardy yellow turnips are highly resistant; Marian swede and some other sorts moderately resistant.

Dry rot or canker (*Phoma lingam*)

This attacks many brassicas, causing concave necrotic areas followed by rotting and collapse of plants. Infection can be seed borne, carried in soil or may arise from brassica waste or farmyard manure. Best seedsmen build up stocks from seed given hot water treatment or fungicidal soak. Sow clean or treated seed. Sanitation as for clubroot.

Beet cyst eelworm (*Heterodera schachtii*)

Brassicas act as host, but though rarely damaged, are treated as 'excluded crops' in rotations containing sugar beet (*see* p. 172 Sugar beet, rotations).

Brassica cyst eelworm (*Heterodera cruciferae*)

Adopt long rotation between brassica crops.

Other general pests and diseases

Numerous pests and diseases attack brassicas; the following are most important.

Flea beetle (*Phyllotreta spp.*)

Dress seed of *all* brassica crops with combined insecticide-fungicide and inspect regularly until well into rough leaf stage; apply insecticide as soon as symptoms of attack appear.

Cabbage aphid (*Brevicoryne brassicae*)

Feeds on leaves during summer and then infests developing brussels sprouts; even with small infestations crop becomes unmarketable. Routine inspection and treatment of sprout and other crops with approved aphicides are essential. Destroy overwintering hosts before May, where practicable.

Cabbage root fly (*Erioischia brassicae*)

Larvae feed on main roots, leading to wilt and death. Causes serious damage in horticultural brassicas, i.e. seedbeds, transplanted and direct drilled crops and early drilled swedes or cattle cabbage. Later generations may seriously disfigure market swedes and brussels sprouts at base of stem. Protect seedbeds of horticultural brassicas, transplants and direct drilled crops with approved granular insecticide, e.g. carbofuran, chlorfenvinphos and brussels sprout bottoms with trichlorphon if needed. Applicators are necessary.

Caterpillars (*several spp.*)

Damage occurs mostly from July to September. Inspect crops regularly and apply approved insecticide when required. Failure to do so may result in serious damage.

Wood pigeons (*Columba palumbus*)

Wood pigeons can strip young crops of brassicas and severely damage and soil older crops with droppings. Avoid isolated fields surrounded by woods. Regular shooting useful.

Slugs (*Agriolimax reticulatus*)

Slugs can strip young seedlings, seriously disfigure brussels sprout crops and spoil cabbage. Broadcast slug pellets as soon as damage is noticed.

BRASSICAS GROWN PRIMARILY AS VEGETABLES

Crops of brussels sprouts, cabbages and cauliflowers may be:

(1) raised in a plant bed, cellular trays or individual peat blocks or pots and then transplanted, or
(2) the seed is drilled directly into its final cropping position in the field (*see* Crop establishment, p. 96).

Brussels sprouts

Brussels sprouts are grown for marketing fresh or freezing. Crops are planted specifically for picking over, which caters mainly for the fresh market or for single harvest, which includes processing and prepacking.

Picking over Sprouts are picked as they mature at intervals of three to five weeks, beginning at bottom of stem. Season starts with earliest varieties, raised under protection, picked in late August or early September and continued until late March. Period of availability of a given variety is much longer than with a single harvest.

Single harvest All sprouts are taken from stem at same time; stems cut in field and stripped or removed elsewhere.

Varieties

Sprouts produced commercially now come almost entirely from F_1 or other hybrids, whose sprouts are much denser and of better quality than from older open pollinated varieties. Seed is very expensive. Development of new varieties is rapid, resulting in previous varieties being outdated quickly. Current varieties are listed in: Vegetable Growers Leaflet No. 3, *List of Brussels Sprouts* (NIAB).

Varieties overlap considerably in maturity, forming a continuous series to provide a season-long succession. For picking over: *earlies*: late August or early September to end November or early December. *Early midseason*: mid September or early October to late December or early January. *Midseason*: mid October or early November to late January or early February. *Late*: November to end of March. Shorter season for single pick, from about early October to late January. Many varieties are suited to both single pick harvest and picking over. When selecting varieties, attention should be paid to height and susceptibility to lodging, ease of picking or deleafing, quality and disease resistance.

Climate and soil

Commercial sprout production can be successfully undertaken in many areas in England and Wales on land below 275 m altitude, especially in the eastern counties. In south west England protect against ringspot (*Mycosphaerella brassicicola*). Soil must be well drained, with pH 6.5 or above. Deep moisture retentive mineral soils are best, notably well structured calcareous clays and silts. Avoid peaty soils and shallow or sandy soils liable to dry out.

Cultural methods for transplanted crops

Prefer cellular trays or peat blocks for early crops sown under protection. Bare rooted or peg transplants are satisfactory for maincrops (*see* Plant beds, p. 92).

Sowing and *transplant* dates depend on variety and harvest date.

Variety	Early	Early midseason
Sow	Mid–end Feb (under protection)	First week March in open (under protection in cold areas)
Transplant	Beginning–mid May	Mid May
Single pick harvest	Late Aug–end Sept	Late Sept–end Oct

Variety	Midseason	Late
Sow	First week March, in open (under protection in cold areas)–early April	End March–mid April
Transplant	Mid May–first week June	First week June
Single pick harvest	End Oct–end Jan	Late Dec–early March

Seedrate for seedbed drilling in rows

Use formula:

$$\text{Seedrate (kg/ha) of seedbed} = \frac{1000 \times \text{required population/m}^2}{\text{no. of seeds/g} \times \text{field factor} \times \text{germination \%}}$$

Use field factor:

	Feb	Mar	Apr (1–14)	Apr (15–30)	May
Frames					
Good soil conditions	0.8	0.8	0.9	–	–
Open ground					
Good tilth	–	0.6	0.7	0.8	0.8
Lumpy tilth	–	0.5	0.6	0.7	0.7
Irregular compaction	–	0.5	0.6	0.6	0.6

Aim for population of 40 plants/run of row

$$\text{Number of plants/m run of row} = \frac{\text{row width (mm)} \times \text{plant population/m}^2}{1000}$$

or

$$\text{Plant population/m}^2 = \frac{\text{no. of plants/m run of row} \times 1000}{\text{row width (mm)}}$$

At 40 plants/m of row, the following row widths will give plant populations/m^2 as follows:

Row width (mm)	Population/m^2
150	270
200	200
250	160
300	135

$$\text{Spacing between seeds within row (mm)} = \frac{1000}{\text{no. seeds/m run of row}}$$

$$\text{Number of seeds to sow/m run of row} = \frac{\text{row width (mm)} \times \text{plant population/m}}{10 \times \text{field factor} \times \text{germination \%}}$$

Transplanting

Plough land in early winter to ensure stale furrow and well weathered, settled soil. Provide fine firm moist uniform tilth, not compacted underneath, soft enough on surface to allow penetration of transplanting machine. Plants pulled from seedbed should be placed in shallow trays and protected immediately from sun; plant at once. Insecticides for cabbage root fly (*Erioschia brassicae*) control should be applied at transplanting. Replace any failure within 7–14 d of transplanting. Irrigation immediately after transplanting ensures rapid root development; essential with plants raised in peak blocks but not with cellular transplants with moist soil. Apply 25 mm of irrigation at 25 mm moisture deficit. Glasshouse raised plants are brittle and need very careful handling.

Cultural methods for direct drilled crops

Widely used for close spaced crops, occasionally at wider spacings. Plough at least 250 mm deep before end of December, having previously subsoiled if necessary, when soil was dry enough to shatter readily.

In spring prepare a fine firm level seedbed with minimal working and compaction for accurate drilling. Precision drills are most satisfactory. Sow groups of three spaced seeds centred at required final spacing 25–35 mm within row or

Table 5.3 Number of plants required/100 m^2 (1/100 ha) according to spacing

Planting distance cm (i.e. mm/10)	10	15	20	25	30	40	50	60	70	80	90	100	125	150
10	10000	6666	5000	4000	3333	2500	2000	1666	1428	1250	1111	1000	800	666
15	6666	4444	3333	2666	2222	1666	1333	1111	942	833	740	666	533	444
20	5000	3333	2500	2000	1666	1250	1000	833	714	625	555	500	400	333
25	4000	2666	2000	1600	1333	1000	800	666	571	500	444	400	319	266
30	3333	2222	1666	1333	1111	833	666	555	476	416	370	333	266	222
35	2857	1905	1428	1143	952	714	571	476	408	357	317	286	228	190
40	2500	1666	1250	1000	833	625	500	416	357	312	277	250	200	166
45	2222	1481	1111	889	740	556	444	370	317	278	247	222	178	148
50	2000	1333	1000	800	666	500	400	333	285	250	222	200	160	133
55	1818	1212	909	727	606	454	363	303	259	227	202	182	145	121
60	1666	1111	833	666	555	416	333	277	238	208	185	166	133	111
65	1538	1026	769	615	513	385	308	256	220	192	171	154	124	103
70	1428	942	714	571	476	357	285	238	204	178	158	142	114	95
75	1333	889	667	548	444	333	267	222	190	167	148	133	107	89
80	1250	833	625	500	416	312	250	208	178	156	138	125	100	83
90	1111	740	555	444	370	277	222	185	158	138	123	111	88	74
100	1000	666	500	400	333	250	200	166	142	125	111	100	80	66
120	833	556	417	333	278	208	167	139	119	104	92	83	67	56
125	800	533	400	319	266	200	160	133	114	100	88	80	64	53
150	666	444	333	266	222	166	133	111	95	83	74	66	53	44

abreast 25 mm apart 20 mm deep, removing unwanted plants later.

Drilling date

This depends on required date of harvest and variety. Very high populations, 44 000 plants/ha and over mature about a month later than those between 30 000–36 000 plants/ha.

Type of variety	Drill	Single pick harvest
Early	1st week Apr	Late Aug–end Sept
Early mid-season	1st week Apr	Late Sept–end Oct
Mid-season	Throughout Apr	End Oct–end Jan
Late	End Apr–mid May	Late Dec–mid Mar

Plant population

Increasing plant population reduces sprout size, improves uniformity of sprout development and delays maturity; length of stem is increased and sprouts are formed further apart on the stem. Level of plant population is more important than precise arrangement. Choice of population depends on market and size of sprout required, method of harvest and variety. Populations may vary from 12 000 to 50 000 plants/ha although populations over 36 000 plants/ha do not necessarily give greater yields of sprouts. Some non hybrid types are unsuitable for close spacing, while many hybrids are unsuitable for traditional wide spacings of 12 000 plants/ha (900 × 900 mm spacing). As an approximate guide aim for 42 000–45 000 plants/ha (450 × 500 mm) for small sprouts for freezing and 26 000–28 000 plants/ha (600 × 600 mm) for picking over, according to variety. For single pick prepack sprouts populations range from 36 000 plants/ha for early shorter varieties to 28 500 for later varieties. For spacing and plant population *see Table 5.3*.

Manuring (*Table 5.4*)

Fertilisers are applied during preparation of direct drilled seedbeds or while soil is being worked prior to transplanting. Where N exceeds 100 kg/ha, excess should be top dressed within one month of crop emergence or transplanting on light soils. Top dressing of further N may also be necessary when rainfall substantially exceeds transpiration rate within two months of base application. Excessive applications of N may result in loose sprouts.

Stopping

Stopping is done to improve uniformity of sprouts in crops to be taken for single harvest up to end of November, but not for later crops. Apical bud is given a sharp downward blow with soft rubber headed hammer or a piece 30 mm diameter, including apical bud, is removed from four to ten weeks before harvest, when 50% of basal buds reach 12 mm diameter. Stopping too early may cause 'blowing' of upper sprouts.

Market and saleable yield

Sprouts for *fresh market* must be clean, sound, tight and fresh in appearance, and free from aphids or pest damage, graded into EEC classes I, II or III and packed into nets, holding 10 kg or boxes holding 6–14 kg or in small prepacks of 0.5–1 kg.

Yield
Poor   12.5 t/ha
Average   20–22.5 t/ha
Good   30 t/ha or more

Sprouts for freezing must be clean, tight and blemish free and are closely graded into sizes: small, standard and large. Usually packed in 250 kg returnable bulk bins for delivery to freezing plant. Being smaller than the fresh pick market, yield is much lower.

Yield
Poor   5 t/ha
Average   7.5–10 t/ha
Good   12.50 t/ha

Sample grading of 11 t crop:

Small: 13–29 mm   10%
Standard: 29–35 mm   85%
Large: 35+ mm   5%

Cabbages

Cabbages are now grown primarily for human consumption. Traditional cattle cabbages, once widely grown, are now of little importance. Stockfeed from cabbages comes largely from unsold crops and waste, although large headed white varieties can be grown for livestock. Cabbages must be regarded as a group of crops rather than a single crop. Types include: spring cabbage, summer and autumn cabbage, winter white cabbage, savoys, January Kings, January King or savoy × winter white hybrids and red cabbage. With the

Table 5.4 Manuring of brussels sprouts for fresh picking or processing (kg/ha)

N, P, K or Mg index	N		P_2O_5	K_2O	Mg	
	Silt and brickearth soils	Other soils	All soils		Sands and light loams	Other soils
0	200	300	175	200	90	60
1	150	250	125	175	60	30
2	100	200	75	125	Nil	Nil
3			50	60		
4			25	Nil		

ADAS figures.

exception of spring cabbage and some January Kings, practically all commercially grown varieties are F_1 hybrids. Seed costs many times that of open pollinated sorts but heads are much more dense, heavier, very uniform and mature more evenly. Standing period without splitting in field of many varieties is excellent.

Calculation of seed requirements for all types, see brussels sprouts p. 101.

Spring cabbage

Crop may be grown as spring 'greens' or 'collards', which are unhearted or as semi-hearted or as fully hearted cabbage. Nowadays 'greens' are the most favoured type of production, often grown as a packhouse filler to supplement other vegetables, when 500 g packs of three to four pieces of clean leafy cabbage are required; they must not be stemmy or bolting. Production can now be obtained all the year round from suitable areas: SW England, December–March; Kent, February–April and Lincolnshire for summer and autumn. Prices are not usually high except when severe winter weather limits marketing and then only favoured sites can profit. Crop may be unsaleable if there is a surplus of other vegetables or a glut in a mild winter. The same crop may be used for two purposes, hearted cabbage being cut selectively as they heart; it is wise to cut some crops before maturity if prices are very high or growth is sappy and liable to frost damage.

Removal of field heat is generally essential before marketing (*see* Cool chain, p. 98).

Varieties

All varieties have pointed hearts. Choose varieties according to purpose, i.e. greens only, greens and hearted cabbage, hearted cabbage. Varieties are listed in Vegetable Growers Leaflet No. 2, *List of Cabbages* (NIAB).

Soil, cultivations and establishment

Choose a free draining soil of light to medium texture which warms up early in spring but retains adequate moisture. Minimum pH 6.0 but preferably 6.5–7.0. Prepare fine firm seedbed with good tilth in adequate time for sowing between late July and September, taking care to conserve moisture. If irrigation is needed apply after completion of cultivations and allow soil to become dry enough to drill. To avoid compaction and improve uniformity many growers use the bed system. Nowadays spring cabbages and greens are almost entirely direct drilled (i.e. seed is drilled directly into cropping position); raising in plantbeds and transplanting only used on very small areas.

For spring greens aim for 28–40 plants/m^2 precision drilled, 20 mm deep using seed dressed with an approved fungicide/insecticide, in 350 mm rows to produce plants 70–100 mm apart; hearted spring cabbage requires 10 plants/m^2 in rows 400 mm apart drilled to produce plants 250 mm apart.

Manuring

Seedbed/base applications, *see Table 5.5*. Crop requires a steady supply of N throughout growing period but excess ruins winter hardiness. For spring harvested crops base dressing should not exceed 75 kg/ha N but for greens harvested before Christmas increase N to 125 kg/ha. Rate of spring N top dressing depends on expected size of crop; fully mature cabbage harvested in late spring can use up to 380 kg/ha whereas lighter crops of greens need less than half this quantity. Dressings of 100–200 kg/ha N at any one time are normally adequate. Timing depends on weather conditions and potential marketing period. To avoid scorch and subsequent infection by grey mould fungus (*Botrytis cinerea*) top dress when crop is dry or very wet to minimise adhesion to foliage.

Cutting date

Determined by sowing date and variety.
Spring greens: according to district.
Hearted cabbage: April to early June.

Yield

Extremely variable and difficult to quantify, income/ha main consideration. Crops may be cut completely for unhearted greens, cut over partly for greens and residue left to heart or cut over hearted, according to market. Average: 12.5 t/ha hearted cabbage.

Table 5.5 Manuring of cabbages for human consumption (kg/ha)

N, P, K or Mg index	N					P_2O_5	K_2O	Mg	
	Spring, hearted and greens	Summer and autumn	Winter white for storage	Winter and savoy		All cabbage types			
				Pre-Christmas cutting	Post-Christmas cutting			Sands and light loams	Other soils
0	75	300	250	300	150	200	300	90	60
1	50	250	200	250	125	125	250	60	30
2	25	200	150	200	100	75	175	Nil	Nil
3	–	–	–	–	–	50	75	Nil	Nil
Over 3	–	–	–	–	–	25	Nil	Nil	Nil
N top dressing	200–400			0–75					

ADAS figures.

Table 5.6 Summer and autumn cabbage – types and times of sowing, transplanting and cutting

Type of cabbage	Sow	Method of raising	Plant out[1]	Shape of head	Cutting period	Direct drilled
Summer–early	late February	P	mid April	Pointed, e.g. Hispi F_1[2] or round, e.g. Golden Acre	Late May–late June	–
Summer–midseason	late March to mid April	P	late May	Round[2]	July–August	–
Summer–late	April	O or P	[1]	Round[2]	September ⎫	⎫
Autumn	late May	O or P	mid July	Round[2]	September–late October ⎬ May to mid June

[1] For spring and early summer sowings six week old protected plants and eight week old open ground plants give better yield and quality than older plants.
[2] F_1 hybrid varieties usually produce more uniform crops than open pollinated varieties or selections but seed is dearer.
P = raised under protection: O = raised in open.

Summer and autumn cabbage

Grown to produce a succession of hearted cabbage to market between late May and late October, for which sequential sowings and a range of varieties are necessary. Varieties are listed in Vegetable Growers Leaflet No. 2, *List of Cabbages* (NIAB). Sowing in February and March for late May, June and July harvesting are made under protection with heat; April and May sowings can be made outdoors. Growers often prefer to continue sowing under protection for harvests from July to October to ensure continuity of supply. Varietal types, and periods of sowing, transplanting and cutting are given in *Table 5.6*. Sowings made in April and May can also be drilled direct into final cropping position, where suitable.

Soil, cultivations and establishment

Soil must have stable structure to permit extensive root development with at least 600 mm effective rooting depth; good drainage essential, especially autumn cabbage. Earliest summer crops come from sandy loams, loams or peats. For later crops water retentive medium to heavy soils and silts are suitable.

Crops respond well to irrigation; on loamy sands it is usually essential. Soil pH should be 6.5 on mineral soils, 6.0 on peats.

Prepare a fine level tilth but not overworked or compacted. Subsoil if required and plough during previous autumn. Spring ploughing is unsatisfactory on heavier soils; allow ample time for working and weathering before planting. Direct drilled cabbage require rather more working than transplanted crops; with the latter a single pass with a spring tine cultivator should be adequate. For direct drilling use only the best soils, ploughed in autumn.

Plant density and spacing

This depends on maturity and size of cabbage required. Summer cabbages for market allow nine plants/m², rows 380 mm apart and 300 mm between plants; for autumn cabbages allow three to four plants/m² rows 600 mm and plants 400–600 mm apart, according to size required.

Manuring

Fertiliser is thoroughly worked into soil while preparing tilth, except where N topdressing is appropriate (*Table 5.5*).

Where heavy dressings of P and K are required (e.g. index 0) part should be applied before ploughing.

Yield

Variable. Good growers achieve 20–30 t/ha.

Winter cabbage and savoys

The group includes hearted cabbage grown for marketing from October until hearted spring cabbage are ready and winter white cabbages for storage.

Types and varieties

Winter white cabbages These are round, dense, hard cabbages, white inside ready for cutting up to end of November or early December. Varieties are suitable for the fresh market, storage or both. Type is not hardy enough to overwinter outdoors, except in mildest coastal areas. White cabbages for storage must be cut and stored *before* there is any frost damage and can then be marketed in late winter onwards.

January King varieties Early selections may be cut before Christmas but are valued for January–February; best sorts are very frost hardy. Few outer leaves, with purple cast, prominent veins and producing heads rather under 1 kg. Leaves somewhat wrinkled and classed as a savoy elsewhere in Europe.

Savoys These may be distinguished by their deeply wrinkled or 'blistered' foliage and are often considered to have a stronger flavour than ordinary cabbage. Varieties have a range of maturity from about 90–200 d from transplanting; savoys can provide crops for cutting for the fresh market from late summer to the following March, according to variety. Winter varieties are very frost hardy and especially suitable for colder areas.

Savoy or January King × winter white hybrids Head density intermediate between the two parents; high yielding with good frost hardiness. Grown as January Kings. Varieties are listed in Vegetable Growers Leaflet No. 2, *List of Cabbages* (NIAB).

Red cabbage Similar in form to winter white with hard

dense heads. Cultivation and storage identical. Small retail market; main use pickling and processing. Best varieties F_1 hybrids.

Soil, cultivations and establishment

Winter cabbages are grown on a wide range of soils but soil must be free draining. Deep soils are best. Avoid light sands or shallow soils. Plant on early, medium and late fields to secure succession. Tilth as for previous types but allow ample time for preparation after clearing previous crop. Plant into moist soil. If soil remains dry for a few days after transplanting, irrigation is essential on all but the most moisture retentive soils. Plants raised in modules can give increased uniformity, particularly with high density crops planted in dry conditions.

Plant population exerts a major influence on head size and maturity date. High populations are necessary to obtain the small 'supermarket size' of cabbage, not more than 1 kg each, now required: aim for populations of 40 000–60 000 plants/ha (*see Table 5.3*). Processors require cabbages of 3–4 kg each, so reduce to 24 000–28 000 plants/ha. Total fresh yield remains constant over the normal range of populations as specified.

Manuring (Table 5.5)

Work in base dressing during tilth preparation, including all P_2O_5 and K_2O. Adequate N is essential for healthy growth but excess damages winter hardiness of crops of savoy and winter cabbage to be left after Christmas; excess also impairs quality of storage cabbage. Direct drilled crops and transplants on light soils should not receive more than 150 kg/ha in seedbed; top dress remainder within one month of crop emergence or transplanting. Additional N may be required as top dressing where rainfall greatly exceeds transpiration within two months of base application.

Storage cabbage

Winter white cabbage is now available all the year round for fresh market, shredding for summer salads and coleslaw manufacture; the latter requires top quality cabbage which cannot be produced from the field until early September, so long-term storage is essential to meet summer demand. Suitable storage facilities require considerable capital expenditure. Winter white cabbage may be drawn from the field during September, October and November but is not frost hardy and must be cleared by late November. White cabbage for fresh market is often left to overwinter in the field in south-west England, but this is risky, especially in hard winters.

Barn store:	Dec–Mid Mar
Cool store:	Early Mar–early July
Controlled atmosphere store:	July–early Sept

Cut early in the day to minimise field heat in the heads and cool quickly. When growing ensure efficient production and programmed market or contractual arrangements.

Marketing from field

The earliest crops of winter cabbage and savoy sometimes overlap late crops of summer and autumn cabbage. Winter cabbage and savoy mature over a long period and regular cutting secures firm, compact heads of uniform size and quality. Modern F_1 hybrids mature more evenly.

Cut into tractor mounted boxes for transport to farm packing shed or to a packhouse for trimming, wrapping and packing for supermarkets. Indoor packing ensures a far higher standard of grading than is possible in the field, especially in winter.

Yield
Storage cabbage:

Average	30–32 t/ha
Good	45 t/ha

Winter hardy cabbage/savoy:

Average	20–22 t/ha
Good	30 t/ha

Cattle cabbage

Cattle cabbage produce large-hearted cabbages for autumn and winter use but are of little importance nowadays, more suited to small farms where massive productivity from small areas is essential. Under really fertile conditions yields of 150 t/ha are obtained but 50–75 t/ha is more common. Grown as a maincrop, it has a high labour demand and its feed value is well below that of marrow stem kale.

Devon Flatpoll, with purple leaves, is the main variety. Seed is usually sown in nursery beds in mild areas in August or early September at 1.25 kg seed, dressed against flea beetle, 12 mm deep in 230–300 mm rows on 0.125 ha. This provides plants for 1 ha of crop. Transplant to final cropping position in April or early May in rows 760 mm–910 mm apart, 910 mm between plants.

Cattle cabbages need very fertile soils and do well on deep,

Table 5.7 Winter cabbage, types and times of sowing, transplanting and cutting

Type of cabbage	Sow	Method of raising	Plant out	Cutting period	Direct drilled
Winter white cabbages	late April	O	early July	Should be cut and stored by November	mid May
January King and savoy × winter white hybrids	late May	O	early July	November onwards	June–early July
Savoys	late April–mid May	O	late June	early varieties September to December, late varieties January to March	mid–end May

O = raised in open.

rich medium-to-heavy soils, liberally manured with farmyard manure 50–60 t/ha ploughed in. Work 125–200 kg/ha N, 125 kg/ha each of P_2O_5 and K_2O into the seedbed, giving the lower rate of N after applying a heavy dressing of dung. The seedbed must be deep and loose enough for the transplanter or planting by hand with a dibber. Cabbages must be firmly planted.

Cauliflowers

Cauliflowers, which are produced all the year round from different areas in the UK, may be divided into two main groups:

(1) *winter cauliflowers*, marketed from early November to late May or June according to variety; and
(2) *summer and autumn cauliflowers*, cut from early May to November or even early December in mild open years.

A comprehensive list of varieties of all types suited to culture in England and Wales is issued by the National Institute of Agriculture Botany – Vegetable Growers Leaflet No. 1, *List of Cauliflowers*. The two groups are quite distinct in their varieties and cultural methods and must be treated as separate crops.

Winter cauliflowers

Grown as farm crops in relatively frost-free areas with mildest winters: south west England, the south coast, South Wales and the Channel Islands. To avoid frost damage to curds, varieties heading in mid winter should be grown only in most favoured situations, usually close by sea coast. Only varieties resistant to ring spot (*Mycosphaerella brassicicola*) of, or derived from, Roscoff type (as recommended by the NIAB) should be grown in the south west. More winter hardy sorts, heading from March to June include varieties of Angers, English Winter and Walcheren types. These are only half grown when winter sets in and are more widely cultivated; especially important in Lincolnshire.

Soil

Choose free draining friable soil, pH 6.5–7.0 on arable farms, pH 6.5 on holdings with livestock. Crop highly susceptible to clubroot (*P. brassicae*) and must not be grown more than once in five years except on strongly alkaline light soils near Penzance in west Cornwall; for intensive production aim for soil pH 8.0 and ensure drainage. Maintain good structure with regular use of leys or farmyard manure. *Never* catchcrop with brassicas.

Plant bed preparation and sowing

Winter cauliflowers are unsuitable for direct drilling and are nearly always raised in plant beds (for plant bed management *see* Plant beds, p. 93).

Sow Sow Roscoff types during first week in May. Sow others late May and during June, all to be ready for transplanting in July. Sowing too early results in subsequent losses from transplants being too large, buttoning, and powdery mildew (*Peronospora parasitica*).

Allow 280–420 g of seed to produce enough transplants for 1 ha of crop. Drill in beds, rows 225–250 mm apart, seed 20 mm deep or cut into soil moisture. Seedrate in bed 28 g/91 m of row to give 33 plants/m of row. Place dressed seed 18 mm apart with precision drill. Plants are lifted from seedbed when required; protect against sun and transplant immediately. Select sturdy plants 180–230 mm height with good root systems. Reject weak, deformed or blind plants, or those damaged mechanically or by pest or disease. Use routine calomel root dip in areas where clubroot is prevalent. (Use 49 g 100% calomel in 1 litre water dipping roots in bunches of 25–30 plants; this will treat 600–1000 plants.) Protect transplants against aphids and cabbage root fly.

Field preparation and transplanting

Plough before end of March at latest to allow adequate time for preparation of well settled fine moist tilth to receive transplants. Always have land waiting for plants, not other way round. Mark out land with spaced tine implement across line of transplanting to ensure adequate population.

Transplant late varieties in first half of July and earliest during third week July; too early can cause bracting (i.e. small leaves growing through curd). Finish planting by end of July at very latest. Complete gapping up failed plants or misses within one week of planting. *Banking up* plants with earth is traditional in windy areas but must be completed early to avoid severe damage to roots.

Plant population and spacing

Aim for 23 500 plants/ha for varieties heading after early March and 20 000 plants/ha for earlier varieties, which require row widths 650–750 mm and interplant spacings 550–600 mm.

Irrigation

Irrigate or water in when transplanting during a dry period. Plant *firmly*.

Manuring

Base fertiliser, including all P and K is worked in during soil preparation (*Table 5.8*). Roscoff types cut in February in frost free areas are top dressed in late autumn; crops for cutting after mid February are top dressed in mid February. Other types for spring cutting are top dressed between January and March. Avoid excess N applications.

Cutting

Inspect crop frequently and once production starts cut every other day to minimise losses from frosting or yellowing of curds.

Cut every good head *before* it has reached full size. During frosty periods cut only in middle of day, when temperature is rising. Cutting at other times damages growing crop and keeping quality of heads; also cut harder during periods of sustained frost although heads will be smaller. Cut into tractor mounted bins, one man usually taking three rows. Cutters trim while harvesting, leaving sufficient leaves and length of leaf above top of curd to protect during transit.

Quality, packing and yield

Heads when cut should be firm, compact and of good white colour. Once seriously yellowed by sunlight or blown by

Table 5.8 Manuring of cauliflower (kg/ha)

N, P, K or Mg index	N			P_2O_5	K_2O	Mg	
	Early summer, late summer, early autumn, late autumn	Winter		All cauliflower types			
		Roscoff types	Winter hardy types			Sands and light loams	Other soils
0	250	75	75	175	300	90	60
1	200	40	40	125	200	60	30
2	125	Nil	Nil	75	125	Nil	Nil
3	–			50	60		
4	–			25	Nil		
N top-dressing[2]		60–125[3]	125–200[4]				

ADAS figures.

[1] When winter cauliflowers follow crop leaving large residues reduce P_2O_5 by half and K_2O by 60 kg/ha.

[2] For transplants on light soils or direct drilled crops top dress N in excess of 100 kg/ha; apply residue within one month of emergence or transplanting.

[3] For February cuttings in frost-free areas, apply top dressing in late autumn. For crops to be cut later than mid-February give top dressing in mid-February.

[4] Apply top dressing January, February or March according to circumstances.

overstanding, they are unwanted. Discard also badly damaged heads. 'Ricey' and 'bracted' curds are of low quality.

All cauliflowers must normally be graded according to EEC regulations (*see* MAFF leaflet EEC Standards for Fresh Cauliflowers) – Extra class and Classes 1, 2 and 3. Most good production usually falls into classes 1 and 2. Extra is too exacting for most commercial winter production; grade 3 is of low quality and usually only justifies local marketing. As with cabbage, pack under cover or send to packhouse for trimming and wrapping for supermarkets. Produce for wholesale markets is packed in non-returnable containers of 12, 16, 24 or 30 heads. Cauliflowers will usually keep for a week in a cool store: only store top quality curds cut as soon or immediately before they are quite ready.

Yield

	(Crates/ha)
High	625
Average	525
Low	425

Summer and autumn cauliflowers

Crops are grown to provide succession of cuttings throughout summer and autumn until November and are grouped as follows.

Early summer cauliflowers Sown in late September or early October, pricked out into frames or Dutch lights and transplanted as 'peg' plants into cropping position in March or early April. Alternatively sown in gentle heat in February or early March and transplanted four to five weeks later, as soon as ground conditions permit. Undue delay involves increased risk of 'buttoning'. Maturity (50% of marketable heads cut – NIAB) varies from approximately 80–110 d after transplanting according to variety. Use of cellular trays, peat blocks or pots minimises root disturbance.

Late summer cauliflowers Sown under protection in March or in open ground in April transplanted mid–late May, maturing about 60–80 d after transplanting.

Early autumn cauliflowers Sown in late April or early May and transplanted mid to late June. Mature 70–85 d after planting.

Late autumn cauliflowers Sown in mid May, transplanted late June or early July, maturing 70–130 d after transplanting.

Soil type and cultivations

These are more specialised crops and require greater care in soil selection and management than winter cauliflowers. *Plants must never receive a check through lack of moisture or nutrients*, or buttoning (i.e. premature formation of tiny worthless curds) is induced.

Choose deep rich moisture retentive soils, such as alluvial silts, medium to heavy loams, brickearths and lower greensands. Avoid soils prone to drying out. Soil should have good structure and be liberally supplied with organic matter, either by leys or heavy dressings of farmyard manure. Cultivations should be deep and thorough as for winter cauliflowers. Allow adequate time after harvesting previous crop for preparation of high quality moist weed free seedbed. Loose soils result in crop failures and low quality curds. Firm with roller but follow with harrows prior to transplanter.

Crop also highly susceptible to clubroot – control pH to 6.5–7.0 or even 8.0 where there are no breeding livestock.

Plant raising

Cellular tray, block or pot raised plants do not suffer root disturbance and are preferred to bare root transplants by many growers. For earliest crops 75 mm pots or 60 mm blocks are superior to smaller cells and bare root transplants but are much more expensive; their use should be restricted to the earliest sites. Normal cells of 9–15 ml volume are otherwise adequate. For plant bed and cellular tray management *see* pp. 93–96. Calculate seedrate from formula and plant requirement in *Table 5.3*. For transplanting *see* p. 94–96. Treat blocks or transplants against cabbage root fly (*Erioischia brassicae*).

Plant population and spacing

This depends on maturity group and size of curd required, head size increasing with lateness of maturity.

Type of crop	Plant popula-tion/ha	Spacing (mm)
Earliest crops (from frame raised transplants)	37 000 to 30 300	600 × 450 to 600 × 550
Late summer and early autumn crops	27 700	600 × 600
Late autumn crops	23 800	600 × 700

Manuring

A steady supply of N is essential. If deficient, plants make poor growth and start to button, whereas excess leads to undesirable features such as browning, scorching, bracting, hollow stems and loose curds. Dressings over 150 kg/ha N should be split; top dress within one month of transplanting. Potash is said to increase head firmness. Very heavy dressings of muriate of potash can damage plant, so apply well before planting or part before ploughing. Some growers prefer sulphate of potash, which is essential under glass (*see Table 5.8*).

Irrigation

Summer and autumn cauliflower make heavy demands on soil moisture; do not allow soil to dry out. In dry periods water immediately after planting, which can suppress buttoning. Soil should be well irrigated in early stage of curd development. Regular watering with 10 mm when needed gives best results.

Quality, packing and yield

Only top quality high yielding crops justify production. Quality and packing as for winter cauliflower. Yield increases with lateness of maturity; late autumn crops tend to have large heads and give high proportion of crates of '12's.

Type of crop (good yield)	Crates/ha
Early summer	1300–1500
Late summer and early autumn	1500–1750
Late autumn	2000–2250

Calabrese

Also known as green sprouting broccoli or broccoli; it is botanically similar to purple and white sprouting broccoli and 'cape broccoli' and is an annual. The edible part consists of a stem terminating in a tight head of green flower buds. The primary spear, which should be large and succulent, is produced first. Secondary or side-stems usually appear after the primary stem is harvested; these are much smaller, of lower quality and it is not usually worth waiting for them to develop.

Varieties

All improved varieties are F_1 hybrids. Maturity varies from around 75–90 d after drilling. There are a number of excellent varieties.

Season

Calabrese can produce heads from May to late November. Earliest crops in May and June can be obtained from glasshouse raised transplants, sown in cells or peat blocks in February and March. Subsequent crops are drilled directly into the field. Late autumn harvested crops are only suitable in the mildest areas of southern England.

Soil

Crop needs to grow rapidly without check to produce top quality heads; it grows well on a wide range of soils. Light soils are excellent for earliest crops but need irrigation later on. If irrigation is not available, choose deep moisture retentive soils. Soil must be well drained and of good structure; soils liable to capping are risky with direct drilled crops.

Plant population and establishment

Aim for 8.5–10 plants/m² according to size of head required for market. Normally drilled in 500 mm rows × 230 mm in rows or in beds 1.68–1.80 m wide containing 4 × 400 mm rows 200 mm apart. Increase spacings within row to 250–300 mm for larger heads, where irrigation is unavailable or on less fertile soils.

Bare rooted transplants are generally highly unsatisfactory, especially large transplants. Transplants raised in peat blocks and cells have given satisfactory results and are suitable for early crops. Plant out as soon as two true leaf stage is reached; delay checks growth and results in reduced yields. Before transplanting soak modules with water or preferably high N liquid feed. If soil is dry, apply irrigation or water in planting furrow from tank on transplanter. Protect from cabbage root fly (*Erioischia brassicae*).

When drilled directly into field, crop is sown by precision drill in groups of three seeds and singled (as with brussels sprouts) or drilled to a stand. Seedrate is calculated by same formula as brussels sprouts (p. 101). Use high germination seed; where soil capping is a problem use anti-capping materials. Ensure a smooth fine firm even seedbed; bed systems, which avoid compaction, are preferable.

Continuity of production

With late spring and summer drillings, crop may mature earlier than from transplants. Use only early varieties for early transplanted crops and sowings from mid July onwards. Sowings maturing in late autumn are only reliable in mildest situations. ADAS (1984) suggest the following programme for continuity of production:

Sowing date	Planting date	Harvest period
Transplants raised in peat blocks/modules		
Mid Feb	Early Apr	Late May
Mid Mar	Mid Apr	Early June
Late Mar	Mid Apr	Mid June

Sowing date	Planting date	Harvest period
Field drilling		
Mid Mar	–	Late June
Late Mar	–	Early July
Mid Apr	–	Mid July
Late Apr	–	Late July
Mid May	–	Early Aug
w/b 28 May	–	Mid Aug
w/b 4 June	–	Late Aug
w/b 10 June	–	Early Sept
w/b 17 June	–	Mid Sept
Late June	–	Late Sept
Mid July	–	Early Oct
Late July	–	Mid Oct
Early Aug	–	Late Oct–early Nov
Mid Aug	–	Mid–late Nov

Manuring

The crop is highly susceptible to clubroot (*Plasmodiophora brassicae*) so maintain soil pH at least 6.5; where brassicas are intensively grown prefer pH 7.5–8.0.

Apply fertilisers as shown in *Table 5.9*. Spread maximum of 100 kg N before drilling; apply the rest after emergence. Where high K is required, apply well before drilling and work in thoroughly.

Table 5.9 Fertiliser requirement for calabrese (kg/ha)

N, P, K or Mg index	N	P_2O_5	K_2O	Mg	
				Sandy soils	Other soils
0	250	150	150	90	60
1	200	75	100	60	30
2	160	60[1]	75[1]	–	–
3	–	60	50	–	–
4	–	30	–	–	–

[1] Apply if no soil analysis available.

Irrigation

Irrigation is essential to ensure timely and consistent supplies of top quality heads. Except on water retentive soils, e.g. silts when 50 mm water may be given 20 d before cutting, apply 25 mm whenever soil moisture deficit (SMD) reaches 25 mm. If necessary, irrigate to bring soil up to field capacity, then drain, prior to drilling to ensure even germination.

Crop protection

Calabrese is subject to the same pests and diseases as other brassicas and requires protection. The number of herbicides specifically recommended for weed control is limited.

Harvesting, cooling and marketing

Primary spears are required for the best trade. Cut with knife while heads are firm with tightly packed buds and no trace of yellow petals or opening buds. The cut stem must be succulent without trace of woodiness.

Preferred size Head diameter: 60–80 mm, maximum 100 mm; length 140 mm, maximum 150 mm; stem base diameter: 15–20 mm, maximum 30 mm. Even F_1 hybrids do not reach optimum stage simultaneously, so crop usually requires cutting over three times.

In summer cut early in the morning and remove cut spears from the field quickly and cool immediately. Wet air cooling with positive ventilation is best; vacuum cooling is suitable. Conventional cool stores are unsatisfactory. Market through the cool chain (*see* p. 98). There is little interest for processing but demand from multiples and supermarkets for cool chain handled produce is good. Before growing check and agree with buyers details of their specifications for prepack weight, quality and trimming. Generally 'posy' packs containing one to four spears, total weight 220–360 g are preferred. Crop is best handled in a packhouse. Yield may vary from 2–12 t/ha but 6 t/ha is usual.

Kohl rabi

The edible part of this vegetable is a turnip-like bulb carried on a short section of the stem just above ground level. Popular on the European mainland, preferred size is 70–90 mm diameter; bulbs must be tender and grown quickly. Taste is similar to a turnip.

Types

White Vienna Pale green skin with white flesh.

Purple Vienna Has a purple skin and is a little later. Leading breeders offer named varieties.

Kohl rabi thrives on a wide range of fertile light to medium and peaty soils but, unlike turnips, to which its general culture is similar, tolerates hot summer conditions. Maintain soil pH at minimum 6.5. Irrigation is essential for top quality crops. Hold soil at near field capacity and apply 25 mm water every time soil moisture deficit reaches 25 mm.

Crops for early summer harvest are raised in cells or peat blocks under glass and planted out in mid March at the three to four leaf stage. Outdoor sowings start in mid March and continue in succession until late June or beginning of July in warm areas. Crops from summer sowings are ready in 9–12 weeks. Drill in beds in 350 mm rows, otherwise 450 mm rows with final in row spacing 80 mm. For seedrate and seed spacing in row use formulae on p. 101. Seed counts usually vary from 280–360/g; rate is usually around 2 kg/ha.

Table 5.10 Fertiliser requirements (kg/ha) of kohl rabi

N, P, K or Mg index	N		P_2O_5	K_2O	Mg	
	Early crop	Main crop			Soil	
					Light	Other
0	150[1]	100	250	250[2]	90	60
1	100	50	200	200	60	30
2	50	0	125[3]	125[3]	Nil	Nil
3	–	–	60	60	–	–
4	–	–	60	60	–	–

ADAS figures.
[1] Apply maximum of 100 N/ha pre-sowing: top dress the rest.
[2] Plough in part of K_2O with heavy dressings.
[3] Apply this rate in absence of soil analysis.
Maintain soil pH to at least 6.5.

Pests and diseases

Kohl rabi is affected by all pests and diseases that attack other brassicas and should be given appropriate seed dressing and other protection. Crop should not be grown more than once in five years. As far as is known herbicides recommended for use on other brassicas are suitable for kohl rabi.

Harvesting

Small areas are pulled by hand, large scale production by a suitable root harvester. Bulbs must be trimmed off close to base of the bulb and usually require washing before packing and marketing.

Yield around 40 t/ha.

BRASSICAS GROWN PRIMARILY FOR FODDER

Fodder radish

Although fodder radish produces sizeable bulbs or roots, it is sown thickly to produce an abundance of foliage and used as a green forage crop. Initial growth rate is extremely rapid and yields of 75 t/ha of green material are recorded, but 37–50 t/ha is more usual. Unfortunately the crop runs to seed in 6–11 weeks according to weather and variety and is then highly unpalatable and neglected by stock. Foliage is not frost hardy and the crop must be cleared by early autumn. Period of use is inflexible and the crop is not popular.

Resistant to mildew, it is virtually immune to clubroot and does not harbour beet cyst nematode (*Heterodera schachtii*), hence it may be of special value on sites heavily infested with clubroot or as a catch crop in a sugar-beet rotation. It is highly sensitive to poor seedbed conditions and adverse climatic conditions, as in the north of England and Scotland.

Varieties

These include early and late cultivars but only late sorts should be sown, e.g. Neris, Slobolt. Early varieties flower much too quickly.

Seeding

Drill 9 kg/ha in 180 mm rows or broadcast 13.5 kg/ha in June or July (August in southern England only) and consume before flowering starts. Seedbed and manuring as forage rape.

Kale

Widely grown, kales are unimportant in Wales and northeast Scotland. Planted mainly as early sown catch crops or late-sown maincrops, they may follow early harvested crops, Italian ryegrass or grass for hay, silage or grazed.

Types and varieties

Marrowstem   Gives heavy yield 60–70% stem, which is very thick, fleshy with low plant densities, when it lodges. Not very frost hardy and unsuited after the new year. Stems become woody quickly and are then neglected by stock. Grown mainly for dairy cows. Green and purple stemmed varieties exist, the latter of little importance. Resistant to clubroot.

Thousand head   Yield mostly as leaf and tender sideshoots arising in late winter and early spring. The stem becomes hard, woody and neglected by stock. It is really frost hardy and well suited to feed in late winter. Both tall (stem height approximately 1.1 m) and dwarf types (0.7–0.9 m) available, the latter, including Canson, most frost hardy and least liable to lodge. Resistant to clubroot. Mainly used for sheep.

Hybrid varieties

These are somewhat similar in appearance to marrowstem, are shorter, less liable to lodging and retain their digestibility longer, stems being slower to become woody. These varieties have largely replaced marrowstem and thousand head, although the latter is still valuable for lambs and broken toothed ewes; use in mixture.

Details of kale varieties are given in Farmers Leaflet No. 1. *Recommended Varieties of Fodder Crops* (NIAB).

Hungry gap and rape kales   See Rape.

Soil, cultivations and sowing

Kale is very tolerant of soil type but avoid heavy or poorly drained soils owing to liability to severe poaching when utilised. Direct drilling (destruction of surface vegetation by herbicide and drilling seed into undisturbed surface with special drill) is highly suitable under the right conditions, as it minimises poaching, annual weeds and loss of soil moisture and requires much less labour. With conventional cultivations ensure a fine free moist seedbed; drying out of soil is a particular hazard when grown as a second crop. For sowing details *see Table 5.11*.

Manuring

Kales require liberal application of N, of which direct drilled crops require an additional 25 kg/ha N. Relatively unresponsive to P and K unless deficiency exists (*see Table 5.12*).

Utilisation

Kales may be fed *in situ* or zero grazed. Early sowings provide excellent part replacement for grass in August and September for cows. Successional sowings of suitable varieties can provide green fodder until mid March in south. Thousand head or winter hybrids may be sown in alternating blocks with swedes. It is unsafe to feed more than 25–28 kg or 18–20 kg/d to large (Friesian, etc.) or small (Guernsey, Ayrshire) breeds of cow, respectively. A phosphorus and iodine rich mineral supplement should be fed. Calcium rich foods such as sugar beet pulp should not be given with kale.

Table 5.11 The kales – dry matter content, D-value, seeding and approximate period of utilisation

Crop	Dry matter content of whole plant (%)	D-value of whole plant	Approximate range of fresh yields (t/ha)	Usual row width (mm)	Seedrate (kg/ha) Drill	Seedrate (kg/ha) Broadcast	Approximate range of sowing dates	Approximate period of utilisation
Marrow stem varieties	12	66–67.5	50–75 Very heavy crops 100 Late sown S. England 25–40	180–350 up to 550 with conventional cultivations	OD 3–4.5 PD 1–2.5	7	N 15 April–31 May	N Aug–Nov
Early hybrid varieties	12	69–70					S 1 April–31 July (according to district)	S Aug–Dec
Frost hardy hybrid varieties	12	69–72.5	25–70 according to sowing date	180–350	OD 2–3.5 PD 1–2.5	5	As marrow stem	N Aug–Dec S Aug–Mar according to variety and sowing date
Thousand head Tall	15	Tall 65–66	Tall 35–50	250–550	OD 1–1.7	2.5–3.0	N May–June	N Oct–Dec (not common)
Dwarf	14	Dwarf 69–70	Dwarf 25–40		PD 0.6–1.0		S June–July successional sowings pointless	S Jan–Mar

Key:
N = Scotland, north of England, hill, upland and other colder areas.
S = South, south western England, west Wales and other areas with mild autumn and winter.
PD = Precision drills—use *graded seed* and check suitability of belt or wheel.
OD = Other drills. Seed may require bulking with *inert* diluent, e.g. mini slug pellets. Do not use fertiliser. Includes direct drills of non precision type.

Table 5.12 Manuring of kales and summer turnips (kg/ha)

N, P or K index	N	P_2O_5	K_2O
0	125	100	100
1	100	75	75
2	75	50	50
3 and over	–	25	50

ADAS figures.
Increase N by 25 kg/ha for direct drilled crops. Above recommendations assume crop is grazed.

Rapes

Rapes grown for fodder include forage rape and so-called rape kales. Oilseed rape is a completely different crop.

Forage rape

Forage rape is a quick grown, green forage, ideal for catch cropping, especially in cereal stubbles. Subject to mildew in dry areas if sown early, it does best in the west, north and Wales. Not really frost hardy, it can only be relied on to provide feed to Christmas, except in mild winters. Clubroot resistant varieties exist but most are susceptible and it is unwise to rely too much on resistant varieties as their resistance can break down, e.g. Nevin. It is inadvisable to catch crop with rape in rotations containing maincrop brassicas, or grow more than once in five years, because of susceptibility to clubroot. Soil should be maintained at pH 6.0–6.5. Initial growth is quicker than the kales but final yield only higher in a short growing season of 8–12 weeks.

Varieties

Varieties of forage rape are now all of the 'Giant' type. Dwarf varieties, which were much lower yielding but were said to give a second feed if not grazed too hard, have disappeared. Continental turnips are far better for spring and early summer sowings, especially in southern England as they are higher yielding and not susceptible to powdery mildew (*Peronospora parasitica*) and mostly, clubroot (*Plasmodiophora brassicae*) resistant. Details of varieties of forage rapes are given in Farmers Leaflet No. 1 (NIAB).

Oilseed rape varieties are unsuitable for growing for forage.

Swedelike or rape kales

These include two well known varieties, *Hungry Gap kale* and *rape kale*, are really rapes and possess many features of rape, notably susceptibility to mildew and clubroot and capacity for rapid initial growth. Cultivated almost entirely in the south, where they are used for stubble sowings for keep in the hungry gap period during the following March and April, both are frost resistant, although the variety rape kale is somewhat hardier but lower yielding. Both varieties lose a large amount of leaf around Christmas; the bulk of the keep is produced from sideshoots which develop in spring.

Sowing

Forage rape and rape kales require a fine firm seedbed, when drilling in close rows is usually more reliable than broadcasting; crops may also be direct drilled, depth 20 mm (*see Table 5.13*).

Table 5.13 Dry matter content, D-value, seeding and approximate period of utilisation of forage rapes

Crop	Whole plant (% DM)	Whole plant D-value	Range of fresh yield (t/ha)	Usual row width (mm)	Seedrate		Approximate range of sowing dates	Approximate period of utilisation
					Broadcast (kg/ha)	Drilled (kg/ha)		
Forage rape	12	68	30–40 (average at 8–12 weeks	180 (or broadcast)	7–8	4.5	N 30 May– 15 July S 15 June– 15 Aug	N 1 Sept– 30 Nov S 1 Sept– 31 Dec
Swede-like kales– rape kale and hungry gap kale	12–13	–	25–40	180 (or broadcast	6	3.5	S only 15 July– 15 Aug	S only late Mar–mid April

N = Scotland, northern England, hill, upland and other colder areas.
S = South and southwest England, western Wales and other areas with a mild autumn and winter.

Table 5.14 Manuring of forage rape and stubble turnips (kg/ha)

N, P or K index	N	P_2O_5	K_2O
0	100[1]	75	100
1	75	50	75
2	50	25	50
3 and over	–	Nil	50

ADAS figures.
[1] Crops sown after mid-August most unlikely to respond to > 75 kg/ha N.

Manuring

Rapes require ample lime and high fertility. They cannot be grown without fertiliser, especially in cereal stubbles (*see Table 5.14*).

Utilisation

Grown largely for sheep, rape is almost invariably fed *in situ*. Stock, especially cattle, must be introduced gradually to it with clean roughage (hay or sweet straw), and with grass, available at all times. Gorging with rape should never be allowed. If properly managed, rape is second to none for fattening lambs in the autumn.

White mustard

Crop now only of localised importance and for feeding purposes is largely superseded by continental turnips or fodder radish. Sometimes grown as a catch crop for green manure or flushing ewes after early potatoes or barefallow. It must be fed or ploughed in before flowering, seven to eight weeks after sowing and is not frost hardy. Drill in 180 mm rows at 11–17 kg/ha or broadcast 23–25 kg/ha on fine firm seedbed mid April to mid August. Fertiliser is unnecessary after well manured potatoes or barefallow. Give 75–90 kg/ha N to stubble sown crops. Green yield 25–50 t/ha.

Turnips and swedes

Swedes may be readily distinguished from turnips by the presence of a 'neck' from which leaves arise; these are usually bluish or dark green and smooth. With turnips the 'neck' is so small that it is almost absent and leaves are coarsely hairy. Both may have white or yellow flesh and are used for (a) human consumption, or (b) animal fodder.

Varieties

Turnips These mature more quickly than swedes, give a higher yield in a shorter growing season and can be sown much later, although they have a lower dry matter content and feeding value. Types of turnip differ widely in time of maturity, keeping quality and frost hardiness.

Human consumption–garden turnips Grown for their small bulbs, white or yellow flesh, in several skin colours. Shape mainly round or flat. Varieties numerous. A very quick growing specialist crop.

Animal fodder

Varieties grown primarily for leaf or forage

White fleshed 'continental stubble turnips' Mainly of Dutch origin, having amazing initial growth capacity and produce heavy crops ready for folding eight to nine weeks after sowing. Under ideal conditions 110 t/ha of fresh material can be produced in 11 weeks. This is a quick-growing, green forage crop rather than a root crop, for 50% of the yield is leaf and the roots deteriorate rapidly. It is essential to consume continental varieties whilst still growing actively, as the bulbs become unpalatable, woolly and finally hollow as they approach maturity. Earliest sowings in late April or May (summer turnips) produce heavy crops capable of fattening lambs weaned off grass from late June onwards; sowings can be made throughout the summer for a succession of feed to early January. There are many varieties but yield differences are relatively small; although all varieties are early maturing there are still marked differences. Earliest varieties include Civasto and Debra, later varieties Ponda, Taronda and Vobra. Many, including the above, show good resistance to clubroot.

Choose varieties with good resistance to bolting for early spring sowing; Civasto, Marco, Taronda.

Hybrid grazing turnips They produce 80% top with only a small bulb. Heavy yield of leaf and will produce regrowth. Root anchorage is superior to continental turnips. Varieties: Appin, Tyfon.

Varieties grown primarily for root production

Traditional British white fleshed varieties These are not resistant to clubroot, make slower initial growth, but produce heavier crops of roots which are sweeter, more palatable and last longer than continental types. Except for

Hardy Green Round, these varieties are not frost hardy and mature early; best varieties include Imperial Green Globe (syn. Green Globe, Norfolk Green Globe), Hampshire Hardy Green Round, Lincolnshire Red Globe, Purple Top Mammouth and Red Tankard.

Yellow fleshed turnips   They are seriously underestimated; easier to grow than swedes, they grow on poor land and can be sown later. *Soft yellow* varieties (Centenary, Fosterton or Dales Hybrid, Grampian) are comparable to white types, being early maturing and less frost hardy. *Hardy yellows* are akin to swedes in hardiness and keeping quality and are preferable where fertility is low. Varieties Aberdeen Green Top Yellow (syn. Green Top Scotch, Green Top Yellow Bullock), Aberdeen Purple Top Yellow (syn. Aberdeen Purple). The Bruce (syn. Tammie Mackie) and The Wallace and Champion Green Top Yellow. Latter three show resistance to clubroot, which is excellent in Brimond and Findlay.

Swedes

Frost hardiness and keeping quality, in which swedes are generally superior to turnips, increase with dry matter content. Few varieties apart, swedes have yellow flesh and are classified by skin colour. There are no distinct varieties for market and livestock feed; market, fodder or dual purpose crops may be grown. *Market* requires swedes 0.5–1.0 kg each with small top and neck, yellow flesh and purple skin. Varieties Acme, Devon Champion, Marian. Manchester market prefers green skinned varieties.

Purple skinned varieties

These are by far the most widely cultivated, are divided into

light and dark skins, the latter having the higher dry matter content and best frost hardiness and are most important.

Bronze skinned varieties

These have comparable characteristics to purple but are unsuited for market sale.

Green skinned varieties

They are rarely grown in England and Wales; roots are smaller but have a higher dry matter content, flesh is very hard, skin tough and pulping is advisable to avoid damaging teeth of stock. Longest keeping, these varieties will store into May in Scotland.

Details of varieties of turnips and swedes are given in Farmers Leaflet No. 1, *Recommended Varieties of Fodder Crops* (NIAB). Sow *only* varieties resistant to clubroot on infected land (e.g. Marian); varieties resistant to powdery mildew are preferable for early sowing, especially in the south.

Climate and soil

Apart from continental and grazing varieties of turnip, which can be grown in most areas, swedes and turnips grown for their bulbs are best suited to cooler moister parts of the UK, including Scotland, Wales and that part of England lying north and west of a line drawn from Hull to the Severn, to the south east of this line they are of little importance, being highly susceptible to dry weather and giving low yields. Market crops of swedes are common in Devon, especially on red soil. Crops are grown on a wide range of soils provided they are well drained but retentive of moisture or where rainfall is ample, soil pH at least 6.0, preferably 6.5.

Table 5.15 Row width, spacing (precision drill), seedrate and range of sowing dates for turnips and swedes

Crop	Usual row widths (mm)	Spacing between seeds (*precision drill*) (mm)	Seed rate Broadcast (kg/ha)	Drilled (kg/ha)	Approximate range of sowing dates
Continental turnips Summer crops					
Bulbs and top (sheep) Tops only (cows) Stubblecrops	180	–	2–3 4.5–7 4.5–7	1.5–2 4.5 2.5–4.5	1 April onwards S only, mid July–mid Aug
White turnips (British) for bulb production	500–550	100–150	–	0.5–1.1	N 20 May–15 June S 20 June–31 July
Soft yellow turnips for bulb production	500–550	100–150	–	0.5–1.1	N 15 May–15 June S 15 June–15 July
Hardy yellow turnips for bulb production	500–550	100–150	0.6 yellow turnips + 23–28 Italian ryegrass	0.5–1.1	N 1–31 May S 1–30 June
Swedes (fodder)	500–550	125–175	–	0.5–1.1	N 20 Apr–31 May S 1 June–25 June
Swedes (market)	375–500	100–150	–	0.5–1.5	S 15 May–5 July

Key:
N = Scotland, northern England, hill, upland and other colder areas.
S = south and south west England, western Wales and other areas with a mild autumn and winter.

Rotation, cultivations and sowing

Turnips for green forage or leaf They are grown as catch crops, either early, as with summer turnips or after a silage cut or, in the south, a winter cereal. Do not grow as catch crops in rotation containing swedes. May be sown in narrow rows, up to 180 mm, or broadcast on fine firm conventional seedbed or direct drilled. Use dressed but ungraded seed (*Table 5.15*).

Swedes and turnips for bulbs These are grown as maincrops or very early sown catch crops. Sowing date according to area and variety. In Scotland and the north swedes and yellow turnips are sown early as traditional root break between cereal crops, whereas in the south they are grown as late sown maincrops after spring fallow or early bite or as an early sown catch crop following early harvested crops, say early potatoes. White turnips are usually grown as catch crops. Clear first crop early enough to allow ample time for seedbed preparation and timely sowing.

Nowadays all crops for bulbs are drilled to a stand on a fine firm level seedbed; singling is obsolete. Precision drill *graded* seed of high germination capacity, treated with combined insecticidal/fungicidal seed dressing, into the soil moisture, usually 10–20 mm deep and leave well firmed. Watch emerging crop for flea beetle attack and treat if required. An appropriate residual herbicide will be necessary. Seed is usually spaced 125–175 mm in row, according to size of bulb required. Numerous small bulbs are wanted for market and are preferable for feeding direct, being less liable to frost, splitting and disease and keep better. For cultural details, *see Table 5.15.*

Manuring

Turnip for green forage or leaf Where a very heavy crop is expected, e.g. summer turnips, apply fertiliser as for kale

(*Table 5.15*), stubble catch crops and others, apply as for rape (*Table 5.14*). Give 25 kg/ha extra for direct drilled crops. Crop is not very responsive to P or K.

Manuring swedes and turnips for bulbs Nitrogen applications depend on level of soil fertility (*Table 5.16*), rainfall and method of cultivation. Crops grown with conventional cultivations should receive low dressings of N especially in high rainfall areas; excess produces large tops and necks with reduced yields of bulbs of poor keeping quality. Swedes and bulb turnips are very responsive to phosphate and potash, especially at low soil levels. In high rainfall areas with soil pH below 6.5 ground mineral phosphate is a suitable alternative to water soluble phosphates. On soils known to be deficient in boron spray solubor or borax.

Table 5.16 Manuring of turnips and swedes for bulb production (kg/ha)

N, P or K index	N	P_2O_5	K_2P
0	100	150	150
1	50	100	125
2	25	50	100
3	–	50	60
Over 3	–	25[1]	Nil

ADAS figures.
[1] Not justified at index 5 or more.

Harvesting and utilisation

Catch crops and turnips grown for leaf are normally eaten *in situ.* Turnips and swedes grown for roots are similarly treated but the practice is wasteful in very heavy crops, where it is generally best to clear lanes at regular intervals for

Table 5.17 Approximate fresh yield, DM content and period of utilisation of turnips and swedes

Crop	Approximate range of DM (%)	Approximate range of fresh yield (t/ha)	Approximate period of utilisation
Continental turnips Summer Stubble	leaf 9–11	50–110 20–50	June to Dec according to sowing date. Later use possible but deterioration can be severe. Allow 50–60 d and 80 d before feeding summer and stubble crops respectively
White turnips (British) for bulb production	bulb 6–8	40–75	N, Aug–Oct S, Oct–Dec
Soft yellow turnips for bulb production including hardy low DM varieties	bulb 7–8	40–75	N, Sept–Nov S, Nov–Dec. Hardy varieties, e.g. Champion Green Top Yellow usable till late Feb
Hardy yellow turnips for bulb production – (mainly grown in Scotland)	bulb 8–9	40–75	N, Oct–Feb S, Dec to Jan or Feb
Swedes – purple and bronze skin varieties	bulb 8–11	N, 75–115 S, 40–50	N, throughout winter, Oct–April, partly stored S, Jan, Feb and Mar
Swedes – purple, market crops		S, 30–37	Oct–Mar
Swedes – green skin varieties (mainly grown in Scotland)	9.5–12	N, 75–105	Late winter, early spring; partly stored

Key:
N = Scotland, north of England, hill, upland and other colder areas.
S = south, south western England, west Wales and other areas with mild autumn and winter.

feeding elsewhere and fold the residue. Harvesting and feeding of roots is fully mechanised; they may be pulped and fed mechanically into mangers, or whole roots may be stored and self-fed indoors.

Yield and dry matter content

See Table 5.17.

BRASSICAS GROWN FOR SEED PROCESSORS

Oilseed rape

A useful break crop for intensive cereal growers, oilseed rape does not harbour pests and diseases of cereals. Once established foliage produces tall dense cover; used in conjunction with suitable herbicides crop gives good control of annual and perennial grassy weeds. Handled by same basic machinery as cereals, it has a low labour requirement and integrates well on intensive cereal farms. An excellent preparation and re-entry for winter wheat, although spring sown rape may be harvested too late in the north.

Uses of oilseed rape

The only outlet for rape seed is for crushing, so it is advisable to grow only on contract to a merchant or cooperative. Crushers are not usually interested in small tonnages. Rape oil is used for:

(1) human consumption, e.g. margarine, cooking oils and shortening, when it is extracted only from varieties with a low erucic acid content,
(2) industrial purposes, e.g. lubricating oils and detergents, when oil with a high erucic acid content is appropriate.

The by-product from crushing, rapeseed meal, is included up to 10% in concentrates for animal feed; the meal must then be low in glucosinolates which are toxic to pigs and poultry.

Varieties

Varieties of oilseed rape behave quite differently from forage rape varieties and the two are not interchangeable. Only the swede rape (*Brassica napus*) is now grown in Europe and UK; turnip rape (*B. campestris*) has been displaced.

A list of varieties, Farmers Leaflet No. 9, *Recommended Varieties of Oilseed Rape*, is issued by NIAB. Varieties are selected for yield of seed (at 9% MC), which is the main characteristic determining cash return and oil content and yield. A high oil content is important, as crushers pay a premium (or make a deduction) of 1.25% for every 1% above or below the standard 40% oil content. This figure can vary but the principle holds good. Selection is also based on the edible qualities of rapeseed, notably erucic acid and linolenic acid contents; the latter may reduce shelf-life of oil and cause undesirable flavours. Varieties with very low erucic acid and glucosinolate contents are referred to as 'double zero' varieties. Information on shortness of stem, resistance to lodging, resistances to stem canker (*P. lingam*), light leaf spot (*P. brassicae*) and downy mildew (*P. parasitica*), earliness of flowering and ripening are given.

Winter rape This is much more widely grown than spring rape and is the major non-cereal break crop grown in UK. It ripens early (late July–early August) and can be relied on in most cereal growing areas, including Scotland. Yield is some 0.5–1.5 t/ha higher than spring rape.

Spring rape This confers benefits of lower production costs and can follow a wider range of crops. Varieties have improved and crop is more popular then previously. Main problems are establishment, lateness of harvest (mid–late September) and lower oil content, which may not reach contract standard of 40%. Unsuitable for Scotland and the north of England.

Rotation

Susceptible to clubroot (*P. brassicae*) and canker (*P. lingam*). Both are soil and residue borne, so rape should not be grown closer than one in five years. As it is an alternate host to beet cyst nematode (*H. schachtii*) rape is best not grown in the same rotation as sugar beet. Rape is subject to Beet Cyst Nematode Order 1977. Winter rape must be drilled in second half of August or early September and normally follows a winter cereal.

Adequate isolation is essential in areas producing seed crops of other brassicas. Avoid land infested with charlock (*S. arvensis*) which cannot be controlled in rape with selective herbicides or contamination and rejection of crop will result.

Climate and soil

Winter rape is suited to most UK cereal areas, including Scotland and the north of England. *Spring rape* is most viable in southern England. Both are suited to most arable soils, provided pH is > 6.0, preferably 6.5, although highest yields are obtained on heavier and moisture retentive soils if well drained. Winter rape can yield well on very light soils, given a high standard of cultivation, but avoid spring rape.

Culture, seedrate and plant population

Winter rape Lack of seedbed moisture is a major problem. Provide fine surface tilth and minimise moisture loss to ensure rapid plant growth. Compacted soils prevent taproot development. Crop responds well to direct drilling and fits well into a sequence of direct drilled cereals; also minimal cultivations, especially on heavy soils. Ploughing with conventional drilling or broadcasting is feasible on light soils.

Sow early to achieve adequate development before winter; in Scotland and the north of England sow mid-August, in southern England up to 5 September in good growing conditions. Later drilling results in reduced yields, if not in poor establishment. Winter rape branches freely. Target population at establishment 100–110/m²; minimum after winter 40/m² on fertile soils. Number of seeds/kg 200 000–220 000. Sow 6–8 kg/ha or 130–160 seeds/m² using higher rate for poor seedbeds; increase to 9 kg/ha for sowing after mid-September.

Drill 15–20 mm deep in rows 90–180 mm apart; these give less lodging, earlier maturity and slightly higher yield than wider rows, which in winter rape are also more susceptible to pigeon damage. Narrow rows provide better support for swathes where windrowing is used at harvest.

Spring rape Plough early for good frost mould to ensure fine firm surface tilth. Prepare tilth with minimal working to avoid compaction and loss of seedbed moisture. Direct drilling is unsuitable. Drill from early March to mid April, but it is essential to wait for good seedbed conditions; heavy soils can be difficult and prefer winter rape. Late sowings are more liable to powdery mildew in dry years.

Spring rape does not branch as freely so 120–130 plants/m^2 are necessary, but seed is smaller; 230 000–240 000 seeds/kg so 6 kg/ha or 140–150 seeds/m^2 is usually adequate. Drill in rows 110–180 mm apart 12–20 mm deep or broadcast.

Pests and diseases

Flea beetle Dress all seed with an approved seed dressing before sowing and be prepared to spray until seedlings are in rough leaf stage.

Incidence of other pests and the need for treatment varies greatly. Some pests are widespread, others more localised in occurrence. Types and numbers vary not only from farm to farm but between different crops on the same farm, so that each crop presents an individual situation and may require different insecticidal treatment.

Main insect pests include:

(i) *Winter rape seed*:
  Cabbage stem flea beetle (*Psylliodes crysocephala*),
  Blossom or pollen beetle (*Meligethes* spp.),
  Cabbage seed weevil (*Ceutorhynchus assimilis*),
  Bladder pod midge (*Dasyneura brassicae*),
  Wood pigeons (*Columba palumbus*);
(ii) *Spring rapeseed*:
  Blossom or pollen beetle (*Meligethes* spp.),
  Cabbage seed weevil (*C. assimilis*),
  Cabbage stem weevil (*Ceutorhynchus quadridens*),
  Bladder pod midge (*D. brassicae*).

To safeguard bees Never spray crops in flower and spray only if pest levels justify it. Use an insecticide of *low toxity* to bees, e.g. endosulphan and phosalone. Apply insecticide *only* in late evening (best) or early morning. Give beekeepers 24 h notice. Dispose of spray washings, excess chemicals and empties safely.

Diseases Clubroot (*P. brassicae*): Avoid infected land. Canker (*P. lingam*): Good stubble sanitation, grow resistant variety and treat seed. Leaf and pod spot (*Alternaria* spp.): spray approved fungicide. Light leaf spot (*Pyrenopeziza brassicae*): treatment benefit unlikely. Other diseases include: downy mildew (*Perenospora parasitica*), powdery mildew (*Erysiphe cruciferarum*), grey mould (*Botrytis cinerea*) and sclerotinia stem rot (*Sclerotinia sclerotiorum*). Diseases are usually not a great problem with spring rapeseed.

Pollination and honey value

The presence of bees does not appear to increase seed set, but they do appear to cause total seed set to be attained more rapidly and hence the crop ripens more evenly.

Oilseed rape is now a major nectar producing flower in the UK, having a very high 'estimated honey potential'; honey is pale, granulates rapidly and is ideal for mixing with darker honeys (Hooper, 1982). Problems are:

(1) Winter oilseed rape flows before bee colonies have normally built up.
(2) Honey crystallises very rapidly and needs special treatment.
(3) Spray poisoning. Rapeseed flowers which they visit in their thousands are extremely attractive to bees and they often fly 2 miles from one field to another. Careless spraying can be quite disastrous to bee colonies for several miles around.

Manuring

Fertiliser recommendations are shown in *Table 5.18*. Unless soil is deficient in P and K (index 0), oilseed rape rarely shows any response to these elements. Applications for indices 1 and 2 are to maintain soil reserves. Ample N is of vital importance. *N top dressing*: winter rape: normally apply all N in February or early March, *Spring rape*: apply all N by early May. Split N dressing on sandy soils to avoid

Table 5.18 Manuring of oilseed rape

N, P or K index	Winter oilseed rape						Spring oilseed rape		
	N				P_2O_5	K_2O	N	P_2O_5	K_2O
	Mineral soils		Humose and organic soils		All soils		Seedbed[4]		
	Seedbed	Spring[1] topdressing	Seedbed	Spring topdressing					
0	50	200[2]	Nil	100	100	90	150	75	75
1	50	175	Nil	50	50	40	125	50	40
2	50	150	Nil	0	50[3]	40[3]	100	50[3]	40[3]
3					50[3]	Nil		50[3]	Nil
Over 3					Nil	Nil		Nil	

ADAS figures.
[1] Where leaching risk is high – i.e. sandy or shallow soils over chalk or lime stone apply ½ dressing at start of spring growth – remainder by early April.
[2] Increase to 240 kg/ha N on crops of high yield potential ($\geqslant$ 3.5 t/ha).
[3] Maintenance dressing to sustain soil reserves yield response unlikely: K_2O not necessary on most clay soils.
[4] With high risk of leaching as on sandy or shallow chalk or limestone soils apply 50 kg/ha N to seedbed and rest by early May.

damaging germination and losses from leaching. All N must be applied by early May.

Harvesting and drying

Crop may be either direct combined or windrowed and combined from swath. Whichever method is used, seed must be fully ripe when threshed from pod. Fit vertical cutter bar. Unevenly ripened crops require windrowing, which is normal for winter rape. Direct combining is quicker, but liable to serious losses in strong wind. Combining from swath gives cleaner, more uniform seed with lower moisture content but persistent rain on windrow will result in sprouting and even loss of entire crop.

For windrowing, cut when seeds in pods at middle of stem are turning brown. Leave a stubble 200 mm long to keep pods clear of soil. Crop is left in windrow for one to two weeks and then combined. *Direct combining*: seed should be black in majority of pods. Use of desiccant, e.g. Diquat is especially valuable in wet harvests, in weedy or late maturing crops, especially spring rape in the north, giving cleaner samples with lower moisture content and reduced sprouting. Spray when majority of pods at centre of stem are yellow and seed is chocolate brown and pliable, usually some three days after windrowing would have occurred. NB desiccants must *only* be used for crop desiccation – they do not hasten crop maturity. Seal all gaps in trailers, lorries or combine to prevent heavy seed losses.

Immediate drying is essential, since seed heats rapidly and oil goes rancid, losing value. Pre-clean trashy crops. Dry to maximum of 8% moisture content for long-term storage. Allow maximum temperature of 66°C, which is safe for oil quality; do not exceed 50°C for crops intended for seed. Use two-stage drying for very wet seed. After drying blow *cold* air through to reduce temperature as low as possible to make conditions unfavourable for development of mites and insects.

Yield

Winter rape   This is a reliable yielder provided a good stand is obtained. Range 2250–4000 kg/ha; up to 3000 kg/ha has been obtained on dry sandy soils.

Spring rape   Reliable yielder, range 1750–2500 kg/ha. Except in seasons of severe litter shortage, rape straw is valueless to the grower and does not burn readily, unless fired as soon as combine has finished and is very dry, so may be best chopped and ploughed in or collected into heaps and burnt; it may have a fuel value if baled dry.

Mustard grown for condiment

Like oilseed rape, mustard is a valuable cereal break crop and an excellent preparation for wheat. Also a useful cleaning crop for wild oats. No special labour or machinery is required. Mustard is grown on arable farms in eastern and south eastern England, mainly in Lincolnshire, Norfolk and Cambridgeshire; also in Essex and Suffolk. Area is small.

Varieties

Two distinct types are grown for manufacturing: brown mustard (*Brassica juncea*) for pungency and white mustard (*Sinapis alba*) for heat. Old established seed stocks were used for 100 years but current varieties were bred by Colmans of Norwich.

Brown mustard   Trowse; Stoke and Newton have yellow seed coats to facilitate processing. Brown mustard must not be grown on charlock (*B. sinapis*) infested soils as their seeds cannot be separated.

White mustard   Bixley, Kirby and Tilney.

Soil, cultivation and sowing

Crop does best on free working medium loams but yields are satisfactory on light or sandy soils if rainfall and fertiliser are adequate. It is often difficult to obtain a good seedbed on heavy soils, so if mustard must be grown, white mustard is rather safer.

A fine firm seedbed is essential. Plough in autumn to give a good frost mould, work to a fine tilth and consolidate with roller if necessary. Use cage wheels to avoid compaction.

Sow as soon as a good seedbed can be obtained in March or early April. Sowing after end of April may result in failure. Precision drill seed in rows 340–510 mm apart; within these limits there are no yield differences. Optimum established interplant spacing 40–60 mm. To allow for normal minor losses sow 0.84 brown and 2.47 kg/ha whiteseed. Do *not* sow thicker and try to thin. Cover seed only lightly unless conditions are very dry; never drill brown seed deeper than 30 mm and white deeper than 35 mm. Leave rolled if soil is dry.

Start tractor hoes as soon as rows are visible from tractor seat. One, or occasionally two hoeings are adequate, as mustard has a strong smothering effect. Where required *only* approved herbicides may be used (*see* Chapter 8).

Manuring

Adequate supplies of N are essential for high yields. Apply:

N index	Light soils and sands	Silts	Peats
0	250	200	150
1–2	150	125	75

(Colmans of Norwich figures)

Modern varieties are very stiff strawed and lodging is unlikely. Early sowings are most responsive to N; reduce rates for sowing after mid April. All N is applied at seedbed stage; top dressing is only considered after a wet spell and heavy leaching when an *additional* dressing may be of value. There is no response to K and only brown mustard benefits from P at indices below 2, when 60 kg/ha P_2O_5 should be broadcast during seedbed preparation. Mustard tolerates pH down to 5.6 but to discourage clubroot it is wiser to maintain pH 6.5 or above. Where magnesium deficiency is a problem apply (kg/ha):

Mg index	Sandy soils	Other
0	90	60
1	60	30
2 and above	Nil	Nil

(Extrapolated from ADAS figures)

Harvesting

Modern varieties of mustard stand well in all conditions of fertility. The tops bind together to form a dense near horizontal layer and hold their seed well without shedding in all but extreme wind and hail conditions, even after full ripeness.

Provided crop is clean, direct combining is cheapest and best method. Mustard is usually fit to combine in early September, when seed in all pods is fully coloured and hard to bite. Straw brittle below lowest pod, although it may still be green at base. Ripe mustard dries out very quickly while standing; allow adequate drying before combining. Seed moisture content usually: brown mustard 8–12%, white 10–15%. They can rarely be combined efficiently above 16–18% MC.

Swath *only* if there is much green in crop as seed may be discoloured and heavy losses can occur. Desiccant sprays not recommended unless crop is very weedy and *never* to accelerate maturity. Combine leaning crops one way (i.e. with the lean). Gaps in trailers, combine or lorries should be sealed to prevent loss.

Drying

Seed above 14% MC heats up rapidly. If seed is insufficiently dressed by combine use farm dresser or precleaner to remove cosh, weed seed and rubbish. Unless seed is *reliably known* to be below 14% MC when it will keep for a few days in cool airy conditions, dry or deliver to buyer *without delay*. Most farm moisture meters are *only* accurate to within ± 1% MC; also sample must be representative of the whole parcel. Maximum temperature of drier airflow:

Moisture content of seed	Maximum temperature (°C)
< 16%	66
16–18%	60
18–20%	54
20–22%	49

Temperature in seed should never exceed 52°C.
Finish with prolonged blowing of air to cool seed.

Dry down to 10% MC or as directed by buyer. Seed below 10% MC is not subject to normal storage pests but above 10% and in dirty seed infestations soon develop.

Pollination

Brown mustard is self fertile and also wind pollinated, while white is mainly wind pollinated and pollination is normally 100%. Presence of beehives is thus of no value to the crop. Both types are an excellent source of pollen and nectar with a long flowering period and extremely popular with beekeepers.

Pests and diseases

Flea beetle Seed is dressed with insecticide; attacks after emergence controlled with γ HCH.

Brown mustard: inspect daily from bud formation, in last half of May, to start of flowering in early June for pollen beetle, swede seed weevil and stem weevil and treat if infestation develops. Give second application, if necessary, before flowers open; do not treat later to avoid killing bees.

White mustard does not need treatment but is attacked by mustard beetle (*Phaedon cochleariae*). Treatment similar to pollen beetle.

Clubroot (*P. brassicae*) incidence is spasmodic; do not grow on infected or poorly drained land.

Yield of seed

Average crops give around 2300 kg/ha, best crops up to 3500 kg/ha. Yields are expressed at standard 15% MC, as are contractual prices; there are premiums for seed of < 15% MC. Crop is only grown on contract, condiment manufacture being the only outlet.

Further information: Crops Department, Colmans, Carrow, Norwich NR1 2DD, UK.

Graminae – cereals, maize, cereal and herbage seed production; also linseed and miscellaneous combinable break crops

The group includes:

(1) major cereals grown primarily from grain; wheat and durum wheat, barley, oats, triticale;
(2) rye, grown for early bite and, in UK on very limited scale for grain;
(3) silage maize; UK climate is too cold for grain production; sweet corn;
(4) seed production of cereals and the major grasses; clover seed is included for convenience of treatment;
(5) miscellaneous combine harvested break crops: borage, evening primrose, linseed, sunflowers.

CEREALS GROWN PRIMARILY FOR GRAIN

General

Choice of seed and seed treatment

Always sow high quality certified seed of high vigour. Seed should be treated with combined fungicide/insecticide to protect seedlings against certain soil-borne fungi and to control certain seed-borne diseases. While surface acting fungicides are adequate for surface-borne diseases, systemic fungicides are necessary for diseases carried inside grain (*see*

Chapter 9). Seed dressings are best applied by merchant, as application is much more even and thus effective. Recent failures to dress seed have resulted in the reappearance of certain seed-borne diseases that have been controlled by seed dressing for many years, e.g. Bunt or stinking smut of wheat (*Ushlago tritici*). Do *not* feed surplus treated seed to livestock, as it is dangerous.

High vigour seed is especially important when it is sown in adverse conditions or if such conditions occur shortly after sowing, e.g. early sown spring crops, cold wet soil. Improvements in plant establishment, early growth, ear size and yield of grain can then be very substantial.

Choice of variety

Sow only varieties included in Farmers Leaflet No. 8, *Recommended Varieties of Cereals* issued annually by NIAB. Choose varieties suitable for the purpose for which they are being grown. The following features are common to all cereals.

Winter and spring varieties

Winter varieties of wheat, barley and oats usually give higher yields than spring sowings as they have a longer growing season and get their roots down before the onset of dry weather. On heavy land adverse soil conditions usually preclude drilling of spring crops early enough for maximum yield. Winter cereals, especially barley, ripen well before spring varieties, thus allowing an earlier start to harvesting, when weather conditions are good, and better utilisation of combine capacity. Earlier maturity of winter barley allows more time for stubble cleaning and makes crop especially suitable to precede winter oilseed rape or stubble catch crops which must all be drilled early for best results.

Winter hardiness

Winter wheat and most winter barleys are generally frost hardy enough for UK conditions. Winter oats are not normally grown north of the Humber owing to insufficient hardiness.

Vernalisation requirement

Winter wheat and winter barley require adequate cold to cause them to come into ear. Sowing too late in spring gives insufficient cold for vernalisation and crop stays in vegetative state. For winter wheat complete sowings before date given in NIAB leaflet No. 8 for latest safe sowing date for each variety. Winter oats may be sown up to 7 March but thereafter yield is reduced and maturity is later than that of spring oats. Winter varieties of barley *must not* be sown in spring.

Earliness of ripening

Choose early ripening varieties where earliness of harvesting is important, e.g. with late sowing of spring crops, in cold late areas or where cereal is to be followed by catch crop or late summer sown ley.

Disease resistance

Choosing resistant varieties is the most economical way of reducing crop losses. Ascertain which diseases are most troublesome in your area and select varieties which have a high degree of resistance (i.e. low risk); most diseases have a characteristic geographical distribution (*see* NIAB Farmers Leaflet No. 8). Avoid growing large areas of a single variety, as it may become vulnerable to new races of disease, e.g. mildew in barley and yellow rust in wheat. Grow several varieties and use Farmers Leaflet No. 8, *Variety Diversification Schemes* (NIAB). There are no varieties of wheat or barley resistance to take-all (*Ophiobolus graminae*) while the incidence of Eyespot (*Pseudocercosporella herpotrichoides*) depends on intensity of cereal growing. Choose resistant varieties for second or third crops and continuous wheat growing. Loose smut can be controlled by seed treatment.

Thousand grain weight

This is a measure of grain size and thus of the genetic potential of a variety but is influenced to a very high degree by growing conditions; it must be ascertained where it is desired to control plant population by seed numbers.

Specific weight

This is a measure of the weight of a given volume of grain, stated as kg/hl (formerly bushel weight). Affected by variety but growing conditions exert a large influence.

Lodging

Lodging (i.e. collapse of straw) can occur early or late; the former may result in near total loss, the latter is not usually serious if combine can get under ears. Most likely in crops with long straw, in exposed or windy positions and wet summers on high fertility soils; also occurs with eyespot infections. Barley is more susceptible to lodging than wheat so prefer wheat on highly fertile sites and select shortest stiffest strawed varieties. Avoid excessive seedrates and applications of N. Adjust N level to ADAS N index. Ensure timing of application suits soil fertility. Where there is serious risk of lodging or heavy dressings of N are given, use recommended straw shortener (e.g. chlormequat on wheat and oats) at appropriate growth stages. Spring grazing of proud winter cereals reduces risk of lodging. Grow eyespot resistant varieties. Modern varieties of wheat and barley are much more resistant to lodging than old cultivars.

Straw shorteners (growth regulators, e.g. chlormequat)

These reduce effect of lodging by increasing internode diameter and reducing straw length. Yield is increased where lodging is a problem; straw shorteners should be used:

(1) where lodging is a problem, and
(2) to allow maximum use of N without lodging, especially on longer or weaker strawed types or varieties. Very stiff strawed varieties of wheat are not worth treating.

Hardly worthwhile on lower yielding crops. Check for which cereals a particular straw shortener is suited; also recommended growth stage of crop before applying product.

Grazing winter cereals

Spring grazing may be obtained from winter cereals:

(1) specially grown for the purpose, usually forage rye,
(2) planned dual purpose cereals, mainly winter barley, very occasionally grown for this purpose in mild south west England to provide extra sheep keep, and

(3) to remove excess vegetative growth from winter proud crops, which is only a problem with early sowings in mild winters and districts; also in case of feed shortage.

Grazing shortens straw, reducing incidence of lodging and perhaps disease but *only* winter proud crops benefit; grazing frequently reduces grain yield and is now rarely practised. Growth stage for grazing is critical and *must be completed before* growing point at centre of ensheathing leaves has started to elongate. Avoid grazing on wet ground, as soiling of crop results in rejection and consequently uneven ripening. Top dress with N immediately after grazing is complete, not before. Apply normal N recommendations for particular cereal.

Growth stages of cereals

Wherever practicable all recommendations for stage of crop growth at which N fertilisers, growth regulators, herbicides, fungicides and insecticides are applied to cereals are now given in terms of Zadok's decimal code for the growth stages of cereals (*see Table 5.19*).

Production systems

Highest grain yields do not necessarily give maximum profits. Aiming for highest yields frequently increases total cost of inputs such as fertilisers, fungicides and pesticides without generating a corresponding increase in cash return. Before embarking on high cost production be sure that quality of site, especially soil type, depth, structure and drainage, has potential to achieve high yield; if not, it is much safer to reduce costs with lower inputs. Various 'blue print' systems, mainly of continental origin, have been tried which use high plant densities and high inputs of fertiliser and fungicide but these are unnecessarily rigid. Although such systems frequently give very high grain yields, profit is often well below those obtained from lower cost production. Each individual cost increase must thus be justified on its own merit on each particular site.

Highest yields

Highest yields, e.g. 10 t/ha depend not only on a suitable site and factors such as high fertility levels, adequate plant population, right variety and efficient disease control, but *on careful attention to detail in all factors*. Numerous apparently less important items can each produce a *small* yield reduction *but added together they can reduce yield considerably*.

Rotation

Cereals are grown under many different systems, which may be broadly divided into:

(1) *intensive cereals*, i.e. 50% or more of arable land in cereals, including continuous cereals or cereal sequences with occasional breakcrops, or
(2) *mixed systems* with under 50% of arable land in break crops.

Situations include mixed cropping, where cereal is used to cash in on built up fertility, e.g. ley or permanent grass or as a change crop from more valuable non cereal crops.

Wheat (mainly winter) is traditionally grown as first crop after a break crop, e.g. potatoes, beans or ley or up to four successive crops after fertile old grass. Lucerne is an ideal precursor for two crops of wheat. Many wheat crops are now included in intensive cereal rotations, when highest N applications are needed. *Barley and oats* are most accommodating and can be grown in any position in rotation but

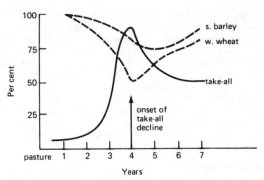

Figure 5.3 Yield and take-all in continuous cereals. (After Lester, E. Disease problems in continuous cereal growing, *Agricultural Progress*, 44, 1969. Reproduced by courtesy of the Editor and publishers)

prefer wheat wherever fertility is high, it is less liable to lodge and cash returns are better. *Oats* is very susceptible to cereal cyst nematode (*Heterodera avenae*); laboratory cyst count should be taken by ADAS before inclusion in intensive cereal rotations.

Disease

Only soil- or residue-borne pests and diseases are controllable by rotational means, e.g. Eyespot (*Pseudocercosporella herpotrichoides*), Take-All (*Ophoibolus graminae*) and cereal cyst eelworm (*Heterodera avenae*). First two can heavily damage wheat and to lesser extent, barley; heavy infestations of cereal cyst nematode (*H. avenae*) can damage all three but oats most susceptible. Use resistant varieties and fungicides to control leaf-borne diseases (*see* Chapter 9).

Soil-borne diseases and pests are mainly a problem with intensive cereal situations, although some grasses can act as alternative hosts and serious outbreaks can occur after old pasture, e.g. take-all (*O. graminae*), Eyespot (*P. herpotrichoides*), primarily attacks wheat, and is worst with early sowing, mild winters and wet springs. Use resistant varieties with continuous cereals or second and subsequent white straw crops. There are no varieties resistant to take-all (*O. graminae*); level of infestation increases to around fourth successive crop of wheat or barley (*Figure 5.3*) and then declines; yield decreases until maximum infestation is reached then rises and may attain higher level than at first – phenomenon known as 'Take-All (TA) Decline'. Intensive cereal grower has two major choices for containing effect of take-all (*O. graminae*):

(1) to go through TA decline, minimising losses culturally and then to continue to grow cereals; grow barley in most susceptible years, or
(2) to introduce non-susceptible break-crops, e.g. beans, oilseed rape, mustard, oats, but these should be grown in two *consecutive* years.

Good soil conditions, especially drainage, structure and lack of compaction, ample N and stubble sanitation are valuable for reducing effects of take-all (*O. graminae*). Although not controllable by rotation, leaf diseases can build up to high levels in intensive cereal crops; crop sanitation and protective treatment are very important.

Weeds

Broad leaved weeds, annual and perennial, are most readily controlled by suitable modern herbicides in intensive cereal rotations (*see* Chapter 8). A few, e.g. forgetmenot (*Myosotis*

Table 5.19 Zadok's decimal code for the growth stages of cereals

Code		Code	
0	GERMINATION	5	EAR EMERGENCE
00	Dry seed	50	–
01	Start of imbibition	51	First spikelet of ear just visible
02	–	52	–
03	Imbibition complete	53	$\frac{1}{4}$ of ear emerged
04	–	54	–
05	Radicle emerged from seed coat	55	$\frac{1}{2}$ of ear emerged
06	–	56	–
07	Coleoptile emerged from seed coat	57	$\frac{3}{4}$ of ear emerged
08	–	58	–
09	Leaf just at coleoptile tip	59	Emergence of ear completed
1	SEEDLING GROWTH	6	FLOWERING
10	First leaf through coleoptile	60	–
11	First leaf unfolded	61	Beginning of flowering (not easily detectable in barley)
12	2 leaves unfolded		
13	3 leaves unfolded	62	–
14	4 leaves unfolded	63	–
15	5 leaves unfolded	64	–
16	6 leaves unfolded	65	Flowering half-way
17	7 leaves unfolded	66	–
18	8 leaves unfolded	67	–
19	9 or more leaves unfolded	68	–
		69	Flowering complete
2	TILLERING		
20	Main shoot only	7	MILK DEVELOPMENT
21	Main shoot and 1 tiller	70	–
22	Main shoot and 2 tillers	71	Seed coat water ripe
23	Main shoot and 3 tillers	72	–
24	Main shoot and 4 tillers	73	Early milk
25	Main shoot and 5 tillers	74	–
26	Main shoot and 6 tillers	75	Medium milk
27	Main shoot and 7 tillers	76	–
28	Main shoot and 8 tillers	77	Late milk
29	Main shoot and 9 or more tillers	78	–
		79	–

75 Medium milk ⎫
76 – ⎬ Increase in solids of liquid
77 Late milk ⎪ endosperm visible when crushing
78 – ⎪ the seed between fingers
79 – ⎭

Code		Code	
3	STEM ELONGATION		
30	Pseudostem erection (winter cereals only) or stem elongation	8	DOUGH DEVELOPMENT
		80	–
31	1st node detectable	81	–
32	2nd node detectable	82	–
33	3rd node detectable	83	Early dough
34	4th node detectable	84	–
35	5th node detectable	85	Soft dough (Finger-nail impression not held)
36	6th node detectable	86	–
37	Flag leaf just visible	87	Hard dough (Finger-nail impression held, head losing chlorophyll)
38	–		
39	Flat leaf ligule just visible	88	–
		89	–
4	BOOTING		
40	–	9	RIPENING
41	Flag leaf sheath extending	90	–
42	–	91	Seed coat hard (difficult to divide by thumb-nail)
43	Boot just visibly swollen	92	Seed coat hard (can no longer be dented by thumb-nail)
44	–		
45	Boot swollen	93	Seed coat loosening in daytime
46	–	94	Over-ripe, straw dead and collapsing
47	Flat leaf sheath opening	95	Seed dormant
48	–	96	Viable seed giving 50 per cent germination
49	First awns visible	97	Seed not dormant
		98	Secondary dormancy induced
		99	Secondary dormancy lost

arvensis), are difficult. Grassy weeds readily build up to infestation level, especially when a run of winter crops is taken; proper preventive measures, according to species, are essential and include stubble treatment and use of herbicides. Modifying rotation, e.g. following winter barley with spring barley allows ample time for stubble cleaning to control couch; spring crops are much less subject to autumn germinating grassy weeds, e.g. blackgrass (*Alopecurus myosuroides*) but yield is lower.

Cultivations

All cereals may be grown by *conventional cultivations*, with or without the mouldboard plough, to varying depths or by *direct drilling*. The latter involves placing seed in soil without prior cultivation or, where necessary, with minimum precultivation, using special drills and coulters of various designs. Provided there are no pans or compaction, deep seedbeds are not necessary and working to 150 mm deep is normally adequate. Subsoil where necessary to break up pans or compaction. With omission of ploughing and especially with direct drilling, presowing control of surface vegetation and elimination of trash, which is mainly straw, is essential.

Suitability of cultural system depends on cropping sequence, condition of the field or site and depth and texture of the topsoil. In a mixed arable cropping system, including crops such as sugar beet, potatoes and large seeded legumes ploughing or at least deep cultivation is essential to secure a suitable tilth and the right conditions for root expansion and crop development. Only continuous cereals with oilseed rape are appropriate for sequential direct drilling or minimal (i.e. shallow) cultivations and then only if soil type and condition of the site is suitable, i.e. it is free from trash (p. 123).

In the late 1970s reduced cultivation systems (i.e. direct drilling and minimum cultivations) increased greatly in popularity, as tractor hours and power requirement for seedbed preparation were fractional compared with those needed for ploughing and conventional cultivations. On many farms in the main UK cereal producing areas, especially in the eastern counties, there was neither home use nor sale for the huge excess of straw, which was burned in the field. A good burn provides clean trash-free conditions which are ideal for tilth preparation and subsequent germination and growth of the cereal crop. Burning also reduces numbers of grassy weed seedings, e.g. barren brome (*Bromus sterilis*), and blackgrass (*Alopecurus myosuroides*) but where regular burning and direct drilling or surface cultivations are practised, the build-up of carbon in the soil surface inactivates or reduces the efficiency of residual (soil applied) herbicides. Thorough incorporation of the ash is necessary. Where burning is not practised, direct drilling or shallow cultivations result in the build-up of excessive populations of grassy weeds and there is much to be said for a return to the plough.

The prime problems with stubble burning, especially when improperly done, are the damage, road accidents and inconvenience caused to others, resulting in public outcry. The practice is being controlled by legal means with increasing severity and may well be totally banned, at least in sensitive areas. Burning should *only* be carried out *if there is no other alternative* and then *only* if the guidelines in the latest *NFU Straw Burning Code*, local byelaws and other legislation in force at the time *are strictly adhered to*. It is also essential to be adequately covered by public liability insurance.

Straw utilisation and disposal (*see* Butterworth, 1985)

UK straw production is currently some 7 million tonnes in excess of requirements. Straw is bulky and difficult to handle economically, because of high transport, storage and processing costs even when using today's technology. If it could only be tapped economically, the energy and other potential is tremendous. Physical properties, chemical composition and feeding value are extremely variable. For on-farm use, untreated straw has a very low feed value and is suitable for use as bedding or coarse roughage. When properly treated, e.g. with sodium hydroxide or ammonium hydroxide, feeding value of straw is much improved and it can, with appropriate adjustment to the rest of the ration, be used in place of conserved grass, especially on mixed livestock/arable farms. If suitable plant is available, straw can also be used as a fuel. Possibilities for sale depend on the local situation and include: interfarm sales, fuel for glasshouses, board and packing tray manufacture. Potential uses for building and as a chemical feedstock appear to be a long way off.

Straw frequently contains large numbers of weed seeds, especially wild oats (*Avena* spp.); these species have often been introduced onto clean farms in purchased straw.

When growing cereals, choice of cultural system must depend on (i) whether the straw is cleared from the field (i.e. by baling or burning); or (ii) is to be left on the field, requiring incorporation. In case (i) the choice is wide open to the grower and depends on the constraints of cropping system and soil type. With (ii) substantial cultivations (i.e. some form of 'conventional' cultivation) are essential for thorough straw incorporation, which is greatly aided if a straw chopper is fitted to the combine (which then requires extra power), and the straw is spread. Alternatively a trailed straw chopper may be used, but this involves an extra pass. Incorporation to a depth of 150 mm is generally considered to be adequate.

There are numerous approaches to the problem of straw disposal, which depend substantially on soil type. Ploughing is widely regarded as the best solution but some pre-cultivation is frequently advisable to ensure adequate straw incorporation and avoid dense layers of straw under the ploughing.

Inadequate incorporation can lead to serious damage to subsequent crops. Straw can shelter slugs and snails. If it breaks down anaerobically in wet conditions various toxins are produced which can seriously affect the germination of the seed.

Conventional cultivations

Winter cereals Prepare seedbed early, firm but not compacted with ample crumb to receive seed but not too fine on top, i.e. some surface clod to reduce capping. Delay drilling until some rain falls if excessively dry but do not work final tilth too far ahead of drilling; in wet autumns drill close behind minimal working. Late wet cloddy seedbeds impair plant establishment. Mouldboard plough has now been superseded on many clayland and other farms by heavy cultivators or chisel ploughs where straw can be baled or burned and trash is not a problem. On heavy soil start shallow and work gradually deeper or tilth will be excessively cloddy. Herbicide may be required to control grass weed with no ploughing, especially in wet autumns.

Spring cereals

(1) Early sown crops (say up to late March) plough before Christmas for well settled frosted tilth, especially on heavy soils. Do not travel on land before it is fit, whether minimally cultivated or ploughed. With earliest sowings use minimal number of passes, consider cage wheels and do not leave to fine a surface.

(2) Late sowings (early April onwards). When sowing cereals after late cleared crop (e.g. folded roots, winter cauliflower) wait until land is dry enough to plough a crumbly furrow slice then work down *quickly* tight behind plough to secure fine seedbed and conserve moisture. Leave crop cambridge rolled after drilling; loose seedbeds are fatal to crop. Except earliest sown crops, aim for fine firm even and level seedbed, especially malting barley and long leys undersown.

Direct drilling

All cereals may be direct drilled either occasionally or sequentially; long-term direct drilled sequences may include suitable combine harvested break crops, e.g. ideal for oilseed rape.

Soil type and conditions are critical for technique, especially sequential direct drilling. Major problem areas include:

(1) *drainage*: install proper drainage system, mole drain and/or subsoil before starting; ponding must not occur on surface;

(2) *lack of tilth*: localised tilth is required to cover seed and allow penetration of soil by rootlets; drilling into compacted soil fails to cover seeds, gives smearing by coulters, causes ponding around seeds and heavy losses. Crop establishment is better after stubble burning, which improves surface tilth, keeps soil drier for a while and disposes of trash which can produce toxic residues; predrilling cultivations and tine coulters on drill also improve tilth;

(3) *topsoil compaction is produced by any heavy machinery* including combine harvesters and grain trailers or tractors if soil is wet, when structural strength is minimal; keep off land or if essential cut travelling to minimum on wet soil, reduce implement weight and use wide or flotation tyres. Problem is reduced on quick draining soils with high organic matter content and surface mulching. Soil texture is most important, e.g. non-calcareous clays, silts, sandy clays and low organic matter sands showing little resistance to topsoil compaction.

Classification of soils according to suitability for sequential direct drilling of cereals

Group 1

Soils give similar yields from direct drilling and conventional cultivations for winter and spring cereals. These are well drained permeable soils of good structure; loams, including some clays with high natural lime and humus rich soils.

Group 2

Similar yields from direct drilling and conventional cultivations, with good management, for winter cereals only. Spring cereals risky. Imperfectly or moderately well drained soils but relatively impermeable layer at moderate depth. Correctable drainage.

Group 3

Risk of loss of yield from direct drilling, especially spring sown crops. Low organic matter sands, silts, wet alluvial soils, clayey soils and clay loams which return to field capacity before 1 November.

Soils in groups 2 and 3 may be direct drilled with cereals provided: drainage is improved, shallow predrilling cultivations are given, excess traffic is avoided and *only* winter cereals are planted, drill before mid October.

General precautions for direct drilling cereals

Ensure clean weed and trash free site, bale straw if practicable or burn stubble if situation permits, drill in good time into friable tilth and adopt high standard of husbandry throughout. Maintain adequate pH in top 25 mm of soil, give extra N, drill slug pellets and watch for slug damage. Do *not* drill in adverse soil conditions or too deeply and use suitable coulters. Check on soil group, drain and precultivate if necessary. Some annual grassy weeds are much more troublesome in sequentially direct drilled cereals than in conventionally cultivated crops where small seed is buried by the plough, so ensure appropriate herbicidal control in direct drilled crops; a good stubble burn is helpful with some species, if feasible.

Plant population, yield and seedrate

Yield of grain is the product of number of ears (fertile tillers) × number of grains/ear (ear size) × weight of individual grains. (Thousand Grain Weight, i.e. TGW.) Ears arise from vegetative tillers which in turn arise from individual plants, each produced by a single grain on germination. It is therefore necessary to ensure sowing an adequate number of seeds to secure a population large enough to utilise fully the resources of the area of land they occupy.

Cereal types differ greatly in their tillering capacity. Barley and winter wheat tiller profusely, while spring oats and spring wheat produce relatively few tillers. Types such as winter wheat and winter barley, which are capable of creating high tiller numbers are especially flexible or plastic in their ability to compensate for low plant densities and give similar yields of grain from widely differing plant and ear populations. Large varietal differences in tiller production have also been noted. If initial population is inadequate, yield potential is not reached; once plant numbers are adequate for the site, yield remains on a broad plateau until it declines through overcrowding (*Figure 5.4*) quite apart from excessive seed cost; ear size is also reduced. Initial plant density and seedrate are not critical in high tillering winter wheat and winter barley varieties, but control of seedrate is an essential first step in helping to give a dense enough plant population to provide a high yield crop structure. Drill a seedrate high enough to ensure an adequate margin of safety – there is no point in exceeding this.

Plant establishment and subsequent development are highly dependent on soil conditions, weather and management. Early sown crops on good seedbeds, especially in autumn, tiller vigorously, so less seed is then required. With crops drilled late in cold, wet soil (e.g. November), seedling mortality is high and more seed is needed but this does not compensate for retarded crop development, especially winter barley. Ear numbers and size are determined by many other factors. Farmer can control

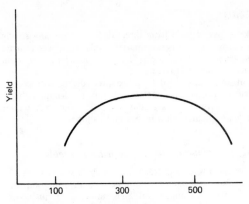

Yield

Plant population (established plants/m²)

Figure 5.4 A diagrammatic representation of the relationship between yield and plant population. (After Hayward, P.R. *et al.*, *Developments in the Business and Practice of Cereal Seed Trading and Technology*, p. 149, 1978. RHM Gavin Press. Reproduced by courtesy of the publishers)

choice of variety, quality of seed and N application; he cannot control emergence factor, winter survival nor summer rainfall, which much affects N utilisation and consequent crop performance.

Always sow

(1) modern high yielding variety recommended by NIAB and suited to your circumstances,
(2) high vigour seed,
(3) seed treated with appropriate fungicidal/insecticidal dressing,
(4) an appropriate seedrate, and
(5) drill winter cereals early, spring crops as soon as soil conditions permit.

Drilling a standard seedrate or weight of seed is a poor way of controlling number of seeds sown, as it takes no account of differences in seed size and hence number of seeds/kg between varieties and different lots of seed; where required population and TGW are known, it is more accurate to calculate seedrate as follows:

$$\text{Seedrate (kg/ha)} = \frac{\text{target population/m}^2 \times \text{TGW (g)}}{\text{expected \% establishment}}$$

e.g. target population = 250 plants/m² established, TGW 50 g, expected establishment 80%, e.g.

$$\text{Seedrate (kg/ha)} = \frac{250 \times 50}{80} = 156 \text{ kg/ha}$$

Formula applies only to high laboratory germination, 95% and over, when allowance for percentage germination makes little practical difference; *avoid sowing low germination seed*. Have home grown seedlots tested for germination by Official Seed Testing Station, NIAB. Field establishment is *very difficult to forecast* and can vary from 60% or less with winter crops sown in bad conditions, about 80% in good conditions and up to 90% in spring grown crops in very good conditions. Standard seedrates and target populations, as appropriate, are given with individual crops.

Drilling and plant distribution

Well distributed populations are much better placed to compensate for low plant numbers than are patchy stands; plant counts can then be unreliable. Operate drill carefully, avoid high speed drilling, using wider or an additional drill if necessary. Buy a reliable drill and maintain it properly. Sow seeds as shallowly as covering, tilth and cutting into soil moisture permit, usually 25–30 mm.

'Tramlines'

These are paired continuous plant free strips, designed to allow passage of tractor with sprayers and fertiliser distributors without damaging standing crop and may be formed by:

(1) shutting off coulters at drilling, or
(2) making *accurate* continuous wheelings in emerged crop.

Tramlines are an integral part of modern high input cereal production (*cf.* bed system in vegetable crops) and necessary to allow numerous passes through standing crops, especially at later stages of crop growth, e.g. fertiliser application, and spraying fungicides and aphicides. It is essential to match widths of drill, fertiliser distributor spread and spray boom.

Manuring of all cereals grown for grain production, P_2O_5 and K_2O

Liberal applications of P, and on many sites, especially lighter soils, of K, are necessary on newly reclaimed deficient soils; also lime. On cultivated land, including well managed temporary grassland, past applications of P and K have built up reserves of these nutrients, so that cereals rarely give an increased yield from fresh P or K fertiliser. Only where soil indices are low (e.g. K on light and chalky soils) or where little P and K has previously been applied, are large responses likely. Responses are rare where P and K indices are over 2 (*Table 5.20*). It is advisable to maintain soil

Table 5.20 Phosphate and potash manuring for all cereals for grain production (kg/ha)[1]

	0	*1*	*2*	*3*	*>3*
Phosphate: P_2O_5					
Yield level 5.0 t/ha	90	40	40	40	Nil
7.5 t/ha	110	60	60	60	Nil
10 t/ha	130	80	80	80	Nil
Potash: K_2O					
Straw ploughed in or burnt					
Yield level 5.0 t/ha	80	30	30[2]	Nil	Nil
7.5 t/ha	95	45	45[2]	Nil	Nil
10.0 t/ha	110	60	60[2]	Nil	Nil
Straw removed					
Yield level 5.0 t/ha	110	60	60[2]	Nil	Nil
7.5 t/ha	140	90	90[3]	Nil	Nil
10.0 t/ha	170	120	120[3]	Nil	Nil

[1] At index 0 large amounts of P and K are recommended to raise soil index over a number of years. At indices 2 and 3 dressings are to maintain soil reserves rather than to give yield increase.
[2] Not needed on most clay soils.
[3] Less may be used on most clay soils.

reserves of P and K, which are applied annually, or, with continuous cereal production, P may be given at regular intervals during the rotation. Levels of maintenance dressing depend on yield, as heavier crops remove more nutrients. Allow for straw removal which removes some K but little P. Light soils tend to be low in K and even at index 2 maintenance dressings of K are wise. Most soils with high clay content can release enough K and maintenance dressings are unnecessary at index 2. Be sure to maintain adequate levels when rotation includes P or K demanding crops.

Combine driling slows down work considerably and such drills are more expensive and laborious to maintain than ordinary drills. P and K should be combine drilled at index 0; if broadcast, increase each by 30 kg/ha. There is no response from combine drilling at high P and K levels.

Nitrogen manuring

See individual cereal crops.

WHEAT

Common ('bread') wheat (*Triticum aestivum*)

Grain quality

Grain of suitable quality is essential for human consumption, when it is milled and blended into a wide range of flours for bread, confectionery or biscuit making; remainder is used for animal feed, mainly by feedstuff compounders. Suitability of sample for milling and bread or biscuit making is essentially a varietal feature but modified by growing, harvesting and storage conditions.

Milling quality

Varieties may be grouped as 'hard' or 'soft' in milling texture. In good milling (i.e. hard) varieties of wheat, bran is readily separated from endosperm or flour; the latter is granular and so is free running, passing through sieves quickly without blockage. Conversely, with poor (i.e. soft) varieties, bran does not come away cleanly and flour is irregular and lumpy, so milling has to be slowed down to prevent blockage of sieves. Extraction rate also varies, plump well filled samples giving highest percentage extraction. Other than plumpness of sample, these are essentially varietal characteristics, little affected by cultural methods.

Bread making

'Hard' wheats are best for bread making because they give a flour which absorbs more water in the dough, making more loaves from a given weight of flour and which take longer to go 'stale'. Good dough is also elastic, thereby trapping the carbon dioxide bubbles from the yeast giving large, soft, well piled loaves. Suitable quality is indicated by:

(1) *Hagberg Falling Number* (*FN*) *test*, which indicates alpha amylase activity. A high FN is essential, showing that the grain has a low activity of the enzyme alpha amylase. A high activity, indicated by a low FN, results in degradation of starch during baking, giving a sticky crumb structure of loaf; loaves are then difficult to slive mechanically. Where crop is to be sold for bread making, choose only varieties with a high Hagberg FN in NIAB

recommended list and good resistance to sprouting; once germination starts, alpha amylase is produced and Hagberg FN is reduced.

(2) *Protein* (*i.e. gluten*) *quality*, which determines elasticity of dough. A 'strong' gluten gives a loaf of large volume which is light and spongy in texture but effect is also determined by *quantity* of gluten present. Gluten *quality* is largely a varietal feature, unaffected by husbandry techniques.

(3) *Protein content* is largely determined by environmental conditions, especially late application of N; an extra 40 kg/ha N may be considered for application between flag leaf and ear emergence (GS 37–39 to 51) to increase protein content of grain.

Biscuit making

Good biscuit flour produces a weak dough which combines maximum extensibility with minimum elasticity, which can be rolled into a thin, even sheet; the biscuits cut from it retain their shape. Good biscuit varieties are soft milling in texture, yet reasonably free milling and have suitable protein quality. Samples of intermediate protein content are required.

Varieties

Choose varieties suitable for the required purpose. For *bread making* select varieties with good scores for bread making and ease of milling, with hard endosperm texture. Minimise likelihood of rejection by selecting those with good scores for Hagberg falling number and resistance to sprouting. Prefer varieties with naturally high grain protein content. For biscuit making choose high yielding soft wheats of suitable quality. For crops grown only for feed, high yield is the main criterion.

Climate and soil

Wheat is widely grown but is more reliable in drier climates of south and eastern Britain. Crop is suited to a varying range of soil types but deep rich moisture retentive soils with good structure are essential for highest yields. Well drained fertile medium heavy loams, clay loams and alluvial soils are especially suitable. Avoid thin, infertile and badly drained soils, as yields are usually low.

Sowing date

Winter wheat

Best yields are usually obtained from sowing during first two weeks of October; complete drilling before end of month, as November sowings invariably give reduced yields. Drilling should begin in September on large areas where a later start may result in November completion. September sowing is also beneficial in cold areas where winter begins early. Spring sowing of winter varieties should be finished before latest safe sowing date recommended by NIAB, Farmers Leaflet No. 8.

Spring wheat

Sow as soon as soil is fit to drill, the earlier the better. Range 1 February–15 April.

Plant population and seedrate

Winter wheat

Winter wheat is extremely flexible in its ability to compensate for variations in plant density, similar grain yields being obtained over a very wide range of initial plant and ear populations. Experiments have shown that on evenly distributed stands, yield increases only by 5% when population is increased from 100–200 plants/m^2; established populations of 200–300 plants/m^2 are adequate for highest yields. When calculating seedrate aim for 250 plants/m^2 to allow for high safety margin and to produce minimum of 500 ears/m^2. Thresholds below which yield declines are probably around 125 plants/m^2 on good seedbeds on heavy soils, 150 plants/m^2 on light soils. Evidence suggests that provided thin crops can be liberally top dressed with N, it is not worth redrilling with spring barley unless established density is below 50–60 plants/m^2. There are large differences between some varieties in tillering capacity. When calculating seedrate, allow 80% establishment for early sowing, good seedbed, reduce to about 60% for late sowing, adverse conditions; these allowances are only approximations.

If TGW is not known, sow 140 kg/ha for early sowings, choice seedbed; 160 kg/ha average conditions and 180–200 kg/ha November–December sowing, cold rough seedbeds. High seedrates for early sowing on good tilth may reduce yield, e.g. by favouring disease growth in resultant dense crop.

Spring wheat

Tiller production is low and high seedrates are essential. Sow 190–220 kg/ha, according to TGW and seedbed conditions, average 200 kg/ha.

Manuring

Nitrogen (P and K, see Table 5.20)

Nitrogen supply is a prime factor in production of high grain yields but timing of applications is most important.

Total application of N fertiliser This depends largely on soil index; *Table 5.21* provides an *approximate* guide to rates for winter and spring wheat. Increase or decrease according to condition of crop and soil and expected economic response on site.

Winter wheat

Timing of N application

Nitrogen stimulates tillering and subsequent vegetative growth and encourages tiller survival and duration of green leaf after flowering. *Adequate but not excess* N is essential for each phase of crop development but amounts required vary greatly. Nitrogen uptake is small from sowing until tillering is completed about end of March. Deficiency during this phase results in a crop whose frame is inadequate to achieve high yield; it cannot be remedied by later applications. After completion of tillering, vegetative growth and N uptake of winter wheat accelerate rapidly, the latter being maintained until about end of June. Nitrogen requirements are heavy during this period but N uptake and N response can be seriously reduced by drought.

Seedbed applications

No response with conventional cultivations; give 30 kg/ha N to direct drilled crops only, in addition to main recommendation.

Early top dressing (GS 20–23) About mid March, during tillering. Split dressings, giving a maximum of 40 kg/ha of total N dressing at this stage, are advisable in the following cases:

(1) direct drilled crops,
(2) crops on lighter soils, N index 0,
(3) crops with low populations, weak tillering or poor soil conditions, and
(4) if main top dressing is delayed.

Do not apply N to vigorous crops on N rich soils at this stage, e.g. after lucerne, intensively managed grass or on peaty soils or yield may be reduced.

Main top dressing (GS 30/31) Mid–late April. Nitrogen uptake is now rapid and restricted N supply adversely affects dry matter production. Translocation of N to ear is more effective now, as opposed to earlier application. Spread remainder of total N recommendation now and normally total recommendation for soil N indices 1 and 2.

Very late N applications (GS 37–39 to 51) Flag leaf emergence and later. Applications during this period increase N content of grain but usually fail to give economic

Table 5.21 Nitrogen manuring of wheat (kg/ha N)

	Winter wheat														Spring wheat				
Yield level	Up to 7 t/ha						7–9 t/ha				> 9 t/ha				All yield levels				
Soil type	1	2	3	4	5	6	1	2	3	4	1	2	3	4	1	2/3	4	5	6
N index																			
0	175	150	150	150	50	90	225	200	200	200	275	250	250	250	150	125	125	40	70
1	150	50	75	100[1]	Nil	4[1]	200	100[1]	125	150[1]	250	150	175	200	100	50[1]	75	Nil	35
2	75	Nil	Nil	50	Nil	Nil	125	Nil	Nil	100	175	50	50	150	50	Nil	30	Nil	Nil

ADAS figures.

Soil types: 1 = sandy soils or shallow soils over chalk or limestone; 2 = deep silty soils; 3 = clays; 4 = other mineral soils; 5 = organic soils; 6 = humose soils.
[1] Increase N by 25 kg/ha where harvesting of previous crop has damaged structure.

response in yield; 40 kg/ha N may be considered to raise N content of grain for sale for baking or possibly on sites of highest potential yield when soil moisture is adequate and late application of fungicide is given. Given as an *addition* to main application and is an integral part of some continental wheat systems.

Spring wheat

Split applications of N may increase yield on light chalky and sandy soils, when half is worked into seedbed in compound fertiliser and rest is topdressed at GS 30, mainly with early sowings. Elsewhere all N is incorporated in seedbed. Do not combine drill more than 75 kg/ha N. Crop is less responsive to N than winter wheat (*Table 5.21*).

Harvesting

Wheat must be dead ripe when combined. Given a choice, harvest wheat in preference to barley, as it is slower to dry out in wet weather. Combine wheat crops for breadmaking, first, then biscuits and finally feeding if fit.

Drying and storage

Dry wet crops at once. Temperatures should not exceed 43°C for seed wheat and 65°C for milling wheat, when excessive temperatures 'perish' gluten and loaves will not rise. Dry grain to 16% moisture content (MC) for sack and 14% MC for bulk storage.

Yield

Grain

Winter wheats normally outyield spring sowings considerably. Well managed winter crops give around 7 t/ha, best reaching 9 t/ha or a little over. Average to moderate crops and spring sowings usually range from 3.75–5.0 t/ha but badly grown or drought stricken crops can yield less.

Straw

Yield 2.5–3.75 t/ha.

Specific weight

Average 77 kg/ha.

Durum wheat (*Triticum durum*)

Durum or semolina wheat is quite distinct from the common (or bread) wheat traditionally grown in the UK and is distinguishable by awns on the glumes and hard amber coloured near translucent grain. Used to produce semolina for manufacturing pasta products (e.g. macaroni, spaghetti, lasagne) and for some proprietary 'breakfast cereals'. Traditionally confined to Mediterranean Basin and Great Plains of North America it is now widely grown. Introduction of new varieties and growing techniques have enabled rapid expansion of area grown in France and now in the UK, where good quality grain can be produced.

Manufacturers want high yield of golden semolina after milling to produce pasta of preferred colour, texture and cooking characteristics; it is wise to grow on contract. Buyers requirements are: (i) *Vitreous (non-floury) grain* ≥ 70%

which when milled gives 'gritty' flour that sieves easily and gives good product. High % of mitadine (floury) grain gives voluminous flour, difficult to process and a pasta with dull unattractive appearance. (ii) *High Hagberg Falling No.* (HFN), ≥ 200. Low HFN denotes high alpha amylase activity, causing serious frothing during cooking and softness of cooked pasta. (iii) *Colour*. Best pasta is bright golden or yellow, largely determined by variety but improved by higher protein levels. (iv) *Resistance of product to shattering and good cooking quality*. (v) *Hectolitre wt* ≥ 76 kg/hl, and (vi) nil ergot in 2 kg grain.

Durum production is critically different from common wheat in respect of variety, harvesting and drying.

Variety

Grow specialised variety with high vitreous grain production (low mitadine content), naturally high HFN, good colour and cooking quality.

General husbandry

Climate

At first grown only in eastern counties, now more widely grown in southern England. Avoid high rainfall areas because of risk of harvesting problems; also situations of very low winter temperatures (winter hardiness only, *cf.* Maris Otter (winter barley).

Soil and rotation

Grows on wide range of soils except most marginal types, provided good seedbed can be obtained. More susceptible than common wheat to impeded drainage. Select uniform fields, to ensure even ripening, and in middle yield range, where yield may equal feed wheats but with grain selling at 50% premium. May follow as second or third wheat crop or after a break crop but avoid following winter barley with risk of serious contamination from volunteer barley.

Sowing

Drill between 20 September and 15 October to ensure well established crop, depth 25–50 mm; 100 mm often preferable to 175 mm rows. Durum wheats produce stronger main shoot with smaller less effective tillers than common wheat, so aim for higher initial population of 300–350 plants/m². Assuming TGW 50 and 80% establishment, this requires 185 kg/ha; for late drilling, poor seedbeds or heavier soils, allow 215 kg/ha. Minimum post-winter population for reasonable yield 100 plants/m² provided stand is even and extra N is given.

Fertiliser

Seedbed P and K applications as for common wheat. With low N situations, e.g. in run of cereal crops, give 15 kg/ha N in seedbed. For spring N top dressing, *see Table 5.21* but in assessing yield potential of durum wheat assume two-thirds of yield of common wheat at similar position in rotation. To help ensure high protein levels and to delay onset of mitadine grain formation, give extra 40 kg/ha at Zadok index 37 (*see Table 5.19*). Give routine application of straw shortener.

Harvesting and drying

Direct combine at high MC, 20–22% (during August 7–10 d before feed wheats) and dry carefully down to 15%;

maximum grain temperature 66°C, seed crops 50°C. Allowing re-absorption of moisture (NB damp weather) encourages development of mitadine grain; durum is also susceptible to sprouting of grain and delay reduces HFN. Different grain structure allows much easier drying than common wheat. For seed production combine at 18% MC and thresh gently.

Disease control

Monitor crop carefully; durum wheats are susceptible to eyespot (*Pseudocercosporella herpotrichoides*) so give routine protective spray; also broad spectrum fungicide application at flowing. Take measures to prevent ergot (*Claviceps purpurea*).

Yield

Grain Average 4–5 t/ha; best crops 6–7 t/ha. Crop only moderately reliable: yield in poor conditions as low as 2 t/ha.

BARLEY

Major use of barley is as an animal feedstuff, either consumed at home or sold to animal feed compounders. Top quality barley used for malting, 'green grain' alcohol production, a small market for pearling and other miscellaneous uses, e.g. coffee substitute.

Malting

A good malting sample has germination as near 100% as possible, grain plump, even, a bright pale lemon colour, finely wrinkled skin, total absence of moulds or mustiness and free from broken grains or impurities. In cross section it is white and mealy, not grey and steely, the latter denoting high nitrogen content. A good sample will not contain over 1.5% N, although up to 1.6% is acceptable by maltsters; more in bad seasons. Malting quality is dependent on choice of variety and environment; a dry season with enough moisture at the end to mellow the grain, a fine, firm, level seedbed, avoidance of lodging from excessive seed rates or too high N application, harvesting when fully mature, not drying above 43°C and dry storage. A late top dressing increases the N content of the grain.

Feeding

Good feeding barley should be clean, dry (*see* storage) and free from mould. Plump well filled sample is best, as thin poorly filled grain with high fibre or husk content has lower feeding value. Fibre content is largely determined by growing conditions, although six row barleys do have a somewhat higher fibre level than two rowed sorts. High protein content of grain is beneficial, as it reduces the amount of protein concentrate needed to balance a ration.

Varieties

Recommended varieties are listed in Farmers Leaflet No. 8, *Recommended Varieties of Cereals* (NIAB). Winter varieties include two and six row types, the latter being very high yielders but unsuitable for malting. For malting select variety with high score for malting grade, although this is no guarantee of a malting sample. For feeding go for high yield provided variety shows adequate resistance to foliar diseases

which are troublesome locally. Rhychosporium is most frequent in wet seasons and around south coast so choose resistant variety for winter and early spring sowings; brown rust can be especially serious in the south west. Mildew is wide spread. Use the NIAB Varietal Diversification Scheme to reduce spread of mildew in spring barley and consider routine fungicidal treatment for winter and late sown spring crops; it is essential for susceptible varieties. Choose earliest maturing varieties for late spring sowings. Winter varieties are totally unsuitable for sowing after mid February.

Climate and soil

Barley is grown in most areas. Traditionally regarded as a light land crop, it is well suited to a wide range of soils. Good drainage and soil pH 6.0 or over are essential; usually pH 6.5 is safer.

Sowing date

Winter barley

Early drilling necessary for top yields, especially in colder more exposed areas, e.g. Cotswolds. Start mid September and complete by mid October, particularly direct drilled crops. Early November drilling seriously reduces yield if seedbed is cold and wet.

Spring barley

Range 15 Jan–30 April, according to soil and rainfall. The soil should be dry and friable and it is better to wait for the right conditions than force a tilth. Sow as soon as conditions permit but generally sow in March and in the southwest, mid March to mid April.

Plant population and seedrate

Winter barley

Given even distribution and adequate density, high yields can be achieved over a wide range of ear populations. Precise level is unimportant. For safety aim to establish 250–300 plants/m^2 by early spring with general target of 900 ears/m^2 at harvest; with six row varieties 800 ears/m^2 is adequate, two row varieties 1000 ears/m^2 is preferable. Optimal seedrate depends on TGW, expected percentage establishment, date of sowing and condition of soil, assuming seed of high percentage germination. With early sowing, good seedbed expect 80% establishment; November sowing, poor seedbed, 60% or less and poorly developed late tillered crop. Calculate seedrate or, if TGW is not known, usual rate is 160 kg/ha, range 140–190 kg/ha, according to above conditions.

Spring barley

Spring barley tillers vigorously and high seedrates are unnecessary. Usual range 125–160 kg/ha, former for really good seedbed conditions. Reduce to 100 kg/ha if undersowing with slow establishing long ley.

Manuring

Nitrogen (P and K, see Table 5.20)

Total N recommendations are shown in *Table 5.22*, winter barley and *Table 5.23*, spring barley. *Do not* reduce total amount of N for *malting barley* as this reduces yield and there is no guarantee of suitable sample or compensatory

Table 5.22 Nitrogen manuring of winter barley (kg/ha N)

N index	Sandy soils, shallow soils over chalk or limestone	Other mineral soils	Organic soils	Humose soils
0	160	160	50	90
1	125	100	Nil	45
2	75	40	Nil	Nil

ADAS figures.

price. Late topdressing increases N content and reduces size of grain, so for malting apply all N early in life of crop, according to type, i.e. winter or spring barleys.

Time of application

Winter barley

Development of crop is parallel to that of winter wheat but stages occur earlier; principles are similar and N is again key factor for high yields.

Seedbed Nitrogen is usually not worthwhile except for direct drilled crops, when 25 kg/ha N should be given, in addition to recommended dressing. All P and K applied to seedbed.

Spring top dressing All N is normally applied in spring; finish application to winter barleys in March for malting crops to minimise adverse effects of N fertiliser on grain size and N content. All N may be applied as a single dressing in March but split application of two or three dressings may be given and appear to give best response, i.e. (a) February and again in late March or early April, or (b) early February, mid March and again late March–early April, or (c) all N applied in first week of April. Flexibility of approach to N application is essential and rate and timing should be governed by plant populations, state of crop and soil and weather conditions: e.g. N requirements may be less after exceptionally dry winters.

Spring barley

Apply all N to seedbed for malting barley; otherwise N may be applied at any time from shortly before sowing to GS 13 (three leaf stage). Split applications are usually only worthwhile for very early sown crops on light sandy soils liable to leaching.

Harvesting and storage

Barley should be combined when dead ripe and with the ear turned right down. Further delay results in shedding and necking, to which some varieties are very prone. The grain should be dried down 16% MC for storage in sacks or 14% in bulk, the latter cooled with dry, cold air to prevent weevil infestation. Maximum drying temperature for seed and malting crops: MC 22% and over 43°C, MC under 22% 49°C. Grain for stock feed needs no drying if treated with propionic acid or stored in airtight silos.

Yield

Grain

Winter barleys outyield spring sowings. Heavy crops exceed 6 t/ha, highest yields 8–10 t/ha. 'Average'–moderate yields 4.0–5.0 t/ha.

Spring barleys good over 5.0 t/ha, average 3.75–5.0 t/ha, poor under 3.75 t/ha.

Straw

Yield 1.8–2.3 t/ha, longer strawed winter varieties may give more.

Specific weight

Average 760 kg/hl.

OATS

Importance of oats has declined greatly in last 25 years. Grain is grown mainly as a livestock feed, especially for horses. Market is limited as feed compounders are not interested on account of high husk content. Minor market for best samples for production of oatmeal and porridge oats. Traditionally some crops have been cut immature and fed unthreshed in wetter parts of Scotland and north west England. Many years ago, when oats were cut immature by binder, oat straw was considered equivalent to average hay but today mature combined oat straw is of less value than that of barley. Oats are now of minor importance.

Varieties

Suitable varieties are listed in Farmers Leaflet No. 8, *Recommended Varieties of Cereals* (NIAB).

Table 5.23 Nitrogen manuring of spring barley, winter and spring oats and rye for grain (kg/ha N)

N index	Spring barley					Winter and spring oats and rye for grain			
	Sandy soils	Shallow soils over chalk or limestone	Other mineral soils	Organic soils	Humose soils	Sandy soils and shallow soils over chalk or limestone	Other mineral soils	Organic soils	Humose soils
0	125	150	150	40	70	125	100	40	70
1	100	125	100	Nil	35	100	60	Nil	35
2	50	50	40	Nil	Nil	50	30	Nil	Nil

ADAS figures.

Climate and soil

Oats require ample moisture for straw production and grain filling and cannot tolerate drought. Crop is best suited to cooler and moister regions of Scotland, Wales and north and western England. Oats may be grown in south and eastern England but winter oats are more reliable and then only on moisture retentive soils. Good drainage is essential. Crop may be grown on a wide range of soils provided they are not liable to drying out. Oats tolerate some degree of acidity but are extremely susceptible to manganese deficiency (grey speck) on overlimed soils and peats of pH 6.0 and over.

Rotation

Oats are not susceptible to take-all (*Ophiobolus graminae*) or eyespot (*Pseudocercosporella herpotrichoides*) and are used as breakcrops by some cereal growers. As they are extremely susceptible to cereal cyst nematode (*Heterodera avenae*), which wheat and barley can carry to a much higher level without showing adverse effects, cyst (soil) counts by ADAS are necessary before inclusion in an intensive cereal rotation. Oats can also carry stem and bulb nematode (*Ditylenchus dipsaci*) which affects several other unrelated crops. Oats will tolerate any position in rotation but choose shortest stiffest strawed varieties for rich conditions, e.g. after heavily stocked ley or permanent grass.

Sowing date

Winter oats

These can be sown from 1 October to early November, optimum first half of October. Complete direct drilling by mid October. Not only are winter oats higher yielding but they are less affected by frit fly and dry summers, have a higher kernel content and ripen earlier than spring oats. May be sown in early spring up to 7 March; thereafter yield is reduced and maturity is very late.

Spring oats

These are often less reliable than winter oats, with sowing dates 15 January–15 April, optimum as soon as soil is fit in late February, early March. Late sowings are more susceptible to frit fly damage, so barley is frequently preferred for April sowing.

Seedrate

Sow according to seedsize, sowing date and quality of tilth, using lower rates for good conditions.
Winter oats 150–170 kg/ha.
Spring oats 180–200 kg/ha, on account of low tillering capacity.

Manuring

Total N recommendations are shown in *Table 5.23*, P and K recommendations in *Table 5.20*.

Winter oats

On cultivated land all N is applied as a top dressing, usually in April when crop is 150–200 mm high; pale crops may require some earlier N. Direct drilled crops should receive 20 kg/ha N at drilling time in addition to total recommendation and may require an early top dressing. If grazed, crop should be top dressed immediately *after* grazing. Recommended straw shortener should be used if lodging is likely, as crop is susceptible in wet areas.

Spring oats

Nitrogen is usually applied to seedbed in compound fertiliser.

Harvesting

Grain is extremely susceptible to heating and moulding in store if at all immature or damp. Crop *must* be fully ripe, i.e. straw completely dead and all grains hard, when it is combined and dried immediately. Drying and storage as for barley.

Yield

Grain 2.5–5.0 t/ha, according to soil and season; yield seriously reduced if drought occurs when grain is filling. Good winter crops can yield considerably more.
Straw 2.5–3.8 t.ha.

Specific weight

Average 52.4 kg/hl.

RYE

Rye may be grown as a green forage crop or for grain; each must be treated as a distinct crop, requiring totally different varieties and management.

Forage rye

Rye starts growth at lower soil temperatures than Italian ryegrass; so is especially suitable for earliest spring bite for dairy cows, ewes and lambs. Well drained soil, which warms up early in spring, is essential. Crop grows away quickly, soon becoming stemmy and unpalatable; therefore start grazing early, before stem formation or waste is excessive. Start cows when rye is 150 mm high, sheep earlier.

Varieties

Use leafy long strawed forage varieties (e.g. Lovaszpatonai, Rheidol); do not leave grazed crops to set grain as yields are usually very low.

Sowing

Early sowing, August in the North and late August to mid September in the south, is essential; later sowings do not give the massive flush of early spring growth from many tillers. Rye may be sown after grass or other cereals and direct drilled or drilled conventionally on a fine firm seedbed. Graze in early November only if growth is excessive. Drill 220 kg/ha seed for pure stands. Where several grazings are required, reduce rye to 100–125 kg/ha and include 25 kg/ha leafy tetraploid Westerworths (milder areas) or hardy Italian ryegrass, whichever is cheapest.

Manuring

Rye is almost insensitive to acidity. Give 40–50 kg/ha N, 40–50 kg/ha P_2O_5 and 40–50 kg/ha K_2O in the seedbed. Top dress with 75–90 kg/ha N as soon as spring growth shows signs of starting; give two dressings of 50–65 kg/ha N if two shorter grazings are taken.

Grain rye

Grain rye suits poor, light, sandy, acid soils and is grown as second or subsequent straw crop or used for reclamation on acid land. Grown on contract for rye crispbread and straw for thatch, it is of little importance compared to the other cereals.

Varieties

Spring and winter varieties occur but winter varieties only are sown in the UK. Recommended varieties are listed in Farmers Leaflet No. 8, *Recommended Varieties of Cereals* (NIAB).

Sowing

Direct drilled or drilled on conventionally prepared moderately fine, firm seedbed during September and as late as mid October according to district. Sown after mid February, the crop may never come into ear. Drill 150–180 kg/ha of seed (around 600 seeds/m^2) 50 mm deep with the higher rate for late sowing.

Manuring

Nitrogen recommendations are shown in *Table 5.23* and PK recommendations in *Table 5.20*.

Harvesting

Rye required for thatch is cut by binder when a little green under ear and grain can be broken cleanly over the thumbnail. Stook for 7–10 d and cart. Rye for combine harvester is direct combined and should be dead ripe. Ready mid August. Straw is very tough and durable and unsuitable for feeding.

Yield

Grain 2500–5000 kg/ha; straw 3750–5000 kg/ha. Safe storage MC 14–15%.

Specific weight

Average = 68.7 kg/hl.

TRITICALE

Triticale is a manmade plant obtained by artificially cross pollinating wheat (*Triticum* spp.) with rye (*Secale* spp.) to combine high yield of wheat with rugged hardiness of rye. It is thus well adapted to the poorer drought prone soils and cold winter conditions. At present used for livestock feed and is potential replacement for feed barley and longer strawed feed wheats on UK grade 3 and 4 soils or as third or fourth cereal on grade 2 soils, being fairly resistant to take-all (*O. graminae*). Varieties have been bred worldwide for either grain or forage production and differ greatly in appearance, habit and tillering capacity. Newer grain varieties (e.g. Lasko, Newton, Salvo) have high yield potential, having consistently outyielded Maris Huntsman wheat in trials.

Culture comparable to wheat. On light soils sow mid September, others 1–15 October preferred or to end of month. With low vernalisation requirement can be sown until March. Seedrate 90–150 kg/ha or 275–325 seeds/m^2 according to TGW and tillering capacity of variety. Optimum plant no. at establishment 200–300/m^2. Minimum after winter 150/m^2. Fertiliser dressings similar to wheat and responds to 120–150 kg/ha N or even 200 kg/ha N at index 0. Use stem shortener containing ethephon with high N applications. Top dressing occurs late March–early April GS 30. Harvesting in August, drying and storage as for wheat but slower and harder to thresh. Grain subject to sprouting.

Yield

Very variable. 3–9 t/ha according to soil. At farm yields ≥ 8 t/ha crop becomes uncompetitive with feed wheat.

MAIZE

Maize is grown world wide in subtropical areas and warmest parts of temperature zone for grain, silage, green forage, sweet corn and other uses. There are numerous types. Sweet corn (corn on the cob) is a luxury vegetable requiring distinct varieties and specialised production and marketing. Maize has been grown from grain in warmest parts of south eastern England but even there climate is neither warm nor reliable enough, so grain is not fit for combining until early winter, if at all. Maize has been grown for provision of green fodder in early autumn but wastage of 25% has been experienced with grazing *in situ*; kale is cheaper and more reliable.

Forage maize

Only forage or silage maize can be grown in the UK with any degree of reliability and then its use must be confined to low lying, warm, sheltered or south facing sites, mostly south east of a line drawn from the Wash to the Severn; even then, crop may not reach desired state of maturity before it is killed by first sharp frost of autumn. Earliest maturing varieties may be considered for most favoured sites north of this line. Maize is not a difficult crop to grow but a very high standard of cultivation and attention to detail is essential.

For highest yield and dry matter content of silage maize, grain must reach cheesy ripe stage at cutting, usually late September to mid October. To succeed crop needs long warm growing season and no exposure to strong winds. Very early sowing is impracticable as germinating seeds cannot tolerate cold soil and many rot. Growth is terminated by onset of cold weather on first sharp frost in autumn. Crop cannot reach necessary growth stage or yield with late springs and cold short summers.

Variety

Suitable varieties, listed in Farmers Leaflet No. 1, *Recommended Varieties of Fodder Crops* (NIAB) are essential. Avoid late varieties for all sites in UK; although potentially high yielding they fail to reach required degree of maturity. Medium and medium late varieties (NIAB maturity classes 5 and 6) have highest yield potential but only achieve it on good sites in warm summers. For greatest reliability, choose early or medium early (NIAB maturity classes 9 to 7) and always on less favoured sites. High scores for early vigour indicate varieties with rather better tolerance of colder spring conditions and ability to compete with any weed growth in early life. Best varieties have good resistance to lodging.

Soil

Maize is *totally dependent* on good soil conditions, requiring first class seedbed on fairly deep well drained and structured topsoil. Also moisture retentive subsoil, as crop is particularly susceptible to drought during tasselling and silking period (i.e. male and female flowers). Deep, free working loams are best; avoid shallow or infertile soils, those subject to drying out, i.e. light sandy soils or soils overlying gravel or shale and thin chalks. Cold heavy clays are late to warm up, give poor seedbeds and result in difficult harvesting in wet autumns.

Seedbed preparation

Although maize seed is large it is highly sensitive to poor seedbed conditions. Rough seedbeds give stunted growth, immature cobs and failure of soil-applied herbicides to control weeds. Land should be ploughed before Christmas to obtain a good frost mould. Fine seedbed, loose enough to allow minimal working. Drill marks help rooks to find the seed and should be removed by rolling or chain harrowing immediately after drilling. Atrazine should be sprayed on to the soil surface when work is completed but shallow incorporation of herbicide improves weed control in dry periods.

Sowing date

Sowing period is short and critical. Drill as soon as soil temperature reaches 10°C, but not before; occasionally in late April on best sites, more often early to mid May. Sowings made from late May onwards germinate and grow rapidly in warm soil, producing tall vigorous dark green plants but these are very deficient in cob. Result is an extremely immature crop at harvest which gives only a low quality watery silage.

Plant population and seedrate

Maximum yield of dry matter compatible with high proportion of cob, one cob per plant, is required. Excessive populations reduce proportion of cob while inadequate plant numbers give lower yields of dry matter. Aim for 110 000 plants/ha (i.e. 11 plants/m^2). Seed size varies greatly between and even within varieties. Usually 2600–4000 seeds/kg, so it is normally sold in packs calculated to give 110 000 plants/ha after allowing 10% for losses after drilling. Crop should be precision drilled 50 mm deep. Merchants usually state number of seeds/kg and size grade of seed, indicating belt or wheel number required; suitable belts or wheels should be ordered well in advance of drilling. Drilling in 760 mm rows requires interseed spacing of about 100 mm to achieve requisite population. Check that seed has been dressed with appropriate fungicidal–insecticidal seed dressing.

Bird damage

Rooks can devastate a promising crop in hours just at emergence and are difficult to control. Phorate granules, used to protect crops against frit fly, are useful repellents. Heavy duty black nylon thread, strung about 1 m above ground on canes 14–18 m apart in each direction, is valuable on smaller areas, together with banger sound scarers. Dead rooks shot on dawn patrols should be strung up.

Manuring

Maize gives a poor response to fertiliser applications and recommendations for P and K shown in *Table 5.24* are largely for maintenance; much more K is usually removed: e.g. a 12 t/ha dry matter crop will remove around 150 kg/ha K_2O which must be replaced during rotation unless regular applications of slurry are being given. When liberal dressing of slurry is given prior to sowing, there is no need to apply P or K for a maize crop. Some 40 kg/ha N should still be applied unless dressing was very heavy (i.e. 50 m^3/ha or more of undiluted cow slurry) and spread shortly before sowing. All N should be applied to seedbed; avoid top dressing as fertiliser granules are trapped in leaf funnels, resulting in scorch. Never combine drill, as it may seriously damage germination.

Table 5.24 Manuring of silage maize (kg/ha)

N, P or K index	N	$P_2O_5^l$	K_2O^l
0	60	80	180
1	40	60	150
2	40	40	120
Over 2	–	Nil	Nil

ADAS figures.
[1] Yield responses to P and K are small; recommendations are for maintenance of reserves. Plough down most in autumn or apply on rotational basis.

Harvesting

Feeding value of the whole plant stays constant over harvest period, with average D value of around 70 and crude protein content 9%. Aim to harvest when dry matter content of whole plant is between 20 and 25% for clamp and 25–30% for sealed towers. Crop is fit to harvest up to 25% MC, the grain being the consistency of hard dough and tough to handle; in the south this stage is reached by early and medium–early varieties in late September or early October. The later stage (around 30%) is reached when grains are hard and difficult to crack with the finger nail and husks and lower leaves are papery.

Frosting does not harm mature crops but increases dry matter content. Harvesting must be within 10 d of frosting or deterioration sets in, and it is essential to use precision chop harvester, either a special single row machine or one with one or two row adaptors. Unadapted flail mowers result in much waste of cob, uneven chopping and poor fermentation, and animals pick out best parts. Properly chopped maize ensiles without difficulty and is rich in soluble carbohydrates; additives are unnecessary.

Yield

Well grown crops of silage maize usually produce 11–14 t/ha of dry matter or 52–62 t/ha of fresh material at 21% dry matter, which should produce from 44 to 53 t/ha of silage.

SWEET CORN

Sweet corn or corn on the cob consists of the core (cob), the edible kernels (seeds) and the husks (enveloping leaves). A marginal crop in UK, it is grown almost entirely for the fresh market from July to October; canned and frozen sweet corn is almost entirely imported. It must be really fresh or it loses its sweetness and is best marketed through the 'cool chain'.

Varieties and pollination

Although a member of the maize family, sweet corn is entirely distinct from maize grown for grain or fodder, with its own varieties. In sweet corn carbohydrate is laid down in the kernels as sugar, which like garden peas, is only converted into starch as the crop matures; in grain and forage varieties conversion to starch occurs quickly and the kernels are unsweet. Maize and sweet corn readily cross-pollinate; the gene for sweetness is recessive, so cross pollina-

tion with maize results in unsweet cobs. Isolate from all other maize crops by at least 200 m, preferably 400 m.

Varieties listed in Vegetable Growers Leaflet No. 10, *Sweet Corn* (NIAB) are classified by maturity as early, early midseason, midseason, late midseason, late and very late. Producers aim for a succession of pickings, but midseason and later varieties are only suitable for the most favoured sites.

'*Supersweet*' varieties, which are much sweeter than 'normal' varieties are gaining popularity. The gene controlling the characteristic of 'supersweet' is a double recessive, so cross pollination with 'normal' varieties masks sweetness. Grow at least 100 m from 'normal' varieties or ensure at least 14 d difference in flowering from other varieties. Supersweet seeds have smaller food reserves, so raise in cells or blocks, provide clear plastic mulch protection or sow later.

Pollen is shed from the tassels or male flowers for about one week. Efficient pollination is essential for top quality well filled cobs; never plant in single lines, always grow each variety in squares.

Climate and soil

Like other forms of maize, sweet corn is very sensitive to temperature and soil conditions. Sweet corn is primarily suitable to low lying sites (maximum 100 m above sea level)

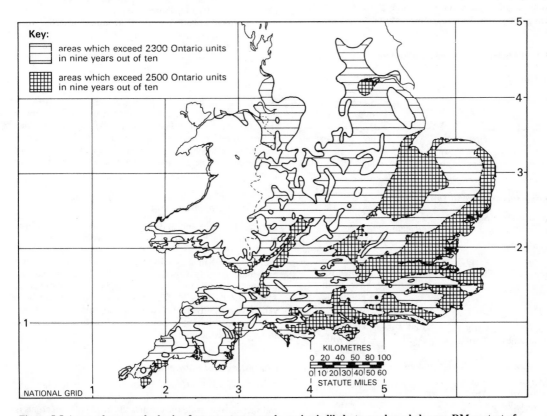

Figure 5.5 Areas where, on the basis of temperature records, maize is likely to reach a whole crop DM content of 25% (2300 Ontario units) or 30% (2500 Ontario units) in at least nine years out of ten. Most early sweet corn varieties are suitable for picking after 2500 Ontario units have been accumulated and the earliest can be picked after 2300 Ontario units. (Source: Meteorological Office, Bracknell)

and production is mostly concentrated in SE England; possible areas occur mostly south-east of a Wash–Severn line (*Figure 5.5*). On less favoured sites choose earliest varieties and use surface plastic mulches (p. 97). Consider raising in modules.

Sweet corn is supersensitive to compaction and poor tilths; as with silage maize plough early for good frost mould and sow with precision drill in fine smooth seedbed with minimal working to conserve moisture. Minimum soil pH 6.5. Deep water retentive but well drained medium loams or silts are best; avoid shallow soils, those liable to dry out or poorly structured or compacted soils.

Irrigation

Prevent water stress during vegetative growth; maximum requirement occurs in July and August, starting with tassel elongation (male flowers) and silking (female flowers). Water stress reduces crop height, delays tasselling and silking and results in incomplete pollination with poorly filled ears of low value or even unsaleable ears. Irrigate, especially low rainfall areas, with 25 mm when SMD reaches 50 mm or with limited water, 50 mm at tasselling.

Manuring

All fertiliser is applied to the seedbed; should it be necessary to apply N later, always give as a *side dressing* to avoid scorching crop. Fertiliser recommendations are shown in *Table 5.25*.

Table 5.25 Manuring of sweet corn (kg/ha)

N, P, K or Mg index	N	P_2O_5	K_2O	Mg	
				Sandy soils	Other soils
0	100	75	75	90	60
1	75	50	50	60	30
2	50	25	25	Nil	Nil

Planning succession and sowing date

Required harvest period	Sow	Transplant modules	Situation
Protected crops			
Early July Early varieties	Early April	End April	Raised and grown to harvest in polythene tunnels
Late July Early varieties	Early April	End April	Grow in field with plastic cover
End July–early August Early varieties	End April–early May	—	Drill directly into field with plastic cover
Mid August Mid season varieties	Second week in May	—	

Unprotected crops			
Late August Early varieties	Late April or early May	—	Field
End August and September Early and mid season varieties	Second half of May	—	Field
October Mid season or late varieties to mature in succession	Late May or early June	—	Field

Succession of harvests may be achieved sowing a limited number of varieties every 10–14 d or by sowing some six to eight varieties on the same date to mature sequentially.

Transplants

Use only in conjunction with clear film protection, otherwise advantage is lost. Transplant cell or block plants at one and a half to two leaf stage into furrow 100 mm deep to avoid damage from plastic cover. Once established, make small slit in plastic cover above each plant to allow emergence but leave cover in place to photo-degrade.

Crops drilled directly into the field should never be sown before soil temperature is at least 10°C or poor germination and stunted growth follows. Pneumatic seeders cope best with sweet corn seed.

Protected field drilled crops

Drill in two row beds 1.5–1.8 m wide with 750 mm between rows. Work in fertiliser during seedbed preparation; apply herbicide and any insecticide *before* covering. Drilling and covering is best done as a single operation with a combined drill/covering machine. Accurately placed slits for seedling emergence must be provided; otherwise slits must be made manually.

Spacing and plant population

Ear size decreases and maturity tends to be delayed at high plant densities. Best grown in beds to avoid compaction.

Maincrops   1.8 m beds with two rows 760 mm apart; within row spacing: early varieties 200 mm; mid-season and late varieties 250 mm to give required field populations of 44 000–56 000 plants/ha. Variation in seed size between both varieties and individual seed-lots is extremely large (3500–6500/kg), so obtain plant count and calculate seedrate as for all large seeded legumes (p. 142).

Harvesting and marketing

Crop is usually ready four to six weeks after fertilisation, which occurs when the silks first appear. These are brown and shrivelled outside the ear and the inside of the kernels is thick and creamy. As an initial test, if several kernels are pressed by the thumb nail and the contents squirt out like thick cream, the crop is ready to pick. Once the 'dough' stage is reached, the crop is no longer tender. Picking is done by hand and it takes some experience to judge the correct stage; it is most inadvisable to allow 'pick your own' customers to harvest this crop. Two or three pickings are usually necessary to clear the crop.

Picking at the right stage and speed of handling to market are vital. Once picked the sugars in the kernels soon turn to starch; this process speeds up as temperature increases. Remove picked ears from the field immediately and put straight into cooling plant. Wet air cooling with positive ventilation is ideal as it can remove relatively high field heat with minimal weight loss. Market crop quickly; cool chain (p. 98) handling is essential to maintain flavour and quality and is required by supermarkets.

Pests and diseases

Pests

Frit fly (Oscinella frit) This is by far the most important pest of sweet corn. It lays eggs on or near the seedling and the hatched larvae bore into the main shoot causing death or extensive tillering, with loss of yield and ear quality. Un-protected drilled crops are most at risk and preventive treatment with approved granular insecticide is advisable. Transplanted protected crops are past the vulnerable stage and covered drilled crops are not often attacked.

Cereal cyst nematode (Heteroda avenae) This pest can stunt the crop, which should not be grown where this pest is present.

Bird cherry aphid (Rhopalosiphum padi) Bird cherry aphid and other cereal aphids can enter the ear and cause crop rejection. Regular precautionary spraying required but allow manufacturer's full recommended time interval between the final spraying and picking. Protect against birds (*see* Silage maize, p. 132).

Diseases

Seed-borne   Pythium can cause serious damage to seedlings, especially if slits in covers of protected crops cause localised waterlogging. Dress seed with approved fungicide.

Stalk rot (Fusarium spp.) Symptoms visible about one month after flowering: leaves wilt, turn yellow and then brown. Stem bases of affected plants have watersoaked appearance and then lodging occurs. *Fusarium* spp. causing stalk rot can attack other cereals. Do not grow sweet corn after cereals or maize or grow successively on the same land: allow three years between susceptible crops. Risk of attack is much greater where compaction, poor soil structure or pans cause wet surface conditions. Grow *only* on freely drained soils in good condition.

Yield

Yields from crops grown under surface mulches can be 30% higher than for unprotected crops. Range of ear size 200–300 g. If yield 2500 × 12 boxes, weight would be around 7 t/ha.

SEED PRODUCTION

Seed certification

The aim of seed certification is clean, healthy stocks of seed, free from weed seed, pests, diseases, alien varieties and other cultivated species. Seed must produce a crop true to type as produced by breeders, of the stated variety, guaranteed at a minimal standard and of high germination. All seed offered for sale in EEC must be certified, which includes seed

(1) produced under statutory certification scheme,
(2) certified seed imported from EEC,
(3) from third country having equivalence as certified seed.

For details of producers and requirements for producing seed for certification in England and Wales, consult MAFF or elsewhere in UK, Department of Agriculture, Scotland or Ministry of Agriculture, Northern Ireland. *National lists* of varieties are published by National Institute of Agricultural Botany, who also publish a range of seed growers leaflets.

Production of certified seed is carried out exclusively under the control of the appropriate certifying authorities. Only authentic seed of varieties included in the *national list* for each crop or group of crops (e.g. cereals, grasses, potatoes) is eligible for certification; to qualify for this a variety must be distinct, uniform, stable and of acceptable agricultural value. *Recommended lists* have no statutory status. All crops being grown for certified seed are inspected solely by *licensed inspectors* and must reach minimal standard for each grade of seed. Processing and sale of seed is also controlled; only *licensed seed merchants* may sell seed, only *licensed processors* may treat, clean and label basic or certified seed while samples for processor must be taken by a *licensed sampler*.

Seed production and marketing is a highly complex business which may be divided into three stages: growing, storage and processing. The latter is entirely outside the province of the grower.

Seed growing

Production of certified seed is a specialised job requiring excellent husbandry and management. Meticulous care, good planning, and very high standards of training and supervision of staff are indispensable. Farmers who cannot satisfy these essentials should not attempt seed growing.

It is necessary to meet the following basic requirements.

Authentic seed

Crops eligible for certification may only be planted with authentic seed of a grade of high standard and of a variety approved by the certifying authority; this normally includes foundation seed in sealed sacks straight from breeder and other seed with an appropriate certificate. Be sure there is adequate seed to finish the field: once-grown seed must not be used.

Adequate isolation

Isolation from crops of the same species prevents cross pollination and contamination of the crop with disease or unwanted seeds of other varieties. Cereals, normally self pollinated (i.e. wheat, oats, barley), need a physical barrier, hedge, fence, or an area of fallow or non-cereal crop at least 2 m to prevent admixture. A distance of 50 m from potential sources of loose smut infection (e.g. crops not planted with certified seed) is recommended to prevent cross infection; six row barleys are prone to cross fertilising and should be similarly isolated from two row barleys. Crops normally cross fertilised require much larger minimum distances, e.g. rye: 250 m. Grass seeds, same spp. or IRG and PRG: fields up to 2 ha 100 m, fields over 2 ha 50 m. Leys and permanent pasture are included in minimum isolation distance: if they adjoin seed stand, keep topped or grazed to prevent heading during flowering periods of seed crop, unless they contain *only* same variety. Be careful in selecting fields on farm boundary or adjoining grass verges of roads. Brassicas require at least 200–1000 m according to species and grade of seed.

Clean land

Seeds likely to contaminate seed crops lie dormant in the soil for a long time so a cleaning rotation is necessary. Cereal crops for seed should be preceded by two years of non-cereals, unless the previous crop was the same variety and grade. Similarly it is necessary to use the same grass or clover variety in ordinary leys as that grown for seed. It is also desirable, if possible, to keep the same field for the same variety and species of grass. The field must be free from noxious weeds or those likely to contaminate the crop. Do not feed hay on seed-producing leys or apply fresh farmyard manure to land used for seed production.

On arrival at the buyer's premises seed must be cleaned. Modern seed cleaning machinery removes impurities provided they are substantially different in size, weight or shape from seed crop. The closer the form of the impurity to the seed crop, the more difficult to remove and the greater the amount of crop seed removed with the impurity during cleaning. If seeds of crop and impurity are identical or very similar, separation is impossible and the contaminated crop is worthless. Thus, some weeds and other crop plants often disregarded in normal crops are highly undesirable in a seed crop.

Clean machinery and equipment

These are essential to prevent contamination. Drills or broadcasters must be thoroughly cleaned on waste land. Combine harvesters should be run empty and then thoroughly cleaned out with a compressed air line and industrial vacuum cleaner. Use clean sacks. Trailers, augers, elevators, conveyors and bins must be emptied and thoroughly cleaned before use. To avoid admixture of stocks, produce of each crop should be kept separate and be identifiable by variety, reference number of stock, field OS number and name. Bins should be kept locked and sacks clearly labelled, with stacks of sacks separated by a solid partition.

Farm suitability

Cereals and herbage seed production are best suited to large farms, in a ring fence. Small farms or those spread out find it impossible to exert control over adjoining cropping needed to obtain the required isolation. It is advantageous for farmers to group together to plan cropping of boundary fields, especially for grass seed crops. Large cereal farms have the advantage of complete cereal growing equipment, combines and drier under their own control. Contractors usually fail to clean out combines between farms and fields; contamination from other crops is soon transferred to the sample. Seed production is thus better suited to larger farms.

CEREALS

Cultivation for seed crops is dealt with under the appropriate cereal crops. Inspection by a qualified inspector should be preceded by roguing when other species, varieties or weeds can be removed from the crop; dirty crops cannot be rogued. Avoid herbicides which stunt but do not kill all wild oat plants (*Avena* spp.). Drying needs care, as high temperatures damage germination capacity; drying to 49°C is normally safe for grain up to 24% MC. Dry down to a maximum of 15% MC and store in a cool, dry place, protecting from vermin. (*See also* Technical Leaflet No. 1, *Seed Quality* (NIAB) and Seedgrowers Leaflet No. 1, *Growing Cereals for Seed*.)

HERBAGE SEED PRODUCTION

Leys for herbage seed production are valuable break crops; the roots restore structure of worn out arable soils and provide large amount of organic matter. In addition, they provide out-of-season grazing and additional livestock can be kept. The stover or straw, often sold as threshed hay is useful fill-belly. Principal crops include grasses (perennial and Italian ryegrass, cocksfoot, timothy, meadow and red fescues) and clovers (red and white clover).

Crop establishment – general

Seedbed

Plough early for good frost mould. Fine, firm, level weed-free seedbed is essential; knobbly seedbeds give weak plants. Shallow drilling 12–18 mm on rolled surface is best in dry areas, otherwise broadcast on rolled surface. Place seed in soil moisture for rapid germination and vigorous growth, drilling as shallowly as possible. Small seeded species (white clover, timothy) 8–12 mm, larger seeded sorts (ryegrasses, cocksfoot, meadow fescue 20–25 mm deep).

Use of cereal cover crop

Undersowing or sowing seeds without a cover crop after cereal harvest gives income from cereal during year of establishment, but in some slower developing species, notably timothy, tall fescue and cocksfoot, yield in first harvest year is heavily reduced (20–70%). Sowing bare (without cover crop) eliminates this loss but it is necessary to balance loss of income from cereal crop against this increase. Practice depends on species and circumstances; major risk of undersowing is total loss of seed crop under lodged or dense cereal. To minimise adverse effects, drill grass and cereal on same day, reduce cereal seedrates and N applications (*see* section on lodging in cereals, p. 119). For sensitive species grown in wide rows and undersowing either

(1) block off every second or third row in cereal drill, according to grass crop row width, and drill seed crop after cereal emergence, or
(2) fit dividers into seed box and sow cereals and grass in same operation.

Row width

Two major methods include

(1) drilling in narrow rows, 180 mm or less, or broadcasting, and
(2) drilling in wide rows 450–600 mm apart.

Method 1 is perfectly satisfactory for all ryegrasses, red and white clover but in cocksfoot, timothy and meadow fescue results in excessively dense tiller populations and consequent reduction in seed yield, especially in high tillering pasture varieties and in second harvest year.

Sowing date

Undersown crops are sown simultaneously with cereal in March or early April. Early drilling of *all* crops ensures adequate soil moisture for establishment in dry areas as well as full tiller development in time for vernalisation the following winter. If sowings without a cover crop are intended late in the year, speed of establishment and tillering rate are determining factors in deciding how late a species can be sown with safety; it also depends on season and soil temperature. In mild southern areas Italian ryegrass, hybrid ryegrasses and early flowering perennials can be sown up to mid September, perhaps even up to end of September but the later the sowing the greater the risk of failure. It is usually inadvisable to delay sowing other grasses beyond end of July at latest. Red clover performs very poorly from late summer or autumn sowings.

Seedrate

Recommendations depend on soil conditions. With choice seedbed and consequent high tillering, low seedrates are best; increase if conditions are not so good. Excessive seedrates result in very high tiller densities, which suppress seed yields. Aim for high quality seedbeds with reduced seedrates, to give low population of vigorous plants. Increase seedrates with closer spacing or broadcasting, according to seed size and establishment rate of species. Ideal population appears to be one mature plant/150 mm of drill run, as a general rule.

Establishment of individual species

Perennial ryegrass

Prefer to undersow in spring cereal but can be sown bare after early harvested winter cereal. Drill in rows 100–200 mm apart, diploid varieties 9 kg/ha, tetraploids 17 kg/ha or broadcast 13 kg/ha diploids, 22 kg/ha tetraploids; 0.5–1 kg/ha wild white clover may be added to crops of pasture varieties to improve grazing; large leaved white clovers may cause difficulty in harvesting.

Italian ryegrass

This grass establishes rapidly from seed and may be undersown in a spring cereal or drilled bare after cereal is removed up to mid September. Sow in drills 120–170 mm apart, diploid varieties 13–16 kg/ha, tetraploids 17–22 kg/ha.

Tetraploid hybrid ryegrasses

As for Italian ryegrass but drill 17–22 kg/ha.

Cocksfoot

For highest seed yields in first harvest year drill in spring without cover crop in rows 400–600 mm apart. Sow 2–5 kg/ha, lower amount for superfine seedbed and good conditions.

Timothy

Drill bare in spring or up to late July in rows 500–600 mm apart, 2–4 kg/ha. If undersowing, reduce row width to 350 mm and increase seed to 5–8 kg/ha.

Meadow fescue

Drill in spring under cereal crop in rows 150 mm apart with 12–18 kg/ha seed.

Tall fescue

Drill in spring without cover crop in wide rows 600 mm apart, seedrate 5–7 kg/ha, after cleaning crop.

Red fescue

Drill in spring under cereal crop in rows 100–200 mm apart at 10–15 kg/ha.

White clover

Broadcast, or in dry areas drill in narrow rows, under cereal crop in early spring. Large and small leaved varieties may be sown pure at 2–4 kg/ha or mixed with 3–5 kg meadow fescue or pasture perennial ryegrass, whose seeds are easily cleaned from white clover; avoid timothy which is difficult to separate. Inclusion of grass improves grazing; sward then forms pasture when seed production is finished.

Red clover

Undersown in cereal crop in spring; avoid late summer sowings as establishment is then unreliable. Prefer drilling in narrow rows in dry areas. Best sown alone, diploid varieties 8–10 kg/ha, tetraploids 11–13 kg/ha. Some growers mix 3–5 kg/ha timothy or perennial ryegrass in broadcast stands.

Row indictor plants

Grasses grown in wide rows germinate slowly and cannot be seen early enough for inter-row cleaning. Include 0.6 kg/ha mustard, or on charlock (*Sinapsis arvensis*) land 0.3 kg/ha lettuce with the crop. These seeds emerge early, indicating the rows and are subsequently killed by herbicides.

Manuring

Establishment year

Total applications shown in *Table 5.26*.

Direct sown grasses Phosphorus and potassium may be applied in previous autumn or to seedbed; if crop is cut for forage give extra 40 kg/ha P_2O_5 and K_2O. Backward crops may need extra N but control weeds first.

Table 5.26 Manuring of herbage seed crops establishment year (kg/ha)

N, P and K index	Direct (bare) sown grasses		
	N	P_2O_5	K_2O
0	60	120	120
1	60	80	80
2	60	50	50
3	–	30	Nil

ADAS figures.

Undersown grasses and clovers If there is a P or K deficiency, broadcast fertiliser for cereal and increase these nutrients; do not combine drill or stripping occurs. Otherwise all additional P and K required is applied immediately after cereal harvest, with N; clovers do not require N. For backward grass seed crops. NB cocksfoot, give up to 50 kg/ha N; at soil P or K index 0 apply an extra 50 kg/ha P_2O_5 or K_2O, all immediately after cereal harvest for grasses and clovers.

Production years

Annual applications of P and K are shown in *Table 5.27*, N in *Table 5.28*. Phosphorus and potassium may be applied in spring, according to N index. Spring N may be given in one or two dressings and timed according to species and variety. For Italian ryegrass, early perennial ryegrass, cocksfoot and meadow fescue, apply in early spring; apply N to medium and late perennial ryegrasses and timothy in late spring. Rate of application also depends on weather, management and condition of sward, older, weak or backward swards need rather more N. Conversely, excessive N application, e.g. direct combined late varieties of perennial ryegrass, can cause considerable harvesting difficulties from lodging.

Table 5.27 Manuring of herbage seed crops production years P and K requirements – all crops and soils

P and K index	P_2O_5	K_2O
0	100	150
1	60	120
2	30	90
3	30	Nil

ADAS figures.
Apply in spring at indices 0 or 1.

Autumn applications, of N, given as compound fertiliser immediately after harvest are designed to produce vigorous leaf growth of grasses but amount is dependent on age and condition of sward. Up to 50 kg/ha N may be needed for backward crops, especially cocksfoot and red fescue. Also apply up to 50 kg/ha N if sward is to be grazed or cut for forage in autumn.

Sward management and defoliation

Herbage seeds are grown under such a wide range of soil, climatic and farming conditions that it is impossible to lay down precise details of sward management. Some crops are grown on stock farms but most are grown on large predominantly arable farms, many with no stock at all. Surplus grass may be removed by stock or mechanically.

Appropriate extent and timing of removal of excess herbage vary not only between species but between types and varieties within same species. Obtain appropriate advice when growing a new variety for first time. Appropriate defoliation technique is also modified greatly by amount and distribution of rainfall and water holding capacity of soil. Species differ greatly in their dates of flower formation and head emergence; in general, grazing should be completed well before critical stage of flower formation; defoliation at this stage or later reduces seed yields drastically in most species. The earlier flower formation occurs, the earlier must grazing be completed; late varieties or species can be grazed later. Effect of spring grazing is much influenced by soil moisture. In dry years and areas spring grazing where appropriate should be early and light, omission is often preferable as late grazing in these circumstances can reduce yield heavily. Always aim to complete grazing *quickly* – then move on.

Perennials and Italian ryegrasses

These produce the bulk of their fertile tillers in spring. Autumn or early winter grazing, October to end January, is beneficial to all varieties, as it encourages tillering and removes excess herbage, which suppresses tillering and causes winter burn. Value of spring grazing, other than supplying feed, is much less certain and can reduce seed yield; properly carried out, grazing shortens straw and helps to reduce lodging. Adequate N is essential to prevent loss in yield.

Finish grazing early varieties of perennial ryegrass, e.g. S24 by first week in April, intermediate, e.g. S321 by second week in April and late varieties, e.g. S23 by third week in April; the latter allow heavier grazing and need higher N levels.

Italian ryegrass

Spring grazing shortens straw and reduces lodging risk. Crop produces ample fertile tillers in spring; provided adequate N is given in early March, crop can be grazed up to end of April, according to variety, without loss of seed yield. If crop is not grazed reduce N. Additional N does not mitigate effects of grazing too late.

Hybrid ryegrasses

Growth pattern permits considerable flexibility of management, which is similar to that of Italian ryegrass.

Table 5.28 Manuring of herbage seed crops. Production years (kg/ha N)

Time of N application	Soil type	Ryegrasses	Cocksfoot	Timothy, fescues
Autumn	All soils	0–50	0–50	0–50
Spring	Sandy/shallow soils over chalk and limestone	150	170	100
Spring	Other soils	120	120	80

ADAS figures.

Removal of flowering tillers in some varieties does not reduce yields, as they are replaced, provided grazing is not too late and ample N is given. Check on varietal requirements.

Cocksfoot, timothy and tall fescue

Majority of seed bearing tillers are produced in autumn and different grazing pattern is required. In general graze after autumn growth has stopped but before spring growth starts, quickly and once only. Removal of excessive leaf, which restricts tiller growth is beneficial but avoid grazing too tightly.

Cocksfoot best grazed leniently in December; danger of heavy yield suppression if grazed early February onwards. Do not graze backward crops in establishment year. Least susceptible species to lodging; generous N application increases number of fertile tillers and size of inflorescence. *Timothy*, management similar to cocksfoot. Graze leniently once only in period October–December; late grazing disastrous. *Meadow fescue*, graze once during autumn or early winter and never later than early March. Very susceptible to lodging. *Tall and red fescues*, graze once in autumn or early winter and complete by mid February.

Clovers

Grazing after removal of cover crop is beneficial in first autumn; graze thin crops *lightly* with sheep, lush crops may be grazed fairly hard to produce short even sward; do not overgraze or bare tight.

Table 5.29 Harvesting of herbage seed crops

Species	Type/variety E: early, M: medium, L: late flowering	Ready for swathing	Usual/best method of harvesting	Target yield (kg/ha)	No. of harvest years
Italian ryegrass Hybrid rye grasses	—	Very variable—mid July to mid Aug	Swathed, then combined. Can be direct combined but only advisable in badly laid crops	800–2000	1
Perennial ryegrass	E	Early to mid July		750–1000	2
	M	Mid to late July		800–1250	2
	L	Late July (good years), mid Aug (late wet summers)		600–1000	2
Cocksfoot	E	End June to early July— 14 d later in bad year	Best swathed, then combined. Stands well but only direct combined in emergency	450–1000	
	L	Mid July to 14 d later in bad year			2[1]
Meadow fescue	E	Early to mid July	Swath and combine or direct combine if lodged, with little greenery	525–750	2[1]
	L	Mid July		350–500	
Timothy	E	Mid Aug	According to variety: swathing double- or triple-combining for varieties difficult to thresh, e.g. S48	375–650	3
	L	Late Aug to mid Sept		375–650	
Red fescue		Early July	Swath, then combine. Does not shed readily	375–500	4–6[2] or even longer
Red clover[3]	Broad red varieties		Seed does not shed readily when ripe. Direct combining. Green crops require chemical desiccant	< 500	1
	Late flowering red varieties	Sept onwards according to season			
White clover[3]	Giant White, varieties	Late July to early Sept according to weather	Tricky except in dry weather; choose dry spell. Swath crop and combine from swath: put into windrows first with light crops	375	2[1]
	Wild White, varieties			< 375 old pastures	2[1]
				80	indefinite if old pastures

[1] Three crops can be taken but two are usually more economic.
[2] Sward may be sold profitably as fine turf at end of its life with landlord's agreement.
[3] Yields highly dependent on weather.

Red clover

Broad red clover is traditionally first cut for silage or hay and the seed crop taken from the aftermath to prevent heavy leaf and straw production. Late or heavy cuts reduce yield severely; do not take more than a light early cut of silage before laying up. *Late flowering red clover* should be left untouched or a *very short* silage cut (10 cm) taken in spring. Research indicates that highest seed yields may be obtained by omitting early cutting and or grazing and taking seed cut first. A recommended desiccant can deal with heavy greenery if properly applied.

White clover

Swards sown with companion grass should be grazed *hard* when grasses are growing actively in early spring, to encourage clovery sward. Clover is grazed in late spring to remove excess growth which interferes with flower production and gives harvesting difficulties. Cease grazing *before* significant number of flower buds are taken (i.e. when first flower buds appear): medium to large leaved varieties finish by mid May; small leaved or wild white varieties, cease end of May or first week of June. In dry areas or seasons, finish two weeks earlier, under wetter conditions grazing may be extended by a week.

Post harvest treatment

Mow over after harvest and remove dead stubble and trash; stubble removal is most important if grazing is required, otherwise stock reject grass. Cocksfoot stubbles should be burned if cocksfoot moth (*Glyphipterix simpliciella*) is a problem. Red fescue swards are best gang mown. Apply fertiliser to cleared stubble. If inter-row cultivating wide drilled crops do not work too close to drills.

Harvesting

The binder, ideal for cocksfoot, has largely disappeared.

Crops cut loose may be put on *tripods*, a safer practice, producing better samples in wet seasons but too laborious and now little used. Alternatively cut and cure in the swath and thresh with a combine and pick-up attachment. Reasonably long stubble is left, according to species. Moving weathered crops around to dry results in heavy seed losses: if grass has grown through swaths, undercut with mower, allow to dry, and combine. Swathing is best suited to unevenly ripened crops, those containing greenery and late flowering or pasture varieties which ripen unevenly. *Direct combining* is most suitable for early ripening grass varieties and red clover: the latter may require use of chemical desiccant. Crop must be ripened evenly and free from green material. This method has lowest labour requirement but shedding is very serious if combining a fully ripe crop is delayed in wet windy weather.

Double combining

This is when crop is threshed and then re-threshed after an interval of a few days, is suitable for unevenly ripened crops, especially timothy.

Drying, conditioning and storage

Damp or trashy parcels must be dried, immediately or, if impracticable, spread thinly on a clean airy floor or hessian; never leave in bags, even overnight, for rapid heating ruins germination. Maximum air temperature reaching the seed should not exceed:

	Normal seed (°C)	Very wet seed (°C)
Timothy and clovers	37	26
Other grasses	48	32

Store in cool, dry, weatherproof permanent building and periodically check for condition. Method and time of harvesting, target yield and duration of the stand are given in *Table 5.29*.

Miscellaneous non-leguminous combinable cereal break crops

BORAGE

Borage (*Borage officinalis* L) is a herb whose seed has a high gamma-linolenic acid content (GLA) similar to evening primrose and is used for pharmaceutical purposes. It is considered to be indigenous to Britain as it has been cultivated and grown wild for centuries. It is fast growing and aggressive and will smother weeds rapidly at high densities. Flowering is indeterminate (i.e. occurs over a prolonged period) and seed shedding at maturity is the major problem. Like evening primrose, it is essentially a speculative crop for the specialist, not a cereal break crop for the average farmer and should only be grown on contract. Borage is a new combinable crop which has received very little development work.

Climate and soil

Most suitable in low rainfall areas for easier harvesting. Prefers medium loams; avoid light sands and acid soils.

Culture

Drill March–April on a fine firm seedbed, similar to that for spring barley, 20–25 mm deep in rows 120 mm apart for ease of swathing or up to 500 mm apart to allow inter-row hoeing. Suggested seedrate 22 kg/ha to give 115 seeds/m^2.

Manuring

Apply kg N/ha: 75 N; 40 P_2O_5; 40 K_2O during seedbed preparation. No top dressing required.

Harvesting

Harvesting in late July–early August is the really critical period; borage is very sensitive to weather conditions and shedding is the major problem. The indeterminate flowering pattern makes judging the optimum time for swathing extremely difficult: swath as soon as the first seeds drop and combine 5–10 d afterwards. Seed is likely to have 20–25% moisture content at harvest and must be dried down to 10% at once.

Yield

Seed yield is extremely variable: range 250–750 kg/ha. Higher yields can be very profitable.

Further details from Bio Crops Ltd, West Street, Coggeshall, Colchester, Essex, UK.

EVENING PRIMROSE

Evening primrose (*Demothera biennis*) is indigenous to North America. It is grown for its seeds, whose oil has a high gamma-linolenic acid content (GLA). Evening primrose oil is marketed as a dietary supplement and is under investigation for a wide range of pharmaceutical uses. Outlets for the crop are extremely limited and it must be stressed that it is a specialist crop to be grown *only on contract* by a limited number of growers rather than as an alternative cereal break crop for the average farmer. Cash returns can be very good, but the crop must be regarded as highly speculative and is in an early stage of development. Critical stages are establishment and harvesting, which can be very difficult in a wet autumn or, at worst, impossible.

Climate and soil

Crop is best suited to the southern half of England owing to warmer and earlier harvesting conditions and is safest on light-medium free working well drained arable soils; avoid soils on which a good tilth cannot be readily obtained or do not drain freely.

Culture

Sowing must be completed by the end of July in the year before harvest. A clean burned barley stubble is ideal. Prepare a fine firm tilth with minimum cultivations and broadcast 2.5 kg/ha seed followed by heavy rolling or precision drill 1.0–1.5 kg/ha seed in 40–60 cm rows as shallowly as possible, 8–10 mm deep. The seed is minute. A combination of warmth and moisture is essential for successful establishment, so irrigation is usually required. Optimum established plant population 30–50 plants/m^2; minimum 5–10 plants/m^2. Transplants raised in modules have also been used, planted at 30 plants/m^2.

Manuring

Response to phosphate and potash is small and on soils in good condition no seedbed fertiliser is needed. If soil reserves are low give (kg/ha): 25 N; 50 P$_2$O$_5$; 50 K$_2$).

Spring top dressing

Broadcast 50 kg N/ha; there is no economic response to more.

Growth pattern and harvesting

After germination the plant grows to a flat rosette of 150–300 mm in diameter for overwintering. In late April it throws up a strong central shoot and tillers, which grow to a height of 0.9–1.5 m. Flowering begins in late June and continues until August. Lower pods start ripening in late August and ripening continues during September. Crop is usually ready for desiccation with approved desiccant approximately 21–30 September; allow 7–10 d and then direct combine with minimum draught 1–14 October. Seed is liable to shedding.

Indeterminate flower growth makes it difficult to estimate precise stage to combine for maximum yield.

Sample usually contains 60–80% seed, with moisture content 30% or more. Seed requires immediate drying after combining or it will deteriorate rapidly. Dry down to 10%. Seed then requires cleaning to the necessary purity standard. Yield is extremely variable: range 300–700 kg/ha.

Further details, contracts and seed from: Evening Primrose Oil Co. Ltd, Unit 1, Jubilee Drive, Loughborough, Leics, LE11 0FL, UK.

LINSEED

While botanically identical, arising from the same parent plant (*Linum usitatissium*), two quite distinct groups of varieties of (a) flax (producing straw for fibre manufacture (i.e. linen)) and (b) linseed (for oilseed production) have been selected over the centuries. In the UK areas of both crops are currently relatively small. There is now increasing interest in both; in flax, because new methods of harvest and processing are being developed and in linseed for import saving and crop diversification. Under the CAP of the EEC, subsidies are available for both crops (contact Intervention Board for Agricultural Produce, Fountain House, 2 Queens Walk, Reading, Berks, UK).

UK requirements for linseed and linseed oil are currently largely met by imports (about 95%). Linseed is crushed to produce drying oil, used in paint and linoleum manufacture; oil is of low edible quality owing to high linolenic acid content. The crushed by-product, linseed meal, is a protein rich animal feed. Straw is usually burned in the field, as it is not absorbent enough for stock bedding; if baled it is ideal for straw burners, giving off more heat than wheat straw. Some growers sell straw to fibreboard manufacturers. Linseed is a useful cereal break crop.

New shorter strawed higher yielding varieties for better standing and suitable for direct combining are now available. For good germination obtain seed from reliable source.

Linseed grows well in cool humid climates but harvesting can be difficult in cold late seasons. Grows well on most soils from chalk to clay but avoid thin soils likely to dry out badly, e.g. gravels and lightest sands. Plough in autumn for good frost mould and a firm fine moist tilth, avoid compaction; drill, preferably in narrow 100 mm rows, from mid March or as soon as land is fit. A seedrate of 75–90 kg/ha is usually recommended to give populations of 750–1000 plants/m^2; in recent trials populations of 350–450 plants/m^2 have given highest yields. Take account of grain weight of different varieties; calculate seedrate from formula on p. 92 if number of seeds/g and germination are known. Calibrate drill carefully as the flat slippery seed runs very freely.

Apply lime if pH is less than 6.5. Broadcast all fertiliser during seedbed preparation; do not combine drill. Response to P and K is poor; excess N causes extended flowering period, late ripening, lodging and reduced oil content. Apply kg/ha:

	Soil index		
	0	1	2
N	75–100[1]	60	40
P$_2$O$_5$ K$_2$O	75	60	40

[1] Up to 120 kg N/ha may be needed after a run of cereal crops: apply 40 kg N to seedbed; topdress remainder shortly after emergence.

Good weed control at seedling stage is vital as crop does not compete well with weeds (*see* Chapter 8 *Tables 8.5* and *8.6*). Any substantial weed growth will also slow harvesting and with weed seeds retard throughput of crop driers.

Linseed is late ripening and usually the last crop to be combined. It is ready from late August to mid September, when capsules turn golden brown and rattle; leaves turn brown and drop but stems remain green and tough. Use a dessiccant (e.g. diquat) when most of the heads are ready and allow *at least* 10 d before attempting to combine. Crop is dense on top, so thorough spraying is essential.

Do not try to rush combining. Wait until the afternoon when crop is thoroughly dry. Ensure *sharp* knives and slow forward speed to avoid blockage. Take precautions against seed leakage (NB trailers, conveyors and driers). Probable harvest moisture content 12–16%; safe storage MC 9%. Limited information on drying suggests ⩽17% MC 65°C; >17% MC and seed crops 50°C. For very wet lots allow two passes. Seed is very resistant to airflow, so use shallow layers in bulk driers.

Yield of seed is variable, range usually 1.5–2.5, average 1.75 t/ha. Very good crops of new varieties may exceed 3.0 t/ha.

SUNFLOWERS

Grown for their oil for human consumption (uses include high grade cooking oil and margarine manufacture); residual meal is used for compounding. Widely grown in France where it has spread north rapidly with breeding of early maturing varieties; these are unsuitable for UK conditions. The crop is likely to remain non-viable in the UK until suitable early maturing varieties have been developed and tested. The main problem is conditions at harvesting, when a warm dry September–early October is required to dry the seeds. Crop requires desiccation and is direct combined with maize headers; combine needs adaptation to deal with the tough stems. A typical harvest moisture content is 25%.

Leguminous crops

This group includes:

(1) Beans *Vicia faba* type, i.e. field and broad beans. *Phaseolus* spp., i.e. dwarf or french beans, runner beans, navy beans and soya beans.
(2) Peas
(3) Other legumes including lupins, lucerne, sainfoin and vetches (tares).

Grown variously for mature and immature grain or pods for human and animal consumption, according to variety; lucerne, sainfoin, vetches and forage peas grown for ruminants.

Rotation

All are valuable break crops in cereal rotations, although amount of atmospheric N they 'fix' varies greatly. Lucerne and sainfoin leys, down for several years leave rich residues (ADAS N index 2) but peas and *V. faba* beans leave less (ADAS N index 1). *Phaseolus* spp. do not fix their own N unless inoculated with appropriate *Rhizobium* culture.

For reasons of pest and disease control, peas, field, broad, dwarf and runner beans, lupins, vetches and oilseed rape should be treated as a single crop. Peas, broad and field beans, lupins and vetches are host to pea cyst nematode (*Heterodera goetingiana*) and are also hosts to some soil-borne diseases attacking peas and dwarf beans, such as *Sclerotinia*, which also attacks oilseed rape; other brassicas can be affected but not, as yet, severely.

None of these crops should be grown more frequently than once in five years; if previous crops have been attacked, a longer interval is advisable.

Calculation of seedrate of all large seeded legumes

Either of the following two formulae may be used according to seed weights available

$$\text{Seedrate (kg/ha)} = \frac{10\,000 \times \text{required population/m}^2}{\text{no. of seeds/kg}}$$

$$\times \frac{100}{\%\ \text{germination}} \times \frac{100}{100 - \%\ \text{expected field loss}}$$

or

$$\frac{\text{Required population/m}^2 \times \text{TGW}^1 \times 100}{\%\ \text{germination} \times (100 - \%\ \text{expected field loss})}$$

[1] Thousand Grain or Seed Weight (g)

Appropriate population/m² and expected field losses are shown under individual crops.

LEGUMINOUS SEEDCROPS FOR HUMAN AND/OR ANIMAL CONSUMPTION

Beans

Field beans

Field beans are a valuable break crop in cereal rotations as cereal diseases are not perpetuated; it has been estimated that the crop is capable of fixing 200 kg/ha N from atmospheric nitrogen, leaving good reserves for the next crop. Only normal cereal equipment is needed and labour requirement is low.

Varieties

Winter and spring varieties are used. The latter includes (a) *tick beans*, small seeds (weight/1000 grains or seeds, i.e. TGW 280–560 g) and (b) horse beans (TGW 560–840 g) are flatter and similar in size to winter beans, which are all of the horse bean type. Winter beans ripen two to four weeks earlier than spring beans and usually give about 500–1000 kg/ha more grain. Recommended varieties are listed in Farmers Leaflet No. 10, *Varieties of Field Peas and Field Beans* (NIAB).

Climate and soil

Winter beans are more frost tolerant than spring varieties (the most frost resistant will stand up to − 15°C) but are rarely grown in Scotland and the north. Both types do best in areas where relatively mild winters are followed by warm dry springs and medium summer rainfall. Yield of spring beans is badly reduced by prolonged summer drought. Still sunny weather during flowering increases yield.

Soil pH should be above 6.5. Soils should be well drained but not liable to drying out in summer, to which spring beans are especially susceptible. Clay or chalk/clays are ideal. Avoid very light soils or those high in humus. Spring beans are grown on a wide range of soils; light soils permit early drilling.

Sowing

Beans require a seedbed similar to wheat, of adequate depth but not too fine. Complete *winter* sowings by the end of October, so that plants have three to four leaves before onset of frost.

Spring beans should be sown as soon after 15 February as possible to ensure rapid development before dry soil conditions occur. Drilling after mid March usually results in serious yield loss. Drill at least 75 mm deep if simazine is which must be applied within 7 d of sowing. Alternatively broadcast seed and plough in *shallowly*. Seed germinates satisfactorily at only 25 mm depth but herbicide choice is limited and risk of bird damage is increased.

Row width may vary from 150 to 500 mm and has little effect on yield of beans; some growers prefer 360 mm drills as this permits tractor working if inter-row cultivation or post emergence sprays are needed.

Prefer high germination seed (minimum 90%) which is tested for leaf and pod spot (*Ascochyta fabae*) and stem nematode (*Ditylenchus dipsaci*). Always sow seed dressed with an approved fungicide seed dressing.

Plant population and seedrate

Winter beans Optimum at establishment: 20–22 plants/m²; > 15/m² satisfactory if well distributed. Minimum after winter: > 10/m² if well distributed. Seedrate: 180–250 kg/ha according to seedsize to sow 25 seeds/m². Probable field losses %: average conditions 15% poor conditions, late sowing 25% or more.

Spring beans Optimum at establishment: 35–40 plants/m². Seedrate: tick beans: 180–220 kg/ha to sow 40–50 seeds/m². Horse beans: 230–250 kg/ha to sow 40–50 seeds/m². Probable field losses: average conditions 20%, poor conditions, very early sowing 30%.

Manuring

Beans of *V. faba* species, give no response to N, which should not be given in any circumstances, even with poor seedbeds. Rates of application for P_2O_5 and K_2O are given in *Table 5.30*. Broadcast dressings should be given after ploughing and cultivated deeply into the seedbed. Do not combine drill more than 50 kg/ha K_2O; K is unnecessary at index 2 on most clay soils.

Harvesting

Field beans are direct combined when the pods are black and dry and the haulm shrivelled. Shedding is not a problem; the

Table 5.30 Manuring of field beans (kg/ha)

P or K index	P_2O_5	K_2O
0	75	120
1	50	50
2	40[1]	40[1]
3	40[1]	Nil

ADAS figures.
[1] To maintain soil reserves; little if any yield response.

crop combines easily. Desiccate with diquat where crops for animal or pigeon feed are weed infested, unevenly ripened or spring crops are still green in late September. Use slow drum speed to avoid damage. Dry at low temperature (< 40°C) and ensure proper drying to below 15% MC before storage.

Diseases and pests

Winter beans are susceptible to chocolate spot (*Botrytis fabae* and *B. cinerea*). Spray with benomyl as soon as first symptoms appear. *Leaf spot* (*Aschochyta fabae*) exhibits similar symptoms – control by using only healthy seed checked for absence of disease.

Black bean aphid (*Aphid fabae*) is a serious pest of beans and can ruin spring sown crops. Inspect regularly before and during flowering. Spray with recommended aphicide as soon as first aphids are seen. Do not spray crops in flower, to avoid injury to bees.

Pollination

Yield of field beans is improved if visited by adequate numbers of bees during flowering. Wild bees are normally adequate for pollinating areas up to 4 ha, but for larger fields allow a further one hive/5 ha if bees are available. Take spraying precautions as given for oilseed rape (p. 116).

Yield

Very variable: winter beans, range 2.5–5.0 t/ha, spring beans, range 2.0–4.0 t/ha.

Markets

Compounding UK compounders now use the home grown crop; only a small proportion is exported.

Pigeons There is a good home and export market (especially Belgium) for tic beans only.

Export for human consumption Dry beans are a staple protein diet in the Middle and Far East, with a steady trade for top quality beans only; sent in 18 t containers.

Broad beans

Broad beans are grown for picking green for market or for processing, i.e. freezing or canning. As with peas, picking green is relatively unimportant. Harvest of broad beans for processing must be integrated with other crops in freezing and canning programmes. Maximum yields are obtained from early sowings but sowing dates must be arranged so that broad bean crops are ready between pea and dwarf bean harvest or before pea harvest.

Types and varieties

Uses of broad beans may be differentiated by their flower colours. Varieties with pigmented flowers include:

(1) *garden varieties*, winter sorts – Seville Longpod and Aquadulce Claudia and spring sorts – Windsor and Longpod varieties. Large seeded and used for fresh market or slicing young;
(2) *freezing varieties*, Ipro, Minica, Pax, Primeur and Primo. Dual purpose varieties have non pigmented (i.e. white) flowers and are suitable for *freezing* or *canning* and include: Polar (Autumn), Archie, Beryl, Feligreen, Rowena and Threefold white.

Flower colour is important, as coloured flowers indicate presence of leucoanthocyanidins, which cause discoloration of beans on canning or cooking. Winter sown varieties of broad beans tend to outyield spring sown sorts, but their simultaneous maturation with vining pea crops presents problems to processors unless they are geared to harvesting and processing as a separate operation. Varietal selection to suit agronomic and marketing requirements is thus of prime importance.

Distribution, soil, seedbed and sowing date

Apart from crops for picking green, must be in easy range of factory. Avoid extreme soil types. After ploughing use minimum cultivations necessary to produce an 'open' type of seedbed. Avoid very fine overworked seedbeds and compaction. Drill seed into soil moisture, usually 50 mm deep, but increase depth to 75 mm to avoid damage if simazine is used. Owing to large seed, not all drills are suitable. Consult manufacturer.

Broad beans are sown to reach required maturity immediately after last crops of vining peas, except where overwintered crops are harvested first. For processors requiring young freezing beans start sowing spring varieties in late April; for more mature crops for canning start sowing in early April. Successional sowings will then give supply of beans of correct maturity for rest of season. Overwintered varieties come to harvest just before vining peas. Sow all winter varieties in first half of October and Spring varieties for picking green in late February and throughout March, according to soil conditions. Seed should be dressed with approved fungicide against pre-emergence damping off.

Plant population and seedrate

Optimum population is 18 plants/m^2, drilled in rows 450 mm apart and 125 mm between seeds; may be reduced to 14 plants/m^2 if seed is very dear or on fertile silt or organic soils which are likely to produce extra strong vegetative growth. Use higher density on poorer soils, e.g. sands. Seed varies greatly in size, so it is essential to use formula on p. 142 to determine seedrate. Field loss; processing: Feb 7%, March–April 5%. Picking crops; autumn and early sown crops: allow same rate of field loss as field beans.

Manuring

Broad beans largely fix their own N and response is only obtained to N fertiliser on N deficient soils; crop very responsive to P and K in deficient situations (*Table 5.31*).

Table 5.31 Manuring of broad beans for marketing and processing (kg/ha)

N, P, K or Mg index	N	P$_2$O$_5$	K$_2$O	Mg Sand and light loams	Other soils
0	60	250	250	90	60
1	25	200	150	60	30
2	Nil	150	100	Nil	Nil
3		50	50		

ADAS figures.

Harvesting

Beans are windrowed with pea cutter, usually wilted for 12–14 h to ease threshing; then vined with mobile viner. Efficient vining and cleaning with minimal delay before processing is essential.

Pests and diseases

Largely similar to those of field beans.

Yield

Canning and freezing 3.0–5.0 t/ha.

Dwarf beans (green or french beans)

Dwarf beans are now a large scale farm crop grown on contract for processing, i.e. freezing, canning or artificial drying. Before introduction of efficient mechanical harvesters they were grown solely as a market garden crop and picked by hand for fresh sale.

Varieties

'Stringless' varieties, slender, smooth skinned and straight, now predominate. Varieties are numerous. Choice of variety, which is a most important factor influencing yield, quality of produce and efficiency of mechanical harvesting, is usually made by processor.

Seed

Seed is extremely fragile; skin is tender and easily cracked and attachment between embryo and cotyledons is slender and easily broken. Avoid rough handling of seed, e.g. dropping bags on hard floor, or germination will be adversely affected.

Climate, soil and cultivations

Dwarf beans are best suited to warmer areas of south and east England; further north only warm sheltered south facing slopes are suitable. Crop is very sensitive to soil conditions and best grown on deep loamy soils with good structure and ample organic matter. Avoid heavy, wet, or overdrained soils. Do not overcompact soil when preparing seedbed. To conserve moisture use minimum cultivations necessary and leave level surface for efficient harvesting. Drilling is delayed until soil temperature has risen to 10°C and is usually completed in first half of May.

Sowing

Rows 140 mm apart give the highest yields but row width is

determined by type of harvester available, many machines working on 450 or 600 mm row widths. Different harvesters are required for narrow rows. Optimal plant densities are 36 and 42 plants/m^2 for wide and narrow rows respectively, but varied according to cost of seed.

To calculate seedrate use formula on p. 142. Field loss usually 10%.

Seed dressed with an approved fungicide is drilled into the soil moisture; 25–40 mm is usually adequate but a depth of 50 m may be necessary in some situations.

Irrigation

Irrigation before flowering has little effect on yield, but water applied during flowering and early pod growth increases yield substantially.

Manuring

Present position is unsatisfactory, as adequate experimental evidence is lacking. Dwarf beans may now be inoculated with N fixing Rhizobium; unless this is done, heavy N applications are required, especially at N index 0, as crop does not fix its own N. However, P level can be increased at index 2 to build up soil P reserves (*Table 5.32*). All fertiliser is usually applied to seedbed, but a top dressing of 25 kg/ha N may help if N deficiency develops during growth.

Table 5.32 Manuring of dwarf beans for market and processing (kg/ha)

N, P, K or Mg index	N[1]	P$_2$O$_5$	K$_2$O	Mg Sands and light loams	Other soils
0	150	250	275	90	60
1	100	200	175	60	30
2	75	150	100	Nil	Nil
3	–	50	50		

ADAS figures.
[1] Reduce N application by 90 kg/ha N if Rhizobium inoculation is used.

Pests and diseases

Aphids (Aphis spp.) These cause distortion and stunting of plants and pods; treat with approved aphicide.

Anthracnose (Colletotrichum lindemuthianum) Produces angular brown spots in which holes appear on leaves, and sunken pink bordered brown lesion on pods. Most severe disease attacking dwarf beans. Apply approved systemic fungicide at full flower stage.

Grey mould fungus (Botrytis cinerea) Thrives in high humidity attacking dying or damaged tissue, e.g. scratched tips of pods touching soil. To prevent apply approved systemic fungicide at full flower stage of crop.

Halo blight (Pseudomonas phaseolicola) Produces brown spots in centre of large yellow spots on leaves, the halo, spots on pods and wilt and death of plants in wet humid weather. Remove affected plants in bags from field, burn and apply copper oxychloride or colloidal copper spray at once and at 10 d intervals until cured.

Yield

Average crop 7.5 t/ha.
Good crop 12.5 t/ha or more.

Runner beans

Runner beans are the most popular type of fresh green bean with consumers in the UK and are preferred to french beans as soon as they are available. Quality is all important: they must be fresh, straight and long with tiny seeds just starting to develop but not swelling, free from stringiness and well presented. Labour requirements for picking are high (total 400–450 man hours/ha) and they soon deteriorate. They are popular on 'pick your own' units. For wider marketing to be achieved the use of 'cool chain' facilities is required.

Choice of site

Choose a sheltered site, as runner beans are extremely vulnerable to wind when grown on string, wires or poles; growing behind shelterbelts or in the lee of woods or copses, provided they do not restrict the sunshine (*see* shelter-belts and windbreaks Chapter 7, p. 235). Not only does wind severely damage the crop but it also restricts the activity of bees, which are essential for pollination; smaller insects are unable to 'trip' the flowers. Failure to pollinate reduces the crop; close proximity of beehives is thus most advantageous.

Runner beans are mostly grown in southern England and Wales; in the north only the most favoured sites are appropriate. South-facing slopes are ideal.

Soil Crop is very sensitive to compaction, poor soil structure and waterlogging, so soil must be well drained and free-working. Light to medium textured soils with a high water-holding capacity are ideal.

Soil preparation, cultivations and manuring

Subsoil if necessary to ensure that roots can penetrate deeply. A heavy dressing of well rotted farmyard manure at rates up to 75 t/ha is well worthwhile. Prepare a fairly fine friable tilth. Ensure soil is at least pH 6.5 (5.8 on peat soils); apply magnesian limestone if both soil pH and magnesium levels are low. Apply fertiliser as shown in *Table 5.33*. Like french beans, runner beans do not fix their own nitrogen, which must be supplied in the fertiliser. If crop is pale or of poor appearance at early picking stage give 75 kg/ha N as a side dressing.

Table 5.33 Manuring of runner beans for picking green (kg/ha)

N, P, K or Mg index	N	P$_2$O$_5$	K$_2$O	Mg Sandy soils	Other soils
0	150	250	275	90	60
1	100	200	175	60	30
2	75	150	100	Nil	Nil
3	–	50	50	Nil	Nil

ADAS figures.

Varieties

Most varieties have red flowers but pink and white flowering varieties exist. Market requires long straight smooth deep green pods free from stringiness; 'stringless' varieties are now available but these must still be picked while tender. Yield must be high. Short podded varieties are ready to harvest 10–14 d before the long podded sorts, but are only suitable for early and 'pinched' crops (*see* Systems of Production, Early Crops below).

Systems of production

Early crops These are usually sown with short podded varieties and are usually 'dwarfed' by pinching out the growing point ('pinched crops') 0.4 m above ground level and subsequently as required. Such crops are earlier than supported crops but yield is lower. The beans, hanging near the ground, may become bent and dirtied by soil. Early pinched crops are usually cleared by early August, allowing time for another crop to be planted.

Maincrops These are the most important and only long podded varieties are suitable. Maincrops are grown on supports. There are various systems, which rely on either canes, wires or strings. Some training is required in the initial stages of growth to ensure that the stems wind round the supports; once they have done so they look after themselves. The mainstems require some 'stopping' once the growing points are 0.5 m above the top of the supports to encourage the plants to produce lateral stems lower down.

Sowing

For *earliest* and *protected* crops plants may be raised under glass in 30–40 mm blocks or containers. Plant one seed per block. Plants are ready for transplanting when the first two true leaves are fully expanded, but ensure plants are properly hardened off and protect against frost or wait till danger is past, usually late April on sheltered sites in southern England, May elsewhere. Seed may also be 'chitted' for 2 d in layers of damp peat and then sown, taking care not to damage the growing point.

Maincrops When sowing directly into the field without protection, wait until soil temperature at 100 mm deep has reached 10–12°C, usually during early May according to site. Before this, germination and growth is often slow and patchy and protection is advisable.

Field protection is particularly valuable for enhancing the earliness of runner beans. Low continuous tunnels give a slightly earlier crop than surface mulches. Both are suitable for pinched and supported crops. Low continuous tunnels cover one row of pinched beans spaced 1 m apart; for supported crops a wider tunnel is used for a double row of beans. Ventilation is gradually increased until the crop has fully emerged as flowering starts; the tunnel is slit along the top and plants allowed to emerge. The cover is removed shortly afterward to allow watering and avoid damage. Completion of production under protection is very difficult, as protection restricts the access of bees for pollination.

Plant population and arrangement

Pinched crops are sown 50–70 mm deep in rows 0.75–1.2 m apart, either by hand with dibber or by planting machine which opens a furrow. Some makes of drill will cope. Seeds are placed 50 mm apart for earliest crops, up to 150 mm later.

Supported crops are often planted after canes or strings have been erected and seeds are then placed to coincide with supports. Yield tends to increase with plant density but support beyond a certain limit is costly and uneconomic. With *cane supports* low density crops, 84 000 plants/ha give the best return: allow 0.9 m paths with canes in double rows 0.6 m apart and 0.3 m apart in row, with two plants/cane. If there is no restriction on area available, cane spacing within row is increased to 0.6 m giving 42 000 plants/ha.

Systems using posts with string or wire supports are cheaper and justify high density crops of around 100 000 plants/ha, e.g. use twin rows 0.3 m apart at 1.5 m centres and plants spaced 130 mm in row; paths are 1.2 m wide.

Irrigation

Runner beans respond strongly to watering at the flowering and fruiting stages of growth. Give the first application of water just before the colour of the first flower petals is exposed, returning soil to field capacity. Rate of application depends on soil type (ADAS, 1982):

Soil classification	Critical SMD[1]	Start SMD	Apply water (mm)
(A) Max. 40 mm available water/300 mm depth	25	20	25
(B) > 40–60 mm available water/300 mm depth	50	40	50
(C) > 60 mm available water/300 mm depth	75	65	50

(ADAS figures).
[1] Soil moisture deficit

On retentive soils (Classes B and C) leave a small part of deficit unsatisfied in case rain falls soon after. Pinched crops are irrigated by sprayline or rotary sprinklers set on risers 0.45 m above crop; supported crops are irrigated by rotary sprinklers set on supporting posts or on risers 0.45 m above top of supports.

Picking and marketing

Crops are picked by hand, the harvest period lasting 10–14 weeks. Regular picking (three to four times/week) is essential to maintain quality. Remove and discard all overstood beans, as these retard the production of new beans; regular picking is thus essential, even if the market is weak. Beans from early crops must be at least 125 mm long, maincrops 175 mm, with a maximum of 350 mm; they must be tender and not stringy or coarse. Select for quality as picking proceeds; reject and discard any low quality beans or those missed in a previous pick. Beans must be clean and attractively presented, all facing one way. Early beans may be packed in 3.6–5.5 kg cardboard boxes; maincrops in larger packs 12.7–13.6 kg or in small weighed packs as specified by supermarkets. Remove from field, cool quickly and forward by cool chain. Beans may be kept in cool store at air temperature 4.4°C and relative humidity 90–95% for up to 7 d.

Yield

Yields are extremely variable. Range: early crops 7–17 t/ha; maincrops 24–35 t/ha.

Peas

Vining peas are grown in large blocks in the eastern counties, within easy range of the processing plants. Combining peas are also important eastern counties crops but with the encouragement of the EEC. Animal Feed and Human Consumption Schemes are now widely grown in other arable areas. Forage peas are unimportant. Peas, which leave behind considerable N residues, are an excellent cereal break crop.

Types and varieties

There are at least three quite distinct ranges of product: 'green' peas, combining peas and forage peas.

Green or 'garden' peas

Peas in this group are harvested in an immature state and must be fresh and sweet. As with sweet corn, the immature seeds contain a high proportion of sugars when freshly harvested at the correct stage; if harvest is delayed beyond this, the skins become increasingly tough, the core starchy and sweetness disappears. Delay in marketing or processing has a similar effect.

Garden peas for picking green for market

Nowadays much the least important outlet. Area has declined to a small fraction of that thirty or more years ago, owing to high quality, reliability and 'ready to cook' packs of frozen peas avoiding the labour and waste of shelling. There is little demand for 'fresh' picked peas, as quality is often unreliable due to staleness, flats, maggots and previous use of low quality early round seeded varieties.

Pods are carried in groups up the stem, maturing sequentially from the lower portion of the stem upwards and are usually ready to pick over a period of 10 days. Varieties have white flowers with either round or wrinkled seed. The latter are of far higher quality and sweet; round seeded varieties are usually unsweet. Pods should be picked before they are fully filled and seeds young and tender. Problem is to maintain freshness. Cool chain could help where demand existed. They can also be grown for 'pick your own'. Some growers produce seed crops for the retail garden packet trade.

Varieties

Varieties may be divided into first earlies (a) round seeded types, which are hardier and can be sown in autumn or very early spring, although quality is very poor (b) wrinkle seeded types, less hardy but very high quality. *Second earlies*, e.g. Early Onward. *Maincrops*: Onward, Hurst Greenshaft; the latter two types are wrinkle seeded.

Contract growing for vining Shelled green by mobile viners while young and tender and transported rapidly to factory for processing into frozen, canned or dehydrated *garden peas*. This type of pea is entirely confined to the eastern counties, as all processing plants are in the east. Processors require minimum blocks of 120 ha for vining. Factories require an even flow of high quality peas for a long period during growing season and consequently contract with numerous growers to exploit climate and soil type with a succession of sowing dates of varieties of differing maturity.

Special varieties of garden type with white flowers and wrinkled seed are used but pods are carried at top of plant and mature simultaneously. For high quality, peas must be harvested at correct stage of maturity, which is decided by means of a *tenderometer*. For freezing a tenderometer reading (tr), 100–105 is usual, but canning peas are harvested more mature, (tr) about 110–115; yield is then slightly higher. The ideal variety is of upright habit with haulm of medium length (about 0.6 m), resistant to diseases such as downy mildew (*Peronospora viciae*) and fusarium wilt (*Fusarium oxysporum*) and herbicides to be used. Pods must mature uniformly and be thin to allow shelling without excessive beating and damage to peas. Peas should not be prone to splitting and be of uniform size, moderately dark but bright green in colour, with good fresh pea flavour. Texture should be smooth, even, skin not tough. Varietal lists are published by Processors and Growers Research Organisation (PGRO), The Research Station, Thornhaugh, Peterborough, together with other technical literature.

Varieties may be divided into: *first earlies* which include earliest peas that can be successfully grown and harvested. Short straw, relatively low yielding but allow early start to vining. Early maincrops and maincrops form the bulk of the vining pea area; late varieties are used largely to extend the drilling season. Other varieties are grown to produce *petits pois* (very small green peas). Pea haulm makes good silage if it is clean and allowed to wilt; otherwise it can be ploughed in as green manure.

Edible podded peas These are special varieties where the whole pod is cooked and eaten. Interest and popularity is increasing. '*Flat podded*' varieties which are picked before pod swelling are very suitable for the UK fresh market; they are also available in frozen form.

'*Full podded*' or *sugar snap* varieties have thick fleshy pods which are picked at the full pod stage but pod and seeds are still sweet and tender. This type is potentially suitable for machine harvesting. Stringless and semi-stringless varieties exist. Cool chain handling will be necessary to exploit the fresh market potential of all edible podded peas.

Combining peas, which are grown to maturity and threshed with a combine harvester, include all peas harvested dry, whether they are 'dried' peas grown for human consumption or 'field', 'feeding' or the so-called 'protein' peas grown for inclusion in compound animal feeds. The actual use to which many 'combined' pea crops will be put cannot be guaranteed at sowing time, as it depends on the quality of the crop and the state of the market; frequently samples of types and varieties suitable for human consumption are sent for compounding owing to staining, poor quality, damage or oversupply of a particular market.

Varieties can be divided into two main groups according to potential use. *White flowered varieties* are suitable for human consumption if of high enough quality. Varieties with *purple flowers* are only used for animal consumption.

Marrowfats Large seed (TGW 310–360) seed coat blue-green flat round dimpled and dented. Medium straw length (60–80 cm) and suited to a wide range of soil types. Relatively late maturing. Varieties in this group are by far the most important for human consumption, being used for dry packet trade and canning as 'processed' peas; their high yield also makes them suitable for growing for compounding.

White peas (TGW 250–350) seed coat white-yellow, round and smooth. Medium length straw (50–70 cm) and give high yields on a wide range of soil types. Most varieties mature about 7 d before marrowfats. Primarily grown for inclusion in compound animal feeds, but also used as split peas, used in the manufacture of soups and prepared meals, canned 'pease pudding' and mushy peas.

Blue peas have blue-green seed coats. *Small blues* (TGW 200–225), early maturing with short-medium length straw, short strawed varieties being most suited to fertile soils. Good quality samples suitable for canning as small processed peas; otherwise used for animal feed. *Large blues* (TGW 290–310), round and smooth. Mature 5–7 d after marrowfats. Short strawed and mainly restricted to fertile soils. Small market for dried packet trade and splitting; main outlet for stockfeed.

Purple flowered peas, marrowfat type. Large dimpled grain (TGW 380–400), colour olive green to brown. Similar in yield, straw length and maturity to white flowered marrowfats. Use: compounding.

Recommended varieties are listed in Farmers Leaflet No. 10, *Varieties of Field Peas and Field Beans* (NIAB).

Maple peas Seed round, dark brown and mottled, e.g. Minerva. A traditional long strawed variety, only really suitable to light soils, which tend to reduce straw length. Can then produce acceptable yields. Main market for pigeon feed. Traditionally also cut green for fodder.

Choice of variety for combining Consider the following points: (i) *Market*. Not all varieties are acceptable for human consumption; those that are suitable often do not yield as heavily but carry a useful premium if quality is high enough. Only assume such a sample will be saleable for human consumption if you have a contract. (ii) *Straw length*. Select variety to suit your soil, although varietal behaviour is modified by rainfall, soil moisture content and fertility. Inherently short strawed varieties are at their best on deep fertile soils, but can be difficult to combine if lodged on stony soils. In wet seasons, longer strawed heavy haulmed varieties produce excessive growth, lodge early and the haulm usually rots. (iii) *Standing capacity* is particularly important in a wet harvest. Not all semi-leafless varieties stand well while others do and are easy to combine. Standing capacity is of equal importance to leafless or semi-leaflessness. (iv) *Maturity*. Select early varieties for north of England, notably small blues and early whites; marrowfats are usually too late maturing.

Reduced foliage peas Conventional varieties of peas have leaflets, stipules and small tendrils. Their large leaf area is more than enough for satisfactory seed production and has the disadvantages of increasing lodging, reducing air circulation through the crop with attendant disease and pale seed risk and increasing harvest difficulties, especially in a wet year. 'Leafless' varieties have had their leaflets eliminated by breeding and replaced by increased tendril growth and stipules reduced in size; in 'semi-leafless' varieties the leaflets have been replaced by larger tendrils. There are now high yielding varieties of semi-leafless pea which stand well and are easy to harvest.

Forage peas Include varieties selected for their ability to produce a high yield of haulm. Widely cultivated in Europe, have been grown commercially in UK since 1975. Flowers variegated, pink, maroon, red or mixed with white, seed mottled, or brown. May be sown alone or with cereals.

Suggested as a nurse crop for Italian ryegrass, but may form a dense 'smother' and kill young seeds before harvest under 'growthy' conditions. There are numerous varieties. Early varieties ready for cutting in 12 weeks, late varieties take 15 weeks, giving higher yield in longer growing season.

Seed quality

Certified seed has a minimum level of 80% germination, but there is no minimum standard for infection with the seedborne disease, leaf and pod spot (*Ascochyta pisi*) and (*Micosphaerella pinodes*), except for seed passing through the NIAB Quality Pea Seed Scheme. Always have seed tested and if level of leaf and pod spot (*A. pisi* and *M. pinodes*) is ≥ 5% treat with a prescribed fungicidal seed protectant. PGRO (Processors and Growers Research Organisation), The Research Station, Thornhaugh, Peterborough, PE8 6HJ, UK provides a complete seed testing service for members including tests for germination, presence of *Ascochyta pisi* and associated organisms and seed vigour (Electrical Conductivity Test for Vining Pea Seed). Vining peas are especially susceptible to the effects of low vigour. For early sowing use only high vigour seed (up to 24μ). Medium vigour seed ($25–29 \mu$) is less reliable for early drilling but fully satisfactory for later sowings; low vigour seed ($30–34 \mu$) is unsuitable for early sowing and may fail in cold wet conditions. Very low vigour seed ($> 43 \mu$) should never be sown. The seed of small and large blues and marrowfats seldom suffer from seedbed losses due to low vigour, so for these types the test is considered to be irrelevant.

Soil and cultivation

Peas suit a wide range of soils provided they are well drained, of good structure and texture, friable, and with depth of mould to allow easy root penetration. Cold wet conditions, raw soil, capping or compaction result in failure. Avoid very light sands and heavy wet clay. A good mouldy seedbed is produced by ploughing early, giving maximum weathering by frost; prepare seed bed with minimal working – a single stroke of the harrows is frequently adequate. Use of pre-emergence residual herbicides requires reasonably fine seedbed.

Sowing

Sowing date Round seeded varieties for *picking green* may be sown in October or February; other sowings are made until end of April, according to variety. Sowing date of *vining peas* is decided by processor, in order to achieve succession of crops for harvesting on planned dates; product deteriorates rapidly if 'overstood'. Combining peas must be sown early for maximum yield. Sow late February or by end of first week in March; thereafter yield falls by 125 kg/ha per week of delay so do not wait for ideal conditions. Early sown peas suffer less from pests and mature earlier, usually in better weather.

Seedrate It is essential to achieve the right population. Optimum depends on type of pea and varies according to cost of seed and value of produce; high value produce and low cost seed justify higher plant populations. Lower densities can be used where vigorous vegetative growth is likely; for poor growth, increase density. Aim for the following target populations: (plants/m^2).

Vining peas Average 90; range 77–100 according to seed cost and value of produce.

Combining peas

	Normal	Lower priced seed
Marrowfats	55	65
Large blues or whites	60	70
Small blues	85	95

Use formula on p. 142 to calculate seedrate. Expected field losses are given in *Table 5.34*.

Drilling and rolling

Seed should be covered with at least 30 mm of settled soil after rolling. It is essential to roll in order to press any stones into the soil surface to avoid damage to harvesting machinery and to break down clods; a smooth surface is also essential for effective pre-emergence weed control. Where possible, rolling should be done shortly after sowing; where this is impracticable roll when the crop is over 5 cm high – the stems are too brittle before this stage. Most cereal drills are suitable for peas.

Peas for livestock feeding

Sowing date Forage peas are best sown in late March or early April. July sowings give faster initial growth but poor autumn bulk, low dry matter and are not recommended. Sow peas for grain as early as seedbed conditions permit.

Seedrate If requisite information is available, use formula for peas for human consumption, otherwise *drill forage peas*: small seeded varieties 150 kg/ha, large seeded varieties 175 kg/ha *or* 90 kg/ha peas with 60 kg/ha oats or barley; do *not* exceed given rate for cereal *grain*. Drill 125–160 kg/ha according to size of pea grain. Drill both types 30–50 mm deep with corn drill in rows 200 mm apart or less; do *not* broadcast.

Manuring

Peas do not respond to N, which is unnecessary even on poor seedbeds. Rates of application for P_2O_5 and K_2O are shown in *Table 5.35*. Broadcast fertiliser on top of ploughing and work in deeply. Do not combine drill more than 50 kg/ha K_2O, K is unnecessary on most clay soils at index 2.

Table 5.35 Manuring of peas for vining, drying and forage (kg/ha)

P or K index	P_2O_5	K_2O
0	50	150
1	25	50
2	Nil	40[1]
> 2	Nil	Nil

PGRO figures
[1] Combine drilled only.

Harvesting

Vining peas

Date of cutting is decided by contracting firm on tenderometer reading. Peas are harvested by minimum intake viner; previously cut into windrows with pea windrower and shelled in the field by mobile viner; the peas are then rushed to the freezing plant, since they perish rapidly once shelled. Peas must be free from maggots and impurities like poppy heads and mayweed, which cannot be removed, and toxic berries such as black nightshade and bryony. Efficient weed control is thus essential.

Combining peas

Peas for animal feed Provided that the crop is not mouldy, quality is of no importance to compounders; cash returns are largely determined by yield.

Peas for human consumption Much more care must be taken when harvesting peas for the packet and canning trades. The price paid for a parcel depends largely on its colour and freedom from defects. Peas to be sold loose or in packets should be dark green while those for canning as 'processed peas' should be of a lighter but even colour. Quality is much affected by weather at harvest. Lodged crops become stained, while value of the crop is reduced for the packet trade if bleached in the sun.

Peas are normally direct combined when moisture content of the mature seed is < 25%. Higher quality in peas for human consumption is often obtained by combining at 20–25% MC followed by careful drying. At low moisture contents peas are inclined to split; this does not matter in samples destined for animal consumption, which can be combined at < 20% MC to reduce drying costs. *Peas for human consumption* must not be contaminated with dust or

Table 5.34 Expected field loss % in vining and combining peas – sowing time

Type of pea	Very early (Feb) Heavy/poorly drained	Free draining	Early (March) Heavy/poorly drained	Free draining	Mid-season (April)	Late (May–June)
Vining and small blues	25	20	20	15	10[1]	5[1]
Whites and large blues	23	18	18	13	[2]	[2]
Marrowfats	20	15	15	10	[2]	[2]

[1] Vining peas only – too late to sow small blues
[2] Too late to sow

soil. Avoid combining when there is surface moisture on the haulm and ensure the combine is clean. Careful setting of the combine with drum speed reduced as low as possible is necessary to avoid damaging the crop. Choose appropriate screen size for the size of peas. Combine in the opposite direction of the lodging so that the tips of the plants enter the combine first.

Desiccation

An approved chemical, e.g. diquat, is applied to the crop to kill green weeds and crop foliage and accelerate maturity. Combining can take place 2–7 d after application providing desiccation is complete and according to weather. Apply when moisture content of the pea seeds has fallen to 40–45% and the crop has turned yellow, with the lowest pods brown and dry, the upper ones yellow. The seeds are easily detached from their stalks and rubbery. *Do not* desiccate too early or yield and quality may be reduced. Desiccation should be reserved for weedy situations and difficult harvest weather; direct combining without a desiccant is preferable wherever practicable.

When *drying* great care is needed to avoid damage. Do not exceed the following drier temperatures:

	Moisture content of peas	
	< 24%	*24% and over*
Seed peas	43°C	38°C
Consumption peas	49°C	43°C

When double drying allow at least 2 d between dryings for MC to even out. For storage in bulk or closely stacked bags dry down to 15% MC for winter storage; 17% MC is adequate for storage up to four weeks; 1% higher MC is permissible for storage in bags fully exposed to air, in bulk ventilated by forced draught or turned frequently.

Forage peas

Cut early varieties at podding, when dry matter yield reaches peak (about 12 weeks) or senescence and leaf loss occur, reducing yield and quality. Late varieties flower and produce pods over a longer period and do not suddenly senesce; date of cutting is less critical. Lodging likely by harvest; avoid setting cutter too low or soil contamination of produce may result.

Pests, diseases and disorders

Pea weevil (*Sitona lineatus*) needs treatment if large numbers appear when plants are small on cloddy seedbeds in slow growing conditions. *Pea aphids* (*Acyrthosiphon pisum*) do severe damage in large numbers; control before build up in warm weather.

Pea moth (*Cydia nigricana*) causes maggoty peas; larvae feed on peas in pod and severely reduce value. Use 'Oecos pheromone' pea moth trapping system to assess need and timing of treatment. Consult ADAS for further information.

Fungus diseases include pea wilt (*Fusarium oxysporum*) and downy mildew (*Peronospora viciae*); sow resistant varieties. Foot rot (*Aschochyta pinodella*) worst on heavy soils. Improve drainage, avoid compaction. Leaf and pod spot (*A. pisi* and *Mycosphaerella pinodes*): sow disease free seed.

Marsh spot caused by deficiency or non availability of manganese on organic or alkaline soils, yellowing around leaf margins and between veins appears. If untreated, brown spots appear in centres of peas, reducing value. Apply 11.4 kg/ha manganese sulphate HV with wetting agent on appearance of symptoms, again if they reappear: to prevent spot formation apply also at full flower and again 7 d later.

Yield

	t/ha
Picking green (market) Early	7.5
Vining	
First early	3 to 6
Second early	3.5 to 6
Early maincrop	4 to 9
Maincrop	3.5 to 7
Dry	
Marrowfat	2 to 6
Small blue	2 to 4
Large blue	2 to 6
Whites	3 to 6
Forage peas (cut green)	
Dry matter	3.5 to 6
Maple and other	2.5 to 3

Lupins

Lupins are extensively grown for the production of their protein rich grain in Australia and the USSR but are still in an early stage of commerical development in the UK. Blue lupins have been used for green manuring in the UK but as a grain crop lupins are new. Crop is included in EEC subsidy scheme with peas and beans. Lupins are most suited to sheltered south facing situations in southern England (*cf.* forage maize). Appropriate soil type depends on species.

Types and varieties

Only varieties with a low alkaloid content in the seed are suitable for grain production, i.e. sweet lupins, as alkaloids are toxic. The first alkaloid free varieties were bred some 50 years ago. *White lupin* (*Lupinus albus*) is the preferred species having large protein rich seeds and requiring fertile loamy soils with good drainage and pH below 7.5. *Yellow lupin* (*L. luteus*) is better suited to less fertile, more acid soils. Lower yield but has a higher oil content and is earlier than white lupin. *Blue or narrow-leaved lupin* (*L. augustifolius*) gives lower yield and prefers moderately acid to neutral soils.

Culture   Seed requires inoculation with appropriate type of *B. radicicola* before sowing. Plough in good time to prepare a friable tilth, ensuring absence of compaction. Sow late March–early April with cereal drill 25–35 mm deep in rows 120–170 mm apart. Optimum plant population at establishment 50–60/m^2, minimum 30–40/m^2. Sow 185–300 kg/ha according to seedsize which varies greatly, to give 70 seeds/m^2. Lupins are not responsive to fertiliser. N is never applied and only 40 kg/ha each of P_2O_5 and K_2O is recommended as a maintenance dressing.

Harvesting   Lupins are direct combined but are not ready until late September–early October, when conditions are often difficult. Desiccate, as for combining peas, only when essential but not before pods are opaque. Threshing similar to beans. Probable harvest MC 20%; safe storage MC 14–15%.

Market and yield Lupins are suitable for compounding for livestock feed, but a contract is essential. Composition: must contain < 5% bitter grains; typical protein 32–35%, oil 9%. Yield is extremely variable, range 1–4 t/ha.

Miscellaneous combinable legumes

Fenugreek (*Trigonella foenum graecum*) An annual traditionally grown in the Mediterranean, Middle and Far Eastern regions. Variously used for production of spice, crushing for vegetable oil, protein and raw steroid for the pharmaceutical industry. Has not as yet achieved any commercial production in the UK and development work appears to have ceased.

Navy beans The dried beans are used to manufacture 'baked beans'; the entire UK requirement is imported, only high quality samples being acceptable to the processors. As yet there are no varieties with sufficient cold tolerance for UK conditions and the crop cannot compete with imports for quality or cheapness.

Soya beans These are the major protein crop traded on the world market. All UK requirements are imported. Attempts have been made to grow this crop in the UK but it is quite unsuitable as it is extremely sensitive to cold in the seedling stage and requires a warmer climate, at least comparable to grain maize. Development of sufficiently cold tolerant varieties appears to be unlikely. Combining peas and beans are the most appropriate protein crops for culture in the UK.

LEGUMES GROWN FOR FODDER ONLY

Lucerne

A native of Asia Minor, lucerne is now cultivated world wide; known as alfalfa in the USA. A deep taprooted legume, it is highly drought resistant and fixes large amounts of atmospheric nitrogen to provide heavy yields of protein rich fodder and rich nitrogenous residues; well managed crops yield up to 10.5 t/ha or even more of dry matter. A perennial crop, it normally occupies the land for four years.

Essentially a cutting crop, it has been grown widely for drying and makes excellent silage if it is wilted first. Difficult to make into hay by conventional methods as stem is difficult to dry and leaf shatters readily; well suited to barn hay drying. Grazing only taken on a limited basis. Lucerne requires very different treatment from grass and clover crops. Management is rather specialised and new growers should take advice first. A sun loving plant, it is especially suited to drier warmer summers of southern and eastern counties but withstands hard winters well. May grow satisfactorily in higher rainfall areas but grass is easier to manage. The crop tolerates soil types ranging from light sand to clay but cannot stand poor drainage or acidity, requiring a pH of 6.5–7.0 in the top soil and a subsoil minimum pH of 6.0 to at least 30 cm deep. Soil profile must allow deep root penetration; useless on unfissured rock or where pans exist but excellent on chalky soils.

Varieties and seeds mixtures

The once popular midseason type *Provence* is now superseded by varieties of the early or *Flanders* type, which give the earliest spring growth, longest growing season and highest yield. Variety Europe has consistently given highest yields but where either *verticillium* or bacterial wilts or stem nematode (*Ditylenchus* spp.) are known to occur, resistant varieties should be chosen, even if of lower yielding capacity. Varieties are listed in Farmers Leaflet No. 4, *Varieties of Herbage Legumes* (NIAB). Lucerne varieties are not mixed together in seeds mixtures.

Lucerne may be grown pure or sown with a companion grass; not grown with clovers. Pure stands are preferable for drying or where grassy weeds are controlled chemically. Otherwise the inclusion of a companion grass helps to suppress weeds, gives variety to the herbage and is said to make hay and silage making easier. Lucerne does not thrive in a dense grass sward so the grass chosen should be a minimum space demander and a similar growth rhythm to lucerne. Cocksfoot is only suitable in close drilled stands on light dry soils or dry areas and requires hard winter grazing to control it. Meadow fescue is preferable for general use. Do not mix with lucerne seed but cross drill to avoid close competition with the young lucerne plants. Drill 13 kg/ha lucerne seed 12 mm deep in rows 100–180 mm apart; cross drill 1.2 kg cocksfoot or 3.5–4.5 kg meadow fescue if required.

Seed inoculation

Lucerne fixes atmospheric nitrogen only if effective nodules essential for vigorous growth are formed on the roots by the appropriate nodule bacterium *Rhizobium meliloti*; this is usually not present where lucerne has not been grown before. Inoculation of seed with correct nodule bacterium is then essential and may only safely be omitted if field has recently grown lucerne. Methods include a slurry peat inoculum in water applied on the day of sowing; seed is allowed to dry before drilling. Inoculation may also be done by pelleting up to two weeks before drilling.

Establishment and seedling management

Lucerne is best drilled without a cover crop on a superfine firm level seedbed after soil has warmed up and several crops of weed seedlings killed by harrowing, between late April and mid July; midsummer sowings are usually best, as growth is very rapid in warm soil. Avoid early spring sowings, as germination and growth are slow and weed competition severe. Crops may also be undersown in cereals, when a silage crop removed in July is preferable to a grain crop; use reduced seedrates of still strawed barley variety, 100 kg/ha.

Young plants each develop initially from a slender single stem and do not form a cover dense enough to suppress vigorous early weed growth, so treatment with a suitable herbicide is normally advisable. If weeds are mown, set cutter *above* growing tips of young lucerne. Allow flowering to begin before removing first crop or allow growth to die back in autumn.

Management of established stands

Allow a large bulk to develop before cutting. First cut, about mid May, should be taken as soon as flower buds appear, subsequent cuts not later than opening of first flowers. Further delay seriously reduces crop digestibility and retards next cut. Cutting too early weakens stand; overfrequent cuts kill it. On strong stands second or third cuts can be taken before recommended stage without damage in favoured

Table 5.36 Manuring of lucerne

	N index			P index				K index			
	0	1	2	0	1	2	3	0	1	2	3
Seedbed	25	Nil	Nil	120	80	50	30	120	80	50	Nil
Conservation											
P_2O_5 per cut				100	80	50	30				
K_2O first cut[1]								150	120	90	30
K_2O second[1] and subsequent cuts								120	90	60	30

ADAS figures.

[1] On some K-rich clay soils, K application may be reduced by 50 kg/ha K_2O *per annum* at K index 2 and omitted at K index 3.

areas. If grazed, *under no circumstances* must young regrowth be eaten; always use a back fence close to the grazing.

Autumn management is critical. Rest for eight weeks in autumn from about late August to end of October to allow build up of root reserves. Crops may then be grazed or cut while green or allowed to die back. Companion grasses are grazed heavily in winter, especially cocksfoot, but remove stock before spring growth starts. If needed mechanical treatments loosen a compacted surface but complete while lucerne is dormant.

Herbicides are more reliable for controlling heavy growth of weed grasses. Paraquat is effective for controlling creeping grasses and checking green growth.

Manuring

Check soil pH and apply lime if necessary.

Seedbed dressings (*Table 5.36*)

This should be worked in prior to sowing. Nitrogen is only worthwhile for grass mixtures or soils of ADAS N index 0. If recent analysis is not available use highest rates of P_2O_5 and K_2O.

Maintenance dressings (*Table 5.36*)

Lucerne gives no response to fertiliser N once established but demands a liberal supply of K_2O. Deficiency must not be allowed to develop for it is difficult to cure. A crescent of white or yellow spots around tips of leaflets, yellowing of lower leaves and stunting of the plants, often in patches, indicates K_2O deficiency. A single cut of young lucerne, 19 t/ha fresh material, removes some 120 kg/ha K_2O, while a crop of total production 10.5 t/ha dry matter removes 190–225 kg/ha K_2O.

Sainfoin

Also known as St Foin, Cockshead or Holy Grass, sainfoin was once grown extensively on chalk and limestone soils, especially in the Cotswolds, Hampshire and around Newmarket; now rarely grown. Like lucerne, sainfoin has a deep taproot, is highly drought resistant and leaves valuable N residues for subsequent cereal crops. Digestibility, palatability, protein content and high feeding value, long appreciated, have now been confirmed by modern research. Well made sainfoin hay has been much in demand for race horses:

aftermath is invaluable for fattening lambs. Yield of British stocks, about 7500 kg/ha dry matter is well below that obtainable from lucerne; some European stocks of sainfoin appear to do considerably better. Cut for seed, 625–750 kg/ha of seed in husk is obtained.

Sainfoin is very suitable for hay or silage making, with aftermath grazing, on light dry soils, provided they are well supplied with lime and have pH 6.5. Primarily suited to well drained soils in warmer dry climates of the south, sainfoin fails utterly on cold wet soils.

Varieties

There are two main types, *Common* and *Giant*. Common sainfoin is truly perennial and once remained down 15–20 years, later shortened to about four years according to weed incidence. Flowering in late May or June gives a single cut of hay but remains prostrate thereafter and is best suited to aftermath grazing. Giant sainfoin establishes quickly but only persists for about two years, giving two cuts annually.

Seedbed, sowing and seedrate

Sainfoin replaces red clover in the rotation and is undersown in spring barley. Clean land and fine firm seedbed are essential. Drill at right angles across the rows of barley in rows 100–180 mm apart and 12 mm deep from March to early May; deep drilling leads to poor establishment; ground must be rolled after drilling.

Seed may be obtained in husk (unmilled) or husk removed (milled); the latter contains fewer hard seeds, giving better germination and is free from empty husks and weed seeds. Only fresh coloured light brown plump seed of high germination should be bought; black shrivelled seed is old or harvested badly.

Seedrate 65 kg/ha for milled and 125 kg/ha for seed in husk. Giant sainfoin is sown as a pure stand while common may be sown pure or mixed with grass, the latter giving higher yields and less weed. Either 7 kg/ha meadow fescue or 3.5 kg/ha timothy may be included, each with 1 kg/ha giant white clover. Cocksfoot at 1 kg/ha may be considered on very poor soils but is very aggressive unless heavily grazed in winter.

Manuring

Little UK experimental evidence exists for recommendations. Overseas work indicates that sainfoin is able to utilise P_2O_5 of very low availability and does not respond to applications of either P_2O_5 or K_2O. Up to 125 kg/ha N given in first harvest year has been shown to increase dry matter yield. As a precaution, give 60 kg/ha each of P_2O_5 and K_2O in seedbed and an annual maintenance dressing of 40 kg of each. Where the sward contains substantial proportions of grass or clover, higher applications of K_2O appear desirable. A little N early in the life of the plant increases nodulation.

Management and conservation

Cut at flowering bud stage or quality is seriously reduced. Like lucerne, sainfoin hay requires careful handling in later stages and is well suited to 'barn drying'. Crimping immediately after cutting bruises the fleshy stems and speeds up drying. Start grazing *before* flower buds appear. Quick defoliation is best, as in folding; use a back fence and avoid overgrazing. Allow adequate autumn regrowth for plants to build up carbohydrate reserves in roots.

Vetches or tares

Fodder

Winter and spring types are listed but little is known of differences between them; winter varieties are hardier. Winter varieties are sown in September or early October, spring varieties from February to April, using 190 kg/ha of seed for pure stands. Seedbed and manuring are as for peas. Winter varieties are mixed with rye for early spring sheep feed or oats or beans and oats for silage, including 35–70 kg/ha in mixture. Of less importance than previously since the seed is expensive, dry matter production per unit area is less than well manured cereals. Vetches grown for silage rot in the base if left too long in wet weather. However, they produce heavy yields of green protein rich material.

Seed production

Market is very small and crop must be grown on contract; mainly for seed, bird food and fishing bait. Suitable for most arable soils except the very heavy or poorly drained. Sow 150 kg/ha winter vetches in late October with cereal drill in 120–200 mm rows 40 mm deep. Manure as for peas. Crop is direct combined as for peas. Yield varies from 1500–2500 kg/ha. Pigeons are the major problem.

Root and bulb crops: miscellaneous vegetable crops

This group includes:

(1) *Grown primarily for human consumption, direct or processed*: beetroot, carrots, leeks, onions (dry bulb), parsnips, potatoes and sugar beet.
(2) *Grown for livestock feeding*: fodder beet, mangels.

BEETROOT (RED BEET)

Beetroot is grown for human consumption. Market has changed in recent years. Principal outlet is for *processing*, mainly in glass jars. 'Baby beet' or 'baby picklers' graded 25–45 mm diameter are required. Larger beet are sliced or diced. Retail trade is mostly for *prepacked cooked beet*. Fresh market is small. Early crops are bunched with leaves left on, later crops are sold as topped loose beet.

Varieties

Varieties described in Vegetable Growers Leaflet No. 7, *Varieties of Beetroot, Calabrese, etc.* (NIAB) all have globe shaped (Detroit type) roots for which the main demand exists. Types with other root shapes include: *Long roots* (Cheltenham or long beet); these store well but are now little grown, *Flat roots* (Egyptian type) and cylindrical roots (Continental type).

Only varieties with *high bolting resistance* should be used for early production (as with sugar beet); all varieties are suitable for maincrop work. Other important features include good root uniformity, good internal quality and skin smoothness.

Soil type and rotation

Quality of root is best on sandy loams, silts and well drained loams which have been well manured for the previous crop. As beetroot are closely allied to sugar beet, mangels and fodder beet, and are hence susceptible to beet cyst nematode (*Heterodera schactii*), none of these crops should be grown more than once in five years. Do not grow within two years of ploughing out turf owing to susceptibility to attack by larvae of pests, e.g. wireworm (*Agrotese* spp.).

Seedbed preparation

Treat couch grass (*A. repens*, etc.) with suitable herbicide and subsoil if required. Medium loams are best ploughed 250 mm deep in winter, light soils may be ploughed and pressed in spring. Methods of seedbed preparation as for sugar beet (see also Irrigation, pp. 154 and 172).

Manuring

If soil pH is below 6.2 on mineral soils or 5.8 on peat, apply lime in autumn previous to sowing. Do not overlime as it may induce trace element deficiencies, especially manganese and boron and also cause scabbing. Apply fertilisers as shown in *Table 5.37*.

Table 5.37 Manuring of beetroot (kg/ha)

N, P, K or Mg index	N^1	P_2O_5	K_2O^2	Mg — Sands and light loams	Other soils
0	250	100	300	90	60
1	200	100	200	60	30
2	150	100	200	Nil	Nil
3	–	50	100	Nil	Nil

ADAS figures

[1] Do not apply more than 50 kg N/ha in the seedbed, especially with high K; apply the rest soon after emergence.
[2] At ADAS Index 2, all P and K can be ploughed in. At lower indices apply 60 kg of each in seedbed; plough the rest in.

If deficiencies occur: *manganese* – apply a foliar spray of 6–9 kg/ha manganese sulphate in 200–300 litres/ha water when plants are 100–200 mm high; if severe two or three sprays may be required; *boron* – give pre-sowing Spray Solubor 11 kg/ha or 22 kg/ha borax and work into seedbed.

Beetroot seed, like the natural 'seed' of mangels and fodder beet, is a corky fruit composed of several segments, each containing a seed; it also contains a natural germination inhibitor, which affects the germination of some seeds. Seed may be also supplied rubbed and graded, which increases the proportion of monogerm seeds and improves crop evenness and precision drill performance. Quality of genetic monogerm varieties is as yet inferior. Beet seed may be treated by suppliers with 'thiram soak' which controls seed-borne canker (*Phoma betae*) and blackleg (*Pleospora betae*) and also washes out the inhibitor, resulting in much more even germination.

Plant population, spacing and seed requirement

Optimum plant population depends on crop use (*Table 5.38*).

Table 5.38 Recommended plant population and approximate seed requirement

Type of crop	Plant population (plants/m^2)	Approximate seed requirement (kg/ha)
Early bunching	54	5
Fresh/maincrop	110	10
Baby beet	160	15

ADAS figures

When calculating seedrate, especially of natural beet seed, it is necessary to know the number of clusters (i.e. true seed) per gram and the germination/100 clusters. Modified formula is thus used for beetroot (other than genetic monogerm).

$$\text{Seedrate (kg/ha)} = \frac{1000 \times \text{required population/m}^2}{\begin{array}{c}(\text{No. of clusters/g}) \times \\ (\text{No. of seedlings/100 clusters}) \times \\ \text{Field factor}\end{array}}$$

Field factor: allow 0.5 to 0.6 for poor conditions; 0.7 for average conditions; 0.8 for ideal conditions.

Inter-row spacing is decided by size of beet required and harvesting method.

Early bunching: single rows 300–380 mm apart.

Fresh maincrop: single rows 380 mm, although yield does not decline in rows up to 500 mm apart. Alternatively *beds* of 4–6 rows 280 mm apart are suitable.

Baby beet Very close rows are essential to restrict size. Arrangement must fit harvesting machinery. Mini-beds of four to six rows 50 mm apart on 510 or 710 mm centres are appropriate.

Production methods

Early bunching beet

Propagation under protection in 14 ml cellular trays or equivalent can produce a crop two to three weeks earlier than field sowing. Sow 5 M-P grade seeds/cell of bolting resistant variety 14–21 February and germinate at 15°C; maintain glasshouse or tunnel at 10°C until cotyledon expansion and then reduce to ambient but give frost protection. Apply a liquid feed 100 mg/litre N: 200 mg/litre

K$_2$O to drench trays once after cotyledon expansion; if planting is delayed give a second feed. Plant out, allowing 10 blocks/m^2, equivalent to 100 000 blocks/ha, end of March–early April when plants have two true leaves and are fully hardened off. When harvesting it is impracticable to pull individual plants selectively, so evenness of development is essential. Herbicide may be incorporated preplanting in modules or applied post planting after plants are fully hardened off.

Floating plastic covers, using perforated film, can accelerate the growth of transplants; planted out late March on a bed system. Ridges of soil 50–80 mm high are drawn along the outside of the bed to prevent the sheet from damaging the outer rows of the crop in windy weather. Owing to the high cost and risk with early crops, only areas small enough to be cleared before the main field sowings are ready in late June should be grown and confined to a single sowing.

Unprotected field production Earliest crops for bunching are drilled from end March–early April. *Always use bolting resistant varieties.*

Maincrop beet

This is sown from April onwards, the main sowings being made about mid-May. Later sowings for smaller or later beet may be made in early July. Drill 35–38 mm deep.

Irrigation

Adequate soil moisture at drilling time is essential for even germination. Cultivations for seedbed preparation should be of minimal depth and carried out just before drilling to conserve moisture. Unless immediate rain is expected, especially in May and June, a pre-drilling watering is advisable. Plentiful watering helps to disperse germination inhibitors. Apply 25 mm in April, up to 50 mm in May and June and drill as soon as soil is dry enough. Do not irrigate post drilling or during emergence to avoid serious capping of the soil.

Globe varieties are shallow rooted and readily wilt in hot weather. They respond well to irrigation from April to the end of August, according to lifting time. Bunched crops should be watered with 25 mm pre-harvest. On class A soils (low water holding capacity) apply 25 mm at 25 mm soil moisture deficit (SMD) and on class B and C soils (medium to high water holding capacity) apply 25 mm at 50 mm SMD as required during the growing season. Excessive irrigation of maincrops can lead to large coarse roots.

Pests and diseases

Pests

Pests are largely those which attack the sugar beet crop and will require similar rotational and control measures. Particular precautions should be taken against pests causing loss of plant, or damage to the root, e.g. cutworms (*Agrotis segetum*) or to the young plant, e.g. beet leaf miner (*Pegomya betae*).

Diseases

Scab (*Actinomyces scabies*) disfigures the roots and is most likely on alkaline soils with young crops in dry conditions. *Violet root rot* (*Helicobasidium purpurea*) attacks many root vegetables and is only controlled rotationally; do not grow beet or carrots closer than one in five years.

Harvesting and storage

The bunching crop is lifted in June and July. Foliage is trimmed and beet are bunched, usually in sixes, and packed in trays. Be careful not to bruise young beet. Maincrop beet is lifted from July to mid-November. Timing of lifting depends on size required and whether crop is to be stored. If lifted too early or immature, especially in warm weather, beet soon heats up unless marketed at once. If unduly delayed, beet bulks up rapidly in autumn and roots become too large.

With mechanical harvesting top pulling harvesters do least damage but some varieties are unsuitable and crop must be drilled in rows or narrow bands, minimum centres 380–400 mm. Avoid cutting crowns of beet during topping or they will rot. For high density crops or if top lifter is not available, remove tops with flail mower or forage harvester leaving adequate petiole above crown and lift with elevator digger.

Storage

Although sophisticated methods of storage give good results, the relatively low value of the stored beet crop limits their use. *Outdoors*, beet may be clamped like potatoes on a 1.5 m base to about 1.4 m high and covered with 150 mm loose straw and then 300 mm soil or protected by straw bales; alternatively, they may be stored in bulk boxes with slatted bottoms and protected with straw bales.

Indoors With unventilated stores stacking should not exceed 1.8–2.0 m wide and 1.5 m high. Even then, weldmesh ventilation ducts are advisable to increase convective ventilation. Cover with 150 mm straw. With proper management *fan ventilated stores*, using ambient air, will keep beet satisfactorily until the end of March or even April; thereafter refrigerated stores are required. The low value of the beet crop makes it most unlikely that the special erection of such a store can be financially justified; this method can only be considered where a refrigerated store is required for other purposes (ADAS, 1983).

Yield

There is considerable variation. 'Average total commercial yield' 37 t/ha including ware 32 t/ha + baby beet 5 t/ha. Total trial yields (NVRS) have reached about 60 t/ha. Total yield is largely unaffected within normal commercial densities. Grading is strongly affected by plant density: 150 plants/m² has achieved almost 50% baby beet (25–42.5 mm diameter).

CARROTS

Carrots are a specialised crop, grown on contract for processing, i.e. canning small whole or sliced, dicing, freezing or dehydration or grown for market and prepacking as earlies or maincrops. Only clean straight undamaged roots of required diameter and type are suitable; reject the rest and

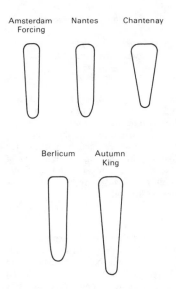

Figure 5.6 Typical root shape for each group of carrots

feed to livestock. Crops are usually washed and must have deep orange flesh and core colour, be free from green shoulders, be well graded and of bright attractive appearance.

Types and varieties

Varieties may be grouped into the following types (*see Figure 5.6*).

Amsterdam Forcing

Early maturing small tops, roots small to medium size, slender cylindrical stump rooted. Grown for early 'bunching' market and prepacks.

Autumn King ('Flakkee' type)

Late maturity, large vigorous tops, roots very large, taper stump rooted. Grown mainly for late-season fresh market and dicing.

Berlicum ('Berlikum')

Rather late maturity, medium foliage, roots large, cylindrical, stump rooted. Grown particularly for fresh market and prepacks, quality good. Also slicing and dehydration.

Chantenay

Late maturity, medium foliage, roots medium size, conical stump rooted, good core and flesh colour. Widely grown for fresh market and processing. Crop often graded: small roots canned whole, medium roots sliced or fresh market and large roots diced for processing or for fresh market.

Nantes

Medium maturity, medium foliage and root size, cylindrical

stump rooted shape. Widely grown for prepacks, also market and canning.

The latest recommended varieties are listed in Vegetable Growers Leaflet No. 5, *Varieties of Carrots* (NIAB).

Soil

Shape is very important; misshapen roots go for stockfeed. Soils must be of such a texture and structure as to allow easy root penetration and even unrestricted root expansion in lateral and vertical directions, to ensure good shape and easy harvesting. Best carrot soils are well drained, deep, stone free sands or light and loamy peats, having minimum water holding capacity of 38 mm/300 mm topsoil and overlying moisture retentive subsoil. Avoid soils with high silt or clay contents or overlying gravel subsoil.

Irrigation

Carrots can be grown without irrigation on sandy soils in years with normal rainfall. If irrigating wait until four-leaf stage and then apply water at 25 mm soil moisture deficit on sand overlying sand, if drying out. If large deficit is allowed to build up, subsequent irrigation encourages splitting, normally only at low plant densities. Irrigation before seedbed preparation may be worthwhile to help rapid uniform emergence and early growth.

Rotation

Carrots may follow cereals but ensure that soil is free from couch grass (*A. repens et al.*) and other perennial weeds. Stubble clean and/or apply recommended herbicide pre-sowing. Frequent cropping with carrots leads to a build up of carrot cyst eelworm (*Heterodera carotae*) and violet root rot (*Helicobasidium purpurea*) in the soil, which can only be controlled by rotational means; ensure a minimum of five years between carrot crops, preferably longer.

Manuring

Ideal pH is 5.8 on peats and 6.5 on sands. Avoid overliming, as crop is sensitive to deficiencies of boron, manganese and copper on sands and manganese and copper on peats. (Treat respectively with: Bo – incorporate 22 kg/ha borax or 11 kg/ha solubor pre-sowing, boron index 0–1; Mn – foliar spray 9 kg/ha manganese sulphate; Cu – foliar spray 2.2 kg/ha copper oxide or oxychloride or incorporate 60 kg/ha copper sulphate pre-sowing.)

Response to N is minimal and N is only applied on deficient soils (*Table 5.39*). Crop responds well to P and K. On sandy soils 400 kg/ha salt (150 kg/ha Na) should be ploughed in or worked deeply into the soil before drilling. All fertiliser should be applied and worked in at least one month before sowing. Use cage or double wheels and wheel track eliminators.

Seedbed preparation and cultural methods

Ploughing may be done at any time during the winter, but many growers on light soils prefer to plough and press immediately before seedbed preparation begins; this ensures a firm surface. Prepare a fine firm level, clod-free seedbed with minimal working. Good carrot soils work freely and seedbed preparation is easy but take care to avoid moisture loss by overworking or badly timed cultivations.

Use of a bed system avoids any tractor wheelings in the soil where the roots are to grow; all passes for working, drilling and spraying occupy the same wheelings, thus eliminating compaction in the bed itself and consequent misshapen roots. Ideally a single pass method of seedbed preparation and drilling is used, where the drill is attached to the seedbed preparation equipment working directly on the ploughed land. With late drilling the 'stale seedbed' technique may be used: the seedbed is prepared two weeks before drilling and paraquat is applied immediately before sowing to eliminate weed seedlings; technique also reduces moisture loss and gives more even moisture distribution in the seedbed. A full herbicidal weed control programme is essential to give protection throughout the life of the crop (*see* Chapter 8).

Soil separation

Stones and clods in the soil during crop growth cause serious malformation of the roots and damage during lifting with some harvesters. Stone windrowing machines (or separators)

Table 5.39 Manuring of carrots (kg/ha)

N, P, K or Mg index	Maincrop and processing				Early bunching			Early and maincrop	
	N		P_2O_5	K_2O^1	N	P_2O_5	K_2O	Mg	
	Fen peats	Other soils	All soils					Sands and light loams	Other soils
0	Nil	60	250	250	60	300	250	60	60
1	Nil	25	150	150	25	250	150	30	Nil
2	Nil	Nil	125	125	Nil	175	125	Nil	Nil
3			50	50		100	50	Nil	Nil
4			25	Nil		75	Nil	Nil	Nil

ADAS figures
[1] When salt is applied reduce K_2O by 60 kg/ha.

remove stones and clods and place them in a trench or furrow at the side of the bed that they have formed; the carrots are then drilled in the bed. This technique, originally developed for potatoes, greatly reduces misshapenness and damage to roots from stones and clods.

In the *absence* of a bed system, e.g. when growing single rows at wide spacing (say 380 mm centres) for large processing carrots, control wheelings and/or use cage or double wheels or flotation tyres to minimise compaction.

Plant population and distribution

Within a given variety, size of produce is decided jointly by plant density and sowing date; size of root is also affected by water supply (i.e. timing and amount of rainfall and irrigation) and soil fertility. It is thus impossible to do more than predict that a particular combination of sowing date and density is *likely* to give a good proportion of roots within the required grade (*Table 5.40*). Precise density will be varied to suit variety and major grade required. Crops are frequently graded into several classes for market and processing.

Each market or processor normally requires a particular size of root; these usually are:

Use	Shoulder diameter (mm)
Small roots for canning whole	19–32
Medium roots for canning whole	32–45
Prepacking	20–50
Loose packs for market	25–70
Freezing	> 25
Large roots for slicing or dicing	> 45

Many row arrangements and widths may be used but correct choice depends on type and width of harvester, grade of carrot required and soil texture. Systems vary from single, double or triple rows on 380–500 mm centres to scatter rows and multi-row beds with at least 460–500 mm for wheel spacings between them; width of bed is determined by width of share on the harvester. A wide range of row widths and arrangements is possible within the bed. Particular advantages of the bed system are the elimination of soil compaction and improved yields from higher plant densities at close row spacings. Choice of row width within beds is limited by the proximity it is possible to achieve between drill units; double or treble scatter rows from each unit are popular solutions (ADAS, 1980 and Hardy and Watson, 1982).

Sowing date

At 325–375 plants/m^2 maximal yield and proportion of canning size occurs some 16–20 weeks after sowing; thus, for small carrots, sow in March, April, May early June for harvesting end of July, end of August, end of September and October and later, respectively. Irrigation should be available for crops sown on sands after late May; complete all sowings by 20 June at latest. Crops grown at low densities for large roots are best sown in April and May.

Drilling

Drill as shallowly as possible but place seed in moist soil at

Table 5.40 Relationship between yield, population and root size for two types of carrots

Anticipated yield (t/ha)		Root size (mm diameter) required					
		20–25	25–30	30–35	35–40	40–45	45–50
Stump rooted	Cylindrical rooted	Plant population (plants/m^2) at harvest[1]					
20	34	130	70	43	27	17	13
25	42	170	90	53	33	21	16
30	51	200	110	64	40	25	19
35	59	230	125	75	47	29	23
40	68	270	140	85	53	33	26
45	76	300	160	96	60	38	29
50	85	330	180	110	67	42	32
55	93	370	200	120	73	46	36
60	102	400	210	130	80	50	39
65	–	430	230	140	87	54	42
70	–	470	250	150	93	58	45
75	–	500	270	160	100	63	48
80	–	530	290	170	110	67	52
85	–	570	300	180	110	71	55
90	–	600	320	190	120	75	58
95	–	630	340	200	130	79	61
100	–	670	360	210	130	83	65

ADAS, 1980.
[1] The figures in this table are for the required plant population at harvest. Plant losses between emergence and harvest commonly number around 10–30% of those emerged.

en even depth. Seed must be dressed; it is *essential* at drilling to apply a recommended insecticide to the soil to protect against first generation carrot fly (*Psila rosae*).

Seedrate

This must be adjusted to obtain plant density required. Rate for a given density depends on seedsize, laboratory germination and field conditions, i.e. field factor. Use formula on p. 92. *Number of seeds/g varies from 550–1600. Field factor should be applied as follows: cold soil and poor tilth 0.5; average conditions 0.6; good conditions 0.7; ideal conditions 0.8.*

Floating plastic film production

Earliest field carrots are now produced under floating plastic film covers which give about a 14 d advantage over unprotected crops. An early F_1 Nantes variety is drilled in beds and row arrangement must fit within width of plastic cover. Seed is drilled in furrows 30 mm deep (*Fig. 5.2d*) so that young seedlings are not lifted out of the soil by adhesion to the cover before they have an adequate root system. Apply a residual herbicide immediately after drilling and lay cover as soon as possible. Remove cover at appropriate growth stage – usually seven true leaves. If removal is too early growth is lost; if too late, growth rate falls off. Apply irrigation immediately after removal of cover and again as soon as soil moisture deficit reaches 25 mm. Choice of cover is important. Those with least perforation produce the fastest growth, but need to be removed earlier than those with more perforations.

Earliest crops are sown early–mid October and overwintered under film covers (ADAS, 1984c). Target seed rate: enough to produce 100 plants/m^2 at harvest. Overwintered losses about 30% of seedlings.

Pest control

Carrot fly (Psila rosae)

Presents the main problem. First generation must be controlled by soil application of persistent approved insecticide at drilling. Crops to be lifted *after* end of September *must* be treated for control of second generation carrot fly, according to manufacturers instructions. Untreated crops are almost invariably heavily damaged and unmarketable. It is also necessary to control *carrot willow aphid* (*Cavariella aegopodii*) with approved insecticides.

Harvesting

Carrots are harvested with a top lifting harvester or an elevator digger type. With the former the roots are loosened with a small share and twin spring-loaded belts grip the tops and lift the roots, which are topped on the machine. It gives high rates of work, and soil, stones and tops are all left behind in the field. The machines are limited to row widths and cannot be used after tops have died down.

With elevator digger types it is generally necessary to remove tops before harvesting begins. They can operate after the tops have died down. Elevator diggers are also suited to a wider range of growing systems but dirt and stones may be carried over with the roots unless sorted on the machine. System and row width must be selected to suit the harvester.

Storage

Canning carrots are generally cleared early. Carrots may be stored in the field by the following methods:

(1) Earthing over: the rows, minimum width 510 mm are covered with at least 150 mm of earth by plough (or ridger) but this hardly practicable in wide beds. Used for crops lifted from January onwards.
(2) Strawing: straw may be laid loose 300 mm deep in late November to early December. Very costly; only justified with low value straw and high crop returns.
(3) Clamping in the field: carrots store satisfactorily in clamps till March.
(4) In mildest areas carrots may be left as grown in field but there is serious risk of total loss in hard winters.

Controlled temperature storage is only justified if higher prices are obtained to cover the extra cost. This type of storage can be very profitable when lifting is impossible or clamps cannot be opened in hard frosty or snowy weather. Optimum storage temperature is 1°C at 95% relative humidity.

Washing and grading

Processors inform growers of their requirements. When selling in the fresh market carrots must be washed, graded and packed according to market or supermarket requirements; damaged and misshapen carrots usually go for livestock feed. Considerable capital outlay is needed for a washing plant and grading and packing line. It is either a large scale 'on farm' operation or is undertaken at a packhouse. For details see ADAS, 1980.

Total yield

Yield varies greatly according to soil, weather and other variables.

Average: mineral soils 40 t/ha, peaty soils 48 t/ha.

Some high density systems can produce 50–100 t/ha; exceptional crops can produce yields in excess of 100 t/ha.

LEEKS

Leeks are primarily a winter vegetable, although some production is available as early as July. The UK is entirely self sufficient; popularity is increasing. High grade market requires long straight shaft with good length of blanch, well presented and clean. Size, washing and dressing requirement depends on particular market.

Labour requirement is high, especially with transplanted leeks in June and early July (about 380 h/ha) and during the harvest period from September to April (about 900 h/ha).

With some field trimming, leeks leave large organic residues in the field and ADAS N index 2 can usually be assumed for the succeeding crop.

Types and varieties

Many 'varieties' are selections of a particular type; there are also distinct varieties, describe in Vegetable Growers Leaflet No. 7, *Varieties of Beetroot, Calabrese, Courgettes, Leeks, Salad Onions, Sweet Corn and Parsnips*. Some varieties are

Table 5.41

Type	Harvest period	Stem/shaft length (mm)	Remarks
Copenhagen Market	Sept	long > 250 mm	Large, early. Not frost hardy
Genvilliers	Sept, Oct	long > 250 mm	Similar to above, but stouter stems
Swiss Giant	Oct–Dec	med–long approx. 200 mm	Clear by Christmas; hardiest varieties only end of January
Autumn Mammoth (or Giant)	Oct–Feb	medium approx. 160 mm	Vigorous and frost hardy in all but hardest winter
Giant Winter	Jan–March	med–short approx. 150 mm	Continental type of winter leek Dark green. Very frost hardy
Bluegreen (Blauwgroene) Winter	March–early May	short approx. 130 mm	Dark green turning blue in cold weather. Very frost hardy. Make much of their growth in February and March and should only be harvested at end of season
Empire and Blue Solaise types similar} to Bluegreen Winter			

Note: Availability will depend on weather conditions; length of shaft on variety and cultural conditions

available for harving over a very long period, e.g. Snowstar, Argenta. The main types are given in *Table 5.41.*

Soils

Leeks grow well on a wide range of soils provided they are well drained and have an adequate pH, minimum 6.5 but preferably 7.0. Deep water retentive loams and peats are best. Avoid heavy soils, on which leeks are difficult to harvest during the winter and also may not give good tilths; also coarse sandy soils, as the coarse particles penetrate the leaf sheaths and are strongly disliked by the consumer.

Rotation

Treat as onions, as they are susceptible to white rot (*Sclerotium cepivorum*) and do not grow any susceptible crops, i.e. leeks, bulb or salad onions, shallots or garlic more than once in six years or on land affected with white rot (*S. cepivorum*).

Manuring

Leeks respond well to a heavy dressing of well rotted farmyard manure (60–70 t/ha) which should be ploughed in. Fertiliser requirements are shown in *Table 5.42.* For each 10 t FYM reduce fertiliser application by 15 kg/ha N; 20 kg/

ha P_2O_5; 40 kg/ha K_2O; and 8 kg/ha Mg, but *never* apply less than 50 kg/ha P_2O_5 and 50 kg/ha K_2O.

Leeks are grown in three ways:

(1) *Raised in a bed*, either under protection (for earliest crops) or in the open and then transplanted into their final cropping position. This method has a very high labour requirement at planting compared with (3) direct drilling, but has several major advantages: stems are longer and blanch (white part of stem) very much longer, so that quality and value is higher. A high degree of precision in spacing is possible, giving a very uniform crop, especially with a dibbing machine. Crop establishment is much less chancy, giving greater scope for higher yields, weed control is easier and costs are lower, saving up to £200/ha on herbicides. This is the traditional method and is suitable for early and high quality crops and for intensive horticultural situations, increasing the time available for the growth of the previous crop.
(2) *Raised in cellular trays* and then transplanted. Leeks can be single seeded into small (about 9 ml) volume cells or multiseeded into larger (14 ml) volume cells or multiseeded into larger (14 ml) cells, when the aim is to establish three plants/cell. More than four plants per cell results in leeks have curved unacceptable shafts. Quality of leeks from multiseeded cells is lower than from single

Table 5.42 Manuring of leeks (kg/ha)

N, P, K or Mg index	Other soils		Fen peats	P_2O_5	K_2O	Mg	
	Base dressing	Top[1] dressing		All soils		Sands and light loams	Other soils
0	150		60	300	275	90	60
1	100	Up to 100	30	250	150	60	30
2	60		Nil	150	125	Nil	Nil
3	–		–	50	50	–	–

ADAS figures
[1] If required

seeded cells. Method especially suitable for earliest crops; single seeded cells can be very cost effective compared to protected beds.

(3) *Direct drilled crops* have the lowest labour requirement at planting but a very high standard of seedbed preparation is essential and establishment can be risky. Crop occupies the land for a longer time. Quality and shaft length are inferior to (1). Method mainly suitable for midseason and late crops.

Production management

Transplanted crops

Sowing dates, method of raising, transplanting and corresponding harvest periods are shown in *Table 5.43*. On bare rooted transplants trimming root length to 45 mm to ease planting is common practice but can reduce yield. Prepare deep well worked tilth for transplanting. There are three methods of transplanting: (1) hand dibber, suitable for small areas only; (2) multi-row machine dibber. This method makes holes but leeks still have to be dropped into them by hand. Earthing up is not necessary, as a good length of blanch is produced on the leek shaft. Soil should not be too loose for this machine, so it is preceded by a light rolling; (3) a conventional brassica transplanter, fitted to a tractor with a reduction gearbox is common, but planting is not as deep as with dibbing; length of blanch and shape is not as good.

Transplanted leeks are best grown on a bed system, e.g. five to six rows 250–300 mm apart per bed, at a spacing of 120–150 mm between plants, although actual spacing depends on machinery used. With a tractor mounted transplanter, row widths may vary from 450–600 mm and interplant spacing 75–100 mm is suitable. Crop may then be earthed up to increase length of blanch.

Plant beds

Protected beds Production is costly so high density is required. Aim to obtain 800–1200 plants/m^2; drilled in rows 50–150 mm apart. Calculate required inter-seed spacing from formulae on this page. Seedrate varies from 3–6 g/m^2.

Open air beds Drill in rows 150–300 mm apart to obtain populations of 450–900 plants/m^2. Calculate required interseed spacing from formulae below and seedrate as formula on p. 92.

Drilled crops

Crop may be drilled in: (i) beds of five to six rows 250–300 mm apart. This method results in a very short blanch, as earthing up is impracticable; (ii) rows 450–500 mm apart to allow earthing up in late summer or early autumn.

Method is unreliable for early crops, i.e. late August–early September harvest. Drill maincrops late March or early April and latest crops in mid April; any later results in reduced yields.

Drill seed 12 mm deep. Seedrate depends on plant population required; calculate using formula on p. 92.

For precision drilling, requisite number of seeds/m of row

$$= \frac{\text{mean row width (mm)} \times \text{plant population/m}^2}{10 \times \text{field factor} \times \% \text{ laboratory germination}}$$

$$\text{Inter-seed spacing (mm)} = \frac{1000}{\text{no. seeds/m of row}}$$

Seed counts vary between varieties and season, but normally fall in the range 280–444/g. Minimum germination 65%.

Field factor: Allow 0.5 for early drilling in March; 0.7 *average* conditions in late March; 0.8 in early April.

Approximate guide to seed requirements: 1–2 kg/ha.

Pelleted seed of many varieties is available; obtain seed count from merchant.

Plant population

Plant density varies from 20–50 plants/m^2 according to size of leek required. Higher densities are appropriate for small leeks for supermarket prepacks and direct drilled crops, lower densities for traditional markets, although there is an increasing preference for smaller leeks. Yield declines at densities below 20 plants/m^2. It is difficult to transplant much more than 25 plants/m^2.

Irrigation

May be necessary for direct drilled crops in April, applied presowing. For transplanted crops apply immediately after

Table 5.43 Sowing, transplanting and harvesting periods for leeks

Crop type	Sow	Method of raising	Transplant	Harvest period
Extra early	Jan	Heated glasshouse Single seeded cells	April	Mid July onwards
Early	Jan–Feb	Heated glasshouse Single seeded cells or beds Also tunnels	April– May	Aug–Sept
Early	March	Unheated tunnel/cold frames/cloches in beds	May	Sept onwards
Mid-season and late	Mid March– early April	Unprotected plant bed	June– early July	Oct–May
Late crops, reduced yield of smaller leeks	Early– mid April	Unprotected plant bed	Late July	April–May

planting to aid establishment. Water along row or give 25 mm overall application. Additional application during crop growth until the end of August may be worthwhile (ADAS, 1982).

Pests and diseases

The following may cause problems:

Pests

Onion fly (*Delia antiqua*), stem eelworm (*Ditylenchus* spp.), onion thrip (*Thripstabaci*), and in some years, cutworms (*Agrotis segetum*).

Diseases

White rot of onions (*Sclerotium cepivorum*) also affects leeks and is most severe on seedlings, Rust (*Puccinia allii*) best controlled by preventive fungicidal sprays, and white tip disease (*Phytophthora porri*).

Harvesting

Most leeks are lifted by hand after mechanical undercutting. The vibrating share, driven by tractor PTO, is much preferable to the fixed blade, making a thorough job of loosening the soil around the roots. Simple elevator diggers have been used but these leave the leeks on the ground and crop is likely to be badly soiled. When purchasing larger equipment, give consideration to the needs of other crops. For larger scale work where the crop is destined for washing and grading in the packhouse, complete harvesters (i.e. top lifting or elevator digger types) may be used on light friable or organic soils. The former gives a cleaner crop.

Preparation for market

To meet high quality market requirements, leeks are best handled in the packhouse, although rough trimming often occurs in the field. Leeks are skinned (i.e. removal of outer leaf sheath) and roots and leaves trimmed; they are then washed, drained and packed.

Statutory grading applies to leeks. Class I must be of good quality and 'blanch' must equal at least one-third of total length or half the sheathed part. In Class II blanch must be at least one-quarter of total length or one-third of sheathed part. Amount of trimming and tightness of grading depends on market requirements. For 500 g supermarket packs containing three to eight stems all root is removed and leaves very tightly trimmed, to give stems 200–300 mm long with maximum 20% leaf. English wholesale market requires that leaves are trimmed to half the length of the stem and roots trimmed to 12 mm, packed loose in trays 4–8 kg. In Scotland half stem and half leaf is adequate.

Yield

Variable, depending on weight of crop, method of production and extent of dressing 20–28 t/ha. Transplanted crops generally give a higher yield.

DRY BULB ONIONS

Onions are grown as (a) *salad* or *spring* onions, which are specialist crops, usually grown on a market garden scale to provide a succession for bunching, and (b) *dry bulb onions*.

These may be grown for ware for fresh household use or by caterers (small ware bulbs 25–40 mm diameter; other ware 40 mm and over. Caterers prefer bulbs over 45 mm), for processing for soups, canning and dehydration or for pickling when bulbs under 25 mm are required.

Good appearance and quality of bulbs is essential to command top prices and compete with high quality imports, especially Spanish. Skin bright golden brown, may vary from straw colour to brown, *must not* be stained. Necks thin and bone dry; bull necked onions must be removed as they will not dry. Shape varies from flat to globe, according to variety; globes fetch best prices.

Types and varieties

Onions for dry bulb production are classified as

(1) *autumn sown*, which stand the winter and are harvested in June and July, and
(2) *spring grown*, harvested in September.

Both open pollinated (normal pollination between selected plants) and F_1 hybrids are available.

Varieties for autumn sowing

Japanese varieties   Only bolting resistant varieties mainly of Japanese type or bred from these in Europe, are suitable. Spring varieties must never be sown in autumn as they invariably bolt. Open pollinated and F_1 hybrid varieties are available. Maturing in June and July, bulbs are generally rather flat or irregular.

Appearance is much inferior to Rijnsburger types and bulbs do not keep well; crop should be sold by 31 August at latest. Valuable for earliest crops.

Varieties for spring sowing

Rijnsburger varieties   Globe shaped with skin colour varying from pale straw to brown. Good keepers and croppers. Favoured by large scale growers, e.g. Robusta, Balstora.

North European F_1 hybrid varieties   Globe shaped, appearance very similar to Rijnsburger varieties from which they are usually bred, but more uniform and usually higher yielding, maturity mainly midseason. Good keeping quality, e.g. Hyper, Hygro Hydeal. Over 90% of UK spring grown crop is drilled with Rijnsburger and North European F_1 hybrids.

North American F_1 hybrid varieties   Early maturing but low yielding. Bulbs high shouldered with dark copper coloured skins, e.g. Elba Globe, Granada.

Other UK varieties   Generally outclassed by Rijnsburger and North European hybrids, which are superior in yield and far better keeping quality, e.g. Ailsa Craig, Bedfordshire Champion.

Pickling varieties   Are sown thickly to produce small onions; they include brown pickling varieties and smaller white skinned varieties, e.g. Barletta and Paris silverskin.

Climate and distribution

Warm sunny climates are best to promote quality of produce and easy harvesting. Mainly grown in eastern counties.

Soil and cultivations

Choose only best soils on the farm, which must be well drained, deep, free working and not subject to capping after heavy rain. Select moisture retentive soils holding more than 38 mm/300 mm depth. Avoid stony and heavy soils and those likely to dry out. Onions are very sensitive to acidity; soil pH should be between 6.3 and 7.0 on mineral soils and 5.5 to 7.0 on peaty soils. Best soils include well drained silts, brick earths and medium loams. Peats are particularly suitable. Sandy soils are used extensively and are especially suitable for autumn sown crops as they are very free draining; yields are lower with spring sown crops.

Seedbed must be fine, firm, level and clod free. Plough in autumn to allow weathering and settling. To avoid compaction during seedbed preparation use cage or double wheels, with minimum number of passes. Bed system of cultivation is highly suitable. Prepare seedbed on day of drilling.

Rotation

Previous crop must be cleared in ample time for seedbed preparation for autumn sown onions and for autumn ploughing for spring sown crops. Onions and leeks should never be grown closer than one year in six as both are susceptible to white rot (*Sclerotium cepivorum*), which is cheaply controlled by rotational means; crops are also attacked by stem eelworm (*Ditylenchus dipsaci*) whose wide range of host crops includes peas, beans, clover and other legumes, oats, carrots, parsnips and sugar beet.

Brassicas, potatoes, wheat and barley are not affected and are suitable for rotations which include onions.

Manuring

Onions are highly responsive to P and K, but response to N is usually poor (*Table 5.44*). Keeping quality is improved by K but reduced by N. Autumn sown onions require top dressing in spring; apply 100 kg/ha N (50 kg/ha N on fen peats) in late February. After exceptionally wet winters top dressings of 150 kg/ha N may be necessary (100 kg/ha N on fen peats) but split dressing between January and normal time. Spring sown crops may require a top dressing of 50 kg/ha N on mineral soils. With heavy applications of K, part should be applied before ploughing. Avoid overliming peaty soils or manganese deficiency is likely. If lower reaches of soil are acid, plough in some lime.

Plant population seedrate and drilling

Total yield of bulbs increases with plant density but size of individual bulbs is reduced. Optimum density depends on size of onion required: *Ware* allow 70–80 plants/m^2 (range 65–85 plants/m^2); *picklers* allow 320 plants/m^2.

When calculating seedrate, seed size (range 240–290 seeds/g), percentage germination and field factor (0.5 cold soil, poor tilth; 0.6 fair conditions; 0.7 good conditions; 0.8 ideal conditions) must be taken into account. Use formula on p. 92.

No. of seed/m in row =

$$\frac{\text{mean row width (mm)} \times \text{plant population/m}^2}{10 \times \text{field factor} \times \text{\% laboratory germination}}$$

Interseed spacing (mm) = 1000 no. seeds/m of row.

Row width and arrangement

This must fit share width of harvester to be used. Ideal mean row width is 300 mm, using single rows or paired rows 50–75 mm apart; the latter can only be obtained by certain makes of drill. Row width should never exceed 450 mm for ware and 350 mm apart and 1500 mm wide have proved satisfactory. Drill into soil moisture, just sufficient to cover seed, 12–25 mm deep; autumn sowings may require slightly deeper sowing to contact soil moisture.

Irrigation

Autumn sown crops may require watering at sowing time to ensure rapid germination, as timing is critical for success. Otherwise it is preferable to grow onions on soils where irrigation is not needed, as uneven application results in uneven ripening. Irrigation is usually only required for spring sown crops on sandy soils in a hot dry summer: apply 2.5 cm in the post crook stage *before* mid July. Multi-seeded peat block crops require irrigation immediately after planting out.

Table 5.44 Manuring of dry bulb onions (kg/ha)

N, P, K or Mg index	N				P$_2$O$_5$	K$_2$O	Mg	
	Pre-sowing				All soils		Sands and light loams	Soils
	Autumn sown		Spring sown					
	Fen peats	Other soils	Fen peats	Other soils				
0	Nil	50	30	90	300	275	90	60
1	Nil	25	Nil	60	250	150	60	30
2	Nil	Nil	Nil	30	150	125	Nil	Nil
3					50	50		
4					25	Nil		

ADAS figures

Methods of culture and sowing dates

Onions may be grown commercially in three distinct ways:

(1) *overwintered crops*, sown in August,
(2) *spring sown crops*, and
(3) sown in *multi-seeded cellular trays*, raised under cover and planted out. These have substantially superseded multi-seeded peat blocks.

Onion 'sets' are of little commercial importance to growers.

Overwintered crops

Use only appropriate varieties. Sowing date is critical, as it is essential to obtain a strong plant to enter winter, with several lateral roots to prevent frost lift, without bolting. Sowing too early causes bolting, while late sowing results in weak plants and serious losses. Sow 9–15 August in north England, 15–31 August in south England. A really free draining site is essential, as is efficient chemical weed control. Quick marketing is important as potential quality of Japanese types does not justify expensive conditioning techniques. Improved varieties are available.

Spring sown crops

The major part of the UK onion area is of this type. Sow as soon as a fine seedbed can be obtained in late February or early March; if not, it is better to delay until conditions are suitable but complete as early as possible. Sow by mid April at latest, as delay results in serious loss of yield. Crop must be harvested as early as possible in September.

Multi-seeded cellular trays and peat blocks

All too frequently soil conditions do not permit early spring sowing, with resulting loss of yield or even of crop and lower quality; problem is particularly serious in north England. With multi-seeded cellular trays or peat blocks a crop is assured, with larger bulbs, higher yield and improved quality; also harvesting occurs in August, when good conditions are more likely. Succession is obtained by sowing range of varieties; Japanese and N. European F_1 hybrids preferred to open pollinated Rijnsburger varieties. Problem is greatly increased costs, but these can be justified by much higher yields and prices obtainable with crops harvested in late summer.

Other advantages include easier weed control; crop is well past the one leaf stage, giving a wider choice of herbicide with lower cost. With greater precision in production, it is easier to predict required populations and bulb size.

Method Using multicellular trays (cell volume range 9–15 ml) sow six to eight seeds/cell and cover level to top of cells with inert material (e.g. silver sand or perlite). Alternatively 27–38 mm cubed peat blocks may be used. Germinate at 21°C (emergence takes 4–5 d) and then remove to glasshouse with some heat 10–15°C (*see* operating cellular tray system, p. 95) aiming to have plants with one and a half to two true leaves at planting. During period in glasshouse plants will require two or three liquid feeds according to growth and colour of plants: N 100 g/litre, K_2O 200 mg/litre give first feed at one true leaf stage. Immediately prior to planting out soak cells with high N feed: N 200 mg/litre: K_2O 200 mg/litre.

Table 5.45 Within-row spacing required for multi-seed cellular tray or block onions (mm)

Bed width (m)	1.42	1.52	1.68	1.83
No. of rows/bed	4	4	4	5
Field population cells or blocks 10^3/ha				
85	330	310	280	320
100	280	260	240	270
125	230	210	190	220

Field population (cells or blocks/ha) recommendation:

Proportion in bulb sizes		*Blocks/ha (10^3)*
75% 60–80 mm diameter		80–85
50% 40–60 mm + 50% 60–80 mm diameter		100 (standard)
75% 40–60 mm diameter		125

ADAS, 1982.

Sow seed in trays or blocks 25 January–15 February and plant out end March or early April in four to five row beds (*see Table 5.45*).

Pre-harvest treatments

Apply sprout suppressant, maleic hydrazide, while tops are still erect; it is not absorbed once tops have fallen over. Effect is to kill sprouts when dormancy breaks, usually in March but also in wet autumns. Use foliar desiccant only as a last resort and then only in conjunction with sprout suppressant.

Harvesting

Peak yields are attained in first half of September. Harvest as soon as tops have gone down. If too early, yield is lost; if too late quality suffers and skins split.

Traditional methods

Crop is undercut and windrowed as soon as tops go down and then left to dry in field for a maximum of 7–10 d in dry weather. Crop may be loaded into drying room earlier if weather is difficult, but ensure adequate airflow or crop will be ruined. Wet weather can result in skin discoloration and also root growth in crops untreated with sprout suppressant. Lift and load into store or sell on completion of drying period; stack up to 3 m deep, not more or crop will be damaged by compression.

Do not lift while dew is on crop or excessive soil will be brought into store. Removal soil and small, bull-necked or diseased onions before storage.

Direct harvesting methods

No field drying period is allowed. As soon as tops have gone down they are removed by flail machine, leaving 80 mm length of neck. Some 3 h are allowed for stem debris to dry off. Crop is then lifted mechanically into trailers and put into store up to 3 m deep on same day that tops are removed. Dry with heat immediately. Surface moisture must be removed within 3 d of harvesting. Direct harvesting results in improved skin colour and retention, but choose varieties with retentive skins and long keeping characteristcs.

Drying

Process is carried out in two stages.

Removal of external moisture

Start blowing air heated to 10–15°C above ambient as soon as store loading begins. Ensure rapid uniform continuous drying for 3 d or longer. Even a short period of wetness results in discoloration and moulds. Stage one is complete when onions on top of heap are dry. Skins should then be bright golden brown.

Removal of neck moisture

Blow intermittently until *all* other batches have completed stage one. Then blow continuously at reduced airflow, using heat if required. Stage two is complete when necks of onions on top are raffia straw dry, usually middle or even end of October. Do not cool until necks are quite dry, then reduce temperature as low as possible without freezing.

Normal ventilated storage may be used until end of March. Cold air is blown at night, using differential thermostat, when ambient is 3°C below store. Frost guard set at − 2°C. *Refrigerated storage* is necessary for storing onions from April to June.

Disease

Neck rot (*Botrytis allii*)

This can cause serious losses in stored onions if not controlled. Sow only seed treated with recommended fungicide. Previous onion crops and dumps are a serious source of infection, so grow new crops well away from these areas. Handle onions carefully at harvest.

White rot (*Sclerotium cepivorum*)

Disease is soil-borne and best controlled by rotational means (*see* Rotation); also by fungicidal treatment. Do not dump packhouse waste on fields. Less important diseases include *downy mildew* (*Peronospora destructor*) and *leaf spot* (*Botrytis squamosa*; *B. byssoidea*; *B. cinerea*) which are controlled by appropriate protective spraying.

Pests

Stem eelworm (*Ditylenchus dipsaci*) is the most serious pest of onions (*see* Rotation). Onion fly (*Delia antiqua*) is much less serious and can be controlled with an appropriate seed dressing.

Yield

	t/ha
Mineral soils	
Poor	30 or lower
Very good	40
Peat	37
Multi-seeded	40–45
cellular trays	
or peat blocks	

PARSNIPS

Parsnips have now become almost an all the year round crop. The earliest crops come to market in July but the main fresh market demand is during the winter from November to April, once they have been 'frosted', which improves flavour and sweetness. The market demands washed clean white straight roots without fangs, bruising or other damage. Prepackers require small to medium roots 35–65 mm diameter at shoulder and 100–180 mm long for the supermarket trade. Traditional, e.g. greengrocer markets, will take medium to large roots, 45–75 mm diameter and 150–230 mm long, while processors require large roots of 150 mm diameter.

Types and varieties

Shape (*Figure 5.7*)

'Bulbous' and 'wedge' shapes are much easier to harvest than the long pointed 'bayonet' type which sits deeply in the ground and is very liable to snap off when large. Prepackers tend to prefer 'wedge' and 'bayonet' shapes with a smooth skin; choose suitable varieties.

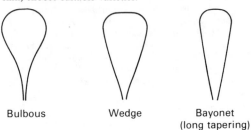

Bulbous   Wedge   Bayonet
                                 (long tapering)

Figure 5.7 Root shapes of parsnips

Susceptibility to bruising

Varieties exhibit substantial differences in this character, varying from good resistance to susceptible.

Canker

Varieties differ markedly in their susceptibility. This disease, which takes two forms, causes the shoulder and crown of the parsnip to rot during winter. Orange brown canker can be caused by three different organisms (*Itersonilia* spp.). Black canker (*Phoma* and *Mycocentrospora* spp.) occurs after injury when various weak pathogens invade the roots. Varieties of parsnip are listed in Vegetable Growers Leaflet No. 7, *Varieties of Beetroot, Calabrese, Courgettes, Leeks, Salad Onions, Sweet Corn and Parsnips* (NIAB).

Soils

The best quality crops are produced on deep sandy well drained soils. Ideally, soil should be stone free; on stony soils the problem can be reduced with the use of a stone windrower, as with carrots and potatoes. On peaty soils high levels on canker organisms can occur and the peat can be difficult to wash from corrugations in the roots. Avoid heavy soils as the roots become fanged and winter harvesting is extremely difficult.

Cultivations and seedbed preparation

The medium to heavier parsnip soils should be subsoiled while the soil is dry and then deep ploughed in the autumn (about 200–250 mm) and left rough for frosting. Do not apply dung as this causes fanging. Light soils are best ploughed after Christmas or ploughed and pressed immediately before sowing. Parsnips are *extremely sensitive* to compaction, so prefer bed systems if possible; otherwise use controlled wheelings, cage or double wheels or flotation tyres. In all cases prepare a fine seedbed with minimum number of passes.

Manuring

Parsnips require a fairly low N fertiliser application (*Table 5.46*). Except for ADAS indices 0–1, all P and K should be ploughed in. Nitrogen is then given as a top dressing after crop emergence; increasing N above the recommended level only increases tops without a corresponding growth of roots. Parsnips are susceptible to manganese deficiency, especially on organic soils. Apply one or more sprays of manganese sulphate at 9 kg/ha with wetting agent in 225 or more litres/ha as soon as symptoms appear. Soil pH should be ⩾ 6.5 on mineral soils and ⩾ 5.8 on fen peats.

Table 5.46 Manuring of parsnips

N, P or K index	N		P_2O_5	K_2O	Mg	
	Fen peats	Other soils	All soils		Sands and light loams	Other soils
0	60	100	175	225	90	60
1	40	75	100	150	60	30
2	Nil	Nil	75	150	Nil	Nil
3	–	–	60	75	Nil	Nil

ADAS figures

Plant population and sowing

Older large rooted varieties require wider spacing than the newer more canker resistant varieties. For small prepack parsnips allow 40–60 plants/m², average 45. For traditional fresh market and processing allow 20–30 plants/m², average 27, according to size of root required. Row spacing depends on cultural and harvesting system. Allow bed system four rows on 1.86 m bed; top lifter harvester 400–450 mm between rows and for digger-elevator drill in twin rows on 500 mm centres. Traditional row spacing varies from 380–460 mm depending on the size of the parsnip.

Where possible precision drill to a stand. Use only seed with high percentage germination and vigour. Natural or pelleted seeds are available but although the latter is easier to drill, natural seed usually germinates more quickly and gives a more uniform stand. Use tightly graded seed and check drill settings carefully, as the seed is flat and does not flow easily; the addition of french chalk helps seed flow. Do not drill seed deeper than 15 mm.

Time of drilling On favoured sites drilling starts in January or February for early crops, as soon as soil conditions permit. Maincrops are drilled in the first half of March. For small parsnips drilling continues until early May.

Irrigation

There is no experimental evidence on the effects of irrigation although it is likely to be useful for the establishment of sowings made after mid-April. If soil is dry, give 25 mm presowing. Otherwise, marginally economic. May be useful to easy lifting of early crops in very dry soil conditions.

Pests and diseases

Carrot fly (*Psila rosea*)

Young parsnips can be killed by larvae mining the tap root. Damage similar to that on carrots. *Control*: grow as far as possible from overwintered carrots and where other *umbelliferae* (carrots, celery and parsley) are grown on the same farm, adopt a wide rotation. Insecticidal treatment as for carrots.

Canker (*Itersonilia* spp. *et al.*)

Roots are rendered completely unsaleable. Sow canker resistant varieties (e.g. Avonresister). Late sowings for small roots in April and early May are also much less susceptible to damage.

Harvesting

Top lifter harvesters may be used before leaves have died down but after this only digger-elevator machines are suitable. Great care is required to prevent bruising and damage, to which some varieties are especially susceptible. For high quality blemish free pre-packaged roots many growers lift with an elevator digger and pick into plastic bags for dispatch to the washing plant. Special shares are normally required for lifting parsnips. These consist of two angled prongs which lift and squeeze the roots from the ground; one pair is required per row. Parsnip roots show good frost resistance, but cannot be lifted from frozen ground. During mid-winter it is thus wise to have up to three weeks' supply lifted and stored in an unheated but frost proof barn.

Yield

Range 15–30 t/ha. Average 23 t/ha.

POTATOES

Potatoes are grown for many purposes, each market having its own requirements and needing a different type of tuber. Hence there are several distinct types of production.

Maincrop 'ware' or table production

By far the largest outlet, requiring heavy yield of saleable, medium to large sized potatoes; larger sizes preferred by chip fryers. Crop is lifted at full maturity and earliness is unimportant, provided crop is ready for lifting by early October. Tubers for pre-packing must be free from blemishes or losses in grading are heavy. Local markets and demands vary.

Earlies

Tubers are lifted in immature state to provide a succession of new potatoes retaining firm waxy or soapy texture after cooking through the summer. First, then second early varieties are used. Earliness is of prime importance for earliest crops; they are planted very early and lifted as soon as marketable. Earliest areas like Cornwall and Pembroke start lifting about 20 May. Price falls sharply as other districts start lifting and yield increases. Later planted first and second earlies, giving higher yields, continue to provide a supply of new potatoes until maincrops are ready.

Potatoes for processing

Generally grown on contract and used to manufacture wide range of products. *Potato crisps* require uniform medium-sized tubers with shallow eyes, dry matter content over 20% and reducing sugar content below 0.25%. *Chips or french fry* require medium to large tubers, shallow eyes, regular shape, white creamy flesh, dry matter over 20% and reducing sugar under 0.4%. With *dehydrated* potato for instant mash, maximum yield of dry matter is needed at minimum cost. Requirements are large, uniform, mature tubers with minimal peeling loss, over 20% dry matter, reducing sugars under 0.5%, white creamy flesh, free from taints and after-cooking discoloration, the starch granules and cells separating readily during processing. *Canned new potatoes* should give high yield of small uniform tubers 18–38 mm with creamy flesh colour, dry matter under 19%, remaining waxy and solid after steam peeling and canning, with clear, sludge-free brine. *Canned potatoes* for dicing need large blocky tubers over 50 mm, white flesh, waxy texture, below 10% dry matter; cubes must retain shape after cutting.

Seed production

Objective is to grow high yield of small tubers within range 32–60 mm, producing crops true to type, free from undesirable variations, other varieties, virus and certain other diseases. Seed production is a specialised job mainly carried out in cool areas of low aphid incidence such as north and east Scotland, Northern Ireland, designated areas in north, north west and west England and Wales. Many early growers produce their own seed for planting earliest crops.

General quality requirements for cooking potatoes

Smooth even tubers with shallow eyes for minimal wastage and labour in peeling, without coarse or misshapen tubers. Damage from second growth (glassy, cracked tubers), greening, pests (slugs, wireworm), disease (blight, *Phytophthora infestans*, scab, *Streptomyces scabies*) must be absent whilst mechanical damage during lifting, storing, grading and transportation must be minimal. Cooking quality is partly a varietal feature but is substantially modified by soil type, rainfall and soil moisture, manuring and storage temperature. Well matured tubers lifted in early autumn before soil has cooled down and become wet give best quality; wounds also heal more readily.

Boiling tubers

These must not disintegrate during boiling nor discolour or blacken after cooking. Maincrop tubers may be waxy or floury after cooking.

Roasting or frying

Product must be golden brown outside, not dark brown or black due to presence of reducing sugars. Chips should be crisp, floury inside and absorb minimal fat.

Varieties

Essential varietal features include ability to produce commercial yield, saleability, good cooking quality and absence of deep eyes and poor shape, with adequate resistance or at least lack of extreme susceptibility to diseases (e.g. late blight, *P. infestans*, dry rot, *Fusarium caerulum*). Varieties differ much more than in many other crops in suitability to cultural and climatic conditions and response to management. Particular attention is needed for varietal requirements, including storage temperature, sprouting regime, seed rate and spacing.

Varieties are classified according to maturity group.

First earlies

Produce the earliest crop of new potatoes. They only produce dwarf foliage, thus give lower yields than maincrop if left to maturity.

Second earlies

Are slower bulking than first earlies but give high yields of new potatoes for follow on digs.

Maincrop varieties

Tubers bulk later than previous types, so larger foliage is built up before growth is slowed down by tuber production, hence heavier yields. Except for a few crops for very late new potatoes, these varieties are only lifted when mature in October.

Recommended varieties are listed and described in Farmers Leaflet No. 3, *Recommended Varieties of Potatoes* (NIAB).

Seed certification

It is only possible to maintain satisfactory yields of potatoes when seed is free from severe virus and certain other pests and diseases, e.g. potato cyst nematode (*Globodera rostochiensi* and *G. pallida*) and blackleg (*Erwinia carotovora* var. *atroseptica*). It is essential that seed is of variety stated, free from rogues, undesirable variations such as bolters and semi-bolters, producing a crop true to type. Certified seed is produced under supervision and control of Agricultural Departments in the UK. Certified seed is only produced from approved parent material and growing crop inspected by officers of the appropriate certifying authority.

Grades of seed

Certificates are issued by the Agricultural Departments in the UK, i.e. *England and Wales* – Ministry of Agriculture, Fisheries and Food (MAFF); *Scotland* – Department of Agriculture (DAS); and *Northern Ireland* – Ministry of Agriculture (MANI). Certificates refer only to health and purity of crops at time of inspection. Grades fall broadly into two groups

(1) basic seed intended primarily for multiplication for seed: VTSC (Virus Tested Stem Cuttings), the highest grade raised initially in aphid proof greenhouses. SE (Super Elite) derived from VTSC or SE stocks. A number 1–3 indicates number of years' multiplication since VTSC. E (Elite) may only be grown in the protected area (Scotland and parts of Cumbria and Northumberland) and in former SS grade areas. AA for producing CC or ware crops.

(2) seed for production of ware crops only:

Certified Seed CC may not be planted for further certified seed production or grown in the protected area.

The letters NI after the grade indicate that variety is non-immune to wart disease (*Synchitrium endobioticum*). Particulars of certificates will be supplied by Department of appropriate country on application.

Source of seed

Growers may

(1) purchase total requirement every year of basic or certified seed from seed producers, or
(2) they may purchase enough basic seed every year to grow on to produce enough once-grown seed to plant the whole crop the following year. Purchased seed is grown in isolation, given appropriate insecticidal treatment and rogued:
(3) seed may be taken from the ware crop and new seed purchased every other year or as required. Practice varies according to type of production.

Earlies

'Once-grown' seed produces an early crop for lifting 10–14 d earlier than one grown from imported seed. Early growers produce their own once-grown seed from grade SE or E imported seed, or purchased seed from a grower producing seed of identical and predictable performance, which involves controlling planting date of the seed crop.

Maincrops

In areas of high aphid incidence it may be necessary to purchase new certified seed every year. Elsewhere, stocks remain healthy for longer periods, depending on the number of aphids, date of migration, and freedom from virus of surrounding crops. Inexperienced growers should change stocks annually, but experienced growers will know how long to keep stock before renewal.

Seed size

Seed size is of less importance than seed rate. As seed size increases, the number of mainstems (or individual plants) produced from the seed tubers increases, so large seed produces more stems and tubers than small seed, but average tuber size is smaller. This effect is modified by widening spacing as size of seed increases, up to certain limits, i.e. by modifying seed rate. Small seed gives better yields than large seed planted at the same seed rate but optimum size is determined by type of production and variety.

First earlies

Small seed under 30 mm with less food reserves is slower bulking, tubers are lifted later and show poor recovery from frosting. Seed must be size graded and planted at intervals from 200 to 350 mm according to size; do not use chats.

Maincrops

Modern planters work better with closely graded seed. If not already size graded, this should be done on arrival. Seed graded 30 × 60 mm should be split into two or sometimes three grades according to size, distribution and shape of consignment. All sizes between 30 and 60 mm are satisfactory if planted at appropriate spacing and similar seedrate. Tubers over 50 mm are less efficient especially with varieties or on sites where numerous small tubers are formed. Chats (i.e. under 30 mm) from healthy stock are prefectly satisfactory if planted at appropriate seed rates.

Seed production

Chats under 30 mm are unsuitable for producing seed potatoes; otherwise, all sizes may be used if planted at appropriate seed rates. Seed producers in Scotland plant ware tubers over 50–60 mm, 230–300 mm apart, as they are worth less than seed size and give a good yield of seed. Chats may be used to produce ware for re-planting for seed.

Seedrate and spacing within row

Seedrate is determined by the number of seed tubers planted per hectare (i.e. spacing within row × row width) and average weight of individual seed tubers. The optimum seedrate varies greatly, according to type of production (earliness and grade of tuber required), variety, conditions of growth (soil type, fertility and water supply) and ratio cost of seed: expected value of ware.

Table 5.47 Optimum seed tuber populations and seedrates for some maincrop potato varieties

Variety and grade (mm)	No. of sets/50 kg	Assuming ratio cost of seed: value of ware = 2:1	
		Optimum population (10^3/ha)	Optimum seedrate (t/ha)
Varieties tending to produce undersized tubers:			
King Edward			
30–50	850	37.5	2.2
30–60	700	33.8	2.4
Maris Piper			
30–50	850	36.2	2.1
30–60	700	32.3	2.3
Record			
30–50	750	38.7	2.6
30–60	600	34.8	2.9
Varieties tending to produce over-sized tubers			
Desire			
30–50	750	48.8	3.3
30–60	600	42.6	3.5
Pentland Crown			
30–50	750	48.8	3.3
30–60	600	42.6	3.5
Pentland Squire			
30–50	750	57.5	3.8
30–60	550	47.0	4.3

ADAS, 1982d.

Table 5.48 Guide to seedrates for potatoes

Type of production	Seedrate (kg/ha)	Remarks
Earlies	3200–4500	Single sprouted seed required for earliest crops; multiple sprouted seed satisfactory for later lifting and second earlies. First early seed should be tightly graded and planted 200–360 mm apart according to size
Canning	Optimum 7500 Economic range 2500–5000	High stem density 160–220/m^2 required. Use multisprouted seed spaced 150–230 mm apart according to size. Size grade seed. Optimum seedrate depends on cost of seed and value of produce
Seed production Home produced seed, e.g. Scotland	5000–7500	High seed rates are fully justified where cheap seed is available; home produced ware size, which is of lower value than seed, is planted
Purchased seed	3800–5000	High seed rates are very expensive if seed is bought at high price, e.g. early growers producing once grown seed. Use lower rate and multisprouted seed

NB When growing new variety check on suitable seedrate – there are large differences

Maincrops The best measure of plant population is the number of main stems per unit area. In practice it is difficult to achieve a precise control over sprout and hence mainstem numbers and it is therefore necessary to rely on planting an appropriate number of tubers. As seedsize increases, so does the number of mainstems it produces. Seed tuber populations are thus adjusted to maintain as constant a main stem density as possible, according to the average weight/set (expressed as number of tubers/50 kg).

Variety Some varieties (e.g. King Edward, Maris Piper and Record) tend to produce large numbers of undersized tubers at too high population and seedrate, especially on low fertility soils. Others (e.g. Cara Desiree and Pentland Squire) produce oversized tubers at low populations and seedrates, especially on fertile soils. Price of seed varies but is usually at the ratio of 2:1 to the value of the produce; as the ratio widens, the optimum seedrate is reduced. Selected examples of optimum seed tuber populations and seedrates for 'average' conditions are shown in *Table 5.47* but growers will

need to adjust them according to their growing conditions and grade of tuber required by their market. If tubers are too large, increase the seedrate; if they are too small, reduce it.

Seedrates may range from 2.5–4.5 t/ha for varieties producing large tubers and 1.8–2.4 t/ha for King Edward. Where the production of undersized tubers from King Edward is a problem, chats (i.e. < 30 mm) may be planted but never try to grow King Edwards on a poor soil.

Seedrates for earlies, canning and seed population are shown in *Table 5.48* and appropriate spacings in *Table 5.49*.

Row width

In high rainfall areas and where tubers are lifted immature (i.e. first earlies) close rows 540–660 mm give more uniform distribution and higher yields: Maincrops have traditionally been grown in 710 mm rows but wider rows, 760, 800 or 900 mm allow more room for tractor tyres and more soil for a ridge, resulting in similar yields with few green tubers. Canning potatoes and seed are grown in 660–760 mm rows.

Table 5.49 Spacing of potatoes according to seed rate

(t/ha)	Required seed rate												
	17.5	20.0	22.5	25.0	27.5	30.0	32.5	35.0	37.5	40.0	45.0	50.0	75.0
No. of sets/ 50 kg	Spacing between sets (mm × 10) *in 760 mm rows*												
400	94	84	74	66	61	56	51	48	43	41	37	33	21
500	76	66	58	53	48	43	41	38	35	33	29	26	17
600	63	56	48	43	41	38	33	30	30	28	24	21	15
700	53	48	43	38	35	30	28	28	25	23	21	19	12
800	48	41	38	33	30	28	25	23	23	20	19	16	11
900	43	38	33	30	28	25	23	20	20	18	16	15	10
1000	38	33	30	28	25	23	20	18	18	18	15	14	9
1100	35	30	28	25	23	20	18	18	15	15	14	12	7
1200	30	28	25	23	20	18	18	15	15	13	12	11	6

Seed treatment

Pre-sprouting or chitting

Seed is placed in chitting trays, four trays per 50 kg for earlies, three and a half trays for maincrops; these are stacked in greenhouses or suitably lit buildings to produce sprouts 12–18 mm long before planting. Pre-sprouting accelerates cycle of growth; emergence, tuber initiation, bulking and maturity are brought forward 10–14 d. Effect on tuber yield depends on circumstances. Sprouted seed invariably increases yield of early crops, preferably always use sprouted seed. With maincrops increases from sprouting are less certain, depending on conditions of growth. With late planting, or in very cold late areas, well sprouted seed gives up to 5 t/ha more ware, but when unsprouted seed is planted early in warmer areas and blight is controlled, there is usually no increase in yield. Even so, sprouting some seed allows an earlier start on harvesting and seed which does not sprout properly can be picked out; misses from skin spot (*Oospora pustulans*) are also eliminated. Full sprouting is costly and heavy on labour, and as yet sprouted seed is unsuitable for bulk handling or in fully automatic planters.

Sprout size

This is controlled by varying storage temperature: below 4°C no sprout growth; 5.5°C fast sprouting varieties grow; 7°C all varieties start, fast sprouting varieties grow rapidly; 10°C all varieties grow rapidly, fast sprouting varieties grow excessively. Effect of temperature depends on level and duration; thus 7°C maintained over long periods is too high for a fast sprouting variety. Provided the store is frostproof, the difficulty is to keep the temperature of fast sprouting varieties low enough to prevent excessive growth. A fan and ducting fitted with differential thermostat for blowing cold (but not frosty) night air into the store to control temperature is usually necessary; refrigeration equipment may be needed in mild areas but is expensive. Ideally sprout length should be restricted to 19–25 mm for all types of production. Varieties differ in speed of sprout growth and need different management; two differing varieties should not be sprouted in the same store, e.g. Home Guard (fast), Ulster Prince (slow). Sprout growth must be watched and adjusted accordingly; to speed up sprout growth, raise temperature; to slow down, lower temperature of the store.

Number of sprouts

This is controlled by varying time at which sprouting starts. If tubers are sprouted at 16°C in the early autumn, a single sprout develops which suppresses growth of all other sprouts; excessive growth is prevented by lowering temperature once the single sprout is formed.

Multi-sprouted seed is produced by delaying start of sprouting until the New Year or later, or removing the first sprout and resprouting; the number of sprouts produced increases with size of seed but small seed still produces relatively few sprouts. Single-sprouted seed is essential for earliest crops while seed for canning and seed crops need as many sprouts as possible.

Mini-chitting

Seed is managed to produce mini sprouts 2–3 mm long at planting. Labour required is low, trouble from skin spot (*Oospora pustulans*) eliminated and possible increase from sprouting maincrops largely obtained; cold storage facilities are essential. Seed may be treated in bags, boxes of up to 500 kg capacity or in bulk. On arrival seed is cured at 13–16°C for 10 d to heal wounds; temperature is then reduced and held at 3°C to stop sprout growth and then raised to 7–10°C to obtain the required sprout development three weeks before planting; management must be precise and is critical.

Storage of unsprouted seed

Where cold store facilities are not available bags should be stacked, maximum 1.8 m high, on boards or straw with ample airspace between in a frostproof building; the shallower the better, for heating is less likely. If necessary, cover sacks with straw in frosty weather but keep as cool as possible as spring approaches by opening doors.

Cutting seed

Small whole tubers always give higher yields than cut pieces of the same weight. Cutting is now rarely done and is only worthwhile with very large tubers and expensive seed. Unsprouted seed should be cut lengthwise and then heaped and covered by moist bags at 13–16°C for 10 d to allow healing. Cut surfaces must not dry out nor be treated with disinfectant. Lime and soot are worthless. Seed for sprouting may be cut for two-thirds of way from apical end towards heel, the two sides left pressed together until planting, then pulled apart; each side sprouts well.

Date of planting

First earlies

These when fully sprouted, are planted as soon as soil conditions permit. In earliest areas, e.g. Cornwall and Pembroke, the crop is planted in February; other areas follow in early March.

Unsprouted maincrop

This seed is planted as early as practicable, certainly by mid April. In the south, planting occurs in late March and first week in April; in the north, early April. Automatic planters speed planting and hence may increase yield of unsprouted seed.

Sprouted maincrop

This seed may be planted later, without loss of yield; if possible plant by mid April in the south, end of April in the north. Precise results of late planting depends on season; in dry years heavy losses result and tops of some varieties may never meet across the rows. By contrast, with ample later rainfall there is relatively little effect provided blight is controlled and planting is not excessively late.

Floating plastic film covers for first earlies

Films advance lifting by 7–10 d compared with unprotected crops, competing with imports rather than home grown produce. Returns are greatly improved but with such a high cost system it is essential to market the crop while prices are still high. Choose sheltered fields, as covers are very susceptible to wind damage. Avoid frost pockets, as foliage coming into contact with film freezes; mist irrigation has been shown to be beneficial. Best responses are obtained in years with poor growing conditions.

Herbicides must be applied before covering. Once this is done, film should be laid as soon as possible after planting, although delays of up to 20 d post planting do not reduce yield significantly; thereafter yield reductions are increasingly serious. Timing of cover removal depends on type of cover material: i.e. whether it is ventilated by slits or perforations; timing of the removal of the latter is more critical. There is little response to keeping covers on for longer than two weeks after 95% emergence; if covers are left on more than four weeks after 95% emergence yield is reduced. Generally total removal rather than slitting longitudinally is best, although precise effect depends on weather conditions. When using plastic covers, physiological age of seed tubers needs to be less than for unprotected plantings.

Climate

Earliest crops

These are grown in areas close to the sea where winters are mild and very early planting is possible, e.g. Cornwall, Pembroke. As climate and site become progressively colder, early lifting is increasingly delayed, late first earlies merging into second early crops.

Maincrops

Maincrops are grown over a very wide range of climatic conditions; for these soil texture and structure are more important than climate.

Soil

Good potato soils are well drained, deep and allow deep cultivation. Deep loams, sandy loams or alluvial soils are admirable but avoid still clays, harvesting being very difficult in wet autumns. Peat soils give heavy yields but tuber dry matter is usually lower. Potatoes are almost insensitive to acidity and lime is only necessary on the most acid reclaimed land, after bracken.

First earlies

These require light soils, light loams or sandy loams with southerly aspect, which warm up early.

Maincrops

They are best on deep moisture retentive soils; avoid soils prone to drying out unless irrigation is freely available.

Cultivations and seedbed

Deep friable moist mould clod free seedbed is required. Plough flat broken furrow slice 200–300 mm deep, according to topsoil depth, with reversible plough well before Christmas to allow weathering and settling. To avoid compaction and clod production use cage or double wheels and minimum number of passes in spring; power harrows are especially useful for reducing number of passes. Most planting is done mechanically; where done by hand, ridges are opened and closed quickly to avoid moisture loss. In dry areas ridges may be rolled down after planting.

Stone and clod separation (soil separation or stone windrowing)

Clods and stones can bruise potatoes severely during harvesting and slow down work-rate. Conventional cultivations cannot remove stones and frequently produce clod. A stone windrower separates out stones and clod and passes them via a lateral web into trench by the side of the bed it has created. Medium to heavier soils are ploughed in the autumn and the bed is then set up; sandy soils may be ploughed in the spring and then destoned. Twin rows of potatoes are then planted on each bed. If properly managed the technique results in good seedbeds and minimal damage to tubers.

After cultivations

Where weeds are controlled by cultivation, ridges may be pulled down with chain harrows, followed by periodic inter-row cultivations after crop emergence. This needs doing carefully, not running too close to the plants to avoid damage and loss of yield. Crop should be finally ridged up when 200–250 mm high and before roots have started spreading between rows. Weed control in early crops and high rainfall is best effected by use of chemicals; elsewhere practice varies (see Chapter 8).

Prevention of potato blight (Phytophthora infestans)

At least one spraying with a MAFF approved fungicide must be given to all maincrops before haulm meets across rows. Thereafter, spraying frequency depends on weather. In warm, humid, *blight* weather, i.e. in the south west or in wet years, weekly spraying is necessary when using contact type of fungicide. In dry periods and years (other than in the south west) a second spraying can be delayed until blight warning; spraying is then advisable at 7–10 d intervals. Burning off the haulm in autumn with MAFF approved desiccant prevents spread of blight to tubers. Essential preventive measures include good earthing up, destroying groundkeepers and old clamps and avoiding very susceptible varieties.

Manuring

Dung, when used, should be well rotted and ploughed in during autumn. Fertilisers ploughed in give inferior results to spring applications. Recommendations in *Tables 5.50* and *5.51* are for broadcasting on the flat, in front of the planter. Placing fertiliser close to but not in contact with the seed gives more efficient utilisation of nutrients but there is risk of damage if fertiliser comes too close to the seed. Risk is greatest with sprouted seed, high rates of fertiliser and on soils liable to dry out. When fertiliser is broadcast over ridges or band placed, reduce recommendations by 25%. Fertiliser attachments to planters slow down rate of working. Lime is *only* applied before potatoes on *extremely acid* soils such as bracken reclamation land in upland areas and then only in small quantities.

Nitrogen

Raising level of N application increases tuber yields up to a certain point but reduces dry matter or specific gravity. Excess N delays tuber initiation, and maturity of foliage and tubers, which being immature are watery and of low cooking quality. Susceptibility of haulm and tubers to blight is also increased. Level of N should be reduced for very fertile soils

Table 5.50 Manuring of maincrop potatoes (kg/ha)

N, P, K or Mg index	All mineral soils except warp soils			Organic soils except moss soil			Humose, moss and warp soils			All soils
	N	P_2O_5	K_2O	N	P_2O_5	K_2O	N	P_2O_5	K_2O	Mg
0	220	350	350	130	350	350	180	350	350	100
1	160	300	300	90	300	300	130	300	300	50
2	100	250	250	50	250	250	80	250	250	Nil
3	–	200	150	–	200	150	–	200	150	Nil
Over 3	–	100	100	–	200	100	–	200	100	Nil

ADAS figures

Table 5.51 Manuring of early, seed and canning potatoes (kg/ha)

N, P, K or Mg index	Earlies			Seed and canning			All crops
	N	P_2O_5	K_2O	N	P_2O_5	K_2O	Mg
0	180	350	180	180	350	250	100
1	130	300	150	130	300	200	50
2	80	250	120	80	250	150	Nil
3	–	250	60	–	250	75	Nil
Over 3	–	200	60	–	200	75	Nil

ADAS figures

(e.g. after heavy applications of dung or rich turf or in areas of high summer rainfall) whereas an increase is necessary for low rainfall areas with arable rotations.

Phosphorus

This encourages early tuber growth, important for early crops, as well as earlier maturity, harder skins and lower blight incidence. Potatoes respond well to P in terms of tuber yield but level of response is lower at high P indices. Use only water soluble forms of P. First early potatoes grown on slightly acid soils of the west give very large responses to P.

Potash

The more mature the crop at lifting, the better the response. Maincrops and second earlies are highly responsive to K, first early crop much less so. The use of sulphate of potash instead of muriate (chloride) increases tuber dry matter somewhat and may reduce incidence of after cooking blackening; sulphate form never used for new potatoes. Heavy dressings of muriate form reduce dry matter content of tubers and consequently reduce susceptibility to internal bruising.

Harvesting and storage

Ware and seed crops

Crops must be completely mature, haulm dead, tubers with hard skins, before lifting; crops should never be lifted with immature or blighty foliage. Allow at least 10 d for skins to harden after burning off foliage, around 21 d for crops with large vigorous foliage. Certified crops for seed are burnt off by order of certifying authority by specified date, to prevent aphid infestation and introduction of virus; burning off also reduces size of seed tubers. Relatively early harvest, from mid September to mid October, gives drier conditions, cleaner, easier lifting, less dirt, better healing of wounds, better cooking quality. Careful implement setting and handling are essential to prevent damage during harvesting and transport.

Storage

Potatoes are now mostly stored in buildings. Indoor storage gives independence of weather conditions for sorting and grading which can then be carried out when other work stops, so farm organisation is improved. Working under comfortable conditions results in better quality grading and 60–100% higher output. Rotting is minimised and temperature control improved. Against this, should rotting occur, many tons of tubers may need moving quickly to reach the trouble, which could prove disastrous if not immediately checked. Capital cost need not be high and a wide range of buildings can be adapted cheaply. General purpose umbrella buildings are ideal, provided tubers are properly protected.

Indoor storage

Clean (no loose earth), dry, sound tubers are required, with minimal damage or disease. Buildings should have:

(1) strong walls, reinforced if necessary, as a stack of potatoes exerts side thrust and weak walls collapse; temporary walls of straw bales should be well reinforced;
(2) proper insulation to combat frost and weather;
(3) adequate ventilation, including ducting to let air through the stack, and gable end ventilation with ample headroom to prevent condensation and keep the potatoes dry, with up to 600 mm of loose straw on top of heap;
(4) proper siting for ease of access;
(5) correct management is essential.

Potatoes may be stacked up to 1.8 m with only convection ventilation; stacking beyond this is likely to require forced draught ventilation.

Sprout suppressants

Potatoes keep satisfactorily until January or February, according to district, with only normal ventilation. Thereafter they show increasing tendency to rapid sprouting as temperature rises. Refrigeration is unsatisfactory and uneconomic for ware. Late storage thus requires a sprout suppressant approved by MAFF to be used. Follow maker's instructions closely.

Yield

Earlies 7.5–10 t/ha according to current price; up to 25–30 t/ha according to earliness of lifting.

Canning potatoes 7.5–10 t/ha or tubers 18–38 mm with up to 10 t/ha oversized tubers.

Maincrops average 25–30 t/ha, good crops 45–50 t/ha: exceptional crops may exceed this.

SUGAR BEET, MANGELS AND FODDER BEET

Rotation

Sugar beet, mangels and fodder beet make excellent cereal break crops, especially on deep light or medium soils; they should *not* follow potatoes owing to danger of ground-keepers. Also avoid following crops where persistent herbicides were used. In order to control beet cyst nematode (*Heterodera schachtii*) sugar beet, fodder beet, mangels, red beet, spinach beet or any brassica crop, including brassica catch crops, should not be grown in the two years preceding the beet crop. A longer break (e.g. six years) between beet crops and/or brassicas is *much* safer, especially where weed beet has appeared (*see* Chapter 8). A sound rotation is essential, as it is the only satisfactory way of controlling beet cyst nematode (*H. schachtii*).

Sugar beet

Grown only on contract to British Sugar plc, who have a monopoly of seed supplies and purchase all roots. Sugar beet is confined mainly to eastern counties and east and west midlands, within easy reach of factories; elsewhere the area of beet is small.

Varieties

Suitable varieties are listed in Farmers Leaflet No. 5, *Recommended Varieties of Sugar Beet* (NIAB); trial results are published annually in summer editions of *British Sugar Beet Review*. When selecting choose at least two varieties; one of them may suit individual farm conditions better than the other. Also buy bolting resistant varieties for March sowings; sow tolerant or resistant varieties in areas where virus yellows or downy mildew are a problem. Choose smaller topped varieties for Fenland or high fertility situations, larger topped varieties for livestock feeding or where obtaining full ground cover is a problem.

Soil and cultivations

Beet requires deep well drained stone free soil that is not acid. Shallow soils, plough or chemical pans or acidity result in fangy roots, making lifting difficult and with high dirt tare. Stony soils cause excessive wear on implements. In practice beet is grown on a wide range of soils from sand to better structured clays or marls but good medium textured soils regularly give the highest yields. A high standard of management is required to provide a well structured soil free from compaction.

A smooth level seedbed ensures even drilling. An excessively fine surface should be avoided, to prevent capping after rain but seed must be surrounded by fine firm moist soil to ensure rapid even germination. Compaction prevents root penetration and seriously reduces yield. Seedbed preparation starts in autumn, with stubble cleaning and subsoiling if required. Dung and certain nutrients, according to soil type and analysis (*see* manuring), are applied prior to ploughing to avoid compaction.

Medium and heavier soils

Plough a coarsely broken level furrow slice. Deep ploughing is unnecessary on most soils; 200–250 mm is usually adequate. Late ploughing of medium or heavy soils when wet causes pans and poor seedbeds. Autumn or winter post ploughing cultivations for levelling allow the production of the spring seedbed with a single pass, but must *only* be undertaken when the soil is dry or well frosted. Chisel ploughing is used on some farms. Type of ploughing and winter work have a profound effect on quality of spring seedbed. Cage or double wheels on tractor should be used wherever possible to avoid compaction.

Wait until soil is dry enough before starting spring seedbed preparation. Work must be shallow with minimum number of passes to avoid loss of moisture and compaction. Deep working brings up lumps and gives loose cloddy seedbeds. To conserve moisture in dry weather, avoid working down a larger area than can be sown the same morning or afternoon. Precise choice of method depends on soil type and growers experience. Always select harrows with *straight* tines to prevent bringing up clod and avoid over-working.

Very light or sandy soils

These are frequently ploughed and pressed in the spring and drilled immediately across the direction of the ploughing. This leaves a coarse surface between the rows which helps to prevent 'blowing' (i.e. wind erosion). The lightness of the soil, however, allows the seed to be surrounded with fine moist soil. FYM and other appropriate fertilisers are also applied before ploughing (see fertiliser application p. 174).

Prevention of compaction

Compaction results in fanging (formation of multiple tap roots), increased dirt tare and loss of yield through poor growth and is mainly caused by tractor wheelings. Bed systems are appropriate (*see* p. 96) as they eliminate wheelings under the rows; four-row beds are normal but these require a four-row harvester. Alternatively, various systems of 'controlled wheelings' may be used, combined with double or cage-wheels and wide or flotation tyres. Ploughing in fertiliser also helps to reduce traffic on the seedbed.

Sowing

Early sowing is essential for maximum yields of sugar. Drilling should start as soon as possible after 20 March and be completed by 10 April, subject to satisfactory ground conditions. After this date yield of sugar declines; in dry years there is inadequate soil moisture for satisfactory plant establishment and activity of residual herbicides. Very early sowing, i.e. before 20 March, is inadvisable as it may give a marked increase in the number of bolters (*see also* Weed beet in Chapter 8) and poor establishment and uneven stands; germination is slow and erratic at soil temperatures below 5°C.

Full evenly distributed plant population is needed for maximum yield of sugar. Aim for a final stand of at least 75 000 plants/ha. Up to 100 000 plants/ha produce no harvesting difficulties. If under 62 000 plants/ha, yield is seriously affected. Irregular distribution may reduce yield, seriously at low populations, and increases harvesting difficulties. Closer seed spacing does not compensate for irregular establishment, as topping mechanisms cannot cope adequately with large variations in size of beet and many small beet remain unharvested.

Row width may need to be adjusted to suit other crops but yield is reduced if row width excess 500 mm.

Only monogerm seed of high germination capacity and in pelleted form, usually graded to 3.40–4.75 mm diameter, is now supplied by BSC to growers; this gives an emergence rate at least equivalent to the former rubbed and graded seed. A standard inter-seed spacing of 175 mm is normally satisfactory, provided that seedbed conditions and standards of drill operation are good enough to achieve an establishment rate of at least 70%. In practice many growers fail to achieve this level. For the earliest sowings, difficult conditions or where problems have been experienced in obtaining satisfactory populations, inter-seed spacings down to 155 mm should be considered.

Failure to achieve satisfactory standards of maintenance and operation of precision drills is frequently the prime cause of poor establishment. Coulters must be sharp, depth of sowing should be checked and sowing mechanisms serviced well before drilling. Store units under cover. In operation do not exceed forward speed of drill of 5 km/h, when each unit should cover 1 ha in 8 h. For faster ground coverage, increase number of drill units, not forward speed. Match number and spacing of drill units to other operations, e.g. spraying, tractor hoeing and harvesting.

Manuring

Adequate N is essential for rapid foliage development and establishment of full leaf canopy but all too frequently excess is applied, increasing leaf growth with very little extra weight of root. Sugar content and consequently financial returns are reduced (*Table 5.52*).

Beet does not respond well to P and yield increases are not likely above ADAS index 1. Applications suggested for index 2 and over are solely to maintain soil P reserves and may be given at any stage of rotation.

Sodium

Sodium is an essential nutrient for sugar beet and should always be given except on fen silts and peaty soils; it reduces need for K and application of 150 kg/ha elemental sodium (Na)(400 kg/ha agricultural salt) is assumed in K recommendations in *Table 5.52*. If salt is not applied, increase potash applications by at least 100 kg/ha K_2O except on fen silts and peaty soils; this treatment is more expensive than a combination of sodium and potassium. When tops are carted off soil is depleted by about 150 kg/ha K_2O, which should be applied subsequently in rotation.

Magnesium deficiency

Magnesium deficiency, to which crop is highly susceptible, is common in light soils where farmyard manure is not given. Kieserite at 600 kg/ha will supply 100 kg/ha Mg and is more available than calcined magnesite but is less persistent in soil. If lime is also needed, use magnesium limestone where available. Deficiency symptoms cannot be controlled by foliar application.

Manganese deficiency

Manganese deficiency, causing speckled yellows, may occur on peaty soils pH over 6.0 and on sandy soils pH over 6.5. Avoid overliming and when symptoms occur spray with 10 kg/ha manganese sulphate in low volume. Up to three applications may be required. On land regularly affected, consider use of sugar beet seed pelleted with manganous oxide, which delays need for foliar applications.

Boron deficiency

Boron deficiency, causing heart rot occurs on alkaline sandy soils. Apply 10–40 kg/ha borax or 7–10 kg/ha solubor to soil; solubor may also be mixed with aphicide and applied to leaves.

Table 5.52 Manuring of sugar beet (kg/ha)

N, P or K index	N					P_2O_5	K_2O	Mg
	Sandy and shallow chalk soils	Lincoln-shire lime-stone soils	All other mineral soils	Humose soils	Organic soils	All soils		
0	125	140	100	75	50	100	200	100
1	100	100	75	50	25	75	100	50(M)
2	75	75	50	25	Nil	50	75	Nil
3	–	–	–	–	–	50	75	Nil
Over 3	–	–	–	–	–	Nil	75	Nil

ADAS figures
M = a maintenance dressing to prevent depletion of reserves. Yield increase unlikely.

Fertiliser application

Germination can be seriously damaged by N, potash and Na if applied in quantity shortly before drilling. Agricultural salt (sodium) is best applied before ploughing in autumn except on very light soils where leaching may occur; it is then applied on top of ploughing. All P_2O_5, K_2O and Mg may also be ploughed in during autumn except where soil analysis shows one or more to be deficient, when part should be applied on top of ploughing at least two weeks before drilling, i.e. ADAS index 0 or 1. Nitrogen is always applied in spring but if applied too early may be lost by leaching; it can also damage germination. Top dress 40 kg/ha N immediately after drilling; broadcast the rest post crop emergence at two to three leaf stage.

Lime

Aim for pH 6.5–7.0 except on sandy (pH 6.0–6.5) or peaty (pH 6.0) soils. Lime requires adequate time to penetrate soil and is best applied to crop which *precedes* beet. If lime is applied just before beet crop, plough in half and apply remainder after ploughing. Application of whole dressing after ploughing acid soil results in acid layer about 100 mm below surface and consequent fanging. On sandy soils apply lime little and often to avoid trace element deficiencies. In areas near beet factories, use waste factory lime.

Pests and diseases

Apart from beet cyst eelworm (*H. schachtii*) which can only be controlled satisfactorily by sound rotation and epidemic attacks of virus yellows, pests and diseases do not usually reduce beet yields seriously.

Seedlings are so small that damage is often fatal. Millipedes (*Class Diplopoda*), symphalids (*Class Symphyla*), spring tails (*Order Collembola-Onychiurus* spp.) and pigmy mangel beetle (*Atomaria linearis*) can all cause serious loss of plant. Routine treatment with approved granular insecticides, other than for control of early aphid attacks and docking disorder caused by *Trichodorus* and *Longidorus* spp., is only worthwhile where soil-borne pests have previously caused serious losses. Mere 'blanket' treatment of all crops puts up cost without increasing yield, may kill beneficial species and induce resistance to chemical in some species. Slugs (*Derocreras reticulatum*) and leatherjackets (*Tipula* spp.) can also cause severe damage. Wireworm (*Agriotes* spp.) can be serious when beet closely follows grass in the rotation; prior treatment with an approved insecticide is essential.

Field mice (*Apodemus sylvaticus*), which live on insects and weed seed, can cause loss of beet seed. Ensure all seed is well covered or mice will eat seed on surface and then work along rows by smell. Inspect field the day after drilling and place approved poison bait (under cover, e.g. section of drainpipe) at three points/ha, according to makers' instructions.

Virus yellows (beet virus yellows BVY or beet mild yellowing virus BMYV) carried and spread by peach potato or green aphid (*Myzus persicae*), is the most damaging disease of beet. Extent of losses depend on earliness and severity of attack. As a rough guide, 3% of potential yield may be lost per week of infection before mid October. Losses are greatest in East Anglia and in bad years may be as high as 40–50%. Reduce incidence and severity of disease by controlling aphids: deal with overwintering sites of aphids,

e.g. residues of beet crop, ground keepers, plants at cleaner loader sites, overwintered green crops and steckling beds (beds of young beet plants for planting out for seed crop). Local gardens and allotments are often a serious source of aphids. Early sowing gives quicker total ground cover which is less attractive to aphids than partially bare ground. Routine seed furrow applications of approved insecticide are of value in areas of highest infestation in East Anglia but are of questionable economic value elsewhere. Routine foliar treatment, according to incidence of aphids is advisable. First foliar application of approved aphicide should be given as soon as an average of one aphid can be found on four plants or on receipt of BSC spray warning card. A second application may be economically viable in bad years. Aphids have now built up resistance to some organophosphorus insecticides and use of others, e.g. carbamates, is often wise.

Powdery mildew (*Erysiphe betae*) can reduce yield in dry years from August onwards. Spray with wettable sulphur in early stages of disease and not later than end of August.

Irrigation

Sugar beet ultimately produces a deep tap root, about 1.8 m deep by September but before this it is vulnerable to water shortage, especially in early stages of growth. Crops on sands and loamy sands suffer most. Irrigation in a dry June, applying 250 mm as water deficit approaches 500 mm, is particularly beneficial in helping to accelerate root extension and early completion of leaf canopy. Irrigation is also worthwhile in dry weather in July and August to prevent wilting, when closure of stomata stops sugar production. Prevent deficit exceeding 500 mm; irrigate to near field capacity. There is no response to irrigation in September, as the tap root has penetrated deep enough to draw adequate water from the soil.

Harvesting

Correct setting and operation of harvester is essential to minimise field losses of beet. Aim to finish by early December. When clamping ensure that beet are clean, correctly topped and free from trash. Concrete rafts are well worth their cost and provide by far the best clamping site. To allow maximum ventilation and reduce heating, use long narrow clamps in early portion of harvesting period. Later on, when damage from frost is likely, use wider clamps. Do not exceed height of 2.5 m. Protect beet from frost as factory will reject frosted beet once thawed, as it cannot be processed and may bring refining to a halt. Frosting also reduces sugar content seriously. Use baled straw to protect tidy well made clamp in frosty weather; in mild weather covering impedes ventilation and allows temperature to rise excessively. Use the thermometers at side and in centre of clamp to monitor temperature. Keep clamp as cool as possible without freezing, i.e. below 5°C. Respiration and consequent sugar losses rise sharply above 10°C; temperatures over 15°C are too high. Load beet for factory in same sequence as it is harvested.

Payment for beet

All sugar beet is grown under contract to British Sugar plc. Terms and conditions are laid down in the 'Inter Professional Agreement' between British Sugar plc and National Farmers Union (NFU). Payment is based on the weight of washed and correctly topped beet delivered to the factory at a sugar content of 16%, with additions or deductions for every 0.1% sugar $\pm 16\%$. Other additions include early and late delivery and transport allowances; deductions include levies for research and to pay NFU representative at factory. All beet deliveries are subject to EEC quotas; payment varies according to quota classification. For copies of the above agreement apply to: British Sugar plc, PO Box 26, Oundle Road, Peterborough, PE2 9QU or NFU, Agricultural House, Knightsbridge, London SW1X 7NJ or your area sugar factory or local NFU branch.

Yield

Yield is extremely variable, some growers achieving very poor results.

Washed roots 30–40 t/ha, very good crops 50 t/ha. 'Good average' attainable yield of sugar 5.5 t/ha.

Yield of sugar/ha =

$$\frac{(\text{weight of washed beet/ha} \times \% \text{ sugar})}{100}$$

By-products

In addition to cash return, growers benefit from by-products.

Tops (leaves and crowns) Tops give fresh yield of 20–30 t/ha and are equal in feeding value to marrow stem kale; 1 ha beet tops gives yield equivalent to 0.5 ha kale at 50 t/ha. Tops must be kept free from soil and wilted for a few days before feeding, since fresh tops contain oxalic acid; 1 ha of tops should last 250 ewes one week. Dairy cows should be limited to 19 kg/d. Tops may be ensiled if free from dirt; harvesters should be specially chosen if tops are to be saved.

Pulp Pulp, of which an allocation is made to growers, is the residue left from sliced roots after sugar extraction; it is available as beet pulp or molassed beet pulp. Dry pulp or nuts are equivalent in feeding value to oats.

Mangels and fodder beet

These crops are capable of giving very heavy yields of highly digestible dry matter. When succulent roots are included in right proportion in a ration, they can increase the total dry matter intake of cattle and sheep. Provided they are protected against frost, roots may be stored over a very long period, sometimes until early June. Mangels are an invaluable feed for ewes penned at lambing time or in times of hard frost.

Varieties

These form a continuous series from mangel to fodder beet but may be grouped as follows:

Mangels, low dry matter (10–12% DM)

Include traditional English varieties, globe or intermediate in shape, with a red, orange or yellow skin. Varieties include Prizewinner, Wintergold, Yellow and Red Intermediates. These varieties sit well on top of the ground and are easy to lift and clean by hand; they give the highest yield of roots, over 200 t/ha having been recorded. Flesh is soft and much

less likely to damage teeth of stock than roots of high DM fodder beet. Tops are small, so yield of DM/ha is some 10% below that of fodder beet, although fresh root yields are 20–30% higher.

Mangels, medium dry matter (12–15% DM)

Mainly of continental origin, usually giving higher yields of dry matter, they have larger tops but roots are smaller.

Fodder beet, medium dry matter (15–17% DM)

Mainly of continental origin, roots are intermediate to pear shaped, sitting reasonably well on top of the ground and fairly easy to lift with swede harvester or by hand. Tops large, with high yield of DM. Suitable for sheep and cattle.

Fodder beet, high dry matter (17–20% DM)

Of continental origin, roots grow mainly below ground level and are very similar to sugar beet in shape; require lifting with sugar beet harvester. Yields of fresh roots are low but these varieties can give very high yields of DM; however, if the yield of DM is to be utilised fully, it is necessary to feed the tops, which constitute some 25% of total DM yield. Excessive dirt, especially where fanging occurs, causes serious problems with feeding.

Varieties of mangel and fodder beet are listed in Farmers Leaflet No. 6, *Varieties of Fodder Root Crops* (NIAB).

Seed

Plant breeders have developed a range of monogerm varieties, mostly in the fodder beet group, which produce over 90% of single plants. Cultivation of these types can be readily mechanised, as with sugar beet, so that handwork is eliminated. Traditional varieties of mangel and fodder beet are multigerm, giving two or more plants per seed; some handwork is still necessary.

Climate and soil

Mangels and fodder beet require warm, sunny climates and produce heaviest yields in the southern half of Britain; they are little grown in Scotland. Crops are grown on a wide range of soils provided they are well drained and of high fertility. Heavy soils need careful management to produce a tilth and cause problems at harvesting.

Cultivations and seedbed

Mangels and fodder beet usually follow cereals. Stubbles should be cleaned and, with mangels and medium DM fodder beet, receive a heavy dressing of well rotted farmyard manure, 50 t/ha or more together with 400 kg/ha agricultural

Table 5.53 Manuring of mangels and fodder beet (kg/ha)

N, P or K index	N	P_2O_5	K_2O
0	125	100	200
1	100	75	100
2	75	50	75
3	–	50	75
Over 3	–	Nil	Nil

ADAS figures

salt (150 kg/ha Na) before ploughing. P and K may also be applied before ploughing, on similar conditions to sugar beet. Remaining fertiliser is applied during spring seedbed preparation, so that it is well incorporated into soil (*Table 5.53*). Where salt is omitted, increase K dressing by 100 kg/ha K_2O except on fen silts and peats.

Land must be ploughed before Christmas to allow production of a good frost mould, which secures the *stale* seedbed required and gives a fine, firm, level tilth containing adequate moisture. These pre-conditions are essential for precision drilled crops.

Sowing, plant population and spacing

Drill in second half of April in the south of England, as soon as weather permits in the north. Later sowings may yield well, but potential DM yield is reduced and drier soil conditions give poor results from soil applied herbicides. Varieties susceptible to bolting (Farmers Leaflet No. 1, *Fodder Crops* (NIAB)) should not be sown very early, especially in long-day conditions of the north.

Most crops are now sown with a precision drill; monogerm varieties are drilled to a stand, multigerm varieties may be chopped out subsequently. Aim to achieve final population of 75 000 plants/ha. Cut seed into soil moisture 12–18 mm deep in rows 500–600 mm apart. Drill pelleted monogerm 175–200 mm apart rubbed and graded precision seed (2.75–4.25 mm) 100–125 mm apart in rows respectively.

Harvesting

Mangels must mature or they will not keep. Maturity occurs in late October to November, according to district, evidenced by withering of outer leaves. Root or bulb must not be cut during harvesting and varieties with low dm need gentle handling.

Fodder beet is harvested like sugar beet with tops kept clean for feeding. Both crops can be harvested mechanically with an adapted swede harvester (not long types of fodder beet), an adapted sugar beet harvester, or multi-purpose harvesters. The crop is drilled in rows wide enough to accommodate tractor tyres; 600 mm width is the minimum required for mangels lifted by swede harvesters pulled by tractors with standard tyres. Stored roots need protection from frost by at least 0.75 m of straw on fine hedge trimmings and should be earthed up or stored under cover in cold situations.

Yields

These vary considerably.

Mangels average 75–115 t/ha of roots to 7.5–11.5 t/ha DM. Stored mangels should not be fed until after Christmas.

Fodder beet 50–75 t/ha of roots to 7.5–11.5 t/ha DM. Tops 30–50 t/ha, 5–8 t/ha DM, which must be fed to achieve full crop potential; wilt before feeding.

Further reading

ADAS (1979) *Brussels Sprouts.* MAFF Reference Book 323, 68 pp. London: HMSO

ADAS (1980) *Carrots.* MAFF Booklet 2371. 24 pp. London: HMSO

ADAS (1981) *Propagating and Transplanting Vegetables.* MAFF Reference Book No. 344, 58 pp. London: HMSO

ADAS (1982) *Bulb Onions.* ADAS/MAFF Reference Book No. 348, Grower Books No. 348, 85 pp. London: HMSO

ADAS (1982a) *Cauliflowers.* ADAS/MAFF Reference Book No. 131, Grower Books, 87 pp. London: HMSO

ADAS (1982b) *Irrigation.* ADAS/MAFF Reference Book No. 138, 122 pp. London: HMSO

ADAS (1982c) *Leeks.* ADAS/MAFF Booklet 2069, 25 pp. London: HMSO

ADAS (1982d) *Seed Rate for Potatoes Grown as Maincrop.* MAFF Leaflet 653, 18 pp. London: HMSO

ADAS (1983) *Beetroot.* ADAS/MAFF Booklet 2444, 40 pp. London: HMSO

ADAS (1984) *Calabrese.* ADAS/MAFF Booklet 2371, 24 pp. London: HMSO

ADAS (1984a) *Early Carrot Production under Plastic Film Covers.* MAFF Leaflet 888, 5 pp. London: HMSO

ADAS (1984b) *Parsnips.* ADAS/MAFF Booklet 2395, 15 pp. London: HMSO

ADAS (1984c) *Plastic Film Covers for Vegetable Production.* ADAS/MAFF Booklet 2434, 40 pp. London: HMSO

ADAS (1984d) *Rapid Cooling of Horticultural Produce.* MAFF Leaflet 860, 8 pp. London: HMSO

ADAS (1984e) *Sweet Corn.* MAFF Leaflet 297, 13 pp. London: HMSO

ADAS (1985) *Vegetable Propagation in Cellular Trays.* MAFF Leaflet 909, 12 pp. London: HMSO

ADAS/MAFF Publications: in addition to the references cited above, ADAS issue a very wide range of publications covering most field and vegetable crops. Lists of priced and unpriced ADAS/MAFF publications obtainable from: MAFF (Publications), Lion House, Willowburn Estate, Alnwick, Northumberland, NE66 2PF, UK. Unpriced individual publications obtainable from local ADAS advisors (*see* Telephone Directory)

BUTTERWORTH, W. (1985) *The Straw Manual.* E. & F.N. Spon, London: 212 pp

CAMBRIDGE AGRICULTURAL PUBLISHING (1982) *Oilseed Rape Book*, The Stag, Cornish Hall End, Finchingfield, Essex, 160 pp

GRAY, D. and SALTER, P.J. (1980) *Fluid Drilling Grower Guide.* No. 13. 27 pp. London: Gower Books

HARDY, F. S. and WATSON, G.D. (1982) *The Complete Guide to Commercial Vegetable Growing.* 294 pp. London: F. Muller

HOOPER, W. E. J. (1982) *Oilseed Rape as a Honey Bee Crop.* 4 pp. Chelmsford, Essex: Agricultural College, Writtle

RICKARD, P. C. (1979) *Plastic Mulches for Vegetable Production.* Grower Guide No. 7. 43 pp. London: Gower Books.

SALTER, P. J., BLEASDALE, J. K. A. *et. al.* (1980) *Know and Grow Vegetables.* 212 pp. Oxford: Oxford University Press

SUGAR BEET RESEARCH AND EDUCATION COMMITTEE (1985) *Sugar Beet – a Grower's Guide.* 81 pp. Obtainable from Library Brooms Barn Experimental Station, Higham, Bury St Edmunds, IP28 6NP, UK. Priced + postage

SUGAR BEET RESEARCH AND EDUCATION COMMITTEE and BRITISH SUGAR plc publish jointly British Sugar Beet Review quarterly

6

Grassland

J.S. Brockman

Grassland occupies a major part of the agricultural land of the UK, as shown in *Table 6.1*.

Rough grazing

Rough grazing can be defined as uncultivated grassland found as unenclosed or relatively large enclosures on hills, uplands, moorland, heaths and downlands. Thirty-four per cent of all agricultural land is in rough grazing and nearly two-thirds of this area is in Scotland. As shown in *Table 6.2*, the contribution of the rough grazing area to national ruminant nutrition is only 5% and this low figure indicates the low level of production associated with rough grazing. The natural vegetation has low yield and nutritional characteristics: the period of summer growth and the total yield are restricted by a combination of altitude, poor soil fertility, high rainfall, difficult access and the consequent poor control of defoliation management.

The main types of rough grazing are as follows.

Mountain grazing

This is land over 615 m in altitude, where high rainfall, low soil temperatures and extreme acidity restrict the plant species that will grow and limit their growth. Winters are very cold in these areas, farms are isolated and the stock carrying capacity of the land is very low.

Table 6.2 Contribution of grass to energy input of UK ruminants

	%
Permanent and rotational grass	52
Rough grazing	5
Forage crops	4
Other feeds (mainly cereals)	39

Source: NEDO (1974) with permission.

Moors and heaths

Moors and heaths are found at lower altitudes and are often dominated by heathers (*Calluna* spp.) in relatively dry areas and purple moor grass (*Molinia caerulea*) or moor mat grass (*Nardus stricta*) in wet areas. Heather can be utilised by sheep and these areas can be improved by liming, selective burning and application of phosphates, grass will be encouraged, often the indigenous sheep's fescue (*Festuca ovina*).

Stock will eat moor mat grass in spring, but as the season progresses it becomes unpalatable, especially to sheep. Most natural species occur in patches and clumps, their occurrence reflecting the environment and the selective grazing of

Table 6.1 Grassland in the UK (10^6 ha)

	England and Wales	Scotland	N. Ireland	Total UK	%
Grass					
rough grazing	1.7	4.2	0.2	6.1	34
permanent	4.0	0.6	0.5	5.1	28 ⎫38 ⎫72
rotational	1.2	0.4	0.2	1.8	10 ⎭ ⎭
Cereals	3.5	0.5	0.1	4.1	22
All other crops	1.0	0.1	0.02	1.1	6
Total	11.4	5.8	1.0	18.2	100

animals. Normally cattle will graze moor mat grass and purple moor grass more readily than sheep, and some improvement in overall production can be encouraged by regular grazing together with the use of lime and phosphate. Under these circumstances, encouragement will be given to bents (*Agrostis* spp.), sheep's fescue and red fescue (*Festuca rubra*). Although these three grass species have a low production potential compared with grass species that are sown in lowlands, they have a greater value than *Molinia* and *Nardus*.

Some rough grazings occur on lime-rich soils, such as the downs in the south of England, where the soil is too shallow or too steep for cultivations: here sheep's fescue and red fescue dominate the sward.

Permanent grassland

Permanent grassland can be defined as grassland in fields or relatively small enclosures and not in an arable rotation; 28% of all agricultural land is in permanent grassland and four-fifths of this area is in England and Wales. The 5.1 million ha of permanent grassland represent many different types of grassland, and as the above definition suggests, much of the area is on land not suited to arable cropping.

It has been estimated that 30% of permanent grass in the UK suffers from serious physical limitations that impede use of machines: the main factors are rough and steep terrain, stones and boulders on the surface and very poor drainage. Farmers can improve drainage and such a step can lead to a marked increase in output from this type of land. Much useful information on permanent grass is given by Hopkins and Green (1978).

Several surveys have classified permanent grass on the basis of its botanical composition, an account of the classification being given by Davies (1960). In simple terms, swards can be placed in order of probable agricultural value as follows (bearing in mind that grassland is rarely homogeneous in composition, and most natural species occur in irregular patches in the field):

rough grazing

— heather (*Molinia, Nardus*)   poor
— areas with red fescue   quality
— *Agrostis* with red fescue
   (with rushes and sedges on wet soils)

permanent grass

— mainly *Agrostis*
— *Agrostis* with a little perennial ryegrass
   (PRG) (less than 15% PRG, often
   with rough-stalked meadow grass, meadow
   foxtail, cocksfoot and timothy)
— second-grade PRG (15–30% PRG with
   Agrostis) (often with crested dogstail,
   sheep's fescue, cocksfoot and tufted hair grass)
— first-grade PRG (over 30% PRG) (often   good
   with white clover, rough-stalked meadow   quality
   grass, *Agrostis*, cocksfoot, sheep's fescue
   and timothy).

Table 6.3 Classification of permanent grass on botanical composition

	% of fields surveyed containing	
	(*Baker, 1962*)	(*Green, 1974*)
15% or more of PRG	20	26
Mainly *Agrostis*	59	56
Fescue dominant	7	10
Others	14	8

Table 6.3 gives a summary of surveys conducted by Baker (1962) and Green (1974): it shows that *Agrostis* dominates the permanent grasslands of UK, although the proportion of ryegrass swards increased slightly from 1962 to 1973; for further information *see* Hopkins (1979). From 1971 to 1980 a special Permanent Pasture Group studied permanent grassland (Forbes *et al.*, 1980). Subsequently, work on this subject has been based at the Permanent Pasture Division of the Institute for Grassland and Animal Production at North Wyke in Devon.

Rotational grass

Rotational grass (temporary grass) can be defined as grass within an arable rotation. Ten per cent of all agricultural land is in rotational grass and it has one or more definite functions, quite apart from its role in supporting an animal enterprise. These functions include:

(1) essential part of rotation to
    (a) break disease and/or pest cycle,
    (b) improve fertility,
    (c) break weed cycle,
    (d) improve soil structure,
(2) grow grass species best suited to animal enterprise, giving
    (a) better fertiliser response than indigenous grass,
    (b) high yields of grass and animal product,
    (c) good seasonal distribution of grass production.

It is estimated that about 600 000 ha of grassland are sown annually, of which 450 000 ha are in rotational systems and 150 000 ha are in areas defined as permanent grass.

DISTRIBUTION AND PURPOSE OF GRASSLAND IN THE UK

Grassland is not distributed evenly over the UK, although every county within the UK contains a good proportion of farmed grass. For example, predominantly arable counties such as Norfolk and Kent have 18% and 42% respectively of their agricultural land in grass. Most grassland occurs in the western half of Great Britain and in Northern Ireland, where rainfall is high and the soils are heavier, and so less suited to arable cropping, than in the east. (For more details of crop distribution *see* Coppock, 1976.)

In some mainly arable counties grass is grown either as a break from regular arable cropping or on land that is unsuited for other crops. In the truly grassland areas, grass is grown as a highly specialised crop, often on farms growing little but grass.

Most of the grass grown in the UK is eaten by the ruminant animals, dairy cows, beef cattle and sheep, and the proportion of total individual farm income derived from these three animal enterprises varies considerably, ranging from 100% down to a very low proportion indeed. It follows that there is a great diversity in the way in which grass is farmed. This chapter outlines general principles that underlie economic grassland production, irrespective of the type of grassland farming being practised.

Use of grass

Grass as such is rarely a cash crop, unless sold as hay, dried grass, or big-bale silage. On nearly every farm, grass is processed by ruminant animals which are adapted to digesting the cellulose found in grass and also synthesising proteins from simple nitrogen compounds (*see* Chapter 13). Whilst it is possible to measure grassland production as weight of grass produced (and research results are often expressed in this way), it is important to realise that growing grass is only a part of the production process. Apart from producing grass, the farmer is concerned about the proportion of the grass that is grown which is eaten by stock, i.e. its degree of utilisation. Also, ruminants are fed material of non-grass origin, e.g. cereals. *Figure 6.1* illustrates the essential steps.

Although the ruminant can extract most of the energy and protein it requires from grass, it has been estimated that in the UK some 40% of the energy used by ruminants comes from cereal feeds, as shown in *Table 6.2* (NEDO, 1974). In pastoral countries, such as New Zealand, where animal products have a low cash value, virtually all ruminant production is derived from grass: this illustrates that the relative proportion of grass and other feeds in the ruminant diet is related to the economic value of the animal product. Thus it is possible that grass could contribute more significantly to ruminant nutrition in the UK in the future; such a move would depend on improvements in grass production and utilisation, or other feeds becoming relatively more expensive. On the other hand it would be possible for grass to become even less significant than now if alternative feeds became more attractive – either in monetary terms or convenience to grow and feed.

GRASSLAND IMPROVEMENT

The botanical composition of permanent grass and older rotational grass is related to the nature of the environmental factors operating there. The main constraints are as follows.

Within farmers' control

Drainage
Compaction
Acidity
Low fertility
Weed problems
Low stocking rate
Poor grazing control

Outside farmers' control

Steepness
Altitude
Climate
Rough or stony terrain
Aspect

If grassland improvement is being considered, the above factors can be used as a checklist. In fields where drainage, acidity and low fertility are not major problems, grassland production can often be improved by better defoliation management. The following section gives a brief account of basic grass physiology and shows how the farmers' management of grass can have a marked influence on the type of sward produced.

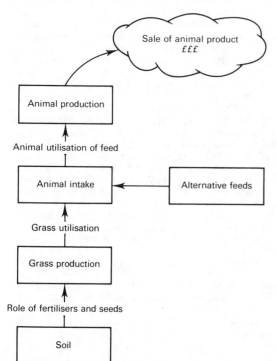

Figure 6.1 Factors affecting cash value derived from grassland

Nature of grass growth

Figure 6.2 illustrates a plant of perennial ryegrass (PRG) in a vegetative state. All the while the plant is not producing seed heads, the region of active plant growth (or stem apex) remains close to the ground and below cutting or grazing height. Growth of the plant continues in two possible ways:

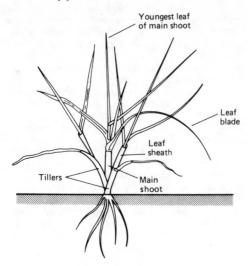

Figure 6.2 Vegetative development in grasses

(1) Increase in size of existing tillers. As each leaf grows (*Figure 6.3*), cells close to the stem apex divide and elongate, so that the youngest part of the leaf is at its base and the oldest at the tip. As each leaf develops, its base surrounds the apex, so the stem apex is continually sheathed by developing leaves. This method of vegetative growth maintains the stem apex at ground level and protects it from damage caused when the oldest part of the leaves may be removed during cutting or grazing.
(2) Tiller production. Each tiller bud contains a replica stem apex and further tillers will develop as the plant grows, giving rise to vegetative reproduction. These young tillers are also well-protected from damage and rapidly develop a root system and an identity of their own – to further continue the process of vegetative growth and development.

When grasses have the ability to tiller freely, and are encouraged to do so, they can spread laterally over the ground surface and form a thick or 'dense' sward, characterised by a large number of grass tillers per unit area.

During inflorescence development (sexual reproduction) the pattern of grass growth is quite different in two important ways. When inflorescence development is triggered, by a mechanism based on day length (and slightly modified by temperature), two changes alter the normal vegetative growth:

(1) new tiller development is suppressed and existing tiller buds start to form the inflorescence,
(2) the stems of these inflorescence buds elongate rapidly, to carry the developing ear well above the ground.

When ear-bearing stems are defoliated, they are incapable of regrowth as the inflorescence was their apex, and plant regrowth can only occur from further developing tiller buds at the base of the plant.

When a grass plant bears visible ears, much of its impetus for vegetative growth is lost, and the longer the plant is left bearing ears, the slower will be the subsequent development

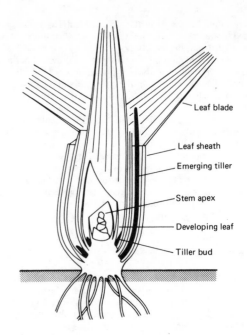

Figure 6.3 Position of stem apex and development of leaves and tillers in grass

of vegetative organs. Thus removal of a heavy, tall grass crop with ears emerged will leave a thin or 'lax' sward, with few tillers per unit area and a slow propensity to regrow. However if regular frequent defoliation is then practised, the sward will become more dense, particularly if there is a high population of PRG, as this species tillers freely in the vegetative state. The effect of cutting date is well-illustrated in *Table 6.4*, where a delay in taking the first silage cut gave a high silage yield but seriously delayed speed of regrowth.

Tillers and plants are in a state of continual change, the life of an undefoliated leaf in PRG being about 35 d in summer and up to 75 d in winter. If grass is left undefoliated, there is a rapid burst of vegetative (tiller) development in spring which is suppressed in May/June by tall inflorescence-bearing stems that restrict further vegetative development until autumn, followed by drastic tiller restriction during winter. Frequent defoliation in spring will restrict ear development, maintain tiller formation and lead to a dense sward. On soils of low–medium fertility, application of fertiliser N will further encourage overall plant growth and tiller production.

The timing and frequency of defoliations will have a marked influence on the seasonal distribution of grass

Table 6.4 Effect of first cut date on regrowth (S23/Melle)

Date cut	DM (t/ha)	Delay in regrowth (d)	Days to next 2.5 t/ha	Total yield up to 20 July
May 20	4.2	8	32	9.7
May 30	5.4	11	35	9.5
June 10	6.6	15	39	9.2
June 20	7.8	20	44	8.9

Based on Wolton, 1980.

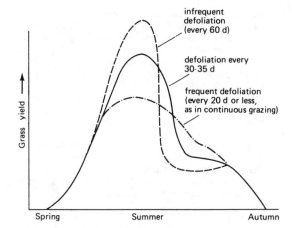

Figure 6.4 Effect of defoliation frequency on seasonal distribution

growth, as shown in *Figure 6.4*. It is important to realise that there is always an uneven pattern of grass growth during the growing season, but this pattern is exaggerated when spring defoliations are infrequent.

Most varieties of PRG can tiller freely compared with less desirable 'weed' species of grass, so that regular frequent defoliation enables PRG to compete favourably for space in the sward, and become dominant. Also most varieties of PRG are comparatively prostrate in growth habit compared with weed grasses, and so defoliation close to the ground will further encourage development of a PRG dominant sward.

The practical important of the basic physiology described above was demonstrated 50 years ago in classic experiments by Jones (1933). He showed on a relatively fertile lowland site that provided there was at least 15% PRG in the sward, an *Agrostis* dominant sward could become PRG dominant in two years if subjected to regular, tight grazing. Similarly, he showed that a newly-sown PRG sward would rapidly degenerate in composition if subjected to irregular high defoliation.

Some other plants, like white clover and creeping bent (*Agrostis stolonifera*), can propagate vegetatively by stolons: under some circumstances close defoliation will favour these plants as well.

Why improve grassland?

Any method used to improve the botanical composition of grassland will cost something and clearly it is of no value to improve for cosmetic reasons; improvement must lead to an increased profit. On most farms, it has been found that a combination of some of the following factors are most likely to lead to profitable improvement:

(1) increased output (usually from a higher stocking rate being carried *or* higher yields of silage or hay),
(2) improved herbage quality giving better animal production,
(3) longer growing season,
(4) quicker regrowth,
(5) bigger response to fertiliser N,
(6) more palatable grass,
(7) repair of areas damaged by poaching or drought.

Methods of improvement

The farmer has three improvement strategies available:

(1) reseed – plough or chemically destroy the previous crop and reseed;
(2) renovate – introduce new species into the old sward, with or without partial chemical destruction;
(3) retain – improved defoliation technique, often with attention to drainage, pH and fertilisers.

Which improvement method?

	Reseed	*Renovate*	*Retain*
speed of action	very rapid	quick	slow
cost	expensive	moderate	very cheap
stability	can revert easily	adjustable to acceptable level	improvement based on farmer's ability
special consider-ations	other limita-tions must be corrected	topography may limit	no real physical limits

Methods of grassland establishment

Grassland is established by sowing seeds either as part of a replacement/renovation procedure where grass follows grass or as part of a rotation involving other crops. Time of sowing will vary, but there are two six-week periods in the year when sowing should take place. These are: mid-March – April *and* August – mid-September.

Sowing outside these periods is risky and can be unsuccessful due to summer drought or winter cold. The timing and method of sowing the seeds will be influenced by the purpose of the crop, soil texture and structure, climate and farm type. Whatever method of sowing is used, it must be remembered that the grass seed has a relatively small food reserve and so must have an adequate supply of available nutrients near the surface of the soil. The biggest single demand is for phosphate and except on high phosphate soils a fertiliser supplying 80 kg/ha of P_2O_5 should be applied at seeding time. Also grass seeds should be sown no deeper than 25 mm below the soil surface; where sowing is preceded by cultivations care should be taken to ensure that the seedbed is level and fine. After sowing, grass establishment is enhanced greatly if the seedbed is well-consolidated; the firmer the better, provided soil structure is not impaired or surface percolation rate reduced.

Undersowing

In 1980/81 FBC Limited estimated that 33% of reseeding used this method. It can be a very successful technique provided a few simple rules are observed. The most important rule is to prevent the 'nurse' crop, usually spring cereals, from overshadowing the establishing grasses and thus depleting the density of the sward before the nurse crop is removed. Where cereals are undersown, an early-maturing, stiff-strawed variety should be used. Whilst farmers successfully undersow wheat, barley and oats, and the choice often depends on rotational considerations, barley is the most suitable, having short straw and an early harvest. Seedrate should not exceed 125 kg/ha and fertiliser N rate should be about two-thirds that which would have been used had the crop not been undersown, to avoid lodging. The seeds mixture may be broadcast or drilled, with drilling always being preferred if conditions are likely to be dry. The grasses may be sown immediately after the spring cereals are sown and the ground then harrowed and rolled: alternatively, some farmers prefer to wait until the cereal is established before sowing the grass seeds. If the cereals are sown in February, it is a good idea to delay sowing the grasses until March. When the cereals are taken through to harvest, the straw must be removed as soon as possible and often the grasses will benefit from a dressing of complete fertiliser, say 400 kg/ha of a 15:15:15 compound, followed by grazing three to four weeks later to encourage tillering.

Some farmers do not take the cereal to harvest, but cut the crop for silage when the grain is still soft: others grow cereal/legume mixtures for use in this manner. Removal of the crop at this earlier stage benefits grass establishment and is good practice, provided there is a use for the arable cereal silage that is made.

Whenever crops are undersown, it must be remembered that herbicide use on the cover crop may be restricted, particularly if the seeds mixture contains clover.

Seeding without a cover crop

Here the seed mixture is sown as a crop in its own right, the nature and timing of operations being adjusted to the needs of the grass crop. In the past this technique has been known as *direct seeding* but this term is not encouraged now as it can be confused with the modern (and different) *direct drilling*. It is advisable to sow without a cover crop in either spring (mid-March to end of April) or late summer (August to mid-September).

Where grass is sown without a cover crop, establishment must be rapid both to minimise weed competition and reduce the period that land is unproductive. Thus adequate fertiliser must be applied to the seedbed; 80 kg/ha of both P_2O_5 and K_2O and 60–80 kg/ha of N if no clover is sown. If clover is in the mixture, it is important that no more than 25 kg/ha of N is used, otherwise clover establishment will be seriously impaired. An early grazing will benefit the establishment by encouraging tillering, but a September sowing should not be grazed before winter as it will drastically reduce the sward's ability to build up root reserves during the winter.

Cocksfoot and meadow fescue establish more slowly than ryegrass, with tall fescue the slowest of all. Seeding without a cover crop is best suited to ryegrass mixtures, particularly straight Italian and hybrid mixtures with no clover.

Direct drilling

Grass and clover seeds can be direct-drilled successfully provided:

(1) the drilling date is within the periods given in the preceding section,
(2) the previous crop has been thoroughly killed off using a herbicide such as paraquat (if annual weeds are present) or glyphosate (if perennial weeds exist). If there was a dense mat of previous vegetation and paraquat is used then two half-rate sprays at three to four week intervals

should be made. After spraying with either chemical at least two weeks should elapse before sowing,
(3) the seeds should be placed 10–20 mm deep in mineral soil. Where there has been a previous mat of vegetation and/or where the ground is rough from poaching by grazing animals, it is not easy to ensure all the seeds are sown where they are required. Direct-drilling grass into grass can result in failure, principal reasons being insufficient time between spraying and sowing seeds into the surface mat and not into the soil proper.

Normal direct-drilling practice includes application of fertiliser at similar rates to the preceding section and use of 8 kg/ha of mini slug bait. Autumn sown crops are particularly susceptible to damage from frit fly and in some localities routine use of an insecticide is advised.

Slit-seeding

This technique is used for sward renovation, the principle being that at least some of the original sward is allowed to survive.

Three basic rules are:

(1) ensure aggressive weeds in the old sward are killed (particularly stoloniferous species), otherwise they will 'strangle' the newly emerging young plants;
(2) use the narrowest width of drill feasible and cross-drill if possible, to get the most rapid knitting together of the sown species;
(3) sow at time of year to give seedlings ample time to grow.

If the old sward contains a reasonable proportion of PRG, but is lax and beginning to fill with annual meadow grass (*Poa annua*) then slit-seeding may be the quickest and cheapest way of renovating the sward and boosting production. Also some farmers use this technique to introduce Italian ryegrass into old swards in an attempt to extend the grazing season and increase production. Where there are no real problems from either grassy or broadleaved weeds, then no herbicide treatment is needed. Otherwise, paraquat or glyphosate will have to be bandsprayed ahead of the slit-seeding of the grass.

There are several types of slit-seeder available, but the basic technique is to drill rows of grass seed into an existing sward with no previous cultivation. Some machines can apply a band of herbicide ahead of the drill; some cultivate the narrow band of the drill before the seed is sown; some can apply fertiliser and slug bait at time of sowing.

The fertiliser rates and sowing periods are similar to the preceding sections. In some cases, slit-seeding may be tried on very old swards and on fields not suited to normal cultivations. It is essential to check soil pH and also ensure that bad drainage will not cancel out any good that may arise from sowing new grasses.

Scatter and tread

On an opportunist basis, it is possible to renovate grassland by simply scattering the seeds on the soil surface under moist conditions. This should be followed immediately by turning in some grazing stock (preferably sheep) for 2–3 d to tread in the seeds. This technique is a gamble, as its success depends on a thorough, quick tread and then some two weeks of continuing moist, but not wet, weather to allow the seeds to establish. The technique is cheap, and can be used to patch up worn areas in grazed paddocks. Also, some farmers report successful grass establishment following mixing of grass seeds in a slurry tanker and application with the slurry.

CHARACTERISTICS OF AGRICULTURAL GRASSES

In this section, the important characteristics of the commonly-used grass species are reviewed, as a precursor to information on the formulation of seeds mixtures. All the information on the performance of grass species and on seeds mixtures is based on information from NIAB and ADAS. An essential publication is NIAB Farmers' Leaflet No. 4 (revised annually).

The important herbage grasses used in the UK are shown in *Table 6.5*.

Table 6.5 Main grass species used in agricultural mixtures in UK

	% by weight of seed sold
Perennial ryegrass	62
Italian ryegrass	18
Hybrid ryegrass (PRG × IRG)	10
Westerwolds ryegrass	1
Timothy	6
Cocksfoot	2
Meadow fescue	less than 1
Tall fescue	less than 1
Others	approx 1

Perennial ryegrass (PRG)

Perennial ryegrasses are very successful under British conditions; not only do they account for over half the grass seeds sown, but also the best permanent pastures have an appreciable PRG content. Provided they are defoliated regularly and grown under conditions of medium-to-high fertility, they will give high yields of digestible herbage over a long season. PRG has the following valuable characteristics:

(1) tiller well and so form dense swards under good management;
(2) fairly prostrate and ideal for grazing (but must be protected from prolonged overshadowing by taller species on cut swards);
(3) recover well from defoliation, both because of high tiller numbers and good rate of replacement of root reserves;
(4) persist well and are suited to long-term grassland.

PRG is classified into three main groups, based on date of ear emergence in spring. The following heading dates are for central-southern England:

Very early and early group	10–25 May
Intermediate group	26 May–1 June
Late and very late group	2–15 June

These heading dates are approximately 12 d later in northern Scotland: if a ruler is placed from Swindon to Aberdeen, then each twelfth of the distance represents one day later in heading.

In general, when early heading varieties of PRG are compared with late heading varieties, the early heading varieties:

(1) grow earlier in spring, but not as much earlier as the difference in heading date: for example a three week earlier heading date would give 7–10 d earlier spring growth;
(2) are more erect;
(3) tiller less freely;
(4) are easier to cut for conservation, both because of erect growth and good stem elongation at heading;
(5) do not grow well in mid-season.

Compared with early heading varieties, late heading varieties are:

(1) more prostrate;
(2) tiller well and are more persistent;
(3) give good mid-season growth (June–August).

PRG has a limited temperature tolerance compared with other grasses, being unable to withstand very hot summers where temperatures exceed 35 °C (not found in the UK but can occur in some temperate Continental areas) or very cold winters where temperatures persist below about −10 °C. Lack of winter hardiness is a problem with PRG in the north of Scotland, where only the most winter-hardy varieties are recommended. Winter survival in PRG is enhanced if little or no autumn nitrogen is applied and the plants are defoliated to at least 100 mm before the onset of winter.

Recommended varieties of PRG come from both British and continental breeders, with little difference in general characteristics, except that continental varieties can be slightly more winter hardy.

Breeders have produced *tetraploid* varieties of PRG. Compared with diploid varieties of the *same heading date*, tetraploid varieties:

(1) appear higher yielding and may have a greater yield above, say 80 mm, but total DM yields are generally the same. Cattle may be able to graze tetraploids more easily under some conditions;
(2) have a higher soluble carbohydrate content which may enhance palatability when grazed;
(3) are more erect, with leaf of higher moisture content; the higher moisture content can render them less suitable for conservation;
(4) tiller less freely and are less persistent;
(5) more winter hardy and more drought resistant;
(6) have larger seeds and so require a higher seed rate.

Italian ryegrass (IRG)

These are grasses with an expected duration of two to three years and heading dates in the period 10–30 May (in southern England). They are very erect and although capable of tillering, they rarely tiller enough to prevent a steady decline in sward density from establishment onwards. IRG will grow for a long season, starting growth in spring before PRG and continuing well into the autumn under conditions of high fertility. However, the bulk of growth from IRG is in the April–June period, making them particularly useful for early spring grazing and a heavy conservation cut. During the summer they tend to run into seedhead every 35–40 d and this can lower their feeding value at this time. Most varieties of IRG are even less winter hardy than PRG. The spring heading date of IRG coincides with the early group of PRG. IRG is specially useful as a rotation grass, where its two to three year duration is not a disadvantage. *Tetraploid* varieties are important, as tetraploidy emphasises the natural advantages of IRG in terms of erectness, leaf size and soluble carbohydrate content.

Hybrid ryegrass (HRG)

During recent years plant breeders have combined the vigorous pattern of growth found in IRG with the greater tiller density and persistence found in PRG. The New Zealand hybrid Grassland Manawa (formerly known as H1) has been available for many years, but is only suited to mild areas as it is not winter hardy. The first of a new series of hybrid ryegrasses coming from the Welsh Plant Breeding Station was recommended by NIAB in 1974 and more varieties have come forward. In 1968 only 2% of grass seed used in UK agriculture were HRG compared with 10% in 1985. All of the new HRG varieties are tetraploid, and the varieties at present available are best regarded as slightly more persistent types of IRG that are also rather more leafy in summer than IRG and can head in early June, rather than May.

It is likely that other hybrid grasses, not necessarily involving a ryegrass parent, will become available as breeders perfect new techniques for developing useful crosses between grass species.

Westerwolds ryegrass

These are annuals. When sown in the spring or summer they flower in the year of sowing and do not persist over winter. They can be sown in autumn in mild areas and can then provide very early spring growth. They are only of real value where high yields of grass are required for a short period and they are best regarded as catch crops, otherwise IRG should be used, as it will give equivalent yields and longer duration for a similar or lower seed cost.

Timothy

Timothy is a persistent and winter hardy grass that will grow well in cooler, wetter parts of Britain. Also, it has survived well in drought years in other areas. Varieties of timothy head 1–25 June – as late and later than the latest heading of the PRG varieties. Timothy is very palatable and grows well in summer and autumn, but when grown alone it does not give a dense sward or very high yields. It is an ideal companion grass in seed mixtures and historically was included with meadow fescue and clover. Today it is used with PRG and is found in most mixtures sown in cool, wet parts of the country. Timothy seed is some two and a half times lighter than ryegrass seed, so its percentage use by weight of 6% shown in *Table 6.5* is an underestimate of its real importance. Timothy is very winter hardy and is widely used in Scotland, where sometimes it is grown alone for heavy hay cuts followed by sheep grazing.

Cocksfoot

Cocksfoot is a native grass that is very responsive to nitrogen and, with a very well developed rooting system, is useful on light soils in low rainfall areas, where it shows drought resistance. It grows and regrows rapidly, is erect in habit and has relatively broad leaves for a grass, the leaves also being coarse and hairy. Cocksfoot heads in early to mid May and has an aggressive growth habit that makes it competitive in swards that are not closely defoliated during the main growing season. Over recent years cocksfoot has declined in popularity as it was found that its digestibility was below that of PRG and IRG. Breeders are producing new varieties that will correct this deficiency, and also are seeking to breed varieties that are less coarse and hairy, thus increasing its palatability.

Meadow fescue

This is a very adaptable grass that tolerates a wider range of climatic conditions than ryegrass. It heads mid to late May. It does not tiller as freely as PRG and so forms an open sward that is more prone to weed invasion, particularly as it is not aggressive. Meadow fescue makes an ideal companion grass in mixtures, for example with timothy and/or cocksfoot: also, it remains leafy during summer under conditions of fairly low fertility, unlike PRG which demands higher fertility if it is to remain leafy. As fertiliser use on grass has increased over the last 30 years, the importance of PRG has risen and the popularity of meadow fescue has declined, except where the winters are too severe for PRG – for example in Scandinavia. Meadow fescue could become more important again if there is a resurgence in the use of white clover, where timothy/meadow fescue/white clover swards could become very important.

Tall fescue

This is one of the earliest grasses to grow in spring, it is winter hardy, drought resistant and very persistent under regular cutting (which PRG is not). It is very slow to establish, taking up to one season to become fully productive. Also it has a very rough, stiff leaf that is extremely unpalatable to grazing stock, although this unpalatability does not apply to silage made from the grass. Tall fescue's main role is on grass drying farms in eastern England, although a few farmers on light land have used tall fescue in areas they cut regularly for silage.

HERBAGE LEGUMES

Legumes have root nodules that can be inhabited by rhizobial bacteria which can fix atmospheric nitrogen (N) and make it available to the host plant. Herbage legumes are grown either on their own ('straight') or mixed with grasses

Table 6.6 Main forage legumes used in UK

	% by weight seed sold
White clover	46
Red clover	35
Alsike	5
Lucerne	6
Others	8

where their ability to acquire fixed N can also benefit associated grasses.

The important herbage legumes used in UK are shown in *Table 6.6.*

Red clover (RC)

There are three main types of red clover:

(1) *Single cut* Traditionally this was used for one hay cut in rotational farming where the clover break lasted one year. It is slow to grow in spring and lower yielding than other types.
(2) *Early red* (also Double-cut or Broad Red) These give early growth and high yields over two cuts and the possibility of autumn grazing. It is not very persistent particularly when grazed regularly and is best grown for two years only, either straight or in mixtures with erect high-yielding grasses such as IRG for cutting.
(3) *Late red* These flower two to three weeks later than Early Red and are more persistent, being suited to medium-term leys where a red clover constituent is required. Such leys should be cut periodically as red clover does not persist well if regularly defoliated more frequently than every 35 d.

Single cut clover is scarcely used now and NIAB do not recommend varieties. In both other groups there are recommendations for both diploid and tetraploid varieties. Tetraploid red clovers do not necessarily give a higher yield than diploid counterparts and choice of variety can depend on selection for disease resistance. Red clover can suffer severely from two diseases:

(1) clover rot (*Sclerotinia trifoliorum*),
(2) clover stem eelworm (*Ditylenchus dipsacci*).

Both diseases can be problems in areas where red clover has been commonly grown, e.g. in the arable areas of East Anglia. Where either disease is present, resistant varieties must be grown. Tetraploids give low seed yields and have bigger seeds than diploids, so tetraploid seed in expensive.

White clover (WC)

White clover is scarcely ever grown straight in the UK: even white clover seed is produced from suitable defoliation management on a grass/clover sward. Unlike red clover, white clover is persistent under grazing, often developing well when associated grass competition for light is reduced by constant defoliation. White clovers are classified according to their leaf size:

(1) *Small leaved* These are very prostrate and have been selected from cultivars existing in old pastures (e.g. Kent Wild White). They are very hardy, but always remain small even when under favourable growing conditions: thus they do not stand up to grass competition when fertiliser N is used or when a conservation cut is taken. Their real value is on sheep grazed swards, conditions of low fertility and at high altitudes.
(2) *Medium leaved* Leaf size is not a consistent characteristic in the field: if a medium-large clover is used in a sward that is regularly and tightly grazed, its leaf size will be quite small. Also, there is a general tendency for persistence to decrease as leaf size increases.
(3) *Large leaved* In recent years breeders have produced bigger types of WC that can persist better under high N usage and in conserved swards. Even so, large-leaved WC does not persist for as many years as medium or small leaved WC at lower levels of N use.

It is important to select the most appropriate type of white clover for the expected management conditions: often it is advisable to sow a range of types in the mixture.

Alsike

This is rather more persistent than red clover under grazing and is also more tolerant of wet acid conditions and is more winter hardy and disease tolerant. It is often used with late red clover in mixtures.

Trifolium (crimson clover)

An annual, its main use is to provide a heavy crop of green material in May following late summer sowing. It can be sown into cereal stubbles, either straight, or with rye (corn) or IRG. Seed rates are 25 kg/ha if alone, or 10 kg/ha with 125 kg/ha of rye, or 15 kg/ha with 10 kg/ha IRG.

Trefoil (black medick)

Very useful for heavy yields of catch crop on thin calcareous soils. A mixture of 6 kg/ha with 20 kg/ha IRG sown in August will give excellent production the following spring for early grazing or conservation.

Sanfoin

See Chapter 5.

Lucerne

See Chapter 5.

BASIS FOR SEEDS MIXTURES

Seeds mixtures should be made up from specific varieties that are recommended for their individual merits. There are very many seeds mixtures available to farmers in the UK and the following outlines a way in which the suitability of mixtures can be examined and appropriate mixtures formulated. A golden rule is to keep the mixture as simple as possible commensurate with its function, also ensuring that sufficient of each ingredient is present to confer its expected characteristic.

Purpose of mixtures

Spread growth pattern

A single variety will exaggerate the spring peak in production and use of several varieties with varying heading dates can spread this peak to some extent, e.g. a mixture of early, intermediate and late PRG.

Speed up regrowth

When grasses are defoliated for conservation, regrowth is slow from grass plants that were fully headed. If a mixture is used it is possible that some grass plants are relatively vegetative and will give rapid growth from tillers when the crop is cut. *However*, the date of cutting such mixtures is critical, as if cutting is delayed, not only will the later heading varieties start to head, but also the feeding value of the early-heading varieties will fall rapidly.

As a general rule:

(1) *for cutting areas*, have grasses with well-matched heading dates;
(2) *for grazing areas*, a greater spread of heading dates can give a greater spread of growth over the season than a single cultivar.

Easier grazing management

With a mixture the timing of grazing is less critical than with single-variety swards. Also, if a range of varietal types is present, there is a good chance that varieties well-suited to the particular management will develop. However there is always a danger of incursion of weed species if this management is *too* lax, or if some of the varieties in the mixture are too short-lived for the expected duration of the ley.

Cheaper seed cost per hectare

This is possible as merchants can buy seed under contract or in bulk. The individual components of a mixture should be checked, as sometimes mixtures include non-recommended and unsatisfactory varieties.

Lessen disease risk

Grasses can suffer from a number of diseases and pests such as crown rust and ryegrass mosaic virus, but so far research has not shown the economic merit of controlling these. It is likely that some varieties are less susceptible to pests and diseases than others and a mixture spreads the risk.

Types of mixture

Mixtures can be classified into 12 possible groups, based on expected duration and intended use, as follows:

very short term (1–2 years)
short term (2–3 years)
medium term (3–5 years)
long term (over 5 years)

} × {
mainly for cutting
mainly for grazing
general purpose
}

Table 6.7 gives an outline of the basis for formulating seeds mixtures in ten of these groups and suggests the other two areas are not appropriate.

Seeds rates

The correct seed rate will vary to some extent with the climatic and soil conditions at sowing and expected during the establishment period. The most common single problem is that seeds are sown too deep, either becuase of bad drilling practice or when broadcast on to a rough seedbed. Under *ideal* sowing conditions, experiments have shown that a PRG seed rate of 4 kg/ha will establish a good sward. Farmers always use much higher rates than this because field conditions are not ideal, but even so it should be remembered that although a very high seed rate may give a rapid green cover, it will not lead to higher yields of grass as plant competition kills out many of the young plants within three months of sowing.

The following rates are guidelines for seed rates of straight species under good seedbed conditions:

IRG (diploid)	35 kg/ha	PRG (diploid)	25 kg/ha
IRG (tetra-ploid)	40 kg/ha	PRG (tetra-ploid)	30 kg/ha
Meadow fescue	25 kg/ha	Cocksfoot	25 kg/ha
Timothy	15 kg/ha[1]	Tall fescue	40kg/ha
White clover	2 (+ grass)	Red clover	10 (with IRG)
			4 (in leys)

[1] But rarely sown alone.

For mixtures, the appropriate mean can be taken, *for example*:

20% of mixture tetraploid IRG, 20% of 40 = 8 IRG
60% of mixture diploid PRG,   60% of 25 = 15 PRG
20% of mixture timothy,   20% of 15 = 3 timothy
                                         ‾‾‾‾
                                         26 kg/ha

For difficult seedbeds, add 20% to all the above seed rates.

Check-list for studying mixtures

The main points to note are:

(1) rate/ha of individual components, at least 4 kg/ha for grass (2 kg/ha for timothy), at least 2 kg/ha for clover;
(2) use of recommended variety, if not recommended, is there a reason? e.g. suited to locality, cheapness, personal preference;
(3) diploid or tetraploid, bearing in mind characteristics of each;
(4) heading dates, generally little spread for spring cutting swards;
(5) date at which the grass is at 67-D (note the silage made from this grass would have a D-value of 64–65);

Table 6.7 Basis for seeds mixtures

Main use	Very short term (1–2 years)	Short term (2–3 years)	Medium term (3–5 years)	Long term (over 5 years)
Cutting	35 of 1a (diploid) or 40 of 1a/b (tet) or 40 of Westerwolds or 25 of 1a/b + 8 of 8a	40 of 1a/b or 20 of 1a/b + 15 of 2a	10 of 1b + 10 of 2a + 10 of 2b or 10 of 2a + 10 of 2b + 5 of 2c + 3 of 7c	not applicable except 35 of 6a for grass drying
Grazing	40 of 1a/b (tet) or 20 of 1a/b + 15 of 2a (tet)	20 of 1a/b + 15 of 2a/b or 15 of 2a (tet) + 15 of 2b + 3 of 7c	8 of 2a + ⎫ dip 8 of 2b + ⎬ and 8 of 2c + ⎭ tet 4 of 3a/b + 3 of 7b/c or 20 of 5a + 5 of 3a/b + 3 of 7b/c	12 of 2b + 12 of 2c + 4 of 3b + 3 of 7a/b or 24 of 2c + 4 of 3b + 3 of 7b (some 7a for sheep)
General purpose	not appropriate	not appropriate	8 of 2a + 8 of 2b + 8 of 2c + 4 of 3b + 3 of 7b/c or 8 of 2a/b + 4 of 3b + 8 of 5a + (4 of 4a +) 2 of 7b/c + 2 of 8b	10 of 2b + 10 of 2c + 4 of 3b + 3 of 7a/b or 10 of 2b/c + 4 of 3b + 4 of 5a + (4 of 4b +) 2 of 7b + 2 of 8c

All rates are in kg/ha: several varieties may be used within a species group, but no variety should be used at less than 4 kg/ha (1.5 kg/ha for white clover, red clover and timothy varieties)

Key to species groups:

1a = Italian ryegrass	3a = Early timothy	7a = small white clover
1b = Hybrid ryegrass	3b = Late timothy	7b = medium white clover
2a = Early PRG	4a = Early cocksfoot	7c = large white clover
2b = Intermediate PRG	4b = Late cocksfoot	8a = early red clover
2c = Late PRG	5a = Meadow fescue	8b = late red clover
	6a = Tall fescue	

(6) persistence, particularly important in medium- and long-term mixtures.

Table 6.8 illustrates the way in which a seeds mixture can be analysed. Points to note are:

(1) the presence of IRG in a mixture sown to last over three years is undesirable: as the IRG dies out there is a serious risk weeds will take its place;
(2) although the use of tetraploid PRG can be an asset under cow grazing, both are present in insufficient quantity;
(3) there is a reasonable spread of HD in the PRG varieties;
(4) the mixture uses two outclassed varieties and one unrecommended variety;
(5) the mixture could be cut about 30 May for grass at 67-D;
(6) the total seed rate is for ideal conditions and should be raised to 30 kg/ha for other conditions.

Table 6.8 An analysis of a seeds mixture available in 1987. (Name of mixture, 4–5 year ley; purpose, mainly for cow grazing)

kg per acre	Species	Variety	Ploidy	NIAB recc	HD Date 67D	Pers score
5	IRG	Wilo	tet	O	47 56	$2\frac{1}{2}$
7	PRG	Cropper	dip	G	43 54	$6\frac{1}{2}$
2	PRG	Reveille	tet	O	48 58	5
5	PRG	Melle	dip	G	74 65	8
2	PRG	Meltra	tet	G	65 65	6
2	Tim	Climax	dip	X	70? ?	?
2	WC	Huia	dip	G		
25						

See text for comments

FERTILISERS FOR GRASSLAND

As with any crop, the soil is able to supply some or all of the essential nutrients needed for growth. Fertilisers are used to make up the difference between those which the soil can supply and the quantity required by the particular crop. In grassland the nitrogen (N) supply is so important that it has the effect of controlling the level of grassland yield and the requirement for other nutrients is related to the yield of grass produced and hence the N supply. Because grass has to be utilised through an animal system to achieve monetary value, not all farmers seek the highest possible grass yields, but rather to balance the level of grass yield with the requirements of their particular stock enterprises. The skill of economic grassland production lies in balancing production with requirement very carefully.

It follows that farmers are justified in operating at a wide range of N inputs on their grassland, depending on the yields needed. As the next section shows, the quantity of N available from non-fertiliser sources can vary, and so the amounts of fertiliser N used to 'top-up' N supply will range from nil up to more than 400 kg/ha of N (equivalent to 1160 kg/ha of ammonium nitrate fertiliser). It is important the rate of fertiliser N applied is adjusted for the level of production required and the other sources of available N.

Sources of available nitrogen

Soil nitrogen

Very large quantities of N are present in the organic matter in the soil, for example a value of 10 000–12 000 kg/ha can be expected under old grassland, but most of this N is present either in living matter or complex organic compounds and is not available to plant roots. However, biochemical activity does release some N in available forms each year, the quantity depending on:

(1) organic matter (OM) content of the soil,
(2) biochemical activity of that organic matter,
(3) length of time in grass.

There can be a very wide range in available N supply from the soil, from 20 kg/ha on soils low in OM (e.g. old arable and light-textured soils) to 150 kg/ha on soils with large quantities of active OM (e.g. old grassland). In the UK there is no accepted method of soil analysis for predicting available soil N under grass and estimates have to be based on the above factors.

For example, assume the following annual values:

30 kg/ha for a young ley;
100 kg/ha for a fairly productive grazed sward over three years old;
150 kg/ha for a highly productive permanent pasture.

Animal recirculated N

Grazing animals retain only some 5–10% of the N they eat in herbage, voiding about 70% of returned N in urine and 30% in dung.

Potentially, grazing animals can recirculate a high proportion of herbage N, but both dung and urine occur in relatively small patches in the field, affecting only a small proportion of the sward in one grazing season. Recirculation of N by animals is not very effective in the first two years of grazing and after that the effect is related to stocking rate (SR). For grazed swards over two years old, assume the following values for animal recirculated N:
nil if grazing season SR less than 2 'livestock units' (LSUs)/ha.
20 kg/ha if grazing season SR 2–2.5
40 kg/ha if grazing season SR 2.5–3.5
60 kg/ha if grazing season SR 3.5–5.0
90 kg/ha if grazing season SR over 5 LSUs/ha

Where slurry (and farmyard manure) is applied to grassland there is an effective recirculation of animal N, the value depending on the rate of application and the N-value of the material (*see* Chapter 4).

Clover nitrogen

Rhizobial bacteria in clover root nodules will fix atmospheric N and this N is used by the clover plant. As the clover plants grow, old roots die and some fixed N in these roots becomes available to grass roots: this is termed *underground N transfer*. Also clover foliage contains fixed N and if the clover is grazed, then some of this N is returned to the soil in dung and urine: this is termed *overground transfer*.

The value of clover N depends on the quantity of clover in the sward. Where there is a very large amount of clover, likely to average 30% or more of the weight of herbage grown in the year, then clover N can be equivalent to 150 kg/ha of fertiliser N (and even 200 kg/ha under some favourable conditions). At the present time, it is rare to find a grass/white clover sward that contains enough clover to average 30% over the year. *Figure 6.5* shows the differing seasonal growth patterns of grass and white clover. The best time to assess clover content is in July/August, when clover content must exceed 60% if the sward is to average 30% over the year. Scoring clover content in summer can give a guide to the value of clover in grazed swards. If 60% clover, assume 150 kg/ha N, 30% clover assume 75 kg/ha, and so on.

Even with skilful management the application of fertiliser N will decrease the clover content, so that fertiliser N and clover N are not additive. As a rule of thumb, application of every 2 kg/ha of fertiliser N will decrease clover N by 1 kg/ha. This means that if 300 kg/ha of fertiliser N is applied over the

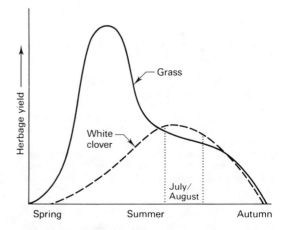

Figure 6.5 Relative seasonal production of grass and white clover in grass/clover swards

season to a grass/clover sward, then clover N contribution will be nil (and *pro rata* for lower fertiliser N rates).

Fertiliser nitrogen

This should be applied to make up the difference between the total available N supply from other sources and the required N level (*see Table 6.11*).

Efficient use of fertiliser N

The target annual rate of fertiliser N application can be obtained from the above. The seasonal use of this N can have a marked influence on the quantity of grass produced and the economic value of the N.

As shown in *Figure 6.6* grass takes up N much more rapidly than it grows in response to that N. Not only is N uptake always more rapid than growth response, but also N uptake can occur when temperature and moisture limit growth. The greenness of grass is associated with N content (i.e. N uptake) and not grass growth, and it does not follow that very green grass following N application is growing rapidly. Research has shown a definite relationship between the quantity of N grass can use for growth and the number of days of active growth available to the grass. From this work, a *maximum* N rate of 2.5 kg/ha for every day of active growth is suggested. Thus *maximum* individual dressings of fertiliser would be:

for 21 d paddock grazing, 21 × 2.5 = 52.5 kg/ha

monthly application in continuous
grazing, 30 × 2.5 = 75 kg/ha

for silage grown 40–50 d,
40–50 × 2.5 = 100–125 kg/ha

Because grass takes up N so rapidly and completely, there is very little residual N left following defoliation: where high annual rates of fertiliser are needed, some N should be used for each growth period rather than relying on infrequent heavy dressings and a season of 200 d of active growth would justify a maximum of 200 × 2.5 = 500 kg/ha of N.

Grass grows most rapidly in spring (May and June) and it gives the biggest responses to N in this period. However very high rates of N application in the spring may deplete tiller production and limit valuable mid-season growth. It is suggested that in grazing systems the maximum daily N rate should never exceed 2.5 kg/ha, even in spring. Where silage cuts are taken, a lack of big N responses in summer should be recognised and a guide is 120 kg/ha of N for spring silage, 100 for July cuts and 80 for September cuts. From September onwards the day-length is decreasing rapidly and grass growth slows appreciably: it is recommended that no N dressing should exceed 40 kg/ha after mid-September.

N rate and site class

As a result of an analysis of the National Manuring Trials, it was found that the major factor determining grass response to fertiliser N was soil available water content (AWC) (Morrison, Jackson and Sparrow, 1984). In situations where AWC was reasonably good throughout the growing season the growth of grass could be maintained by adequate N application. Where AWC fell to low levels during the summer grass growth was restricted by lack of water and failed to respond fully to N application.

AWC can be estimated from two easily assessed factors:

(1) water holding capacity of the soil, based on the combination of soil texture and rooting depth;
(2) summer rainfall.

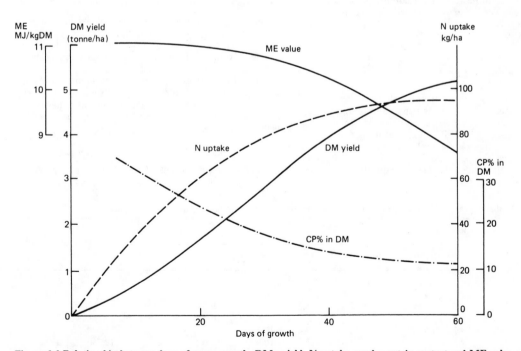

Figure 6.6 Relationship between days of grass growth, DM yield, N uptake, crude protein content and ME value

Table 6.9 Estimation of site class

Soil type	April to September rainfall (mm)		
	over 400	300–400	under 300
Clay loam and heavier (moisture retentive)	1	2	3
Loams, deep soils (intermediate)	2	3	4
Sandy and shallow (dry out easily)	3	4	5

If field is over 330 m above sea level, add 1 to above score

Table 6.9 shows how the use of the two factors can lead to an appropriate *site class number* being given to a farm (or part of farm).

It was also found that altitude had some effect on grass growth (mainly via temperature) and so it is recommended that 1 is added to the value derived from *Table 6.9* if the altitude is over 330 m above sea level.

It follows that fields with a high site class number have a more limited yield potential than low site class sites: thus it is logical that the *maximum* fertiliser N rate and total annual DM yield will vary with site class (*see Table 6.10*).

Table 6.10 gives a refinement to the '2.5 kg/ha/d of growth' rule for N outlined earlier. Two hundred days of growth (giving a maximum N rate of 500 kg/ha) would be found only on Class 1 sites and all other sites, with less growing days, would justify less N over the season. Normally, the reduction in fertiliser N use in these other sites should come from decreasing the number of fertiliser applications and *not* by diluting the 2.5 kg/d rule in the first part of the season.

Phosphate and potash

The soil can supply some or all of the phosphate and potash needed by grass, fertiliser being required to bridge the gap between supply and demand. As a guide, total phosphate demand is 60–80 kg/ha of P_2O_5 and that for potash is 150–300 K_2O, depending on the level of grass production (and hence available N supply). For both phosphate and potash soil analysis can be a useful guide to the quantity of fertiliser needed (*see Table 6.11*).

Grazing animals

These return a substantial proportion of both the phosphate and potash eaten, virtually all the phosphate being in dung and most of the potash in urine. Dung is very unevenly distributed over the field and broken down very slowly to release the phosphate: as a result animal recirculation of phosphate is not assumed. Urine covers a much greater area of the sward than dung and all the potash in urine is in a readily available state: thus allowance is made for potash recirculation on grazed swards.

Plant uptake of phosphate is steady and is well-related to plant needs: it follows that the annual fertiliser requirement for phosphate can be applied in one annual application if desired, without causing nutrient imbalance. However, uptake of potash by the plant is related to the quantity of potash available to the roots and *not* to the plant needs: thus grass can take up in a single growth period as much as twice the potash it needs for adequate growth. This excessive uptake is called 'luxury uptake' and can have two harmful consequences:

(1) Mg content is inversely related to K content in grass, so the unnecessarily high level of K leads to very low Mg content and a greater risk of hypomagnesaemia in stock eating the herbage, particularly in spring and autumn when herbage Mg contents in grass are naturally low;
(2) growth following defoliation of this herbage may be restricted by lack of available potash.

Where high annual rates of potash are needed, some potash should be applied for each growth period, except none should be applied in spring to swards that will be grazed.

Fertilisers for grassland establishment

Young developing grass and clover plants have a high demand for phosphate and potash, particularly phosphate. As the young plant has a very small root system, it is necessary to ensure that fertiliser supplies large quantities of available phosphate and potash.

For undersowing use normal cereal recommendation for phosphate and potash at sowing: if soil is index 0 or 1, apply 50 kg/ha of P_2O_5 and/or K_2O immediately after harvest.

For other methods of establishment apply the following before or at time of sowing:

	kg/ha of P_2O_5				kg/ha of K_2O			
Soil index	0	1	2	3+	0	1	2	3+
	100	80	60	40	100	80	40	nil

Table 6.10 Probable grass DM yields (t/ha) at range of fertiliser N rates and maximum N rate to apply

Site class	Fertiliser N (kg/ha)										Maximum to apply to apply
	0	50	100	150	200	250	300	350	400	450	
6	1.4	2.5	3.8	4.9	6.1	7.2	7.5 at				270
5	1.6	2.9	4.2	5.4	6.6	7.7	8.4 at				300
4	2.0	3.3	4.6	5.9	7.2	8.2	9.1	9.5 at			330
3	2.4	3.8	5.1	6.4	7.7	8.7	9.6	10.3	10.5 at		370
2	2.8	4.2	5.6	7.0	8.3	9.4	10.3	11.0	11.5	11.6 at	410
1	3.2	4.7	6.1	7.5	8.9	10.0	10.9	11.6	12.2	12.7 at	450

Table 6.11 Basis for fertiliser recommendations on grassland

Stocking rate (LSUs/ha)	Amount of nutrient to apply (kg/ha)												
	N		P_2O_5 soil index				K_2O soil index					
	High clover and/or soil N	Low clover and/or soil N	0	1	2	3^2	Utilisation	0	1	2	3
1.0–1.3	none	50–100	60	40	20	0^2	cut	100	80	40	0
1.3–1.7	40–100	100–160					grazed	80	40	20	0
1.7–2.2	$80–150^1$	160–300	80	60	30	0^2	cut	200	120	90	45
over 2.2	$150–300^1$	over 300					grazed	80	60	30	0

[1] High clover N unlikely at these stocking rates
[2] Some may be needed every three to four years to maintain level

Young grass also benefits from some fertiliser N, particularly when sown in spring. However, little or no seedbed N should be used if good clover establishment is expected.

	kg/ha of N in seedbed	
	spring sowing	autumn sowing
IRG, no clover	80	60
PRG types, no clover	60	50
PRG types, with clover	0–30 (max)	0

Sulphur

Application of fertiliser N to grass increases the synthesis of S-containing proteins, so the S requirement of grass is related to N use. At low N levels the S requirements are modest but high N grass needs the relatively large amount of about 40 kg/ha of S over a growing season. Few fertilisers contain much S and in recent years most available S has come from atmospheric pollution. As the pollution is reduced, more S is needed from other sources to meet the requirements of high N grass.

Table 6.12 gives results from 23 sites of an ADAS experiment studying grass grown for silage. The results show two important aspects.

Table 6.12 Grass response to sulphur under regular cutting (t/ha of DM)

	N rate (kg/ha)			
	0	200	400	600
Nil sulphur	5.0	10.3	12.2	12.5
50 kg/ha of S	5.2	10.6	13.2	13.6

% increase in DM yield from each cut due to S (mean of N rates 400 and 600 only)

cut 1	0
cut 2	10
cut 3	20
cut 4	18

(1) S response interacted with N, with the biggest yield response to added S at the highest N rate;
(2) major S responses occur after utilisation of the first growth in the season, as S deposition over winter is often adequate to support spring growth.

Calcium

Calcium is applied in lime and maintenance of a pH of 5.8 or above will ensure that acidity does not limit the growth of grasses and clovers *and* that adequate calcium is present in the soil.

Magnesium

Magnesium deficiency can occur in grazing animals as hypomagnesaemia. This is a problem that cannot be cured reliably by applying magnesium to the soil; the best prevention of the disease is to place magnesium compounds in the animals' food or water. However, where soil magnesium is low, it is sound practice to use a magnesium-containing form of lime when lime is applied.

Other nutrients

Grassland in the UK suffers rarely if ever from other nutrient deficiencies, although occasionally grazing stock may suffer from lack of minerals such as copper. The safest rule is to supply the deficient element directly to the animal rather than through the herbage: this not only ensures that the animals concerned receive a correct dosage but also ensures that areas of grassland with adequate levels of minerals are not enhanced to reach toxic levels.

PATTERNS OF GRASSLAND PRODUCTION

The basic pattern of grassland production is shown in *Figure 6.7* and shows that grass can be available for grazing over a

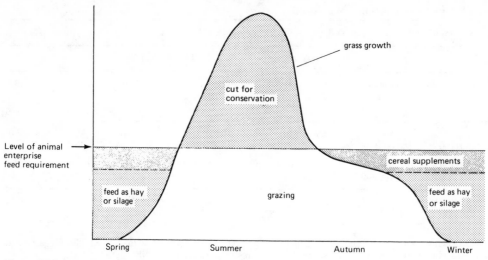

Figure 6.7 Pattern of grass production and use

long period of the year, with conservation as hay or silage removing surplus grass and making this grass available for winter feeding. As conserved grass has a lower feed-value than fresh grass, other feeds have to be used with hay and silage, normally a cereal-based concentrate.

Generally over 50% of total annual grass production has occurred by the end of May and some management practices can emphasise the spring peak even more than this. For example, heavy use of N during spring with much lower rates later in the year can result in 70% of annual production by the end of May: where one very heavy conservation cut is taken in early June, this cut can account for as much as 85% of annual production.

Spring peak in production will be *emphasised* by:

(1) very high N use in April–May, particularly as this can limit July–September production below that which would otherwise be obtained;
(2) taking one very heavy cut in spring, particularly as this will deplete tiller numbers and subsequent regrowth;
(3) use of a single grass variety, particularly an early-heading one;
(4) low summer rainfall reducing production from June onwards.

Spring peak will be *minimised* by:

(1) restricting N use in April–May, although this will lower annual grass yield as grass is most responsive to N during this period;
(2) very frequent defoliation in the May–June period;
(3) using a mixture of grasses with different spring growth periods.

Effect of site class

Figure 6.8 shows the effect of site class on the seasonal growth pattern of grass receiving optimum N supply throughout the season, and indicates there are two parts to the growing season:

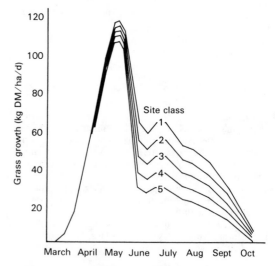

Figure 6.8 Seasonal pattern of DM production from a perennial ryegrass sward at five site classes

(1) up to the end of May, when site class has little effect on yield;
(2) from June onwards, when differences in site class can give a 100% variation in expected grass yields.

Farmers have to take into account such differences in grass growth when planning their grassland management strategy.

For example, if it is assumed grazing animals require the same amount of grass every day throughout the season and the accumulated 'surplus' can be cut periodically for silage, then it might be possible to take two to three cuts on site class 1 or 2 farms. On site class 5 only one spring cut will be possible and often the overall stocking rate will be lower as well.

Estimation of site class is based in part on average rainfall.

If, in a particular year, rainfall is *above* average, the grass growth will be better than expected and surplus grass can be ensiled as an effective way of maintaining good grass utilisation. In years when rainfall is *below* average, less silage will be made and on site class 4 or 5 the growth of grass may be below even that required for grazing: here other forage may be needed, e.g. 'buffer feeding' of silage.

EXPRESSION OF GRASSLAND OUTPUT ON THE FARM

Farmers seldom weigh grass and so have little idea of grassland output as 't/ha'. Also, it is not the yield of grass *grown* that is important but the quantity of grass *utilised* by the livestock systems. There are two general ways in which the output of grassland can be assessed. There is a 'quick' way based on types of stock and overall stocking rate and a more detailed method based on utilised metabolisable energy (UME).

Stocking rate method

The forage intake of ruminant stock is closely related to their liveweight and it is convenient to take a standard 'livestock unit' (LSU) of 550 kg and express all other stock on the basis of this LSU (sometimes called 'cow equivalent'). *For example* at a particular time in the season, a farm has 250 ewes of average weight 60 kg *plus* 360 lambs of average weight 20 kg *plus* 40 heifers of average 200 kg, then the total LSU would be

$$\left(250 \times \frac{60}{550}\right) + \left(360 \times \frac{20}{550}\right) + \left(40 \times \frac{200}{550}\right)$$

$$= 54.9\,\text{LSU}$$

Not all animal types have the same appetite, and more precise LSU equivalents are given in *Table 6.13*. Baker (1964) found that forage intake could be related to animal liveweight by

$$y = 0.0234x + 0.32$$

where y = DM intake in kg/d and x = liveweight of animal.

This is a good guide for individual classes of stock, and would suggest the 550 kg LSU has an appetite of 13.2 kg DM/d.

Growing animals change weight over the year, and some animals are sold and others bought, so that the annual expression of stocking rate as LSUs depends on:

(1) conversion of all stock to LSUs;
(2) allowance for changing stock weight and/or numbers.

This latter point is best assessed on a monthly basis.
If in the previous example, the ewes were kept on the total grass area of 35 ha for the whole year, the lambs grew from 10 to 40 kg liveweight over seven months and the heifers increased in weight over the year from 160 to 350 kg. Then the animal LSU carried on the area is:

Table 6.13 Cow equivalents of other ruminant stock

Live-weight (kg)	CE		Live-weight (kg)	CE	
10	0.04	growing sheep			
20	0.06				
30	0.08				
40	0.10		40	0.08	ewes
60	0.13		60	0.10	
80	0.17		80	0.13	
100	0.20	growing cattle			
150	0.29				
200	0.38	heifers			
250	0.47				
300	0.56	Channel Island cows			
350	0.64				
400	0.73		400	0.56	beef cows
450	0.82		450	0.61	
500	0.91	Friesian dairy cows	500	0.68	
550	1.00		550	0.75	
600	1.09				
650	1.18				

From Forbes *et al.*, 1980

ewes 250, weighting 60 kg, for whole year $= 250 \times \dfrac{60}{550}$

$$= 27.3$$

lambs 360, average weight 25 kg for 7 months

$$= 360 \times \frac{7}{12} \times \frac{25}{550} = 9.5$$

heifers 40, average weight 255 kg, for whole year

$$= 40 \times \frac{255}{550} = 18.5$$

$$\text{total} = \overline{55.3}$$

So stocking rate $= \dfrac{55.3}{35} = 1.58\,\text{LSUs/ha}.$

Utilised metabolisable energy (UME)

The total energy value of the food eaten by the animal is its *gross energy* (GE). Some of this energy is passed out in faeces and the remainder is *digestible energy* (DE). Some of the DE is lost as methane from the rumen and some is passed out in urine; that remaining is termed *metabolisable energy* (ME). Some ME is lost as heat and the remainder, *net energy* (NE), is used for maintenance and production. Whilst NE is the nearest assessment to the production potential of a food, it has been found that ME is well-correlated with production and is much easier to assess on a routine basis than NE.

Energy is expressed as mega Joules (MJ) or giga Joules (GJ), where 1000 MJ = 1 GJ.

Animals extract useful (metabolisable) energy from their feed to

(1) maintain their body functions;
(2) increase body weight;
(3) lactate (where appropriate);
(4) provide nutrition for any fetus carried.

The energy required to fulfil the above functions in all classes of stock is known and so it is possible to estimate the total quantity of ME utilised by animals (this can be for an individual animal or a livestock unit *or* a whole farm, depending on the most convenient way to calculate). The ME value of all commonly used supplementary feeds is also known, so that where grass is used along with other feeds, it is possible to produce a balance sheet that leads to an estimate of UME from grass, as shown in the example below. Note this example illustrates the UME method of estimating output from grass; the exact UME standards for size and type of animal, its weight change and ME value of milk of specific quality can be obtained from published tables (and some organisations offer computerised methods).

Example 100-cow Friesian herd, averaging 5500 litres of milk and using 1.3 t/cow of concentrates per year in addition to grazing and silage made from 50 ha of grassland.

Maintenance	365 d at 60 MJ/d × 100 cows	= 2 190 000 MJ
Weight change	none	
Calf	1 calf produced/cow 100 × 2400 MJ	= 240 000 MJ
Milk	5500 litres × 5.3 MJ × 100 cows	= 2 915 000 MJ
	energy utilised	5 345 000 MJ
Concentrates	supply 13 MJ/kg DM (or 11 MJ/kg @ 85% DM) 1300 kg × 11 × 100 cows	= 1 430 000 MJ
Other feeds	none	
Utilised from forage		3 915 000 MJ or 3915 GJ
	3915 GJ from 50 ha	= 78.3 GJ/ha

Efficiency of grassland use

The UME figure gives a value for energy used; if the total quantity of ME available were known, then the % utilisation of the energy could be calculated. It is very difficult to measure grass yield on a farm, but some organisations use a prediction based on information similar to that given in *Table 6.10*.

Example The farm above has a site class of 2 and average fertiliser N use is 300 kg/ha. It is assumed the average energy value of grass (grazing and silage) over the year is 10.5 MJ/kg DM.

Grass growth	estimated at	= 103 00 kg DM/ha
ME produced	10300 × 10.5	108 150 MJ/ha or 108.15 GJ/ha

$$\text{utilisation of ME} = \frac{78.3}{108.15} \times \frac{100}{1} = 72\%$$

But note all such calculations are based on assumed yields and so are only a guide.

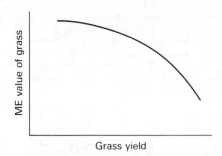

Figure 6.9 General relationship between yield and quality in grass

Grass digestibility

In the 1950s workers at the Grassland Research Institute found that the *digestibility* of grass (i.e. the percentage of grass eaten that was digested by the animal) was well-correlated with the animal intake of grass and with the subsequent animal production from it. Since that time much advice and literature refers to digestibility. By convention, the term 'D-value' (i.e. the use of the capital letter D) should refer specifically to *the digestible organic matter as a percentage of the dry matter*. There is a relationship between D-values and ME in grass, such that between D-values of 60–70 ME (MJ/kg of DM) = 0.16 D.

Figure 6.6 shows the effect of increasing grass growth period (maturity) on yield, N uptake, CP content and ME value. As grass matures it increases in yield but decreases in % CP and ME value: it is important to define the quality of grass required and defoliate it at the appropriate stage of growth. Generally high grass quality is associated with low yields and *vice versa* (*see Figure 6.9* and *Table 6.14*). Not only does a feed low in ME provide a low feed value, but also the animals eat less of such a feed. *Table 6.15* shows the double-action of low ME value on intake and production (Brockman and Gwynn, 1961).

Clearly the target ME value of grass must depend on the productive potential of the animals concerned and on whether the grass is grazed or conserved: the latter point being influenced by the inevitable loss of feed value when grass is conserved and the need for at least a moderate yield to justify the costs of conservation. Guidelines for ME levels for good production from grass are:

Utilisation	Minimum ME (MJ/kg DM)	Optimum M.
Dairy cow grazing	10.5	11.5
Grazing other growing stock	10.0	11.0
Grazing dry cows	10.0	10.5
Grazing dry ewes	9.5	9.5
Silage and barn-dried hay	10.0	10.5
Field hay	9.0	10.0

OUTPUT FROM GRAZING ANIMALS

Grazing experiments with dairy cows, beef cattle and sheep have all shown that if the quantity of herbage available to the grazing animal is reduced below a certain value, then intake

Table 6.14 Effect of cutting frequency on grass yield and quality

Three cut system				Two cut system		
Date	DM yield (t/ha)	ME (MJ/kg DM)		Date	DM yield (t/ha)	ME (MJ/kg DM)
(a) *Field data*						
late May	4.6	10.6		early June	7.8	9.6
early July	3.2	9.8		mid-August	3.7	9.0
mid-August	1.8	9.6				
Total	9.6	Mean 10.0		Total	11.5	Mean 9.2
(b) *Feeding data*						
concentrate feeding	low	high			low	high
concentrate intake (kg/d)	5.0	9.3			5.0	9.3
silage intake (kg/d DM)	11.2	9.0			10.3	8.4
milk yield (litres/d)	19.9	20.8			17.3	19.7
land required for 180-d winter (ha/cow)	0.25	0.21			0.20	0.16

Source: Leaver and Moisey, 1980.

Table 6.15 Effect of grass ME value on intake and production in dairy cows

Grass ME value (MJ/kg DM)	Grass DM intake (kg/d)	ME intake (MJ/d)	Milk yield (litres/d)
11.3	12.8	145	14.6
9.5	12.0	114	11.8
8.4	10.6	89	10.0

per animal falls below appetite, and production per animal is below potential.

Figure 6.10 summarises the data and shows that if herbage on offer is less than double that required at any grazing occasion, then intake will decline. This means that during a specific grazing period herbage utilisation should not exceed 50%. From this conclusion, two major practical points arise:

(1) can utilisation in grazing systems exceed 50% without loss of production?
(2) is it possible to monitor grazed swards to ensure animals have sufficient grass to reach appetite and thus potential production?

The remainder of this section deals with these two questions.

Utilisation

Over a season a grazed field is grazed on a number of occasions (from, say, 5–6 in a rotational system to 80–100 in a 'continuous' system). Whilst on no single occasion the utilisation should be allowed to exceed 50%, the key to good overall utilisation is to ensure that most of the unutilised herbage on one grazing occasion is available and utilised at a subsequent one.

In this way, it is feasible to achieve 70% (or even 80%) utilisation over a grazing season. Indeed, the successful use of any of the grazing methods outlined in the next section is to ensure that:

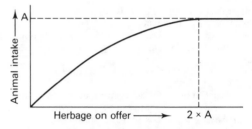

Figure 6.10 Relationship between herbage on offer at grazing and animal intake

(1) animals are not forced to eat too tightly;
(2) grass uneaten early in the season is available for utilisation later.

Monitoring herbage available

Many experiments within the Institute for Grassland and Animal Production (IGAP) have shown that sward height is a useful practical guide to the availability of herbage.

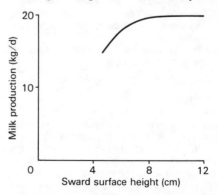

Figure 6.11 The influence of sward surface height on milk yield: average grazing season yields for spring-calving cows

For example, work with spring calving dairy cows has given the relationship shown in *Figure 6.11*. Similar relationships have been found for other classes of stock and critical sward heights have been established, which are:

Type of stock	Minimum sward height (cm)
ewes and lambs (normal)	4–5
flushing ewes, finishing lambs	6
heifers, stores, dry cows	6–8
sucklers with calves	7–9
finishing beef	7–9
lactating dairy cows	8–10

Note that as animals become larger, so the critical height is higher.

Measuring sward surface height

Place a ruler (or similar marker) vertically in the sward, with the lower edge just touching the ground. Find the height at which a finger descending the ruler first touches a green leaf. Points to watch are:

(1) ensure the ruler is vertical and is not pressed into the soil or standing in a depression;
(2) 30–40 readings are taken in the field, avoiding bias and including a fair proportion of grazed and ungrazed areas;
(3) contact with stems or seed heads is ignored.

In rotational systems the assessment of sward surface height is a guide for moving the stock.

In continuous grazing it is suggested readings are taken every one to two weeks, so that adjustments can be made to stock numbers and/or grazing area.

(For more on the practical application of sward height, *see* Hodgson, Mackie and Parker, 1986.)

GRAZING SYSTEMS

Many areas of grass, particularly those used for beef and sheep, are grazed on a very extensive basis that is not systematic in nature. On the other hand most dairy farmers have a very definite basis to their grazing policy and an increasing number of the more successful beef and sheep farmers are adopting grazing strategies similar in principle to dairy systems. This section will deal with the major methods used to graze dairy cows; some of the methods outlined are appropriate for other stock.

Whatever the grazing method or stock used, the principles outlined in the previous section apply.

Two sward system

Here one area is regularly cut for conservation and the other regularly grazed. It is difficult to accommodate the seasonal growth pattern in the grazed area, but the system has advantages where part of the grass area is inaccessible to grazing (e.g. on a split-farm) or where the cutting grass is in an arable rotation and does not justify fencing and a water supply.

Otherwise the integration of cutting with grazing is a powerful tool in smoothing out the availability of grass for grazing, because whereas only one-third of the total area may be needed for grazing in May, two-thirds may be needed in June–July and the whole area from August onwards.

For	Against
useful if part of area cannot be grazed	risk of surplus grass in grazed area in spring and insufficient later

Set-stocking

This implies a given number of stock on a fixed area for a long period, often the whole season. The term *set-stocking* is sometimes used erroneously for *continuous (or full) grazing* (*see below*). Set-stocking is advised under very extensive conditions only at low stocking rates, as there is a tendency for undergrazing in spring, leading to poor quality mature herbage, followed by overgrazing in late summer and autumn resulting in low animal performance. One advantage is that provided the perimeter of the area is stockproof, fencing costs and water supply problems are minimal.

For	Against
simple and cheap	does not match grass growth
good for large groups	possible disease build-up

Continuous grazing (or full grazing)

This refers to systems where the stock are allowed to graze over a large area for a fairly long period, say two to three months. Many farmers use this system for dairy cows and associate it with a high stocking rate, high and regular fertiliser N usage and a willingness to bring in other feeds if grass supplies become inadequate. As rate of grass growth slackens during the season, farmers can both reduce stock numbers as cows dry off and increase the grazing area by adding some silage aftermaths. Thus part of the area may be grazed for the whole season, but neither the total area nor the stock numbers are fixed, and other feed is used as a buffer against periods of low grass availability. This system depends on grass growing as rapidly as possible over the whole season and is best associated with high N application, with monthly applications of about 70 kg/ha of N from March to August and site class 1 or 2.

Some farmers have a 'rapid rotation', where the area is divided into three or four blocks and each is grazed for about one day at a time. In other situations there is one area for day grazing and another for night grazing. Because of the frequent defoliation that occurs, these are variants of continuous grazing and are not types of rotational or block grazing.

Overall there should be 1 ha of available grazing for every three cows: at turnout in spring the whole area may be needed but as growth picks up the area for spring grazing will be about 1 ha per five cows; following spring silage cuts the area may need extension to 1 ha per four cows, with the whole area in use from August onwards, depending on the season.

For	Against
reduces poaching as stock dispersed	no visible assessment of grass growth (but sward height can be measured)
saves cost in fencing and water	regular check on performance essential
increases sward density and clover	cow collection can take time
eliminates daily decision on grazing area allocation	willingness to supplement feed at grass essential

Three field system

This is one that has much to commend it if the whole grass area can be grazed conveniently (this is often difficult with cows that have to walk to and from milking and for this reason this method is more often used in beet and sheep systems). Basically this method formalises a continuous grazing system into three periods in the season and in a very simple way adjusts to the expected pattern of grass production. The grassland area should be split by stock-proof fencing into two areas, one area being approximately double the size of the other. In spring the stock graze the smaller area and the larger area is cut, preferably for silage as this will provide more reliable aftermaths than where hay is taken. As soon as the aftermaths are available, the stock are switched and the smaller area is taken for a conservation cut. Once the smaller area has available aftermaths, the whole area is grazed.

For	Against
simple self-adjusting to seasonal growth	inflexible
suitable for large animal groups	based on *expected*, not *actual* growth
good parasite control 'clean' grass	often unsuited to dairy grazing

Block grazing systems

These are ones in which the grazing area is divided into a number of fairly large blocks with the aim of grazing on a rotational basis. Thus one block of grass is grazed with a large number of animals for a short period, usually 1–7 d, and then they are moved to other blocks whilst the grazed block regrows. After about 20–28 d, the block can be grazed again. Blocks that are surplus to grazing are cut for conservation. This basic principle is used for all classes of grazing stock, with the following common amendments.

For dairy cows Farmers often like cows to have a fresh allocation of herbage each day, so an electric fence can be used to ration the grass. This is formalised in the *Wye College* system where cows are given one-seventh of a block each day and one block lasts one week.

For ewes and lambs The fence separating adjacent blocks

can have spaces wide enough for the lambs to pass but not the ewes. Thus lambs can obtain the pick of the next block to be grazed by the ewes (*forward creep grazing*) or creep into an area never grazed by ewes (*sideways creep grazing*).

For cattle rearing Young stock may be grazed one paddock ahead of older cattle so that the younger ones obtain the pick of the grass and the older cattle clean up, ensuring good utilisation. This system is called *leader follower grazing* and has been used particularly in heifer rearing.

For dairy cows Occasionally farmers split their herd and do a leader/follower with the cows, putting the high yielders in the leader group, or putting dry and nearly-dry cows as followers behind the main herd.

For	Against
efficient grass utilisation	good fencing round each paddock essential
good parasite control for young stock	regular stock-moving decisions needed
areas large enough for conservation	many water points needed

Paddock grazing

This represents a very formal method of rotational grazing, where the grazing area is divided into some 21–28 permanently-fenced and equal-sized paddocks. The aim is that one paddock is grazed each day and the rotation around the area is completed as soon as the first paddock is ready for grazing again. Surplus paddocks can be taken out for conservation but usually the operation is restricted by the small size of the paddocks. Paddock size must be related to the number of grazing animals, as the stock density must be sufficient to ensure that the grass is efficiently utilised during the short grazing period. A guide is to allow 100–125 cows/ha daily, depending on stocking rate.

Thus for a high-stocked herd of 100 cows and where 25 paddocks are selected, each paddock should be about 1 ha in size. Every paddock must have a water point and independent access on to a track to avoid poaching.

For	Against
easiest rotational system to manage	high fencing and water costs
gives objectivity to grazing plan	small areas for conservation some wasted land in trackways

Strip-grazing

This involves the use of an electric fence to give a fresh strip of herbage once or twice daily. Ideally, stock should be confined to the daily strip only by use of a regularly moved back fence to prevent the regrazing of young regrowth, but this is done rarely, because of problems with animal access and water supply. Although this is the most sensitive system to allow adjustment for fluctuations in grass growth, it is not common because:

(1) it requires daily labour to decide on area allocation and move the fence;
(2) there is a tendency for grass to become over-mature ahead of grazing in large fields;
(3) there is a risk of serious poaching along the fence line in wet weather.

Many of the best features of strip-grazing are achieved more easily in block grazing.

Zero grazing

This implies that grass at grazing stage is cut and transported to stock that are either housed or kept in a 'sacrifice' area. Zero-grazing is practised widely in some countries where grass is inaccessible for grazing (e.g. in 'strip field' systems) or where a range of crops needs to be grown to ensure continuity of supplies (e.g. in semi-arid areas). In temperate climates there is research evidence (Marsh, 1975) that stocking rates can be higher under zero grazing because:

(1) the sward does not suffer the deleterious physical effects of treading and selective grazing,
(2) herbage can be harvested at the ideal stage for the stock concerned, and
(3) utilisation can be high as there is no field refusal of grass (although this implies the stock eat most of the grass carted to them).

Both milk and meat production per hectare have been shown to be greater from zero grazing than from other grazing systems. However, zero grazing is not used commonly in the UK except for specific opportunist reasons such as:

(1) reduce poaching in early spring and autumn;
(2) provide 'grazing' from distant or inaccessible fields;
(3) use other crops during crisis periods, as in drought.

The reason for the lack of popularity of zero grazing despite its technical excellence for high production per hectare are:

(1) high labour and machinery costs for feeding *and* for removing slurry;
(2) complete dependence on machinery, with risk of breakdowns;
(3) output per animal tends to be depressed (but high SR gives high output per hectare);
(4) problems of refused grass at feeding face, particularly as wet grass heats rapidly when heaped.

Buffer grazing and feeding

This term does not refer to a grazing system but to the provision of additional feed if and when that available from the grazing area is insufficient. The buffer (or additional) material can come from

(1) another (often adjacent) grass area,
(2) a specifically grown forage crop,

(3) a bulk forage such as silage, hay or straw,
(4) a dry feed such as cereals, concentrates or dried grass.

CONSERVATION OF GRASS

Conserved grass is used as the basis of ruminant feeding when grass for grazing is not available. Grass for conservation is either grown specifically for this purpose or taken as a surplus in a grazed area: often both situations apply in any one year on an individual farm. Where grass is grown for conservation, it is possible to tailor the crop to fit the particular requirements of the unit, in terms of herbage varieties, yield and feeding quality. Where grass arises as a grazing surplus, an over-riding factor is the rapid removal of the crop in order to allow regrowth for a further grazing (*see Table 6.4*).

When green crops are cut biochemcial changes occur and if these are allowed to continue unchecked, degradation of the material will take place, releasing heat and effluent (water that contains much of the soluble cell contents). There are four methods by which this progressive degradation can be restrained, but at present only the first two methods are used in commercial practice and only these two methods will be discussed in detail.

Acidification

Degradation ceases when the pH of the material falls below 4.2–5.0 (depending on the moisture content) *provided* the material is anaerobic (oxygen-free). This material is *silage*, and can be made by heaping herbage in as near oxygen-free conditions as possible and allowing the natural process of biochemical change to produce acids which can effectively 'pickle' the material. Additives can be used to both accelerate the process and give a more reliable end-product.

Dehydration

Degradation ceases when the material reaches 85% DM (grass in the field averages about 20% DM). This means the production of each tonne of hay at 85% DM necessitates the loss of 3.25 t of water.

Dehydration can take place under natural conditions in the field and is *haymaking*. Sometimes hay that is almost dry is placed in a building and subjected to forced-draught ventilation to complete dehydration and is *barn hay drying*. On some specialised units, grass at or near field moisture is carted to a high-temperature drier and dehydrated very rapidly; this is *green crop drying*.

Preservation

Some chemicals can prevent the process of degradation and so preserve the material in almost the same chemical state as at cutting.

At present research is seeking suitable chemicals that will preserve herbage economically and yet not restrict animal intake, affect rumen fermentation or be carried into the animal products.

Freezing

Freezing will inhibit completely chemical breakdown in the material.

For many years research institutes have been using frozen grass in feeding experiments as a normal means of preserving fresh grass and all its intrinsic value. At present the cost of freezing and storing herbage on a farmscale is prohibitive.

Silage making

The process of ensilage consists of preserving green forage crops under acid conditions in a succulent state. When such green material is heaped, it respires until all the oxygen in the matrix is exhausted: during respiration carbohydrates are oxidised to carbon dioxide with evolution of heat. Continued availability of oxygen, as in a small outside heap of grass, will lead to enhanced oxidation with resultant decomposition and overheating.

Assuming the oxygen supply is restricted, bacteria can control the fermentation process. These bacteria, present on the crop, the machinery and in soil contamination, fall into two categories – desirable and undesirable.

The desirable bacteria are ones which can convert carbohydrate into lactic acid and are mainly of the *Lactobacillus* and *Streptococcus* species. These are anaerobic bacteria and their even distribution and activity throughout the grass in the silo is encouraged by mechanical chopping of the grass, rapid consolidation and exclusion of air. Lactic acid is a relatively strong organic acid and its rapid production within the ensiled grass leads to a low pH and conditions which inhibit the lactic acid producing bacteria and *all* other bacteria as well. The pH at which this 'pickling' occurs depends on the moisture content of the grass: the wetter the grass the lower the pH needed and the greater the quantity of lactic acid that has to be produced.

Silage DM%	*pH for stable silage*
18	3.8
20	4.0
22	4.2
26	4.4
30	4.6
35	4.8

Silage with a good lactic acid content is light brown in colour, has a sharp taste and little smell: it is very stable and can be kept for years if necessary provided nothing is done to permit oxygen to enter the material.

The undesirable bacteria are:

(1) aerobic species of bacteria that can oxidise carbohydrate to carbon dioxide and water;
(2) obligate anaerobes of the *Clostridium* species that can ferment carbohydrate and lactic acid to form butyric acid; *Clostridium sporogenes* can break down amino acids to ammonia and amines, some of the latter being toxic to stock.

An indicator of clostridial activity is the ammonium-N content of silage, as the ammonia produced by these bacteria is retained in the silage.

	Fermentation quality		
	Good	Moderate	Bad
NH_4-N as % of total N	0–10	10–15	over 15

Butyric silage is olive-green in colour, has a rancid smell and is unpalatable to stock. Also it has a higher pH than lactic silage, is unstable and will not keep for more than a few months.

Clostridial activity in silage can be inhibited by:

(1) reducing moisture content of the grass, as this will lessen the quantity of acid needed to prevent decomposition;
(2) ensuring adequate carbohydrate is present for lactic acid bacteria, or applying an acid to assist in lowering pH;
(3) adding inoculants of live *Lactobacilli* (although this is not fully proven);
(4) avoiding contamination from soil or animal manure, both of which contain large quantities of Clostridia.

Intrusion of air *during* fermentation will delay or even prevent the achievement of a stable pH and will lead to an excessive amount of carbohydrate being used, thus lowering the nutritional value of the silage. Intrusion of air *after* the silage has reached a stable condition will lead to secondary respiration and a further progressive loss of carbohydrate. Secondary respiration can shorten the storage life of well-made silage and can occur when the silo is opened for use if the exposed feeding face is too large for the rate of silage removal.

Silage-making can result in considerable loss of material during the wilting, fermentation and feeding periods. Field losses can range from 0–10% of the DM yield depending on degree of wilting in the field: fermentation losses are inevitable and commonly range from 10–20% of the DM yield: effluent can give a loss of up to 5%: not all of the material present in the silo is suitable for feeding, due mainly to side and top waste, and in clamp systems this 'visible' loss can be 5–15% of the DM yield. Thus in a really good silage system, only 80% of the weight of grass cut will be available for feeding as silage and often the figure is as low as 65–70%.

For the best fermentation the crop should have a high carbohydrate content (to provide ample substrate for fermentation) and a low moisture content (to reduce the volume of material requiring acidification). For these reasons emphasis is given to cutting crops when their carbohydrate (soluble sugar) content is high and when they can be wilted quickly. ADAS advice is that sugar content should be at least 3% at time of cutting and grass should be wilted from its normal moisture content of about 80% down to 70–75%. Often these two desirable conditions cannot be met and farmers can use additives to help alleviate the problem.

Additives can be used to:

(1) *Increase water soluble carbohydrate* (WSC)
By adding extra WSC (e.g. molasses) the *lactobacilli* are better able to produce lactic acid. (Note also that wilting decreases the quantity of lactic acid needed for effective pickling.)
Also *enzymes* such as cellulase and hemicellulase can be used to convert some of the non-soluble material in the plant into WSC. Note there is no starch in grass and so amylase is not effective.

(2) *Increase effective bacteria*

Inoculants containing *Lactobacillus* and *Streptococcus* can be used to ensure lack of suitable bacteria does not limit the rate of lactic acid production.

(3) *Add acid*

In some countries inorganic acids are used at high rates in the silo at time of filling to create a pickled effect. At present in the UK acids such as formic and sulphuric are applied at 3–5 litres/t of grass as the grass is picked up from the field. These acids reduce the quantity of lactic acid needed to reach a stable pH and consequently reduce the time taken for stability to be reached in the silo.

(4) *Use preservatives*

Some acid additives also contain chemicals that should suppress unwanted biochemical reactions: examples of such chemicals are formalin and sodium metabisulphite. *Table 6.16* summarises the Liscombe Star system that can be used to warn of possible fermentation problems that might be alleviated by additive use.

Storage of silage is in bulk silos (clamp or tower) or big bales (contained in sealed polythene bags).

Clamps

Clamps are found in a variety of forms, e.g. walled or unwalled, roofed or open, on the surface or in pits.

Towers

Towers are made of concrete or galvanised-vitreous enamelled steel: material for ensiling is always added to the top but, depending on type, silage is removed from either the top or the bottom.

Big bales

Big bales can be stored on level sites around the farm. They must be protected from wind which can damage the polythene, and from rodents who can eat holes in the bags.

Making silage in clamps

Assuming the material to be ensiled is either high in sugars and low in moisture or having an additive applied, the main principle during the filling process is to eliminate as much air from the matrix of herbage as possible and keep the material airtight. Polythene sheeting is an essential feature of silage-making and is available for this purpose in 300 or 500 gauge and in widths up to 10 m. The use of polythene sheets can be taken to the ultimate in the production of *vacuum silage*, where the crop is stacked on a sheet laid on the floor of the silo: then another sheet is placed over the heap and the two sheets joined together at ground level by a plastic 'strip-seal', after which the whole mass is evacuated by a vacuum pump. Such complete removal of oxygen leads to excellent

Table 6.16 The Liscombe star system[1]

		Your score
Grass variety	Timothy/Meadow fescue	*
	Perennial ryegrass	**
	Italian ryegrass	***
Growth stage	Leafy silage	O
	Stemmy mature	*
Fertiliser Nitrogen	Heavy (125 kg/ha +)	– *
	Average (40–125 kg/ha)	O
	Light (below 40 kg/ha)	*
Weather conditions (over several days)	Dull, wet (less than 2% sugar)	– *
	Dry, clear (2½% sugar)	O
	Brilliant, sunny (3% sugar or more)	*
Wilting	None (15% DM)	– *
	Light (20% DM)	O
	Good (25% DM)	*
	Heavy (30% DM)	**
Chopping and/ or bruising	Flail harvester or forage wagon	*
	Double chop	**
	Meter/twin chop	***

[1] A total of 5 stars needed for a good fermentation. Consider each factor, such as variety, growth stage etc, and add up the stars (*).

HOW THE STAR SYSTEM WORKS

For example a perennial ryegrass sward (**), in leafy silage growth stage (O) and heavily manured (– *), being ensiled in dry weather (O) and only lightly wilted (O) with pick-up double chop (**), gives a total score of *** and will show a benefit from additive use.

In comparison an Italian ryegrass sward (***), in a leafy stage (O), which is heavily manured (– *), in average weather (O) but well wilted, 25% DM (*) and meter chopped (***) will give a total score of ****** and not require additive.

5 Stars – no additive needed
4 Stars – use additive at recommended rate
3 Stars – use additive at recommended rate
2 Stars – use additive at higher recommended rate
1 Star – use additive at higher recommended rate
0 Stars – unsuitable conditions for making silage.

fermentation, but the process is laborious and difficult to carry out on a large scale. Polythene sheeting is easily punctured and on some farms polythene in any form is nibbled by rodents.

Farmers have found that many of the advantages of vacuum and similar techniques can be obtained more simply by making silage in a walled pit using a *wedge-filling* principle (sometimes called *Dorset wedge*).

On the first day of filling the cut crop is stacked at one end of the silo, against an end wall that is either solid or has a polythene sheet lining. If the side walls are not solid, they too must have a polythene lining. The material is normally put into the silo using a push-off buckrake and the buckraking tractor maintains the slope at the steepest reasonable angle. When the material has reached the maximum intended height the slope is progressed forwards, leaving a fixed height of material. One principal objective of the wedge system is to prevent warm air rising out of the silo, as this will encourage oxygen-rich cold air to come in at the bottom and sides: so a polythene sheet is placed over the grass each night, and when one section of filling is complete, the sheet is left in place, so that the silo is gradually wrapped in polythene sheets. If the technique is carried out correctly, the oxygen in the air in the silo is soon used up, lactic acid fermentation proceeds, and the crop consolidates under its own weight, often resulting in a drop in crop height to two-thirds to three-quarters of the original. Because of this shrinkage the polythene sheet should not be fixed rigidly, but rather covered with a flexible and convenient material such as old tyres, sand bags or even a net. If the sheet is not held down tightly it will flap in the wind, allowing in more air and eventually tearing the sheet.

It is essential to fill a silo rapidly: a rate of 100–200 t/d, depending on silo width, should be the minimum.

Long material can be made into good silage in clamps, but chopped material is easier to handle and consolidates better, with less oxygen trapped in the matrix. Fairly wet grass can be placed in a clamp, but it will produce effluent. ADAS figures show that grass ensiled at 20% DM will release an average of 200 litres of effluent per tonne of grass ensiled, and the quantity of effluent decreases progressively until material ensiled at 28% DM should give no effluent.

Silage effluent

This is a real problem if it enters a water course, as it has a high Biological Oxygen Demand (BOD), and will kill many oxygen-demanding organisms in the water, including fish. Many farmers wilt their grass simply to avoid or minimise the effluent problem. As a precaution it is advisable to construct an effluent tank adjacent to the silo, taking care to ensure that *only* silage effluent can enter it (i.e. no surface or rainwater). Silage effluent is very acid and all effluent-conducting channels and ducts must be coated with acid-resisting material. Silage effluent contains some plant nutrients (say 2, 1 and 1.5 kg/1000 litres of N, P_2O_5 and K_2O respectively) but as it is very acid it is very phytotoxic. It can be applied to arable land by tanker or a slurry irrigation system, but care is needed to ensure it does not get into land drains and hence to a water course.

Silage effluent can be given to stock, normally either cows or pigs. Care must be taken in collecting and storing the material to ensure there is no seepage. Also effluent deteriorates on storage and becomes unacceptable to stock, so a preservative such as formalin should be added.

Making silage in towers

The crop has to be blown into the top of the tower and at feeding time the silage is removed by mechanical means: thus it is vital that the ensiled material is well-chopped and sufficiently dry to remain friable after compaction in the tower during storage. For this reason, grass going into a tower must be above 35% DM and many tower operators prefer 45–55% DM (sometimes called *haylage*). Very wet material must *never* be placed in a tower as the effluent from it will cause very serious corrosion to the tower structure leading to possible collapse of the tower. Because the material going into a tower must be well-chopped, and because a tower is almost airtight, compaction in a tower is good and excellent fermentation is assured. Towers have a deserved high reputation for producing good silage with minimum in-silo losses of 5–10%. (But field losses during the necessary prolonged wilting period will be 10–15%.) Some farmers have shown that application of 'tower' techniques in terms of wilting and chopping the grass, but then placing the material in a polythene-lined clamp produce silage equivalent to the tower: even so the tower is a first-class starting point for mechanised feeding and on some farms is justified for this reason alone.

Making big-bale silage

Precut grass can be picked up by special big balers. Unwrapped bales can be stacked very close together and tightly covered by a polythene sheet. Generally, they are either placed in individual polythene bags after baling or wrapped in polythene. Whatever system is used, it is essential that the minimum volume of air is trapped with the grass and that no further air gets near the silage until feeding-time. If this practice is followed fermentation will proceed rapidly and well, giving a well-preserved and very edible silage. This rapid fermentation may produce enough CO_2 to cause well-sealed bags to swell up: it is important to ensure that bags do not flap loosely in the wind once this distention of the bags has passed.

Effluent is very undesirable in big bags and so normally grass is wilted to at least 28% DM and some farmers prefer 35% DM. Compared with clamp silage, big bags have the following features:

Advantages	Disadvantages
effective way to deal with small quantities of grass when baled by contractor, very low capital expenditure	rupture of bag can lead to spoilage of whole contents
bales can be stored in field and moved when needed for feeding	bags can be damaged by wind, rodents and birds
can have very efficient fermentation (low losses)	feeding systems need careful planning
excellent when small amounts of silage needed (e.g. buffer feeding)	

Feeding clamp silage

Feeding methods can be split into two types – self-feeding and mechanical feeding.

Self-feeding

This takes place when the animals are allowed to help themselves to the silage. The settled height of the silage must not be higher than the animals can reach and usually the animals' access to the face is restricted by some physical means, such as an electric fence, so that they cannot climb on the silage or pull out big lumps of silage and waste it by treading it under foot. Where animals are given 24 h access to the face, then some 20 cm of face width should be allowed per cow equivalent. If access time is limited, the width per beast must be greater. The rate of feeding can be controlled by the distance the electric fence is moved each day, and by this means the silage can be rationed to last the whole winter. Self-feeding is a very cheap and effective method of feeding silage and is very popular with farmers. The only real justification for using other feeding techniques for clamp silage are where several groups of animals are to be fed from one silo face, or where a balanced feeding programme is planned in which silage must be premixed with other feeds.

Mechanical removal

Mechanical removal of silage from clamps can be done by fore-loader, grab, block-cutter or a cylindrical silage-cutter. Where the silage is greatly loosened during removal, it must be eaten within 12 h or it will deteriorate from re-fermentation: in this respect the block-cutter is good as it removes blocks weighing some 0.6–0.7 t without losing much of the original density of the silage.

Feeding big-bale silage

Normally big bales are moved using a single protruding tine fixed to a fore-end loader, preferably with a levelling control and push-off device. Big bales are heavy and care is needed to ensure safe movement of bales in this way. ADAS state that the weight of a big bale can be estimated by:

wt of bale (kg) $= 725 - (7 \times$ DM% of grass)

e.g. if 30% DM, then guide weight $= 515$ kg

With each bale holding about 0.5 t of silage, the feeding method must be based on mechanical placement of the bale. Where big-bale silage is used to supplement other forage, then ring feeders placed in the animals' exercise area can be very effective in allowing easy access for moving bales. If the big-bale silage is the major feed, and if this must be fed in a building, then there must be easy access for the tractor carrying the bale and proper control of feeding to prevent animals climbing on to the bale.

Crops for ensilage

Many green crops can be ensiled and also some arable by-products such as *sugar beet tops* and *pea haulms*. By far the most common crop is *grass* (including *grass/clover herbage*) and grass should be ensiled when its digestibility is at least 65, and for high quality silage, the D-value at cutting must be 67–70.

The crop will lose about 2 units of D during the ensilage period even where there is a good fermentation, and as much as 5 units of D can be lost if fermentation is poor.

Whole-crop cereals

These can be made into silage, the cereals being cut about two weeks after full ear emergence. Yields of 8–10 t/ha of DM can be obtained from this crop, but digestibility is often low at around 62–63 D, so that animal intake and performance are fairly low: also it is possible for whole-crop barley to have an intake below expectation due to the physical dislike of the barley awns in the silage.

Special arable mixtures

These are sometimes grown for arable silage, usually based on the traditional oats–legume combination: this has the advantage of combining the high protein value of the legume with the good carbohydrate content of the cereal. Also this type of mixture requires less fertiliser N than straight cereals or grasses. Examples of such mixtures are:

	125 kg/ha oats	125 kg/ha oats
	35 kg/ha vetches	45 kg/ha forage peas
or	35 kg/ha beans	

Legume silage

This, usually based on red clover or lucerne, can have a high protein content, but it is more difficult to get a good fermentation than with grass, cereals or maize because of the low sugar content in legumes: an additive is very often essential. Also legumes tend to have fibrous stems when cut for silage and the material compacts much better in the silo if a precision-chop harvester is used.

Maize

Maize can make excellent silage provided it is well-chopped before ensiling. A crop of maize with grain at the 'pasty' stage has a high carbohydrate and low protein content, a D-value of 63–65 and a moisture content of about 70% as it stands in the field. In the silo, this crop ferments well without additives and it is very well suited to mechanical handling. Provided its low protein value is recognised at feeding time, it is excellent material for silage and a good contributor to stock nutrition. The main snag with maize is the relatively low yield obtained in the UK coupled with the late harvesting in October or even November. A good crop of maize should produce 30 t/ha of 30% DM material in mild, sunny parts of southern England, but in practice yield of only 16–20 t/ha have been recorded.

Silage – facts and figures

Yields

High yields of crop are essential for silage to justify the machinery labour and fuel costs involved. *Table 6.17* is a guide to the fertiliser rates and expected yields of grass silage cut at three different times in the season.

Density of silage

clamp silo	unchopped	20–25% DM $= 720$–800 kg/m^3
	chopped	20–25% DM $= 800$–850 kg/m^3
tower	chopped	35–45% DM $= 500$ kg/m^3

Table 6.17 Good yields from grass cut for silage (t/ha)

Cutting period	Fertiliser to apply (kg/ha)	Site class 1			Site class 3			Site class 5		
		DM cut	Grass cut @ 25% DM	Silage @ 25% DM	DM cut	Grass cut @ 25% DM	Silage @ 25% DM	DM cut	Grass cut @ 25% DM	Silage @ 25% DM
May	125 N (+ 30 P_2O_5 and 50 K_2O on index 0 and 1 soils)	6	24	20	6	24	20	6	24	20
July	100 N (+ at least 50 K_2O on soils below index 3)	4	16	12	3	12	10	second cut likely in wet years only		
Aug/Sept	80 N (+ at least 40 K_2O on soils below index 3)	2½	10	8	third cut likely in wet years only					
Total over season		12½	50	40	9	36	30	6	24	20
Expected number of cuts			3			2			1	

Feeding values

Typical values for a range of silages are:

	DM%	ME (MJ/kg DM)	D-value	Crude protein (% in DM)	DCP (g/kg)
Grass silage					
excellent	25	11.2	70	18	120
good	25	10.7	67	16	105
average	25	10.2	64	14	95
poor	25	9.8	61	12	75
Whole-crop cereals					
barley	30	9.3	62	10	50
oats	30	8.6	57	9	60
wheat	30	8.4	55	8	35
Oats and vetches	27	9.6	60	16	95
Lucerne	27	9.4	62	20	160
Forage peas	22	9.4	62	20	140
Maize	24	10.6	65	10	70

Haymaking

Haymaking is still a popular method of conservation, involving the reduction of moisture from fresh grass at 80% to about 20% when the product can be stored. Successful haymaking should follow these principles:

(1) Grass should be at the correct stage of growth when cut. As grass is allowed to mature its total yield increases and its moisture content falls, so it may be tempting to allow a very heavy, mature crop to develop before cutting. *But* digestibility falls at a rate of one-third to one-half of a D-value unit/d once the seedheads have formed; mature hay has low feed value and low intake characteristics, even though it may be well made.

(2) Losses should be kept to a minimum. Losses can arise in the following ways:
(a) respiration losses will occur in the field as the herbage continues to respire after cutting. Rapid drying will minimise these losses, which can amount to 1–5% of the DM yield.
(b) Mechanical losses occur if herbage is fragmented by machinery into pieces too small to be picked when the hay is collected for storage. Mechanical losses tend to become greater as the herbage becomes drier and when the action of the machines is abrasive. A mechanical loss of 10% is acceptable.
(c) Leaching of nutrients can take place when the cut material is exposed to rain. Ideally hay is made and removed from the field without rainfall, when this loss is zero, but long periods of heavy rain can lead to a DM loss of up to 15% and a soluble nutrient loss far in excess of this.
(d) Some of the hay may itself be inedible due to dust and mould. Both these factors arise when hay is made under adverse conditions: the loss can be up to 15% of the DM and it emphasises the importance of making hay under good climatic conditions.

When grass is cut and a wide area of swath exposed to wind and sun, there is a rapid initial loss of moisture, as external moisture is lost and water from the outer cells of the leaves and stems. At this stage drying will take place without much sun provided the atmosphere has a low relative humidity. Also at this stage of drying the material can be treated quite roughly by machines. As drying proceeds it becomes progressively more difficult to remove water and the rate of moisture loss declines, together with an increasing risk of mechanical damage. For continued drying sun is necessary to provide heat and wind is valuable for removing water vapour from within the swath. The rate of drying can be speeded up if the grass is cut with a flail mower or passed through a crimper/conditioner immediately after cutting: however, grass treated in this way suffers more in bad weather, and flail mown material can suffer mechanical losses as high as 50% if subjected to prolonged wet weather.

Field drying rate is maximised if the cutting machine leaves the largest possible leaf area exposed to sun and wind, and not tight swaths. If bad weather threatens during the drying process, the material should be windrowed to present the smallest area to the rain. Also it is sound practice to swath-up the grass at night to minimise the effect of dew, ensuring the ground and surface of the swath are dry before the material is again spread next morning.

The rate of drying depends on the weather, proper use of machinery, rates of N used, varieties in the sward, stage of maturity at cutting, bulk of grass present, time of year and desired moisture content at transporting from the field. Most hay is baled and it should not be baled until it is fit for storage, i.e. with a moisture content of about 20%. Frequently hay is baled at a higher moisture content than this and then the bales are left in the field to 'cure'. Very little further moisture loss can occur in the field from the bale, and yet the bales will acquire considerable moisture if rain falls. As a general rule, baling should be regarded as the first step in the process of transport into storage, and grass should not be baled until the herbage is dry and transport into storage is organised.

Barn hay drying

This is a very useful technique for reducing the risk of bad weather, minimising losses and making hay from younger material (at a higher feed value) or where higher N rates have been used. The material is cured as far as possible in the field, certainly down to at least 30% moisture. It can then be baled, taking care not to over-compress in the bale, and the bales carefully placed over a grid or ducts through which air can be blown. There must be no gaps between the bales, otherwise air will take the line of least resistance and fail to pass through the bales. If the outside humidity is low and only some 5–8% of moisture needs to be removed, then cold air blown for up to a week will suffice, often ceasing to blow at night when humidity might be high. During periods of sustained high humidity or when considerable moisture must be removed, then the air must be heated. The air can be heated by either a flame or electric heaters, but in any event the heating of air is *very expensive*. Also, barn-drying installations cannot deal with a sudden large volume of material to cure. Barn-drying is best regarded as a means of producing a limited quantity of quality hay.

Success in barn-drying depends on:

(1) justifying the extra costs involved, rather by the higher quality of the hay than by a salvage operation on a mediocre crop;
(2) allowing moisture content to fall to below 30% (and certainly 35%) before baling;
(3) using bales at a low-moderate packed density;
(4) stacking the bales carefully to avoid cracks through which the majority of the air can pass;
(5) having sufficient fan capacity to obtain a good flow of air, with heating available if humidity is high or bales are wet;
(6) having a dry secondary store, so that dried bales can be moved to allow more bales to be dried.

Hay additives

Hay additives are based on either propionic acid or ammonium bispropionate. Both chemicals have a strong antifungal activity and will reduce the rate of decomposition of hay (and control heating) when the material is still too wet for immediate storage. However there are four snags to hay additives:

(1) It is very difficult to apply these chemicals evenly to the herbage. Attempts to apply within the bale chamber or spray on the swaths before baling have both failed to obtain the necessary even application (in contrast to the addition of silage additives in a forage harvester).
(2) Although both the chemicals mentioned give good initial control of decomposition, they are broken down gradually and give protection for some two to three weeks only. Thus the hay must be dehydrated to a proper storage moisture content soon after baling and storage.
(3) Under warm conditions at baling some 50–75% of the additive may be lost by volatilisation.
(4) Rate of application should depend on the moisture content of the hay – and this will vary from swath to swath and throughout the day, for example:

% moisture	kg of additive/t of hay
28	7
30	9
32	11
34	13

Development of a successful hay additive is proving difficult and at present it should not be regarded as a means of storing wet hay but rather as an aid to delaying the need to get to a safe storage moisture content.

Hay facts and figures

Quality in hay can be judged by its colour, which should be bright green/yellow, by its sweet smell and an absence of dust. Also the feeding value of hay can be determined by its analysis: typical figures for a range of types of hay are given:

	DM%	ME (MJ/kg DM)	D-value	Crude protein (% in DM)	DCP (g/kg)
Grass hay					
excellent[1]	85	10.7	67	13	100
good	85	9.6	64	11	60
average	85	9.0	60	10	50
poor	85	8.4	56	9	40
Grass/clover hay	85	10.0	64	15	110
Lucerne	85	8.8	55	20	140
Sanfoin	85	9.3	58	16	120

[1] Barn dried samples can have higher values for ME, D and protein.

Yields

Yields of hay vary considerably, and high N rates should not be used to grow a hay crop because it aggravates the curing problem: a maximum N rate of 80 kg/ha for the growth period is recommended.

light crop	= 2–3 t/ha of made hay
medium crop	= 4.5 t/ha of made hay
heavy crop	= 5 t/ha and over

A standard bale of hay measures approximately 0.9 m × 0.45 m × 0.35 m and weighs 20–30 kg depending on type of material and density (33–50 bales/t). A big round bale of hay measures about 1.5 m high and 1.8 m in diameter and contains some 25 times more material than a standard bale, weighing 500—600 kg. A big rectangular bale is approximately 1.5 m × 1.5 m × 2.4 m with the same weight as the round bale.

Approximate storage volumes

	m^3/t
loose medium-length hay	12–15
standard bales	7–10
big rectangular bales	10

Green crop drying

When green crops are passed through an efficient high temperature drier, they are rapidly reduced to a stable moisture content with little or no loss of nutrient value. Also artificially dried green crops are very palatable to stock. In the UK the majority of green crops dried are either grass or lucerne; both are traded under the general term 'dried grass'. Less than 1% of the grassland area of the UK is used for dried grass production, and most of the drying is done in very specialised large units where the size of operation justifies the use of the large, very efficient high-temperature triple pass drier, the smallest of which can produce some 3000 t of product per year, needing about 400 ha of adjacent land to provide material. Most green crop driers are members of the British Association of Green Crop Driers, who supply specialist information to members on a variety of topics related to the industry. Although dried grass has a very high reputation as a supplement to silage in cattle and cow rations, it is unlikely that green crop drying will expand in the foreseeable future because it demands a high input of fossil fuel – equivalent to about 200 litres of oil/t of material produced, quite apart from the large equipment needed to cut and transport the grass to the drier and the power needed by the mill and cuber to package the material after drying.

In an efficient grass-drying unit, dry-matter losses are only about 3–5%, and in this respect the technique is far superior to hay and silage making. Some feed analysis figures for dried grass and lucerne are given:

	DM%	ME (MJ/kg DM)	D-value	Crude protein (% in DM)	DCP (g/kg)
Dried grass					
leafy grass	90	11.2	70	16	110
average	90	9.6	64	15	95
lucerne	90	9.6	60	24	170

GRASSLAND FARMING AND THE ENVIRONMENT

As grassland is the most widely occurring crop in the UK it is inevitable that grassland farming has a broad interface with many aspects of the environment.

Main points to consider are as follows:

Pollution of water courses and drinking water

Animal wastes and manures should never be allowed to run directly into drains and water courses: a barrier ditch system may help to alleviate the problem but most of the material is best applied to the land, where the maximum recommended rate is 25 000 litres/ha of material in *any* one year.

Silage effluent is a very potent pollution hazard. The best single measure of the harmful effect of such materials is the amount of oxygen taken out of the water, which is measured as BOD (Biological Oxygen Demand). As the following figures show, silage effluent is 168 times more demanding than domestic sewage and only milk is worse, being twice as demanding as silage effluent.

	BOD (mg/litre)
raw domestic sewage	400
dairy washings	2000
cow slurry	35 000
silage effluent	67 000
milk	140 000

Other farm wastes can cause problems, notably used chemical containers (from dairy hygiene, animal health and crop protection materials). Guidelines for disposal are:

(1) empty the container completely at time of use for its intended purpose (do not wash out and let water down drain);
(2) destroy containers as soon as possible and do not use them for any other purpose;
(3) *combustible containers* should be burnt well away from habitations, stock and edible crops;
(4) *metal containers* should be flattened and buried at least 450 mm deep well away from drains. The spot should be marked and recorded for future reference;
(5) *glass containers* should be broken in a sack and buried as (4).

Fertiliser N does not leach rapidly from grassland if it is applied in the February–September period. There is a substantial leaching of nitrate from all grassland in the autumn and this leaching will be greater from high N grassland than from low N grass or grass/clover swards. Ploughing of grassland gives very substantial losses of nitrate over the next few months.

Effect on botanical composition of swards and soil

As shown by Jones (1933), the present composition of a farmed area of grassland is a reflection of its immediate past management. If the yield and speed of regrowth of desirable species is encouraged, then other species in the sward will suffer from enhanced competition. Thus productive grassland will contain a narrower range of plant species than non-productive land.

Grassland encourages the build-up of good soil structure and an increase in soil organisms. If an area of permanent grass is productive enough to support 2 LSUs/ha, it has been estimated that the total weight of organisms below ground is equivalent to 20 LSUs/ha.

The following table (based on Garwood and Gilbey, 1985) shows the interactive effect of fertiliser N and drainage on earthworms and the harmful effect of reseeding.

	Earthworm biomass (kg/ha)	
	drained	*undrained*
permanent pasture		
200 kg/ha N	165	100
400 kg/ha N	285	125
reseed		
400 kg/ha N	60	105

Other factors

Good grassland farming should not give grounds for serious environmental concern. Care should be taken to avoid offence from annoying odour caused by animal units and farm waste. Also, the appeal of the countryside can be marred by such items as scattered polythene sheets, untidy fencing and restricted access to public footpaths.

GLOSSARY OF GRASSLAND TERMS

Based on Hodgson (1979) and Thomas (1980), with additions.

Anthesis: flower opening.
Biomass: weight of living plant and/or animal material.
Browsing: the defoliation by animals of the above-ground parts of shrubs and trees.
Buffer feeding (or grazing): provision of additional forage when quantity in grazing area is insufficient.
Canopy: the sward canopy as it intercepts or absorbs light.
Canopy structure: the distribution and arrangement of the components of the canopy.
Closed canopy: a canopy which either has achieved complete cover or intercepts 95% of visible light.
Cover: the proportion of the ground area covered by the canopy when viewed vertically.
Cow equivalent (CE): an aggregated liveweight of animals equivalent to a standard cow of 550 kg (synonymous with LSU).
Crop growth rate (CGR): the rate of increase in dry weight per unit area of all or part of a sward.
Crop (standing): the herbage growing in the field before it is harvested.
Crown: the top of the tap-root bearing buds from which the basal leaf rosette and shoots arise (appropriate to clovers but *not* grasses).
Culm: the extended stem of a grass tiller bearing the inflorescence.

D-value: digestibility of organic matter in the DM eaten (as %).
Date of heading (or date of ear (influorescence) emergence): for a sward, the date on which 50% of the ears in fertile tillers have emerged.
Defoliation: the severing and removal of part or all of the herbage by grazing animals or cutting machines.
Density: the number of items (e.g. plants or tillers) per unit area.
Foliage: a collective term for the leaves of a plant or community.
Forage: any plant material, except in concentrated feeds, used as a food for domestic herbivores.
Forage feeding: the practice of cutting herbage from a sward (or foliage from other crops) for feeding fresh to animals.
Grassland: the type of plant community, natural or sown, dominated by herbaceous species such as grasses and clovers.
Grazing: defoliation by animals.
Grazing cycle: the length of time between the beginning of one grazing period and the beginning of the next.
Grazing period: the length of time for which a particular area of land is grazed.
Grazing pressure: the number of animals of a specified class per unit weight of herbage at a point of time.
Grazing systems:
　Continuous stocking: the practice of allowing animals unrestricted access to an area of land for the whole or a substantial part of a grazing season.
　Set stocking: the practice of allowing a fixed number of animals unrestricted access to a fixed area of land for a substantial part of a grazing season.
　Rotational grazing: the practice of imposing a regular sequence of grazing and rest from grazing upon a series of grazing areas.
　Creep grazing: the practice of allowing young animals (lambs or calves) to graze an area which their dams cannot reach.
　Mixed grazing: the use of cattle and sheep in a common grazing system whether or not the two species graze the same area of land at the same time.
Harvesting: defoliation by machines.
Harvest year: first full harvest year is the calendar year following the seeding year.
Herbage: the above-ground parts of a sward viewed as an accumulation of plant material with characteristics of mass and nutritive value.
Herbage allowance: the weight of herbage per unit of live weight at a point in time.
Herbage consumed: the herbage mass once it has been removed by grazing animals.
Herbage cut: the stratum of material above cutting height.
Herbage growth: the increase in weight of herbage per unit area over a given time interval due to the production of new material.
Herbage mass: the weight per unit area of the standing crop of herbage above the defoliation height.
Herbage residual: the herbage remaining after defoliation.
Inflorescence emergence: the first appearance of the tip of a grass inflorescence at the mouth of the sheath of the flag-leaf.
Leaf: in grasses = lamina + ligule + sheath; in clovers = lamina + petiole + stipule. Note leaf is *not* lamina.
Leaf area index (LAI): the area of green leaf (one side) per

unit area of ground; *Critical LAI* is LAI at which 95% visible light is intercepted: *Maximum LAI* is the greatest LAI produced by a sward during a growth period: *Optimum LAI* is the LAI at which maximum crop growth rate is achieved.

Leaf area ratio: total lamina area divided by total plant weight: *specific leaf area* is the lamina area divided by lamina weight.

Leaf burn (or scorch): damage to leaves caused by severe weather conditions, herbicides, etc. This contrasts with *leaf senescence* which is a genetically predetermined process where leaves age and die, usually involving degradation of the chlorophyll in the leaves.

Leaf emergence: a leaf in grass is fully emerged when its ligule is visible or when the lamina adopts an angle to the sheath. In clovers the leaf is emerged when the leaflets have unfolded along the midrib and are almost flat.

Flag leaf: in grass this is the final leaf produced on the flowering stem.

Leaf length to weight ratio: lamina weight divided by lamina length.

Livestock Unit (LSU): *see* CE.

Net assimilation rate (NAR): defined as $1/A \cdot dW/dt$ where A = total leaf area, W = total plant weight, t = time, units = $(g\,m^2)/d$.

Palatable: pleasant to taste.

Plastochron: the interval between the initiation of successive leaf primordia on a stem axis.

Preference: the discrimination exerted by animals between areas of a sward or components of a sward canopy, or between species in cut herbage. *Preference ranking* is based on the relative intake of herbage samples when the animals have a completely free choice of the materials.

Pseudostem: the concentric leaf sheaths of a grass tiller which perform the supporting function of a stem.

Regrowth: the production of new material above the height of defoliation after defoliation, often with initial regrowth at the expense of reserves stored in the stubble.

Rest period: the length of time between the end of one grazing and the start of the next on a particular area.

Seed ripeness: the stage at which the seed can be harvested successfully.

Seeding year: the calendar year in which the seed is sown.

Seedling emergence: a seedling has emerged when the shoot first appears above the ground.

Selection: (by animals) the removal of some components of a sward rather than others.

Shoot bases: the part of the sward below the anticipated height of defoliation. This becomes *stubble* after defoliation.

Sod: a piece of turf lifted from the sward, either by machine or grazing animals.

Spaced plant: a plant grown in a row so that its canopy does not touch or overlap that of any other plant.

Standing crop: the herbage growing in the field before it is harvested.

Stem: the main axis of a shoot, bearing leaves.

Stocking density: the number of animals of a specified class per unit area of land actually being grazed at a point in time.

Stocking rate: number or weight of animals kept on a unit area of land for a long period (preferably for 12 months, so including grazed area *and* conserved area).

Sward: an area of grassland with a short continuous foliage

cover, including both above- and below-ground parts, but not any woody plants.

Sward establishment: the growth and development of a sward in the seeding year. *A primary sward* is one that has never been defoliated. *A mixed sward* is one that contains more than one variety or species. *A pure sward* is one that contains a single stated variety or species.

Simulated sward: an assemblage of plants intended to represent in convenient form a 'normal' sward.

Sward height: height above ground at which a descending finger first touches vegetative material (ignoring stems and seed heads).

Tiller: an aerial shoot of a grass plant, arising from a leaf axil, normally at the base of an older tiller. *An aerial tiller* is one that develops from a node of an extended stem.

Tiller appearance rate (TAR): the rate at which tillers become apparent to the eye without dissection of the plant.

Tiller base: the part of a growing tiller below the height of defoliation.

Tiller stub: the part of a tiller left after defoliation.

Turf: the part of the sward which comprises the shoot system plus the uppermost layer of roots and soil.

Utilised metabolisable energy (UME): the quantity of ME accounted for in animal production. *If UME of grass*, then UME after deduction of ME supplied from non-grassland sources.

Winter burn: leaf burn in winter.

Winter hardiness: the general ability of a variety or species to withstand the winter.

Winter kill: death of plants in winter.

References

BAKER, H. K. (1962). *Proceedings of the Sixth Weed Control Conference, Brighton* **1**, 23–30.

BAKER, R. D. (1964). *Journal of the British Grassland Society* **19**, 149–155

BROCKMAN, J. S. and GWYNN, P. E. J. (1961). *Journal of the British Grassland Society* **16**, 201–202

COPPOCK, J. T. (1976). *Agricultural Atlas of England and Wales.* London: Faber & Faber

DAVIES, W. (1960). *The Grass Crop.* London: Spon

FORBES, T. J., DIBB, C., GREEN, J. O., HOPKINS, A. and PEEL, S. (1980). *Factors affecting the Production of Permanent Grass.* Hurley: Grassland Research Institute

GARWOOD, E. A. and GILBEY, J. (1985) Grassland Manuring Occasional Symposium No 20, 94–96. Hurley: British Grassland Society

GREEN, J. O. (1974). *Preliminary Report on a Sample Survey of Grassland in England and Wales, Report 310.* Hurley: Grassland Research Institute

HODGSON, J. (1979). *Grass and Forage Science* **34**, 11–18

HODGSON, J., MACKIE, C. K. and PARKER, J. W. G. (1986). British Grassland Society: Grass Farmer No. 24, 5–10

HOPKINS, A. (1979). *Journal of the Royal Agricultural Society of England*, 140–150

HOPKINS, A. and GREEN, J. O. (1978). Changes in Sward Composition and Productivity, Occasional Symposium No 10, 115–129. Hurley: British Grassland Society

JEWISS, O. R. (1979). British Grassland Society: Grass Farmer No. 4, 3–7

JONES, M. G. (1933). *Empire Journal of Experimental Agriculture* **1**, 43–47; 122–128; 361–367

LEAVER, J. D. and MOISEY, F. R. (1980). British Grassland Society: Grass Farmer No. 7, 9–11

MARSH, R. (1975). *Pasture Utilisation and the Grazing Animal, Occasional Symposium No. 8,* 119–128. Hurley: British Grassland Society

MORRISON, J., JACKSON, M. V. and SPARROW, P. E. (1980). *Report of joint GRI/ADAS Grassland Manuring Trial.* Hurley: Grassland Research Institute Report No. 27

NEDO (1974). *Grass and Grass Products.* London: National Economics Development Office

THOMAS, H. (1980). *Grass and Forage Science,* **35,** 13–23

THOMAS, C. and YOUNG, J. W. O. (1982). *Milk from Grass.* Billingham: ICI Agricultural Division

WOLTON, K. M. (1980). AgTec, Summer 1980. Felixstowe: Fisons Fertilizer Division

Further reading

HOLMES, W. (1980). *Grass – its Production and Utilisation.-* Oxford: Blackwells Scientific Publications

JOLLANS, J. L. (1981). Grassland in the British Economy. Reading University: Centre for Agricultural Strategy, Paper No. 10

WILKINSON, J. M. (1984). *Milk and Meat from Grass.* London: Granada Publishing Ltd

7

Trees on the farm

R.D. Toosey

Trees grown for timber have an extremely important part to play in rural land management as they achieve high productivity on land such as steep banks, waste places and on the poorer soils, with little or no agricultural value. Managed woodlands, even where small, can also be a valuable source of farm timber, fencing posts and firewood. Trees for shelter are often neglected; windbreaks are usually essential for horticultural production, while shelter belts increase productivity on exposed farms. Well managed woodlands also improve amenity, landscape and sporting value of the countryside.

CHOICE OF SITE AND SPECIES

Once planted, a tree crop is unlikely to be changed for 50–100 years unless it fails completely. Also it takes many years for mistakes to become apparent, so that a wrong choice of tree can result in wasted time and serious financial loss. As it is essential to marry the tree successfully to the site, ecological factors override all other considerations, e.g. degree of exposure, frost risk, rainfall, surface soil and underlying rock formation, drainage, susceptibility of the site to drought, local topography, elevation and aspect. Only when species of trees suited to the particular site have been determined can a final selection be made on the basis of volume production, marketability and other considerations.

When selecting species to plant it is advisable to consult local experts and inspect woodlands on similar nearby sites. Natural flora, occurring either as single species or plant associations (e.g. grass/herbs, grass/rush) can give a good guide to probable site conditions but are not infallible. To give a reliable guide a species must occur in *quantity* and be *thriving* (*Table 7.1*).

Careful inspection of the soil and subsoil is required and the nature of the underlying rock should be ascertained. Compacted soils do not permit satisfactory root penetration and ploughing or other preplanting cultivation will usually be required. Depth of soil and the state of other drainage should also be noted. Soil analysis is only of limited value. Species differ greatly in water requirements, and consequently suitability to high or low rainfall areas, and in their ability

to withstand exposure; the latter is more important than altitude. *Every ecological factor must be examined carefully as the neglect of only one can result in failure.* On fertile ground, especially old or long established woodland, there is a wide choice of species. Conversely, on infertile or exposed sites only one or two species may be suitable.

Most common timber trees are fully resistant to winter cold, but new spring growth of some species, e.g. Douglas fir, is particularly susceptible to damage by late spring frosts. Only frost hardy species should be planted in hollows where cold air collects, i.e. frost pockets. *Light demanding* trees can only be grown in full light and require earlier thinning. *Shade tolerant* trees thrive initially in partial shade cast by a tall pole crop of larger trees and are suitable for underplanting, which is particularly advantageous on frost prone sites.

Coniferous trees (softwoods) generally grow most rapidly, produce the highest volume of timber, need the shortest rotation (i.e. number of years from planting to clearfelling) and, except for larch, are evergreen. Most prefer acid soils and some tolerate exposed hill sites. As a group they give the highest volume of timber and financial return.

Broadleaved trees (hardwoods) are mostly slower growers, lower volume producers (*see* Estimation of growth rate, p. 211) and better suited to more fertile lowland conditions. With lower volume production and low priced thinnings they are much less profitable than conifers but have greater amenity value and provide useful firebreaks; well grown hardwood timber is scarce and valuable. With improved establishment techniques and planting grants, broadleaves are now more attractive financially.

The main characteristics and site requirements of tree species for planting for timber and shelter are shown in *Tables 7.2; 7.3; 7.4* and *7.5*. For further details and identification consult Forestry Commission (1978, 1984, 1985a, 1985b) and Caborn (1965).

MEASUREMENT OF TIMBER

With the occasional exception of some low value felled material (e.g. firewood, pulpwood which may be weighed) timber is sold and measured by volume (m³). All methods of

Table 7.1 Assessment of site conditions and quality from vegetation present, i.e. indicator plants

Indicator species	Site conditions indicated						Quality of site	Remarks
	Natural fertility		Wetness of soil		Presence of lime (Ca)			
	High	Low	Wet	Dry	Lime present	Acid		
Alder, black (*Alnus glutinosa* L.)	√		√		M		2–3	Lowlands, by rivers, marshy sites (D)
Ash (*Fraxinus excelsior* L.)	√		M	M	M		1	
Beech (*Fagus sylvatica* L.)				√	M		1–2	Soil can be acid
Bilberry (*Vaccinium* spp.)		√		√		√	4	
Bluebell (*Endymion non-scriptus*)	√			√	√	√	1	Ancient woodland
Bracken (*Pteridium aquilinum*)				√		√	2–3	Covers wide fertility range; good indicator of aeration
Cotton grass (*Eriophorum* spp.)		√	√			√	5	Wet uplands (D)
Deer grass (*Trichophorum caespitosum*)		√	√			√	5	Wet uplands (D)
Dog's mercury (*Mercurialis perennis*)	√			√	√		1	
Dog rose (*Rosa* spp.)	√			√	M		1	Usually occurs on neutral to alkaline sites
Hazel (*Corinus avellana*)	√			√			1	Often found on stiff clays pH 7 or more
Heaths: (*Erica* spp.)								
(i) Bell Heather		√		√		√	4	
(ii) Cross Leaved Heath		√	√	√		√	4	
Heather or ling (*Calluna vulgaris*)		√		√		√	4	Also moist soils
Honeysuckle (*Lonicera periclymenum*)	√	√		√			1–3	Often found on soils pH 7 or more
Hornbeam (*Carpinus betulus*)	√		M	M			1	Moist soils
Horsetails (*Equisetum* spp.)	√		√				1–3	Wet subsoil
Maple, field (*Acer campestre*)	√			√			1	
Meadowsweet (*Filipendula ulmaria*)	√		√					Wet lowland situations (D)
Moor mat grass (*Nardus stricta*)		√	M			√	4–5	
Oaks (*Quercus* spp.)								
Sessile oak (*Q. petraea*)	√			√	M	M	1–2	When showing strong growth
Pedunculate oak (*Q. robur*)	M	M	M		M	M	1–2	
Primrose (*Primula vulgaris*)	√		M		M	M	1	Ancient woodland
Purple moor grass (*Molinia caerulea*)		√	√			√	3–4	When in heavy tussocks
Rhododendron (*Rhododendron ponticum*)						√	1–3	
Rushes: (*Juncus* spp.)								(D)
(i) Hard rush			√		√		1–2	
(ii) Soft rush			√			M	2–4	

Table 7.1 Continued

Indicator species	Site conditions indicated						Quality of site	Remarks
	Natural fertility		Wetness of soil		Presence of lime (Ca)			
	High	Low	Wet	Dry	Lime present	Acid		
(iii) Heath rush		✓	✓			✓	4	
Sedges: (*Carex* spp.)	✓		✓				1–2	(D)
Sphagnum moss (*Sphagnum* spp.)		✓	✓			✓	4–5	Unplantable unless drained (D)
Wild garlic or ramson (*Allium ursinum*)	✓		M	M	✓		1	
Wild privet (*Ligustrum vulgare*)	✓				✓		1	
Wild raspberry (*Rubus idaeus*)	✓				✓		1	Good forest soil, wide choice of
Willows (*Salix* spp.) including			✓				variable	(D)
(i) Goat willow			✓					
(ii) Sallow			✓					

Key: *Quality of site indicated:*
(1) Very good, wide choice of species to plant
(2) Good
(3) Moderate
(4) Poor–very poor
(5) Poorest sites
D = Drainage normally essential before planting or plant water tolerant species if appropriate, i.e. Alder, Poplar or Willow
M = Indicates fertility, wetness or acidity only to moderate or variable degree.

measuring unprocessed or partly processed timber are of necessity estimates and vary greatly in their degree of precision; the greater the precision required, the higher the cost per unit of volume. Measurable volume is described as ≥ 70 mm overbark (Forestry Commission measures all diameters in centimetres (cm); minimum diameter is thus 7 cm) or, as in broadleaves to where no mainstem is visible (i.e. to junction of mainstem) with lowest crown branch. Suitability of method depends on: (a) whether timber is felled or standing; (b) purpose of measurement (e.g. sale or woodland management requirements); (c) value of product. High value felled logs (e.g. veneer, top quality sawlogs) require very precise measurement; for low value material less precision is usually acceptable; (d) quantity of timber; (e) availability of facilities (e.g. weighbridge); (f) acceptability to each party concerned.

Estimation of growth rate and productivity of trees

For practical management purposes it is necessary to measure and predict the rate of growth of a plantation, i.e. the annual volume increment of timber produced per unit area of land, expressed as cubic metres per hectare (m^3/ha). *Current Annual Increment* (CAI) is the volume added in any single year, while *Mean Annual Increment* (MAI) is the *total* volume produced to date divided by the *age of the stand*, i.e. number of growing seasons since planting.

Yield class is a classification of growth rate in terms of the potential maximum MAI of volume greater than 70 mm top diameter, irrespective of tree species or age of culmination.

These measurements only apply to even aged stands (Forestry Commission, 1971). The MAI and therefore the yield class (or maximum potential MAI) are determined jointly (a) by the species and (b) by the quality of the site. As shown in *Tables 7.2* and *7.3* the yield class varies greatly from species to species and at the same time there is a wide range within species, according to the quality of the particular site. A high volume producing species on a good site will thus produce highest yields while a low volume producer on a poor site will produce the lowest (*Tables 7.2* and *7.3*).

There is a good correlation between top height of largest diameter at breast height (dbh) 1.3 m above ground level and total volume production of a stand of timber. This relationship is used to avoid detailed measurement of total volume production. In any even aged plantation it is thus possible to obtain the General Yield Class (GYC) which is adequate for most purposes, by calculating the top height of a small representative sample of largest diameter trees at dbh and plotting it against the age of the stand in years on the GYC curves (Forestry Commission, 1971). These graphs also indicate the appropriate ages of first thinning and maximum MAI for each species according to GYC.

Estimating volume of standing and felled timber

For detailed account of recognised procedures for timber mensuration, measurement conventions and conversion tables see Forestry Commission (1975). The following are most suitable for farm woodlands:

(Continued on p. 224)

Table 7.2 Characteristics and site requirements of major timber species and species planted for timber or shelter – The Conifers

Species and origin	Approximate length of rotation (years)	GYC (m³/ha/annum) Range	GYC (m³/ha/annum) Average	Specific site conditions Tolerance to Dry site	Exposure	Shade in early life	Spring frosts	General site requirements Suitable situations	Situations to avoid	Timber quality, uses strength class and general remarks
Douglas fir (*Pseudotsuga menziesii* (Franco) mink) (Western North America)	45–55	8–24	14	2	3	2	3	Valley slopes, well drained sheltered, moderately fertile soils.	Exposed or frosty sites, heather, ground, wet, soft or shalloww soils	Strong construction timber BSC Sr; ND. High strength: weight ratio Imported timber = Oregon or Columbian pine
The Larches (*Larix* spp.)	35–45	4–16	8	3	3	3	3			Grade for grade all larches have similar timber quality BSC1S1, MD. Wide range of outdoor work
European larch (*L. decidua* Mill		Very susceptible to larch canker; now rarely planted.						Thrives over wide range of conditions	Dry, poorly drained or v. exposed sites. Rainfall ≤750 mm frost hollows	Best quality boatbuilding. Used as 'nurse' in establishing hardwoods. 'Corkscrews' on very fertile sites. Runs out of top-growth quickly so thin in early life (good stand year 13).
Japanese larch (*L. kaempferi* (Lambert) Carr) (Japan)								–		
Hybrid larch (*Larix × eurolepis* Henry) (First raised in Scotland)								Outgrows Japanese larch on marginal sites		Plant first generation (F₁) hybrids
The Pines (*Pinus* spp.)										Imported timber known as Redwood, red or yellow deal, Baltic redwood BSC S2; ND
Scots pine (*P. sylvestris* (L)) (British Isles and Northern Europe)	65–80	4–14	8	1	2	3	1	Heather sites, poor gravel or sandy soils, low rainfall areas	Wet or soft ground, chalk, limestone high rainfall moorlands	General purpose softwood Best quality for furniture. Low growth rate and volume production. Useful as nurse for hardwoods. Shelter belt tree
Corsican pine (*P. nigra* var. *maritima* (Ait) Melville) (Corsica)	60–70	6–20	12	1	2	3	1	Low rainfall areas and elevations. Succeeds on sand and clay soils, especially near sea	High elevations. Wet moorland. Plant only in southern England	Timber similar to Scots pine but coarser and a little weaker. Higher volume production than Scots pine but more difficult to establish; seedlings in small containers give good results. Useful shelterbelt tree

Species								Soils	Sites	Remarks
Lodgepole pine (*P. contorta* Dougl. ex Loud.) (Western North America)	65–80	4–14	8	1	1	3	1	After suitable soil preparation of poorest health, peat or sand dunes	All but poorest sites where no other tree will grow	Timber similar to Scots. Very slow growing, low volume production. Useful pioneer and shelter tree
The Silver Firs (*Abies* spp.) European silver fir (*A. alba* Mill.)										Should not be planted in UK as heavily damaged by insect *Adelges nüsslini*
Grand fir (*A. grandis* Lindl.)	55	14–30	3	3	3	1	3	Well drained moist deep soils	All poor soils; dry, frosty or exposed sites	Soft white timber of only moderate quality. Produces large volume in short time. Useful for underplanting
Noble fir (*A. procera* Rehd.)	70	10–22	15	2	1	1	2	Best on well drained deep moist soils tolerates acidity	Poor dry soils	Timber as grand fir. Withstands exposure well. Tolerates drier sites than sitka space. A useful shelter belt tree in Scotland
The Spruces (*Picea* spp.)										Imported timber known as whitewood, white deal or Baltic whitewood. BSC S3; ND
Norway spruce (*P. abies* L. Karst.) (Europe)	55–60	6–22	12	3	2	3	2	Moist grassy or rushy sites, most reasonably fertile soils and fairly heavy clays.	All dry exposed or poor conditions. Prefer sitka spruce on exposed wet uplands.	Good general purpose timber, works and nails well. Stable in changing humidity conditions so suitable for building. Best quality joinery and whitewood furniture. Planted to produce Christmas trees.
Sitka spruce (*P. sitchensis* Carr.) (Western North America)	55–60	6–24	14	3	1	3	2	Wet exposed uplands in N. and W. of British Isles. Thrives on peats and grasslands in high rainfall areas	All sites liable to dry out and rainfall areas ≤1000 mm. Not E or SE England	Timber similar to Norway but too coarse for joinery work. Not an amenity tree but stands severe exposure and ideal for upland shelterbelts
Western hemlock (*Tsuga heterophylla* (Raf.) Sarg.) (Western North America)	45–55	12–24	14	1–3	2	1	1	Tolerant to high and low rainfall areas, acid mineral soils and the better peats	Very subject to butt rot (*Fomes annosus*) so avoid previous conifer sites if suspected. Slow to establish on heather and open heaths.	General purpose building timber; if graded, use for joinery. BSC. S3; ND. Establishes best if given some shade.

| Western red cedar. (*Thuja plicata* D. Don) | 45–55 | 12–24 | 14 | 2–3 | 2–3 | 1 | 1 | Moderately fertile soils, shallow or clays. | Exposed, poor or very acid sites. | Lightweight timber. BSC S3; D. Estate work, portable buildings, greenhouses and cedar shingles. Amenity and single row screens. Not hedges as subject to Keithia disease. |

Key:
Specific site conditions, tolerance to:
Dry site
1 Suited to dry soils and situations
2 Avoid sites liable to dry out badly
3 Moisture demanding species but site must be well drained
4 Thrives on poorly drained soil sites with fluctuating watertable or along streams
Exposure
1 Stands severe exposure
2 Stands moderate exposure
3 Avoid all exposed sites or give adequate side shelter
Shade in early life
1 Strong shade bearer and suitable for underplanting
2 Moderate shade bearer or tolerates side shade only, which may be beneficial for establishment – e.g. group selection system
3 Strong light demander, clearfell before replanting. Thin early in life or good sites
Spring frosts
1 Tolerant
2 May be damaged in severe cases
3 Highly sensitive – avoid all frost pockets
BSC = *Building Strength Class*, S1 strongest see BSCP 112
Durability of untreated heartwood
ND = Perishable
  D = Durable
MD = Moderately durable
NB. Sapwood is non-durable and must be treated for outdoor use

Table 7.3 Characteristics and site requirements of major timber species and species planted for timber or shelter – The Broadleaves

Species and origin	Approximate length of rotation (years)	GYC (m³/ha/annum) Range	Average	Specific site requirements — Tolerance to: Dry site	Exposure	Shade in early life	Spring frosts	General site requirements Suitable situations	Situations to avoid	Timber qualities, uses, strength class and general remarks
Ash (*Fraxinus excelsior* L.) (British Isles and Europe)	50–70	4–10	5	3	2	3	3	Good timber ash must grow quickly. Very exacting on site – deep fertile moist soils; also deep chalk and limestone, pH > 5.5. Wild garlic a good indicator. Give side shelter during establishment.	Dry, shallow heavy clay or badly drained soils, heaths and moorlands	Springy timber with high shock resistance. Top grade for sports goods, oars, tool handles furniture, panelling. Branches very good firewood. Usually grown in mixture with beech, oak, gean, sycamore or larch. Thin regularly to encourage large crowns.
Beech (*Fagus sylvatica* L. (Southern England and Europe)	80–100	4–10	6	2	1	1	3	Light well drained soils on chalk or limestone, deep sands and acid brown earths	Cold, wet poorly drained soils, dry infertile sands	Timber strong fine even texture; clear timber good for furniture; bends, stains and polishes well. Fine amenity tree. Good shelter but old stands become open, draughty and difficult to regenerate. Best mixed with oak, ash and gean. Very subject to squirrel damage. Valuable hedging plant.

Table 7.3 Continued

Species and origin	Approximate length of rotation (years)	GYC (m³/ha/annum) Range	Average	Specific site conditions — Tolerance to: Dry site	Exposure	Shade in early life	Spring frosts	General site requirements: Suitable situations	Situations to avoid	Timber qualities, uses, strength class and general remarks
Native oaks. (*Quercus* spp.) (British Isles and Europe) Pedunculate oak (*Q. robur* L.)	120	2–8	4	3	3	2 then 3	3	Deep clay loams with ample moisture; tolerates heavy clays.	All exposed areas. Shallow, infertile or poorly drained soils.	Timber quality very erratic, varies from prime to rubbish. Good prime oak scarce; is used for veneer, furniture, panelling, joinery. Lower grades fencing, gates, mining timber; all grades boat building. Do not try to establish in open or low 'toffee apple' crowns produced.
Sessile oak (*Q. petraea* (Matt.) Lieb.) The two species have intercrossed freely and many trees are hybrids.								Best on deep porous brown earths. Tolerates clay soils.	As above and wet clays or fluctuating water table.	Use tree shelters and give side shelter to draw stems up and suppress epicormic branches. Side prune. Regenerate with group selection system. Sessile oak less likely to produce epicormics. Timber grown on lighter soils liable to 'shake'.
Poplars (*Populus* spp.) Black poplar (*P. robusta*) and black hybrids. (*Populus* × *euramericana* Dode Guinier) (Europe) (Vars: *Eugenii, Gelrica, Heidemiz* T-78, *Serotina*)		4–14		4	3	3	1	Base rich loamy soil water table 1–1.5 m below surface in summer. Will also grow on wide range of soils if moist.	Exposed acid dry or infertile sites	Demand for match timber much reduced. Used for pallets, chip baskets. Useful pulpwood. Poplar production for timber is a very specialised job. Poplars are only suitable for perimeter horticultural windbreaks; they are too vigorous for use inside orchards.

Species (origin)	Rotation (years) / method	Yield class		a	b	c	d	Suitable sites	Unsuitable sites	Remarks
Balsam poplars, (*P. trichocarpa* (Var. Fritzi Pauley and Scott Pauley) *P. tacamahaca* × *trichocarpa*-Var 32 (North America)				4	3	3	1	Tolerate more acid soils than black hybrids and more suited to cooler wetter parts of Britain.	As above.	Balsam poplars subject to bacterial canker; plant only resistant clones. Preferred to black hybrids as they are fastigiate and trim more easily. Hybrid *P. txt* Clone 32 most commonly used. Other poplars include grey and white spp.
Southern beech (*Nothofagus* spp.)		10–18						Tolerant of wide range of soils: heavy clay to deep sands	Badly drained exposed or frosty sites. Frost causes early life kill and stem cankers.	Timber of both spp. similar to native beech, but 20% less bending strength. Immature timber of lower quality. *N. procera* best timber form. Both spp. fast growing. Prefer *N. procera* in wetter west country, *N. obliqua* in drier east. Light demanding spp. but start under thin canopy. Thin early in life.
N. procera				3	2 then 3	3				
N. obliqua				2	2 then 3	3				
Sweet chestnut (*Castanea sativa* Mill.) (Mediterranean)	*High forest* 50–70 *Coppice* commonly 12–16, range 10–20 walking sticks (2)–3	4–10	6	2	3	3	3	Deep fertile soil in warm climate NB southern England. Warm sunny acid sandy loam banks ideal (pH 4.0–5.0)	Cold wet badly drained, exposed or infertile sites. Chalk, limestone or alkaline soils. Not N. England or Scotland	*High forest* Timber subject to ring, also star, shake esp. large trees circa 80 years. Sawn timber furniture, also coffins. *Coppice* cleft fencing, hop poles, stakes. Firewood must be very dry to burn. 90% of area occurs in Kent and Sussex
Sycamore (*Acer pseudo-platanus*) (Central Europe)	50–70	4–12	5	2–3	1	2	2	Exposed uplands where broadleaves are mixed in conifer shelterbelts	Dry, shallow ill-drained or heavy clay soils. Not an amenity tree. Clear area of grey squirrels	White timber used for turnery and where in contact with food, e.g. butchers' blocks. Figured sycamore valuable for veneer and furniture

Table 7.3 Continued

Species and origin	Approximate length of rotation (years)	GYC (m³/ha/annum)		Specific site conditions — Tolerance to				General site requirements		Timber qualities, uses, strength class and general remarks
		Range	Average	Dry site	Exposure	Shade in early life	Spring frosts	Suitable situations	Situations to avoid	
Walnuts (*Juglandaceae* spp.) Common walnut (*J. regia* L.) (Turkey and Asia) Black or American walnut (*J. nigra*) (USA)	80–100			2	3	3	3	Require warm summers; only South and Central Britain. Sheltered mid slope situation on fertile deep well drained medium texture soils. Optimum pH 6 to 7. Chalk or limestone ≥600 mm of overlying soil	Only planted on carefully selected sites which meet requirements	Only common walnut yields edible fruit. Valuable veneer timber, often dug out and taken to mill with main roots as 'figuring' occurs in base. Also valuable furniture and gun stocks. Branches good firewood. Grown as 'orchard' spaced crop. Can be ready in 40 years on best sites.
Wild cherry or Gean (*Prunus avium* L.) (Britain and Europe)	50–60			2	3	3	3	Timber: Deep fertile well drained forest soils, especially over chalk. Less exacting requirements for amenity	Heavy soils, depressions and all infertile sites	Widespread natural occurrence. Heartwood rich reddish brown; veneers, furniture, decorative panelling, turnery. Used on margins of woods or roads; sometimes mixed with ash, beech or pedunculate oak. Not attacked by squirrels
Cricket bat willow (*Salix alba* var. *coerulea* Sm) (England)	14–15							Only margins of flowing streams or rivers with highly fertile soil	Useless elsewhere	Cricket bats; timber must be knot free. Planted as sets 10.5–12 m apart; all sideshoots removed on bottom 3.05 m of stem. Very specialised production. (Forestry Commision 1986.) Also planted for shelter

Key: see Table 7.2, p. 214

Table 7.4 Characteristics and site requirements of species planted primarily for shelter or amenity. Timber of secondary or nil importance – The Conifers

Species and origin	Specific site conditions Tolerance to				General site requirements		Remarks
	Dry site	Exposure	Shade in early life	Spring frosts	Suitable situations	Situations to avoid	
The Cypresses Lawsons cypress (*Chamaecyparis lawsoniana* Parlat)	2	2	1	2	Requirements not exacting but best on deep fertile soils	Dry infertile sites and heather ground	Amenity, single row screens and hedges only. Liable to snow break
Leyland cypress (*Cupressocyparis leylandii Dall*)	2	2		1	Does well under wide range of conditions	As Lawsons cypress	Hybrid (Nootka × Monterey cypresses). One of the fastest growing cypresses. Grown only from cuttings. Plant *small* trees to root before wind catches them. Can grow to 20 m or more. Popular for single row screens. May have some timber value. Usually too fast a grower for hedges
Monterey cypress (*Cupressus macrocarpa* Hart.)					Do not plant		Has been planted extensively in SW England but dies back, is not frost hardy and superseded by Leyland cypress
Other Pines (*Pinus* spp.) Austrian pine (*P. nigra*)	1	1	3	1	Shelterbelts near sea, smoky conditions, dry limestone soils	North and West Britain, wet soils	Prefer Corsican, Scots or Lodgepole pines wherever suitable. Coarse, knotty and low timber value. Only worth planting for shelter where site conditions rule out other spp.
Mountain pine (*P. mugo* Turra)	1	1	3	1	Only plant on poorest sites		Useless for timber. Stands great exposure in poorest situations. Possible for very exposed shelterbelt margins. Suffers from leaf diseases
Radiata or Monterey pine (*P. radiata* D. Don)	1	1	3	3	Near south west coast only. Not frost hardy; stands salt laden winds	All wet hollows and frosty sites	Major timber species in New Zealand but not as yet in UK. Fast grower, mature tree in 30 years. For timber is grown in plantation; requires early and continuous side pruning otherwise timber is coarse and knotty. May have future on Devon, Cornwall and south coast drylands. For shelter use only where quick growth is required; plant other spp. to take over. Plant *small* containerised seedlings on exposed sites, *never* larger transplants.

Table 7.4 Continued

Species and origin	Specific site conditions Tolerance to				General site requirements		Remarks
	Dry site	Exposure	Shade in early life	Spring frosts	Suitable situations	Situations to avoid	
The Redwoods (*Sequoia* spp.)							
Californian redwood (*S. sempervirens* (D. Don) Endl.	3	3	1	3	Only deep fertile soils in higher rainfall areas	Dry exposed infertile or frosty sites	Valuable timber tree in USA but British timber has low density and strength. Durable, slow establishing. Plant under tall cover. Only for amenity in UK. GYC 16–30. Very long lived
Wellingtonia (*Sequoia dendron giganteum* (Lindley)) Buch.	2	2	2	1	Tolerates drier more acid sites than Cal. redwood. Prefer in Scotland		Timber as Cal. redwood. Only planted as amenity tree in UK. parks, avenues, etc. Hardy and fairly windfirm. GYC 16–30

Key:
Specific site conditions, tolerance to:
Dry site
1 Suited to dry soils and situations
2 Avoid sites liable to dry out badly
3 Moisture demanding species but site must be well drained
4 Thrives on poorly drained soil sites with fluctuating watertable or along streams.
Exposure
1 Stands severe exposure
2 Stands moderate exposure
3 Avoid all exposed sites or give adequate side shelter.
Shade in early life
1 Strong shade bearer and suitable for underplanting
2 Moderate shade bearer or tolerates side shade only, which may be beneficial for establishment – e.g. group selection system
3 Strong light demander, clearfell before replanting. Thin early in life or good sites.
Spring frosts
1 Tolerant
2 May be damaged in severe cases
3 Highly sensitive – avoid all frost pockets.
BSC = *Building Strength Class*, S1 strongest see BSCP 112
Durability of untreated heartwood
ND = Perishable
D = Durable
MD = Moderately durable.
NB. Sapwood is non-durable and must be treated for outdoor use.

Table 7.5 Characteristics and site requirements of species planted primarily for shelter or amenity. Timber of secondary or nil importance – The Broadleaves

Species	Specific site conditions Tolerance to				General site requirements		Remarks
	Dry site	Exposure	Shade	Frost	Suitable situations	Situations to avoid	
The Alders (*Alnus* spp.)							
Black or Common Alder. (*A. glutinosa*)	4	1	3	1	River and stream banks and wet marshy places in lowland and uplands	All dry sites. For very acid sites prefer grey alder	When of timber size dries red brown, good for turning, staining and polishing; also bending. Grows rapidly for first 25 years. Coppices vigorously. Grown on 15–20 year rotation can give 8–10 t/ha DM/annum. Only burns if air-dried. GYC 4–12 Good shelterbelt tree. Windfirm and withstands salt spray in coastal belts. Can be planted on pulverised fuel ash. Valuable for retaining river banks. Usually found in pure stands. Mixes with birch and ash and is a good nurse for oak on wet clay soils. Fixes atmospheric N in root nodules. Acts as host for 90 insect spp.
Grey alder (*A. incana*)	2–3	1	3	1	Requires drier situation than black alder but tolerates poorer sites	Dry infertile situations, heathlands, shallow and thin chalky soils	Timber worthless. Widely planted for windbreaks, shelter and as a pioneer spp. for reclaiming derelict land. Suckers freely. Also used for horticultural internal windbreaks, especially fruit. Extremely hardy. Prefer to black alder on very acid sites
Italian or cordate leaved alder (*A. cordata* Desf.)	2–3	2	2	2	Adaptable. Also drier chalk and limestone soils of S England	Avoid dry thin acid or infertile soils	Not timber. Useful pioneer species. Used for single row windbreaks, fruit etc. Fast growing, handsome
Red alder (*A. rubra*)							Native of N. America, natural range Alaska–California. Not yet fully tested in UK but shows promise, especially upland situations
The Birches (*Betula* spp.)							
Downy birch (*B. pubescens*)	2–4	2	2	1	Well drained sandy loams but tolerates wet and acid soils, lakesides, peat	Do not plant unless wanted for silvicultural or amenity reasons	Not timber. Pioneer spp. for land reclamation; on difficult sites use containerised plants. Also nurse, planted or natural for frost tender conifers, oak or beech. Cut out before crowns of timber spp. are damaged. Widespread natural occurrence, seeds and coppices readily. Popular amenity tree, supports numerous insect spp.

Table 7.5 Continued

Species	Specific site conditions Tolerance to				General site requirements		Remarks
	Dry site	Exposure	Shade	Frost	Suitable situations	Situations to avoid	
Silver or warty birch (*B. pendula*)	1–3	2	3	1	Adaptable drier soils than downy birch: brown earths, podsols, sands, gravels	Not planted unless wanted for silvicultural or amenity reasons	An underestimated spp. Not timber in UK but grown in Scandinavia with improved cultivars. Pioneer spp; light open crown gives suitability for underplanting with shade tolerant spp. Excellent visual amenity and supports numerous insect spp. Seeds and coppices strongly
Elms (*Ulmus* spp.)							⎰Unsafe to plant until cultivars resistant to Dutch Elm disease have been bred and fully tested. Prefer ⎱other spp.
Hawthorn, May or Quickthorn (*Crataegus monogyna*)			3		All but poorest soils	Proximity to apple and pear orchards as susceptible to fireblight bacterium *Erwinia amylovora*	Widely planted for stockproof hedges but requires regular layering or cutting or it becomes open and leggy. Stands severe cold and salt laden winds. Quick growing valuable windward margin for shelterbelts. Flowers and fruit valuable to many spp. of insects and birds
Hornbeam (*Carpinus betulus*)	2–3	1	1	1	Moist soils, chalk, limestone and acid brown earths	Thin, infertile dry and very acid sites. Control grey squirrels before planting	Hard heavy tough timber giving very smooth finish. The 'white beech' of the continent. A substitute for beech on clay soils and where high frost resistance is required. Especially valuable as margin or underwood for shelterbelts and for hedges; also under oak to control epicormics. Normally used as understorey, not pure stand
Horse chestnut (*Aesculus hippocastanum*)	2–3	3	1	1	Best on deep moist fertile loams	All infertile sites	Soft white timber turns and machines well. Pulpwood. Not good firewood. Fast growing seldom lives beyond 80 years. Valued for its snowy white flowers. Widely used in parks and avenues. Also mixed woodland where it keeps oak stems free from epicormics and shades out thorn. Gives deep leaf litter
Small-leaved lime (*Tilia cordata*) **Common lime** (*T. × vulgaris*) is hybrid (*T. cordata* × *T. platyphyllos*)	2	2	2	1	Fertile well drained soils. Occurs naturally on chalk and limestone soils	All infertile soils	Good all-round timber: turnery and carving. Stains and polishes well. Productive species to replace elm. Valuable for amenity, lowland shelterbelts and conservation. Flowers produce abundant nectar for bees. Little squirrel damage seen. Grow in mixed woodland, rotation 50–60 years. Small-leaved lime grows from cuttings. Should be more widely grown. Casts dense shade

Species					Soil preference	Sites to avoid	Remarks
The Maples (*Acer* spp.) Field maple (*A. campestre*)	1	2	2	1	Well drained clays on chalk and limestone, tolerates shallow soils	Poorly drained very acid or light sands	Rarely available in large sizes so mostly firewood. Useful in hedgerows, edges of plantations and outside rows of shelterbelts. Good autumn colour, good source of food for small mammals and insects. Easy to establish, coppices readily, tolerates clippings. Understorey to oak on calcareous soils
Norway maple (*A. platanoides*)	2	2	2	1	Moist deep free rooting soils with high base status	All infertile situations	Timber as sycamore; flooring, furniture, turnery, veneer. Good firewood. Grown in mixture with ash, gean and larch on 45–60 years' rotation. Good amenity tree, bees like flowers. Suited for screens and mixed shelterbelts. Heavily damaged by squirrels
Red oak (*Quercus borealis* Michx.)	2			1	Fertile slightly acid sandy soils	Very infertile sites	General purpose hardwood only; no substitute for English oaks. Brilliant red autumn leaves. Grown for amenity only — little planted. Heavily damaged by squirrels
Rowan or mountain ash (*Sorbus aucuparia*)	1	1	2	1	Adaptable. Prefers light, acid brown earth	Waterlogged sites	Not timber but a beautiful amenity and conservation tree. Suited to shelterbelts and as understorey in woods and as a pioneer spp. Severely broused by deer
The Willows (*Salix* spp.) White willow (*S. alba* L.)	3–4	2	3	2	Deep alluvial loams, fertile boulder clays with high pH; proximity to running water	Dry or waterlogged sites with stagnant water	Not timber. Selected Dutch varieties make ideal single row perimeter windbreaks with better winter protection than poplars and easier to trim. Cheap and easy to establish with unrooted rods at 0.3 m spacing. Insect pollinated and attract insects to orchards
Goat willow (*S. caprea* L.) Other willow spp. of bushy habit	3–4	1	3	1	Very adaptable. Tolerates waterlogged peats	Dry situations	Mainly shrub up to 3 m tall, very hardy and adaptable. Valuable as edging for high elevation shelterbelts and as pioneer. Selected clones and hybrids for osier production and low horticultural windbreaks

Key: See Table 7.4, p. 220

Standing timber   For small numbers or scattered groups of trees the volumes of individual trees are estimated. This method is of only moderate precision and unduly expensive for large numbers of trees (Forestry Commission, 1975). Requires measurement of dbh (diameter breast height 1.3 m above ground level with girthing tape) and height of tree with altimeter. For larger numbers of trees other methods are appropriate (Forestry Commission, 1975).

Estimating volume of felled timber

(1) *Assessment of volume from length (m) and mid-diameter (cm) measured overbark* is the traditional method of assessing volume and the most accurate, as it does not assume a standard taper. Best for small lots, valuable produce and long log lengths. Measure and round down length (m) on logs up to 20 m long (logs ⩽ 10 m long to nearest 0.1 m; > 10 m to nearest 1.0 m) and measure diameter at mid point with girthing tape. Consult Forestry Commission booklets 1975 or 1978b for detailed instructions and conversion tables. Logs > 20 m long are estimated in *two* sections. Hardwood logs of varying quality are measured in individual sections.

Volume is derived from the formula:

$$V = \frac{\pi d^2}{40\,000} \times L$$

Where: V = volume (m³); L = length (m); d = mid-diameter (cm); π = 3.1416.

(2) *Assessment of volume of conifer sawlogs from length (m) and top diameter (cm) measured underbark* is less accurate than (1) but quicker and cheaper for stacked batches of short logs crosscut to uniform length. Standard taper assumed 1:120. Measure standard length (to lowest 0.1 m) and top diameter (smallest end) of each log individually with rule to lowest whole centimetre. Consult Forestry Commission booklets 1970 or 1975 for detailed instructions and conversion tables.

(3) *Assessment of smallwood volume from length (m) and top diameter (cm)*   For small lots of small diameter material (e.g. stakewood) when other methods are impracticable. Assumed taper 1:84 measure individual lengths (to lowest 0.1 m); top diameters measured with rule to lowest whole centimetre. Consult Forestry Commission booklet 1975 for detailed instructions and conversion tables.

(4) *Measurement of stacked timber*   Method used for cut uniform length smaller diameter material (e.g. pulpwood). Length of stack equals length of cut pieces. Stretch tape parallel to base of longside of stack (width) (*Figure 7.1*), read off and measure height with rule, first point 0.5 m from start and at every 1.0 m interval to give average height, all measurements in metres. Use formula

$$\text{Volume of stack (m}^3) = \frac{(h_1 + h_2 + h_3 \ldots + h_n)}{\text{no. of readings of } h} \times W \times L$$

Where h = height, W = width × L = length   of pieces. Multiply × 0.70 or 0.75 to convert volume of stacked to solid timber. For full details consult Forestry Commission booklet 1975 (see *Figure 7.1*).

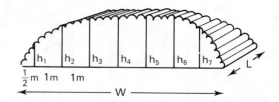

Figure 7.1 Method of measuring stacked even length small timber

LENGTH OF ROTATION

Length of rotation may be defined as the number of years which elapse between the planting of the tree crop and the point when the mature crop is felled; the area is then replanted. Maturity is not easy to define.

Physical maturity indicates the age at which a stand is physically mature for felling – the best saw timber needs large trees from a long rotation while trees for small poles, chipboard and pulpwood only need a short rotation or are taken from thinnings. *Financial maturity* is the age at which a stand will give the best financial return, which in turn depends on the value of the timber at each stage of growth or maturity.

The main financial problem in forestry is the long period during which capital is locked up and consequently high accumulated interest charges. Major costs such as land clearance, fencing and planting occur at the beginning of the rotation, and attract interest from the day they are incurred, but it is many years before a cash return is available for repayment of capital and accumulated interest. In order to reduce the interest liability it is necessary to delay all expenditure until the latest feasible stage, speed up the process and obtain a quicker return by planting fast growing species and by shortening the rotation where possible (*Tables 7.2 and 7.3*).

Although the larger trees produced in a long traditional rotation will be more valuable and may give a higher income per hectare than those in a shorter rotation, there is so much capital locked up that the percentage return is seriously reduced. Thus, even though a plantation in a shorter rotation may yield less cash when clearfelled, it is usually more profitable on account of the lower interest charges, and because the rate of growth of timber falls off once maximum MAI is reached; the best financial return is often given by felling just before this but in practice it is usually recommended to clearfell ± five years from the point of maximum MAI shown in the Forestry Commission GYC curves for the appropriate species (Forestry Commission, 1971).

The precise timing of felling depends on current timber demand and value. The new crop following replanting will give a better growth rate and return than the ageing stand it replaced.

There are no hard and fast rules on length of rotation, which is largely determined by the rate of growth, being shorter with fast growing conifers and on good sites, longer with slower growing species and on poor sites. Very slow growing broadleaved trees such as oak and beech require such a long rotation that the interest charge alone becomes excessive.

TIMBER QUALITY

The best quality timber is grown under forest woodland conditions. Quality is determined jointly by species, provenance (see Plant Supply, p. 230), site conditions and silvicultural management, especially planting distance and thinning regime. Hedgerow and parkland trees are generally unsatisfactory for quality timber, as the trunks tend to be short and the open conditions allow heavy branching with the consequent development of large knots. The lower part of the trunk frequently conceals old staples, nails and wire, and has to be discarded.

Good quality saw timber comes from trees which are long, straight, knot free and with the minimum taper. Knots, wide rings and cross grain all tend to cause a rough surface finish. Large knots and rings of knots cause weakness. A variable growth rate of the tree gives unevenly spaced rings, which may result in cupping or twisting of the sawn timber. Homegrown timber is frequently inferior to the imported article in these respects.

Wood is used for a very wide range of purposes. The higher the price, the more stringent are the quality requirements. Only the pick of the clean, straight, knot-free large diameter logs are suitable for veneer, which is a highly specialised job. This is followed by trades where a fine finish is required, notably furniture manufacture, high-grade joinery and boat-building. Good sawn building timber used for structural purposes needs to be strong in bending and in compression – Douglas fir is ideal in this respect.

Although building timber does not need to be as fine as the joinery grade, it should be clean, straight, even grained and free from large knots. Softwoods to be used for structural purposes may also be stress graded to BS 4978 into five categories of strength class (numbered SC1–SC5 in descending order of strength) and according to species; timber may be visually or machine graded. Thinnings and the lower grades are used for a wide variety of work such as turnery, fencing, fencing stakes, garden furniture, rustic work, general estate work, mine timber, chipboard, fibreboard, wood wool, packing cases and pulpwood, but command lower prices than sawlogs, minimum top diameter 18 cm. There are also various local markets for timber. Timber quality can be ruined by inefficient felling, cross cutting and processing, so harvesting, extraction and milling are jobs for highly skilled specialists only.

SYSTEMS OF WOODLAND MANAGEMENT

Woodland may be managed either as 'coppice' or 'high forest'.

Coppice

Coppice is a broadleaved woodland crop managed on a regrowth system, the new crop being raised from the stumps (stools) of the previous crop, cut near ground level. The new shoots arise from either dormant buds on the side of the stool, adventitious buds at the edge of the cut surface or with hazel, birch and willow from root buds near the stump.

Initially shoots make very rapid growth as they are supported by a well developed root system and food reserves but growth rate gradually slows down; old unworked coppice fails to regrow vigorously and usually needs replanting. Only broadleaved species that regrow rapidly after cutting are suitable, e.g. sweet chestnut (*Castanea sativa*), pedunculate and sessile oaks (*Quercus* spp.), hazel (*Corylus avellana*), ash (*Fraxinus excelsior*), hornbeam (*Carpinus betulus*) and willows (*Salix* spp.); others do not regrow vigorously or only from small young stumps, e.g. beech (*Fagus sylvatica*), birch (*Betulus* spp.), some poplar species (*Populus* spp.) and gean or wild cherry (*Prunus avium*). The latter and some poplars produce suckers freely. All conifers (except Coast redwood (*Sequoia sempervirens*) and monkey puzzle (*Araucaria araucana*) do not regrow. *Coppicing* is repeated many times on the same stand, some stands being hundreds of years old. The practice is ideal for rapid production of small roundwood but value of produce (e.g. firewood) is relatively low while handling and conversion costs are usually high and financial returns are far below those obtainable from well managed high forest. Suitability of the system to farm woodland conditions depends largely on margin of returns over production and conversion costs.

Coppice products were of prime importance to the rural economy until the industrial revolution, when coal superseded wood as a fuel and sawn timber was required; labour was cheap and the skills for conversion were readily available. Decline was rapid after the 1920s and large areas of coppice, e.g. hazel (thatching spars, sheep cribs and hurdles) and oak (charcoal and bark for tannin) fell into disuse as demand for products disappeared. Only sweet chestnut (*C. sativa*) (split palings and stakes, hop poles) is now of any significance in coppice worked for profit. Much of the old coppice has lost its vigour and requires replanting; it is usually best converted to high forest. Vigorous coppice can be used to provide a wide variety of products, e.g. turnery wood, poles, pea sticks, decorative foliage and firewood, but profitable markets are limited.

Types of coppice

Simple coppice   Consists solely of coppice all worked on the same cycle (even aged); it may be of one (pure) or several (mixed) species. Most sweet chestnut (*C. sativa*) is worked as pure simple coppice.

Coppice with standards   Two storey woodland with coppice underwood, e.g. hazel (*C. avellana*) with scattered standards, usually oak (*Quercus* spp.) being allowed to reach timber size through several coppice cycles. Obsolete and usually unsatisfactory, as standards are short in the bole and partly shade the coppice below, resulting in poor growth.

Short rotation coppice   Worked on a cycle of less than ten years to produce stick sized material for rural crafts, e.g. osiers (willows) for wickerwork, hazel for thatching spars, sweet chestnut for walking sticks. Also new willow varieties for biomass production on disadvantaged wet lands.

Pollards   Regenerate exactly like coppice but are cut off 2–3 m above ground to prevent browsing of regrowth by livestock; an obsolete wood-pasture system (e.g. riverside willows). Appearance is similar in trees (e.g. alder, *Alnus* spp.; poplar, *Populus* spp.) topped for shelter or screens.

Storing up   Each coppice stool is reduced to the best single stem which is then allowed to grow on to timber size. Initial growth is rapid but ultimately trees from stored coppice are much smaller and of lower quality than maidens, having a typical swept (curved) butt.

Conversion of coppice to high forest   Much low quality oak woodland in Somerset and Devon is stored coppice. Storing is a cheap method of converting coppice to high forest but is only a wise practice on sites with good quality mixed species. Pure hazel and worn out or low quality coppice should be replanted or enriched with transplants of good timber provenance or certified parents of good timber quality.

Production of firewood and biomass

In remote areas or where supplies of other fuels are expensive, growing coppice for fuelwood may be attractive, either for (a) firewood or (b) biomass.

Firewood   Firewood is neither a cheap nor an easy alternative to fossil fuels. Its use presupposes skilled labour to harvest and convert it, ample covered storage for drying and an adequate area of coppice. Only burn air dried wood as wood with high moisture content gives little heat and causes tar deposits in stoves and flues.

With wood as sole heat source for a three bedroomed house, about 7 t air dried wood/annum are needed occupying 4 m^3 covered space. Coppiced woodland may give about 5 t/ha green wood/annum $\equiv$ 12.5 t/ha/annum air dried wood, requiring about 3.0 ha coppiced woodland cutting and processing 0.2 ha/annum managed on a 15 year cutting cycle.

Only slower growing broadleaved species should be used in open fires; mixed air dried wood is best – sweet chestnut (*C. sativa*) is poor. Conifers, willow (*Salix* spp.) and poplar (*Populus* spp.) spit and are only suitable for closed stoves. Waste from timber felling is valuable. Ash (*F. excelsior*), beech (*F. sylvatica*), birch (*Betula* spp.), hazel (*C. avellana*), holly (*Ilex aquifolium*) and old fruit trees burn well. Oak (*Quercus* spp.) must be well dried, cut small and is best mixed with other woods. Elm (*U. procera*) and black alder (*A. glutinosa*) burn well only if very dry. Machines are now available to convert small material and lop and top to chip size for burning in furnaces but capital costs for machinery and installation are high and can only be justified for large scale operations.

Biomass production   Coppiced willows have proved highly productive on wet disadvantaged land at Castle Archdale Experimental Station in Northern Ireland (surface water, gley soils), Sweden and Newfoundland. In Northern Ireland the following mean annual yields of DM have been obtained from *Salix aquatica gigantea* (t/ha): one year rods 10.7, three year rods 15.8, compared with perennial ryegrass 9.4 and improved permanent pasture 5.7. Mechanical harvesting is essential for economic production. Crop used for pulp, processed cattle feed and in Sweden, petrol manufacture. Northern Ireland probably has most potential in the UK. Only appropriate for wet soils.

High forest

High forest is grown to produce mature sawmill timber and is mainly established from transplants; some stands arise from natural regeneration or stored coppice. There are two main systems (a) the plantation or clearfell system, and (b) the selection system or continuous cover forest.

Plantation forest

This is the normal system of management in the UK. The forest is divided into even aged blocks which are clearfelled and then replanted. This system is less complex to manage and felling and extraction are easier and cheaper, as each harvest is confined to a single area; it also permits the use of large extraction machinery.

Unless carefully managed this system can cause serious soil erosion, especially on step slopes in high rainfall areas when conifers have produced dense shade, precluding the development of ground cover. Destruction of the litter layer, and leaching and compaction from heavy equipment can also occur.

Plantation forestry is frequently identified with large areas of even aged conifer monoculture, e.g. sitka spruce, which produces a dreary featureless landscape. Best modern plantation practice now varies species and ages to fit the landscape and give a varied and pleasing appearance; whenever possible irregular planting boundaries are followed. Large areas should not be clearfelled, as this ruins the appearance of the landscape and silviculturally destroys the woodland environment, making it much more difficult to establish species requiring some side shelter. Clearfelling should be confined to small areas, preferably not more than 2–3 ha in one block to shelter establishing transplants and varied uneven aged woodland. Boundaries and rides can also be softened by using broadleaved species, which should also be planted within 30 m of streams to avoid acidification of the water.

Selection system

In continuous cover forest felling and regeneration are not concentrated on certain clearly defined areas but spread all over the forest, so that each compartment contains an uneven aged mixture of all sizes. There are two basic approaches: (a) the single tree system, where single trees are removed, and (b) the group selection system, where groups of trees are felled and the area replanted; this is comparable to plantation management on a small scale but retains the woodland environment.

This system is widely used on the continent, as it is especially suitable where soil erosion is likely (e.g. mountainous areas with steep slopes) and suits small scale working (e.g. Germany, Switzerland). This system encourages a stable ecological balance, and preserves amenity, especially in sensitive landscapes. Silviculturally it greatly reduces liability to windthrow and late frost damage and is suited to production of large high quality timber. Disadvantages include complex management and working. Very experienced foresters are required, considerable skill is needed in felling and handling to avoid damage to adjacent trees and extraction is much more costly. The single tree system only suits shade tolerant species, western red cedar (*T. plicata*), western hemlock (*T. heterophylla*), beech (*F. sylvatica*) etc., group selection system also suits species demanding more light but needing side shelter and is ideal for regenerating worn out old oak woodland provided cleared areas are large enough: minimum 0.2 ha, preferably 0.5 ha. Prefer a 'lozenge' shape.

Natural regeneration may come from seed shed by nearby trees or from stored coppice regrowth. With no plant or planting costs it looks superficially attractive but there are snags. Regeneration is limited to the quality of tree and species already present; without transplants new species and improved timber provenances cannot be introduced. For natural regeneration to be successful the following basic requirements must be met:

(1) Accept regeneration only from top quality timber trees of seed bearing age – these are absent in most farm woodlands. All low quality trees must be removed before seeding begins.
(2) Regular 'mast' (fertile seed production) years – (beech (*F. sylvatica*) only masts every five to ten years).
(3) Adequate seed supply: ash, sycamore and conifers set seed abundantly.
(4) Absence of weed growth: lowland woodlands, especially broadleaves, produce vigorous undergrowth as soon as top cover is opened (e.g. bramble (*Rubus* spp.) thickets); such situations are hopeless for natural regeneration (tree shelters plus herbicide are then appropriate).
(5) There are good soil and seedbed conditions, not compacted.
(6) Adequate light is essential for seedling establishment; no species will grow under dense cover.

Even with good management success is far from certain. The mast may be slow in coming and regeneration may be patchy or so dense that respacing (i.e. thinning to normal planting distance) is costly and difficult. Policy should be to hope for it, allow for it and foster it if it comes but *not to wait* for it.

Woodland improvement and regeneration of worn-out woodlands

The most suitable management for woodland improvement is usually a compromise, as it depends on many factors: e.g. owners' requirements (timber, firewood, game, amenity) and capital available; condition of the woodland and its variability; soil, climate, topography and degree of exposure; value and marketability of constituent trees; area of woodland concerned; management skills, labour and machinery available; visual effect, sensitivity of landscape and official planning constraints; weed growth and likelihood of further weed regeneration. As a result several methods of improvement may be used simultaneously in a single wood.

Complete clearance and replanting (i.e. clearfelling)

This suffers from all the disadvantages already listed but is essential with infestations of rhododendron (*Rhododendron ponticum*), laurel (*Prunus* spp.), blackthorn (*Prunus spinosa*) and briar (*Rosa* spp.) which should all be cut down. Herbicidal treatment is usually advisable on either actively growing foliage or on the stumps after cutting. Failure to treat stumps may result in heavy regrowth. When replanting with light demanding species (e.g. larches, pines, spruces) clearfelling is usually necessary. On these sites do not use small transplants; prefer those 300–460 mm height.

Group selection

This is silviculturally the best method for regenerating broadleaved woodland composed of large mature trees, especially when replanting with oak or ash, as it provides essential side shelter, helping to prevent formation of low crowns. Damage to young trees from felling is avoided. Method is also suitable for amenity situations and sensitive landscapes.

Underplanting

Underplanting where the new crop is planted under existing cover, is only suitable under large trees where all are to be cleared subsequently at the same time, otherwise the young crop suffers heavy damage during felling and extraction; for these cases group selection is much safer.

Underplanting a thinned pole crop, which is retained to provide *temporary* cover, is a useful method of establishment for shade tolerant species on sites subject to late spring frosts, heavy soils where total removal of the cover results in surface waterlogging or where exposure is a problem. Weeding costs are reduced and protection from heat, cold and drying winds gives better establishment with fewer losses. Beech (*F. sylvatica*), hornbeam (*C. betulus*), western hemlock (*T. heterophylla*), western red cedar (*T. plicata*), grand fir (*A. grandis*) and noble fir (*A. procera*) are all suitable. Douglas fir (*P. taxifolia menziesii*) is less safe as it does not like top shade.

To prepare a site for planting, cut out all rubbish and all large or heavily branched trees, thinning to leave a pole crop 4.5–6.0 m high, with all stems brashed to 1.5 m above ground level; this should leave a dappled shade on the forest floor. Silver birch, giving a light shade, is ideal; also hazel, oak, ash or chestnut coppice. Timeliness of removal of cover is critical or the young crop will be ruined. Removal usually occurs in two stages, half after two years and the rest after four years; six years for final removal is the absolute maximum. Further delay seriously restricts growth and reduces yield. Cover trees may be felled or, in areas of 'non-sensitive landscape' may be 'killed on their feet' by ring barking or appropriate chemical means (Chapter 8).

Enrichment

This refers to interplanting to fill an imperfectly stocked stand in a wood with a reasonable number of useful trees, which may be patchy or even absent in some areas. Existing trees have usually arisen from natural regeneration or coppice regrowth and may include a wide range of ages from old standards, trees not yet ready to fell and to young trees which will form part of the next rotation. This system maintains the woodland environment and retains most of the useful trees; it is expensive in terms of individual plantings but is cheaper over the whole area. Before planting cut out rubbish, poor quality or old trees and treat stumps with herbicide to prevent regrowth. Choose species on basis of their silvicultural suitability. Tree shelters (*see* Tree Shelters p. 228 and *Figure 7.2*) which speed up growth of the transplants, are ideal for use in enrichment situations. Wild cherry (*Prunus avium*) is especially suitable for wood margins.

METHODS OF ESTABLISHMENT FOR BROADLEAVES

Broadleaved trees, especially oak, produce much lower financial returns than well managed conifers; growth is slow, volume production relatively low and earlier thinnings are of low value, e.g. firewood. If the final crop is of low quality, e.g. timber is stained, shaky or of poor form, interest and production costs far outstrip revenue. The aim must therefore be to speed up returns and improve timber quality. Various methods of broadleaf establishment have been tried, some with scant success. Traditional close plantings of pure broadleaves at 0.9×0.9 or 1.2×1.2 m spacings are very wasteful of plants, excessively costly and obsolete.

Mixtures of species

Mixtures of species are notoriously difficult to manage, especially so if they contain a number of species and should be avoided by all except experienced foresters. Many attempts with mixtures have been made where the final crop species has been thought unlikely to provide satisfactory results over the whole rotation, e.g. pure broadleaves give little or no return for many years after planting.

Mixtures of conifers with broadleaves

This approach is only feasible where landscape considerations are not of prime importance. Conifers are included to improve returns in the earlier years of the rotation, as well as to 'draw up' the stems of the broadleaves. The main danger is that the conifer will compete unduly, so reducing the growth of the broadleaves or even suppressing them. It is essential to remember the following:

(1) *Choose species with compatible initial growth rate*, e.g. with beech (*F. sylvatica*) plant Japanese larch (*L. leptolepis*), western red cedar (*T. plicata*) or Scots pine (*P. sylvestris*), although growth rate of conifer can become excessive on some sites, e.g. Norway spruce has proved compatible with oak in low rainfall areas but in wetter south-western England is likely to make excessive growth. Avoid fast growers, e.g. Douglas fir (*P. taxifolia menziesii*).
(2) While the mixture should look after itself into the thicket stage, remove conifers immediately adjacent to the broadleaves early enough and before the latter are suppressed.
(3) It is usually safest to avoid intimate mixtures and plant small groups of broadleaves (oak, *Quercus* spp., in tree shelters) in a matrix of conifers.

Tree shelters (Figure 7.2)

These are translucent open-ended tubes which are placed over each tree and secured to a stake. They act as individual 'tree greenhouses' giving protection and shelter with a warm moist environment inside. The result is very rapid initial growth, which slows after the tree has emerged. Survival rate of trees is also much improved. Hardwoods respond best, oak (pedunculate *Q. robur* and sessile *Q. petraea*) being

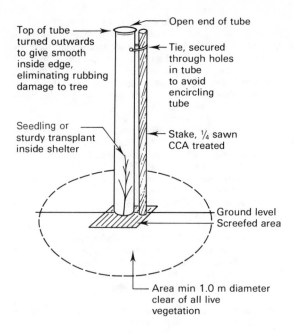

Top of tube turned outwards to give smooth inside edge, eliminating rubbing damage to tree

Open end of tube

Tie, secured through holes in tube to avoid encircling tube

Seedling or sturdy transplant inside shelter

Stake, ¼ sawn CCA treated

Ground level
Screefed area

Area min 1.0 m diameter clear of all live vegetation

Figure 7.2 Circular tube type of tree shelter

especially responsive; other species to respond include ash (*F. excelsior*), alder (*A. glutinosa*), black walnut (*J. nigra*), hornbeam (*C. betulus*), limes (*Tilia* spp.) and Norway maple (*A. platanoides*). Response of southern beech (*Nothofagus* spp.) is variable. Results to date with beech (*F. sylvatica*), common walnut (*J. regia*) and horse chestnut (*S. hippocastanum*) are disappointing; conifers rarely justify treatment. Shelters protect young trees against rabbits and deer but not cattle: required height: rabbits 0.6 m, deer: roe 1.2 m, fallow, sika and red 1.8 m; sheep 1.2 m; cattle and horses 2.0 m. Shelters also protect young trees against close application of herbicide (glyphosate) which is essential for first two years after planting especially with tall weeds (e.g. bramble, *Rubus* spp.; bracken, *P. aquilinum*); also grass which may halve growth rate of young trees. Also control coppice regrowth close to trees. (For subsequent management *see* Thinning.) Shelters are *not* a substitute for weeding–keep an area of minimum diameter 1 m clear of all grass and weeds for at least two years after planting.

To erect shelter screef cut and remove a turf from the planting site, drive stake and plant tree; cover with tube pushing base tightly onto ground and fix tube securely to stake. Quarter sawn round CCA (copper chrome arsenate) treated (external) stakes are stronger than small ones, do not break across knots and are more resistant to deer rubbing.

Shelters are especially suitable for the enrichment of understocked woodlands, establishment of standards in poor coppice (after cutting holes in the crop), in group selection and for speeding up growth of some broadleaves planted in a matrix of suitable conifers. On very exposed sites initial growth is good but may be constantly cut back as they try to emerge from shelter; slower growing trees in the open are not so affected. There may then be a case for using short shelters; also for protecting conifers under rabbit attack. For overall planting on open sites space trees 3 m apart; 4 m is satisfactory in clearfelled woodland areas where intermediate coppice growth is accepted provided regrowth close to the trees is controlled. Alternatively, in cases of enrichment of heavily thinned woodland, groups of three oaks may be planted in 1.2 m shelters 8 m apart *provided* that intermediate broadleaves are allowed to grow up. In year 3 or 4 the best of the three is given a 2 m tube. Before making a final choice of spacing, check with the Forestry Commission whether a grant is available. There are many designs of shelter but all give comparable initial results. Choose a type that will last at least five or six years to prevent deer rubbing, where applicable. Coloured shelters are less obtrusive in the landscape than white and less obvious to vandals. Always use transplants from best timber quality parent trees. Although the tree shelters themselves involve considerable additional outlay, *cf.* unprotected trees, there is a large saving in number of trees required in planting the extra trees and making good losses.

PREPARATION OF SITE

Trees must form a canopy quickly to suppress all unwanted vegetation early in the life of the crop. Thorough preparation of the site is thus essential to provide an environment in which the young trees can establish and grow rapidly.

The work requires provision of drainage to control surface and spring water, fencing to prevent damage by farm livestock, rabbits and deer, clearance of unwanted vegetation, cultivation if required, the control of fungal diseases and insect pests and ensuring adequate plant nutrients are available.

Drainage

The prime function of woodland drains is to remove surface water and prevent surface waterlogging, which kills trees and results in serious windthrow later on. Once fully established, the trees themselves draw water from the soil, which they transpire and this forms a very efficient drainage system.

Open drains, i.e. ditches or ploughed furrows, are the rule in forest or woodland, as tree roots invariably block unsealed land drains, which they enter at the joints. Salt glazed, solid plastic, concrete or similar pipes, where the joints can be sealed, are necessary for farm drains passing through or near tree roots. Except in high rainfall areas, most of the water comes from higher lying land above a plantation or from springs within it.

Before starting a drainage scheme *within* the plantation, it is necessary to intercept any external water entering the plantation from above and lead it away in long *cut off* or interceptor ditches. These run across the slope, appropriately at right-angles to the water flow. Once external water is controlled, drainage within the plantation may proceed.

Drainage is expensive and one should not overdrain. The system must last for the whole rotation. Choice of method depends on the circumstances – soil type, rainfall, topography and the size of the catchment area. Plantations on heavier soils and in high rainfall areas require more drains which are drawn closer than elsewhere. Wet peats and impacted or panned podsols need overall ploughing; in the latter case, busting the pan allows vertical percolation of the water and an immediate improvement; some soils then drain

quite freely. Springs should be picked up and led into the drainage system.

On *unploughed land* water falling within the plantation is collected by contour drains, with a small fall running across the slope, at right-angles to the flow. Short feeder drains may be herringboned in. Each drain leads eventually to a main stream or, on agricultural land, a main ditch. Drains running downhill on moderate or steep slopes collect little water and scour badly.

The layout, spacing and size of drains should vary with soil and local conditions. A common size of drain is 0.3–0.4 m deep, 0.6 m wide and 0.2 m across the bottom. Main ditches are usually 1.0 m deep and 0.4 m across the bottom.

On low lying land with little fall, deep ditches may be dug. Trees are planted on the spoil thrown on to the banks and as rooting depth is increased the water table is lowered.

A plan of the drainage system should be drawn to scale and kept in the office for reference. The system should be regularly maintained. Operations such as brashing and thinning are especially likely to result in blockages. When clearfelling takes place, restitution of the drains should be written into the sale contract. *See also* Forestry Commission leaflet No. 72 *Drainage schemes*.

Fencing

Fencing is expensive and for productive forest large square plantations, which give a relatively low cost per unit area, are the general rule, but for small farm woods and shelterbelts, fencing forms a very high proportion of the total cost of establishment. Sound stock-proof fences are necessary where livestock come into contact with woodland but rabbit and deer-proof fences are only essential where these are a problem. In some cases individual tree guards may be cheaper.

Specifications depend on needs, e.g. rabbits, sheep, cattle or a combination of them. Deer fences should be 1.8 m high with at least eight strands of wire barbed and plain; sheep netting must be used at the bottom against roe deer. Specifications are obtainable from the Forestry Commission (Forestry Commission, 1986 and Leaflet No. 87, *Forest Fencing*). CCA treated timber and stakes should always be used.

Clearing unwanted vegetation and cultivation

It is essential to provide the young trees with satisfactory soil conditions for rapid root development and to control vegetation immediately around them. Failure to do so results in the suppression of the young tree, which is robbed of water, light and nutrients or even smothered by dense vegetation. Grass competition is especially serious, even when the grass is kept mown; it is best killed. Site clearance is generally expensive, the amount of work necessary depending on the quantity and type of vegetation or dead material present and on whether it is likely to hinder subsequent operations. The approach depends on whether the land to be planted is bare land (i.e. land which has not carried trees previously) or former woodland.

Bare land

Handwork is slow and expensive and is largely confined to small areas and steep or difficult terrain. In the absence of herbicidal treatment on grass or heather ground each tree site is '*screefed*' shortly before planting, i.e. a thin square of surface vegetation 300–450 mm is removed with a spade or mattock to expose bare soil and temporarily suppress unwanted growth.

Where planting and establishment conditions are difficult, e.g. with heavy wet clay or where screefing would leave a hollow in which water accumulates, *turf planting* may be considered. A square of turf with sides 300–450 mm and 150 mm thick is cut and replaced upside down in the original hole to receive the young tree. *Raised turves* are preferable on wet ground where an elevated planting site is necessary. Again, turves 300–400 mm × 150 mm thick are cut and inverted, but are laid alongside the original hole. Alternatively, on steep slopes, *hinged turves*, cut only on three sides, may be swung out from the original hole to prevent them from sliding or washing downhill. The rotting faces of the inverted turves are believed to provide some plant nutrients, but cut and invert the turves well in advance of planting to allow time for settlement and rotting, or failure will result. On water-logged sites it is also necessary to provide drainage channels.

Ploughing

Wherever practicable ploughing should be used when cultivation is required, as it is quick and relatively cheap. Forestry ploughs, which can work up to 0.9 m deep, have two types of mouldboard assembly:

(1) for digging or disturbing the soil and breaking up pans, or
(2) drainage ploughs for producing a clean drainage furrow.

Ploughing is *essential* on soils with a pan or compacted layer and on waterlogged soils where it can make all the difference between success and failure. On compacted heaths or on soils where iron pans exist ploughing opens up drainage passages in the soil, allowing water to reach the lower levels and young roots to penetrate. On waterlogged soils, notably peats and peat gleys, it is essential to plant on a site raised above the original soil level, so that the soil around the roots of the young trees is well drained. The drainage plough produces a strip of inverted raised turf for planting. On old pasture or in relatively fertile situations ploughing is not usually necessary provided appropriate steps are taken to control vegetation and there is no compaction.

Where ploughing is not necessary, the use of preplanting total herbicide treatments should be considered (*see* Chapter 8). Overall or strip applications 1.0 m wide, in the centre of which the young trees are to be planted, not only removes direct competition but also results in the decomposition of unwanted vegetation which provides the new crop with additional nutrients.

PLANTING

Plant supply

For production of good quality timber it is essential that transplants are the progeny of selected parents of good timber form with good crowns and of high productivity. They should also be suited to the climatic conditions in which they are to be planted and must therefore be raised from seed of a suitable '*provenance*', defined as 'the geographical source of place of origin from which a given lot of seed or plants was collected; the material from such a source or origin; often restricted to imply material from a specified race or species'. Suitable provenances for each species are recommended by the Forestry Commission, e.g. Queen Charlotte Islands sitka spruce. The use of seed of unsuitable provenances or sources usually gives disastrous results.

Unless a large area is to be planted annually, the highly skilled job of raising trees from seed or even bought in seedlings is not worthwhile. Plants should be purchased from a professional *forest tree* nurseryman. Names and addresses can be obtained from the Royal Forestry Society of England, Wales and Northern Ireland. Under the Forest Reproductive Material Regulations cuttings and plants of the listed species intended for the production of timber may not be sold unless obtained from sources approved and registered by the Forestry Commission in Great Britain or other appropriate authority elsewhere in the EEC.

Bare rooted transplants are used for most forest or woodland planting. Sturdy transplants with well developed roots should always be bought. Small trees recover more quickly from transplanting than large ones, but are more likely to be smothered by weeds or grass if not kept clean. Larger transplants are preferable where rapid weed growth is likely but such sites are usually fertile and relatively sheltered and can support a larger tree. Smaller trees are much safer on poor or exposed sites as they are less liable to damage from wind-rocking and easier to protect. Trees over 600 mm require staking. Age is normally described by number of years in the seedbed and transplant lines respectively, e.g. a 1 + 1 tree has spent one year in each. Trees may be undercut in the nursery bed, i.e. mechanical severance of the taproot 75–100 mm below the surface, to encourage vigorous development of fibrous roots. Hardwoods generally benefit most.

Plants may be offered straight from the seedbed but transplants or a similar size of undercut plant are normally used, owing to their balanced top and root development, with *conifers* 2 + 2 and 2 + 1 are suitable, but 1 + 1 or a similar size in undercut plants is most popular as these are cheaper. Plants 300–380 mm tall are suitable for average conditions but smaller plants, 150–250 mm in size are preferable for exposed situations, wet peat or dry heath provided initial weed competition is eliminated by ploughing or a suitable herbicide. Fertile sheltered positions and sites subject to strong weed growth need larger plants, up to 600 mm tall.

With most *hardwoods*, two-year seedlings, undercut in the nursery bed or 2 + 1 transplants are best. Well grown one year seedlings may also be used. Poplars and willows are supplied either as unrooted or rooted two to three year cuttings or sets.

Conifer seedlings grown in containers (i.e. tubed or paper pot planting stock) and raised in controlled conditions in polyhouses are now used to plant considerable areas (Forestry Commission leaflet No. 61, *Tubed Seedlings*). From seed to planting in the forest usually takes from 12 to 16 weeks. The plants are small and it is essential to have a well prepared weed free site. A special tool is necessary for planting paper pot stock. This technique is especially suitable for species which are difficult to establish, e.g. Corsican (*P. nigra var. calabrica*) and Monterey (*P. radiata*) pines.

Amenity plants may also be purchased in polythene containers which must be removed before planting. These are many times more expensive than transplants and cannot be justified for plantings for timber. Suitable for trees difficult to establish as transplants, valuable amenity trees and late spring or summer plantings. Care should be taken in planting and the plants should be watered as needed during the establishment period.

Plant requirement

Spacing

Optimum spacing, usually on the square, depends on initial growth rate of the species and on quality of the site. Fast growing species on a good site need relatively wide spacing while slower growing species on poor or very exposed sites may be set more closely. Spacing also depends on length and type of bole required, length of rotation and thinning technique. Wider spacings than those used traditionally are now employed, as plant quality, soil preparation and weed control have all improved greatly in recent years. As a general guide approximately 2×2 m is suitable for conifers and mixed stands in most situations. Closer settings may be preferred on poor or very exposed sites but it is questionable whether the extra cost is justified. Pure stands of hardwood are no longer planted at the traditional very close spacings. Poplars and willows are not thinned and are planted at their final cropping positions. *Table 7.6* gives an approximate guide to spacing; *Table 7.7* shows the number of trees required.

Time of planting

Bare rooted transplants are planted only when the plants are dormant, from October until early April. Sites in colder areas and at higher altitudes may be planted with cold stored (i.e. refrigerated) plants in late May.

Spring planting is commonest, as autumn planted trees suffer from winter gales and hard frosts. Late planting frequently leads to failures. Planting starts as early as possible, provided there is no frost. Larch, which flushes early, is planted in autumn or first in spring, as also are the hardwoods. Other conifers, notably western red cedar (*T. plicata*) and Lawson cypress (*C. lawsoniana*) are probably best planted in the spring. Excellent establishments have been obtained in mild areas by planting in early autumn while the soil is warm, notably with paper pot transplants of Corsican (*P. nigra var. calabrica*) and Monterey pines (*P. radiata*).

Ordering

Trees should be ordered well in advance. Nurseries normally pack plants in polythene bags. On no account must bags be exposed to direct sunlight as trees soon heat up and wither; newly arrived trees should be protected against frost. If there is to be any delay in planting or trees are not in polythene bags, heel in at once. A trench should be dug in well prepared ground before the arrival of plants, the roots placed in the trench, covered and firmed. The soil should be well packed around the plants to prevent damage by drying winds or frost. In frosty weather the plants should be covered with straw or bracken.

Planting

When trees are taken out for planting the roots must not be allowed to dry out – even a few minutes' exposure to hot sun or wind can kill. When planting in drying wind, the roots should first be dipped in water. Trees are best carried in plastic or plastic lined bags slung on the planter's shoulder and withdrawn as required. On no account should the trees be 'lined out' prior to planting.

Planting at regular intervals in straight lines is essential to expedite subsequent operations. The requisite inter-tree spacing may be obtained initially by using a planting stick cut to the appropriate length, but more experienced planters pace the distance. On *unploughed land* straight lines are achieved by furnishing each planter with two long straight sighting sticks, one for each end of the planting area. The planter aims for his stick at the opposite end, reaches it and then offsets it to the required row width. He then aims back to his other stick. With more than one planter, each man has his own colour of cloth tied to the top of his sticks. On *ploughed land* the ridge or furrow, as appropriate, are followed to plant each tree in a similar position.

Table 7.6 Spacing of trees

Species and situation	Approx. spacing (m) (planting on the square)
Conifers	
Christmas trees	0.75–1.00
Slow growers (e.g. Scots, Lodgepole pines on poor exposed sites)	1.5–1.75
All conifers on average to good sites	2.0
Hardwoods	
Tree shelters	
in open	3.0
overall planting in woodland	4.0
enrichment of woodland up to	10.0
Other hardwoods	2.0–4.0
Poplars	
Lombardy type	5.0
Bushy top types	6.0–10.0
Willows	10.0

Table 7.7 Tree requirements (number of trees required/ha planted on the square)

Spacing (m)	No. trees/ha	Spacing (m)	No. trees/ha
0.75	17 778	2.75	1322
1.00	10 000	3.00	1111
1.25	6400	4.00	625
1.25	4444	4.50	494
1.75	3265	5.00	400
2.00	2500	6.00	277
2.25	1975	10.00	100
2.50	1600		

Methods of planting

Notch planting

Notch planting is normal for most forest and woodland planting except on loose ploughed ground. Although it can be done with a garden spade, a straight bladed notching spade is quicker; on rocky ground a mattock is best. Drive the tool to full depth, withdraw it and make a notch at right angles to the first. The pattern is like a letter T or L. Lever the ground up for a crack to open. Insert the tree into this crack, draw the tree through until the old ground level mark (or collar) of the tree coincides with the surface of the disturbed ground. Withdraw the tool and allow both earth and tree to fall back, level with the surrounding ground. Finally, stamp everything firm with the heel.

On ploughed ground choice of position on ridge or in the furrow depends on soil type and conditions: planting in the furrow bottom for dry sites but in heavier soil on the side of the furrow. On sheltered wet ground planting on the top of the ridge is suitable but not on peaty ground as the peat tends to shrink or crack, but rather on the side. Planting on the side of the ridge away from the prevailing wind should be practised on the wettest sites with more than 900 mm rainfall.

Turf planting

Allow time for turf or furrow to settle before planting. Side notches are inclined to dry out and split open, exposing the young roots. Cutting a T notch in the centre without breaking the side of the turf is safest although a slower operation than side notching. The roots lie between the two surfaces which rapidly decompose; air pockets can cause failures.

Pit planting

This is used only for poplars, willows and valuable specimen trees. A hole large enough to take the spread-out roots is dug, the tree held at the right level, crumbled earth returned around it and finally stamped firm. Be sure that the tree is planted at the same depth as it grew in the nursery; this can be determined by the soil mark at the base of the stem. Trees of 600 mm height or greater require staking and tying; ties must be inspected annually and adjusted to allow for stem expansion. In grass, turves are returned upside down to the bottom of the pit, a mulch of leaves, cut grass or farmyard manure used as filling, this method being applicable on stiff soils lacking aeration, and on old woodland sites. Not used for normal forest planting.

Beating up

This is the replacement of young trees that failed on newly planted land, a few trees invariably failing from drought or severe frost. In woodland and wide shelterbelts beating up is not profitable unless failures exceed 10–15% in a pure stand, provided they are evenly distributed. Beating up is advisable where failures occur in groups, a wide spacing has been used or, in a mixed stand, where numerous trees of the species which is to give the final crop have failed. For shelterbelts, the deciding factor is whether losses are likely to reduce effectiveness. Failures must be replaced in narrow belts and single-row windbreaks. Beating up must be carried out the year after planting to be fully effective, best quality plants used, at least as large as the originals, and the work done carefully. Slower growing species are likely to be suppressed by the original planting.

FERTILISER APPLICATION

Trees are far less responsive to fertilisers than agricultural crops. When planting on good soils fertiliser is not usually given. Poor, acid, heathland, wet and peaty soils are usually deficient in one or more of the major elements, nitrogen, phosphorus and potassium. Lime is rarely, if ever, applied and is undesirable for conifers. Unless the necessary fertilisers are applied the trees may 'take' but then remain in check, i.e. make little or no growth for many years or permanently. Phosphorus is by far the most frequently deficient element, followed occasionally by potassium. Nitrogen is rarely required. Where in doubt, a dressing of P should be given.

On all acid soils phosphorus must be applied as URP (Unground Rock Phosphate) which is less dusty and much pleasanter to handle than GMP (Ground Mineral Phosphate); there is little to choose between them in cost and effectiveness. Water soluble phosphates, i.e. Superphosphate and Triple superphosphate are dearer and soon become unavailable to plants on acid soils.

On non-acid soils in low rainfall areas water soluble phosphates are suitable. The spruces are especially sensitive to phosphate deficiency; phosphate is of prime importance in the early stages of tree development.

Potassium is applied as muriate of potash (KCl) and N as prilled urea. Applications may be made as follows:

Plant nutrient	kg/ha	Fertiliser	kg/ha
P_2O_5	175	URP (GMP) (29–30%)	600

and only if required:

K_2O	120	Muriate of potash (60%)	200
N	160	Prilled urea (46%)	350

Fertiliser is normally applied as a top dressing soon after planting. Large areas, especially on rough or hilly country, and tall crops are most cheaply treated from the air but staff must be trained to cooperate in loading and marking the area to be covered. Small plantations are top dressed by hand in the area immediately surrounding each tree.

PLANTATION AFTERCARE

Weeding means the suppression of vegetation in the immediate vicinity of the young tree and is essential for three to four years after planting. Neglect allows weeds to rob the young trees of nutrients, water and light, checking growth severely; *grass* is especially damaging around young trees and is best controlled by herbicide, e.g. glyphosate, but protect the young tree. Treat an area of *at least* 1 m diameter around tree and keep clear of vegetation for two to three years after planting. Applies to all species. If weed growth is allowed to collapse on the young trees in the autumn they may be smothered. Vigorous weed growth may follow cultivations which often encourage germination of weed seeds of unexpected species, e.g. gorse (*Ulex* spp.), spearthistle (*Cirsium lanceolatum*).

Cleaning refers to the removal of woody growth, which includes:

(1) fast growing wild tree species, e.g. birch (*Betula* spp.), sallow or goat willows (*Salix caprea* L.) and coppice shoots from hardwood stumps;
(2) harmful climbers, e.g. old man's beard (*Clematis vitalba*), ivy (*Hedera helix*) and honeysuckle (*Lonicera periclymenum*), which are especially damaging in broadleaves; and
(3) other species, e.g. brambles (*Rubus* spp.).

Most of these can restrict growth rate or even suppress the planted species if not controlled up to the thicket stage. Weed control may be effected by manual, mechanical or chemical means or by a combination of methods (*see* Chapter 8).

Preparation for thinning

It is common practice to cut inspection racks or paths during the thicket stage to allow the state of the plantation to be assessed; these should be adequate in number and are made by cutting off the lower branches of two adjacent rows with a pruning saw or lightweight chain saw, tight to the trunks.

Brashing is the removal of the branches on the lower 2 m of the trunks with a *sharp* saw: either a pruning saw or lightweight chain saw, but almost flush with the trunk without damaging the bark. Traditionally whole plantations were brashed but the work is expensive. Today, with selective thinning, only enough trees are brashed to allow inspection and marking of those to be felled. With line thinning, brashing is largely eliminated. Species differ in the ease with which their lower branches are removed: spruces are very tough, larch branches break off readily, while pines are intermediate between the two.

THINNING

The aim of commercial woodland management is to exploit each site to its maximum financial return. What is feasible depends on site quality. On good sites the best returns come from large final crop trees of high timber quality into which management aims to divert maximum resources from an early age, i.e. by early and hard thinning. On poorer sites less is achievable: e.g. on wet exposed sites highly susceptible to windthrow, usually planted with conifers, a 'no thinning' policy may be adopted and the whole crop of relatively small trees is clearfelled at an earlier age for lower quality material, e.g. pulpwood, chipboard. Alternatively, 'oceanic' forestry is occasionally practised with conifers, when an early 'non-commercial' thinning is given (the thinnings being left to rot where they fall); the remainder are left to reach maturity. Whatever the system, it is essential to have a clearly defined management objective from the outset.

Requirements for top quality sawlogs are:

(1) *Sufficient length of clean knot free stem or bole*, obtained by 'drawing up' the tree and suppressing 'side branch' development by competition from adjacent trees, so eliminating knots, or at least large knots.

(2) *A large girth*, with increased growth rate, is only obtainable from a large well developed crown. These requirements are only met by reducing inter-tree competition at an early stage by hard thinning to concentrate growth and girth production on the best trees. This approach improves timber quality, gives earlier marketability and so improves financial returns.
(3) *Evenness of ring growth* Thinning only affects *subsequent* ring growth and not the wide ringed juvenile core; density and strength increase with growth after the juvenile period. Timely thinning gives an even decline in ring growth.

While there is little difference in the total *physical* output of thinned and unthinned crops, quality and hence financial return are greatly affected by thinning.

Undue delay in thinning seriously affects crown and root development. Crown size and with it growth of stem girth can be seriously reduced. In larch, a strong light demander which requires early thinning, the lower branches are killed leaving only a small inadequate crown. The crowns never recover, as larch ceases stem elongation early and the underdeveloped stems may be no better than 'whips' in severe cases. Ash is also badly affected by delayed thinning. The restricted root development of dense late thinned stands makes them very prone to windthrow once they are opened up; on sites prone to windthrow thinning should be done early or not at all. Delayed or infrequent thinning also results in uneven ring growth. As competition within the stand increases, ring growth (i.e. distance between rings) declines; with seriously delayed thinning to a low level. Thinning revitalises ring growth and wider rings are immediately produced, so much so that the point of thinning can be ascertained from the rings; the position of adjacent tree competition can also be determined by ring inspection.

Although basic principles are similar for conifers and broadleaves, there are considerable differences in practical management.

Thinning conifers

Thinning management is the normal method of controlling timber quality in conifers (Monterey pine, *P. radiata*, is an exception – *see* Pruning). The crop is initially planted at close spacing (about 2500 trees/ha) to draw the stems up straight, suppress side branching and hence the development of large knots; this also provides a wide selection for the final crop and gives rapid ground cover. Age at which the *first thinning* is taken is determined jointly by species and quality of site. In general, the higher the yield class (GYC) the earlier the first thinning should occur. Species differ considerably. On a good site Japanese and hybrid larches may need thinning 13–14 years after planting, whereas other species may wait until 16–20 years. Poorer sites are thinned later.

When selecting trees to be taken out at the first thinning the following must be included for removal (*Figure 7.3*):

Wolf trees Vigorous but deformed trees of planted species which suppress the growth of good adjacent trees unless removed.

Whips Trees with long whiplike stems, which flay about in wind and damage the tops of adjacent trees. *Suppressed* trees should be removed where economically worthwhile.

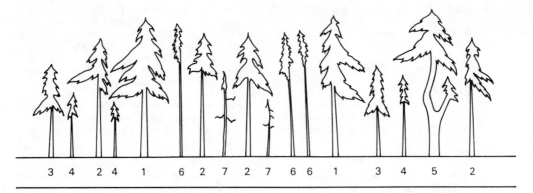

Figure 7.3 Thinning: classification of types of tree likely to be found in a crop. 1, dominants (tallest); 2, co-dominants; 3, sub-dominants; 4, suppressed trees; 5, wolf trees; 6, whips; 7, leaning, dead or dying trees

Detailed information on the date of first thinning, according to GYC and the volume to be removed is given for the main timber species in Forest Management Tables (Forestry Commission, 1971).

Some 50–60% of the total volume production and 80–90% of tree numbers are removed as thinnings. A single thinning cannot be treated in isolation, as each forms an integral part of management of the plantation. Frequently the first thinning is unprofitable or may even make a loss, but it must still be carried out on time or subsequent tree development will be impaired. The overall plan and type of management of sequential thinning operations in a plantation is known as the '*thinning regime*' which is determined by

(1) thinning intensity,
(2) thinning cycle, and
(3) thinning type, which are all interdependent.

There are many possible permutations.

The *thinning intensity* is the annual yield from thinnings or the rate at which volume is taken as thinnings from a plantation, i.e. V/C where V = volume removed at a single thinning, and C = number of years in the thinning cycle.

Total volume production remains unaffected over a very wide range of thinning intensities. Increasing the intensity removes a greater volume and allows increased development of the remaining trees, so that the stem diameter of final crop trees is greater and a high price and marketability are reached earlier. The maximum volume which can be removed without leaving too few trees and causing a subsequent volume reduction is known as the Marginal Thinning Intensity (MTI); this is the most profitable rate of thinning and is adopted as the standard in Forest Management Tables (Forestry Commission, 1971). MTI approximates to 70% of maximum mean annual increment (i.e. 70% of General Yield Class – GYC), e.g. for a stand of GYC 10 the annual yield of thinnings will be 7 m³. This criterion applies only to the period during which annual thinning yields are applicable and varies with species.

Thinning cycle is the frequency or number of years between thinnings and depends jointly on yield class, species and local conditions. As a general guide fast growing conifers on a good site (i.e. with high GYC), notably Japanese and hybrid larches (*Larix* spp.), Douglas fir (*P.*

menziesii) and sitka spruce (*P. sitchensis*) may require a three year cycle; the majority require four to six years while for slow growers an eight, nine or even ten year cycle may be appropriate. Thinning cycle and thinning intensity are interdependent. A longer thinning cycle with more volume removed is less costly in labour but may result in increased windthrow. With light thinning on a long cycle individual trees suffer excessive competition between thinnings; heavy or excessive thinning results in an understocked crop and loss of volume.

Thinning type refers to the dominance class at thinning (*Figure 7.3*) denoting the position in the canopy from which the thinning is taken. *Low thinnings* are taken from the lower canopy and include subdominant and suppressed trees, especially at the first thinning.

In a *crown thinning* some of the dominant and co-dominant trees in the upper storey of the canopy are removed to make room for the full development of top quality final crop trees. An *intermediate thinning* is predominantly low but includes removal of some competing dominants and co-dominant trees. Thinning type has little effect on volume yield but it is essential to leave enough vigorous trees for final selection. A guide to the appropriate number of trees to fell at each thinning is given in Forest Management Tables (Forestry Commission, 1971).

Marking

Sound selection of trees needs considerable experience and is made on the basis of stem form and crown. Potential final crop trees should be the best in the wood, having long clean straight stems with a large diameter, little taper and free from heavy branches or forking. The crowns should be large and well developed. Remove trees which are coarse, leaning or have a wide spread, a weak unhealthy crown or spiral bark, which frequently denotes spiral grain in the wood. When marking, canopy and crowns must be continuously observed so that excessive gaps in the canopy do not occur after thinning; thus when the removal of a poor tree is likely to create an excessive gap in the canopy, it may be spared until the next thinning. In later thinnings density of crop canopy is a good guide to what is required, as it indicates the degree to which the crop has responded to previous thinning. It is important to ensure individual trees have enough room to maintain satisfactory girth increment. It is worth considering marking the potential final crop and final thinning trees initially so that the best trees are favoured from the outset. This course makes working easier later but see that there are ample trees for the final choice – allow at least two to three times the number required.

Line thinning

Conifers may be thinned by the traditional *selective* system or thinned *systematically*. With selective thinning trees are removed or retained on their individual merit. This method ensures retention of all the best trees but financial returns at the first thinning are low and may even result in a loss, as extraction costs are high in relation to the sale value of the timber. Trees cut but with tops caught up in dense thicket can be very difficult to extract. Extraction costs may be reduced greatly by the adoption of systematic methods of thinning, collectively referred to as line thinning, when trees are removed on a systematic basis in lines or a series of interconnected lines, regardless of their individual merit. Such thinnings may be taken as *row thinnings*, when thinning follows the planting lines and a regular sequence of rows is removed, as *strip or corridor thinnings*, where the cuts run across the rows, or as a combination of row and strip thinning, e.g. chevron and staggered chevron thinnings. Line thinnings are normally confined to the first thinning and are occasionally used in the second thinning as well but never later. Subsequent thinnings are always given selectively. Line thinning should only be practised where the risk of windthrow is small, never on sites subject to windthrow, e.g. exposed windy sites on soft soils (Forestry Commission Leaflet No. 77, *Line Thinning*).

The simplest form is single row thinning, when 1:2, 1:3, or 1:4 lines (i.e. 50, 33, or 25% respectively) may be removed. Where rows are closer than 2 m or with wider machinery, removing pairs of rows (2:5 or 2:4, i.e. 40 or 50%) may be considered but this does result in some loss of future volume production and increases liability to windthrow. Removal of three consecutive rows is rarely if ever justifiable. Only the trees in the rows adjacent to the cleared rows respond to the thinning; for 1:4 row thinnings an early selective thinning is desirable for the remaining centre row.

Chevron and staggered chevron designs (*Figure 7.4*) are appropriate for reducing visual impact and liability to windthrow, especially the staggered chevron, where the junction of the side racks with the main rack are staggered.

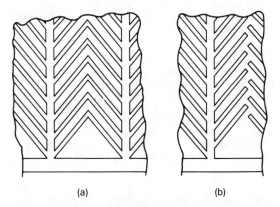

(a)   (b)

Figure 7.4 Patterns of line thinning (a) chevron, (b) staggered chevron (After Forestry Commission, 1978)

Main racks (4 m wide) usually occur every 20 rows, two rows being removed; side racks (strips) are 8 m apart and 2 m wide. The side racks are angled towards the entrance to ease extraction.

Thinning broadleaves

Stem form of the final crop trees is all important in broadleaves, as these provide the bulk of the financial return in a rotation; for this thinning is essential. To make top prices for veneer, furniture and other high grade uses, the bole must be of good girth, clean, straight, and free from branches alive or dead. Owing to lack of apical dominance in many broadleaved species, it is not usually possible to obtain a bole more than 6–8 m long below the bottom branch of the crown. Internal faults, which result in severe downgrading (e.g. rots, staining and shake) are checked at stump. As thinnings are generally of very low value, the prime objective of thinning is to improve yield and quality of the final crop trees; total yield/ha, including thinnings, is of relatively little importance. 'No thinning' regimes are inapplicable.

Make an early selection of potential final crop trees, marking three times the number required to ensure sufficient choice. The stand is thinned hard early in order to allow the selected trees room for full development. Thinning begins when trees are about 8 m tall. Except for first thinnings of densely stocked stands and row-mixtures with conifers, all thinnings are selective. The ideal is to thin little and often to obtain even ring growth. The need for thinning is judged in the summer when the density of the leaf canopy can be judged, according to the degree of crown overlap and those trees barely making canopy. The decision on *which* trees to remove is made during winter, when the extent of individual crowns and quality of upper stem form is clearly visible. The criteria, in order of importance, are

(1) good lower stem form, 7 m long if possible,
(2) a large crown but no forking of main branches,
(3) good vigour,
(4) freedom from defects in upper stem and crown, e.g. disease, squirrel damage,
(5) little or no epicormic branching (*see* Epicormic branches p. 236),
(6) proximity of other selected trees.

There are several problems in thinning broadleaves. Frequently there are few trees of good form and it is often necessary to thin heavily and then enrich with a new planting in tree shelters, especially oak. Some old woods are so poor that there is nothing worth leaving. Species differ greatly in their requirements and response (Forestry Commission, 1984); some species readily produce epicormic branches, especially oak; NB pedunculate oak (*Q. robur*).

Present day plantings, often in tree shelters, are spaced much wider than traditional dense plantings. Side shelter is still essential to 'draw up' the stems, together with growth of adjacent coppice and planted trees. Quality of final crop trees is enhanced by side and high pruning and, in English oaks (*Q. rubur* and *Q. petraea*) adoption of free growth principles.

Pruning

Side and high pruning is the progressive removal of small branches on the lower stem before they can develop and form knots and eliminates fungal infection of the stem via dead branches. Work is expensive and is largely confined to potential final crop trees in broadleaves with the express object of improving timber quality. Pruning is essential for producing veneer and other top quality hardwood timber; also essential if veneer and high quality timber is to be produced from the fast growing pine species, e.g. Monterey pine (*P. radiata*), as practised in New Zealand (Everard and Fourt, 1974), but is rarely used in other conifers. Early pruning (cut before diameter at base of branch is > 5 cm) ensures that no knotty branch core is formed, no heartwood is exposed and that healing is rapid. Use *sharp* pruning saw, with extensible handle for high pruning, cutting not quite flush with bark but slightly proud to avoid bark damage. Pruning 5–6 m high is normally adequate: prune 3 m up stem by first thinning and the rest by the second thinning. Prune most broadleaved species in late winter or early spring but not when flushing; wild cherry from June to August to minimise disease infection. Pruning of fast growing pines (*P. radiata*) is still in experimental stage in UK; work starts in year 4 or 5 to remove two lower branches. Delay results in heavy branching and low value timber.

Singling refers to the singling of dual stems to improve stem quality and upgrade marginal timber.

Epicormic branches usually arise from dormant buds, occasionally from adventitious buds on a pruning scar, on the stem and are a major problem in pedunculate (*Q. rubur*) oak, less so in sessile oak (*Q. petraea*). If allowed to persist for longer than one year a knot is formed in the wood. Epicormics can develop into large side branches in ample light but in woodland usually remain as semi-moribund shoots. Any knots must be completely absent in veneer and only small knots are permissible in high quality plank timber. Emergence is best prevented by avoiding sudden changes in stand conditions, e.g. by adopting frequent light thinnings; they usually arise after a thinning. Regular pruning is necessary and the shoots should be rubbed off at least annually in mid season. Trees badly infested with epicormics are best removed.

Free growth in oak Growing oak in the open, e.g. hedgerows results in short stems, wide crowns and branches lower down the stem but rapid growth of stem diameter. Nowadays the 'free growth' technique in woodland involves thinner planting, allowing side competition to draw the stems up and then thinning when the crop is 20–40 years old, more than 8 m mean height, to allow free development of the crowns of the chosen trees. This accelerates stem diameter growth, resulting in a much earlier harvest of high quality stems (Forestry Commission, 1977, 1984). Epicormics must be rigorously controlled. Thinning occurs every five years.

FELLING AND MARKETING

Before felling growing timber it is necessary to obtain a licence from the Forestry Commission under the Forestry Act 1967; there are listed exceptions, including a small quarterly allowance which an owner may fell, mostly for his own use. For details apply to Forestry Commission, local private woodlands officer. Application must be made by the owner, or the tenant if he is entitled, not the merchant even if he is felling the trees. The merchant must satisfy himself before felling that a valid licence has been issued, as he is liable to prosecution and a heavy fine if none exists. Applica-

tion forms together with map and proposals must be submitted at least three months before felling to allow time for field inspection by Forestry Commission Officer.

Licences cover a specified number of years and must be renewed if not used. Special permission from other bodies may also be required, as in a Conservation area, Site of Special Scientific Interest (SSSI) or trees covered by a Tree Preservation Order (TPO).

Modern timber harvesting, i.e. felling and extraction, which includes the operations of thinning and clear felling, is essentially a specialist's job, requiring efficient planning and organisation coupled with a high degree of skill on the part of the work force. Chain saws and other forest machinery are extremely dangerous in untrained hands and much money can be lost by relying on an inexperienced work force. Extraction equipment, e.g. skidders and timber forwarders, requires heavy capital investment and can only be justified where continuous usage is involved. Timber harvesting gangs are normally employed full time on this work. The work may be undertaken by:

(1) a timber merchant who has bought the standing trees,
(2) specialist timber handling contractors, or
(3) on large estates, the timber owners' own gang and equipment, if these can be justified.

When selling timber, the further it is processed the more the vendor receives but work must be done skillfully. Owners of small woods or where no skilled labour exists do best to sell the timber *standing*, although the price of quality trees is normally agreed *at stump* after felling. Selling *at roadside* implies the vendor has skilled labour and equipment, i.e. skidder or forwarder, for felling and extraction, while *delivery* to buyers' works also requires road transport and is normally done by a merchant or a management agency. It is essential to have a buyer before felling.

Methods of sale

Negotiation

Price and conditions of sale are agreed between buyer and seller and a suitable contract is then drawn up, possibly several years in succession. It is essential that the seller has an accurate knowledge of timber quality, prices and operation costs, otherwise it is prudent to employ a consultant. Negotiation is especially suitable for small lots and scattered trees.

Tender

This method gives a true reflection of current market prices. Tenders may be invited either from selected merchants or by trade press advertisement. Decide on *precise* terms before advertising. Copies are sent to interested parties who then know the precise terms before tendering; acceptance of tender automatically concludes acceptance of contract on published conditions. The problem is that a keen buyer who might pay more offers just below the best bid and may fail.

Auction

Auction usually obtains the maximum possible but is most appropriate to larger quantities. It may be possible to enter for Forestry Commission or other auctions.

The contract

Everything must be laid down precisely in legal format; once signed by both parties and witnessed it is legally binding. The more conditions the vendor imposes, the less the buyer is usually willing to pay. Contract should include detailed description of parcel(s), price, period of contract – date of entry, expiry and removal of equipment, logging requirements, details of working sites, claims, fire precautions, use of subcontractors, penalties and arrangements for arbitration; also making good damage, methods and limitation of extraction routes and clearing of lop and top.

The most profitable outlet for poles and timber grown on a farm or estate is generally some use on the property itself. Usually there is a surplus of material or it is of unwanted size or some of it is of special quality. Larger estates of landowners with experience of the trade and their own equipment, saw mills transport and a skilled labour force can profitably process their timber to some finished form, except for the highest grades such as logs for veneer or furniture, where the buyers require to do their own processing. Otherwise the best course is to sell the timber to a merchant, who will have the skilled labour to handle it and can place *all* grades, each on a suitable market, unlike many manufacturers who will handle only the best.

Conversion of timber on small saw-benches driven by a power take-off is an excellent way of meeting small home needs, but such equipment cannot compete in accuracy of sawing and speed of output, with more specialised plant.

SHELTERBELTS AND WINDBREAKS

Types of wind

Strongest wind causes serious and obvious damage to buildings, structures, e.g. glasshouses, plastic tunnels and coverings, caravans and to plant life, e.g. windthrow of trees de-leafing, shredding and flattening of crop plants. Results can be catastrophic. *Damaging winds* do serious but not always visible damage to plants and animals at a sensitive stage, e.g. fruit blossom and newborn lambs. *Prevailing wind* refers to the direction from which wind is commonest; it is not necessarily damaging.

Shelter for crops

Effects of wind on plants

Wind removes water or vapour film around leaves, often causing excessive transpiration, when conducting tissues cannot transport water quickly enough; wilting or worse follows. The height to which plants grow is limited by their capacity to transport water upwards. The taller they grow, the worse they suffer; thus trees and shrubs are dwarfed on the windward side, giving a 'leaning' or 'windswept' appearance. Wind accentuates the bad effects of cold (chill factor), heat and salt spray. Chilled roots and stems, especially with frozen soil, preclude water transport, so that in a freezing wind plants are killed (e.g. winter beans) when they would be unharmed in still air at the same temperature. Hot scorching winds with dry soil can de-leaf trees and hedgerows. Even before wilting, serious water loss checks photosynthesis and reduces yields. Cold winds delay or damage germination, growth, flowering, pollination and ripening and reduce crop yield and quality. Fruit and vegetables are especially at risk. Wind also causes serious heat loss from glasshouses and soil erosion.

Well designed windbreaks improve growing conditions, raising early spring soil temperatures (average $+3°C$), reducing seedbed moisture loss and subsequent risk of wilting, improving habitat for pollinating insects and allowing more spraying opportunities. The result is earlier crops and better yields.

Design of windbreak

Badly designed, sited or maintained windbreaks can be useless or even harmful. Major climatic features cannot be altered. With a crop or orchard windbreak the objective is to reduce windspeed by acting as a filter, not to eliminate it (*Figure 7.5a*). Solid or near solid windbreaks cause an upward deflection of the airstream, producing an area of low pressure to leeward of the barrier, which results in intense turbulence, more damaging than the original wind (*Figure 7.5b*). A windbreak of 50% permeability reduces windspeed up to 30 times height but maximum benefit is obtained only up to ten times height of barrier. Single row windbreaks, occasionally double rows, are the rule on valuable horticultural land; wider barriers only use more land with no increase in benefit. A series of windbreaks is most effective, e.g. a series with height 7.5–9.0 m spaced every 90 m should give a good response with minimal land use; on very windy sites allow about 45 m. If windbreak is too short, wind is deflected around its ends; minimum length:height ratio is 12:1.

When siting allow for influence of contours on surface wind. When wind funnels down a valley, place windbreaks across valley at right angles. Windbreaks running across a slope should only be solid when sited *above* area to be protected to deflect wind; solid barriers below the protected area dam up frosty air and should always be kept open at their base (*Figure 7.5c*). The siting and number of windbreaks depends on local circumstances. With fruit or long-term crops establish windbreak several years beforehand if possible. The young windbreak itself may need some temporary shelter, usually using manufactured materials.

It is easy to create problems with living windbreaks. Roots, especially of poplars, severely damage foundations and block drains. Keep well away from buildings and if drains pass through windbreak at least 5 m should be of solid pipe with sealed joints. Prevent 'root robbery' (i.e. water and nutrients) by running subsoiler 1.2 m away from windbreak to prune roots annually. Trim regularly to prevent shading of crops and glasshouses. Attend to gaps to prevent funnelling. Choose species which do not act as alternative hosts to pests and diseases of crop, e.g. hawthorn (*Crataegus* spp.) harbours fireblight (*Erwinia amylovora*) of apple and pear. Dense windbreaks provide habitats for pests such as pigeons and bullfinches and prevent normal drying, favouring fungal diseases, e.g. grey mould (*Botrytis cinerea*).

Windbreaks made from manufactured material are suited for special purposes, e.g. intensive horticulture, protecting container stock, also establishing young living windbreaks. They give immediate protection, can be moved about, give constant permeability and do not offer root competition. Living windbreaks should be cheap to establish, quick to become effective, easy to maintain and not act as hosts to

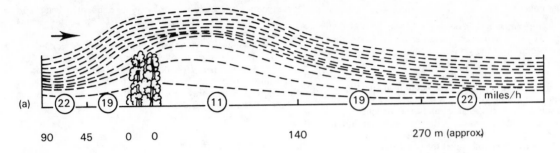

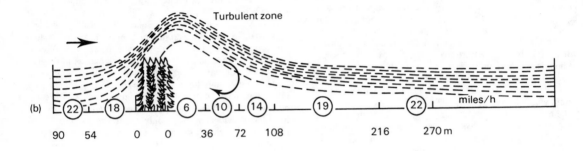

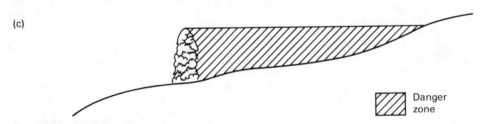

Figure 7.5 Effect of (a) moderately penetrable and (b) dense windbreaks on the flow of wind. Figures circled denote approximate wind speeds in the various sections. Horizontal distances (in metres) refer to windbreaks 9 m tall (after Kuhlewind *et al.*). (c) Frost pocket created by a dense windbreak

pests. The main species for orchard and field crop protection are as follows.

Poplars

Poplars (*Populus* spp.) grow readily from cuttings. Keep far apart from buildings and only use for boundaries, not internal windbreaks, as branches are too large; poplars are greedy feeders and tend to throw suckers. Choose variety according to purpose. Balsam hybrid (*P. tacamahaca × trichocarpa*) is most generally useful. Some species are readily damaged by trimming; poplars are less supple than willows.

Willows

Willows (*Salix* spp.) grow tall but are more supple than poplars, can be planted closer, giving more winter protection and easier to trim. Less susceptible to diseases, they are cheap and easy to establish, growing rapidly from unrooted cuttings. Choose finer branching types, e.g. cricket bat willow or Drakenburg.

Alders

Alders (*Alnus* spp.) are extensively used for internal windbreaks between rows of fruit. Growing only moderately tall, they are easily trimmed, thrive on a wide range of soils and fix atmospheric N but can be damaged by orchard sprays. *Grey alder* (*A. incana*) suits a wide range of soils, retaining adequate vigour on drier sites but throws suckers. *Black alder* (*A. glutinosa*) is a wiser choice than poplar or willow on wet heavy clays but loses vigour on drier soils.

Italian or cordate alder (*A. cordata*) is a handsome, adaptable, more vigorous species but relatively expensive; flushes later in the spring but retains leaf well into autumn.

For single row windbreaks allow *spacings*: alders and birches 1.5 m; poplars (balsam hybrids) 2.0 m, unrooted cuttings of *P. robusta* 1.0 m; willow, selections of *S. alba* (rooted) 1.0–2.0 m, unrooted 0.5–1.0 m; conifers Leyland cypress (*C. leylandii*) 1.5–2.5 m; Lawsons cypress (*C. lawsoniana*) and western red cedar (*T. plicata*) 1.0–2.0 m.

Conifers

Conifers are only suited where a solid evergreen screen is required for privacy or screening motorway lights, etc. Do not use as solid tall windbreaks for crops or fruit.

Planting

Windbreaks require as much attention in establishment as the crops they are to shelter, needing careful planning, site preparation, adequate fertiliser application and weed control. Mulch roots with leaves, straw or black polythene and keep well watered during year of establishment. Trim as required. For support tie to WG 10 plain wire carried on 7.5 cm × 150 cm posts, 3 m apart. Do *not* use barbed wire. On exposed sites stake individually. Guard against rabbits.

Shelter for livestock

Applies especially to exposed upland and mountainous areas, where prolonged winters are cold, wet and windy. Lowland areas are often less exposed and better supplied with buildings and other shelter. Land is much less valuable in marginal upland areas, so wider shelterbelts can be used jointly to enhance productivity of hill land and for timber production.

Livestock suffer severely from exposure to wind, enduring serious heat loss, and use more food just to maintain or even lose condition. Still cold air usually presents no problem but high wind with severe cold or driving rain results in severe chilling and even death. Young or thin coated animals, e.g. lambs and dairy cattle are at much greater risk than adult long coated hill breeds of sheep and cattle, although these too can benefit from shelter. Shelter is particularly valuable around the homestead and the 'in bye' land, where the length of the growing season and grassland productivity can be improved by the better growing conditions provided by shelterbelts, thus increasing stocking rate and livestock output.

The situation is different from that of crops. Animals are mobile and can pack tightly into the small still area in the lee of a belt in a gale; dense belts are thus suitable. As the belt becomes taller, the area it will shelter becomes larger but the lower branches die back leaving the bottom open and increasingly draughty; old beech stands are usually the worst. A staggered 'strip' system of felling and replanting should be adopted, so that belts can be reasonably wide, productive and being of mixed ages, closed at the bottom in part of their width. The use of low growing shrubs at the windward margins should also be considered (Caborn, 1965).

Conifers are usually most suited to upland situations, choice depending on the ecological situation. Most suitable species include sitka (*P. sitchensis*) spruce for quickest growth; Scots (*P. sylvestris*) and lodgepole (*P. contorta*) pines are slow. Where timber production is unimportant

Austrian pine (*P. nigra var austriaca*) or noble fir (*A. procera*) are possible. Corsican (*P. nigra var calabrica*) and Monterey (*P. radiata*) pines for south of England only. Birches (*Betula* spp.), grey alder (*A. incana*), Norway maple (*A. platanoides*) and sycamore (*A. pseudoplatanus*) may be included in mixed upland plantings. (*See Tables 7.2–7.5 for characteristics of species.*)

Establish as for conifer timber production on bare land. Stagger planting lines for maximum wind interception, using 1.5 m × 1.5 m spacing on poorer soils and very exposed situations, otherwise 2.0 m × 2.0 m. Beat up at end of first season if required. Fence to keep out farm livestock, deer and rabbits as required.

Shape, size and siting

There are no hard and fast rules but avoid long straggling shelterbelts and site shelter according to lie of the land. Use baffle systems for 'wind funnel' valleys, while on gentle slopes wind follows contours and needs diverting. See that there is ample shelter around homestead and provide sheltered paddocks. Belts must be large enough or effect is inadequate and see to drainage on leeward side or severe poaching occurs. Arrange shape, fencing and stock control so that stock can be moved around according to wind direction and to avoid severe overgrazing to leeward, to prevent occurrence of poaching and fouling and heavy worm burdens in stock. It is essential that shelter is properly maintained and its advantages are fully exploited with a high standard of livestock and pasture management.

TIMBER PRESERVATION

A wide range of preservatives against fungal and insect attack both indoors and outdoors is available. Use treated timber for construction work, but the processes are highly specialised and expert advice should be sought., Only treatment of farm or estate timber is considered here.

Seldom is timber durable when exposed to alternating wet and dry conditions favouring growth of wood rotting fungi. The heartwoods only of oak, sweet chestnut, larch, western red cedar, Lawson cypress and yew possess natural durability; small logs and all *sapwood* do not. Untreated non-durable wood used for fencing stakes lasts barely four to five years whereas properly treated timber has an almost indefinite life.

Moisture must be removed from timber before it can be successfully treated with preservative either by kilning or more slowly by 'seasoning', which involves storing in free moving air, under cover to lower the water content to at least 30%, preferably 25%, according to species. Logs or stakes are peeled, stacked criss-cross or with spacers to allow the passage of air; supports are used to keep the bottom layer off the ground. Two to six months' seasoning is usually necessary, depending on the month of felling. Seasoning is not a preservative treatment. Home grown plants and scantlings required for indoor work need seasoning for six months in a dry airy building, small sticks between them allowing air circulation. Seasoned timber is lighter, slightly stronger and much less liable to shrinkage or change in shape.

Preservatives

The effect of any wood preservative depends on concentration, amount retained in the wood and penetration achieved.

Factory treatment, though more expensive, gives far better control and a really durable product.

Brush applications

These are ineffective for stakes and are only appropriate for thin boards up to 12 mm thick, above ground level and reached on both sides as in a board fence.

The CCA process (copper chrome arsenate)

This has superseded creosote. Water soluble at the time of treatment, it gives a high degree of penetration and forms a totally insoluble chemical complex with the conifer wood resins which remain permanently in the timber.

The process requires a specialist plant operated under factory conditions, some local sawmills provide the service. Stakes require de-barking, pointing and kiln drying before treatment.

Factory treated CCA timber has a life of 30 years or more and it is a once and for all treatment. CCA treated stakes and timber for outside use replace all others.

PROTECTING WOODLANDS

Fire risk

The danger of fire is always present, but degree of risk varies greatly. On the average wooded estate, where varied kinds and ages of tree are intermixed and broken up by fields, it is never so high as in a large expanse of young conifer plantations. Everyone owning woods should take simple precautions:

(1) Have available telephone numbers and addresses of fire brigade and police; if in doubt dial 999.
(2) Inform fire brigade of location and type of woods.
(3) Make sure all parts of woods are accessible for fire fighting; locked gates, broken culverts or fallen trees can cause delay, allowing minor outbreaks to become major disasters.
(4) Have adequate, if simple, fire fighting tools at strategic points; including spades, buckets, axes, bill hooks and birch brooms. Only birch brooms can safely be left out in the woods; site stacks in full view near gateways.
(5) Clear away inflammable vegetation from points of high risk; for example, clear gorse along roads or railways.
(6) Brash young conifers early, especially near roads or footpaths.
(7) Ensure that all staff and contractors observe fire precautions and are trained in fire drill.
(8) Insure against fire risk.

Few private owners have woods big enough to merit a regular patrol, but in dry weather both owner and woodmen should keep alert for signs of fire. The most dangerous period is usually from March to May; afternoon and early evening are the most dangerous times of day. Fires may arise at any time when herbage is dry.

Wind damage

Avoidance of serious damage by gales is mainly a matter of correct choice of species for site, proper maintenance of drains to prevent soft ground conditions and proper and timely thinning. In particular, the spruces, Douglas fir and larch are very susceptible on shallow clay soils; shallow peats affect all species. Avoid long periods between thinnings. Line thinned plantations are more susceptible. With best management, however, no forester can stop the occasional gale or gust from bringing down groups of trees. Windblows are rarely a total loss; any tree big enough to be blown down is also big enough to sell. *Windsnaps*, where the tree breaks off part way up the trunk, cause considerable loss, as split wood has to be discarded, and lengths are shortened.

INSECT PESTS AND FUNGI

Insects

When clearfelled conifers are replanted or conifers are planted in close proximity to existing conifer plantations, the young trees are liable to be attacked by either the large *pine weevil* (*Hylobius abietis*) or the *black pine beetle* (*Hylastes* spp.).

Pine weevils breed in conifer stumps. The adults emerge and feed on the bark of all species of newly planted conifers, which die as soon as they are ring barked by the weevils. The traditional method of control was to leave the site bare for three or four years after felling and trap the insects, but this method wasted time and allowed heavy weed infestations to develop. Control at planting time may be achieved by dipping the aerial parts of the bundles into a liquid dip of methrin. If control is incomplete or has not been undertaken prior to planting, a topspray of methrin is given to the young trees. Alternatively, chlorpyrifos may be used. Neglect of control measures can result in total loss of the planting.

Black pine beetles also breed in conifer stumps and then attack the young transplants, but below the region of the collar. Pines are especially liable to damage, also spruces. Root dipping of the bundles in liquid methrin dip as far as the collar gives good control. Alternatively, control of both pests may be achieved from a total dip of the bundles.

A number of other pests may attack woodlands and forests, some causing considerable damage, but they usually appear sporadically and are very localised in their occurrence.

Fungus diseases

Trees are subject to attack by many kinds of fungi. The most serious for the ordinary grower are those causing heart rot or butt rot. The fungi concerned, such as *Fomes annosus*, gain entry to the heartwood of the tree either through some deep wound, such as the breaking of a branch, or through the roots. They cause decay in the centre of the trunk, which may be slight or so serious as to make it both dangerous to passers by and worthless as timber. Unfortunately, only when infection is well advanced is it revealed by fructifications – toadstools or brackets – usually at the foot of the tree, but sometimes higher up the trunk.

Control of heart rot in hedgerow or woodland trees is seldom possible once there is cause to believe trees are affected. Some should be felled and the trunks examined. If rot is prevalent, it may be advisable to clear the crop before sound trees deteriorate in quality and value, but the occasional unsound tree need cause no general alarm. Few stands of maturing timber are wholly free, while the timber merchant can usually cut sound lengths of timber from all but the worst affected logs.

To prevent spread of butt rot timber fellers should carry cans of a concentrated solution of urea coloured by a green indicator and flood the newly cut surface of each stump immediately after felling or at latest within 15 minutes. Douglas fir (*P. menziesii*) is fairly resistant and a good choice for replanting. Avoid very susceptible species, e.g. Western Hemlock (*T. heterophylla*).

Honey fungus (*Armellaria media*) can be a problem in old broadleaved woodlands, affecting both broadleaves and conifers. It spreads from old stumps through the soil and kills young trees. Losses are sporadic and rarely affect a whole wood. There is no economic treatment. Other problem diseases include Dutch elm disease (*Ceratocystis ulmi*) and beech bark disease (*Nectria coccinea*), both carried by insects.

Further details of insect pests and fungus diseases and their control may be obtained from the Forestry Commission, bulletin (1986) and appropriate leaflets, listed therein.

AGENCIES FOR ASSISTING WOODLAND OWNERS

To employ a full time forester ensures maximum returns in the case of several hundred hectares of woodland, but a common difficulty with small acreages is that need for thinning, felling or replanting arises only at intervals of several years. The help of a forestry adviser, in constant touch with local conditions, particularly markets and ruling prices, is worthwhile. Numerous landowners now let out the management of their woodlands to contract. The *Forestry Commission* gives free technical advice to wood-owners.

Cooperative forestry societies

These are independent associates operating throughout Scotland and in many districts of England and Wales. In return for a small subscription, members have the services of a professional forester for advice on planting or marketing of produce, and most societies undertake the actual operations for a fee or commission. Societies also arrange bulk buying of young trees, tools, and materials at favourable rates.

Planting contractors

The planting and subsequent maintenance of woods for a stated period are undertaken by firms of planting contractors for an agreed charge. Some timber merchants maintain a planting service which can be employed to re-stock land cleared by felling.

Forestry consultants

These are professionally qualified men who advise on ˌwoodland management, or actually undertake it, for an agreed fee. Many land agency firms offer a similar service and can advise on taxation and other legal problems appertaining to woodlands.

Forestry societies

The Royal Forestry Society of England, Wales and Northern Ireland, 102 High Street, Tring, Herts HP23 4AH, UK and The Royal Scottish Forestry Society, 26 Rutland Square, Edinburgh, EH1 2BU, issue quarterly journals, organise excursions and conferences at which forestry problems are discussed and can advise on legal and other matters.

GRANTS FOR WOODLAND OWNERS AND FARMERS

Forestry Commission

The Dedication Scheme Basis III and Small Woodland Schemes closed to new applicants from 1 July 1981. These schemes continue to operate for existing recipients.

The Woodland Grant Scheme (effective from April 1988)

This scheme replaces the *Forestry Grant Scheme* (effective from 1 October 1981) and the *Broadleaved Woodland Grant Scheme* (effective from 1 October 1985); both schemes were withdrawn on 15 March 1988. The Woodland Grant Scheme aims to encourage expansion of private forestry in a way which achieves a reasonble balance between the needs of the environment, increasing timber production, providing increased rural employment from new woodlands and enhancing the landscape, amenity and wildlife conservation. It is also designed to encourage restocking and rehabilitation of existing woodlands by planting or by natural regeneration.

Application for grant may be made by the landowner or tenant provided all parties concerned join in the application. Grant is available under this scheme is respect of areas $\geqslant 0.25$ ha; it will not be acceptable to aggregate areas < 0.25 ha, unless, with the Forestry Commission's agreement.

The scheme applies to the establishment of conifer, broad-leaved and mixed woodlands established by planting or natural regeneration. Applicants must observe the following provisions: the broadleaved component of existing woodlands must be maintained or enhanced; ancient woodland must be planted or regenerated with appropriate broadleaves or mixture of conifers; native pinewood must be replaced with Scots pine; conifer planting consisting mainly of one species must allow for a proportion of other conifers or broadleaves; ancient monuments must be protected; the legal requirements regarding SSSIs or NMRs must be satisfied before planting; all public rights of way must be safeguarded and the 'Water Guidelines', as published by the Forestry Commission must be observed during all operations affecting drainage of the site, domestic water supplies and other related matters. Full particulars of the Woodland Grant Scheme and scales of grant (much higher for broadleaves) can be obtained from the Forestry Commission Headquarters, 231 Corstorphine Road, Edinburgh, EH12-7AT, or from your local FC Conservancy Office listed in the Telephone Directory.

The Farm Woodland Scheme

This scheme is expected to come into operation in Autumn 1988 for an initial period of three years. The primary sources of advice will be ADAS in England and Wales and the Scottish Agricultural Colleges advisory service in Scotland. The Forestry Commission will be responsible for queries and applications in respect of grant. The main aims of the scheme will be: to divert land from agricultural production and assist reduction of agricultural surpluses; to enhance the landscape, create new wildlife habitats, encourage recreational use, including sport and tourism; to contribute to supporting farm income and rural employment; to increase interest in timber production from farms and in the longer term to contribute to national timber requirements. Leaflets may be obtained from your local ADAS or Scottish College Office.

Ministry of Agriculture, Fisheries and Food (MAFF)

Agriculture Improvement Scheme (AIS) includes grants for shelterbelts or shelter hedges to protect crops or livestock, using native broadleaved species, wherever practicable and trees for shading livestock if planted singly. Grants for hedges apply to new plantings, replanting extensive sections in 'gappy' hedges, temporary protective fencing and hedge laying, i.e. complete reconditioning of hedge in which main stems are cut, laid over and secured to make stockproof barrier to last up to 20 years with routine maintenance. Consult AIS leaflets *Farm Environment and Energy-Saving Grants* Nos. 1 and 4, and *Conservation of the Countryside*, No. 1S, all obtainable from your local MAFF Divisional Office.

Countryside Commission and County Councils or District Councils

Grants are available for *approved* amenity plantings of areas < 0.25 ha financed by Countryside Commission; scheme is operated by Local Authorities. Contact your appropriate Local Authority Woodlands Officer or his equivalent.

Further reading

CABORN, J. M. (1965). *Shelterbelts and Windbreaks*. London: Faber & Faber, 288 pp.

EVERARD, J. E. and FOURT, D. F. (1974). 'Monterey and Bishop Pine as Plantation Trees in Southern Britain *Quart. J. Forest.* **68**, (2), 111–125.

FORESTRY COMMISSION (1970). *Metric Top Diameter Sawlog Tables*. Booklet No. 31, 23 pp.

FORESTRY COMMISSION (1971). *Forest Management Tables (Metric)*. Booklet No. 34. 201 pp.

FORESTRY COMMISSION (1972). *Nursery Practice*. Bulletin No. 43. 184 pp.

FORESTRY COMMISSION (1975). *Forest Mensuration Handbook*. Booklet No. 34. 274 pp.

FORESTRY COMMISSION (1977). *Free Growth of Oak*. Forest Record No. 113. 16 pp.

FORESTRY COMMISSION (1978). *Forestry Practice*. Bulletin No. 14. 138 pp.

FORESTRY COMMISSION (1978a). *Managing Small Woodlands*. Booklet No. 46. 40 pp.

FORESTRY COMMISSION (1978b). *Volume Ready Reckoner for Round Timber*. Booklet No. 26. 81 pp.

FORESTRY COMMISSION (1984). *Silviculture of Broadleaved Woodland*. Bulletin No. 62. 232 pp.

FORESTRY COMMISSION (1985a). *Broadleaves*. Booklet No. 20. 104 pp.

FORESTRY COMMISSION (1985b). *Conifers*. Booklet No. 15. 67 pp.

FORESTRY COMMISSION (1986). *Forestry Practice*. Bulletin No. 14, 10th Edn. 104 pp.

A complete list of all Forestry Commission publications may be obtained from Alice Holt Lodge, Wrecclesham, Farnham, Surrey, GU10 4LH, UK (0420 22255).

8

Weed control

R. D. Toosey

Weeds are injurious or harmful to growing crops, and if not controlled reduced yields from crops and grass result, harvesting and other operations are hindered, produce contaminated or taints imparted and the product rendered unfit for sale. A number of weeds are poisonous.

Most weeds of significance are *annuals* or *perennials*. *Annuals* complete their life cycle within a year and reproduce only from seed, of which they set an abundance. Growth is rapid, some species producing two or more generations in a year. *Perennials* also produce seed but do not die after seeding and, as they also reproduce from vegetative storage organs, serious infestations can build up rapidly if not controlled. *Biennials* only live two years. They make vegetative growth in the first and set abundant seed in the second, after which they die; they reproduce only from seed. Few are important, e.g. ragwort (*Senecio jacobea*) and spear thistle (*Cirsium lanceolatus*), which infest badly managed grassland. For convenience of weed control, weeds may also be classified as grasses, broadleaves, rushes and sedges and woody weeds.

OCCURRENCE OF WEED PROBLEMS

Grassy weeds

Annuals

A serious problem in cereals, especially wild oats (*Avena* spp.), blackgrass (*Alopercurus myosuroides*) and *sterile brome* (*Bromus sterilis*), but can be controlled chemically; not usually a serious problem in broadleaved crops. Only annual meadow grass (*Poa annua*) occurs in grassland.

Perennials

A widespread problem in grassland and they occur on poorly managed arable land, especially cereals, where they cannot be controlled chemically while crop is growing actively; stubble treatment or spraying shortly before harvest is usually necessary. Good chemical control in young trees, no problem later.

Broadleaves (non-woody)

Annuals

Except for mouse eared chickweed (*Cerastium vulgatum*) these occur almost entirely on arable land, where there are good chemical controls in most crops. Elsewhere only a problem following soil disturbance, i.e. on new leys or poached grassland.

Perennials

Sprayed cereals are an excellent cleaning crop for these, where they are readily controlled by translocated herbicides. Cannot usually be controlled chemically in broadleaved arable crops. Some are difficult to control chemically in grassland, e.g. docks (*Rumex* spp.) unless ploughed prior to treatment. Adequate controls in young trees.

Rushes and sedges (perennial)

Found on poorly drained and managed grassland – sign of low fertility. Should not be an arable weed.

Woody weeds

Mainly a problem in young woodland. Can be readily controlled by chemicals in establishing conifers. May establish in undergrazed grassland, e.g. gorse (*Ulex* spp.), brambles (*Rubus* spp.), etc.

GENERAL CONTROL MEASURES – APPLICABLE TO ALL SYSTEMS

Prevention of spread of weeds by seed

Preventive sanitation is necessary to stop the spread of seed of a wide range of weeds, e.g. wild oats (*Avena* spp.), docks (*Rumex* spp.) and gorse (*Ulex* spp.). Sow clean seed, burn straw infested with weeds and avoid buying weedy hay or straw or transporting onto land destined for seed production. Allow farmyard manure to heat up in a neat pile before spreading. Prevent spread of weed seeds from hedgerows or banks, e.g. sterile brome (*Bromus sterilis*) and waste places, e.g. docks (*Rumex* spp.), ragwort (*Senecio jacobea*), and seeds and vegetative organs from headlands.

Cultural and rotational methods

See also major cropping systems – Arable Crops (Chapter 5), Grassland (Chapter 6) and Woodland (Chapter 7).

PRINCIPLES OF CHEMICAL (HERBICIDAL) WEED CONTROL

The effect of any herbicide or chemical weedkiller depends on the manner and circumstances in which it is used, namely the herbicidal treatment. If recommendations are to be successful they must be given in relation to the whole treatment. The three basic components of herbicidal treatment are the crop, type of weed, and type and dosage rate of herbicide. Herbicidal treatments can be:

(1) *total or non-selective*. Applied with the object of killing all vegetation, either before planting the crop or in non-crop situations, e.g. paths, around buildings and waste land;
(2) *selective*. Designed to suppress weeds without damaging the crop. However, all treatments become total if a large excess of herbicide is applied.

Herbicides may be applied either to the foliage (foliar application) or to the soil (residual application).

Foliar application

Foliar applied herbicides may be of two types.

Contact types

These kill only parts of the plant they contact. Single applications are suitable only for use against annual weeds.

Translocated types

These are transported within the plant to parts remote from the point of application. They can be used for perennial weeds with well protected storage organs.
Applications may be made at these stages of crop growth:

(1) *pre-sowing* or *pre-planting* before the crop is sown or planted;
(2) *pre-emergence*, after sowing but before emergence;
(3) *post-emergence*, after crop emergence.

Each may consist of contact, translocated or residual applications or a combination of two or more of these.
Herbicides may be applied *overall*, when the whole crop area or soil is covered. Most spraying for fallows, cereals or grassland is so treated.
Alternatively the application may be *directed*, where only part of the area is sprayed. Special equipment is needed, hence the acreage must justify the capital cost. With *band spraying*, where a band of chemical about 180 mm wide is applied along the rows, say for sugar beet (pre- or post-emergence), some two-thirds of the cost of herbicide is saved. Weeds between the rows are controlled by rowcrop tackle. Chemical hoeing, where the herbicide is applied to weeds or soil between the rows of an emerged crop, the crop being protected by metal guards on either side of the row, is used only on a limited scale. Directed applications may also be made from *tractor-mounted spraylances, knapsack sprayers* or *granule applicators*. Directed applications are widely used in woodlands and forests.

Selective treatments

Contact applications

These are used almost entirely for controlling annual weeds but kill only emerged weeds, being applied to the seedling or young plant. Contact herbicides usually have little or no residual effect and are thus ideal for application immediately before sowing. A high degree of operator efficiency is necessary to achieve adequate spray penetration and coverage. Higher pressures and smaller droplets are required.

Uses

(1) *Pre-sowing or pre-planting*. For (a) control of emerged annual weed seedlings, a 'stale seedbed' technique; (b) sward desiccation prior to direct drilling or ploughing (paraquat); (c) stubble cleaning, desiccation of stubble for burning and couch control (paraquat). Can be combined with a residual herbicide to control weed seedlings not emerged at the time of application. Several residual herbicides have some contact action.
(2) *Post-emergence*. Specific herbicides used on brassicas, sugar beet, mangolds and similar crops. Generally used in mixture with a translocated herbicide for cereals.

Translocated applications

These are used for controlling annual and perennial weeds. These usually act more slowly than contact herbicides but a much lower degree of cover is necessary, provided there are no misses. Damage to susceptible crops from drift is a serious hazard. High pressures and small droplets should not be used. Usually applied in low or low–medium volume. Many translocated herbicides need 6 h of fine weather but thereafter unaffected by rain; others need longer.

Uses

(1) *Pre-sowing*. This is for (a) control of couch and other grassy perennials in stubbles or fallows; (b) control of late starting perennials in stubbles and seedlings germinating after a first spraying; (c) control of perennials for direct drilling, provided chemical has little residual effect.
(2) *Post-emergence or post-planting*. By far the greatest proportion of land sprayed is treated in this way, including most of the cereals treated for broadleaved weeds and all permanent and temporary grassland where selective weed control is required. Also used on clovers, lucerne, peas and in forestry and some other crops.

Soil or residual applications

These remain active in the soil for a variable period, depending on rate of application and speed with which herbicide is dissipated from the soil by rainfall and bacterial activity. Cropping restrictions exist according to the individual herbicide and dosage rate and the time necessary between application and sowing a susceptible crop. Degree of activity, effectiveness and requisite dosage rate of residual herbicides depend on soil type. Some are of limited value in certain soils, e.g. organic soils and peats.

Uses

(1) *Pre-sowing*. These herbicides are usually incorporated into the soil and those with high volatility need immediate incorporation. The ideal tool is a rotary cultivator and some are fitted with spray bars. Alternatively, two passes of heavy harrows at right-angles to each other may be used. Some herbicides are also formulated as granules, obviating need for incorporation. They control difficult perennial weeds like couch, or, if applied shortly before sowing, annual broadleaved weeds, wild oats or blackgrass in a wide range of annual crops, cereals, peas, brassicas and sugar beet.
(2) *Pre-emergence*. Treatment is highly dependent on adequate moisture and a smooth, high-quality tilth, and is almost useless in dry conditions and rough seedbeds. Some herbicides in this group are mixed with a contact herbicide to give immediate control of annual weed seedlings. Several have some contact properties of their own used to control annual grassy and broadleaved weeds in a wide range of broadleaved and cereal crops.
(3) *Post-emergence*. Largely used for directed application to perennial crops, orchards, soft fruit, and forestry; of little importance in farm situations.

A number of herbicides are used for crop desiccation prior to harvesting, e.g. diquat for the desiccation of leafy red clover seed crops, potato haulm and green material in a cereal crop ready for harvesting. They may be used for root and shoot destruction on potato and mangold clamps or old clamp sites.

CHOICE OF HERBICIDE

Choice of herbicide is governed by the following factors.

Crop

Only crops and varieties for which a herbicide is specifically recommended should be treated; others are liable to be severely damaged or even destroyed. Even where several herbicides are recommended for a particular crop, they may vary in their degree of crop safety.

Weeds to be controlled

The occurrence of weed species and the nature of weed problems vary greatly between different types of crop and from farm to farm; species which are very troublesome in one situation may not even be present nearby. In some cases a weed infestation may consist of a single species, which may be dealt with by a simple herbicide, but the problem is usually much more complex, involving a wide spectrum of weeds, some of which are resistant to the simple herbicide. Successful weed control, at reasonable cost, involves a combination of the following approaches.

Broad spectrum

These treatments generally contain a mixture of two or more (sometimes four) herbicides given as a single application. These can be very successful for the control of a wide range of weeds, which have reached simultaneously an appropriate growth stage at the time of application, especially for the late spring foliar treatment of broadleaved weeds in cereals. Herbicides must be miscible. The same priciple is used also for residual applications.

Herbicidal programmes

Single applications, even when using broad-spectrum mixtures, are only partially effective on mixed weed floras which develop over a relatively long period, e.g. winter cereals, where the crop may first be infested by autumn germinating annual grassy and broadleaved weeds, and then by spring germinating annual grass and by a wide range of broadleaved weeds. An autumn application, including a residual herbicide, may be highly effective for much of the winter but may lack persistency to give adequate control of spring germinating wild oats and broadleaved weeds. Conversely, if application is delayed until the spring, the autumn weed germination may well have reached infestation proportions. A *sequence* of applications, designed to deal specifically with the weeds being encountered, is therefore given at appropriate times in the life of a particular crop and is referred to as *a herbicidal programme*. The principle is now used in a very wide range of crops – sugar beet, root and vegetables, cereals, perennial crops and forestry.

Integration of herbicide use with those used on other crops

Crops differ greatly in the ease with which certain weed types may be controlled selectively during the life of the crop and in the post-harvest opportunities which they offer for chemical or mechanical cleaning. Cereals are an excellent crop for cleaning broadleaved weeds with herbicides – a sequence of two or three cereal crops sprayed with translocated herbicides will control most, if not all, perennial broadleaved weeds.

A winter cereal stubble allows ample time for cleaning grassy perennials before a spring sown crop. Conversely, it is not often possible to control selectively broadleaved perennials in broadleaved crops such as sugar beet and potatoes; these weeds usually present serious problems if not controlled in the preceding cereal crop. Grassy annuals, which are frequently expensive to control in cereals, especially early sown winter varieties, can be readily controlled in spring sown broadleaved crops, where they should not present any problem. Integration of herbicidal treatments over a run of crops pays dividends.

Stage of growth of crop and weed

Treatment at the wrong stage of crop growth is likely to result in damage to the crop, which can be very severe or even lethal, while application of a herbicide to weeds at an unsuitable stage results in unsatisfactory control. Foliar applied contact herbicides used to control annual weeds need very early application while translocated types, especially for control of perennials, need later treatment, when there is an adequate leaf area to absorb enough herbicide. A wide degree of crop or weed tolerance is especially valuable when climatic or weather conditions restrict the period of application (e.g. late spring applications to winter cereals) when a more expensive herbicide or one with a narrower spectrum may be justified. Avoid, where possible, herbicides with very narrow tolerances of the stages of growth of crop or weed between which application must be made.

Suiting the environment

Wherever possible herbicides least harmful to animal life should be selected. Highly toxic substances may only be justified where the situation demands such use and no satisfactory alternative is available.

Herbicides differ in their reaction to low temperatures. Hence suitability must be checked before applications are made very early or late in the season. The same basic herbicide is frequently formulated as a number of different compounds differing in type of activity. Choose the most suitable for the purpose in hand. Residual herbicides should not be used on unsuitable soils.

Farming system and cropping programme

Use of sophisticated herbicidal treatments requires considerable expertise and equipment, both generally available on intensive arable farms. On many livestock farms, however, neither the equipment nor the experience exist and simple treatments, often applied by a contractor, are preferable. Herbicidal treatments should be planned in conjunction with the cropping programme and must take probable residual effects into account.

Cost of treatment

This must be considered in relation to type and degree of weed infestation controlled. The value of crop and saving in labour and cultivations must also be taken into account. Heavy infestations of wild oats, blackgrass or couch grass in wet conditions justify expensive herbicidal treatment. Light infestations or conditions where cultivations are cheap and effective seldom do. Herbicides which result in complete mechanisation of row crops or permit use of narrow rows giving higher yields, as with carrots, are well worth while. High value crops like sugar beet, potatoes or carrots permit the use of sophisticated herbicidal treatment. Less valuable fodder crops seldom justify heavy expenditure on herbicides. Where labour is freed by the use of herbicides it must be saved or used for other profitable enterprises. When labour is in short supply use of herbicidal treatment makes the difference between growing a crop or not growing it.

See also 'Food and Environment Protection Act 1985 – The Control of Pesticides Regulations 1986.' (p. 247).

Spray application

Success or failure of herbicidal treatment depends largely on efficient application. A high standard of maintenance is required and sprayers should be overhauled and worn parts replaced before winter storage. Sprayers should be calibrated before use and output checked periodically. Use correct pressure volume rate and nozzle type for each job. Good marking and correct height of spray bar are essential; fans or cones of spray should meet a few centimetres above the top of the crop being sprayed.

The manufacturer's latest instructions on dosage rate, mixing, application, and use of protective clothing must be followed. Use clean tap water, for dirty water blocks filters. Establish a proper spraying routine using gloves, face shield, and protective clothing when handling dangerous concentrates. Empty drums must be rinsed out into a spray tank, the empty drum being disposed of properly or burned to avoid contaminating water supplies or streams. Wash out the sprayer thoroughly, using a synthetic detergent, after each day's work or when changing types of chemical or crop, on waste land away from water courses. *Even small traces of herbicide can severely damage susceptible crops.*

Field conditions, stage of crop growth and weed must be right, for spraying at the wrong stages gives poor control or severely damages the crop. Best results are obtained when weeds are growing vigorously, not in cold weather or droughts. Spraying immediately before heavy rain is expected should be avoided. *Drift* is a major hazard and does irreparable damage to susceptible crops. Even slight traces of growth regulator herbicides ruin glasshouse, fruit and many market garden crops. Spray only in calm weather, never up-wind with susceptible crops, and avoid unnecessarily high pressure or small droplets.

Always possess adequate insurance cover against third-party claims from damage by spray drift. Cover needs to be substantial for glasshouse crops involve very large sums of money.

Nearly all herbicides are applied as a spray, water being the normal diluent. The volume of water used may be:

Volume	litre/ha
very low	< 90
low	90–200
medium	201–700
high (rarely used)	> 700

In ULV (ultra low volume) water is *not* used. Special formulations required (mainly forest application) 5–20 litres/ha.

Low or low-medium volume applications of 280 litres/ha or under require less labour and a larger acreage can be sprayed in a given time.

Tank mixes

Chemicals are mixed in the sprayer tank as opposed to buying a ready made mixture. Ingredients may include herbicides, fungicides, insecticides, aphicides, trace elements, wetting agent and growth regulating substances. The practice is common. NB Tank mixes *must only be applied within label recommendations or in accordance with Pesticides 1986 or subsequent revision of every product in the mix.* It is also essential to ascertain (especially where mixing additional herbicides to broaden the weed control spectrum) that varietal, stage of growth of crop and weed and any other restrictions do not differ from that of the main or initial herbicide or other ingredient.

The method and order of mixing *must* conform to mixing instructions provided and *only* done in that way.

Rope wick applicators consist of a 'rope wick' soaked in a strong solution of total herbicide (glyphosate) which is wiped over the leaf surface of the weed to be killed. Application relies for selectivity on weed being taller than crop to be treated, e.g. bolters and flowering weed beet in a sugar beet crop or developed rushes in grazed pasture. Applicators vary from wide boom type machines for field use to small hand held units where the wick is mounted on a cross piece (*cf.* broomhead) and the herbicide is contained in a hollow handle. The operator then wipes the target foliage. The latter models are suitable for small areas such as spot treatment, parks, gardens and in forest work.

Food and Environment Protection Act 1985. The *Control of Pesticides Regulations 1986* came into operation on 6 October 1986. The term pesticide includes chemical substances and certain micro-organisms prepared or used to destroy pests, which include plants as well as animals. Herbicides are therefore pesticides and their advertisement, storage, sale or supply and application are controlled. Only herbicides or mixtures of herbicides registered for each product by the manufacturer and currently listed in Pesticides 1986 (or appropriate subsequent revision) Part A 1.1. *Herbicides* (Reference book 500, Ministry of Agriculture, Fisheries and Food (MAFF), Health and Safety Executive), may be applied to cropped or uncropped land; this book is obtainable from MAFF (Publications) Lion house, Alnwick, Northumberland, NE66 2PF, UK.

Purchase and use of herbicides

Before purchasing a herbicide or herbicidal product, it is essential to ensure that it is entirely suitable for the purpose for which it is intended. Before mixing and application, it is equally essential to *read the manufacturer's instructions carefully and to follow them exactly,* taking all recommended precautions for the safety of operator, livestock, wildlife and beneficial insects and to avoid drift and pollution of streams or watercourses. The herbicidal product should *only* be used for purposes and applied only by methods and in circumstances which the manufacturers specifically recommend.

Herbicide recommendations for a wide range of field and vegetable crops are published in booklet form by ADAS and revised regularly; obtainable (priced) from MAFF (Publications).

The purpose of this chapter is to outline the main principles and methods of cultural, rotational and chemical weed control. Individual herbicides listed are given as examples, their inclusion does *not constitute a field recommendation for use* nor does the omission of any herbicide necessarily suggest it to be inferior in any way. While every reasonable precaution has been taken to ensure the correctness of this chapter, neither the author, editor nor the publishers will be held responsible for any losses arising from its use.

ARABLE CROPS AND SEEDLING LEYS

Once established, a heavy leaf canopy suppresses further weed growth but individual crops differ in density and duration of the canopy. Crops like kale produce a heavy, lasting canopy but cereals and potatoes may 'open up' late in the season allowing weed growth to re-start. The problem of weed control is thus twofold:

(1) To establish a dense leaf canopy as quickly as possible and maintain it throughout the life of the crop. All factors which promote rapid growth and increase yield, such as free drainage, good soil structure, adequate soil moisture, ample fertiliser, high quality seed and freedom from pests and diseases, allow crops to form a thick, well maintained leaf canopy. Conversely, drought or poor plant population favour weed growth.

(2) Control measures to suppress weeds until a complete leaf canopy is formed or where the crop canopy is poor. These measures include preventive sanitation, rotations, cultivations and chemical weed control.

Rotations

A well balanced rotation prevents build up of certain weeds by changing the environment from one year to another. If one crop is grown continuously, management and timing of cultivations remain constant and particular types of weed increase, e.g. blackgrass in winter cereals, wild oats in all cereals. Rotations including three year leys and roots give much less trouble. The alternation of cereals with broad-leaved crops also allows herbicides to be rotated and used much more efficiently. Build up of couch and other perennial grasses can be prevented by routine stubble cultivations.

The rotation may also be adapted to deal with other weed situations. Wild oats die out if left undisturbed under a long ley (seven to eight years), while two to three year perennial ryegrass leys, if *grazed hard* and not mown, usually eradicate couch.

Cultivations

Annual weeds

(1) *Preparation of 'false seedbed' prior to sowing crop.* The seedbed is prepared early, but not too fine, to encourage weed seeds to germinate. They are then killed by harrowing, another crop of weeds is allowed to grow and then killed; the process may be repeated several times. Method results in a very clean seedbed and is suitable where time allows, e.g. summer seedbeds for lucerne, direct sown leys, swedes or kale, for general annual weed control. Only suitable for the control of weeds whose seed will germinate at the time of year cultivations are carried out.

(2) Thus, the similar technique, *stubble cleaning* or early autumn cultivation of cereal stubbles for annual weed control is highly effective against autumn germinating weeds, e.g. blackgrass, (*Alopecurus myosuroides*) or sterile brome (*Bromus sterilis*), but is useless against weeds which germinate mainly in spring, e.g. spring wild oat (*Avena fatua*). The latter may be killed by delaying the sowing of spring crops to allow harrowing in March and April.

(3) *Delayed sowing of crop,* combined with repeated harrowing, is an effective method of controlling some weeds, e.g. delaying sowing of winter cereal until 5 November gives good control of blackgrass and seedling

grasses but usually reduces yield; also loss of yield from delaying spring cereal sowings.

(4) *Inter-row cultivations for crops grown in wide rows*, e.g. sugar beet. Hand hoeing obsolete. Weeds within rows most effectively controlled by herbicides but overall application expensive. Use of band sprays (directed band of spray 90 mm each side of row), with inter-row cultivations saves two-thirds of chemical cost at each application.

(5) *Green smother crops*, e.g. cereal silage may be grown to control wild oats, which ensile satisfactorily and prevents seeding. May be followed with a catch crop or undersown if not planted too thickly.

Perennial weeds

Cultural methods for controlling perennial weeds are nowadays restricted to grassy species, couch (*Agropyron repens* and *Agrostis gigantea*) and others, in cereal stubbles in hot dry weather. Broadleaved perennials are most economically controlled in a run of sprayed cereal crops.

Fallows may be used but where possible stubble cleaning is cheaper. Costs are minimal where routine stubble cleaning is employed as a *preventive* measure. Regular attention should be given to the *headlands* where infestations frequently start.

There are two approaches to controlling infestations:

Desiccation

This process relies on hot dry weather, when it is cheapest and very effective. The ground is loosened by chisel plough or heavy cultivator, the couch rhizomes worked to the surface and desiccated in the sun by frequent shallow cultivations. The rhizomes should not be ploughed in during autumn and the seedbed for the following cereal should be prepared by shallow cultivations.

Exhaustion

Normal growing weather is ideal. Soil and rhizomes are broken up, usually with a rotary cultivator, to encourage fresh growth. This is then destroyed by further cultivations when shoots are 50 mm long or have an average of one and a half to two leaves each. The process is repeated several times to exhaust the rhizomes and buds. A *bare fallow*, which covers a full year, is rarely used. The *bastard fallow*, the land being broken up with heavy cultivators or ploughed in late summer and then baked in the clod, is excellent for dirty, worn-out leys or old pastures prior to a winter cereal.

General chemical weed control

Pre-sowing sward desiccation

Used prior to direct drilling, minimal cultivations or to kill grass swards before ploughing to give grass free seedbed. Only herbicides with minimal residual effect suitable. Use (1) *paraquat* (foliar applied contact) for general purpose where couch grass (*Agropyron repens*) and rhizomatous grasses and perennial broadleaved weeds are absent, or (2) *glyphosate* (foliar applied translocated) for couch and other rhizomatous grasses and perennial broadleaved weeds.

Pre-sowing control of grassy perennials

Suitable herbicides are listed in *Table 8.1*. Herbicides are more reliable than cultural methods for heavy infestations in wet weather. Routine annual treatments around headlands should be considered.

WEED CONTROL IN CEREALS

Cereals not undersown

Winter cereals

These occupy the land for some ten months of the year and when they follow another winter cereal little time is available for cleaning. Crops are now sown earlier to increase yield (September and early October) and are then more exposed to infestations of autumn germinating *annual and other weed*

Table 8.1 Presowing of perennial weeds – stubble cleaning and elsewhere

Herbicide	Type of activity	Species controlled	Limitations of herbicide
Aminotriazole	Foliar: translocated	Couch (*A. repens*) and other grassy and broadleaved weeds	Specified time interval before sowing crop, 125 mm regrowth required before spraying
Dalapon	Foliar: translocated	Couch, perennial and annual grasses only	Specified time interval before sowing crop, 125 mm regrowth required before spraying
Glyphosate	Foliar: translocated	Couch, most grasses and perennial broadleaved weeds	No residual effect. 125 mm regrowth required before spraying
Paraquat	Foliar: contact	Annual and stoloniferous perennial grasses. Couch only if repeated applications are given. NOT a single application	No residual effect except on trash; then allow 10 days for photochemical degradation. 75 mm regrowth required
TCA	Residual (soil)	Couch and other grasses	Specified time interval before sowing next crop. Requires incorporation in soil

grasses, including blackgrass (*Alopecurus myosuroides*), sterile brome (*Bromus sterilis*), annual and rough stalked meadow grasses (*Poa* spp.) and winter wild oat (*Avena ludoviciana*).

Cultural and rotational controls include: prevention of seeding from banks and hedgerows and control on headlands where infestations of sterile brome (*B. sterilis*) often start, stubble cultivations and delaying sowing until 5 November, which usually reduces yield somewhat, or changing to a spring sown crop (incomplete control of *A. ludoviciana*).

Herbicidal treatments

Those for autumn application are shown in *Table 8.2*. If needed, failure to treat at this stage results in a heavy infesta-tion by spring. Some pre-emergence herbicides used for grass seedlings also control several species of broadleaved annual weeds, but may not control wild oats and do not persist long enough to control spring germinating broadleaves; post-emergence applications for the latter two situations are frequently necessary. A full programme for an early sown winter cereal crop may thus include as many as four applica-tions:

(1) stubble application for perennial grasses (*Table 8.1*);
(2) autumn applied residual for annual grasses and broad-leaved weeds (*Table 8.2*);
(3) spring post-emergence application for wild oats (*Table 8.2*);
(4) spring post-emergence application for broadleaved weeds (*Table 8.3*).

Table 8.2 Selective weed control in cereals: annual/seedling grasses

Herbicide	Weeds controlled[1]						Suitable crops[2]	Remarks
	Type of activity	Barren brome (*B. sterilis*)	Blackgrass (*A. myosuroides*)	Meadow grasses (*Poa* spp.)	Wild oat (*Avena* spp.)	Some annual broadleaved weeds		
Benzoylprop ethyl	Tr				Yes		Wheat	
Bifenox mixtures	R		Yes	Yes		Yes	Barley, winter wheat	
Chlorsulfuron mixture	CoR		Yes			Yes		
Diclofop-methyl	Fo				Yes		Barley, wheat	Also controls Awned Canary grass
Difenzoquat	Tr				Yes		Barley, wheat, winter rye	Also crops undersown with ryegrass/clover
Flamprop-M-isopropyl	Tr				Yes		Barley, wheat	Also crops undersown with ryegrass/clover
Flamprop-methyl	Tr				Yes		Wheat	Also crops undersown with ryegrass/clover
Isoproturon	Tr R	(Yes)	Yes			Yes		Barren brome controlled only after tri-allate
Methabenzothiazuron	Tr R		Yes	Yes		Yes	Winter barley, winter oats, winter wheat	
Metoxuron	Tr R	(Yes)	Yes			Yes	Winter barley, winter wheat	Barren brome controlled by double application or after tri-allate, also controls mayweed
Pendimethalin	R		Yes	Yes		Yes	Winter barley, winter rye, triticale, winter and durum wheat	
Terbutryn	R		Yes			Yes	Winter wheat, winter barley	
Tri-allate	R		Yes	Yes	Yes		Barley, wheat	
Trifluralin	R		Yes			Yes	Winter barley, winter wheat	Main purpose broadleaved weed control and annual grasses/seedlings

Key to Table 8.2:
Type of application: Co = contact foliar; Tr = translocated foliar; R = residual or soil acting; Fo = foliar.
[1] Weeds controlled depends on mixture (where applicable).
[2] Suitability to crop depends on mixture (where applicable).

Table 8.3 Selective control of annual broadleaved weeds in cereals *not* undersown

Herbicide or predominant herbicide in mixtures (M) = mixtures only suitable/available	Type of activity	General weed control	Combinations difficult to control with simple herbicides[1]								Suitable crops[2]	Remarks
			Black bindweed (P. convolvulus) Redshank (P. persicaria)	Common chickweed (S. media)	Cleavers (G. aparine)	Corn marigold (C. segetum)	Knotgrass (P. aviculare)	Mayweed spp. (Anthemis etc.)	Speedwell spp. (Veronica spp.)			
Bentazone M	Co	Yes		Yes	Yes	Yes	Yes	Yes	Yes	Barley, winter wheat	also field pansy, meadow grasses
Bifenox M	R	Yes		Yes					Yes		
Bromoxynil M	Co	Yes	Yes			Yes	Yes	Yes	Yes	All cereals	
Chlorsulfuron M	Co R	Yes						Yes	Yes	Winter wheat	
Clopyralid M	Tr	Yes	Yes	Yes	Yes	Yes	Yes	Yes		Barley, oats, wheat	
Cyanazine M	Fo R	Yes	Yes	Yes	Yes		Yes	Yes	Yes	Barley, oats, wheat	
2,4-D	Tr	Yes								All cereals exc. spring oats[3]	
Dicamba M	Tr	Yes	Yes	Yes	Yes		Yes	Yes		All cereals	
Diclorprop and mixtures	Tr	Yes	Yes	Yes						Barley, oats, wheat	Not rye
Fluroxypyr and mixtures	Tr	Yes		Yes	Yes		Yes		Yes	Wheat, barley, durum wheat, rye, trificale	*also* hemp nettle
Ioxynil M	Co	Yes	Yes	Yes	Yes	Yes		Yes	Yes	All cereals	
Isoproturon M	Tr R	Yes								Winter barley, winter wheat, rye	
Linuron	Co R	Yes			Yes					Spring cereals	
Linuron + trifluralin	Co R	Yes	Yes	Yes			Yes			Winter barley, winter wheat	May require spring treatment for perennials
MCPA	Tr	Yes								All cereals[3]	
Mecoprop and mixtures	Tr	Yes	Yes	Yes	Yes					All cereals exc. rye[3]	
Pendimethalin	R	Yes								Winter barley, winter wheat, triticale	
TBA mixtures	Tr	Yes	Yes	Yes	Yes		Yes	Yes		All cereals	

Key to Table 8.3:
Type of activity: Co = contact foliar; Tr = translocated foliar; R = residual or soil acting; Fo = foliar acting.
[1] Weeds controlled depends on mixture (where applicable).
[2] Suitability to crop depends on mixture (where applicable).
[3] Cheap general purpose translocated herbicide for controlling broadleaved perennial weeds and alternating with broad-spectrum herbicides (as required).

One of the problems of spring applications of broadleaved herbicides to winter cereals is to complete the job before the cereals have passed out of a safe growth stage. Select a herbicide with a *wide application tolerance* – several broad spectrum herbicides are now available for winter cereals which can be applied up to the two node stage (Zadok scale 32, *see Table 5.19*, p. 121). Autumn germinating grassy weeds are unlikely in late sown winter cereals, but these crops are subject to spring germinating wild oats and broadleaved weeds.

Spring cereals

Spring cereals (sown some six months after winter crops) allow ample time for stubble cleaning; residual effects from herbicides applied in the previous season are much less likely. The weed situation is much simpler and, in the absence of wild oats, only a single application for broadleaves is normally necessary.

Herbicides for broadleaved weeds

Perennials require treatment with *translocated* herbicides – contact types are useless. The 'simple' herbicides, e.g. mecoprop, MCPA and 2,4-D give satisfactory control in cereals provided adequate leaf area is allowed to develop before spraying. Do not spray too early – wait as long as possible. Control is enhanced by a series of sprayed cereal crops. More complex mixtures containing translocated

Table 8.4 Selective control of annual broadleaved weeds in undersown cereals; seedling leys (grass and clover, lucerne and sainfoin) (a) Established legumes and seed crops

Herbicide or predominant herbicide in mixtures and type of action. (M = mixtures only suitable or available)	General weed control	Combinations difficult to control with simple herbicides							Suitable crops	Limitations of herbicide
		Black bindweed (*P. convolvulus*) and Redshank (*P. persicaria*)	Common chickweed (*S. media*)	Cleavers (*G. aparine*)	Corn marigold (*C. segetum*)	Knotgrass (*P. aviculare*)	Mayweed spp. (*Anthemis*, etc.)	Speedwell spp. (*Veronica spp.*)		
Benazolin M Tr	Yes	Yes	Yes	Yes		Yes			Undersown cereals, direct sown leys. *Not* lucerne or sainfoin	
Bentazone M Co	Yes		Yes	Yes	Yes		Yes		Undersown cereals; *not* undersown lucerne	
Bromoxynil M Co	Yes	Yes			Yes	Yes	Yes		Undersown cereals	Annual weeds only
2,4-DB Tr	Yes	Yes				Yes			Lucerne – undersown cereals, direct sown leys	
2,4-DB + MCPA Tr	Yes	Yes				Yes			Seedling red and white clover, undersown cereals, direct sown leys not lucerne	Not established clovers
2,4-DB + Bromoxynil + Ioxynil Co Tr	Yes	Yes				Yes			Undersown cereals, seedling leys not lucerne	
MCPB Tr Herbicides used only in MCPB + MCPA above mixtures: 2,4-D Ioxynil MCPA	Yes Yes	Yes Yes							Sainfoin, seed clovers undersown leys, *not* lucerne Sainfoin, leys and undersown cereals, *not* lucerne	Not alsike for seed or yellow trefoil

For control of wild oats (*Avena* spp.) in some cereals undersown with grass and clover *see* difenzoquat, flamprop-M-isopropyl and flamprop-methyl in *Table 8.2.*
Key: Type A activity; Co = contact foliar; Tr = translocated foliar; M = mixtures only; R = residual or soil acting.

herbicides are only necessary where the presence of other difficult annual weeds requires them.

A wide range of annuals is controlled by simple herbicides but some species are resistant and more complex herbicides are necessary (*Table 8.3*). Choose according to crop and weeds present. Check stages of crop and weed growth, etc. suitability from manufacturers' literature.

The more expensive 'broad spectrum' herbicides are not usually necessary every year – cheaper simple herbicides or those with a more limited weed spectrum may be considered for intervening years, but in no circumstances is it advisable to omit the annual cereal spray for *broadleaved* weeds – even if only MCPA is used – otherwise, docks and thistles soon develop.

Undersown cereals and seedling reseeds containing clover

Use only herbicides to which clover is tolerant. General purpose herbicides such as MCPB or MCPB + MCPA control many weeds but a number are resistant (*see Table 8.4*). Selection should be made on this basis. Sprays which control chickweed should always be chosen where the weed is known to be a problem, as it can rapidly suffocate young leys. *All* undersown and seedling leys should be sprayed as dock seedlings, which are usually present and inconspicuous, are highly susceptible in the seedling stage to all translocated herbicides – adult docks are much more difficult to control.

Table 8.4 (b) Selective weed control in established legumes and herbage seed crops

Herbicide and type of action	Purpose	Suitable crops	Limitation of herbicide
Legumes			
Carbetamide Fo R	Winter control of annual grasses, volunteer cereals, chickweed and speedwells	Clover, lucerne and sainfoin	Soil type limitation
Dalapon *Glyphosate* *Propyzamide* *TCA* }	Couch grass; also other grassy weeds	Lucerne, pure stands and winter application only	Soil type limitation, stands at least 1 year old
Grass for seed/pure grass stands			
Bifenox + mecoprop Tr R	Broad spectrum control, especially cleavers, field pansy, speedwell	Ryegrasses	
Difenzoquat	Wild oats *see Table 8.2*	Ryegrasses,	
Ethofumesate Tr R	Meadow grasses, wild oats	Ryegrasses, Tall fescue. Wide range of pure grasses	Soil type limitation
Methabenzthiazuron Also 2,4-D, dicamba mixtures	Meadow grasses *see Table 8.2*	Perennial ryegrass	
MCPA, *mecoprop*, TBA mixtures	Broadleaved annual and perennial weeds – *see Table 8.3*	All pure grass swards	

Key:
Type of activity; Co = contact foliar; Tr = translocated foliar; R = residual or soil acting.
Italics denote herbicide also or solely used by itself.
Desiccation of clover seed crops; diquat.

ROOT CROPS, ROW CROPS AND OTHER BROADLEAVED CROPS

The principles of weed control are similar in all these crops – the main differences lie in the herbicides used.

Weed problems

Unlike the situation in cereals, perennial weeds are generally difficult or impossible to control selectively during the growing period of these crops. Grassy perennials (couch *Agropyron repens* and others) are best controlled in cereal stubbles or pre-sowing by the herbicides listed in *Table 8.1*. Exceptions include couch in *maize* – use atrazine (high rate) but crop must be grown for two successive years. *Potato* – EPTC, worked in pre-sowing.

Broadleaved perennials

These should be controlled by translocated herbicides applied to preceding cereal crops wherever possible or, if feasible, by pre-sowing application of glyphosate. Exceptions include creeping thistles (*Cirsium arvense*) in *peas* – apply MCPB post-emergence (NOT beans or MCPB + MCPA) and *sugar beet* – clopyralid.

Grassy annuals

Grassy annuals (e.g. blackgrass *Alopecurus myosuroides* and wild oats *Avena* spp.) may be controlled by herbicides listed in *Table 8.5*.

Broadleaved annuals

For broadleaved annuals consult *Table 8.6*. Selection should be made according to the weeds present or expected to be present and with residual herbicides on soil type; on peaty soils the range is very limited.

Herbicidal programmes

These are necessary in broadleaved crops which germinate slowly and take a long time to form a complete leaf canopy. Such crops are extremely vulnerable to weed competition when there are successive germinations of different weed species. Sugar beet is an excellent example, where programmes, which are constantly changing, have become sophisticated and costly.

WEEDS FROM SHED CROP SEED OR GROUNDKEEPERS

Problems arise from weed beet, shed oilseed rape, potato and beet groundkeepers and, when oilseed rape or stubble brassicas are to follow, shed cereal grains. Wherever possible prevention is better than cure; methods include preventive sanitation, adequate length of rotation, cultural controls as for annual weeds and herbicides, see *Table 8.7*.

Weed beet

Weed beet is an annual form of beet which produces no harvestable root and sheds large quantities of seed in the year of germination. Although indistinguishable from sown beet in the seedling stage, weed beet may be detected as they occur in patches. Any beet out of place between or within sown rows must be treated as weed beet; they grow from seed shed by bolters in beet crops, which may come from varieties

Table 8.5 Selective control of grassy weeds in broadleaved field and vegetable crops

Crop (W = winter sown) (S = spring sown)	Weeds to be controlled and appropriate herbicides					Remarks
	Annual grasses and grass seedlings	Barren brome (B. sterilis)	Blackgrass (A. myosuroides)	Couch grass (A. repens)	Wild oat (Avena spp.)	
Field crops						
Beans W		1	1, 4, 13, 15	1	1, 3, 4, 8, 13, 15	
S		1	1, 15	1	1, 3, 8, 14, 15	
Brassicas						
Oilseed rape W	4, 7, 11, 13, 14, 15	11	4, 11, 13, 14, 15	7, 11, 14	3, 4, 8, 11, 15	
S	7, 14, 15		14, 15	7, 14	3, 8, 14, 15	
Seed crops + mustard			4*, 13*, 15	14	3†, 4*, 14, 15	*not mustard †mustard only
Fodder brassicas (rape, kale, swede, turnip)		1	1, 15	1, 14	1, 14, 15	1. swede and turnip only
Linseed	15		15		8, 15	
Maize			15	2†, 10*	9, 10*, 15	*with protectant †also with spray additive
Potato	1			1, 7*, 10	1, 8, 10, 14	1, not annual meadow grass *pre-emergence/harvest
Sugar beet	1		1, 5, 6, 11, 13, 15		1, 5, 6, 8, 11, 14, 15	
Fodder beet and mangel	1, 11*		1, 5, 6, 11*, 15		1, 5*, 6, 11*, 15	*not mangel
Field vegetable crops						
Bean, broad	1*		1, 15	1	1, 8, 15	*not meadow grass
french, runner			15		8*, 15	*french only
Beetroot	1*		1, 6, 15	1	1, 6, 15	*not meadow grass
Brassicas						
Spring cabbage			4, 13, 15		4, 8, 14, 15	
Calabrese			15		15	
Other			15			
Carrot	15*	1†	1, 15†	1, 7, 14	1, 8, 14, 15	*also linuron + metoxuron †also metoxuron
Leek		(1)	(1), 15	(1)	(1), 15	() drilled crops only
Onion		1	1, 15	1	1, 8, 15	
Parsnip	15	1	1, 15	1	1, 8, 15	
Pea	1†		1, 15	1, 14*	1, 8, 14, 15	*dry peas pre-harvest †also prometryn and terbutryn mixtures
Sweet corn				2, 10*	10*	*with protectant

Key:
Herbicides

1 Alloxydim-sodium	Tr	9 Difenzoquat	Tr	F = foliar		
2 Atrazine	R	10 EPTC	R	R = residual (soil)		
3 Benzoylprop-ethyl	Tr	11 Fluazifop-P-butyl	Tr	Tr = translocated		
4 Carbetamide	F, R	12 Metoxuron	Tr R			
5 Chloridazon mixtures	R	13 Propyzamide	R			
6 Cycloate + lenacil	R	14 TCA	R			
7 Dalapon	Tr	15 Tri-allate	R			
8 Diclofop-methyl	F					

with low bolting resistance and weed beet or groundkeepers elsewhere. Bolting must be prevented as sugar beet varieties are genetically unstable when allowed to cross pollinate freely in an uncontrolled environment and soon degenerate into the annual weed beet form.

A single bolter is capable of producing up to 2000 viable seeds in a single season; if not removed it can soon build up an infestation.

Control

Good isolation and tight control of cross pollination by seed growers is essential. Root growers should walk their fields before row cropping, looking for patches of seedlings outside rows or in cereal crops and identify the problem. *Prevention of seeding is better than cure.*

Measures

Pull *all* bolters before flowering (i.e. yellow pollen visible); if pulled after flowering remove from field and destroy to prevent shedding of viable seed. Destroy all weed beet at cleaner loader sites or on wasteland to prevent cross pollination with bolters in root crops. If hand labour is not available

Table 8.6 Selective control of broadleaved weeds in broadleaved field and field vegetable crops

Crop	Suitable herbicides		Creeping thistle (C. arvense)	Mayweed (Anthemis spp.)
	General weed control – annuals (see manufacturers' literature for weeds controlled)			
Field crops	*Basic herbicide*	*mixture(s)*		
Beans (field/tick)	chlorpropham	mixtures		
	simazine	—		
	terbutryn	mixtures		
	trietazine	mixtures		
	trifluralin	—		
Winter beans only	carbetamide			
	propyzamide			
Brassicas			clopyralid	clopyralid
Oilseed rape	chlorthal-dimethyl	with propachlor		mixtures
	clopyralid	mixtures		
	propachlor			
	trifluralin			
Winter oilseed rape only	carbetamide			propyzamide
	napropamide	mixture		mixtures
	propyzamide	and mixtures		
	tebutam with TCA			
Other brassica seed crops	carbetamide			not mustard
	chlorthal-dimethyl	with propachlor		
	propachlor			
	propyzamide			not mustard
	trifluralin			
	also: paraquat	and mixtures		pre-emergence contact
Fodder brassicas	chlorthal-dimethyl	with propachlor	clopyralid	clopyralid
	desmetryn			fat hen in kale and rape
	propachlor			
	sodium monochloracetate			kale
	tebutam	with propachlor		
	tebutam	with trifluralin		swede, turnip
	trifluralin			kale, swede, turnip, contact
	also: paraquat	and mixtures		pre-emergence
Linseed	Bentazone			
	MCPA			
Maize	atrazine			
	atrazine	with spray additives		
	cyanazine			
	EPTC	with protectant		
Peas	*see* field vegetable crops			
Potato	chlorbromuron			
	cyanazine	mixtures		
	linuron			
	metribuzin			
	monolinuron	and mixtures		
	prometryn			
	terbutryn	mixtures		
	trietazine	mixtures		
	also: paraquat	and mixtures		contact pre-emergence

Table 8.6 Continued

Crop	Suitable herbicides		Creeping thistle (C. arvense)	Mayweed (Anthemis spp.)	
	General weed control – annuals (see manufacturers' literature for weeds controlled)				
Sugar beet	carbetamide				Sugar beet stecklings all used in various mixtures and programmes
fodder beet and mangel (Proprietary and tank mixes – numerous)	chloridazon	and mixtures	clopyralid	cloyralid	
	cycloate	and mixtures			
	ethofumesate	and mixtures			
	lenacil	and mixtures			
	metamitron	and mixtures			
	phenmedipham	and mixtures			
	propham	and mixtures			
	trifluralin				
Field vegetable crops					
Beans, broad	chlorpropham	mixtures			
	simazine				
	terbutryne	mixtures			
	trietazine	mixtures			
	trifluralin				
Beans, french and runner	bentazone, also	with spray additives*			* french beans only
	chlorthal dimethyl	with diphenamid*			* runner beans only
	diphenamid				
	monolinuron	and mixtures*			* french beans only
	trifluralin				
Beetroot	cycloate	with lenacil	clopyralid	clopyralid	
	ethofumesate	with lenacil or propham mixtures			
	lenacil	with tri-allate			
	metamitron	or with spray additives			
	phenmedipham				
	propham	mixtures			
Brassicas General (except calabrese)	chlorthal-dimethyl	with propachlor or propachlor + paraquat	clopyralid	clopyralid	
	propachlor				
	tebutam	with propachlor			
	trifluralin				
Brussels sprout and cabbage only	aziprotryne				fat hen
	desmetryn				
	sodium monochloracetate				
Spring cabbage only	carbetamide				
Calabrese	propyzamide				
	tebutam	with propachlor	clopyralid	clopyralid	
	trifluralin				
Carrot	chlorbromuron				
	chlorpropham				
	linuron				
	metoxuron				
	prometryn				
	trifluralin				
Leek	chlorpropham				
	propyzamide				
	trifluralin				
	also paraquat	and mixtures			contact pre-emergence

Table 8.6 continued

Crop	Suitable herbicides		Creeping thistle (*C. arvense*)	Mayweed (*Anthemis spp.*)
	General weed control – annuals (*see manufacturers' literature for weeds controlled*)			
Field crop Onion	*Basic herbicide* aziprotryne chloridazon chlorpropham chlorthal-dimethyl	*mixtures* mixtures and mixtures with propachlor or propachlor and paraquat	clopyralid	clopyralid
	cyanazine ioxynil propachlor sodium monochloracetate also paraquat	and mixtures		pre-emergence contact
Parsnip	chlorbromuron linuron trifluralin also paraquat	and mixtures		pre-emergence contact
Peas	aziprotryne bentazone chlorpropham cyanazine MCPB prometryn terbutryn trietazine	and mixtures mixtures and with MCPB/MCPA mixtures mixtures		
Sweet corn	atrazine cyanazine EPTC	or with spray additives with protectant		

Table 8.7 Control of volunteers or groundkeepers in some crops

Volunteer or groundkeeper	Affected crop	Suitable herbicides	Limitations
Selective controls			
Beet, weed	Cereals	bromoxynil mixtures	
	Sugar beet, fodder beet, mangel	glyphosate	Rope wick application only
Cereal volunteers	Brassicas		
	Cabbage (S)	propyzamide	
	Forage crops	TCA	
	Seed crops, some	carbetamide, propyzamide	Not mustard
	Rape oilseed		
	(W) only	carbetamide, fluazifop-P-butyl, propyzamide	
	(W) + (S)	dalapon, TCA	
	Seed crops, some	carbetamide, propyzamide	
	Legumes		
	Clover, lucerne, sainfoin	carbetamide	
	Sugar beet, fodder beet	fluazifop-P-butyl	Not mangels
Oilseed rape	Cereals	bromoxynil mixtures, mecroprop	
Non-selective controls			
Beet, cereals and oilseed rape	Pre-sowing	paraquat	
Potatoes	Pre-sowing	glyphosate	

Table 8.8 Control of weeds in established grassland

Weed	Situations where especially troublesome	Preventive cultural and management controls	Most effective herbicidal treatments and herbicides
Bent, creeping (*Agrostis stolonifera*)	Poorly managed lowland pastures, overgrazed in winter and spring, undergrazed in summer	G, E, F, G	Asulam used for docks gives some useful effect
Bracken (*Pteridium aquilinum*)	Heavily understocked sites, upland, moor and hill lands, poisonous	B, C, I, K, L	Asulam
Buttercup, bulbous (*Ranunculus bulbosus*)	⎧ Common in most permanent grassland. Poisonous except in hay. *R. bulbosus* occurs in drier conditions	⎧ If very serious N otherwise CE	QR (2, 4-D[1]) (MCPA[1])
Buttercup, creeping (*R. repens*)		A, C, E	P2, 4-D[1] MCPA[1] MCPB and mixtures P MCPA[1]
Buttercup, tall (*R. acris*)	⎭	⎩ A, C, E	MCPB and mixtures
Chickweed (*Stellaria media*)	Poached grassland along with other annual weeds	D	*Benazolin mixtures* mecoprop[2]
Daisy (*Bellis perennis*)	Overgrazed poor permanent grass and lawns	C, E R if necessary	2, 4-D[1] MCPA[1]
Dandelion (*Taraxcum officinale*)	Common in poorly managed grassland	C, E, F	2, 4-D[1] MCPA[1]
Docks, broad and curly leaved (*Rumex obtusifolius* and *R. crispus*)	Universal on land stocked only with cattle; also treated with slurry and mown	J, M, N	O *Asulam Dicamba mixtures*[1] or *Triclopyr*[1] and mixtures
Gorse (*Ulex* spp.)	Grossly undergrazed acid permanent grass and waste places	B, C, D, E, G, H, J, M	*Triclopyr* and mixtures[1]
Horsetails (*Equisetum* spp.)	Soils with wet subsoil; remains poisonous in hay and silage	A, C, E, F, G	RT(2, 4-D)[1] (MCPA[1]) (MCPB mixtures)
Meadow grass, annual (*Poa annua*)	Very common in reseeds and poor, open or poached grassland. May suddenly appear/disappear	D, E	Ethofumesate[1] *if* cost is justified
Mouse-eared chickweed (*Cerastium holosteoides*)	Universal in grassland	C, D	(2, 4-D[1]) (MCPA[1]) (MCPB mixtures)
Ragwort (*Senecio jacobea*)	Neglected and overgrazed old pastures and wasteland, hedgerows. Remains poisonous in hay and silage	C, D, E, J, M Undersow new leys	2, 4-D[1] MCPA[1] Once sprayed keep stock out until ragwort has disappeared
Rush, hard. (*Juncus inflexus*) Rush, soft, (*Juncus effusus*)	Badly drained wet land of low fertility; establish from seed in poached or open sward	⎧ A, C, D, E, G, I Reseed new leys *without* cover crop A, B, C, D, G, E, I Reseed new leys ⎩ *without* cover crop	Resistant to selective herbicides P, 2,4-D[1] MCPA[1]. Cut 4 weeks after spraying
Sorrel common (*Rumex acetosa*)	Poor permanent grass, usually damp	B, C, G	(2,4-D[1]) (MCPA[1])
Sorrel, sheeps (*Rumex acetosella*)	Indicates acidity on poor permanent grass	B, C, G	(2,4-D[1]) (MCPA[1])
Stinging nettle, great (*Urtica dioica*)	Mainly on grazed grassland, often with loose structure; encouraged by surface litter	E, H, J, K N if serious	⎰ PQRS, *Triclopyr* and ⎱ mixtures[1]
Thistle, creeping (*Cirsium arvense*)	Universal in undergrazed grassland — rarely seeds	C, E, F, G N if serious	P QR 2,4-D[1] MCPA[1] *MCPB mixtures*, Clopyralid[1]
Thistle, spear (*C. vulgare*)	Universal. Establishes readily from blown seed	M	O 2,4-D[1] MCPA[1] MCPB mixtures
Tussock grass (*Deschampsia caespitosa*)	Badly managed wet grazings, establishes from seed	A, C, D, E, H, M	No selective chemical control
Yorkshire fog (*Holcus lanatus*)	Wet, acid or badly managed lowland grassland	A, B, C, E, F, G	Asulam used for docks gives useful effect

[1] Clovers killed or severely checked. Prefer 2,4-D to MCPA if suitable.
[2] Mecoprop is effective against chickweed but kills clover. Use only on grass seed crops, pure grass swards and in emergency, when chickweed (*S. media*) is likely to smother sward or oversowing.
( ) brackets denote limited effect.
Italics denote first choice herbicide.

Table 8.8 Continued

Key to Table 8.8.

A   Improve drainage.
B   Apply lime.
C   Increase fertility. Apply N, P and K as required; a base dressing of 190 kg/ha insoluble P_2O_5 given as basic slag or ground mineral phosphate on phosphate-deficient pastures in high rainfall areas or wet situations.
D   Avoid poaching; keep cattle off grassland in winter.
E   Avoid overgrazing in winter, spring or early summer.
F   Avoid undergrazing in summer.
G   Increase stocking rate, mow or top over after grazing.
H   Change management to include close cutting, using forage harvester to cut silage.
I   Cut twice annually for two or three years.
J   Heavy grazing with sheep reduces infestation.
K   Heavy trampling with cattle but unstock cattle on bracken to avoid bracken poisoning.
L   On ploughable land plough bracken deeply with digger plough when fronds are three-quarters open in spring or early summer. Cut up rhizomes with several passes of disc harrows and consolidate. Grow pioneer crop, e.g. rape or turnips, graze in autumn, disc harrow in December; do not plough. Follow with re-seeded ley or potatoes; re-seed must be intensively stocked and managed or bracken returns.

M   Prevent flowering and seed production by spudding, cutting, flail or swipe mower.
N   Several perennial weeds are difficult to control with inexpensive herbicides in established grassland but are easily controlled by such herbicides in the presence of cultivations. Where practicable, ploughing and cultivations, followed by either (a) re-seeding with Italian ryegrass and spraying with MCPA, 2, 4-D or mecoprop as soon as possible, followed by two further sprayings during the summer, or (b) growing a run of two or more sprayed cereal crops.
O   Undersown leys and reseeds must be sprayed with an MCPB or 2, 4-DB mixture (*see Table 8.4*) to kill perennial broad leaved weeds in the susceptible seedling stage. Once established these weeds become much more resistant to herbicidal treatment.
P   Spray in spring or early summer before flowering.
Q   Spray in autumn.
R   Spray in at least two successive years.
S   Spot treatment — wet clumps to run-off point.
T   Spray when plants have made maximum growth.
     Note. When pasture contains a very large amount of weed and little or no valuable grass or clover, reseeding should be considered, especially if intensive stocking is intended.

control bolters by mechanical cutting at least twice in July. Obtain advice from the British Sugar Corporation field staff as to correct stage for cutting. Drill only after 20 March to avoid bolting, use only *non-bolting varieties* (NIAB Farmers leaflet No. 5).

Additional measures (*once weed beet are identified*); lengthen rotation with non-beet crops and use effective herbicides in other crops (e.g. bromoxynil mixtures in cereals). Spread beet over whole farm. Tractor hoe frequently, close to rows to remove about 90% of weed beet seedings and use trifluralin for late inter-row weed control. Alternatively use rope wick with glyphosate when beet is above crop. If seed of weed beet has been shed, leave it on the surface; avoid ploughing in, which preserves it until brought to the surface by later ploughings. Use direct drilling or minimal cultivations for cereals on suitable soils. Only relatively few fields are as yet seriously affected but weed beet poses a serious menace to the whole British home grown sugar industry; it could equally become a major problem with fodder beet.

WEED CONTROL IN ESTABLISHED GRASSLAND

The presence of infestations of weeds or weed grasses is generally an indication that all is not well with management and growing conditions. Weed infestations can also arise in leys, where the species originally sown lacked persistency and faded out, leaving the field to be colonised by weeds.

The first step is to decide whether the sward justifies retention and subsequent improvement or whether to kill all vegetation and reseed or grow a run of one or more arable crops before reseeding. Retention is often best on moist fertile soils or in wetter situations as a more poaching resistant sward is maintained; costs are lower and no grass production is lost. Conversely, where a temporary sward is worn out, in dry situations or where a poor sward is infested with perennial weeds which are difficult to control by herbicides while the field remains down to grass, destruction of the sward has much to commend it. A herbicide which gives a total kill if properly used (e.g. glyphosate) is a valuable first step. The majority of grassland weeds will not withstand well managed cultivations and arable crops for

long, especially where the arable break starts with cereal crops sprayed with translocated herbicides.

The main weeds of grassland and their appropriate controls are listed and summarised in *Table 8.8*. Where weedy grassland is to be retained, improvements in management, growing conditions and fertility levels are essential. Herbicides are often an essential part of the programme but cannot maintain any improvement by themselves.

WEED CONTROL IN FOREST AND WOODLANDS

Adequate weed control is essential for the satisfactory establishment of young trees and their subsequent growth. The nature of the problem and weed spectrum is much more variable than with normal agricultural crops where much of the land is reasonably level and the weed spectrum is confined largely to weed grasses and broadleaved herbaceous species. In woodland a much wider weed spectrum occurs and terrain is often difficult. The situation varies from sites where little or no weeding is necessary to those where weed growth is vigorous and rapid and may cause total failure of the young trees; the commonest cause of smothering is the collapse of tall vegetation on top of them in the winter. Frequently the control of one type of weed allows another to develop, e.g. grasses often follow cutting down of brambles (*Rubus* spp.) while ploughing temporarily controls some grass species but may result in heavy growth of thistles (*Cirsium* spp.) or gorse (*Ulex* spp.). It is necessary to be prepared for these changes. Total eradication of weeds over the whole site is unlikely to repay the cost; generally the suppression of weed growth in the immediate vicinity of the young tree is adequate, provided that tall woody broadleaves are not allowed to develop. Considerable time and money can be saved by knowing when to intervene and when intervention is unnecessary.

Weed types

For practical purposes weeds may be grouped as follows:

(1) 'soft' or 'fine' grasses: bents (*Agrostis* spp.), fine leaved

fescues (*Festuca* spp.), Yorkshire fog and creeping soft grass (*Holcus lanatus* and *H. mollis*) and meadow grasses (*Poa* spp.);

(2) coarse grasses and rushes: oat grasses (*Arrhenatherum* spp.), cocksfoot (*Dactylis glomerata*), purple moor grass (*Molinia caerulea*) and rushes (*Juncus* spp.);

(3) mixed herbaceous broadleaved weeds and grasses;

(4) bracken (*Pteridium aquilinum*);

(5) heaths (*Erica* spp.) and heather or ling (*Calluna vulgaris*);

(6) low growing woody weeds, e.g. gorse, furze or whin (*Ulex* spp.);

(7) taller woody broadleaved weeds. These include:

    (i) trees and large shrubs or bushes: birches (*Betula* spp.), sallow and goat willows (*Salix* spp.), rhododendron (*Rhododendron ponticum*), laurel (*Prunus* spp.), blackthorn (*Prunus spinosa*) *et al.*;

    (ii) coppice regrowth – most broadleaves; sweet chestnut (*Castanea sativa*) is especially vigorous;

    (iii) trailers: blackberry (*Rubus* spp.), dog rose (*Rosa* spp.);

    (iv) climbers: honeysuckle (*Lonicera periclynemum*) and ivy (*Hedera helix*).

Weeding and cleaning

Low growing weeds, i.e. groups 1–6, suppress the growth of young trees and may even kill them unless controlled by *weeding*. This is usually carried out during the first three to four years after planting, when the young trees should have made enough growth (about 2 m high) to suppress any further low weed growth. Grass, even if kept tightly cut, suppresses growth badly, as its all pervading root system robs the young trees of water and nitrogen (N). If not controlled, grass and bracken can also collapse on top of the young trees in winter and smother them. The best methods of controlling grass in the immediate vicinity of young trees are (i) ploughing before planting, or (ii) the application of a grass killing herbicide (glyphosate or paraquat) pre-planting in strips 1 m wide along the proposed planting lines, or (iii) spraying around the transplants with a knapsack sprayer with its lance fitted with a tree guard.

Cleaning, i.e. the removal of unwanted growth, is carried out in the pre-thicket stage, usually 6–12 years after planting according to the state and type of the growth present; if growth is substantial, give an early 'low cleaning' with a second 'high cleaning' later.

Unwanted types include: fast growing broadleaved 'weed' trees, especially birches (*Betula* spp.), sallow or goat willow (*Salix caprea*), which can seriously damage or even suppress a young plantation of conifers as well as broadleaved crops. These can be removed by companion saws or application of herbicide.

Harmful climbers, e.g. honeysuckle (*L. periclynemum*) must be removed from broadleaves as they wind round the trunk and produce serious distortion. Unwanted coppice, laurel or rhododendron regrowth is best prevented by treating the old stumps with herbicide *immediately* after cutting (*see Table 8.9*). Where retained to draw up broad-leaved crop stems, e.g. between tree shelters, regrowth is cut out as and when it starts to compete with the crop.

Methods of control

Control measures must be carefully integrated to suit the weed flora, topography and general conditions of the site. Each operation should be carried out as part of the overall establishment plan and not in isolation. Methods can be grouped into the following categories: (a) hand, (b) mechanical, and (c) herbicidal.

Handwork

Weeding

Weeding with hand tools, e.g. grass hooks, is slow, laborious and expensive, especially if two cuts are wanted in a single season; it has largely been superseded by ploughing, mechanical aids and herbicides. On some ploughed upland sites no weeding may be required for two to three years after planting, sometimes not at all. The effect of handwork is only temporary and may not last a whole season; it does not prevent grass from reducing the growth rate of young trees. Handwork also creates a summer labour peak when on the farm there are heavy labour demands.

The traditional method is to walk down a line of plants, trimming away vegetation around each plant for a sufficient distance to prevent smothering, using a curved grass or reaping hook. *Do not make any cuts until the plant is located.* A light stick is carried to push weed growth away and if of the same length as the planting distance within the row, location of small trees is much quicker. Work is much easier where rows are quite straight and trees are evenly spaced.

Cleaning

If no mechanical facilities are available, handwork is suitable for dealing with light infestations or small patches, e.g. laurel or rhododendron. It is essential for dealing with injurious climbers on broadleaves unless a hand held rope wick herbicide applicator is available.

Mechanical weeding and cleaning

Machines can be tractor mounted, pedestrian controlled or hand held. Tractor mounted machines with a suitable attachment, e.g. rotating flails or chains, can give a high output at low cost on soft weed or deal with considerable woody growth, but cannot operate on very rough or steep terrain. A good scarifier can deal with lop and top if not too large and produce a surface tilth for planting. The output of hand machines is low, but they are very useful on steep land, in confined spaces, for small patches and light infestations. String trimmers ('strimmers') are suitable for soft growth but avoid damaging the bark of young trees with the plastic 'string'. Companion saws are invaluable for most hand cleaning work, except injurious climbers. Machines are useful on very mixed weed floras, where the vegetation is out of hand or where it is impracticable to use herbicides, e.g. around reservoir catchment areas.

Herbicidal treatments

These have now displaced handwork on large areas of forest as they are normally cheaper, last for a whole season or more and save labour. A wide range of well tried herbicides for most weed situations is now available. Some methods require no diluent (e.g. ULV and granules) and are well suited to steep or rough sites or where water supply is difficult. The main problems lie in persuading some woodland owners to use herbicides. Under no circumstances may the use of herbicides be entrusted to anyone but fully trained and properly instructed and supervised operators. (Consult 'The Control of Pesticides Regulations 1986 SI 1986/1510 and the Consents to sale, supply, storage and use of pesticides, also the MAFF leaflet: *Pesticides: Guide to the New Controls* UL 79).

No single herbicidal treatment can be relied on to control the very broad spectrum of weeds that may develop in the early years of a plantation – some weeds are resistant and different herbicides may be needed later. Some treatments have a short period of use in a growing crop, while timing in relation to growth stages of trees and weeds is important; if wrong, weed control may fail or the crop may be damaged. Careful forward planning is therefore essential to integrate the herbicidal programme with all other operations during the establishment period – from the initial preparation of the site until the trees have formed a dense thicket.

Herbicidal application

Equipment

Steep or rough terrain and tree growth largely preclude the use of normal farm equipment for woodland. The use of lances or tractor mounted sprayers may be feasible in some situations. The following are suitable, according to herbicide, species and type of application.

Knapsack MV sprayers

Directed applications around base of young trees – some herbicides require use of *tree guards* (i.e. to cover tree or spray jet). NB glyphosate, paraquat. Also stump and basal bark treatments.

Mistblowers LV

Knapsack or tractor mounted for overall foliar applications. The latter machine only of limited value in lowland forest.

Ultralow volume (ULV)

Overall foliar application by incremental spraying.

Distributors for granular herbicides

Airflow type distributors are essential for accurate granule distribution. Handwork is too laborious, inaccurate and wasteful of expensive herbicides; overdosing can occur. Suitable for overall or strips 1–2 m wide; flow can be cut off with wide tree spacings for spot application. ULV and granules require no diluent and are ideal for areas where water supply or carriage is difficult.

Tree injectors

These are for injecting undiluted herbicide into standing trees at (a) breast/waist height – cut made with axe, or (b) base of tree with chisel bit for penetration.

Rope wick application

Handheld applicators are suitable as a replacement for handwork on climbing or tall growing weeds or climbers.

Types of treatment

Pre-planting

These treatments are given to control unwanted vegetation prior to planting, either overall or in strips, in the centre of which trees are subsequently planted; selectivity at this stage is not needed but some herbicides require an appropriate interval between application and planting. Avoid drift onto adjacent susceptible crops.

Post-planting

Post-planting treatments if given *overall must be selective* in the species concerned, as determined by age, size and growth stage of the trees, time of year and herbicidal activity; otherwise *directed* applications must be given, using *tree guards* where necessary (e.g. glyphosate or paraquat applications in susceptible spp.). *Foliar* applications are dependent for success on the active growth of *weeds* at the time of application, e.g. with broadleaved woody weeds apply before leaf senescence. *Selectivity* is often obtained by application when the crop is not making rapid growth or is dormant. Timing is all important for control of weeds and crop safety, as application periods are frequently *very limited*. *Surface* or soil applications, granular or liquid depends on adequate rainfall for activity. Restrictions exist on timing of applications.

TREATMENT OF INDIVIDUAL LARGE TREES

Stump treatment

This treatment is given to freshly cut stumps (within 24 h of felling) of broadleaved trees or woody weeds to prevent coppice regrowth. Give as a *pre-planting* treatment. Use triclopyr, saturating cut surface and remaining bark. Use paraffin or diesel oil as diluent. Hawthorn (*Crataegus* spp.), laurel (*Prunus laurocerasus*) and rhododendron (*Rhododendron ponticum*) require high dosage rates.

Unwanted standing trees

Basal bark spray

Saturate bottom 300 mm (thin bark) or 450 mm (thick bark) of whole circumference of trunk to run off; use triclopyr in paraffin or diesel oil. Application in conifer crops should be delayed until growth extension has ceased and resting buds formed.

Frill girdling

A ring of overlapping downward cuts encircling the trunk close to ground level is first made with a light axe or bill hook, to penetrate the cambium and preferably the sapwood; herbicide is then sprayed onto bark just above cuts and runs into them. Use triclopyr. Also suitable for ammonium sulphamate (AMS) for trees less than 150–200 mm diameter breast height (1.3 m above ground level).

Notching

Notching is for applying solid AMS crystals to larger trees; cut a ring of steps with an axe, maximum 10 cm edge to edge apart at base of tree, floors of steps sloping slightly inward to retain AMS crystals, 15 g per step.

Tree injection

Tree injection requiring a special tool for injecting undiluted herbicides into the translocation systems of a tree, may be carried out safely at any time of the year.

Table 8.9 Selective weed control in forests, woodland, shelterbelts and other tree situations (except fruit and ornamentals)

Weed situation	Herbicide and type of activity	Method of application	Apply pre- and/or post planting	Months of application	Remarks
Predominantly grasses	Atrazine R	FS	2	*Feb–Apr* (May)	Not organic peaty soils. Molinia, Calamagrostis and Deschampsia spp. resistant
	Atrazine + dalapon } RTr	FS	1 or 2	*Mar–May* (June–Aug)	Not organic peaty soils
	Glyphosate Tr	FS	1 or 2	*July–Sept* but any period of active foliage growth	Post planting; protect actively growing trees; some conifers can be sprayed overall when dormant
	Hexazinone RTr	FS	1 or 2	*Mar–May* (June–Aug)	Limited range of spp. spruces and pines only
	Paraquat C	FS	2	*Feb–May* and *Aug–Oct* but any other time to green foliage	All spp. susceptible and must be protected. Not rhizomatous grasses
	Propyzamide R	Granules or FS	1 or 2	*Oct–Dec* only exc. Jan in upland Britain	Reduced activity on organic soils, max peat depth 10 cm *Dactylis, H. mollis* and Calamagrostis show some resistance
Grasses with substantial proportion of broadleaved herbs	Glyphosate Tr	FS	1 or 2	*July–Sept* but any period of active foliage growth	Post planning; protect actively growing trees; some conifers can be sprayed overall when dormant
Bracken (*Pteridium aquilinum*)	Asulam Tr	FS	1 or 2	*Late June–Aug* Best effects finds unfurled and canopy almost complete	Best control but only controls bracken. Conifer tolerance is high, except Western Hemlock (*T. heterophylla*)
	Glyphosate Tr	FS	1 or 2	*July–Aug* but before senescence starts	Post planting: see above. Wider weed spectrum than asulam
Heather (predominantly ling, *Calluna vulgaris*)	2,4-D Ester Tr	FS	2	*June–Aug* before mid July only as directed spray with knapsack sprayer	Never on broadleaves; larches and lodgepole pine sensitive. Some crop damage usually occurs. Toxicity to crop increases in hot weather. Water catchment area limitations
	Glyphosate Tr	FS	1 or 2	*Late Aug–end Sept* after conifer growth has hardened; timing varies with season	Spruces and pines tolerant but avoid leaders
Gorse (*Ulex spp*) and broom (*Sarothamnus scoparius*)	Triclopyr Tr	FS	1 or 2	*July–Sept*	Post planting: larch and broadleaves very susceptible to damage. Spruces will tolerate overall sprays if leader growth has hardened. Pines more sensitive. Do *not* apply VLV or ULV. Wear face shield
Rhododendron (*Rhododendron ponticum*)	Ammonium sulphamate } Tr R	FS	1 only	Any time of year dry	Allow 3 months between application all species can be heavily damaged
	Glyphosate Tr	FS	1–2 only if trees are protected	*June–Sept*	All species heavily damaged if spray comes in contact

Table 8.9 Continued

Weed situation	Herbicide and type of activity	Method of application	Apply pre- and/or post planting	Months of application	Remarks
	Triclopyr Tr	FS	1–2 only if trees are protected	*June–Sept*	Not VLV or ULV. Wear face shield. All spp. can be heavily damaged

NB All Rhododendron treatments are best given pre-planting

Weed situation	Herbicide and type of activity	Method of application	Apply pre- and/or post planting	Months of application	Remarks
Woody weeds including brambles (*Rubus* spp.)	Fosamine Ammonium } Tr	FS	1 or 2	*July–Sept* prior to leaf senescence	Only deciduous broad leaved woody spp. controlled. Rhododendron gorse and broom resistant. Protect wanted broadleaved species. Conifer sensitivity varies
Foliar applications	Glyphosate Tr	FS	1 or 2	*June–Aug* when weeds are growing actively	Post planting: Protect all wanted broadleaves. Only spray suitable conifers overall after leader growth has hardened. Not larches
Tree shelters (Protected broadleaf crop) Control of all unwanted spp.	Glyphosate Tr	FS	2	*June–Aug* when weeds are growing actively	Use only glyphosate. Direct spray at vegetation in ring round each shelter, minimum diameter 1.2 m for 2 years after planting
Woody weeds Stem treatments	Ammonium sulphamate } Tr R (Water soluble crystals)	Frill girdle or notch	1 preferable	Any time of year in dry weather	Corrosive; only apply by recommended plastic applicator. Allow minimum 3 months before planting. Especially suitable for rhododendron
	2,4-D Amine Tr	Tree injection only	1	Any time of year	Allow at least 1 month before planting
	Glyphosate Tr	Notch	1 or 2	Any time of year except sap rise period Mar–late April	Unwanted individual stems can be safely treated in any spp. Cut 2 notches for larger trees
Woody weeds Cut stump treatment	Ammonium sulphamate Tr R (Water soluble crystals)	dry crystals or spray	1 only	Any time of year on a dry day	All spp. can be heavily damaged by direct contact or root uptake. Allow *at least* 3 months before planting
	Glyphosate Tr	Knapsack sprayer, paint brush or clearing saw with spray attachment	1 or 2	*Oct–Apr* before spring sap flow	Use 20% solution in water with red dye added. Planting can occur immediately after treatment. Post planting; safe in any spp. but avoid foliage contact

Key for Table 8.9.
Type of activity. C = contact (foliar)
     Tr = translocated (foliar)
     R = residual (soil)

Pre- or post-planting: 1 pre-planting
          2 post-planting

Method of application FS = foliar spray
Note: Detailed recommendations for all suitable herbicides and treatments are given in Forestry Commission (1986) and subsequent revisions; consult this before application

Table 8.10 Weed control in non-crop situations

Herbicide, type of activity and persistency of initial dose in soil	Main species/types affected when used for initial knockdown of foliage	Availability in mixtures with very persistent residual herbicides; also other leaf applied herbicides. Yes/No. Remarks
Foliage absorbed herbicides		
Aminotriazole. Tr, LP	Rhizomatous grasses	Yes. Also affect broadleaved herbaceous weeds
Dalapon Tr, MPR	thizomatous grasses	Yes. Broadleaves resistant to straight herbicide
Paraquat C None	All green foliage	Yes. Straight herbicide only burns tops of perennials
Triclopyr Tr LP	Wood spp.	No. Also docks (*Rumex* spp.) and nettles (*Urtica* spp.)
Foliage and/or root absorbed or vapourising herbicides		
Chlorthiamid VP ⎫	Bracken (*P. aquilinum*) Docks (*Rumex* spp.) grasses and general weed control	No + dalapon only ⎱ Granules only. Do not use near glasshouses, hops or susceptible crops
Dichlobenil VP ⎭		
Picloram Tr VP	Ragwort (*S. jacobea*)	Yes. Straight herbicide general broadleaved weeds
Sodium chlorate Tr MP2	General weed control	Yes. Straight herbicide highly inflammable. Fire risk. Do not apply near roots of trees or shrubs of value
Root absorbed herbicides (residual)		
Atrazine VP ⎫	Used for	Yes
Bromacil VP ⎪	long-term	Yes. Do not use near trees or shrubs of value
Diuron VP ⎬	residual weed	Yes. Do not use near trees or shrubs of value
Simazine VP ⎭	control	Yes

Key to Table 8.10
Type of activity C = contact.
　　　　　　　　Tr= translocated.
Persistency
Persistency of chemicals is determined to a large extent by the dosage rate (increased dosage rate = longer persistency) soil type and rainfall. The following times are *only approximate* and apply to the *straight herbicide only*. Additives may increase persistency greatly:
VP　(Very persistent) a season or longer
P　　(persistent) a season
MP^1　(moderately persistent) 3 months to a season
MP^2　(rather less persistent than MP^1) 3 to 4 months
LP　Low persistency. Up to 8 weeks
None　No persistency—sowing can follow soon after treatment if required.
NB There is a need for great care when using herbicides in the proximity of glasshouses, orchards, ornamentals and any susceptible crops. Avoid drift at all times. Many of the above herbicides are quite unsuited for amateur use.

Choice of herbicidal treatment

A sequence of treatments is likely to be necessary in the early life of the crop; the treatments for various weed situations are summarised in *Table 8.9*. Particular attention should be paid to the requirements and limitations of each herbicidal treatment.

WEED CONTROL (UNSELECTIVE) IN NON-CROP SITUATIONS

The object is usually to keep the ground free from vegetation for a prolonged period, as with yards, drives, paths and roadways. If temporary control only is needed and cropping is to follow, select only herbicides with short residual effects.

It is not possible to keep land permanently free from vegetation by herbicidal application unless there is an 'ongoing' programme. There are three distinct aspects to the problem: (a) The '*initial knockdown*' to kill existing vegetation. A translocated foliar acting herbicide is usually preferred, e.g. glyphosate. Contact types alone, e.g. paraquat, only kill annuals; they simply defoliate perennials. (b) *Prevention* of immediate re-establishment with a very persistent residual herbicide, e.g. bromacil, diuron. Established perennials require a heavy dose for a complete kill.

(a) and (b) may either be given as separate applications or, where a particular mixture is suitable, are given as a single combined application, e.g. glyphosate + simazine; paraquat + diuron. (c) *Maintain freedom from weeds*. If (a) and (b) are fully effective, only a low rate–low cost application will be required in the following spring for maintenance; if (a) and (b) are not fully effective, a translocated (foliar) application will be required later.

Persistency of herbicides is temporary and depends on (i) *herbicide*: in some e.g. paraquat and glyphosate, there is little persistency; others, e.g. diuron, and bromacil are very persistent; (ii) *dosage rate*: persistency and cost of residual herbicides increases with dosage rate; (iii) *rate of leaching or decomposition* of herbicide is accelerated by high rainfall on soil types which leach readily (e.g. sand) and with increasing soil temperature, resulting in higher microbial activity. Adsorptive surfaces, e.g. peat, clay or ashes require higher dosage rates.

When choosing a herbicide, particular references must be made to surrounding crops, gardens and glasshouses. Spray drift dangers apart, trees can be killed by uptake of herbicides through their roots, especially with sodium chlorate; glasshouses are very vulnerable to vapour given off by chlorthiamid and dichlobenil. Always ensure that drift run off or leaching will not affect adjacent or local sites. Obtain advice before treating slopes or hard impenetrable surfaces where run-off may affect watercourses, streams or wanted vegetation. A selection of suitable herbicides is given in *Table 8.10*.

References

FORESTRY COMMISSION (1986). *The Use of Herbicides in the Forest*. Booklet No. 51, 122 pp. Edinburgh: Forestry Commission

MINISTRY OF AGRICULTURE, FISHERIES AND FOOD (1986, superseded by subsequent editions). *Pesticides 1986*. Pesticides approved under the control of Pesticides Regulations, 253 pp. London: HMSO

ROBERTS, H. A. (1982). *Weed Control Handbook – Principles*. 7th Edn, 253 pp. British Crop Protection Council. Oxford: Blackwell Scientific Publications

Further reading

BRITISH SUGAR PLC publish herbicide recommendations (for herbicidal programmes) annually: 'Herbicides recommendations for post emergence low volume sprays for annual broadleaved weed control'

MINISTRY OF AGRICULTURE, FISHERIES AND FOOD (ADAS) publish a range of priced booklets on weed control in crops. Obtainable from MAFF Publications, Lion House, Willowburn Estate, Alnwick, Northumberland, NE66 2PF

PROCESSORS AND GROWERS RESEARCH ORGANISATION (PGRO), Thornhaugh, Peterborough PE8 6HJ publish leaflets (free to members) on choice of herbicides in peas and varietal susceptibility to herbicides; also associated leguminous crops

9

Diseases of crops

G. Moule

Recent changes in farming practice have resulted in changes in the disease problems of crops. The increase and intensification of the cereals and oilseed rape cropping in recent years, coupled with the trend towards increased autumn sowings, has resulted in the widespread incidence of many air-borne foliar diseases. Many seed-borne diseases however, are now quite rare due to the general use of the various seed treatments available. Similarly, disease problems in the future are only likely to change rather than disappear totally.

The cost effectiveness of any crop protection treatment is dependent upon many factors which may vary widely with the region, season, the chemical used and the individual crop and variety concerned. In the short term the cost of the treatment has to be set against the increased value of the crop and there are likely to be large differences between the treatment of high value crops such as field scale vegetables compared with fodder crops for livestock. In the long term it is likely to be preferable to undertake uneconomic measures with an initial infection of a persistent problem (e.g. long-lived soil-borne disease or pest) to prevent or reduce recurring losses in future crops. Many problems of this nature are best controlled by strict preventative measures in the first instance.

Crop diseases can be divided into the following general groups:

(1) air-borne,
(2) soil-borne,
(3) seed-borne,
(4) vector-borne diseases.

Many diseases have two or more distinct phases of attack and may belong to more than one of these groups, e.g. canker (*Leptosphaeria maculans*) of brassicas may be seed- and soil-borne as well as having a very distinct and destructive air-borne phase. However, each disease has been allocated to the most appropriate group for control measures.

AIR-BORNE DISEASES

These quickly establish in the crop and mostly attack the foliage and stems producing an overall blanket-typed field infection, i.e. virtually all plants are affected to the same degree. They are all of fungal or bacterial origin and produce spores which are easily dispersed by wind and air currents. Spore production may be vast and rapid during suitable environmental conditions and long distance spread can occur in a short period of time. Many are associated with wet weather conditions particularly when the air temperatures are above 10°C, i.e. April to September. Important diseases included in this group are potato blight (*Phytophthora infestans*), the yellow rusts (*Puccinia striiformis*) of wheat and barley, *Septoria* spp. of wheat, *Rhynchosporium* leaf blotch of barley, rye and ryegrasses, crown rust (*Puccinia coronata*) of oats, canker (*Leptosphaeria maculans*) of brassicas and various leaf and pod spots (*Aschochyta* and *Botrytis* spp.) of peas and beans. Their activity is often curtailed by hot dry weather. Others tend to be more prevalent during the warm dry spells of summer and early autumn, e.g. powdery mildews (*Erysiphe* spp.) and brown rust *Puccinia* spp.) of various crops. The powder mildews, however, are more independent of specific climatic conditions for spread than most other diseases.

Many air-borne diseases are very host specific and produce large numbers of asexual spores giving rise to specific races of the fungus with the ability to attack only a very few species or even several varieties (cultivars) of one crop. This extreme specificity can be utilised to advantage when choosing varieties of certain cereals (*see* National Institute of Agricultural Botany Diversification Schemes).

Control

Disease epidemics are only likely to result if

(1) a susceptible host is sown,
(2) a virulent race of the pathogen occurs, and
(3) the environment is favourable for disease attack and spread.

If any of these three factors is limiting in any way crop damage is not likely to be severe. Wherever possible all forms of control measures should be used to give a fully integrated

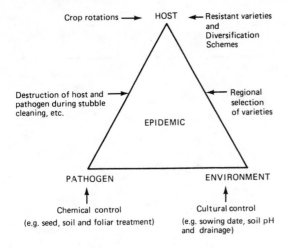

Figure 9.1 A hypothetical epidemic and its integrated control

control programme *(Figure 9.1)*. If, however, a resistant variety is used or climatic conditions have not been favourable for disease spread, chemical control can safely be reduced or omitted on economic grounds.

Cultural

Crop rotations are not very useful in controlling air-borne diseases as they cannot prevent wind-blown spores coming in from neighbouring infected areas. Similarly, efficient stubble clearing after harvest drastically reduces carryover of spores onto the following crops but at best, like crop rotation, is

only likely to delay subsequent re-infection. Both are very much more useful for soil-borne diseases and some types of weed control. Burning of stubble followed by deep ploughing to bury the remainder is preferable to the use of heavy cultivators with their poor burial ability. The date of sowing may affect the length of the intercrop period considerably and the subsequent disease carryover. Many diseases are very common on late tillers and volunteers, consequently early sown autumn crops are particularly prone to disease carryover from the previous crop. The seedlings of these crops often become greatly infected with various seed-borne diseases from the later germinating shed corn after stubble cleaning *(Figure 9.2)*.

September sowings of winter cereals are often very high yielding but seldom give as high returns as October sown crops due to the increased costs incurred for the early autumn disease, pest and weed control required. For these reasons spring sown crops may be useful as cleaning crops in a sequence of autumn sown crops, although returns tend to be considerably lower.

Air-borne diseases are best controlled by the use of resistant varieties. Where these do not exist or break down due to new pathogen races, chemical control should be used. All variety choice should be made after consultation with the National Institute of Agricultural Botany (NIAB) recommended leaflets paying particular attention to those diseases common in the region. Where such schemes exist, use should also be made of the NIAB Variety Diversification Schemes to help reduce the spread and severity of those air-borne diseases which exist in many physiological races.

Farmers should grow, preferably, several varieties chosen from different diversification groups and not just one or two popular varieties that may often belong to the same diversification group and thus the same pattern of susceptibility. Diversification in time is also useful, e.g. where wheat follows wheat it is advisable if the second wheat crop variety is chosen from a different diversification group from the first wheat variety.

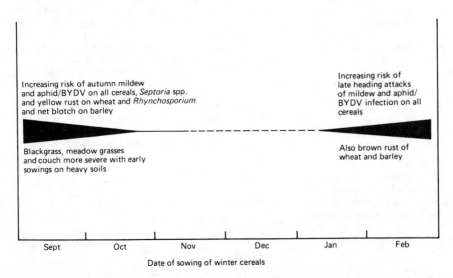

Figure 9.2 The relationship between the date of sowing of winter cereals and associated disease, pest and weed problems (BYDV – barley yellow dwarf virus)

Chemical control

Seed treatments

These are useful in controlling many seed- and soil-borne diseases and give good control during the first half of the plant's life. Few are persistent enough to give effective control of most air-borne diseases of later critical stages of growth such a flag leaf emergence and heading of cereals. Those that are persistent are often more expensive than similar foliar treatments and, in years of relatively low disease levels, are unnecessary. Seed treatments for foliar diseases are best restricted to very susceptible varieties only.

Foliar treatment

This treatment is considerably more flexible than seed treatment particularly with regard to the number and choice of chemicals available. The correct chemical(s) can be chosen accurately for the specific disease(s) as and when they occur. Protectant fungicides with little or no eradicant action must always be applied before disease build-up. As a general rule even when using systemic fungicides with good eradicant activity the best economic responses are obtained when applications are made at the first sign of disease build-up particularly if this coincides with weather conditions that favour disease development *(Figure 9.3)*.

It is normally more cost effective to use a fungicide in high risk situations, for example

(1) susceptible varieties,
(2) high disease levels in nearby crops,
(3) suitable weather conditions for disease spread,
(4) pre-disposing factor(s) operative.

Broad-spectrum fungicides generally give better responses than specific types particularly if the variety is susceptible to more than one disease and/or two or more diseases are likely to attack the crop. With any crops there are often critical growth stages that benefit from good disease protection, e.g. flag leaf/ear emergence in cereals and rapid bulking up of potatoes in August. Generally crops should not be sprayed with any crop protection chemical if they are suffering from stress. Fungicides used during periods of drought often give rise to quite significant yield reductions.

FUNGICIDE RESISTANCE

In recent years, as a result of the overuse of fungicides on certain crops, e.g. eyespot and powdery mildew control on cereals, insensitive/resistant strains of the fungal population

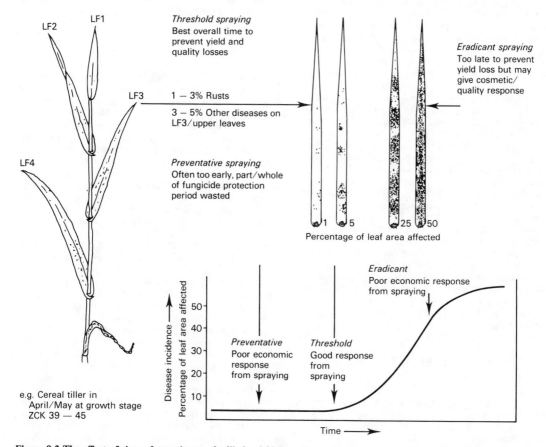

Figure 9.3 The effect of time of spraying on the likely yield response

have been artificially selected out and now dominate the population. This can result in a lack of effective disease control. To minimise the problem fungicides with different modes of action are mixed, alternated and/or rotated to prevent selection and build up of resistant strains. The ADAS fungicide grouping scheme for cereal disease control contained in Booklet 2257 'Use of Fungicides and Insecticides on Cereals' should be consulted for further information.

SOIL-BORNE DISEASES

The main symptoms usually occur on the roots and stem bases which often give rise to wilting and stunted plants. Affected plants occur in patches of varying sizes in the field and may be confused with poor drainage, shallow soil depth and various mineral deficiencies. These diseases generally have very limited powers of mobility and are commonly dispersed in contaminated soil on machinery wheels, animals' feet, clothing and footwear. Severe wind and water erosion of soil may be significant in certain areas also.

The organisms generally exist in a limited number of races and either attack a wide range of crops and are short lived without a host, e.g. 'take-all' *(Gaeumannomyces graminis)* of cereals and grass or possess a narrower host range coupled with long-term survival in the soil as a resting spore, e.g. potato wart *(Synchitrium endobioticum)*, onion white rot *(Scerotium cepivorum)* and pea wilt *(Fusarium oxysporum)*. Club root *(Plasmodiophora brassicae)* of brassicas is a particularly difficult disease as it has a wide host range including most agricultural/horticultural brassicas and many cruciferous weeds such as charlock *(Sinapis arvensis)*, shepherds purse *(Capsella bursa-pastoris)* and field pennycress *(Thlaspi arvense)*. Its soil-borne resting spore is also capable of surviving for at least eight years between susceptible hosts. Once established, eradication of these diseases is extremely difficult and often quite costly, therefore the main form of control measure should be preventative using a sensible cropping sequence. Overcropping should be avoided and extreme care should be taken with contaminated fields to prevent further spread.

Often the most severe effects of the disease can be alleviated by various cultural measures such as generous fertiliser application, drainage and liming where appropriate. Some specific crops can be sown on infected soils provided resistant varieties are used. As a result of legislation (The Wart Disease of Potatoes (Great Britain) order 1973) requiring the use of immune varieties to the UK races of potato wart since the 1920s this particular disease has been virtually eliminated. Chemical control is considerably more difficult, expensive and generally less effective than with air-borne diseases. Often it is only worthwhile with high value arable crops.

SEED-BORNE AND INFLORESCENCE DISEASES

Symptoms are most likely to be seen in the seedling and young plant stages followed later by further major attack at the end of the crop's life on the flowers, seed pods and seeds in general. They may often cause a systemic infection of the plant causing no visible external symptoms during the vegetative phase and then appear quite dramatically on the ears, e.g. loose and covered smuts *(Ustilago spp.)* of cereals, bunt *(Tilletia caries)* of wheat and maize smut *(Ustilago maydis)*. Others cause severe attack on the seedlings and foliage during the early stages of growth and relatively little damage until the heading stage, e.g. various seedling blights of cereals caused by *Fusarium*, *Pyrenophora* and *Septoria* spp. This latter group is often associated with untreated seed such as shed corn as a result of poor stubble cleaning. The diseases develop on the volunteer plants then infect the newly sown autumn crops. This is particularly common with early sown winter cereals.

Seed-borne diseases are potentially very serious indeed and can cause considerable yield reduction. In the UK most of them are controlled by the general use of healthy seed produced in various Seed Certification Schemes and such fungicidal seed dressings as organomercury, benomyl, carboxin, iprodione and thiram. Seed treatments are an inexpensive form of disease control and should always be used where the health of a seed sample is in doubt. Bunt of wheat and the covered smuts of barley and oats are now exceedingly rare as a result of the standard organomercury seed dressings used today. Partly as a result of this effective treatment research into breeding resistant varieties has been limited in recent years. However, strains of seedling blight on oats *(Pyrenophora avenae)* have developed which are resistant to organomercury dressings and more research may be necessary in the future.

VECTOR-BORNE DISEASES

The important diseases in this group are all viruses and the main vectors are either aphids or nematodes. Aphid-borne diseases may show a superficially similar blanket pattern of distribution to air-borne diseases in the field. However not all plants are infected and the appearance and severity of symptoms is more variable. Crop attack and appearance of symptoms are directly dependent upon aphid activity initially and therefore seldom occur much before May and continue until October. Yield reductions are related to the age of the plant at the time of infection. Seedling infection, as often occurs with early sown autumn and late sown spring crops, can cause severe reductions in yield in individual plants whereas late attacks on mature plants may cause negligible loss. The yield loss is therefore directly related to the numbers of individual plants affected and their age at infection. Most true seed is naturally virus-free at planting and will give rise to healthy seed despite being infected during vegetative growth. However, where vegetative 'seed' is used, e.g. potato seed tubers, once infected these will give rise to infected seed. Consequently to maintain virus-free stock for commercial growers, potato seed tubers are produced in areas such as Scotland and Northern Ireland with low aphid populations and sold subject to statutory restrictions (Prevention of Spread of Pests (Seed Potatoes) (Great Britain) order 1974 and Seed Potatoes Regulations 1978) on purity and health. Similar schemes also exist in horticulture for production and sale of virus free stocks of strawberries, raspberries and top fruit that are typically reproduced by vegetative means.

Nematode-borne viruses occur in patches in the field similar to most other soil-borne pathogens. Attack by these free-living types of nematodes and subsequent virus infection is much more common on lighter sandy soils where nematode populations are naturally higher. Many crops and weeds may be attacked by these nematodes but potatoes, sugar beet and raspberries are at greatest risk.

Once infected with viruses, plant yields can be severely reduced and chemical control is not possible. Control should be aimed at preventing or delaying infection for as long as possible by the use of healthy seed, isolation from infected sources of material and then chemical control against the vectors concerned. The control of aphid vectors tends to be generally less expensive than for nematodes but prevention of virus infection is variable and dependent upon virus type.

Lastly it should not be forgotten that plants are often attacked by several organisms simultaneously and virus infection may occur alongside or be confused with fungal diseases and in particular mineral deficiencies.

DISEASES OF WHEAT

Leaf and stem diseases

Powdery mildew (Erysiphe graminis)

Superficial grey-white fungal pustules on leaves, stems and ears, particularly at heading. Pustules darken with age. Attacks wheat, barley, oats, rye and grasses but cross-infection unlikely. Common in May/June (ears) and Oct/Nov (seedlings) and favoured by warm dry conditions. Air-borne disease surviving on stubble, late tillers, volunteers and early sown winter wheat.

Cultural control

(1) Use NIAB Farmers leaflet No. 8 for variety choice and Diversification Schemes.
(2) Destroy stubble and preferably plough soon after harvest.
(3) Avoid early autumn and late spring grown crops.
(4) Avoid close proximity of winter and spring crops and wheat after wheat.

Chemical control

Foliar treatment Use of one of the following fungicides alone or in mixtures at first sign of disease threshold in spring and up to full ear emergence (ZCK 59) especially on NIAB rated susceptible varieties.

ethirimol	prochloraz
fenpropidin	propiconazole
fenpropimorph	triadimefon
flutriafol	triadimenol
nuarimol	tridemorph

Seldom economic to apply fungicides in autumn or more than one well timed application at start of epidemic in spring/summer months especially on NIAB rated resistant varieties. The numbers of applications best minimised to one or two per season to prevent fungicide resistance.

Yellow rust (Puccinia striiformis)

Orange-yellow pustules occurring in stripes on mature leaves or groups on leaves of young plants. Air-borne diseases favoured by cool moist conditions in May/June and wet summers in general. Survives on volunteers, late tillers and early sown winter wheat. Occurs also on barley and rye but cross-infection from barley is unlikely. Very common in Eastern England.

Cultural control

(1) Use NIAB Farmers leaflet No. 8 for variety choice and Diversification Schemes.
(2) Destroy stubble and preferably plough soon after harvest.
(3) Avoid early sowing of winter wheat.
(4) Avoid close proximity of winter and spring crops and wheat after wheat.

Chemical control

Foliar treatment Use of one of the following fungicides alone or in mixtures at first sign of disease threshold in spring and up to full ear emergence (ZCK 59) especially on NIAB rated susceptible varieties

benodanil	triadimefon
fenopropimorph	tridemorph and 'Polyram'
flutriafol	triadimenol
propiconazole	triforine

Brown rust (Puccinia recondita)

Orange-brown pustules randomly scattered or grouped in patches on leaves. Air-borne disease favoured by hot dry conditions in June and July. Seldom severe until after ear emergence, therefore less common or as damaging to yield as yellow rust. Survives on stubble, late tillers, volunteers and early sown winter wheat in mild winters. Also occurs on rye and barley but a different species is involved in barley and, therefore, cross-infection is not possible. Generally of infrequent occurrence.

Cultural control

(1) Use NIAB Farmers leaflet No. 8 for variety choice.
(2) Destroy stubble and preferably plough soon after harvest.
(3) Avoid late sown and late maturing crops.
(4) Avoid close proximity of winter and spring crops and wheat and spring crops and wheat after wheat.

Chemical control

Foliar treatment Use of one of the following fungicides alone or in mixtures at first sign(s) of disease threshold in spring and up to full ear emergence (ZCK 59) 3 especially on NIAB rated susceptible varieties

benodanil	triadimefon
fenpropimorph	tridemorph and 'Polyram'
flutriafol	triforine

Glume blotch (*Leptosphaeria nodorum* syn. *Septoria nodorum*) and leaf spot (*Mycosphaerella graminicola* syn. *Septoria tritici*)

Septoria spp. cause brown, often irregular shaped lesions on leaves and purple brown glume blotch phase on glumes at heading. Difficult to diagnose unless leaves and head are still green. Associated with high rainfall areas, wet summers and high humidity at heading. Rain splash/air-borne disease surviving on infected stubble and seed.

Cutural control

(1) Use NIAB Farmers leaflet No. 8 for variety choice.
(2) Destroy stubble and preferably plough soon after harvest.
(3) Avoid early sown winter and wheat after wheat.

Chemical control

Seed treatments containing the following will give partial control of *Septoria* spp.
  benomyl + thiram   guazatine
  fuberidazole   organomercury

Foliar treatment Most cost effective when applied to protect flag leaf and ear (ZCK 39-59) at first sign of disease threshold especially on NIAB rated susceptible varieties.
  benomyl
  captafol (up to ZCK 49 only)
  carbendazim ($\pm$ maneb. manocozeb)
  chlorothalonil   prochloraz
  flutriafol   propiconazole
  iprodione

The number of applications best minimised to one or two per season to prevent fungicide resistance.

Barley yellow dwarf virus (*vectors – cereal aphids*)

Seedling infection can result in stunted plants of wheat, barley, oats, rye and grasses. Infection in early summer results in purple-red leaves in wheat, bright canary yellow in barley and pink leaves in oats. Spread by two main cereal aphids, bird-cherry *(Rhopalosiphum padi)* and grain aphid *(Macrosiphum avenae)*, during warm dry weather in autumn and May/June. Most common in early sown winter and late sown spring crops. Survives in late tillers, volunteers and grasses.

Cultural control

(1) Winter wheat most tolerant, spring barley most susceptible to yield reductions.
(2) Destroy stubble and preferably plough soon after harvest.
(3) Avoid early sown winter and late sown spring crops.
(4) Avoid close proximity of cereals and grasses.

Chemical control

Foliar treatment

(1) Delay infection to as late in the plant's life as possible by avoiding high aphid populations.
(2) September and May sown cereal crops are high risk for aphid/BYDV and mildew infection. Dual low cost treatment of both may be worthwhile.

Root and stem base diseases

Eyespot (*Pseudocercosporella herpotrichoides*)

Dull brown indefinite eyespot at base of stem eventually causing lodging. Attacks wheat, barley, oats and rye. Common in intensive cereals on heavy damp soils. Soil- and stubble-borne surviving two to four years between susceptible crops. Favoured by long wet cold periods in winter and spring.

Cultural control

(1) Use NIAB Farmers leaflet No. 8 for variety choice.
(2) Deeply plough stubble soon after harvest.
(3) Avoid three or more years of continuous wheat and/or barley especially on heavy land.
(4) Use three year break from wheat/barley for disease reduction.
(5) Avoid early sown lush crops.

Chemical control

Foliar treatment Use of one of the following fungicides alone or in mixtures at late tillering stage (ZCK 30-31).
  benomyl
  carbendazim
  propiconazole
  prochloraz (especially if MBC fungicide resistance suspected)
  thiophanate-methyl
One application only per season required.

Take-all (*Gaeumannomyces graminis*)

Dead, stunted and thin patches of varying sizes in fields of wheat and barley. Black fungus kills roots causing empty bleached 'wheat heads' and premature ripening. 'White heads' often attacked by sooty moulds at harvest. Oats and rye are usually resistant to wheat/barley race of fungus. Disease is favoured by light alkaline soils and above average rainfall in winter and spring. Survives on couch *(Elymus repens)* and stubble for two years. Common in intensive cereals.

Cultural control

(1) Use two year break from wheat and/or barley for eradication.
(2) Prevent build-up by avoiding more than three years of continuous wheat or barley.
(3) Avoid wheat after barley.
(4) Direct drilled crops often less affected than traditional sown crops.
(5) Oats, maize and grasses are effectively resistant in most commercial situations.
(6) No resistant varieties of wheat or barley.

Chemical control

(1) No economical chemical control.
(2) 40 kg/ha N extra top dressing at first sign of disease in spring may alleviate most damage.

Sharp eyespot *(Pellicularia filamentosa syn. Rhizoctonia cerealis)*

Attacks stems and stem bases causing numerous and clearly defined 'eyespot' lesions. Soil-borne disease with wide host range (including barley, oats and rye) and, therefore, not easily controlled by rotation. Most common on light sharp soils of neutral/acid nature during cold dry conditions. More prevalent in rotation with grass leys, peas and root crops. Less prevalent in intensive cereals.

Cultural control

(1) No resistant varieties, oats and rye most susceptible, barley least so.
(2) Avoid late sown winter and early sown spring crops.

Chemical control

No economic chemical control but prochloraz ± carbendazim has given some disease suppression.

Brown foot rot and ear blight *(Fusarium spp.)*

Ill-defined brown rotting of stem base, roots and seedlings especially on cold, heavy, poorly drained acid soils in wet autumns and winters. Also occurs on other cereals.

Cultural control

(1) No varietal resistance known at present.
(2) Avoid late sown winter and early sown spring crops in cold soils.

Chemical control

Seed treatment Use of a seed dressing containing one of the following will give partial control.
 benomyl + thiram
 guazatine organomercury

Foliar treatment Use of one of the following alone or in mixtures may give good disease suppression when applied at late tillering (ZCK 30-31).
 benomyl thiophanate-methyl
 carbendazim
 prochloraz

Ear diseases

Loose smut *(Ustilago nuda)*

Ears only visibly affected. All grains destroyed and replaced by black spores. Conspicuous in early June, at the start of heading. Not serious in UK. Seed-borne disease.

Cultural control

(1) Use of NIAB Farmers leaflet No. 8 for variety choice.
(2) Use clean certified seed.

Chemical control

Seed treatment Use of seed dressing containing one of the following may be necessary for many seed certification schemes.
 benomyl
 carboxin
 triadimenol + fuberidazole

Bunt *(Tilletia caries)*

Ears small and stunted. Internal contents of grain replaced by black fishy-smelling spores but seed coat intact initially. Now very rare in UK. Seed-borne disease.

Cultural control

(1) Use clean certified seed.
(2) Avoid untreated home saved seed.

Chemical control

Seed treatment Use of a seed dressing containing one of the following
 benomyl guazatine
 carboxin organomercury
 triadimenol +
  fuberidazole

Black/sooty moulds (Cladosporium herbarum/Alternaria spp.)

Black sooty appearance on overipe/late harvested crops in wet seasons and/or on prematurely ripened grains as a result of other disease attacks, e.g. Take-all, foot rots. Cosmetic damage only.

Cultural control

(1) Avoid late harvesting of crops where possible.

Chemical control

Foliar treatment Chemicals used for *Septoria* glume blotch control at heading will give incidental control of sooty moulds. Otherwise not economic to spray.

DISEASES OF BARLEY

Leaf and stem diseases

Powdery mildew (Erysiphe graminis)

Superficial grey-white fungal pustules on leaves and stems particularly during stem elongation. Pustules darken with age. Attacks wheat, barley, oats, rye and grasses but cross-infection unlikely. Common in May/June (adult plants) and Oct/Nov (seedlings and shed corn). Favoured by warm dry conditions. Air-borne disease surviving on stubble, late tillers, volunteers and early sown winter barley. Very common and serious on spring barley.

Cultural control

(1) Use NIAB Farmers leaflet No. 8 for variety choice and Diversification Schemes.
(2) Destroy stubble and preferably plough soon after harvest.
(3) Avoid early sown winter and late sown spring barley crops.
(4) Avoid close proximity of winter spring crops and barley after barley.

Chemical control

Seed treatment Use of a seed dressing containing one of the following especially on NIAB rated susceptible varieties.

ethirimol	triadimenol
flutriafol	triforine

Foliar treatment Use of one of the following alone or in mixtures at first sign of disease threshold in spring and up to full ear emergence (ZCK 59) on susceptible varieties.

fenpropidin	pyrazophos
fenpropimorph	triadimefon
flutriafol	triadimenol
nuarimol	tridemorph
prochloraz	triforine
propiconazole	

Seldom economic to apply fungicides in autumn or more than one well-timed application at start of epidemic in spring/summer months especially on NIAB rated resistant varieties. The number of applications also best minimised to prevent fungicide resistance.

Yellow rust (Puccinia striiformis)

Distinct form from that on wheat and cross-infection unlikely. Very similar in all other respects and full account given under 'Wheat'. Control similar to that on wheat with the addition of triadimenol and flutrifol seed dressing mixtures which may control infections on young plants. A later foliar application is also likely to be needed with these particular seed dressings.

Brown rust (Puccinia hordei)

Distinct species from that on wheat and cross-infection is not possible. Very much more common than wheat brown rust and likely to cause greater crop damage. Yields and quality are likely to be reduced. Most common in low rainfall areas of Southern and Central England during July at heading time, especially in hot, dry seasons. Contorl is similar to that of brown rust of wheat with addition of triadimenol and flutriafol seed dressing mixtures which may give control on young plants although these early infections tend to be of irregular occurrence. Seed dressing most useful and economic on late sown spring barley only.

Leaf blotch (Rhynchosporium secalis)

The disease causes blotches with purple-brown borders on the leaves and stems of barley, IRG, PRG, and rye. Blotches initially pale grey-green, often diamond shaped or at base of leaf blade. Air-borne foliar disease often very severe in high rainfall areas, coastal regions and generally widespread in wet summers. Yield and quality reduction can be very severe when weather is cool and wet in May/June particularly. Survives on stubble, late tillers, volunteers and early sown winter barley crops. Also sometimes seed-borne.

Cultural control

(1) Use NIAB Farmers leaflet No. 8 for variety choice.
(2) Destroy stubble and preferably plough soon after harvest.
(3) Avoid early sowings of both winter and spring barley.
(4) Avoid close proximity of winter/spring crops and infected rye grasses.
(5) Avoid barley after barley.

Chemical control

Seed treatment Seed dressings may give control of early infections.

> flutriafol + ethirimol + thiabendazole
> triadimenol + fuberidazole

Foliar treatment At first sign of disease threshold in spring and up to full ear emergence (ZCK 59).

benomyl	prochloraz
captafol (up to ZCK 49 only)	
carbendazim	propiconazole
chlorothalonil	thiophanate-methyl
fenpropimorph	triademifon
flutriafol	tridimenol

Barley yellow dwarf virus (vectors – cereal aphids)

Extremely damaging on yield where infection occurs in seedling stage on early sown winter or late sown spring barley. Later infection at heading time in May/June has minor yield effect. Barley and oats are much less tolerant of attack than wheat. Control measures are similar for those on wheat. Aphicidal control is likely to be more cost-effective on barley particularly when combined with powdery mildew control which often accompanies BYDV infection. Aphicide control after symptoms appear on plants will only reduce further losses occurring and will not affect initial infection damage. A full account is given under 'Wheat'. Use NIAB resistant varieties in high risk situations.

Net blotch (Pyrenophora teres)

The disease causes dark brown net-like blotches on seedlings of volunteers in particular. Symptoms spread to young crops in autumn and attack mature plants in June. Favoured by warm wet conditions in Sept/Oct and May/June. Disease may be seed-borne or from crop debris and volunteeers left in field.

Cultural control

(1) Avoid untreated seed.
(2) Avoid early sown winter barley.
(3) Avoid winter barley after spring barley.
(4) Destroy stubble and preferably plough soon after harvest.

Chemical control

Seed treatment (partial control only)
> carboxin + thiabendazole + imazalil
> flutriafol + ethirimol + thiabendazole
> guazatine + imazalil
> organomercury
> triadimenol + fuberiadazole
> (± imazalil)

Foliar treatment Spray to protect flag leaf in May/June with one of the following

fenpropimorph	propiconazole
flutriafol	pyrazophos
iprodione	triforine + mancozeb
prochloraz	

Leaf stripe (Pyrenophora graminea)

Long brown stripes on leaves usually running entire length. Leaves often split and shred later. Most common on untreated seed in cool wet weather.

Cultural control

(1) Avoid untreated seed.
(2) Avoid early sowings.
(3) Avoid barley after barley.

Chemical control

Seed treatment Easily controlled by most seed dressings.

Halo spot (Selenophoma donacis)

Small angular spots with dark margins and pale centres. Most common on top leaves and lawns in May/June during cool moist conditions. Seldom damaging in UK. Seed- and stubble-borne but may also survive on wheat, rye, cocksfoot and timothy.

Cultural control

(1) Avoid untreated seed.
(2) Avoid barley after barley.

Chemical control

Foliar treatment If symptoms are severe spraying with one of the following may be beneficial

benomyl	thiophanate-methyl
dichlofluanid	zineb
prochloraz	

Root and stem base diseases

Eyespot (Pseudocercosporella herpotrichoides)

Similar but generally less important than that on wheat. Cross-infection is likely and taken into account for crop planning. Chemical control measures are unlikely to be as necessary or as economic compared with disease control on wheat. A full account is given under 'Wheat'.

Take-all whiteheads (Gaeumannomyces graminis)

Similar but generally less important than that on wheat. Cross-infection occurs but barley is more tolerant of attack and there is less time for the disease to spread in the soil with spring sown barley. Spring barley usually shows a lower incidence of attack than winter barley due to the longer intercrop period. A full account is given under 'Wheat'.

Sharp eyespot (Pellicularia filamentosa) syn. Rhizoctonia cerealis

Rarely economic on barley. *See* under 'Wheat'.

Brown foot rot and ear blight (Fusarium spp.)

Seldom recorded on barley probably due to the majority of barley being spring sown and on lighter free draining soils which discourage disease development. *See* under 'Wheat'.

Ear diseases

Loose smut (Ustilago nuda)

This disease is a distinct form from that on wheat and cross-infection does not occur. Otherwise very similar in all other respects and for further information *see* under 'Wheat' plus the addition of seed treatments containing flutriafol and fenfuram.

Covered smut (Ustilago hordei)

Very similar to bunt of wheat, for control *see* under this disease plus the addition of seed treatments containing flutriafol.

Black/sooty moulds (Cladosporium herbarum/Alternaria spp.)

Identical to those on wheat causing cosmetic damage to diseased and/or late harvested crops. Chemicals used for *Septoria* glume blotch control on wheat are likely to be uneconomic for sooty mould control solely. Many are, however, used for control of various leaf diseases of barley also.

DISEASES OF OAT

Leaf and stem diseases

Powdery mildew (Erysiphe graminis)

Grey-white superficial fungal pustules on leaves and stems. Pustules darken with age. Often very severe on late sown spring oats during warm dry weather in May/June and on winter oat seedlings in Oct/Nov. Also occurs on wheat, barley, rye and various grasses but cross-infection unlikely.

Cultural control

(1) Use NIAB Farmers leaflet No. 8 for variety choice.
(2) Avoid early sown winter and late sown spring oat crops.
(3) Destroy stubble and preferably plough soon after harvest.
(4) Avoid close proximity of winter/spring oat crops.
(5) Avoid oats after oats.

Chemical control

Seed treatment
    Triadimenol + fuberidazole

Foliar treatment Use of one of the following at first sign of disease threshold in spring and up to full ear emergence (ZCK 59).
    fenpropimorph   triadimefon
    nuarimol   triadimenol
    propiconazole   tridemorph

Crown rust (Puccinia coronata)

Orange-brown pustules on leaves and sometimes stems. Attack cultivated and wild oat species as well as several important grasses (PRG and IRG). Cross-infection from grasses is unlikely. Air-borne disease favoured by cool wet conditions especially during May/June and Oct/Nov. Survives on stubble, late tillers, volunteers and early sown winter oats. Most severe in high rainfall areas.

Cultural control

(1) Use NIAB Farmers leaflet No. 8 for variety choice.
(2) Avoid early sown winter and late sown spring oat crops.
(3) Destroy stubble and preferably plough soon after harvest.
(4) Avoid close proximity of winter/spring oats.
(5) Avoid oats after oats.

Chemical control

Seed treatment
   triadimenol + fuberidazole

Foliar treatment Use of one of the following at first sign of disease threshold in spring and up to full ear emergence (ZCK 59).
   fenpropimorph
   triademifon
   triadimenol

Leaf spot and seedling blight (Pyrenophora avenae)

Seedlings discoloured and stunted, adult lower leaves with short purple-brown stripes and upper leaves with spots. Seed- and stubble-borne disease, most damage occurs on seedlings during cold wet weather.

Cultural control

(1) Avoid untreated seed.
(2) Destroy stubble and preferably plough soon after harvest.
(3) Avoid very early autumn sowings of winter oats.
(4) Avoid sowing in cold wet soils.
(5) Avoid oats after oats.

Chemical control

Seed treatment Resistant strains to organomercurial seed dressings have become common. Therefore where thought necessary use one of the following seed dressings.
   carboxin + thiabendazole + imazalil
   guazatine + imazalil
   triadimenol + fuberidazole

Foliar treatment Very seldom necessary to use fungicides on adult infections.

Barley yellow dwarf virus (or red leaf)

Similar to virus infection on other cereals. For further details *see* under 'Wheat'.

Oat mosaic virus (fungal vector)

Plants stunted in patches in field. Few tillers formed and leaves have dark green mosaic appearance. Fungus vector on *Polymyxa graminis* present in infected soil causes virus infection. Most common in winter oats during cool wet springs and summers. Infested soils remain so for long periods and thus crop rotations are not very effective. No chemical treatment economic on a farm scale.

Root diseases

Take-all (Gaeumannomyces graminis)

Almost immune to 'Take-all' that affects wheat and barley and often used as a break crop where appropriate. However it is susceptible to a distinct oat strain occurring in several Western and Northern parts of the UK.
   Control measures for oat strain of Take-all are similar to those used for wheat and barley.

Ear diseases

Oat smuts (Ustilago hordei and Ustilago avenae)

Similar to those on wheat and barley. Rarely economic on the oat crop. For more details *see* under 'Wheat'. Most seed treatments are effective including organomercurials which are likely to be the cheapest available.

DISEASES OF RYE/TRITICALE

See under 'Wheat diseases' for the following diseases of rye: powdery mildew, yellow and brown rust, eyespot, sharp eyespot, take-all, brown foot rot, bunt and sooty moulds.
   See under 'Barley diseases' for leaf blotch of rye and barley.

DISEASES OF FORAGE MAIZE AND SWEET CORN

Stalk rot (Gibberella/Fusarium **spp.***)*

Base of stalks rot, plant wilts and often lodge especially in July/Aug. Some strains may also attack wheat. Control includes the use of resistant varieties found in NIAB Farmers Leaflet No. 7 and prevention of lodging by avoiding windy exposed sites and late harvesting. Captan and thiram seed treatments may give partial control.

Take-all (Gaeumannomyces graminis)

Roots attacked mainly but seldom damaging in UK. May act as an alternative host of the disease for following crops of wheat or barley. No recommendations for control of the disease on maize. Further details *see* under 'Wheat'.

Maise smut (Ustilago maydis)

Typically causes black galls on the cob or elsewhere. Spectacular in appearance but of little economic importance at present in the UK. Soil- as well as seed-borne survival. No chemical control measures can be recommended. Growers should avoid maize growing more than one year in four and avoid known infected sites.

Damping-off (Pythium spp.)

Poor emergence and slow early growth especially under cold wet conditions. Control measures include the use of treated seed and avoidance of cold early sowing especially in poorly drained and/or low lying fields. Chemical seed treatment with thiram is recommended.

DISEASES OF POTATO

Leaf and stem diseases

Potato blight (Phytophthora infestans)

Grey-brown leaf and stem lesions spreading rapidly to kill haulm and infect tubers. Very common and serious on main crop potatoes after long periods of wet windy weather in July, August and September. Very common in high rainfall areas and the main potato growing regions of UK. Yield and quality losses with first and second earlies may be slight but the plants may act as a source of infection for nearby maincrops. Infected tubers may cause very high storage losses as a result of secondary bacterial infection.

Cultural control

(1) Use NIAB Farmers leaflet No. 3 for variety choice.
(2) Use clean pre-sprouted seed tubers.
(3) Avoid late sowings.
(4) Avoid close proximity of main crops with earlies or previous potato fields.
(5) Maintain stable and well earthed up ridges to reduce tuber blight.
(6) Destroy groundkeepers and potatoes on dumps and in clamps.
(7) Isolate potato dumps from growing areas.
(8) Defoliate and harvest early.

Chemical control

Foliar treatment
Protectant fungicides
(1) Start spraying maincrops approx. every ten days just before haulm meets in rows or at first sign of disease if earlier.
(2) Vary spraying interval rate and chemical used depending upon incidence of wet weather and age of crop.

Young plants	
Non-systemic	Systemic
captafol	Max five applications per
chlorothalonil	season up till mid August
cufraneb	benalaxyl + mancozeb
mancozeb	cymoxanil + mancozeb
maneb	cyprofuram + mancozeb
'Polyram'	metalaxyl + mancozeb
propineb	ofurace +
zineb	bisdithiocarbanmate
	oxadixyl + mancozeb

Mature plants and tuber blight control	
captafol	fentin acetate + maneb
chlorothalonil	fentin hydroxide

Restrictions on systemic mixtures designed to prevent fungicide resistance. If an acceptable yield has been formed stop spraying and kill haulm where 5% disease level occurs on varieties susceptible to tuber blight. Haulm destruction may be delayed in those varieties showing good tuber resistance. If the crop is required for long-term storage do not lift for at least ten days after the haulm is completely dead.

Potato leaf roll virus (main aphid vector – peach potato aphid – Myzus persicae)

Symptoms vary with variety of potato, strain of virus and time of infection. Plants stunted, leaflets with margins rolled upwards and inwards often with purple tinges. Foliage is hard and leathery; tubers are smaller, fewer in number and yield is much reduced. Mother tuber seldom rots during season and remains hard till harvest. Infection after flowing produces no visible symptoms till after planting next year. Very common in southern England with home saved seed.

Cultural control

(1) Use certified virus free VTSC seed.
(2) Use NIAB Farmers leaflet No. 3 for variety choice.
(3) Isolate new seed from home saved seed.
(4) Isolate potato dumps and clamps from growing fields.
(5) Destroy groundkeepers and potatoes on dumps.
(6) Isolate growing fields from allotments, market gardens and housing areas.
(7) Rogue virus infected plants early in season.

Chemical control (aphid vectors)

Prevent aphid infestations at all times in chitting houses, stores, potato fields and dumps using appropriate aphicides. Spread in field can be drastically reduced by efficient use of one of the following.

aldicarb	oxydemeton-methyl
demeton-s-methyl	phorate
dimethoate	pirimicarb
disulfoton	thiofanox
malathion	thiometon

Potato virus Y (severe mosaic and leaf drop streak – main aphid vector – Myzus persicae)

Symptoms vary with variety of potato, strain of virus and the time of infection. Plants stunted, leaves small, crinkled, showing mosaic, mottling and necrosis. Some varieties show leaf drop streak symptoms with certain strains of Y. Mosaic symptoms usually more severe than Virus X.

Cultural control

As for leaf roll virus.

Chemical control

Prevent aphid infestations at all times in chitting houses, stores, potato fields and clamps. Spread in the field is *not* easily reduced by aphicides due to non-persistant nature of the virus in the aphid.

Potato virus X (mild mosaic)

Typically symptomless or a very mild mosaic. No reduction in size of leaflets and yield losses usually slight but may accentuate effects of other viruses. No vector involved, mechanically transmitted by machinery, footwear, clothing and general physical contact of contaminated with healthy plants. Physical method of spread coupled with general lack of symptoms makes roguing difficult and spread can be rapid in susceptible crops. No visible symptoms in tubers. NB Several other viruses of generally minor importance can also cause mild mosaic symptoms. Best controlled by the use of certified virus free VTSC seed.

Cultural control

(1) Use certified virus free VTSC seed.
(2) Use NIAB Farmers leaflet No. 3 for variety choice.
(3) Isolate new seed stocks from home saved seed.
(4) Destroy groundkeepers.

Chemical control

None

Black leg (Erwinia carotovora var. atroseptica)

Soft black bacterial decay on lower parts of stem and heel end of tuber. Plants may wilt and die and stems are easily pulled out. Storage losses greater than in field. More common under wet conditions in field and store. Mainly seed-borne.

Cultural control

(1) Certified virus-free VTSC seed is generally safer than home saved seed.
(2) Avoid cutting or damaging seed at planting.
(3) Avoid poorly drained fields.
(4) Avoid putting wet tubers in store.
(5) Lift early under clean dry conditions thus reducing tuber damage.

Chemical control

Storage fumigation using a bactericide such as dichlorophen or an organo-iodine complex may reduce infection in seed tubers.

Spraing (tobacco rattle virus – vectors – nematodes)

Leaves show yellow mottle, lines or rings with distorted leaf margins. Locally common on light dry sandy soils with high populations of free living nematodes. Tubers unmarketable due to brown concentric rings or arcs internally. Symptoms not usually visible till tuber cut. Yield reductions negligible compared with loss of tuber quality. Most of the progeny from spraing infected tubers are virus free.

Cultural control

(1) Use NIAB Farmers leaflet No. 3 for variety choice.
(2) Efficient weed control in preceding crops may be beneficial in lowering nematode populations.
(3) Avoid home saved seed in virus prone areas.

Chemical control

Use of one of the following nematicides may be useful on known infected sites.
    aldicarb
    carbofuran
    oxamyl
Also phorate when used for wireworm control.

Spraing (potato mop top virus – vector – powder scab fungus)

Symptoms and control similar to spraing caused by tobacco rattle virus. The powdery scab fungus vector *(Spongospora subteranea)* is most common in the wetter west and northern parts of the UK. Most top spraing is not associated with any particular soil unlike tobacco rattle spraing. Mop top virus is more readily transmitted in infected tubers than tobacco rattle so that they should not be used for seed purposes.

Tuber and storage diseases

Gangrene (Phoma exigua var. foveata)

An increasingly troublesome disease of storage maincrop and seed potatoes especially from northern Scotland and Northern Ireland. Dark coloured round or oval shallow depressions in tubers one to two months after lifting. Soil-borne fungus enters damaged areas at lifting especially under cold late conditions. Often secondary invasion by dry rot and black leg organisms occur also. Large dry hollow cavities usually form in the centre of the tuber in store. Well sprouted but infected tubers usually produce normal plants.

Cultural control

(1) Use of NIAB Farmers leaflet No. 3 for variety choice.
(2) Early lifting of maincrops (i.e. before 10 October).
(3) Avoid damage at lifting/riddling time.
(4) Curing period of 10 d at 13–16° C after lifting and riddling will help wound healing.

Chemical control

(1) Fumigation with 2-aminobutane 14–21 d after lifting controls gangrene and skin spot.
(2) Thiabendazole dips, dusts and ULV mists at harvest or within three weeks after also controls gangrene, silver scurf and skin spot.

Dry rot (Fusarium solani – f.sp. caeruleum)

Soil- and seed-borne disease causing wrinkled concentric rings with pink, white or blue fungal pustules on tubers within one to two months after storage. Eventually tuber dries out, shrinks and mummifies. Bacterial wet rots may also invade under damp storage conditions. Less important than gangrene.

Cultural control

(1) Use of NIAB Farmers leaflet No. 3 for variety choice.
(2) Early lifting of maincrops (i.e. before 10 October).
(3) Avoid damage at lifting and riddling time.
(4) Curing period as for gangrene.

Chemical control

(1) Use of sprout suppressant tecnazene (TCNB) dust at lifting.
(2) Thiabendazole treatments as under gangrene.

Common scab (Streptomyces scabies)

Very common superficial skin blemishing disease especially on light/gravelly alkaline soils low in organic matter. Seed-/soil-borne disease affecting selling quality but not yield.

Cultural control

(1) Use of NIAB Farmers leaflet No. 3 for variety choice.
(2) Avoid liming potato or preceding crop.
(3) Avoid low organic matter soils.
(4) Irrigate crop especially during June.

Chemical control

    quintozene

Powder scab (Spongospora subterranea)

Very uncommon except on wetter soils of the west and north. Most common in wet seasons and low lying areas of field. Long-lived soil and seed-borne disease.

Cultural control

(1) Use NIAB Farmers leaflet No. 3 for variety choice.
(2) Avoid use of known infected sites for at least five years after last potato crop.

Chemical control

No chemical control available.

Silver scurf (Helminthosporium solani)

Very common superficial skin blemishing disease affecting appearance but not yield. Silvery grey lesions develop during storage under high temperature and humidity conditions, causing loss of fresh weight in storage. Mostly seed-borne.

Cultural control

(1) Avoid planting infected tubers.

Chemical control

(1) Treatment of seed tubers before planting with benomyl or thiabendazole.
(2) Post-harvest treatment with thiabendazole or benomyl mists and dusts.

Skin spot (Polyscythalum pustulans)

Superficial skin blemishing disease common and important. Purple spots appear during storage and can affect crop emergence if used for seed. Most common on cold dry or heavy loam soils.

Cultural control

(1) Presprout seed before planting and discard affected tubers.
(2) Avoid late cold wet harvesting.
(3) Store tubers in boxes under dry ventilated conditons.

Chemical control

(1) Treatment of seed tubers before planting as for silver scurf.
(2) Fumigation of seed tuber after lifting with 2-aminobutane.
(3) Post-harvest treatment with thiabendazole within 14 d of lifting.

Pink rot and watery wound rot (Phytophthora and Pythium spp.)

Soil-borne diseases causing wilting in field and pink rot and/or rapid rot in store. Very sporadic occurrence. Locally common in hot summers on heavy badly drained soils. Commonly spreads in store.

Cultural control

(1) Avoid damaging tubers at lifting/riddling time especially under damp soil conditions.
(2) Extend period between potato crops to eight years or more and prevent infection of clean fields.
(3) Improve drainage.

Chemical control

No chemical control available.

Wart disease (Synchytrium endobioticum)

Large external warts and deformities on tubers and stolons. Now very rare due to government legislation and use of field immune varieties. Common in Europe. Soil-borne disease. More common in wet north and west.

Cultural control

(1) Use of NIAB leaflet No. 3 for selection of immune varieties.
(2) Avoid spread on infected implements and dung fed with infected tubers.
(3) Rest field for 30 years. Outbreaks must be reported to MAFF (Notifiable disease).

Chemical control

Not permitted.

Black scurf and stem canker (Thanatephorus cucumeris)

Superficial black scurfy skin blemishes and brown or white girdling of stem bases. Black patches easily removed from skin. Young sprouts destroyed on seed tubers causing delayed emergence. Soft leaf rolling and wilting may occur (cf. virus leaf roll). Control includes the use of well-sprouted healthy seed. Avoid early/deep plantings in cold dry conditions and it is generally less severe if the crop is harvested early.

Chemical control

Apply tolclofos-methyl dust on seed tubers during planting.

DISEASES OF BRASSICAS

Leaf and stem diseases

Powdery mildew (Erysiphe cruciferarum)

Silvery white patches on most brassicas causing eventual defoliation. Very common on swedes, turnips and brussel sprouts in dry summers. Air-borne disease.

Cultural control

(1) Use of NIAB Farmers leaflets Nos. 2, 6 and 9 and Vegetable Growers Leaflets Nos. 1, 2 and 3 for variety choice.
(2) Delayed sowing of spring and summer sown brassicas may be beneficial.

Chemical control

Foliar treament   At first sign of disease use one of the following

benomyl	tridemorph
chlorothalonil	triadimefon
dinocap (Brussel sprouts also)	
fluotrimazole	triadimenol
sulphur	

Canker (Leptosphaeria maculans/Phoma lingam)

Beige leaf spotting in autumn followed by stem cankers and lodging in spring and eventually pod and seed infection. Very important disease of oilseed rape and all brassica seed crops may be affected. Seed-, stubble- and debris-borne disease favoured by wet weather.

Cultural control

(1) Use of NIAB Farmers Leaflets Nos. 2, 6, and 9 and Vegetable Growers Leaflets Nos. 1, 2 and 3 for variety choice.
(2) Isolate crop from other brassica crops and previous oilseed rape fields.
(3) Use treated seed.
(4) Chop/burn and deeply plough oilseed rape stubble soon after harvest.
(5) Use at least four year break between brassica crops.

Chemical control

Seed treatment   Use seed treated with one of the following

benomyl + thiram	iprodone
carbendazim + thiram	thiabendazole

Foliar applications   If infection occurs in autumn or spring on foliage use of one of the following may be beneficial. Maximum of three applications per crop to prevent fungicide resistance.

benomyl	thiabendazole
carbendazim	thiophanate-methyl
chlorothalonil	vinclozolin
prochloraz	

Light leaf spot (Pyrenopeziza brassicae/Cylindrosporium concentricum)

Pale green/bleached areas on leaves and passing onto stems and inflorescences especially during wet weather in May/June. Common on most brassicas especially seed crops. Seed- and debris-borne.

Cultural control

(1) Isolate crops from other brassica crops and previous stubble.
(2) Use treated seed.
(3) Chop/burn and deeply plough stubble soon after harvest.
(4) Use at least four year break between brassica crops.

Chemical control

At 20% of the leaf area affected use of one of the following may be beneficial

benomyl	prochloraz
carbendazim	thiabendazole
chlorothalonil	thiophanate-methyl
iprodione	vinclozolin

Dark leaf spot (Alternaria spp.)

Small dark leaf spots on foliage and seed pods especially during wet weather. Very common on all brassica seed crops and rape and stubble turnips in general. Seed- and debris-borne.

Cultural control

(1) Isolate crop from other brassica crops and previous stubble.
(2) Use treated seed.
(3) Chop/burn and deeply plough stubble soon after harvest.
(4) Avoid early sowing (before 20 August).

Chemical control

Use of thiram seed soak treatment or dry seed treatment with iprodione.

Foliar treatment (*Botrytis* spp. control also)   From early flowering onwards one or two applications to prevent pod infection.

Downy mildew (*Peronospora parasitica*)

Yellowing of lower leaves with white fungal growth on undersurfaces especially on autumn sown seedlings and young plants during winter months, and wet weather. Not usually important on mature plants or during summer period. Air-borne disease.

Cultural control

(1) Isolate crop from other brassica crops and previous stubble.
(2) Chop/burn and deeply plough stubble soon after harvest.
(3) Avoid very thick plant populations and low lying cold wet fields.

Chemical control

Not normally necessary but foliar applications on seedlings and young plants with one of the following may be beneficial at times.
  chlorothalonil
  dichlofluanid
  metalaxyl + mancozeb
  ofurace + bisdithiocarbamate
  zineb

Cabbage ringspot (*Mycosphaerella brassicola*)

Brown/black concentric ringspots on foliage of brassicas especially cauliflowers, brussel sprouts, kale and cabbages in south west England during winter months. Seed- and stubble-borne.

Cultural control

(1) Isolate crop from all other brassica crops and previous stubble.
(2) Chop and deeply plough stubble soon after harvest.

Chemical control

Seed infection is deep-seeded requiring hot water treatment and thiram dressing.

Foliar treatment One of the following at first sign of infection during Nov–Feb
  benomyl   dichlofluanid
  carbendazim   mancozeb
  chlorothalonil   prochloraz

Cauliflower mosaic virus (*aphid vectors – Myzus persicae and Brevicoryne brassicae*)

Mottling and vein clearing of leaves, plants stunted and yields severely reduced. Very common on all brassicas especially cauliflowers where curd production is affected.

Cultural control

(1) Isolate crops from all other brassica crops and previous stubble.

Chemical control

Control aphid populations at all times and delay virus infection to late in plant's life.

Foliar treatment Use of one of the following aphicides at first sign of aphid attack will reduce spread.
  demeton-s-methyl   mevinphos
  dimethoate   thiometon
  malathion

Clubroot (*Plasmodiophora brassicae*)

Stunting of plants as a result of tumour growths on and later rotting of crop roots. All brassica crops may be affected but particularly severe on summer crops though rarely damaging on kale. Swedes and turnips on cold wet acid soils may be very severely affected. Soil-borne disease occurring in patches in fields. Resistant spores may survive for more than eight years between susceptible crops in affected soils. Potentially very serious on oilseed rape crop.

Cultural control

(1) Use of NIAB Farmers Leaflets Nos. 2, 6 and 9 and Vegetable Growers Leaflets Nos. 1, 2 and 3, for variety choice.
(2) Avoid overcropping of brassicas (not more than one year in five as a preventative measure).
(3) Avoid transport of infected soil on boots, etc.
(4) Avoid feeding of infected roots on fields planned for future brassica production.
(5) Liming to at least pH 6.5 will reduce severity of attack.
(6) Improved drainage may reduce severity of attack.
(7) Use a break of at least eight years from brassicas after infection.

Chemical control

Seed treatment Sterilization of seedbed for high value brassica crops with dazomet.

Transplant application Pre-plant dip treatments with one of the following
  calomel
  benomyl
  carbendazim
  thiophanate-methyl

Sclerotinia stemrot (Sclerotinia scelerotiorum)

Bleached areas on the stem from May onwards with hard block sclerotia within cavity of affected stem area resulting in lodging. Sclerotinia affects a wide range of plants – beans, brassicas, beets, carrots, celery, peas, potatoes and many weeds. Sclerotinia contaminate seed, persist in soil for eight years or more and may produce air-borne spores in May to infect through wet petals shed on to leaves and stems. Disease favoured by wet weather during flowering and petal fall.

Cultural control

(1) Avoid over cropping of brassicas and/or other host crops (not more than one year in five as a preventative measure).
(2) Use sclerotinia-free seed and clean equipment.
(3) Chop or burn sclerotinia infected stems and stubble followed by deep ploughing of remaining debris.

Chemical control

An application at first petal fall with one of the following will control *Sclerotinia*, *Botrytis* spp. and *Alternaria* spp. infections.
 iprodione
 prochloraz
 thiophanate-methyl
 vinclozolin

DISEASES OF SUGAR BEET/FODDER BEET/MANGOLD

Virus yellows (aphid vectors – peach-potato – Myzus persicae – black bean aphid – Aphis fabae)

Leaves turn yellow and may be more susceptible to fungal attack. Yield and quality reduced depending upon earliness of infection. Aphids spread the two viruses concerned after feeding on infection plants. Beet yellow virus (BYV) carried only for a few hours in aphids but beet mild yellowing virus (BMYV) carried by aphids for most of its life. Reduce spread in spring to young crops from clamps, seed crops and any over-wintered beet, plants or remnants left in field. Many aphicides now less effective due to aphid resistance in south and east.

Cultural control

(1) Use of NIAB Leaflets Nos. 5 and 6 for variety choice.
(2) Sow early.
(3) Avoid very low plant populations.
(4) Use aphicides and delay virus infection to as late as possible in the season.
(5) Avoid close proximity of root crops to steckling beds and previous crops.
(6) Inspect crops regularly and spray if one plant in four has aphids.

Chemical control (aphid vectors)

Seed-furrow treatment　Using one of the following:
 aldicarb   oxamyl
 carbofuran thiofanox
followed by:
Foliar treatments　of any one of
 acephate   pirimicarb
 dimethoate phorate
 demeton-s-methyl thiofanox
 disulfoton thiometon

Downy mildew (Peronospora farinosa)

Lower leaves yellow with white mycelium underneath especially during winter six months on seed crops and early sown root crops in spring. Root yields and juice quality can be seriously affected in wetter seasons and low lying areas.

Cultural control

(1) Use of NIAB Farmers Leaflets Nos. 5 and 6 for variety choice.
(2) Avoid close proximity of seed and root crops.
(3) Avoid early sowing of root crops on heavy soils.

Chemical control

Foliar treatment　Spray autumn and spring seedling plants with one of the following
 copper oxychloride
 maneb
 zineb

Powdery mildew (Erysiphe spp.)

Powdery white mycelium on foliage especially during dry weather in late summer. Sugar yield may be significantly reduced in south and east. Spray foliage with wettable sulphur or triadimefon at first sign of disease in summer.

Blackleg (Pleospora bjoerlingii)

Important fungus causing damping off and poor seedling growth leading to low plant populations. Seed-borne disease more important in crops drilled to a stand. Use seed treated with EMP (ethylmercury phosphate).

Beet rust (Uromyces betae)

Orange-brown pustules randomly scattered or grouped in patches on leaves. Air-borne disease favoured by hot dry conditions July to Sept. Spraying at first sign of disease with triadimenol will give good disease control.

Leaf spots *(Ramularia beticola, Phoma betae)*

Spray *seed crops only* at first sign of disease(s) during summer months. There should be a maxium of three applications at 14–21 d intervals with fentin hydroxide.

DISEASES OF PEAS AND BEANS (FIELD, BROAD/DWARF, NAVY)

Damping off, foot rot and seed-borne diseases *(Pythium phytophthora, Fusarium, Ascochyta and Colletotrichum spp. but not halo blight)*

Poor germination and early growth.

Cultural control

(1) Use disease free or treated seed.
(2) Avoid sowing into cold wet soils.
(3) Chop/burn and preferably plough stubble soon after harvest.
(4) Avoid poorly drained or low lying fields.
(5) Use appropriate fungicides on growing crops.

Chemical control

Seed treatment

Peas
  drazoxolon   fosetyl-Al + captan
  (± thiabendazole) metalaxyl mixtures
  thiram
Field /broad beans
  benomyl   thiram
  drazoxdon
Dwarf beans
  drazoxolon   thiram

Leaf, stem and pod spots *(Ascochyta pisi – peas, Ascochyta fabae – field and broad beans, Botrytis fabae – chocolate spot of field and broad beans, Collectotrichum lindemuthianum – anthraconose of dwarf/navy beans)*

Various types and sizes of brown/grey coloured spots of leaves, stems and pods. Cause partial defoliation, collapse of stem and discoloration of pods and seeds especially during wet periods in May/June, July. Chocolate spot is most severe on winter beans; seed- and stubble-borne.

Cultural control

As above for seed-borne diseases.

Chemical control

Foliar treatment At first sign of disease build-up in spring spray with benomyl, carbendazim iprodione thiophanate – methyl or vinclozolin rrepeated three weeks later if necesssary. OR a single application at flowering may be sufficient.

Pea wilt *(Fusarium oxysporum)*

Rapid wilting of plants within patches in fields during late May and June. Foliage turns grey then finally goes yellow starting at base of plant then upwards. Very persistent soil-borne disease severely affecting yields in the affected patches. May be seed-borne. Mostly controlled by rotation and resistant varieties.

Cultural control

(1) Use disease free or treated seed.
(2) Use NIAB Pea Leaflet for variety choice.
(3) Use at least a four year break after a *healthy* crop of peas *or* beans. Longer if unhealthy.

Chemical control

No chemical control available.

Downy mildew *(Peronospora viciae) of peas and field/broad beans*

Lower leaves yellow with white mycelium on undersurface especially during wet periods/seasons. If required spray at first sign of disease build-up with metalaxyl + mancozeb or chlorothalonil on field/broad beans.

Halo blight *(dwarf, navy beans) (Pseudomonas phaseolicola)*

Small lesions surrounded by a yellow halo and greasy spots on pods. A bacterial disease favoured by wet windy weather causing reduction of pod quality in the main. Seed- and stubble-borne.

Cultural control

(1) Use disease free seed.
(2) Isolate crops from other dwarf bean crops and previous stubble.
(3) Chop/burn and deeply plough stubble soon after harvest.
(4) Avoid poorly drained or low lying fields.

Chemical control

Foliar treatment If occurrence is likely spray every 10–14 d from emergence to pod set with copper oxychloride.

Grey mould/pod rot of peas/beans (*Botrytris spp.*)

Lesions with grey fungal growth on pods especially those damaged or in contact with soil. Very common and severe in wet seasons causing serious reduction in quality.

Cultural control

(1) Chop/burn and deeply plough stubble soon after harvest.
(2) Avoid poorly drained or low lying soils.

Chemical control

Foliar treatment Spray at flowering with one of the following to protect pods.

benomyl	thiophanate-methyl
carbendazin	vinclozolin
chlorothalonil	

Broad bean rust (*Uromyces fabae*) of field/broad beans

Orange-brown pustules randomly scattered or grouped in patches on leaves. Air-borne foliar disease favoured by hot dry conditions July/August. Spraying at first sign of disease with fenpropimorph will give good disease control.

DISEASES OF CARROTS

Damping-off and leaf blight (*Alternaria dauci*)

Causes damping-off and leaf blight on wet soils and during wet seasons. Seed- and soil-borne.

Cultural control

(1) Use disease-free or treated seed.
(2) Avoid poorly drained or low lying cold fields.
(3) Avoid early sowings.

Chemical control

Seed treatment Thiram seed soak for 24 h.

Violet root rot (*Helicobasidium purpureum*)

Roots covered with purple fungal growth at lifting time especially on cold poorly drained soils. Wide host range including sugar beet, beetroot, parsnips, potatoes and weeds such as docks and dandelions. Brassicas are resistant. Soil- and stubble-borne disease. Yield reduced in store and quality markedly reduced.

Cultural control

(1) Avoid susceptible root crops being grown more than one year in five.
(2) Chop and deeply plough crop debris soon after harvest.
(3) Practise good weed control.

Chemical control

No chemical control available.

Black rot (*Stemphylium radicinum*)

Large black sunken lesions on mature roots and rotting in store. Also considerable seed losses can occur in seed crops and damping-off of seedlings in the field. Seed-borne disease mainly.

Cultural control

(1) Use disease free or treated seed.
(2) Chop and deeply plough crop debris soon after harvest.

Chemical control

Seed treatment Thiram seed soak for 24 h.

Storage treatment Benomyl on roots before storage/prepacking.

Carrot motley dwarf virus (*vector – willow carrot aphid*)

Yellow mottling of leaves and stunting. Also affects parsnips and celery.

Cultural control

(1) Isolate carrot field from other aphid hosts and previous cropped fields.
(2) Large fields are less prone to overall attack than small fields.
(3) Reduce aphid populations at all times.

Chemical control

Seedbed treatment Carbofuran, disulfoton or phorate granules at crop emergence.

Foliar application At first sign of aphids use one of the following

demeton-S-methyl	oxydemeton-methyl
dimethoate	pirimicarb
malathion	thiometon

Storage rots (Sclerotinia sclerotiorum, Botrytis cineria)

Grey and white fungal mycelium on roots in store especially when damaged and stored under damp conditions.

Cultural control

(1) Avoid late lifting under cold wet conditions.
(2) Avoid damaging roots at harvesting.
(3) Provide adequate ventilation to keep roots dry and cool in storage.

Chemical control

Use of benomyl on roots before storage.

DISEASES OF ONIONS AND LEEKS

Downy/mildew (Peronospora destructor)

Pale oval lesions on leaves and die back on tips. Spreads extensively within field under cool wet conditions. Fungus overwinters in bulbs and in soil for many years, causing further infection. Common on autumn sown onions during winter months. Attacks all onions and shallots.

Cultural control

(1) Isolate crops from other onion crops and previous onion stubble.
(2) Avoid using known infected fields for at least five years.
(3) Avoid low lying cold or poorly drained sites.
(4) Practise good weed control to help air circulation within crop.

Chemical control

Foliar treatment Apply zineb or chlorothalonil at first sign of infection and repeat every 14 d to a maximum of six applications.

Smut (Urocystis cepulae)

Leaves bliser and rot to release black powdery fungal spore-masses. Can be very common and serious on seedlings and young plants of onions, leeks, shallots, chives and garlic. Fungus penetrates in seedling stage only. Therefore if raised in disease-free seedbed transplants cannot become infected later. Fungal spores survive in soil for at least ten years.

Cultural control

(1) Avoid contamination soil with infected debris, or soil on machinery, boots, wheels, etc.
(2) Burn all infected plants immediately. Do *not* bury in soil.
(3) Encourage fast germination and early growth. Avoid sowing early in cold wet soils.
(4) Avoid using infected fields for at least ten years.

Chemical control

Seed treatment
(1) For each sowing treat seed with thiram + methyl cellulose sticker for dry powder applications.
(2) For each sowing treat seed furrow with 40% formalin solution while drilling.
Thiram treatment more suitable under high rainfall conditions at sowing and/or lightly infected soils.

White/rot (Sclerotium cepivorum)

Plants yellow, stunted/wilted with rotten base and covered with white fungal growth. Soil-borne fungus surviving in soil for many years.

Cultural control

(1) Avoid contaminating soil with infected debris or soil on machinery, boots, wheels, etc.
(2) Burn all infected plants immediately. Do *not* bury in soil.
(3) Avoid any infected soils for at least eight years.

Chemical control

Seed treatment Use of one of the following will give some control
| benomyl | carbendazim |
| calomel | iprodione |

Foliar application Spray overwintered crops in March, or spring sown 14 d after emergence with iprodione or vinclozolin. Followed by further applications as necessary at three week intervals.

Neck and storage rots (Botrytis allii)

Onions soften and rot internally while in store. Discoloration and rotting of neck occurs after several weeks in store. Very common in December and January. Mostly seed-borne but crop debris and onion dumps can be an important source of infection. Infected seedlings and plants appear healthy in field.

Cultural control

(1) Use disease free or treated seed. Care is needed in using home saved seed.
(2) Deeply bury crop debris soon after harvest.
(3) Completely cover old onion dumps with soil.
(4) Avoid damage at harvest and ensure adequate curing/drying occurs.

Chemical control

Seed treatment   Dry or slurry applications of benomyl + thiram or iprodine.

Foliar application   Several applications of one of the following to the plant foliage at three to four week intervals may be beneficial particularly to machine harvested crops during wet season.

benomyl	iprodione
carbendazim	vinclozolin
chlorothalonil	

Leek rust (Puccinia allii)

Orange-brown pustules randomly scattered or grouped in pustules on leaves. Air-borne foliar disease favoured by hot dry conditions Aug/Sept. Spraying at first sign of disease with fenpropimorph or triadimefon will give good disease control.

DISEASES OF GRASSES AND HERBAGE LEGUMES

Crown rust of rye grasses (Puccinia coronata)

Crown rust causes orange fungal pustules on the leaves of rye grasses, fescues and cultivated oats, but cross-infection is unlikely. Most common on late summer/autumn silage cuts of rye grasses (especially IRG). Associated with hot weather and cool dewy nights coupled with low nitrogen application. Reduces yield, palatability and digestibility of forage. Disease is more severe in a pure stand of rye grass cultivars especially early heading types. Very common in south and west England. Less frequent in north. Air-borne disease surviving on established leys.

Cultural control

(1) Use NIAB Farmers Leaflet No. 16 for variety choice.
(2) Increase defoliation by cutting or grazing more frequently.
(3) Change management to all grazing in late summer through to early winter.
(4) Increase nitrogen levels to 250 kg/ha or more.

Chemical control

No chemical treatment economically feasible at present but the following will give good control at low disease levels.
benodanil
triadimefon
triadimenol

Leaf blotch of rye grasses (Rhyncosporium secalis and Rhyncosporium orthosporum)

Leaf blotch causes dark brown blotches with light centres on barley, cocksfoot, couch, timothy and rye grasses (especially IRG). Cross-infection is possible but generally restricted. Most common under cool moist conditions on spring IRG silage crops. Yield, palatability and digestibility are reduced. Air-borne disease surviving on established plants.

Cultural control

(1) Use NIAB Farmers Leaflet No. 16 for variety choice.
(2) Increase defoliation by cutting or grazing more frequently.
(3) Change management to all grazing in early spring and summer months.

Chemical control

No chemical control economically feasible at present but the following chemicals will give good control at low disease levels.

benomyl	triadimefon
carbendazim	triadimenol
thiophanate-methyl	

Powdery mildew (Erysiphe graminis)

Grey/white superficial fungal growth on leaf surface. Attacks a wide range of grasses/cereals but cross-infection unlikely. Common during and after dry periods. More conspicuous under high soil nitrogen conditions. Reduces yield, palatability and digestibility but generally less damaging than crown rust and leaf blotch. Air-borne disease surviving on established plants.

Control

Use NIAB Farmers Leaflet No. 16 for variety choice. Increase defoliation by cutting or grazing more frequently. No chemical control economically feasible at present but tridemorph, triadimefon, triadimenol will give good control at low disease levels.

Barley yellow dwarf virus

Attacks a wide range of grasses mostly without showing symptoms. Very important disease of cereals and although grasses can be severely affected main importance in rye grasses is as a reservoir of infection for neighbouring cereals. The virus is transmitted by various cereal/grasses aphid species.

Cultural control

(1) Isolate cereal and grass crops from each other, particularly early sown winter barley.
(2) Destroy cereal stubble and volunteer as soon as possible after harvest.
(3) Avoid early sown winter cereals.

Chemical control

Chemical control is unlikely to be economically feasible on grass crops alone. Aphicides may be worthwhile on undersown cereals and grass seed crops.

Rye grass mosaic virus (mite vector – abacus hystix)

Causes mottling and streaking in mild strains and a dark brown leaf necrosis with severe strains. Disease is widespread and severe in the south in seed crops and conservation and grazed leyes (especially IRG). Undersown spring sown crops are more severely affected than straight autumn grass reseeds. Yield, palatability and digestibility can all be seriously affected.

Control

Hard autumn grazing reduces mite populations and the following spring virus infection. Chemical control not economically feasible.

Clover rots (Sclerotinia trifoliorum)

Very serious disease affecting trefoil, lucerne, sainfoin, white and particularly red clover. Soil-borne disease killing large patches of plants in high rainfall areas during autumn and winter months. Often damaging on first year lucerne crops but once established lucerne is generally resistant. Responsible for clover sickness with continuous clover cropping.

Cultural control

(1) Use NIAB Farmers Leaflet No. 4 for variety choice.
(2) Use healthy seed free from fungal sclerotia.
(3) Maintain sward in short condition during autumn but do not poach in winter months.
(4) Use at least five year break between susceptible crops.

Chemical control

Foliar treatment Fungicide treatments using benomyl and quintozene have been partially effective but remain uneconomic.

Verticillium wilt (Verticillium albo-atrum and Verticillium dahliae)

Yellowing and wilting of lucerne followed by poor re-growth, stunting and finally death of large patches in crop. Plants shrivel from the base upwards particularly in late summer and autumn. Very serious disease of lucerne after three to four years cropping on infected soil. Seed- and soil-borne surviving for long periods in soil.

Cultural control

(1) Use NIAB Farmers Leaflet No. 4 for variety choice.
(2) Restrict lucerne crops to maximum of three years duration.
(3) Cut/graze healthy crops first.
(4) Use as long a break as possible on infected sites.

Chemical control

Seed treatment
(1) Thiram seed treatment to prevent disease introduction on new sites.
(2) No foliar treatment possible.

10

Pests of crops

D. J. Iley

Farm crops are subject throughout their growth to attack by pests belonging to various animal groups, the most important being insects, mites, nematodes, slugs, birds and mammals. Different pest species vary in their mobility, host-specificity, period of peak abundance, regularity of occurrence and response to climatic conditions.

The rapid expansion in areas devoted to 'new' crops such as oilseed rape, other oilseed crops and field grown vegetables, and the development of trends in crop production such as the early sowing of winter cereals, have created new pest problems, some of which have been only partially solved.

It is well known that problems such as the development of resistance in pests, the killing of beneficial organisms and the entry of persistent chemicals into food chains have resulted from the use of pesticides. Because of these problems, and the increasing cost of chemical control, a great deal of effort is now being directed towards the development of techniques aimed at improving the effectiveness of pesticide usage so that smaller quantities may be used in a more selective manner. Examples of such techniques are:

(1) establishing economic thresholds so that growers can be advised more precisely on pest population levels at which control is justified;
(2) forecasting the abundance of pests so that growers can be advised whether or not preventive control measures are necessary;
(3) employing biological and ecological information so that other methods of control can be used to replace or enhance chemical control;
(4) employing equipment which is capable of applying pesticides accurately in smaller quantities.

A growing body of knowledge related to these is now available and is incorporated in the text where appropriate.

There may be legal constraints or recommended codes of practice associated with the control of a pest or the use of a pesticide. All the factors mentioned above should be considered when contemplating prevention or control of a pest and additional information should be sought, if necessary, from appropriate sources such as ADAS,

growers' organisations, technical representatives and other professional advisers. The way in which such information is obtained and disseminated is changing rapidly and much advice which was recently freely available through ADAS and other sources may now have to be paid for. Information technology is rapidly being applied to pest control and systems such as Prestel-Farmlink and Agviser point the way to further developments in the near future.

In this chapter the pests are dealt with under 'host-crop' headings but it must be remembered that many of the general feeders are associated with a wide range of crops. It is not possible to include every pest which might injure a crop and those covered have been selected somewhat arbitrarily on the basis of their actual or potential economic importance or on their frequency of occurrence. Chemicals named are given as examples. The omission of an insecticide, nematicide, acaricide, molluscicide or other crop protection chemical does not necessarily imply ineffectiveness.

CEREAL PESTS

Many pests whose normal hosts are grasses infest cereals either directly via mobile winged adults or indirectly when the feeding stages migrate from ploughed or desiccated grass to a following cereal crop. Direct drilled cereals are especially prone to the latter.

Wheat bulb fly (*Delia coarctata*)

A serious pest of winter wheat in eastern Britain; winter barley, triticale, rye and early sown spring wheat or barley may also be attacked. Adult fly lays eggs in cracks and crevices in bare soil in July–September, eggs hatch and larvae invade susceptible cereal plants in following January–March; hatching delayed by cold conditions. Larvae feed inside base of main shoot or tillers and migrate, as they grow, to infest more shoots.

Predisposing circumstances

Host crops drilled in fields which were fallow or sparsely covered in previous July, e.g. roots. Late sown winter wheat, untillered, is especially susceptible. ADAS forecasts, based on egg counts of soil from sample fields, are issued.

Identification

Patches of dead or damaged plants in January–March, expanding in April–May. Damage shows as 'deadhearts', centre leaf dies and turns yellow, outer leaves remain green. The larva, a typical legless fly maggot blunt at posterior end, is revealed on peeling outer leaves away from base of shoot.

Cultural control

Sow winter wheat before end of October at a shallow depth. Sow spring wheat and barley after mid-March.

Chemical control

(1) Seed treatment of chlorfenvinphos, carbophenothion or fonofos. These are unlikely to be beneficial on wheat sown before late October and must not be used after 31st December because of the risk to wild birds. Control less effective on deep drilled seed.
(2) Fonofos granules or spray of chlorfenvinphos, chlorpyrifos or fonofos mixed into top 50 mm of soil at, or immediately before, drilling give better protection but cost more. Use when high risk is forecast.
(3) Sprays of chlorpyrifos, chlorfenvinphos, triazophos or pirimiphos methyl at egg hatch where soil conditions permit; advice on timing from local ADAS.
(4) Where earlier treatments have been omitted, or have failed to stem an attack, spray with dimethoate, formothion or omethoate if damage is observed; not later than mid-March.

Note. Some of the above treatments are approved only for use on winter wheat.

Yellow cereal fly (*Opomyza florum*)

A pest of early sown winter wheat and occasionally winter barley; spring wheat and barley unlikely to suffer. Eggs laid in soil near base of host plants in October and November. Start of egg hatch at any time between late January and March, actual time varying from year to year. Newly hatched larvae climb the shoots, enter them between the outer leaves and burrow downwards to the growing points which they destroy.

Predisposing circumstances

Wheat drilled before mid-October in districts where this species has caused damage previously is especially at risk.

Identification

Untillered plants and tillers developing 'deadhearts' and then dying; distinguish from wheat bulb fly damage by distinct thin brown line circling or spiralling down the central shoot. Larvae small, pointed at both ends, complete their development inside one shoot.

Cultural control

Early sown crops, well advanced at egg hatch, compensate well for tiller damage and chemical control is unlikely to give an economic return.

Chemical control

Probably justified only on slow growing winter wheat crops in high risk situations, which have not already been treated with a synthetic pyrethroid insecticide against BYDV. Sprays, including chlorfenvinphos, fonofos, omethoate, triazophos and various synthetic pyrethroids, should be applied at egg hatch (timing from ADAS) or following ADAS warnings.

Frit fly (*Oscinella frit*)

Grasses, especially ryegrass, are the natural host plants of the frit fly whose larvae feed inside the bases of the shoots. There are three generations, sometimes four in southern counties, of flies in a year and larvae overwinter in the shoots of their host plants. Cereals may be attacked in two different ways.

Migration from sward

Predisposing circumstances

Autumn sown wheat, barley, oats and rye drilled over ploughed grass or grassy stubbles, or directly drilled into desiccated swards may be invaded by larvae migrating from the decomposing grass.

Identification

Slight angular bend above coleoptile where larva has penetrated at single shoot stage of growth, followed by 'deadheart', i.e. central leaf dies and turns yellow. Legless, transparent larvae 2–5 mm long, inside base of shoot. Young plants usually killed.

Cultural control

Leave an interval of 4–6 weeks between ploughing/desiccating and drilling. Late sown crops suffer less because lower soil temperatures reduce migration.

Chemical control

Spray with chlorpyrifos, fonofos, omethoate, pirimiphos methyl or triazophos

(1) at drilling in high risk situations on farms where damage has occurred frequently.
(2) at crop emergence following local ADAS warnings.
(3) when centre leaves of 10% of plants at one-leaf to two-leaf stage detach easily on being gently pulled, and show a small brown feeding scar at the base.

To be effective the treatment should be given before obvious damage is seen.

Direct oviposition

Female flies lay eggs directly on spring oat and maize plants, the eggs hatch after a few days and the larvae enter the shoots.

Predisposing circumstances

Later sowings of spring oats are at greater risk of attack by first generation flies because the latter are attracted only to plants or tillers with fewer than five leaves. Maize is often at risk because of its late sowing date but attacks are unpredictable.

Identification

First generation attack results in 'deadhearts' and death of young oat and maize plants or excessive tillering of older plants. Second generation flies lay eggs in spring oat spikelets, the larvae feeding producing blindness or shrivelled and damaged grain.

Cultural control

Sow spring oats before mid-March in south, late March in Wales and north Britain.

Chemical control

For spring oats, spray with chlorpyrifos if crop has not reached the four-leaf stage by first half of May in south, late May in Wales and north. Or apply triazophos immediately damage is observed. Economic yield response likely to be small unless attack is severe. For maize, granules of phorate or carbofuran in the seed row as a precaution. Or spray with chlorpyrifos, fenitrothion, pirimiphos methyl or triazophos at crop emergence.

Wireworms (*Agriotes* spp.)

The larvae of click-beetles; natural habitat is the soil under permanent grass where they feed on the underground parts of plants.

Predisposing circumstances

Only likely to be a problem in the first three years after permanent grass. All cereals are affected but wheat and oats are most susceptible.

Identification

Main damage occurs in autumn and spring when wireworms feed on underground stems and hypocotyls of cereals. Seedlings may be completely severed and turn yellow and die while still remaining upright in the ground. Several plants in a row may show progressive symptoms. Damage often occurs in patches where other factors are contributing to poor growth, e.g. disease, poor drainage. Careful excavation of soil around injured plants reveals stiff, smooth, yellow larvae up to 20 mm long.

Cultural control

Increase seed rate in high risk situations. Assist recovery by rolling and nitrogen top dressing if soil conditions are suitable.

Chemical control

Seed treatments based on gamma HCH are cheap and are recommended when wireworm population is low. To reduce the risk of plant injury the seed should be sown soon after treatment, moisture content should not exceed 16% and the dressing should be applied uniformly at the prescribed rate. For higher populations apply gamma HCH* as a spray to the soil when the seedbed is being prepared. Fonofos treatment for wheat bulb fly control will also provide some degree of protection against wireworms.

Leatherjackets (*Tipula* spp.)

The larvae of the daddy longlegs or craneflies. The flies lay eggs in grassland mainly in September. The eggs hatch in 10–14 d and the leatherjackets feed in the soil on the underground parts of plants but will come to the surface in dull, moist and relatively warm conditions to feed at ground level. Activity is reduced in cold and dry conditions.

Predisposing circumstances

Cereals following grass or grassy stubbles are at risk especially when preceding September–October has been wet. All cereals suffer; winter cereals may be attacked in early winter but the greatest damage normally occurs on spring cereals in April and May. ADAS monitors populations in most regions and forecasts are issued or are available on request. Rooks are attracted to cultivated fields with large numbers of leatherjackets and may be seen searching for them by turning over surface clods.

Identification

Young plants may be chewed below or at ground level and spring cereals in particular may be completely severed or roughly grazed down. Leaves with ragged holes. Damage often in patches coinciding with wet areas of fields. Confirm by examining soil in vicinity of affected plants for presence of legless, grey-brown, fleshy grubs up to 50 mm long, with tough, wrinkled skins and a number of small, pointed protuberances at tail end.

Cultural control

Plough or desiccate grassland before September; kill grassy stubbles immediately after harvest. Rolling and top dressing with nitrogen when soil conditions permit will encourage the crop to grow away from an attack.

Chemical control

Examine crop daily for signs of attack in risk situations because damage can occur very rapidly and insecticide must be applied immediately to avoid excessive loss. Chemical control is justified when a *total* of 15 or more leatherjackets are found on examining ten separate 30 cm lengths of row selected at random diagonally across the field. The figure applies when the rows are 17.5 cm apart, at narrower spacings a total of ten or more is the critical level. The soil within 3–4 cm of the plants should be carefully searched.

Insecticides may be applied in these circumstances as sprays, granules or poison baits, the latter normally giving best results. In all cases the application is best made when the leatherjackets are active on the surface, i.e. in the evening in damp, warm weather.

---

* Gamma HCH should not be used in this way if potatoes or carrots are to be planted within 18 months because of risk of taint.

Sprays Chlorpyrifos, gamma HCH*, quinalphos and triazophos. These may also be applied during the preparation of the seedbed when leatherjacket populations are known to be high.

Granules Apply chlorpyrifos using an applicator capable of distributing small quantities/ha.

Poison baits Mix fenitrothion or gamma HCH* with bran moistened to a crumbly consistency. Thorough mixing to ensure even distribution is essential, and the bait should then be distributed as evenly as possible over the field. Bran plus gamma HCH pellets are available commercially.

Field slug (*Deroceras reticulatum*)

This is the most important of several species of slugs which may feed on cereals, and is the only one which regularly feeds above or close to the soil surface. Slug numbers vary considerably according to soil type, previous cropping and climate; their activity is also affected by temperature and humidity but, in general, their numbers are greatest in autumn and late spring.

Predisposing circumstances

Autumn sown cereals, especially wheat, are most likely to suffer economic loss; damage may be seen on spring cereals but this is not normally important. Populations are highest in undisturbed soil, especially if it has good moisture retaining properties and where there is dense ground cover. Direct drilled cereals or cereals following grass, oilseed rape, peas, beans and other crops with bulky residues on silt or clay soils are most at risk.

Identification

Winter wheat, rye and barley may be severely damaged by slugs feeding on the germ of the seed shortly after drilling, giving the impression of total or partial seed failure. Young shoots of all cereals may be eaten as they germinate, or grazed at ground level and their growing points destroyed. The leaves of older plants may be shredded longitudinally but this damage, though conspicuous, is not normally of great importance.

Cultural control

Prepare a fine, well consolidated seed bed which restricts movement of slugs through soil.

Chemical control

Treat with metaldehyde or methiocarb baits when seedbed conditions and cropping history indicate a high risk. Test baiting is no longer recommended for forecasting slug damage but is useful in indicating when slugs are active on the surface and therefore amenable to chemical control. Best results are obtained with baits applied to the surface before drilling, preferably following rain, and left undisturbed for 3–4 d. Failing this, post-drilling applications or mixing and drilling slug pellets with the seed may be carried out. A spray of copper sulphate + aluminium sulphate + borax

---

* Gamma HCH should not be used in this way if potatoes or carrots are to be planted within 18 months.

('Nobble') applied just prior to or just after drilling may also be used.

Cereal cyst nematode (*Heterodera avenae*)

Mainly a problem on spring oats. Winter oats, maize, wheat, barley and rye are progressively more resistant to damage. Present in small numbers in grassland which increase rapidly when cereals are grown frequently. Small, less than 1 mm long, lemon-shaped cysts, the bodies of dead females, containing approximately 400 eggs when young may persist in the soil for many years their contents declining slowly in the absence of suitable hosts. Minute larvae emerge from the cysts in spring to response to substances exuding from growing cereal roots and invade and feed inside the rootlets. Females swell and form young cysts in July which may be seen attached to roots of affected plants, white at first and darkening as they age.

Predisposing circumstances

Mainly confined to chalky and light soils in southern England. Oats following several successive crops of wheat and/or barley are most at risk but other cereals may suffer when populations are high. Suspect fields can be sampled to determine population levels.

Identification

Damage normally shows up as patches of pale, stunted plants whose roots are short and much branched in comparison with those of healthy plants.

Cultural control

Reduce high populations by sowing a ley. Avoid growing oats after several successive crops of wheat or barley. Resistant varieties may be grown where cyst counts are high and are useful in reducing nematode populations; the spring barleys Tyra and Tintern, the winter oat Panema and the spring oat Trafalgar all exhibit resistance to the two principal pathotypes (races) of cereal cyst nematode.

Chemical control

Chemical control is not economically justified at present.

Cereal aphids

Three species are important; like all aphids they feed by sucking plant sap causing direct damage to their host plants, they may also transmit cereal virus diseases. Correct identification, and careful observation of population trends wherever possible, is important because different species may require different treatment. Special care must be taken when deciding whether or not control measures are justified because spraying is costly in both economic and ecological terms on a crop which occupies such large areas of land.

Grain aphid (*Sitobion avenae*)

Predisposing circumstances

Mainly a problem on winter wheat causing direct damage in summer, but may occur on all cereals. Winged females migrate from grasses in late May–early June in southern Britain, later in the north. Numbers build up in hot, humid conditions. Losses greatest when heavy infestations occur when flowering heads are developing. Also transmits BYDV (*see* Chapter 9) from grasses to cereals and then within the cereal crop, especially in the north.

Identification

Largish aphids, 2–3 mm long, colour varying between reddish brown and green, two black tubes (siphunculi) projecting from upper surface at rear of body. Early arrivals feed on leaves but transfer to flowering heads as latter emerge and feed on rachilla and developing grain; resulting grain is light and shrivelled.

Cultural control

Direct damage in summer – none recommended. BYDV transmission – see under Bird cherry aphid.

Chemical control

Direct damage in summer – spray winter wheat when, at the beginning of flowering (GS 61), an average of five or more aphids per ear is found on at least 50 ears selected at random from all parts of the field except the headlands, and when weather is conducive to aphid build up. Chlorpyrifos, demeton-S-methyl, dimethoate, heptenophos, oxydemeton-methyl, phosalone, pirimicarb and thiometon are all effective. Local beekeepers should be warned of impending spraying and, where there is a risk to bees, pirimicarb should be the insecticide of choice.

BYDV transmission – see under Bird cherry aphid.

Rose grain aphid (*Metapolophium dirhodum*)

Predisposing circumstances

Winged females disperse from brambles and wild roses to cereals and grasses late May–early June. Rapid build up in warm, humid conditions.

Identification

Light green aphids, 2.25–3.0 mm long, with darker green stripe down middle of back. Normally on undersurfaces of lower leaves but will colonise upper leaves and surfaces and stems as numbers build up. Only likely to cause economic loss when flag leaves are heavily infested.

Cultural control

No cultural control recommended.

Chemical control

In winter wheat and spring barley spray when average number of aphids per flag leaf exceeds 30 at any time between flowering and milky ripe stage (GS 75). Insecticides and precautions as for grain aphid.

Bird cherry aphid (*Rhopalosiphum padi*)

Mainly important as a transmitter of BYDV (*see* Chapter 9) from grasses to cereals and then within the cereal crop. Overwinters naturally as eggs on bird cherry but, in areas with a mild winter climate, can also overwinter as small colonies of wingless adults and juveniles on grasses and early

sown winter cereals; may then build up and spread rapidly in the spring.

Predisposing circumstances

Main risk of virus transmission is in southern England and Wales on early sown, September–early October, winter cereals. Late sown spring cereals may also become infected by aphids migrating in May. Cereals following grass, in close proximity to grassland, and in fields with high hedges are more likely to suffer.

Identification

Small, 1.5–2.3 mm long, brown to greenish-brown aphids with rust red patches at rear of body; found on all parts of the plants. Cereals infected in the autumn do not normally show BYDV symptoms until the following spring when patches of affected plants can be seen. Infected plants in spring sown crops normally become conspicuous at ear emergence.

Cultural control

Careful ploughing under of grass before drilling cereals.

Chemical control

(1) BYDV may be introduced into winter cereals up to mid-October by winged aphids migrating from infected grasses, the risk increasing in proportion to the earliness of drilling, the severity of previous infections and the percentage of migrating aphids which are carrying the virus (Infectivity Index). General recommendations on spraying are summarised in *Table 10.1*.

Sprays include alphamethrin, cypermethrin, cyfluthrin, demeton-S-methyl, deltamethrin, fenvalerate and permethrin.

(2) BYDV may also be transmitted by aphids which survive when cereals follow grass. This is best prevented by killing the sward with paraquat 7–10 d prior to cultivation and 14 d before sowing.

Table 10.1 Spraying to control BYVD in winter cereals

Sowing date *BYDV*	*Before mid-* *September*	*Mid–late* *September*	*Early–mid* *October*
Severe in past	A	C	D
Little previous history	B	D	D

A, spray in mid-October
B, spray in mid-October only if aphids present in crop or following ADAS warnings
C, spray in late October or early November
D, spray only if aphids present in crop or following ADAS warnings

POTATO PESTS

Pests of potatoes fall into two main categories – those that reduce the productivity of the plants by direct feeding damage or by transmitting virus diseases, and those that disfigure the tubers thereby reducing their market value. In both cases the most effective control measures are normally applied before or at the time of planting.

both cases the most effective control measures are normally applied beofre or at the time of planting.

Peach potato aphid (*Myzus persicae*)

Overwinters as eggs on peach trees and as small colonies of adult and juvenile females on a wide range of host plants especially in mild winters and in protected situations such as glass houses and chitting houses. Winged females arrive on potato plants in spring and feed by sucking the sap, colonies of wingless aphids build up rapidly in hot dry conditions, movement within the crop is responsible for transmission of virus from infected plants. Mainly important as a vector of leaf roll virus and rugose mosaic (virus Y) into and within home grown 'seed' crops and in the chitting house. Early migration from overwintered colonies and rapid build up in early summer may create a virus problem in main crop potatoes in some years.

Predisposing circumstances

More likely to be a problem following a mild winter and early spring and when calm, warm weather prevails in summer.

Identification

Small scattered colonies of green to pinkish aphids, relatively inactive even when disturbed.

Cultural control

Destroy volunteers, discards, clamp site debris and other potential sources of potato viruses. Plant certified healthy 'seed'. Grow 'seed' crops in isolated sites and rogue virus infected plants as soon as symptoms appear. Burn off foliage early.

Chemical control

Routine control is recommended only for home grown 'seed' production and is more effective against leaf roll virus which, in contrast to virus Y, is not transmitted instantly by aphids. Granular applications of aldicarb, disulfoton, phorate and thiofanox to the seedbed or planting furrow will generally protect until mid-June, thereafter one or more sprays should be applied until haulm destruction. Alternatively, spray at 80% plant emergence using demeton-S-methyl, deltamethrin + heptenophos, dimethoate, oxydemeton-methyl, pirimicarb or thiometon and repeat at 14 d intervals.

On ware potatoes, direct damage may be caused by this and other aphids including the potato aphid (*Macrosiphum euphorbiae*) a larger species which may form dense colonies in hot, dry summers. Spraying is justified only when an average of 3–5 aphids per true leaf from a sample of equal numbers of upper, middle and lower leaves is found in July. Do not spray if aphid predators (ladybird and hoverfly larvae) and parasites (indicated by swollen, brown, 'mummified' aphids) are present. In chitting houses and seed stores pirimicarb smokes used as required from the end of November, will kill aphids on the sprouts.

Chemical control is complicated by the increasing prevalence of strains of aphids which are resistant in varying degrees to some of the insecticides listed above. This pattern of resistance is constantly changing and growers are advised to obtain local advice on the best choice of insecticide.

Potato cyst nematodes (*Globodera rostochiensis* and *G. pallida*)

The most important pests of potatoes in the UK. *G. rostochiensis*, which forms yellow cysts on the roots is present in all potato growing districts; *G. pallida*, with white cysts, is more restricted in its known distribution but is probably present at low population densities even in areas where it is not considered to be a threat. Life cycles are similar to that of the cereal cyst nematode except that potatoes are the only important field hosts; tomatoes are also susceptible.

Predisposing circumstances

Yields are reduced when potatoes are grown in soils containing viable cysts, and cyst numbers increase rapidly when potatoes are grown frequently in infected fields. The number of eggs/g of soil can be estimated by advisers and the information should then be used to plan a cropping and control programme in consultation with specialist advisers.

Identification

Infestations normally show up first as patches of stunted plants which wilt readily. Their root systems are shortened and much branched and small spherical cysts may be seen attached to the rootlets.

Cultural control

The main objective of control is to contain the cyst/egg population to an acceptable level and is best achieved by combining lengthened rotations, resistant varieties and nematicide treatment. EEC regulations require that potatoes sold as 'seed' can only be grown in fields which prove to be eelworm free following a statutory test. Where initial egg counts are high potatoes should not be grown for several years; thereafter the rotation can be shortened. Destroy volunteer plants.

The effectiveness of resistant cultivars is lessened by the presence in the UK of one pathotype of *G. rostochiensis* (RO1) and four of *G. pallida* (PA1, PA2, PA3 and PA4). RO1 predominates in field populations in east and south England, Scotland and Ireland. PA3 predominates in the Humber basin and the Channel Islands. In other regions these two main pathotypes are randomly intermingled and field infestations of PA1, PA2 and PA4 are also present. The first early varieties Aminca, Pentland Javelin and Ukama, and the maincrop varieties Cara, Kingston and Maris Piper are resistant to pathotype RO1 only. They tolerate invasion by the nematodes and alter the sex ratio so that few new cysts are formed. It is unfortunate that where RO1 resistant varieties are grown frequently, populations of *G. pallida* will almost certainly increase. The new varieties Cromwell and Sante, combining RO1 resistance with partial resistance to *G. pallida*, may lead the way to more effective control.

Chemical control

Incorporate granules of aldicarb, carbofuran or oxamyl in the seedbed. Rates of application may vary according to soil type and whether an early or maincrop variety is to be grown. Effectiveness depends on thorough mixing with the soil to the correct depth. Autumn application of dazomet, metham-sodium or dichloropropene may also be used, special applicators may be required.

Free living nematodes (*Trichodorus* and *Paratrichodorus* spp.)

Minute, less than 1.00 mm long, soil inhabiting nematodes. Migrate through the soil and feed externally, sucking the contents of surface cells of tubers and roots.

Predisposing circumstances

Associated with coarse, sandy soils.

Identification

Loss of yield negligible. Important as vectors of tobacco rattle virus (TRV) causing spraing which renders tubers unmarketable (*see* under potato disease, Chapter 9).

Cultural control

The first early variety Aminca and the main-crop Record, used only for crisping, are resistant to spraing.

Chemical control

Aldicarb or oxamyl granules incorporated into the seedbed. Phorate granules used primarily against wireworms also bring about some control.

Garden slug (*Arion hortensis*)

The principal slug species affecting potatoes, it lives almost entirely underground only coming to the surface in numbers during rainy weather in July and August. The keeled slug, *Milax budapestensis*, may also contribute to tuber damage.

Predisposing circumstances

Numerous in wet, heavy soils, especially those with high organic matter content. Damage likely to be greatest during a wet autumn following a mild, wet summer; also in irrigated crops.

Identification

Slugs penetrate tubers through relatively small, rounded entrance holes and excavate wider tunnels and cavities thereby reducing the marketability of the crop. The garden slug is black with a yellow sole. Millipedes may be present as secondary feeders.

Cultural control

Improve drainage. Harvest as soon as possible. Maris Piper and Cara are highly susceptible and should not be grown in slug risk situations. Pentland Dell and Pentland Ivory are among the least susceptible varieties.

Chemical control

Surface baits containing methiocarb can be useful when applied in wet conditions in late July and August when the slugs come to the surface. A number of marked plants of Maris Piper within the crop can be examined regularly to provide an early warning of an attack.

Wireworms (*Agriotes* spp.)

See wireworms in cereals. Relatively low populations can cause serious economic loss.

Predisposing circumstances

Numbers are likely to be high in the two years following permanent grass but damaging populations, above 75 000/ha, may persist even in arable land. Numbers are estimated by soil sampling. Wireworm activity increases in autumn. Unlikely to be a problem on early lifted varieties.

Identification

Small round holes in the tuber leading into tunnels of the same diameter, wireworms may be found on cutting tubers open. Entry of slugs, millipedes and other soil pests is encouraged. Loss in yield is negligible but crop may be rejected or market value greatly reduced.

Cultural control

Lift maincrops as soon as possible after maturation.

Chemical control

Work aldrin dust or spray into the soil prior to planting when population levels merit chemical control; higher rates are needed on peaty soils. Alternatively, apply phorate granules in the furrow at planting, this will also reduce spraing and early aphid attack. On no account should gamma HCH be used.

Cutworms

The caterpillars of a number of related species of nocturnal moths, which damage a wide range of crops. The most important species in recent years has been the turnip moth (*Agrotis segetum*). Eggs are laid on foliage and stems of many crop and weed plants from May–July. The small caterpillars hatch and feed on the leaves for two to three weeks, during which time they are highly susceptible to wet conditions, and then descend into the soil feeding just below ground level although they may come to the surface at night.

Predisposing circumstances

Prevalent when weather has been warm and dry during early feeding stages, especially on light, well drained soils and weedy fields. Irrigated crops are unlikely to suffer.

Identification

The fleshy, smooth, greenish-brown caterpillars, up to 5 cm long when fully grown, excavate irregular shallow pits in the surface of the tubers, especially those close to the surface.

Cultural control

Control weeds.

Chemical control

Spray with chlorpyrifos, triazophos or cypermethrin when the newly hatched caterpillars are feeding on foliage prior to descending into the soil.

SUGAR BEET PESTS

The seedling is the most vulnerable growth stage of any plant, a fact of particular relevance to the sugar beet crop which is now almost entirely sown to a stand with pelleted monogerm seed. This results in a low seedling population and soil pests feeding on seed or seedlings can damage or destroy a high proportion of the plants very quickly even though they themselves are present in relatively small numbers. Effective weed control increases the problem in the case of general feeders. It is essential, therefore, that protection should be applied in advance against predictable soil pests and that the crop should be examined every day from sowing until it is well established so that control measures can be applied in time to be effective.

Seed furrow treatment with granular insecticides, extensively used against aphids and free-living nematodes, will also control or reduce the injurious effects of most soil pests. In general, they are preferable on ecological grounds to overall seedbed or post-emergent treatments against soil pests which may kill important predators and parasites of aphids.

Sugar beet pests also affect fodder beet and mangels but the cost of control on these crops is only justified in exceptional circumstances.

Soil pest complex

This includes millipedes, symphylids and springtails and it is convenient to deal with them as a group. Main damage is to the roots and underground stems of seedlings which may result in death or permanent stunting of growth. Tolerance to attack normally develops between the two and four true leaf stage of development.

Millipedes

Body composed of many similar segments, examination with a hand lens reveals the presence of two pairs of legs per segment. The spotted snake millipede (*Blaniulus guttulatus*), cylindrical in section with a line of reddish-brown spots along each side, is the most injurious but flat millipedes including *Polydesmus angustus* may also contribute to the damage.

Seedling roots and stems tunnelled or rasped causing collapse of plants; large numbers may congregate around individual seedlings in May, extending the damage caused by other soil pests. Damage often patchy.

Symphylids (*Scutigerella immaculata*)

White, up to 15 mm long, 12 pairs of legs when fully grown. Become active in May and June, may be seen running rapidly over surface of disturbed soil.

Springtails (*Onychiurus spp.*)

Minute, up to 2 mm long, white insects with three pairs of legs. Active in the soil at lower temperatures than other soil pests.

Predisposing circumstances

Damaging populations likely only in soils with a good structure containing many minute fissures and macropores which the pests can move through to congregate rapidly around seedlings. Silts and chalky soils generally most heavily infested. Damage by symphylids is normally confined to crops in silty soils, whereas millipedes and springtails are present in a wider range of soil types. Sandy soils carry the lowest populations. Field populations show enormous variation even between similar neighbouring fields and predictions of damage have to be on a field basis.

Cultural control

None.

Chemical control

At present, there are no accurate guidelines as to when control is justified and decisions on preventive treatment are based on soil type and especially on the previous history of damage in a field. To this end, it is desirable that small portions of a field should be left untreated to help in assessing the effect and value of any control measure. Investigations have shown that an assessment of springtail population, which can be made at lower temperatures prior to drilling, will give an indication of the overall risk of attack; the use of baited traps for this purpose is under development. The main objective of any control measure is to bring about a uniform stand of between 75 000 and 100 000 plants/ha. Current recommendations are:

Seed furrow applications of granular formulations of aldicarb + HCH, bendiocarb, benfuracarb, carbofuran or carbosulfan;
or
Spray soil with gamma HCH and work into the seedbed (not on some peat soils) – but see introductory remarks.

Woodmouse (*Apodemus sylvaticus*)

The mice live in burrows in the open field hence their alternative name of long-tailed field mice. They dig up and break open the husks and feed on the ungerminated seed, working their way along rows. Each individual has a large foraging area. Populations vary enormously from year to year during the relatively short time that seed is at risk.

Predisposing circumstances

Damage is greatest on early sowings in dry soils and where some seed is exposed on the surface. BS field staff monitor populations during the spring.

Identification

Soil disturbed along rows. Seed pellets fragmented.

Cultural control

Avoid spillage and exposure of seed.

Chemical control

Place proprietary mouse baits or breakback traps at a density of three to four/ha in the open fields away from hedgerows as soon as damage is observed. Baits and traps should be placed in pipes or other suitable containers to protect non-target mammals and birds. Traps must be reset and baits replenished until the seed germinates.

Pygmy beetle (*Atomaria linearis*)

Small, 2 mm long beetles which disperse, by flying when temperatures exceed 15°C and wind speeds are below 7 km/h, during mid-April–June from previous year's beet fields into fields carrying the current year's crop where they feed on the seedlings. Populations build up during the year but little harm is caused to plants beyond the seedling stage. Adults overwinter in soil and under clods and plant debris on the surface.

Predisposing circumstances

Numerous in intensive beet growing areas.

Identification

Blackened feeding pits in the hypocotyl below, or just above the soil surface which may bring about complete severance and death. At a slightly later stage beetles feed in the young heart leaves which become very distorted and misshapen as they unfurl and grow.

Cultural control

None.

Chemical control

Methiocarb, normally present in the seed pellet, gives some protection against early attack. In areas where damage occurs regularly control as for soil pests, thiofanox used for aphid control is also effective.

Wireworms (*Agriotes spp.*)

See under cereal pests, p. 290. On sugar beet, relatively low populations can lead to poor establishment.

Predisposing circumstances

Most likely in the second or third year after permanent grass, but damaging populations may persist longer. Damage more serious on early drilled crops.

Identification

Hypocotyl chewed below ground level, seedlings die.

Cultural control

None.

Chemical control

Seed furrow treatments of aldicarb, bendiocarb, carbosulfan, carbofuran and thiofanox (but not oxamyl) used against other pests give reasonable control. Otherwise, work gamma HCH* into the seedbed, using higher rates on soils with a high organic content.

Beet flea beetle (*Chaetocnema concinna*)

Small, 2.5–3.5 mm long, shiny bronze beetles active in dry sunny weather. Adults overwinter in hedgerows and other sheltered situations close to the fields in which they have been feeding. Emerge on first warm days of spring and disperse to fields of seedling beet to feed on cotyledons and leaves. Seedlings beyond the cotyledon stage are normally able to withstand an attack.

Predisposing circumstances

Outbreaks sporadic, more likely in dry areas with abundant shelter and in cold dry springs when beetles are active but seedling growth is slow.

Identification

Small circular pits, later developing into holes on the cotyledons – 'shothole' damage. In bad attacks the cotyledons are completely consumed and the seedlings killed. Beetles may be seen on the cotyledons, jumping when disturbed.

Cultural control

Early sown crops are more likely to be past the susceptible stage at the time of attack.

Chemical control

Spray rows with carbaryl or gamma HCH* only when absolutely necessary, because of danger to predators and parasites of aphids.

Leatherjackets (*Tipula ssp.*)

See under cereal pests, p. 290.

Predisposing circumstances

A problem in wetter, western beet growing areas and on fields with a high water table; also when beet is grown after grass or weedy stubbles. Much lower populations are injurious to beet than to cereals.

Identification

Leatherjackets feed in spring, cutting off the seedlings at or just below ground level. Grubs in soil near damaged plants.

Cultural control

Plough grassland before September.

Chemical control

Apply gamma HCH as a spray or bait* as soon as damage is seen. If slugs are also likely to be a problem use methiocarb as a surface bait.

---

* Do not plant potatoes or carrots within 18 months following this treatment.

Cutworms

See under potato pests, p. 294. Sporadic attacks by a number of species occur on sugar beet. In the fens the caterpillars of the garden dart moth (*Euxoa nigricans*) cause occasional damage to beet seedlings, grazing and severing them at ground level between April and mid-June. Control, when damage is observed, by band sprays of triazophos applied in large volumes of water in the late afternoon in moist conditions when cutworms are more likely to come to the surface. Effectiveness is increased if the insecticide is then hoed into the surface soil alongside the seedlings. Other species may feed later in the season on the tap roots but control is not normally justified.

Skylark (*Alauda arvensis*)

Birds of the open fields, skylarks have emerged as seedling pests since the rapid increase in sowing to a stand. They feed on the cotyledons and leaves of the plants throughout the spring and early summer and the defoliation can cause death or reduction in vigour. Other birds, notably partridges, may cause similar damage. Control is difficult and uncertain but aldicarb, used as a seed furrow treatment against other pests, may give some protection. Commercially available hawk-shaped silhouettes suspended from 25 ft poles could be tried.

Peach potato aphid (*Myzus persicae*)

See also under potato pests, p. 294. Sugar beet is one of the numerous host plants of this species which is of major importance as a vector of beet yellow virus (BYV) and beet mild yellowing virus (BMYV) – *see* section on sugar beet diseases, p. 282.

Viruses are not transmitted via the true seed of sugar beet, so every plant starts life free from infection but may become infected by aphids migrating into the crop after having fed on other infected sources, further spread then occurs within the crop as aphid populations increase and disperse.

Predisposing circumstances

Early migration, and therefore, early virus infection follows a mild winter. Crops in which a lot of bare soil is exposed are more attractive to aphids. The Virus Yellow Warning Scheme provides early warning of the likelihood of infection in time for preventive seed furrow treatments to be carried out, and updates the information as the season progresses. The highest risk is normally in the south of the beet growing area.

Identification

Green aphids may be seen on the plants.

Cultural control

Sow early and encourage rapid early growth. Sow virus resistant varieties in recognised 'yellows' areas. Break the cycle of virus transmission by eliminating overwintering sources of virus; plough in plant residues on previous years beet fields, destroy debris on cleaner-loader and mangel and fodder beet clamp sites before aphid migration gets under way in April. The seed crop may be an important overwintering 'bridge' for virus and aphids, stecklings should be grown in isolation and/or under cover crops.

Chemical control

The development of resistance in peach potato aphid populations to many widely used insecticides, especially in areas where virus yellows is a recurrent problem, is a major threat to the beet industry. It is, therefore, very important that insecticides are used only when absolutely necessary in order to prolong their active field life.

Preventive seed furrow treatments, specifically directed against virus transmission, should only be used in high risk 'Yellows' areas, and following receipt of early 'Yellows' warnings. Materials include aldicarb and thiofanox granules which should protect up to the six to eight leaf stage. Carbofuran, carbosulfan and benfuracarb, used against soil pests, provide shorter lived protection.

Sprays containing demeton-S-methyl, pirimicarb, delta-methrin + heptenophos, or dimethoate may be used on crops which are otherwise unprotected following later warnings of aphid migration or when an average of one wingless green aphid per four plants is observed. Sprays may have to be repeated in years when aphid migrations persists, but not after the plants have reached the 15–20 leaf stage of growth.

Black bean aphid (*Aphis fabae*)

This aphid can seriously reduce the yield and quality of beet crops as a result of direct feeding damage. It can transmit BYV but is relatively unimportant in this respect. Overwintering occurs almost entirely as eggs on spindle trees (*Euonymus europaeus*) from which winged females migrate to summer hosts, including beet, in May and June on which they do not normally become numerous until July. Dense colonies of wingless females grow rapidly in hot weather and more plants become infested as the season progresses. Numbers decline in September as a result of emigration and build up of predators and parasites.

Predisposing circumstances

Injurious in hot dry seasons especially on late sown crops and when plant growth is retarded by drought. Loss in yield occurs when average number of aphids per leaf exceeds two.

Identification

Conspicuous black aphids often co-existing with green peach potato aphids. Damage most important on heart leaves which become distorted and brown.

Cultural control

Early sowing.

Chemical control

Normally controlled by sprays used against peach potato aphid. Where specific action is required high volume applications (450 litres/ha) give better coverage of the larger plants. Pirimicarb, because of its selective action enables aphid predators and parasites to persist and 'mop up' surviving aphids.

Beet cyst nematode (*Heterodera schachtii*)

The life cycle is similar to that of the potato and cereal cyst nematode, dealt with under their respective headings. There are, however, some special features which are significant when considering control measures. The development time from the emergence of the larvae from the cysts to the formation of the next generation of cysts is short and up to three generations may be completed in a single growing season, resulting in a much greater increase in cyst population when a host crop is grown. This species also has a wide range of hosts including all beets, various weeds and most brassicas including oilseed rape; the latter, although efficient hosts producing many cysts, are relatively unaffected by the invading larvae.

The Beet Cyst Nematode Order of 1977 gives MAFF the power to prevent host crops being grown on infected land, but these powers have not yet been used. The clause governing length of rotation in British Sugar contracts was removed in 1983 and since then growers have been able to follow rotations with no formal constraints. In recent years, rotations in the main beet areas have tended to become shorter and surveys have shown that numbers of infected fields and levels of infection have risen, especially in the Fens. Significant yield reductions may be expected if this trend continues.

Predisposing circumstances

Injurious populations are more likely to occur in light sandy or peaty soils. Frequent cropping with sugar beet, fodder beet, red beet, spinach, mangel and brassicas (except radish and fodder radish) will lead to a rapid increase of cyst populations in fields which are infected.

Identification

Causes 'beet sickness' in fields where no rotational control is carried out. Patches of stunted, unhealthy plants whose outer leaves wilt readily in dry conditions, and turn yellow and die prematurely. Tap roots poorly developed with numerous fibrous side roots. Lemon shaped cysts visible on roots.

Cultural control

The only reliable control is to adopt long rotations covering all host crops, combined with weed control and the prompt destruction of all host crop residues. Length of rotation is influenced by the likelihood of damage and the soil type and growers should carefully examine roots from suspect crops for the presence of cysts, or have the soil sampled to ensure early detection of infestations. The following guidelines are suggested:

Field category	Maximum host crop frequency*
(1) No known infection	1 year in 3
(2) Infected, but no heavy yield loss	1 year in 4
(3) Yields affected	Rest 4 years then follow a long rotation

* 1 year longer on mineral or high organic soils.

Chemical control

Too costly and unreliable at present. Soil fumigation may be justified on some soils where potato cyst nematodes are also a problem.

Free living nematodes (Docking disorder) (*Trichodorus* and *Longidorus* spp.)

Similar to the species which attack potatoes.

Predisposing circumstances

A problem on light, sandy soils only, particularly when plants are under stress from other causes.

Identification

Patchy damage in crop, associated with sandier areas. Individual healthy plants often conspicuous among surrounding stunted plants. Attacked roots distorted, fangy or with lateral extensions.

Cultural control

Provide good growing conditions.

Chemical control

Seed furrow applications of aldicarb, carbofuran, carbosulfan or oxamyl in high risk situations and where there is a history of infection.

OILSEED RAPE PESTS

This relatively new crop has raised a number of new pest problems and as the area and intensity of cropping increases it is probable that more will arise. Pests which affect the developing pods, flowers and the flowering stems are of particular importance and in this connection it is necessary to distinguish between winter and spring sown crops because their susceptible growth stages occur at different times and, since the pests normally have relatively fixed times of maximum abundance, they are subject to a different range of pests.

Bees are strongly attracted to rape flowers and, although they make little impact on seed set in this self-pollinating crop, sprays should not be applied to crops in flower. This is not normally necessary, but if it should be, e.g. when flowering is uneven, the sprays should be applied in the early morning or the evening using pesticides which are *relatively* safe to bees. Local beekeepers should be given adequate warning of any intention to spray.

Spray booms should be set high to promote even distribution when advanced crops are being treated.

Routine insecticide treatments are not recommended for any pest; populations should be carefully monitored and sprays applied only when threshold levels, if these have been established, are reached.

Field slug (*Deroceras reticulatum*)

See also under cereal pests, p. 291.

Predisposing circumstances

A problem mainly in the autumn and early winter on winter rape directly drilled into cereal stubbles on wet soils and in cool conditions which retard germination and plant growth.

Identification

Leaves shredded, seedlings grazed.

Cultural control

None.

Chemical control

Drill slug pellets containing metaldehyde or methiocarb with the seed *or* apply pellets as soon as damage is seen.

Brassica flea beetles (*Phyllotreta* spp.)

Small, 2.5 mm long, dark, shiny beetles often with a yellow stripe running down each wing case. Life-cycle and injurious effects similar to beet flea beetle. Mainly a problem in spring sowings. Most seed is dressed with gamma HCH as a protection but it may be necessary to band spray the seedling rows with gamma HCH in the face of sustained attacks when seedling growth is slow. Potatoes and carrots should not be grown for 18 months after such treatment.

Cabbage stem flea beetle (*Psylliodes chrysocephala*)

A pest of winter rape at present mainly in parts of the East Midlands, East Anglia and south eastern England, but is widely distributed and is a potential pest in other areas. Adults migrate in August or September from fields of the current year's crops of rape and other brassicas into fields where autumn sown seedlings are emerging. The adults cause shothole damage on the seedlings but this is normally of little importance. Females lay eggs in the soil in October–November, these hatch at temperatures above 5°C. The larvae burrow into the stems of the plants via the petioles, mainly between October and January, and accumulate there as hatching continues. When feeding is completed they emerge to pupate in the soil in April and May giving rise to adults in May and June.

Predisposing circumstances

The presence of large numbers of the small, 4 mm, dark metallic blue beetles with thickened hind legs in the current year's crop and/or shothole damage in autumn seedlings indicate that damage is likely to occur during the winter until early spring. Crops growing next to previously infected crops are especially at risk.

Identification

Attacked plants collapse as their stems are hollowed. Small, white grubs with dark heads and six small legs may be seen on opening affected stems. Distinguish from cabbage leaf miner larvae, found widely in petioles, which is a typical fly maggot with no legs and no obvious head.

Cultural control

None.

Chemical control

Only when, in autumn or early winter, an average of more than five larvae per plant are found in the petioles on a random sample of at least 20 plants. Later treatment of larger plants is not likely to give an economic response. In general, the last time for treatment is in November or early December, or following official warnings. Treatments include granules of carbofuran or fonofos or sprays of cyfluthrin, cypermethrin, deltamethrin, fenvalerate, gamma HCH, permethrin, pirimiphos-methyl or alphamethrin. Synthetic pyrethroid sprays may also be applied earlier if large numbers of adults are seen; the crop should still be monitored later for larval damage.

Rape winter stem weevil (*Ceutorhynchus picitarsis*)

A pest of winter oilseed rape; a rare species until 1981 since when localised outbreaks leading to economic damage have occurred in eastern England south of the Humber. Adults assemble in oilseed rape fields late September–October. Eggs laid, usually in batches of up to 12 in punctures and small cracks at base of leaf stalks. These hatch during the winter and the larvae burrow into the plant and move towards the crown, which may be hollowed out; and leave to pupate in the soil in March. Young plants may be killed outright or die later in the winter; surviving plants develop secondary shoots and are stunted with a rosette-like habit of growth leading to uneven ripening and reduced yield.

Predisposing circumstances

No obvious pattern as yet; attacks likely to become more frequent in intensive rape growing areas.

Identification

Adults 3–4 mm long, black with minute yellow scales. Typical weevil snout, long and curved. Larvae creamy white, up to 6 mm long, distinct brown head, no legs, body curved and wrinkled.

Cultural control

None.

Chemical control

No thresholds established. Control as for cabbage stem flea beetle.

Pigeon (*Columba palumbus*)

Winter crops are attractive to flocks of pigeons which can cause extensive damage as a result of their grazing activities especially on late sown crops of small plants with a high proportion of bare soil. Control by sowing early and encouraging rapid germination and growth. Dalapon and TCA applied to control grass weeds and volunteer cereals render the crop unpalatable for a few weeks. The deterrent effect of scaring devices is usually short lived and their use is best restricted to the period mid-January to mid-March when most of the serious damage is caused. A combination of several devices, moved around if necessary, is likely to give the best results. These include ground based mechanical scarecrows and the inflatable man, and aerial devices such as hawk-shaped silhouettes suspended from poles, and kites and balloons. The latter must not be used within 5 km of an airfield and must not be flown more than 60 m above ground level.

Cabbage stem weevil (*Ceutorhynchus quadridens*)

Affects spring crops. Adults migrate from fields of previous year's crops, feed on seedling leaves and lay eggs. Larvae hatch and tunnel in petioles and stems in May and June causing parts of the plants above where they are feeding to wilt and collapse. Larvae white with darker heads and legless. Control is not normally directed specifically against this pest; the sprays applied to control pollen beetles and seed weevils in spring will control it.

Cabbage aphid (*Brevicoryne brassicae*)

Occasionally a problem, especially on spring crops in hot, dry seasons when dense colonies of the mealy grey aphids form on the flowering stems. Specific control is necessary only when an appreciable proportion of the stems is colonised, care must be taken to avoid killing bees if insecticides are applied. Pirimicarb is the least harmful to bees and natural enemies of aphids.

Pollen beetles (*Meligethes* spp.)

Also knows as blossom beetles. A problem mainly on spring crops but may affect backward winter crops. The adult beetles hibernate in sheltered situations such as hedges and copses and fly to rape crops, assembling in increasing numbers during hot, sunny periods in April and May. Eggs are laid in the early green flower buds and the larvae, which hatch after a few days, feed on and destroy the reproductive parts of the flowers. Feeding by adults and larvae continues as the flowers open. Large numbers of beetles cause a significant reduction in the number of pods formed.

Predisposing circumstances

Common in most districts especially where there is an abundance of winter shelter. Most damage occurs when peak numbers coincide with the green bud stage of development.

Identification

Look for small 1.5 mm long, oval, shiny black beetles with conspicuous knobs on the ends of their antennae. Fly readily in summer weather when disturbed.

Cultural control

None.

Chemical control

Because of the damage to bees and other beneficial insects, and the potential waste of money, it is essential that sprays are applied only before the late yellow bud stage and only when pest population levels exceed the threshold value. It is well worth while to make careful counts at the appropriate stage of crop growth. This is done by selecting 20 plants at random throughout the field, avoiding headlands, on a sunny day when beetles are on upper parts of plants; move carefully to avoid disturbance. Carefully bend and shake the flower heads over a tray or stiff card approximately 30 × 25 cm and count the beetles dislodged.

(1) On spring rape this must be done at the early green bud stage and treatments given if the *average* population exceeds three per plant. Counting and treatment may have to be repeated in the event of re-infestation but only up to the yellow bud stage.
(2) On winter rape, count during green to yellow bud stage and treat when average per plant reaches 15–20.

Sprays include azinphos methyl + demeton-S-methyl sulphone, endosulfan, gamma HCH* and malathion.

Cabbage seed weevil (*Ceutorhynchus assimilis*)

A major pest in both winter and spring crops. The adult weevils overwinter in hedgerows and other sheltered situations and migrate in spring to rape crops. They fly actively in hot, calm weather and rapidly colonise the fields. Some time is spent feeding on the leaves and other parts of the plants but this causes little or no loss. Eggs are laid only in the soft young green pods shortly after flowering. Normally only one egg is laid per pod and the damage is caused by the larvae feeding on and destroying the young seeds. The larvae emerge when fully fed and pupate in the soil, giving rise to a second generation of adults which move to other brassica crops in the area in July–August.

This pest is often associated with the brassica pod midge which exploits the holes made in the pods by the weevil as egg-laying sites.

Predisposing circumstances

Highest populations develop in areas of intensive oilseed rape production where the crop has been grown for many years. High levels of parasitism occur, especially in unsprayed crops and many larvae do not complete development.

Identification

Small, 2–3 mm long, grey beetles with long thin snouts are seen on the plants, flying actively in sunny weather, often in association with pollen beetles. White, legless grubs with brown heads in the seed pods, normally only one per pod.

Cultural control

None.

---

* Do not plant potatoes or carrots until at least 18 months after application.

Chemical control

Because of the danger to bees and other beneficial insects sprays must never be applied when crop is in flower.

(1) On spring crop maximum populations usually assemble on the plants before flowering and the same sprays used against blossom beetle, plus phosalone, should be applied at the late yellow bud stage when an average of one or more adult weevils is found per plant. Use the same counting technique as for pollen beetles.
(2) On winter crops, which are usually more advanced, peak populations often coincide with flowering. In this case it is necessary to make the counts when the crop is in flower and then wait until after petal fall, when the crop presents an overall green appearance, before spraying with phosalone or triazophos if the threshold was exceeded at any time during flowering. Phosalone should be used if late flowers are persisting in the crop. Local beekeepers should be warned at least 48 h before spraying and the application should be made preferably in the evening after bees have ceased flying actively.

Brassica pod midge (*Dasyneura brassicae*)

Sometimes known as the bladder pod midge because of the swellings characteristic of pods which may contain up to forty minute white maggots feeding on and weakening the inner walls of the pods. Similar, but pink, maggots are found sometimes but these are fungus feeders and can be disregarded. Affected pods ripen and shed seed prematurely. The midge is associated with the seed weevil because the adults, which are fragile and weak, make use of the weevil's oviposition punctures and exit holes to penetrate the pod for egg laying. Control of seed weevil usually protects the crop against the midge but when exceptionally large populations occur a specific treatment using phosalone or triazophos, on winter rape only, may be required. In many instances, because of the weak flight of the midge, only the headlands need treatment.

On spring rape the midge is controlled by sprays directed against seed weevils pre-flowering.

KALE PESTS

The main pests of kale, grown as animal fodder, are those which affect the establishment of the seedlings.

Brassica flea beetles (*Phyllotreta* spp.)

Life cycle and injurious effects similar to beet flea beetles.

Predisposing circumstances

Mainly a problem in hot, dry conditions in April and May on spring sowings when the seedlings are in the cotyledon stage and growth is slow. Plants are much more tolerant of injury after the first true leaves develop.

Identification

'Shothole' damage to cotyledons and stems, leading to complete loss in severe attacks. Small, 2–3 mm, beetles some with a yellow stripe down each wing case, jumping actively when disturbed.

Cultural control

Sow early and encourage rapid early growth by providing a fine, well fertilised seedbed.

Chemical control

Dress seed with gamma HCH, usually combined with a fungicide. This may have to be supplemented by a spray or dust of gamma HCH in the face of sustained attacks. Potatoes and carrots should not be planted within 18 months of this treatment.

Cabbage stem flea beetle (*Psylliodes chrysocephala*)

See under oilseed rape pests, p. 299. Mainly a problem on late sown overwintering kale crops. Early sown, well grown crops are more tolerant of injury. Chemical control as on oilseed rape.

PEA AND BEAN PESTS

Apart from the general pests which disrupt the normal growth of the crops, growers of peas in particular have to contend with a number of pests which feed within the pods and affect the appearance of the peas. A relatively small percentage of damaged peas can reduce the market value of the crop and may even lead to rejection by the processor.

Pea and bean weevil (*Sitona lineatus*)

Hibernating adults become active from mid-April onwards and move into pea and bean fields, where they feed on the leaves. In May and June they lay eggs in the soil and the larvae burrow and feed on the root nodules.

Predisposing circumstances

Leaf injury is only likely to be important when the leading shoots of young plants are heavily attacked at a time when plant growth is slow. Extensive leaf-notching, however, indicates that more serious nodule damage is likely to occur later.

Identification

Characteristic U-shaped notches in the leaf edges or on the folds of unopened leaves.

Cultural control

Encourage rapid early growth.

Chemical control

(1) *On peas* Foliar sprays containing azinphos methyl + demeton-S-methyl sulphone, carbaryl, cypermethrin, fenitrothion, fenvalerate, permethrin or triazophos, when significant leaf damage is seen are more likely to lead to an economic yield response. Aldicarb, oxamyl or

phorate granules applied pre-drilling or at sowing against other pests will give some control.

(2) *On beans* Spring sown crops are most at risk and treatment with foliar sprays containing cypermethrin, fenvalerate, permethrin or triazophos when significant leaf notching occurs are more likely to be economic.

Pea cyst nematode (*Heterodera goettingiana*)

Life cycle similar to cyst nematodes of cereals, potatoes and sugar beet but with peas, beans and vetches as hosts. Peas are seriously damaged but beans, although efficient hosts, are little affected by larval invasion. Important in parts of eastern England where peas are grown intensively, occurs locally in other regions.

Predisposing circumstances

Cyst numbers build up rapidly in fields cropped too frequently with peas and beans.

Identification

Patches of stunted, prematurely senile pea plants are normally the first indication of trouble, lemon shaped cysts can be seen on the short, fibrous roots which bear few root nodules.

Cultural control

Lengthen the interval between host crops. Cysts are persistent and expert advice on the length of the interval should be sought.

Chemical control

Reduce damage by mixing oxamyl granules in the seedbed before drilling. This must be considered to be a supplementary treatment and is not an acceptable alternative to lengthening the rotation.

Pea aphid (*Acyrthosiphum pisum*)

Winged females migrate from a wide range of leguminous plants into pea crops during the summer months. These found colonies of wingless aphids and further spread occurs within the crop. Seldom a problem on beans.

Predisposing circumstances

Colonies build up rapidly in hot, settled weather reaching a maximum in June–July.

Identification

Colonies of large, long-legged green aphids on the growing points and spreading to the pods when numerous. Growing points yellow, leaves and pods distorted. Plants may be covered with sticky honeydew.

Cultural control

None.

Chemical control

Foliar sprays applied at high volume to assist penetration into the crop on peas grown for human consumption when colonies are easily seen and numbers are increasing. On peas for animal consumption spray when 50% of plants are infected between flowering and four trusses set. The choice of aphicide may be limited if the crop is to be harvested shortly after spraying. Active ingredients include: azinphos methyl + demeton-S-methyl sulphone, cypermethrin*, deltamethrin*, demeton-S-methyl, dimethoate, fenitrothion, fenvalerate*, heptenophos*, malathion, mevinphos*, nicotine*, oxydemeton-methyl, pirimicarb* and thiometon.

Pea moth (*Cydia nigricana*)

The moths emerge from the previous year's pea fields and congregate in the current year's crop, the flight period lasting from early June to mid-August with a peak round about mid-July. Eggs are laid on the leaves and the minute caterpillars crawl to and penetrate the soft young pods and feed inside for three weeks. They emerge after this and overwinter in the soil.

Predisposing circumstances

Common in all pea growing areas and particularly prevalent in East Anglia and Kent. Especially important on peas for processing because even a light infestation may result in rejection. All peas which flower in the main flight period are at risk.

Identification

Peas burrowed into and fouled by the excrement of the small, pale caterpillars with black heads.

Cultural control

Early picking peas which flower before the main flight period escape serious attack.

Chemical control

Commercially available pheromone traps set up in individual fields should be used in all fresh pea and vining crops to indicate the arrival of the adult moths. The crop must be sprayed if any moths are caught up to full flowering, treatment may have to be repeated after 10 d. In peas to be combined for human consumption, pheromone traps may also be used to determine the need to spray and the timing of the treatment. Peas grown for animal feed need no treatment. The following insecticides can be used: azinphos methyl + demeton-S-methyl sulphone, carbaryl, cypermethrin, deltamethrin, fenitrothion, fenvalerate, permethrin and triazophos.

Pea midge (*Contarinia pisi*)

Mainly a pest of vining peas. Midges migrate from previous year's pea fields and lay eggs near the developing flower

---

* Short minimum harvest interval, *see* manufacturers' labels.

heads; the small pale, jumping maggots feed in the developing buds, on the leaves and sometimes inside the pods. Attacks localised and sporadic.

Predisposing circumstances

Rain stimulates emergence from previous year's pea fields. Small, fragile, yellowish flies concealed within the terminal clusters may be found by careful examination when the midges assemble about a week before flowering.

Identification

Flower heads sterile; terminal parts of flowering stems shortened and leaves distorted to form 'nettle-heads'.

Cultural control

None.

Chemical control

In vining peas, in areas where there is a regular occurrence of damage, crops should be sprayed when the oldest green buds are about 6 mm long, about 7 d before flowering. In slow growing crops a second application may be required 6–10 d later. In other crops the leading shoots should be examined as they become susceptible and sprayed if adult midges are found in about 15% of the plants. ADAS or PGRO may issue spray warnings. Suitable insecticides are: azinphos methyl + demeton-S-methyl sulphone, carbaryl, demeton-S-methyl, dimethoate, fenitrothion, and triazophos.

Black bean aphid (*Aphis fabae*)

See also under sugar beet pests, p. 297. Mainly a problem on spring sown field beans.

Predisposing circumstances

Serious infestations more likely on late sown spring crops and in southern England. Colonies build up rapidly in hot, dry weather.

Identification

Dense colonies of black aphids on stems and growing points. Pods distorted, poorly filled, contaminated by honeydew and moulds. Bean virus diseases transmitted.

Cultural control

Sow early and encourage rapid early growth.

Chemical control

Beans are bee pollinated and aphicides should be applied just before flowering when forecasts issued by ADAS in conjunction with Imperial College suggest that this is necessary. In areas where the risk is lower, treatment may be restricted to the headlands.

Foliar applications of granules of disulfoton or phorate, or sprays of demeton-S-methyl, dimethoate, malathion, mevinphos, nicotine, oxydemeton-methyl, pirimicarb or thiometon are recommended. Granular formulations and pirimicarb sprays are the least harmful to bees and should be used if it becomes absolutely essential to treat field beans which are being worked by bees, if this is the case the application should be made in the evening.

FIELD VEGETABLE PESTS

Uniformity of appearance and freedom from blemish are the main requirements of vegetables grown for human consumption and very low pest populations, which have a negligible effect on yield, may therefore be a cause of considerable financial loss. Standards of control must be exceptionally high and often involve routine and intensive use of insecticides. This has created problems of increasing resistance to many of the chemicals which have been successfully used in the past. The choice of insecticide may be further limited by the short interval between final treatment and harvest, and other factors such as soil type.

Anyone growing such crops is recommended to take full advantage of forecasts, spray warnings and advice provided by the local ADAS and commercial organisations. The principal pests of the main vegetable crops are:

(1) brassicas – flea beetles, cabbage root fly, cabbage caterpillars (various species), cabbage aphid;
(2) broad and french beans – black bean aphid, bean seed flies;
(3) carrots – carrot fly, willow-carrot aphid;
(4) onions – onion fly, stem and bulb nematode.

SCHEMES, REGULATIONS AND ACTS RELATING TO THE USE OF INSECTICIDES

Virtually all the insecticides in use are potentially hazardous, some more than others, to users, consumers, wildlife and beneficial organisms and there are many schemes and legal requirements governing their purchase, use, application, storage and disposal. The principal schemes and Acts are as follows:

(1) The Food and Environment Protection Act 1985 – and the Control of Pesticides Regulations 1986 (made under the Act).
(2) The Control of Pollution Act 1974.
(3) The Health and Safety at Work Act 1974 – and the Poisonous Substances in Agriculture Regulations 1984 (made under the Act).
(4) The Poisons Act 1972 and the related Poisons Rules.
(5) Wildlife and Countryside Act 1981.

11

Grain preservation and storage

P. H. Bomford

Grain is stored after harvest so that it can be marketed, or used, in an orderly manner. Prices are generally at their lowest at harvest time, and can be expected to rise with length of storage, until the next harvest approaches. If livestock are to be fed with home-produced grain, careful storage can ensure a year-round supply of high quality feed.

A successful storage system will preserve those qualities of the product which are important for its proposed end-use. These qualities may include a high germination percentage, good baking or malting properties, ease of extraction of constituents such as starch, oils or sugars, freedom from impurities, discoloration or taint, the nutritive value of the grain to livestock, or the conditions of moisture content and temperature necessary to meet Intervention standards.

Grain storage alternatives

Grain is generally preserved in good condition by controlling its moisture content and temperature so that the organisms which cause deterioration cannot develop. Grain stored in this way can be sold into any appropriate market, or fed to livestock on the farm.

CONDITIONS FOR THE SAFE STORAGE OF LIVING GRAIN

The main agencies which cause deterioration in stored grain are moulds, insects and mites. Some of these organisms occur naturally on cereal crops, and so are likely to be brought into the store with the crop. Residual populations of insects and mites may be present in crevices or crop residues in the 'empty' grain store, but thorough cleaning and insecticide/miticide treatment of the store before it is filled can greatly reduce this hazard.

The grain is protected from attack by a tough outer skin or pericarp and impermeable seed coat or testa. Unfortunately, the severe threshing treatment to which the crop is subjected at harvest can damage these layers, allowing easy access by mould organisms to the starchy endosperm or the embryo. In one study between one-fifth and one-third of all

wheat grains were found to have threshing injuries to the skin of the embryo area. The embryo contains sugars, fats and proteins and is thus an excellent source of nutrients for the moulds. Damage to the embryo will impair the viability of the seed.

The relationship between grain condition and pests (over a 35-week period) and storage conditions is illustrated clearly in the classic diagram (*Figure 11.1*) which has been produced by research workers at the MAFF Slough Laboratory.

In the figure, *line A* defines the conditions under which mites will breed, and therefore infestations can build up. Breeding can be prevented by keeping the temperature below 2.5°C, or the grain moisture content below 12%. In fact, protection from most major mite species is achieved if grain is dried below 14% moisture content (MC) (Wilkin, 1975). Careful cleaning and acaricide treatment of the store when empty will minimise the risk of mite populations persisting from one season to the next.

Line B covers the risk of loss of germination due to embryo damage, and also the deterioration of baking (wheat) and malting (barley) qualities. Grain stored under conditions to the left of the line may be expected to retain all these qualities and thus be suitable for sale into premium markets.

Line C defines the conditions under which moulds will grow on and subsequently in the grain. In addition to the 'invisible' forms of damage mentioned in the previous paragraph, mould growth can lead to a loss of crop weight as the mould organisms feed on the grain, discoloration due to visible mould growth, and the unpleasant health hazards of mycotoxins (poisons produced by the mould organism which can kill or debilitate livestock or even humans who eat contaminated foodstuffs) and mould spores (a very fine dust which can cause farmers' lung, a pneumonia-like affliction, when inhaled). It is essential to wear an adequate *organic dust respirator* if working with grain (or hay) which is suspected of being mouldy.

Moulds, like other living organisms, produce *heat* as they break down food with oxygen, liberating *water* and carbon dioxide. Thus, any large concentration of moulds (or insects or mites) will make the grain around it warmer and more

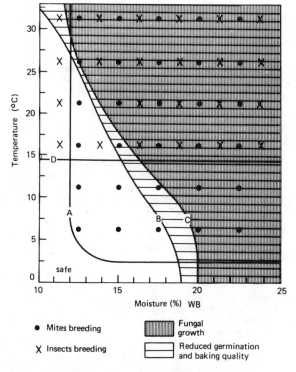

● Mites breeding   ▦ Fungal growth

X Insects breeding   ▢ Reduced germination and baking quality

Figure 11.1 Combined effects of temperature and moisture content on grain condition and grain pests over a 35 week storage period (Crown copyright. Reproduced with permission) (After Burges and Burrell, 1964)

moist. As can be seen from the diagram, this will have the effect of making conditions even more favourable for their own development, and may also encourage the development of other pests or even allow the grain to germinate, which will happen if the local moisture content increases to 25–30%.

Severe mould infestation, if not detected and controlled, will result in the grain being caked together, particularly at or just below the surface and against the walls of the store, and sprouted at the surface.

This can reduce or eliminate the market value of the corn, and may also render it hazardous to livestock and to the farmer.

Line D shows that insects such as grain weevils and beetles can generally be prevented from becoming active and breeding if the temperature of the grain is kept below 14.5°C. In fact, the cooler the grain, the less active the insects and the greater the margin of safety, even down to freezing point. Reducing grain moisture content has no effect in controlling insect activity.

Insect activity in the grain will lead to loss of dry matter and reduced germination, contamination with insect remains and faeces, and as above, to an increase in temperature and moisture levels which can lead to moulding and even sprouting of the grain.

Where an insect problem has been found to exist and it is not possible to control it by cooling the grain, the insects may be controlled by fumigation of the grain store with substances such as methyl bromide or phosphine. These

substances are toxic and their successful use involves sealing the grain store and the deployment of specialised control and safety equipment. This should be left to specialist pest control contractors.

The grain may also be treated with an approved insecticide as it is conveyed into store, if insect problems are expected.

As is the case with mites, the best defence against insect infestations is 'good housekeeping', in cleaning out dust pockets and crop residues in which populations can survive in the empty store, and treatment with an approved insecticide. During storage, the extent of insect populations should be assessed by trapping.

It can be seen from *Figure 11.1* that (ignoring mites) the preservation of living grain in top condition for a 35-week storage period may be achieved by keeping conditions within the area below line D and to the left of line B. If the grain is intended for use or sale as livestock feed, then the right hand boundary moves out to line C, allowing a higher moisture content to be accepted.

Generally, the crop is safest when temperature and moisture content are low, and at increasing risk as either factor rises.

THE TIME FACTOR IN GRAIN STORAGE

As may be expected, insect, mould or mite infestations take time to build up. This means that grain, too moist or too warm to be stored safely for seven months without drying or cooling, can be held for several weeks with little or no harm. The length of time such a parcel of grain can be held without loss of quality decreases with increasing temperatures or moisture content. *Table 11.1* shows the maximum recommended moisture content for bulk grain stored at various temperatures and for various periods. (It must be remembered that storage for any extended period at temperatures over 14.5°C entails the risk of insect build-up *however dry the grain*.)

Table 11.1 Estimated maximum number of weeks of mould-free storage of barley at a range of moisture contents and temperatures (ADAS, 1985)

MC (%)	Temperature °C				
	5	10	15	20	25
16	120	50	30	12	5
17	80	28	10	4	2
18	22	9	4	2	1
19	14	6	3	1.5	0.5
20	10	4.5	2	1	0.5
22	7	3	2	1	0.5
24	4	2	1	0.5	0.2
26	2.5	1	0.5	0.2	0.2

GRAIN MOISTURE CONTENT AND ITS MEASUREMENT

Grain moisture content is expressed as wet-base (WB) percentage, or the ratio of the weight of water contained in a sample of the grain, to the total weight of the sample, including the water.

In a laboratory, the moisture content of a sample of grain is found by drying the weighed sample in an oven until no water remains, and then re-weighing to find the weight of the water which has been driven off. The moisture content (wet base) is found from the expression:

$$\text{MC\% (WB)} = \frac{\text{Loss of weight on drying}}{\text{Original weight of sample}} \times 100$$

This is the only completely accurate method of measuring the moisture content of a sample of grain, and is always used in scientific research, or when moisture meters are to be calibrated. However, the complete oven-drying of even a milled sample of grain will take $1\frac{1}{2}$ to 4 h. Faster moisture measurement systems have been produced to satisfy the need for an 'immediate' evaluation of grain moisture content.

Most grain moisture meters measure an electrical property of the grain such as its resistance or capacitance, which have been found to vary with moisture content. A good quality instrument, well maintained, can be expected to give a reading which is correct to within $\pm 1\%$ of the true moisture content of the sample, and this reading can be available within minutes of the sample being taken.

Some instruments require the grain to be ground, which increases the time needed to make a measurement. However, a whole-kernel instrument may give false results with grain that has just passed through a drier, or with very cold grain since in these cases the condition of the exterior of the grain may not reflect the internal situation.

When measuring the moisture content of a sample of grain, it is most important to ensure that the sample is truly representative of the bulk of grain from which it has been drawn. This can be best achieved by systematically extracting several small samples from different points and depths in the load or store, mixing them together and then taking the final sample from this mixture.

Without great care in the taking of samples, even the best of moisture meters and the most careful of operators will fail to produce satisfactory results.

DRYING GRAIN

Where incoming grain is at too high a moisture content for long-term storage, it must be dried at the beginning of the storage period. In the process of drying, heat is provided to evaporate the moisture from the grain, and a stream of air picks up the water vapour and carries it away into the atmosphere.

The more a batch of grain has to be dried, due either to a high initial moisture content or to a low final moisture content, the greater will be the cost both in terms of energy usage and drying equipment operating time.

When grain is dried the resulting product will weigh less than it did before drying, by the amount of water which has been removed. It is often important to know what this weight loss will be, either for the purpose of estimating drying costs or for converting harvested grain weights to equivalent weights of dry grain. The following formulae are helpful.

Where

X is the weight loss on drying,

M_1 and M_2 are initial and final moisture contents (% WB)

W_1 and W_2 are weights of grain before and after drying.

$$X = \frac{W_1(M_1 - M_2)}{100 - M_2}, \qquad X = \frac{W_2(M_1 - M_2)}{100 - M_1}$$

In addition to the weight loss on drying, a further minimum weight loss of 1–1.5% can be expected over the storage period, due to respiration of the grain itself, and to the activities of storage pest organisms.

Drying systems range from those which are capable of drying grain within 2 or 3 h of its arrival from the field to those where complete drying can take two weeks or more.

Because of the air temperatures which are used, high-speed drying systems may be classified as 'high-temperature driers', while the slowest are known as 'low-temperature', or 'storage' driers, the latter term because drying takes place after the grain has been put into the store, and the store does double duty as a drier.

High-temperature driers

High-temperature driers can extract moisture from grain very rapidly because a high temperature allows moisture to diffuse rapidly out from the interior of each grain, and the very hot air which is blown through the grain has a very great attraction for water vapour and a very large water vapour carrying capacity.

As the hot air passes through the grain, it gives up some of its heat to evaporate water, and is thus cooled as it picks up water vapour. The distance that the air travels through moist grain must be limited so that there is no risk of it being cooled to the point where it becomes saturated (i.e. can no longer carry all its water, and water is deposited back on the grain by condensation). In high-temperature driers the thickness of the grain layer rarely exceeds 250 mm, while in low-temperature storage driers it may be as much as 3 m.

The higher the temperature at which a given drier is operated, the faster is the rate of moisture extraction, the higher is the grain throughput of the drier, and in many cases the lower is the energy consumption per unit of water removed. Hence, operating a drier at a higher temperature can reduce drying costs in terms of both overhead and fixed costs per tonne dried.

If the temperature of the drying air is too high, grain temperature will rise to a level high enough to cause damage to germination, baking or malting properties, extraction of

Table 11.2 Recommended maximum air temperatures for high-temperature grain driers

	Moisture content (% WB)	Drying air temperature (°C)
Malting barley and seed grain	$\leqslant 24$	49
	> 24	43
Milling wheat	$\leqslant 25$	66
	> 25	60
Grain for stockfeed		82–104

From Blakeman, 1982.
These recommendations are based on research carried out in the 1930s. New recommendations are likely to be based on grain temperature rather than the temperature of the drying air.

oils or starch or even to the livestock feeding value of the grain. *Table 11.2* shows the recommended maximum air temperatures at which to dry grain.

Unless the grain is completely dry, grain temperature will always be lower than air temperature due to the cooling effect of evaporating water; the greatest cooling effect will be where the grain is moist, and the evaporation rate is high. Even at the end of the drying process, average grain temperatures will still be 10–20°C lower than drying air temperature. If in doubt, drier manufacturers' recommendations should be followed.

High-temperature driers may be classified as 'continuous', where there is a constant flow of grain through the machine and into store, or 'batch', where a quantity of grain is loaded into the machine where it remains while it is dried, and is then unloaded into store.

Continuous flow driers

Most continuous grain driers are of the crossflow type, where the grain moves in a layer 200–300 mm thick while the drying, and later cooling, air is directed through the grain at right angles to its direction of travel. In order to give an equal drying effect to all grains, the grain should be turned or mixed several times as it passes through the drier.

The crossflow principle is popular due to its simplicity, but other airflow systems can offer advantages, and a few have appeared on the market. In a counterflow drier system, drying air flows in the opposite direction to the grain, so that the hottest and driest air meets the driest corn. This gives faster and more efficient drying, but there is a greater risk of damaging the grain by overheating, and this method is in fact only used in the cooling section of some driers where the temperature gradient is in the reverse direction and this system gives a more gentle temperature reduction and the advantage of maximum heat transfer can be fully utilised.

In a concurrent flow drier, grain and drying air move in the same direction although the air moves at a much faster rate. Thus, the hottest air always meets the wettest corn, where the cooling effect of evaporation will be greatest. By the time the air reaches the driest corn, its temperature will have been much reduced, minimising the temperature stress on the grain. This may permit air temperatures above the recommended maximum, without excess grain temperatures. There is a faster throughput from a given drier size, and more efficient use of heat (Brooker *et al*, 1974).

The mixed-flow drier combines features of all three types. Heat transfer is efficient, and high air temperatures can be used without risk of overheating the grain. At a plenum temperature of 120°C and when drying grain from 20 to 15% MC, grain temperature only reached 50°C (Bruce and Nellist, 1986).

Not all the drying is done in the drying section of the drier; a final 1 or 1.5 percentage points of moisture are removed while the grain is being cooled, utilising the heat energy remaining in the grain from its passage through the drying section. The cooling section occupies one-third to one-quarter of the drier, and is generally capable of cooling grain to near ambient temperature before it is discharged from the drier. The greatest demand on the cooling section is when only a small amount of moisture needs to be removed from the grain, and drier throughput is higher than rated.

If there is cooling available in the store, drier capacity can be increased since the whole of the drier can be used for drying the grain, which is cooled in the bulk store later. A refinement of this system, where the grain is held, still hot, in a sealed bin for a 12-h period of 'tempering' before being cooled by low-volume aeration, is known as 'dryeration'. 'Tempering' will improve efficiency, as better use is made of the residual heat in the grain (Bakker-Arkema, 1984).

The drying capacity, or 'throughput' of a continuous flow drier is expressed as the number of tonnes/h of grain at 20% MC which can be dried to 15% at a specified drying-air temperature (often 65°C). Output of the drier will be increased if a higher air temperature can be (safely) used, or if the grain requires drying by less than five percentage units, and vice versa.

The amount of drying done to each grain depends on the length of time for which that grain is exposed to the stream of hot air in the drying section of the drier, and this is controlled by the rate at which grain moves through the machine. This rate is adjustable, and adjustments are made according to the moisture content of the incoming grain and the resulting final moisture content of the dried product. Both of these values should be checked regularly. Several manufacturers offer automatic drier control units, which can monitor final grain moisture content (or grain temperature at the end of the drying section which is a related value) and regulate the flow of grain accordingly.

It should be remembered that grain can take 3 h or more to pass right through a continuous-flow drier, so that any adjustment made to the machine will take this long to come into full effect.

Energy consumption by continuous driers

Continuous-flow and other high-temperature driers, which evaporate water at a very rapid rate, consequently consume energy at a rapid rate (*Table 11.3*).

Such a high heat requirement clearly rules out many sources, and at present the available types are fired by diesel (gas oil) or LP gas (propane). Potential alternative fuels are natural gas (if a piped supply is available) or coal.

To convert 1 kg of water into vapour at 20°C, 2.45 megajoules (MJ) of heat energy are required. Tests at NIAE have shown an average energy use of 6.7 MJ/kg of water evaporated, for a range of traditional cross-flow continuous driers. Clearly, significant amounts of energy are used up in other ways. Apart from the energy used for the operation of fans and conveyors, and the proportion which leaves the drier in the form of latent heat locked up in water vapour, most of the 'wasted' energy is lost in the form of heated, unsaturated air either through the lower part of the drying section or from the cooling section of the drier. Recycling of this exhaust air to exploit the energy it still carries can give energy savings of up to 40% (Bakker-Arkema, 1984).

Several designs of mixed flow driers can achieve specific energy figures of less than 4.0 MJ/kg under test conditions,

Table 11.3 Furnace capacity of continuous driers

t/h of wet corn to be dried from 20–15% MC at 65°C air temperature	*Drying capacity*					
	2.5	*5.0*	*10.0*	*20*	*30*	*60*
Evaporation rate (kg water/h)	150	300	600	1200	1800	3600
Furnace capacity (kW)	220	450	900	1860	2700	5300

Table 11.4 Crop weights and volumes

	kg/m^3	m^3/t
Wheat	785	1.28
Barley	689	1.42
Oats	513	1.95
Beans	817	1.20

Table 11.5 Round bin capacities (wheat)

Bin diameter (m)	Volume (m^3 m)	Weight (t/m)
2.4	4.5	3.5
3	7.1	5.5
3.7	10.8	8.4
4.3	14.5	11.3
4.9	18.9	14.8

although they are not often attained under farm conditions.

In operating the drier to save energy, the most important factor is to *avoid over-drying*. Every kg of water that is removed from the grain adds to the cost of the operation and slows output. Reducing the drying air temperature will *not* save energy, and will certainly increase the cost of drying a tonne of grain.

On farms, the major component in drying costs is the overhead cost of the equipment, and any cost-saving exercise should address this aspect first, before moving on to matters of energy consumption.

High-temperature batch driers

The high-temperature batch drier carries out the same functions as the continuous drier, but it operates in a series of separate steps instead of a flow process. The sequence of events can be controlled manually, or automatically so that the drier can be left to run unattended. Each batch spends 2–4 h in the drier.

Batch driers can be installed on a permanent base, or some

models can be moved on their own wheels from one site or one farm to another. Portable driers can be easily re-sold. The unit may be rated by the amount of grain in a batch, or by the number of tonnes of grain that can be dried by five percentage points in 24 h operation. Capacities range from 38–135 t/24 h. The capacity of a drier in 24 h should be equal to the daily harvesting capacity.

Storage, or low-temperature, driers

Where grain is to be stored on the farm after it has been dried, it is possible to reduce costs by combining the functions of drying and storage. *Tables 11.4* and *11.5* refer to the storage volumes of various grain crops. Capital cost is generally less than for a separate drier and a store, and typical energy consumption is 3.5 MJ/kg of water evaporated. Against this, the system gives a slow drying rate, 0.5%/d or less, and its successful operation demands a high level of management.

Measurement and control of air relative humidity

The operation of any low-temperature drier relies on the fact that there is an equilibrium relationship between the moisture content of grain and the relative humidity (RH) of the air which surrounds it (*Table 11.6*). Grain is brought to a safe moisture content by ventilating it for two weeks or more with air at the corresponding equilibrium relative humidity. Any deviation in air quality will affect both the rate of drying and the final moisture content, so it is clear that the accurate measurement and control of relative humidity must be a prime requirement for successful drier management.

Relative humidity is measured with a wet and dry bulb hygrometer, which consists of a pair of identical thermometers. The bulb of one thermometer is covered with a wet cloth sleeve. After exposure to the air stream until readings stabilise, the two thermometers are read and the relative humidity is found from a slide rule or chart supplied with the instrument. All other instruments for measuring or controlling relative humidity must be calibrated regularly against the basic, and inexpensive, wet and dry bulb instrument.

If the relative humidity of the air is too high, it may be

Table 11.6 Equilibrium moisture contents of a range of seeds at 25°C (% WB)

Species	Air relative humidity (%)						
	40	50	60	70	75	80	90
Wheat	10.7	12.0	13.7	15.6	16.6	17.6	23.0
Barley	9.8	11.5	13.2	15.1	16.1	17.3	23.1
Oats	9.5	11.0	12.5	14.3	15.4	16.7	22.0
Oilseed rape	5.7	6.6	7.4	8.8	10.2	11.9	17.4
Linseed	6.3	7.3	8.5	10.0	11.0	12.5	18.2
Field peas	9.5	11.2	13.0	15.8	18.0	20.9	28.6
Perennial ryegrass	9.2	10.6	11.9	13.9	15.3	17.4	24.4

After McLean, 1980
Note: As temperature rises, the equilibrium shifts so that RH increases in relation to MC, and the air gains moisture from the grain. The reverse happens if temperature falls (Pixton and Warburton, 1971).

reduced by heating the air. Equipment which will raise the air temperature by 6°C is adequate to deal with any weather situation. Electrical heater banks are commonly used, because of their ease of installation, operation and control, but oil or gas fired heaters are available, or a stove may be fuelled with wood or straw. Diesel-engine driven fans apply the waste heat of the engine, which can raise the air temperature by up to 4.5°C. A heat input of 1.1 kW is needed to raise the temperature of 1 m³/s of air by 1°C, and thus reduce the relative humidity by 4.5 units.

Heater output must be adjusted with every change in ambient conditions, if the maximum drying rate is to be maintained without unnecessary use of heat. An automatic humidity control (humidistat) which will regulate the output of a series of electric heaters to maintain the drying air at the desired relative humidity will soon pay for itself in energy savings and reduced overdrying.

An alternative method of controlling air relative humidity is by means of a heat pump, which cools the air to condense moisture, and then re-heats the dehumidified air to further reduce RH. A unit of adequate size can give a satisfactory performance at lower energy cost than heaters, but at a very much greater capital cost (Brook, 1986).

The low-temperature drying process

As the drying air moves upwards through the grain mass it picks up water and is cooled. Grain at the bottom of the store dries first, and a 'drying front' then moves slowly upward through the grain until all the grain is dry after two weeks or longer. During drying, the grain above the drying front remains as wet (or wetter) than when it was harvested. This portion of the grain is preserved before drying by the cooling effect of the air which has dried the lower layers of grain.

If grain at a moisture content above 22% is to be dried, it must not be loaded to a depth of more than 1 m. This reduces the drying time, increases drying efficiency and it also prevents the soft grains being squashed by the weight of a full depth of corn. If grains at the bottom were crushed, this would reduce the air spaces between them and restrict the flow of drying air.

Drying rate and efficiency are improved for grain of any moisture content if the grain store is filled in layers as drying proceeds, rather than piling the grain in one portion of the store first (Sharp, 1984).

Heat should not be used for drying grain above 18% MC. Initial drying should be carried out with unheated air, with heat being applied only in the final stages of drying. Use of heat at an earlier stage will result in re-condensation of moisture in the upper layers of grain.

Progress of the drying front towards the grain surface can be followed by extracting samples from different depths with a sampling spear. The depth of the transition between dry and wet grain indicates the position of the drying front.

Air supply and distribution

The drying air is distributed over the base of the grain mass by means of a system of ducts, or through a fully perforated floor. An adequate quantity of air must be provided to make the drying period as short as possible. An airflow rate of 0.1 (m³s)/m² of floor area, or 0.05 (m³s)/m² of grain, whichever is greater, is recommended (ADAS, 1982). Since not all the grain in the store is being dried at once, the recommendations are applied to one-half to three-quarters of the store

according to geographical region. Adequate fan capacity is a cornerstone of successful crop drying.

FANS FOR GRAIN STORES

A fan moves air by creating a pressure behind it. The air output of the fan decreases as the back pressure or static pressure against which it is working increases. Back pressure builds up as the air is forced along ducts, round bends and through the grain. The greater the air speed, the higher the back pressure.

The pressure generated by a fan is measured with a water manometer (WG = water gauge), and fan output can be determined from this reading by reference to manufacturers' performance curves. Working pressures for crop drying fans are 60–150 mm WG.

A fan is 40–85% efficient. The remaining 60–15% of its power consumption is lost as heat, which warms the air by up to 2.5°C (Sharp, 1982). this may be enough heat to dry a crop, but it is a disadvantage when the fan is used for cooling. Cooling systems sometimes operate under suction with the fan at the outlet; this way, the fan's heat does not enter the crop.

Some fans absorb more power as back pressure drops and air output increases, and the motor may become overloaded and overheat. Only non-overloading fans are suitable for crop drying.

Centrifugal fans

Centrifugal fans develop pressure by accelerating the air in the rotor, and then converting its kinetic energy into pressure energy in the diverging housing or volute. They are quiet in operation, and bulky. A form of centrifugal fan with backward-curved (b-c) blades on the rotor is non-overloading, and is frequently used in grain stores. Pressures up to 300 mm WG are possible, according to rotor diameter and speed.

Axial-flow fans

Pressure is created by the rotation of an airscrew-like impeller in a close fitting circular housing. Air passes in a straight line through the fan, giving high peak efficiencies. The fan alone can only produce a pressure of 80–120 mm WG, but stationary intake guide-vanes can increase this by 20–60%. Two fans contra-rotating in the same tubular housing can increase the pressure up to three times. The fan is non-overloading, and more compact than a centrifugal fan of the same output. It is noisy, particularly when guide-vanes are fitted. Silencers are available, or the fan can be shielded with bales to absorb the noise.

Table 11.7 shows the output of various crop drying fans against a range of static pressures.

Ducts and drying floors

After being pressurised by the fan, and heated if necessary, air reaches the grain via a main duct and a series of lateral ducts spaced at 1–1.2 m. To limit the build-up of back

Table 11.7 The output of various crop drying fans against a range of static pressures (manufacturers' figures)

Type and power of fan	Static pressure (mm WG)					
	50	75	100	125	150	175
	Air output (m³/s)					
3.8 kW b–c	4.6	3.9	3.0	—	—	—
7.5 kW b–c	8.2	7.7	7.0	5.9	4.4	1.8
18.7 kW b–c	13.7	13.1	12.4	11.5	10.5	9.4
30 kW axial	10.0	8.8	7.1	4.7	1.8	—
52 kW axial	18.8	17.7	15.9	14.1	11.8	9.4

pressure, ducts should have a minimum cross-sectional area of $1 \, m^2$ for every $10 \, m^3/s$ of air that passes. This will give a maximum air speed of $10 \, m/s$ in the ducts. Similarly, the size of the openings through which the air passes from the ducts into the grain should ensure that air speed at this point does not exceed $0.15 \, m/s$. Lateral ducts do not normally exceed $10 \, m$ in length because the output of air along the duct becomes uneven beyond this length. A better distribution of air to the grain lying between the ducts is achieved when air escapes from the sides rather than the top of the duct.

Above-floor ducts convert any sound, level concrete floor into a drying floor, and can be removed if the building is needed for some other use. However, the operations of filling and emptying the store are made more difficult by the presence of above-floor ducts. Many stores are equipped with permanent below-floor ducts which give a flat floor and are strong enough to support the weight of trailers or bulk lorries. Although the flexibility of use of the store is curtailed, and cost is increased, moving the grain in and out of store is greatly facilitated.

Completely even distribution of drying air over the floor area can be achieved if the entire floor is perforated, and air passes up from a plenum chamber below the floor, through the perforations and into the grain. This system is used in some grain bins. A further refinement to this type of floor is an arrangement of louvres which direct jets of air to 'sweep' the last few tonnes of grain out through an unloading chute using the power of the drying fan. This eliminates the dusty task of sweeping or shovelling the last corn out of a flat-bottomed bin.

Uniform air flow throughout the grain mass also demands that the grain itself be uniform. Any differences in depth or compaction, or any local concentrations of rubbish will affect uniformity of air flow. Pre-cleaning the grain before it enters the store will remove most of the rubbish. Even layers of grain should be built up over the whole floor area by the careful use of conveyors, spreaders or grain throwers, to prevent rubbish, or a particular batch of wet grain, from being concentrated in one spot.

If air flow to any part of the store is restricted, the grain in that part of the store will dry slowly or not at all. Air flow through the grain can be measured at many points on the surface using an inverted funnel type of airflow meter, and any shortfall can be corrected. The instrument costs little more than the price of a tonne of corn, and is simple and quick to use.

To ensure that moisture-laden air can escape easily from the roof space above the grain store, exhaust openings of at least $1 \, m^2/(2.5 \, m^3 s)$ of air flow must be provided at eaves, ridge or gable ends.

GRAIN CLEANERS

It is desirable to remove chaff, dust and impurities by pre-cleaning grain before it is dried or put into store. This avoids the expense of drying valueless material, and reduces the risk of fire in high-temperature driers. The presence of impurities in a bulk store can block the spaces between the grains, and inhibit air flow. Concentrations of such impurities can build up under the discharge of stationary conveyors.

Grain may also be cleaned or graded when it is removed from store prior to sale; the ability to do this is particularly valuable where the grain is to be sold at a premium for seed, malting or baking, or in a situation of over supply where it is necessary to produce a good-looking sample.

Dual-purpose grain cleaners are available from many sources. For pre-cleaning, machines can accept 4–40 t/h; this output is halved for cleaning after storage when more time is available and a more thorough job is required. Most machines combine one or two aspirations (separation by airblast) with a series of vibrating sieves. Air blast and sieve sizes are adjusted to suit the crop being cleaned. Inexpensive aspirator pre-cleaners are available with outputs to 12 t/h.

Cleaners with reciprocating sieves should be mounted on a rigid base so that no damage will result from their constant vibration.

MANAGEMENT OF GRAIN IN STORE

Monitoring grain temperature

After the grain has been dried and placed in store, careful management and monitoring of its condition is essential if the full value of the crop is to be retained. Grain temperature is normally monitored, since heating is a symptom of most of the problems which affect stored grain. Any local increase in grain temperature must be detected and the cause corrected before there is time for grain damage to occur.

Many manufacturers offer permanently installed systems of grain temperature sensors, with a central monitoring station where all the measurement points can be checked at least weekly. Some systems will themselves perform the checking, and sound an alarm if a pre-set temperature is exceeded. At the other end of the scale, a temperature sensing spear can be pushed into the grain and the temperature read off in a number of positions and depths in the grain store to build up a composite picture of grain temperature.

Aeration

Any grain store more than 20 t in capacity should be fitted with aeration equipment. Storage driers will have this equipment anyway, but where the grain has been dried elsewhere and then put into store, small fans and air distribution ducts must be provided. The required airflow is 0.5 to $0.7 \, m^3/min$ for each m^2 of floor area, or $10 \, m^3/(h \, t)$ of grain (ADAS, 1985). A $0.375 \, kW$ fan unit can aerate up to 50 t of grain at a time. To give even air distribution, ducts should not be spaced more widely apart than twice the depth of the grain.

Aeration helps to maintain grain quality in the following three ways.

General cooling

The cooler the grain, even down to freezing point, the better it will keep (*see Figure 11.1*). Without refrigeration, grain can only be cooled to ambient temperature. The ability to force cold air into the grain as the season cools, and to take advantage of cold nights in order to reduce the temperature of the grain is a valuable aid to the management of the store.

Where aeration equipment is available in the store, the output of a continuous drier can be uprated by converting its cooling section into additional drying space, and cooling the dry grain in the store.

In addition to the conditioning of dry grain, aeration has been used to preserve undried grain of up to 21% MC which is to be used on the farm. The success of this system is dependent on the season; cold nights during harvest, and a cool dry autumn are ideal; a mild, moist autumn might not permit the grain to be cooled quickly enough to prevent mould development. As with other low-cost storage systems, careful management and a thorough understanding of the principles involved are essential for success.

Elimination of hot-spots

When localised heating of the grain is detected the traditional solution is to 'turn' the grain to break up the hot spot and allow it to cool down by exposure to the air. When aeration equipment is fitted, cool air can be blown through the grain to reduce spot temperatures without the need to move the grain.

Maintaining an even temperature throughout the grain

When the grain goes into store, it is all at approximately the same temperature as the outside air. As the season progresses, the layers of grain in contact with the walls of the store will cool, while the grain in the centre of the store will remain at its original temperature. Sunshine will heat the grain at the south side of a store. Any temperature inequalities of this sort will set up convection currents in the air which fills the spaces between the grains. This will result in moisture removal from the warmer grains and condensation on the cooler grains, particularly just below the surface at the centre of the store. This process can result in mould development, which may be followed by heating, insect activity and even sprouting.

If the grain is cooled in line with the cooling of the season, temperature will be uniform throughout the grain mass and moisture movement will not occur. This eliminates the need for costly over-drying which gives a 'safety margin' to allow for the effects of moisture movement. As well as minimising any losses due to deterioration of the grain, the correct use of aeration equipment will permit drying costs to be reduced.

HANDLING GRAIN

Most grain storage installations rely on conveyors for moving the grain. The characteristics of the common types of handling equipment are summarised below.

Auger conveyors

The grain is moved by a screw, or 'flight' which rotates inside a steel tube. Grain is drawn in at one end, and is normally discharged at the other end although intermediate discharge points are possible.

Augers are made in diameters of 90, 115, 150, 200 and 300 mm. Outputs of dry wheat, when the conveyor is inclined at 45 degrees to the vertical are as follows: 114 mm–25 t/h; 150 mm–35 t/h; 200 mm–58 t/h; 300 mm–120 t/h. Output is reduced with other grains, or if the grain is damp, or if the angle is steeper, and is increased if the angle is flatter. Portable augers are made in lengths up to 15 m: large units have a wheeled tripod stand.

The auger is a versatile conveyor, which will operate in many positions without adjustment. There is considerable churning and abrasion of the conveyed material, particularly if the auger is only partially full. The churning effect may be used to advantage in mixing a preservative with grain, but augers must not be used for conveying fragile materials such as rolled barley or pelleted feeds, because of the risk of damage.

An auger may be used to meter the flow of, for example, an ingredient into a mixer by regulating its speed of rotation. This gives very accurate control of output.

The flight of the auger projects beyond the tube at the intake end. This portion must be guarded at all times with a welded-mesh screen to avoid the risk of serious personal injury.

Chain and flight conveyors

The grain is moved along a smooth wooden or steel trough by an endless chain fitted with horizontal scraper or 'flights'. The grain can be fed into the trough at any point, and is discharged at the end of the conveyor or at intermediate discharge points if these are fitted. Some models are reversible.

Chain and flight conveyors are normally horizontal, although some can work at small slopes. They are quiet in operation, long-lasting, economical in power, and do little damage to the grain. Lengths up to 60 m are available; outputs range from 15 t/h to more than 100 t/h. They are widely used as top conveyors in bin-type grain stores. 'Double-flow' models can supply grain to a continuous drier, and also take excess grain back to store.

Belt conveyors

The grain lies on an endless fabric-reinforced rubber belt which slides along a flat bed. Grain can be loaded onto the conveyor at any point, through a hopper which concentrates the grain in the centre of the belt. Discharge is over the end of the belt, or at a movable discharge point where the belt passes in an S-shape over a pair of rollers. The grain shoots over the upper roller and is deflected into a side discharge trough to right or left.

A motorised discharge can move slowly back and forth along a conveyor distributing the grain evenly along the whole length. In conjunction with a grain thrower to project the grain across the store, a level layer of grain can be built up over the complete area.

The belt conveyor is quiet, and very gentle with the material being conveyed. It is found as top and bottom conveyors in bin-type grain stores. Conveyors with cleated belts can operate at steep angles, but the typical grain version has a smooth or textured surface without cleats, and operates horizontally.

Belt conveyors are available in lengths up to 45 m and capacities from 20 to more than 50 t/h. Power requirement is greater than for the chain and flight machine.

The bucket (or belt) elevator

This is a device for raising grain vertically. A series of steel or plastic scoops or buckets are bolted to a flat endless belt which moves vertically between a pair of pulleys. The motor and drive pulley are at the top, while the belt passes round a tensioning idler pulley at the bottom. Grain is fed into the conveyor through a regulating orifice up to 1 m above its lowest point, and is discharged from a spout up to 1 m below its highest point. For this reason, the elevator usually stands in a hole at least 1 m deeper than the bottom of the receiving pit, and is often subject to flooding in winter when the water table is high. The top of the conveyor is generally accommodated in a small extension of the grain store roof where access for maintenance may be less than good.

No damage can occur to the grain while it is in the buckets, but careful operation of the intake slide, and correct design of intake and discharge points are necessary if damage is to be avoided entirely.

Bucket elevators are available in capacities from 5 to 120 t/h, and heights to 25 m. 'Double-leg' models combine two separate conveyors with a single drive unit. The two legs may be used separately; for example, one leg may load wet grain into a continuous drier while the other takes dry grain from the drier and raises it to the top conveyor of the store. When a high throughput is required, for example when loading a lorry, both legs can be used to raise grain from the bottom conveyor into the reversed top conveyor which carries it to a discharge chute outside the store.

Multiple valving devices can be used to direct the grain from the conveyor's discharge to alternative destinations. Controls at ground level simplify the operation of such valves.

Pneumatic conveyors

The grain is carried along a closed tube 130–210 mm in diameter by a stream of air moving at 70–100 km/h. The air stream is produced by a narrow, large-diameter fan driven by electric motor or tractor power take-off.

Intake may be by a flexible suction spout, which is very convenient for emptying flat bottomed bins or clearing floor stores. Alternatively, the grain may be injected into the air stream through a metering hopper. Discharge is from the far end of the system, which may be up to 200 m from the intake if sufficient power is applied, at outputs of 5–40 t/h. The grain does not pass through the fan but considerable abrasion can occur as it moves along the pipes and round bends; this shows itself in the large amount of dust which is generally produced at the discharge end.

This is a very versatile conveying system, as the pipework can be joined together to accommodate any storage arrangement, and the power unit is generally portable and can be used at several locations during its season of work. Against this must be set a high initial cost, and the power consumption of 1–2.5 kW/(t h) is high, particularly for those models

with a suction intake. Fragile materials such as rolled grain or pelleted diets should not be handled in this way.

The grain thrower

A floor store can be filled in a series of flat layers if a grain thrower is available. The grain is metered onto a fast moving conveyor belt or a rotating paddle-wheel which throws it up to 12 m horizontally, and to a height of up to 5 m. Larger models stand on the floor, smaller units may be mounted at the discharge of other types of conveyors. Outputs range from 5–120 t/h, at a power consumption of 1 kW per 4–7 t/h. There is little damage to the grain, although some dust is produced.

A further advantage of loading a floor store in this way is that any rubbish present in the grain is distributed evenly throughout the store and will not have any local effect on the drying process.

Grain weighers

A very important instrument in the grain producer's armoury is the weigher. It permits him to determine yields at harvest time, to control his stock of grain, and to weigh the grain as it comes out of store for sale or for livestock feed on the farm.

The weigher consists of a counterbalanced hopper which is positioned in the flow path of the grain. When the hopper has received a pre-set weight of grain, the flow is cut off, one increment of weight is recorded on a counter, the hopper is discharged and the cycle is repeated. Very accurate weighing is possible at rates from 6 to more than 100 t/h according to model.

Loading lorries

If the crop is to be sold in bulk, it may have to be discharged into lorries of 25 t or more. If conveying equipment of 20 t/h is used, which is adequate for most other functions in the grain store, the lorry will be tied up for more than an hour and this may make the product less attractive to a potential customer. High capacity conveyors are available, but would only be fully utilised for a few hours per year.

To overcome this problem some grain stores are equipped with covered V-bottomed hoppers of 30 t capacity with several large discharge ports, mounted high enough for a bulk lorry to be driven underneath. The hopper can be filled slowly with the store's existing conveyors, allowing plenty of time for the grain to be passed through cleaning equipment if required, and yet the lorry can be filled in 10 minutes or so with the minimum of effort or dust. In the likely event of there being a buyer's market for grain, it is this type of detail, plus a quality product, that will ensure a sale in a competitive situation.

STORAGE OF GRAIN IN SEALED CONTAINERS

Where grain is to be processed into livestock feed on the farm, there are many advantages to storing it without drying. Apart from the fact that there is no drying cost, the grain can be handled into store at a very fast rate so there are no harvest bottlenecks. Moist grain can be rolled easily, and produces little dust when ground; this can improve palatability to stock and improve working conditions for the stockman.

When stored under airtight conditions, the oxygen concentration is quickly reduced to less than 1% by the respiration of the grain and other aerobic organisms such as moulds. The grain is killed and most destructive organisms die or become dormant (Hyde, 1965). Some fermentation may take place if the moisture content is above 24% and this will taint the grain.

As long as the container remains sealed, the condition of the stored grain is stable. When grain is removed from the store, care must be taken to allow only the minimum of air to enter, and to reseal as soon as possible.

Once exposed to the air, the moist dead grain will deteriorate rapidly especially if the weather is warm. Normally, only enough for 1 or 2 d use should be withdrawn at a time.

The container for airtight storage may be a plastic or butyl bag, or the more durable enamelled steel tower silo. Capacities are from 15 to 800 t/silo. Unloading is by means of augers, a sweep arm which can traverse the circular bottom of the silo, and a fixed arm which brings the grain from the centre of the silo to discharge outside through a resealable spout.

If grain above 22% MC is to be stored in a tower silo the lower layers are likely to be compressed by the weight above, and may bridge and be difficult to unload. A few tonnes of dry corn at the bottom of the silo will facilitate this initial stage of unloading, as will the use of a gentle method of filling such as a bucket elevator rather than the more convenient but damaging blower.

USE OF CHEMICAL PRESERVATIVES TO STORE GRAIN

An alternative method of controlling moulds and pests in undried grain is to treat the grain with a chemical preservative such as propionic acid, at a rate which increases with the moisture content of the grain.

After treatment, no special storage structure is required; the grain must be kept dry to avoid dilution or leaching of the preservative so floor, walls and roof must be waterproof. Since propionic acid is highly corrosive, even to galvanised steel, the materials of the store must resist corrosion, or be coated with plastic or a chlorinated rubber paint for their protection. Any equipment which handles either the preservative, or grain which has been freshly treated, must be thoroughly washed out with water after use, to minimise corrosion.

The preservative liquid or its fumes can cause irritation or injury to the skin, eyes, mouth or nose. Protective garments should be worn, and a supply of water kept close by in order to wash off splashes. The operator should stand upwind of working application machines or piles of recently treated grain.

The applicator generally consists of a pump, flow regulator and nozzle which draw the chemical from a container and spray it onto the crop at the intake of a short (1.5 m) length of auger. The tumbling effect of the auger is very effective in distributing the chemical over every grain, an important requirement of this system. Charts are provided so that the chemical flow rate can be adjusted to match the grain moisture content (up to 30% and more) and the (measured) throughput of the auger.

The cost of the chemical is relatively high, so in order to be viable the handling and storage parts of the system must be inexpensive.

Since the grain is killed and tainted by the preservative, there is no market for the grain other than for stockfeed. Weed seeds are also killed, which may be an advantage.

Because the chemical is absorbed into the grain, protection continues after the grain is removed from store, so quite large batches of feed can be prepared at a time even in warm weather, without risk of deterioration.

References

ADAS (1985). Booklet 2497. *Low-rate aeration of grain.* Alnwick: MAFF Publications

ADAS (1982). Booklet 2416. *Bulk grain driers* Alnwick: MAFF Publications

BAKKER-ARKEMA, F. W. (1984). Selected aspects of crop processing and storage: a review. *J. agric. Engng. Res.* **30,** 1–22

BLAKEMAN, R. (1982). Fuel saving in grain drying. *Agr. Engr.* **37,** 60–67

BROOK, R. C., (1986). Heat pumps for near ambient grain drying: a performance and feasibility study. *Agr. Engr.* **41,** 52–57

BROOKER, D. B., BAKKER-ARKEMA, F. W., and HALL, C. W. (1974). *Drying cereal grains.* Westport, CT, USA: Avi Publishing Co.

BRUCE, D. M. and NELLIST, M. E. (1986). Simulation of continuous-flow drying. Proc. BSRAE Association Members'Day 26–27 February

BURGES, H.D. and BURRELL, N. J., (1964). Cooling bulk grain in the British climate to control storage insects and to improve keeping quality. *J. Sci. Fd Agric.* **15,** 32–50

HYDE, MARY B., (1965). Principles of wet grain conservation. *Agr. Engr.* **21,** 75–82

McLEAN, K. A., (1980). *Drying and Storing Combinable Crops.* Ipswich: Farming Press Ltd.

PIXTON, S. W. and WARBURTON, SYLVIA, (1971). Moisture content/relative humidity equilibrium of some cereal grains at different temperatures. *J. stored Prod. Res.* **6,** 283–293

SHARP, J. R., (1984). The design and management of low-temperature grain driers in England — a simulation study. *J. agric. Engng. Res.* **29,** 123–131

SHARP, J. R., (1982). A review of low-temperature drying simulation models. *J. agric. Engng. Res.* **27,** 169–190

WILKIN, D. R., (1975). The effects of mechanical handling and the admixture of acaricides on mites in farm-stored barley. *J. stored Prod. Res.* **11,** 87–95

Part 2
Animal production

12

Animal physiology and nutrition

R. A. Bourne and R. M. Orr

Animal production involves the conversion of feedstuffs, mostly of vegetable origin, to animal products in the form of meat, milk, eggs and wool. In this chapter those physiological and metabolic processes which are especially relevant to this conversion are considered. The chapter is divided into eight sections:

Regulation of body function
Chemical composition of feedstuffs and animals
Digestion
Metabolism
Voluntary food intake
Reproduction
Lactation
Growth
Environmental physiology and behaviour

Since animal production involves various species of farm animal an attempt is made throughout to include reference to the comparative aspects of the subject. In addition, since animal productionists are concerned with the efficiency of the conversion of feedstuffs into products, reference is likewise made where appropriate to this important aspect of livestock science.

REGULATION OF BODY FUNCTION

In order that animals can survive their environment and changes in that environment they must possess the necessary coordinated mechanisms to maintain their own internal environment in a steady state. These physiological reactions that maintain the steady states of the body have been designated 'homeostatic' and are achieved by the combined action of the various organ systems of the body. This coordinated action is carried out by the nervous and endocrine systems. These systems must initiate the necessary adjustments that enable animals to respond to external environmental changes such as the availability of food, temperature and light.

The nervous system

The nervous system in mammals is an extremely complex collection of nerve cells which plays an essential part in the functioning and behaviour of the animal. The basic units of the nervous system are the nerve cells or neurons and the associated receptors.

The ability of animals to maintain homeostatis is dependent on the nervous system receiving and responding to information from specialised receptor cells and to this end mammals have evolved receptors that are sensitive to a wide range of physical (e.g. temperature), chemical (e.g. specific components of the blood) and mechanical (e.g. muscle stretch) stimuli. When any response to information received at receptors is necessary the information is carried from receptors to the cell or tissue (effector) where appropriate adjustments are made. This is the function of the neurons.

There are three different types of neurons:

(1) sensory or afferent neurons which carry information from receptors to the spinal cord and brain;
(2) motor or efferent neurons which carry information from the spinal cord and brain to the effector organ; and
(3) interneurons or association neurons which make connections between sensory and motor neurons within the spinal cord and brain.

Neurons consist essentially of a cell body with an elongated projection, the axon, and numerous shorter projections, the dendrites (*Figure 12.1*). For our purposes the cell body can be thought of as performing two functions: firstly, it receives information from the axons of other cells both directly onto its surface membrane and also via the dendrites and, secondly, it acts to sum all the effects of inputs, which may be either excitatory or inhibitory, that it receives. Should this sum of inputs exceed a critical level then an impulse is triggered which is then carried along the axon to the synaptic region which in turn passes the neuronal signal to the next stage in the nervous system, either another neuron or, alternatively, a muscle or gland. Propagation of signals along an axon is dependent on changes in the electrical potential between the inside and the outside of the

317

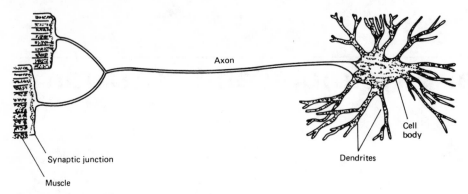

Figure 12.1 A nerve cell

axon membrane which results from changes in the permeability of the membrane to sodium, potassium and chlorine ions. Passage of neuronal signals at synaptic regions involves the release of a chemical transmitter substance into the small gap between the synaptic region and the membrane of the next neuron. The nature of the transmitter substance determines whether the influence on the next neuron is excitatory (e.g. acetylcholine and noradrenaline) or inhibitory (e.g. γ-amino butyric acid). These transmitter substances released by the synapse are received by specialised sites on the dendrites of the next neuron which have the effect of changing the electrical potential of the neuron.

It is the nerve cells – the basic building blocks – from which the various components of the nervous system of mammals are built (*Figure 12.2*).

Division of the nervous system between central and peripheral is made simply on the basis of whether the fibres or cell bodies lie outside the spinal cord. The peripheral nervous system can be further subdivided into the somatic nervous system and the autonomic nervous system. The former is involved in the control of the skeletal muscles in the body whereas the latter controls specific target organs in the body such as the heart, lungs, blood vessels and intestines. The autonomic nervous system has two subdivisions, the sympathetic and the parasympathetic. In many cases these two divisions have opposite effects on the target organs and glands. The sympathetic nervous system acts to increase the level of activity or secretion in an organ whereas stimulation of the parasympathetic will have the opposite effect.

The central nervous system consists of the spinal cord and brain. The spinal cord is contained in a continuous channel within the vertebrae, which form the backbone. At each vertebra there are two openings in the base through which nerves can pass in and out of the spinal cord. Sensory nerves

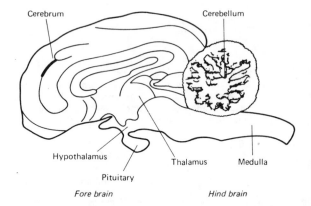

Figure 12.3 Midline section of the brain showing the location of major structures

enter through openings on the dorsal surface, motor nerves leave through openings on the ventral surface. The brain itself is a vastly complex organ consisting of billions of nerve cells, interconnected by neurons. However, certain areas of the brain (*Figure 12.3*) have been found to have specific functions:

(1) the hind brain: this consists of the medulla and the cerebellum. The medulla is involved in the regulation of the heart, breathing, blood flow and posture. The cerebellum is involved in the coordination of muscle activity,

(2) the fore brain: this consists of the thalamus, the hypothalamus and the cerebrum. The thalamus plays an important role in the analysis and transmission of sensory information between the spinal cord and the cerebral cortex. Lying directly beneath the thalamus is the hypothalamus which, despite its small size has been shown to be involved in the control of basic behaviours such as hunger, thirst and sexual behaviour. The hypothalamus also has close connections with the pituitary – a hormone secreting gland – which, under the control of the hypothalamus, secretes hormones into the blood stream. In mammals, the cerebrum constitutes by far the major part of the brain. Definite areas of the cerebrum have been shown to control specific motor and sensory function.

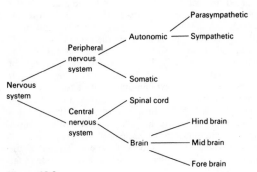

Figure 12.2

The endocrine system

The endocrine system, like the nervous system, acts as a means of communication in the regulation of the physiological and biochemical functions of the body. It differs from the nervous system in that its actions are slower and more generalised. The endocrine system employs chemical messengers known as hormones. These substances act in an integrated fashion to regulate activities such as growth, reproduction, milk production, intermediary metabolism and adaptation to external environmental factors including light, temperature and stressors (stimuli that evoke a stress response).

The endocrine system consists of various endocrine glands, the location of which are shown in *Figure 12.4*. These glands produce two types of hormone: steroids (which are fat-soluble substances formed from cholesterol) and protein hormones (which are water-soluble). Hormones are secreted directly into the blood to be carried to target tissues or organs where they exert their effects, which may be either excitatory or inhibitory. The pituitary gland is said to be the 'master' endocrine gland because of the control it exerts over many of the other endocrine glands through the action of its trophic hormones, such as thyroid-stimulating hormone (TSH), adrenocorticotrophic hormone (ACTH), and the gonadotrophins, follicle-stimulating hormone (FSH) and luteinising hormone (LH). The activities of the pituitary are, in turn, controlled by the region of the mid-brain known as the hypothalamus. Hypothalamic releasing factors are responsible for stimulating or inhibiting the release of specific anterior pituitary hormones into the blood circulation. A complete list of the hormones produced by the various endocrine organs, their chemical nature, target tissues and main physiological effects are shown in *Table 12.1*.

The blood concentration of a given hormone will depend on its rate of secretion and rate of removal from the blood. Normally, hormones are continuously being secreted in small amounts and inactivated by the liver and kidney. The rate of inactivation tends to remain the same, therefore changes in blood concentrations of hormones are mainly due to changes in secretion rate, which are either controlled by the nervous system or by the composition of hormones and other chemicals in the blood. Hormones control their own secretion rates or those of other hormones by means of a feedback mechanism which is either positive or, more commonly, negative feedback.

The action of hormones on their target tissues is not well understood. The initial interaction of a hormone with its target cell is for it to become attached to a specific receptor either at the cell membrane or within the cytoplasm. The number of receptor sites and the proportion of those that are occupied is thought to determine the degree of responsiveness of the target tissue.

Hormones are thought to act on cells in one of two ways: (i) by changing the activity of critical enzymes in the cells' metabolic pathways; (ii) by changing the transportation of substances across the cell membrane (for example insulin increases the uptake of glucose into the cell).

Since the endocrine system plays such a fundamental role in the control of animal functions, there has been considerable attention in recent years to the use of natural and synthetic hormone preparations in animal science, particularly in the areas of promoting increased growth and the regulation of reproductive processes.

CHEMICAL COMPOSITION OF THE ANIMAL AND ITS FOOD

The animal and its food contain the same types of chemical substances. These constituents may be conveniently illustrated by the following diagram (*Figure 12.5*). Comparisons of the chemical composition of the animal and its food (*Table 12.2*) show that the predominant fraction in the dry matter of most foods is the carbohydrate fraction, although in some foods protein predominates. In contrast, the animal body contains very little carbohydrate and is largely composed of protein and fat. Thus, in simplistic terms, animal production involves the conversion of inputs that are mainly in the form of carbohydrates with lesser quantities of protein into outputs that are mainly composed of fat and proteins. The minerals (ash) and vitamins are present in relatively small proportions and are collectively termed micronutrients.

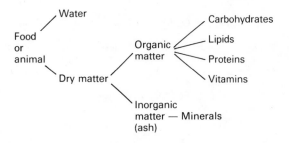

Figure 12.5 The composition of the animal and its food

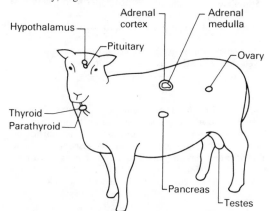

Figure 12.4 Location of major endocrine glands

Proteins

One of the most characteristic chemical features of all organisms is their content of proteins. In both plants and animals proteins perform the important function of *enzymes* which are responsible for metabolic processes. In the animal

Table 12.1 Endocrine structures and hormones

Endocrine structure	Hormone	Chemical nature	Major activity
Hypothalamus	Growth hormone releasing hormone (GHRH)	Polypeptide	All are neurohormones regulating the release of the pituitary hormone indicated
	Prolactin inhibiting hormone (PIH)	Polypeptide	
	Luteinising hormone releasing hormone (LHRH)	Polypeptide	
	Follicle stimulating hormone releasing hormone (FSHRH)	Polypeptide	
	Thyrotropin hormone releasing hormone (TRH)	Peptide	
Anterior pituitary	Growth hormone (GH)	Protein	Stimulates growth
	Adrenocorticotrophic hormone (ACTH)	Protein	Stimulates release of hormones of the adrenal cortex
	Thyroid stimulating hormone (TSH)	Protein	Stimulates release of thyroxine
	Follicle stimulating hormone (FSH)	Protein	Regulates development of ovarian follicles in females and spermatogenesis in the male
	Luteinising hormone (LH)	Protein	Triggers ovulation and stimulates progesterone and testosterone release
	Prolactin	Protein	Stimulates production of milk by mammary gland
Posterior pituitary	Oxytocin	Peptide	Stimulates uterine contraction and involved in milk ejection
Adrenal gland	Vasopressin	Peptide	Reabsorption of water by the kidney tubules
Cortex	Aldosterone	Steroid	Regulates water and electrolyte balance
	Cortisone and corticosterone	Steroid	Regulator of carbohydrate metabolism
Medulla	Adrenaline and noradrenaline	Amino acids	Regulator of carbohydrate metabolism in muscle and liver, constriction of peripheral vessels and contraction of smooth muscles
Thyroid	Thyroxine	Amino acid	Stimulates oxidative metabolism
	Calcitonin	Polypeptide	Regulates calcium levels in body fluids ($\downarrow$)
Parathyroid	Parathyroid hormone	Polypeptide	Regulates calcium levels in body fluids ($\uparrow$)
Pancreas	Glucagon	Polypeptide	Raises blood glucose levels
	Insulin	Polypeptide	Lowers blood glucose levels
Digestive tract			
Duodenum	Secretin	Polypeptide	Stimulates release of pancreatic enzymes
	Cholecystokinin	Polypeptide	Stimulates release of pancreatic enzymes and regulates the gall bladder
Stomach	Gastrin	Peptide	Stimulates secretion of HCl and pepsin by gastric mucosa
Testes	Androgens (e.g. testosterone)	Steroids	Stimulates development of secondary sex characteristic in male; maintains accessory sex organs
Ovary	Oestrogens	Steroids	Stimulates development and maintenance of female secondary sexual characteristics
	Progesterone	Steroid	Stimulates uterus in preparation for ovum implantation; development of mammary gland for lactation
	Relaxin	Polypeptide	Relaxation of pelvic tissues at parturition

they act as the structural proteins of the skin (collagen and keratin), connective tissues, tendons and bones (collagen) and muscle (actin and myosin), as hormones (e.g. insulin), as *antibodies* and as substances performing transport and osmotic functions in the blood (e.g. haemoglobin, albumin).

The fundamental units from which all proteins are constructed are amino acids. All amino acids contain the elements carbon, hydrogen, oxygen and nitrogen. The nitrogen content of most proteins varies between 15–17% with an average of 16%. Measurement of protein content of foods is based on estimation of nitrogen content. Traditionally protein content is expressed as crude protein (CP) where

$CP = N \times (100/16)$. Three amino acids, cystine, cysteine and methionine also contain the element sulphur. Thus the sulphur content of proteins is a function of the proportion of sulphur containing amino acids present, e.g. wool has a high sulphur content reflecting its high cystine content. Proteins are synthesised from a pool of 20 different amino acids (*Table 12.3*). These conform to the general formula

$$H_2N-\underset{\underset{H}{|}}{\overset{\overset{R}{|}}{C}}-COOH$$

Table 12.2 The composition of the animal and feedstuffs (%)

	Water	Carbohydrate	Protein	Fat	Ash
Pig, 30 kg	60	< 1	13	24	2.5
Adult cow	58	< 1	16	20	4
Hen	56	< 1	21	19	3
Grass – leafy	80	12	5	1	2
Wheat straw	10	75	3	2	9
Wheat grain	13	72	12	2	2
Turnips	90	7	< 1	1	1
Soyabean meal	12	35	45	2	6

in which a central carbon atom has attached to it an amino group ($-NH_2$), a carboxylic acid group ($-COOH$), a hydrogen atom and a variable side chain designated here by the letter R. The condensation of the amino group of one amino acid with the carboxyl group of the next, forming peptide bonds is the mechanism by which proteins are produced. Proteins generally contain several hundred amino acid residues and are thus large molecules with molecular weights ranging from 35 000 up to several hundred thousand. Proteins are synthesised in plant and animal cells where the cell nucleus contains genetic material which determines the nature of the newly synthesized protein. The physical and chemical characteristics of proteins are altered by the different proportions of amino acids, the sequence in which they occur in the protein, the degree of cross-linking that occurs between different parts of the molecule and by the presence of other compounds in their structure. For example, some proteins contain lipids (lipoproteins), carbohydrates (glycoproteins) or mineral elements (haemoglobin, casein). From the viewpoint of the nutritionist the proportion of the different amino acids present in a protein and those characteristics of the protein which may influence their availability to the animal are of primary importance.

Whereas plants and many micro-organisms are capable of synthesising all of the different amino acids found in proteins provided they have an adequate supply of inorganic nitrogen and organic compounds capable of supplying the other elemental components, higher animals are not capable of synthesising all amino acids required by the various tissues. Thus, some amino acids are required in the diet of most animals and are referred to as *essential* amino acids. Those

Table 12.3 Major amino acids occurring in proteins

Essential	*Non-essential*
Arginine	Alanine
Histidine	Aspartic acid
Isoleucine	Citrulline
Leucine	Cystine
Lysine	Glutamic acid[1]
Methionine	Glycine
Phenylalanine	Hydroxyproline
Threonine	Proline[1]
Tryptophan	Serine
Valine	Tyrosine

[1] Amino acids also essential to the chick

not specifically required in the diet are called non-essential amino acids (*see Table 12.3*).

Thus, the ability of a food protein to furnish the animal with essential amino acids is a parameter that particularly influences its value to the animal. In practical animal nutrition the amino acids most likely to be limiting in their supply are lysine, methionine and tryptophan. Ruminants do not require dietary amino acids to the same extent as monogastric species such as pigs and poultry. This is because the microflora in their digestive tracts are capable of synthesising amino acids (both essential and non-essential) from the simple organic compounds found in the rumen. This supply of amino acids is, however, inadequate to meet the demand of high producing animals.

The ease with which food proteins may be hydrolysed during digestion may influence their nutritional value. This is influenced by the solubility characteristics of the protein. Feed processing techniques (e.g. heat or formaldehyde treatment) which bring about denaturation of the physical structure of a protein may influence its solubility characteristics.

In ruminant nutrition the proportion of dietary protein hydrolysed by the rumen microbial population is referred to as its *degradability*. Protein hydrolysed by rumen micro-organisms is termed rumen degradable protein (RDP) and that which is resistant to hydrolysis is undegradable protein (UDP).

Comparisons of the crude protein, RDP, UDP and essential amino acid content of animal feeds is given in *Table 13.3* pp. 374–375.

Carbohydrates

Carbohydrates are the main product of photosynthetic activity in plants and contribute up to 80% of the dry matter in many feedstuffs. The nutritive value of these carbohydrates to the animal is variable and dependent on both the chemical structure of the carbohydrate and the digestive capacity of the animal. Generally speaking the variety of carbohydrate types in plants is far wider than in animals where it accounts for only a very small proportion of the body composition.

Carbohydrates are composed of the elements carbon, hydrogen and oxygen. The general formula $C_n(H_2O)_n$ may be used to represent any carbohydrate and indicates that the ratio of hydrogen to oxygen atoms is always 2:1, as in water. A classification of commonly occurring carbohydrates is given in *Figure 12.6* and the carbohydrate content of a range of plant tissues is shown in *Table 12.4*.

From the point of view of the nutritionist a distinction may be made between those carbohydrates that occur within plant cells, either as simple sugars or storage reserve compounds (e.g. starch, sucrose and fructans) and that carbohydrate which occurs in the cell wall performing a structural role (e.g. pectins, cellulose, hemicellulose). Such a distinction is based on the digestibility characteristics of carbohydrates, the former being readily hydrolysed whereas the latter are relatively resistant, their degradation being largely a function of microbial activity in the digestive tract.

The non-structural carbohydrates of plant material can be further categorised in terms of their cold water solubility. The term 'water-soluble carbohydrates' refers to the monosaccharides, oligosaccharides and some polysaccharides, principally fructans and distinguishes them from the starches

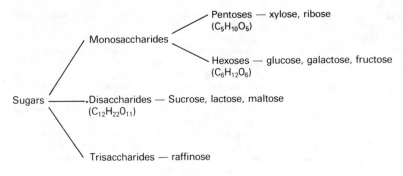

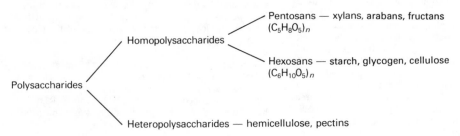

Figure 12.6 Classification of carbohydrates

that are the principal component of most seeds. Cereal grains are the major source of starch to farm animals.

Although small amounts of various free monosaccharides may be detected in feeds (e.g. glucose, fructose) most are of sufficiently low concentrations to be of little importance in nutrition. Sucrose is the main sugar in the sap of plants and in the case of root crops serves as the primary form of energy storage. Many plants convert sucrose into other forms for storage. Temperate grasses store fructans in leaves and stems and starch in the seed. There are two general types of fructans, the levans that occur in grasses and inulins that are characteristic of the Compositae, e.g. the Jerusalem artichoke. Fructan content may account for up to 25% of dry matter in perennial ryegrasses, especially in cool growing seasons.

The water soluble carbohydrate of forages is especially important in the ensilage process (*see* p. 199) and may also influence their palatability. The water soluble carbohydrate content is chiefly influenced by the physiological conditions of growth. High light intensity and photosynthetic rate increase water soluble carbohydrate content. Hence marked diurnal variation in the water soluble carbohydrate content occurs in the living plant. Respiration of cut and drying forage may markedly reduce sugar content.

Starch is the most important storage carbohydrate in plants. Two types of starch exist, amylose and amylopectin. Both have crystalline structures which are disrupted by heating. The temperature at which this occurs is called the gelatinisation temperature and these changes in the starch structures upon moist heating are partly responsible for the improvement in utilisation resulting from steaming, flaking, micronising and pelleting. Physical processing of the gelatinised starch is often required to prevent recrystallisation (retrogradation).

Excessive heat treatment may cause caramellisation of the carbohydrates which has a detrimental effect on utilisation. The conflicting results that have been obtained in feeding studies of processed grains are most likely due to these interacting effects of gelatinisation, retrogradation and caramellisation.

Of the structural carbohydrates the most abundant are the celluloses along with lesser quantities of hemicelluloses, prin-

Table 12.4 Carbohydrate content of plant tissues, % of DM

Component	Tropical grasses	Temperate grasses	Cereal seeds	Alfalfa	Green vegetables
Sugars	5	10	negligible	5–15	20
Fructans	0	1–25	0	0	–
Starch	1–5	0	80	1–7	low
Pectin	1–2	1–2	negligible	5–10	10–20
Cellulose	30–40	20–40	2–5	20–35	20
Hemicellulose	30–40	15–25	7–15	8–10	low

cipally xylans. The former are polymers of glucose, the latter polymers of xylose. The proportion of these carbohydrates in the plant increases as it matures. Their nutritional availability to the animal varies from total indigestibility to complete digestibility and depends on the animal to which it is fed and on the degree of lignification. Differences between animals in their ability to utilise these carbohydrates is largely a function of microbial activity in their digestive tracts – ruminants and to a lesser extent other herbivores (e.g. horses, rabbits) having a greater capacity than monogastric species.

Lignin is a complex substance, a phenylpropanoid polymer of high molecular weight which associates with the structural carbohydrates in the cell wall to form an amorphous matrix. It is particularly resistant to degradation and is the main factor limiting digestibility of forages. The lignin content of plant cell walls increases with maturity and accounts for the reduction in digestibility of herbage as it matures. Cereal straws are examples of highly lignified feeds. Chemical treatment of cereal straws with alkalis such as sodium and ammonium hydroxide breaks the mainly ester bonding between lignin and the structural carbohydrates in the cell wall and increases the susceptibility of the cell wall to degradation in the digestive tract of animals. The efficiency of the treatment is dependent upon the proportion of lignin–carbohydrate bonds that are broken.

In animal tissues the main carbohydrates represented are glucose, which features in the energy metabolism of animals and glycogen which is synthesised in muscle and liver from glucose and acts as a readily available form of energy storage. The amount of glycogen and its rate of metabolism is of particular relevance to changes which occur in the muscles of meat animals after slaughter. Lactose is a disaccharide synthesised in the cells of the mammary gland representing over 95% of the total carbohydrate present in milk.

Lipids

The lipids are a diverse group of substances which share the common property of being relatively insoluble in water and readily soluble in organic solvents such as ether or chloroform. Most animal feeds contain up to 5% of lipid in the dry matter whilst the animal body may contain up to 40% of lipids. A variety of different types of lipids are found in plant and animal tissues performing a range of important biochemical or physiological functions. A chemical classification of lipids and some of their functions is given in *Figure 12.7*.

Fats and oils are constituents of both plants and animals and account for about 98% of all naturally occurring lipids. They have the same chemical structure and properties differing in that oils occur in liquid form in plants and fats chiefly in the solid form in animal tissues. Structurally, they are mainly triglycerides (neutral fats) in which three fatty acids are joined by ester linkages to glycerol. The fatty acids have the general formula $CH_3(CH_2)_nCOOH$. As well as differing in chain length – naturally occurring fatty acids are mainly of chain length 4–18 carbons – they also differ in their degree of unsaturation (*Table 12.5*).

In most triglycerides there is a mixture of fatty acids and in both fats and oils there is a mixture of triglyceride types. Animal body fats are characterised by a high proportion of saturated and mono-unsaturated fatty acids – particularly

Table 12.5 Commonly occurring fatty acids

	Abbreviated designation[1]
Saturated fatty acids	
Acetic	C2:0
Propionic	C3:0
Butyric	C4:0
Caproic	C6:0
Caprylic	C8:0
Capric	C10:0
Lauric	C12:0
Myristic	C14:0
Palmitic	C16:0
Stearic	C18:0
Arachidic	C20:0
Unsaturated fatty acids	
Palmitoleic	C16:1
Oleic	C18:1
Linoleic	C18:2
Linolenic	C18:3
Arachidonic	C20:4

[1] The first number after C indicates the number of carbon atoms and the second number indicates the number of double bonds present.

stearic and oleic – plant and fish oils by a high proportion of poly-unsaturated fatty acids (*Table 12.6*). Three of the fatty acids, linoleic, linolenic and arachidonic acids are known as the essential fatty acids since they cannot be synthesised by the animal body and must be provided in the diet from vegetable sources. However, arachidonic acid can be formed from dietary linoleic acid. The essential fatty acids are precursors of the prostaglandins.

The energy content of triglycerides is considerably greater than that of other nutrients – whereas 1 g of carbohydrate has a gross energy content of about 16 kJ, 1 g of triglyceride has a gross energy content of about 40 kJ. Thus fat may be included in animal diets as a means of increasing their energy density. Triglyceride, unlike the glycogen of liver and muscles, can be laid down in virtually unlimited amounts in adipose tissue (white fat) and serves as a more economical means of energy storage. Much of the white adipose tissue of the body is deposited under the skin (subcutaneous) with smaller proportions of the body's fat to be found around the kidneys, between muscles (intermuscular fat), and between and within muscle fibres. Species and strain of animals are important factors determining the distribution of fat depots. For example, 'beef' breeds of cattle have lower proportions of internal fat depots than 'dairy' breeds of cattle. The composition of the body fat in simple stomach species may be influenced by the type and amount of dietary fat consumed.

The membranes of cells and intracellular organelles in animals are largely comprised of a group of lipids termed phospholipids characterised by their phosphoric acid component. The most commonly occurring are the lecithins which also contain the water soluble vitamin choline. The phospholipids are higher in unsaturated fatty acids than the triglycerides of adipose tissue. Arachidonic acid is especially prominent and acts as a depot of this metabolically important fatty acid. The phospholipids are, along with another group of compound lipids the sphingolipids, major components of nervous tissue. As substances with emulsifying properties phospholipids fulfil important functions in

Table 12.6 Fatty acid composition (%) of some common fats and oils

Fatty acid	Palm oil	Soyabean oil	Rapeseed oil	Butter	Beef tallow	Lard	Fish oil
C14 and less	1	1	–	15	2	1	5
C16:0	45	12	5	23	35	32	13
C18:0	4	4	2	9	16	8	4
C18:1	40	27	56	35	44	48	22
C18:2	10	50	25	3	2	10	6
C18:3	–	6	9	–	0.4	1	4
C20 and above	–	–	3	–	–	–	37

lipid transport in the blood in which they combine with simple fats, cholesterol and proteins to form lipoproteins. Blood lipoproteins are classified on the basis of their density which is, in turn, a reflection of the proportions of their balance of lipid and protein.

Whereas triglycerides are the major lipid components of concentrate feeds and seeds the lipids in roughages are mainly in the form of glycolipids which in addition to containing fatty acids, chiefly linoleic and linolenic, have a carbohydrate component in the form of galactose.

The steroids are a large group of physiologically important compounds in plants and animals. All are derivatives of cyclic alcohols, the parent compound being the sterol nucleus.

Sterol nucleus

In animal tissues the sterol cholesterol is the most common, occurring in cell membranes, nervous tissue and in the blood. It is important as a precursor of such substances as the bile acids, sex hormones (oestrogens, androgens and progestins), hormones of the adrenal cortex (cortisol, corticosterone, aldosterone) and vitamin D. In recent years cholesterol has received much attention in the context of levels in the blood being associated with the condition of coronary heart disease in humans.

Micronutrients

Micronutrients are dietary components that need to be present in only relatively small quantities compared with carbohydrates, lipids and proteins. There are two broad groups of micronutrients; *the minerals*, which are elements other than carbon, hydrogen, oxygen and nitrogen required by the animal, and *the vitamins* which are organic

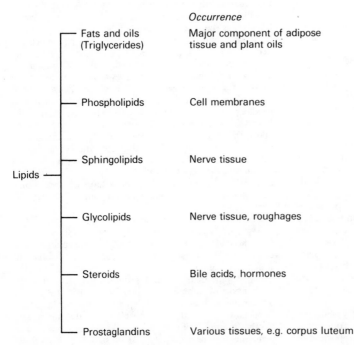

		Occurrence
	Fats and oils (Triglycerides)	Major component of adipose tissue and plant oils
	Phospholipids	Cell membranes
Lipids	Sphingolipids	Nerve tissue
	Glycolipids	Nerve tissue, roughages
	Steroids	Bile acids, hormones
	Prostaglandins	Various tissues, e.g. corpus luteum

Figure 12.7 Chemical classification of lipids

compounds required for particular body functions. Details of the occurrence of individual micronutrients in feeds and their role in the animal are given later (p. 333).

Water

Water is contained in all plant and animal cells. In foods the water content varies widely from over 90% in certain roots to 10–15% in dried feeds such as cereal grains and hay. Such variation in the water content of feeds makes it essential that comparisons of the nutritive value are made on a dry matter basis. The water content of the animal body is also highly variable. Since muscle tissue contains some 75% of its weight as water, whereas fat contains only 12–15%, the water content of the whole animal varies with the proportions of these two tissues, being highest in young lean animals and least in mature, fat animals.

Water is obtained by drinking and from food. Additionally, water is produced by metabolic processes during the oxidation of nutrients within the body. This is termed metabolic water. The oxidation of 1 kg of carbohydrate yields 0.6 kg water whilst for every kg of fat oxidised 1.1 kg water is produced. Metabolic water production is usually of the order of about 5–10% of the total water consumed. Animals lose water through faeces and urine, water vapour in expired air and through skin. Milk represents a major loss in lactating animals, whilst the pregnant animal deposits large quantities of water within the uterus particularly in late pregnancy.

Within the animal, body water is the major component of all body fluids and internal secretions. In the ruminant especially large amounts of water are secreted into the digestive tract through saliva and digestive juices. Although the bulk of this water is reabsorbed from the digestive tract faecal water losses are considerably higher in ruminants than in other species. The high specific heat and latent heat of evaporation of water aids in the regulation of body temperature, whilst chemically water takes part in hydrolytic reactions and in the absorption of digested nutrients.

Water intake and requirements are influenced by dietary and environmental factors. Thus water intake is related to dry matter intake and to the composition of the dry matter, especially in relation to the mineral salt content. The levels of sodium chloride and to a lesser extent the protein content of the diet influence urinary excretion and therefore requirements for water. Increases in the environmental temperature which increase water losses through respiration and sweating result in greater water consumption. All stock need adequate supplies of water at all times; insufficient water means reduced DM intake with resultant depression in productivity.

DIGESTION

The process of digestion breaks down complex food materials in the diet to simple products which can be readily absorbed from the digestive tract into the blood and lymph.

Anatomy of the digestive tract

Farm mammals may be divided into either simple-stomached (monogastric) animals, such as the pig, or ruminants which possess a compound stomach of four regions, cattle and sheep being the main examples.

The monogastric digestive tract

The arrangement of the monogastric digestive tract is shown in *Figure 12.8*. Food is digested by physical and chemical means while in the digestive system. The process begins in the mouth with the physical breakdown of food by the teeth. Saliva from the salivary glands helps the mastication and swallowing of the food. Movement of food through the digestive tract is brought about by waves of alternate contraction and relaxation of the muscular layers of the digestive tract wall, an action which is termed peristalsis.

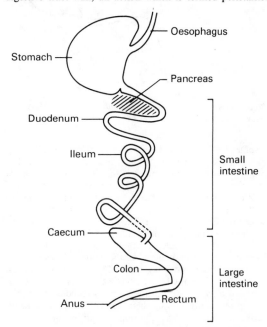

Figure 12.8 Diagrammatic representation of a monogastric digestive tract

Chemical digestion in the stomach begins with the secretion of acidic gastric juices. The digesta passes from here into the duodenum and the rest of the small intestine, where it is digested further by the actions of bile, pancreatic juice and intestinal juice. A more detailed description of the chemical processes involved is given later in this section. Digestive products are absorbed mainly in the small intestine. The large intestine is chiefly concerned with absorption of large quantities of water from the digestive waste, or faeces, before it is voided from the body (defaecation).

The ruminant digestive tract

The ruminant stomach, unlike that of the monogastric animal, is divided into four compartments: rumen, reticulum, omasum and abomasum. The first three of these are known as the forestomachs and as such have no digestive glands. The rumen is the largest of these compartments, comprising of more than 60% of total stomach capacity. The fourth stomach, the abomasum, closely resembles the structure and function of the stomach of the monogastric animal. The passage of food through the ruminant stomachs

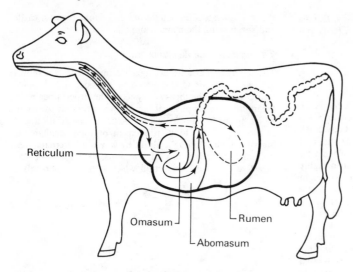

Figure 12.9 Diagrammatic representation of the pathway of food through the ruminant digestive system of the cow

is somewhat complicated (*see Figure 12.9*). After swallowing, food enters the rumen to be mechanically mixed and subjected to fermentation by the host population of microbes. Food may spend up to 30 h in the rumen. From time to time, coarse material is regurgitated into the mouth to be rechewed (chewing the cud) and then returned to the rumen for further digestion. The more fibrous the diet, the greater the rumination time, which may be as much as 8 h a day. In total, a cow may regurgitate 50 kg of food per day. Eventually the finer material in the rumen passes into the omasum, where water is extracted, and then on to the abomasum. Here any bacterial activity within the digesta is inhibited by the acidity of the gastric secretions. Microbial protoplasm is digested by the proteolytic enzymes present. From this stage onward the ruminant digestion resembles that of the pig.

The pre-ruminant digestive tract

At birth the rumen and reticulum are very much underdeveloped compared with the adult, thus the animal is said to be at a preruminant stage. Consequently, little or no rumen fermentation takes place in the newborn animals. Instead the liquid milk diet bypasses the rumen-reticulum and goes directly to the abomasum, where milk proteins are clotted and partially digested. The rumen bypass is made possible by the presence of an oesophageal groove running from the oesophagus to the omasum. Suckling or the drinking of a liquid diet stimulates the groove to close over into a muscular tube so that the contents are prevented from entering the rumen or reticulum. As the young animal is slowly introduced to a more solid diet, the reflex closure mechanism gradually diminishes, thus allowing solid material to enter the rumen so that fermentation can commence (*see* description of calf rearing in Chapter 14).

Biochemistry of digestion

Few of the dietary nutrients present in an animal's food can be directly absorbed by the animal and must therefore be

modified to make them available. This is accomplished in the digestive tract by enzymes either produced and secreted into the digestive tract by the host animal or of microbial origin, the micro-organisms living in symbiotic association with the animal. The former is most characteristic of monogastric species whereas the latter is characteristic of the ruminant. Neither monogastrics nor ruminants are, however, solely reliant on either host-produced or microbial enzymes for nutrient breakdown.

Carbohydrate digestion

Monogastric

Only non-structural plant carbohydrate such as starch and the various water-soluble carbohydrates are subject to hydrolysis by digestive secretions. There are no enzymes present in digestive secretions capable of hydrolysing cellulose and hemicellulose.

Saliva contains significant quantites of only one enzyme α-amylase and like pancreatic amylase this enzyme initiates the breakdown of starch into a mixture of maltose and dextrins. With an optimum pH of 7 salivary amylase activity is short-lived since as the food reaches the stomach the pH falls to a value of around 2. In the duodenum the food mixes with pancreatic and duodenal secretions and bile, the pH rises and conditions become suitable for the action of pancreatic amylase which continues the breakdown of starch principally to maltose.

The mucosal cells of the small intestine produce a number of carbohydrates. These are located on the brush border of the mature cells and include maltases, lactase and sucrase. Lactase activity is maximal in the suckling animal and declines as milk makes a declining contribution to the diet. These carbohydrases can hydrolyse the appropriate simple sugars to their monosaccharide constituents which are then readily absorbed from the small intestines, mainly by 'active transport' mechanisms. Thus the main end product of carbohydrate digestion is glucose.

Cell wall carbohydrates pass through the small intestine of pigs and poultry and reach the caecum where they are substrates for the microbial population that exists in this region. This microbial population can ferment cellulose and hemicellulose (and any starch evading breakdown in the small intestine) to volatile fatty acids in a similar fashion to ruminal fermentation (*see* next section) but it is quantitatively limited. These volatile fatty acids are absorbed from the rear gut and have in the case of the pig been estimated to supply between 10–20% of the absorbed energy.

Ruminant

All the dietary carbohydrates are subject to some fermentative degradation in the rumen. The major end-products of this fermentation are volatile fatty acids (VFA) mainly in the form of acetic, propionic and butyric acids along with the gases carbon dioxide and methane. VFA production represents up to three-quarters of the effective energy value of the diet. The rates at which dietary carbohydrates are fermented to these end products is dependent on type, soluble carbohydrate being fermented more rapidly than starches which in turn are fermented more rapidly than cell wall carbohydrates.

The concentration of VFA in the rumen is a function of their rates of production and rates of absorption. Absorption of VFA is in the free form without active transport. The acids produced during fermentation are partially neutralised by the buffers present in the saliva. Where diets containing high quantities of readily fermentable carbohydrates are fed acid conditions may result. Such lowered pH conditions can interfere with rumen fermentation and may lead to acidosis in the host (*see* p. 340). Where such diets are being fed violent fluctuations in rumen pH may be prevented by more frequent feeding, the inclusion of certain minimum amounts of long roughage to induce greater production of saliva and the inclusion of agents such as sodium bicarbonate in the diet.

The proportion of the various types of VFA produced by the fermentation of carbohydrate is dependent on diet. Typical values for a range of diets are given in *Table 12.7*. The acetic + butyric/propionic (non-glucogenic VFA/glucogenic VFA) is of particular relevance to the efficiency of dietary energy utilisation and efforts to manipulate this ratio is a feature of ruminant production. For example, the inclusion of ionophore-type and other antibiotics such as monensin and avoparcin to feeds can increase the proportion of propionic and thus narrow the ratio, whereas increasing

Table 12.7 The effect of diet on the molar proportions of VFA in rumen liquor

Diet	Molar % of VFA			
	Acetic	Propionic	Butyric	Higher[1]
Grass silage	76	13	7	4
Fresh grass	63	20	13	4
60% hay:40% concentrates	60	25	12	3
40% hay:60% concentrates	55	30	10	5
20% hay:80% concentrates	51	36	10	3
10% hay:90% concentrates	45	39	11	5

[1] Higher VFA include valeric, isovaleric and caproic acids.

the frequency of feeding promotes an increase in the proportion of acetic and thus a wider ratio. The former change is of benefit to fattening animals, the latter of benefit to lactating animals. Differences in the proportions of VFA end products is a function of the type of microbial population present in the rumen in that different microbial species use different nutrients as their principal substrates. Thus, for example, whereas *B. amylophilus* utilises starch *B. succinogenes* is a principal utiliser of cellulose.

Under some circumstances a portion of the fermentable carbohydrates may escape rumen fermentation. This is mainly in the form of structural carbohydrate which is sufficiently lignified to prevent microbial degradation. Further fermentation of this fraction may take place in the caecum leading to a further source of VFA to the animal but, in practice, the extent of this is very limited. Under some circumstances dietary starch may also escape rumen fermentation and reach the lower gut. This is a possibility when high cereal diets processed in particular ways are fed (e.g. finely ground maize). This starch can be digested in the small intestine in the same way as occurs in monogastric species with glucose as the principal end-product. Although this starch is used more efficiently the small intestine of the ruminant is more limited in its carbohydrase activity than the monogastric.

Gas production resulting from the fermentation of carbohydrates is lost by eructation and in energy terms accounts for about 7% of the food energy going into the rumen. This is in the form of methane and results from the reduction of carbon dioxide by hydrogen. A wide range of compounds have been shown to be capable of depressing methane production with the aim of increasing the efficiency of the fermentation. None of these are at present being used commercially.

Protein digestion

Monogastric

The enzymes concerned with protein digestion may be considered either as exopeptidases (enzymes which hydrolyse peptide bonds at the ends of polypeptides) or endopeptidases (those which hydrolyse peptide bonds within a polypeptide). The exopeptidases are either carboxypeptidases or aminopeptidases, the endopeptidases include pepsin, trypsin, chymotrypsin and dipeptidases. With the exception of pepsin, which operates in the acid environment of the stomach, the protein digesting enzymes are to be found in the small intestine. Trypsin and chymotrypsin are secreted in the pancreatic juices, the other enzymes are secreted from cells within the mucosa of the small intestine. A feature of these proteolytic enzymes is their specificity for peptide bonds involving particular amino acid types. For example, chymotrypsin has a preference for peptide bonds involving the carboxyl group of aromatic amino acids.

The end result of such proteolytic activity is the hydrolysis of dietary protein to free amino acids. These amino acids are absorbed by cells lining the small intestine and subsequently enter the portal blood and are transported to the liver. Small quantities of short peptides may also enter the absorptive cells of the small intestine where they are hydrolysed to allow free amino acids to enter the blood.

During the passage of food through the gut considerable quantities of endogenous protein are added in the form of digestive secretions and desquamated mucosal cells. This protein is itself digested and absorbed like dietary protein

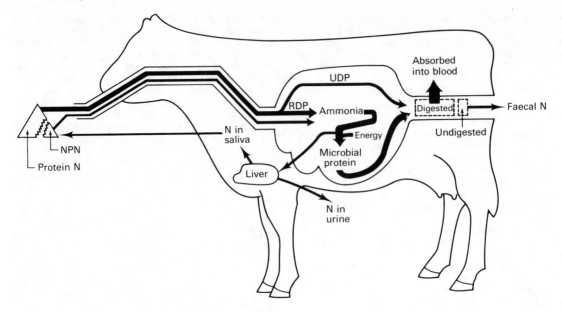

Figure 12.10 Summary of nitrogen utilisation by the ruminant

and complicates the assessment of dietary protein require-ments.

Ruminant

In the ruminant animal consideration must include not only the digestive utilisation of dietary protein but also the utilisa-tion of other sources of dietary nitrogen, termed non-protein nitrogen (NPN)(*Figure 12.10*).

Within the reticulo-rumen microbial activity results in the degradation of large but variable amounts of dietary protein. That which is broken down is termed rumen degradable protein (RDP), the protein that resists breakdown by bacteria is termed undegradable protein (UDP). Typically the ratio of RDP/UDP is about 70:30 but this may vary according to the protein sources in the diet.

Dietary RDP is initially hydrolysed by bacterial processes to yield free amino acids. These may be subsequently utilised by micro-organisms to synthesise microbial protein but the bulk of these free amino acids are further degraded to produce ammonia, organic acids and carbon dioxide. This ammonia is the major source of nitrogen for the synthesis of microbial protein. The rate of microbial protein synthesis and the efficiency with which this form of nitrogen is converted to microbial protein is mainly a function of the energy available in the rumen. It has been estimated that 7.8 g of microbial protein are synthesised per megajoule of metabolisable energy provided ammonia is not limited.

Thus readily fermentable carbohydrate sources lead to more efficient conversion of ammonia into microbial protein. When rates of ammonia production are greater than rates of utilisation ammonia concentrations build up in the rumen and may result in absorption of ammonia into the blood. On reaching the liver it is converted to urea and mainly excreted in the urine, although a small quantity ($\sim 20\%$) is recycled to the rumen via the saliva.

Ammonia needed by rumen micro-organisms may be derived from NPN sources, the most commonly used

commercial source being urea. This is rapidly hydrolysed by bacterial ureases to yield ammonia. Where urea is included in the diet it is essential that there be sufficient readily fer-mentable carbohydrate to achieve efficient conversion to microbial protein. Failure to balance rates of ammonia production with energy availability in the rumen may lead to ammonia toxicity in the animal (*see* p. 340).

In addition to RDP and carbohydrates the rumen micro-organisms also require sulphur and phosphorus for protein synthesis. Sulphur is required for the synthesis of the sul-phur-containing amino acids, a ratio of N:S of 10:1 is considered optimal, and phosphorus is a component of the nucleic acids which are involved in protein biosynthesis.

The microbial protein resulting from ruminal biosynthesis reaches the abomasum and small intestine in the form of micro-organisms carried by fluid out of the rumen once they lose their attachments to food particles. They pass into the abomasum and small intestine together with any dietary UDP. The processes of digestion and absorption of these proteins in the ruminant abomasum and small intestine are little different from those occurring in the monogastric animal. Thus the mixture of amino acids available for absorption from the small intestine is comprised of those that make up microbial protein and those present in UDP. It is thus markedly different in composition from dietary protein.

The most advantageous feature of the digestion of nitroge-nous compounds in the ruminant is that the microbial population can provide the animal with a source of relatively high quality protein-containing essential amino acids – from NPN and poor quality protein in the diet. The major disad-vantage of the system is that the ammonia produced in the rumen cannot be converted into microbial protein without some losses into the blood being incurred. Most efficient use of the system can be obtained when the amount of RDP in the diet matches the available energy in the rumen and should the resulting microbial protein be quantitatively inadequate to meet the requirements of the animal the shortfall is made up by the appropriate amount of UDP.

In recent years manipulation of the degradability of the dietary protein through either feeding an appropriate mixture of protein types or through chemical or physical treatment of proteins has been used to improve the efficiency of protein utilisation. Chemical treatment with formaldehyde and physical treatment with heat can reduce the solubility and therefore degradability of feed proteins. At the present time research is directed towards providing information that will allow quantification of the individual essential amino acids available for absorption from the intestine when particular protein sources are fed. Such information will allow nutritionists to formulate rations for ruminants that will supply appropriate amounts of essential amino acids to meet the animal's requirements for these nutrients. On silage based diets methionine has been shown to be the limiting amino acid for milk production. By protecting methionine from rumen decomposition inside a coat of hydrogenated lipid it is possible to supplement the animal.

A feature of the nitrogen content of micro-organisms reaching the small intestine is that a proportion (20%) of it is in the form of nucleic acids. The pancreatic juices of ruminants contain high activity of nucleases which break down the nucleic acids. The phosphorus released in this process is absorbed and recycled to the rumen via the saliva but the other end products are of no value to the animal and represent a further source of nitrogen loss in the digestive utilisation of dietary protein by the ruminant.

Lipid digestion

The digestion of lipids requires that they become miscible in water before they can be absorbed through the villi of the intestine. In this respect fat digestion and absorption differs from that of carbohydrate and protein.

Monogastric

The small intestine is the site of fat digestion and absorption. As it enters the small intestine fat becomes mixed with bile salts from the gall bladder in the form of taurocholic and glycocholic acids which have emulsifying properties reducing the size of fat particles and giving an increased surface area for digestive hydrolysis.

The major enzymes with lipolytic activity are to be found in the duodenum and are secreted in the pancreatic fluids. Pancreatic lipase hydrolyses triglycerides to a mixture of mono-, di- and triglycerides which, in the presence of bile, result in dietary fat forming micelles which after disruption on contact with the microvilli can enter the mucosal cells. Other lipid components in the form of fat-soluble vitamins and sterols are likewise components of these micelles. Within the mucosal cells the various lipid fragments absorbed are resynthesised into triglycerides and phospholipids and along with sterols and protein combine to form particles called chylomicrons. These pass into the lymph and then into the general circulation.

Differences in the efficiency with which different fats are digested and absorbed exist. Unsaturated fatty acids are digested better than saturated ones and digestibility decreases with chain length. Synergistic effects between fatty acids exist such that absorption of saturated fatty acids is greater in the presence of unsaturated fatty acids.

Ruminant

Ruminant diets normally contain only some 3–5% lipid either as triglyceride in concentrate feeds or galactolipids in forages but may contain up to 10%. The fatty acid composition can vary widely but in forages and cereals high proportions of linoleic and linolenic acid are present. In the rumen, microbial lipases hydrolyse a high proportion of these fats releasing fatty acids. Of the polyunsaturated fatty acids some 80–90% are rapidly hydrogenated to saturated and mono-unsaturated fatty acids. In addition this may also result in the production of fatty acids in the *trans* configuration. Ruminal micro-organisms also synthesise long-chain fatty acids many of which are of the odd-numbered and branched chain type. These processes account for the presence of these unusual fatty acids in both ruminant body and milk fat.

Unlike VFA the long chain fatty acids resulting from dietary and bacterial fat are not absorbed until the digesta reaches the small intestine where mechanisms of absorption are similar to the monogastric.

Although the modification of fat in the rumen is restricted to hydrolysis and hydrogenation, dietary fats themselves may inhibit fermentation with consequent effects on the extent of digestion in the rumen. The digestion of cellulose is particularly affected and may result in a 'high propionate type' fermentation. This occurs when the fat content of the diet is around 7–10% but is dependent on the type of fat and manner of incorporation in the diet. For example, unsaturated fat types have a greater effect on the fermentation than saturated fats and free fat a more deleterious effect than fat present in 'whole' oilseed meals.

In recent years systems of 'protecting' fat by coating the fat with undegradable protein have been developed. The inclusion of additional fat is of greatest relevance to dairy cows where there is often the need for more energy dense diets, although it may also be used to manipulate the amount and composition of milk fat. Appropriate processing of whole oilseeds may produce a similar 'protected-fat' system.

METABOLISM

Metabolism is the name given to the sequence of chemical processes that take place in the tissues and organs of the animal. Some of these processes involve the breakdown of compounds (catabolic), others involve the synthesis of substances (anabolic). Catabolic processes are frequently oxidative in character and are primarily concerned with generating energy for mechanical work and for the chemical work of synthetic processes. Thus the anabolic processes of carbohydrate, protein and fat synthesis in the body are inextricably linked to the catabolic processes. The common currency of energy production and energy utilisation in catabolic and anabolic processes is the substance adenosine triphosphate (ATP).

Energy from catabolism → / ADP \ → Energy for work, anabolism, etc. \ ATP /

The starting point of metabolism may be looked upon as the substances absorbed from the digestive tract.

Dietary nutrient	Major end-product of digestion	
	Ruminant	*Monogastric*
Carbohydrate	VFA-acetic, propionic, butyric acids	Glucose
Proteins	Amino acids	Amino acids
Fats	Mono, -di- and triglycerides, fatty acids	Mono, -di- triglycerides, fatty acids

As can be seen these end-products are different in ruminant and non-ruminant species and give rise to differences in metabolism between these animal groupings.

In the coverage of metabolism given here only a general outline of the processes involved is given. Further detail of particular pathways may be obtained by referring to appropriate textbooks.

Glucose metabolism

The major pathways by which glucose is catabolised to yield energy as ATP involves firstly the glycolytic pathway which results in the production of pyruvate and secondly the tricarboxylic acid cycle which results in the complete oxidation of pyruvate to carbon dioxide and water (*Figure 12.11*).

Glucose for these pathways is mainly obtained in the monogastric through absorption from the digestive tract. Glucose that is absorbed but not immediately catabolised may be stored as glycogen in muscle and liver cells. The amount of glucose that can be stored as glycogen is relatively limited and the bulk of glucose that is not catabolised is stored in the animal's adipose tissue. The concentration of glucose in the general circulation is kept within fairly narrow limits through the action of a variety of hormones but

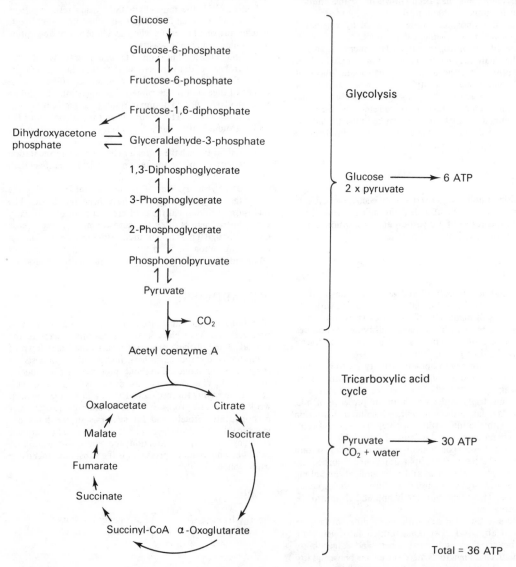

Figure 12.11 The oxidation of glucose via glycolytic and tricarboxylic acid pathways

especially insulin and glucogon. Thus, although the rates of absorption of glucose from the digestive tract fluctuate with meal patterns the levels of glucose in the blood are kept relatively constant. The storage of surplus glucose as glycogen and the ability to reconvert glycogen, especially that in the liver, back to glucose is one means by which blood glucose levels are regulated. Liver cells are also able to synthesise glucose from certain metabolites (e.g. amino acids, oxaloacetic acid) in order to maintain blood glucose levels. This is termed gluconeogenesis. Body fat reserves may also be called upon as an energy source in order to spare glucose catabolism when rates of absorption are lower than rates of utilisation.

In addition to acting as a principal energy source for the organs and tissues of the animal glucose is also used as the precursor for lactose synthesis in the mammary gland of lactating animals and is the principal energy supplying substrate to cross the placenta and be used by the growing fetus.

For the complete oxidation of glucose the presence of oxygen in cells is necessary. Under anaerobic conditions the pyruvate produced by glycolysis is converted to lactic acid. This is of relevence in the muscle cells of the animal post-slaughter where the catabolism of cell glycogen to lactate produces a fall in the pH of the muscle. The amount of lactic acid produced and the resulting pH can thus be influenced by the reserves of glycogen at slaughter. When animals have low muscle glycogen reserves at slaughter only a relatively small fall in pH results. The meat from such animals (most commonly bulls) is usually characterised as being dark in colour and firm and dry in texture (DFD meat), both of which are undesirable characteristics from the viewpoint of the consumer. Under some circumstances in animals that show excessive adrenalin release at slaughter, most commonly pigs with particular genetic characteristics, glycogen is quickly metabolised to lactic acid giving a rapid fall in muscle pH while the carcass is still warm. This tends to give rise to meat which is of pale appearance and soft, exudative in texture (PSE). Again these are undesirable characteristics to the consumer.

VFA metabolism

The volatile fatty acids are absorbed across the rumen wall down a concentration gradient. Rates of absorption are thus mainly dependent on rates of production in the rumen although other factors such as rumen pH may have an influence. Although fluctuations in the rates of absorption exist and are dependent on feeding regime and ease with which dietary carbohydrates are fermented the diurnal variation in absorption of energy supplying metabolites is considerably less in the ruminant than the fluctuations in glucose absorption in the monogastric.

Of the three main VFA produced in the rumen only acetic appears in the peripheral circulation. This is because propionic is rapidly converted to glucose by the liver and butyric to 3-hydroxybutyric acid during passage across the rumen wall.

Acetic acid is the major VFA absorbed and as such is catabolised via the tricarboxylic acid cycle by a variety of tissues to provide energy. When acetic acid is absorbed in amounts surplus to immediate energy needs the surplus is synthesised into long chain fatty acids and stored in the adipose tissue. Although the acetic acid acts as the precursor for fat synthesis the energy required for this anabolic process is supplied by glucose. Likewise in the lactating ruminant acetic acid is the precursor for milk fat synthesis within the mammary gland.

Although little glucose is absorbed from the ruminant digestive tract levels of glucose in ruminant peripheral circulation are only slightly lower than those found in the monogastric. A major source of this glucose is propionic acid although gluconeogenesis from amino acids also makes a significant contribution. Whilst glucose is not used as a principal energy source in the ruminant it is utilised in many ways as in the monogastric. Thus, for example, it is the main source of energy for the nervous tissue and the fetus, for glycogen and lactose synthesis and as an energy supplier for fat and protein biosynthesis.

Although usually absorbed in somewhat lesser quantities than acetic acid and propionic acid, butyric acid can be used as an energy source by a variety of tissues after being converted to 3-hydroxybutyrate in its passage through the rumen wall. Like acetic it may also serve as a precursor for both body and milk fat synthesis.

The proportions of the different VFA absorbed into the peripheral circulation can have an influence on the endocrine balance within the animal such that the partition of nutrients into different body processes may be affected. Thus high concentrate, low forage diets which favour the production of propionate in the rumen depress milk fat secretion in the lactating ruminant and promote lipogenesis in adipose tissues. It is thought that the stimulation of insulin secretion by propionate is the primary factor influencing partition.

Fat metabolism

Body fat metabolism is used as a means of regulating the energy metabolism of the animal. Thus dietary energy consumed in excess of immediate requirements is deposited as fat in adipose tissue and is released from adipose tissue as free fatty acids when the dietary supply of energy is inadequate. Such fluxes in fat reserves are characteristic of lactating and pregnant animals which have to call on fat as an energy source when food intake is inadequate to meet requirements and replace this fat at other times in their production cycle. Even on a diurnal basis, particularly in monogastric animals being fed discrete meals, a cycle of deposition and mobilisation of fat occurs. Fat is a considerably better energy store than either glycogen or glucose because not only does it have more than twice the energy content per unit weight than these nutrients it has also very little water associated with it ($\sim 12\%$ in fat) whereas glycogen and protein are heavily hydrated ($\sim 75\%$). Thus the energy stored per kg of tissue is about 32 MJ for fat and 4 MJ for glycogen and protein. Thus fat can supply a starved animal with energy for several months whereas the energy reserves in glycogen are sufficient for only about a day.

A principal site of fat metabolism is the liver. Both catabolism and anabolism of fatty acids is located in this organ. Fatty acids may be completely oxidised by the processes of β-oxidation and the tricarboxylic acid cycle to carbon dioxide and water and release energy. Such catabolism also takes place in muscle. Glucose and acetate are the major precursors for fatty acid synthesis. The relative importance of these two substances depends on species, the former being more important in monogastrics, the latter more important in ruminants. Adipose tissue can also synthesise fatty acids and is the major site of fatty acid synthesis in some species, including pigs and ruminants.

Fat circulates in the blood in a variety of forms. Fat resulting from the digestion and absorption of dietary fat is in the form of particles called chylomicrons. These particles consist mainly of triglyceride with small amounts of protein. Mostly they are processed in the liver but may release their fatty acids to adipose tissue or mammary tissue in the lactating animal, and thus contribute to body fat and milk fat. It is in this way that the fatty acid composition of dietary fat may influence that of the body and milk fat. Fatty acids that have been synthesised in the liver are transferred for storage in adipose tissue in the form of lipoproteins. It is also in this form that liver-synthesised fatty acids are transferred to the mammary gland for direct inclusion into milk fat. The other form in which fat circulates in the plasma is as free fatty acids (FFA), also termed as non-esterified fatty acids (NEFA). These are mostly transported in the plasma bound to albumin. Although the amount of NEFA in the plasma is very small this fraction is important metabolically because lipid is released in this form from adipose tissue and transported to the liver or other tissue to supply energy in times of need. The levels of NEFA are typically elevated in the undernourished animals. These NEFA may also be taken up by the lactating cell and be incorporated into milk fat.

Amino acid metabolism

Amino acids resulting from the digestion of dietary protein are absorbed from the small intestine into the portal blood. These amino acids contribute to the metabolic 'pool' of amino acids in the blood and tissues. In the body amino acids are also constantly being liberated into the blood from the breakdown of tissue proteins. This pool serves as a source of amino acids, some of which are used to build nitrogenous substances such as muscle or milk proteins or certain hormones whilst the vast majority are catabolised, their nitrogen being excreted as urea (*Figure 12.12*). The process of tissue protein synthesis and degradation is referred to as the 'turnover' of body proteins. Clearly in the growing animal synthesis exceeds degradation.

Muscle accounts for up to 60% of body protein. Thus the growth of animals involves a substantial uptake by muscle tissue of amino acids from the pool for protein synthesis. The efficiency with which amino acids in the pool are utilised is dependent on the proportions of the different amino acids that make up the pool. The proportions of amino acids in the pool are mainly influenced by the profile of amino acids absorbed from the intestine. The idealised profile is where the amounts of all essential and non-essential amino acids absorbed from the intestine is the same as that required for tissue protein synthesis. Only the total of non-essential amino acids supplied need be considered since there exists in the liver a capacity to synthesise non-essential amino acids that are in deficit from others that are in surplus by the process of transamination. Such an idealised profile does not exist in practice and some imbalance in the supply of amino acids exists (*Figure 12.13*). In this example lysine is the amino acid in greatest shortage, termed the 'limiting' amino acid and is responsible for the restriction in protein synthesis.

In monogastric species the profile of amino acids absorbed from the intestine is a reflection of the amino acid composition of dietary protein. In pig diets the amino acid most likely to be limiting is lysine, levels of which are especially low in cereals. Hence considerable attention is given to ensuring adequate lysine levels in practical diets through the inclusion of protein feeds high in lysine or of synthetic lysine.

In ruminant species the profile of absorbed amino acids is influenced by the composition of the microbial protein produced from dietary RDP and any UDP present in the diet. In many production situations the supply of amino acids from microbial protein is sufficient to meet the animal's

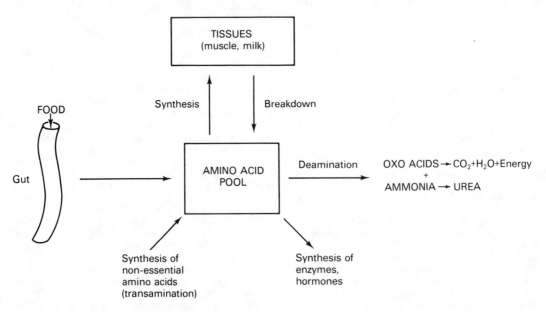

Figure 12.12 Representation of amino acid metabolism

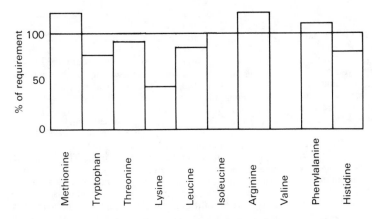

Figure 12.13 Diagrammatic representation of amino acid balance

requirements. In certain situations of high productivity, however, in particular the high yielding cow, the supply of particular essential amino acids may be limiting. In grass silage based diets methionine is most likely to be limiting whereas in corn silage based diets lysine may be limiting. Through the inclusion of UDP of appropriate amino acid composition or synthetic amino acids protected from rumen degradation an appropriate balance may be obtained.

Amino acids which are surplus to requirements for protein synthesis are converted by the liver in a process known as deamination to oxo acids and ammonia. The ammonia is rapidly synthesised to urea which is subsequently excreted in the urine. The oxo acids may be used as energy sources mostly by way of the tricarboxylic acid cycle. Surpluses of amino acids occur when an excess of protein is fed and/or where the profile of absorbed amino acids differ significantly from requirements. A further situation where the extent of deamination may be elevated is in starvation where tissue protein may be catabolised to supply energy and amino acids used to maintain blood glucose levels.

Mineral nutrition and metabolism

The essential mineral elements are designated as either macro- (or major) or micro- (or trace elements) depending on their concentration within the animal body (*Table 12.8*). Although animal tissues and feeds contain about 45 mineral elements only about half have been shown to have an essential function. Three general functions may be identified for minerals:

(1) a structural function in bones, e.g. Ca, P, Mg and fluorine:
(2) as electrolytes in body fluids maintaining acid/base balance, osmotic pressure and inducing excitation of nerves and muscles, e.g. Na, K, Cl and Ca;
(3) as integral components of enzymes and other biologically important compounds. Trace elements especially function in this capacity.

Animals obtain most of their mineral requirements from the feed they consume. Animal feeds vary widely in their mineral composition and individual feeds may vary in their mineral content according to the soil, fertiliser and other environmental influences on its production. The trace mineral composition of feeds tends to vary more widely than that of the major elements. As well as the absolute amount of individual mineral present in feed the availability of the element in terms of the proportion absorbed and utilised is also important. The availability of individual mineral elements may be influenced by other dietary components. Thus, for example, a surplus of phosphate may reduce the availability of calcium through the formation of insoluble salts. An excess of molybdate may likewise reduce the availability of copper. In some instances, availability of certain ions may be increased through the formation of soluble chelates. Such interactions present difficulties in the determination of requirements and in some instances the identification of particular deficiencies and imbalances.

In situations where feedstuffs fail to provide the animal with sufficient amounts of particular minerals to meet requirements mineral supplementation may be given in concentrated forms such as finely divided powders that may be efficiently distributed through feeds, suitable licks containing the deficient elements or through 'bullets' and 'needles' that lodge in the reticulum and slowly release the mineral throughout the animal's lifetime. The latter are particularly useful for the supply of trace elements and as a means of giving supplementary minerals to animals kept under extensive systems of production.

Table 12.8 The essential mineral elements

Major	Trace
Calcium	Iron
Phosphorus	Zinc
Magnesium	Copper
Sodium	Cobalt
Potassium	Manganese
Chloride	Molybdenum
Sulphur	Selenium
	Iodine

Calcium, phosphorus and magnesium

These three elements are found together in the animal's skeletal tissue. The calcium and phosphorus exist mainly in the form of crystalline hydroxyapatite $[Ca_{10}(PO_4)_6(OH)_2]$ which mineralises the organic matrix. About one-third of the magnesium in bones is bound to phosphate, the remainder being absorbed on the surface of the mineral structures. The main component of the organic matrix is protein in the form of collagen. Some 98% of the body calcium, 80% of the body phosphorus and 65% of the body magnesium are present in bone tissue. The remaining amounts of these elements are present in the blood either as free ions or in complexed forms and in the soft tissues. In these tissues they have diverse functions. Thus calcium participates in neuro-muscular activity and has a role in blood clotting, phosphorus functions in energy metabolism as a component of ATP, is present in the phospholipids of cell membranes and certain proteins whilst magnesium, like calcium, has an involvement in the excitation of nerves and muscles, and acts as an enzyme cofactor in metabolism.

The overall metabolism of these elements is illustrated in *Figure 12.14*. An important element of this metabolism is the maintenance of constant levels of calcium and phosphorus in the blood. This is achieved through the interaction of the two hormones, parathyroid hormone and calcitonin and the active metabolite of vitamin D_3, 1,25-dihydroxycholecal-ciferol($1,25(OH)_2D_3$) which control the absorption from the digestive tract, influence the balance of resorption and deposition in bone and influence the extent of excretion in faeces and urine. Parathyroid hormone (PTH) is secreted in response to a fall in blood calcium and its action in raising blood calcium through increasing absorption efficiency from the gut, increasing resorption relative to deposition in bone

and depressing urinary excretion is mediated through $(1,25(OH)_2D_3)$. The effects of calcitonin are antagonistic to PTH but do not involve $1,25(OH)_2D_3$. PTH and $1,25(OH)_2D_3$ are likewise involved in the regulation of plasma phosphate concentration. The amounts of ionised magnesium in blood serum are normally in the range of 20–40 mg/litre and although there is constant exchange of magnesium ions between serum and bone surfaces there do not appear to be the same type of homeostatic regulatory mechanisms of the type that exist for calcium.

Simple deficiencies of calcium, phosphorus or of vitamin D result in bone abnormalities such as rickets in the young and osteomalacia in the mature animal. The metabolic disease, milk fever, most commonly found in dairy cows just after parturition, is the main problem associated with a breakdown in the regulation of calcium metabolism (*see* p. 339). Because of the animal's lesser ability to mobilise bone phosphorus low levels of blood phosphate may readily arise when intakes of phosphorus are inadequate. Low levels of blood phosphate have been associated with poor fertility in animals. Low levels of serum magnesium (hypomagnesae-mia) may arise through dietary insufficiency but especially in ruminants low coefficients of absorption may be a contribut-ory factor. Absorption which occurs from the reticulorumen may be adversely affected by high levels of ammonia, potassium and phosphates. Outbreaks either of chronic or acute hypomagnesaemia (grass staggers) occur in ruminant livestock (*see* p. 340).

The amounts of calcium and phosphorus found in different feeds varies widely whilst most of the commonly fed diets for farm animals contain sufficient magnesium to meet body needs. As a generalisation, green forages are relatively high in calcium but low in phosphorus whilst animals on diets high in concentrates (cereals and oilseeds) are likely to

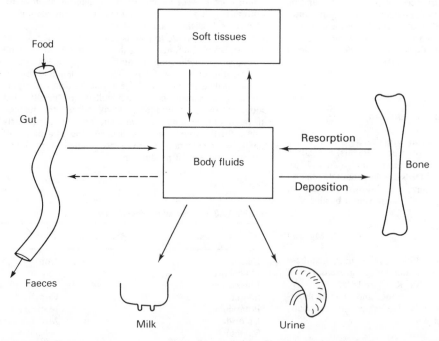

Figure 12.14 Diagrammatic representation of the calcium, phosphorus and magnesium metabolism

be receiving insufficient calcium but sufficient phosphorus. A number of supplementary sources of both minerals are available to correct problems of either insufficiency or imbalance in the dietary supply of calcium and phosphorus, e.g. ground limestone, mono-, di- and tricalcium phosphates, sodium phosphate and bone flour. These supplementary sources supply different amounts of the two elements and thus particular supplements may be used in given situations. Phosphates used as supplements must be defluorinated before feeding since fluorine is toxic.

Sodium, potassium and chloride

These three minerals are found mainly in the soft tissues and body fluids and together are largely responsible for the maintenance of osmotic pressure, fluid balance and acid-base balance and also have, along with calcium and magnesium, an important function in neuromuscular activity. Sodium is the major cation in body fluids, almost exclusively in the extracellular fluid and blood, whereas most of the potassium is found intracellularly. Chloride is the major anionic constituent of body fluids. It is a constituent of hydrochloric acid secreted in the gastric juice.

Deficiency of these elements is unlikely since the potassium content of most foods is high and, although the supply of sodium and, to a lesser extent, chloride are not always present in sufficient amounts, the practice of including common salt or sodium bicarbonate in the diet means that these elements are supplied in sufficient amounts. The intake of potassium is frequently much higher than that of sodium but a balance between the two is maintained by the hormone aldosterone which ensures the excretion of excess potassium and reabsorption of sodium by the kidney. Excessive intake of potassium may, however, limit the absorption of magnesium in the ruminant.

In situations of inadequate water intake or where either an excess or imbalance of sodium and chloride in the diet exists animals may be unable to regulate osmotic and acid-base balance allowing the development of either an alkalosis or acidosis. This is more likely in monogastric species and in young animals. There is now evidence that since these elements are all involved in the maintenance of acid/base and water balance that an optimum balance must be achieved between them and that imbalances, even in the absence of deficiency or toxicity, can adversely affect production. For chickens it has been suggested that the sum of $Na + K - Cl$ should be between 250 and 300 mmol kg^{-1} of feed.

Sulphur

Sulphur present in the body and food is largely present in the form of protein since sulphur is a constituent of the amino acids cysteine, cystine and methionine. Wool is particularly rich in cystine and contains about 4% of sulphur. Some sulphur is contained in the vitamins biotin and thiamine. In monogastrics sulphur deficiency is reflected in a shortage of these essential sulphur containing amino acids which are often limiting in tissue protein synthesis. In the ruminant the micro-organisms utilise elemental sulphur or sulphate to synthesise these sulphur containing amino acids. Deficiency in the ruminant is unlikely in most situations but there is evidence that animals receiving a high proportion of their nitrogen from urea or animals grazing pastures growing on soils low in sulphur may benefit from supplementation.

Iron

More than half the iron (60–70%) of the animal body is present as a constituent of the protein haemoglobin found in red blood cells and functioning in the transport of oxygen from the lungs to the tissues. Small amounts are to be found in myoglobin which serves as an oxygen store in muscle and in certain enzymes but the remainder of body iron is present in the form of storage compounds ferritin and haemosiderin. These are to be found in the liver, spleen, kidneys and bone marrow.

Although iron is poorly absorbed from the digestive tract – a coefficient of absorption of about 5–10% pertains in the adult – that which is absorbed is efficiently retained. Thus iron liberated from the breakdown of red blood cells and haemoglobin is efficiently recycled in the resynthesis of haemoglobin in the bone marrow and only small amounts are lost through excretion in the faeces (as components of bile) and in urine.

Iron requirements in the adult are generally low and easily met by dietary sources but young suckling animals are at risk due to the low level of iron in milk and the poor reserves of the newborn. The pale colour of the meat in veal calves is due to the low levels of iron-containing pigments. Suckling piglets are especially vulnerable to deficiency and should be routinely given supplementary sources either orally or by intramuscular injection. Piglets reared out of doors acquire sufficient iron by rooting in the soil. Supplementing pregnant females with iron compounds may have the effect of increasing the iron reserves of the newborn, but dietary intake of iron has no influence on the level of iron in the milk of lactating animals.

The main consequence of iron deficiency is anaemia, characterised by low blood haemoglobin, subnormal growth and diarrhoea. Low levels of haemoglobin may arise for reasons other than iron deficiency. For example parasitic infection, insufficient dietary protein and the presence of the unusual amino acid S-methyl-cysteine sulphoxide, which is present in brassicas, may contribute to the development of anaemias.

Cobalt

The chief role of cobalt is as a constituent of the vitamin B$_{12}$ molecule. In the ruminant micro-organisms synthesise this vitamin provided they have a source of dietary cobalt. Vitamin B$_{12}$ is an essential cofactor in the metabolism of propionic acid in the liver.

Cobalt deficiency occurs in grazing areas where the levels and uptake of soil cobalt is low. The condition is given various names, for example pine, and is characterised by a general unthriftiness, poor appetite and anaemia. Prevention of deficiency depends on oral administration of cobalt either as salts in the diet or as 'bullets'. The latter are small pellets that contain cobalt oxide which remain in the reticulum and release cobalt slowly.

Copper

Copper is widely distributed in the body and functions as a cofactor in several enzyme systems. It is involved in the formation of haemoglobin, is present in certain pigments found in the hair and has an involvement in the production of characteristic physical properties (crimp) of wool. The liver has an ability to concentrate copper and thus acts as a storage organ.

Various conditions arise, especially in ruminant stock when there is a shortage of copper. These range from anaemia resulting from copper being necessary for the utilisation of iron in haemoglobin synthesis, bone disorders resulting from copper being required to produce the structural integrity of the organic matrix, pigmentation failure, diarrhoea and impaired fertility. A condition in lambs known as swayback is associated with copper metabolism. Affected animals are unable to walk properly, suffering incoordination of the muscles and rear limbs. Postmortem examination reveals damage to the spinal cord which arises through inadequate synthesis of the myelin sheath in which copper has a role.

Such conditions may arise from a simple deficiency due to grazing areas with low levels of soil copper but frequently are due to poor utilisation of ingested copper. The absorbability of copper is influenced by a number of other dietary components. The presence of molybdenum and sulphate may, for example, make copper unavailable through the formation of insoluble copper thiomolybdate in the rumen. Such interactions influence the requirement of the animal for copper.

Whilst copper is an essential element it may also be highly toxic when consumed in quantity. In this context there is considerable species variation. Pigs can tolerate high levels of copper, it is routinely included at up to 150 ppm to stimulate growth, whereas sheep and to a lesser extent cattle are particularly susceptible to copper toxicity. This toxicity is due to the accumulation of copper in the liver and its subsequent liberation into the blood where it causes haemolysis.

Molybdenum

This is an element required in traces for certain enzyme systems in both the animal and ruminal micro-organisms but which is toxic at higher levels. In practice, deficiency is rarely encountered. Toxicity is related to its interaction with copper (*see above*) and may occur when animals graze herbage with a high molybdenum content. Molybdenum uptake by herbage is influenced by soil pH being greater at higher pH, such that molybdeosis can follow overliming of pastures. Control is through oral administration of copper sulphate.

Zinc

Zinc is distributed widely in the animal body with higher concentrations in bones, skin, hair and wool and the testes. Its major role within the animal is as an integral constituent of certain enzyme systems. Most diets contain sufficient zinc for farm animals but insufficiency may arise through inadequate absorption from the small intestine. Excesses of calcium and copper in the diet may depress zinc absorption as may also the phytic acid present in cereals and oilseeds. Deficiency initially results in skin lesions, termed parakeratosis in growing pigs, a condition readily alleviated through the addition of zinc to the diet.

Manganese

This element occurs in most tissues but is present in higher concentrations in bone, liver, kidney and pancreas. It is a component of enzymes involved in the synthesis of the mucopolysaccharides that occur in bone, in pyruvate carboxylase an enzyme central to fat and carbohydrate metabolism it has a role along with the vitamin biotin and is a cofactor for enzymes involved in the synthesis of cholesterol and urea. Deficiency is rare in farm livestock but when it occurs it can cause retarded growth, skeletal deformity and reduced fertility in breeding animals. The element is widely distributed in feeds but only some 5–10% is absorbed. The presence of excessive calcium and phosphorus may further depress this coefficient of absorption.

Manganese tends to be concentrated in the exterior layers of cereal grains such that bran is a good dietary source.

Iodine

The sole role of iodine in the body is as a constituent of thyroxine, a hormone secreted by the thyroid gland and unique in having an inorganic constituent. Thyroxine is primarily involved in the regulation of energy metabolism, but may also influence development and fertility. The secretion of thyroxine is regulated by hypothalamic (thyroid releasing factor, TRF) and pituitary (thyroid stimulating hormone, TSH) factors which are subject themselves to negative feedback. A deficiency of iodine leads to the production of insufficient thyroxine to prevent TSH from continually stimulating the thyroid leading to its enlargement, a condition known as goitre.

Iodine deficiency is generally a problem in regions with soils containing low levels of iodine. Where such situations exist supplementation with iodised salts is necessary taking care to avoid excessive levels. Some iodine deficiencies can arise in situations of apparently adequate intake due to the presence of dietary constituents termed goitrogens. These occur most commonly in brassicas in the form of thiocyanates and thiocarbamides and interfere with thyroxine synthesis. Whereas the goitrogenic effects of the thiocyanates can be overcome through the addition of extra iodine to the diet the effects of the thiocarbamides can only be partially overcome by this measure.

Selenium

The role of selenium as an essential nutrient is closely related to that of vitamin E in that both micronutrients are involved in the prevention of oxidation of tissue lipids. As a component of the enzyme glutathione peroxidase, selenium is involved in the destruction of peroxides produced by the oxidation of unsaturated fatty acids before they have an adverse effect on cell membranes. It is also necessary for the production of pancreatic lipase.

A deficiency of selenium causes a variety of syndromes in farm animals ranging from the nutritional muscular dystrophy (white muscle disease) found most commonly in calves and lambs and characterised by degeneration of skeletal muscle; mulberry heart disease found in piglets in which the heart muscle is affected and sudden death may occur; and the exudative diathesis found in chicks in which damage to the cell walls leads to leakage of fluids from cells and oedema of the breast. As well as being cured by inclusion of trace amounts of selenium to the diet these conditions also respond to vitamin E supplementation.

The margin between sufficiency and toxicity levels of selenium in the diet of farm animals is a relatively fine one and considerable care is required in providing supplementary sources. Selenium in foods exists as either the inorganic form of selenites or in organic forms bound to sulphur containing amino acids, e.g. selenomethionine. The organic forms are more available than the inorganic forms. The amounts of these different forms in feeds can vary widely and feeds originating from areas with low soil selenium may be deficient.

Vitamins in metabolism and nutrition

Vitamins are organic compounds required in extremely small quantities for the normal function, health and productivity of animals. As a generalisation animal cells are unable to synthesise these substances and must therefore obtain them from exogenous sources, either the diet or in some instances through microbial synthesis within the digestive tract. Individual vitamins are required for specific metabolic roles, frequently as integral parts of various enzyme systems. Thus deficiencies reveal a variety of disorders and symptoms that are frequently non-specific especially in the marginal deficiencies that are more likely to pertain in farm livestock.

Vitamins encompass a variety of chemical structures but may be readily classified according to their solubility characteristics into fat soluble and water soluble vitamins (*Table 12.9*). All the vitamins have chemical names but many continue to be known by letters of the alphabet by which they were designated prior to knowledge of their chemical identity.

The solubility characteristics of vitamins have implications for their absorption and storage. Fat soluble vitamins are absorbed along with fats in micelles and may be stored in fat containing tissues whereas water soluble vitamins are absorbed mainly by passive diffusion and there is little or no capacity for body storage necessitating frequent intake in the diet.

Most foodstuffs contain some vitamins or in some cases the precursors from which the animal derives vitamins (provitamins) but the amounts of individual vitamins in feeds varies widely. Monogastric animals and young ruminants must obtain most of their vitamin requirements from feedstuffs but in the case of the ruminant the microbial population in the rumen synthesises the B vitamins and also vitamin K which subsequently become available to the animal. In most instances the supply of these vitamins is sufficient to meet the ruminant's requirement for them but in situations of high productivity there may also be a need for dietary supplementation. Microbial synthesis of these vitamins also occurs in the rear gut of animals, both ruminants and monogastrics, but very little is absorbed and is only of real value to animals which practise coprophagy.

Vitamin A

All animals have a dietary requirement for vitamin A. In considering the supply of this vitamin two groups of compounds are of interest. One group are the carotenoids, the most important of which is β-carotene. Carotenoids are principally found in the leaf tissue of plants, to a lesser extent in seeds, and are precursors of vitamin A. The other group of compounds are forms of vitamin A itself which are only found in animal products and thus, from the viewpoint of farm animals, are present as supplementary rather than naturally occurring components of the feed.

The utilisation of dietary carotenes differs between species. In the pig, sheep and goat they are largely converted to vitamin A in the intestinal mucosa prior to absorption whereas in cattle and poultry some carotene escapes conversion and appears in the blood. Such absorbed carotenoids can be converted to the active vitamin in the liver and kidney but are also evident in the pigmentation of egg yolks, poultry carcasses and the milk and fat of cattle. The efficiency of conversion of carotene to vitamin A varies from almost 100% in poultry to less than 30% in ruminants. In all animals the conversion efficiency declines with increasing intake of either carotenes or vitamin A. Surpluses of vitamin A in the body may be stored in the liver and protect the animal during periods of vitamin A insufficiency.

Vitamin A performs a variety of functions in the body but many of these relate to the maintenance of the integrity of epithelial tissues. Thus in vitamin A deficiency, keratinisation of epithelia in the respiratory tract, genitourinary tract, alimentary tract and cornea occurs. Such damage to the membranes in these tissues not only leads to poor absorption from the digestive tract and also respiratory and reproductive problems but also allows ready entrance of bacteria such that secondary infections may arise.

Vitamin A has an important role in the formation of bones, especially in the formative stages. Deficiency causes retardation of growth, bones being shorter and thickened. The synthesis of certain glycoproteins, a major constituent of the organic matter, is depressed.

The animal normally receives its vitamin A from carotene in plant materials. The carotene content is high in green crops, especially in young leafy material, whilst roots and cereals are generally poor sources. The carotene content of feeds can be influenced by harvesting and storage processes.

Both enzymic and non-enzymatic oxidation may lead in some instances, e.g. haymaking and storage, to up to 80% destruction of carotene in feeds. When vitamin A supplements are added to animal feeds it is usually protected in a gelatin–carbohydrate coating containing antioxidant to prevent loss in activity during feed storage.

Vitamin D

Two forms of vitamin D are of practical importance, namely D_2 and D_3. These active forms of the vitamin, chemically known as ergocalciferol and cholecalciferol, are formed from the effects of ultraviolet irradiation on the steroids ergosterol and 7-dehydrocholesterol which are the respective provitamins. Ergosterol occurs commonly in plants and is transformed during the sun-curing of forages to the active

Table 12.9 Vitamins of importance in animal nutrition

Vitamin	Chemical name
Fat soluble vitamins	
A	Retinol
D_2	Ergocalciferol
D_3	Cholecalciferol
E	Tocopherol
K	Phylloquinone
Water soluble vitamins	
B_1	Thiamine
B_2	Riboflavin
–	Nicotinamide
–	Pyridoxine
B_6	Pantothenic acid
–	Biotin
–	Folic acid
–	Choline
B_{12}	Cyanocobalamin
C	Ascorbic acid

form of the vitamin. The provitamin of D_3 is synthesised in animal tissues, especially skin. Vitamins D_2 and D_3 have similar effectiveness for mammals but D_2 is virtually inactive in poultry.

The role of vitamin D is chiefly concerned with calcium and phosphorus metabolism and the mineralisation of bone (*see* p. 334). It is now clear that vitamin D undergoes metabolic change in the liver (to 25-hydroxvitamin D_3) and the kidney (to 1,25-dihydroxyvitamin D_3) before it can perform its role of influencing the intestinal absorption, bone mobilisation and urinary excretion of calcium and phosphorus. The mode of action of $1,25(OH)_2D_3$ is similar to that of steroid hormones through inducing the production of messenger RNA.

Intensively kept livestock that do not receive direct access to sunlight and which receive little of the vitamin from cereal based diets require supplementation whereas grazing animals receive adequate amounts from irradiation. Synthetic metabolites of vitamin D have been used in the prevention of milk fever.

Vitamin E

Vitamin E refers to a group of substances called tocopherols, the most important of which is α-tocopherol. Although first identified to have a role in reproduction it is now known to function primarily as an antioxidant preventing peroxide damage to cell membranes resulting from the oxidation of polyunsaturated fatty acids. In this role it is supported by the selenium containing enzyme glutathione peroxidase. The conditions that result from selenium deficiency (*see* p. 336) may also arise if vitamin E is deficient.

Green foods and cereal grains are good sources of the vitamin although some of the activity may be lost during storage, e.g. high moisture storage of cereal grains. Where supplementation is required it is usually provided as α-tocopherol acetate. The need for supplementation is increased if the diet has increased levels of polyunsaturated fatty acids as, for example, when calves are turned out to grass in spring.

Vitamin K

Vitamin K is required for normal blood clotting through its involvement in the synthesis of prothrombin. Deficiency of vitamin K is rare in mammals since it is synthesised by the micro-organisms of the rumen in ruminants and in pigs hind gut synthesis combined with a degree of coprophagy is usually sufficient to meet requirements, although young piglets housed on slatted floors require supplementation. Likewise, poultry are routinely supplemented with vitamin K concentrates or with feeds such as lucerne meal that are high in vitamin K. The dietary requirement is increased during periods of treatment with antibiotics which reduce intestinal synthesis.

A relative vitamin K deficiency can occur when its action is blocked through antagonists such as dicoumarol which is found in mouldy clover hay. It can be prevented by giving extra vitamin K. Agents which similarly block the action of vitamin K, e.g. warfarin, have found use as rat poisons and in the treatment of thrombosis in humans.

B vitamins

The group of vitamins referred to as B vitamins function primarily as components of enzyme systems involved in carbohydrate, fat and protein metabolism. Ruminant animals usually obtain their requirements from microbial synthesis in the digestive tract and since the B vitamins are present in significant quantities in a wide variety of feeds it was formerly thought that only in particular situations did pigs and poultry require supplementation. The need for supplementation, however, is constantly being increased through the improved performances demanded of animals such that not only monogastrics but even high producing ruminants may now benefit.

Some of the metabolic roles and symptoms arising in deficiency are given in *Table 12.10*. Identification of problems resulting from B vitamin insufficiency is frequently difficult because they are often associated with non-specific

Table 12.10 The metabolic roles and deficiency symptoms associated with the B-vitamins

Vitamin	Metabolic role	Deficiency symptoms
Thiamine (B_1)	Decarboxylation in carbo-hydrate metabolism	Nervous disorders — cerebro-cortical necrosis (CCN), polyneuritis
Riboflavin (B_2) Niacin	Hydrogen transfer in energy metabolism	Non-specific; curled toe paralysis (chick)
Pantothenic acid	Part of Coenzyme A	Non-specific; 'goose-stepping' (pigs)
Pyridoxine (B_6)	Amino acid metabolism	Non-specific; skin lesions, anaemia
Biotin	Fatty acid metabolism	Skin and hoof lesions, reduced fertility
Choline	Phospholipids, methionine metabolism	Fatty liver, perosis (birds)
Folic acid	Nucleic acid synthesis	Non-specific; anaemia, reproductive problems
Cobalamin (B_{12})	Nucleic acid synthesis Propionate metabolism	Non-specific; dermititis (pigs), poor feathering (poultry)

symptoms such as reduced appetite, poor performance, diarrhoea and general poor condition.

In meeting the requirements for individual vitamins not only is the vitamin content of feedstuffs relevant but also the availability and stability of the vitamin. Thus, for example, thiamine deficiency in ruminants results not from a lack of thiamine in the digestive tract but from a combination of thiamine destruction by thiaminases and reduction in availability through the action of antagonists. Antivitamin factors have likewise been associated with biotin (streptavidine), folic acid (sulphonamide), niacin and pyridoxine. As far as stability of vitamins in feeds is concerned the effect of heat and light during processing and storage can lead to loss of thiamine, pyridoxine and riboflavin activity. In the case of biotin, levels in feed do not relate to activity of the vitamin. Thus, for example, in wheat and barley, although biotin is present in significant amounts, it is almost totally unavailable whereas in feeds such as soya and fish meal all of the biotin is available.

Dietary supplementation using commercial forms of vitamins is primarily dependent on the ability of dietary raw materials to supply the vitamin but also on the cost of the vitamin. Thus since excess of water-soluble vitamins is not harmful supplementation may in some instances (e.g. riboflavin) be carried out merely as a precautionary measure in cases where the vitamin is inexpensive, but in the case of a highly expensive vitamin such as biotin there is greater need to justify and be precise in supplementation. A further important parameter in deciding whether supplementation is necessary is the degree of confidence that can be placed in tables of composition and requirements.

Vitamin C

It is often considered that since farm animals can synthesise this vitamin there is no need for its consideration in the diet. Some evidence exists, however, that extra dietary vitamin C may help animals deal with stress resulting from adverse environmental conditions such as temperature and ill health. Such an effect of vitamin C may result from its role in the synthesis of corticosteroids. Beneficial effects of supplementation have been reported for poultry kept at high temperatures and for early weaned piglets.

Vitamin C is also known to function in collagen synthesis, as an antioxidant and as an enhancer of iron absorption from the gut.

Metabolic disorders in livestock

A number of disorders occur in farm animals which appear to be related to an imbalance in the dietary supply of and demand for nutrients. These disorders tend to occur in animals with a high demand for particular nutrients and are sometimes referred to as 'production diseases'. They occur where the classic physiological mechanisms which normally deal with fluctuations in supply and demand and partition of nutrients are unable to cope or adapt quickly enough thus leading to metabolic imbalances.

Ketosis

This condition occurs both in the cow and ewe. Typically it occurs in cows (bovine ketosis/acetonaemia) in early lactation when milk yield is at a peak and in ewes (pregnancy toxaemia/twin lamb disease) in late pregnancy when carrying two or more lambs. The metabolic origins of the conditions are an imbalance in the animal's energy supply and demand. The energy intake in these conditions is limited in the case of the pregnant ewe by the growing fetus restricting rumen volume whilst in the dairy cow maximum voluntary food intake is not achieved until some time after peak yield. In both ewes and cows, animals that are overfat tend to have more restricted food intake.

Glucose is the energy supplying metabolite particularly in demand in the pregnant animal for the metabolism of the fetus and in the lactating cow for milk synthesis. Thus a characteristic of the condition is low blood glucose levels. Further to this the imbalance in energy supply and demand leads to the metabolism of body fat reserves. The metabolic consequence of this is that in its attempts to maintain glucose supplies through gluconeogenesis the mobilised body fat is incompletely oxidised and leads to the formation of the ketone bodies that are characteristic of the condition. The accumulation of ketones can be recognised through blood analysis, the use of the Rothera test on milk samples, or the presence of the sweet odour of acetone on the animal's breath. As well as ketone bodies being produced, fatty acids mobilised from adipose tissue may lead to fat deposition in the liver ('fatty liver' syndrome). This restricts gluconeogenesis and further exacerbates the problem. The low availability of glucose eventually leads to nervousness, inappetance and in the cow a drop in milk yield. The drop in milk yield in the cow can lead to spontaneous recovery as the supply and demand for energy gets back in balance but in sheep the condition is often fatal unless the fetuses are aborted.

Prevention of the condition involves taking measures which allow the animal to keep energy metabolism in balance. Avoidance of overfatness, maintenance of an appropriate energy intake through ration formulation and avoidance of sudden changes in diet that may affect the rumen function are essential. The inclusion of appropriate amounts of protein, particularly UDP, has also been shown to be an important preventative measure.

Milk fever (parturient paresis)

Milk fever is a metabolic disturbance affecting high producing dairy cows normally within two to four days of calving. It is less frequently found in sheep in late pregnancy.

The condition arises from an imbalance in the animal's calcium metabolism which leads to a dramatic fall in blood calcium from the normal of 2.5 mmol/litre to less than 1.5 mmol/litre. The clinical signs include initial excitement and change in muscle tone developing into muscle tremor, stiffness of gait and eventual recumbency and coma. The condition arises through the inability of the mechanisms which normally regulate calcium metabolism to adjust to the sudden increase in demand for calcium for milk synthesis at the onset of lactation. Under normal circumstances any imbalance between the supply of calcium in the diet and demand for calcium is taken care of through the mobilisation and resorption of calcium from and into skeletal reserves. Such regulation is mainly brought about through the actions of parathyroid hormone, calcitonin and vitamin D. Higher yielding and older cows are more susceptible, the latter being due to the lesser ability to mobilise calcium from the skeleton.

Preventative measures suggested include the feeding of low calcium diets in late pregnancy to stimulate the regulatory mechanisms into action. In practice, it is difficult to feed diets with sufficiently low levels of calcium. Alternatively, treatment with vitamin D_3 and its analogues at appropriate times pre-calving can lead to the mobilisation of bone calcium but the timing of such treatment is critical. Therapy involves the intravenous administration of calcium borogluconate solution. Other compounds such as calcium hypophosphate and magnesium sulphate are necessary in certain cases. Recovery from therapeutic treatment is usually rapid although relapses may occur.

Grass staggers (hypomagnesaemic tetany)

This is a disorder of cattle and sheep and is characterised by low levels of blood magnesium resulting from an inadequate absorption of magnesium from the digestive tract. In its acute form, which most commonly occurs in lactating cows shortly after 'turn-out' in spring and less commonly during periods of rapid growth in autumn, animals show progressive hyperirritability leading to staggering gait, tetany, violent convulsions and death. A chronic form is found particularly in beef cattle being fed poor quality winter feeds and in outwintered animals.

The inadequate absorption of magnesium from the gut may be due to a restricted food intake, low magnesium content in the diet or low absorption efficiency. Very often a combination of these factors is involved. Unlike the situation with calcium metabolism the animal has little ability to buffer fluctuations in supply and demand for magnesium since there is little skeletal reserve and this is largely immobile. A steady supply of magnesium in the diet is therefore essential. Amongst a variety of factors that have been implicated in influencing the efficiency of magnesium absorption are rates of nitrogen and potassium fertiliser application to grass, rumen pH and the stress sudden adverse weather conditions can cause.

It has been well demonstrated that daily administration of magnesium oxide (calcined magnesite) in the concentrate part of the feed will prevent grass tetany during seasons when it is prevalent. Dosage levels of $50 \, g/d$ for cattle and $7 \, g/d$ for sheep are adequate. The use of magnesium bullets which lodge in the rumen and slowly release magnesium are a useful alternative in more extensively managed stock.

Rumen related metabolic disorders

The health of ruminant animals may be adversely affected by imbalances in rumen metabolism. Such imbalance may occur through the excessive or sudden ingestion of concentrates leading to *acidosis* and when excessive non-protein nitrogen and rumen degradable protein is fed leading to *ammonia toxicity*.

The problem of acidosis results from the production of very high levels of lactic acid and low rumen pH. This may influence gut function causing rumen stasis and epithelial damage or alternatively the rapid entry of lactic acid into the blood may so upset the animal's acid/base that hypotension and respiratory failure ensue.

Sudden ingestion of concentrates brings about the acute form of the condition but chronic forms may be seen in cattle given very high concentrate diets with little or no forage. Such diets are associated with characteristic changes in the microbial population of the rumen, Gram-negative organisms being replaced by Gram-positive organisms such as lactobacilli.

Ammonia concentration in the rumen is a function of its rate of production from dietary RDP and NPN and its rate of utilisation for microbial protein synthesis. The latter is mainly dependent on energy availability in the rumen. An imbalance in the production and utilisation of ammonia in the rumen leads to accumulation of ammonia in rumen liquor. This eventually spills over into the blood upsetting the animal's acid/base balance and producing signs of hypertension and in extreme cases respiratory failure.

Prevention of both these conditions is dependent on care in ration formulation and feeding regimes. Observation of appropriate ratios of forage to concentrate and of degradable protein and NPN to available energy are essential. The mixing of ingredients and frequency of feeding need also be considered.

Blood chemistry and nutritional status

Under normal circumstances the concentration of blood metabolites is kept within well defined limits and it is only in circumstances of metabolic imbalance that deviations from the 'normal' occur. Identifying deviations from normal in the composition of blood allow the diagnosis and correction of subclinical conditions which may adversely affect production. The analytical details of levels of metabolites present in blood samples taken from animals is referred to as its 'metabolic profile'.

Metabolic profile testing has to date been mainly used to monitor the nutritional status of dairy herds. Animals considered to be representative of high yielding, medium yielding and non-lactating groups in the herd are sampled and the blood analysed for glucose, 3-hydroxybutyrate, free fatty acids (indicators of energy status), albumin, globulin, urea and haemoglobin (all indicators of protein status), packed cell volume (an indicator of blood dilution) and various minerals (e.g. Ca, P, Mg). After analysis of samples the results are processed by computer and presented as a histogram allowing comparison of levels of individual metabolites with 'normal' values. Normal values are considered to be the mean for the population ± 2 s.d. from the mean.

Full interpretation of the data requires information on both the feeding regimes and performance of animals. Only then may recommendations on changing the nutritional inputs be made.

VOLUNTARY FOOD INTAKE

Voluntary food intake may be defined as the weight of food eaten by an animal per unit of time when given free access to food. Its importance as a parameter influencing both the biological and economic efficiency of production has been increasingly realised in recent years. The level of intake dictates the rate of production, the proportion of the intake going towards production and thus the efficiency of food conversion. The level of intake may also influence the products of production. For example, when intake is excessive then excessive fat deposition may occur in growing and lactating animals. Thus an aim in production is to match intake to the required level and type of production.

An ability to predict and manipulate the voluntary food intake of farm animals requires a knowledge of those factors that influence it. Like other animals, farm animals appear to control their food intake primarily by monitoring their consumption of available energy (energostasis). Thus if animals are fed a range of diets of differing energy concentration they will alter their dry matter consumption through altering either meal frequency or meal size in order to achieve energy balance. The precise mechanisms by which such regulation occurs is still unknown but involves centres in the hypothalamus of the brain that receive neural and endocrine information pertaining to energy balance and thus regulate the 'drive' to eat.

Mechanisms of energostasis

The mechanisms by which animals are capable of regulating their energy balance are imprecisely understood. Many of the early theories on the control of food intake have proposed single factors. These include such classic ideas as the *chemostatic* theory in which it is proposed that levels of primary blood metabolites, such as glucose in the monogastric and volatile fatty acids in the ruminant provide the feedback information; the *thermostatic theory* that proposes appetite to be linked to the animal's thermoregulatory mechanisms; and the *lipostatic control theory* which proposes that the feeding drive is in some way controlled by the body's fat reserves. It is unlikely, however, that food intake is regulated by any single mechanism and that whilst some aspects of these theories are involved, the centres in the brain concerned with the control of feeding are more likely to receive a variety of feedback signals which they integrate to determine feeding behaviour. As well as the involvement of primary blood metabolites it is likely that metabolic hormones, such as growth hormone, insulin and glucogen, and gut peptides, such as cholecystokinin, have a part to play in energostasis.

Limits to energostasis

Although the underlying mechanism by which animals regulate their intake is related to energy balance a number of factors may limit the capacity of animals to achieve 'energostasis'.

The ultimate limiting factor to the intake of food must clearly be the physical capacity of the digestive tract, the rates of digestion and absorption and the rate of passage of food through the digestive tract. Thus, although animals may seek to control their energy consumption through phy-siological mechanisms they may fail to achieve this when fed bulky and/or indigestible feedstuffs. In practice the extent of dietary dilution necessary to produce limitations to gastric distension is unlikely to pertain in the monogastric and is more likely to arise in the ruminant animal. Evidence for the physical limitation of intake comes from observations relating intake to the available energy concentration of the diet. This is illustrated in *Figure 12.15*. The range of energy concentrations over which energostasis may be accomplished or physical limitations to intake prevail is dependent upon the physiological status of the animal and characteristics of the feed other than simply energy concentration. Thus, for example, in pregnant or overfat animals where the developing uterus or internal fat deposits restrict rumen volume physical limitation occurs at a lower dietary energy concentration. As far as feed characteristics are concerned these features which influence the rate of passage of food through the gut are relevant. Thus digestibility and rate of digestion and those features of a feed such as chemical composition, level of processing and water content which may influence these parameters are influential. The extent of gut fill is monitored by stretch receptors in the gut wall and this information is relayed to the brain via the nervous system where it influences feeding behaviour.

The prediction of intake

Since a knowledge of the dry matter intake which an animal will eat is essential for the precise and economic formulation of diets the ability to predict intake is of the utmost importance. For ruminants, particularly dairy cows, many relationships have been proposed. The simplest is that proposed by MAFF and widely used in advisory work.

Dry matter intake (kg/d) $= 0.025\,W \times 0.1\,Y$

where W is body weight in kg and Y is daily milk yield in kg. Such a simplistic prediction equation has obvious drawbacks in that characteristics of the food, such as physical form, palatability, acidity and chemical position and characteristics of the animal other than bodyweight and milk yield (e.g. pregnancy, degree of fatness, liveweight change and stage of lactation) all of which are known to influence intake are not taken into account. A number of multiregression equations relating some of these factors to intake have been derived but incorporation of many of the factors is limited by the lack of quantitative data. Such prediction equations tend to be highly complex and to date have not been widely used in advisory work.

Prediction of intake for pigs has received much less attention since the practice in pig feeding has been to restrict intake in order to achieve appropriate carcass composition at slaughter. In prediction equations derived from poultry a parameter with a considerable influence is environmental temperature.

Manipulation of intake

The objectives in the manipulation of intake are dependent on the species involved. For pigs attempts to manipulate voluntary intake have concentrated on limiting energy intake whilst allowing *ad libitum* feeding and thus saving labour. At present no practical way appears open to achieve this objective and in the long term is more likely to be

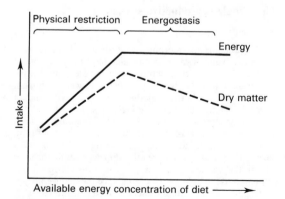

Figure 12.15 Representation of the relationship between the energy concentration of the diet and energy and dry matter intake

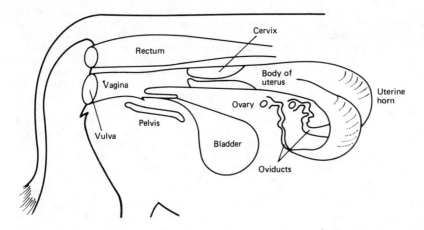

Figure 12.16 Reproductive tract of cow

attained through the selection of animals which can feed *ad libitum* without becoming overfat.

With ruminant animals production is frequently limited by physical restraints on the intake of forage and the objective is to increase total dry matter intake, especially the proportion of forage in the diet.

The greatest scope for manipulating intake at present lies in altering the characteristics associated with the feed which affect intake. Thus the processing of foodstuffs can have a major effect on the intake of feeds. The grinding and pelleting of forage, the chemical treatment of forages with alkali, and the processing of cereal grains either physically or through techniques such as micronisation that gelatinise the starch may all influence intake through their effects on the rate of passage and/or digestibility of feeds. The method of feeding may likewise influence total daily intake. The mixing of the various dietary components to produce a 'complete diet' may increase intake as may the method of feeding employed in feeding systems involving the separate allocation of compound and roughage. Thus the more frequent feeding of compound through the use of 'out-of-parlour' feeders and the mechanised feeding as opposed to self-feeding of silage may increase intake.

In the grazing animal there is now a conceptual basis for understanding the influence of plant morphology and sward structure on herbage intake in terms of the effect of such parameters as bulk density, sward height and shearing strength. Quantification of such relationships so that sward variables may be objectively manipulated are now required.

The facility to manipulate the animal in ways which will influence intake is more limited. Since body fatness has a clear influence on intake, attaining appropriate body condition targets in the production cycle of pregnant and lactating animals is important. Social interaction can influence feeding behaviour such that group size, pecking order and social facilitation of feeding are characteristics with potential for manipulation. Feeding behaviour and intake is affected by photoperiod and although the mechanisms involved differ in different species it may be possible to manipulate this parameter to advantage. The recent observation that Brahman cattle consume considerably more forage dry matter in relation to body weight when compared with Friesians has led to the suggestion that

selection for the characteristic of rumen volume and/or giving animals early experience of very bulky, low quality diets may be worthwhile.

REPRODUCTION

Reproduction is of crucial importance to all animal populations, both for herd or flock replacement and as a means of genetic improvement. Reproductive physiology is the science of reproductive functions within the animal. As such it holds the key to increased reproductive efficiency. It may also be able to provide explanations when things go wrong – such as subfertility or infertility in breeding stock. The manipulation of reproduction through the application of reproductive technology is now standard practice in most modern livestock enterprises. The following section seeks to provide a basic outline of the physiology of reproductive processes and to review the more recent advances in the technology of reproduction.

The female reproductive system

Anatomy

In all mammals the reproductive system of both sexes is bilaterally symmetrical, which means that organs consist of right and left paired structures. The female reproductive system is comprised of two ovaries, two oviducts or fallopian tubes, two uterine horns, a uterus, a cervix, a vagina and an external opening, the vulva (*see Figure 12.16*). Although the female reproductive anatomy is essentially similar between farm species, there exists a distinction in the degree of fusion of the uterine horns and their relative length with respect to the body of the uterus. This relationship depends on the litter size of the species. Thus the pig has long uterine horns compared to the body of the uterus since the many embryos of the sow develop within these horns. In contrast, the cow normally has a single offspring which develops in the uterine body, therefore the uterine horns are relatively much shorter. The ewe reproductive tracts falls somewhere between these two extremes.

Sex determination

The sex of an offspring is determined by its genetic makeup or chromosomes which are donated by its parents. A female mammal has a pair of identical sex chromosomes, designated XX, whilst those of the male are non-identical, XY. The primary reproductive organ, the ovary, produces female gametes or eggs which carry only half the chromosome complement of the nuclei of the parent cells. Therefore each egg cell can carry only one sex chromosome and this must be an X chromosome. Male gametes or sperm cells on the other hand can be either X- or Y-bearing. Thus, at fertilisation, the sex of the resulting embryo is determined by the 'sex' of the fertilising sperm cell. Within ejaculated semen there are approximately equal numbers of X- and Y-bearing sperm, thus the chances of a fertilised egg being male or female is 50:50. This is known as the primary sex ratio. The secondary sex ratio is the number of male:female offspring at birth.

The genetic sex of the female is determined at fertilisation. The phenotypic expression of that sex, or more simply, the appearance of 'femaleness', is largely determined during embryonic development. Of primary importance in this respect are the embryonic ovaries which secrete the female sex hormone, oestrogen. This hormone determines that the embryo reproductive tract shall develop into a female one rather than a male reproductive system.

Impairment of normal embryonic sexual development sometimes occurs, as for instance in the case of the freemartin heifer. In the small proportion of twins that occur in cattle, when a male and female embryo share the uterus, the production of male hormone, testosterone, from the testes of the male embryo, causes the reproductive system of the female embryo to be effectively 'masculinised'. The result is a female calf, or freemartin, which is usually infertile.

Sexual development

The ovaries are the site of production of the female gametes or oocytes. Their production is known as oogenesis. At birth there are something like 200 000 primary oocytes already present in the ovaries, far in excess of those needed for the animal's entire reproductive life. During their development into mature oocytes, female gametes go through a stage of meiotic division which results in a halving of the chromosome number of each egg cell. Oocytes are surrounded by layers of ovarian cells and the whole structure is termed a follicle (*see Figure 12.17*). Follicle development is controlled by the gonadotrophins from the anterior

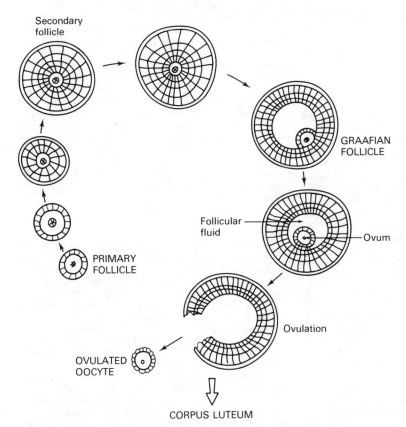

Figure 12.17 Diagrammatic representation of the sequence of follicle development, ovulation and corpus luteum formation during the ovarian cycle

Table 12.11 Some data on normal female reproduction

(a)

Animal	Onset of puberty	Age at first service	Length of cycle	Duration of oestrus
Cow	8–12 months	15–18 months	21 d (18–24 d)	18 h (14–26 h)
Ewe	4–12 months (first autumn)	First autumn or second autumn	17 d (14–20 d)	36 h (24–48 h)
Sow	4½–10 months	7–10 months	21 d (18–24 d)	48 h (24–72 h)

(b)

Animal	Time of ovulation	No. of ova shed	Length of gestation	Optimum time for service
Cow	10–15 h after end of oestrus	1	282 d (277–290 d)	Mid to end of oestrus
Ewe	Near end of oestrus	1–4	149 d (144–152 d)	16–24 h after onset of oestrus
Sow	Middle of oestrus	10–20	114 d (111–116 d)	15–30 h after onset of oestrus

pituitary, chiefly FSH. Maturing follicles produce increasing quantities of steroid hormone, oestrogen, which influences both growth of the reproductive tract and the sexual behaviour of the animal.

Follicle development does not reach completion until puberty occurs.

Puberty

This signals the beginning of the female's active reproductive life when the animal commences breeding activity. Thus puberty is indicated by the first heat or oestrus during which the female is receptive to the male. The age at which puberty is reached is dependent on the animal's body maturity which is related to its bodyweight and liveweight gain. Onset of puberty coincides with the point of inflexion on the animal's growth curve (*see Figure 12.34*). Breed size influences the onset of puberty; smaller, faster maturing breeds reach puberty before larger breeds. Other factors are nutrition, health, season of birth and the presence or absence of a mature male. The introduction of a mature boar to late prepubertal gilts advances their onset of puberty. Data on the age at puberty for farm species is given in *Table 12.11*.

Oestrus cycles

All females of farm species, once puberty has been attained, exhibit sequences of reproductive activity known as oestrous cycles. They may be continuous throughout the year as in cattle and pigs or may be restricted to a specific breeding season as in the case of sheep, goats and deer, all of which are autumn breeders. The non-breeding time of the year is known as anoestrus. In sheep seasonal anoestrus is much longer for temperate or northern breeds, such as Scottish Blackface, than it is for breeds of more equatorial origin such as Merinos, which tend to breed for most of the year. For

seasonal breeders, daylength or photoperiod is the cue to onset of breeding activity. Photoperiodic changes influence the activity of the pineal gland, via the optic pathway. The pineal regulates seasonal changes in reproduction through its control of gonadotrophin release from the anterior pituitary. Melatonin, a hormone secreted by the pineal, is produced in increased amounts during darkness. This hormone has an important role as a mediator between environmental light patterns and seasonal differences in gonadotrophin secretion (*see* later section on melatonin and out-of-season breeding).

The sequence of events in the oestrous cycle of the sow, ewe and cow are essentially similar, although their precise timing and duration do differ. (For a more detailed comparison *see Table 12.11*.) The oestrous cycle may be divided into four main stages: oestrus, met-oestrus, di-oestrus and pro-oestrus.

Oestrus

This is the stage of the oestrous cycle when the female comes into heat and is characterised by physiological and behavioural changes, such as reddening of the vulva; increased mucus secretion of the cervix and vagina; increased irritability and vocalisation; mutual grooming activity, and most importantly, a willingness to stand for the male. In cattle, in particular, in the absence of a male, the oestrous female will allow mounting by other females, who themselves are often in oestrus or approaching it. For a more detailed coverage of behaviour and detection in oestrus, see chapters on cattle, sheep and pigs (14, 15 and 16).

Duration of oestrus and length of the oestrous cycle differ between individual animals and may be influenced by such factors as age of animal, its standing in the social hierarchy, stage of the breeding season, time of year, climate, nutritional status, health and stress.

In pigs and sheep, egg release (ovulation) occurs at the end of oestrus. In contrast, ovulation in the cow occurs some 10–12 h after the end of behavioural oestrus. There is evidence that the interval between onset of heat and time of ovulation is significantly shortened when females are naturally mated compared with those artificially inseminated.

Met-oestrus

This is the stage immediately post-oestrus when the corpus luteum forms in the ovary from the cells of the post-ovulatory Graafian follicle (*see Figure 12.17*).

Di-oestrus

This is the longest stage of the oestrous cycle and coincides with the maturation of the corpus luteum. It is otherwise referred to as the luteal phase.

Pro-oestrus

This is of relatively short duration when one or more follicles undergo the final stage of maturation. Animals in pro-oestrus may show a reddening of the vulva and increased sexual activity, although they will not stand to be mounted.

Hormonal control

The oestrous cycle is regulated by pituitary and ovarian hormones. Generally speaking the concentrations of these hormones in the blood reflect the physiological changes occurring in the animal (*see Figure 12.18a* and *12.18b*). The endocrine events in the oestrous cycle of the cow have been assumed as typical of all farm mammals.

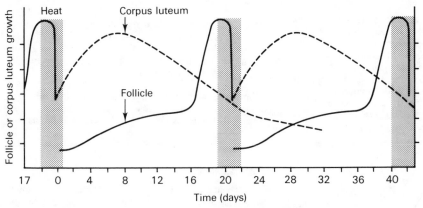

Figure 12.18(a) Follicle and corpus luteum development during the oestrous cycle

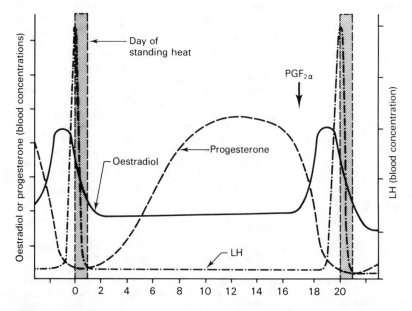

Figure 12.18(b) Changes in concentrations of the various hormones associated with the oestrous cycle

If the beginning of oestrus is day 0, then this is the stage at which Graafian follicles reach maturity under the influence of the gonadotrophins FSH and LH. In the cow, ovulation occurs shortly after the end of oestrus as a result of the increases in the levels of the luteinising hormone, LH. Thus a mature egg is released into the oviduct and the follicle cells within the ovary form the corpus luteum. The corpus luteum gradually matures during di-oestrus, secreting progesterone, whose function is to prolong the luteal phase of the oestrous cycle until the fate of the ovulated egg has been decided. Progesterone, through its effects on the pituitary, inhibits the release of FSH and thus prevents maturation of further follicles. Progesterone, in conjunction with the low levels of circulating oestrogen at this time, helps prepare the uterus lining or endometrium for the possible arrival of a fertilised egg. Sometimes animals become acyclic due to the presence of a persistent corpus luteum, or luteal cyst as it is generally termed. Thus a return to oestrus is indefinitely delayed.

Normally, however, in the absence of fertilisation, the corpus luteum naturally regresses at about day 16 or 17 of the cycle. Luteal regression is accompanied by a fall in progesterone. The cause of these events is the release by the uterus of a hormone known as prostaglandin.

With the end of the luteal phase, follicle development accelerates. Maturing follicles are the major source of oestrogen production, the chief oestrogen in the cow being oestradiol. During this pro-oestrus stage, blood concentrations of oestrogen are therefore greatly elevated. Oestrogen is the hormone responsible for the physiological and behavioural changes taking place during pro-oestrus and oestrus.

The number of follicles that mature and are eventually released by the ovaries is termed the ovulation rate. Although influenced by gonadotrophins, ultimately ovulation rate is genetically determined. Thus some species and breeds have a greater ovulatory capacity than others. Other factors influencing ovulation rate are age, stage of the breeding season and nutritional status.

The practice of flushing, where the level of nutrition is increased shortly prior to mating, usually results in an improved ovulation rate.

Synchronisation of oestrus

Synchronisation of oestrus, or controlled breeding, is possible in farm species through the manipulation of oestrus cycles by means of hormone administration. This can be achieved in two ways:

(1) by premature regression of the corpus luteum (luteolysis) using prostaglandins;
(2) by prolongation of corpus luteum activity beyond normal length of the luteal phase, using progesterone; followed by sudden withdrawal of treatment.

In both cases animals return to heat shortly after treatment, in a synchronised fashion. *See Figure 12.19* for a summary of hormonal changes associated with oestrus synchronisation.

Prostaglandins

Prostaglandin $F_{2\alpha}$ or its synthetic analogues, when administered to a breeding female with an active corpus luteum, causes an effective shortening of the oestrous cycle. For instance, in cattle, the injection of prostaglandin causes the corpus luteum to immediately regress and oestrus normally follows two to three days later. Insemination can then be carried out three to four days after prostaglandin treatment. However, not all cows respond to prostaglandin in the same manner. Animals between days 1–5 and days 17–21 of the oestrous cycle are unaffected by prostaglandin administration. Between days 1–5, the corpus luteum is immature and therefore unresponsive to prostaglandin and cows between days 17–21 have no corpus luteum present. The solution to this problem has been either to use two injections 11 days apart (when all animals should then be between days 5–17 of the cycle) or to use one injection and combine this with oestrus detection of unaffected animals. The use of prostaglandins to synchronise oestrus in pigs is largely ineffectual since the corpora lutea of the sow will only regress if treated between days 12–15 of the 21-day cycle.

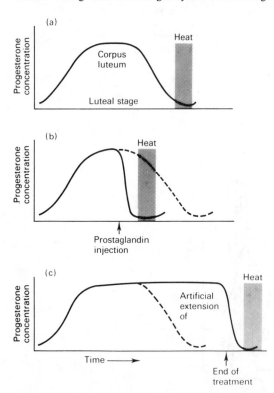

Figure 12.19 Diagrammatic representation of the hormonal events of the oestrous cycle in (a) a normal cycle, (b) after prostaglandin injection, and (c) after progestagen treatment

Progesterone

Use of progesterone or its synthetic analogues (collectively referred to as progestagens) simulates the activity of a corpus luteum thereby artificially prolonging the luteal phase of the oestrous cycle. Thus once progestagen treatment is ceased, the animal returns to oestrus shortly afterwards. Greater synchrony of oestrus is achieved by either combining the progestagen with oestrogen or by $PGF_{2\alpha}$ injection following progestagen withdrawal.

In cattle, progestagens are administered intravaginally with oestradiol, in the form of a coil known as a PRID (progesterone-releasing intravaginal device). PRIDs have an advantage over prostaglandins in that their effectiveness is not dependent on the presence of a corpus luteum and unlike $PGF_{2\alpha}$ inadvertent use on pregnant animals is not likely to cause abortion. In sheep, progestagen-impregnated vaginal sponges are used in a similar fashion to PRIDs. However, immediately following sponge removal, PMSG may be injected to stimulate ovulation rate when synchronising sheep at the beginning of their breeding season. (For further details of oestrus synchronisation procedures in sheep *see* Chapter 15.)

The success of the above techniques depends on the degree of oestrus synchrony that can be achieved. Unfortunately, there is an inevitable variability in animals' responses to treatment, which means that in practice, synchronisation is never perfect. Pregnancy rates following synchronisation are generally similar to those of untreated animals.

The male reproductive system

Anatomy

The male reproductive system is comprised of paired primary and secondary sex organs, together with accessory sex glands (*see Figure 12.20*). The primary sex organ is the testis whose function is to produce sperm and male sex hormones. The testes are situated outside the abdominal cavity in the scrotal sac. The secondary sex organs consist of the reproductive tract, extending from the testis to the urethra. The chief role of the reproductive tract is to transport sperm. The urethra, in addition, carries urine from the bladder. Accessory sex organs (the seminal vesicles, the prostate gland and the bulbo-urethral glands) are situated at the base of the penis and are responsible for seminal fluid production (*see* later section).

Sexual development

The possession of a Y chromosome by the male fetus causes the undifferentiated gonad to develop into a testis rather than an ovary. The production of male hormone, testosterone, by the fetal testis regulates the subsequent development of the male reproductive tract. The entire reproductive system is thus fully differentiated prior to birth and by then the testes have normally descended into the scrotal sacs via an opening in the ventral abdominal wall known as the inguinal canal. In a small proportion of males, the testes may remain within the abdomen, a condition known as cryptorchidism. Since sperm production depends on the testes being maintained at 2–4°C lower than body temperature, such animals are invariably infertile.

After birth, significant further development is limited until puberty.

Puberty

Puberty in the male is characterised by both physical and behavioural changes. The marked increases in testosterone output at puberty stimulate sperm production and the development of secondary sex characteristics associated with the mature male. These characteristics include further growth of the reproductive tract, an increase in muscle formation, development of a masculine voice and body odour, together with an increase in sexual and aggressive

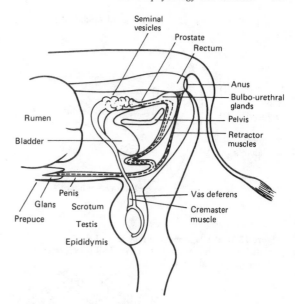

Figure 12.20 Reproductive tract of bull

behaviour. The exact onset of puberty, in contrast to the female, cannot easily be determined. Thus puberty in the male may be said to have occurred when enough sperm can be produced to successfully impregnate a female. For example, in the bull, the ejaculate should contain a minimum of 5×10^6 spermatozoa with at least 10% motility. Such a definition, however, takes no account of sexual behaviour of the animal. Sexual activity may reach full intensity considerably later than puberty. Thus puberty should not be confused with full sexual maturity. For bulls, puberty usually occurs at between seven and nine months of age, whilst sexual maturity may not be attained until four to five years of age.

Male sexual behaviour has two components: libido (or sex drive) and the ability to copulate (mating behaviour). Whilst libido is largely genetically determined, mating behaviour may depend on the social conditions in which the male is reared.

Castration, which is the removal or destruction of the testes, results in a loss of both sperm production and libido. Males castrated soon after birth fail to develop characteristic male appearance. Growth is also affected by castration (*see* section on growth). It is important not to confuse castration with vasectomy which is the sectioning of the vas deferens to render the male infertile. Unlike castration, vasectomy has no apparent effect on sexual behaviour or the animal's ability to sexually arouse the female.

Sperm production

Sperm is produced by the seminiferous tubules of the testis (*see Figure 12.21a and b*). The germinal epithelium only begins significant sperm output at puberty. Spermatogenesis is temperature-dependent, the temperature of the testes being lower than the body. This is achieved by countercurrent heat exchange between the blood of the spermatic artery and vein supplying the testes. Maintenance of testes temperature is also helped by the action of the cremaster muscle which in cold conditions draws the testes closer to the body and in hot conditions moves then further away.

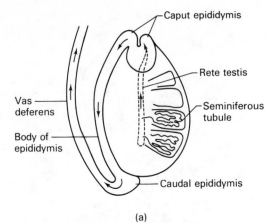

(a)

Figure 12.21(a) Diagrammatic representation of a longitudinal section of the testis of the bull (→ represents passage of spermatozoa)

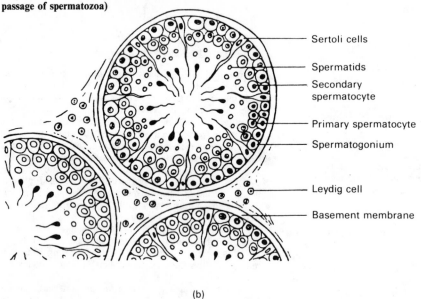

(b)

Figure 12.21(b) Cross-section of the seminiferous tubules of the testis

matogenesis. The process of spermatogenesis is outlined in *Figure 12.22*.

Spermatogenesis is characterised by a massive cell proliferation accompanied by the eventual halving of the chromosome complement (meiosis) of each spermatozoon. By the time spermatozoa have collected in the lumen of the seminiferous tubule, they have attained their distinctive appearance, as shown in *Figure 12.23*. They do not, however, become fully motile until they have been transported to, and resided in, the caudal epididymis. The complete sequence of spermatogenesis takes between 40 and 60 days in farm species.

Testis diameter is a useful criterion for sperm production. The testis of the ram noticeably increases in size as the breeding season approaches.

Seminal fluid is added to spermatozoa prior to ejaculation

Sperm production is seasonal in wild species. In farm species, seasonal breeding in the male is not so clearly defined. However, the ram shows marked changes in sperm production and testis size depending on the season of the year. There are, however, breed differences and this should be borne in mind when choosing the ram for out-of-season breeding of ewes. Sperm production, as well as being daylength dependent, is also affected by environmental temperatures. For example, hot weather may seriously affect sperm production capacity in the boar.

Spermatogenesis

Sperm production by the seminiferous tubules is controlled by FSH and LH. The Leydig cells are stimulated by LH to produce testosterone, which in turn stimulates sper-

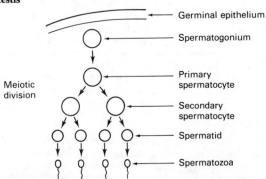

Figure 12.22 Spermatogenesis (courtesy of Peters, A. R. and Ball, P. J. H. (1986) *Reproduction in Cattle.* **London: Butterworths)**

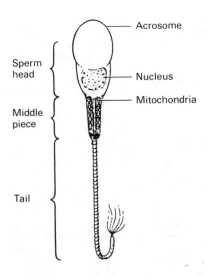

- Acrosome

Sperm head {

- Nucleus

- Mitochondria

Middle piece {

Tail {

Figure 12.23 Structure of a spermatozoon (or sperm cell) (courtesy of Peters, A. R. and Ball, P. J. H. (1986) *Reproduction in Cattle*. London: Butterworths)

by the prostate, seminal vesicles and bulbo-urethral glands. A comparison of sperm and semen output in farm species is given in *Table 12.12*.

Artificial insemination (AI)

The technique of artificial insemination permits the semen from a genetically superior male to inseminate a large number of females. The number of inseminations that can be performed using the semen from a single ejaculate depends on the extent to which the semen can be diluted without impairing its fertilising capacity. As many as 1000 cows can be inseminated with the diluted semen from one ejaculate of a bull. This figure is far less for the boar and the ram (*see Table 12.12*). Another factor limiting the widespread use of AI in pigs and sheep is the difficulty of long-term preservation of semen. Unlike cattle semen, which can be stored

deep-frozen, boar and ram semen do not survive the process without some reduction in their fertilising capacities. Thus AI in pigs and sheep has largely been confined to the use of fresh semen.

Artificial insemination involves a number of distinct stages: semen collection; examination and evaluation of semen; semen dilution and storage, and, lastly, techniques of insemination.

Semen collection

One commonly used method in cattle is the use of a teaser female or a dummy, with the ejaculate collected into an artificial vagina. With pigs, a dummy mounting stool is used and the penis is grasped manually. A technique which is especially successful in the ram is that of electro-ejaculation, where a rectal probe provides a low voltage stimulation of the musculature of the reproductive tract.

Examination and evaluation of semen

During collection and examination every effort is made to avoid temperature shock to the semen, which might result in a drastic reduction in sperm viability. Semen examination involves an evaluation of a number of aspects of sperm quality: sperm numbers; morphological appearance of the spermatozoa; and sperm motility. The sperm count is important in that it determines the extent to which the semen can be diluted, which in turn dictates the number of inseminations possible from a single collection. Morphological examination assesses the quantities of abnormal or immature spermatozoa present. For instance, in the case of a young male or a sire that has been over-used, a high proportion of immature spermatozoa may be present. In these instances the semen would not be used for commercial inseminations. Sperm with poor motility would also be rejected on the basis of its probably low fertilising capacity.

Semen dilution and storage

Semen is diluted with chemical 'extenders' so chosen because of their ability to preserve the life of the sperm and at the same time have a negligible effect on the fertilising ability of spermatozoa. Such diluents contain nutrients, a buffer to protect against pH changes, and in the case of deep-frozen

Table 12.12 Data on sperm and semen production in farm animals

Species	Volume of ejaculate (ml)	Expected sperm concentration ($\times 10^6$/ml)	Total number of sperm per ejaculation ($\times 10^9$)	Number of collections per week	Site of semen deposition at — mating	Site of semen deposition at — AI	Potential number of inseminations per ejaculate	Volume of inseminate after dilution (ml)	Number of motile sperm ($\times 10^6$)
Bull	4–8	1200–1800	4–14	3–4	Anterior vagina	Cervix and/or uterus	400	0.25–1.0	5–15
Ram	0.8–1.2	2000–3000	2–4	12–20	Anterior vagina	External cervical os	40–60	0.05–0.2	50
Boar	150–500[1]	200–300	40–50	2–4	Uterus[2]	Uterus[2]	15–30	50–100	2000
Stallion	30–>150	100–250	3–15	2–6	Uterus[2]	Uterus[2]	5	20–50	1500[3]

[1] Includes gelatinous secretion of the bulbo-urethral glands.
[2] The ejaculate makes passing contact with the cervical canal.
[3] Total number deposited in three inseminations during the prolonged period of oestrus.
Source: courtesy of Hunter, R. H. F. (1980). *Physiology and Technology of Reproduction in Female Domestic Animals*. London: Academic Press.

semen, a chemical such as glycerol to protect against the effects of chilling. Antibiotics are also included to prevent bacterial contamination. In pigs and sheep, where fresh semen is used, such chemical extenders may prolong its effectiveness for up to three days. Once diluted, bull semen is stored in liquid nitrogen at $-196°C$ in plastic straws. Each straw contains sufficient diluted semen for a single insemination.

Insemination techniques

Insemination is performed using specially designed catheters or insemination 'guns'. Such devices in the case of cattle and sheep permit the semen to be deposited further up the female reproductive tract than would be achieved by natural insemination. This allows a more dilute semen to be used without adverse effects on conception rates. It is important to note that with any artificial insemination procedure, stressful circumstances surrounding the event may have detrimental effects on its successful outcome. The timing of insemination is also vital relative to the stage of oestrus of the female and this is dealt with more fully in Chapter 14.

Factors affecting male fertility

In summary there are a number of factors which affect optimum fertility in the male:

(1) Genetics – there are considerable individual and breed differences in male fertility.
(2) Age of the animal – performance increases with age from puberty.
(3) Nutritional status – e.g. rams should be in good body condition at mating to sustain performance throughout the mating season.
(4) Environment – daylength and ambient temperature affect level of sperm production.
(5) Frequency of use – both overuse and underuse affect fertilising capacity of the male (a mature boar may be used 2–3 times per week and a bull 3–4 times per week).
(6) Health – e.g. physical damage to the penis may cause sufficient pain to affect libido. Damaged legs or feet may hinder mating ability.

Mating

Mating, the process which ensures that spermatozoa are deposited in the reproductive tract of the oestrus female, represents the culmination of a complex sequence of behavioural and physiological events. Prior to actual mating the male and female seek each other out by a combination of visual and olfactory signals. For example, the bull and the ram both display the characteristic Flehmen response which involves curling the upper lip and sniffing the air. This enables the male to detect the sexual pheromones given off by the oestrous female. In a similar way, the female may be sexually aroused by the odours of the mature male. Pheromones are believed to play a part in individual animal recognition. That a male may show a clear preference for some females over others is important with regard to mating management of herds or flocks. For example, running several rams with a flock or using males in rotation takes account of this fact.

Mating itself consists of mounting the female, thrusting, intromission and ejaculation. The exact site of deposition of sperm in the female reproductive tract depends on the species. In the cow and the ewe, sperm is deposited in the vagina, whilst in the sow, sperm is placed beyond the cervix into the uterus. Once in the female reproductive tract, sperm need to make their way to the site of fertilisation, the oviduct. This may take as little as 2 h in the case of the pig but a much longer time may be necessary in cattle and sheep. Only a few hundred of the billions of sperm in the ejaculate manage to reach the oviduct. Fertilisation of the egg is still not possible until the spermatozoa have completed the process of sperm capacitation. This term refers to a minimum period of 'acclimatisation' to the uterine conditions which is necessary before sperm have the capacity to successfully fertilise an egg.

Spermatozoa and eggs undergo the effects of ageing while in the female reproductive tract. For sperm the effective fertilisable life-span is 1–2 d and for the egg it is 10–12 h from time of ovulation. Given the rapid ageing of an ovulated egg and the delay in time between insemination and when spermatozoa are able to effect fertilisation, it is clear that mating or artificial insemination must precede ovulation by an appropriate amount of time if conception rates are to be optimised (*see* Chapter 14 for advice on timing of AI in cattle).

Fertilisation

Once spermatozoa have become fully capacitated, the acrosmal portion of the sperm head releases enzymes which enables the sperm to penetrate the protective membrane, the zona pellucida, which envelops the egg. There then follows a fusion of nuclear contents of the egg and sperm cells. The zona pellucida undergoes a subsequent chemical change which prevents entry of further sperm. The fused egg and sperm is termed an embryo or zygote.

Pregnancy

Embryonic development

Whilst in the oviduct the embryo begins to divide mitotically thus doubling its cell mass at each successive division. The embryo, by the time it has reached the uterus after about 3–4 d, has developed into a hollow ball of cells.

Implantation

It is generally considered that pregnancy only becomes established once the embryo has successfully attached itself to the lining of the uterus wall or endometrium. This process is known as implantation.

Fetal development

An embryo may be regarded as having attained the status of a fetus once recognisable organ development is apparent; heart, limb buds, liver, spinal cord are all present at an early stage of pregnancy. The majority of growth taking place in the first half of pregnancy, however, is mainly that of the fetal membranes and the placenta. Most of the growth of the fetus itself is confined to the latter stages of pregnancy (*see* Figure 15.7, p. 418).

A summary of the sequence of the major events of pregnancy is shown in *Figure 12.24*.

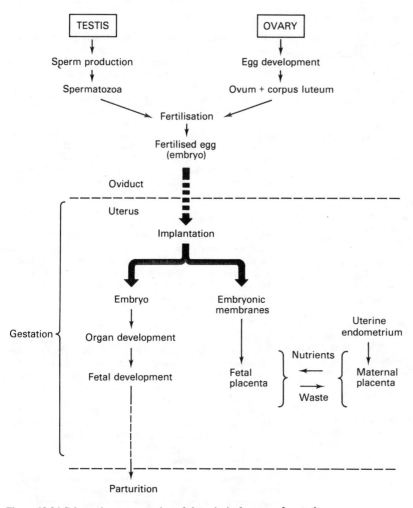

Figure 12.24 Schematic representation of the principal events of gestation

Maintenance of pregnancy

Hormones, especially progesterone, play an important part in the establishment and maintenance of pregnancy following mating and fertilisation. In the cow, for example (*see Figure 12.25*), following ovulation and service, the blood and milk progesterone levels remain identical up to day 17 or 18 whether the egg has been fertilised or not. After this time, in the non-pregnant animal, progesterone levels fall, as described earlier (*Figure 12.25*). In the pregnant animal, however, progesterone levels remain elevated due to the continued presence of a corpus luteum. It is thought that the embryo itself plays a major role in preventing the prostaglandin $F_{2\alpha}$-induced regression of the corpus luteum that occurs if fertilisation has not taken place. Thus the continuing production of progesterone after day 18 in the pregnant animal ensures both the maintenance of a uterine endometrium to receive the fertilised egg and also acts on the hypothalamus/anterior pituitary to inhibit gonadotrophin release. Progesterone therefore prevents further oestrous cycles. The high circulating levels of progesterone charac-

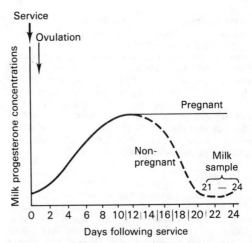

Figure 12.25 Milk progesterone profiles in a pregnant and non-pregnant cow following service and ovulation

teristic of pregnancy eventually fall at the end of pregnancy, shortly before parturition.

The source of progesterone through pregnancy differs in farm species. In the sow, the corpus luteum persists throughout the whole of pregnancy and is the main source of progesterone. In the cow and the ewe, in contrast, the responsibility for progesterone production switches from the corpus luteum to the placenta approximately halfway through pregnancy. There are obvious practical implications for prostaglandin treatment. Administration of prostaglandin to pregnant animals with an active corpus luteum would result in an abortion.

Embryonic and fetal mortality

It is estimated that 30–40% losses in potential offspring occur between fertilisation and parturition in farm species. In cattle, where normally only one egg is ovulated, the death of an embryo or fetus means a failed pregnancy. In contrast, in pigs, where there is a much higher ovulation rate, success or failure may be calculated in terms of the proportion of ovulated eggs (as measured by the total number of corpora lutea on both ovaries) surviving to term.

Some two-thirds of all pre-natal deaths can be put down to embryonic losses. Much of this loss tends to occur either in the first few days following fertilisation or at around the time of implantation. Genetic abnormalities, hormonal insufficiency or failure due to an aged egg are thought to be among the chief causes. Any nutritional problems or imbalance occurring at this stage will only serve to exacerbate such losses. If embryo mortality is largely attributed to failure of the fertilised egg, then fetal losses are much more likely to be the result of some deficiency on the maternal side, such as placental inadequacy or uterine infection. A dead fetus may result in an abortion or in some cases it may be retained within the uterus in a mummified condition, where the fetal fluids have become reabsorbed by the uterus lining.

Pregnancy diagnosis

An accurate diagnosis of pregnancy is vital to the reproductive efficiency of breeding livestock. Any pregnancy diagnosis service should also be early, fast and reliable so that animals not successfully inseminated may be re-bred as soon as possible. In addition, as in the case of sheep, a determination of the number of fetuses present allows the farmer to feed the mother accordingly. Techniques which can offer early diagnosis suffer the disadvantage of the high rates of early embryo mortality and are therefore less able to give an accurate estimate of viable offspring at the end of pregnancy. Some of the more commonly employed pregnancy diagnosis techniques are outlined below.

Hormone assay

This involves the measurement of hormones found in the blood or milk of pregnant animals, by means of a highly sensitive, laboratory-based procedure known as radio-immunoassay. Hormone assay techniques have been confined commercially to dairy cattle. There are two types of hormone diagnosis available: an early progesterone test and a later oestrone sulphate test.

Progesterone test

This method allows pregnancy diagnosis to be made as early as 21–24 d after ovulation. Samples of blood or, more conveniently, milk, are taken at this time and the levels of progesterone present are determined.

Pregnant animals will have high progesterone values, while low values will indicate that the animal has returned to oestrus (*see Figure 12.25*). Although the laboratory procedure is highly accurate, the technique does produce a proportion of false positive diagnoses, which may be due either to cows which have abnormal length oestrous cycles or to insemination at the wrong time of the cycle. There is a further drawback implicit in such an early diagnosis, in that an animal correctly diagnosed as pregnant may subsequently lose the embryo.

Oestrone sulphate test

This is a much later test (at around 15 weeks) which can confirm pregnancy with almost 100% reliability. This is because the method, unlike the progesterone test, measures a hormone, oestrone sulphate, which is derived from the fetus itself. The test is of limited practical use because of the lateness of the diagnosis.

Rectal palpation

The technique involves the insertion of an arm into the rectum of an animal in order to manually detect (palpate) the presence of a fetus within the underlying uterine horns. This procedure is largely limited to cattle and needs to be performed by a trained operator. The fetus can be palpated from around day 40 up to the fifth month of pregnancy. After this time palpation becomes impossible until after month seven because the uterine horns become greatly enlarged and 'disappear' below the pelvic brim.

Ultrasonic techniques

Such procedures consist of ultrasonic detection of fetal or placental structures. A variety of techniques have been developed, all of which involve the same basic principle. A probe is applied to the surface of the animal's abdomen which transmits and receives an ultrasonic beam. The reflected signal, which is able to detect the presence of the fetus and associated tissues, is then converted into an auditory or visual output. Recently, real-time ultrasound scanners, used routinely in human medicine, have been developed for farm use. This device allows an image of the uterine contents, including the fetus, to be displayed on a screen. This method has been commercially used in sheep to detect fetal lambs between 50 and 100 d of pregnancy (*see* Chapter 15 on Sheep for a fuller description). With all the available ultrasonic methods, a skilled operator is needed to correctly interpret the results obtained. Commercially, ultrasonic pregnancy diagnosis has been restricted to pigs and sheep.

Parturition

Gestation length

The onset of parturition or birth is determined by the length of gestation. Gestation length is largely determined by the fetus. Whilst it is the mother that decides the hour of birth, it is the fetus that decides the month, week and day. Other factors which influence gestation length include species, breed, litter size, sex of the fetus, age of the dam and the genotype of the sire.

Initiation of parturition

As stated above, it is the fetus that is chiefly responsible for initiating birth. Recent evidence has shown that once the fetus achieves its maximum growth, hormonal changes cause progesterone production to be 'switched off'. Thus the withdrawal from circulation of the hormone responsible for maintaining pregnancy, signals the beginning of parturition. Increases in oestrogen and prostaglandin $F_{2\alpha}$ follow, with a resultant release of oxytocin, which is responsible for causing uterine contractions (i.e. labour). A diagrammatic summary of events in the sheep is shown in *Figure 12.26*.

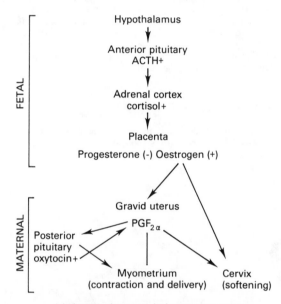

Sheep

(+) denotes increase in hormone output
(-) denotes decrease in hormone output

Figure 12.26 A postulated sequence of hormonal events controlling the onset of parturition in the sheep (courtesy of First, N. L. (1979) Mechanisms controlling parturition in farm animals. In *Animal Production*, pp. 215–257. Ed. by H. Hawk. Monclair, NJ: Allanheld Osmun)

Stages of labour

The process of parturition may be divided into three stages:

The first stage – the preparatory stage involving dilatation of the cervix and relaxation of the pelvic ligaments.
The second stage – the expulsion of the fetus through the pelvic canal.
The final stage – the expulsion of fetal membranes and the initiation of uterine involution.

Any difficulty or prolongation of the birth process is given in term dystocia.

Induction of parturition

Simulation of the events which initiate parturition (as outlined in *Figure 12.26*) can be brought about by exogenous hormone treatment of late pregnant animals. Corticosteroids, prostaglandin $F_{2\alpha}$ and their synthetic analogues, when administered to animals nearing the end of their natural gestation, will induce parturition within two to three days of treatment. The method is not without its problems however. For example, in cattle such inductions of parturition have been associated with increased calf mortality and greater incidence of retained placentas. Used commercially the procedure allows a more exact timing of birth to fit in which management and labour requirements. (For a detailed description of induction in lambing *see* Chapter 15.)

Lactational anoestrus

This describes the situation, following parturition, when active lactation is accompanied by minimum ovarian activity. Behavioural oestrus and complete resumption of breeding cycles are not observed for several weeks in cattle, sheep and pigs, depending on such factors as milk yield, nutritional status and suckling intensity.

Recent advances in reproductive technology

This section will attempt to review some of the more recent technological developments not already mentioned earlier in this chapter.

Superovulation

This process involves the injection of hormones such as FSH and pregnant mare serum gonadotrophin (PMSG) to increase the number of ovulatory follicles and hence the number of eggs released by the ovaries. Superovulation has been used in association with embryo transfer to allow the genetic potential of superior breeding females to be more fully exploited.

Immunisation against ovarian steroids

Immunisation against ovarian steroids is a newly developed technique which improves the ovulation rate without the risk of excessively large litters than sometimes results from superovulation. The female is injected with ovarian steroid and the immune system is stimulated to produce antibodies against the hormone. Normally within the animal, ovarian steroids regulate, by a process of negative feedback, the gonadotrophins that are responsible for follicle development and ovulation. Immunisation is believed to interrupt this negative feedback and thereby produces an increase in ovulation rate. The ovarian steroid androstenedione, commercially available as Fecundin, has been successful in improving lambing percentages. Two injections are required, at eight and four weeks prior to mating.

Melatonin and out-of-season breeding

The pineal gland and its secretion, melatonin, have a central role in the photoperiodic control of reproduction in seasonal breeders. Melatonin, when fed in the afternoon to out-of-season sheep, has the effect of creating a chemical 'darkness' which convinces the animal that it is experiencing a shortened daylength. This results in an advance of the breeding season. Melatonin, at the time of writing, has not yet been cleared for commercial use.

Embryo transfer

The commercial application of the technique of embryo transplantation has been largely confined to cattle. A number of eggs are produced by superovulation in a donor cow and, after fertilisation, non-surgically collected and transplanted into the uteri of recipient cows. In this way a greater number of calves can be produced from a genetically valuable female than would be achieved normally during its reproductive life-time. It is possible to freeze embryos for long-term storage and still achieve acceptable pregnancy rates with the thawed embryos. This facility therefore offers a useful extension to the scope of embryo transfer.

Sex determination

This can be achieved either by sexing of embryos or by attempting to separate the X- and Y-bearing spermatozoa. Although laboratory techniques exist to do this, no commercial method is yet available for use in animal production. The Y chromosome of the male produces a protein known as the HY antigen, which it is possible to detect by immunological means. Alternatively, it is possible to karyotype the cells of the embryo by a procedure which examines and identifies the sex chromosomes. A successful method of sex determination would have enormous benefits

for the AI industry, given that half the number of dairy calves produced consist of males that may be unwanted.

A rapid milk progesterone test

A rapid, highly sensitive progesterone assay is now available commercially, employing ELISA procedures (enzyme-linked immunosorbent assay), which can be used 'on-farm' by untrained operators. A bulk milk sample is obtained between 21 and 24 d after service in the same way as for the laboratory-based radioimmunoassay. The ELISA method shares the same disadvantage as its more established counterpart in the number of false positives that are produced (*see* section on pregnancy diagnosis).

LACTATION

The growth and development of the mammary gland and the subsequent manufacture and secretion of milk represents an important phase in the reproductive cycle of mammals. This section describes the structure and development of mammary tissue, the processes involved in the synthesis, secretion and release of milk, together with the underlying hormonal control of these events.

Anatomy of the mammary gland

The structure of the adult mammary gland is shown in *Figure 12.27*. The majority of the gland consists of closely-

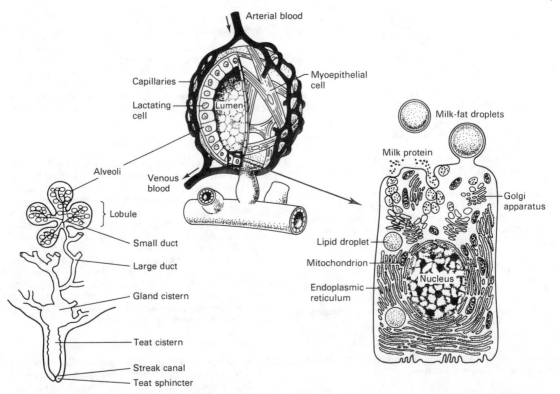

Figure 12.27 Diagram showing structure of the mammary gland of the cow

packed lobules, each made up of numerous clusters of alveoli. It is the epithelial cells lining the lumen of each alveolus, the secretory cells, which are the site of milk synthesis. The secreted milk is stored in the alveolar lumen prior to passing into the small ducts leading from each alveolus. The arrangement of the alveoli and ducts is rather similar to a large bunch of grapes, with the alveoli being the grapes and the ducts the stem branches. The ducts merge into ever larger ducts until the largest of these, the collecting ducts, empty into the gland cistern. In the cow, prior to milking, 40–50% of the milk is held in the gland cistern and larger ducts, while the rest is retained within the small ducts and alveolar lumen. In contrast, in the sow, approximately 90% of the milk is held in these latter structures.

Continuous with the gland cistern is the teat cistern and the streak canal through which the milk is eventually secreted. The streak canal is surrounded by a sphincter composed of circular smooth muscle fibres. In cows which tend to leak milk, this sphincter is not tight enough, whilst the opposite is the case with slow milkers. The mammary gland of the sow differs in that each teat is served by two

separate streak canals, each leading to the teat cistern and gland cistern.

The sow normally has seven pairs of glands distributed bilaterally on the ventral surface of the body from thorax to inguinal region, a single teat serving each gland. The cow has two pairs of glands, found in the inguinal region, the rear two accounting for approximately 60% of total milk production. The ewe has a single pair of glands, inguinally situated.

The udder of the cow may weigh as much as 15–30 kg and as much again when full of milk. Support for the udder is therefore very important. This is achieved by means of suspensory ligaments which are attached to the body wall and pelvic girdle by tendons (*see Figure 12.28*). The lateral suspensory ligament is fibrous and inelastic while the median suspensory ligament is more elastic; thus when the udder fills with milk, the median ligament stretches to allow the udder to expand.

Mammary development

At birth there is little mammary development present apart from teats, connective tissue and a rudimentary duct system. There is no secretory tissue, its place being taken by fat. At puberty there is an increase in ductal development, but it is only when the animal is pregnant that significant development of the duct and lobule-alveolar system commences.

Mammary gland differentiation is under the control of a complex battery of hormones from the anterior pituitary, ovary and adrenal gland. A summary of their effects on mammary development is given in *Figure 12.29*. Lactogenesis, which is the initiation of milk secretion, occurs towards the end of pregnancy and at parturition. It is thought that the trigger for this event may be the decline in progesterone concentrations at parturition, although prostaglandin $F_{2\alpha}$ may also be involved. Following parturition there is a proliferation in the secretory cell population which,

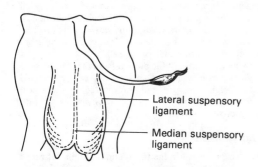

Figure 12.28 Support structures for the udder of the cow

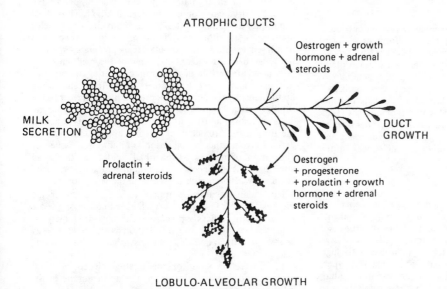

Figure 12.29 Summary of hormonal control of mammary developments (courtesy of Peters, A. R. and Ball, P. J. H. (1986) *Reproduction in Cattle*. London: Butterworths)

together with an increased output of each secretory cell, is responsible for the dramatic mammary secretion at this time. Suckling activity, which is itself related to the number of offspring, encourages greater secretory response. Conversely, a decline in suckling or milking leads to a diminished milk output. At weaning or 'drying off', the mammary gland regresses or involutes and this is marked by a dramatic reduction in the secretory cell population.

Milk biosynthesis

Whilst the composition (*Table 12.13*) and quantity of milk secreted by the dams of the different domestic species varies considerably, the milk from all species has the same basic qualitative composition. Milk essentially consists of two phases; an aqueous phase, in which are partitioned proteins in colloidal suspension and lactose, minerals and water-soluble vitamins in solution; and a lipid phase containing triglycerides, phospholipids, sterols and fat-soluble vitamins. The lipid in milk is about 99% triglyceride and exists in the form of globules ranging from 3 to $5\,\mu$m in diameter surrounded by a complex membrane, comprised of the other lipid components consisting of phospholipids, cholesterol and fat-soluble vitamins and which is acquired at the time of secretion from the lactating cell.

Table 12.13 Approximate average composition of milk of different species (% by weight)

	Total solids	Fat	Protein	Lactose	Minerals/ water soluble vitamins
Cow	12.5	3.7	3.4	4.7	0.7
Sow	20.0	8.0	6.0	5.2	0.8
Ewe	17.0	5.5	6.2	4.5	0.8
Rabbit	35.0	18.3	14.0	2.0	0.7
Human	12.4	3.8	1.0	7.0	0.6

The protein fraction in milk consists of two major protein groups, the caseins, which exist in the form of colloidal particles called micelles and of which there are four types – α_s, β, γ, and κ-caseins – and the serum (whey) proteins consisting of principally β-lactoglobulin and α-lactalbumin.

Because of the commercial importance of cow's milk as a food more is known about it than about the milk of other mammals and the mechanisms of synthesis of milk have been much investigated. In this section, therefore, synthesis will be described with reference to synthesis in the ruminant and attention drawn where appropriate to differences in the monogastric.

Site of biosynthesis

The site of biosynthesis is the vast number of epithelial cells which line the alveoli, each cell in an alveolus discharging its milk into the lumen or hollow part of the structure (*see Figure 12.27*). A feature of the secretory cells is the large number of mitochondria present, being indicative of the high energy demand of biosynthesis and also the marked hypertrophy of the Golgi apparatus and endoplasmic reticulum which occurs at the onset of lactation, these being the sites of synthesis of milk constituents. The alveoli of the mammary gland are well supplied with blood capillaries and it is from the arterial supply that the metabolites used in milk synthesis are drawn. Arterio-venous (AV) difference studies have shown that acetate, 3-hydroxybutyrate, free fatty acids, glucose, amino acids and proteins are withdrawn from the blood as it passes through the mammary tissue and act as precursors in milk synthesis. The origins of these precursors are illustrated in *Figure 12.30*.

Milk protein synthesis

A-V difference studies have shown that the total free amino acids withdrawn from blood are apparently sufficient for synthesis but since the balance of absorbed amino acids is not the same as that present in the milk proteins a certain amount of amino acid synthesis must occur. Of the amino acids incorporated into milk protein the provision of the essential amino acids for the protein synthesis is of greatest concern since the provision of non-essential amino acids can be achieved by synthesis within the lactating cell. A number of studies have indicated that under the dietary circumstances encountered by dairy cows in the UK, methionine is likely to be the first limiting amino acid for milk production and that there may be value in giving 'protected' methionine supplements. Protein synthesis involves the assembly of amino acids into a polypeptide chain through peptide bonds. The sequence of the amino acids in the various milk proteins is dictated by the usual method of transcription by ribonucleic acid on the ribosome of the cell.

Following the release of the protein from the endoplasmic reticulum they are 'packaged' in secretory vesicles derived from the Golgi apparatus and transported to the apical membrane of the cell before being secreted by exocytosis into the lumen of the alveolus.

Milk fat synthesis

The fatty acids present in the triglycerides in milk fat range from C_4–C_{18} in chain length and in comparison with other fat sources are characterised by having a relatively high proportion of short chain fatty acids. The origins of the different fatty acid groups vary (*Figure 12.31*). Thus short chain fatty acids and most of the medium chain fatty acids are synthesised within the lactating cell using acetic acid and to a lesser extent 3-hydroxybutyrate as precursors whereas the long chain fatty acids and some of the medium chain fatty acids are derived directly from circulating lipoproteins in the plasma. Fatty acids are removed from circulating lipoproteins by the enzyme lipoprotein lipase present in the capillary membrane. Whereas acetic and 3-hydroxybutyric acids are mainly derived through absorption from the rumen the fatty acids present in circulating lipoproteins may be derived from the mobilisation of adipose tissue or absorption of dietary or microbial lipid from the small intestine. The presence of odd-numbered and branched chain fatty acids in milk fat is due to the latter source.

Non-ruminant animals appear to use glucose for the synthesis of milk fat. In all species the metabolism of glucose by the lactating cell is the precursor of the glycerol moiety for triglyceride synthesis and in addition is the main energy source for both fat and also protein synthesis in the lactating cell.

In the secretion of fat from the lactating cell droplets of fat coalesce and migrate towards the apical membrane of the cell where they are enveloped in membrane as they push their way into the lumen of the alveoli. The true nature of the process remains obscure.

357

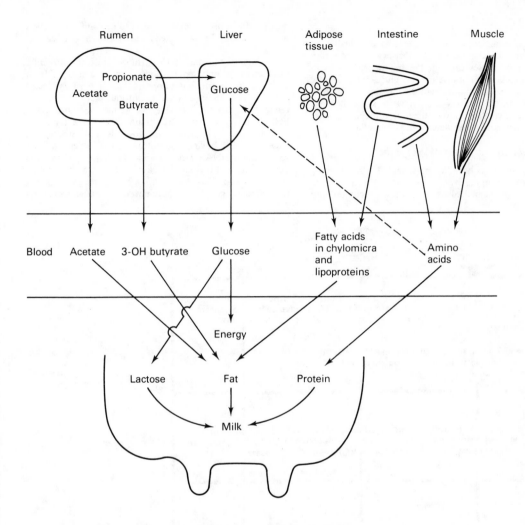

Figure 12.30 Diagram showing precursors for milk synthesis in the ruminant

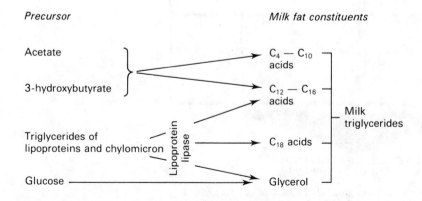

Figure 12.31 The origins of milk fat triglycerides

Lactose synthesis

Of the major milk constituents lactose is chemically the simplest. Synthesis takes place from glucose withdrawn from the blood as it passes through the mammary tissue. In ruminant animals this utilisation of glucose for lactose synthesis represents a major drain on the animal's glucose reserves. Lactose secretion would appear to share the same secretory route, via vesicles, with the milk proteins, together with water, ions and other water-soluble materials.

Nutritional influences on milk biosynthesis

The rates of synthesis of milk components by the lactating cells is dependent on the availability of milk precursors and the ability of the animal to partition these precursors into

milk production. The latter appears to be mainly a function of the genetic make-up of the animal but it is also influenced by diet.

The ability of animals to partition nutrients into milk production as opposed to body gain is controlled by a variety of hormones, but it appears that the most important are growth hormone (GH), insulin and glucagon. Recent research has shown that GH, also termed somatotrophin, has a major role in partitioning nutrients away from deposition in body tissues and towards milk production. Research using bovine somatotrophin (BST) derived from recombinant-DNA (rDNA) technology has shown that the daily injection of BST can increase yields in cows by between 10–40%. The BST is thought to operate through supplying the mammary gland with more nutrients through increasing

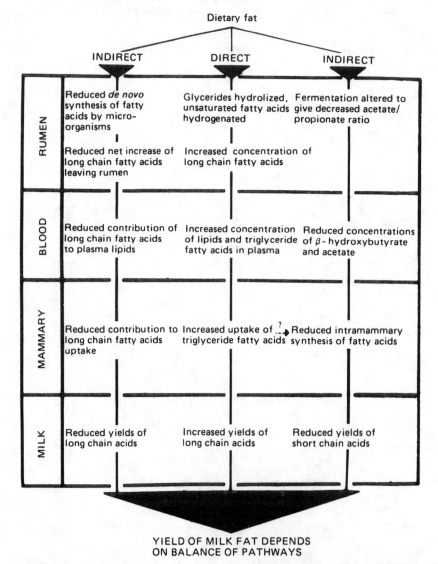

Figure 12.32 Effects of dietary lipid on aspects of metabolism related to milk fat synthesis (courtesy of Storry, J. E. (1981) *Recent Advances in Animal Nutrition.* **London: Butterworths**

the rate of body fat mobilisation and hence the levels of plasma free fatty acids, acetoacetic acid, 3-hydroxybutyric acid and lipoproteins available to lactating cells and perhaps also through increased uptake of glucose and amino acids by the mammary gland.

The greatest effects of diet on the secretion rates of milk constituents are caused by the form of dietary carbohydrate and the fat content of the diet. Dietary protein, within normal limits, has much less effect.

It has long been recognised that the ratio of acetic + butyric:propionic acid in the rumen, which is largely a function of type of dietary carbohydrate (*see* p. 327), has an influence on milk fat content. Thus increasing the roughage: concentrate ratio increases the yield of milk fat and vice versa. This is due to the fact that propionic acid derived from the fermentation of concentrates in the rumen is gluconeo-genic and elevates plasma insulin. As a consequence there is reduced mobilisation of fatty acids from body fat and milk fat secretion falls due to a reduced supply of precursors for milk fat synthesis. In contrast to the effect that propionic acid has on milk fat, increased levels of propionate absorbed from the rumen increase milk protein synthesis.

The effects of dietary fat on milk composition are complex and are mediated through direct effects on the availability of diet derived fatty acids in the plasma and the indirect effects of dietary fat on rumen fermentation (*see* p. 327), and on the *de novo* synthesis of short chain fatty acids in the rumen (*Figure 12.32*). Thus the yield of milk fat depends on a balance of these effects. Not only total yield but also the composition of the milk fat can be influenced through the inclusion of dietary fat. The use of 'protected' fat systems can allow the incorporation of dietary fat without some of the indirect effects of dietary fat being incurred.

The milk ejection mechanisms

As a result of secretion, milk accumulates in the alveolia and ducts of the mammary gland. This milk is ejected from the alveoli and ducts by contraction of the smooth muscle fibres (myoepithelial cells) which surround the alveoli. This process of milk 'let down' or, more accurately, milk ejection, is the result of a neurohormonal reflex (*see Figure 12.33*). The reflex is made up of a neural sensory input and a hormonal output. Tactile stimulation of the teats (through the butting and suckling of the offspring) provides the stimulus which evokes the release of oxytocin from the posterior pituitary. Oxytocin then acts on the myoepithelial cells to cause their contraction, and hence milk ejection. The time from initial stimulation to the start of alveolar contraction is less than 1 min; in the cow, the effective duration of milk ejection is 2–5 min and in the sow, it is about 10–30 s.

Because a proportion of milk is stored in the gland cistern and larger ducts, initially milk is passively withdrawn from the gland, to be followed by the active milk-ejection phase. The completeness with which milk is removed from the gland depends on the amount of oxytocin released from the posterior pituitary, which itself depends on the strength of the initial stimulation. Natural suckling is more efficient in this respect than either hand or machine milking. Once the pituitary store of oxytocin has been depleted, there is a substantial delay while more milk is synthesised and the store of oxytocin is replenished.

In machine-milked dairy animals the normal routine prior to milking is sufficient to trigger a milk ejection reflex. This

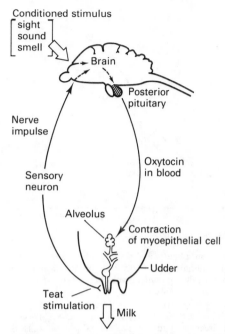

Figure 12.33 Diagrammatic representation of the milk ejection reflex

is a good example of what is termed a conditioned reflex. The milk ejection reflex can be inhibited by emotional disturbance. It is likely that this is due to a release of adrenaline which, by its vasoconstrictive action on the blood supply to the udder, reduces oxytocin availability to the gland.

GROWTH

A knowledge of the processes involved in animal growth and an ability to manipulate these processes may be used to influence the efficiency of production. In the production of animals for meat a knowledge of mechanisms involved in the control of muscle, fat and bone deposition is relevant such that we may efficiently produce animals with a high proportion of muscle and an optimal amount of fat at market weight. Growth is also a characteristic of animals that is related to reproductive performance (*see* earlier section).

The nature of growth

Growth has two aspects. The first is the increase in mass (weight) per unit time of the whole animal or part of the animal. The second involves changes in form and composition resulting from differential growth of the component parts of the body. In studies of meat animals we are primarily concerned with the growth of the major tissues of the carcass which are muscle, fat and bone and with the proportions of these three major tissues in the carcass. A growth curve for weight or size in the whole animal follows the pattern shown in *Figure 12.34*. From birth the animal will typically grow along a sigmoidal curve showing

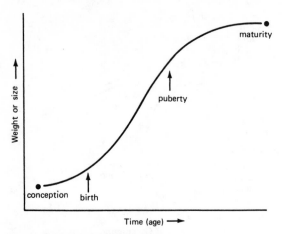

Figure 12.34 The sigmoid growth curve

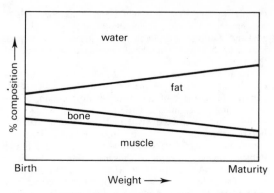

Figure 12.36 Diagrammatic representation of changes in carcass composition

acceleration about puberty and slowing down as maturity is approached. Animals are usually slaughtered around the pubertal phase of their growth curve at between one-half and two-thirds of mature body size.

The liveweight growth of farm animals is the gross expression of combined changes of the carcass tissues, organs and viscera and gut fill. The animal productionist is primarily interested in the growth of the carcass tissues comprising the muscle, bone and fat. The carcass weight of the animal is comprised of the live weight less the viscera, gut fill, blood, hide, head and feet and may be influenced by a variety of biological factors, particularly level of fatness and diet type and by carcass handling procedures.

The carcass is an extremely variable commodity reflecting the temporal growth of its component tissues. Although each of the carcass components follows a sigmoid growth curve similar to that of liveweight the different growth curves are not in phase one with another. This is due mainly to the fact that the different tissues vary in priority for the available nutrients. The bone tissue reaches maximal growth rate prior to maximal muscle growth with adipose tissue being the latest of the body tissues to attain peak growth intensity (*Figure 12.35*). Thus carcass composition in terms of the proportions of muscle, fat and bone changes as an animal grows (*Figure 12.36*).

The chemical composition of the body changes with age in a way which clearly substantiates the phasic development of the tissues. The most marked changes with age are the decrease in the proportion of water in the body and the increase in the proportion of lipid. The almost inverse relationship between the water and fat content of the body reflects the lower water content of fat (10%) compared with that of muscle tissue (75%). The percentages of protein and mineral matter decline only slowly as growth proceeds mainly as a result of changes in lipid content; in fact if chemical composition is expressed on a fat free basis the chemical composition of the body remains remarkably constant during growth reflecting that the bones and muscles of the limbs and trunk remain in proportion to one another.

Tissue growth

The growth of individual tissues is a reflection of the increase in both the size and number of cells. An increase in number of cells is called *hyperplasia*, an increase in cell size is termed *hypertrophy*. Hyperplasia and hypertrophy of all body cells occur during embryonic life.

Bones grow during both prenatal and postnatal periods through the interaction of three different cell types: chondrocytes, osteoblasts and osteoclasts. *Chondrocytes* are cells that produce cartilage, *osteoblasts* produce bone collagen and other bone components, and *osteoclasts* break down

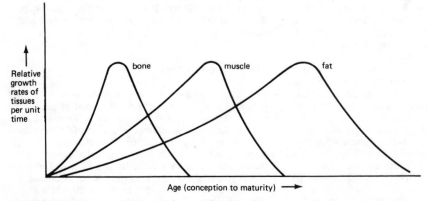

Figure 12.35 Diagram indicating relative growth rates of body tissues from conception to maturity

bone during the process of resorption. Bone grows in length by ossification of the epiphysial cartilage. Once this is complete the bone will stop growing in length. In mature animals bone is composed of approximately 50% mineral [$Ca_{10}(PO_4)_6(OH)_2$] and 50% organic matter and water, on a weight basis.

In muscle tissue development the number of muscle cells is fixed just prior to birth. Thus the growth of muscle that follows birth is due to hypertrophy of cells through protein deposition. Deposition of protein in muscle represents the net effect of two processes; protein synthesis and protein degradation. Control of protein deposition may be exerted through either or both of these processes although growth clearly involves synthesis exceeding degradation. The supply of nutrients, especially amino acids and energy, and the hormonal status in the animal, particularly the balance of insulin, growth hormone, the glucocorticoids and the sex steroids are influential.

Fat consists of fat cells (adipocytes) and supporting connective tissue. The period during which hyperplasia of adipocytes occurs is unclear but some postnatal as well as prenatal development does occur; in pigs up to five months of age; in sheep up to 11 months of age; and in cattle up to 14 months of age. Hypertrophy of adipose tissue occurs through the net deposition of fatty acids in adipocytes. Adipocyte size may vary considerably according to the animal's energy balance, reducing in size in times of negative energy balance and increasing in times of energy surplus. There are two types of adipose tissue which are referred to as *white fat* and *brown fat*. Most body fat is white and it functions as a depot of stored energy. Brown fat is very active metabolically and can be used to maintain body temperature, which is especially important in neonatal animals (*see* p. 364).

Factors influencing growth

The patterns of tissue growth and consequent changes in chemical composition of the body are influenced by several interrelated environmental and genetic factors.

Genetic effects

Genetic differences occur in that animals of the same species vary in their mature size and weight, a feature which is reflected in differences in carcass composition. Generally animals which reach the asymptote of the growth curve at an early age have lower mature body weights. The converse also applies. This is exemplified in the Angus and Friesian breeds of cattle and largely reflects differences in the timing of adipose tissue growth. Differences in mature weight are, however, probably not solely responsible for differences in composition between breeds. Thus, for example, Soay sheep have been shown consistently to contain less fat than Down breeds at the same stage of maturity. Although it is often implied that animals which have large mature body weights are more efficient because of their lower maintenance energy requirement per unit of body weight, there is little evidence that this is the case and the overriding factor governing efficiency within a species is the choice of slaughter weight.

In addition to these differences in total body size and body composition there are genetic influences on the proportions of muscle and bone tissue. These are most noticeable in the comparisons of dairy and beef breeds of cattle, the latter generally having more muscle relative to bone when comparisons are made at equal levels of fatness and stage of maturity. In cattle and sheep, although not pigs, there are clear breed differences in the distribution of body fat, especially between the carcass (subcutaneous and intramuscular) and intra-abdominal (perinephric fat, omental fat and mesenteric fat) depots. Thus 'dairy' breeds of cattle and sheep breeds such as the Finnish Landrace tend to deposit a higher proportion of fat intra-abdominally than 'beef' breeds of cattle and sheep such as Suffolks and Hampshires.

Differences in the growth and composition of body tissues occur between the sexes. In cattle and sheep, if comparisons of body composition are made at equal body weights or ages, entire males have less fat than castrates and castrates less fat than females. The situation differs slightly in pigs where gilts have less fat than castrates at equal body weight. Differences in mature weight between males and females explain some of the differences in composition attributable to sex. The rest is most likely due to differences in the levels of testosterone found in males, females and castrates. The effect of androgens is specifically shown in the more pronounced development of the forequarter muscles of the bull and boar compared with the female or castrate. The influence of sex on muscle to bone ratio is unclear but at the same level of fatness it is usual that entire males have higher muscle:bone ratios than females illustrating further that the impetus for fattening supersedes the impetus for muscle growth at lighter weights in females than in males.

Nutritional effects

Nutrition is generally the dominant factor influencing the expression of growth potential in farm animals. Both quantitative (plane of nutrition) and qualitative (diet composition) variation in nutrition influence growth. The general influence of plane of nutrition on growth develops from the priorities that exist for available nutrients as an animal grows (*Figure 12.35*). On the high plane of nutrition the growth curves are telescoped together, whereas on a low plane the sequence is extended. The most profound effects are on the deposition of fat. This is illustrated in *Figure 12.37* and shows that not only does plane of nutrition above maintenance influence the proportion of the different tissues synthesised but also that a plane of nutrition below maintenance can lead to 'negative' growth, especially of fat tissue in the first instance.

The main feature of diet composition which can influence growth is the energy and protein content of the diet and the ratio of energy:protein in the diet. Fat growth relative to muscle and bone growth is dependent on the level of energy intake. Increasing the level of energy intake at a specified protein intake level increases the deposition of fat relative to muscle. Additionally with any specified level of energy intake animals increase in fatness if their muscle growth is restricted by the amount, and in the case of non-ruminants, the quality of protein in the diet. Thus a balance of energy and protein is necessary for the production of carcasses with the desirable level of fatness. The plane of nutrition can be used to adjust the growth rate in different stages of the growing/fattening period. Thus, for example, by restricting the energy intake of pigs the amount of fat in the carcass at slaughter may be limited and the time to slaughter in cattle may be manipulated through adjusting plane of nutrition in order to make best use of price fluctuations.

A further nutritional influence that requires mention here is that of 'compensatory growth'. Animals whose liveweight

Rate of gain/loss

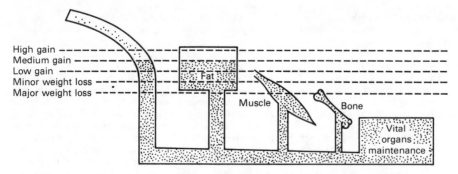

Figure 12.37 A model of priorities for nutrients during growth (after Berg, R. T. and Butterfield, R. M. (1976) *New Concepts in Cattle Growth.* **Sydney University Press)**

growth has been retarded by restriction of feeding exhibit this phenomenon which is characterised by more rapid than normal growth when introduced to a high plane of nutrition. Compensatory growth is characterised by rapid deposition of cellular protein so that the maximum protein/DNA ratio in the cell is achieved. Although in biological terms the issues involved in compensatory growth are complex, in economic terms the phenomenon can be put to good use. Under UK conditions the growth rate of cattle may be restricted during the winter (store period) with 'compensatory growth' being achieved on summer grass.

The hormonal control and manipulation of growth

The endocrine system is a major regulator of animal metabolism and probably all hormones either directly or indirectly influence animal growth. Those considered to have specific effects on growth are the hormones of the anterior pituitary, the pancreas, thyroid, adrenals and gonads. These hormones affect body growth mostly through their effects on nitrogen retention and protein deposition. The precise modes of action of the individual hormones is only partly understood and is complicated by the fact that in some cases at least their effects depend on the sex, species of animal and balance of other hormones. When workers have attempted to correlate the concentration of hormones in the blood and growth no clear relationships have emerged.

Although there is little evidence of an association between blood growth hormone (GH) levels and growth in livestock some workers have shown there to be greater growth hormone secretory activity in genetically superior cattle and pigs and more recently it has been shown that exogenous growth hormone may produce an anabolic response in farm livestock. The anabolic response of GH appears to be mediated through a group of peptide hormones termed somatomedins and that not only do they increase growth rate but also produce leaner carcasses. The secretion of GH from the pituitary is controlled by the balance of an inhibitory hypothalamic peptide, termed somatostatin and a stimulatory GH releasing factor (GHRF) which is also a hypothalamic peptide. Much recent research has looked at the possibilities of either removing or neutralising the effects of somatostatin by active immunisation techniques and at

supplementing the effect of GHRF as a means of manipulating growth.

It is now generally accepted that the natural steroid hormones – oestrogens, androgens, progestins and glucocorticoids – exert an effect on growth. The difference in growth rate and mature body size of male, female and castrate animals suggest that the sex hormones play an important role in the control of growth. There can be no doubt that these differences are in the main androgen-dependent. Deprived of the anabolic effect of androgens, castrates divert energy intake into the synthesis of fat rather than muscle and are thus less efficient in food conversion efficiency terms. Castration can only be justified on the basis of its effects on sexual and aggressive behaviour and in some situations the possible risk of carcass taint. Such characteristics only reveal themselves in the period close to slaughter and the technique of late castration using immunological techniques is at present being considered by research workers as a means by which the maximum anabolic advantage of the testes may be obtained. Immunological castration involves the active immunisation of animals against gonadotrophin releasing hormone (GnRH). This has the effect of inhibiting the secretion of follicle-stimulating hormone (FSH) and luteinising hormone (LH) such that spermatogenesis ceases and testosterone secretion declines sharply.

Several synthetic compounds have been used to promote growth and in many cases increase the protein content of the carcass. Classically, androgenic compounds would be expected to increase protein synthesis and hence protein deposition in muscle but the synthetic androgen, trenbolone acetate, has been shown to decrease protein synthesis and exerts its growth promoting effect through decreasing protein degradation in muscle. This it appears to do through depressing the catabolic effects of the glucocorticoids. As with the androgen, trenbolone acetate, the oestrogenic anabolic agent zeranol, which has been widely used commercially, would also appear to promote growth through reducing the rate of protein turnover rather than through stimulating muscle protein synthesis. It has also been suggested that oestrogens stimulate circulating GH concentration and change thyroid hormone status and that this accounts in part for their growth stimulatory effects.

Exogenous anabolic agents to stimulate the growth of animals, especially cattle, were widely used in production practice until 1986 when EEC legislation banned their use on the grounds of their possible risk to human health. The main substances used prior to the ban were trenbolone acetate (a synthetic androgen) and the oestrogenic substances hexoestrol and zeranol. The ability of these agents to stimulate growth is dependent on their producing an optimum balance of androgens and oestrogens in the animal. Thus response to implants is determined by the sex of the animal; best responses have been obtained by the exogenous administration of androgens to females, oestrogens to entire males and a combination of androgenic and oestrogenic agents to castrates.

One of the major concerns of the meat producing industry in recent years has been the excessive amount of fat deposited in carcasses. The origins of this concern have been primarily due to public concern about the health risks associated with the consumption of animal fats although excessive fat production also represents a source of inefficiency in livestock production. Although leaner carcasses may be achieved in the short term by slaughtering at lighter body weights and in the long term by appropriate selection programmes an alternative presently being investigated is the feeding of a group of substances termed β-agonists. Agents such as clenbutarol and cimeterol which are substituted catecholamines have been shown to reduce fat deposition and increase the protein content of the carcass. Their mode of action is not fully understood but influences on fat deposition would appear to be due to a reduction in lipogenesis and an increase in lipolysis whilst the effects on protein content appear to be by a combination of reduced protein degradation in muscle and an increase in the rates of protein synthesis. As yet, however, β-agonists have not been approved for commercial use.

ENVIRONMENTAL PHYSIOLOGY

Throughout their evolution animals have been subjected to environmental influences and farm animals, like their ancestors, have to employ a range of structural and functional strategies to adjust to changes in their environment. The term environmental physiology describes those physiological mechanisms involved in the process of homeostasis and the external stimuli, such as temperature, daylength and stocking rate, that necessitate such responses.

The nature and importance of photoperiod in regulating seasonal activities such as reproduction has already been mentioned both in this chapter and in Chapter 15. Perhaps the most important environmental influence, particularly with regard to animal productivity, is that of temperature. This governs the amount of heat energy lost from the animal, which in turn determines the amount of dietary energy retained by the animal.

Maintaining thermal balance

Farm animals maintain a high (37.5–39°C) and relatively constant body temperature despite large fluctuations in environmental temperature. The combination of physiological and behavioural changes which regulate body temperature are controlled by the hypothalamus.

The maintenance of a constant body temperature requires that there is a balance of heat production within the body and heat loss from the body. Heat is produced as a by-

product of the metabolism of ingested nutrients used for maintenance and the various productive processes. Heat transfer between the body and the environment occurs by means of conduction, convection, radiation and evaporation. Since generally the animal body is at a higher temperature than the environment, heat may be lost by all these methods. The partition of losses between the various routes of heat exchange varies according to the air temperature, with evaporative losses being high at high air temperature, whereas radiation and convection are the main avenues of loss at low and intermediate temperatures. In addition to air temperature, other interacting environmental factors – wind, precipitation, humidity and sunshine – are determinants of rates of heat transfer. Wind, precipitation and humidity particularly affect evaporative losses; wind also affects convection losses by increasing air movements around the animal's body, thus destroying the insulating layer of air trapped in the animal's hair or wool, and sunshine especially affects the net radiation exchange between the animal and its environment in animals kept outdoors.

The mechanisms by which animals can regulate body temperature fall into two categories: alteration in the rate of heat production and the rate of heat loss. The animal has a number of mechanisms by which the rate of heat loss can be altered – by varying the degree of vasodilation, sweating, piloerection (erection of body hair), and numerous behavioural responses such as change in posture, activity and food intake – and it is these mechanisms which come into play first of all in the regulation of body temperature. The range of temperature within which body temperature can be regulated without changes in metabolic heat production is termed the thermoneutral zone. The air temperature at which these mechanisms are no longer successful in dissipating sufficient heat to prevent a rise in body temperature is termed the upper critical temperature; the temperature below which the animal must increase its rate of heat production is termed the lower critical temperature (*see Figure 12.38*). These critical temperatures are not fixed and

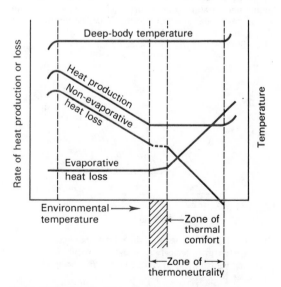

Figure 12.38 Graph of thermoregulatory responses to a range of environmental temperatures

are especially dependent on the animal's metabolic body size and level of production.

Environmental temperature and animal production

To date much of the precise work on the effects of the environment on the animal, in temperate regions, has been concerned with intensively-kept pigs and poultry and has primarily studied their interaction with the nutrition of the animal. Energy is the main nutrient affected, higher environmental temperatures releasing more food energy for productive purposes but having secondary effects on protein metabolism, food intake and carcass composition.

Of all the farm species the pig is perhaps the most vulnerable to the effects of environmental temperature. It has little body hair for insulation and few sweat glands to help with evaporative cooling. However, pigs do make certain behaviour adaptations to temperature, such as huddling together in cold weather and wallowing in hot conditions. The lower critical temperature of pigs changes markedly with age and body size. Small pigs with a larger surface:body mass ratio will lose relatively more heat from their skin surfaces than their older and larger counterparts. Thus a newborn pig is particularly vulnerable to a cold environment. *Figure 12.39* shows how the lower critical temperature of a pig improves with increase in body weight. Apart from the level of dietary energy, lower critical temperature is also affected by factors such as ventilation rate, wetness and type of flooring, and provision of straw.

In the light of these effects it is clear that where quantitative data on environmental effects are available, they may be taken into account in the planning of feeding programmes, the design of buildings, and optimising of investment in insulation and supplementary heating.

In the outdoor situation experienced in the UK, most interest in environmental effects centres on the combined effects of extreme cold, wind and rain. The effects on adult animals are primarily discomfort and loss of production but in the young and newborn, such weather conditions are a threat to life. Surveys have shown that perinatal lamb mortality can vary from 5 to 45% on hill farms and that, at the higher mortality rates, extreme cold is the main contributor. In extreme conditions, shivering thermogenesis is often insufficient to maintain body temperature in the neonate. These animals may increase heat production by non-shivering thermogenesis, a process mediated by the effects of cold on the sympathethic nervous system and particularly by the action of noradrenaline on brown adipose tissue. The breakdown of this special type of adipose tissue results in much higher levels of heat production than are produced from other catabolic processes and serves to help prevent hypothermia in the neonates of many species. It appears likely that variations in the amounts of brown adipose tissue present in the body at birth may explain in part the genetic variation in the ability of the newborn lambs to survive adverse conditions.

Environmental physiology and animal welfare

In recent years the welfare of farm livestock has come under increasing scrutiny, particularly the more intensive methods of animal production. In such systems, an animal may be housed in a largely artificial environment. The extent to which animals are able to adapt to their surroundings will depend mainly on the design of the accommodation and the husbandy skills of the stockman.

Failure to interpret correctly the animal's physiological and behavioural needs may have serious consequences for both animal productivity and animal welfare. Animals that are physiologically or behaviourally deprived may suffer stress, a response which in the long term can result in an increased incidence of disease through a lowered resistance to infection and reduced growth rates. Even off the farm, it has been shown that pre-slaughter stress results in a loss of carcass quality and weight as in the case of dark-cutting meat of beef animals and the pale, soft, exudative meat of pigs stressed immediately prior to slaughter. Thus attention to the conditions during transport, auction, lairage and at slaughter may have benefits for animal productivity as well as animal welfare.

The Farm Animal Welfare Council (FAWC) in their codes of welfare recommendations offer guidance to the farmer on the physiological and behavioural needs of farm animals that will help safeguard their welfare.

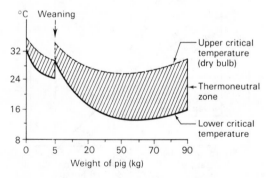

Figure 12.39 Graph showing the relationship between upper and lower critical temperatures and bodyweight in a growing pig (courtesy of ADAS (1982) *Pig Environment*, Booklet 2410

Further reading

COLE, D. J. A. and HARESIGN, W. (1985). *Recent Advances in Pig Nutrition.* London: Butterworths

FRANDSON, R. D. (1981). *Anatomy and Physiology of Farm Animals.* Philadelphia: Lea and Febiger

HARESIGN, W. and COLE, D. J. A. (1981). *Recent Developments in Ruminant Nutriton.* London: Butterworths

HUNTER, R. H. F. (1982). *Reproduction in Farm Animals.* London: Longman

McDONALD, P., EDWARDS, R. A. and GREENHALGH, J. F. D. (1981) *Animal Nutrition.* London: Longmans

ROOK, J. A. F. and THOMAS, P. (1983). *Nutritional Physiology of Farm Animals.* London: Longmans

WHITTEMORE, C. T. (1980). *Lactation of the Dairy Cow. (Handbooks in Agriculture)* London: Longmans

13

Livestock feeds and feeding

R. M. Orr

Feed represents the major cost to animal production. Thus, the efficiency of its use can have a considerable impact on the performance of an enterprise.

In general terms efficient utilisation of feeds is dependent on the following elements:

(1) a knowledge of the nutrient value of the range of feedstuffs available;
(2) a knowledge of the nutrient requirements of animals;
(3) an ability to formulate diets and rations from a mixture of feedstuffs which will meet nutrient requirements within the potential voluntary food intake of animals. Clearly, an important element of this process of formulation is that of meeting nutrient requirements most economically.

In this chapter the procedures used in determining the first two elements are given attention as are characteristics of the major feeds, in terms of feeding value and limitations to their use, that are used in diet formulation. The aspect of feeding particular classes of livestock is dealt with in the chapters on the individual species.

NUTRIENT EVALUATION OF FEEDS

The value of a feed is dependent on how much of particular nutrients in the feed the animal is able to use to meet the requirements of the various body processes. In the utilisation of feed there are inevitable losses of nutrients due to inefficiencies in digestion, absorption and metabolism. Thus, the available nutrients in a food are the difference between the potential value of a feed as measured by chemical analysis and these inevitable losses of nutrients. Losses through incomplete digestion are a major factor in determining feeding values and digestibility measurements are in some instances used to give an indication of feeding value – this is particularly true in the evaluation of roughages. For the purpose of diet formulation, however, the nutritive value of foods is determined in terms of their ability to furnish the animal primarily with energy and protein and, to a lesser extent, with minerals and vitamins.

Digestibility measurements

The digestibility of a food is defined as the proportion of the food which does not appear in the faeces and is assumed to be absorbed. This is strictly termed 'apparent' digestibility because a very small fraction of faecal excretion is in the form of fragments of gut mucosa and therefore not of dietary origin but in practice this is not taken into account. The apparent digestibility of a feedstuff may be determined in animals restrained in metabolism crates which allow the precise measurement of food intake and of faecal excretion over a specified period of time. From this and the analysis of feed and faeces, digestibility coefficients for the food dry matter and also for individual food nutrients may be calculated, i.e. organic matter, protein, energy.

The percentage of digestible organic matter (OM) in the dry matter of a feed is termed the D-value. The D-value is widely used to describe the nutritive value (especially energy potential) of roughage feeds, particularly hay and silage.

Since digestibility assessments using live animals are tedious to conduct and unsuitable for routine estimations of large numbers of samples, alternative laboratory techniques (*in vitro* methods) have been developed. The earliest *in vitro* methods were developed for roughages at the Grassland Research Institute and aim to simulate ruminant digestive processes. A two-stage process is carried out: the first involves incubation of a sample with rumen liquor to simulate rumen fermentation, the residue from this being further incubated with an acid pepsin solution to simulate the digestion of protein in the lower gut. More recently, methods based on the use of fungal cellulases have been used successfully to predict the digestibility of forages and other feeds. Following boiling with neutral detergent solution which hydrolyses the cell contents, the remaining cell wall or neutral detergent fibre fraction (NDF) is incubated with a cellulase solution, the residue from which can be ashed to determine the digestibility of the organic matter in the dry matter. This is referred to as the cellulase digestible organic matter (NCD).

The apparent digestibility of feedstuffs may vary between animals due to such factors as type of animal (e.g. ruminant *versus* non-ruminant; sheep *versus* cattle). The main characteristics of foods that affect digestibility are its chemical composition, particularly the fibre, protein and lignin content. Thus, for foodstuffs such as cereal grains and protein concentrates that are fairly constant in chemical composition digestibility values are fairly constant but for forages where composition varies widely according to growth stage wide variation in digestibility values may be found.

The other main characteristic that can have an effect is the method of feed preparation and processing involved. The effects of processing on the digestibility of cereal grains is complicated by different responses in the different species of farm animal but such techniques as pelleting, rolling, milling and cooking through micronisation and steaming can be of use in particular situations. Likewise, chemical processing through alkali treatment can be used to effect improvements in digestibility of both grains and forages. Physical processing of forages, especially grinding, decreases digestbility through increasing rate of passage. The decreases in digestibility of foods observed when the level of intake is increased are likewise largely due to rate of passage effects. Because the level of intake affects digestibility, measurements are usually made at the maintenance level of feeding.

Measurement of feed energy

Animals use energy for a variety of body functions ranging from energy required for essential muscular activity and maintenance of body temperature to energy required for the synthesis of body tissues. It is primarily energy intake which determines the level of performance that an animal can sustain.

The energy content of feeds may be expressed in the following ways: Gross energy (GE), Digestible energy (DE), Metabolisable energy (ME), Net energy (NE) and to understand the meaning of the terms involves a recognition of the ways in which food energy is utilised by the animal.

The large number of chemical bonds that make up the organic compounds present in food contain the energy present in the food. This chemical energy, termed gross energy or heat of combustion, is determined by the technique of bomb calorimetry in which a known weight of food is completely oxidised and the heat liberated measured. Of this gross energy in a food only a portion is eventually available to provide for the animal's energy requirements. In the utilisation of the gross energy there are several routes of energy loss, namely:

(1) losses of indigestible energy in the faeces,
(2) energy losses in the urine, mainly in the form of nitrogen-containing substances such as urea,
(3) losses of energy as methane produced during the fermentation of dietary energy in the rumen and to a much lesser extent in the large intestine,
(4) heat produced from inefficiencies in metabolic processes of the animal and dissipated from the body surfaces and to a lesser extent heat produced from inefficiencies in the metabolism of microbial cells in the digestive tract (heat of fermentation). The sum of these forms of heat is termed the Heat Increment of Metabolism.

The energy remaining after these losses are accounted for represent the energy the animal utilises for maintenance and

production. This may be represented by the following diagram:

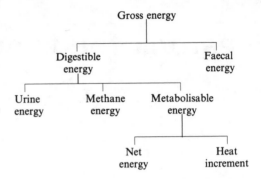

The unit of measurement used in the assessment of food energy is the joule (J). More typically it is customary to use the kilojoule (kJ) or the megajoule (MJ).

$$1\,MJ\ =\ 1000\,kJ\ =\ 1\,000\,000\,J$$

Gross energy value

This is measured by burning a known weight of the food and recording the total heat produced. The gross energy value of a food is dependent on the proportions of carbohydrate, protein, fat and ash that are present. The approximate energy values of these components are: carbohydrate 17.5 MJ kg^{-1} DM, protein 23.5 MJ kg^{-1} DM, fat 39 MJ kg^{-1} DM and ash 0 MJ kg^{-1} DM. Thus the fat and the ash content of a food have potentially the greatest influence on gross energy values. In practice, most animal foods have a low and fairly constant amount of these component, mainly consisting of carbohydrate with lesser amounts of protein, such that the gross energy of a wide range of foods is fairly consistently around 18 MJ kg^{-1} dry matter. Although gross energy tells us the potential energy in a feed it is the least satisfactory assessment of energy value since no account is taken of the availability of that energy to the animal.

Digestible energy

Digestible energy (DE) is a measure of the gross energy of the food minus that part of the food energy lost in the faeces. Strictly speaking this is the 'apparent' DE of a food and is determined in digestibility trials in which the intake and faecal excretion of gross energy is measured. The DE value of foods is dependent on the gross energy value and the factors that influence the digestibility of foods.

Metabolisable energy

The metabolisable energy (ME) of a food is the digestible energy less the energy lost in the urine and as methane. Typically about 8–10% of the digestible energy is lost in urine. About 10–12% of the digestible energy is lost as methane in ruminants whereas in the pig losses are negligible. Thus, for pigs metabolisable energy values are approximately 90% of DE values and for ruminants are approximately 80% of DE values.

Metabolisable energy values are determined by feedings trials in which animals are kept in metabolism cages within respiratory chambers that allow the accurate measurement of intake, faecal and urinary excretion and methane production. ME values are primarily influenced by the digestibility of the feed. Although the proportion of digestible energy lost in urine and methane is relatively consistent urine losses are affected by the protein content of the diet.

Net energy value (NE)

The metabolisable energy supplied by the diet may be looked upon as the absorbed nutrients available for the various metabolic processes in the tissues. Thus, in the ruminant, volatile fatty acids, and in the monogastric, glucose, can be looked upon as some of the main components of ME. The efficiency of the metabolic pathways into which these and other metabolites are directed so that forms of energy (e.g. ATP) that are useful to the animal may be created, are not 100% efficient, some of the energy being converted into heat which is termed the heat increment (HI). Some heat is also produced from the work of digestion whilst in the ruminant heat arises from the inefficiencies in the metabolism of rumen microbes (heat of fermentation). The HI of a food is measured by the technique of animal calorimetry, and when subtracted from the ME gives the NE value. Typically the HI is greater in the ruminant (35–65% of ME for mixed diets) than in the monogastric (10–40% of ME for mixed diets.)

In the monogastric considerably less HI results from the metabolism of fat (9%), than carbohydrates ($\simeq 17\%$) or proteins ($\simeq 25\%$) but in contrast to the ruminant the efficiency with which ME is used for the different body processes, such as maintenance and growth, is very similar. The NE value of a food to the monogastric is, therefore, a function of its ME value and the nutrients comprising the ME, especially the proportion of fat present.

In the ruminant the extent of heat production in the metabolism of ME and therefore the NE value of a feed is dependent on the purpose for which the ME is utilised. This is because the products of digestion (the constituents that make up the energy of ME) are used at different levels of efficiency for maintenance or conversion to the various productive functions of lactation, growth and reproduction. The situation is further complicated by the fact that the principal energy containing end-products of digestion, the VFA – acetic, propionic, butyric – are used with differing efficiencies. Thus, diet composition, which influences the molar proportions of VFA, as well as physiological function, have an effect on the efficiency of utilisation of ME. This is illustrated in *Figure 13.1* and means that individual feeds cannot be assigned single NE values.

Since the efficiency of utilisation of ME for maintenance does not vary widely it is currently the practice to assume that little error is involved through using an average efficiency factor of 0.72 to convert ME values to NE values for maintenance (NE_m) thus:

$$NE_m = ME \times 0.72 \tag{1}$$

Likewise for lactation little error is introduced by using an average efficiency of utilisation of ME for lactation (k_l) such that:

$$NE_l = ME \times 0.62 \tag{2}$$

For growth and fattening the efficiency of utilisation of ME varies widely according to diet such that for growing ruminants an efficiency factor for the particular diet in question must be used. The efficiency of utilisation of ME for gain (k_g) varies according to the equation:

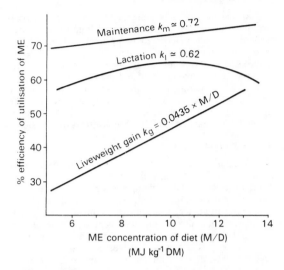

Figure 13.1 Effect of diet and production function on the efficiency of utilisation of ME

$$k_g = 0.0435 \times ME \text{ value of the diet} \tag{3}$$

Thus the NE value of a diet used for growth (ME_g) is:

$$NE_g = ME \times k_g$$

Thus for a diet (A) having an ME value of $8 \text{ MJ kg}^{-1} \text{DM}$:

$$k_g = 0.0435 \times 8$$

$$= 0.345$$

and

$$NE_g = 8 \times 0.345$$

$$= 2.78 \text{ MJ kg}^{-1} \text{DM}$$

and for a diet (B) having an ME value of $12 \text{ MJ kg}^{-1} \text{DM}$:

$$k_g = 0.0435 \times 12$$

$$= 0.522$$

and

$$NE_g = 12 \times 0.522$$

$$= 6.26 \text{ MJ kg}^{-1} \text{DM}$$

Although diet B has 50% more $ME \text{ kg}^{-1} \text{DM}$ than diet A it has 125% more $NE \text{ kg}^{-1}$ when fed to growing/fattening ruminants. In other words the relative value of foods for maintenance is not the same as their relative values for growth and fattening.

Protein evaluation of feeds

Animals require protein to provide them with amino acids. These amino acids are used for the synthesis of body tissue proteins and proteins in products such as milk and wool which leave the body. Simple stomached animals obtain their supply of amino acids from the digestion and absorption of dietary protein whereas in the ruminant the situation is complicated by the presence of the rumen which means that the supply of amino acids comes from a combination of microbial protein produced from dietary rumen degradable protein (RDP) or non-protein nitrogen (NPN) and dietary undegradable protein (UDP). Thus, different approaches to the evaluation of foods as protein sources are necessary for the monogastric and the ruminant.

Before considering the more recent and more precise methods of evaluating foods as sources of protein it is worth mentioning methods of evaluation that have been in use for a number of years and are of relevance to current systems.

Crude protein (CP)

Proteins in food are the main nitrogen containing components. Thus by measuring the nitrogen content of a food using the Kjeldahl technique it is possible to derive its protein content. Two assumptions are made in calculating crude protein content from nitrogen content. Firstly, it is assumed that all nitrogen in a food is present in protein. In reality this is incorrect in that some nitrogen is present in such compounds as amides. For most feeds this nitrogen is less than 5% of the total but for root crops it may be considerable. Secondly, it assumes that all food protein contains 16% nitrogen although, in practice, a range of 15–18% for different protein types exists. Because of the inaccuracy of these assumptions the protein content of feeds is expressed as crude protein thus:

$$\% \, CP = \% \, N \times (100/16)$$

or more commonly

$$\% \, CP = \% \, N \times 6.25$$

The crude protein content gives no indication of how efficiently the protein is utilised.

Digestible crude protein (DCP)

The DCP content of a diet is determined from the results of a digestibility trial where it is assumed that the difference between the nitrogen in the food and faeces represents the quantity of nitrogen absorbed. Since in the ruminant some nitrogen may be absorbed from the digestive tract as ammonia rather than amino acids, and that in all animals some of the faecal nitrogen is of metabolic origin (e.g. digestive enzymes) such an assumption is technically incorrect. The latter of these criticisms is of less importance since the loss of nitrogen of faecal origin is inevitable and through not being accounted for gives a more realistic measure of nutritive value but the former criticism means that the use of DCP as a system to evaluate the ability of foods to supply protein to the ruminant is fundamentally flawed and has led to the development of more relevant methods of evaluation.

In pig feeding, it is thought that DCP values are not significantly better than CP values since the degree of digestibility of protein in a variety of feeds is considered fairly constant.

Measures of protein quality

Monogastric

The usefulness of a food as a protein source is not only dependent on the amount of CP or DCP present but also on the proportions of the different amino acids that make up the DCP content compared with the balance of amino acids required by the monogastric for tissue synthesis. The amino acid make-up of the absorbed protein influences its efficiency of utilisation. There are a number of methods of assessing protein quality but an initial assessment may be given by its apparent biological value (ABV). This is determined by nitrogen balance experiments from which the proportion of the nitrogen absorbed that is retained is calculated from the nitrogen intake and nitrogen excretion in the faeces and urine. Thus:

$$ABV = \frac{N \text{ retained}}{N \text{ absorbed}} \times 100$$

$$= \frac{N \text{ intake} - (\text{faecal N} + \text{urinary N})}{N \text{ intake} - \text{faecal N}} \times 100$$

A perfect match between the balance of amino acids absorbed from the feed and amino acids needed is expressed as a biological value of 100. Knowledge of BV in pig nutrition is relevant in that it may affect the cost effectiveness of diet formulation. Thus it may be economically sound to use more of a cheaper, poorer quality protein source than lesser amounts of a more expensive high quality protein source.

In practice, the efficiency with which the digested protein is utilised by monogastrics is largely dependent on the supply of essential amino acids; that amino acid that is present in least supply in relation to requirement determines the extent of protein synthesis (the limiting amino acid). In the largely cereal based diets that are used in pig and poultry production the limiting amino acids are normally lysine and methionine. Thus the quality of protein in foods to supplement the cereal component of the diet is largely dependent on the available lysine and methionine content of the food. Routine techniques for the determination of available lysine in feeds exist and this information may be used in diet formulation to ensure that the lysine content of the diet meets the requirements of the animal. The availability of synthetic amino acids at economical prices has, to some extent, simplified the procedure of balancing the amino acid composition of monogastric diets.

Ruminant

Although ruminant animals, like monogastrics, must necessarily have requirements for specific amounts of amino acids the nature of the digestive system ensures that the amino acid composition of absorbed protein is less variable and less influenced by the amino acid composition of dietary protein. Thus at the present time the ruminant nutritionist is more concerned with wanting to predict the total amount of protein reaching the small intestine and being available for absorption rather than with the amino acid composition of that protein. It is worth noting, however, that future developments in this area may allow prediction of the amino acid composition of the protein as well as the total amount of protein being absorbed.

The amount of protein reaching the small intestine in the ruminant is dependent on the contribution of microbial protein which is in turn a function of the rumen degradable protein (RDP) and energy content of the diet and of undegradable protein (UDP) contained in the diet (*see* p. 328). Thus, a crucial element in the calculation of protein supply to the lower gut is knowledge of the degradability (dg) of food proteins. Simply:

$$CP = RDP + UDP$$

and

$$RDP = CP \times dg$$

The degradability of dietary protein in the rumen is currently assessed either by the so-called *in situ* technique or by *in vitro* methods. The former is the most widely used and involves suspending raw materials in a polyester (Dacron) bag in the rumen of fistulated animals. The degradable fraction may be calculated from the proportion of nitrogen lost from the bag during the incubation period. The point of greatest contention in this technique is the length of the incubation period since the length of time that feed particles remain in the rumen is highly variable and dependent on such factors as particle size, whether the diet is pelleted and the characteristics of the forage component of the diet. *In vitro* methods involve incubation of protein sources with rumen fluid in the laboratory rather than in the rumen itself.

Published values (*see Table 13.3*) using these techniques have until now been highly variable and this is the major source of error in protein rationing systems based on RDP and UDP. As well as the rate of passage effect mentioned above it is clear that the degradability values for certain proteins are modified by the treatment to which they are subjected. Thus although the protein in fresh grass is already highly degradable the conservation of grass as silage further increases this degradability whilst the addition of formalin to grass at ensilage reduces the degradability of the protein. The protein concentrate feeds likewise vary markedly in degradability according to their treatment during processing. Thus, the extent of heat treatment in the drying of fish meals and in the roasting of soya to inactivate trypsin inhibitors can affect degradability values.

Micronutrients

The mineral composition of goods is usually described in terms of the total quantity of the individual elements present. Since availability of minerals in food varies greatly and is influenced by a number of factors there is no attempt to include this parameter in evaluation.

Vitamin concentration of foods is usually expressed as milligrams or micrograms per kilogram. However, the levels of fat soluble vitamins are normally expressed as International Units (IU).

1 IU Vitamin A = 0.3 mg crystalline Vitamin A alcohol

1 IU Vitamin D = 0.025 mg crystalline Vitamin D_3

1 IU Vitamin E = 1 mg α-tocopherol acetate

Chemical analysis and the prediction of feed value

The most widely used routine methods of evaluating the nutrient content of feedstuffs are based on chemical procedures. Such analyses are only of value, however, if the feed analysed is representative of the entire batch and to this end it is important that appropriate sampling procedures be followed. Using appropriate sampling equipment a number of core samples should be taken from different areas of the feedstuff concerned, these samples thoroughly mixed together and a suitable size of subsample taken for analysis. It is important that samples for analysis are appropriately packaged, labelled and stored prior to analysis. For example, silage samples not analysed immediately should be frozen.

Current methods of analysis of concentrate feeds frequently rely on the Weende or proximate method of analysis whilst the analysis of forage feeds is largely based on the methods developed from Van Soest. From elements of these analytical procedures a number of equations have been developed to predict feeding values that may be used in the processes of ration formulation.

Concentrate and compound feeds

Current methods of analysis are still largely based on the Weende or proximate system in which feeds are broken down into six fractions (*Table 13.1*). The system has many criticisms, particularly with respect to its failure to completely separate the carbohydrate fraction into fibrous and non-fibrous carbohydrates and its overestimation of true protein. It is likely that the use of certain components of the proximate analysis will decline in the future as new techniques are used more widely but the contents of crude protein, oil, crude fibre and ash are still required by the Feedingstuffs Regulations 1983 to be declared on bags of both compounds and straight feeding stuffs.

Since the proximate analysis fails to adequately describe the carbohydrate fraction of feeds several methods have been derived to measure the starch and sugar content but the precision of these methods requires improvement. Although not required by law a statement of starch and sugar levels is occasionally given on bags of feedingstuffs. A further development is that the fibre fraction in certain raw materials may be characterised according to the Van Soest scheme of analysis (*Figure 13.2*) into neutral detergent fibre (NDF) which corresponds to the sum of cellulose, hemicellulose and lignin, and acid detergent fibre (ADF) which consists of cellulose and lignin.

Table 13.1 The fractions of proximate analysis

Fraction	Components	Procedure
(1) Moisture	Water and any volatile compounds	Heat sample to constant weight in oven
(2) Ash	Mineral elements	Burn at 500°C
(3) Crude protein (= N × 6.25)	Proteins, amino acids, non-protein nitrogen	Determine nitrogen by Kjeldahl method
(4) Ether extract (oil)	Fats, oils, waxes	Extraction with petroleum spirit
(5) Crude fibre	Cellulose, hemi-cellulose, lignin	Residue after boiling with weak acid and weak alkali
(6) Nitrogen-free extractives	Starch, sugars and some cellulose, hemicellulose and lignin	Calculation (100 minus sum of other fractions)

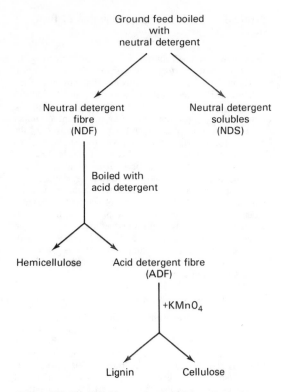

Hemicellulose

Lignin   Cellulose

Figure 13.2 The Van Soest scheme of forage analysis

The use of crude protein overestimates the protein content of feeds because of some of the nitrogen being derived from non-protein sources but where NPN sources such as urea are used in the formulation of ruminant compound feeds a statutory statement of the protein equivalent of the feed derived from NPN sources must be given. Such a statement must also accompany the declaration of protein content of compound feeds for non-ruminants if greater than 1% of protein equivalent is derived from NPN sources. For ruminant feeds a further description of the protein that can be but is not required to be given is the proportion of rumen degradable (RDP) and undegradable protein (UDP) present, for which *in vitro* laboratory methods of determination exist.

In recent years a considerable amount of research effort has been put into the development of equations which relate some of the above aspects of chemical composition to the energy value of feeds. This is especially useful for compound feeds. In some instances equations have been put forward as suitable for voluntary declaration by feed manufacturers with a view to possible future legislation. Although a vast amount of work has gone into such equations it is important at the present time to realise the limitations of these equations as far as their precision is concerned. Examples of such equations are:

For ruminant compound feeds an equation derived from Rowett Research Institute data is:

$$\text{ME (MJ k}^{-1}\text{DM)} = 11.56 - 2.38\,\text{oil}\,\% + 0.03\,\text{oil}\,\%^2$$
$$+ (0.03\,\%\,\text{oil} \times \text{NCD}\,\%)$$
$$- 0.034\,\text{ash}\,\%$$

where NCD is Cellulase Digestible Organic Matter (*see* p. 365).

An alternative equation which relates proximate analysis to ME but which has a lower degree of precision is:

$$\text{ME (MJ kg}^{-1}\text{DM)} = 11.78 + 0.0654\,\text{CP}\,\%$$
$$+ 0.0665\,\text{oil}\,\%^2 - (0.0414\,\text{oil}\,\%$$
$$\times\,\text{CF}\,\%) - 0.118\,\text{ash}\,\%$$

For compounded pig diets an equation derived by the Edinburgh School of Agriculture is:

$$\text{DE (MJ kg}^{-1}\text{DM)} = 17.49 + 0.157\,\text{oil}\,\% + 0.078\,\text{CP}\,\%$$
$$- 0.325\,\text{ash}\,\% - 0.149\,\text{NDF}\,\%$$

and for poultry an equation derived by the Poultry Research Centre to predict the metabolisable energy is:

$$\text{ME (MJ kg}^{-1}\text{DM)} = 5.39 + 0.113\,\text{CP}\,\% + 0.281\,\text{oil}\,\%$$
$$+ 0.113\,\text{starch}\,\% - 0.136\,\text{CF}\,\%$$

Analytical descriptions of forage feeds

Forage feeds are characteristically much more variable in their nutritive value than concentrates and must be routinely evaluated prior to their use in ration formulation. The methods of forage analysis developed by Van Soest in the USA (*see Figure 13.2*) have served as the basis of forage evaluation in recent years. From such measurements he devised the following regression equation for estimating digestibility of forages:

$$\text{Digestibility} = 0.98\,\text{NDS}$$
$$+ [1.473 - 0.789\log_{10}\text{lignin}]\,\text{NDF}$$

where NDS and NDF are percentages of the forage dry matter and lignin is expressed as a percentage of the ADF fraction.

Such analysis according to the Van Soest method is both too costly and time consuming for routine analysis but Modification of the Acid Detergent Fibre (MADF) determination serves as the basis of the routine analysis carried out by MAFF laboratories. Modified acid detergent fibre (MADF) is obtained by refluxing the dried ground forage with an acid detergent solution (cetyl trimethyl-ammonium bromide) for 2 h and recovering the residue that is resistant to hydrolysis. Using appropriate regression equations it is possible to predict the DOMD (Digestible Organic Matter Digestibility) and ME value (MJ kg⁻¹DM) of forages:
e.g. for grass silages

$$\text{DOMD-value} = 94.51 - (0.912 \times \text{MADF})$$

$$\text{ME (MJ kg}^{-1}\text{DM)} = 15.33 - (0.152 \times \text{MADF})$$

where MADF is calculated on a 'corrected' % of DM basis. The DM correction factor involves the addition of 1.9 % units to 'oven-dried' values and corrects for the average loss of volatiles during drying.
For grass hays

$$\text{DOMD-value} = 104.8 - (1.27 \times \text{MADF})$$

$$\text{ME (MJ kg}^{-1}\text{DM)} = 17.1 - (0.221 \times \text{MADF})$$

where MADF is calculated as % of oven DM.

Further routine information on forage composition and quality is also given by the following chemical procedures:

(1) dry matter or moisture percentage,
(2) crude protein,
(3) pH in silages,
(4) ammonia–nitrogen in silages.

The latter two measurements are indicative of the type of fermentation (*see* Chapter 6).

Near infrared reflectance analysis

Near infrared reflectance analysis is the most up-to-date, rapid and precise routine method of analysis currently available. Illumination of a feedstuff in the near infrared region of the electromagnetic spectrum (1100–2500 nm) causes certain molecular changes to occur which can be quantified such that the amount of light absorbed at a particular wavelength is proportional to the concentration of particular organic components. By selecting particular wavelengths and using computer interpretation of the spectra rapid measurement of constitutents, such as moisture, starch, protein, oil and fibre can be made in straight concentrate, forage and compounded feeds.

From such information an assessment of the energy status of the feed may be made. Although this method of analysis is only as precise as the traditional 'wet' chemical analysis of materials that are used to calibrate the NIR analysis equipment it allows analysis of feeds in as little as 10 minutes compared with the 24–48 h period required for 'wet' chemical analysis. For feed compounders it provides an opportunity to analyse routinely each consignment of feed entering the mill and in the light of this information to update rapidly feed formulations. In addition, it allows a check on the analysis of compounded feeds before they leave the mill giving an opportunity to prevent release to farms if they do not conform to statutory or company specifications.

RAW MATERIALS FOR DIET AND RATION FORMULATION

A great variety of feedstuffs is used for the formulation of animal diets and rations. Whilst pig and poultry diets consist almost entirely of concentrated feedstuffs the requirements of ruminant livestock are met by a combination of roughages and concentrates, the contribution of concentrates to total requirements being greatest in dairy cow feeding and least in the feeding of sheep and suckler cows.

In considering feedstuffs it is useful to distinguish between the feedstuffs which are relatively high in fibre and/or water content (bulky feeds or roughages) and those which are low in fibre and water (concentrate feeds). *Figure 13.3* indicates into which of these major groupings different feedstuffs fall.

The following terms that are frequently used in animal feedingstuffs terminology are also worth defining at this stage:

Compound feeds: a number of different ingredients (including minerals, vitamins, additives) mixed and blended in appropriate proportions, to provide a properly balanced complete diet or, in the case of ruminants, a technically designed supplement to natural

Table 13.2 Compound feed formulations Great Britain – 1986, June. % of feeds

	Cattle		Pigs			Poultry		
	Standard dairy	Summer grazing	Pregnant/ lactating sow	Rear- ing	Finish- ing	Broiler starter	Broiler finisher	Layer
Cereals								
Wheat	14	13	36	44	35	60	61	51
Barley	6	4	5	9	11	–	–	3
Maize	–	–	–	–	–	1	2	2
Other[a]	1	–	1	1	–	–	1	1
Total	21	17	42	54	46	61	64	57
Cereal By-products								
Wheat offals	11	10	23	8	13	–	3	2
Dried grains	4	4	–	–	–	–	–	–
Maize gluten	7	4	–	–	–	–	–	–
Other[b]	12	21	2	1	2	1	1	6
Total	34	39	25	9	15	1	4	8
Animal/vegetable proteins								
Soya	2	3	12	22	23	21	17	12
Rape	9	8	–	–	–	–	–	–
Other[c]	11	10	9	10	7	12	–	11
Total	22	21	21	32	30	33	27	23
Miscellaneous[d]	23	23	12	6	9	6	6	12

Source: MAFF statistics
[a] Oats, rye, sorghum, etc.
[b] Nutritionally improved straw, rice bran, biscuit meal, malt culms, oatfeed meal
[c] Sunflower meal, peas, beans, olive residue, cottonseed
[d] Molasses, sugar beet pulp, citrus pulp, oils and fats, minerals and vitamins, manioc, limestone (especially in layers rations)

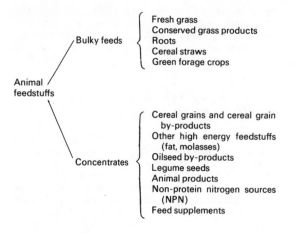

Bulky feeds
{
Fresh grass
Conserved grass products
Roots
Cereal straws
Green forage crops
}

Animal feedstuffs

Concentrates
{
Cereal grains and cereal grain by-products
Other high energy feedstuffs (fat, molasses)
Oilseed by-products
Legume seeds
Animal products
Non-protein nitrogen sources (NPN)
Feed supplements
}

Figure 13.3 A classification of animal feeds

foods (e.g. grass or roughages). An indication of the composition of compounds manufactured in Britain for the different classes of livestock in 1986 is given in *Table 13.2*.

Straights: single feedingstuffs which may or may not have undergone some form of processing before purchase, e.g. wheat, barley, flaked maize, soyabean meal, fish meal.

Protein concentrates: products specially designed for further mixing before feeding, at an inclusion rate of < 5%, with planned proportions of cereals and other feedingstuffs either 'on farm' or by a feed compounder. These usually contain protein rich ingredients such as fish meal, meat meal, soya which may be supplemented with minerals, vitamins.

Additives: substances added to a compound or a protein concentrate for some specific purpose other than as a direct source of nutrient, e.g. coccidiostats, antibiotics, flavourings.

Supplements: products used at < 5% of the total ration and designed to supply specific amounts of trace minerals, vitamins and non-nutrient additives.

In animal feeding a basic consideration is the nutritional value of available feeds. This information has been accumulated over many years and is being continually updated as more precise measures of nutritive values are obtained. Details on composition and nutritive values of the more widely used feedstuffs for the various classes of livestock are given in *Table 13.3*. More comprehensive information may be obtained from the following publications: *Feed Composition UK Tables of Feed Composition and Nutritive Value for Ruminants* (1986) MAFF; *Energy Allowances and Feeding Systems for Ruminants* (1984) RB433 MAFF; *Nutritional Allowances for Pigs* (1978) *Advisory Paper No. 7* (LPR22); Bolton, W. and Blair, R. (1977), *Poultry Nutrition*. MAFF, Bulletin 174, London: HMSO.

The value of a feedstuff is not solely described by its compositional analysis or nutritive value. Limits to the use of a feed may be imposed by the presence of antinutritional or toxic factors or by particular organoleptic or physical properties. In the consideration of individual feeds the possibility of such factors operating will be given attention.

Grazed herbage

Grazed herbage provides a major source of nutrients for ruminant animals. As a feedstuff, however, it has the disadvantage of being variable in nutritive value and in addition its efficiency of utilisation is difficult to control. Aspects of this are covered in Chapter 6, the salient points being that whilst a number of factors such as climate, soil fertility and botanical composition of the sward may affect nutritive value, the most important factor is stage of growth (*Figure 13.4*).

Early in the growing season grass has a high water, organic acid and protein content and a low content of structural carbohydrates and lignin making it highly digestible to the animal. As the plant matures and the yield of forage increases there is an increase in the proportion of of stem tissue and a decrease in leaf tissue. This is reflected in increased levels of structural carbohydrates and lignin, decreased protein and thus declining digestibility and metabolisable energy values. A further consequence of these changes in nutritive value that occur is that the animal's voluntary intake of herbage declines as it matures. Thus, systems of utilisation of grassland must take into account those variables together with animal production targets and other management constraints, e.g. amount of herbage required for winter use. A relatively high degree of control of these factors may be exerted on lowland pastures and has led to the development of systems which allow the measurement of the yield and efficiency of utilisation of ME from grass (*see* Chapter 6). In the case of hill pastures the restricted growing season and difficulties in controlling herbage utilisation leads to a very cyclical pattern of nutrition in grazing animals. Although during the growing season (May–October) grass in hill pastures will be in the range of 65–55 D, during the winter months D-values may be as low as 40 with little or no protein. Such low levels of protein restrict the rate of herbage breakdown in the rumen and consequently adversely affect the voluntary intake of the animal.

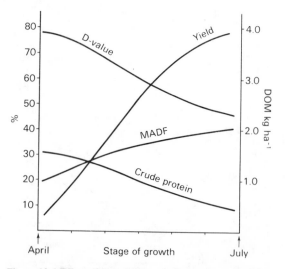

Figure 13.4 Effect of increasing maturity on the composition and yield of digestible organic matter (DOM) of a typical ryegrass sward

The presence of clover in pasture increases the protein supply to the animal and since the effects of stage of growth on the feeding value of clover are less obvious, due to the lower degree of lignification of stem tissue, the inclusion of clover in a sward results in a reduced rate of decline of nutritive value with advancing maturity. In the grazing of pastures with a high clover component a degree of care must be taken to guard against 'bloat' which may occur due to the stimulation of foam production in the rumen by certain cytoplasmic proteins.

Conserved grass

A feature of all conserved grass is that the nutritive value is dependent on the quality of grass used and the efficiency of the conservation process. Thus, the aim of all conservation processes is to minimise the losses of nutrients from the grass utilised and to produce a feedstuff that fulfils the production objectives of the livestock to which it is to be fed.

The main methods of conserving grass are as hay, silage and artificially dried grass. Most grass in the UK is now conserved as silage. Because of the costs of production artificially dried grass is produced on a relatively small scale for specialized purposes. Details of these processes and factors affecting the efficiency of the processes are given in Chapter 6, Grassland.

As the nutritive value of conserved hays and silages can vary considerably it is advisable that appropriate samples be routinely analysed so that most prudent use of both conserved forage and supplementary concentrate feeds may be made. Since grass is generally cut at an earlier growth stage for silage than for hay the nutritive value is generally higher. Thus when compared with the best quality hays good quality silages have approximately 10% more ME and up to 50% more CP. Retention of the vitamins and minerals in grass during hay and silage making processes is variable but in general the mineral content of both materials reflect levels in the crop unless effluent losses or leaching losses are high; silages are good sources of carotene but contain negligible amounts of vitamin D whereas the opposite situation pertains with sun-cured hays.

During haymaking losses of nutrients through oxidation of water soluble carbohydrates, protein breakdown, leaf shatter and the possibility of leaching losses when the crop is subjected to continuous wetting and re-drying combine to increase the proportion of cell wall material and thus reduce the digestibility when compared with the original grass. The calcium and phosphorus content of hay is frequently only half of values for grass. Absence of green colour is a good indication that carotene has been destroyed.

Some reductions in nutrient losses may be achieved through baling at lower dry matters (50–60%) and resorting to some form of barn drying. Failure to achieve dry matters of 85% in store may result in fungal spoilage which represents a hazard to both human health since it causes farmers' lung, and to animals due to the fungal toxins being a cause of abortion in cattle.

The feeding value of silage whilst primarily dependent on the stage of growth of grass used may be further influenced by the extent of nutrient losses incurred during the ensilage process. The digestibility of silages are, however, usually very similar to those of the original grass except where there have been high losses through continued respiration either during wilting in the field or prior to anaerobic conditions being achieved in the clamp. Restricting the extent of the fermentation in the clamp through the use of either acidifying or inhibitor type additives results in greater retention of water-soluble carbohydrates (WSC). These give the silage a higher energy value through their more efficient fermentation in the rumen compared with the fermentation of the organic acid end products that result from their fermentation in the clamp. Although nutrients such as WSC, proteins and minerals may be lost in the effluent of silages made from unwilted grass recent experimental evidence indicates that greater milk production may be obtained from high moisture silages. This is probably the result of an influence on rumen fermentation altering the partition of nutrients in the animal.

A further factor that may influence the feeding value of silages is the extent of nutrient losses that are incurred due to the exposure of silage to air prior to its consumption. Oxidation of any residual WSC and organic acids by microorganisms is especially likely and the management of the clamp face and the method of feeding should aim to restrict these losses.

The ensilage of whole crop cereals, particularly forage maize, gives a product of higher energy value than grass silage, but lower levels of CP. Such silages may be usefully supplemented with NPN compounds.

Whilst the nutritive properties of conserved forages may be defined to an extent by chemical analysis one of the main problems in feeding roughages when they are a major component of the diet is to predict how much animals will consume. This is more of a problem with silages where factors such as digestibility, dry matter content, chop length and the fermentation pattern are operating. Intakes tend to be enhanced by high dry matter, high digestibility and reduced chop length. Whilst intakes of silages with high pH and high ammonia levels resulting from poor fermentation are low, at the other end of the scale high levels of acidity, especially in maize silages, may also restrict intake. At present there is insufficient information to predict accurately roughage intake by different classes of livestock.

Typical artificially dried grass is a product which, by virtue of its relatively low fibre content, occupies an intermediate position between the bulky feeds and the concentrates. Due to the rapid removal of water at high temperatures dried grass has a similar feeding value to the original grass when expressed on a dry matter basis. Since artificial drying is an expensive process only young leafy grass of high D-value and high crude protein content is used. An effect of high temperature drying is to decrease the degradability of protein. Intakes of dried grass do not present a problem since not only is it highly digestible but also its cost dictates that its use is limited to a supplemental role.

Cereal straw

The gross energy content of cereal straw is similar to that of grains but because of its low digestibility even to ruminants much of this energy is lost in the faeces and makes no contribution to the animal. A further constraint to the use of straw is that a low rate of digestion and therefore rate of passage through the gut limits the intake.

Since the levels and digestibility of the protein in straw are negligible and since there are very few vitamins and minerals present the only nutrient straw can supply is energy. The

Table 13.3 Guide to the chemical composition and nutritive values of some commonly used feedstuffs

Feedstuff name	*Dry matter content* (%)	*Chemical composition* *CP* (% of feedstuff DM)	*EE*	*CF*	*NFE*	*Ash*	*Nutritive value – energy* *Metabolisable energy* *Ruminants* (MJ/kg DM)	*Poultry* (MJ/kg A–F)*	*Digestible energy* *Pigs* (MJ/kg A–F)
(1)[2] *Pasture grass* (average)	20.6	13.9	2.2	24.2		7.6	11.2	–	–
(2)[2] *Conserved grass products*									
Silage (average)	28.9	15.9	3.7	27.1		8.7	10.7	–	–
Hay (average)	85.8	12.2	1.8	30.1		7.8	9.2	–	–
Dried grass (average)	89.0	19.7	3.8	22.4		11.4	9.5	–	8.90
Maize silage	30.4	8.8	3.0	17.2		4.9	11.2	–	–
(3) *Roots*									
Cassava (dehydr. ground)	88.0	2.8	0.3	8.7	84.8	3.4	12.8	14.43	15.3
Fodder beet	18.0	6.8	0.3	5.9	79.1	7.7	11.9	–	2.9
Mangels	10.0	9.2	0.8	6.2	76.9	6.9	12.4	–	1.97
Swedes	10.5	10.8	1.7	10.0	71.7	5.8	13.9	–	1.7
Turnips	10.0	12.2	2.2	11.1	66.6	7.8	12.7	–	1.4
Sugar beet pulp (dried)	90.0	9.9	0.7	20.3	65.7	3.4	12.7	–	12.9
(4) *Cereal straws*									
Barley (spring)	86.0	3.8	2.1	39.4	49.3	5.3	7.3	–	–
Barley (winter)	86.0	3.7	1.6	48.8	39.2	6.6	5.8	–	–
Oat (spring)	86.0	3.4	2.2	39.4	49.3	5.7	6.7	–	–
Wheat (winter)	86.0	2.4	1.5	42.6	47.3	6.2	5.7	–	–
(5) *Green forage crops*									
Kale–marrowstem	14.0	15.7	3.6	17.9	49.3	13.6	11.0	–	–
Rape	14.0	20.0	5.7	25.0	40.0	9.3	9.5	–	–
(6) *Cereal grains and cereal by-products*									
Barley	86.0	10.8	1.7	5.3	79.5	2.6	12.8	11.13	13.2
Wheat	86.0	12.4	1.9	2.6	81.0	2.1	13.5	12.18	14.3
Oats	86.0	10.9	4.9	12.1	68.8	3.3	12.0	11.05	11.4
Maize	86.0	9.8	4.2	2.4	82.3	1.3	13.8	13.22	14.9
Maize (flaked)	90.0	11.0	4.9	1.7	81.4	1.0	15.0	–	15.3
Sorghum	86.0	10.8	4.3	2.1	80.1	2.7	13.4	12.98	14.3
Brewers' grains (dried)	90.0	20.4	7.1	16.9	51.2	4.3	10.4	7.91	7.6
Distillers' grains (dried)	90.0	30.1	12.6	11.0	44.3	2.0	12.1	9.37	–
Bran	88.0	17.0	4.5	11.4	60.3	6.7	10.1	8.45	8.9
Fine wheat middlings	88.0	17.6	4.1	8.6	65.0	4.7	11.9	11.80	11.8
(7) *Other high energy feeds*									
Fat and oil (FGAF)	98.0	–	100	–	–	–	–	35.5	33.0
Molasses	75.0	4.5	0	0	84.1	6.9	12.5	7.91	–
(8) *Oilseed by-products*									
Ex soyabean	90.0	50.3	1.7	5.8	36.0	6.2	13.4	10.67	13.6
Ex rape seed	90.0	41.3	3.4	10.4	36.6	8.2	12.0	7.36	12.2
Ex. groundnut (decort.)	90.0	50.4	2.1	27.3	31.6	4.7	9.2	11.4	13.7
Ex. sunflower	90.0	42.3	1.1	18.1	31.2	7.2	10.4	7.5	12.6
(9) *Legume seeds*									
Field beans									
– spring	86.0	31.4	1.5	8.0	55.1	4.0	12.8	10.42	12.2
– winter	86.0	26.5	1.5	9.0	59.1	4.0	12.8	10.00	12.2
Peas	86.9	26.1	1.4	6.0		3.2	13.8	10.86	14.34
(10) *Animal products*									
Fish meal (white)	90.0	70.1	4.0	0	1.8	24.1	14.2	11.46	11.8
Herring meal	90.0	76.2	9.1	0	4.4	10.2	–	13.35	13.5
Meat and bone meal (medium protein)	90.0	52.7	4.4	0	1.7	41.2	7.9	–	9.6
Dried skim milk	95.0	37.2	1.1	0	53.2	8.5	14.1	12.26	15.6

* A–F = as fed

Sources of information

Bolton, W., and Blair, R., (1974). *Poultry Nutrition.* MAFF Bulletin 174 (HMSO London).

Ensminger, M.F., and Olentine, C. G., (1978). *Feeds and Nutrition complete* Clovis, California, Ensminger Publishing Co.

McDonald, P., Edwards, R. A., and Greenhalgh, J. F. D. (1981) *Animal Nutrition*, Longmans, London.

Nutrient Allowances and Composition of Feedingstuffs for Ruminants (1976) MAFF. LGR21.

Nutritive value – protein					Amino acid composition						Mineral composition	
Rumen degradability of CP[1]		Digestible crude protein			Glycine + serine (% of total feed A–F)	Lysine	Methionine + cysteine	Iso-leucine	Threonine	Trypto-phan	Calcium (% of total feed DM)	Phosphorus
RDP (g/kg DM)	UDP	Ruminants (g/kg DM)	Pigs (% A–F)	Poultry (% A–F)								
84	55	93	–	–	–	–	–	–	–	–	0.50	0.29
127	32	105	–	–	–	–	–	–	–	–	0.62	0.30
97	25	73	–	–	–	–	–	–	–	–	0.55	0.28
118	79	127	8.95	–	–	0.75	0.44	0.68	0.68	–	0.92	0.33
70	18	70	–	–	–	–	–	–	–	–	0.39	0.18
–	–	20.0	1.7	–	–	0.09	0.03	0.05	0.05	0.01	–	0.03
54	14	50.0	0.7	–	–	0.035	0.02	0.024	0.036	0.02	0.22	0.22
74	18	80.0	0.8	–	–	–	–	–	–	–	0.22	0.22
86	22	64.0	0.9	–	–	–	–	–	–	–	0.42	0.33
98	24	70.0	0.9	–	–	–	–	–	–	–	0.44	0.33
60	39	66.0	3.1	–	–	0.69	–	0.37	0.37	0.09	0.96	0.09
30	8	9.0	–	–	–	–	–	–	–	–	0.42	0.08
30	7	8.0	–	–	–	–	–	–	–	–	0.39	0.08
27	7	11.0	–	–	–	–	–	–	–	–	0.24	0.09
19	5	1.0	–	–	–	–	–	–	–	–	–	–
94	63	123.0	–	–	–	–	–	–	–	–	2.14	0.31
120	80	144.0	–	–	–	–	–	–	–	–	0.93	0.42
86	22	82	7.6	9.0	0.81	0.35	0.34	0.35	0.30	0.15	0.05	0.38
99	25	105	9.8	8.8	1.00	0.31	0.39	0.40	0.31	0.12	0.03	0.40
87	22	84	7.3	8.5	1.00	0.38	0.34	0.36	0.34	0.12	0.09	0.37
59	39	69	6.7	6.7	0.78	0.26	0.29	0.32	0.32	0.08	0.02	0.27
66	44	106	9.0	–	–	0.29	0.33	0.36	0.36	0.10	–	0.29
65	43	87	7.2	8.4	0.65	0.22	0.31	0.38	0.30	0.08	0.03	0.68
122	82	149	13.5	14.1	1.80	0.52	0.55	0.94	0.55	0.21	0.32	0.78
181	120	214	–	–	–	–	–	–	–	–	0.31	0.33
136	34	126	10.4	11.1	1.50	0.60	0.50	0.55	0.45	0.25	0.16	0.84
141	35	129	12.2	15.0	1.43	0.64	0.40	0.49	0.49	0.22	0.13	0.91
–	–	–	–	–	–	–	–	–	–	–	–	–
27	18	31	–	–	–	–	–	–	–	–	–	–
302	201	453	39.0	42.8	3.94	2.94	1.35	2.52	1.81	0.60	0.23	1.02
248	165	343	31.0	26.1	3.32	2.04	1.26	1.36	1.58	0.44	0.59	0.94
404	100	316	46.2	37.8		1.67	1.25	1.67	1.36	0.52	0.12	0.51
254	169	381	31.0	30.3	2.91	1.60	1.49	1.71	1.56	0.50	0.41	1.33
251	63	248	21.6	21.1	2.39	1.78	0.52	1.25	1.05	0.25	0.16	0.66
212	53	209	18.2	17.5	2.08	1.50	0.36	0.93	0.84	0.21	0.19	0.67
131	87	218	19.4	–	1.05	1.6	0.59	0.97	0.87	0.20	0.09	0.83
421	280	631	56.8	59.0	7.12	4.40	1.26	2.46	2.46	0.61	7.93	4.37
457	305	–	64.4	66.6	–	5.62	1.62	3.10	2.91	0.58	3.26	2.22
211	316	411	42.2	–	8.00	2.61	1.48	1.26	1.55	0.29	11.44	6.00
223	149	350	34.7	27.5	2.82	2.63	1.26	2.42	1.58	0.47	1.17	0.96

Nutrient Allowances for Pigs (1978) MAFF Advisory Paper No. 7(LPR 22)

Whittemore, C. T. and Elsley, F. W. H. (1976). *Practical Pig Nutrition.* Ipswich: Farming Press

[1] Values provisional for rumen degradable protein (RDP) and undegradable protein (UDP).

[2] Values highly variable and samples require to be routinely analysed.

Feed Composition UK Tables of Feed Composition and Nutritional Values for Ruminants (1986) MAFF Chalcombe Publications.

availability of this energy is limited by the very high levels of lignin protecting the fibrous carbohydrates from microbial breakdown. In addition when straw is a major component of the diet the low levels of nitrogen available to the rumen micro-organisms may be a further factor limiting its digestibility. It has, however, been observed that considerable differences in nutritive values exist not only between the different cereal types, oat straw being superior to barley, which in turn is of higher quality than wheat, but also between varieties. Typically, spring varieties give higher nutritive values than winter varieties but even comparisons of spring barley varieties have shown that digestibility values can vary from as low as 40 for varieties such as Delta and Golden Promise to nearly 60 for varieties such as Doublet and Corgi. Likewise for wheat straws digestibilities ranging from 30 to 50 have been found for different varieties. It should be noted that growing conditions may also influence nutritive value although the influences of fertiliser treatment, season and daylight are largely unknown.

In recent years a great deal of research has been carried out with the objective of improving the nutritional value of straw. Chemical treatments with alkali in the form of sodium and ammonium hydroxide leads to improvements in nutritive value and dry matter intake. The extent of this improvement is dependent on the quality of the original straw, greater improvements are obtained with poor quality materials, and the method of treatment chosen. A number of 'on-farm' methods of treatment have been devised ranging from the low cost stack methods to high capital investment oven treatment methods. In the production of nutritionally improved straw (NIS) by feed compounders the straw is additionally ground and pelleted. As with the feeding of untreated straw, however, it is important that the remainder of the ration supplements the low levels of protein and nitrogen present. The use of ammonium hydroxide as the alkali helps overcome this problem. Due to their increased digestibility the intake by animals of treated straw is much higher than that of untreated straw allowing much wider inclusion in the diets of ruminant animals.

Both treated and untreated straw may be used to effect in devising feeding strategies. Thus, in the feeding of dairy cows it may be used in the case of dry cows to save silage; in the case of freshly calved cows alkali treated straw may be used as a chemical 'buffer' to increase appetite; as a buffer feed for cows at grass as a means of maintaining milk fat content; or as a means of ensuring adequate rumination and so prevent displaced abomasum in animals being fed very high density diets. Traditionally straw has formed an important part of the ration for suckler cows and 'store' animals but for situations where higher levels of performance are required from growing beef animals the level of inclusion of straw in the ration and therefore the potential benefit from treatment are severely limited. Some farmers have used both treated and untreated straw to effect as an alternative roughage source in the feeding of ewes.

The economic advantage from using treated straw is difficult to assess but is largely dependent on the cost of the original straw and the price of alternative energy sources, such as barley.

Roots and tubers

The root crops, turnips, swedes, mangolds and fodder and sugar beet, are characterised by having a high water content (75–90 %) and low crude fibre (5–10 % of DM). The main component of the dry matter is sugar. Thus roots are highly digestible and the principal nutrient they supply is energy.

The crude protein content is in the range of 50–90 g kg^{-1} DM, a fairly high proportion of this being in the form of NPN. As suppliers of vitamins and minerals (potassium excepted) roots are poor and rations containing a high proportion of roots require appropriate supplementation.

Whilst roots can be fed and are palatable to pigs their high water content and bulkiness limits levels of incorporation for young growing pigs although diets for sows may contain up to 50 % of their DM as roots. Roots are more commonly used in the diets of ruminant animals either being fed *in situ* to sheep or carted and fed to housed animals. Although roots are capable of producing very high yields of energy per hectare the costs and difficulty of harvesting represent a considerable problem as far as the feeding of housed animals is concerned. A further problem is that where large quantities of roots are fed to housed animals recognition should be made of the large quantity of urine that is excreted.

Sugar beet is not primarily grown for animal consumption but sugar beet pulp, the residue left after sugar has been extracted at the factory, is widely used in ruminant diets where in energy terms it can substitute for barley. The higher fibre content compared with barley makes it a useful component of the concentrate diet when cows go out to grass in spring but account must be taken of its low protein value when substituted for cereals. This product is sold either directly to farmers as fresh 'pressed pulp' with about 28 % DM and is therefore a perishable commodity or as dried pulp which may be readily stored for many months. Dried pulp is available in shredded form, or as pellets or nuts. Its feeding value is frequently modified prior to sale by the addition of molasses, magnesium, minerals and vitamins.

The feedstuff variously termed tapioca, manioc or cassava, is a raw material which acts as one of the main cereal substitutes. It is extracted from the tropical root crop cassava and became widely used by compounders in the 1970s and early 1980s when it often made up around 10–20 % of pig, ruminant and pig compound feeds. More recently the reduction in price advantage has led to lower rates of inclusion. Manioc is principally comprised of starch with low levels of protein (2.5 %) and oil (0.5 %). The energy content is similar to wheat and barley but when used as a replacer for these cereals requires greater protein supplementation. Raw manioc contains cyanogenic glucosides which can be enzymically hydrolysed to hydrocyanic acid. Processing must eliminate these glucosides since the hydrocyanic acid has the effect of depleting the body reserves of the essential sulphur containing amino acids methionine and cystine. A further quality control characteristic that must be assessed is the ash content since on occasion it may acquire excessive levels of silica from the soils on which it is grown. This reduces the energy value. As ground manioc is very dusty it is normally included in pelleted feeds rather than meals.

Green forage crops

The main green forage crops fed are kale, forage rape and stubble turnips. These brassica crops can provide a useful source of succulent feed during autumn and winter. Composition varies between the different brassica species and varieties, being particularly affected by the leafiness. As a generalisation they compare favourably with the feeding value of good quality silage in energy terms and are rich in protein, minerals and vitamins.

These crops are generally fed *in situ*, kale principally to dairy cows and rape and stubble turnips mainly to fattening lambs. For the latter they may be fed without cereal supplementation, ·but cereals can usefully complement the high protein supply of these crops to give increased liveweight gain.

In feeding these brassica crops account has to be taken of certain agents that may prove harmful when brassicas are fed in excess. The presence of goitrogens which affect iodine utilization (*see* p. 336) and of the chemical S-methyl-cysteine sulphoxide (SMCO) which can produce haemolysis and consequently anaemia mean that their use in diets should be limited. The latter problem is minimised if the crop is grazed before it is too mature and before secondary leaf growth occurs. For dairy cows it is recommended that intake of these brassica forage crops be limited to a maximum of 30 % of the total dry matter intake and particular attention paid to the health of animals. Additionally, it should be remembered that the high levels of calcium present in brassicas may upset the Ca:P ratio in the diet, necessitating phosphorus supplementation.

A further problem is that rape can cause a skin condition termed yellowsis in white faced lambs. In this condition the face and ears become sensitive to light.

Cereal grain and cereal grain by-products

Cereal grains and their by-products are the main ingredient of rations for pigs and poultry and provide the major source of energy in compound feeds fed to ruminant animals. The main cereal grains used in the UK are barley, wheat and oats which are mainly homegrown along with imported maize and sorghum.

All cereal grains are rich in carbohydrate which is mainly in the form of starch. Starch accounts for about 70 % of the seed, varying between grain types. In raw material terms the crude protein of cereals is rather low and is the most variable item ranging from 6 to 14 %. This crude protein is of relatively poor quality being low in the essential amino acids lysine, methionine and tryptophan and containing 10–15 % of the nitrogenous compounds in the seed as NPN.

The crude fibre levels in cereals varies with species, being lowest in maize and highest in oats. Its level has a direct bearing on the digestibility and therefore energy value of the whole grain. The oil content also varies with species; oats and maize having higher levels than other cereals. Cereal oils are unsaturated, the main acids being linoleic and oleic which leads them to become rancid fairly quickly after processing.

Cereals are relatively good sources of phosphorus but much of this is present as phytates which adversely affect its availability to livestock. In general, cereals are deficient in calcium and contain varying levels of other trace minerals. Of the vitamins most cereals are good sources of vitamin E, although under moist storage conditions much of this may be destroyed. Except for yellow maize cereals are low in carotene and vitamin A and are deficient in vitamin D and most of the B vitamins, thiamine excepted.

The feeding value of cereal grains is relatively constant during prolonged storage but occasionally poor harvest and storage conditions can lead to the presence of fungal toxins such as aflatoxin, zearalenone, vomitoxin and ochratoxin. *Aspergillus, Penicillium* and *Fusarium* moulds are particularly responsible for the production of these toxins. Affected grains are usually thin and shrivelled with a pinkish colour. In the case of vomitoxin maximum concentrations in feed should be < 1 ppm for pigs and dairy cows and < 5 ppm for poultry and other cattle. Cocktails of mycotoxin in contaminated feed are, however, potent at lower levels than are individual mycotoxins.

Maize

Of the commonly used cereals yellow maize has the highest metabolisable energy content. With a metabolisable energy value of $16\,MJ\,kg^{-1}DM$ it is especially useful for broilers where yellow carcass pigmentation resulting from its high carotene content is desirable. For laying hens it not only supplies xanthophylls that enhance the colour of egg yolks but it also supplies linoleic acid which is a necessary dietary component for birds to produce eggs of a satisfactory size. The high proportion of unsaturated fatty acids means that its inclusion in pig diets must be restricted to 35 % of the diet otherwise there is the likelihood of soft fat depots high in polyunsaturated fatty acids. The yellow pigmentation in body fat resulting from the feeding of yellow maize is considered undesirable in pig and ruminant carcasses. White maize has all the attributes of yellow maize without the problems of carcass fat pigmentation.

Wheat

Although not appropriate for the production of yellow broilers, wheat is suitable to all other feeding situations. It frequently forms up to 70 % of diets for poultry, can be included at high levels for pigs provided that care is taken to control the fineness of grinding whilst for ruminants inclusion levels of up to 30–40% in compound feeds may be used provided the remainder of the diet contains sufficient fibre. The gluten content of wheat can have a beneficial effect on the quality of heat processed extruded cubes or pellets.

Barley

Traditionally barley is considered to be particularly suitable for pig feeding having an appropriate amount of both fibre and also of oil which is associated with the production of saturated carcass fat. The higher levels of fibre than are found in wheat or maize means that the upper inclusion rate for poultry must be limited to about 30% in many cases. High levels of barley in broiler diets have been associated with wet droppings.

Oats

Oats are normally only used for ruminants and horses. Their low energy and high fibre means that they are seldom used for pigs or poultry. In recent years varieties of 'naked' oats have received some attention as being especially suitable for young piglets, since without the husk the grain is of high energy value due to its comparatively high oil content and the protein quality is somewhat better than in other cereals.

Sorghum

From a nutritional viewpoint the grain sorghums (milo, kaffir and hybrids) resemble wheat. They have a low fibre content and in comparison with maize contain more protein and less oil. Certain varieties have a high content of phenolic compounds including tannins which not only influence palatability but also lower protein digestibility and reduce energy values for pigs and poultry.

With respect to other cereals rye can only be tolerated at low inclusion rates because of its content of B-glucans and phenolic compounds which reduce performance and result in wet droppings in poultry and are toxic in quantity. Rice and millet may be used as an alternative to wheat once the husk has been removed. Triticale, a cross between durum wheat and rye, is similar to wheat but with a higher protein content.

Cereal preparation and processing

The aim in all cereal processing methods is primarily to increase the efficiency of utilisation of the nutrients. Such improvement may result simply from an increased nutrient availability but other factors such as changes in palatability and nutrient density may also contribute towards improvements in performance.

Most grain processing methods have as their main objective improvement in the availability of the starch present. Some of the techniques involve solely physical change, others chemical and some a combination of both physical and chemical; in addition, some processes are carried out 'wet', others 'dry'; some involving heat treatment, others under cold conditions. Mechanical alterations of the grain are the most widely used and involve physical disruption of the grain such that the starch is made more available. Grinding, rolling and crushing are the most common processes employed. It is worth noting, however, that such treatments have little or no effect on the nutritive value of barley or wheat offered to sheep.

Processing procedures which bring about chemical changes through the gelatinisation of the starch include such techniques as steam flaking, micronising, popping and pelleting. The 'flaking' process has long been applied to produce 'flaked maize' in a process involving steaming of the grain either at atmospheric pressure or in a pressure chamber followed by rolling. Micronising and popping are both 'dry heat' processes. In the former grain is passed under gas-fired ceramic tiles, the radiant heat from which produces rapid heating within the grain causing it to soften and swell. It is then crushed in a roller mill which prevents reversal of the gelatinisation process. Popping is the exploding of grain through the rapid application of dry heat. Popped grain is usually rolled or ground prior to feeding. Popping has been shown to be particularly effective in processing sorghum grain.

When feeding concentrate mixtures to certain livestock classes it is common practice to produce it in the form of pellets. From a management point of view it reduces wastage, prevents selection of ingredients and makes for easier storage and handling. Additionally, there are in some instances nutritional advantages to pelleting, this being due in the main to improvements in the available energy content.

Recent studies involving the treatment of grain for ruminant consumption with alkali appear to indicate potential as a means of increasing the availability of energy.

Cereal by-products

When cereal grains are processed for human consumption a number of by-products are produced which are used extensively as livestock feeds. The main sources of by-products are the flour milling industry, the brewing and distilling industries and, imported from the USA, by-products from maize processing.

Wheat by-products

When wheat is milled to produce flour for human consumption about 28% of the grain becomes available as by-products of the process. Wheat millfeeds are usually classified and named on the basis of decreasing fibre as bran, coarse middlings and fine middlings or shorts. These arise from the removal of the outer layers of the kernel. Wheat bran is comprised of the coarser fraction (about 50 % of wheatfeed) and is usually fed to ruminant species. The finer fractions are lower in fibre and widely used in formulating diets for pigs and poultry. Since the protein in wheat grain is concentrated mainly in the outer layers, crude protein levels in these residues are higher than the whole grain. Likewise wheatfeeds are relatively good sources of most of the water-soluble vitamins, except niacin.

Brewing/distillery by-products

The main by-products arising from the brewing and distilling industries are brewers' grains and distillers' grains (draff), which are essentially the part of the grain that remains after the starch has been removed in the malting and mashing processes. They may be purchased without being dried and fed either fresh or after ensiling or alternatively may be purchased after being dried. Since both feeds have a relatively high fibre content their use is limited to ruminant rations. As well as the fibre fraction being more concentrated by the loss of the starch so also are the crude protein and oil fractions. In the case of distillers' grains the relatively high lipid content (80–90 g kg^{-1} DM) is known to interfere with the cellulolytic action of the rumen microflora but this can be overcome to an extent by the addition of suitable amounts of calcium which results in the formation of insoluble calcium soaps.

Other by-products of these industries, namely malt culms, dried brewers' yeast and dried distillers' solubles may also be fed to livestock.

Rice bran

In recent years rice bran has become increasingly important in animal feeds and up to 10 % may be included in compound feeds for ruminants. It is also used in sow diets. Rice bran consists mainly of the bran and outer part of the grain and is obtained as a by-product when rice is 'polished' for human consumption. It tends to be a variable product but good quality rice bran contains very little of the less nutritious hull fragments and is rich in oil (14–15 %) making it susceptible to rancidity. This oil is usually extracted either by solvent leaving less than 1 % oil in the product or by expeller press in which case up to 10 % oil may remain. Solvent extracted brans have ME values of the order 6.5–7.5 MJ kg^{-1} DM whereas those with higher oil content can have ME values of 11 MJ kg^{-1} DM. Such differences emphasise the need for users to ascertain the composition of rice bran before purchase or use in a formulation.

Maize by-product feeds

The wet milling of maize is used in the USA for the production of starch, sugar and syrup. A number of by-products result from this process including bran, germ meal and gluten. The latter is the most important and large amounts of maize gluten are exported from the USA into the EEC.

Maize gluten is a good source of energy and protein for both dairy and beef rations. Energy levels are similar to barley with crude protein of the order of 20 %. Up to 20 % of the total DM may be included in ruminant rations and it may also be included on a more limited scale in pig and poultry diets. The quality of maize gluten from different sources can very considerably such that it is essential to monitor the quality closely.

Other high energy feeds

Whilst feed grains are the main energy supplying concentrate feeds, other feeds are routinely used to supply energy to livestock.

Fats and oils

These are of particular value in increasing the energy density of the ration since its energy value is more than twice that of digestible carbohydrate. Fats may also improve rations by reducing dustiness and increasing palatability. The inclusion of fats in animal diets has found greatest application in milk replacers for suckling animals and in the diets of pigs and poultry where energy density is a factor controlling total energy intake. In recent years fats have also been included in dairy cow and sow diets. Energy costs can represent up to 75 % of total formulation costs for animal feeds and fat is frequently the cheapest source.

A major source of fat is feed grade animal fat (FGAF), often referred to erroneously as tallow. This is the fat rendered from meat and bone meal and in fatty acid compositional terms is typical of animal fats being high in palmitic, stearic and oleic acids. In recent years FGAF has been increasingly blended with vegetable oil by-products to produce blended fats that have fatty acid profiles appropriate to the class of livestock to which they are fed. The vegetable oil by-products come mainly from edible oil refining and consist of a mixture of neutral oil and free fatty acid. A further source of vegetable oil is oil recovered from such processes as potato crisp manufacture. The blending of oils to produce an appropriate fatty acid profile is of relevance in that this parameter can influence the digestion, absorption and consequently metabolisable energy value. Synergistic effects between fats frequently exist as evidenced by mixtures of tallow and soyabean oil having higher metabolisable energy values than either of the individual sources. In nutritional terms the most important aspect is the ratio of saturated to unsaturated fat and the specific requirements for polyunsaturated fatty acids such as linoleic acid for pigs and poultry.

Current opinion is that hard fats based on palm or FGAF are most suited to ruminants and high levels of free fatty acids which are unacceptable to monogastrics are thought to be beneficial to energy values. The use of 'protected fat' systems (*see* p. 329) either through the mixing of fat with a carrier such as Vermiculite or with formaldehyde-treated protein sources allows the incorporation of extra fat into ruminant diets without the adverse effects on cellulose digestion normally associated with supplementary raw fat. It also presents an opportunity to manipulate milk fat output and through the incorporation of polyunsaturated fatty acids the fatty acid composition of milk fat and carcass fat.

These protected fats are usually free-flowing powders which are easily incorporated into the diet without special equipment.

For poultry diets a typical profile for supplementary fat is 35 % saturated fatty acids, 20 % linoleic acid and less than 50 % of the total fatty acids as free fatty acids. Relatively soft fats are also recommended for pig rations provided it is within the limitation of producing soft carcass fat. It is frequently observed that the utilization of the non-fat components of the non-ruminant diet are improved by adding fat to the ration. This is probably due to an effect of additional fat on rate of passage of food through the gut.

A further source of dietary fat that has received attention in recent years is the inclusion of full fat oilseeds in the diet. Most attention has been given to full fat soyabeans. When used in the diet of non-ruminants there is the need for severe physical processing in order to make the oil fully available. Extrusion techniques and to a lesser extent micronisation are most effective. Appropriate processing of whole oilseeds also provide a means of supplying additional fat to dairy cows within a 'protected' sytem. Extensive heat treatment, considered to be overheating in the preparation of whole oilseeds for monogastrics is recommended in the processing of oilseeds for dairy cows.

Molasses

This is a by-product of sugar refining. It is very low in protein, the main constituent being sugar, giving it an energy value of about 85 % of the value of cereal grain. It is mainly used in beef and dairy rations and also sow diets where it may be of particular value as a pellet binder or as a component of feeds which include NPN sources. The limiting factor to its inclusion, usually 5–10 %, is the difficulty of mixing it into concentrated feeds. This requires specialised equipment making it very difficult to use 'on-farm'.

Citrus pulp

Dried citrus pulp is prepared from the residue resulting from the manufacture of citrus juices and consists of a mixture of pulp, peel, seeds and cull fruits. Since a variety of fruits go into it, it has variable composition and requires routine analysis before use. It is similar in feed value to dried sugar beet pulp with a lower protein content (5–8 %). It is mainly used for dairy and beef rations and quantities up to 50 % of the total DM in the diet may be used if desired. At such levels it may produce taint in milk. This product is not very palatable to monogastrics although up to 10 % may be included in sow diets.

Oilseed residues

Several oil-bearing seeds are grown to produce vegetable oils for human consumption and industrial processes. The residues that remain after the extraction of the oil are rich in protein and of great value as livestock feeds.

Among such high protein feeds are soyabean meal, rape seed meal, sunflower seed meal, cottonseed meal, groundnut meal, palm kernel meal and sesame meal.

Oil is extracted from these seeds by hydraulic pressure (expelled) or solvent (extracted). Most oilseeds are now subjected to the latter treatment, the efficiency of oil extraction being much higher, leaving little oil (< 1 %) in the residue whereas expeller methods leave up to 6 % of the oil in the residue. The amount of oil left in the residue affects the energy value of the feed.

Fibre levels will also affect the energy value of the feed and in some cases, e.g. groundnuts, fibre levels will vary according to whether the seeds have been decorticated (removal of husk) prior to processing.

Of those oilseeds used widely, soya has the best quality protein followed closely by sunflower and rapeseed. Soya is slightly deficient in methionone, whereas sunflower and rapeseed, whilst being slightly higher in methionine, are deficient in lysine. Groundnut meal, although high in crude protein content, has very low levels of methionine and is low in both lysine and tryptophan.

In addition to nutritive values a number of anti-nutritional factors have to be taken into account when feeding certain oilseed residues.

In rapeseed meal the presence of glucosinolates, which under the action of myrosinase produce compounds (isothiocyanates and oxazolidinethione) that are goitrogenic (*see* p. 336), limits its inclusion in animal feeds. In recent years newer varieties with lower levels of glucosides have been produced. These newer varieties contain less than 1 % W/W of glucosinolate but whilst they are common in Canada and given the name canola they are less widely grown in the EEC. Further problems with rapeseed meal can occur when it is fed in excess of 5 % to laying poultry in that it can increase the incidence of haemorrhagic fatty liver and through the presence of sinapine promotes the accumulation of trimethylamine which can cause a fishy taint in eggs. In some instances the presence of anti-nutritional factors can be nullified by heat during their processing. This is the case with the trypsin and urease inhibitors present in raw soya and with the yellow-pigment, gossypol, which is present in cottonseed meal. Such heat processing reduces the solubility of the protein and in addition reduces the quality of the protein through involving amino acids such as lysine in browning reactions with carbohydrates. A particular problem associated with groundnut meals is that they are prone to infestation by the mould *Aspergillus flavus* which results in their subsequent contamination with the mycotoxin called aflatoxin. Current regulations prohibit the importation of groundnut meal containing > 0.05 ppm of aflatoxin B_1.

Legume seeds

The seeds of legumes such as peas, beans and lupins are useful sources of both energy and protein although it is for the latter nutrient that they are primarily included in rations.

The protein in field beans is particularly rich in lysine but the low levels of methionine restrict the extent of its inclusion in the diets of pigs and poultry. Varietal differences in the digestibility of the protein present exist and is related to the amounts of tannins, vicin and convicin present. Similar varietal differences in the digestibility of peas exist. As a consequence of their only moderate protein content the extent of their use tends to be restricted, especially in the diets of young growing stock.

Lupins have a significantly higher protein content than peas and beans but they tend to be deficient in lysine, the sulphur containing amino acids and tryptophan. For pigs and poultry the poorer protein quality restricts the amount that may be included as an alternative to soyabean meal to about 10 % of the diet. Further limiting factors to their use

at present are the levels of the alkaloids, lupenine and sparteine, which by imparting a bitter taste depress feed intake and also the presence of α-galactoside sugars, which are fermented in the rear gut of monogastrics causing flatulence.

Animal protein supplements

Feedstuffs of animal origin other than rendered fats are principally included in diets as supplemental sources of high quality protein which can remedy deficiencies in the essential amino acid composition of the rest of the diet. Although this role has long been recognised in the feeding of pigs and poultry current thinking on the protein nutrition of ruminants indicates that high quality protein, which is resistant to degradation in the rumen, may be of value.

Protein supplements of animal origin are derived from slaughterhouse wastes as meat meal, meat and bone meal, blood meal and feather meal; from milk and processed milk chiefly in the form of skimmed milk powders and wheys; and from fish and processed fish as fish meals. Although the bulk of these by-products are simply available as dried ground meals of the raw materials a recent innovation is the production of protein hydrolysate of these waste products. These are produced through the action of proteolytic enzymes and have the advantage of being virtually odour free and readily produced in a variety of formats to meet market demand, e.g. in extruded admixtures with grain or grain by-products. These protein hydrolysates are especially useful for inclusion in poultry and pig starter rations and with their high degree of solubility may also be included as an ingredient of low antigenicity in milk replacer formulae for calves. A further attraction of protein hydrolysis is that it allows less biodegradable materials such as poultry feathers to become useful protein sources.

A problem with the major animal meals is that their composition tends to be variable according to the composition of the raw materials used. This is especially true of their oil and ash content which is dependent in the case of slaughterhouse waste on the efficiency of rendering and on the proportion of bone present, and in the case of fish meals, on the species of fish used. A further variable factor is that during processing overheating of these products can markedly reduce their digestibility to monogastrics and also reduce the available lysine content. Variations in processing temperatures can also cause variability in the RDP:UDP ratios in these products.

Meat and bone meal

It is available in two types – as a low fat meal (~ 6 % oil, 48–50 % protein) or as a high fat meal (12–14 % oil, 45 % protein). The cost of the latter is often attractive because of its high energy value but the high oil content can present handling difficulties. The ash content within meat and bone meals constitutes a valuable source of available phosphorus as well as calcium.

Meat meal

This is produced from slaughterhouse waste from which all bone, hoof and horn have been excluded and consequently has a higher protein level (50–55 % CP). As with meat and bone meal it is relatively low in methionine and tryptophan when compared with fish meals.

An important aspect of the production of meals from slaughterhouse wastes is that adequate sterilisation must occur to prevent the infection of the feed with salmonella.

Fish meals

These are excellent sources of protein being especially high in lysine and methionine and also of minerals. In the UK over 60 % of fish meals go into poultry rations, 20 % into pigs with the remainder going into ruminant and fish rations. The composition of fish meals is dependent on the type of fish used, demersal species, such as cod, producing meals with a low oil content (2–6 %) whilst pelagic species, such as herring, produce meals with fairly high oil levels (7–13 %). Such variation in oil contents influence the metabolisable energy values but the use of high oil content meals is more limited because of the possibility of their imparting fishy taints or 'soft' fat in the product. As a component of the diets of early weaned pigs fish meal has certain advantages over a protein source such as soyabean meal in that it appears to have a relatively lower antigenicity whilst for ruminants it is a particularly useful source of UDP rich in lysine and methionine. The latter amino acid is usually first limiting in milk protein synthesis, especially in cows on silage-based diets.

Milk by-products

The main product that can be used in compounded animal feedstuffs is skimmed milk powder. The extent of its use is largely dependent on EEC Regulations designed to reduce intervention stocks. Such powders have been extensively denatured such that protein digestibility to monogastrics is reduced. If included in pelleted feeds the level of inclusion needs to be limited to avoid pellets that are too hard. Liquid milk by-products, such as wheys and ultrafiltrates from cheese-making operations, are frequently used in liquid feeding systems for pigs but levels of inclusion have to be monitored because of their high lactose and mineral contents.

Non-protein nitrogen sources (NPN)

Feedstuffs which contain nitrogen in a form other than protein are termed non-protein nitrogen. Such compounds can serve as useful components of ruminant diets in that they provide a source of nitrogen for rumen micro-organisms to synthesise microbial protein. Although a wide variety of compounds can be used as NPN sources the market is dominated by urea. When urea is fed it is initially broken down to ammonia and carbon dioxide by microbial urease. This ammonia may then be utilised along with appropriate oxo-acids in the synthesis of microbial protein. The efficiency of urea utilisation is particularly dependent on the availability of oxo-acids and of energy to meet the needs of protein synthesis. These are affected by the amount and type of dietary carbohydrate. Starch appears to be the best source. Failure of micro-organisms to incorporate ammonia rapidly into microbial protein leads to a loss of nitrogen through urinary excretion and in extreme cases of ammonia production outstripping utilisation, toxic levels of ammonia in the blood may result. A variety of factors must be considered in utilising urea in feeds. These may be summarised as follows:

(1) The diet to which urea is being added must be suitable in terms of its energy, protein and mineral status to allow efficient use of NPN.
(2) Diets containing urea must be introduced gradually to allow rumen micro-organisms to adapt.
(3) Urea should not be used in pre-ruminant diets or where the level of NPN in the diet of adult ruminants is already fairly high, e.g. silage.
(4) Levels of inclusion should be appropriate to the class of stock being fed. For example, levels exceeding 1.25 % in the concentrate ration will affect production in dairy cows, particularly in early lactation whilst for beef cattle and suckler cows levels should not exceed 2.5 % with restricted feeding or 3 % with *ad libitum* feeding.

Feed supplements (nutrients)

In formulating a ration the primary aim is to fulfil the animals' requirements for energy and protein and sometimes fibre. Should the ration prove to be lacking in micronutrients then additions of the appropriate minerals, vitamins or amino acids may be made. Nutrient supplements are commercially available which allow the addition of small amounts of specific nutrients without changing the general make-up of the initial formulation.

Supplementation of rations with specific amino acids is mainly of concern to non-ruminants where cereal based diets may be sub-optimal in such amino acids as lysine, methionine and tryptophan. In the feeding of high yielding dairy cows there is evidence that supplementation of silage based diets with 'protected' methionine may give economic responses.

Mineral supplementations may be provided in the form of licks or feeding blocks which allow the animal free access to a suitable combination of minerals or alternatively those specific minerals in deficit in a ration may be added to the ration in the form of a powder. In using mineral supplements the interrelationships among minerals must be recognised since excessive amounts of one mineral can cause a deficiency of another. Many proprietary supplements containing mixtures of macro- and trace minerals, appropriate to particular production situations are available commercially.

As with minerals, any vitamins in deficit in a formulation must be made good by supplementation. For ruminants fat-soluble vitamins are the major consideration whilst for pigs and poultry both the water and fat-soluble vitamin content of the diet may require supplementation. A variety of balanced vitamin premixes formulated for particular circumstances and usually containing the vitamins in a chemically pure form such that only very small amounts are required are available commercially. In assessing the need for vitamin supplementation it is important to recognise the variability in the vitamin content of feedstuffs and also that vitamins are easily destroyed by agents such as heat, light and oxidation.

NUTRIENT REQUIREMENTS

The requirements of animals for nutrients are initially derived in net terms and subsequently converted to and expressed in the same terms and units of measurement that are used to describe the nutrient content of foods in the rationing process. Thus, for example, the energy requirements for maintenance of a ruminant are determined in net energy terms then converted to and expressed in metabolisable energy terms. Nutrient requirements are a measure of what the average animal requires for a particular function. Tabulated data on requirements generally include a safety margin over and above what the average animal requires in order that animals with requirements higher than the average are adequately fed. Such values are referred to as recommended nutrients allowances.

Recommended nutrient allowances may be expressed in two ways, either as a quantity or as a proportion of the diet. Thus the lysine allowances of a growing pig may be expressed either as $12\,g\,d^{-1}$ or $8\,g\,kg^{-1}$ of the diet, on the assumption that the pig is consuming $1.5\,kg\,d^{-1}$ of diet. In practice, the allowances for ruminants are expressed in quantitative terms on a daily basis whereas those for pigs and poultry are expressed as a dietary concentration of the nutrient in the feed.

The remainder of this chapter is devoted to a survey of how requirements are derived and of the factors influencing the requirements of animals. The total requirement for particular nutrients of animals is arrived at factorially by adding together the requirements for maintenance and the various production functions of lactation, growth and reproduction. For ruminants, the nutrient allowances for maintenance and the productive functions are usually given separately whilst for pigs and poultry nutrient allowances for maintenance are usually combined with the appropriate production function. Tabulated data on nutrient requirments and allowances are contained in the following publications: *The Nutrient Requirements of Farm Livestock. No. 1 Poultry* (1975) ARC; London; *The Nutrient Requirements of Ruminant Livestock* (1986), Commonwealth Agricultural Bureaux, Slough; *The Nutrient Requirements of Pigs* (1981), Commonwealth Agricultural Bureaux, Slough; *Nutrient Allowances for Pigs* (1982), MAFF/ADAS, Booklet 2089; *Nutrient Allowances for Sheep and Cattle* (1986), MAFF/ADAS, P2087 (formerly Booklet 2087).

In this section energy and protein requirements will be considered for each of the body processes. Micronutrient requirements will be considered at the end of the section.

Energy and protein requirements for maintenance

The maintenance requirements of an animal refers to the amounts of nutrients an animal requires to keep its body composition and weight constant when it is in a non-productive state.

Energy

Since the energy used for maintenace must leave the body through heat the net energy requirement for maintenance may be determined by measuring the heat produced by an animal when kept in the fasted state (fasting metabolism, FM). Such measurements may be made in animal calorimeters and show that energy requirements for maintenance are roughly proportional to the surface area of the animal or more closely, and of more practical value, related to a function of body weight (W), namely $W^{0.73}$. This is termed the metabolic weight of the animal. Typical fasting

metabolism values for different species are:

	Cow	Ewe	Pig
Body weight (W) (kg)	500	50	70
Metabolic weight ($W^{0.73}$) (kg)	93.4	17.4	22.2
FM per kg $W^{0.73}$ (MJ d^{-1})	0.36	0.24	0.30
FM per animal (MJ d^{-1})	33.6	4.2	6.7

Thus although the body weight of the ewe is only one-tenth that of the cow its fasting metabolism or net energy requirement for maintenance is approximately one-eighth that of the cow.

A number of factors other than body weight may influence energy requirements for maintenance. Those with greatest influence are the activity of the animal and environmental conditions although lesser effects related to breed, sex and age have also been identified. Clearly animals which have to forage for food expend more energy than stall fed animals. Likewise animals kept under adverse environmental conditions expend additional energy. These additional energy expenditures are debited to the maintenance requirement.

Environmental effects are chiefly mediated through the interacting effects of temperature, wind speed and rain. Within certain limits (the thermoneutral zone) the animal can maintain thermoregulation by limiting heat loss/gain through altering peripheral blood circulation and behavioural mechanisms. Under more extreme conditions, however, the animal may have to use energy to maintain thermoregulation either to produce heat or to dissipate heat. The environmental temperature at which the animal is forced to produce heat to maintain thermoregulation is referred to as the *lower critical temperature* and the environmental temperature at which it is required to use energy to dissipate heat is the *upper critical temperature*. Neither of these values are constant and are influenced by wind speed, rain and the effective insulation of the animal as influenced by coat or fleece and subcutaneous fat.

To convert net energy requirements for maintenance to the same units of measurements used in the energy evaluation of feeds the efficiency with which food is used to meet net energy requirements, the addition of allowances for activity ($+10\%$) and a safety factor ($+5\%$) must be taken into account. Thus, for example, in cattle where fasting metabolism is approximately $0.36\,MJ\,kg^{-1}\,W^{0.73}$ and the efficiency with which metabolisable energy (k_m) is used to meet net energy (NE) requirements for maintenance is 0.72, the addition of a 10% activity allowance and a 5% safety factor gives:

ME required for maintenance (M_m)(MJ d^{-1}) =

$$\frac{0.36 \times W^{0.73}}{0.72} \times 1.1 \times 1.05$$

or more simply

$$M_m (MJ\,d^{-1}) = 8.3 + 0.091\,W$$

Protein

Proteins of body tissues are constantly being renewed. During this *turnover* of body tissues the amino acids resulting from breakdown are not re-utilised with 100%

efficiency for synthesis into new protein. Thus the diet must provide the animal with sufficient protein to remain in nitrogen balance. This *net requirement* of protein for maintenance can be determined by measuring the faecal and urinary nitrogen losses from the animal when fed a nitrogen-free ration. The net requirements for protein for maintenance are like those for energy more related to metabolic weight rather than body weight. For example, *see Table 13.4.*

Table 13.4 Tissue protein requirements for maintenance in cattle

Liveweight	Requirement (g d^{-1})
200	49
250	53
300	57
350	61
400	64
450	67
500	69
550	72
600	74
650	76

In the pig it has been shown that the losses from protein 'turnover' are of the order of 6 % of the protein cycled. As an animal grows from birth to slaughter weight it has been shown that the amount of protein turnover decreases from 13 to 6 % of the protein mass of the animal. Since protein mass is usually about 15 % of liveweight the net amount of protein needed daily to supply the requirements of turnover (maintenance) varies from 0.12 % of body weight soon after birth (6 % of 13 % of 15 %) to 0.05 % of body weight towards slaughter (6 % of 6 %, 15 %), (*Figure 13.5*). The amount of protein required in the diet further depends on the

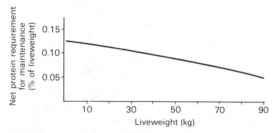

Figure 13.5 Relationship between the net protein requirements for maintenance (% of liveweight) and liveweight in the pig

efficiency with which food protein is digested and on the efficiency of utilisation of digested protein (the biological value, BV). Thus if BV of dietary protein = 65 and digestibility of dietary protein is 80 % the dietary crude protein requirements for maintenance vary from 0.23 − 0.10 % of body weight as the animal grows.

Energy and protein requirements for lactation

The nutrient requirements of the animal for milk production are dependent on the yield of milk being produced and the composition of that milk.

Energy

The energy value of milk, which is a measure of the net energy required for milk production is dependent on its compositional analysis. It may be determined by measurement of its gross energy using a bomb calorimeter or more usually by prediction equation using information on its compositional analysis. The energy yielding components of milk, i.e. the fat, protein and lactose are considered to have energy concentrations of 38.5, 24.5 and 16.5 MJ kg^{-1} respectively. However, on a routine basis the constituents of milk commonly determined are fat and protein. With a knowledge of these values, fairly accurate predictions of milk energy value can be made using the equation:

$$EV_1(MJ\ kg^{-1}) = (0.376 \times \%\ fat)$$
$$+ (0.209 \times \%\ protein) + 0.946$$

Taking into account the efficiency with which metabolisable energy is used for milk production (k_1) to be 0.62 and the addition of a 5 % safety factor

$$ME\ required\ for\ lactation\ (ME_1)(MJ\ kg^{-1}) = \frac{EV_1}{0.62} \times 1.05$$

$$or\ ME_1(MJ\ kg^{-1}) = 1.694 \times EV_1$$

Provision of dietary metabolisable energy to lactating dairy cows is complicated by the relationships between milk yield, the partition of dietary energy and the voluntary food intake of the animal through the lactation. The situation is illustrated in *Figure 13.6.*

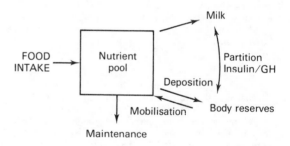

Figure 13.6 The partition of nutrients in the cow

In the early part of lactation it is usually the case that cows are unable to consume sufficient energy to meet energy requirements for milk consumption and call on body reserves, chiefly fat, resulting in a loss of body weight.

The amount of dietary ME for milk production which this weight loss is equivalent to may be calculated as follows:

Energy value of 1 kg tissue = 20 MJ

Efficiency of use of body tissue energy for milk synthesis = 0.82. Therefore 20 × 0.82 = 16.4 MJ dietary NE for milk.

Since the efficiency with which dietary ME is converted to milk energy is 0.62 and including the 5 % safety margin the dietary ME equivalent for milk production of 1 kg tissue = (16.4/0.62) × 1.05 = 28 MJ. Thus the mobilisation of 1 kg body tissue saves 28 MJ from the ration and allows dairy cows of high potential to meet their energy requirements for milk production when appetite is limiting. The tissue that is mobilised in early lactation must be replaced in later lactation through the inclusion of energy in the ration over and above that required for maintenance and milk production. The amount of dietary energy required for body gain may be calculated as follows:

Energy value of 1 kg tissue = 20 MJ

Efficiency of utilisation of ME for body gain in the lactating animal = 0.62 and inclusion of a 5 % safety factor gives:

$$\text{ME required/kg gain} = \frac{20}{0.62} \times 1.05 = 34\,\text{MJ}$$

Whilst these calculations are appropriate for rationing purposes it is worth pointing out that the partitioning of dietary energy between milk production and body tissue is an individual characteristic of the cow, related in physiological terms to the hormonal – especially the growth hormone/insulin – balance of the animal. In addition the composition of the diet may affect partition, diets with a high energy concentration encouraging the deposition of energy in body tissue at the expense of milk production.

As with the cow the energy requirements for milk production in the sow may be obtained from food or from body fat. The amount of energy being obtained from the latter source depends on the feed intake of the sow, a characteristic which itself may be influenced by level of body fatness; sows fed well in pregnancy and carrying more fat at the start of lactation frequently eat less food and lose weight faster than sows fed less well during pregnancy. The efficiency factor involved in converting the metabolisable energy in food into milk energy is about 65 %; when food metabolisable energy is converted to body fat ($\sim 75\%$ efficiency) and that body fat then converted into milk ($\sim 85\%$ efficiency) the overall conversion efficiency is likewise about 65 % (85 % of 75 %).

Protein

The net protein requirements for milk production are dependent on the protein content of the milk. Average cows' milk contains 34 g protein/kg milk. The net protein requirement for maintenance and milk production in the dairy cow must be met from a combination of microbial protein and undegradable protein (UDP). The provision of microbial protein is dependent primarily on energy intake which in turn influences the rumen degradable protein (RDP) requirements. Thus dairy cows have separate requirements for RDP and UDP. This is best illustrated by the following calculation for a 600 kg cow yielding 35 kg milk (38 g kg^{-1} fat; 34 g kg^{-1} protein) and losing 0.5 kg of body weight.

Total ME allowance   = 232 MJ d^{-1} (*see* pp. 382–384)

Tissue protein
required for maintenance = 74 g d^{-1} (*Table 13.4*)

Tissue protein
required for milk product = 34 × 35 = 1190 g d^{-1}

Total net tissue
protein requirement   = 1264 g d^{-1}

Assuming 7.8 g microbial protein synthesised/MJ of ME (*see* p. 328)

Microbial protein synthesised = 232 × 7.8

                              = 1809.6 g d^{-1}

Therefore RDP requirement   = 1809.6 g d^{-1}

In the utilisation of microbial protein the following efficiency factors may be assumed:

Proportion of 'True Protein'
in microbial protein   = 0.8 (*see* p. 329)

Efficiency of digestion of
protein reaching small intestine = 0.7

Efficiency of utilisation of
absorbed amino acids   = 0.75

Then:

Microbial protein available
to tissues   = 1809.6 × 0.8 × 0.75 × 0.7

                             = 760 g d^{-1}

Using the above assumptions on the efficiency of digestion and utilisation of protein reaching the small intestine then:

UDP requirement = (1264 − 760) × (1/0.7) × (1/0.75)

                 = 504 × 1.91

                 = 962.6 g d^{-1}

Therefore the protein requirements of this cow may be met by a diet containing 2772 g CP with a degradability of 65 %.

In the case of lactating sows net protein requirements are converted to a crude protein requirement using an efficiency of utilisation of absorbed protein as measured by biological value (0.70 on average) and digestibility of the protein (0.80 on average). From this the proportion of protein required in the diet at a particular level of feeding may be calculated.

Energy and protein requirements for reproduction

The nutrient requirements for reproduction may be considered from the following viewpoints:

(1) how the nutrient intake of animals may influence reproductive potential in terms of its effect on such parameters as age at puberty, ovulation rate, conception and re-breeding; and
(2) the determination of the nutrient requirements of the pregnant female animal for fetal and extra-uterine growth in preparation for lactation.

In general the requirements of pregnancy increase exponentially from insignificant levels in early pregnancy and it is only in the last trimester of pregnancy that it becomes necessary to make special dietary provision (*Figure 13.7*).

Energy

The plane of energy nutrition during the *rearing* period can influence the age at which puberty is achieved in cattle and sheep. Whilst earlier puberty may be achieved by increased levels of feeding caution must be taken to ensure that excessive fat deposition does not occur. In practice, the level of energy fed and its effect on age at puberty is determined by the requirements of the production system chosen. In pigs, plane of nutrition would appear not to have such a marked effect on age at puberty.

The *period before mating* is considered important since ovulation rate may be influenced by the condition of the

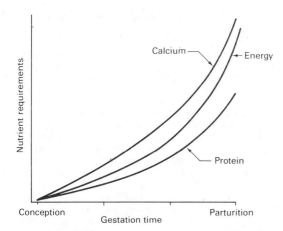

Figure 13.7 Effect of stage of gestation on the relative requirements for nutrients during pregnancy

animal. In ewes, improving the body condition can have a particularly marked effect on the fertility of hill breeds of sheep but more prolific breeds and crosses of these with hill breeds respond less dramatically. The more short-term practice of 'flushing' ewes prior to mating would also appear to improve fertility especially of ewes with lower condition scores. There is also reason to suppose that a similar relationship operates in the sow. From the viewpoint of feeding, the most practical advice is to ration energy according to body condition taking account not only of those animals that require an improvement in body condition but also of overfat animals which tend to have reduced fertility. In the dairy cow mating must occur during early lactation at which time body weight loss is frequently occurring. Excessive weight loss is associated with infertility and the aim in feeding over this period is to restrict the extent of weight loss.

The period immediately *post-conception*, when the developing embryo is nourished by direct absorption from its fluid environment, is critical in that faulty or inadequate nutrition at this time can jeopardise the successful implantation of the fertilised ova in the uterine wall. Adequate nutrition must therefore be continued over this period although subsequently there is little extra requirements for energy until the last trimester of pregnancy when accelerating fetal growth increases energy demand. Although severe energy undernutrition in early pregnancy can affect placental growth some loss in body condition can occur in ewes without adverse effects on fetal development.

Just as an awareness of body condition governs feeding in the premating period so also it is an important parameter to be monitored as pregnancy progresses. Whilst over-thin sows and ewes will produce weak, underweight young with reduced survival rate the development of over-fatness, especially in mid-pregnancy can lead to pregnancy toxaemia and problems at parturition. Overconditioning in late pregnancy may also lead to reduced voluntary food intake in the subsequent lactation. In the feeding of ewes during the last six weeks of pregnancy it is important to take into account the likely number of fetuses being carried. In this context the use of pregnancy scanning techniques has greatly improved the precision of rationing.

With dairy cows it is practice to terminate a lactation some eight weeks before the birth of the next calf. Restoration of body tissue converted to milk in early lactation should take place prior to 'drying off' since the conversion of feed energy into liveweight gain is more efficient in the lactating than the non-lactating state. Provision of extra energy during the dry period is not generally recommended because not only is it less efficiently used but there is the risk of excessive fattening and it has little influence on the birth weight of the calf.

Additional energy requirements need to be satisfied in pregnant animals that are themselves still growing. Sows, for example, do not reach their mature body weight until their fourth parity whilst cows' mature weight is not reached until their third lactation. There is still considerable uncertainty about the quantitative aspects of requirements for this aspect of pregnancy, because of the impact of pregnancy itself on energy utilisaton. There is some evidence that the maintenance requirements of the pregnant animal are reduced in early pregnancy whilst later in pregnancy the so called 'heat of gestation' increases energy needs for maintenance.

Protein

In the pregnant animal the net protein requirement is represented by not only the products of conception, fetal and uterine growth, mammary tissue growth and protein turnover losses, but also lean tissue growth in the maternal body. Clearly these various requirements are not easily separated out one from another and further to this precise estimation of requirements is complicated by changing efficiency of utilisation during pregnancy. Especially towards the end of pregnancy there appears to be more efficient utilisation of protein, termed pregnancy anabolism, probably resulting from the animal's changed hormonal status. Requirements are clearly at their greatest during the latter part of pregnancy when uterine and fetal growth predominate.

In sows the quality of protein in the diet will influence the amount fed whilst in ruminants the need for protein supplementation is dependent on the dietary balance of energy, RDP and UDP. In the cow energy intakes in late lactation and the dry period ensure sufficient amino acid supply to the growing fetus but in the ewe an energy and protein gap between intake and needs may exist in late pregnancy, particularly if poor quality hay is fed. This can affect the amount of microbial protein synthesised and necessitates an increase in the level of concentrate feeding and level of protein in the concentrate. Some workers have claimed benefits from the feeding of protein sources with a high proportion of UDP such as fish meal in late pregnancy.

In dairy cows it has been suggested that excessive dietary protein in early lactation may decrease reproductive efficiency and that this is related to the absorption of ammonia-nitrogen from the rumen in animals fed excessive rumen degradable protein (RDP). Conversely improved conception rates have been observed when the ratio of RDP/UDP in the diet is decreased.

Energy and protein requirements for growth

The commercial measure of growth in farm animals is liveweight gain. In determining the nutrient requirements for liveweight gain the major problem to be faced is that the composition of liveweight gain is not constant. The major influences on the composition of liveweight gain are:

(1) the physiological age or weight of the animal, and

(2) the rate of liveweight gain.

Detailed coverage of these influences on the composition of liveweight gain is given in Chapter 12 (*see* p. 361).

Energy

The net energy content of the LWG made by an animal is dependent on the proportions of protein, fat, water and ash present. Thus the energy content is represented by the protein and fat fractions. In general the gain made by young animals has a low energy content since it has a high proportion of water (in lean) and ash (in bone) and little fat. As animals grow and mature the energy content of gain increases, this being due to the declining proportions of water present and the increasing proportions of fat.

The rate of LWG influences the net energy content of the gain since the greater the rate of gain the higher the proportion of fat produced.

In specifying energy requirements for growth it is necessary, therefore, to indicate the rate of growth anticipated. This is especially the case in beef production where it is desirable to achieve target LWG at each stage of the particular production system (*see* Chapter 14). In some situations the target LWG is the maximum growth potential for some or all of the production cycle. This is normally achieved in practice by allowing *ad libitum* intake of high energy feeds. Some restriction of energy intake is frequently desirable as animals approach slaughter weight as a means of reducing the fat content of the carcass, a principle applied in the production of pigs for bacon where overfatness of carcasses is penalised.

In the rationing of growing ruminants a particular problem arises in that energy requirements are not simply affected by the energy content of the gain but also by the metabolisable energy concentration (total ME in diet/total DM in diet, M/D) of the diet which has a marked effect on the efficiency of utilisation of metabolisable energy. What it means in practice is that a higher proportion of the metabolisable energy in a largely concentrate based diet is laid down as body tissue than in a largely roughage based diet. The origins of such differences lie in the types of rumen fermentation and particularly the VFA end products that different diets produce. Thus increasing the proportion of propionate decreases the heat increment of metabolism. This extra dimension to assessing the energy requirements for growth has led to the development of a Net Energy System sometimes referred to as the Variable Net Energy System which takes account not only of the effects of the weight of the animal and rate of gain but also the M/D of the diet. Details of the Variable Net Energy System are given in Chapter 14.

Whilst energy requirements for gain are mainly influenced by the factors considered above, the breed and sex of the animal also have an influence. Thus for cattle later maturing breeds require less energy per unit of gain at equal body weights than early maturing breeds; the energy content of gain at a particular weight is greater in heifers than in steers which in turn is greater than in bulls.

Protein

Growth rate in animals is primarily determined by the energy intake. The same principles which govern the net energy requirements for gain also affect the net protein requirements for gain. In reality the requirements are not so much for protein but more specifically for the various amino acids that contribute to growth.

For both pigs and poultry the amounts of dietary protein (as CP) required for growth are dependent on the digestibility and biological value of the dietary protein. A reduction in the biological value can be countered by an increase in the CP requirement. In practice, because certain essential amino acids are likely to be limiting to the efficiency of utilisation of absorbed amino acids, recommended DCP allowances are also accompanied by recommendations for limiting amino acids such as lysine, threonine, methionine and tryptophan. These requirements are determined by monitoring the growth response or nitrogen retention resulting from the feeding of graded levels of the particular amino acids, a technique subject to the criticism that it does not take account of the interactions that occur between essential amino acids, e.g. methionine and cystine.

For ruminants the make up of the dietary protein in terms of RDP and UDP required to support a particular level of gain is dependent not only on the protein content of the gain and the efficiency with which protein reaching the small intestine is utilised for gain but also on the energy intake of the animal. In practice the protein requirements of growing animals can usually be met in full by the supply of microbial protein to the small intestine except where high growth rates are demanded in which case an appropriate supply of UDP must be fed.

Micronutrient requirements

Micronutrients have a vast number of roles in the animal body and failure to meet the animal's requirements for particular micronutrients can at least impair the performance and at most produce clinical conditions. The precision of tabulated data or allowances for micronutrients is limited by the methodology of determination and by the fact that numerous interactions occur between individual micronutrients. Such interaction can influence availability in terms of absorption and their functional activity in a metabolic pathway. In addition, a variety of mechanisms exist in the animal to provide a relatively constant micronutrient milieu for metabolic activity under circumstances of variable dietary intake, thus ensuring the normal physiological functioning of the animal. Nevertheless, such homeostatic mechanisms can break down under more extreme dietary supply situations leading to impaired physiological function either through an inadequacy or an accumulation of micronutrients.

Allowances for individual mineral elements have been determined through either factorial methods or by feeding trial techniques. In the former method the net requirement for the element is determined by adding together the endogenous losses of the element that occur from the animal when fed a diet free of the element – equivalent to maintenance requirement – to the amount of element that is present in the product of the animal. For a product such as milk this is relatively easy but for estimation of the mineral composition of liveweight gain it is a somewhat more laborious procedure. To convert the net requirement to a dietary requirement the availability of the particular element in terms of absorbability has to be taken into account. In the performance or feeding trial method of estimation diets containing different amounts of the element are fed and their influence on performance monitored. The main difficulty is in setting the parameters that relate to the optimum intake of the element. For example, in the case of an element such as calcium, maximum growth rate may be obtained at a level of intake that is inadequate in terms of bone strength. The ability of the animal to store some mineral elements further complicates the situation.

Estimation of requirements for vitamins are obtained from feeding trials. The same criticisms made for the use of this method for estimating mineral requirements also apply for vitamins.

In practice, levels of supplementary inclusion in particular situations are arrived at empirically, taking into account any conditions such as feed composition and health status of the herd which may influence requirements.

Further reading

McDONALD, P., EDWARDS, R. A. and GREENHALGH, J. F. D. (1981). *Animal Nutrition*. London: Longmans

WISEMAN, J. (Ed.) (1987). *Feeding of Non-Ruminant Livestock*. London: Butterworths

CHURCH, D. C. (1984). *Livestock Feeds and Feeding*. Corvallis: O & B Books

WHITTEMORE, C. T. and ELSLEY, F. W. H. (1976). *Practical Pig Nutrition*. Ipswich: Farming Press

14

Cattle

J. Kirk

Domesticated cattle are nearly all descended from two major species: *Bos taurus*, which includes the European types, and *Bos indicus* to which the Zebu cattle belong. Selection from these has led to the development of a number of well defined breeds. These breeds vary from types used primarily for milk production to those developed for beef production.

In some breeds an attempt has been made to combine the desirable qualities of both dairy and beef cattle to provide dual-purpose cattle.

Since the 1950s a number of exotic breeds have been imported into the UK, mainly from Europe. The rationale behind this importation of breeding stock has been to improve beef production although some are dual-purpose breeds in their country of origin. Most of these immigrants have been large bodied, fast growing cattle producing lean carcasses (i.e. Charolais, Limousin, Simmental), characteristics particularly valuable for satisfying consumer demand for lean meat. Similarly, the dairy sector has benefited from the importation of Holstein cattle from Canada, the USA, Denmark and New Zealand.

Table 14.1 UK agricultural output 1984

	£ million	% Total
Livestock		
Fat cattle and calves	1922	16.0
Fat sheep and lambs	579	4.8
Fat pigs	1000	8.4
Poultry	664	5.5
Others	94	0.8
Total livestock	4258	35.5
Livestock products		
Milk	2293	19.1
Eggs	537	4.5
Wool	37	0.3
Others	26	0.2
Total livestock products	2893	24.1
Total crops	3541	29.6
Total horticulture	1241	10.4
Total agricultural output	11933	99.6

Fat cattle, calves and milk are an important sector of total UK agricultural output (*Table 14.1*).

The last two decades have seen a dramatic change in the UK dairy herd structure. The number of registered milk producers has fallen by over 60%, whilst dairy cow numbers have fallen by about 20%. This has resulted in a rapid increase in the average size of dairy herd (*Table 14.2*).

Milk yield per cow has also risen. This coupled with a decline in liquid milk consumption has resulted in more milk being used for manufacture (*Table 14.3*).

Milk sold for processing commands a lower price than that in the liquid milk market. Milk price as paid to the producers has to take into account the quantity and lower price of milk sold for manufacture (*Figure 14.1*).

A similar pattern is seen when looking at milk utilisation between the liquid consumption and manufacturing sectors within the EEC (*Table 14.4*).

Milk product surpluses arose because the increased amount of milk being produced and reduced consumption. In 1986 it was estimated that accumulated intervention stocks stood at: butter 1 500 000 t and milk powder 1 100 000 t compared with beef at 620 000 t. Associated with this was the massive cost of the Common Agricultural Policy for the initial purchase of these and other products and their storage. It was for this reason that in March 1984 the EEC Agricultural Ministers agreed on a policy for curbing the

Table 14.2 Cattle numbers in the UK (× 10)

Year	Number of producers	Number of dairy cows	Average herd size	Average milk yield (litres)
1955	175 000	3 972 000	18	3065
1965	124 700	3 186 000	26	3545
1976	71 200	3 242 000	43	4300
1982	52 221	3 246 000	62	4745
1983	51 726	3 328 000	64	5055
1984	50 625	3 278 000	65	4940
1985	48 827	3 147 000	64	4770
1986	47 927	3 141 000	66	4880

After: MMB (1986) *UK Dairy Facts and Figures.*

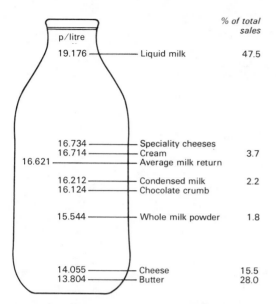

Figure 14.1 Average return from milk for each product –
England and Wales 1985–86 (After MMB (1986) *UK Dairy
Facts and Figures*)

Table 14.3 Utilisation of milk in the UK

	1977	1979	1985
Liquid market	6800 (60%)	6533 (51%)	6049 (48%)
Manufacture	4648 (40%)	6206 (49%)	6627 (52%)
Average consumption (litres/head/week)	2.76	2.66	2.36

After: MMB (1986) *UK Dairy Facts and Figures.*

Figure 14.2 Sources of home produced beef – 1985 (After
MLC (1986) *Beef Yearbook*, reproduced with permission)

Table 14.4 Milk utilisation in the EEC (in %)

	Liquid	Cream	Butter	Cheese	Milk powder	Total utilisation ($10^{-3} \times$ t)
Belgium	16	5	53	6	7	4242
Denmark	8	6	39	27	11	5205
France	9	4	37	23	4	34 787
Germany	13	9	42	14	3	26 142
Irish Republic	10	3	59	12	3	6480
Italy	28	4	14	42	–	13 408
Luxembourg	11	9	56	3	–	320
Netherlands	6	3	42	28	11	12 701
UK	41	3	26	15	2	17 680
Total	17	5	36	22	4	120 965

After: MMB (1986) *EEC Dairy Facts and Figures.*

increase in EEC milk production, by the introduction of a
system of quotas. The agreement allowed for a superlevy to
be charged on all milk produced above a specific quantity or
'quota'.

In the UK there is a close relationship between milk
production and beef production. Dairy herds commonly use

beef bulls for crossing and the crossbred calves, together
with pure bred bull calves from dairy herds are fattened for
beef. A number of heifers from the dairy herd are reared as
replacements for suckler beef herds (*Figure 14.2*).

After reaching a peak in 1984 beef and veal production in
the EEC fell in both 1985 and 1986 (*Figure 14.3*).

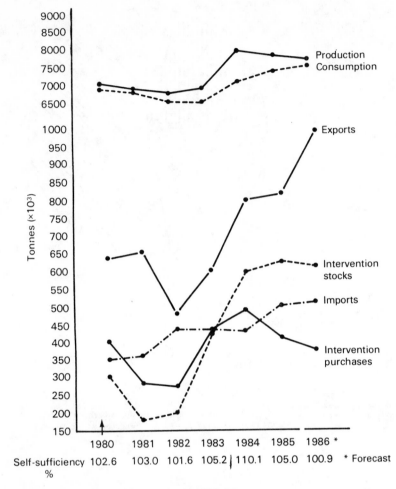

Figure 14.3 EEC beef and veal production 1980–86

Table 14.5 Estimated meat consumption (kg/head) in the UK 1981–85

	1982	1983	1984	1985
Beef and veal	18.3	18.5	18.3	19.9
Mutton and lamb	7.3	7.1	6.8	6.8
Pork	12.9	13.1	12.6	12.9
Bacon and ham	8.4	8.4	8.3	8.2
Poultry	14.4	14.7	15.6	16.3
Offal	2.2	2.1	2.0	2.0
Total	63.5	63.9	63.6	64.1

After: MLC (1986) *Beef Yearbook* (Reproduced with permission)

Consumption of beef is increasing in all EEC member states except Portugal. *Table 14.5* shows the pattern of meat consumption in the UK.

Intervention purchases of beef are declining but sales from intervention are also small. By the end of 1985 intervention stocks were 633 000 t. Imports into the EEC, in particular imports of fresh or chilled beef into Italy from Yugoslavia

and Austria and live cattle from Eastern Europe into Italy are significant. However, the EEC retained its position of a major net exporter of beef and veal. Self-sufficiency in beef and veal has declined from 110.1% in 1984, it is expected that this decline will continue in 1987 (*Table 14.6*).

DEFINITIONS OF COMMON CATTLE TERMINOLOGY

At birth
    male – bull calf, bullock calf if castrated
    female – heifer calf, cow calf

First year
    male – yearling, year old bull
    female – yearling heifer

Second year
    male – two year old bull, steer, ox, bullock
    female – two year old heifer

Third year
    male – three year old bull, steer, ox, bullock
    female – three year old heifer, becomes cow on bearing calf

Table 14.6 Cattle population 1985 (10^3 head)

Country	Total cattle	Total cows	Dairy cows	Others cows	% Dairy cows in total cow population
Belgium	2960	1130	946	184	83.7
Denmark	2623	972	913	59	93.9
France	22 803	9763	6506	3257	66.6
Germany	15 627	5625	5451	173	96.9
Greece	776	344	219	126	63.7
Irish Republic	5779	1942	1528	415	78.7
Italy	9010	3485	3120	365	89.5
Luxembourg	220	86	70	16	81.4
Netherlands	5076	2333	2333	–	100.0
UK	12 695	4579	3257	1322	71.1
Total	77 569	30 259	24 343	5917	80.4

Source: MMB (1986) *EEC Dairy Facts and Figures.*

Heifer – usually applied to a female over one year old which has not calved. An unmated animal is known as a maiden heifer and a pregnant one as an in-calf heifer. In some areas the term first-calf heifer is used until after the birth of a second calf. A barren cow is either barren, cild or farrow and when a cow stops milking she is said to be yeld or dry.

Stirk – limited to males and females under two years in Scotland. It is usually applied to females only in England, the males being steers.

Store cattle – stores, are animals kept usually on a low level of growth for fattening later.

Veal – is the flesh from calves reared especially for this purpose and normally slaughtered at about 16 weeks of age.

Bobby veal, slink veal – flesh from calves slaughtered at an early age, often only a week or two old. These animals tend to be of extreme dairy type, thus making them undesirable for rearing as beef cattle.

Cow beef – beef from unwanted cows, often used in processing.

Bull beef – beef from entire male animals, the majority produced as an end product but some as a by-product from redundant breeding stock.

Ageing

The development of the incisor teeth can be used as an indication of the age of cattle.

The dental formula for a full mouth is:

Permanent molars	Temporary molars	Incisors	Temporary molars	Permanent molars	
3	3	0 0	3	3	
–	–	–	–	–	= 32
3	3	4 4	3	3	

There are four pairs of incisor teeth. These are found in the lower jaw, the upper jaw has no incisors but is a hard mass of fibrous tissue known as the dental pad. Starting in the middle of the jaw the pairs are known as centrals, medials, laterals and corners. The times at which the temporary

Temporary incisors

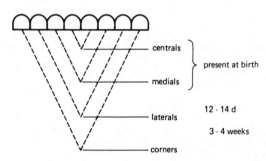

- centrals
- medials — } present at birth
- laterals — 12 - 14 d
- corners — 3 - 4 weeks

Permanent incisors

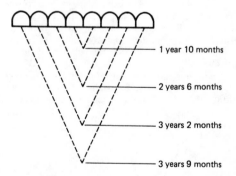

- 1 year 10 months
- 2 years 6 months
- 3 years 2 months
- 3 years 9 months

incisors are shed and replaced by the permanent incisors are important.

Individual animals will vary from these figures. Breed, management and feeding all have an influence.

CALF REARING

The foundations of the future health and well being of the calves are laid by good feeding and management throughout the first three months of age.

Table 14.7 Composition of colostrum (first 24 h after calving) and of milk

	Colostrum	*Milk*
Fat (%)	3.6	3.5
Non-fatty solids (%)	18.5	8.6
Protein (%)	14.3	3.25
Casein (%)	5.2	2.6
Albumin (%)	1.5	0.47
Immune globulin (%)	6.0	0.09
Ash (%)	0.97	0.75
Calcium (%)	0.26	0.13
Magnesium (%)	0.04	0.01
Phosphorus (%)	0.24	0.11
Iron (%)	0.20	0.04
Carotenoids (μg/g fat)	25–45	7.0
Vitamin A (μg/g fat)	42–48	8.0
D (μg/g fat)	23–45	15.0
E (μg/g fat)	100–150	20.0
B (μg/g fat)	10–50	5.0

After: Roy (1980)

The rectal temperature of a healthy calf is 38.5–39.5 C. Common to all the systems of rearing is the need for an adequate supply of colostrum (the secretion drawn from the udder at the time of parturition).

At birth the calf is virtually free of bacteria but quickly becomes infected with organisms from its surroundings. The blood of the newborn calf contains no antibodies until the calf has received colostrum.

As can be seen in *Table 14.7* colostrum contains a high percentage of proteins especially immune globulins with their attendant antibodies, minerals, vitamins (especially the fat soluble A, D and E) as well as carotene which is the cause of the yellow coloration.

It is vitally important that the calf receives colostrum during the first 24 h of life. The calf's ability to absorb the antibody protein is greatest during the first 6–12 h of life. This is due to the calf's stomach wall changing and becoming impermeable to the immunoglobulins (*Figure 14.4*) Hence, to obtain the maximum protection the calf should be fed colostrum during the first 6 h of life.

Colostrum feeding should continue for 4 d. Some farmers achieve this by leaving the calf to suckle, others remove the calf at birth. If the calf is removed at birth it is easier to teach

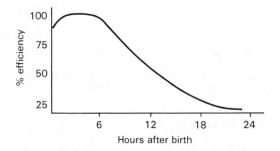

Figure 14.4 Effective uptake of antibodies from colostrum in the young calf

it to drink from a bucket. Frequently, a greater quantity of colostrum is produced by the cow than is needed by the newborn calf. Any excess colostrum can be diluted with water at the rate of two parts of colostrum to one of water and fed to older calves in place of milk or milk substitute. Alternatively colostrum may be frozen and kept in case of an emergency when it may be fed to newborn calves. The composition of colostrum changes to milk during the first 4 d milking. If pre-calving milking is practised this change may take place before the calf is born.

The zinc sulphate turbidity test (ZST) enables the concentration of circulating antibodies in the calf's blood to be determined. There is a high correlation between the results of this test and calf health, showing that it is essential calves get adequate amounts of colostrum. This is undoubtedly one of the major factors involved in ensuring that disease and ill health are kept under control (*Table 14.8*).

Although the newborn calf has the same four stomach compartments, rumen, reticulum, omasum and abomasum, as that of the adult they vary in size (*Table 14.9*) and development.

The same relative proportions would also apply to the volume of the various compartments.

The stomach of the newborn calf is similar to that of a monogastric animal. The oesophageal groove conveys liquids or semiliquids directly to the omasum and from there they pass to the abomasum. It is in the abomasum where true gastric digestion takes place.

The age at which transition to ruminant digestion takes place depends on the diet the young calf is fed. The longer a calf is fed milk the less is the urge to consume other feeds and the later will be the development of the rumen.

Ideally calves should be kept for at least 7 d on the farm on which they were born. Newly purchased calves should be allowed to rest after a journey. If the calves have only had a

Table 14.8 Relationship of colostral status to calf performance at five weeks

	Colostrum rating		
	Low	*Medium*	*High*
No. of calves	87	158	182
% Treated for illness			
No treatment	17.2	37.3	48.9
Treated once	27.6	41.2	30.2
Treated twice	33.3	12.0	13.8
Treated more than twice	21.9	9.5	7.1

Source: Thickett *et al.* (1979)

Table 14.9 Proportion of tissues by weight of a calf's stomach

	Age (weeks)			
	0	*4*	*12*	*20–26*
Stomach compartment				
Rumen-reticulum	38.0	52.0	64.0	64.0
Omasum	13.0	12.0	14.0	22.0
Abomasum	49.0	36.0	22.0	14.0

Source: Warner and Flatt (1965)

short journey and were recently fed, additional carbohydrate, e.g. glucose, may predispose the animals to diarrhoea. Generally calves arriving before noon should be left to rest in a dry, draught free, well bedded area and then fed in the evening with a drink of an electrolyte/glucose solution. Calves arriving mid to late afternoon should be left until the following morning before being fed the same solution. Proprietary electrolyte solutions are available or a solution of glucose (22 g) with the addition of sodium chloride, common salt (4.5 g), per litre of water can be used

First drink — 1.5 litres electrolyte/glucose solution
Second drink — 1.5 litres electrolyte/glucose solution
Third drink — 50:50 milk powder/electrolyte solution
Fourth drink — milk powder alone at standard concentration.

Numerically the most common method of rearing calves is 'by hand'. Artificial milk replacers are used instead of cow's milk because the cost per unit of milk replacer is less than the equivalent amount of milk. Calves destined as replacements in the dairy herd are almost always reared by hand. This method of rearing requires specialist accommodation and is more exacting in terms of labour but enables economies in food costs to be achieved. Milk substitutes basically consist of skim milk powder and added fat, protein, vitamins and minerals; whey powder is sometimes included. Most artificially reared calves are 'early' weaned.

Early weaning

In this method the calf is weaned onto a diet of dry food by five weeks of age. The early introduction of concentrates, hay and water by day 7 after birth encourages the development of the rumen. To achieve the desired intake of concentrates it is essential that they be palatable and fresh. Milk replacer is restricted, again to encourage the consumption of hay and concentrates. Calves should be eating 0.75–1.00 kg of concentrates daily with a liveweight of 65 kg by weaning at five weeks. The advantages of early weaning are that concentrates are much cheaper per kg of gain than milk replacer.

'Milk' feeding can be practised in several forms. Once or twice daily feeding of milk substitutes is the common practice on farms.

Ad libitum feeding of milk substitute was first made possible with the advent of high fat milk substitutes and automatic dispensing machines. The introduction of 'acid' milk replacers has enabled *ad libitum* feeding of cold milk to be practised without the need for sophisticated machines. Acid milk replacers are classified according to two types:

(1) Medium acid based on skimmed milk. These have a pH of 5.5–5.8 and a protein content of 24–26% with 17–18% fat.
(2) High acid usually based on whey from cheese manufacture. These have a pH of 4.4–5.8 with 19–20% protein and 12–15% fat.

The acidity has a positive effect in helping to reduce the incidence of digestive upsets. The pH of the abomasum prior to feeding is 2.0–2.8 but after feeding conventional milk replacer of pH 6.2–6.5 the abomasum pH rises to between

4.5 and 6.2. It then declines to pre-feeding levels after 3–5 h. Satisfactory digestion depends on enzyme action and the optimum pH for this to occur is between 3 and 4. The conclusion therefore is that by feeding high acid milk replacer the abomasal pH is kept near optimum levels for enzyme activity and below the pH at which most organisms can survive.

Milk replacers may be prepared for 3 d feed supply at a time and the formulation aids the normal digestive process of the calf.

The suckling action by the calf ensures correct closure of the oesophageal groove allowing milk to enter the abomasum without spilling into the developing rumen, where it can ferment and possibly cause digestive troubles. The argument that cold feeding is bad for calves probably grew out of the bucket feeding system where a calf may suffer a physiological shock when consuming large amounts of cold milk in a short period of time. With an *ad libitum* feeding system the 'little and often' effect of food consumption ensures that the small amount of milk taken in at any one time is rapidly warmed up to body temperature.

There are three important elements if the use of *ad libitum* feeding of acidified milk is to be applied correctly:

(1) Ensuring that calves are consuming enough milk. One of the easiest ways of achieving this is to feed individually the calves for the first 5–10 d by means of a bucket and teat. Any initial difficulty with calves not drinking may be overcome if the milk is fed warm, later the calves readily consume the milk replacer if it is fed at progressively cooler temperatures until cold feeding is practised.
(2) Calves should remain on cold acid milk *ad libitum* for three weeks. At no time should the supply of milk be allowed to run dry as excess consumption may take place when replenished and this may lead to scours.
(3) After three weeks the intake of milk may be reduced by substituting cold water for replacer during the night. This encourages the consumption of concentrate food. Calves must have access to palatable concentrates at all times. These should preferably be sited near the teats where milk is drunk, thus encouraging consumption.

Criticism of *ad libitum* feeding is usually made of the over-consumption of milk, linked to the cost of feeding greater quantities of milk replacer. Commercial farmers claim that although it may cost more to rear a calf, the animal experiences less stress and fewer health problems and as a result is a better calf. Added to this is a reduction in labour coupled with a more flexible work routine.

When reared in groups and machine-fed care should be taken to ensure that individuals know how to suck and that there is no bullying so all have an opportunity to feed. Evenly matched batches and close observation for the first few days are essential.

A pen 1.8 m long and 1 m wide will accommodate an individual calf up to eight weeks old. Less pen space is needed by calves reared in groups, an area of 1.1 m² per calf should be sufficient up to eight weeks, this being increased to 1.5 m² per calf by 12 weeks.

Whatever system of housing is used, warmth, a dry bed and particularly, prevention of draughts are important to the well being of young calves.

The three most commonly practised systems of natural rearing are single suckling, double suckling and multiple suckling.

Single suckling

This is by far the most popular and is carried out mainly on hill and marginal farms, by pedigree beef breeders, on lowland farms with inaccessible grass and, on some arable farms utilising grass as a break crop. In this method the cow rears her own calf with milk being the main source of nutrients. The calf remains with the dam until weaning at approximately six months of age. Cows are normally calved in either autumn or spring. The advantage of autumn calving is that calves are old enough to make full use of the grass in spring and are heavier at weaning in the following autumn. The disadvantage is that cows and calves often have to be housed during some of the winter period. Spring calving has the advantage that cows can be overwintered outdoors, the disadvantages are that the young calf cannot make as good use of spring grass, and spring grass may cause a flush of milk in the cows resulting in the calves scouring. Concentrates should be fed before weaning so eliminating any loss in condition later.

Double suckling

The objective is for each cow to rear two calves together. After calving a second calf should be introduced, and allowed to suckle with the cow's own calf. Particular care should be practised at first until the cow willingly accepts the new calf. Success of the system depends on a supply of newborn calves as required, cows having an adequate supply of milk (often a cull dairy cow or a Friesian cross cow) and good management.

Multiple suckling

This necessitates nurse cows yielding a suitable quantity of milk, say 4000–5000 litres, and a supply of suitable calves. Each suckling period usually lasts about ten weeks. In the first ten weeks of lactation four calves are suckled, in the second ten weeks three new calves are suckled, in the third period another three and for the last period two calves, giving a total of 12 calves reared during the lactation. Cows are often removed from the calves between feeds.

Veal production

For veal production calves capable of high rates of liveweight gain are required, Friesian bull calves normally being used. These were traditionally reared on an all milk diet and slaughtered between 140–160 kg liveweight at 14–18 weeks of age. It should be noted that the recently published welfare codes recommend that calves have access to roughage feed. Friesian bull calves will have a killing-out percentage of 55–60%.

Correct feeding is critical if an adequate return on capital is to be achieved. Milk replacers of the high fat type with at least 15% and up to 25% fat are required for maximum gains. The aim is for a daily liveweight gain to slaughter of 1 kg/d or more. Good husbandry with particular attention being paid to hygiene and observation of animals for ill health is of paramount importance.

Calves were normally housed in buildings with a controlled environment and kept in individual pens on slatted floors. New systems of rearing veal calves involve loose housing in barns with Yorkshire boarding sides, floors bedded in straw and the animals fed *ad libitum* replacer from machines.

MANAGEMENT OF BREEDING STOCK REPLACEMENTS

Rearing policy from weaning depends on two main factors, the season in which the replacement is born (autumn or spring) and the age at which it is to be calved. Age at calving is important as it is related to conception, dystokia, milk yield in first lactation, overall lifetime milk production and the herd calving pattern.

Conception is largely influenced by liveweight, oestrus being associated with weight. *Table 14.10* gives the target liveweights for various breeds.

Table 14.10 Target liveweights (kg)

Breed	Weight at service	Calving weight
Jersey	230	340
Guernsey	260	390
Ayrshire	280	420
Friesian	330	500
Hereford × Friesian	320	500
Aberdeen Angus × Friesian	290	430

Dystokia problems require particular consideration and help to determine the appropriate weight and age at first calving. Calving problems are greater in younger heifers, especially if calved before 22 months of age. Both the size of the calf and of the heifer are important and both are influenced by the feeding of the heifer. Condition scoring is a valuable management aid and heifers should score between 3 and 3.5, six to eight weeks prior to parturition. If the condition of heifers is correct at this time restricted feeding during the last weeks of pregnancy will not affect heifer size but will help to reduce calving difficulties by minimising the growth of the calf.

Choice of bull affects calving difficulties, but the choice is also influenced by the values of the calves as herd replacements. Sires from large beef breeds cause the greatest problems in Friesian heifers. As Friesian bulls tend to cause more problems than either of the two smaller beef breeds, Aberdeen Angus and Hereford, their use on heifers is not recommended unless there is need for a large number of dairy herd replacements. Another advantage in favour of the Hereford and the Angus is that both colour-mark their calves thus adding value. Whatever the choice of breed individual variation within breeds has a great effect on the incidence of difficult calvings.

First lactation yield is lower in heifers calved early, milk yield being closely related to liveweight at calving (*Table 14.11*). In subsequent locations differences in milk yield are minimal and, evidence suggests, because early calved heifers (two years) are kept in the herd to the same age as later calved (three years) they average one lactation more and thus their lifetime yield is increased, as well as providing the extra calf.

Table 14.11 Milk production according to age at first calving

	Age at first calving (months)				
	23–25	26–28	29–31	32–34	35–37
Herd life (lactations)	4.00	4.03	3.84	3.81	3.78
Lifetime yield (kg)	18 747	18 730	17 964	17 991	17 657

Source: Wood (1972)

Table 14.12 Target weights for two year calving

	Weight (kg)
Autumn born heifer	
Birth	35
Turnout (6 months)	150–170
Yarding (6 months)	275–300
Service (15 months)	330
Turnout (18 months)	375–400
Calving (24 months)	500–520
Spring born heifer	
Birth	35
Turnout (3 months)	80–100
Yarding (6 months)	140–160
Turnout (12 months)	250–280
Service (15 months)	330
Yarding (18 months)	400–420
Calving (24 months)	500–520

Source: MLC/MMB joint publication *Rearing Replacments for Beef and Dairy Herds* (Reproduced with permission)

Heifer rearing should be planned so that replacements enter the herd to fit the intended calving pattern. This is one of the main determinants in maintaining a system of block calving. Once the age and the month at which the heifer is to calve have been decided then growth rates to achieve the necessary target liveweights at service and calving can be calculated.

Differences between growth rates for autumn and spring born heifers are largely due to the higher weight gains achieved at grass (*Table 14.12*). Maximum use of grass means economical rearing and the advantage of compensating growth.

Autumn born heifer calving at two years old

Autumn born calves should be weaned at five weeks of age when consuming at least 0.75 kg of concentrates. The concentrate should be palatable and contain 17% crude protein. Between five and 12 weeks calves should be fed concentrates containing 15% crude protein *ad libitum* up to a maximum of 2 kg/d and hay to appetite. Silage as a partial substitute for hay can be fed from six weeks of age. From three months to turnout in the spring the concentrates fed can be cheapened by reducing the crude protein to 14%. Hay or silage should be fed *ad libitum*. By six months each calf will

have consumed 50 kg of early weaning concentrates, about 300 kg of rearing concentrates and 330 kg of hay.

To achieve the desired gains during the first grazing season (6–12 months of age) a continuous supply of good quality grass and the control of parasitic worms are essential. Calves should be vaccinated against husk (lungworm) before turnout unless clean pastures are available; clean pastures being those that have been free of cattle since the previous mid summer. Supplementary feeding of concentrates (1–2 kg/d) after turnout prevents a check in growth which might otherwise occur. A change to clean silage aftermath after dosing against stomach worms in mid summer is generally recommended. At this time cereal feeding (9% protein plus vitamins and minerals) may be introduced when grass becomes scarce or very wet and lush.

The grazing system should be integrated with the conservation area; 0.25 ha/animal can be divided into three sections. One section is grazed until the end of June, whilst the other two are cut. After this time the two conserved areas are grazed and the other one cut, finally all three areas are grazed.

The second winter is best sub-divided into two halves. From yarding until service at 15 months and from service until turnout at 18 months. Cattle benefit from dosing against internal parasites at yarding and from the use of an insecticide against warble fly. A daily liveweight gain of 0.6 kg/d is important to ensure a target service weight of 330 kg and a good conception rate. Conserved forages form the basis of the ration (25 kg silage or 6 kg hay/d) and this is supplemented by 2.5 kg of concentrates (12% protein). The concentrates can be based on barley and the protein content and quantity of concentrate adjusted according to the quality of roughage.

Identification of bulling heifers is often found to be a problem. Careful observation for oestrus, and heat detection devices may prove to be valuable aids. After service the liveweight gain may be reduced to 0.5 kg/d, this allows the concentrate level to be reduced. The overall concentrate use during 12–18 months should be about 250 kg of concentrates and 4–5 t of silage or 1.25 t of hay.

During the second grazing season target weight gain should be about 0.7 kg/d. With good grassland management no concentrates are necessary until steaming-up (generous feeding of cow pre-calving) takes place in late summer. Excess steaming-up should be avoided to prevent overstocking of the udder prior to calving. The best guide to the level of feeding during late summer is body condition score. This should be between 3 and 3.5 six weeks before calving.

Spring born heifer calving at two years old

The management and feeding of the newborn calf until weaning is the same as for autumn born calves. As young calves are too small to make efficient use of grass, concentrate (16% crude protein) feeding should be continued after turnout. Hay should also be available during this time (*Table 14.13*). If growth rate from grass alone falls below 0.5 kg/d concentrate feeding should be restarted. The target stocking rate should be about ten calves/ha. This may be achieved by a similar system of grassland management as for autumn born calves. It is important to ensure that grazing is clean, as very young claves are extremely susceptible to parasites. At yarding calves should weigh between 140 and

Table 14.13 Approximate feed quantities used

	Autumn born	Spring born
Milk substitute (kg)	13	13
Concentrates (kg)	700	920
Silage (t)	5.5	7.25
and hay (kg)	100	
or hay (t)	1.9	2.0

160 kg and have consumed 100 kg of concentrates and 125 kg of hay from weaning.

From yarding at six months to turnout in the following spring at 12 months a growth rate of 0.6 kg/d should be maintained. At yarding animals should be dosed against internal parasites and may be dressed against warble fly during the winter. The basis of the ration will be either silage or hay, the amounts fed being 18 kg or 5 kg/d, respectively. The level of concentrate feeding depends on the quality and quantity of the conserved forages, a guide being 2–3 kg/d of 16% crude protein. If the growth rates are not maintained the target service weight of 330 kg at 15 months of age in the spring will not be achieved and conception rates will be poor.

The second grazing season growth rates from turnout to service will be 0.8 kg/d. After service, target growth rates can be reduced to 0.6 kg/d. Supplementation with mineralised cereals (barley) should commence in early autumn.

During the second winter the aim should be to produce a Friesian heifer calving down at about 500 kg. This can be achieved by feeding 2 kg/d of concentrates and up to 30 kg of silage. This will take approximately 4.5 t of silage or 1.25 t of hay and 360 kg of concentrates. Steaming-up prior to calving can be practised and the heifer's body condition score should be between 3 and 3.5 six weeks prior to calving.

Heifer replacement rearing enterprises compete for resources with milk producing animals. The two major economies that can be made in the resources required in the production of replacement heifers are, the reduction in the number required and reduction of the age at which they calve. Both these enable considerable economies to be made in land, labour and capital invested in livestock and buildings. However, young heifers grown well enough to calve at two years of age need a high plane of nutrition. This necessitates the feeding of greater quantities of concentrates thereby increasing the cost of concentrate feed for animals calving at two rather than three years of age. As far as total feed cost is concerned heifers calving around two years of age cost nearly as much as those calving a year older. Thus intensification of heifer rearing with a greater reliance on concentrates has important repercussions in that the land and labour saved can be made available for milk production or other more profitable enterprises.

In addition to these direct savings in resources the heifer calving at the younger age will have a longer herd life with a greater total lifetime milk production (*Table 14.14*).

These higher lifetime yields make up for a slightly lower first lactation yield; 4300 kg for a two year old heifer compared with 4500 kg for a three year old.

As far as total feed is concerned it appears that heifers calving at two years of age costs nearly as much as those calving a year older. The financial saving in grazing and forages for the younger calving animals is substantially eroded by the need for a higher concentrate input to maintain growth rates. This is of especial importance during

Table 14.14 Effect of age at calving on calf growth (Percentage difference in 400 d weights compared with calf out of a five year old cow)

		Percentage difference
Heifer		
First calving at 2 years		−8
	2½ years	−5
Cow		
Second calf		−3
Third calf		−2
Fourth calf		0

Source: MLC/MMB joint publication *Rearing Replacments for Beef and Dairy Herds* (Reproduced with permission)

the winter periods unless the diet comprises ample good quality forage.

Even if there is no great saving in the cost of producing younger calving heifers there is a saving in capital investment in young stock because of the quicker turnover in animals and a reduction in the number of followers kept. Indirect benefits may result from the successful adoption of such a system of rearing, the intensification required having repercussions throughout the whole dairy enterprise, especially with respect to better grass production, conservation and utilisation. The effects that the more rigorous discipline involved in rearing heifers to calve at two years imposes are likely to extend right across the rearing enterprise and benefit the whole farm economy.

In beef suckler herds calves out of heifers have a slightly slower growth rate than calves out of five year old cows (*Table 14.14*). These slower calf growth rates are largely due to the early calved beef heifers having a slightly lower milk yield. Over their lifetime beef heifers calved first at two years of age produce more calves over their lifetime than those calving at three years.

FEEDING DAIRY AND BEEF CATTLE

Details of the characteristics of the main feedstuffs used by cattle may be found in Chapter 13, and discussion of the theory of animal nutrition in Chapter 12. This section outlines the compilation of rations for dairy and beef cattle.

Breeding stock

One of the main constraints involved in the rationing of dairy cows is Dry Matter Intake (DMI). A simple, yet relatively accurate estimate of DMI is given by the equation

$$DMI = 0.025W + 0.1 \text{ Yield}$$

where W = bodyweight in kg.

For cows in the first six weeks of lactation this prediction should be reduced by 2–3 kg. A more realistic prediction is given by:

$$DMI = [135 \text{ g/kgW}^{0.75} + (0.2 \times [Y - 16.4])] \times CF$$

CF is a conversion factor which takes account of the variation in appetite as lactation progresses, and is taken as approximately 0.8, 0.95, 1.05, 1.08 and 1.10 for each of the first five months of lactation. Intake will be increased by up to 5% by feeding a variety of feeds, and by feeding the concentrate component on a 'little and often' basis, and will be decreased by 5% by silage with an ammonia-N content above 10%, and by 10% if ammonia-N is above 15%.

Energy requirements depend on liveweight, milk yield, milk quality and stage of pregnancy. Requirement for maintenance is given by the formula $ME_M = 8.3 + (0.091 \, W)$ and for production of 1 kg milk by $ME_L = [(0.376 \, BF) + (0.209 \, Prot.) + 0.946] \times 1.69$.

A summary of requirements is given in *Tables 14.15* and *14.16*. In early lactation it is often impossible to meet total energy requirements within the constraints imposed by limited DMI. In such circumstances backfat may be utilised. Each 1 kg of fat used replaces 28 MJ of dietary energy. Provided the cow is in good body condition at calving (condition score 3–3.5), the loss of up to 30 kg backfat over this period is acceptable. Utilisation of backfat in this way allows an increase in the forage:concentrate ratio of diet and may well be economically desirable. At some stage it will be necessary to replace any backfat utilised. This is best done during late lactation rather than in the dry period and will require an input of 34 MJ for each 1 kg backfat gained (see Chapter 13). Fetal growth increases dramatically during the later stages of pregnancy: requirement for fetal growth is calculated as $M + 1.13 \, e^{0.0106t}$ (where t = number of days pregnant and $e = 2.718$). In practice, little accuracy is lost if additions of 10, 15 and 20 MJ/d are made for months 7, 8 and 9 respectively.

Protein requirements for breeding females may be

Table 14.15 Computed maintenance allowances for dairy and beef cattle (MJ/d)

Liveweight (kg)	MJ
100	17.4
200	26.5
300	35.6
400	44.7
500	53.8
600	62.9
700	72.0

(Based on $8.3 + 0.091 \, W$)

Table 14.16 Computed requirements (MJ) for 1 litre of milk of varying composition

Protein (g/kg)	Fat content (g/kg)						
	32	34	36	38	40	42	44
28	4.61	4.74	4.87	5.00	5.13	5.26	5.39
30	4.68	4.81	4.94	5.07	5.20	5.33	5.46
32	4.75	4.88	5.01	5.14	5.27	5.40	5.53
34	4.82	4.95	5.08	5.21	5.34	5.47	5.60
36	4.89	5.02	5.15	5.28	5.41	5.54	5.67
38	4.96	5.09	5.22	5.35	5.48	5.61	5.74

(Based on $1.694 \, (0.0376 \, BF + 0.0209 \, Prot. + 0.946)$)

Table 14.17 Daily DCP requirements for lactating cows (g)

Liveweight (kg)	Milk yield (kg)				
	0	10	20	30	40
350	225	725	1225	1725	2225
400	245	745	1245	1745	2245
450	265	765	1265	1765	2265
500	285	785	1285	1785	2285
550	305	805	1305	1805	2305
600	325	825	1325	1825	2325
650	345	845	1345	1845	2345
700	365	865	1365	1865	2365

expressed either in terms of Digestible Crude Protein (DCP) or Rumen Degradable and Undegradable Protein (RDP and UDP). These requirements are based on different concepts and are not interchangeable. For most circumstances DCP requirement is based on a maintenance requirement of some 300–350 g/d plus 50 g/kg milk produced (*Table 14.17*). An additional amount is added for fetal growth. This should be approximately 40, 90 and 230 g/d for months 7, 8 and 9 of pregnancy.

Calculation of requirements for RDP and UDP are less straight forward, depending as they do on the amount of energy entering the rumen. RDP requirement will be 7.8 × ME of the diet. This is the maximum amount of RDP which the micro-organisms of the rumen can capture. Any RDP or non-protein nitrogen (NPN) in excess of this amount will be lost, passing into the bloodstream in the form of ammonia and possibly causing metabolic disturbance. An excess of more than 250 g/d RDP should be avoided. All animals have a requirement for Tissue Nitrogen (TN). This may be met either from microbial sources (i.e. from captured RDP) or from dietary UDP. Requirements for TN average 12 g/d for maintenance and 4.8 g/litre of milk (5.7/litre for Jerseys). Where backfat is being utilised 16 g TN/kg fat are saved: to replace fat increases demand by 24 g TN/kg. Total TN requirement is compared with the ability of microbes to meet that requirement. Microbial Nitrogen to the tissues (TMN) is given as, 0.53 × ME of diet: where TN < TMN there will be no demand for dietary UDP. Where TN > TMN there will be a demand for dietary UDP, calculated as $UDP = [(1.91 \, TN) - ME] \times 6.25$.

From the above it is evident that there is a significant interaction between energy input and protein requirement. Where, for example, backfat is being utilised as an energy source in early lactation, there is a reduced input of dietary energy. This reduces the rumen microflora's ability to capture RDP and increases demand for dietary UDP. Flat rate feeding of concentrates, especially at low levels of input, also make a response of dietary UDP more likely.

Having established the requirements of the animal, and knowing the analyses of all available foodstuffs, a suitable ration may now be formulated. Amounts of ingredients to be included will depend on availability, cost, palatability, form, but most formulations will aim at maximising forage input and minimising use of purchased foods. The overriding constraint in this respect is the relationship between the animal's potential DMI and its dietary energy needs. This relationship is known as the minimum energy density of the diet (minimum M/D = Total dietary energy (ME)/Total DMI).

As minimum M/D goes up so the maximum amount of forage possible in the total diet goes down. In a simple, two ingredient, situation the maximum forage DM input is given as:

$$\text{Max. forage DM} = \frac{\text{DMI (ME of concentrate} - \text{M/D)}}{\text{ME of concentrate} - \text{ME of forage}}$$

It is obvious from this equation that increases in potential DMI, or decreases in dietary energy demand through backfat mobilisation, can significantly increase forage utilisation. Conversely, if concentrate input is restricted and total appetite is limited, for example in the first few weeks of lactation, backfat loss is inevitable. In this regard, few cows are able to eat more than about 12 kg forage DM/d, irrespective of amount of concentrate fed.

Growing/finishing cattle

The metabolisable energy system is usable in formulating rations for mature/lactating cattle because the efficiencies with which such cattle use energy (0.72 for maintenance; 0.62 for production of milk) are constant irrespective of ration. In beef cattle efficiency of energy use for maintenance is constant (0.72), but that of production (k_g) varies with diet according to the formula:

$$k_g = 0.0435 \, \text{M/D}$$

(where M/D = energy concentration of the diet).

In situations where the diet is known, therefore, the ME system may be used to predict the likely performance of animals on that diet. Formulation of a ration requires the use of a different approach, namely the Net Energy (NE) system.

Predicting performance from a known ration

Once the energy (ME) provided by a ration is known, predicting performance depends on calculating:

(1) the energy needed for maintenance (M_m)
(2) the energy used for gain (E_g)
(3) the energy required for 1 kg of gain (EV_g)

The gain predicted (d) will then be the energy used (E_g) divided by the energy required for 1 kg gain (EV_g) Namely

$$M_m = 8.3 + 0.091 \, \text{W} \tag{1}$$

$$E_g = \frac{(\text{Total ME} - M_m \times k_g)}{1.05*} \tag{2}$$

$$EV_g = 6.28 + (0.3 \, E_g + (0.0188 \, \text{W}) \tag{3}$$

$$DLWG = E_g / EV_g \tag{4}$$

*A 5% allowance for safety in prediction may be added.

The diet must then be checked to ensure that it contains sufficient protein to sustain this rate of gain (*Table 14.18*). This prediction takes no account of breed or sex effects and should be taken only as a guide to performance.

Formulating a diet to meet target daily gains

Formulating a ration to meet required daily gains necessitates the use of Net Energy values. This does not demand knowledge of efficiency with which the animal uses each feed for production but does require estimation of the average efficiency of utilisation of a feedstuff for maintenace and gain (k_{mg}). This estimate is derived from a value known as Animal Production Level (APL), which expresses total energy requirement ($E_m + E_g$) proportional to the requirement for maintenance (E_m). As APL increases (i.e. as the proportion of the energy used for gain increases) so k_{mg} will fall. Since the energy value of feedstuffs is expressed in terms of ME, it is necessary to calculate the Net Energy (NE_{mg}) value of each feedstuff separately for each diet and each animal group, and to use NE values in compiling the ration. The equations used in this process are as follows:

$$E_m = 5.67 + 0.061 \, \text{W}$$

$$E_g = 1.05 \times \left[\frac{\text{DLWG} (6.28 + 0.0188 \, \text{W})}{(1 - 0.3 \, \text{DLWG})} \right]$$

$$APL = \frac{E_m + E_g}{E_m}$$

$$k_{mg} = \frac{\text{ME of feed} \times \text{APL}}{1.39 \, \text{ME of feed} + 23 \, (\text{APL} - 1)}$$

$$NE = \text{ME of feed} \times k_{mg}$$

Having calculated all these values it is then possible to compile a ration, using a modification of the equation used for compiling dairy cow rations:

$$\text{Max. forage DM} = \frac{\text{DMI (NE compound} - \text{N/D)}}{\text{NE compound} - \text{NE forage}}$$

Once again these equations do not take account of breed and sex effects and modification to diets may be necessary in the light of these factors. Separate consideration must also be given to requirements for protein, vitamins and minerals.

Table 14.18 Computed values for $E_m + E_g$ (MJ NE)

Gain required (kg/d)	Liveweight (kg)				
	100	*200*	*300*	*400*	*500*
0	12.4	18.8	25.2	31.6	38.0
0.25	14.7	21.6	28.6	35.5	42.5
0.50	17.4	25.0	32.6	40.1	47.7
0.75	20.7	29.0	37.3	45.6	53.9
1.00	24.6	33.9	43.1	52.3	61.5

Table 14.19 Computed values for APL ($E_m + E_g / E_m$)

Gain required (kg/d)	Liveweight (kg)				
	100	*200*	*300*	*400*	*500*
0	1.00	1.00	1.00	1.00	1.00
0.25	1.19	1.15	1.13	1.12	1.12
0.5	1.41	1.33	1.29	1.27	1.26
0.75	1.67	1.54	1.48	1.44	1.42
1.00	1.99	1.80	1.71	1.66	1.62

Table 14.20 Computed NE values (NE$_{mp}$) of feedstuffs

APL	ME of feed (MJ)					
	8.0	9.0	10.0	11.0	12.0	13.0
1	5.8	6.5	7.2	7.9	8.6	9.4
1.2	4.9	5.7	6.5	7.3	8.1	8.9
1.4	4.4	5.2	6.1	6.9	7.8	8.7
1.6	4.1	4.9	5.8	6.7	7.6	8.5
1.8	3.9	4.7	5.6	6.5	7.4	8.3
2.0	3.8	4.6	5.4	6.3	7.3	8.2

(Based on MEF $\times$ k_{mg})

Where $k_{mg} = \dfrac{\text{MEF} \times \text{APL}}{1.39\,\text{MEF} - 23\,(\text{APL} - 1)}$

Appetite (DMI) in growing beef cattle is generally lower than that of mature cows, falling linearly from 0.029 W at 100 kg liveweight to 0.021 W at 500 kg liveweight. Some values from the equations above are given in *Tables 14.19* and *14.20*.

BEEF PRODUCTION

The EEC beef industry is dependent on dairy farming. Agricultural policy restricting milk output will have a profound effect on the beef sector in both the short and the long term. In the short term dairy farmers will react to milk quotas by reducing herd numbers, and as more cows are culled this will increase the supply of cow beef. In addition less young stock will be retained for breeding. In the long term the number of calves available either as surplus bull calves or dairy cross beef calves, will be reduced. It is forecast that the dairy breeding herd will have fallen by 6% or 1.6 million head by the end of 1988.

Recent developments in beef production systems, using imported breeds, better grassland management and more efficient use of concentrate feeds have resulted in greater liveweight gains and a move from grass to yard finishing.

Baker (1975) suggests four factors need to be considered when choosing a system for beef production.

(1) Financial resources. These include cash flow requirements and the capital availability.
(2) Physical resources, e.g. the area and quality of grassland available, field structure and the availability of water.
(3) Date of birth of calves – autumn or spring.
(4) Type of cattle – pure dairy, dual purpose or beef, or crosses.

Growth is usually measured by the change in liveweight of the animal. As this includes changes in the weight of feet, head, hide and the internal organs, including the content of the gut, besides carcass tissue it is not always a reliable indicator of the final amount of saleable beef. Liveweight gain follows a characteristic sigmoid curve (*see* Chapter 12). The rate of growth being influenced by nutrition, breed and sex.

An important phenomenon in some beef systems is compensatory growth. When animals have been fed on low quantities and/or low quality feed, growth will slow down or cease – this is known as a store period – when full levels of feeding are resumed these animals will eventually catch up animals that have been kept on full feeding levels; this is compensatory growth. It is exploited in beef systems so winter feeding can be kept at as low cost as possible and when spring feeding is started compensatory growth takes place, taking advantage of relatively cheap grazed grass.

The development of the animal varies as different tissues mature at different rates; nervous tissue first, followed by bone, muscle and finally fat (for a fuller explanation *see* Chapter 12). If at any time the energy intake of the animal is in excess of that needed for the growth of earlier maturing tissues, bone and muscle, then fat will be deposited. In time the fattening phase is reached when fat grows fastest. Fat is deposited in the body in a certain order: subperitoneal fat (KKCF, kidney knob and channel fat) first, then intermuscular fat, subcutaneous fat and, finally, intramuscular fat (marbling fat).

In the later maturing breeds, which have characteristically high growth rates, the fat deposition phase occurs at a greater age and weight than in the early maturing breeds fed on similar diets. Sex effects are also important – bulls grow faster than steers and these in turn grow faster than heifers. In parallel, bulls are later maturing than steers and steers later than heifers. As a result heifers of early maturing breeds quickly reach an age when they are depositing fat and are slaughtered at a younger age. Bulls of the late maturing breeds need high levels of feeding before fattening commences (*Figure 14.5*). In practice, different types of cattle are managed differently and are slaughtered at different levels of carcass fat cover to suit the requirements of different sections of the meat trade. An understanding of the relationship between production system and cattle type is fundamental in planning production for a particular market.

The amount of fat in the animal influences killing-out percentage. As the animal gets heavier and fatter the killing-out percentage increases, killing-out percentage (sometimes called dressing percentage) being the yield of carcass from a given liveweight. Systems of describing carcasses have been developed in many countries. The purpose of beef carcass classification is to describe carcasses by their commercially

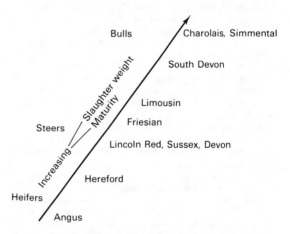

Figure 14.5 Ranking of cattle of different breeds and types in terms of slaughter weight and rate of maturity. (From MLC (1982) Reproduced with permission)

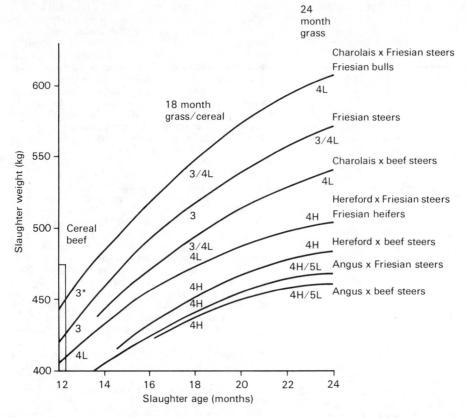

* Fat class on scale 1 (leanest), 2, 3, 4L, 4H, 5L, 5H (fattest)

Figure 14.6 Average liveweight, slaughter age and carcass fat for different breeds and types of cattle. (After MLC (1982) Reproduced with permission)

important characteristics. The development of the classification system sought to improve the efficiency of marketing throughout the whole industry. The description enables wholesalers and retailers to define their requirements. Producers should be able to obtain higher returns by producing animals to match these market requirements. The result is a flow of information from the consumer through the retailer to the farmer. Producers can then assess the economics of providing one type of carcass from others and plan their systems of management accordingly (*Figure 14.6*).

The scheme is based on a grid system which describes fatness on the horizontal scale and conformation on the vertical scale (*Table 14.21*). *Table 14.22* gives the average composition of carcasses in each of the EEC fat classes.

Fatness and conformation are determined by visual appraisal. Conformation relates to shape and takes into account carcass thickness, blockiness and the fullness of the round. Fatness is always referred to before conformation, e.g. a carcass falling in fat class 2 and conformation R would be described as 2R. The scheme also includes information on weight, sex type and sometimes age. The MLC undertake many demonstrations of the system throughout the UK every year and anyone interested should contact the nearest MLC Regional Fatstock Offices.

Selecting cattle for slaughter

The two most important aids to the selection of cattle for slaughter are weight and fat cover (finish). Knowledge of the weight of animals and their previous performance means that decisions can be based on finish. Once animals reach their minimum weight the decision to sell should be based on finish. Excess fat is wasteful and has to be trimmed off carcasses besides which feed turned into unwanted fat is wasted and is very expensive. The best way of assessing carcass fat cover is to handle the cattle.

The diagram (*Figure 14.7*) shows the five key points for handling:

(1) over the ribs nearest to the hindquarters
(2) the transverse proceses of the spine
(3) over the pin bones on either side of the tailhead and the tailhead itself
(4) the shoulder blade
(5) the cod in a steer or the udder in a heifer.

If producers are doubtful about the fat levels in animals they have selected for slaughter they should check their classification reports or follow carcasses through the abbattoir.

Table 14.21 Percentage distribution of Great Britain clean beef carcasses in the EEC classification grid (Jan–Dec 1985)

| | | Fat class | | | | | | | |
| | | Leanest | | | | | Fattest | | |
		1	2	3	4L	4H	5L	5H	Overall
Very good	E			0.1	0.1	0.1			0.3
C O N F O R M A T I O N	U+		0.1	0.7	1.4	0.9	0.3		3.4
	−U		0.5	2.5	5.3	3.6	0.9	0.1	12.9
	R		1.0	7.4	17.2	10.7	2.2	0.3	38.8
	O+	0.1	1.3	7.9	15.2	8.1	1.6	0.3	34.5
	−O	0.1	0.8	2.8	3.4	1.3	0.3	0.1	8.8
	P+	0.1	0.2	0.3	0.2	0.1			0.9
Very poor	−P	0.1	0.1	0.1	0.1				0.4
overall		0.4	4.0	21.8	42.9	24.8	5.3	0.8	

Source: MLC (1986) *Beef Yearbook*. (Reproduced with permission)

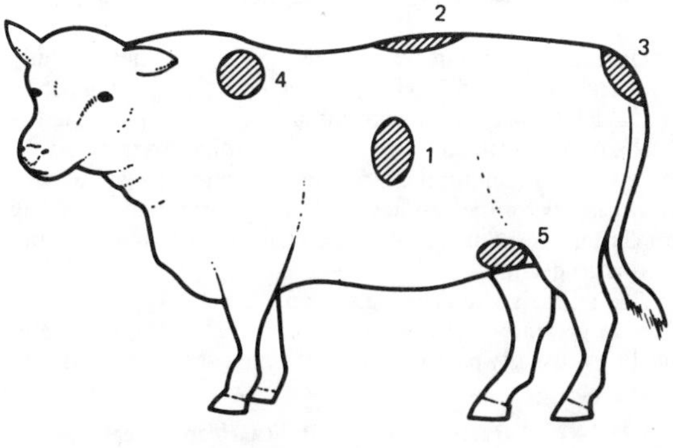

Figure 14.7 Key handling points

An EEC directive banning the sale of carcasses, from animals in which hormone growth promoters were used, commenced 1 January 1988. The British Goverment implemented a ban on the use of hormone growth promoters from 1 December 1986. This ensures that Britain will be in line with other member states when the EEC ban takes effect.

At the time of writing the use of feed additive growth promoters, e.g. Romensin, Avotan, is still permitted.

BEEF PRODUCTION SYSTEMS

Cereal beef

Cereal beef, more commonly known as barley beef, is so named because of the system of production. Dairy bred animals are fed on an all concentrate ration – usually based

Table 14.22 Average composition and saleable meat yield of R conformation[a] carcasses by fat class

| Carcass composition | Fat class | | | | |
| | 1 and 2 | 3 | 4L | 4H | 5L |
	Percentage of carcass weight				
Lean	69	64	61	58	53
Fat	12	18	22	25	31
Bone	19	18	17	17	16
Saleable meat					
Meat yield	74	72	71	70	69
Fat and lean trim	7	10	12	13	15

[a] Most common conformation class; it accounts for about 40% of classified carcasses.
Source: MLC (1982) *Selecting Cattle for Slaughter*. (Reproduced with permission)

on barley, this encourages rapid liveweight gain and animals are slaughtered at 10–12 months of age weighing 430–470 kg. Killing-out percentage is usually 54–56% thus yielding a carcass of 230–265 kg.

Calves are reared on an early weaning system (*see* earlier, p. 393). Weaning takes place at five weeks of age onto a concentrate ration containing 16% crude protein. At 10–12 weeks of age a diet usually based on rolled barley supplemented with protein, vitamins and minerals giving a mix of 145 crude protein is fed *ad libitum*. From six to seven months the protein level of the mix can be reduced to 12% thus reducing costs (*Table 14.23*).

Table 14.23 Liveweight and protein levels (% CP) of diet

Period	Liveweight (kg)	% CP in diet
5–12 weeks	100	16
3–6 months	250	14
6 months–slaughter	430–470	12

The system is best suited to late maturing animals as early maturing animals become overfat at light weights. The Friesian is most commonly used because of its ready supply from the dairy herd. As bulls are later maturing than steers and have better food conversion efficiencies (FCE) there has been an increasing number of entires kept on this system. Keeping bulls is facilitated by animals being housed throughout. Their faster and leaner growth, compared with steers, means that bulls can be slaughtered at the same age (10–12 months) weighing 470 kg. For a comparision of bulls and steers *see Table 14.24*.

Practical problems that may be encountered include respiratory diseases and bloat. With animals housed throughout building design is of particular importance both to the handling and physical management of stock and to reducing respiratory diseases. Good ventilation and draught-free buildings significantly reduce the incidence of pneumonia. Bloat or rumen tympany is a greater problem when animals are kept on slats rather than in bedded pens where they can consume roughage. If 1 kg/d of hay or barley straw is fed this normally prevents bloat.

Table 14.24 Cereal beef targets (Friesians)

Period	Gain (kg/d)		Economy of gain kg feed:kg gain	
	Bulls	*Steers*	*Bulls*	*Steers*
0–5 weeks	0.45	0.45		
6–12 weeks	1.0	0.9	2.7	2.8
3–6 months	1.3	1.2	4.0	4.3
6– slaughter	1.4	1.3	6.1	6.6
Overall	1.2	1.1	4.8	5.5
Slaughter age (d)	338	345		
Slaughter weight (kg)	445	404		
Carcass weight (kg)	231	211		
Feed inputs (kg)				
Milk powder	13			
Calf concentrate	155			
Protein supplement	225			
Barley	1500			

Source: Allen and Kilkenny (1980)

Cereal beef production is sensitive to the relative prices of calves, barley and beef. Its main advantages are that it makes no direct use of land as all feeds can be bought-in and it is not seasonal, hence an even cash flow can be established once a regular throughput of animals is established.

Maize silage beef (*Tables 14.25 and 14.26*).

Continental producers have developed this system of fattening dairy bred calves and it is now a well established system for bull beef production. British interest has been aroused as maize silage systems are fully mechanised and the crop provides a useful arable break crop.

Calves are reared to 12 weeks in the same way as for cereal beef. Maize silage is then introduced and fed *ad libitum*. A protein supplement must be fed to bring the overall crude protein of the diet to 16%. The main problem with maize silage being its low protein content.

As with the cereal beef the system favours animals of late maturing type.

Maize silage can also be used for finishing suckled calves. It is important to ensure that the silage has a high dry matter content (i.e. 25% DM). Silages with low dry matter levels will result in lower feed intake and lower liveweight gains.

Table 14.25 Maize silage beef production (Friesians)

	Bulls	*Steers*
Gain (kg/d)	1.0	0.9
Slaughter weight (kg)	490	445
Slaughter age (months)	14	14
kg feed DM/kg gain	5.6	6.0
Protein concentrate (kg)	500	500
Maize silage DM (t)	1.75	1.65
Stocking rate (cattle/ha assuming 10 t DM/ha)	5.7	6.1

Source: MLC (1978) *Beef Improvement Services* (Reproduced with permission)

Table 14.26 Slaughter weights for maize silage beef

	Slaughter weight (kg)	
	Bulls	*Steers*
Friesian	480–500	430–450
Charolais, Simmental, Blonde d'Aquitaine, Limousin, South Devon × Friesian	510–540	460–480
Devon, Lincoln Red, and Sussex × Friesian	480–500	430–450
Hereford × Friesian	450–480	400–430
Aberdeen Angus × Friesian	420–450	380–410

Source: Alan and Kilkenny (1980)

18 month grass/cereal beef (*Tables 14.27 and 14.28*)

Autumn born calves are reared on a conventional early weaning system.

Friesian and Hereford × Friesian calves are commonly used. The autumn born steer should weigh 180–190 kg when turned out in the spring. Calves born during the winter and early spring weigh less and have weight gains at grass. The daily gain of heifer calves is 20% poorer than steers and they finish at lighter weights. Bulls grow more rapidly and produce carcasses about 10% heavier than steers (Baker, 1975).

During the grazing season the aim is to achieve a daily liveweight gain of 0.7–0.8 kg. To achieve this grass must be managed to provide a continuous supply of high quality herbage. Supplementary feeding with rolled barley may be practised to maintain the target growth rate, this will usually be necessary for the first few weeks after turnout in the spring and after late August when herbage quality and quantity begins to decline.

The animals should weigh about 320–350 kg at yarding. The target rate of liveweight gain is 0.9–1.0 kg/d. Winter feeding is usually based on silage which is supplemented with

Table 14.27 Targets for 18 month grass/cereal beef

	Performance		Feeds	
	Weight (kg)	*Gain (kg/d)*	*Concentrates (kg)*	*Silage (t)*
At start	50		406	1.0
		0.75		
At turnout	197		76	–
		0.77		
At yarding	336		456	4.1
		0.95		
At slaughter	494			
Stocking				
N fertiliser (kg/ha)	264			
Stocking rate (cattle/ha)	4.32			
Grazing gain (kg/ha)	1098			

After: MLC (1986) *Beef Yearbook* (Reproduced with permission)

Table 14.28 Comparison of Friesian steers, Hereford × Friesian steers and Hereford × Friesian heifers in an 18 month beef system

	Friesian steers	Hereford × Friesian	
		Steers	Heifers
Weight (kg)			
At start	50	50	41
At turnout	208	212	213
At yarding	357	352	330
At slaughter	522	484	410
Daily gain (kg)			
First winter	0.78	0.69	0.66
Summer grazing	0.82	0.77	0.66
Second winter	0.97	1.09	0.65
Feeds			
Concentrates (kg)			
First winter	400	380	310
Summer grazing	80	72	50
Second winter	625	340	190
Silage (t)			
First winter	1.1	1.5	1.3
Second winter	4.5	3.7	2.1

After: MLC (1986) *Beef Yearbook*. (Reproduced with permission)

mineralised rolled barley. The amount of barley fed will depend on the quality of the silage and the desired weight gains. Cattle are slaughtered at 15–20 months of age weighing 400–520 kg.

Grass silage beef

Grass silage beef is also known as The Rosemaund Beef System or Storage Beef. The system normally uses dairy bred calves – Hereford × Friesian or Friesian bull calves – fed on grass silage to appetite with rationed compound feeds and housing the animal throughout. Animals are slaughtered at 11–15 months of age.

Table 14.29 Feeding bulls from weaning to sale (storage beef)

Time from calf arrival (kg)	Liveweight (kg)	Feeding
5–6	65–75	Weaned: *ad libitum* early weaning concentrate
6–8	75–85	Calf concentrate replaced by rearing compound fed to appetite
8–12	85–105	Silage first offered to appetite
12–16	105–130	Compound gradually reduced to 2 kg/d
16–sale	130–slaughter	Silage offered to appetite plus 2 kg compound/d rising to 4 kg if needed

Source: Hardy and Meadowcroft (1986)

Calves are reared on an early weaning system, and are gradually changed from the early weaning ration to a rearing compound. High quality silage is introduced at this time. At 12 weeks the rearing compound is restricted to 2 kg/animal/d and this level of compound is fed until slaughter. If silage quality is less good the level of compound may be increased to 4 kg/animal/d (*see Table 14.29* for further details).

The system depends on the ability to obtain high growth rates from young animals; it is therefore important that the quality and quantity of silage is adequate. Rosemaund results show that when grass is cut at a 'D'-value of 70 or over satisfactory liveweight gains can be achieved.

A comparison of results from units using Hereford × Friesian and Friesian calves is given in *Table 14.30*.

Stocking rates are dependent on silage yields and tonnes of silage used per animal. High stocking rates lead to a high working capital requirement and hence high interest charges.

Table 14.30 Performance of Hereford × Friesian bulls, Friesian bulls and Friesian steers on grass silage beef system

	Hereford × Friesian bulls	Friesian bulls	Friesian steers
Performance			
Days to slaughter	393	364	395
Mortality	3	3	2
Weight at start (kg)	105	110	111
Weight at slaughter (kg)	489	483	498
Daily gain from 12 weeks (kg)	0.98	1.05	0.98
Feeds			
Concentrates (kg)	824	954	976
Silage (t)	5.4	5.7	5.5
Stocking rate (head/ha)	7.87	7.44	6.88

After: MLC (1986) *Beef Yearbook*. (Reproduced with permission)

Suckled calf production

Suckled calf production is practised under a variety of environmental conditions, from hill land to lowland. Beef suckler cows vary from the larger and milkier types to the smaller and hardier types.

The number of purebred beef calves has declined, crossbreds becoming more prevalent, and this introduces the heterosis effects of greater fertility and calf viability.

The choice of season of calving is affected by the availability of buildings and winter grazing, the bulk of calvings taking place in either the autumn or late winter/early spring. A short calving season enables the cows to be fed as one group without over- or under-feeding of individuals and also gives a more uniform batch of calves to be sold or fattened.

The use of condition scoring has added some precision to the management of suckler herds. The body condition of breeding cows at service and at calving is particularly important in reducing the incidence of barren cows.

Herds that have a condition score of above 2, with the optimum being 2.5–3, have the better calving intervals and the greatest number of calves reared (*Tables 14.31* and *14.32*).

Suckler calf production usually produces calves for sale post weaning as stores, this is especially true of hill and

Table 14.31 Relationship between body condition and reproductive performance (beef cows)

Cow scores	Calving interval (d)	Herd average	Calves weaned per 100 cows served
1–2	418	1–2	78
2	382	2	85
2–3	364	2–3	95
3+	358	3+	93

Scores on the scale 1 (very thin)–5 (very fat)

Source: MLC (1978) *Beef Improvement Services*. (Reproduced with permission)

Table 14.32 Body condition score targets

Stage of production	Target score	
	Autumn calving	Spring calving
Mating	2.5	2.5
Mid pregnancy	2.0	3.0
Calving	3.0	2.5

upland producers where supplies of winter feed are scarce. These will then be fattened by lowland farmers. Lowland suckler herds are more likely to carry their calves through to slaughter (*(Table 14.33*).

Table 14.33 Performance of suckler herds 1985

	Lowland	Upland	Hill
Cow			
Calving spread for 90% of calving (weeks)	12	13	11
Calves born live	92	93	90
Calves purchased	2	6	3
Calf mortality	4	5	2
Calves reared	90	94	91
Calves			
Age at sale/transfer (d)	284	314	278
Weight at sale/transfer (kg)	300	307	265
Daily gain	0.92	0.85	0.79
Feeds			
Cow concentrates (kg)	123	117	141
Calf concentrates (kg)	133	141	79
Silage (t)	3.7	5.3	4.9
Feeding straw (t)	0.5	0.3	0.6
Stocking			
Stocking rate (cows/ha)	2.15	1.86	0.82
N fertiliser (kg/ha)	160	141	55

After: MLC (1986) *Beef Yearbook*. (Reproduced with permission)

Season of calving tends to determine the production system. Autumn calving is most popular as calves sold in the autumn store sales are older and heavier. Autumn calving however involves more buildings and a greater requirement for winter feed.

Grass finishing of stores

Suckler bred stores are purchased in the autumn sales and may be overwintered before being finished off grass the following summer (*Table 14.34*).

Alternatively the stores may be finished over winter for sale the following spring (*Table 14.35*).

Table 14.34 Targets for overwintering and grass finishing stores (1984–85)

Performance	
grazing period (d)	139
Weight (kg)	
at purchase	240
at turnout	348
at slaughter	458
Daily gain (kg)	
over winter	0.51
over grazing	0.79
grazing gain (kg/ha)	769
Feeds	
Concentrates (kg)	
over winter	325
at grazing	43
silage (t)	3.1
N fertiliser (kg/ha)	190
Stocking rate (cattle/ha)	4.14

After: MLC (1986) *Beef Yearbook*. (Reproduced with permission)

Table 14.35 Targets for winter finishing stores 1985–86

Performance	
Feeding period (d)	164
Weight at start (kg)	334
Weight at slaughter (kg)	490
Daily gain (kg)	0.95
Feeds	
Concentrates (kg)	296
Silage (t)	3.2
Feeding straw (t)	0.2
Stocking rate (cattle/ha)	10.35

After: MLC (1986) *Beef Yearbook*. (Reproduced with permission)

DAIRYING

Milk production is the largest enterprise in UK agriculture with an annual net sum received by producers of some £2293 million (*Table 14.1*), this accounts for about 19% of the total value of all agricultural output.

Lactation curves

If the milk yield of cows is plotted against time a graph of a lactation curve is produced. The standard lactation is 305 d and the annual cycle can be conveniently split into four distinct parts of early, mid, late lactation and the dry period. (*Figure 14.8*). After calving milk yield will rise for a period of four to ten weeks when peak milk yield will be achieved. The time taken to reach peak yield varies with breed, individual, nutrition and yield.

Once a cow has reached peak yield the subsequent decline is approximately 2.5% per week or 10% per month. Many producers are now beating this performance with milk declining by 2% per week, post peak yield. The decline in heifers is about 7–8% per month. The daily peak yield of cows is approximately 1/200 th of total 305 d yield. Thus a cow giving 30 kg at its peak will have a total yield of approximately (30 × 200) 6000 kg.

The heifers' peak yield will be approximately 1/220 th of total 305 d yield. Thus a heifer giving 18 kg at peak will yield approximately (18 × 220) 3960 kg. A useful 'rule of thumb' guide is that two-thirds of the total yield will be produced in the first half of lactation.

It is clear that the height of the curve at peak milk yield has a great influence on the total lactation yield. The factor that is most likely to limit the level at which cows reach their peak lactation is nutrition.

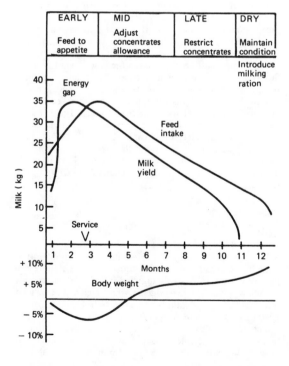

Figure 14.8 Feed intake, milk yield and body weight lactation relationships for a dairy cow producing 35 kg milk daily at peak. (From the Scottish Agricultural Colleges (1979) Reproduced with permission)

Feeding dairy cattle

The nutrition of cattle can be divided into two distinct requirements: first energy and nutrients to provide maintenance (maintenance requirement); secondly, to provide nutrients for growth, the development of the unborn calf and milk (production requirement).

Details of rationing are given earlier in this chapter.

Accurate feeding of dairy cows involves long-term planning as the requirements and allowances at any one time can be influenced by previous nutrition.

The practice of feeding prior to calving is a good example of the nutrition in one period affecting the production in another. During the last two months of pregnancy – the latter part of the dry period – the unborn calf grows rapidly and there is extensive growth and renewal of mammary tissue. The practice of steaming-up (the generous feeding of a cow pre-calving) is designed to fulfil these two needs, as well as providing reserves of body tissue which the cow can catabolise during early lactation and accustom the rumen to consuming increasing quantities of concentrates.

Steaming-up is usually started some six to eight weeks prior to calving, beginning with a small quantity of concentrates which is then increased each week. Traditionally in the week before parturition the level of concentrates fed would be about one-half to three-quarters of the amount that it is expected will be needed at peak lactation. The amount of concentrates fed will depend on the body condition of the animals, the expected level of milk yield at peak lactation and the quality of bulk fodders. For efficient and economic feeding, bulk fodders should be analysed for their nutrient value. Heifers are not usually given more than 3 kg of concentrates/d unless their body condition is very poor.

As the date of calving becomes close the animal may have a temporary loss of appetite, this should have been regained within 3–4 d after calving. As the appetite increases the cow is commonly fed for the quantity of milk being produced and an extra 1 kg/d of concentrates in an attempt to increase future production – this is usually termed 'lead feeding'. However too rapid an increase in concentrate feeding can lead to digestive upsets. It is important to realise that feeding dairy cows should not be viewed entirely in terms of concentrate nutrients as the bulk food part of the ration can contribute a substantial part of the total nutrients. This is increasingly the case since the introduction of milk quotas and at a time of falling returns to milk producers.

Correct feeding pre- and post-calving is essential to prevent excess weight loss in the early part of the lactation. Some weight loss is inevitable as the cow's appetite will not reach its maximum until after peak lactation. This difference in time taken for an animal to reach peak lactation and peak DMI means that a nutrient gap occurs. This deficiency in nutrients is made up by the cow catabolising fat, and a daily liveweight loss of at least 0.5 kg can be expected. Too great a loss in weight can be conducive to a higher incidence of ketosis (acetonaemia). The manipulation of body weight and condition in dairy cows has become necessary to sustain high milk yields. The pattern in *Table 14.36* is commonly suggested.

Frood and Croxton (1978) showed that the ability of a cow to reach a predetermined level of milk was closely related to its condition at calving. Cows whose condition score was below 2 at calving did not achieve their predicted milk yield, those whose score was above 2.5 yielded more

Table 14.36 Liveweight change during lactation of dairy cows

Week no.	Liveweight change (kg/d)	Change during 10 weeks (kg)	Net effect on liveweight (kg)
0–10	−0.5	−35	−35
10–20	0.0	0	−35
20–30	+0.5	+35	0
30–40	+0.5	+35	+35
40–52	+0.75	+63	+98

Source: HMSO (1974) Crown copyright. (Reproduced with permission)

than their predicted yields. Animals with a high condition score at calving, i.e. having ample body reserves, gave a higher earlier peak milk yield than animals in poor condition whose body reserves could not furnish enough nutrients during the time of low appetite and high energy demands of the cows.

The condition at calving and subsequent weight loss can have repercussions on conception rates at service. To ensure good conception rates cows should have a condition score of at least 2.5 at service. Below this score fertility is adversely affected. Cattle that are 'milking off their backs' must have their diet supplemented with protein and minerals additional to that normally included in the diet.

Once peak milk yield has been achieved and production starts to decline concentrate use can be reduced and intake of bulk foods increased. The level of concentrates fed will be determined to some extent by cow condition. Animals that have lost much weight will need liberal feeding until after they are served and are in calf. Concentrate levels should still be fed slightly in excess of milk production in an attempt to prevent the decline in yield. Liveweight during the period of mid-lactation should be stable.

During late lactation cows have a maximum DMI relative to milk yield. The aim is to maintain milk production and restore the body condition of the cow. Maximum use of good quality forage with little or no concentrates should achieve the aims.

Suggested condition scores for dairy cows are as follows:

Calving	3.5
Service	2.5
Drying off	3

It should be stressed that the above deals with obtaining the maximum amount of milk from a cow. Farmers coping with quotas by reducing the yield per cow will usually rely on feeding less concentrates and rely on a greater proportion of milk from forage.

Flat rate feeding

With the traditional method of feeding differing amounts of concentrates are fed according to the stage of lactation and yield. On a flat rate feeding system all lactating cows are fed the same daily amount of concentrates throughout the winter regardless of their individual yields. Bulk food is usually silage because of its better feed value; this must be of good quality (at least 10 MJ/kg DM) and must be fed *ad libitum*. If there is not enough silage available other good

quality bulk foods can be fed in its place. On most farms the optimum level of concentrates per cow will be between 6 and 10 kg/d, this depends on the yield of the cows and the quality of the forage fraction of the diet. The system works as the amount of concentrates fed rather than their pattern of feeding is the important factor.

Flat rate feeding of concentrates is most effective and easier to manage if the herd has a tight calving pattern.

Complete diets

In this system all the dietary ingredients are mixed in such a manner that individual ingredients cannot be selected out from the rest and the mixture is offered *ad libitum*. The whole system can be mechanised and does away with the need for equipment designed to feed concentrates both in and out of the parlour. It is more common with larger dairy herds where cows can be split into at least three groups according to the stage of lactation and appropriate rations can be fed to each group.

Self-fed silage

This system became popular after the Second World War with the advent of loose housing. The silage depth should not exceed 2.25 m for Friesians or they will not be able to reach the highest part of the clamp. If the depth of silage is greater than 2.25 m the upper part will have to be cut and thrown down, otherwise there is a tendency for cows to tunnel into the clamp and animals can be trapped when the overhang collapses, this is especially dangerous if tombstone barriers are used as the cows can have their necks trapped between the uprights of the barriers. The width of the feeding face should be 150 mm/cow. It is unwise to have a greater width than this or the silage will not be consumed quickly enough resulting in spoilage and secondary fermentation.

The commonest barrier to keep the cows from the silage face is either an electric wire or an electrified pipe; the latter being less likely to break than wire and providing a better electrical contact. Consumption can be controlled by the distance between the barrier and the feed face. The layout of buildings is important as cows need ready acess to the silage. The width of the face is relative to the number of animals that will feed from it and one of the problems encountered with such a feed system is that significant herd expansion is difficult because of the inability to increase the width of the silo face.

Milk quality

Milk is a major source of protein, fats, carbohydrates, vitamins and minerals in the human diet. All these are present in milk in an easily digested form. The greater part of milk is water. The remainder is solid which can be divided into two main groups, the milk fat (butterfat) and the solids-not-fats (SNF) (*see Figure 14.9*).

The total solids content of milk varies. Milk fat is a mixture of triglycerides and contains both saturated and unsaturated fatty acids. Casein is the main protein contained in milk.

There is generally an inverse relationship between milk yield and the milk fat and protein percentages. The higher the yield the lower the percentage composition of these components.

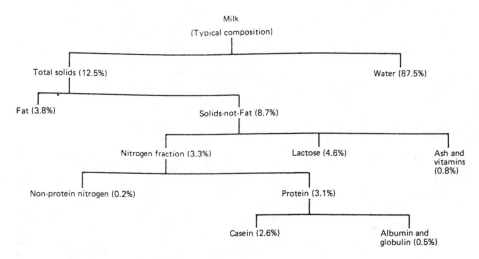

Figure 14.9

The breed of cow is an important factor affecting milk composition. Friesians, one of the heaviest milk yielding breeds, have some of the lowest fat and protein contents. Jerseys on the other hand have low yields but high fat and protein content. Individual variation within a breed can also cause a large effect in milk composition. Old cows tend to produce milk with a lower total solids content because yields tend to increase up to the fourth lactation and udder troubles increase with age, these factors have a depressing effect on fat and solids-not-fats. A regular intake of heifers into the herd will help to maintain milk quality, because of the age structure of the herd.

Day to day variations in milk quality can be caused by incomplete stripping of milk from the udder. The first milk to be drawn from the udder at milking contains low levels of fat when compared with the last milk to be drawn off, solids-not-fats change very little during the milking period. Fat percentage is usually lower after the long period between milkings, part of this is due to the higher udder pressure which causes lower fat secretion.

The fat content of milk often drops when cows are turned out to lush grass in the spring. This is a function of the corresponding increase in yield and because of low fibre levels. The problem can be mitigated to some extent by feeding 2 kg of hay, straw or long roughage before the cows go out to grass and by restricting the grass so that the more fibrous stem fraction is eaten as well as the leaf. Long fibre is necessary for the rumen fermentation to produce acetic acid, acetate being the main precursor of milk fat.

Progressive underfeeding and poor body condition are responsible for cows producing milk with low protein levels. If energy supply is deficient either during the winter or as a result of grass shortage during the summer then solids-not-fats will drop. Subclinical mastitis can be the cause of a reduction in protein levels of milk.

If the above factors are not the cause and there is still a milk quality problem then the answer might be to change the breed or by the use of a progressive breeding programme with bulls selected for with milk quality characteristics.

The milk pricing system is based on:

(1) compositional quality payments – based on the fat, protein and lactose content of the milk produced;
(2) contemporary payments – this means compositional quality payments are based on test results obtained in the month being paid for;
(3) seasonal adjustments.

Current values for each of the above can be found in *Milk Producer* or *Farmers Weekly*.

As milk is valued by the value of its component parts any factors that affect the monthly quality of milk, such as low fat at turnout in spring, can have a marked effect on producers returns. Because of the seasonality adjustments milk price is reduced in the spring and prices are increased in autumn to encourage production in August and September. Producers attempting to alter the balance amongst milk constituents, fat, protein and lactose are restricted by seasonal, lactation and breed factors. Variation in milk constituents with stage of lactation and with season are shown in *Figures 14.10* and *14.11*.

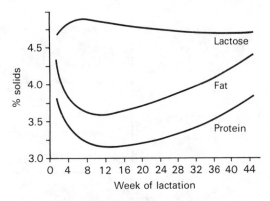

Figure 14.10 Variation in milk constituents with stage of laction (Source MMB (1983) *The new pricing package*)

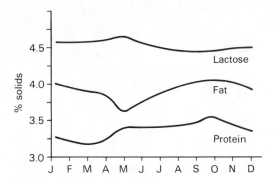

Figure 14.11 Variation in milk constituents with season (Source MMB (1983) *The new pricing package***)**

Oestrus detection

Oestrus (bulling or heat) is the time during which the cow will stand to be mounted by the bull. This period lasts on average 15 ± 4 h. During the winter the period may be greatly reduced, a shorter period of 8 h is also common with heifers. The average interval between oestrus is 20–21 d.

Failure to detect animals on heat can be a major problem, especially with the increasing size of herds. Poor conception rates are often blamed on poor artificial insemination techniques, bull fertility, disease, whilst the actual problem may be poor management.

The first management factor necessary to improve oestrus is good cow identification. There are a number of methods by which cows can by identified clearly, but freeze branding is probably the best. In conjunction with identification should be the keeping of good records. It is essential that simple, accurate and complete records are available whilst cows are being observed for signs of oestrus. Dectecting cows in oestrus is a part of good stockmanship as cow behaviour is the best indication of heat.

Signs indicating oestrus are:

(1) mounting other cows
(2) sore, scuffed tail-head and hip bones
(3) mud on flanks
(4) mucus from vulva
(5) restlessness
(6) steaming animals.

Stockmen must be allowed adequate time for observation of the herd and the most productive times are after the animals are settled after milking and feeding. Mid morning and late evening are particularly productive. Observations should ideally take place three or four times a day and the herd should be watched for at least half an hour at a time, and any sign of oestrus should be written down and not left to memory. The target should be to detect 80% of all cows in oestrus. Various aids may be used, including vasectomised bulls, pads or paint placed on the cows' rumps have all proved useful.

The timing of service is important. If cows are served too early or too late in the heat period poor conception rates may result. Ideally, cows should be served in the middle of the heat period between 9 and 20 h after the start of heat to obtain the best results (*Figure 14.12*).

The aim should be for a calving interval of 365 d. If the cow is to have a lactation of 305 d and a dry period of 60 d, with pregnancy lasting between 278–283 d, cows should conceive 81–86 d after calving. If service is delayed until the first heat after 60 d – a common practice – the result will be a slightly higher conception rate but a longer calving interval.

Oestrus can now be controlled and synchronised by the use of prostoglandins. The technique is useful where a group of heifers needs to be inseminated. The animals will come into heat at the same time and the result will be a tighter

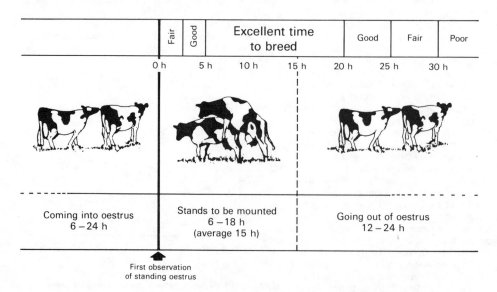

Figure 14.12 Optimum time for insemination. (From HMSO (1984) reproduced with permission)

calving pattern. In older cows the technique can be used on individual animals rather than the herd as a whole. The technique is not a cure for infertility and will only bring an animal on heat that is cycling normally (*see* p. 346).

The practice of calving the herd over a period of about ten weeks in not as common as would be expected despite the potential simplification of management routine. All the main tasks, calving, oestrus detection, service, drying off can occur during a specific time period for the entire herd. Coupled to this is the fact that the feeding management of the herd can be simplified by having uniform groups of animals at similar stages of lactation. Heifer rearing becomes simpler as the animals have a smaller range of age. Not all farmers are attracted to such a routine but the regime can create conditions more conducive to the achievement of better results.

Milk production

Milk production and the premises under which milk is produced are controlled by the Milk and Dairy Regulations 1959, a requirement being that all dairy farms are registered by the Ministry of Agriculture. Premises, stock and methods of production are open for inspection at any time.

The milking process

Milk ejection or 'let-down' is controlled by the hormone oxytocin. Milk let-down is a condition reflex in response to a stimulus. Quiet handling of animals prior to milking is essential; if cows are nervous or frightened milk let-down will be inhibited. The natural stimulus for cattle is the calf suckling, in dairy herds a substitute stimulus in the form of feeding and/or udder washing is used. Besides providing this stimulus udder washing removes dirt which may contaminate the milk as it leaves the teats. Methods of udder washing include sprays, or buckets of water, the warm water often includes an antibacterial agent and cloths or paper towels are used to clean and dry the udder and teats. Disposable paper towels are preferable to cloths as there is less chance of infection being passed from cow to cow.

Before or after udder washing fore-milk (the first milk from the udder) is removed by hand into a strip cup to reveal any signs of mastitis in the form of clots, flakes or watery milk. This is a useful indication of clinical mastitis.

Teat cups should be applied as soon after washing as possible. Maximum rate of milk flow is reached after about 1 min, later the flow rate declines quite rapidly. Once milking has started it should be accomplished as rapidly as possible. Overmilking should be avoided as this may damage the udder, thus predisposing mastitis. Stripping – the removal of the last milk – can be achieved by applying downwards pressure to the teat cup with one hand and the udder massaged downward with the other, the forequarters being done first. In parlours the stripping of milk and automatic removal of teat cups, when the milk flow slows to a predetermined rate, is becoming more common.

Teat disinfection by dipping teats in a cup containing an approved iodophor or hypochlorite solution helps prevent the spread of bacteria causing mastitis.

Efficient milking is largely dependent upon a good milking routine. This should be simple and consistent providing a 'let-down' stimulus for the cow and a routine series of operations which the milker has to perform on each animal.

After milking the milk will be at an ideal temperature (37°C) for the growth and multiplication of most bacteria. As the milk is collected once a day by bulk tankers some must be kept overnight. It is important to cool the milk quickly to prevent its deterioration. Cooling takes place in a refrigerated bulk tank and the Milk Marketing Board requires milk to be cooled to below 4.5°C 30 min after milking. Before milk is allowed into the bulk tank it is passed through a filter.

Milk is routinely tested for keeping quality by direct or indirect tests and penalties are imposed for failure to meet the required standards. Stringent penalties are also incurred if antibiotics are detected in milk, these usually occur from the intra-mammary treatment of mastitis.

Poor keeping quality can be caused by inadequate cleaning of the milking equipment giving a build up of residue. Hand washing is normally used for bucket milking plants. A cold water rinse of approximately 10 litres of water can be drawn through the clusters of each unit, this removes the film of residue left after milking. The equipment is then dismantled and washed once in either detergent or a detergent and sterilant solution at 50°C, then finally rinsed in clean water also containing a sterilising agent.

With pipeline systems where it is impractical to dismantle the equipment, cleaning is done *in situ* by circulation cleaing. Here the success of the operation relies on the properties of the chemicals and heat for the disinfectant effect. Again the basic process starts with a rinse of cold water, this is followed by the circulation of a detergent solution. The initial temperature needs to be fairly high (80–85°C) as the solution will be cooled during the first cycle round the plant. A final rinse with cold water which may contain sodium hypochlorite for sterilisation is circulated.

The commonest sterilising and disinfecting agents are sodium hypochlorite, hypochlorite, bromates and iodophors, they should be used in accordance with the manufacturers' recommendations.

Another method of cleaning equipment *in situ* is by the use of acidified boiling water. Here hot water (96°C) is flushed through the plant for 5–6 min and allowed to run to waste, this pre-rinses and warms up the plant. Nitric or sulphamic acid is mixed with the water, this has no disinfecting effect but removes the milk deposits from the equipment. The aim is to heat the plant to 77°C for at least 2 min to achieve disinfection.

The circulation cleaning method involves the use of expensive chemicals and takes more time (15 min) whereas the acidified boiling water system needs more water (13–18 litres/unit) at a much higher temperature thus using more energy, but only takes 5–6 min.

To prevent a build up of scale on equipment it may be necessary to use a milkstone remover once per month.

Bulk tanks may be cleaned by hand, using long handled brushes, or by a mechanical spray. As tanks are cooling mechanisms and often contain an ice bank at the base of the tank a cold system of cleaning is normally used. This uses an iodophor or bromate cleaning agent at mains water temperature. The tank is rinsed with cold water, immediately after emptying, by hand and the cleaning agents applied. In automatic systems the rinse is sprinkled into the tank; this is then followed by a solution containing the chemicals. The inside is then rinsed before milking.

Particular attention should be paid to the outlet, paddle, dipstick and underneath the lid and bridge of the tank.

CATTLE BREEDING

The aim of selecting breeding animals is to produce a future generation with improved performance.

Improvement is brought about by increasing the frequency of desirable genes and by creating favourable gene combinations. The genotype of an animal is its genetic constitution. The phenotype of an animal is the sum of the characteristics of the animal as it exists and is the result of both genetic and environmental effects. it is possible for animals to have the same genotype but different phenotypes owing to environmentally-produced variation, i.e. how it is fed, housed and managed. Environment does not affect all characters to the same extent. For example, the normal homozygous black coat colour of the Aberdeen Angus is always dominant to the recessive red of the Hereford, thus all Angus × Hereford cattle are black with a white face. Environment has no effect as these characters are controlled by the presence of one pair of genes. The expression of many characters of economic importance – body growth, milk yield, milk quality, fertility – are controlled by many genes and environment also exerts an effect. Characters in which there is a close resemblance between parent and offspring, whatever the environment, are characters of high heritability (*Table 14.37*)

Greater progress can be expected if the breeding programme is concentrating on characters of high heritability.

As the sire often serves many cows a great deal of attention needs to be placed on his selection. The two main methods of evaluating bulls are:

(1) performance testing
(2) progeny testing.

Performance testing involves measuring an animal's individual performance – growth rate, feed conversion

efficiency – and comparing them with animals from a comparable group which have been subjected to similar conditions of feeding and management. The advantage of performance testing is that it is much cheaper and quicker than progeny testing. The disadvantages are that it is only of value where the characters being measured are of a high heritability and thus likely to be passed on to its progeny. It can only be used to measure characters in the live animal, this used to be a disadvantage in assessing carcass composition as that used to necessitate slaughtering, however, new techniques such as ultrasonics have overcome this particular difficulty to a large extent.

Progeny testing involves the examination of an animal's offspring. The characters are measured and compared against the progeny of other sires. Again progeny should be kept under similar environmental conditions. It is particularly useful for testing dairy bulls as milk production is only measurable in the female and the characters in question are often of low heritability. The method is used by the Milk Marketing Board for evaluating its dairy bulls for artificial insemination (AI) purposes. The bull's progeny can be compared with daughters sired by other bulls in the same herds. The records should reflect the differences attributable to the bull's genetic constitution and is referred to as a contemporary comparison. These contemporary comparisons can provide guides when selecting a bull for AI. The higher the contemporary comparison the greater the chances are that the bull will pass on genes resulting in improved daughters. The greater the number of daughters used to evaluate a bull in this way results in a more reliable comparison figure, the number of daughters used is referred to as a 'weighting' which appears with the contemporary comparison figure.

Further reading

ALLEN, D. and KILKENNY, B. (1980). *Planned Beef Production*. London: Granada

BAKER, H. K. (1975). *Livestock Production Science* **2,** 121

FROOD, M. J. and CROXTON (1978). *Animal Production,* **27**(3), 285

HMSO (1984). *Dairy Herd Fertility*. Reference Book 259

HARDY, R. and MEADOWCROFT, S. (1986). *Indoor Beef Production*. Ipswich: Farming Press Ltd

JOHANSSON, I. and RENDEL, J. L. (1968). *Genetics and Animal Breeding*. London: Oliver and Boyd

MLC (1978). *Beef Improvement Services*. Data summaries on beef production and breeding

MLC (1982) *Selecting Cattle for Slaughter*

MLC (1986). *Beef Yearbook*

MLC/MMB JOINT PUBLICATION. *Rearing Replacements for Beef and Dairy Herds*

MMB (1983). *The New Pricing Package*

MMB (1986). *UK Dairy Facts and Figures*

MMB (1986). *EEC Dairy Facts and Figures*

ROY, J. H. B. (1980). In *The Calf*. London: Butterworths

SCOTTISH AGRICULTURAL COLLEGES (1979). *Feeding the Farm Animal – Dairy Cows*. Publication No. 42

WARNER, R. G. and FLATT, W. P. (1965). In *Physiology of Digestion in the Ruminant*. Ed. R. W. Dougherty. London: Butterworths

THICKET, W. S., CUTHERBERT, N. H., BRIGSTOCKE, T. D. A. and WILSON, P. N. (1979). *BSAP Winter Meeting Paper No. 18*

WOOD, P. D. P. (1972). *MMB Better Management* No. 7:1

Table 14.37 Heritability of breeding stock characters

Trait	Heritability
Birth weight	0.4
Weaning weight	0.2
Daily gain, weaning to slaughter	0.3–0.6
Food conversion	0.4
Mature weight	0.4
Wither height	0.5–0.8
Heart girth	0.4–0.6
Dressing percentage	0.6
Points of carcass quality	0.3
Cross-section of eye muscle	0.3
Bone percentage	0.5
Lactation yield	0.2–0.3
Fat content of milk	0.5–0.6

After: Johansson and Rendel (1968) (Reproduced with permission)

15

Sheep and goats

R.A. Cooper

SHEEP

International picture

There were 1122 million sheep in the world in 1985, and in most countries populations are increasing. Details of their distribution are given in *Table 15.1*. Production of sheepmeat by the world's major producers is shown in *Table 15.2*, with EEC output detailed in *Table 15.3*. Mean carcass weight in most situations is 15–16 kg, but in areas where sheep are kept as dual-purpose milk-meat animals (e.g. in Italy and Greece) carcasses may only weigh 7–9 kg as lambs are slaughtered at six to eight weeks of age.

Sheep in the UK

The total ewe population of the UK is 17.2 million head. Data for breeding animals are given in *Table 15.4*, which also gives details of animals eligible for Hill Livestock Compensatory Allowance (HLCA) payments and thus indicates the importance of sheep in hill and upland areas. *Table 15.5* gives a breakdown of the sheep industry by flock size. Sheep numbers increased in all geographic regions by an average of 5% during 1984–85, with increases in flocks of more than 1000 head particularly noticeable.

Table 15.2 Sheepmeat production 1986 (10^3t)

EEC	915	N. Zealand	465
USSR	835	N. America	156
Australia	584		

From *MLC Quarterly Review* with permission

Table 15.3 Production and consumption of sheepmeat in EEC 1986

Production (10^3t)		*Consumption* (10^3t)
EEC total	915	1127
UK	302	369
France	163	245
Spain	146	142
Greece	130	142
Italy	70	85
W. Germany	22	49

From *MLC Production Yearbook*, (1986) (reproduced with permission)

Table 15.1 World distribution of sheep ($\times 10^6$) 1985

Africa 192.7		*N. America* 19.2		*S. America* 102.2		*Asia* 311		*Europe (inc. USSR)* 276.2		*Oceania* 220.3	
South Africa	30.3	USA	10.4	Argentina	29.0	China	108.6	USSR	142.9	Australia	149.7
Ethiopia	23.5	Mexico	6.4	Uruguay	20.6	India	41.3	E. Europe	47.9	N. Zealand	70.6
Sudan	19.0	Guatemala	0.7	Brazil	17.5	Turkey	40.4	UK	23.9		
Algeria	18.0	Cuba	0.4	Peru	13.5	Iran	34.5	Spain	17.5		
Nigeria	13.0					Pakistan	25.0	France	10.8		
Morocco	12.0					Afghanistan	20.0	Italy	9.5		
								Greece	7.9		

FAO Production Yearbook (1986) (reproduced with permission)

Table 15.4 Numbers and distribution of breeding ewes in UK 1985 (10⁶ head)

	England	Wales	Scotland	N. Ireland	UK
Total	7.8	4.8	3.8	0.8	17.2
Of which[1]					
Higher rate	1.7	2.4	2.3	0.3	6.7
Severely					
disadvantaged					
Lower rate	0.9	0.7	1.0	0.2	2.8
Disadvantaged	0.35	0.55		0.1	1.0

[1]Eligible for HLCA payments
(MAFF data)

Table 15.5 Distribution of breeding ewes by flock size (%) 1985

Flock size	England	Wales	Scotland	N. Ireland	UK
1– 99	4	2	2	14	4
100– 299	15	9	11	37	13
300– 499	15	11	9	18	12
500– 699	13	11	8	11	11
700– 999	14	15	11	9	13
1000–1499	15	19	15	6	16
1500–1999	9	12	12	2	10
2000	15	22	31	3	20

From *MLC Production Yearbook* (1986) (reproduced with permission)

The National sheep flock is made up of some 50 'pure' breeds and more than 300 crosses, but few of the pure breeds, and even fewer of the crosses, are of major significance except in localised circumstances. Details of individual breeds may be obtained from the National Sheep Associa-tion publication *British Sheep*. Flock categories, important characteristics and major breeds are given in *Table 15.6*.

Sheep production systems in the UK are many and diverse, because of the varied conditions under which sheep are kept and the multiplicity of breeds and crosses available. They are linked, however, by an interdependence based on a substantial cross-breeding programme. A generalised outline of this programme, which is referred to as stratification, is shown in *Figure 15.1*. Under the very harsh conditions experienced on many hill farms ewes cannot thrive for more than four seasons, thus the hill farm retains many of its ewe lambs as replacement breeding stock while the sale of draft ewes provides an important part of hill farm income (15% of flock gross output at 1986 prices). Drafted into the better conditions on upland and marginal farms these ewes have several years of productive life left, and given better climate and nutrition are capable of lambing levels well beyond the 80–100% they have achieved previously.

On the upland farms to which these ewes go there is often little scope for finishing lambs satisfactorily. Put to a longwool ram these ewes produce a crossbred lamb with the benefits of heterosis evident in terms of fertility, milkiness and general vigour. There is a strong demand for such animals for lowland breeding flocks, where these attributes may be complemented by those of suitable Down rams for the production of prime lamb. Sixty per cent of ewes tupped under lowland conditions are derived from longwool × hill matings. *Table 15.7* gives details of the most important of these crosses.

Factors affecting flock performance

Given the diverse nature of sheep production in the UK it is difficult to discuss flock performance in other than general terms. The standards which flocks recorded by the Meat and Livestock Commission (MLC) are achieving are shown in *Figures 15.2 and 15.3*. In each case target figures are given in

Table 15.6 Breed types, important ewe characteristics and main breeds

Type	Desirable characteristics	Main breeds
Hill Breeds	Regular lambing Milking/mothering ability Ability to rear 100% crop Hardiness Good wool weight and quality	Scotch Blackface Welsh Mountain Swaledale Cheviot Herdwick Gritstone
Upland breeds	Milking ability High fertility	Kerry Hill Clun North Country Cheviot
Longwools	High fertility and prolificacy Milking ability Growth rate	Blue-faced Leicester Border Leicester Teeswater
Down breeds	Growth rate Carcass quality	Suffolk Oxford Down Hampshire Down Texel Dorset Down

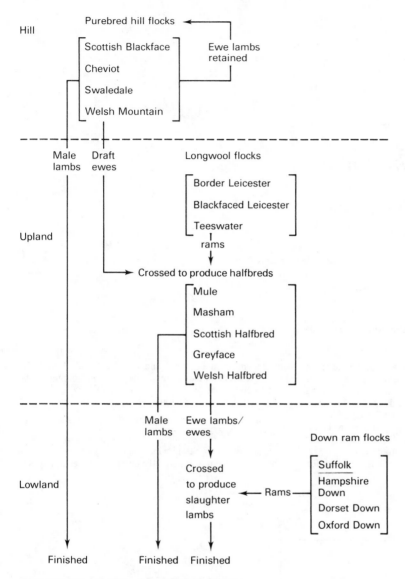

Figure 15.1 Stratification of the UK sheep industry

Table 15.7 Important crossbreeds and their derivation

Cross	Sire breed	Dam breed
Mule	Blue-faced Leicester	Swaledale, Blackface Speckledface, Clun
Greyface	Border Leicester	Scotch Blackface
Masham	Teeswater, Wensleydale	Dalesbred, Rough Fell, Swaledale
Scottish Halfbred	Border Leicester	Cheviot
Welsh Halfbred	Border Leicester	Welsh Mountain

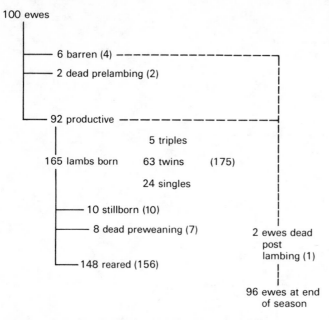

Figure 15.2 Physical performance of lowland flocks (1985 data) (target values in parenthesis)

parenthesis alongside achieved values. Under lowland conditions the top-third producers achieved a gross margin/ha some 47% above average. The two factors contributing most to this superiority in 1985 were stocking rate (40%) and lambing percentage (25%). For hill flocks the superiority of top-third flocks was 36%, these flocks rearing more lambs and selling them for more money. These factors alone accounted for more than 70% of the superiority

It is apparent from these data that there is room for improvement in both hill and lowland flocks. Details on good grassland management can be found in Chapter 6. The improvement of lambing percentage, which should be defined as lambs reared/100 ewes to ram, is discussed below.

Oestrus

The sheep is a species in which there is generally an annual rhythm of breeding activity, the onset of the breeding season

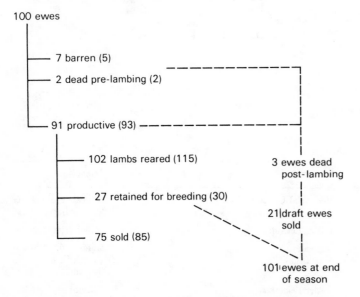

Figure 15.3 Physical performance of hill flocks (1985 data) (target values in parenthesis)

corresponding with a period of decreasing daylength and cessation occurring as daylength increases in spring. There are, however, wide variations within as well as between breeds, and a much reduced cyclicity is evident at lower lattitudes. Some idea of the range of dates involved is given in *Table 15.8*. These values can be taken as no more than guides because of the interactions of many other factors as well as daylength. The response of the ewe to photoperiod is known to be mediated by the production of a hormone, melatonin, by the pineal gland, during the hours of darkness. It has recently been demonstrated that the administration of exogenous melatonin, either in feed, as a series of injections, or in a soluble rumen bolus, may be used to bring forward the onset of breeding activity. Where such a technique is adopted as an adjunct to normal mating dates it may lead to increased conception rates and fecundity.

The sudden introduction of rams during the period immediately before the onset of oestrus may also bring forward breeding activity by a few days. More importantly it can stimulate a considerable degree of synchrony. When associated with the use of vasectomised, or 'teaser', rams the technique can be used to obtain a tighter lambing pattern when 'entire' rams replace the teasers after 14 d.

Ovulation and fertilisation

One or more 'silent' ovulations often precede the onset of behavioural oestrus. Ovulations occur every 17 d (range 14–19 d) throughout the breeding season. Oestrus lasts for about 30 h (range 24–48 h) with ovulation occurring towards the end of this period. One of the prerequisites for a good lambing percentage is a high ovulation rate, and subsequently a high fertilisation rate with minimal embryo mortality. Genotype will have influence here, although nutritional factors may confuse the situation. Within a flock ovulation rate will be influenced by the following factors.

Nutrition

It has been known for many years that putting ewes onto a rising plane of nutrition for two or three weeks prior to tupping has a beneficial effect. Traditionally ewes were weaned onto hard conditions, in part to assist drying off, in part to put them into lean condition before 'flushing'. This increasing body condition, or dynamic effect, gives some 6–8% more twins compared with the performance of ewes mated whilst on a maintenance diet, and 12–16% more twins when compared with ewes mated while in negative energy balance. More recently a 'static' effect has also been recognised, this being dependent on the level of body reserves in the ewe; ewes with greater body reserves having a higher ovulation rate. For ewes on a level plane of nutrition an extra 3.5 kg bodyweight at mating represents a potential 6% extra twins (comparisons within flock).

Recognising the impracticability of weighing ewes on commercial farms, the Hill Farming Research Organisation has devised a simple, rapid and accurate method of judging the body condition of ewes by palpating the spinal column over the loin area. Scoring, on a scale from 0–5, is possible to the nearest 0.5 (see Appendix 1, p. 428). Body condition scoring may be used on sheep of any breed, and at any time of the year, as a guide to the adequacy of nutrition of the flock. It is particularly valuable in the pre-tupping and pre-lambing periods. Pre-tupping it can be used to divide the flock into groups according to their needs in terms of the duration and degree of flushing required. Lowland ewes should have a condition score of at least 3.5 at mating and hill ewes 2.0 to 2.5. In a normal flushing period (approximately three weeks), lowland ewes having an initial score of 3 or above can achieve this target easily. Those scoring 2.5 or less, will need at least six weeks on improved grazing, or supplementary concentrates, if lambing percentage is not to suffer. The relationship between score at tupping and lambing percentage is shown in *Table 15.9*, but many breed differences exist and some of these are illustrated in *Figure 15.4*. A scoring at least six weeks before tupping can identify those animals in need of preferential treatment and a second, three weeks later, can be used to monitor response. An improvement in condition from score 2.5 to score 3.5 is equivalent to a gain of some 6 kg liveweight.

Age

Ewe lambs may attain puberty in their first year. There is a nutrition/date of birth interaction and with lightweight or late-born lambs oestrus may not be attained. Although most animals that attain puberty will rear one lamb at best, there is some evidence that animals which are bred as lambs may be more prolific in later life. Subsequently, mean ovulation rate will continue to rise to five or six years before falling off

Table 15.8 Breeding seasons for some British breeds

Breed	Onset		Cessation	
	Mean	*Range*	*Mean*	*Range*
Blackface	25 Oct	26 Sep–10 Nov	19 Feb	17 Jan–3 Apr
Welsh Mountain	25 Oct	11 Oct–11 Nov	17 Feb	1 Feb–25 Feb
Border Leicester	9 Oct	24 Sep–14 Oct	10 Feb	18 Dec–11 Mar
Romney	4 Oct	16 Sep–19 Oct	2 Mar	20 Jan–23 Mar
Suffolk	3 Oct	12 Sep–23 Oct	17 Mar	7 Feb–20 Apr
Dorset Horn	24 Jul	15 Jun–21 Aug	2 Mar	22 Jan–22 Apr

(Adapted from Hafez, E.S.E. (1952) *J. Agric. Sci.* **42**, 189 with permission)

Table 15.9 Relationship between condition at mating and lambing percentage

Breed	Average score at mating	Live lambs/ 100 ewes to ram
Scotch Halfbred	3.8	184
	2.6	120
Welsh Halfbred	3.5	158
	2.7	133
Masham	3.4	178
	2.2	127
Clun	3.6	156
	2.0	131

After MLC 1980. Sheep Improvement Services, Body condition score of ewes (Reproduced with permission)

Table 15.10 Effect of mating date on mean lambing percentage in lowland flocks

Lambing date	Lambing % MLC data[1]	WSCA[2] data
January	148	137
February	155	145
March 1–15	160·	159
March 16–31	167	
April	168	—

[1]*MLC sheep facts 1978* (Reproduced with permission)
[2]West of Scotland College of Agriculture Research and Development Publication No. 7.

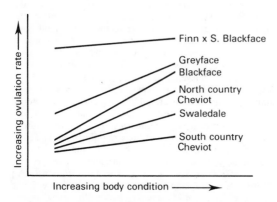

Figure 15.4 Breed effects on the relationship between condition at mating and ovulation rate (From *Science and Quality Lamb Production* (1986). ARC with permission)

again. Hence the draft hill ewe has only just achieved her potential when she is sold.

Season

Many ewes begin the season with a 'silent' oestrus. Mean ovulation rate then rises for three or four cycles and will likewise drop at the end of the season. On average therefore lambing percentages will be lower in early or very late-lambing flocks and are very low in most out-of-season lambings. These problems are exacerbated by higher numbers of barren ewes at such times. The Finnish Landrace, and its crosses, do not appear to be affected in this manner and are thus widely used in 'out-of-season' production systems. The relationship between tupping date (and hence lambing date) and lambing percentage in lowland flocks is shown in *Table 15.10*. Decisions on lambing date are likely to be influenced also by factors such as climate, altitude and politics. Thus, in hill situations the introduction of the ram into the flock will be dictated in part by tradition. Traditional dates in Scotland for example are 10 November for Cheviots and 25 November for Blackface. Then, too, the pattern of guide prices for eligible lamb, giving as it does such low returns in July, August and September, may well influence a decision in favour of earlier lambing so that finished animals are

marketed before the price drops. Under normal conditions all lowland sheep will ovulate and produce at least one ovum; the majority should produce two or more. Genotypically there is no reason why this should not be similar for hill breeds, the lower ovulation rates experienced in hill flocks being largely the result of environmental effects.

If all ova shed are to be fertilised, then rams must be both fertile and active. Low libido (sexual drive) may well be a problem in rams in early-lambing or out-of-season situations. At the beginning of the breeding season all rams should be examined to ensure that they are healthy, sound in feet and legs and without any abnormality, especially of the testes. Microscopic examination of a semen sample, obtained by electro-ejaculation, is an additional check. Correct ewe to ram ratio will depend on circumstance. A strong ram should be able to cover 50–60 ewes adequately, but under extensive conditions the ratio may need to be lower and where an attempt has been made to synchronise oestrus it will need to be no more than 10:1. An inexperienced ram lamb may be used (carefully) in his first year, but only at a low ewe:ram ratio (20:1) and only with mature ewes. It is particularly important that ram lambs do not have to compete with experienced older rams at this time.

The practice of applying ochre to the breastbone of rams (keeling or raddling), or the use of suitable harness and crayon, is a helpful management tool. Rams treated in this way leave a mark on all ewes mated and this allows the monitoring of the progress of the breeding programme and, supplemented by more permanent markings, can form the basis of subdivision of the flock prior to lambing. In larger flocks this can lead to significant savings in concentrate costs. With such a system the colour of the raddle should be changed at least every 16 d, going progressively from lighter to darker colours.

Embryo mortality

A significant percentage of fertilised ova is lost in early pregnancy, especially in the first 30 d. Losses as high as 40% have been suggested but the mean is probably about 25%. Some loss of embryos is to be expected but steps can be taken to minimise it. In particular the avoidance of stress and the adequate nutrition of the ewe are critical; above-maintenance levels of feeding should be maintained for the first month of pregnancy. Embryo mortality also increases with increasing ovulation rate, so that ewes that have been flushed premating are likely to be more sensitive to postmating drops in feed intake. Embryo loss is also greater in early and out of season breeding and in ewe lambs. In many countries a further factor is likely to be heat stress, which can cause very high embryo losses if it occurs at the early cleavage stage.

During the middle part of pregnancy few losses appear to occur. The ewe is able to withstand periods of under-nutrition – hill ewes often losing 5–10% of bodyweight during this period without significant effect. Losses in excess of this, or in the latter stages of pregnancy may lead to fetal death, reduced birthweights and increased perinatal mortality.

Perinatal mortality

Perinatal mortality (including stillbirths and abortions) is a major source of loss in many sheep flocks. Surveys suggest that as many as four million lambs may be lost in the UK every year. Main causes of loss are: abortions and stillbirths 20–40%, starvation and exposure (hypothermia) 35–55%, infectious diseases 5–10%, misadventures 5–10%. Organisms of the *Pasteurella, Toxoplasma, Brucella* and *Rickettsia* groups may be implicated in so-called abortion storms in sheep. Vaccination can help avert the problem. Similarly, appropriate vaccination (e.g. against *Clostridia*), given to the ewe prepartum, can help reduce postnatal lamb mortality.

The nutrition of the ewe in the six weeks prelambing is extremely important. During this period the fetus makes 70% of its growth. Undernutrition of the ewe at this time will lead to reduced birthweights and an increased susceptibility to hypothermia (*see Figure 15.5*), or even to the death of some, or all fetuses *in utero*. Additionally, correctly-fed ewes are much less likely to suffer from pregnancy toxaemia (twin lamb disease), a metabolic disorder which may cause the death of both ewe and lambs. The accuracy with which pregnant ewes can be fed is increased if the number of fetuses carried is known. Pregnancy diagnosis, using real-time ultrasonic scanning and carried out 50–100 d post-mating, can detect pregnancy to an accuracy of 98% and fetus number to 95%. Such knowledge allows appropriate division of ewes, facilitating differential feeding and making lambing management easier. Savings of concentrates of up to 20 kg/barren ewe and 10 kg/single-bearing ewe are possible and will often cover the cost of scanning. Reduced ewe and lamb mortalities may further contribute to the cost-effectiveness of the technique.

Hypothermia

Once the lamb is born, careful shepherding can do much to ensure its survival. Critical points are the lamb's ability to withstand the challenges of climate and infection. A well-planned care routine (*see Table 15.11*) can do much to reduce these problems. Hypothermia is one of the major causes of postnatal mortality. The newborn lamb has a coat with little insulating value, a large surface area:weight ratio, and is wet.

Table 15.11 Intensive care routine at lambing

(1) Check lamb breathing: clear mucus from airways
(2) Treat navel with chloromycetin or suitable alternative
(3) Individually pen ewe and lambs
(4) Draw milk from both teats of ewe
(5) Put lamb to teat and initiate sucking
(6) Use stomach tube to administer colostrum if necessary

It must produce heat as rapidly as it is losing it or it will die. Lambs have only small fat reserves, especially small lambs, and rely heavily on feed energy. The small lamb may quickly burn up body reserves in a cold, wet environment and thence be too weak to suckle properly. Such a situation will lead to death within a few hours of birth. The larger lamb may be able to utilise reserves for up to 24 h but must then replace them if it is to survive.

Body temperature is a good guide to the lamb's status. Workers at the Moredun Institute have developed a

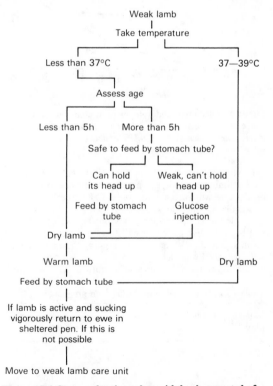

Figure 15.6 Course of action taken with lamb suspected of being hypothermic. (From Eales, F.A. and Small, J. (1986). *Practical Lambing.* **London: Longman, reproduced with permission)**

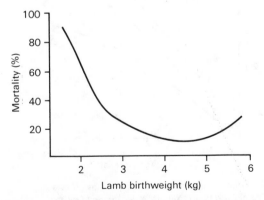

Figure 15.5 Relationship between lamb birthweight and mortality. (From Maund, B. (1974), Drayton EHF results)

programme based on lamb body temperature (*see Figure 15.6*) which, if rigorously applied, should dramatically reduce the number of lambs lost to hypothermia.

Meat production

Growth and nutrition

Under UK conditions the main output of the sheep unit is in the form of lamb – either finished or store. The money obtained for each lamb is a function of weight, either live or carcass, quality and the date of sale. This section discusses the main factors affecting this output.

As can be seen from *Figure 15.7* the development of the components of the conceptus is variable, with most placental development having finished by six weeks prior to lambing while most fetal development is yet to take place (*see Table 15.12*). The importance of ewe nutrition during this period has been discussed in relation to the survival of the newborn lamb, it is also important in terms of the subsequent growth of that lamb, especially in the case of multiple births.

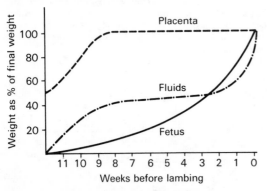

Figure 15.7 Differential development of components of conceptus (From Robinson, J.J. *et al.*, *Journal of Agricultural Science*, 88, 539. Reproduced by courtesy of the editor and publisher)

From a daily requirement of 12–18 megajoules (MJ) of metabolisable energy in late pregnancy, the energy demands of the ewe will increase to 20–30 MJ at peak lactation. Especially in the flock lambing before grass becomes available this energy demand may not be met and the ewe will begin to 'live off her back'. Ideally the ewe will lamb down at condition score 3.5 if she is going to be able to milk satisfactorily. Ewes suckling more than one lamb respond by producing more milk; some 40% more if suckling twins,

Table 15.12 Growth of fetus during twin pregnancy

Stage of pregnancy (days)	Fetus weight (g)
28	1
56	89
84	1 000
112	4 250
140	10 000

After Wallace, L.R. (1948). *Journal of Agricultural Science* **38**, 93

Table 15.13 Average milk yields during first 12 weeks of lactation

Type of ewe	No. of lambs	Yield/d (litres) Month 1	Month 2	Month 3
Hill	1	1.35	1.30	0.95
	2	2.25	1.85	1.25
Lowland	1	1.60	1.55	1.25
	2	2.55	2.10	1.45

50–55% more for triplets, and there may on occasion be a case for running ewes suckling triplets separately and feeding them accordingly. Notwithstanding this extra milk, lambs reared as multiples are unlikely to grow as fast as those reared as singles. Average figures for daily milk yield are given in *Table 15.13*. For the lowland ewe the average of 1.6 kg produced/d will support daily gains of 300 g in the single lamb. When producing 2.54 kg the ewe is supporting some 460–500 g of gain, but this is only 230–250 g/lamb. By the third month the relevant figures are 220 g and 130 g/d for singles and twins respectively. The effects of this on lamb liveweight are demonstrated in *Figure 15.8*. For the single lamb growth rate is almost linear, and with 'milky' ewes the lamb will need no supplementary feed. On the other hand triplet lambs start at lower weights, grow more slowly, become dependent on alternative feeds earlier and when grazing are thus much more prone to roundworm infestation. There are few advantages to be gained by leaving lambs with their dams beyond 12 weeks of age and indeed, in situations where grass is in short supply, there may well be advantage in weaning at ten weeks and allowing the lambs access to the best grazing.

A second factor affecting daily gain beyond the 10–12 week period is the shift, in terms of components of any gain, away from lean and towards fat. The characteristics of growth in the lamb are similar to those for other species, namely that initially bone growth predominantes, followed by lean tissue and finally fat. Animals in which the waves of

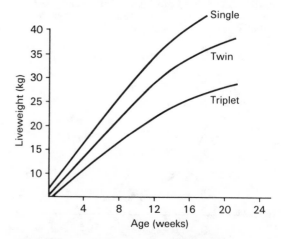

Figure 15.8 Typical growth curves for lambs (From Spedding, C. (1970). *Sheep Production and Grazing Management*. London: Bailliere, reproduced by courtesy of the publishers)

growth for bone, lean and fat are steep and close together are called early maturing. In sheep the tendency is for smaller breeds, such as the Southdown, to be early maturing and to begin to lay down fat at relatively low carcass weights, and for heavier breeds to be later maturing and capable of producing, given time, a heavier carcass at any given level of fatness.

The current demand is for leaner carcasses, but the current EEC sheepmeat regime encourages producers to hold lambs for sale later in the year and thus tends to produce heavier, fatter carcasses. The Meat and Livestock Commission (MLC) have developed a classification scheme for lamb carcasses which can be used to assess and describe any carcass in terms of its conformation (muscling) and fatness (*see Figure 15.9*). On this basis, current demand is for a carcass falling into fat classes 2 and 3L. An indication of the relationship between carcass weight and classification is given in *Table 15.14*. It is possible to predict, with a reasonable degree of accuracy, the likely carcass weight of any breed or cross of lamb, at a fat score of 2–3L, using the formula $(WD/2 + WS/2)/2 \times$ estimated killing out percentage (where WD = mature weight of dam breed; WS = mature weight of sire breed). Thus, for example, using a Suffolk ram (90 kg) on a Scotch Halfbred ewe (90 kg) will produce a 20 kg wether carcass at a killing out percentage of 45. For ewe lambs this weight will be reduced by about 10%. Choice of sire breed is a matter of individual judgement. It is essential that the breed of ram chosen suits the lamb finishing system to be used, and it must be remembered that a heavier lamb will take longer to produce and that this may be inconvenient. For example, lambs aimed at the early market need rapid growth and early maturity, and will tend to be sold at lower carcass weights, while animals to be finished on roots as late lambs or hoggets (animals sold after 1 January) need to be capable of producing a heavy carcass without laying down too much fat.

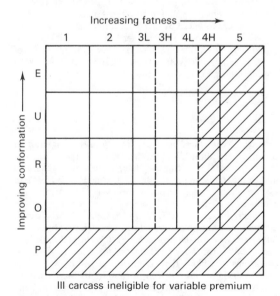

III carcass ineligible for variable premium

Figure 15.9 Sheep carcass classification grid

Systems of lamb production

If lambs are categorised according to date of sale, then in the UK early lambs, sold before the end of June, account for only 16% of total output; lambs finished off grass 35%, lambs finished on forage crops 30% and hoggets 19%. There are many systems of lamb finishing. Important elements of the main ones are as follows.

Early lamb

Aimed particularly at the Easter Market, prices peaking between mid March and late April as hogget numbers decline. Dorset Horns and their crosses are able to lamb regularly in the period September to December, as are selected strains of other breeds. The system is characterised by high lamb prices and high stocking rates for weaned ewes (25 ewes/ha), but these points tend to be offset by lower lambing percentages and higher concentrate costs.

Traditionally Dorset Horn lambs have been run on root crops, when only modest rates of gain are necessary. For later-born lambs the need is for concentrate feeding, with high quality feed (12.5 MJ, 170 g DCP) being fed from three weeks. On such diets an FCE (Feed Conversion Efficiency) of 3.5:1 is possible, with lambs going off at 34 kg liveweight in 12–14 weeks. This type of system fits well onto arable units where grass is limiting, and where 'off-peak' buildings and labour are available.

Lambs off grass

The traditional system of production in many areas, but tending to lose favour because of the low guide price in July, August and September under the current Sheepmeat Regime of the EEC. Such systems should aim for high lambing percentages (170 +), high stocking rates 15–18 ewes/ha and as many lambs sold 'off-the-ewe' as possible (75%). Lambs not sold by weaning may then be finished on silage aftermaths or forage crops, or sold as stores. Performance of lambs on such a system is improved by the adoption of forward-creep-grazing techniques. Two such systems, which have been well documented, are the 'Grasslamb' system developed by the Grassland Research Institute (now known as AGRI) (see GRI Farmers Booklet No. 2), and the Follow-N system of the National Agricultural Centre. Both these systems utilise high levels of Nitrogen (upwards of 250 kg/ha), the main difference being that the GRI system uses rotational grazing and forward creep grazes the lambs, while the NAC set stock and rely on anthelmintics to reduce worm burden. The stocking levels achieved may be compared with average values on more traditional systems of only 10–12 ewes/ha at nitrogen levels below 100 kg.

Table 15.14 Relationship between carcass weight and carcass classification

Carcass wt range (kg)	% carcasses in fat class			
	2	3	4	5
16–17	29	60	7	—
18–19	18	63	15	1
20–21	11	60	24	3
22–23	7	52	33	6
24–25	4	40	41	14

Lambs and hoggets off forage crops

Those lambs which have not been finished by 1 October in any year are generally termed store lambs. Many such animals change hands at this time of year but some will be finished on their home farms. Animals available will be wethers of the hill breeds, weighing 23–30 kg, wethers of the hill crosses (mules, etc.) at 35–35 kg and the remains of fat lamb flocks at 30–40 kg. Feeds for these animals may be grass, silage, catch-crop roots, main crop roots and arable by-products. It is important that lambs are matched to the feeding system proposed and that stocking rates are carefully monitored. For example, main-crop roots are expensive to grow and the aim must be for high yields and high stocking rates. On the other hand if stocking rates are too high individual animal performance may suffer. Details of potential yields and stocking rates for the main crops are given in *Table 15.15*.

Table 15.15 Yields and stocking rates of forage crops

Crop	Season of use	Yield (t/ha)	Stocking (lamb days/ha)
Stubble turnips	Sept–Jan	40–50	2500 (i.e. 60 lambs for six weeks)
Rape	Oct–Feb	30–35	2500
Swedes	Dec–Mar	65–100	8000
Beet tops	Oct–Dec	?	1800
Grass	Oct–Mar	?	600

Grass finishing

Suited to Down crosses for finishing by December, halfbred wethers for February–March sale. Where dairy cow paddocks are being grazed sheep must be off by December to avoid subsequent yield drop; potential silage ground may be grazed until March. Relatively low rates of gain (70–100 g/d) produce carcasses with acceptable fat levels, but Down cross lambs will become fat if held too long.

Silage finishing

Depends on availability of surplus silage of good quality (DM 25–35%, 'D' value 65 +, ME 10.5 MJ +, pH 4, NH$_4$N 10%). Needs lambs of 30 kg liveweight with good frame (e.g. Halfbred wethers). Growth rates are variable around 75 g/d. Where good silage is available (100–120 kg/lamb) it is only necessary to supplement with small quantities (30 g/d) of fishmeal to improve silage intake. For silage of 65 D up to 250 g/d of a 15% protein barley/protein mix will be necessary. This amount will increase to 450 g at 63 D and 750 g at 61 D. These quantities may need increasing further if silage intake is limited (e.g. for long-chop or poorly fermented material). At the same time, in some situations silage intake may be suppressed by concentrate feeding. A 10 cm feed face is required for *ad libitum* silage feeding and 35–40 cm trough for concentrates. One of the problems of this system is that where lambs are held to take advantage of the higher prices in late February/March they may be occupying buildings needed for the breeding flock.

Finishing on roots

The performance of lambs on roots is very variable, between 50 and 200 g/d. Short-term systems, aiming at merely 'finishing' well-grown but lean lambs, will utilise stubble turnips. Long-term systems, usually involving swedes or hardy turnips, aim to have animals putting on lean before finishing and depend on the rise in lamb prices in February/ March for their profitability. One of the main problems with this system is that of balancing wastage and performance. High individual-animal performance is normally associated with high waste (up to 60% in a bad year). Where heavier stocking is practised, to reduce crop waste, lamb growth suffers. The feeding of up to 200 g barley/d can help maintain outputs and reduce variation between animals. Block grazing tends to produce lower wastage rates, without affecting lamb performance, and has lower labour requirement than strip grazing.

Finishing on cereals

The high cost of cereal-based diets means that this system is only applicable to February/March sales. Lambs must be capable of going to 40 kg liveweight without going to fat, but as with root-based finishing the production of over-fat lamb is a major hazard. Ideally lambs are introduced to cereals whilst still at grass, having been vaccinated against pulpy kidney (*C. welchii*), and preferably also against *Pasteurella*, at least two weeks earlier. Lambs may be housed when eating 500 g/d and gradually brought to *ad libitum* feeding, but shy feeders should not be brought in. The ideal cereal for lamb is whole barley, but addition of protein and minerals is then difficult. A compromise, 13% protein ration might be as follows: whole barley 350 kg, rolled barley 300 kg, beet pulp shreds 230 kg, soyabean meal 100 kg, sheep minerals 20 kg. On such a diet lambs will gain 1 kg/week, at an FCE of 5–5.5:1, and will tend to get fat. Regular handling of all animals is necessary if fat class 4 carcasses are to be avoided.

The use of concentrate feeding may also be considered in systems involving 'out-of-season' lambing, when concentrates (12.5 MJ, 17% CP) are introduced at 7 d and fed *ad libitum* to slaughter, or when six-week weaning of one lamb from twins is practised (*see* p. 424). For such concentrate-fed lamb, milk substitute should be offered *ad libitum* for three to four weeks followed by one week restricted access. A feed conversion of 1:1, on a dry matter basis, is possible for the liquid diet, with 4:1 being achieved on concentrates. Gains of 300g/d are possible, with 10 kg milk substitute and 100 kg concentrates necessary per lamb.

Milk production

In recent years there has been an upsurge of interest in milking sheep in the UK. Elsewhere in the world the practice is widespread. Within Europe, Italy, Spain, Portugal, Greece and France have well established industries (*Table 15.16*). In Greece sheep's milk accounts for over 35% of all milk produced.

As may be seen from *Table 15.16*, yields of sheep's milk are extremely variable. In most of Europe, ewes are used as dual-purpose animals, producing both meat and milk, and are suckled for varying lengths of time. Under UK conditions it is likely that milking ewes will be weaned as soon as the lamb has taken colostrum. Many breeds may be milked satisfactorily but yields will vary. 'True' milking breeds, such as the Friesland and British Milksheep, will produce up to 450 litres in a 180–220 d lactation. More 'dual purpose' animals, such as the Clun or Mule, will yield

Table 15.16 Outline of sheep's milk production in Europe

Country	No. of ewes ($\times 10^3$)	Av. flock size	Main breeds	Av. yield (kg)
Italy	5000	10	Sarda	100–250
			Gentile di Puglia	30–35
			Sopravissana	50–55
Spain	2500	60	Churra	120
			Manchega	100
			Castellana	90
Greece	2500	80	Chios	
			Vlahico	90
			Karagunico	
Portugal	1000	60–100	Merino	
			Churra	120
			Bordaliera	
France (Aveyron and Bas Pyrenees)	1000	100–350	Manech	150
			Corse	110
			Lacaune	160

upwards of 150 litres. The lactation curve of a sheep is similar to that of the cow. 'Letdown' occurs in two phases, with an initial flow of milk from the udder cistern followed some 20 s later by that from the alveoli. Udder washing may reduce this delay. Ewe's milk is very rich, having approximately 70 g fat, 50 g lactose and 50 g protein/kg. This creates a high energy demand (7.8 MJ/kg) as well as tending to create difficulties with milkstone deposits on equipment.

As with the dairy cow, the most difficult time in terms of feeding the ewe is in early lactation when appetite is limited. This is especially true of ewes yielding in excess of 3 litres. The nutrition of the ewe is discussed later. One of the major problems likely to be encountered when ewes are lambed in winter is that of low water intake. Ewes may be encouraged to drink by using a feed such as sugar beet pulp presented soaked and covered with water. In some areas too there may be problems with compulsory dipping of sheep for scab. Following dipping there is a compulsory 70 d withdrawal of milk if HCH (BHC) has been used, and a 40 d withdrawal after Diazinon.

Wool

At one time wool was a major source of income for the sheep farmer. Today wool accounts for less than 10% of gross returns. Yields of wool vary greatly between breeds, with longwools such as Teeswater and Devon longwool producing 5–6 kg/head while hill breeds such as the Swaledale and Herdwick will only yield 2 kg.

Wool is produced by follicles in the skin. Two distinct types of follicle are identifiable. Primary follicles, normally associated with an erector muscle and a sweat gland, produce coarser, medullated fibres of hair or kemp. In improved breeds many of these coarser fibres are shed soon after birth. Secondary follicles, associated only with a sebaceous gland, produce wool fibres. The ratio of primary:secondary follicles determines the fineness of the fleece and varies between breeds. For example the fine-woolled Merino may have 25 secondaries per primary, while most British breeds will have no more than eight.

Fineness is a major criterion in wool grading, the standard measure of fineness being the Bradford Count. Bradford Count is the number of hanks of yarn 510 m long that can be spun from 450 g wool. Values will vary from those of the Merino at 80 + to those of the Blackface at 27 to 30. There are also within-fleece differences, with breech wool being coarsest and shoulder wool finest. Breech wool also contains a higher percentage of kemp. A second quality factor in wool is its crimp; that is the number of corrugations per 25 mm. Crimp values vary from 8 to 28 and are closely corrected with fineness.

The growth of wool is photoperiodic and cyclic, 80% of growth taking place between July and November. Wool fibres are almost entirely keratin – a protein with a very high cystine content (12–14%). Notwithstanding this, the main relationship between nutrition and wool growth is in terms of energy input. Wool continues to grown even when the sheep is in negative energy balance. It does however grow more slowly and is finer under such conditions. While wool which is uniformly fine along its length is desirable, the development of a thinner area in an otherwise thicker fibre, a condition known as tenderness, is to be avoided since it reduces the value of the fleece and may even cause premature shedding. Tenderness develops in conditions of underfeeding, for example in ewes suckling triplets, or during bad attacks of parasitic gastroenteritis or liverfluke.

At shearing, care should be taken to preserve the quality of the fleece. Sheep must be dry when shorn, double cuts, which reduce effective staple length, should be avoided, organic matter such as straw and faeces should be kept out of the wool, and the fleece should be carefully packed and stored in the dry. Considerable penalties are incurred for badly presented or marked fleeces. Details of grades, prices and penalties are published annually by the British Wool Marketing Board, who have a monopoly on all wool sales.

No yield advantage accrues from shearing sheep more frequently than once yearly. Sheep are traditionally shorn in late spring or early summer, but recent years have seen the development of winter shearing of housed ewes. This technique allows increased stocking densities (up to 15% more) and tends to produce cleaner fleeces. It also reduces heat stress in the ewes, leading to increased dry matter intakes of the order of 10%. This extra intake is reflected in higher lamb birthweights and lower perinatal mortalities. In the prolific ewe flock at Drayton EHF shearing increased birthweight by 600 g/lamb and reduced mortality from 16 to 7%.

Winter shearing does not fit all situations. To avoid problems the house must be well bedded and draught-free. A minimum of eight weeks must elapse before turnout, to allow some regrowth, and even then it is undesirable to turn shorn ewes out before early March in sheltered areas of the south or early April in the north. Ewes with condition score below 2.5 should not be shorn.

Flock replacements

Flock replacement costs are a major element in the profitability of a sheep enterprise. In 1985 the costs averaged £9.45/ewe in MLC recorded flock.

Keeping these costs down is a question of minimising the number of ewes to be replaced each year and of keeping down the cost of each replacement.

On most hill farms some ewes will be lost and some will need to be culled. The majority of ewes leaving the farm will be drafted out at four to five years of age simply because they are no longer able to cope with conditions. Thus, on average some 25% of the flock will need to be replaced each year. Set against this is the fact that many of these ewes will be sold as draft breeding stock rather than culls, thus making a substantial contribution to the output of the flock (17% for MLC recorded flocks in 1984/85). Traditionally, and of necessity, replacement ewes will be retained from within the lamb crop. This is essential if the ewe lamb is to become a productive member of the flock. Within a hill flock, groups of sheep become territorially organised or 'hefted'. This tendency is encouraged by carefully shepherding, for it reduces the need for fences and assists the survival of the sheep since they know of, and will seek out, areas of shelter and sources of food during bad weather. From a husbandry viewpoint one of the major problems created by the need to keep ewe lambs as replacements is what to do with them over their first winter. If they stay on the hill then they add to feed requirements at a difficult time, and some indeed may not survive the winter. On the other hand, the traditional 'tacking' or 'agistment' of ewe lambs onto lowland farms has been made more difficult by a reduction in the number of farmers willing to cooperate and by escalating costs, both of agistment and of transport. One compromise which may become more common is the provision of housing for these animals on the home farms.

Under lowland conditions most ewes being sold off will be culls. Criteria taken into account when deciding on which animals to cull may include condition of udder, feet and mouth, previous history, condition and temperament. Many of these factors are influenced, to a greater or lesser extent, by management. It could be argued, for example, that culling because of bad udder or bad feet should only occur as an extreme measure, and that if many sheep are involved it may be that flock management is suspect. Culling on teeth condition will depend on two main management factors: stocking rate and winter feed policy. The need for a ewe to be 'sound' in mouth relates to her ability to feed. As stocking rates increase so competition for grass increases and the ewe needs to be able to graze closer to the ground, a facility made more difficult if the ewe is 'broken-mouthed'. Similarly, if

winter feeding involves the use of 'hard' roots, such as swedes or turnips, the broken-mouthed ewe will be unable to compete and will tend to lose condition. Additionally, the feeding of hards roots tends to increase teeth loss. By four and a half years of age up to 75% of ewes may still be full-mouthed if fed on hay and concentrates; in root-fed flocks the figure may be as few as 35%.

In addition to the above factors, opportunity plays a part in determining culling rate. In years when replacements are expensive culling levels tend to be lower than in years when they are relatively cheap. The cost of each replacement can also be influenced by the age of the animal at purchase; ewe lambs being appreciably cheaper than shearlings. Having opted for the cheaper ewe lambs, the farmers must next decide whether to put them to the ram in their first year or not. Puberty in sheep is influenced by age, bodyweight and daylength, older heavier animals being more likely to attain puberty than younger or lighter ones. The advantages of breeding from ewe lambs are that it increases lifetime lamb production per ewe and decreases replacement cost, but against this must be set the likelihood of more mis-mothering problems and difficulties in getting their lambs away fat. In the end the decision may well depend on the source of the lambs. For homebred animals, whose management has been aimed at producing an animal suitable for tupping at seven to nine months, the system can work quite well. For the producer who is using longwool crosses, and who of necessity has to purchase all his replacements, it may be much less viable.

Feeding the ewe

Theory

As has been emphasised, the nutrition of the ewe is a major factor in determining flock performance. Feed requirements vary with liveweight, physiological state and environment. Details of these requirements, and suggested systems for meeting them, are discussed in the MLC publication *Feeding the Ewe*, and in MAFF Reference Book 433. The main points are discussed below.

Voluntary feed intake

Dry matter intake will vary with physiological state and with diet, as well as with bodyweight. Intakes will normally be within the range 60–70 g/kg $W^{0.75}$ (where $W^{0.75}$ metabolic bodyweight), but may be as high as 100 g/kg $W^{0.75}$ for concentrate diets or as low as 45 g/kg$^{0.75}$ on silage-based diets, especially if the silage is acid (pH below 4) or poorly fermented. Appetite is depressed by up to 20% in late pregnancy and very early lactation but recovers quickly to reach a peak some two to three weeks after peak milk yield is achieved. The increase in intake between late pregnancy and peak may be as high as 60%.

Energy requirements

Energy requirements vary according to bodyweight, stage of pregnancy and milk yield, and are reduced by 0.75–1.2 MJ/d in housed ewes. Maintenance requirements, based on $M_m = 1.8 + 0.1 W$, are usually in the range 5–10 MJ/d. During pregnancy requirements increase significantly as fetal growth accelerates in the last six to eight weeks (*see Figure 15.7*), being 20% higher six weeks before lambing and up to 150% higher immediately pre-lambing. During lactation energy requirement varies with milk yield, which is in turn influenced by the number of lambs suckling. Average milk yields are given in *Table 15.13*. Requirements for energy, based on the above considerations, are given in *Tables 15.17, 15.18* and *15.19*.

Protein requirements

Protein requirements also vary according to the physiological status of the ewe. Maintenance levels are generally between 65 and 100 g CP/d. For weight gains (increases in body condition) before tupping these values increase by up to 100% (*Table 15.17*), but during mid pregnancy maintenance levels are again adequate. During the last six weeks of pregnancy requirements increase to between 110 and 235 g CP/d, depending on ewe weight and number of lambs carried (*see Table 15.18*). Throughout this period the provision of 10 g CP/MJ ME in the diet should be adequate, but where ewes are expected to utilise backfat during late pregnancy there will be a need for supplementary protein, preferably in the form of a material of low degradability such as fishmeal or soyabean meal. For the lactating ewe protein requirements depend on stage of lactation and number of lambs suckling (*see Table 15.19*). Here, too, the use of an undegradable protein in advantageous, particularly in situations where ewes are utilising back-fat reserves.

Table 15.17 Mean daily allowances for ewes up to mid pregnancy

Liveweight (kg)					Postmating			
			Premating		Month 1		Months 2 and 3	
			ME (MJ)	CP (g)	ME (MJ)	CP (g)	ME (MJ)	CP (g)
40	M	+0[1]	6	65				
		+0.25[1]	9	95	6	65	5.5	55
		+0.5[1]	12	120				
60	M	+0[1]	8	85				
		+0.25[1]	12	125	9	90	7	70
		+0.5[1]	17	165				
80	M	+0[1]	10	105				
		+0.25[1]	16	155	11	110	9	90
		+0.5[1]	22	210				

[1] Level of body condition change in 28 days premating

Table 15.18 Mean daily allowances for ewes in late pregnancy

Live weight (kg)	No. of fetuses	Weeks before lambing					
		6		4		2	
		ME (MJ)	CP (g)	ME (MJ)	CP (g)	ME (MJ)	CP (g)
40	1	6.7	80	7.4	95	8.2	110
	2	7.1	95	8.2	120	9.5	150
60	1	8.8	110	9.8	125	10.8	140
	2	9.4	130	10.9	155	12.7	185
80	1	10.9	135	12.1	150	13.4	165
	2	11.8	155	13.7	190	15.8	235

Table 15.19 Mean daily allowances for lactating ewes (Based on yield data from *Table 15.13*)

Live weight (kg)	No. of lambs	Month of lactation					
		1		2		3	
		ME (MJ)	CP (g)	ME (MJ)	CP (g)	ME (MJ)	CP (g)
40	1	16.3	200	15.9	200	13.2	150
	2	23.3	300	20.3	240	15.6	190
60	1	20.3	210	19.9	205	16.7	170
	2	27.6	315	24.2	250	19.0	205
80	1	22.3	220	21.9	210	18.7	190
	2	29.6	335	26.2	265	21.0	220

Feeding in practice

In practice it may be neither practical, nor economically desirable, to meet the ewe's requirements in full at all times. Most flocks contain ewes with a range of bodyweights and carrying or suckling varying numbers of lambs. The aim should be to feed accurately at critical times in the production cycle, using the body condition of the ewes as a guide. The need for ewes to be in good body condition, and on a rising plane of nutrition, at tupping has already been discussed. The ewe should not be allowed to drop in condition for the first month of pregnancy, but during the second and third months a moderate degree of underfeeding, leading to weight loss of up to 5%, is acceptable. Forage-only diets, with an energy concentration of 8–9 MJ/kg DM are adequate at this stage. During late lactation it is essential that ewes are brought back to good body condition (condition score 3–3½). Pregnancy scanning can allow grouping of ewes according to number of fetuses and differential feeding. Otherwise lowland ewes should be assumed to be carrying twins and hill ewes singles. Concentrate feeding will normally begin, some six to eight weeks before lambing, at 100 g/head/d. This amount should gradually increase to a maximum of 600 g/d at lambing. With housed ewes and good quality silage concentrates may only be fed for four weeks and to a maximum of 400 g. At all times the body condition of the ewe should dictate concentrate feeding level. Forage will usually be offered *ad libitum*.

In many hill situations such feeding is economically more difficult to justify and practically often very difficult, not only because of the difficulty of reaching the ewes but also because it is undesirable to interrupt the normal grazing behaviour of the flock. In such conditions the judicious use of feed blocks may be considered. Such blocks should be sited at strategic points, allowing one block/30 ewes/week. These blocks should be offered from late January and where possible should be supplemented by small amounts of grain feeding in the last few weeks of pregnancy (up to 200 g/head/d). Additionally storm-feeding provision will often be necessary. Adequate quantities of hay, ideally 1 bale/ewe, should be stored on the hill to be fed when grazings are snow-covered.

For the lactating ewe the amount of hand feeding will depend on date of lambing. Where grazing is immediately available then no supplementary feeds may be needed. In early-lambing flocks root crops may be utilised or forage feeding continued. In these conditions the use of an undegradable protein source will maintain milk yields and lamb growth rates. Under normal circumstances the aim should be

Table 15.20 Suggested rations for 75 kg ewe in late pregnancy (kg/d)

Ration	Silage	Hay	Swedes	Kale	Concen-trates[1]	Barley
1		0.6	4		0.6	
2		0.7		4		0.5
3	4				0.6	
4		1.0			0.6	

[1] 13 MJ/kg DM; 160 CP/kg DM

to restrict concentrate feeding to 45 kg/ewe per year, with about 0.5 t silage or 215 kg hay/ewe being needed during winter housing. Suggested rations for a 75 kg lowland ewe in late pregnancy are given in *Table 15.20*.

Grazing management

Good grazing management is one of the most important factors influencing sheep flock profitability. Variable costs will be affected by the extent to which the flock needs supplementary feeding, and by forage variable costs, while gross margin/ha is very dependent on stocking rate. Examination of MLC figures shows that 40% of the superiority of the 'top-third' flocks is accounted for by better stocking rates and only 25% by better lambing percentages. In 1985, average stocking rate was 12.8 ewes/ha while the top one-third of producers kept 15.1 ewes/ha. These higher stocking rates were achieved at higher levels of nitrogen/ha (184 *versus* 160 kg) but similar levels/ewe.

The use of ewes/ha as a measure of stocking rate can be misleading, since it takes no account of ewe size. In the MLC publication *Prime lamb from grass* an attempt has been made to overcome this problem by interrelating ewe weight, prolificacy and fertiliser usage, rather than ewe numbers. Representative values for average quality land are given in *Table 15.21*, but individual farm circumstances may markedly affect these values. Although there is a trend towards higher fertiliser N levels leading to higher stocking rates (+ 10 kg N = 0.3 ewes/ha), many farmers fail to fully exploit this potential. A second problem associated with the recording of grassland usage, especially on mixed farms, is that of allocating variable costs. The simplest way of standardising figures is to use the livestock units system (*see* Chapter 6).

The times of the year when problems of high stocking

Table 15.21 Relationship between ewe weight, nitrogen use and stocking rate on average land

Targets	Nitrogen (kg/ha)		
	0–75	75–150	150–225
Wt of ewe carried (kg)	750	900	1050
Wt of lamb produced (kg)	600	750	850
Stocking rates to achieve above targets (ewes/ha)			
Welsh halfbred	13	16	18
Mule	11	13.5	15.5
Scotch halfbred	10	12.5	14.5

After *MLC Prime Lamb from Grass*. (Reproduced with permission)

rates are likely to be most apparent are in early spring and in the latter half of the grazing season. In early spring the provision of adequate grazing for the lactating ewe is important in terms of milk yield and lamb growth rates. It is easier to obtain this early grazing if the flock can be kept off the grass over winter. Where sheep are run at grass over winter the effect on subsequent performance is variable and will depend on season and timing. Up to January there should be no effect on yield, but grazing in January and February will reduce early growth by 20%, or even more in a wet year. On the other hand, most swards have recovered from such an early grazing by silage or hay making. Wherever possible therefore sheep should be outwintered on fields destined for conservation.

In-wintering of sheep is becoming more common. More than two-thirds of lowland flocks are lambed indoors but only about 20% are housed throughout the winter. (The question of housing is examined thoroughly in Bryson, 1984.) In-wintering is expensive and unlikely to be justifiable in terms of increased ewe output. It can only be defended on grounds of easier shepherding and if grassland utilisation is improved as a result.

In the latter half of the grazing season the main problem is competition between ewes and lambs, which can lead to reduced rates and increased parasitic gastroenteritis in the lambs. This situation is most acute on all-grass farms where sheep are the only enterprise, since there is less scope for utilising silage aftermaths or alternating grazing between species. In such conditions the forward creep grazing of lambs is worth considering and may give benefit of up to 500 g extra growth/lamb per week.

Increasing lamb numbers per ewe

As was shown in *Figures 15.2 and 15.3*, average flock performance is well below what is theoretically possible. Under most hill farm conditions a 100% lambing is considered satisfactory. Where higher lambing percentages are required, manipulation of ewe nutrition, as discussed earlier, can lead to increased numbers of lambs being born. There is then a need to adopt an alternative management strategy.

Early weaning of hill lambs

The main problems associated with twins from hill ewes are loss in body condition in the ewe, which may be severe enough to cause barrenness, and reduced growth rates in the lambs. Removal of one lamb, leaving a female where possible, can overcome these problems.

Weaning at 24 h, once the lamb has taken sufficient colostrum, allows the ewe to return immediately to the hill. The 'weaned' lamb must be fed on milk substitute and concentrates. A higher labour and feed cost, coupled with generally poor lamb performance, makes this a difficult system to operate sucessfully. Where adequate inby grazing is available, it is more satisfactory to allow both lambs to suckle for five to six weeks before weaning one. The ewe may then be returned to the hill while the weaned lamb will either remain at grass, a technique demanding clean grazing and lamb-proof fencing, or be concentrate-fed indoors. This latter system is more likely to produce a finished lamb, with growth rates in excess of 200 g/d, and can improve the utilisation of buildings used to house the flock through the winter.

Under lowland conditions there is often considerable scope for improving output using conventional management. A number of techniques are however available for those who wish to move further.

Use of more prolific ewes

Prolificacy in sheep is a trait with low heritability (0.1). The development of a flock of highly prolific ewes will generally involve the use of animals with some Finnish Landrace blood (e.g. the Finn–Dorset) or of the relatively new Cambridge breed. Where such animals are used as breeding females then lambing percentages in excess of 250% may be expected. As mean lambing percentage in a flock increases so the number of triplet and quadruplet births increases (*see Table 15.22*). A high level of stockmanship is necessary in

Table 15.22 Relationship between lambing percentage and distribution of litter size

Mean no. born	Litter size distribution (%)				
	1	*2*	*3*	*4*	*5*
1.5	49	50	1		
2.0	12	76	12		
2.5	9	43	41	5	1
2.6	10	36	41	13	2

such flocks. Chief features of the management in these situations are as follows:

(1) careful organisation of mating to facilitate organised lambing;
(2) generous feeding in late pregnancy;
(3) closely supervised lambing, normally indoors in March early April;
(4) weaning by mid-July to allow ewes to regain body condition prior to tupping.

Additionally, the combination of Finnish Landrace blood and the smaller birthweights of triplet/quadruplet lambs means that careful selection of terminal sires is necessary if lambs are to be finished successfully.

Hormonal manipulation

Two broad approaches to the hormonal manipulation of ovulation are possible: direct stimulation (e.g. using PMSG) and reduction of feedback controls (e.g. using Fecundin). An outline of the endocrinological background to these materials is given in Chapter 12. In practice, PMSG is seldom used in flocks lambing at the normal time, due to the unpredictability of the response. Its use is confined to out-of-season lambing (*see later*). Fecundin is used as part of conventional management, and in flocks whose normal performance is between 100 and 180% lambing can give up to 25 extra lambs/100 ewes. This result may be achieved by vaccinating ewes at eight and four weeks pre-tupping in the first year of its use and at four weeks only thereafter. Following Fecundin use it is essential that adequate attention is paid to flock nutrition. In situations where earliness of lambing or lowered genetic potential are the causes of low lambing percentage then Fecundin may prove useful. When poor performance is a reflection of poor overall management then the use of Fecundin is not advisable.

Increased frequency of lambing

Theoretically it is possible for ewes to lamb every six months and some may do this. However, ewes may not recover from one lambing in time to be mated again and the problems of operating such a system are many. In particular low conception rates following the mating of lactating ewes necessitate very early weaning and failure to conceive greatly increases barrener percentage or spreads lambing to an unacceptable degree. More commonly ewes will be lambed every eight months and often two flocks will be run, four months out of synchronisation, so that ewes failing to conceive in one flock may be 'slipped' into the other and so given a second chance (*see Figure 15.10*). In a frequent-lambing flock a variety of breeds may be used, but, although exogenous hormones or daylight manipulation will generally be used, highly prolific ewes having a long breeding season are to be preferred. In this respect ewes of the Finn-Dorset type are probably ideal.

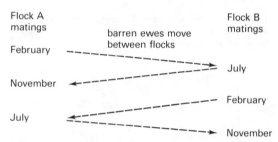

Figure 15.10 Organisation of split flock allowing 'slipping' of animals failing to conceive

Although daylength may be manipulated to increase breeding frequency (*see Figure 15.11*) the method is seldom commercially viable and exogenous hormones will be used. An intravaginal sponge, impregnated with progesterone, is left in place for 12 days. On its withdrawal an intramuscular injection of 500–750 units PMSG is given and ewes will show signs of oestrus 36–72 h later. A detailed calendar describing such a system is given in Appendix 2, p. 429.

Critical to such systems are the management of the rams, the close supervision of lambing, the abrupt weaning of the lambs at one month of age and the careful nutrition of the ewe. Close supervision of lambing may be made easier if lambing is induced. The use of corticosteroid injection (e.g. 20 mg dexamethazone) on the evening of day 142 of pregnancy will result in ewes beginning to lamb some 36 h later. Lamb viability and growth rates, and ewe fertility, are not affected by such treatment.

Diagrammatic representation of the nutrition of a frequently-lambing ewe is given *Figure 15.12*. Of particular importance is the protein input. In early lactation the inclusion of undegradable protein, such as fishmeal, facilitates backfat utilisation and improves milk yield, while the reduction in total protein at the end of week 3 of lactation speeds the drying off process and encourages backfat deposition.

The frequent lambing system is characterised by high variable costs. It has been suggested that an annual production in excess of 2.5 lambs/ewe is necessary for the system to be viable, but the high stocking rates possible with the system (25 ewes/ha) can help compensate for lower margins/ewe. A detailed discussion of this aspect of sheep production may be found in Littlejohn (1977).

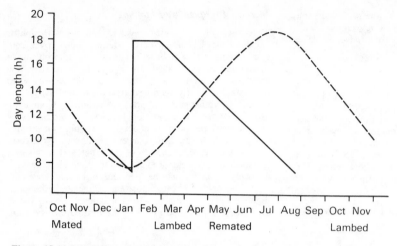

Figure 15.11 Manipulation of daylength to control oestrus. (From Robinson, J.J. *et al.* (1975) *Annals Biol. Anim. Biochem. Biophys.* 15, 345, reproduced by courtesy of the editor and publishers)

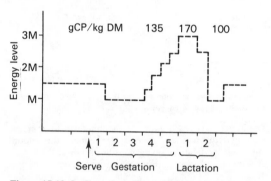

Figure 15.12 Outline of nutrition of frequently-lambing ewe

GOATS

There are more than 460 million goats in the world, largely in Asia and Africa (*Table 15.23*). Within Europe there are regional differences in the uses to which goats are put. In France and the UK the main output is milk and cheese. Here the most important breed is the Saanen (and its crosses) and average yields are 500 litres/year (France) and 800–1000 litres year (UK). In Spain aned Italy approximately 30% of goats are milk-type, with the remainder dual-purpose or meat-producing, whereas in Greece almost all animals are dual-purpose and milk yields are lower. An

outline of main breed types and milk yields is given in *Table 15.24*.

Goat production in the UK is still in its infancy, with approximately 100000 breeding does. The main output is milk, but interest in fibre production is growing. Estimated goat numbers, together with average milk production data, are given in *Table 15.25*.

Milk production

The lactation curve of the goat is similar to that of the cow. Potential lactation yield may be predicted by multiplying

Table 15.24 Types and production of goats in Europe

Country	Main products		Breed(s)	Yield (litres/year)
France	Milk	50%	Alpine	570
	Meat	50%	Saanen	570
Greece	Milk/meat		Not defined	120
Italy	Milk	30%	Sarda	150–200
	Meat	70%	Maltese	450
Spain	Milk	30%	Murciana	500–850
			Malaguena	600
			Delas Mesetas	300
			Canaria	700

Table 15.23 World distribution of goats ($\times 10^6$) (1985)

Africa		Asia		North America		Europe		S. America	
Nigeria	26	India	81	Mexico	10.5	USSR	6.5	Brazil	8
Ethiopia	18	China	63	USA	1.5	Greece	5	Argentina	3
Somalia	16	Pakistan	29	Haiti	1.1	Spain	3	Bolivia	3
Sudan	14	Iran	13			France	1	Peru	2
Kenya	8	Indonesia	11			Italy	1		
Tanzania	6.5	Bangladesh	10						

Header totals: Africa 155, Asia 251, North America 14, Europe 19, S. America 20

FAO Production Yearbook (1986) (reproduced with permission)

Table 15.25 Goat breed distribution and lactation data for UK

Breed	%	Lactation yield (kg)	Fat (%)
Crossbred	53	800 +	3.5
Saanen	17	920	3.6
Anglo-Nubian	12	750	4.5
Toggenberg	12	880	3.3
Alpine	5	920	3.5

Table 15.26 Typical composition of milk of different species

	Lactose (%)	Fat (%)	Protein (%)
Cow	4.5–5.0	3.5–4.0	3.0–3.5
Ewe	5.2–5.5	5.5–11.0	4.5–7.5
Goat	4.0–5.0	3.0–4.5	2.8–3.6

daily yield at peak by 200; beyond peak, yield may be expected to decline at 2–2.5% per week. In the non-pregnant goat, lactation can continue for up to two years. Whilst this will increase lactation yield, annual production is unlikely to equal that of two lactations in the same period.

Table 15.26 shows that the analysis of goat's milk is similar to that of the cow. The fat in goat's milk has a higher content of small globules (28% *versus* 10% $< 1.5\,\mu$) and a higher percentage of short-chain fatty acids (15% *versus* 9%). It is some of these short-chain acids – caproic, caprilic and capric – which give goat milk its characteristic flavour. The protein in goat's milk is characterised by smaller casein micelles and by an increased β-casein content (67% *versus* 43%). The absence of carotene leaves the milk looking very white, although the level of vitamin A is higher than in cow's milk, as is that of nicotinic acid. On the other hand goat's milk has only 10% of the vitamin B_6 level found in cow's milk.

There is a specific demand for goat's milk and goat's milk products from those allergic to cow's milk. Estimates suggest that 7.5% of babies and 2.5% of adults may be allergic to cow's milk and that 60% of these are probably not allergic to goat's milk. There is no quota on goat's milk production and fewer regulations governing its production. The sale of goat's milk for human consumption is covered by Sections 1, 2 and 8 of the Foods Act (1984) which covers the questions of contamination and of health hazards.

Meat production

In many parts of the developing world goats are an important source of meat. Total production exceeds 2 million tonnes. In the EEC production is estimated at 80000 t; Greece accounting for 66% of the total. Goat's meat may come from cull adults, from kids weaned and slaughtered at 6–12 weeks or from intensively-finished animals – usually surplus males.

On intensive systems, goats tend to have higher DMI than lambs of similar weight, but poorer efficiencies. Daily gains

of up to 250 g may be expected at FCRs of 4–6:1. Killing-out percentages are similar to lamb but conformation is poorer and carcass composition is different, with goats having half as much subcutaneous fat but almost twice as much kidney fat. There is some evidence that Angora goats may be fatter than other types.

Fibre production

Most goats have a 'double-coat' fleece with long guard hairs covering a finer under-wool. It is this under-wool which is combed out to produce cashmere. In the Angora goat the under-wool is longer and coarser and is clipped twice yearly to produce mohair.

Cashmere fibres should be less than 6 cm long and 13–16 μm diameter. Yields vary between breed types and range from 50–250 g. Sixty per cent of world cashmere output comes from China. Mohair fleeces can weigh up to 2.5 kg, with fibres up to 15 cm long and a diameter of 20–40 μm. Major mohair producers are Turkey, South Africa and Texas.

Feeding goats

Detailed information on the nutrition of dairy goats is scarce. Dry matter intakes, at 80 g/kg $W^{0.75}$, are higher than for sheep, especially in lactating animals. Appropriate metabolisable energy inputs are of the order of 0.4 MJ/kg $W^{0.75}$ for maintenance and 0.7 MJ/kg $W^{0.75}$ during the last eight weeks of pregnancy. For a 60 kg goat these values equate to 9 MJ and 15 MJ/d respectively. During late pregnancy it is important to monitor body condition to avoid does becoming overfat and thus in danger or suffering from ketosis in early lactation. Mean energy requirement for milk production is 5.1 MJ/litre (6 MJ for Anglo–Nubians). During early lactation it is often impossible to meet energy requirements from the diet. Backfat utilisation can provide 2.8 MJ/100 g backfat loss.

The protein requirement of a goat for maintenance is

Table 15.27 Nutrient requirements of housed dairy goats

Yield (kg/d)	Nutrient	Liveweight (kg) 50	60	70
0	DM (kg)	1.5	1.8	2.1
	ME (MJ)	8.0	9.2	10.3
	DCP (g)	51	59	66
1	DM	1.7	2.0	2.3
	ME	13.1	14.3	15.4
	DCP	106	114	121
2	DM	1.9	2.2	2.5
	ME	18.2	19.4	20.5
	DCP	161	169	176
3	DM	2.1	2.4	2.7
	ME	23.3	24.5	25.6
	DCP	216	224	231

For grazing goats, ME requirements should be increased by 25%.
(Modified from *Nutrient Requirements of Goats*, NRC)

Table 15.28 Nutrient requirements of growing goats

Liveweight (kg)	Nutrient	Liveweight gain (g/d)				
		0	*50*	*100*	*150*	*200*
	DM (kg)	0.45				
10	ME (MJ)	3.0	4.5	6.0	7.5	9.0
	DCP (g)	35	45	55	65	75
	DM	1.30				
30	ME	6.8	8.3	9.8	11.3	12.8
	DCP	50	60	70	80	90
	DM	1.5				
50	ME	10.0	11.5	13.0	14.5	16.0
	DCP	61	71	81	91	101

(Modified from *Nutrient Requirements for Goats* NRC)

approximately 7 g DCP/MJ ME (equal to 60 g for a 60 kg doe). This amount should be increased by 60 g/d for the last eight weeks of pregnancy and by 5 g/litre of milk produced (*see also Tables 15.27 and 15.28*).

In lactating goats water intake is critical. In temperate areas intakes of 140 g/kg $W^{0.75}$ plus 1.4 litres/litre milk (i.e. 4–7 litres/d) may be expected. Goats do not like very cold water and intake may drop significantly in winter, adversely affecting yields. Warm water may need to be provided in such situations. Similarly, in concentrate-fed goats, reduced water intake can lead to the development of urinary calculi (bladder stones), a condition to which Angoras seem especially prone. Addition of 1% salt to the diet (as fed) reduces the danger of this condition developing.

Goats are browsers, but will graze when necessary. In grazing trials comparing sheep and goats, and using grass/clover swards and natural vegetation, goats have been shown to preferentially eat indigenous species such as bent grass (*Molinea caerulea*) and rush (*Juncus* spp.), and to select against clover. This suggest that goats may have an important role in the improvement of hill grazing. On the other hand, problems with internal parasites (anthelmintics may not be used on lactating goats without withdrawing milk from sale) and difficulties with fencing have caused many goat's milk producers to house their animals year-round. The goat's tendency to produce tainted milk if fed on aromatic material is an additional hazard.

The most common feed for housed goats is hay, although good silage may also be used. Sugar beet pulp is a useful additional feed, especially early in lactation, when its palatability stimulates appetite and its digestible fibre helps maintain butterfat levels. Most producers favour concentrate feeds presented as a coarse mix, but there is no evidence that these are any better than pelleted diets; they are certainly more expensive.

Compared with silage, hay is generally lower in energy and much lower in protein. A switch to silage feeding, where possible, will thus lead to a saving in concentrate require-

Table 15.29 Annual feed requirements for a housed dairy goat (kg)

	Silage (10.5 ME)	*Hay* (8.5 ME)
Forage	3500	700
Beet pulp	110	110
Concentrate	200	450

ments. Yearly requirements for a dairy goat are given in *Table 15.29*. In fibre production the most important consideration is avoidance of nutritional stress, particularly in terms of energy input. Feeding at levels equal to those required by a dairy goat of similar size is unlikely to produce problems in this respect.

Reproduction in the goat

Does are seasonally polyoestrous, but less so than ewes. Attainment of puberty depends on date of birth and level of nutrition and is generally at four to six months, with full sexual maturity reached at six to eight months in does and eight to ten months in bucks. Oestrous cycle length is very variable about a 21 d mean. Oestrous lasts 4–40 h, with ovulation 30–36 h after onset. Unmated does will have three to five cycles, with up to seven recorded for Angoras. Does in oestrus become vocal and restless, showing active male-seeking activity. Tail fanning tends to cause a wet area around the tail. Behaviour is more marked in the presence of an odorous male (the odour being produced by secretion from a sebaceous gland at the base of the horns). Problems of oestrus detection in goats are less marked than in cattle because seasonality pre-ordains likely breeding periods and because breeding is unlikely to coincide with peak yield. Techniques for synchronising oestrus, or for stimulating out-of-season breeding, are equally applicable to goats and sheep.

Gestation length varies between breeds, within the range 148–153 d, with Anglo–Nubians tending towards shorter and Saanens towards longer gestations. Litter sizes vary between and within breeds, and are largely a reflection of environment ($h^2 = 0.1$). Mean litter size is approximately 1.75 under UK conditions.

Intersexuality

A well-recognised problem in goats is that of reduced fertility resulting from intersexuality. The condition is related to the presence or absence of horns. Polledness is under the control of a single gene (designated P). Three genotypes and two phenotypes are identifiable:

PP = Homozygous polled
Pp = Heterozygous polled
pp = Homozygous horned

In females, PP genotypes are generally infertile while Pp genotypes have a 50:50 chance of being subfertile: these genotypes are indistinguishable. Fewer males are infertile if they are PP genotypes (up to 50%) but almost all will be subfertile. Careful inspection of the vulva of polled kids can often indicate whether or not they are likely to be infertile, presence of a pea-like protrusion indicating a problem.

APPENDIX 1
BODY CONDITION SCORES

0   Extremely emaciated and on the point of death. Not possible to detect any tissue between skin and bones.
1   Spinous processes prominent and sharp. Fingers pass easily under ends of transverse processes. Possible to feel between each process. Loin muscles shallow with no fat cover.

2 Spinous processes prominent but smooth; individual processes felt as corrugations; transverse processes smooth and rounder, possible to pass fingers under the ends with little pressure. Loin muscle of moderate depth but little fat cover.

3 Spinous processes have little elevation, are smooth and rounded, and individual bones can only be felt with pressure. Firm pressure required to feel under transverse processes. Loin muscles full with moderate fat cover.

4 Spinous processes just detected with pressure, as a hard line; ends of transverse processes cannot be felt. Loin muscles are full with thick covering of fat.

5 Spinous processes cannot be detected, even with firm pressure; depression between layers of fat where processes would normally be felt. Loin muscles are full with very thick fat cover.

APPENDIX 2
CALENDAR FOR FREQUENT-LAMBING FLOCK

July
14 Insert sponges.
27 Withdraw sponges. Inject 750 units PMS.
29 Mating. Allow adequate rams (1:10) and rotate between paddocks every 8 h or use AI.

August
15 Mate repeats.
20 Remove rams.

December
23 Begin lambing.

January
9 'Repeats' lamb.
27 Wean first lambs. Insert sponges into all ewes.

February
10 Withdraw sponges. Inject 750 units PMS. Wean 'repeats'.
12 Mate.
28 Mate 'repeats'.

March
3 Remove rams.

July
9 First ewes lamb.
25 'Repeats' lamb.

September
20 Wean all lambs.

November
5 Introduce rams (normal mating, no hormonal treatment).

April
1 First ewes lamb.

June
14 Wean all lambs.

July
14 Insert sponges.
Etc.

Further reading

BRYSON, T. (1984). *Sheep Housing Handbook*. Ipswich: Farming Press
EALES, F.A. and SMALL, J. (1986). *Practical Lambing* London: Longman
GALL, C.G. (1981). *Goat Production*
LITTLEJOHN (1977). *A study of high lamb output production systems*. Scottish Agricultural Colleges
MLC (1983). *Artificial rearing of lambs*

16

Pig production

M.A. Varley

INTRODUCTION

Pig farming has changed rapidly in recent years both in developed countries and in the less prosperous regions of the world. A principal reason for this is the high biological potential of the pig for converting feedstuffs into meat.

Pigs are omnivorous and can utilise a wide range of food materials including plant proteins, bulky feeds, and human food waste. Generally, modern hybrid pigs grow rapidly and efficiently to yield lean carcasses on energy-dense cereal based feeds balanced with a supplementary protein source such as soyabean meal. Throughout Western Europe the nutrition of pigs is based on cereals such as barley, wheat or maize with soya and canola meal as protein sources. Some countries utilise imported cassava meal as an energy source and this is supplemented with soya protein or canola. In Third World countries, the use of pigs for meat has taken some criticism because the animals compete for available food energy which could be eaten directly by the human population. Much depends on the particular economic development of a country. If a given community has moved off the economic baseline it will demand food in the form of animal protein. The use of pigs is then a viable proposition because of their innate capacity for efficient growth and production.

The reproductive characteristics of the pig are certainly impressive when viewed alongside other species and with today's systems of production it is possible to produce 25 piglets per sow per year. There is still large variation in what is actually achieved on farms and even on the best of farms it is difficult to maintain constant production over a long time span. Research and extension work in recent years have given the industry sound working guidelines for reproductive management and improving performance has resulted.

On a global scale there is now a great degree of uniformity of the methods used in production. Whether pigs are kept in Europe, North America, Australasia or in Africa there has been a trend towards the use of prolific hybrid sows kept in individual stall houses, weaned at between three and four weeks and the piglets reared in flat-deck cages. It seems therefore that the pig industry is following rapidly the path that the poultry industry traversed 20 years ago.

In Western Europe and in Britain particularly another feature of pig farms has been the increasing specialisation of pig units and the continued expansion of existing farms. The economies of scale for labour utilisation dictate that viable units are much bigger than they were ten years ago and this process is not slowing down.

A few years ago an average commercial pig unit was 100 sows. Currently in the UK this figure is nearer 500 sows and herds of 1000 and 2000 sows are not uncommon. This rapid expansion has presented some problems and one of these is the demise of the traditional family farm which was once the backbone of British agriculture. It is still possible to make a living from a smaller unit where the only labour used is family labour but in future the smaller unit will be under increasing pressure from the larger operators with their minimal unit costs of production.

Another new aspect of the European pig industry is the recent concern amongst the general public about the way in which animals are housed and managed. Legislation in some countries prohibits the use of certain practices such as early-weaning or the use of sow tethers. New systems of production which minimise the stress imposed on animals seem likely to have a major impact on pig production in the next ten years or so. It may be that the structural changes seen in the past are gradually reversed as urban populations demand a more active interest in pig farming.

It is the purpose of the ensuing text to give an overview of the pig industry concentrating mainly on the UK. The intention is to cover important technological developments in the fields of genetics, reproduction, nutrition, housing and marketing to give the reader a basic working understanding of the pig industry.

THE STRUCTURE OF THE UK PIG INDUSTRY

The structure of the UK pig industry has altered significantly since the end of the Second World War as a result of social and economic pressures. In the early 1950s, pigs were considered a secondary enterprise as part of a mixed farming pattern. Herds were small and only a small number of specialist producers existed. This scenario worked well in the context of integrated systems of livestock production. Arable farms growing their own cereals could process milled barley and wheat into pig meat, straw was freely available for bedding and manure disposal was not a problem using cereal stubbles in the autumn months.

Today the pig industry is in relatively fewer hands and there is a continuing trend towards specialist production although as pig units have increased in size the large scale arable farmers of the eastern counties of England have found it easier to expand with minimal slurry disposal problems. There has been a gradual shift of production to the East of Britain partly because of the drier climate but mainly because of the logistics involved in the production, processing and transport of the raw materials involved such as corn and straw. In 1964 there were around 75 000 holdings in the UK which had sows on them. By the middle of the 1970s there were only about 22 000 holdings in the UK with pigs and at the present time about 80% of the nation's pigs are in the hands of 2000 producers. These structural changes allied to rapid increases in the average herd size have meant that the cost of pig meat to the consumer has been kept as low as possible and 33% of all meat eaten in Britain at the present time is pig meat.

Economic factors have been largely responsible for the changes and within the European Economic Community (EEC) there is intense competition to secure and hold markets for pig meat. Denmark has traditionally had a large slice of the UK market for bacon and this seems likely to continue because of the vigorous marketing tactics used by the Danes. More recently, significant quantities of pigmeat have been imported from the Netherlands and also from Ireland. In the future as part of a much larger economic community the marketing of pigmeat will be increasingly complex and difficult for the home producer. The EEC itself now contains Greece, Spain and Portugal and these countries also have rapidly developing pig industries which in due course could be major competition for the UK.

One of the facets of pig farming within the EEC is that the industry has not been supported financially by the same system of intervention prices and storage of products that has helped the milk, beef and sheep industries. Pig farmers have therefore been price takers without the insurance of guide prices and subsidy support. As a consequence of this, the industry in general has exhibited increasing technical skills and the uptake of new technology and research has been swift and effective.

A significant component of the financial burden in pig farming has been increasing costs of feedstuffs. Whilst arable farmers have enjoyed strong support from the EEC budget and cereal prices have been buoyant, the price of wheat and barley represents the major raw material cost to the pig farmer. Eighty per cent of the costs of producing pig meat is the cost of feedstuffs. This situation may change in the future and the EEC administrators are changing their stance on the financial support of agricultural production. It may be that with the relaxation of intervention buying, the costs of production in pig farming could fall in real terms.

Another recent problem for the UK farmer in relation to feedstuffs was changes seen in our fishing industry. Surplus fish manufactured into fishmeal was until the 1975 'cod war' with Iceland a relatively cheap and high quality source of protein for most types of pig. At the present time it is difficult to justify the inclusion of fishmeal in the diets of growing pigs or sows because of its very high price. A benefit transpiring

from this situation has been the stimulation of good research and development work in non-ruminant nutrition and this has demonstrated clearly the value of alternative protein sources such as soya and canola meal.

The pig cycle

The so-called pig cycle has existed for many years now and its characteristics are illustrated in *Figure 16.1*. Because the reproductive rate of the pig is so high it is possible for the national herd to expand and contract very quickly over a relatively short time span. At a time of high profitability and confidence in the industry farmers expand and new producers enter the industry. The rapid rise in sow numbers inevitably leads to overproduction and the price of finished pigs falls. Profitability declines, sows are culled, some farmers go out of business and the national production of pigs falls off. The national herd numbers return to baseline again and the cycle begins again.

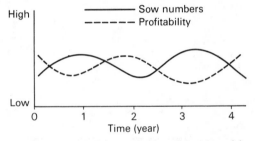

Figure 16.1 A schematic representation of the 'pig cycle'

The amplitude and frequency of the cycle varies considerably but the hope has always been that with large herds committed to steady state production and with heavy investment in buildings and equipment, then the cycle will disappear. This so far has not been the case and the larger producers seem prepared to expand continually. Currently, the indications are that after a very bad year in 1984 producers are being a little more cautious in a reasonably good economic climate.

Summary of the UK industry

In *Table 16.1* are given data illustrating the current position of the UK pig industry. The industry is a significant industry in its contribution to overall agricultural production and, in financial terms, has a turnover around £1 billion per annum. In relation to the rest of the EEC, we have about 9.7% of the total number of pigs within the 12 nations of the EEC. Pig production for the whole European community is also a major force and 43% of all meat produced is pigmeat.

Trends in performance and profitability

In *Table 16.2*, data are presented from the Meat and Livestock Commission's (MLC) Feed Recording Service, showing the trends in physical performance in pig production over the years from 1970 to the present time. There has been steady improvement in the national herd as indicated from this large sample and the process is not

Table 16.1 UK pig industry – 1987

Parameter	
Sow numbers	820 000
Annual production of finished pigs	15 608,000
Bacon and ham production	207 000 t
Pork production	754 000 t
Consumption per capita: pork	13.1 kg
Consumption per capita: bacon	8.2 kg
Bacon production: Self-sufficiency	45%

Table 16.2 Trends in pig performance

	1970	1977	1987
Breeding herd			
Litters/sow/year	1.8	2.0	2.3
Pigs/litter alive	10.2	10.3	10.3
Mortality %	15.2	14.5	11.1
Pigs/sow/year	15.5	18.1	21.0
Feed/sow t	1.6	1.6	1.2
Feed price (£/t)	36	117	139
Finishing pigs			
Food conversion ratio	3.7	3.2	2.7
Daily gain (g/d)	–	–	588
Mortality (%)	–	3.9	2.7

slowing down. What is also evident from the MLC statistics is that a considerable gap exists between the 'average' producer in the recorded herds and the performance of the top 10% of herds as determined by gross margin analysis. For example, in terms of breeding performance in 1987, the average herd achieved 2.3 litters per sow per year, 21 piglets reared per sow per year and a gross margin of £205 per sow. The top producers however achieved 2.4 litters per sow per year, 24.6 piglets per sow per year and a gross margin of £284 per sow (top third of producers). Clearly there is scope for the forward thinking producer through the application of good technical skills to generate a healthy financial return.

PIG HOUSING AND ANIMAL WELFARE

The way in which pigs are housed has changed at a remarkable pace since the move to intensive systems. The rate of change does not appear to diminish and there is a continuous development of new building systems to meet the changing needs of the industry. In the early days of intensification, the criteria were established for the design of systems which could accommodate pigs efficiently, with a gain in productive efficiency for a reasonable life span. Pigs are notorious amongst all the farm species for their capacity to damage buildings and equipment.

As indoor systems became more prevalent and the economic situation changed it became more and more important to operate existing and new buildings as efficiently as possible. It was necessary to build them at minimum cost within the limits of the known environmental requirements of the pig. Consequently stocking densities in pig buildings

have steadily risen. Most successful pig farmers however, are well aware of the relationships which exist between stocking density and the incidence of disease. There is an optimum stocking density for any given building above which production falls off significantly as a direct result of increased bacterial contamination. Unfortunately there are some farmers who increase stocking densities above the optimum and then solve problems with antibiotics. This is, of course, dangerous practice and potentially limits the effective clinical use of the same antibiotics and also in some cases can limit their effectiveness against human diseases.

A more recent phenomenon in the conception of new pig buildings systems is the involvement of public opinion in how animals are managed and housed. It would appear that the consuming public wish to have a say in the care of animals on farms and the industry will have to respond to these wishes. At the same time the general public may want education in order that they may understand why animals are housed and managed as they are. There is at the moment a move to improve the channels of communication between the various interested parties in which central government is playing a role. One aspect which is still a problem area is the economics of welfare. Animals can be produced in less intensive systems using high-welfare management. These systems do cost money, however, and inevitably unit costs of production and therefore the price of meat rises accordingly.

Breeding stock

The package-deal building is the standard housing unit. These are of prefabricated construction and are erected on-site by trained personnel. A very high standard of construction is available in terms of thermal insulation, ventilation, heating control and the internal fittings used. There is scope for small farms building their own facilities using timber and concrete, but for the larger units aiming at high levels of biological efficiency, the package deal building is probably the best option.

Dry sows are housed universally in individual stalls within controlled environment buildings. These include slatted floors and many have completely automatic feeding systems. The labour input is minimal and the management of the sows can be of a high order. Problems arise because of the automation itself. If sows are placed in stalls in early pregnancy and are supplied with all of their needs by mechanical devices then this reduces the necessity to observe animals as individuals on a routine and daily basis. Moreover, if sows and gilts are in stalls allowing only limited exercise for about 100 d at a time, then this may not be conducive to the maintenance of high health status. It is this last point which has brought so much criticism to the stall house and many people believe they should in future be phased out of use. In particular, the use of neck tethers to reduce the constructional costs of the stall has been severely attacked by some pressure groups and in many ways these tethers are difficult to defend.

The well managed stall house does have many advantages, however, and it is hoped that these advantages are considered carefully before stalls are abandoned forever. The stall house evolved in order to fulfil the nutritional, thermal and social needs of all sows. Sows in groups establish social hierarchies after much fighting and physical damage. Animals at the bottom of the social order may be repeatedly attacked and prevented from obtaining any food at all. It is also impossible as animals progress from one phase of the reproductive cycle to the next to avoid constant remixing of groups of animals. Sow stalls therefore prevent such physical damage and allow all animals to receive the correct daily feed allowance. In addition, reproductive management is easier due to the possibility for the timely checking of animals at the different stages.

A porportion of sows are housed in semi-intensive building systems in groups of between ten and 50. These are usually based on straw yards and in the eastern counties of England where they have been popular. The buildings themselves can be multipurpose allowing producers to move in and out of pig production as the pig cycle progresses. There is a renewed interest in these systems and the use of electronic sow feeders may be partly responsible for this. These are devices where pigs can live in a group of say 25 and have access to a special feeding stall. When a sow enters the stall, a gate closes preventing another sow from entering. Each sow wears an electronic transponder/device which contains a sow identification code. This code is read electronically whilst the sow is feeding and the weight of feed eaten by the sow is monitored and recorded automatically by a dedicated computer. This allows group-housed sows to be fed individually and the system can be programmed to allow each sow a set daily allowance.

Many of the development problems of these feeders have now been resolved. Where these feeders are used sows are still housed in a group for most of the day with all the problems this can entail, but the intense competition at feeding is eliminated.

Sows are moved to farrowing crates about 3–7 d before delivery. Farrowing crates are designed to minimise the loss of baby piglets due to overlying by the sow in the early days after farrowing. This is the major cause of death in the neonatal period. Farrowing crates commonly include a 'creep' area where piglets have access to a supplementary dry food and a heat lamp to keep them warm. The crate is necessary to restrain the sow and prevent her rolling on her piglets. Farrowing crates are not popular with some people and it has been suggested that restraining the sow during parturition slows down the process, but this is not yet proven conclusively. Most stockmen and managers prefer the use of farrowing crates as they are used for a relatively short period of time.

There have been innovations in the design of farrowing crates recently. One of the more promising of these is the use of flexible hydraulic bars on either side of the crates. These bars slowly part when the sow lies down. When she stands up, the bars fold inwards so preventing any sudden move downwards, thus protecting the piglets from being crushed. Another development is the use of blower fans around the legs of the sow, activated when the sow stands helping to drive the piglets away from the place of maximum risk and into the creep area.

When sow stall houses were first installed it was usual to have boar pens in the dry sow house and sows and gilts were moved to these pens for heat detection and service. This worked well for some, but the arrangement is not ideal because of the logistical problems in moving animals in and out of stalls or tethers. Boar pens which are remote from weaned sows and unserved gilts do not provide the ideal pheromonal environment. Larger pig units now build specific houses for boars, weaned sows and gilts and these buildings included purpose built pens where mating can take place with the full supervision of the stockman. These service areas contribute to high conception rates and focus management attention on this crucial phase of the cycle of production.

Weaner accommodation

When weaning was carried out at six or eight weeks of age, piglets could be transferred to large straw yards without special environmental controls and there were few problems. With the advent of early-weaning it soon became clear that more careful consideration of the housing of post-weaned pigs was essential. Flat-deck cages were devised as a method of housing small pigs to provide an appropriate environment for minimising the post-weaning growth check and promoting good health. The essence of flat-deck cages is that the completely slatted floor area ensures that piglets are in minimal oral contact with their own faeces. The walls and floors therefore carry less contamination and enteric disease is minimal. The second feature of these houses is that they are well insulated and invariably have a heating system to keep the inside temperature at the optimum throughout the year. For newly weaned three week old piglets this is probably around 25–28°C. Flat-decks have also been improved over the years to provide plastic slotted floors which are kind to the piglets' feet, easy-clean surfaces so that each pen can be completely power-hosed in between batches and better ventilation systems that give air change without draught. The outcome of these improvements is minimal mortality of post-weaned piglets and faster daily gain through to slaughter weight.

Flat deck cages in particular have been criticised on animal welfare grounds. This is partly because they are associated with early-weaning, but also because of the completely slatted floors which do not allow any bedding in the form of straw or wood shavings.

Another popular type of building for weaners is the verandah house. These give piglets access to an outside slatted dunging area and an indoor insulated kennel for groups of 20–40 piglets. These buildings can work well for weaning at four to five weeks but in adverse weather conditions (hot or cold) they can be difficult to manage.

Growing pigs

From 25–30 kg onwards piglets are transferred to the final accommodation where they remain until slaughter. These houses are variable in type and design often depends on the availability of bedding. Many bacon houses are based on package-deal buildings with completely controlled environments and automatic feeding systems. Bedding is not used and most buildings have either a partly slatted floor as a dunging area or in some cases a fully slatted floor. Under these conditions the attainment of lean carcasses from pigs which have grown rapidly and efficiently is routine assuming a good degree of stockmanship.

The main alternative to the completely controlled environment house is a semi-intensive system based on straw yards and insulated kennels. These have become known as Suffolk houses because of the popularity of this type of house in that part of England. Good performance can be achieved with these houses but in the absence of a plentiful supply of cheap straw the system may not be an economic proposition.

Associated with any large pig finishing enterprise is the disposal of large volumes of either slurry or farmyard manure. The straw based systems have the advantage that muck can be stacked in heaps and spread when the land conditions are right. Slatted floor systems can present real problems due to the amounts of low dry matter slurry produced. Most buildings include underground tanks for short-term storage. Vacuum tankers are required to dispose of the slurry in dry conditions on stubbles, grassland and often on sacrifice areas. Because of the vagaries of British weather conditions, regular spreading is not always possible and on many farms there are now large capacity above ground storage tanks. These are emptied at set times in the farming calendar or when weather conditions are suitable. The spreading of the slurry itself is still a major issue in some areas due to potential problems of run-off into waterways and problems of smell from farms adjacent to towns and villages. In some of the high density pig areas of the UK planning authorities are refusing to allow farmers to invest in new slurry-based systems. Currently, the handling and disposal of slurry is a major unresolved problem for the industry. It seems the UK may be treading rapidly down the path of Dutch pig farmers who have been banned from expanding their industry in the south of Netherlands because of persistent slurry problems.

GENETICS AND PIG IMPROVEMENT

The size of the contribution of genetics to the improved performance of modern pigs cannot be over-emphasised. Over the last 20 years, the overt performance characteristics of pigs have altered enormously. A major factor in the success was the foresight of pig breeders in using established principles of quantitative genetic theory and abandoning some of the traditional methodology. To an extent, pig breeders followed a similar course to the poultry industry and adopted the techniques of mass selection for multiple objectives and used large populations to ensure statistical validity in the selection of parents to breed the next generation. A more traditional approach included the selection for traits of no economic merit and the progeny test as a selection tool. The drawback of progeny testing is the length of time before a selection decision can be made as this reduces significantly the rate of annual genetic improvement in a population.

As a result of the intense selection methods deployed over the last 20 years and also the contribution made by national organisations such as the MLC and the hybrid companies, it is easy to see why British pigs are now in demand throughout the world. It is also interesting to note that Denmark, which prides itself on the production of quality pig meat, is now importing and using crossbred sows from a major UK hybrid company.

Breeds

The concept of a discrete breed is fast becoming irrelevant and the use of commercial hybrids has superseded purebreds in commercial meat production. The hybrids themselves are created from the original parent purebreds and these are still maintained within nucleus populations. The bulk of UK slaughter pigs are produced with some degree of relationship to either one or both of the two most important breeds; the Large White and the Landrace. It is believed that the original ancestry of the Landrace breed belongs in Scandinavia and the Large White, related to the American Yorkshire breed, is more indigenous to Great Britain.

It was fortuitous that the Large White and the Landrace breeds came to the fore in Britain at a time when systematic breed improvement was being initiated. Both breeds are extremely prolific and can produce litters of ten to 12 piglets consistently. Furthermore, both breeds produce a very acceptable carcass even as a purebred. They have the length needed for bacon production and also a high lean content and low subcutaneous backfat thickness which is required by today's consumers. More specifically, they both have a white skin and this is needed by the meat trade in the UK industry. Both breeds are also noted for their ability to grow quickly and to show a high efficiency of food conversion. In short the Large White and the British Landrace are good all rounders and are well fitted for meat production. It is easy to understand why the hybrid pig breeding companies initially selected these breeds as the foundation material for their selection programmes.

The other two breeds of any numerical significance in Britain are the Welsh and British Saddleback and these have been given resources over the years in the national improvement programmes. The Welsh breed is similar in some respects to the British Landrace and can be considered as an alternative in a crossing programme. The British Saddleback, however, is a black coated breed with a white 'saddle' just behind the front legs. This breed was established relatively recently by the merging of the herdbooks of a number of similar black saddled breeds such as the Essex. The merits of the Saddleback are in its hardiness and in its mothering ability which some strains of the white breeds lack. British Saddlebacks are used in many parts of southern England for outdoor pig production using grass paddocks or cereal stubbles and portable shelters for accommodation. This type of pig production can be very profitable because of the minimal fixed costs involved and if limited labour is available then the Saddleback is capable of producing strong healthy litters of weaners with the minimum of fuss. Some of the leading hybrid companies now market breeding females specifically designed for outdoor production and most of these have genes from the British Saddleback.

Minor breeds such as the Tamworth, the Large Black, the Gloucester Old Spots and the Middle White are still kept by a few enthusiasts and at rare breed society farms but the commercial performance of these breeds in terms of prolificacy, growth and carcass quality falls way below the breeds mentioned above. It is of incalculable value, however, that these breeds are still maintained both because of their intrinsic qualities as part of our heritage but also because of the need to keep a wide genetic base to allow for changing selection objectives in the future.

There has been periodic interest in imported breeds and some of these are now used in large numbers. The Hampshire breed from North America is used as a top crossing sire in pork production. This is because of the high lean content of these pigs and the very good conformation of the hams. Hampshire pigs also have a good eye-muscle area and a good killing out percentage. The level of prolificacy in Hampshires is low however and as a breeding female they are not a viable proposition. There are also a number of Belgian Pietrain pigs in the UK. These have probably the best eye-muscle and hams of any pig and a high lean content generally. They have been incorporated into sire-lines by the breeding companies to be sold as top-crossing sires for the production of slaughter pigs. There is renewed interest at the moment in the Duroc breed also from North America. These pigs are very hardy compared to our indigenous white pigs and some strains of them can grow extremely quickly. The current interest in Durocs stems from the putative superior quality of the meat. It is said that Durocs have a high percentage of intramuscular fat and as a consequence Duroc meat may be juicier and may have more flavour than other breeds. This remains to be demonstrated under British conditions and to date the results of controlled experiments with Duroc meat have been rather equivocal. Throughout the whole of Europe and further afield there has been an increased use of Durocs as terminal sires both as purebreds and also as crossbreds.

In 1987 after some years of negotiation a sample of Chinese pigs was imported into Britain. The stimulus for this effort was the reports and first hand accounts coming from China to the high prolificacy of these pigs. Some anecdotal accounts suggested that breeds such as the Maishan from central China could regularly produce 20 piglets in one litter. The objective was to incorporate the good genes of the Chinese pigs with British pigs to produce hyperprolific hybrids with an acceptable carcass. This last point may yet prove the stumbling block and research from France, where Chinese pigs have been investigated for some years, suggests that although the prolificacy is good it is not good enough to cancel out the very poor carcasses that crossbred Chinese pigs produce. With the advent of sophisticated genetic engineering techniques it may be possible to identify the specific genes carried by the Maishan and others and then transfer these genes directly into the genetic constitution of our own indigenous pigs.

Other pigs present in the UK in very small numbers include the Vietnamese pot-bellied breed which is kept simply as an ornamental pig. One last breed of interest is the small herd of Mexican Yucatan pigs owned by the University of Leeds. These were principally kept as laboratory mini-pigs because of the similarities (from a physiological point of view) between man and pigs. Recently, it has been discovered that the heart valves of these pigs can be used as replacement valves for defective valves in the hearts of human infants. It may be that in future the pig may serve humans in many more ways than just for food.

Crossbreeding

Systematic crossbreeding has been used in pig production for some years and this seems likely to continue. Part of the problem in any breeding situation is that it is almost impossible to find the perfect blend of characteristics in a single breed. By merging together the good traits of two parent breeds in a crossing scheme it is possible to produce a slaughter generation containing all the good points. In pig breeding the biggest reason why crossbreeding is still so widely used is to exploit hybrid vigour to improve reproductive performance. It was found, in the 1960s, that by crossing together the Large White and the British Landrace, the first cross female expressed a greater prolificacy than either of the two parent breeds. This is an example of heterosis or hybrid vigour and as the carcasses of both parent breeds are good then the use of the first cross female as a breeding sow seemed the obvious choice. The use of LW × LR females has stood the test of time and now a very high percentage of

Table 16.3 Crossbreeding to produce meat pigs

Generation	Male	Female
Grandparent	Purebred LW × Purebred LR	
	Purebred LR × Purebred LW	
	↓	
Parent	Purebred LW × LW × LR	
	Duroc sire-line, etc.	
	↓	
Slaughter pigs	LW × (LW × LR)	

British female pigs are of this genotype. Most hybrid females are based on this same first cross animal. In the context of pig breeding, hybrid is thought of as an 'improved crossbred' where simultaneous genetic improvement is made in the parent lines as well as selection for crossing ability. *Table 16.3* gives a schematic representation of the most popular system of crossbreeding used in Britain.

There are of course a number of variants on the scheme outlined in *Table 16.3* but the majority of producers use bought-in first-cross females and purebred terminal sires from one of the hybrid companies. Some producers use their own gilt replacements and may buy in only a proportion of their annual needs. To gives farmers flexibility in their replacement policies, many breeding companies sell grandparent females directly to commercial farmers who then carry out the crossing programme themselves to generate the bulk of their gilt replacements. Artificial insemination is also used in such schemes to replace the original grandparent gilts and as crossing sires on nucleus purebred females.

As might be expected, there are advantages and disadvantages to the different breeding policies for individual farmers. Those who purchase all of their first-cross gilts every generation from a company or from a pedigree breeder will incur higher replacement costs and they also run a higher risk of introducing disease onto their farms. On the other hand if they are buying in from a progressive company then they will capitalise more quickly on the rapid genetic improvement. If all replacement gilts are bought monthly from one source a simple system is created. In contrast, the management of grandparent gilts can be onerous and badly implemented.

The hybrid companies

There are about ten to 12 major hybrid companies in the UK in the business of selling boars and gilts to commercial farmers. The majority of them use proven scientific principles in their selection programmes allowing for the making of significant advances in the genetic quality of their stock. The origins of individual companies are quite diverse. Some were groups of collaborating pedigree breeders who effectively pooled their resources and expertise to maximise their breeding effort. Others have grown out of the ancillary industries such as the meat processing companies striving to develop products which fitted their own markets. There are also companies which developed by diversification from other breeding operations. All of the successful companies use essentially similar methods. They have large nucleus populations where direct selection is carried out under company control. These nucleus herds are owned predominantly by the companies themselves. Improvement made at the nucleus level is then propagated by multiplying breeders who are usually leading pig farmers operating under contract with and the control of the hybrid company. Gilts are sold off these multiplying farms to the commercial meat producers.

A decision facing commercial farmers is from which company to purchase replacements. It is often difficult to identify real genetic improvement and to separate this from the advertising rhetoric. Perhaps the first thing to decide is what are the traits of most importance in the commercial environment. Some producers may put more emphasis on food conversion ratio. Others may feel that backfat thickness is more important and so on. Once the selection objectives are clearly identified then it is possible to select a particular company who maximises the improvement of those traits in their own schemes. This process was made much easier by the MLC's Commercial Product Evaluation (CPE) exercise. They took samples of gilts from each company to a single farm unit and compared the commercial performance of each company's gilts and the progeny of those gilts in the same environment. The results were published every year and provided very valuable information to meat producers. Unfortunately, the MLC have had to halt this CPE exercise but the reports are still available.

Another major consideration in the selection of breeding replacements is health status. Many of the companies sell stock which are free from certain diseases such as virus pneumonia and atrophic rhinitis. They maintain a high health status by stringent controls at the nucleus level. In-troduction to the nucleus herds may only be carried out via hysterectomy-derived litters which are specific pathogen free. Over long time spans, however, the health status of even the best companies can wax and wane and this requires careful monitoring by the commercial producers.

National involvement in genetic improvement

The UK has been fortunate to have had progressive organisations such as the Pig Industry Development Authority (PIDA) and later the MLC to coordinate our national programme of improvement. Whilst accepting that the role of the hybrid companies has taken over from pedigree breeders it was the original work done by PIDA and MLC that focused the enthusiasm of the pedigree breeders into a collaborative effort.

The cornerstone of the MLC scheme was the rationalisation of the existing structure of breeders into a clearly defined breeding pyramid. The best breeders were designated as nucleus or élite breeders and a second category of units was listed as multiplication units to supply commercial units with breeding stock.

Groups of test pigs were sent from all breeders within the scheme to central pig testing facilities. This breeding pyramid structure is illustrated in *Figure 16.2*.

There were five testing stations throughout the country where large numbers of potential young boars could be performance tested every year. A method of index selection was used on these stations to improve simultaneously a set of multiple selection objectives. These objectives were the important traits such as; food conversion efficiency, growth

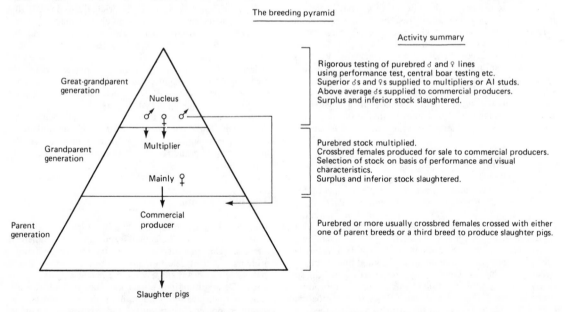

The breeding pyramid

Activity summary

Rigorous testing of purebred ♂ and ♀ lines
using performance test, central boar testing etc.
Superior ♂s and ♀s supplied to multipliers or AI studs.
Above average ♂s supplied to commercial producers.
Surplus and inferior stock slaughtered.

Purebred stock multiplied.
Crossbred females produced for sale to commercial producers.
Selection of stock on basis of performance and visual characteristics.
Surplus and inferior stock slaughtered.

Purebred or more usually crossbred females crossed with either one of parent breeds or a third breed to produce slaughter pigs.

Great-grandparent generation

Nucleus

♂ ♀ ♂

Grandparent generation

Multiplier

Mainly ♀

Commercial producer

Parent generation

Slaughter pigs

Notes (1) Nucleus and multiplier units may be totally independent (e.g. within MLC scheme) cooperating (e.g. in a cooperative breeding company or organisation) or totally integrated and dependent (e.g. in a breeding company organisation).

(2) Multipliers and commercial producers may be either independent, part of a cooperative or integrated (e.g. a commercial producer may be his own multiplier).

Figure 16.2 The breeding pyramid

rate, backfat thickness, killing out percentage, eye muscle area and muscle quality. In other words, the scheme concentrated on traits of high heritability and therefore selected high quality boars passed on their superiority to the next generation.

Each tested boar was given an index score which integrated his total economic value in breeding terms. For ease of understanding, the points score was designed so that an average boar in any given time period scored 100 points. Boars scoring 90 points or less were slaughtered after test. Boars scoring 120 points or more were called approved sires and were returned to the breeders who could then either use the boars themselves or sell them to other breeders. It was in the interest of nucleus breeders to use the highest pointed boars they could in order to stay ahead of the field. All the data were made public and multipliers could use only approved sires. By identifying good genes and disseminating them as widely as possible, rapid genetic progress was made. The hybrid companies also use similar testing methods.

The rewards for this effort and considerable expense have been enormous and it is estimated that over the two decades that the scheme operated the increase in profits to the commercial farmer, were of the order of 50–60 pence annually on every pig sent for slaughter. Nationally therefore the scheme was an investment that has probably paid a tenfold return.

The last testing station in the MLC boar testing scheme closed in 1987 and the remains of the scheme centres around a reduced super-nucleus list of pedigree breeders and the use of on-farm testing. These changes have been made due to the increased presence of the hybrid companies and also to the escalating costs of operating central testing stations.

Artificial insemination

One of the most efficient methods of spreading good genes widely and rapidly is the use of artificial insemination. Pig AI has not been quite the overnight success story that in many ways revolutionised cattle breeding. There are signs now that producers are at last aware of the benefits from using AI in their own breeding programmes. As a percentage of all matings carried out nationally in any one year, about 8% are artificial inseminations and this figure is rising steadily. In some neighbouring countries the figure is much higher. The reasons for not using AI are that there is less risk to human life and limb with boars on a farm compared with bulls. Pig farmers are familiar with the housing and handling of boars with the minimum of problems and costs. Also, and probably more importantly is the variable reproductive performance associated with the use of AI. Surveys in the past have shown that conception rates are down when compared to natural service and in some cases the litter size is reduced. The third reason is that farmers were concerned about the length of time required to carry out the AI procedure. This last point is quite contentious as the supervision of natural matings done properly involves care, attention and therefore time.

The reasons in favour of using AI revolve around the superiority of the genetic material available at the AI studs. In the MLC scheme, any young boar scoring 180 points or more automatically can be held at one of the AI stations and the semen from these boars is available to any commercial producer. In practice, there are boars at stud which have scored 220 points or more and which are available to an average commercial producer at modest cost. The possibility therefore exists of quickly upgrading an average herd using AI.

With regard to the problems of conception there is now a much better understanding of the relationships between the AI technique itself and the probability of failure to hold to service. The MLC give sound and clear guidelines to farmers in the form of publications and visiting consultancies and there are many producers using AI who achieve conception rates around 80–90%.

The principal AI scheme nationally is the one operated by the MLC and this works on a postal delivery service. The AI station is situated at Selby in the centre of the UK and farmers with females in heat or expected in heat telephone the station. Semen is despatched immediately by first class post or by the rail network. Batches of semen therefore arrive within 24 h on the farm. The insemination itself is carried out by the farmer or stockman using a rubber catheter and the process takes about 15–20 min. Many of the larger herds using the scheme place fixed orders on a weekly basis so that semen is available at the time animals are expected in heat.

The timing of insemination is in fact crucial to success and a rigorous heat detection programme must be used to determine the onset of oestrus. The first insemination should be carried out about 24 h after the onset of oestrus (not the first time that oestrus is actually observed) and a second insemination given between 8 and 16 h later.

There are also a number of private pig AI schemes in operation under licence with the Ministry of Agriculture. Some of these are associated with national feed compounders and others with meat processing firms. These companies essentially provide a similar service to the MLC scheme and provide boar semen from their own testing stations.

There have been many attempts to use deep frozen semen in pig AI schemes but all have failed because of poor freezing and thawing characteristics of pig semen. In specialist breeding work there is still a place for the use of frozen semen and research is slowly improving the conception rates achieved with deep frozen semen. New diluents and cryo-preservatives are constantly being tried.

For the meat producer, AI offers the opportunity of replacing the existing herd with superior animals over a period. There are also some meat producers who use AI widely in their main herd to gain from the significant financial advantage of using the progeny of AI boars as meat animals.

SOW AND GILT REPRODUCTION

The breeding sow is the basic production unit for the production of as many weaned piglets as possible in a given time. The profitability of a breeding-finisher unit is intimately linked to maximum sow productivity. Where the statement falls down is that in a changing price situation over long time spans, maximum profits are not always associated with maximum physical output. Furthermore, in the context of the public becoming increasingly interested in animal welfare, the hyper-producing sow may in the future not be compatible with the public's perception and understanding of animal welfare. Within normal bounds, sow management is directed towards farrowing as often as possible in a year and rearing the maximum number of piglets to weaning. It is intended in this section to outline female reproduction from puberty to weaning and illustrate the salient features involved.

A more detailed account of this broad topic is given in *Reproduction in the Pig* (Hughes and Varley, 1980).

Puberty and the gilt

Whatever the source of gilt replacements, it is imperative to introduce them to the main breeding herd as soon as possible. In other words, the sooner they are mated and produce their first litters the lower the overall replacement costs. It was once thought that around a third of all sows a year neeed to be replaced and every time a cull sow was removed from the herd a gilt was mated immediately as a replacement. This is probably not the case and in many herds replacement rates are around 50% per annum. Early weaning is a factor in this but not necessarily because the practice 'wears-out' sows rather quickly but because culling takes place after six parities and with early weaning this point is reached much more quickly. The annual culling rate therefore rises accordingly.

Puberty or first heat in the female pig occurs when the animal is about six months of age at 90 kg liveweight. There is enormous variation around this age and some gilts may attain puberty as early as 130 d and some as late as 280 d. There has been much research carried out to establish the factors controlling puberty, which is now quite well understood. Genotype, nutrition and season of the year are not the least of the factors which influence the time of first heat but the most potent influence is generally referred to as the 'boar effect'. If prepubertal gilts are suddenly exposed to contact with mature boars this stimulates the first heat and normal cyclicity in a high proportion of gilts. There are a number of components of this effect such as sight, sound and tactile with a boar. The biggest component is that of boar odour and the pheromones (airborne hormones) contained within the boar's characteristic smell.

Gilts at 160–180 d of age are introduced to mature boars and given daily contact for at least 20 min/d in the same pen as the boar. It is highly desirable to house the gilts from the time of first introduction to males within the same air space or in the pen adjacent to the boar used for testing the gilts. This maximises the exposure of the gilts to the pheromones. Stimulation with mature boars necessitates a new pen and perhaps relocation within the piggery. Gilts respond to this stimulus and it helps to initiate the first heat in many gilts. It is well known that gilts appearing on a farm from a breeding company exhibit oestrus within 7–10 d of delivery and this is put to good advantage by farmers.

A final stimulus to early puberty is the mixing of groups of gilts into new social groups. This on its own will cause many heats within a few days of mixing. The use of this practice in a management programme may need careful control to avoid the physical trauma resulting from fighting in new social groups.

Despite the abundant data and research on gilts, failure to exhibit oestrus is a perennial and serious problem on some farms. There are now a number of pharmaceutical products which can be very effective as cycle starters. These products are usually based on combinations of the gonadotrophin hormones (FSH and LH) or their analogues which cause follicles to grow and mature on the ovary.

Oestrous cycles

Following first heat, most gilts will have regular oestrous cycles at approximately 21 d intervals. This varies from about 18 d to 23 d in individual animals. In some gilts, however, following first heat a number of so-called silent heats may occur. This is the situation where ovulation takes place and eggs are shed from the ovary under the control of the hormone LH, but the overt signs of oestrus including receptivity to the boar are absent. A similar problem is where gilts show only a weak oestrus response and mating is then tortuous for all concerned.

A major decision is at what stage relative to puberty mating is to be carried out. There is clear evidence that ovulation rate (the number of eggs shed) is very poor at pubertal heat. Mating gilts at this time therefore leads to an unacceptable litter size at the end of pregnancy. Most producers wait until the second or even third heat in order that the ovulation rate increases to promote a reasonable level of prolificacy.

In larger herds where 20 gilts or more are mated in any one week, there may be a role for the use of synthetic materials to control oestrus in groups of gilts. The problem is that because of seasonal variations in the propensity to show first heat, it is difficult to ensure a regular supply of replacements. There is now a product available in the UK and elsewhere based on an analogue of the naturally occurring hormone progesterone. This product (Regumate, Hoechst UK Ltd) can be given as a feed additive and on withdrawal of the product, animals show heat with a high degree of synchronisation. Matings can be planned well in advance to maintain a steady throughput of gilts into the main breeding herd.

Oestrus and mating

When oestrus occurs in pigs, the outward signs are characteristic and clear. Gilts and sows in a group will tend to go off their food and in the pro-oestrus period will mount other females. They will also exhibit vulvul reddening and swelling at this time under the influence of rising levels of the oestrogen hormones. At the onset of oestrus animals will respond to the application of firm hand pressure to their backs and will stand rigidly often holding their ears erect. Once this response has been elicited, it can be extremely difficult to move the animals around. Stockmen often sit astride the backs of sows and gilts as a more vigourous method of heat detection and this is known as the 'riding test'.

In the absence of a boar for heat detection it is possible to pick up only a percentage of animals which are in heat. This figure may be up to 60%. With a boar in the same pen as pro-oestrus gilts or sows there is a maximum chance of detecting heats. Some farms use vasectomised young boars for heat detection and the sterile boars are used as teasers for the stimulation of first heat, without the risk of mating at the wrong time or with the wrong boar.

Oestrus lasts normally for 2–3 d and ovulation takes place about 40 h after the onset of oestrus. The aim as with AI is to have fresh semen in the reproductive tract to await eggs being shed from ripe ovarian follicles. This is important because there are definite physiological processes whereby the sperm cells mature for a number of hours in the females

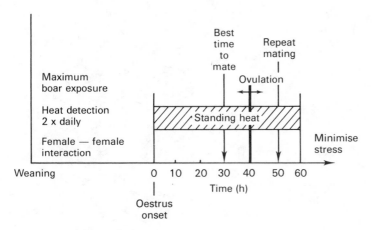

Figure 16.3 Timing of mating

tract before they are able to fertilise viable eggs. If sperm are introduced into the females tract either too early or too late, then the sperm and egg cells are out of phase in their physiological maturity. Conception could therefore fail or the litter size could be extremely small.

With natural service, it is best to try twice a day heat detection to ascertain, as near as possible, the precise onset of oestrus. Service is carried out ideally at 30 h after the beginning of heat and in practice this is on the morning of the first time that oestrus is detected. Mating is then repeated the next day. For sows that were spotted in heat in the afternoon but not on the morning of the same day, it is better to wait until the next morning before the first mating is carried out.

Figure 16.3 illustrates the timing of mating relative to the important reproductive events.

There are now some farmers who allow three matings through the course of the 2 d oestrus period. This does have some merit and evidence has shown that conception rates and litter size are maximised by increasing the chances of placing fresh semen in the female's tract at just the right time. On the debit side, there is an increased requirement for the number of boars needed. If sows or gilts are mated twice each and the optimum services a week for each boar is about four times, then this means that for every 20–25 females in the herd a farmer will need one boar. If triple serving is practised, this changes to around 15 females per boar to avoid the overuse of boars and to allow for the resting of boars periodically.

Feed allowances for gilts

Where gilts are reared on the farm where they are to remain as breeding females, there is more scope to provide the perfect nutritional and social environment. The evidence points to a medium plane of energy intake in the early rearing period to allow gilts to reach the threshold liveweight at an age where they are mature endocrinologically. If they are grown to quickly or too slowly then there is the possibility that puberty will be inhibited. In practice, up to the time gilts are ready for stimulation by boar contact, they are fed about 1.8 kg/d of a conventional breeder diet containing 14–15% CP. The precise amount given may vary from farm

to farm depending on the housing system, group size and so on.

From the time gilts are introduced to mature males, most farmers 'flush' their gilts. Flushing is the abrupt transition to a high plane of energy intake prior to ovulation and gilts are commonly given around 2.5 kg/d or are even fed *ad libitum* on some farms up until the time mating takes place. This practice ensures the maximum number of eggs are shed at ovulation. Immediately after service the feed scale should be adjusted again to a low plane at about 1.8 kg/d. This prevents the excessive loss of fertilised embryos which may happen if gilts are overfed in the first three weeks of pregnancy.

Pregnancy

Sows spend at least two-thirds of their working lives in gestation and the bulk of feed inputs are given during this important period. One of the problems with modern systems is that with automated housing and feeding systems, sows in pregnancy can easily be under managed. If the whole unit is to operate successfully, sows should be monitored as individuals throughout pregnancy.

The period of gestation in the pig lasts about 115 d and the variation around this mean value is generally small although some sows will naturally give birth as early as 110 d after mating or as late as 119 d. The initial days of pregnancy are the most critical in terms of whether the pregnancy will last and also it is in the initial weeks when the final litter size is determined. At first the fertilised eggs are free-living entities within the lumen of the uterus. By day 12, the initial attachment of embryonic membranes begins and this process of implantation contiues until about day 25 of pregnancy. Because implantation is similar in some respects to the host–graft relationship seen in the organ transplantation or in skin grafting then the process is a very delicate one. The developing embryos can be immunologically rejected by the uterus and development stops. Any stress on the sow or gilt at this time causes an imbalance in the hormone status of the sow and more embryos die. Similarly, any difficulty experienced by individual embryos in securing a supply of nutrients across the placental wall will also lead to embryonic death. Once the hurdle of implantation is passed

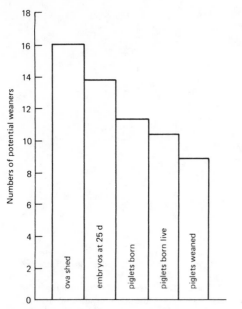

Figure 16.4 Losses of potential weaners through pregnancy and lactation

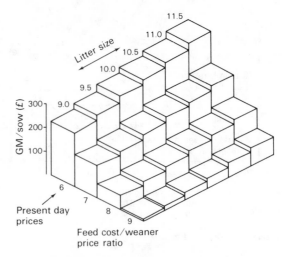

Figure 16.5 Prolificacy and profitability

then the sow is more able to buffer embryos from an adverse external environment.

The final litter size is represented initially by ovulation rate. Throughout gestation the number of potential piglets is reduced by losses at different stages. Under average conditions 25–30% of all fertilised eggs will be lost in the first three weeks of gestation. Another portion of loss occurs in the fetal stages when the developing fetuses compete for available space and nutrients within the uterus. This latter percentage of loss is less significant than early embryos losses. *Figure 16.4* illustrates the general pattern of losses of potential piglets from ovulation onwards.

In the management of sows, the objective is to provide a minimum stress environment, particularly in the first weeks of pregnancy. The understanding of stress has a long way to go before firm recommendations can be given on housing and the ideal social environment. Some interesting work from Australia has shown the importance of social factors on reproductive performance. A novel finding was the impact that social factors in very early life can have on the ultimate lifetime performance of the animal. For example, ten week old piglets which were either reared in isolation or which were subjected to mis-handling by humans in their early lives were especially prone to diminished reproductive performance at maturity. The notion that early stress can have profound influences on the physiology and therefore the physical performance of sows and gilts is an extremely important one. In the future we may see more attention given to the building of minimal stress farms where production systems are good for the animals' general well-being but also good for economical production. The Australian work carried out by Dr Paul Hemsworth and his colleagues includes a study of the way in which humans interact on farms with the animals in their care. This latter work has shown that the personality of the stockman looking after sows has a strong relationship with both litter size and the farrowing rate of individual farms. Where sows

showed particular aversion to humans there was a detrimental effect on reproduction. The selection and training of competent stockmen may therefore turn out to be the biggest factor of all in determining the reproductive capacity of the whole herd. What is certain is that a high litter size is crucial for survival in the modern economic climate. *Figure 16.5* illustrates how in a changing price structure, litter size affects profitability. It can be seen that those units operating below par in a bad year will quickly move into a loss.

The feeding of pregnant sows

Pregnant sows which are gaining body weight from one parity to the next need nutrients for maintenance and growth and also a supply of nutrients to sustain the developing conceptus. The factorial approach has been used in the past to calculate energy and protein requirements for sows and recommendations are based on these calculations along with the results of controlled feeding trials. There are a number of difficulties with this approach. The first is that pregnant sows experience a hormonal environment which is stimulatory to growth in the same way that steroidal growth promoters work in a number of species of farm animals. The steroid hormone progesterone is at high concentrations in the peripheral circulation of pregnant females and this helps them to utilise their food more efficiently than they would otherwise. The second problem is that we have to take what may be an arbitrary decision as to how much weight we wish the sow to gain (or lose) through pregnancy.

Traditional feed scales allowed for high weight gain in pregnancy and much of the gain was catabolised in the ensuing long lactation. This proved an inefficient way to feed sows from one parity to the next. At the present time, sows are offered a plane of energy intake in pregnancy only just above what would be a maintenance intake for a non-pregnant animals. At the end of gestation, the sow has accrued modest weight gains and can then be fed directly in lactation for milk production. In this way, sows gain weight at 15 kg from one parity to the next and do not waste energy in building up body reserves and then breaking them down again in lactation. In current commercial practice sows are offered between 1.8 and 2.2 kg/d throughout pregnancy of diets containing 12.5–13 MJ of digestible energy/kg. This scale of feeding is not altered throughout the whole of pregnancy as there is little merit overall in increasing the scale as parturition approaches. There has been some interest in the short-term increase of the energy concentration in the diets of sows in the last few weeks of gestation. There have been some claims that the average birthweight and the vigour of newborn piglets is improved. The results are often conflicting however and not convincing despite the fact that many feed companies offer high energy diets with an increased inclusion of fats.

Protein requirements for sows have also been well researched and it has been demonstrated that a sow will perform satisfactorily on a diet containing as low as 12% crude protein. This means that a diet composed of high protein barley plus synthetic lysine would be perfectly suitable for a pregnant sow. For sows expected to have a reasonably long productive life these extremely low levels of protein may not be acceptable and a protein level of 14–15% is more commonly used for dry sow diets.

Pregnancy diagnosis

An array of ultrasonic devices is now available which can be used for the determination of pregnancy. The early detection of sows which are non-pregnant is a valuable management tool. Non-pregnant sows or gilts can be moved back to the service area or culled if required without the input of further feed and effort. Diagnosis is possible as early as 28 d after service but repeated testing through pregnancy is necessary to pick up sows which resorbed their fetuses after initially conceiving. There is also a system of pregnancy detection based on a blood test for the hormone progesterone. In the future, this method may be more accurate than ultrasonic machines and if they can be made to work on urine samples then it is likely that they will be used widely.

Parturition

Parturition and the time that sows deliver their offspring is a crucial phase in the reproductive cycle. If things go wrong, piglets will either die before they are expelled from the uterus of they will die in the early hours or days after birth. The hormonal and physiogical events taking place around parturition are now well known and this understanding has given us some valuable tools to use in the control of the process.

Labour is initiated 48 h or so prior to any visible signs. The ovaries respond to a hormone of the prostglandins series produced in the uterine wall and the net result is that the corpora lutea on the ovaries begin to regress and blood progesterone levels fall. Without the support of progesterone, uterine contractions begin and labour commences. A number of other hormones then control the frequency and the strength of the uterine contractions and amongst these are oxytocin, relaxin and oestrogen. Perhaps partly because of the complexity of the hormonal events the whole process is prone to dysfunction. Any stress on the sow prior to parturition can cause delayed farrowing and prolongation of delivery. This in turn leads to oxygen starvation in those piglets born last and they may be stillborn.

Sows should be transferred to their farrowing quarters at least 3 d before the expected time of farrowing. This allows them to acclimatise to the new environment before the litter is delivered. They should ideally be in a group farrowing within a few days of each other in order that they can all farrow in a clean area away from the contaminants associated with older piglets. When a particular sow is about to deliver she will exhibit 'nest building' activity or characteristic pawing with the front legs. Finally she will settle down and at the appearance of a bloody discharge she will begin to deliver piglets. There are no set rules as to the interval between piglets and in some cases it will be a few minutes or even as long as an hour or two. From begining to end, the whole process may be from 1 to 12 h or longer. Above 6 h the percentage of piglets which are stillborn rises exponentially and an obstetric examination may be required.

For a number of years now farmers have used synthetic versions of the naturally occurring prostglandin hormones. These are extremely effective in inducing parturition and are now widely used. A single injection is given 26 h before delivery is required and almost 100% of sows respond to this by going into labour at the prescribed time. It is therefore possible to ensure that night-time farrowings are avoided and also weekend farrowings when stockmen are possibly not around to supervise the deliveries.

The use of products such as prostglandin analogues needs some caution as it is possible to induce farrowing at a time when the fetuses are not mature enough to survive the external environment. Piglets induced 2 d or more before the expected date can hence be lacking in vigour. Postnatal care therefore must be of a high order.

Lactation and weaning

If all has gone well after the rigours of farrowing, lactation is initiated and suckling occurs more or less at hourly intervals. Daily milk production builds up to a peak at the end of the third week of lactation and then declines steadily. *Figure 16.6* shows the pattern of daily milk output for the average sow.

Reproductive activity is at a low level generally throughout lactation and the ovaries are almost completely inhibited by the suckling stimulus. Feral pigs have been observed to show heats whilst they are still suckling but only after quite long periods of lactation. Experimentally it has been demonstrated that the sow can be persuaded to escape the inhibition of the suckling stimulus and can exhibit concurrent pregnancy and lactation. Attempts have been made to exploit this commercially using groups of multi-suckling sows and the introduction of boars to groups of sows and litters at about 12 d after parturition. It has never been possible to achieve conception before about four weeks into lactation, with a high percentage of producers weaning at three weeks and mating at 26–28 d post partum the effort involved in inducing sows to ovulate in lactation does not seem worthwhile.

The age at which piglets are weaned from the sow is a major decision and there is probably no one perfect time to wean which suits everyone. There are now well designed creep feeds for very young piglets and the environmental requirements are well worked out and have been translated into appropriate buildings. There is therefore no technical limitation to the age at which weaning takes place and even systems of weaning at birth have been explored experimentally. About 65% of UK producers now wean at between three and four weeks of age and on average the national

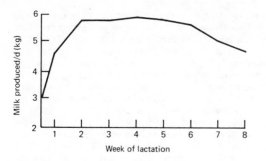

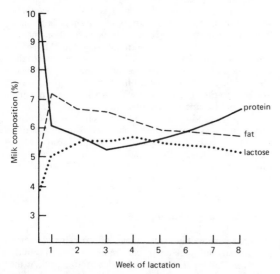

Figure 16.6 Pattern of milk yield and composition in the sow (After Salmon-Legagneur, E. (1961) *Ann. Biol. Anim. Biochem. Biophys.* 1, 295–303. Reproduced by courtesy of the Editor and Publishers)

weaning age has fallen steadily over a long time period. Economic pressure to maximise the annual sow productivity on farms is the biggest single reason why producers have moved in this direction. In theory, by reducing the time each sow spends in lactation, more reproductive cycles are fitted into a 12 month period and annual sow productivity rises. This is illustrated in *Figure 16.7* which shows the components of a sow's normal reproductive life. The faster each sow moves around the cycle, the more output of piglets is achieved.

This relationship is not linear and as we reduce lactation length, the sow's physiology alters so that the litter size falls off and there is a prolonged interval from weaning to remating. In *Table 16.4*, a theoretical analysis is given to illustrate the expected outcome of weaning at different ages together with the real outcome as demonstrated in a number of large studies and surveys.

The data given in *Table 16.4* serve to illustrate the point that by continuing to reduce the lactation length below three weeks, there will be no further advantage in sow productivity because of the precipitous drop in litter size. On individual farms the precise outcome may vary but results in general will follow this same pattern. More recently another factor has come to bear in consideration of weaning age. It is still generally believed that annual sow profitability is directly proportional to annual sow productivity and indeed this may be true in some situations where there is a high ratio between weaner price and feed costs. Through the 1980s we have seen escalating feed costs relative to pig prices and this has shifted the balance.

Early weaning necessitates the use of expensive creep diets and it is this that can severely deplete the profit of the system. On many farms careful consideration of all of the financial aspects of weaning age should be given as well as the likely output in terms of piglets per sow per year.

Feeding the lactating sow

The feeding of the lactating sow should always be assessed in relation to the feed scales adopted in pregnancy, but in

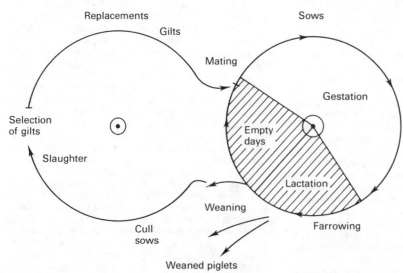

Figure 16.7 The reproductive cycle

Table 16.4 The effects of weaning age on sow productivity

	Lactation length (d)				
	14	21	28	35	42
Theoretical					
Pregnancy (d)	115	115	115	115	115
Lactation (d)	14	21	28	35	42
Weaning to mating (d)	5	5	5	5	5
Farrowing interval (d)	134	141	148	155	162
Litters/sow/year	2.72	2.58	2.46	2.35	2.25
Litter size	10.7	10.7	10.7	10.7	10.7
Piglets/sow/year	29.1	27.6	26.3	25.1	24.0
Actual					
Pregnancy (d)	115	115	115	115	115
Lactation (d)	14	21	28	35	42
Weaning to mating (d)	7.5	6.5	6.0	5.2	5.0
Farrowing interval (d)	137	143	149	155	162
Litter/sow/year	2.66	2.55	2.44	2.35	2.25
Litter size	9.4	10.6	10.7	10.7	10.7
Piglets/sow/year	25.0	27.0	26.1	25.1	24.0

general lactation is a time when relatively high energy intake should be allowed to promote milk production. Until recently most pig units used the same breeder ration in lactation as for pregnancy. The feed compounders now offer high energy diets with a high inclusion of fats to satisfy lactational requirements. In part this move has stemmed from the difficulties faced by gilts in their first lactation. These animals may have been mated relatively young and are still actively growing their first parities. Particularly if they have been overfed in their first period of pregnancy, then they will struggle to consume enough energy in their first lactation for milk production and growth. The net result is that these gilts reach the beginning of their second reproductive cycle in very poor condition and reproductive performance in the second parity will be adversely affected. It is probable that the changing genotype of our pigs has contributed to this problem. By selecting pigs which grow fast and are efficient converters of energy we have also selected pigs with smaller appetites. As a consequence, a number of farmers give lactational diets to sows on an *ad libitum* basis.

For mature sows given a restricted amount of food in lactation the daily allowance offered of a diet containing 13 MJ of digestible energy and 15% CP might be: 1.8 kg plus 0.45 kg for each piglet suckling. A sow suckling a litter of ten piglets would therefore be offered 6.3 kg each day. This allowance would not be offered at the beginning of lactation and from the gestational scale of say 1.8 kg/d an extra increment of 0.45 kg would be added on each day until the set amount is reached.

The weaning to remating interval

After weaning and the removal of the suckling stimulus, the sow under normal circumstances exhibits oestrus and ovulates between 3 and 10 d later. The majority of animals show heat at either day 4 or day 5 after weaning but as shown above the early weaned sow is a little later at 6 or 7 d after weaning.

When things are running smoothly and heats are expressed regularly and predictably then management is

straightforward. All too often however, these short periods in the sow's life can be the most difficult. Anoestrus or the complete absence of any signs of heat at all after weaning is one of the major reasons for culling sows from the herd and even the best herds will have 2–4% of these every cycle. At worst this figure can rise to 10–20% and can lead to the complete disruption of the farrowing programme. The cause of anoestrus is multifactorial but sow body condition, season and health status are often implicated. A thorough heat detection policy is essential to ensure that there are no sows recorded as anoestrus which were simply missed by the stockman.

There is no advantage in the withdrawal of either food or water in the first 24 h after weaning. It was once thought that this practice might help to dry the sow off. It has been shown that this gives no benefit in terms of the time to return to oestrus or the percentage of sows showing oestrus within 10 d of weaning. The objective in feeding the sow during this empty period is to maximise the ovulation rate and therefore a daily feed allowance of 4–6 kg is offered to this end. It is also a time when some sows need to begin to recover from the negative energy balance of lactation and therefore feeding to appetite in this short period is good practice.

WEANER PIGLETS

The management of small piglets has been made more straightforward with the development of flat-deck cages. These building systems help to reduce outbreaks of acute enteric disease which were commonly seen in earlier types of weaner accommodation. The understanding of nutritional and immunological requirements of baby piglets is now well advanced and there are high quality diets available based on complex formulations to ensure minimum problems. Despite the technology which is available, the post-weaned piglet still represents real problems for some farmers.

The ability to withstand disease challenge is a principal attribute which piglets must acquire to survive. They are born devoid of any protective antibodies in their blood stream and therefore are highly susceptible to even the

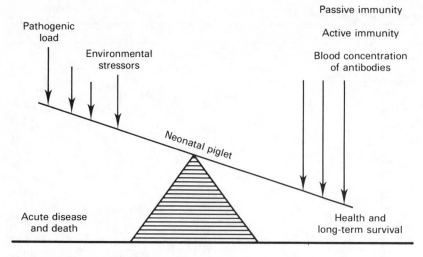

Figure 16.8 The neonatal 'balancing act'

simplest of bacterial types. Each piglet must therefore ingest as much of the sow's first milk or colostrum as possible. Colostrum contains a very high concentration of antibodies or immunoglobulins and immediately after birth these pass straight across the piglet's gut wall into the blood stream. Within hours of birth the piglet's gut is said to close and it loses its capacity to transfer large immunoglobulin molecules straight into the blood stream. It is also well known that piglets do not begin to manufacture their own antibodies in significant amounts until they are about two to three weeks of age. The amount of colostrum ingested in the first 12 h of life determines the piglet's survival prospects for many weeks to come because after birth the blood concentration of antibodies falls at a regular rate. In *Figure 16.8*, a schematic diagram is given of the way in which immunoglobulins are needed to balance out the prevailing pathogenic load. If the balance falls in favour of the bacteria and viruses, then the young animal will succumb to disease and may die.

From three weeks of age onwards the piglet's blood level of immunoglobulins (Igs) rises and the possibility of disease and death diminishes as complete immunocompetence is attained. There may still be some problem piglets which only

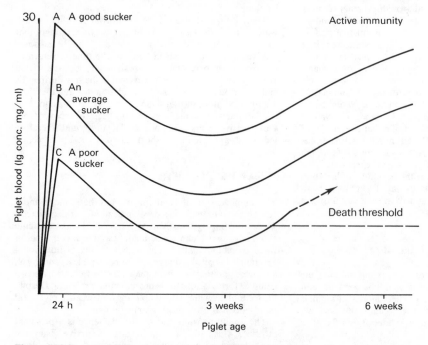

Figure 16.9 Immunoglobulin concentrations after birth

acquire enough colostrum to survive initially. These piglets may fall below a certain threshold of blood antibody at some later stage and they may die. Death could occur many weeks after birth but the time when the problem begins is immediately after birth.

It is important to supervise the postnatal period as much as possible to maximise the suckling performance of all piglets in every litter. Successful suckers within a litter will easily survive but poor suckers are prone to mortality for some time after birth. The relationship between suckling performance and the probability of death is illustrated in *Figure 16.9*.

The use of prostaglandin analogues to induce farrowing during normal working hours may help to facilitate the supervision of suckling. Cross-fostering of piglets between a group of sows which have farrowed at the same time is also a useful technique.

The process of weaning itself is often made an abrupt event rather than a gradual transition from dependence on the dam to complete self-sufficiency. At weaning the piglet adjusts to and finds its new source of food and water. It has also to adjust to the stress of relocation and of being mixed with new individuals from other litters.

These combined stressors often lead to reduced immune competence and this manifests itself as enteric disease. At best, post-weaning scours cause chronic loss in growth and at worst mortality rates are significantly increased. One method to circumvent these problems is to spread the stressors as much as possible. The sow can be moved from the farrowing house leaving the litter of piglets *in situ* for as long as possible. It may then be possible to take away the dividing partitions between two or three litters and allow them to mix freely but still in familiar surroundings. Finally the new group is moved to weaner accommodation where they will find the same food and water delivery system with which they are already familiar.

Water delivery systems for weaned piglets need some care in their selection and operation. Nipple drinking systems have been used for some time and in general they are maintenance free and almost impossible for piglets to foul. Some types may not be easy for piglets to operate particularly post-weaning. It is becoming increasingly obvious that one of the first limiting factors to post-weaning growth may be water intake which in turn limits dry matter intake. There are many sweetening agents which can be added to water supplies to encourage intake.

Another factor which is known to be responsible for a high proportion of the outbreaks of enteric disease seen in the post-weaning phase is the type of protein source used in the feed. Even though a diet contains an ideal blend of ingredients to meet all the requirements for maintenance and growth, some types of protein may cause hypersensitivity reactions in the piglet's gut. Plant proteins are worse than animal proteins in this respect and soya proteins may be amongst the worst. It is easy to formulate a high energy diet with maize meal, wheat and corn oil and to balance this with skimmed milk powder as a protein source. These feed ingredients are highly nutritious and very palatable to young piglets and will cause few problems. The trouble is that skimmed milk powder is a very expensive ingredient and will continue to be so. The inclusion of heat treated soyabean meal in creep diets is highly attractive in keeping the price down. The use of these soya based diets need some caution although it is not impossible to use them. Many of the problems can be avoided by 'tolerising' piglets to the protein source. This entails giving piglets the potentially harmful ingredients in small but regular amounts for as long as possible before weaning. When piglets are subsequently put onto a high level of the same feed ingredient after weaning, their immune system does not overreact with an allergic response because the animal is familiar with the source of protein. Another strategy is to offer no creep feed at all before weaning. The piglet is then presented with dry food for the first time after weaning and the sudden high level of intake of the protein source induces the tolerisation process and the lining of the gut is not damaged. If it works, as it can on some farms, then this latter strategy may cheapen feed costs considerably.

Amongst a variety of methods for minimising post-weaning diarrhoea are the use of orally active vaccines against the *E. coli* organisms which cause the disease. These can be given in the feed and although they may not work on every farm, they give protection against a spectrum of causal organisms. Probiotics or *Lactobaccillus* based agents are also used to minimise enteric diseases. *Lactobaccillus* bacteria are normal inhabitants of the gut and a healthy flora of these organisms tend to crowd out the further colonisation of the gut wall by potentially pathogenic bacteria such as *E. coli*. In addition the *Lactobaccillus* bacteria produce lactic acid which reduces the pH in the gut lumen creating a hostile environment for the proliferation of pathogenic strains of bacteria.

FINISHING PIGS

From weaning to the time when pigs are slaughtered for meat the management is relatively straightforward compared to the problems with sows and piglets. In a well designed finishing house with a carefully controlled environment the process is one of turning feed energy into saleable meat. The investment in terms of fixed equipment and working capital is such that great precision in technical management is needed to make a profit. Currently compared to other animal enterprises, pork and bacon production can give a reasonable return on capital and a cash flow situation allowing regular receipts.

Traditionally, there have been a number of different markets for finished pigs. The first of these is the pork market where the meat is sold as a fresh product. The slaughter liveweight for pork pigs varies from one part of the country to another but is usually in the range 50–90 kg. Jointing techniques for pork pigs also vary from one region to another depending on local customs. Consumer demand shows seasonal trends and the Christmas and Easter trades are the times when demand is the greatest. The smallest demand at one time was seen in the summer months but recently this is not so clearcut because consumers have responded to the competitive price of pork and historical fears about pig meat in hot conditions are now allayed.

Bacon production tends to be the most profitable of all pig meat outlets and large scale producers sell a large proportion of their output into this category. The slaughter weight range is between 80 and 95 kg. Individual bacon companies may specify a much narrower weight band to fulfil their own market requirements. In general, the closer a producer can sell his pigs to the top of the weight range, the closer he is to maximising his profitability. Pigs are therefore weighed carefully every week as they approach slaughter to ensure they attain the correct carcass weight.

As well as pork and bacon weights there is a category of pigs known as cutters. The weight range is between 75 and 100 kg and historically these pigs were processed into manufactured products. There may be a better future for cutter pigs at around 80 kg liveweight. Production at this liveweight can be diverted to either quality bacon or to pork production depending on fluctuations in the markets of each. The hope is to balance one market against the other and to create a more stable price for each.

Heavy hogs or manufacturing pigs are slaughtered at 105 kg or more liveweight and these are destined to be processed into a large number of diverse products such as cooked meats, high quality hams, bacon products, meat pies, sausages and so on. The high slaughter weight means that when the animals are killed, they are laying down body fat at a much greater rate than pigs at lighter weights. This can be somewhat inefficient production in biological terms because of the feed energy required to accrete fat. High quality bacon products sold at a premium price are produced by trimming off the excess backfat to yield rashers and joints with a high lean content. In the future there could be an increasing market for pigs of greater liveweight. This is because of the modern pig's capacity for lean tissue deposition and minimal fat cover. Heavy pigs fed on an *ad libitum* basis might offer better profitability. This strategy has yet to be proven and there is a reluctance on the part of the meat trade to move in this direction due to increased processing costs.

Carcass quality and classification

The pig industry has been aware for some time of the need to generate products which consumers perceive as high quality and not just to produce for expedience. As a consequence, there has been a gradual reduction in the backfat scores of slaughter pigs. It seems that in the UK we are still not capable of producing as consistent a high quality product as do our EEC neighbours in Holland and Denmark. Bacon products from these countries are uniform with a high lean content and the packaging and marketing expertise is of a high order. Part of the apparent differential may be that UK producers sell almost all their produce on the home market. Danish and Dutch processers can select the best pigs for the export trade and the meat sold on their own home markets is a lower quality. The image of British pig products on the home markets needs developing in the future if the industry is to remain competitive even though farmers are capable of producing high quality pigs.

The criteria which determine quality in a finished pig revolve around the fat content of the carcass and in particular the amount of backfat. Most people prefer the taste of lean meat and in the mid 1980s a number of government reports on human nutrition pointed strongly to the need for the nation to reduce the number of calories eaten as saturated animal fats. The population has responded to this vigorously with much advertising effort given to low fat products. As a result there is continuing drive to produce leaner and leaner pigs. One point in favour of pig meat over most other meats is that the level of intramuscular fat is minimal and for a given weight of lean tissue the fat content is significantly lower.

The Meat and Livestock Commission operate a national scheme for the classification of carcasses and a very high percentage of pigs (> 90%) are classified within this scheme.

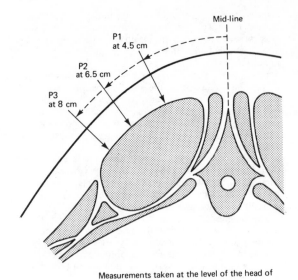

Measurements taken at the level of the head of the last rib

Figure 16.10 Positions for the measurement of backfat thickness on the pig carcass. Note the P1, P2 and P3 probe measurements are taken 4.5, 6.5 and 8.0 cm respectively from the mid-line at the level of the head of the last rib. On carcasses these measurements are taken using an optical probe. On live animals measurements may be taken at similar points using an ultrasonic probe

Classification aims to describe carcasses objectively in a form which can be used by producers, meat wholesalers and retailers as a framework for payment. It is fortuitous that there is such a clear relationship between the depth of subcutaneous backfat on the pig and its saleable meat yield and lean content. This has meant that a relatively simple scheme of carcass measurements give easily defined grading schedules based on what the meat trade and consumers want. The three basic measurements of backfat used in the MLC scheme are taken at the last rib at set points from the mid line. The names given to these positions are P1, P2 and P3 and these are shown in the illustration of a section through a pig carcass given in *Figure 16.10*.

All three measurements are not necessarily used in any one grading scheme. In pork production, for example, the P1 and P3 measurements are taken and added together to give a joint score. This score is then stamped onto the skin of the carcass. The measurements are made by an MLC officer after slaughter on the unsplit carcass. An intrascope device is used for this purpose which is an optical tool pressed in at the appropriate points. The work involved in taking measurements can be automated and computer interfaced systems are available which can identify individual pigs, take and record the necessary information and process the data for grading the pig and paying the farmer.

Grading schemes may be based on three or more discrete categories of carcass based on certain bands of backfat scores. The bands are designed to encourage the farmer to send the maximum number of pigs within the top grade for which the best price is paid. One of the elegant facets of the scheme is that with ultrasonic machines, breeding companies can make the same measurements on live animals and use this information as a basis for selecting carcass characteristics.

Despite the fact that there is considerable variation in the backfat scores of classified pigs, in general pigs are much leaner than ever before. Some wholesale meat companies have built into their grading schedules a penalty for pigs which are foo lean. This is partly because some extremely lean pigs have gone beyond the threshold of consumer acceptability and partly because the texture of the meat is adversely affected and cutting becomes difficult. This seems a rather intractable dilemma and the pig producer is caught in the middle of consumers demanding increasing leanness and the meat trade seemingly reluctant to pay when lean pigs are produced.

The industry has also been aware for some time of the need to carefully monitor meat quality. As pigs become leaner they also tend to show a higher incidence of pale soft exudative meat (PSE). Some breeds and crosses are known to have problems with PSE meat. Most hybrid companies therefore take steps to select against PSE in their breeding programmes and use foundation stock where possible with a low propensity for this problem. Breeds such as the Pietrain and some strains of Landrace pigs have a high incidence of PSE.

Pigs which are prone to PSE are often stress prone and possess the so-called halothane gene and they are extremely sensitive to halothane gas used in anaesthesia. Suseceptibility to the gas is used to identify PSE genes and stress susceptibility. The genes can then be bred out of a population over a period of time. Many of the breeding companies screen their pigs for halothane sensitivity.

Within the EEC, there have been strong pressures on member countries to bring together all classification schemes. There is now an agreed schedule which some nations have adopted. This scheme uses backfat measurements similar to the MLC scheme but a combination of the observations is used with prediction equations to assess the overall lean content of each pig. The UK has not, at the time of writing, subscribed to this scheme but it seems likely in the near future that this will be the case.

The nutrition of growing pigs

The energy and protein requirements of growing and finishing pigs are well established following many research projects over the years. It is not intended to give an exhaustive account of this work here but merely to outline some of the salient features with regard to feeding and management. For comprehensive reports on the nutrition of pigs the reader is referred to the 1981 edition of the AFRC publication *The Nutrient Requirements of Pigs*.

Nutrient requirements are to a large extent determined by the stage of growth or age of the pig. Genotype, sex, thermal environment and social order are other important factors which have a direct bearing on the nutritional needs of the animal. As pigs grow older, because of their changing body composition and increasing propensity for fat deposition, most feeding strategies are based on a change over at around 50 kg or so from a high protein *ad libitum* schedule to a lower protein scale-fed system. Practically it would be more straightforward and convenient to use a single grower diet at about 16% crude protein and to offer this on an *ad libitum* basis through to 90 kg or slaughterable liveweight. This would lead to an increased number of overfat pigs which would fall into the lower categories of most grading schemes. Improved hybrids to have the capability to remain on *ad libitum* feeding to much greater weights without excess fat accretion. The majority of finishing pigs are still fed on a restricted basis in the final weeks up to slaughter to maximise the percentage of pigs achieving the top grade. One of the disadvantages in reducing energy intake in the last few weeks is that overall growth performance, usually expressed as 'average days to bacon', is made worse. As a consequence, the throughput in a given finishing house is reduced. The production of finished pigs is therefore a balancing act with feed inputs, slaughter weight and throughput being the principal variables.

Because of the complexity of the relationships involved in making on-farm decisions, many producers use computer simulation programmes which model their own businesses to optimise inputs and output. These models are valuable tools

Table 16.5 Suggested energy and protein allowances for growing pigs[1]

| Liveweight (kg) | Approximate growth rate | | | | |
| | 600 g/d | | 700 g/d | | |
	Energy requirement (MJ/DE)	Protein requirement[2] (g/Cp)	Energy requirement (MJ/DE)	Protein requirement[2] (g/CP)	
20	12.6[3]	185[3]	12.6[3]	185[3]	
30	18.8	255	18.8[4]	255[4]	
40	22.0	275	24.5	300	
50	23.9	285	26.5	310	
60	25.5	300	28.2	325	
70	27.0	315	29.8	340	
80	28.4	325	31.2	355	
90	29.8	335	32.6	365	

[1] Derived from ADAS (1978) *Nutrient allowances for pigs*. Advisory Paper No. 7, MAFF. Reproduced with permission
[2] Assumes protein digestibility 80% biological value 65. (Equivalent to an amino acid contribution in the protein of lysine 5.1% methionine + cystine 2.8% threonine 3.3% and tryptophan 1.0%)
[3] Growth rate 400–500 g/d
[4] Growth rate 600 g/d

Table 16.6 Suggested vitamin allowances for growing pigs

	Weight of pig		
	Up to 20 kg	*25–55 kg*	*55–120 kg*
Vitamin A (units/kg)	12 000–20 000	10 000–15 000	5000–10 000
Vitamin D (units/kg)	1500–2000	1500–2000	1000–2000
Vitamin E (mg/kg)	12–30	10–15	5–15
Vitamin K (mg/kg)	2–5	1–4	1–4
Thiamine (B1) (mg/kg)	1–3	0–2	0–2
Riboflavin (B2) (mg/kg)	3–6	3–6	3–10
Nicotinic acid (mg/kg)	20–25	10–20	10–15
Pantothenic acid (mg/kg)	10–15	10–15	5–15
Pyridoxine (B6) (mg/kg)	2.5–6	1–4	0–4
Vitamin B12 (μg/kg)	10–30	10–30	10–20
Folic acid (mg/kg)[1]	0.8	0.5	–
Biotin (μg/kg)[1]	200	150	100
Choline (mg/kg)	150	150	100
Vitamin C (mg/kg)[1]	20	10	–

[1] Where only a single value is presented there is a paucity of published evidence on which to suggest likely allowances.

Table 16.7 Suggested dietary allowances of minerals

	Liveweight		
	Up to 20 kg	*20–55 kg*	*55–120 kg*
Calcium (%)	0.6–0.75	0.6–0.90	0.6–1.0
Phosphorus (%)	0.65	0.62	0.60
Salt (NaCl) (%)	0.35–0.60	0.35–0.50	0.35–0.50
Potassium (%)	0.22	0.22	0.22
Iron (mg/kg)	50	50	50
Magnesium (mg/kg)	–	–	–
Zinc (mg/kg)	160	160	160
Copper (mg/kg)	180	180	180
Manganese (mg/kg)	80	80	80
Iodine (mg/kg)	1.5	1.5	1.5
Selenium (mg/kg)	0.2	0.2	0.1

Note: with the exception of calcium, phosphorus and salt values indicate the suggested level of micronutrient supplementation.

in the farm management and as the many variables change it is possible to 'fine tune' the business to maximise profitability. Linear programming techniques are also widely available to farmers using their own microcomputers to enable them to formulate their own pig rations. As different samples of feed ingredients are available and prices change it is possible to produce least cost rations and minimise costs. For larger producers with their own mill and mixing plants, the ability to buy feed ingredients at the right price is a crucial skill. Computers are necessary to evaluable quickly all the possible options before major decisions are taken.

Nutrient requirements for growing pigs are given in *Tables 16.5–16.7*. These requirements can be delivered to pigs in a variety of forms, at different nutrient densities within the final diet and a large number of ingredients can be used depending on regional availability and season. It is impossible therefore to present scales of intake for different liveweights.

The farmers who tend to make the most money from finishing pigs are those with a very flexible feeding system.

They can meet the nutrient requirements of their pigs using a wide variety of ingredients and as the price or availability of an ingredient changes, they are able to respond quickly. This is the basis of the large scale producers around the large conurbations of the UK who have access to large quantities of by-products and waste from the human food industry. They are able to operate with such low feed costs that the quality of their end product is of less consequence. For the majority of farms a blend of barley and soyabean meal with an added package of vitamins and minerals is still the most profitable combination for finishers.

There has been sporadic interest in alternative feedstuffs for pigs such as fodder best, potatoes and even grass silage. A minority of farms may be able to make some use of these materials where they are home grown but for growing pigs appetite is a severe limiting factor. For sows, some of these bulky feeds may offer an economic alternative to conventional rations.

General management of growing pigs

The general management and husbandry of growing and finishing pigs is to a large extent dictated by the type of housing available. In intensive systems the performance of pigs is highly temperature dependent. Normal growth can be sustained within a wide range of ambient temperatures given an adequate supply of nutrients. There may be only a relatively limited range of temperatures where pigs will perform economically. This is because the pigs biology strives to preserve its own internal environment and if external conditions (i.e. house temperature) change then pigs use excess feed energy in temperature regulation. Growth rate and the efficiency of food conversion are adversely affected. The precise house temperature depends on factors such as the use made of straw and the group size. In a modern intensive finishing house the temperature should be maintained at around 70°C. Some of the more sophisticated houses for growing pigs included separate rooms for each batch or group of pigs. Each room has its own control unit and as the pigs reach higher liveweights, the temperature settings can be reduced to provide each pig with as near an ideal environment as possible. These systems offer very high pig performance but at a high building cost.

Growing pigs are accommodated in groups of ten to 20 in any one pen. Above this and it is difficult to monitor the progress of individuals and disturbances in the social order are often seen. As animals approach slaughter weight, they need to be weighed individually on a weekly basis and this is not easy with large groups. As particular groups reach the target liveweight, some individuals get there first and some may take another two or three weeks. In this situation there may be some pens with only one or two pigs housed because they can not be mixed without considerable fighting and trauma. It is possible to use a tranquillising agent to mix group of slow growing individuals together to release some of the available pen spaces.

There are many methods of delivering feedstuffs to groups of finishers but as labour costs have risen, the use of automatic systems has increased. Liquid feeding systems are still popular with some farmers. The blend of ingredients in a meal form is mixed with water into a slurry. The slurry is then pumped around the farm from pen to pen. Some evidence has shown that these wet feeding systems are associated with improved feed conversion efficiency but one of the disadvantages is the need for more trough space within each pen and this tends to reduce the number of pigs housed. Dry feeding systems are usually based on pelleted feeds which reduce dust and waste. These are given on the floor of the lying area of the pen and this allows for increased stocking density and helps to persuade the pigs to dung in the allotted slatted area or dunging passage. There are now many automatic and semi-automatic systems for dispensing dry feeds to groups of pigs and the reliability of these systems is much improved compared to earlier machinery.

The effects of sex on the performance of growing pigs is well documented and it is known that other things being equal, the lean tissue growth potential of boars is better than gilts and in turn gilts are slightly better than castrated males (often called hogs in the UK). In the past, when pigs grew much more slowly, entire males were slaughtered well after sexual maturity. Male pigs, which are sexually mature, produce steroidal products which taint the meat giving it a strong flavour and a characteristic odour. The faster growing pigs of today are slaughtered before a significant problem can be detected although a small proportion of entire males will exhibit boar taint at an early age. There seems little reason to continue the practice of castrating male piglets at about two weeks of age with all the stress this entails. Castration is associated with reduced lean tissue growth and a poor feed conversion ratio and many consider it an unnecessary mutilation. There has been a great reluctance on the part of the meat trade to accept entire males as quality carcasses but in the last few years this has changed and in some regions of the country there are large numbers of entire males produced for meat. With this type of production it is then advantageous to separate males from females and to offer a different feeding regime to each. Many boars can be offered an *ad libitum* scale of feeding through to slaughter with no loss in carcass quality. Females on the other hand still need a period of restricted feeding towards slaughter.

The use of steroidal growth promoters to modify the carcass composition of pigs was once practised throughout Europe and elsewhere. These are now banned in EEC countries as a result of public concern over meat residues. It seems likely that no other products will be used in the future for growing pigs. A possible exception to this may be the future use of the beta-agonist agents now under development for their capacity to divert nutrients into lean tissue growth and away from fat deposition. If consumers continue to demand lean only then this may be an acceptable way of achieving minimal fat production. In the not too distant future it may also be possible, using transgenic animals to design pigs with specific carcass characteristics and growth potential. These 'designer pigs', could give the public meat products which are both nutritious and which can promote good health. They may also be extremely efficient and profitable for producers.

CONCLUSIONS

Those involved in pig production have seen many changes in the methods used in the last few years. The improvements in technical efficiency and management have been of a high order as producers have endeavoured to stay ahead of the economic storms without the protective umbrella of a price support scheme. As a result, the industry is lean and fit and the prospects are encouraging for those who are committed to production and vigorous marketing.

It is still possible to step onto the farming ladder by producing pigs on a small scale but this route is becoming more difficult to follow as the structure of the industry continues to change in favour of the larger units. A major challenge for the future, apart from the production of low fat meats is the accommodation of the views of the groups involved with animal welfare. In view of the industry's record for rapid change and forward thinking it is possible that within a short time the innovators within the industry will have created systems of pig farming which will provide for the welfare needs of the animals whilst still generating profits.

Further reading

BRENT, G. (1986). *Housing the Pig*. Ipswich: Farming Press.

COLE, D. J. A. and HARESIGN, W. (1985). *Recent Developments in Pig Nutrition*. London: Butterworths.

HUGHES, P. E. and VARLEY, M. A. (1980) *Reproduction in the Pig*. London: Butterworths.

MEAT AND LIVESTOCK COMMISSION. Annual Yearbooks.

Nutrient Requirements of Pigs (1981). Commonwealth Agricultural Bureaux, Slough: AFRC.

WHITTEMORE, C. T. and ELSLEY, F. W. H. (1979). *Practical Pig Nutrition*. Ipswich: Farming Press.

17

Poultry

J.I. Portsmouth

Poultry are kept in the UK primarily for meat and egg production. The two sectors are economically different, but naturally have physiological similarities, e.g. meat chickens are hatched from eggs produced by specially bred parent stock and commercial eggs are produced by specialist bred parents. The distinction is thus in the selection of meat or egg production characteristics. The utility fowl of the 1950s/1960s no longer exists. Chickens are specifically bred, managed and marketed to meet a very definite requirement of quality, quantity and price.

MEAT PRODUCTION

For a summary of UK poultry meat production *see Table 17.1* and *Figure 17.1*. Total output increased by 34% in the

ten-year period with turkey meat expanding by 64% and chicken by 12.7%. Waterfowl increased by 32%. A period of further expansion and also consolidation is anticipated over the next ten years as different sectors of the poultry and meat industry concentrate on developing even newer markets for gourmet type products as saturation point approaches for whole carcass sale. Such products are chicken rolls, sausages, portions, meat loaves and many gourmet dishes.

Figures in *Table 17.2* and *Figure 17.2* show how poultry meat consumption has increased by 40% in the ten-year period, an annual increase of some 3.6%. Turkey meat consumption on the other hand has increased by 97% (8.8% per annum). A further increase in the consumption of poultry meat is almost certain to occur at the expense of red meats with the main reasons being price, flexibility of

Table 17.1 UK poultry meat industry (10³ tonnes) 1975–85

	1975	*1980*	*1981*	*1982*	*1983*	*1984*	*1985*
Poultry meat – output	652	754	746	812	804	845	874
Imports	9	27.6	23.7	27.2	51.1	53.0	61.2
Exports	2.0	19.4	17.6	19.9	23.3	27.7	30.9
Consumption/head (kg)	11.4	13.5	13.5	14.4	14.5	15.5	16.0
Broilers – output	584	554.8	555.7	604.7	589.5	632.7	658.0
Slaughterings (10⁶)	324.2	370.9	375.9	401.7	387.7	413.3	428.0
Placings (10⁶)	336.5	402.7	423.9	439.3	434.8	465.1	487.8
Consumption/head (kg)	NA	10.0	10.1	10.9	11.3	11.7	11.9
Producer price (p/lb)	13.5	22.4	23.1	24.7	25.7	26.6	26.0
Turkeys – output	85	122.1	116.7	130.3	136.6	138.5	140.4
Slaughterings (10⁶)	16.9	23.4	22.2	24.7	26.0	26.3	26.7
Placings (10⁶)	18.63	25.51	24.27	27.84	28.26	29.07	29.78
Consumption/head (kg)	NA	2.1	2.2	2.3	2.5	2.7	3.0
Ducks – output	12.8	15.4	15.3	16.5	16.7	16.4	17.2
Slaughterings (10⁶)	5.9	7.1	7.1	7.6	7.7	7.6	7.9
Placings (10⁶)	NA	8.39	7.98	8.48	8.77	9.13	10.67
Geese – output	0.8	0.8	0.8	0.9	0.9	0.8	0.8
Slaughterings (10⁶)	0.2	0.2	0.2	0.2	0.2	0.2	0.2
Hens – output	70.3	60.6	57.0	59.1	59.9	57.0	57.8
Slaughterings (10⁶)	46.6	39.8	36.5	38.3	38.6	36.2	36.9

Source: *Poultry World*, Sep 1986

450

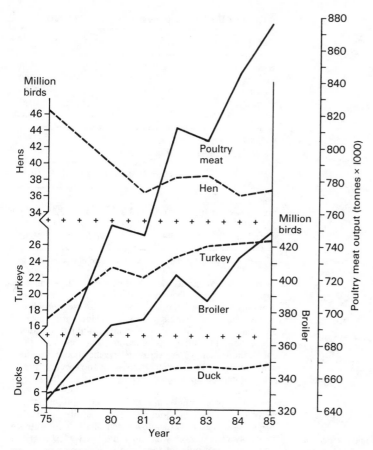

Figure 17.1 UK poultry meat industry 1975–85

product, attractiveness of product, low animal fat content and high protein content. Waterfowl meat consumption has increased modestly over the last five years and with the exception of goose meat, where it was anticipated a small significant increase will occur, the overall situation for waterfowl is one of little change.

Table 17.2 Poultry meat consumption (kg/head/year) 1975–85

Year	Total	Turkey meat	Chicken meat
1975	11.4	1.52	9.88
1976	11.5	1.65	9.85
1977	11.4	1.66	9.74
1978	11.8	1.80	10.00
1979	13.50	1.97	10.00
1980	13.5	2.1	10.0
1981	13.5	2.2	10.0
1982	14.4	2.3	10.9
1983	14.5	2.5	11.3
1984	15.5	2.7	11.7
1985	16.0	3.0	11.9

Source: Poultry Meat Association, reproduced with permission

Broilers

Since the broiler industry began in the early 1950s it has developed rapidly and is now the most integrated form of livestock production. All the major companies, e.g. Buxted, Marshalls, Sun Valley, Padley's and many others produce and market their own nationally advertised brands with total control of live bird growing, breeding, hatching, feeding and processing.

Only a handful of independent producers remain to supply the smaller packing stations with the high quality product demanded by the market.

Of the total meat produced in the UK 75% is derived from broilers. A broiler is a young bird of either sex, marketed between 1.45 kg and 2.75 kg liveweight. It is slaughtered between 35 and 56 d (approximately) depending on the range of body weight needed and the most economical time of production. The broiler is a specifically bred hybrid meat bird derived from breeds and strains originally developed in the USA. Basic leading UK breeding companies are: Cobb (Tyson/Upjohn), Ross (Hillsdown Holdings), Shaver (Cargill), Hubbard (MSD), Marshall (Marshall Foods Ltd), and Arbor Acre (Booker) (name in brackets indicates major shareholding company).

Depending on killing age a broiler will convert food into

(lb)

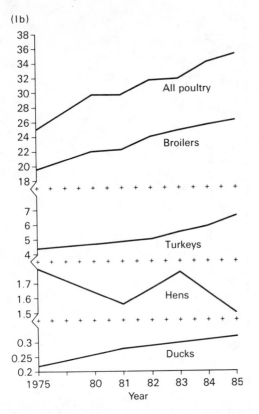

Figure 17.2 UK poultry meat consumption per head

meat with a ratio of 2 to 1. Optimum economic performance depends on correct nutrition and husbandry conditions. Work by ADAS at Gleadthorpe EHF shows that broilers respond optimally to high nutrient density (HND) when housed at 21°C and killed at 47–49 d *Tables 17.3, 17.4, 17.5, 17.6* and *17.7* summarise the nutritional needs of broilers. It is anticipated that by the 1990s the figures quoted in *Table 17.5* will be achieved in 2 d less. By the end of the present decade a 2 kg liveweight broiler will be reached in 40 d.

Anti-coccidial drugs such as monensin and clopidol are an essential part of an overall disease prevention and control

Table 17.4 Broiler feed programmes (kg/1000 birds)

Age of slaughter	*35–45 d*	*46–53 d*	*+ 53 d*
Starter	+ 750	750	500
Grower	1500	1500	1000
Finisher 1	to finish	1500	2250
Finisher 2		750	1000

Source: Peter Hand technical information. Reproduced with permission

Table 17.5 Broiler performance guide

	Days at slaughter				
Age of slaughter	35	42	45	52	59
Av. liveweight (kg)	1.50	1.81	2.00	2.35	2.60
FCR	1.80	1.85	2.00	2.10	2.18
Liveability %	98	97.5	96.5	96.0	95.5

Source: Peter Hand technical information. Reproduced with permission

programme. Without these anti-coccidial drugs modern broiler/turkey production under intensive husbandry methods would be impossible and the disease, coccidiosis, would cause considerable loss. Anti-coccidial drugs such as monensis and halofuginone are an important part of disease prevention in turkey production. The use of these medications is, however, relative to specific disease situations and together with anti-blackhead drugs the specific drug to use will depend upon the degree of disease and the type of management.

Growth promoting drugs such as the antibiotic, bambermycin, are used in about 90% of UK broiler and turkey feeds. It is generally acknowledged that their presence probably enhances growth and improves feed conversion efficiency between 2 and 5%.

COMMERCIAL EGG PRODUCTION

The UK laying flock has been declining since 1968. This is due to a fall in demand for eggs and an increase in egg production per bird due to technological advances. The total number of poultry farms with laying birds has fallen from 125 258 in 1971 to 46 000 in 1985. The egg industry con-

Table 17.3 Recommended broiler ration specifications

Nutrient	*Feed type*			
	Starter	*Grower*	*Finisher 1*	*Finisher 2*
Protein (%)	23	20	20	17.5
Lysine (%)	1.30	1.20	1.05	0.95
Av. lysine (%)	1.18	1.06	0.95	0.95
Methionine (M) (%)	0.62	0.58	0.50	0.40
M + C (%)	0.97	0.90	0.88	0.76
Energy (ME kcal/kg)	3035	3130	3180	3180
Energy (ME MJ/kg)	12.70	13.10	13.30	13.30
Calcium (%)	0.9	0.9	0.9	0.9
Available phosphorus (%)	0.45	0.45	0.45	0.43
Salt (%)	0.34	0.36	0.36	0.36
Linoleic acid (%)	1.50	1.40	1.25	1.10

Source: Peter Hand technical information. Reproduced with permission

Table 17.6 Micronutrients recommended for inclusion in broiler diets

Nutrient	Starter	Finisher
Vitamin A (m i.u.)	13	13
Vitamin D3 (m i.u.)	4	4
Vitamin E (k i.u.)	25	25
Vitamin K3 (g)	2	2
Vitamin B1 (g)	2	2
Vitamin B2 (g)	8	8
Nicotinic acid (g)	30	20
Pantothenic acid (g)	12	12
Vitamin B12 (mg)	15	15
Vitamin B6 (g)	3	2
Choline (g)	250	200
Folic acid (g)	1.5	0.5
Biotin (mg)	250	100
Manganese (g)	80	80
Zinc (g)	70	70
Copper (g)	15	15
Cobalt (g)	0.4	0.4
Iron (g)	30	30
Selenium (g)	0.20	0.20
Iodine (g)	1.0	0.5

Source: Peter Hand technical information. Reproduced with permission

centrates on fewer and larger units with more than 50% of the country's eggs produced from about 400 flocks averaging 20000 or more laying birds each. The trend to fewer and larger units will continue unless intensified action by welfare interests moves the industry away from the battery cage and back to the less intensive management systems where husbandry practice favours the smaller unit.

Free range production

Eggs produced by hens managed under free range conditions account for approximately 5% of the total egg production. The proportion is highest in the southern and home counties of England where the claimed premium is also highest. It is apparent that the saturation point for free range eggs and eggs produced under other less intensive systems is around 10%. Beyond this the premium is eroded by a fall in demand and many free range units which began in the 1980s were out of business by 1987. This was caused partly by poor returns and partly by higher production costs incurred, especially in adverse weather conditions. Under extensive systems of management the need for a high degree of stockmanship is even more important than in the intensive systems.

In the late 1950s and early 1960s the laying flock were made up predominantly of white feathered birds producing white eggs. In 1975, 64% of the flock were brown and producing brown eggs. In 1979 almost 90% were brown egg layers and in 1985 this figure had reached 98%.

The demand for brown eggs (including tinted eggs) has caused breeding companies to intensify efforts to increase food efficiency, which previously favoured white egg layers, because of their smaller body size and therefore lower energy needs for maintenance. *Table 17.8* shows the main 'breeds' and companies involved in breeding commercial laying stock in the UK.

Tables 17.9 and *17.10* show egg yields for hens housed in different husbandry systems and also per capita egg consumption since 1960 (*see also Figure 17.3*). In 1960/1961 birds housed in cages on average produced 18.3 eggs more than the other flocks while in 1978/1979 this widened to 21.5 eggs per bird. The average of 249 eggs per bird is in fact insufficient to cover the production costs of the 1980s. Profitable flocks need to produce 278 eggs on low feed intakes. To obtain high and economic production, certain minimum amounts of nutrients must be provided. These are shown in *Table 17.11*.

Feed programmes for laying birds

Laying hens are invariably fed *ad libitum*. The hens' adjustment of feed intake to change in ration energy level is not precise but sufficiently accurate to make *ad libitum* feeding the most economical system. Restricted feeding, whereby a certain allowance is allocated once or twice daily, is hazardous. Variation in feed consumption within the flock may be as much as 40%, thus any physical restriction of feed penalises the small appetite to the great detriment of egg output and profits. Free choice feeding, whereby cereals are fed whole with a pelleted protein/vitamin/mineral concentrate allows the individual bird to select its own protein (amino acids) needs according to rate of lay, has shown nutritional/physiological advantages. The system needs management perfection before finding practical value.

Feed restriction for replacement pullets

Excess body fat adversely affects subsequent rate of lay. Consequently, replacement pullets and replacement broiler breeding stock are fed controlled amounts of feed to reduce excessive energy intake. With commercial pullets the mid-term restriction during rearing proved to be popular in the early 1980s where it was required to bring them into lay at around 20 weeks of age. It was found that the mid-term restriction was more important than either early or late term restriction. In practice, this meant that the severest form of restriction took place during the period 9–15 weeks. More recently, in the rearing of commercial laying pullets the *ad libitum* feeding system has been favoured as it brings birds into lay at an earlier age and has been shown to produce superior egg production. Such pullets should, however, not be fat and the balance between additional body weight and earlier sexual maturity is a delicate one as it is important to optimise not only egg production but also egg weight.

Feed control with broiler breeders during rearing and laying is vital if excess energy consumption and overfatness is to be prevented. Broiler breeder companies have recommended feeding programmes and these should be followed closely for maximum economic performance. *Table 17.12* details the essential micronutrients such as vitamins and trace elements which are used to supplement normal ingredients of poultry rations.

BROILER BREEDING INDUSTRY

The two major breeding companies in the UK are Cobb and Ross, whilst in the EEC Arbor Acre, Hubbard and Vadette have significant market shares. Both Cobb and Ross are

Table 17.7 Additives commonly used in poultry production without veterinary prescription

Additive	Stock type	Age (weeks)	Level of active ingredient in final feed (ppm)
Growth and egg promoters			
Avoparcin	Broilers	0–16	7.5–15 avoparcin
Virginiamycin	Broilers	0–9	0–20 virginiamycin
Bambermycin	Broilers	0–16	2–5 bambermycin
	Turkeys	0–26	2–5 bambermycin
	Laying hens		2–5 bambermycin
Zinc	Broilers	0–4	5–20 zinc bacitracin
bacitracin	Broilers	4–16	5–20 zinc bacitracin
	Laying hens and breeders		15–100 zinc bacitracin
ZB100	Broilers	0–4	5–50 zinc bacitracin
	Broilers	4–16	5–20 zinc bacitracin
	Turkeys	to 26	5–20 zinc bacitracin
	Layers		15–100 zinc bacitracin
Anti-coccidials[1]			
Monensin	Broilers, pullets	0–16	100–120 monensin
Narasin	Broilers	0–16	70 narasin
Lasalocid	Broilers	0–16	90 lasalocid
pancoxin	Pullets	0–6	amprolium
	Pullets	6–16	65 amprolium
	Pullets	12–16	50 amprolium
Maduramycin	Broilers	6–16	5 maduramycin
Salinomycin	Broilers	0–16	60 salinomycin
Halofuginone	Broilers	0–16	3 halofuginone
	Broilers	0–16	
Halofuginone	Turkeys	Until 5 d before slaughter	3 halofuginone
Nicarbazin	Broilers	Until 7 d before slaughter	125 nicarbazin
Anti-blackhead medications			
Dimetridazole	Turkeys		125 dimetridazole
Nifursol	Turkeys		50 nifursol

[1]Where an additive contains more than one active ingredient only the major constituent is shown.
Source: Peter Hand Technical Information. With permission

Table 17.8 Selection of egg producing strains available in UK

Commercial name	Company
Babcock 380	ISA Poultry Services
Dekalb G. Link	Ross Breeders
Hisex Brown	Euribrid
Golden Comet	Hubbard
Ross Brown	Ross Breeders
Shaver 585	Shaver
ISA Brown	ISA Poultry Services

Source: Peter Hand, Technical information. With permission.

Table 17.9 Egg production/hen per annum by management system

Management system	1960/61	1972/23	1973/74
Free range	166.5	181.2	172.9
Deep litter	187.9	238.5	200.9
Battery cage	206.2	214.0	232.3

Management system	1974/75	1976/77	1978/79	1984/85
Free range	175.7	192.1	196.8	258
Deep litter	207.0	224.5	227.8	264
Battery cage	236.4	245.4	249.3	276

Source: Eggs Authority, with permission

Table 17.10 1960–85 annual egg consumption in the UK (no. eggs/person)

	Total
1960	258
1961	264
1962	264
1963	260
1964	273
1965	271
1966	270
1967	275
1968	273
1969	271
1970	275
1971	273
1972	273
1973	262
1974	256
1975	246
1976	248
1977	248
1978	250
1979	251
1980	236
1981	231
1982	231
1983	226
1984	227
1985	229

Source: Eggs Authority and *Poultry World*, with permission

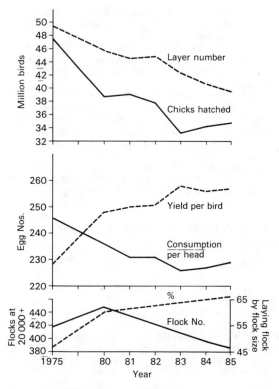

Figure 17.3 UK egg industry 1975–85

Table 17.11 Recommended intake of certain essential nutrients to produce 55 g egg output with house temperature of 21°C

Nutrient	Minimum daily	% Nutrient in ration at specified average daily intake		
		100 g	110 g	120 g
Protein	18.0 g	18	16.4	15
Methionine	420 mg	0.42	0.38	0.35
Methionine + cystine	640 mg	0.64	0.58	0.53
Lysine	860 mg	0.86	0.78	0.72
Calcium	4.2 g	4.2	3.80	3.50
Av. phosphorus	0.33 g	0.33	0.32	0.29
Linoleic acid (EFA)	1.25 g	1.25	1.14	1.04
[1]Xanthophylls	14 mg/kg	14	12.75	11.7
Energy (MJ/d)	1.25	1.25	1.37	1.50

[1]Recommended level of xanthophylls ensures a good yellow-orange yolk colour (fan colour 10)
Source: Peter Hand, technical information, with permission

so-called heavy meat strains yielding superior quantities of high value breast and thigh meat compared with the other strains. Egg yield is, however, inferior and the different markets of the EEC countries assess the economic value of the breeds according to local requirements. Broiler chick production of Cobb and Ross UK strains average between 115 and 128 chicks per hen per 60-week period. The other

Table 17.12 Micronutrient recommendations for commercial layers, replacement and breeding stock

Nutrient	Poultry breeder	Chick grower	Layer
Vitamin A (m i.u.)	16	10	6
Vitamin D3 (m i.u.)	6	3	3
Vitamin E (k i.u.)	30	10	6
Vitamin K3 (g)	2	2	2
Vitamin B1 (g)	1	0.5	0.5
Vitamin B2 (g)	8	5	3
Nicotinic acid (g)	25	20	8
Pantothenic acid (g)	10	8	3
Vitamin B12 (mg)	6	10	10
Vitamin B6 (g)	4	1	1
Choline (g)	250	100	–
Folic acid (g)	2	0.5	–
Biotin (mg)	250	50	–
Manganese (g)	70	65	70
Zinc (g)	60	40	40
Copper (g)	8	8	3
Iodine (g)	1	1	1
Cobalt (g)	0.5	0.5	0.4
Iron (g)	20	15	10
Selenium (g)	0.20	0.15	0.15
Molybdenum (g)	0.5	–	–

Source: Peter Hand, technical information. With permission

strains can produce 10–15 more chicks per hen but the growth rate and meat yield are generally lower. Because broilers are bred to grow fast the replacement parent stock have to be reared under very strict management and nutritional conditions. Control of energy intake is of paramount importance in order to prevent excess body weight occurring.

Special feed programmes must be carefully followed with the emphasis on monitoring a pre-determined body weight for age guide. At 18 weeks increases in both nutrient quantity and quality are synchronised with an increase in the photoperiod. The effect of changing nutrition and day length serves to bring the flock into egg production with body weight still under careful control. Broiler breeding, like other sectors of the modern poultry industry, is highly sophisticated, demanding a great deal of skill and stock sense.

ENVIRONMENT OF LAYERS AND BROILERS

Ventilation

Table 17.13 shows the ventilation rates needed for intensively housed poultry. The optimum temperature for adult layers is 21°C and for broilers, after initial brooding, 20–21°C depending on sex and ratio nutrient density. Ventilation rate is also expressed according to the amount of food consumption. Thus, the optimum minimum for broilers is 2 mstd (cubic metres of air per second per tonne of feed per day). Maximum ventilation is ten times the minimum (that is 20 mstd). Note 1.0 mstd is approximately equal to 1.0 cfm per pound per day (cubic feet of air per minute per pound of feed per day). This standard therefore relates closely to the metric m³/s/tonne/d, 25 cfm/lb/d being equivalent to 25 mstd, and 2 cfm matching 2 mstd.

Table 17.13 Ventilation rate requirements for intensively housed poultry

Stock	Weight (kg)	Max. needed		Min. needed	
		m³/bird	cfm/bird	m³/bird	cfm/bird
Layers	1.2	10	6.0	1.0	0.6
	2.0	12	7.0	1.2	0.75
	2.5	14	8.0	1.5	0.9
	3.0	14	8.0	1.7	1.0
	3.5	15	9.0	2.0	1.2
Broilers	0.05			0.1	0.06
	0.4			0.5	0.3
	0.9			0.8	0.45
	1.4			0.9	0.50
	1.8	10	6.0	1.3	0.75
	2.2	14	8.0	1.7	1.0
Turkeys	0.5	6	3.5	0.7	0.4
	2.0	12	7.0	1.2	0.7
	5.0	15	9.0	1.5	0.9
	7.0	20	12.0	2.0	1.2
	11.0	27	16.0	2.7	1.6

Source: MAFF, Gleadthorpe EHF. Reproduced with permission.

Lighting

Because nearly all modern poultry units employ windowless controlled environment housing, artificial lighting, both in extent and intensity is extremely important. In laying birds changes in day length stimulate or depress the secretion of gonadotrophic hormones. Increasing day length stimulates and vice versa. Change in day length can be used to manipulate age of sexual maturity. Extending the day length in rearing advances sexual maturity. With a laying bird, day length is generally increased from about 8 h at 18 weeks to a maximum of 16 h achieved by weekly increases of 15 min each (*see Figure 17.4*) Day length is then held at this level until the end of laying period. Egg output can be manipulated by so-called ahemeral lighting patterns where day length is more or less than the normal 24 h cycles. With a longer day length, i.e. 26–28 h, egg weight is increased, whilst with cycles less than 24 h the rate of production can be increased at the expense of egg weight. Ahemeral patterns are currently used in practice and may find application in selection of future breeding stock.

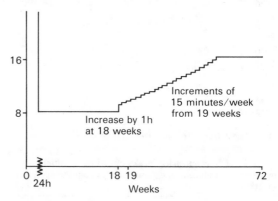

Figure 17.4

In broiler production the most popular lighting system is 24 h for the first 48–72 h followed by a 23 h day to slaughter. The 1 h dark period is used to accustom the bird to darkness in the event of a power failure.

Intermittent lighting programmes are also used. After the continuous 24 h for the first 72 h the programmes may be 2 h light followed by 2 h darkness or 1 h light and 3 h darkness. Intermittent programmes are thought to improve efficiency of digestion through the enforced rest of the dark period and also to reduce the incidence of leg weaknesses which is not uncommon in fast growing broilers.

Light intensity (all poultry)

Under intensive housing systems a low light intensity is essential to avoid feather pecking and cannibalism. Additionally, birds are quieter, easier to manage, less active, thus using less energy for maintenance. Young chicks should be given 20–30 lx for the first 72–96 h to encourage early feeding. Thereafter, 6–7 lx is sufficient to allow normal feeding, optimum activity and ease of operations. For laying birds the optimum intensity is 5–7 lx. In the dark period one-tenth of the optimum figure is recommended i.e. 0.5–0.7 lx.

TURKEYS AND WATERFOWL

Turkeys

Table 17.1 shows turkey production has increased four-fold since 1965. Disastrously low selling prices due to over-production in 1973 saw drastic cutbacks in 1974 which were even more severe in 1975. Since then a better organised industry, with marketing at Easter, Whitsun and August Bank Holiday, has led to an almost speculative expansion. The market prospects to the end of the 1980s are extremely good and, with a demand for high quality white meat, the future for the turkey industry during the 1990s is extremely bright. Predictions like this, however, can be severely affected by the output of our partners in the EEC, particularly Italy and France who have relatively large turkey industries. Both of these countries are potentially large exporters of turkey meat which could seriously interfere with our own expansion plans. In 1971, 2300 flocks produced 14 000 000 turkeys whilst by 1985 this number had fallen dramatically for almost twice the output (28 000 000 birds) thus the trend of fewer and larger is seen also in the turkey industry. *Table 17.14* shows how in 1985 some 92% of all turkeys were produced by units with in excess of 100 000 birds compared with some 4% produced by flocks of 5000–10 000 birds!

Turkeys are into a relatively new phase of marketing with expansion into the 'cut-up' portion, and further processed markets with such products as turkey rolls, sausages, patés, burgers, and a variety of easily prepared convenience foods competing very successfully with other sections of the meat market. This further process market requires a large bird with a high meat to bone ratio. Such birds are normally

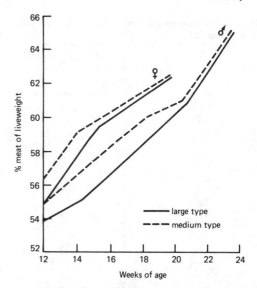

Figure 17.5 Relationship between % meat/liveweight and turkey age (from *Poultry World* 26th June 1980, reproduced by courtesy of the editor and publisher)

20–26 weeks old and may weigh from 12–15 kg compared with the whole bird market which demands a smaller 4–6 kg oven-ready bird. *Table 17.15* shows the carcass analysis of 5.2 kg male turkeys (from MAFF, ADAS, Gleadthorpe, 1980). *Figure 17.5* shows the relationship between percentage meat of liveweight and turkey age. For a 22-week-old

Table 17.14 Number of turkeys in England and Wales and proportion of flock by size of flock

Size of flock	1976 × 10³	%	1977 × 10³	%	1978 × 10³	%	1981 × 10³	%	1984 × 10³	%	1985 × 10³	%
1–25	3.8	0.1	3.3	0.1	3.1	0.1	3.6	0.1	2.7	0.1	2.8	0.1
26–99	6.8	0.1	5.3	0.1	5.6	0.1	6.5	0.1	3.6	0.1	3.9	0.1
100–499	45.0	0.8	32.9	0.8	35.2	0.7	34	0.4	22.3	0.3	19.0	0.2
500–999	53.9	1.0	36.3	0.8	43.7	0.9	38.9	0.5	24.0	0.4	32.7	0.4
1000–4999	371.2	6.6	290.3	6.0	278.4	5.5	238.3	3.0	253.8	3.7	258.2	3.4
5000–9999	425.5	7.6	351.1	7.3	360.3	7.1	340.9	4.3	272.4	4.0	299.6	3.9
100 000 +	4720.2	83.8	4110.8	85.0	4370.1	85.8	7291.2	91.7	6254.7	95.0	6982.3	91.9
Totals	5626.4	100	4830	100	5096.5	100	7953.4	100	6834.4	100	7598.5	100

Source: MAFF statistics. Reproduced with permission

Table 17.15 Turkey carcass analysis

	kg	%
Weight	5.2	100
Breast meat	1.634	31.42
Leg meat	1.31	25.20
Visceral fat	0.007	0.14
Total bone	1.493	28.71
Total meat	3.277	63.02
Total meat/skin	3.70	71.15

Source: MAFF, Gleadthorpe EHF. Reproduced with permission

medium stag some 63% of total liveweight is meat. In the data of *Table 17.15* the 5.2 kg bird is 14 weeks old and such birds have a 55–56% meat yield of liveweight. The larger the bird the greater its meat yield and its usefulness in the cut-up and meat stripping business. Consumption per capita of turkey meat is shown in *Table 17.2*. The average rate of increase is 8.8% per annum. In the last five years this has been due to greater consumption of turkey meat at times other than Christmas, for example Easter and Whitsun. *Table 17.16* shows the growth rate, feed consumption and feed conversion tables for turkeys from day old to 20 weeks of age. These figures are just a guide to performance, strain differences, etc. and can be wide. *Tables 17.17* and *17.18* deal with the nutritional requirements of turkeys at all ages.

Table 17.16 Management standards – turkeys growth rate and food conversion

Age (weeks)	Live weight (kg)	Total food consumed (kg)	Cumulative FCR
1	0.13	0.11	–
2	0.27	0.29	1.07
3	0.45	0.60	1.33
4	0.75	1.13	1.51
5	1.18	1.86	1.57
6	1.68	2.77	1.65
7	2.27	3.86	1.70
8	2.84	5.13	1.80
9	3.47	6.59	1.90
10	4.18	8.20	1.96
11	4.77	10.0	2.09
12	5.54	11.8	2.13
13	6.22	13.8	2.21
14	6.86	15.9	2.31
15	7.5	18.2	2.42
16	8.2	20.7	2.52
17	8.77	23.1	2.63
18	9.32	25.4	2.72
19	10.0	28.2	2.82
20	10.5	31.0	2.95

Source: Peter Hand (GB) Ltd – technical information, with permission.

Table 17.18 Micronutrients recommended for inclusion in turkey diets

Turkey supplements	Turkey starter	Turkey grower/ finisher	Turkey breeder
Vitamin A (m i.u.)	16	12	16
Vitamin D3 (m i.u.)	5	5	4
Vitamin E (g)	20	15	40
Vitamin K3 (g)	4	3	4
Vitamin B1 (g)	2	1	1
Vitamin B2 (g)	10	8	20
Nicotinic acid (g)	60	45	50
Pantothenic acid (g)	15	12	15
Vitamin B12 (mg)	15	10	15
Vitamin B6 (g)	4	2	4
Choline (g)	450	250	200
Folic acid (g)	1	0.5	3
Biotin (mg)	150	100	150
Manganese (g)	80	70	80
Zinc (g)	65	60	70
Copper (g)	8	5	8
Iodine (g)	0.5	0.5	1
Cobalt (g)	0.4	0.4	0.5
Iron (g)	10	10	20
Selenium (g)	0.20	0.15	0.20
Molybdenum (g)	—	—	0.5

Source: Peter Hand, technical information. With permission.

Table 17.17 Nutritional requirements of turkeys

	Pre-starter	Starter	Rearer	Early finisher	Late finisher	Grower	Heavy stag
Protein (%)	30	27	24	21	15	18	14
Methionine (M) (%)	0.86	0.75	0.65	0.45	0.30	0.32	0.26
Lysine (%)	2.0	1.80	1.50	1.20	0.76	0.90	0.70
M + cystine (%)	1.30	1.18	1.00	0.75	0.55	0.60	0.50
ME MJ/kg	12.0	12.0	12.25	12.65	12.25	12.0	11.5
Calcium (%)	1.30	1.20	1.20	1.20	1.25	1.20	1.40
Phosphorus (%)	0.90	0.85	0.80	0.70	0.65	0.75	0.65
Av. phosphorus (%)	0.70	0.65	0.60	0.50	0.45	0.55	0.45
Salt (%)	0.36	0.36	0.36	0.36	0.40	0.42	0.42
Na (%)	0.18	0.17	0.16	0.17	0.17	0.17	0.16

Source: Peter Hand technical information, with permission.

Waterfowl

Ducks

Meat production is small in comparison with turkeys and chickens. *Table 17.19* shows the growth in the duck market from 1979 to 1985 in the UK and other EEC countries. By far the largest output is from France at almost 53% of the total. The UK produces nearly 16% of the EEC duckling production.

Based on 1983 figures consumption per capita is 0.87 kg. Duck meat is still regarded as a luxury. The majority of UK duck production is from the world's largest producer, Cherry Valley Farms Ltd, Lincolnshire. The two main breeds are Aylesbury and Pekin. Plummage of both is white. The Pekin hybrids commonly used in commercial production have better reproductive qualities than Aylesburys, but hybrid crosses are superior to both of these pure breeds. The growth rate of duckling is far superior to chicken and turkeys as shown in *Table 17.20*. For a given age under eight weeks, duckling outgrows all other poultry. Its food conversion, however, is inferior. This is due to the higher feed consumption per unit of bodyweight gain and it is this which makes duckling a relatively expensive meat to produce. The yield of edible cooked meat is lower for duck compared with turkeys and broilers. *Table 17.21* compares the relative cooking losses.

Table 17.19 Duck hatchings ($\times 10^3$) in EEC countries

	1974	1975	1976	1977	1978	1979	1980	1981	1982	1983	1984	1985
West Germany	3710	2301	2943	3779	4176	4103	4300	4200	4100	3600	2900	4630
Netherlands	3519	2672	3314	3019	3141	5915[1]	4700	4300	5000	5400	5700	6100
Belgium/Luxembourg	350	293	268	259	232	256	241	226	216	188	161	115
Italy	2048	2518	3368	4476	4264	4440[1]	5700	5900	3100	4800	5000	6290
France	12988	13888	14085	14797	15492	17538	19100	22600	24300	26500	31700	35800
Denmark	2829	2542	2809	2769	2717	2858	3200	3500	3100	3700	3500	3940
Eire	514	626	1191	1206	1236	1263[1]	1200	1300	1400	1100	NA	NA
UK	7400	6400	6803	7202	7667	7569[1]	8290	7980	8480	8770	9130	10670
Greece	–	–	–	–	–	–	–	–	–	–	–	25
EEC	33358	31232	34780	37507	38924	43942	46831	50006	49696	54058	54011	67570

[1] Jan–Nov
NA, not available
Source: After MAFF and AGRA Europe. Reproduced with permission.

Table 17.20 Comparative growth performance

	Broiler strain	Duckling	Turkey
Age (d)	47	47	47
Live weight (kg)	2.1	3.4	2.25
FCR	2.0	2.35	1.74
Feed consumed per bird (kg)	4.22	8.0	3.93

Source: Peter Hand, technical information. With permission.

Table 17.21 Comparison of cooking losses in ducks, turkeys and broilers

		Duck	Turkey	Broiler
(1)	Live weight (kg)	2.7	3.6	1.8
(2)	Bled/plucked weight of live weight (1)	82%	90%	87%
(3)	Eviscerated weight of (1)	73%	70%	75%
(4)	Cooking loss of (1)	37%	31%	22%
(5)	Field of edible cooked meat of (1)	25%	34%	27%

Source: After Hollows, (1978). *Productivity of Ducks*. Society of Feed Technologists. Reproduced with permission.

Geese

Goose production is the poor relation of the poultry industry but in recent years good attempts have been made to increase demand for the gourmet luxury meat. Few accurate records are available to show how many birds are produced annually. Many are raised on general farms and remain unrecorded. Sales mainly occur at the festive occasions, particularly Christmas. Being extremely good grazers and converters of grass into meat the most profitable management of feeding systems utilises this natural asset. Most popular breeds are the Embden, Toulouse, Roman and various hybrids involving crosses of these major breeds. Weights are dependent on feeding systems and the age of killing with Embdens reaching 14–15 kg by six months,

whilst the smaller Roman may only achieve 6 kg in this time. The Toulouse is slightly smaller than the Embden.

After feeding on a good proprietary chick pellet for three to four weeks after hatching, the growing goose is capable of surviving and growing on good quality grassland and small supplements of concentrate feed. (Recommended micronutrient supplements for both ducks and geese are given in *Table 17.22*.) This can be dispensed with after eight weeks and a satisfactory finish can be achieved on grass provided some supplementary grain is given during the last month of fattening. It is usual to restrict the range area at this time to conserve energy requirements. Penning the geese in straw yards is sometimes used in East Anglia, where straw forms a relatively cheap wall and bedding. The selling of day old or part grown goslings can be a possible side line. Breeding

Table 17.22 Recommended micronutrient levels for duck and geese rations

Supplement	Starter/finisher
Vitamin A (m i.u.)	12
Vitamin D3 (m i.u.)	3
Vitamin E (g)	25
Vitamin K (g)	5
Vitamin B1 (g)	2
Vitamin B2 (g)	12
Vitamin B6 (g)	4
Vitamin B12 (mg)	20
Biotin (mg)	110
Choline (g)	200
Folic (g)	2
Nicotinic acid (g)	80
Pantothenic acid (g)	15
Cobalt (g)	0.5
Copper (g)	6
Iodine (g)	2
Iron (g)	5
Manganese (g)	100
Selenium (g)	0.2
Zinc (g)	100

(Breeding waterfowl should be given a vitamin supplement as specified for 'Poultry Breeder' in *Table 17.12*)

Source: Peter Hand, technical information. With permission.

geese often mate for life and this strong 'pairing bond' is established in the autumn prior to the laying season in early spring. Normally one gander pairs with three to four geese. For best fertility the gander should be one year or older than the geese. Young geese show reproductive improvements with age through to ten years or more. Egg production varies, but may range from 30–75 eggs per goose per year. The eggs are best hatched artificially as geese make indifferent incubators and mothers.

CONCLUSION

The monthly journal *Poultry World* provides an annual update of the UK and EEC poultry industry statistics and is also a very useful source of technical information for the reader interested in keeping up to date.

18

Game conservation

P.B. Beale

INTRODUCTION

Prospects for game conservation in Great Britain in the latter part of this century must be viewed in the context of changes in land use demand by its major user, agriculture. Initiatives proposed by Government in 1987 stress the need for future farming policies to encourage alternative uses of land, greater diversity on farms and to be more responsive to the claims of the environment. Proposals to divert land from agricultural production, by encouraging timber production, recreational and sporting use, and the creation of wildlife habitats and landscape features, can be seen as an opportunity for landowners concerned with conservation of game and who are keen to use or market their shooting.

Shooting is an important source of income in rural areas and not only is the demand by UK residents increasing, but shooting continues to attract greater numbers of participants from abroad. Surveys conducted in 1982 have indicated that nearly 590000 people shoot game and wildfowl in Great Britain. This figure excludes the considerable number of people who only shoot clay pigeons but who do, nevertheless, help to support the rural economy in a number of ways (Cobham, 1983).

Shooting employs the equivalent of 12200 people of whom some 4000 are gamekeepers. Additionally, about 2300 people are employed in trades which are directly associated with shooting. The employment of at least ten times this number is directly dependent on countryside sports and recreation. Another factor to be borne in mind is the annual expenditure on shotgun shooting in the Great Britain of around £200 million (Cobham, 1983).

The willingness of farmers and woodland owners to retain or to plant small woodlands (less than 10 ha) as shooting cover has been shown by Cobham (1983). On the basis of a survey of members of the Timber Growers Organisation (Now The Timber Growers: United Kingdom) 67% of respondents indicated shooting as a reason for the retention of small woods, whereas 56% considered game cover a motive for planting new woods. Only landscape beauty was considered to be a more important reason for the retention and planting of small woods, than for their value as game cover (Cobham, 1983).

In order for woodlands and other habitats on farms to be suitable for game conservation, they require appropriate management. This chapter will try to show that shooting can be a profitable additional enterprise which justifies the costs of management required.

The Dartington Amenity Research Trust (1983) showed that small woods on farms in nine study areas in England and Wales are a badly used (underused) asset, whose function and value is diminishing. The report concluded, however, that given a long-term commitment to management both the environment and the economy would benefit. In particular management can increase the habitat value of woodlands to pheasants, thereby improving the function of these areas.

Where timber production is improved, in a way which is sympathetic to game, then the economic value of two enterprises can be enhanced. In addition, a wide range of wildlife species and landscape beauty also benefit.

Government initiatives announced in the Farming and Rural Enterprise package in 1987, which are designed to encourage alternative farm enterprises, can therefore be related to the contribution which shooting makes to the rural economy and employment. It is evident that, when small woodland planting and encouragement of more extensive farming techniques are linked to the rising demand for shooting, an exciting potential for game conservation, habitat management and rural employment is provided.

This chapter is intended to act as a guide to farmers not already involved with game conservation rather than for owners of estates who already run shooting enterprises. A wealth of excellent advice and practical information is available from the Game Conservancy for those seriously interested in shooting. Consequently, only guideline reference to what can be achieved by a farm shoot, and an indication as to what this involves in terms of habitat and management needs, will be included.

The potential for pheasant shooting on most farms, particularly those most likely to be affected by agricultural policies aimed at reducing levels of food production, is far greater than is the potential for partridges. Greater emphasis will be placed on describing ways in which cover and pheasant numbers can be improved. Where farming in arable areas becomes more extensive, with a new approach to the use of herbicides, fungicides and insecticides, and where the habitat needs of partridges are met more effectively in future, then numbers and the quality of partridge shooting should also increase.

HABITAT REQUIREMENTS

The grey partridge

This bird is native to Britain and is found mainly in open arable farmscapes which substitute for its original temperate grassland habitat of the steppes. The grey partridge was a common bird on farmland prior to mortalities caused by disease in the 1930s and the introduction of modern farming techniques in the 1950s. Use of pesticides, greater use of machinery and increases in field size, with corresponding hedgerow losses, have resulted in a reduction in the population to less than 20% of pre-war numbers.

This species is typical of cereal fields and long grasses and adults depend mainly on grain and seeds of those weeds which are to survive in crops or in stubble. Pastures, especially those with clover, are also used for feeding by adults. Preference for grassy banks or hedge bottoms is shown for breeding sites and chicks feed on insects which, in their turn, feed on weed plants in crops. Greater use of herbicides has reduced many species of broadleaved weeds in arable crops, thereby reducing the food supply of adults and particularly of chicks.

The Game Conservancy's Cereals and Gamebirds Research Project was set up to investigate the benefits to partridges and wildlife of selectively spraying 6 m wide strips (headlands) around the edges of cereal fields, a technique now known as 'conservation headlands'. Controlled use or non-use of herbicides, insecticides and fungicides was monitored. Initial results showed considerable benefits in the form of greatly increased chick survival in those fields with selectively sprayed headlands, compared with fields which are sprayed over their full extent. Survival of pheasant chicks is also greater in areas with reduced pesticide use on the headlands. Other beneficiaries are butterflies, where a doubling in numbers of some species has been noted and uncommon or rare arable weeds have re-appeared after long absences.

Not surprisingly, average grain yields tend to be lower from these lower input headlands compared with those which have been sprayed. However, this does not appear to have an economically significant effect on the *overall* yield of both types of field. Contamination of cereal crops from fields with such headlands with weed seeds can be largely overcome by on-farming cleaning.

Results show that some spraying can be carried out with certain chemicals at specific times of the year without adverse effects on insects or 'beneficial' plant species. These results are made available to subscribers to the Project. So far the results are very encouraging but, as the Game Conservancy points out, the selective spraying of headlands is only part of the method by which partridge and pheasant numbers can be increased. Benefits which derive from reductions in the use of pesticides must be combined with habitat management which aims to improve nesting and holding cover. In addition, effective predator control is essential.

Partridges prefer open farmland with few trees and, in the absence of natural grassland, can only flourish in cereal fields. Grassy-bottomed hedges or grassy banks of 0.5 m or so above field level provide the best nesting cover (*Figure 18.1*).

Since partridges are very territorial during the nesting season, the presence of sufficient hedges to prevent the cock birds seeing each other will increase nesting densities. An average field size of no more than 9 ha, with no more than ten hedgerow trees per linear kilometre, will attract the greatest number of birds (Potts, 1986).

As stated previously, chick survival depends on sufficient numbers of preferred insects in the crop. Wherever possible, the selective use or non-use of herbicides and insecticides, will enable broadleaved weeds and their dependent insects to survive. The adults too will benefit from the seeds shed by these weeds, and by the crop, later in the season.

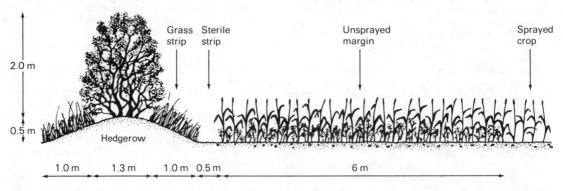

Figure 18.1 Suggested layout of hedge, nesting bank, sterile strip and unsprayed crop margin (headland). The sterile strip will be of little or no benefit to game unless the crop margins (headlands) are unsprayed.

Weeds:

Knotgrass
Fat hen
Mayweed
Chickweed
Annual meadow-grass

Insects:

Leaf beetles
Weevils
Plant bugs
Sawfly larvae
Caterpillars

The pheasant

Despite the fact that it is not native to Britain, the pheasant has filled a niche very effectively. Essentially it is a bird of woodland edges and effective management to increase nesting and holding cover must bear this in mind.

Unlike the partridge, the pheasant is polygamous and the territory of a cock bird may hold several hens. The better the cover in a territory, the more cocks are likely to establish territories.

Farms do not have to be heavily wooded to hold pheasants. Birds can be held effectively on land with 5% of the area devoted to small woodlands and hedges. Woodland is used as cover, particularly during the winter months, and the adult birds move out into cereal fields and grasslands in the spring. During the shooting season pheasants are very dependent on woodland or thick hedges for cover.

DART (1983) has shown that a high proportion of farm woodlands are undermanaged and overgrown. Many farmers do not realise that standing timber can contribute to income, particularly if shooting is incorporated into management strategies. Sporting revenue arising from a well planned woodland can often exceed the value of the standing timber (Robertson, 1987). Pheasants provide a short-term, sustainable crop, compared with the longer-term rotation of most broadleaved trees.

Radiotracking is used by Game Conservancy staff to determine the habitat preferences of pheasants and early results of the Pheasants and Woodlands Project have shown that birds are located mainly within 20 m of the edge of a wood and that hens favour cover which is dense at 1 to 2 m above ground level.

Woodland edges which are critical to pheasants can be increased in a number of ways. A large number of small woodlands of up to 1 ha in size provide for more edge than one large wood. Narrow woodlands provide more edge than a square block, but the former must provide protection from the wind in order to hold birds. A large woodland can be improved by clearing and replanting small glades of up to 0.25–0.5 ha in size. Alternatively, wide rides of 30–50 m wide can be cut. These help timber extraction and are invaluable for feeding the birds in the winter and for positioning guns. Glade and ride cutting lets more light in and thus benefits woodland plants and insects. A woodland which supports a diverse and numerous butterfly population is also likely to prove its value as shooting cover. Great care must be taken in the cutting of rides to avoid creating draughty corridors in a wood. Feeding rides which run in a continuous line to the edge of wood also attract the attention of poachers.

Management of existing woodlands is a first step for farmers who plant to improve or create a pheasant shoot. Thinning, in addition to glade clearance and ride cutting, increases the amount of light able to penetrate into a wood. This, in turn, will encourage the shrub layer which pheasants prefer and will help to reduce draughtiness at ground level. 'Draught-proof' roosting trees are also important.

Planting of conifers such as Western Red Cedar or Lawson's Cypress help to draught-proof woodland edges but plans to plant these need to be clearly identified to the Forestry Commission. There may be amenity objections to conifer planting or rates of grant aid under the Woodland Grant Scheme could be reduced.

The Forestry Commission's Woodland Grant Scheme already provides incentives for their restocking or natural regeneration. Spacing between trees is an important factor where an owner intends to combine timber production and shooting cover. The Commission now accepts 3 m spacings and this allows a greater amount of useful cover to be maintained between the trees. The use of Tuley tubes (*see* Chapter 7) helps to overcome some of the problems caused by weeding, but farmers must recognise the need to weed round the young trees for at least five years after planting. Three metre centres allow the use of machinery to undertake weeding and the possibility of practicability of planting cover crops could also be investigated.

The Farm Woodland Scheme proposals announced in March 1987 are intended to encourage farmers to plant trees on land previously in agricultural production. A minimum area of 1 ha for each woodland planted is envisaged and farmers may be limited to 20 ha over three years in the lowlands or 40 ha elsewhere. Annual payments would be made to farmers in recognition of the land being taken out of agricultural production. In addition, grant aid from the Forestry Commission would be available to help met planting costs. As proposed the scheme will aim at a minimum of 35% broadleaves in the lowlands and 20% in disadvantaged areas. Given the earlier economic return from conifers, farmers are likely to want to plant these in preference to broadleaves. Farmers should be encouraged by grant incentives, which are now available, as well as the prospects of a game crop, to plant as high a proportion of broadleaves as possible.

If the amount of land taken out of agriculture exceeds the amount actually intended for planting, it should be possible to include wide rides or areas of cover between blocks of woodland. Not only would the farmer receive an annual income from the area of woodland, but shooting tenants are prepared to pay up to £600/ha as an incentive for farmers to plant suitable crop covers.

The Game Conservancy have designed a series of game and 'instant' spinneys and the principles involved could be incorporated into a woodland planting programme which comes under the umbrella of the Farm Woodland Scheme and the Woodland Grant Scheme. The potential advantages of using land to provide an *annual* income from timber production as well as income from shooting are evident.

Feeding and holding of pheasants (shooting too) are enhanced by planting of cover crops in areas which adjoin woodlands. Some cover crops have a specific value for pheasants for example, artichokes, canary grass, sunflowers, buckwheat, millet and quinoa. Other crops such as beans, maize, kale and mustard have an agricultural value as well. In addition, pheasants feed on clover-rich grassland, among roots, forage legumes and in cereals.

The provision of cover crops is a specialised topic which is the subject of ongoing research and application. Exciting developments such as the recent introduction of quinoa, a cultivated form of fathen, is providing good cover and feed when combined with kale. The Game Conservancy's booklet *Game and Shooting Crops* should be consulted about the choice of species, siting, minimum widths and other factors.

Cover crops are important in a number of ways since the value of natural woodland, hedgerow and scrub cover can be enhanced and shooting can be improved by increasing the variety of drives.

Cover crops can be used to:

(1) increase nesting and holding cover,
(2) provide extra or return drives,
(3) draw birds into an area where they can be presented to

the guns in a sporting way by using natural topography,
(4) produce supplementary feed in the form of herbage, seeds and insects,
(5) optimise the game 'holding' capacity of a farm.

Where possible, stubbles should be kept over winter or for as long as possible. Since there is a need to reduce overall levels of cereal production within the European Community, stubble retention would be a sound policy. Spring sown cereals not only yield less, but require lower inputs of both nitrogen and pesticides. The value of the game crop would also be increased by stubbles, which are valuable to both grey partridges and pheasants. When combined with cover crops and small woodlands, they provide ideal habitat conditions and shooting cover.

The designation of Environmentally Sensitive Areas which aims to encourage traditional farming practices should also provide the right climate for game conservation, unless more grassland and increased grazing pressure results. There is considerable scope for encouraging farmers to reduce the use of pesticide sprays and to grow spring cereals undersown with legumes to reduce fertiliser needs. The adoption of conservation headlands would benefit game, wild plants and numerous beneficial or benign insects, birds and mammals.

BREEDING, RELEASE AND HOLDING STRATEGIES

Farms in central and eastern Britain, which are naturally suited to wild partridges and pheasants, may not require stocking. In these areas money and labour may be better devoted to improving nesting and holding cover and on predator control than on putting down reared birds. More extensive farming techniques in the drier parts of the country may boost natural stocks of both partridge and pheasant.

In the short term for these areas and in the long term for wetter parts of Britain, rearing and release of pheasants is necessary to ensure sufficient birds for regular shooting. The release of reared birds can, however, create two problems:

(1) mortalities due to predation are very high due to low avoidance behaviour,
(2) restocking allows a rate of shooting which is sustained by the release and not by wild birds.

There are so many questionable features in the release of partridges, that it is not regarded as beneficial to their conservation on the basis of current knowledge (Potts, 1986). Although releasing pheasants can increase the bag it is usually detrimental to the wild population.

Estates which put down 4000–5000 pheasants may find it cost effective to rear eggs taken from wild birds, but many buy in and rear day old chicks. For the farmer with more moderate stocking plans, it is far more satisfactory to buy in six week old poults. They may cost £2.00 each, but the reduction in labour costs and risks makes this worthwhile.

Time and money spent on making sure that release facilities are adequate is vital and a properly designed release pen is essential to give the poults initial security in the 'wild'. Ideally, the pen should be sited within a draught-free wood, with one-third of the pen having trees and shrubs in which the birds can learn to roost, one-third with low ground cover and one-third in the open so birds can sun themselves, dry out and dust bath. Watering points should be provided and feeding by hand will encourage the young birds to come to the feeding area. It is essential to feed at the same time each day, in the morning and early evening and the use of a proprietary food is recommended until the birds are fully fledged. The Game Conservancy's booklet *Pheasant Rearing and Releasing* should be consulted for additional and more practical information.

Once the young birds have dispersed into woodland surrounding the release pen, they will require regular feeding, again at the same times each day, to make sure they are held on the shoot. Rides and glades are invaluable for supplementary feeding and corn fed under straw will keep the birds occupied and will reduce losses to sparrows, squirrels and other opportunists. The Game Conservancy's booklet *Game in Winter; feeding and management* provides a wealth of additional detail.

PREDATOR CONTROL

Farmland is an unstable ecosystem in which they are few natural constraints on either predators or prey. Whereas cover improvements can improve survival chances for birds by reducing predator success, released birds are subject to heavy losses unless predator numbers are kept in check.

Crows and magpies are a serious risk, particularly to nests and chicks and farmers should consider controlling their numbers by nest destruction or by use of cage traps.

Foxes can also take a large number of birds and a successful shoot may depend on effective control. Foxes can be shot at night by lamping or they can be snared. Both require skill and snares demand regular inspection. If there are deer in the area, snares should be filled with stops. Gassing can also be used to destroy foxes in their earths, but care should always be taken to ensure humane death. The Wildlife and Countryside Act 1981 and 1986, imposes controls on the way in which foxes and other predators may be killed. The Act must be consulted to make sure no statutory offence is committed, inadvertently or otherwise.

Mink are also a serious risk in some parts of the country and the Wildlife and Storage Biologists of the MAFF should be consulted for advice and traps if they are present.

Rats and grey squirrels may also cause local problems, by taking eggs and food, which will require their control. In addition, grey squirrels are a serious pest in broadleaved woodlands and any policy aimed at increasing the stock of broadleaved trees will necessitate an active programme of grey squirrel control. The Game Conservancy's booklet *Predator and Squirrel Control* should be consulted.

Badgers, otters and *all* birds of prey are strictly protected and under no circumstances should they be controlled.

KEEPERING

Most farms will not be able to employ keepers unless shooting cooperatives are formed, so the keeper's tasks will devolve onto a member of the farm staff or a keeper employed by a shooting tenant.

Who does the work is not important, so long as predator control, habitat improvements, cover planting, rearing and releasing and winter feeding of birds are carried out effectively. This all requires planning so regular routines are established and adhered to.

OPTIONS FOR SHOOTING

Broadly, the options are as follows:

(1) let the shoot to shooting tenant or syndicate,
(2) develop the shoot and let shooting days,
(3) develop the shoot and let to a syndicate,
(4) operate a cooperative with neighbouring farmers on the basis of (1), (2), (3) above.

Shooting tenants normally undertake their own habitat management and restocking and may pay up to £600/ha to encourage the farmer to grow cover crops. Rentals vary between £2 and £25/ha, but an area of at least 150 ha is necessary. The presence of small woodlands on the farm can boost rentals by 200–400% in comparison with an intensive farm with little cover.

The farmer has more control over shooting which he/she lets as a package of shooting days for the season. Rentals can attract between £12 and £15 per bird shot, but this involves greater expense and organisation. Unless the farmer is very keen to manage his own shoot and is committed to all that this involves, the first option may be a safer bet.

A 'half-way house' lies in the development of a shoot and letting to a syndicate. This involves the cost of putting down birds, managing them and their habitat, but it does reduce the problem of marketing the shoot to individuals.

Some farmers have pooled their resources and developed shoots over much larger areas by forming cooperatives. The terms of the cooperative must be agreed very carefully beforehand and there must be a very high expectation of the farmer being on, and remaining on, good terms with his neighbours. The financial value of a large shoot, so long as cover is equally good, is higher than on individual farms. The chances of holding birds and providing good and sporting shooting are also greatly increased.

ECONOMIC BENEFITS

In addition to the economic benefits mentioned in the previous section, two other benefits need to be mentioned.

If a 160 ha farm with, say, 32 ha of well laid out woodlands in 2 ha blocks were to be let for £12.50/ha, this would yield £2000 per annum for the whole farm. However, standing timber, which is valued at between £2500 and £6500/ha, is limited to 20% of the farm's area.

Successful shooting depends on the woodlands and their sympathetic management for game. Working on a 120 year rotation for the broadleaved trees a sporting rental of £240000 for the farm, or £7500/ha of woodland could be achieved. This compares very favourably with the standing value of timber (Robertson 1987).

Recent farm sales have shown buyer's preference for those which provide good shooting. The capital value of a good shooting farm with a total of 10 ha of woodland may be increased by £250000 over and above its sale value as a farm enterprise (McCall 1987, personal communication).

CONCLUSION

Game conservation provides sport for an increasing number of people and demand in some areas outstrips supply. It provides another, and valuable, function for woodlands, hedges and rough grassland on farms. Wildlife and landscape amenity also benefit which meets some of the increasing demands made on the countryside by a largely urban population.

The need for alternative use of agricultural land could provide the greatest stimulus to game conservation, by re-creating some of the conditions of the halcyon days of shooting which occurred prior to the agricultural revolution in the early 1950s.

Game conservation requires careful planning and ongoing commitment. Given those it could prove a profitable alternative enterprise for many farmers.

Acknowledgements

I would like to acknowledge help and advice received from Drs G.R. Potts, N.W. Sotherton, P.A. Robertson and Messrs I. McCall and M. Swann. They are part of the Game Conservancy's research and advisory teams. The latter are able to provide 'on the spot' advice to farmers, which is tailored to the needs of a particular shoot or its development.

Contacts can be made to The Game Conservancy, Fordingbridge, Hampshire, SP6 IEF, telephone (0425) 52381. Details of subscriptions to the Cereals and Game-birds Research Project can be obtained from the Game Conservancy.

References and further reading

AGRICULTURAL DEVELOPMENT AND ADVISORY SERVICE AND FORESTRY COMMISSION (1986). *Practical Work in Farm Woods* (a series of leaflets) MAFF Publications

BRITISH DEER SOCIETY, GAME CONSERVANCY, FEDERATION OF DEER MANAGEMENT SOCIETIES (1984). *Avoidance of Accidental Snaring of Deer*

BROOKS, A. (1975). *Hedging: a practical conservation handbook*, British Trust for Conservation Volunteers

BROOKS, A. (1980). *Woodlands: a practical conservation handbook*, BTCV

COBHAM RESOURCE CONSULTANTS (1983). *Countryside Sports: their economic significance*. Standing Conference on Countryside Sports

COLES, C. (1984). *The Complete Book of Game Conservation*, Barrie & Jenkins

DARTINGTON AMENITY RESEARCH TRUST (1983). *Small Woods on Farms*, Countryside Commission

EVANS, J. (1984). *Silviculture of Broadleaved Woodland*, Forestry Commission

FORESTRY COMMISSION (1980). *Grey Squirrel Control*

Game Conservancy, advisory booklets:
  2. Game and Shooting Crops (1986)
  4. The Grey Partridge (1986)
  7. Farm Hazards to Game and Wildlife (1981)
  8. Pheasant Rearing and Releasing (1983)
  14. Game in Winter: Feeding and Management (1986)
  15. Woodlands for Pheasants (1981)
  16. Predator and Squirrel Control (1981)

Game Conservancy. Annual Reports.

McCALL, I. (1986). *Your Shoot: gamekeeping and management*. Black

McKELVIE, C.L. (1985). *A Future for Game?* George Allen and Unwin

NIX, J., HILL, P. and WILLIAMS, N. (1987). *Land and Estate Management*. Packard

POTTS, G.R. (1986). *The Partridge, Pesticides, Predation and Conservation*. Collins

ROBERTSON, P.A. (1987). *Pheasant Management in Small Broadleaved Woodlands* in D.C. Jardin (Ed.), *Wildlife Management in Forests*. Proceedings of the Institute of Chartered Foresters discussion meeting 3–5 April 1987

19

Animal health

D. W. B. Sainsbury

INTRODUCTION

Good health is the birthright of every animal and it is the duty of the livestock keeper to do all he can to ensure this.

Considerable advances in the control of animal infections have led to the effective elimination of many of the traditional causes of acute disease and it is no longer necessary to consider disease an inevitable part of a livestock enterprise. This situation has been achieved by a combination of appropriate vaccine usage, good drug therapy, the development of disease-free strains of livestock but above all by improved husbandry. These advances have made it possible to keep animals in much larger groups and more densely housed than hitherto but the results have by no means been a gradual disappearance of infectious diseases altogether.

On the contrary, a number of complex diseases has emerged, difficult to diagnose and induced by a multiplicity of pathogenic agents. Whilst these may cause an apparent or 'clinical' disease it is more likely that the effect will be less obvious and may only reduce the overall productivity of the livestock by, for example, slowing growth and reducing the food conversion efficiency. Animals may not die or even show any symptoms at all so that the farmer may be unaware of what is happening unless he keeps very careful records and uses them with more than the usual degree of skill. It is also a very common phenomenon that intensive production on a livestock unit may start efficiently and effectively but deteriorates in time so gradually that it is not noticed until the consequences have become very serious and control becomes extremely difficult. The nature of these infections is of especial interest and concern because the environmental and housing conditions have a profound effect on their severity. In the field of contagious diseases the new problem of the 'viral strike' is emerging; many apparently new virus diseases, borne by wind or vectors, travel through areas causing some quite devastating effects for a time, especially in the larger livestock units.

A further major problem is that of the *metabolic diseases*. These are a group of diseases that are caused intrinsically by the animals being called to produce an end product faster than the body can process its intake of feed. Enormous efforts are made to provide the right nutrients in an easily assimilated form but as the metabolic disease is rather different from a deficiency disease, this does not necessarily work. In a way the metabolic diseases are the inevitable outcome of the success in conquering most of the acute virulent diseases together with the advances made in improved genetics, nutrition and housing. These improvements have led to much increased growth and productivity but the capability of the animals to keep pace with these has outstripped the normal functioning processes (or metabolism) of the body.

An example of this is provided by considering the growth of table chickens. When the industry started intensively it took about 13 weeks to produce a bird weighing 2 kg, at a food conversion efficiency of approximately 3:1 (that is, 3 kg of feed to produce 1 kg of liveweight). The hazards in growing the birds at that time were innumerable and were largely related to contagious and infectious diseases. Mortalities were high, often up to 30% and the diseases played their part in slowing growth and damaging the food conversion efficiency. Now, some 30 years later, the position is quite different. It takes about half the time for the birds to reach 2 kg, that is about $6\frac{1}{2}$ weeks. Mortalities normally average 3–4% and contagious and infectious disases are controlled largely by good hygiene together with judicious use of vaccines and medicines. But now we have a number of the so-called metabolic or production diseases, for example, birds are affected with excess fatty deposition, causing degeneration of the liver and kidney and heart attacks. The skeletal growth may not keep pace with the rate of muscle development so that locomotor disorders occur, such as twisted, rubbery and broken bones and slipped tendons.

Types of diseases

An important factor influencing the incidence of disease in livestock today is the increasing immaturity of livestock. Improved performance has resulted in animals reaching market weight much earlier, whilst for genetic reasons, breeding animals are also younger on average than previously. Thus, on a modern livestock farm, there is usually a high proportion of young animals that are in a state

Table 19.1 The disease pattern on the farm

Poor husbandry + primary disease agent (often a virus)
 + secondary disease agents (often
  bacteria or parasites)
 = subclinical or overt disease

Examples of bad husbandry

Too many animals on a site
Overcrowding within the buildings
Mixed ages within a house or site
Excessive movement of animals
Poor ventilation and environmental control
Bad drainage and muck disposal
Insufficient bedding
Lack of thermal insulation in construction
Unhygienic or insufficient food and watering equipment
Absence of routine disinfection procedures
Faulty nutrition

of susceptibility to infectious agents whilst they are still developing the ability to resist disease naturally or natural immunity to disease which will normally take place over a prolonged period. This difficult state of affairs is further exacerbated by the considerable size of many livestock units in which the young animals may have originated from various parents of very different backgrounds. In many cases the young or growing stock will have come in from widely separated areas. They may have no resistance to local infections and will therefore be totally susceptible to them and at the same time contributing a new burden of pathogenic micro-organisms to the unit they have entered. Altogether the modern livestock unit may present at any one time a confusing immunological state and the basic design and its management will influence the success or otherwise of disease control (*see Table 19.1*).

There are certain major groups of infections that account for most of these problems. The most widespread are probably respiratory diseases. Many of these are subclinical, have a pronounced debilitating effect on the animals and are caused by a large number of different infective agents even in any one disease incident. They may not respond satisfactorily to vaccines, antisera, antibiotics or any other drugs, so that the fundamental method of approach to their control is by managerial, environmental and hygienic measures.

Another significant group of diseases influenced similarly are the enteric infections. These, also, have many different primary causative agents, ranging from parasites to viruses and bacteria. The reasons for their increasingly harmful effects in recent years are not only those already listed earlier but also the general trend with certain livestock to eliminate the use of bedding, such as straw. The harmful effects of this may often be corrected by the use of good pen design, especially by the use of slatted or slotted floors, but it is nevertheless more difficult to separate animals from their urine and faeces when the flooring is without litter, as the latter has a diluting and absorbent effect on the excreta.

There are, in addition, several important bacterial infections that have tended to increase in large intensive units. Examples of these are *Salmonella* and *Clostridia* bacterial spores, together with *Escherichia coli*, *Pasteurella* and *Campylobacter*. Many forms of these organisms are normal inhabitants of the animal's intestines in small

numbers but excessive 'challenges' causing disease may build up under unhygienic intensive conditions, encouraged by building with poorly constructed surfaces which cannot be cleaned.

Livestock unit size

In addition to the risks of the gradual build-up of disease-causing agents within livestock buildings there are serious dangers of livestock enterprises becoming too large. In all parts of the world where the development of large units has taken place there has emerged a number of viral diseases which tend to 'sweep' like a forest fire through areas with a high livestock population, sometimes leaving a trail of devastation. Often there is likely to be considerable loss over a concentrated period, after which the animal population may develop a natural immunity, at least for a time, or artificial immunities are promoted by the use of vaccines. It is now known that contagious virus and other particles can travel great distances from large infected sites, certainly distances of 50 miles have been virtually proven, but they may well travel much further than this. If livestock enterprises continue to grow in size then the dangers in this respect can only become greater. At the present time it is impossible to give soundly based objective advice on the optimal unit size and in any event the factors that would lead to making proposals are highly complex. However, there is clear evidence that animals thrive less efficiently in large numbers even in the absence of obvious clinical disease. Apart from this there is the difficulty of hygienic disposal of the dung when the unit is of excessive size.

It is normally possible to keep much greater numbers of *adult* animals together than young stock since there is nothing like the same number of contagious disease problems after the difficult growing stage and its immunological uncertainties are passed.

As to the maximum number of livestock that might be kept on a site, it will be appreciated that this will depend on a number of factors, apart from health considerations. Figures have been proposed which attempt to allow for all factors and those suggested are as follows:

Dairy cows 200
Beef cattle up to 1000
Breeding pigs 500
Fattening pigs 3000
Sheep 1500
Breeding poultry 3000
Commercial egg layers 60 000
Broiler chickens 150 000

Such figures can be no more than suggestions made in the light of practical experience. In due time more scientific evidence may be available to establish a better degree of accuracy. It is certainly likely that the figures will require constant adjustment in the light of new developments in husbandry, housing and disease control and especially with the anticipated trend towards the increasing provision of livestock free of specific disease.

Design essentials to minimise the disease challenge

In addition to the size of the livestock unit there are a

number of other basic items which need to be considered in order to provide the bases of good health.

Depopulation

A fundamental concept in maintaining the health of animals is to ensure the periodic depopulation of a building or a site. The benefits of eliminating the animal hosts to potential disease-causing agents are well understood and the virtue of being able to clean, disinfect and fumigate a building when the animals have been cleared is also accepted. Periodic depopulation is especially important for young animals but is less so for groups of older animals which have probably achieved an immunity to many contagious diseases. Much also depends on whether the herd or flock is a 'closed' one with few introductions of fresh animals or an 'open' one with a constant renewal of the animal population. If the latter is the case, then constant depopulation is of greater importance as there is little or no opportunity for natural immunities to develop and the regular removal of the build-up of infection is of great assistance in ensuring the good health of the livestock.

The health status

The policy will also depend on the health status of the stock. At one extreme there are the so-called 'minimal disease' or 'specific pathogen-free' herds which have been developed to be free of most of the common disease-causing agents of that species. Here, depopulation is less critical than the protection of the animals from infections that come in from outside. Since this danger is very serious in most localities it is important to subdivide the animals in a unit into smaller groups, lessening the likelihood of a breakdown and/or enabling isolation and elimination of a group which may become infected. At the other extreme there are those units which have a constant intake of new animals from outside and of a totally unknown health status. In this case there is a high risk, and more usually a near certainty, that some will be either clinically infected with, or carriers of, disease-causing organisms. Design specifications for such units should be quite different from those of the closed herd or flock so that defined areas of the unit should have groups of animals put through them in batches after which the area can be cleared, cleaned and sterilised. Obviously, it is preferable if the whole unit can be so treated since it ensures an absolute 'break' in the possible disease build-up cycle.

Between these two extremes is the more usual case in which a herd or flock is of reasonable health status, though certainly not free of all the common diseases, and in which new livestock are added only occasionally. In such cases the precautions in the housing against disease 'build-up' and spread of infection can be rather more relaxed but there must still be a proper provision for the isolation of incoming and sick animals.

Group size

Animals thrive best in groups of minimal size. If groups are small it is usually easier to match the animals within them for size, weight and age, and it is well-established that growth under these circumstances is likely to be most even and economical. Behavioural abnormalities, such as fighting and bullying, are also kept to a minimum – indeed they may be prevented altogether.

Fighting amongst animals is a highly contagious condition and under the most intensive husbandry systems an almost casual accident that may draw some blood can escalate into a blood-bath. Pens which keep the animals in small groups will tend to reduce the occurrence of such disasters and indeed with good management the removal of an animal that has accidentally injured itself or is off-colour and therefore prone to being bullied will stop the trouble before it ever has a chance to develop seriously.

There is yet another economic advantage in keeping animals in small groups. The farmer will achieve the best economic return if the stock are housed at the densest possible level for optimal productivity. If animals are kept in a house to allow a density such as this, it means that they should spread across the house evenly so they do in effect occupy and use this area. In practice, however, this is very difficult to achieve, especially when large numbers are housed together without any subdivision at all. Hence the birds, or any other livestock as may be at risk, crowd in certain parts of the building which can lead to grossly over-stocked floor areas. If livestock crowd excessively in certain parts of a house, this area is likely to become more polluted with excreta and respiratory exhalations to an abnormal and harmful degree; the humidity becomes high, proper air movement is impeded and the animals soon may become ill. Sick animals feeling cold tend to huddle together more, so the vicious circle is perpetuated and there is seemingly no end to it unless some measures are taken to ensure a better distribution of the stock. When this problem arises under practical conditions it may be impossible to subdivide the animals at once. An immediate trend in the right direction, that is encouraging the animals to spread themselves more uniformly over the house, can often be achieved by introducing some artificial heat. There are excellent gas radiant heaters and oil-fired and electric blower heaters available.

When animals are penned in large numbers the effects of a fright caused by an unusual disturbance can be extremely serious. It is almost impossible to guard against all the extraneous sounds and sights that may affect the stock. The best safeguard, therefore, is to have the animals housed in small groups so that the effect of a panic movement will be more limited and will never build up into highly dangerous proportions.

Floors

The profound effect of the floor surface on the health and well-being of livestock is well established. When bedding was almost invariably used in animal accommodation there were relatively few problems related directly to the flooring. Now that the farmer must frequently use housing systems without bedding, often with slatted or other forms of perforated flooring, to produce economically a comfortable and clean environment for the animals, new problems have emerged. Though we are still far from being able to advise the ideal floor, it is possible by choosing a good combination of surface and bedding where used, to provide the animals with a comfortable, warm, hygienic and well-drained surface.

The best solid flooring is usually based on concrete because when properly made and laid it is hard wearing, hygienic and impervious to fluid. Where an animal lies directly on the floor, without bedding, there should be an area of insulated concrete, most often incorporated by using 100–200 mm of light weight or aerated concrete under the top screed and above a damp-proof course.

The surface of the concrete must not be so smooth that the animals slip or injure themselves yet, on the other hand, if it is too rough it can cause abrasions and injuries. A happy medium is not easy to find but is usually achieved by using a wood float finish or by tamping the floor lightly with a brush. Also, very rarely are the 'falls' on the floor correct; again, there must be a nice balance between being steep enough to drain away liquids yet not so sloping that the animals slip. With some animals the floor is made more comfortable by placing rubber mats on top; these are used with some success with cattle, calves and pigs.

Some of the worst problems of injury, especially with cattle and pigs, have occurred with slatted or other forms of perforated floors. These have been particularly troublesome when the edges have been left too sharp or have worn leaving injurious protrusions. If bedding *can* be used for livestock it is usually better to do so. Good bedding is an insulator and warmer for animals and is probably the cheapest that there is. The air temperature of buildings can be lowered if bedding is used. When the bedding is straw it may form an acceptable, if not an essential, adjunct to the diet of the animal. The fibre may help digestion and create a sense of well-being. It keeps the animal occupied and assists in the prevention of vices.

Bedding also provides just about the safest flooring for the animal, so reducing the danger of injuries to the body, particularly the limbs. It is also much more difficult to keep the floor dry when no bedding is used and if the animal's surface is kept wet there is a further chilling effect created by contact with a wet floor. It is also important to emphasise the health risk from undiluted muck which does not pass through slats but accumulates at some points, and the great risk to the respiratory system of humans and animals from gases rising from slurry tanks or channels below the slats. Not surprisingly, therefore, there has been a trend towards the advocacy of bedding for most animals.

Isolation facilities

There is an urgent need for better consideration to be given by the farmer to the isolation of sick animals. Isolation removes the dangers of contagion to the normal animals. It also makes it more likely that the sick animal will recover without the unwelcome attentions of the other animals who will always tend to act as bullies. In nature the sick animal usually separates itself from the rest of the herd or flock but it is often impossible for the housed animal to do this. Also, under intensive management, more animals suffer from the aggression of their pen-mates and it is essential that such animals are removed to prevent them from suffering or being killed.

It is also easier with an isolated animal to give it such therapy as it requires and to give it any special environmental conditions, such as extra warmth. It is an interesting but important observation, known to most farmers, that if you take 'poor doers', perhaps animals suffering from subclinical disease, out of a group and pen them with more space and extra comfort, they may thrive without any further attention. Nevertheless, with the addition of suitable therapy, the cure may be accelerated and completed.

Health and the disposal of manure

The method used for manure disposal has potentially important effects on health, both human and animal. Whilst the smell from composted solid manure tends to be strong, it rarely travels far or creates a nuisance problem and there is little if any risk to the human or animal population from this form of manure under temperate climatic conditions.

Slurry, however, is quite a different matter. If placed straight on the land from the animal house or after holding in a tank anaerobically it has an extremely offensive smell. Whilst masking agents are possible they are too expensive at present to be considered economic. The worst smell comes from the pipeline and gun spreader, because the droplet size is small and light and particles may carry for considerable distances. Less smell arises from a tanker spreader because the slurry is much thicker and not spread by aerially dispersed small droplets. The most satisfactory way to prevent the slurry from causing offence is to treat it aerobically in some way before spreading. Many human and animal health problems may arise from the spreading of slurry. In surveys it has been found that potentially pathogenic bacteria were able to survive for up to three months in slurry kept under anaerobic conditions. Whilst the particular bacteria studied were *Salmonella* species and *E. coli*, there is little doubt that more resistant organisms, such as *Bacillus anthracis*, *Mycobacterium tuberculosis*, *Clostridium* species and *Leptospira* species could survive as long or probably very much longer. If the slurry should enter a river or stream, the pollution may have far-reaching and infinitely more serious effects. It is thus essential for all enterprises with slurry as the disposal system that either the slurry is placed on land where it cannot be a nuisance or health risk or it is so treated beforehand that the risks are removed.

The dangers from gases in farm buildings

Hazards associated with gases in and around livestock farms have been highlighted recently, particularly in association with gas effusions from slurry channels under perforated floors. Numerous fatalities have been recorded of livestock, and even in humans, due to gas intoxication. There is, in addition, mounting evidence that there may be concentrations of gases in many livestock buildings that may affect production adversely by reducing feed consumption, lowering growth rates and the animals' susceptibility to invasion by pathogenic micro-organisms.

The most serious incidence of gas intoxication arises from areas of manure storage in slurry pits or channels under the stock usually, but not always, associated with forms of perforated floors. The greatest risk arises when the manure is agitated for any reason, usually when it is removed. There is also an ever-present danger if a mechanical system of ventilation fails and this is the only method of moving air in the house. Several cases of poisoning have been reported when sluice gates are opened at the end of slurry channels and the movement of the liquid manure has forced gas up at one end into the building.

High concentrations of gases, chiefly ammonia, may also arise from built-up litter in animal housing. This is most likely in poultry housing since the deep litter system is the most commonly used arrangement with broiler chicken and poultry breeders. The danger has undoubtedly been exacerbated within the past few years, owing to the necessity of maintaining relatively high ambient temperatures in order to reduce food costs, while at the same time there has been good evidence that higher temperature than hitherto should be maintained for optimal productivity. Poultry farmers have often attempted to achieve such temperatures by restricting ventilation in the absence of good thermal insulation of the house surfaces, and the result can be generally harmful if not dangerous.

The most popular form of heating is by gas radiant heaters which are suspended from the ceiling. Well over half of all poultry housing is heated in this way and because the cost of gas as a fuel appears in some respects to be improving in relation to that of other fuels, the system is being used more for pig housing and especially for piglets in the early weaning system. There is a risk that with inexpert use such heaters may be improperly serviced and the house insufficiently ventilated to give complete combustion and that toxic quantities of carbon monoxide may be produced.

DISEASE AND IMMUNITY
Organisms causing infections

There are different groups of organisms causing infectious disease in animals. The smallest of them are viruses, below 300 nm in size, but many are well below this. For example, the foot-and-mouth disease viruses are 10–27 nm ($1 \mu m = 0.001$ mm and 1 nm $= 0.000001$ mm). Viruses cannot usually be seen under the ordinary microscope but can be photographed by the electron microscope. They are simple organisms that can only multiply within living cells and this property distinguishes them from bacteria. Viruses are classified into a number of different groups, e.g. reoviruses, adenoviruses and herpesviruses.

Bacteria are relatively large organisms compared with viruses and can multiply and grow outside living tissues. In size, for example, cocci (round) bacteria measure $0.8–1.2 \mu m$, bacilli (rods) are $0.2–2 \mu m$, and the spiral shaped spirella up to $50 \mu m$. They are visible under the ordinary microscope. Many types, once outside the animal body, may sporulate to form a protective coat so that they can live many years in buildings, soil or elsewhere and can still be capable of infecting animal life. For example, the spores of anthrax and *Clostridium* can live for 20 years or more under favourable circumstances.

Mycoplasma are smaller organisms than bacteria, being about $0.25–0.5 \mu m$ (or $250–500$ nm). They are rather like oversize viruses but, unlike viruses, they can be cultured on artificial media.

Rickettsia are of somewhat similar size to *Mycoplasma*, being $0.25–0.4 \mu m$, but they cannot be grown in ordinary culture media and will only multiply intracellularly, like viruses.

All these organisms are classified as from the vegetable kingdom and can be joined by the yeasts (moulds and fungi) which are larger than bacteria, several varieties of which cause diseases in animals and humans.

There are also many simple organisms from the animal kingdom, known as parasites, that cause disease in farm livestock. These range from the single-celled protozoa, such as the coccidial parasites, to parasites of ever increasing size and complexity, culminating in such relatively complicated organisms as the roundworms (helminths), lice, flies and ticks.

Organisms from these groups are the main cause of infectious disease in farm livestock. The only other agents causing disease and which are not infections, are the metabolic disorders, poisons, deficiencies, excesses and injuries, examples of which are found throughout this chapter.

Immunity

In our understanding of the natural mechanisms that maintain good health it is essential to know the principles of the normal animal's immunological system and the way in which we make use of such products as vaccines and sera to boost it as necessary. This is best understood by studying the cycle of events in any animal's life (*see Figures 19.1*).

When an animal is born it is relatively free of disease organisms but it has what is known as *passive immunity*, which it receives from the mother whilst still *in utero*. If the mother has been vaccinated against a disease or has had an experience of the actual disease, then her immunological system will normally produce the antibodies that are capable of resisting the disease. These antibodies circulate in the blood and will also be transferred to the fetus. The antibodies themselves are effective only for about two to three weeks before they disappear.

The young animal can produce only its own antibodies and thus develop what is known as *active immunity* if it has experience of the organisms that cause the disease in question or has been vaccinated against it. It should be stressed that an animal will have some passive immunity only to those particular infections experienced by the mother. These may be quite limited and 'local'. This is a good argument for rearing an animal in the environment in which it is conceived and a good reason for not transporting young animals too early in life before they have some better ability to resist a 'foreign' disease challenge. Passive immunity in mammals is fortified immediately after birth by the young drawing the first milk (or colostrum) from the dam. Colostrum is especially rich in antibodies and also nutrients.

As the passive immunity the young animal receives from its mother fades it may be replaced by an active immunity if the young animal is vaccinated or has some challenge with the organisms that can cause disease. Under practical conditions the aim is to make sure that any natural challenge of potentially pathogenic organisms is always mild or gradual, not massive and overwhelming, as in the latter case disease will arise.

Sometimes it is perfectly satisfactory to allow the animals to receive a 'natural' challenge from the environment but this

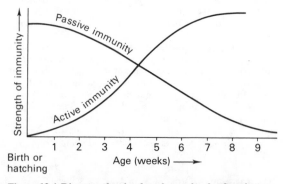

Figure 19.1 Diagram showing how immunity develops in livestock. Curves represent antibody levels (i.e. immunity) produced by active or passive immunisation. Animals derive passive immunity from their dams but it is short-lived. Active immunity comes either from vaccines or a disease challenge – which may be mild

is at the best inexact and at its worst ineffective. Thus, various methods of induced immunity are given, either *passive* or *active*. The passive immunity is induced by injecting into the animal hyperimmune antiserum prepared from the serum of animals which have experienced the actual infection and have recovered. This serum may also be used for the treatment of the disease but it has only a transitory effect lasting about three weeks, similar to the natural immunity received by the young animal from its mother.

For an active immunity vaccines are used. A vaccine is a way of activating the body's defence mechanism by challenging it with the pathogenic organisms so modified that they are active enough to produce an immunity to the disease but not the disease itself. The modification is done in very many ways, by adding chemicals, by growth of the organism in special media or livestock of a different species or by irradiation. Sometimes no modification is required as an organism can be used which is sufficiently closely related to induce immunity but not disease. Occasionally, in actual farming practice, if no vaccine is available for a disease which is almost certainly likely to challenge the animals later in life, the animals are deliberately exposed to the disease causing agent and at such a time when no actual disease will be caused. This is a venture which is hazardous and it is unwise to do it without proper veterinary supervision.

The production of vaccines is a very sophisticated process but can induce an enormously efficient protection for the animal. Vaccines must be used by skilled hands at the right time in the correct dose and repeated as necessary since the length and strength of the immunities developed vary according to the infection. It should be borne in mind that when a vaccine is given the immunity takes time to develop, somewhere in the order of three weeks, so it may be necessary to protect an animal during that time by antisera, medicines or simply by isolation. Some vaccines are of live organisms, others are dead. In general the live vaccine may give a stronger immunity more quickly but it is not necessarily any longer lasting and it may induce more reaction. Dead vaccines given by injection in oil or other special bases can be very effective for long-term immunities and are increasingly widely used. It should be emphasised that when a vaccine is administered to an animal it causes a degree of illness and during the period of immunological development the animal is experiencing stress and is *more* susceptible to infection than at other times. Thus it is necessary to practise even better management at the time of vaccination. Very often this factor is not appreciated and a failure that is unjustifiably ascribed to the vaccine is essentially due to mismanagement of the animals.

Medicines

Vaccines and sera are used to protect or treat animals for specific diseases. They should be used wherever they can be in the prevention of infections but in many cases no serum or vaccine exists and we must deal with illness using one or many of a wide choice of drugs. Also many of the disease conditions we are dealing with are caused by a number of different organisms and the identity of some or all may not be known definitely. Frequently the urgent need for treatment demands that a drug is used before laboratory examinations have elucidated the cause, and we must rely on the experience of the farmer and the clinical skill of the veterinarian to judge the correct drug to use. When the results of the tests are known, an àppropriate change in treatment may be indicated.

Drugs can be given to have an almost immediate action. One administration may have an effect for only a few hours or in some cases a few days but usually, in order to keep the levels up within the blood stream of the animal, there must be a constant boosting of levels. It cannot be emphasised too strongly that the full course of treatment prescribed should be completed and levels maintained correctly or the benefits may be lost and drug resistant organisms produced; this possibility is one of the worst outcomes of the use of drugs. Care must also be taken in the agricultural use of drugs to observe certain conditions of use. For example, with many drugs a compulsory withdrawal period is required before animals can be sent for slaughter, varying from about 5–10 d and in some cases even longer, and milk from cows being treated for mastitis must also not be mixed with normal milk for specified periods.

The majority of drugs used for the prevention and treatment of disease are chemotherapeutic, the term implying the use of chemicals that may have been synthesised, such as the sulphonamides, or produced by the action of living organisms – these being the antibiotics.

The routes by which drugs are given vary from injection by the intravenous, intramuscular or subcutaneous routes to oral dosing and incorporation in feed or drinking water. With so many drugs available and the routes and methods of administration being so variable it is perhaps inevitable that careless and abusive use occurs. One cannot do more than emphasise that the prescribing of drugs requires a high degree of skill and their administration an abundance of diligence and competence.

Growth promoters

For many years the livestock industry has made use of antibiotics and other substances which, when incorporated in small amounts in the ration, may increase the growth rate of livestock and improve food conversion efficiency. In the early days of use of these materials, the same antibiotics were used for growth promotion as for the treatment of disease. Since drugs as growth promoters are used at a much lower level than for therapeutic use, the incorporation of antibiotics in a low level in the feed is unacceptable since drug resistant organisms will soon emerge and not only will the antibiotics no longer be of use as growth promoters but they will also have lost their value for therapeutic purposes. Because of this there is now a clear separation, legally enforced, between the use of antibiotics for these two purposes. Those antibiotics used as growth promoters (e.g. zinc bacitracin, flavomycin and virginiamycin) have no other use and are not used for any therapeutic purposes. The therapeutic antibiotics are used only for the prevention and treatment of disease and are available only on veterinary prescription to be used by or under the control of the veterinarian.

This problem of resistance is not the only one in connection with the use of the feeding of growth-promoting antibiotics. These substances have their greatest effect when the hygiene or husbandry system is less than ideal. Under really good management practices they may do little good at all. Thus they tend to 'mask' bad husbandry and this is a dangerous characteristic since such an effect will not last for ever; the bad practices, however, cannot only last indefinitely but will tend to have a cumulative effect on results.

There is still no absolute certainty as to the means by which growth promoting antibiotics exert their effects but it seems most likely that these effects are due to the favourable action of the antibiotics on the balance of bacteria in the intestines, reducing the harmful ones and promoting the beneficial ones and thereby increasing the efficiency of food utilisation and production.

The importance of disinfection

Livestock are generally housed much more intensively than hitherto. This involves not only the stocking of animals more densely but equally an increased size of units and the more efficient use of building space. It is rare for a livestock pen to be without stock; if there is a break between batches then it is usually a short-lived one. In the days before the use of antibiotic therapy, intensive methods tended to collapse under the burden of disease. Now they are again under threat because an increasing number of disease producing organisms are resistant to drugs and also the cost of almost continuous antibiotic therapy is becoming an economically damaging burden to the stock keeper.

There is one viable answer to this and it is to institute a very efficient programme of cleaning and disinfection of the buildings. Let us be quite clear, however, disinfection needs to be done in a carefully organised way and must not be, as so often is the case, an application of any disinfectant chosen at random in the hope that it will kill the pathogens. Firstly, it is imperative, at least with young livestock, to plan the enterprise in such a way that a building or even a site is totally depopulated of stock before the cleaning and disinfection takes place. This is necessary because the live animal, and it might be literally only one, is potentially a much greater reservoir of infection than the building itself. So the complete emptying of a house can be more important than the disinfection process itself. The disinfection process in sequence is as follows:

(1) Empty the pen and preferably the house of all livestock.
(2) Clean out all 'organic matter' – dung, bedding, old feed, any other material that could contain pathogenic micro-organisms. Remove totally from the area of the buildings.
(3) Remove all portable equipment for cleaning and disinfecting outside the building.
(4) Wash down with heavy duty detergent-disinfectant. Use a power washer wherever possible.
(5) Apply a disinfectant appropriate to the type of infections being dealt with. Usually required is a mixture active against viruses, bacteria, parasites and insects which may carry infection from one batch of livestock to another.
(6) Wherever possible fumigate with formaldehyde or spray with a disinfectant after the equipment has been reassembled.
(7) Dry out and rest for a day or two before restocking.

Signs of health and disease

Early recognition of illness is particularly vital because the risk of spread of anything contagious is very great. Early recognition also increases the opportunity for successful treatment, enables the affected stock to be isolated and speeds any laboratory diagnostic work. It also helps in the good welfare of the animals since sick animals are often maltreated by their neighbours. All in all, it will also generally reduce losses to a minimum, thus being of sound economic sense.

The following are the most important signs to look for.

Appetite

One of the earliest and surest signs of illness is a lack of interest in feed (anorexia) or a capricious appetite, though on occasions it may be the feed itself which is at fault and requires investigation.

Separation from the group

Such animals will usually separate from the others and hide in a corner or bury themselves in bedding, if available.

Excreta

This may be 'abnormal'. Sick animals often have scour (diarrhoea) or, more usually in the earliest stages of disease, they will be constipated, especially if there is a fever. In other circumstances the faeces may contain blood (dysentery) or may be of abnormal colour (which sometimes happens in cases of poisoning, for example).

Urine

This may show abnormalities such as the presence of blood or may be cloudy or yellow (jaundiced). Normal urine is pale and straw-coloured.

Posture

Sick animals may find it impossible to rise, or may hold themselves uncomfortably, and may be lame. The nature of the abnormality of posture may indicate the organ affected.

Appearance

Sick animals will often have a droopy head, with dull eyes and dry muzzle and possibly discharge from the nose and watery or catarrhal eyes. All mucous membranes may be discoloured.

Skin and coat

A healthy coat is clean and glossy. Abnormal coats may be 'hide-bound' which is when dehydration makes it difficult to move the skin over the underlying tissues. Sick animals may have bald, scurfy, staring, 'lousy', mangy or scabby skin, and the animals may scratch or rub. The skin, coat and feathers are very good indicators of disease.

Coughing

Most abnormalities of the respiratory system produce coughing in animals and the nature of the cough will often be diagnostic of certain conditions.

Pain

The presence of pain shows in a number of ways: grunts, groans, grinding of the teeth, squealing or crying, arching of the back.

Mucous membranes

The normally moist membranous linings to the mouth, nose and other external orifices may show abnormalities such as discoloration, dehydration, discharge of a serous, mucous or purulent nature, or haemorrhages.

Temperature

An abnormally high temperature usually indicates an infection; an abnormally low one may be due to a metabolic defect, a poisoning or simply the later or terminal stages of an infection. Normal temperatures and respirations are as follows:

| | Temperatures | | Respirations/ |
	°F	°C	min
Cattle	101.5	38.7	12–20
Sheep	103.0	39.4	12–30
Pig	102.5	39.2	10–18
Fowl	106.5	41.5	12–28

Respirations tend to be accelerated by infectious diseases and slowed by others though the nature of the breathing will also be vastly affected and may change from the deep to the shallow on the occurrence of illness.

Pulse rates

This is normally only studied by the veterinarian but the normal rates are as follows:

	Pulse rate/min
Cattle	45–50
Sheep	70–90
Pig	70–80
Fowl	130–140

Animal welfare and animal health

A recent phenomenon has been the widespread questioning of the humanity of certain methods of rearing livestock. Under critical attack from so-called 'welfare groups' have been systems of housing that greatly restrict the movement of animals, amongst these being cages for chickens, calves housed in crates for the production of veal and sows closely tethered in stalls during pregnancy.

Animal welfare is very pertinent to the question of health since there is no dispute that one of the essentials for the provision of good welfare is the maintenance of health in the animals. It has also become apparent from recent activities in the research field that the eventual goal of establishing what constitutes good welfare in a scientific way is going to be very difficult indeed, if not impossible. Thus it will be necessary to continue to rely, at least in the foreseeable future, on the overall knowledge, expertise and even instinct of the persons concerned with the management and care of the animals, these being the farmers, stock keepers and husbandry and veterinary advisers. It is essential to keep in the forefront of consideration the overriding importance of good welfare and humanity to animals on all occasions. Anyone who has closely observed sick animals can understand what constitutes 'misery' in the animal kingdom and contrast this with the alert, buoyant and inquisitive appearance of an animal in good health.

A feature of this controversial field which has been beneficial is the way in which it has forced us to enquire whether it is necessary to house animals using methods which involve a high degree of restriction on their movement. There has been a vigorous search for alternatives which has been fruitful as it appears there is less need for the extremes of intensification than has been alleged and indeed it may be economically preferable to have systems which are more in harmony with the overall agricultural scene.

Large and intensive livestock enterprises may be unsuccessful for three principal reasons. Firstly, the management of large numbers of *living beings* is not easily organised on the large scale. Stockmanship is partly science and partly art, and art must rely heavily on the individual. And if only one small item goes wrong or is overlooked it can lead to a very serious chain of events on the large scale.

The second reason is that disease is much more likely to have a serious deleterious effect in the large unit in subclinical or clinical forms. The failure of many units can be ascribed to their inability to control disease so that the economic viability of the unit has been totally destroyed.

The third reason is the heavy cost of buying in or bringing in fodder and moving out the muck in such enterprises. It is common for the highly intensive unit to be totally dependent on food from outside the farm or the locality and to be faced with great difficulty in disposing of all the waste products. In smaller units the crops can be used to feed the livestock and the 'muck' they generate is of great value to the land; energy is conserved and this is agriculture of a totally balanced sort.

It may be stated also that the welfare standards that are gradually emerging across the world are based on evidence which, so far as is possible, is not in conflict with economics. For example, if animals are housed at concentrations and intensities above those advised, it is likely that their productivity will fall. Particularly this is so when it comes to stocking density and its relationship to growth, production and the efficiency of food utilisation. Good welfare standards can be used, therefore, as a husbandry standard with some confidence.

The key to good welfare is a high standard of stockmanship and if this is to function then good facilities must be provided for the stockman. Many systems that have been developed in recent years completely ignore this. The essentials are as follows:

(1) plenty of room for the stockman to *see* the stock whenever he needs to;
(2) good lighting and no undue saving on passage space;
(3) facilities for isolation of sick stock;
(4) easy access by the stockman to his animals and, above all, a very high standard of handling facilities so that it is easy to medicate or treat the animals in any way necessary.

Disease legislation

In Great Britain, under the Diseases of Animals Act, there are a substantial number of Orders relating essentially to the State's interest in animal disease control. These regulations cover a wide spectrum of activity. For example, they deal with the importation of animals and animal by-products, the movement of animals, the inspection of livestock and vehicles, and allow for the Ministry of Agriculture, Fisheries and Food to institute a large number of Orders relating essentially to specific diseases.

Under the Diseases of Animals Act certain specified diseases are designated notifiable. Those of current importance are:

Anthrax
Foot-and-mouth disease

Fowl pest
Bovine tuberculosis
Sheep scab
Swine vesicular disease
Aujeszky's disease
Enzootic bovine leucosis
Rabies
Warble fly

There are also others, such as swine fever, which could well occur again as they are widespread elsewhere in Europe.

The most important feature about the notifiable diseases is that the owners or person in charge of an animal suspected as affected by a notifiable disease must report immediately his suspicion to a police officer or an inspector of the Ministry of Agriculture. Once the owner has notified the inspector or police, then steps will be taken to diagnose the condition and measures may be taken to isolate the premises and treat, vaccinate or slaughter the stock. Different diseases have different procedures and these are dealt with later in the chapter under the respective diseases. Also, under the Zoonoses Order, the presence of conditions such as salmonellosis and brucellosis, which infect both animals and humans, must be notified.

THE HEALTH OF CATTLE

In this section health problems are dealt with on a life-cycle basis, dealing firstly with those health problems affecting the calf, continuing with growing stock and concluding with the mature beast and those diseases that have no special age incidence.

The health of the calf

Of the 5–10% mortality that occurs in calves, nearly two-thirds takes place in the first month of life and much of this is due to what is commonly called calf scours. The main cause is ultimately the ubiquitous bacterium *Escherichia coli* following initial environmental stress and usually a virus challenge.

Colostrum and protection against diseases in the calf

It is essential that a calf should have a good drink of colostrum very soon after birth. Colostrum contains the immunoglobulins that are the essential protection to infection. It also contains large quantities of vitamin A (six times as much as in ordinary milk). Immunoglobulins are the proteins which contain the antibodies which protect the calf from disease organisms, whilst vitamin A is the most important vitamin associated with the animal's resistance to infection.

Because calf scour is caused by a number of organisms the use of specific vaccines and sera may have a limited chance of providing either good treatment or prevention, though sometimes this can be the case. More often the disease must be treated with antibiotics with a broad range of activity, such as the synthetic penicillins, together with replacement therapy for the enormous fluid depletion that takes place. This will include electrolytes, glucose, minerals and vitamins in easily assimilated liquid form.

Salmonellosis

The salmonella group of bacteria is a very large one with hundreds of different types, many potentially pathogenic to man and animals. Only relatively few are harmful, the majority being 'exotics' that come in via the feed and are quickly eliminated.

The two most important species in the condition in calves are *Salmonella typhimurium* and *Salmonella dublin*. The symptoms with either of these infections or indeed with other virulent salmonellae can be largely similar. A percentage of calves may die so rapidly that no symptoms are seen; others scour profusely with dysentery, pneumonia, arthritis, jaundice and even nervous symptoms manifesting themselves in some cases. Urgent treatment will be essential to include antibiotics such as ampicillin, chemotherapeutics and fluid therapy, isolation and general nursing, especially warmth. Thorough disinfection of the quarters will be essential and preventive measures may include vaccination and possibly preventive medication. However, the use of the last measure is hazardous as it may generate drug resistance and encourage the establishment of the 'carrier' animal which can excrete salmonella organisms throughout its life.

Whilst salmonellosis is especially a disease of young stock, adult cattle do become infected and may not show symptoms, becoming carriers of great potential danger to man and other livestock. Abortions may be caused by salmonella and salmonella can be a source of infection via milk. In the main the biggest problems from salmonellosis arise from farms where the calves are brought in from outside and infection can easily spread to the adults if they are on the same premises.

In Great Britain salmonellosis is a notifiable infection and official measures can be taken to control the disease as appropriate. In the event of an outbreak in adult cattle, measures that may be taken will include laboratory testing of the cattle to identify the carriers, treatment of animals with antibiotics, such as ampicillin, and appropriate care of the milk to remove the risk of human infection.

Calf diphtheria

The bacterium associated with foul-in-the-foot in cattle, known as *Fusiformis* or *Actinomyces necrophorus*, is also capable of causing a serious disease in calves known as calf diphtheria. The infection attacks very young calves causing inflammation within the mouth, which leads to soreness, ulcers and ultimately necrosis. The calf eventually becomes seriously ill. The condition responds well to treatment. Sulphonamides are effective and also a wide range of broad spectrum antibiotics. It is a disease that is associated with bad hygiene.

Navel-ill or joint-ill

These two terms describe the disease affecting calves where bacterial infection causes inflammation in the region of the joints. It usually occurs because the calves are managed under dirty conditions and are challenged by a serious burden of bacteria before the navel has healed.

Treatment may be undertaken successfully in the early stages of the disease with sulphonamides or broad spectrum antibiotics; local application of dressings to the navel will help to prevent infection.

The health of growing cattle

Internal parasites

There are three major groups of diseases in cattle caused by parasitic worms. There is 'husk' or 'hoose', which is due to worms which live in the lungs; parasitic gastroenteritis, which is caused by roundworms which live in the abomasum and/or intestines of cattle, and finally liverfluke disease, caused by flatworms or flukes, which live in the liver.

Parasitic worms do not multiply within the body of the animal in which they live as adults. The adults produce millions of eggs but none reaches maturity without first passing out of the animal in the faeces. They must then have a period of time outside the animal before they become infective, and this period varies from a few days to many weeks. Thus every worm in the animal has been picked up in the herbage while the animal was grazing. The significance of this knowledge is that it provides for two ways of attack, one within the body and one without. Removal of contact of animals with their own faeces is a vital factor. If the husbandry is good, with clean environment and no over-crowding, the infestation may be so slight as to be unimportant. Mixed grazing is also useful since the parasitic worms are host specific and infect only one species. It is also important to know that it is chiefly the young and growing stock that are vulnerable to infection. Adults should have a naturally produced immunity due to earlier contact with the parasites. Nevertheless, the adults often have a worm burden and will produce some eggs. And even with adults they can become badly affected if they become stressed by poor conditions or nutrition. The worm eggs once voided can remain infective for up to or even more than a year.

Husk or hoose

This is an infection of the bronchial tubes and leads to the cough known as 'husk' or 'hoose'. It is caused by the lungworm *Dictyocaulus viviparus*. The white worms are threadlike, up to 75 mm long and the females lay vast numbers of eggs. After hatching they produce minute larvae which are carried in the mucus up the windpipe and to the mouth. They are then swallowed and pass out onto the pasture with the dung. These larvae have to develop on the ground before they are able to infect cattle who take them in with the grazing. The development takes about 5 d in warm, moist weather and over a month when the weather is cold. Whilst developing in the dung the larvae go through a series of changes and moult their skin twice. The larvae have little resistance to drying and can die off rapidly under these circumstances.

When the infective larvae are swallowed by grazing cattle they penetrate the intestinal wall and pass into the lymphatic system and by this route reach the blood stream and then pass to the lungs. Here they develop into the adult worms and begin to lay eggs after three weeks.

Cattle that have been exposed to lungworm infestation will in due course acquire a resistance. If the level of pasture infestation rises at a moderate rate, the gradual and increasing resistance of the calf will enable it to reject the effect of the worms and no disease will actually occur. But if the pasture infestation rises rapidly, as it may do if the climatic conditions are favourable, disease may occur.

High pasture infestation may occur in a number of circumstancers. In early spring most pastures will be either clean or at least only very slightly infested. Contamination of the pasture by older animals, which have carried over some worms in their lungs from the previous season, may produce more significant levels of infestation on the herbage and may even give rise to dangerous infestation if the season and pasture conditions are suitable. Low levels of herbage infestation may be increased when susceptible calves become infected on the pasture and in consequence increased numbers of larvae are passed on to it in their dung.

Symptoms

The disease may be recognised, or at least suspected, if there is an increase in the respiratory rate, coughing, especially when the animals are moved, and some loss in body condition. If cases become severe then the coughing becomes louder, the respirations shallow with clear pain and distress, and the animal stands uncomfortably with head and neck outstretched and the tongue protruding.

Control

There are a number of useful drugs for lungworms, such as levamisole (Nemicide, Coopers) and fenbendazole (Panacur, Hoechst). Wherever possible it is best to remove the animals from the pasture and house them in good buildings where they can be well fed.

Prevention

It is wise to keep animals grazing their first year off pasture which has had animals on it recently. Strict systems of rotational grazing can always be of help since they will prevent heavy build-up of disease. The calves should be moved twice a week on to ground which has not had calves on it for about four months but even a shorter time would be of help in reducing infection. The vaccine is also a great help in preventing the disease but it is an *adjunct* to good management and not a substitute.

Liverfluke disease

Liverfluke disease tends to go quite undetected or unsuspected until animals start dying. It is admittedly more a disease of sheep than cattle but there is good evidence that liverfluke disease causes a very serious effect on growth and productivity. The liverfluke is a flat worm about the size and shape of a privet leaf, which lives and feeds on the liver of cattle. The flukes lay many thousands of eggs which pass on to the pasture in the dung. The eggs hatch and produce a very small swimming creature which has to enter a particular type of snail if it is to survive. The parasite grows in the liver of the snail, producing many young liverflukes. These look like the flukes which cause disease in the older stock except that they are much smaller. The young flukes leave the snail when the snail is in water and encyst on any herbage which they contact. They lose their tails and surround themselves with a protective covering. There they remain as cysts but are inactive for long periods, up to months on occasion, until eaten by the grazing animal. When eaten the protective coat disappears and the fluke emerges and makes its way into the liver. During the next two to three months the fluke grows to maturity and then lays eggs.

Symptoms

There are two forms which are clinically differentiated. The acute disease occurs when the grazing animal picks up many flukes in a short time. These damage the liver seriously and this can lead to sudden death in the apparently healthy animal. The chronic form is more common and is due to the presence of flukes in the liver over a period of months and affected animals waste progressively.

Control

Certain drugs now exist for treatment, namely oxyclozanide (Zanil, Coopers) and nitroxynil (Trodax PML, RMB). Also recommended is albendazole (Valbazen, Smith, Kline and French). Professional advice should generally be sought before use of these drugs in view of possible side effects, contraindications and withdrawal periods of milk for human consumption.

It is also effective to prevent the disease by controlling the secondary hosts, the snails. Infestation in flukey areas is worse in the months of October to December and hence fluke-free pastures should be grazed in this period wherever possible. Badly drained areas should be properly drained. The snails may also be dealt with by treating the pasture with either copper sulphate at 12 kg/acre or sodium pentachlorphenate at 4 kg/acre. As these are dangerous materials they should be applied with due caution.

Parasitic gastroenteritis

This is a disease, or infestation, caused by the presence in the abomasum or fourth stomach or the anterior part of the small intestine, of a group of small roundworms. The most important species are *Ostertagia ostertagi* and *Trichostrongylus axei* in the abomasum; *Cooperia oncophora, C. punctata, T. vitrinus* and *T. colubriformis* in the small intestine and *Oesophagostomum radiatum* in the large intestine. They cause considerable economic loss from both poor productivity and overt disease, chiefly in young animals.

The life-cycle of the worms is 'direct', there being no intermediate host. Eggs are laid by the adult worms passing out with the dung on to the pasture where they can hatch within about 24 h, producing minute larvae which feed on bacteria in the faeces. Within the next few days they moult twice and only then are they capable of infesting stock who ingest them in the herbage. In order to make the ingestion more certain they leave the dung and pass on to the grass, climbing upwards so that they will readily be consumed by the animals. All this can take not more than 4 d in warm, damp weather but may take weeks in cool conditions.

A limited number of worms do little or no harm to the beasts but a heavy infestation can be lethal. Thus infestation is assisted by overcrowding, moist, warm conditions and long grass in which moisture more readily persists.

Symptoms

These are nearly always indefinite and the usual signs are a progressive loss of condition possibly leading eventually to emaciation and death. There is usually a severe scour but the animals continue to eat voraciously. However, one should be aware of the fact that some animals will not scour and signs will be so indefinite as to go undetected. In every case it is wise to have positive identification by examination of the faeces to determine the degree of infection.

Control

The first course of action is to consider medication and there are a wide variety of materials that can be given. Organophosphorus compounds, such as fenbendazole (Panacur, Hoechst), albendazole (Valbazen, Smith, Kline and French) and thiabendazole (Merck, Sharpe and Dohme); also levamisole (Nilverm, Coopers) and systamex oxfendazole (Autoworm, Coopers).

Preventive measures should consist of the following: overcrowding of the young stock should be avoided. Grass should be kept short as larval worms tend to perish in such grass. Only adult stock should be put on badly infested land. Period treatment should be given to animals after laboratory examination of faeces has established the degree of infection.

Skin conditions and external parasites

Ringworm

This is an extremely common skin disease of cattle which tends to be a problem in winter when the animals are housed. It is caused by fungi belonging to the genus *Trichophyton*, two species of which occur in cattle in the UK and also attack horses, pigs, dogs and humans. *Trichophyton verucosum* is the most common one and is responsible for most cases of ringworm in cattle, farmers, stock keepers and their families.

The fungi that cause ringworm are capable of surviving for very long periods in the farmyard, certainly for over a year and probably much longer. The spores that are picked up by the animals find their way into cracks on the skin and germinate to produce fine filaments which flourish in the surface of skin and hair. These fungal threads grow downwards inside the hair keeping pace with its growth from the hair bulb and weakening it so that it snaps off at the surface of the skin producing the characteristic bald areas. The skin reacts to the infection by inflammation followed by the formation of a grey to yellow white crust. The lesions of ringworm are most often seen on the head and neck, especially in calves, but in adults patches can occur on the flanks, back and rump, wherever rubbing takes place. Animals that are rundown are likely to suffer a severe infestation.

Control

Whilst ringworm will disappear spontaneously in the course of time, it is not advised to ignore treatment since the disease has a serious debilitating effect. Treatment may be effected by removing the crusts with a soft wire brush and applying one of several proprietary preparations containing iodine or quaternary ammonium compounds, also a 2% formalin solution, handling with care, wearing rubber gloves and overalls. A much better treatment, though expensive, is an antibiotic, griseofulvin. It is more vital to practise good hygiene measures by efficient cleansing and disinfection of the buildings and all objects that could harbour the fungal spores.

Mange

Certain forms of mites infest cattle – these are *Sarcoptes scabiei, Chorioptes bovis, Psoroptes communis* and *Demodex bovis*. All are extremely contagious conditions (sometimes called scabies). The result of the disease is to produce very patchy coats and areas of bare skin with an extreme amount of irritation.

The parasitic mites spend their whole lifetime on the animals and only survive a few weeks off them. The symptoms are always at their worst during the final winter months. Diagnosis is made by microscopic examination of skin scrapings. The mites are visible to the naked eye.

Control

Control is rather difficult and requires professional veterinary advice. Any incidence should firstly be identified and once this is done measures are advised to eradicate it from the herd. Amitraz (Taktic, Smith, Kline and French) is applied as a spray and Bromocyclen (Hoechst) as a spray or suspension. All animals should be treated and in order to eradicate the condition it will be necessary to re-treat two or three times at 10 d intervals.

It is also necessary to keep the cattle away from those parts of the building that could harbour the mites and/or disinfect these areas as effectively as possible to prevent re-infestation.

Blackquarter (blackleg)

The disease of blackquarter is similar to so-called blackleg of sheep. It is caused by a bacterium *Clostridium chauvoei*, the spores of which are common inhabitants of the soil in areas where sheep and cattle have long been grazed.

The disease occurs because the organism is able to multiply in the muscles of the leg. The symptoms are stiffness and lameness with swellings appearing in areas such as the loin, buttocks or shoulder. The swellings are hot and painful at first but later become cold and painless. The skin over the affected muscles will become hard and stiff and if pressure is applied there is a papery effect, a dry, crackling noise due to the stiffness of the skin and the movement of the gas underneath. In fatal cases the condition of the animal rapidly deteriorates and it will die quite quickly. Initially there is probably a bruise or injury to the muscle and it is in this area that the clostridia can multiply in the absence of oxygen. Then they produce the toxins which have the serious destructive effect on tissues.

Animals affected with the disease can be treated successfuly with antibiotics, only if the disease is recognised very early. Penicillin preparations are satisfactory. Prevention can be effected by using a vaccine but the best measures of prevention rely on good pasture management, avoiding areas of known susceptibility and draining or cultivating those that must be used.

Coccidiosis in cattle

Coccidiosis is a form of dysentery that primarily affects young stock in the summer and autumn months. It is caused by a microscopic unicellular parasitic organism known as a coccidium. The coccidial parasite forms spores, known as oocysts, that can live outside the animal for many months., Adults are frequently carriers of the coccidial parasites and show no symptoms yet spread the disease to younger animals.

The oocysts are very resistant to destruction and can often remain infective for a year but if they are swallowed by a susceptible animal they hatch and the active forms of the coccidia are liberated. These penetrate the cells lining the intestine where they mature and multiply and break down areas of the mucosa causing diarrhoea together with bleeding.

Animals acquire a resistance to coccidiosis by receiving a light infestation early in life and developing an immunity.

However, if the intake of oocysts is sudden and too great at one time before a resistance has been produced, then disease occurs. This is most liable to happen when animals are densely stocked and the weather conditions are warm and wet. Dirty walls of pens or boxes, as well as the floor, are sources of infection.

The disease is controlled in a number of ways. Affected animals should be treated with a suitable drug, such as sulphonamide, plus plenty of really good nursing, especially fluids. Overstocking must be avoided; good bedding and cleanliness are important and animals should be kept in good thriving conditions so that they can readily resist such infection that occurs.

Bovine virus diarrhoea (mucosal disease)

Bovine virus diarrhoea (BVD) or mucosal disease (MD) is an infectious disease of cattle caused by a virus. It was first recognised in the USA but in recent times has occurred in virtually all parts of the world. Originally the two diseases BVD and MD were distinguished but it is now clear they are caused by the same virus so the best term is probably the BVD–MD complex.

Symptoms

The disease causes ulceration of the mouth and lips accompanied by foul smelling diarrhoea. There is often lameness and extreme scurfiness of the skin. By far the majority of cases occur in cattle between 4–18 months of age. Most cases can recover spontaneously and only exceptionally if there are many young animals affected the death rate can be high.

Treatment and control

There is no specific treatment but supportive therapy in the form of intestinal astringents and electrolytes can be of value.

Infectious bovine rhinotracheitis (IBR)

This is an acute respiratory disease of cattle caused by a virus and is spread by contact or by the airborne route. It is a very contagious disease and can spread through a herd within 7–10 d.

Symptoms

The symptoms are of severe inflammation of the eyes and the upper respiratory tract. The disease can also cause a substantial fall in milk yield, abortion and infertility. There may be a fever, loss of appetite and drooling of the saliva. Sometimes there is a pronounced cough.

Control

This is not too serious a disease in Great Britain. IBR is treated with broad spectrum antibiotics to help reduce pyrexia or the effect of secondary invaders. A vaccine is available to protect the herd.

Respiratory diseases in cattle

The number of pathogenic organisms that can infect the respiratory system of cattle is enormous – there are some 20 bacteria, six fungi, four parasites that *commonly* infect them, not to specify the underlying primary damage that may be wrought by viruses. The wide involvement of so many agents has made the production of satisfactory vaccines most difficult and the use of drugs complicated and expensive. Thus the prime responsibility for dealing with such conditions rests with getting the environment correct.

The general principles are given in *Table 19.2*.

Table 19.2 Essential action to counter respiratory disease in cattle

Expert attention required for correct therapy
Improve ventilation – usually needed is more air
   flow with less draught
Check stocking density – are the cattle overcrowded?
Is more bedding required?
Separate and if possible isolate those cattle that
   are badly affected
Provide warmth for the very sick
Attempt separation of different ages
Reduce movement of animals to a minimum
Consider insulation and other essentials of
   construction relevant to the building
Can serum, vaccine or drugs be used to protect
   the affected animals?

Specific advice for cattle may be summarised as follows:

(1) Depopulate between batches and institute a rigid disinfection programme.
(2) Reduce numbers in one environment to a minimum.
(3) Keep different ages and sizes of animals separate.
(4) Ensure copious ventilation without draught as cattle are hardy animals and do not require cossetting.

Wooden tongue and lumpy jaw

These are two distinct diseases which have rather similar symptoms and are caused by unrelated organisms. They both run a slow course and are characterised by a steady enlargement of the tongue and swellings in various parts of the animal but especially on the head and neck. Their presence may generally interfere with the normal functions of the animal, they can also form abscesses or ulcerate, discharging into or outside the body. Untreated they cause a steady wasting of the animal so that it becomes unproductive and will eventually die.

The cause of the two diseases are quite different. Actinomycosis or lumpy jaw is caused by a ray fungus known as *Actinomycetes bovis*, whereas actinobacillosis or wooden tongue is caused by a bacterium *Actinobacillus ligneresi*. The main difference between the two diseases is that actinobacillosis tends to affect the soft tissues, such as the tongue, and actinomycosis usually causes diseases of the bones, especially those of the jaw.

Symptoms

The tongue is most commonly affected and becomes hard and rather rigid, thus the name 'wooden tongue'. There is a constant dribbling of saliva and food tends to be rejected. The disease will also affect the glands of the neck and swellings appear between the angles of the jaw. If this swelling is allowed to continue it will affect breathing and swallowing also becomes difficult. The lesions may burst and release a characteristic yellow and granular pus. In addition the bones of the jaw are often affected. Pus forms within the bone and these disease areas can eventually break through the skin to form an unpleasant discharging fistula. These are, however, the external signs of the diseases. They may also affect the internal organs and will then cause a progressive wasting.

Control and treatment

Other than in advanced cases, treatment can be very successful. If the disease does seem to have got a firm hold on the animal which is fast losing condition, it is probably best to slaughter it. *Actinobacillus* responds quite well to iodides and sulphonamides and actinomycosis will be suitably treated with penicillin or broad spectrum antibiotics.

Redwater

Redwater, also called bovine piroplasmosis or babesiasis, is a curious condition caused by the presence in the blood of minute parasites called piroplasms. These parasites live inside the red blood corpuscles where they can be seen in the early stages of the disease. It is a condition that is found in certain areas, for example in Wales, southwest England and East Anglia.

The organisms are especially associated with rough scrubland, heathland or woodlands where the ticks *Ixodes ricinus* are present as they transmit the organisms from animal to animal. The tick is infected by sucking the blood of an infected animal. Once the tick is engorged with blood it falls off and lies in the herbage. Later the tick may attach itself to other cattle and when it sucks from these introduces the parasites into the blood. The two worst periods for redwater disease are from March to June and October to November.

Symptoms

The main effects of the disease are to cause loss of condition and a considerable loss of milk yield in the dairy cow, together with blood pigments in the urine. Animals affected also show a high fever and diarrhoea and later constipation. The animal also often becomes jaundiced, the mucous membranes becoming yellow. The heart beat becomes very loud and there are tremors of the muscles of the shoulder and legs. It should be noted that young cattle are less susceptible than adults which have not previously been infected. If contracted early in life the disease is only mild and the animal is then immune against further infection. This indicates one of the great dangers; if adult cattle from a tick-free area are introduced into a redwater area, some of these may contract the infection and die quite rapidly without treatment.

Control

Treatment of infected animals will be twofold. Firstly, the animal will require an early injection with one of a number of chemotherapeutic agents that are lethal to the parasite. Secondly, careful nursing is perhaps even more important. Feed well, house comfortably, do not drive or excite for fear of heart failure and keep the animal warm.

Prevention

The disease could be eradicated if ticks were eliminated. The tick is extremely resistant to most measures of control. In addition it can feed off several species of animals and is not dependent on cattle alone. Sometimes sheep are put on pasture as 'tick collectors'. By placing them alone on the pasture at the worst tick seasons of the year and then adopting a heavy dipping programme, this will reduce numbers considerably. The only effective way of eliminating ticks is to clear the rough ground by cultivation and normal crop rotation.

Wherever this procedure is adopted and a part of the farm is freed from ticks, the importance of exposing young animals to tick infestation must be considered otherwise they will be highly susceptible as adults to a first infection. Great care must be taken not to put them on to tick infested ground for the first time in adulthood.

Diseases of mature cattle

Metabolic disorders

Bloat

In bloat, also known as blown, hoven or tympanitis, the rumen becomes distended with gas and the pressure from this on the diaphragm may lead to the animal dying from asphyxia or shock. Two types of bloat occur: the first is where the gases separate from the contents of the lower part of the rumen, and the second where the gases remain as small bubbles mixed with the contents in a foamy mass, giving rise to what is known as 'frothy bloat'.

Bloat most often occurs in the grazing of lush pastures which contain a high proportion of clover. It may be due to the fact that some kinds of herbage contain substances that may cause paralysis of the rumen muscles and of other organs concerned in the process of rumination. Or it may be due to the fact that more gas is produced than can be got rid of by eructation. Bloat may also be due to the result of foaming in the rumen caused by saponin and protein substances obtained from plants.

Symptoms Symptoms are obvious distension of the abdomen which is particularly pronounced on the left flank between the last rib and the hip bone. In the early stages the animal becomes uneasy, moves from one foot to another, switches its tail, and occasionally kicks at the abdomen. Breathing becomes very distressed and rapid. If the attack is a severe one, any movement intensifies the discomfort and so the animal stands quietly with legs wide apart. If relief is not given very quickly, death from suffocation or exhaustion can follow rapidly.

Control It is advisable to summons a veterinary surgeon as soon as possible. If time allows a stomach tube may be inserted into the rumen to allow the gas out. In acute cases the only remedy is to puncture the rumen on the left flank with the surgical apparatus, the trocar and cannula. In an emergency, where death seems imminent and specialised assistance is not available, the only course of action is to puncture the area with a short pointed knife with a blade which is at least 6 inches long, the blade being plunged into the middle of the swelling on the left side and twisted at right angles to the cut to assist the gas to escape.

Although this treatment will serve to give immediate relief, more treatment is required if the case is one of frothy bloat. Drenching with one of the substances known as silicones, or with an ounce of oil of turpentine in a pint of linseed oil, often helps in the less severe cases and some gentle exercising may help to relieve the pressure of the gas in the rumen. There are also a number of drug treatments available, such as Avlinox, which is an ethylene oxide derivative of ricinoleic acid.

Prevention Efficient pasture management can help to avoid the trouble. After a pasture is closely grazed and then rested, the clover recovers more quickly than the grasses so that when pastures are used intensively the clover becomes dominant in the sward but it is generally considered that if there is less than 50% of clover present, it is usually safe. A variety of seed mixtures has been tried to provide a suitable balance between clover and grass; one which has given promising results contains tall fescue. After it has been closely grazed this species of grass recovers almost as quickly as the clover and maintains a reasonably safe balance betweeen grass and clover.

Another system which has been advocated is to allow the herd to graze a potentially dangerous herbage for a limited period and then turn it on to an old pasture for a time.

Controlled grazing can contribute to the prevention of bloat by ensuring that the grazing animal eats both leaf and stem, whether clover or grass, and that it cannot select clover or clover leaf which is thought to be more dangerous. If an electric fence is moved forward so that the cows have access to a succession of narrow strips of pasture during the course of the day, bloat may be reduced or eliminated but it may reduce the intake of the feed and milk yield.

The second method of controlling grazing is to mow strips of herbage, then allow it to wilt and let the herd graze it as it lies on the ground. Alternatively, the wilted herbage can be carted to the cattle and fed indoors. The feeding of roughage in the form of hay or oat straw shortly before the grazing of dangerous pasture has also been practised for many years. It is only completely reliable if the roughage is fed overnight.

Milk fever

Milk fever occurs principally in cows after calving, usually in dairy breeds of high milk yielding capacity. It is given various synonyms, such as hypocalcaemia, parturient hypocalcaemia, parturient paresis and parturient apoplexy. There is, in fact, no fever and indeed it is more usual for the cows to show subnormal temperatures.

Symptoms Usually within 12–72 h of calving there is a short period of uneasiness with paddling of the hind legs, swishing of the tail and convulsive movements, soon followed by depressed consciousness and paralysis. The animal usually falls on its side with its legs extended, rolling its eyes and breathing heavily. The characteristic picture is a twisting of the neck as it lies on one side, with spasm of the neck muscles, shallow laboured breathing, grunting, cessation of rumination and dryness of the muzzle. In straightforward milk fever the calcium and phosphate levels fall and the magnesium levels rise.

In practice, it is quite difficult to distinguish between the simple hypocalcaemia and the more complicated metabolic disorder where there are a number of deficiencies which have to be corrected.

Background to occurrence   There does not appear to be any real breed incidence but it is usually well fed, high-yielding cows which are affected and the severity of the condition increases with age. The cause is usually ascribed to a temporary failure in the physiological mechanism which controls calcium levels in the blood. The stress of calving and the physical adjustment necessary for lactation will upset the balance of chemical regulators (hormones) produced by various glands and if the cow does not adjust itself quickly enough it may suffer an attack of parturient hypocalcaemia.

Control   The immediate administration of a solution of calcium borogluconate is effective in curing uncomplicated milk fever. It may be given intravenously for a quick response or subcutaneously for a slower effect. In addition, the cow should be propped up in the box by means of straw bales on each side to prevent it from becoming 'blown', and to prevent regurgitation of ruminant contents. Further injections of calcium may be required and in addition, in complicated cases, there will be the need for administration of phosphorus injections and even other minerals. Professional advice is certainly required for all but the simple case of hypocalcaemia.

Injections of vitamin D increase the mobilisation of calcium in the body and thereby may help to prevent levels falling too low.

Ketosis (acetonaemia)

Ketosis is a disease involving ketones which are chemical compounds which may be produced during the metabolism of fat. This condition may occur when an animal's food intake is inadequate for its needs and it is drawing on its own reserves in considerable amounts to make good the deficiency. The usual form of ketosis is ketonaemia which includes an excess of ketones in the blood. Since acetone is the simplest and most characteristic of the ketones the disease is generally termed acetonaemia. It is also called 'post-parturient dyspepsia' as it often occurs soon after calving and is associated with indigestion. It often follows mild attacks of milk fever and almost appears to be a sub-normally slow recovery, hence the traditional term for the disease of 'slow fever'.

Symptoms   The symptoms of ketosis are usually seen within two weeks of calving and in well nourished stock of high milking capacity. But it may also occur much later than this, even several months after calving. It is most common during the tail-end of winter. Affected cows appear listless, a marked loss of condition, dull coat, and a sickly sweet smell of acetone in the breath and also in the urine and milk. The milk production falls away quickly and the milk becomes tainted and undrinkable. Constipation with dark, mucus covered dung is usual. Sometimes there is a licking mania, a rather wild looking appearance and champing of the jaws and salivation, and there may also be hyperexcitability.

Control   Firstly, tests are carried out to confirm the presence of ketones in the blood. It will be usual for the blood sugar to be low and injections of glucose are an early

procedure. Corticosteroids may be administered to give a longer-term beneficial effect. Other useful treatments are glycerine and preparations containing sodium propionate. Molasses and/or molassine meal are also helpful.

Prevention will depend principally on skilful management. The energy provided in the food must keep pace with the draining demands of production, especially in early lactation. In late pregnancy the cow should be kept in good condition but not over-fat. The roughage in the ration should also be carefully selected and kept up to a good level or at least reduced slowly if the cow's appetite and enthusiasm for a balanced diet are to be maintained.

Hypomagnesaemia

The amount of magnesium in the blood of normal animals keeps within a fairly constant range. However, should the magnesium level of the blood drop below this range then the condition known as hypomagnesaemia, or grass tetany, can result. The condition occurs especially in areas of cooler climate where the pasture has been improved and production increased. In adult cattle it is called by various names, such as 'grass tetany', 'grass staggers', 'lactation tetany' and 'Hereford disease', whilst in calves it is known as 'milk tetany' or 'calf tetany'.

Underlying causes   The occurrence of this disease is in the period of sudden change of lactating cows from winter housed conditions to those of the rapidly sown spring grass. But it is now known to occur in cattle under most feeding and management regimes, including winter feeding periods. It is not due to a simple deficiency of magnesium, for sometimes the condition occurs in cattle on pasture which has a very adequate magnesium content. Underfeeding appears to be a predisposing cause, especially in out-wintered beef cattle, but overall the disease is due to physiological dysfunctioning that interferes with the absorption and utilisation of the magnesium which is in the food.

Symptoms   Symptoms vary according to the intensity of the attack. The first signs are often nervousness, restlessness, loss of appetite, twitching of the muscles, especially of the face and eyes, grinding of the teeth and quite soon after, staggering. In less acute cases animals which are normally placid become nervous or even fierce. In severe cases paralysis and convulsions develop very soon after the onset of symptoms and if treatment is not given death can follow very quickly. In milking cows a sudden reduction in milk yield may occur just before an attack. Sometimes the attack comes on so rapidly that no symptoms are seen, merely a dead beast. There is also a chronic form of the disease in which cows show a gradual loss of condition although appetite and even milk yield do not show any drop. This chronic state can last several weeks and then develop into the more acute condition.

Control   The veterinary surgeon will inject magnesium plus other solutions into the affected animal to cure the condition. To prevent the disease supplements of magnesium will be used. Two ounces (60 g) of magnesium oxide/head/d has been used for many years successfully in dosing adults to prevent the condition. The magnesium oxide is usually given as calcined magnesite in a granular form. This must be given daily to cattle under risk or blood magnesium levels can fall quickly.

The use on pastures of magnesium-rich fertilisers has been quite successful in preventing the disease. On some soils top dressing in January or February with 500 kg of calcined magnesite per acre can be a perfect preventative. Even 250 kg may be sufficient. Another method of preventing the disease is the magnesium 'bullet' in the reticulum with its slow release of magnesium and attempts have also been made to give multi-vitamin injections to increase the absorption and utilisation of magnesium.

Calf tetany

Many years ago it was established that calves could not be reared to maturity on milk alone because of the onset of deficiency conditions and in particular hypomagnesaemia or calf tetany during growth. But the condition does also occur under other conditions of feeding and it is an increasing problem in beef calves, especially if they are grazing at the time of the most active growth of spring grass.

The symptoms in calves are similar to those in older animals, hyperexcitability, irritableness and twitching of the muscles. The walk often becomes spastic, the feet being carried well above the usual height. Thereafter in the progress of the disease convulsions may follow and eventually death.

Treatment consists of immediate administration of suitable magnesium injections. Prevention is effected by administering about 15 g ($\frac{1}{2}$ oz) of calcined magnesite daily, or 30 g (1 oz) of magnesium carbonate daily.

Mastitis

Mastitis means literally inflammation of the mammary gland or udder. It has always been an important disease of dairy cattle with potentially disastrous consequences but with the advent of antibiotic treatments there was a general optimism that the days of the disease were numbered. Regrettably this has not been the case, for organisms that cause mastitis have become increasingly resistant to the constant use or, perhaps more correctly, misuse and abuse, of antibiotic therapy. Also there has been a growth of large dairy cattle units with the greater opportunity for disease build-up and cross infection. Cow cleanliness has become less satisfactory and some of the most modern practices lead to dirtier cows and udders, for example, with loose housing and cubicles. There has also been the emergence of different forms of mastitis, called environmental, because of their association with the newer forms of husbandry and rather different balance of organisms involved.

Mastitis may affect as many as 50% of the cows in the National Herd in one form or another, many cases being subclinical and so not easily recognisable. Since a cow with mastitis may *readily* lose a quarter of its milk output there is no doubt of the extreme economic importance of this condition. Whenever mastitis occurs it must be looked upon as a herd problem and should be dealt with urgently and thoroughly. Even a single cow may be a warning of a managemental error and should be carefully considered in this respect.

Subclinical mastitis means that the udder is affected mildly but the cow's health or even the milk yield are not obviously affected. It can be recognised by *cell counts*. If an inflammatory process is proceeding, then the milk will contain more white blood cells than normal. The best way for the farmer to deal with this is to arrange for a regular monitoring of the cells in the milk. The range of 'counts', with the significance given, is as follows:

Cell count ranges (cells/ml)	Estimate of mastitis incidence in a herd	Estimate of milk (litres) production loss per cow
Below 250 000	Negligible	—
250 000–499 000	Slight	200
500 000–749 000	Average	350
750 000–999 000	Bad	720
1 000 000 and over	Very bad	900

The milking machine

A major predisposing cause of mastitis is the misapplication of milk machines, together with bad maintenance and poor hygiene routine during milking. One of the most serious of these is overmilking, i.e. the milking machine is left on much longer than it should be causing considerable damage to the delicate mucous membranes lining the teat canal allowing pathogenic micro-organisms to invade so that mastitis results. Teat liners must be chosen of the right size and of gentle fit. They must never be slack and should be carefully maintained to prevent them from causing damage. Pulsations of the machine must be maintained between 40 and 60/min; if the rate increases squeezing may fail and mastitis is more likely to occur. The ratio between the release and squeeze phase should be between 2:1 and 1:1.

The pump needs to be maintained properly as it 'ages' or a surge of milk may result around the teats, such circumstances also being a predisposing cause of mastitis. The vacuum regulator prevents fluctuation in vacuum which can be very damaging to the udder and teats for, if it fails, a surge of milk results around the teat and clusters may fall off or, if it rises, it damages the teats causing extrusion of the lining of the teat canal.

Types of mastitis

The easily recognisable forms of clinical mastitis are the *acute, sub-acute* and *chronic* forms.

Acute mastitis The cow is obviously ill, feverish, fast breathing, depressed and there is no cudding. Examination of the udder will show that one or more quarters are tense, swollen, painful and possibly discoloured (blue). It may be possible to withdraw from the affected quarter(s) a small quantity of grossly abnormal fluid. Gangrene may set in causing the destruction of part of the udder. This 'acute' mastitis is often the same as 'summer mastitis' which, whilst it is most common in dry cows and heifers in the summer months, can occur in cows at other times. Such mastitis is caused primarily by a bacterium, *Corynebacterium pyogenes*, and must be treated vigorously with antibiotics and sulpha drugs. Because of its common association with 'dry' cows its incidence *may* be reduced by using long-acting antibiotic therapy as soon as the cow is dried off.

Sub-acute mastitis The symptoms are somewhat similar but milder and slower in their progress. The first sign of disease may be the appearance of small clots in the milk with some increased difficulty in extracting it. Pain also gradually increases and the affected quarter(s) swell. The milk eventually becomes yellowish and much decreased in amount.

Chronic mastitis In this form of mastitis there is no pain and any general sickness in the cow is unlikely but there is a gradual hardening of the udder tissue and a decrease in the amount of milk.

Causes of mastitis

At one time the most frequently occurring bacteria was *Streptococcus agalactiae* but this has generally decreased in incidence with the advent of successful antibiotic treatment. Other streptococci, *dysgalactia* and *uberi* are common and also *Staphylococcus* and *Corynebacterium*. There are, however, some serious new forms of mastitis that are of comparatively recent origin and are most difficult to treat.

Escherichia coli is usually associated with the so-called environmental mastitis. Also *Pseudomonas* is a common bacterial cause of mastitis and *Corynebacterium bovis*. Mastitis may also be caused by micro-organisms other than bacteria, such as mycoplasma (*M. agalactiae*) and moulds, the latter being termed mycotic mastitis.

Control of mastitis

The view is now taken that the control of mastitis must depend largely on the application of hygienic criteria of the highest standard. These methods are summarised in *Table 19.3*.

Table 19.3 Summary of measures aimed to prevent mastitis in the dairy cow

Basic design of cow accommodation must be right to ensure cow's bed is dry and clean and minimum chance of injury to udder.

Replace fouled bedding frequently in stalls, yard or cubicles.

Test milking machine regularly, check vacuum pressures, pulsation rates, air bleeds and liners daily

Monitor cell counts in milk. Keep detailed records for frequent reference

Teat-dip always to be used after milking. Use a sanitising mixture with an emollient such as lanolin.

Wash udders before milking with clean running water and dry with disposable towels. Milker should preferably wear smooth rubber gloves.

Ensure early diagnosis by use of 'fore-milk' cup

Treat all cases of mastitis promptly.

Cull chronic cases of mastitis.

Correct laboratory diagnosis of causal organisms assists in taking proper remedial measures.

Treat cows as they dry-off with a long-acting antibiotic.

Disease problems associated with breeding in cattle

Lowered fertility results in lowered production and causes great economic loss. However, infertility is not a disease as such, it is only the symptom of one. It may be due to any condition that prevents a vigorous sperm from fertilising a healthy ovum to produce a robust calf. The process involved is complicated and the different causes of fertility are numerous. Consider the whole physiological process. The bull must produce healthy sperm only but the cow must not only produce a healthy egg but also provide the ideal condition for the fertilisation of the ovum by the sperm and the establishment of the embryo in the uterus. The cow then has to provide the nutrients for the calf and then, when it is mature, she must expel the calf from the uterus.

Infertility has probably become more serious in recent years because it tends to be associated with cows giving an immensely good milk yield.

Causes of infertility

Fundamentally the breeding potential of an animal depends on certain inherited genetic tendencies and the presence of any inherited abnormalities. Such genes may either cause the animals to be infertile or to produce faulty calves. However, some of the most important causes of infertility are bacterial or other infectious agents. Such an agent may be a specific organism producing a condition of which infertility is only one symptom. An example of this is contagious abortion (brucellosis), a disease which has now been virtually eradicated from the UK. Another example is trichomoniasis, the causal organism being *Trichomonas foetus*, which is spread at service by infected bulls which themselves show no obvious evidence of infection. A third example, vibriosis, is due to the organism *Vibrio foetus* and is carried and transmitted by apparently normal bulls. Both trichomonad and vibrio organisms multiply in the uterus and cause the death and abortion of the fetus. Infertility may also be caused by non-specific infections of the genital tract.

Control   When a cow or heifer fails to conceive to a second service it should be examined by a veterinary surgeon who, by carrying out tests, can establish if there is an infectious cause. If there is, it may well be treatable. Prevention depends on a number of factors. The herd should be kept as self-contained as possible and only maiden heifers and unused bulls should come into the herd. There should be no exchange of breeding animals. The diet must be adequate and balanced.

Diseases affecting cattle of all ages

Foot-and-mouth disease

Foot-and-mouth disease is an acutely contagious disease which causes fever in cattle and all cloven hoofed animals, followed by the development of blisters or vesicles, which arise chiefly on the mouth and feet. The disease is an infection, world-wide in incidence, caused by at least seven major variants of the foot-and-mouth virus. Each type produces the same symptoms and they can only be identified by specialist examination. The disease is notoriously contagious, perhaps more so than any other disease, and it is known that it can spread some 50 miles down-wind from one outbreak to another. It has a fairly short incubation period of about 3–6 d so it can spread rapidly throughout a susceptible population.

The disease does not usually cause deaths, except in the young, but it leads to damaging after-effects. It will seriously lower productivity in all animals, especially the milking cow. Mastitis will possibly develop so milk production is permanently impaired and the infection of the hooves may lead to lameness that leads to secondary infections.

Infection with the virus of foot-and-mouth disease can occur in many ways other than by wind-borne spread. People, lorries, wild birds and other animals, wild life such as hedgehogs, markets and so on are just some of the ways in which the disease can spread. In addition, infective material can come in from abroad in imported substances, such as hay, and meat and bones which have not been sterilised.

In this country foot-and-mouth disease is a notifiable disease and has long been dealt with by immediate slaughter of infected animals. This has been a very successful policy and has meant there have been few outbreaks. In fact, the last big outbreak occurred in 1967 and there have been just two small infections since then. The important thing is to be sure of recognising a case as soon as possible so all the official measures can be taken to control the disease. Symptoms that would lead one to suspect foot-and-mouth disease are described below.

Infected cows suffer from fever and anorexia with a sudden drop in milk yield and there are blisters on the upper surface of the tongue and the balls of the heels. Animals prefer to lie down but when they are forced to walk they move painfully, occasionally shaking their feet. The lameness then becomes worse so the animal can barely move. Blisters also develop on the teats so that the cow cannot be milked.

Rabies

Rabies is a notifiable virus disease which affects nearly all animals and is especially dangerous to humans. It causes progressive paralysis and madness in most animals. The virus is present in the saliva so that the animals affected may bite one another, or humans, and in this way cause infection to spread. It is nearly always fatal. The greatest danger presented by this disease is that infection becomes endemic – virtually permanent – in wild animals, particularly foxes but also in badgers, deer, vampire bats, and these infect dogs, cats and farm animals. Thus is the particular danger to humans created.

The disease may be prevented in humans and animals by vaccination but it should be emphasised that no vaccinations are 100% effective and it is difficult to exaggerate the ghastly effects of this disease in the human.

One of the features of this virus is that after an animal becomes infected, perhaps via the saliva by being bitten by another rabid animal, several months may elapse before any signs of the disease are seen.

Symptoms

Symptoms in animals may show great variation, ranging from the classic mad, biting, salivating beast, to 'dumb' forms in which the animals are incoordinate, progressively paralysed and make no noise. Usually the 'dumb' form follows the mad stage.

Control

The British Isles have been free of rabies outside quarantine for 60 years, other than two cases in dogs which were quickly dealt with. Our rigid regulations prohibiting the importation of dogs and cats and certain other mammals without a six month quarantine period which includes vaccination is thoroughly justified. The fear remains that animals incubating the disease may be smuggled in, or could jump off a boat, and thereby rabies could become endemic in the wild animal population. This danger has increased with the spread of rabies in areas of Europe near the UK. In those countries where the disease is endemic, great strides are now possible towards eradicating rabies with the aid of the greatly improved vaccines now available.

Tuberculosis in cattle

Tuberculosis is a chronic infectious disease affecting virtually all species of animal and also humans and birds and is due to a bacterium, *Mycobacterium tuberculosis*. There are three main strains of this organism. There is the human strain which affects primarily the human but can also affect cattle; the cattle (bovine) strain which is most prevalent in cattle but also affects humans, pigs and certain types of wild life; and finally we must be aware of the avian strains which primarily affect birds but also cattle, pigs and other animals at times. All types of cattle may be affected but it is in the dairy cow that the main risk occurs. The milk from an infected dairy cow may contain the organism that causes infection and this may infect calves or human subjects if the milk is not pasteurised or sterilised. Also, in advanced cases of tuberculosis, the uterus may become infected so that the calf is born with the disease.

Tuberculosis is essentially a slow, inflammatory action which produces almost anywhere in the organs of the body nodular swellings, known as tubercles, which are of fibrous tissue with a core of pus-like caseous (cheesy) material. Fortunately it is possible to test an animal for the presence of the disease long before it has reached the contagious stage.

Tuberculosis has been virtually eradicated from the UK though testing continues and some cases do occur. Animals may be infected by humans and wild life and there is, for example, much concern about the infection of cattle by badgers which, in some areas, are serious reservoirs of the disease. Those same measures that have generally been successful in eliminating tuberculosis in the UK are also being applied in a similar fashion elsewhere.

Anthrax

Anthrax is a serious bacterial disease which infects cattle, other animals and humans. It is rare to see symptoms in cattle as infection is so acute that it usually causes sudden death. The cause of the disease is a bacterium, *Bacillus anthracis*, which may persist in the soil of a farm, once present, almost indefinitely. However, few cases occur in cattle in the UK, around 50 a year, because of the means taken to prevent its spread and re-infection.

Anthrax should be suspected if an animal was found dead or there was a very high fever and a swollen neck. At either point professional help should be sought. Diagnosis would be made by the veterinary surgeon taking a very small quantity of blood and examining this under the microscope. If the animal died from anthrax, or is infected with it, the blood will be teeming with the rod-shaped organisms that are distinctive features of this disease.

A major warning: do not allow the carcass to be cut or air to reach the blood since it is then that the organisms form spores which can survive for many years.

Some blood may be oozing from parts of the animal and this must be treated with disinfectant. Anthrax is a notifiable disease and measures that must be taken are mandatory. The carcass must be burned or buried deeply so that it is removed from any opportunity to cause re-infection or for the organisms to form spores. The area around which the animal was found dead should be disinfected and isolated for a short while in case of spillage of other organisms. It is most likely that the source of infection is from a feedstuff, possibly meat or bone meal, which has come from imported infected material.

Enzootic bovine leucosis (EBL)

This is a comparative newcomer to the list of notifiable diseases. It is a virus disease, slow and insidious in its effect, which may have been imported with Canadian Holstein cattle. It causes multiple tumours, known as lympho-sarcomas, and whilst it is certainly not widely present in the UK, its presence has been confirmed. The symptoms in the live and adult animal are chronic ill health, anaemia, weakness and inappetance, which are not diagnostic. Only special laboratory examinations will give a definite diagnosis.

External parasites

Warble fly

There are two species of cattle warble fly that cause trouble, *Hypoderma lineatum* and *Hypoderma bovis*. Both are rather like bumble bees in appearance as their dark bodies are covered with white, yellow and black hairs. They have no mouths and only live for a few days. They have been found almost everywhere in the British Isles and most other parts of the world.

Adult female flies cause substantial losses to cattlemen, being responsible for 'gadding' during the spring and summer when the cattle can be so panicked by the approach of the flies that they may stampede in the fields. Such stampeding lowers milk yield and may cause serious injuries to the beasts and even broken limbs. There is also hide damage and this is by far the worst loss. The maggots of the fly make breathing holes and these can ruin parts of the hide, even after the maggot has left the hide and the hole has healed over with fibrous tissue: this remains a weak area in the leather so that it has a much reduced value. It also damages the meat itself as a yellowish jelly-like substance forms around each maggot and spoils the appearance of the meat.

The life-cycle of the warble fly is as follows. During the short life of the fly the female may lay some 50 or so eggs which are attached to hair on the legs and bellies of cattle, those of *H. bovis* being laid singly (like louse eggs) and those of *H. lineatum* in rows of up to about 20. In 4–5 d the eggs hatch and tiny maggots emerge, crawl down the hair and burrow through the skin of the beast by means of sharp mouth hooks.

The entry of the maggot causes some injury and it shows as a small pimple or scab. Under the skin the maggot moves along, feeding and growing. It takes seven to eight months to reach the skin of the back of the animal. When this point is reached the maggots make breathing holes and become isolated within a small abscess. It stays there for five to seven weeks, casting its skin twice and growing to a length of about 1 inch. The full grown maggot is dark brown, fleshy, barrel-shaped and covered with groups of tiny spines. When fully ripe they squeeze themselves out of their 'warbles' and fall to the ground. Here they slide into shallow crevices in the soil and pass underneath dead vegetation. The skin of the maggot darkens and it then becomes a puparium or chrysalis inside which the warble fly is formed. This usually takes about four weeks.

Control must be achieved by dealing with the maggot. These can be destroyed by treating the animal with one of the more recently produced systemic insecticides.

Warble fly is a notifiable disease and any suspicions must be reported to the divisional veterinary officer of the ministry who will supervise treatment of all cattle over 12 weeks old on the premises. The disease is nearly eradicated from the UK and only 34 cases occurred in 1986. Great care is needed in the use of the systemic insecticides as they are organophosphorus compounds and are given by pouring them on the backs of the animals or by injection. Sick animals must not be treated. Milking cows must be treated immediately after milking to allow at least 6 h before the next milking. The approved 'withdrawal' period – which is on the label of the product used – will specify that no animal may be slaughtered for meat under 14–21 d from treatment.

Lice

Lice infestation is a major problem only in cattle when they are housed. The thick winter coat, lack of sufficient air and sunlight and possibly poor nutrition can contribute to the occurrence of this problem. There are four species of lice infesting cattle, three being sucking lice and the fourth, which is also the commonest, a biting louse. The sucking lice are *Haematopinus eurysternus*, *Linognathus vituli* and *Solenoptes capillatus*, and the biting louse is *Damalinia bovis*. The sucking lice cause the greater damage and irritation as their sharp mouth parts pierce the skin to draw blood on which they feed. The biting louse feeds on the scales of the skin and on the discharge from existing small wounds.

These four types of louse affect only cattle and cannot exist on other farm animals. The entire life-cycle takes place on the skin of the animals where the females lay eggs in small groups. They are attached to the base of the hair by a sticky secretion. There the eggs remain for up to three weeks after which the young live hatch out and very quickly start feeding and reproducing. Reproduction lasts for about five weeks; each female is able to lay about 24 eggs and if conditions are favourable the number of lice can increase very rapidly. Lice cannot survive for more than 3–4 d away from cattle so that transfer of infections is largely by direct contact.

It is not too difficult to recognise an infestation with lice. They tend to congregate on the animal's shoulders, base of the neck, head and root of the tail and examination of these areas will show them to be extremely scurfy and the lice should be obvious. As the infestation worsens more of the body can become affected. The affected animals rub and scratch themselves greatly adding to the damage. The hair goes altogether in some areas. Rubbing thickens the skin which often breaks, so that bacterial infection gets in. All the effects of a bad infestation are ultimately to cause much loss of condition and production and should not be tolerated.

Insecticides must be applied to kill the lice and materials that are satisfactory include organophosphorus preparations. Liquid preparations can be applied either under pressure or with a scrubbing brush. As these preparations have a reasonable residual effect, it is feasible for one good application to kill the lice and the larvae that hatch out after the application.

After treatment, measures must be taken to prevent re-infestation by contact with untreated animals and also by putting in fresh bedding and probably cleaning equipment.

Never under-estimate the damage that can be done by lice both as mechanical irritators and as vectors of any livestock disease.

Foul-in-the-foot

Foul-in-the-foot is an infection with *Actinomyces necrophorus* between the claws of the feet in cattle causing pus formation and necrosis. The organisms enter only through the broken skin, often caused by sharp stones. Cattle become lame and productivity can be severely affected. Treatment consists of trimming, cleaning and dressing the affected feet and giving an injection of a sulpha drug or antibiotic. As a routine preventive where the infection is widespread in a herd the cattle may be routinely walked through a foot-bath containing 5–10% copper sulphate or 5% formalin.

A vaccine is now available for sheep with foot-rot and it is expected to be available soon for cattle with foul-in-the-foot.

THE HEALTH OF PIGS

The sow and piglets around farrowing

Farrowing fever (metritis, mastitis, agalactia syndrome or MMA)

This is a complex condition that occurs in sows during the period of farrowing and which seriously affects the viability of the piglets and the health of the sows. The sow runs an elevated temperature (about 40.5–42°C) and the udder is often hard and tender. There will be little or no milk 'let-down' ('agalactia'), and there is often inflammation of the uterus ('metritis'), which may be associated with retained afterbirths.

It is usually possible to cure the disease by immediate administration of broad spectrum antibiotic injections such as tetracycline or ampicillin, and posterior pituitary hormone (oxytocin) to 'let-down' the milk.

Transmissible gastroenteritis (TGE)

An alarming disease of pigs caused by a virus which causes a severe diarrhoea with many very young piglets rapidly dying of dehydration. Some piglets vomit; all tend to have a very inflamed stomach and intestines. The sows may also be ill and have little milk. Older piglets, above about three to four weeks, will also be affected to varying degrees and for a period will scour and make little progress but should recover. The only treatment is to administer fluids, provide good nursing and inject antibiotics to reduce the likelihood of secondary bacterial infections. (Also *see* p. 488 for general advice against viral infections.)

Piglet anaemia

Every pig which is born in intensive housing requires some extra administration of iron if it is to be ensured that it does not become anaemic. The piglet is born with limited reserves of iron in its liver, iron being an essential constituent of haemoglobin, the oxygen carrying element in the red blood cells. Some supplementary source of assimilable iron must be given. The preferred way of giving the iron is by injection of 200 mg of a soluble iron preparation at not more than 3 d of age and this is sufficient to carry the piglet through to the time it is eating solid food which should be liberally supplemented with iron.

Escherichia coli infection

There are various ways in which this ubiquitous cause of trouble on the livestock farm can do damage. In the young piglet, within the first few days of life, *E. coli* may be a cause of an acute septicaemic condition. Within only a few days of birth a number of pigs in the litter can become very ill with no particular symptoms except extreme illness leading to death.

Neonatal diarrhoea

Those pigs that are infected with *E. coli* but less seriously than are the piglets killed with the septicaemic form of the disease show a most serious profuse watery diarrhoea.

Milk scours and post-weaning diarrhoea

Another form of diarrhoea due to *E. coli* but occurring later is that known as 'milk scours'. A pale, greyish scour affects piglets in the period between one to three weeks of age.

There are many strains of *E. coli* that cause these symptoms and almost certainly many of these strains will be living in perfect harmony within the intestines of the pig and will generally be present around the pigs' quarters. The *E. coli* appear to be able to multiply if there are certain stresses in the management.

Treatment and prevention

The treatment of *E. coli* infections is best achieved either by injections or medication incorporated in the feed or water. There are many antibiotics that can be used for this purpose, such as chlortetracycline, oxytetracycline, streptomycin, amoxycillin, lincospectin or the chemotherapeutic trimethoprim sulphonamide. Orally nitrofurans, framomycin or neomycin may be given. The recovery of pigs suffering with *E. coli* diarrhoea will also be helped by administering fluids fortified with minerals, electrolytes and vitamins.

Prevention may be achieved by using certain antibiotics in the feed. There are a number of alternatives to choose from including furazolidone, organic arsenicals, chlortetracycline or carbadox. An alternative approach is to attempt to create an active or passive immunity by the use of serum or vaccine. The former course makes use of *E. coli* hyperimmune sera. In the case of the vaccines, *E. coli* vaccine may be given to the sow during pregnancy.

Diseases of young and growing pigs

Bowel oedema

Also caused by *E. coli*, the condition invariably occurs about 10 d after weaning, The first indication of an outbreak is usually the appearance, in a pen of weaners, of one dead pig, often the biggest of the bunch. Others may show 'nervous trouble', staggering, incoordination, blindness, loss of balance, falling about when roused. As the disease progresses in these pigs they will lie on their side and paddle their legs and they then go into a coma and die within 1 d. In bowel oedema such pigs usually show swollen (oedematous) eyelids, nose and ears and may have a moist squeal resembling a gurgle. A post mortem, if carried out speedily, shows oedema of the stomach and in the folds of the colon and in the larynx.

The occurrence of this disease seems to be associated with

the stress of weaning, with the addition of *ad libitum* dry feeding.

Treatment is rarely successful. To prevent the spread of the disease all dry foods should be withdrawn and replaced with a limited wet diet.

Streptococcal infections

In recent years streptococcal infections of young piglets have beome very common. *Streptococcus suis Type 1* infects the sow and she then transfers the infection to the piglets soon after birth. The organisms, after getting into the piglet, may cause a general infection.

The disease usually occurs at about 10 d of age. Affected pigs run a temperature, show painful arthritis, have muscular tremors, appear blind and cannot coordinate their movements. Some die suddenly with heart inflammation. Typically, some 20–30% of a litter are affected.

Affected pigs may be treated successfully with injections of suitable antibiotics such as penicillin or with sulphonamides such as trimethoprim.

The same organism, *Streptococcus suis Type 1*, can also cause serious disease in older pigs. Here, what seems to happen is that young pigs, already carriers of this disease, are mixed with older pigs around the early fattening stages, say 8–12 weeks. The carriers infect the pigs they are mixed with and after a few days of incubation the affected pigs show similar symptoms as the younger pigs.

Clostridial infections

Whilst the effects of clostridial infections on pigs are in no way comparable with those which affect sheep or even cattle, there are problems from time to time, particularly in outdoor reared pigs on 'pig-sick' land. For example, *Clostridium welchii (perfringens)* can produce a fatal haemorrhagic enteritis in pigs up to about a week old. The disease has a very dramatic effect, with profuse dysentery, becoming dark from the profusion of blood within it. Usually pigs so affected die very rapidly though a few take a more chronic turn. Most die within 24 h of the symptoms being noted. The litter affected should be treated with an antibiotic such as ampicillin, and those under risk of infection can be given antiserum. A more permanent answer may be needed and the sows can be given a vaccine for the infection, this being administered twice during pregnancy.

Piglets are also infected with *Clostridium tetani* in the same way as lambs by penetration of the organism from dirty conditions, possibly through the navel or by the injury caused during castration.

Greasy pig disease (exudative epidermitis; Marmite disease)

This is an acute skin inflammation of pigs two to eight weeks old with the production of excessive amounts of sebaceous secretion and exudation from the skin which does not cause irritation. It tends to be most common in hot weather.

For treatment inject a broad spectrum antibiotic such as ampicillin or lincomycin. It is vital then to make sure that the cause is removed, in particular the best procedure is to provide plenty of soft bedding.

Swine influenza

The swine influenza virus causes serious losses amongst pigs in all parts of the world. In the normal way, if the animals are well housed and are not infected with other respiratory invaders, the course of the disease is acute but the animals quickly recover.

Atrophic rhinitis

Atrophic rhinitis is a respiratory disease caused principally by a bacterium, *Bordetella bronchiseptica* and other organisms, notably a virus – the inclusion body rhinitis virus. Atrophic rhinitis affects primarily the membranes lining the delicate bones of the nose and leads to inflammation, nose bleeding, sneezing and eventually general degeneration of the bones of the snout. This does profound damage to the growth of the pig and it also makes the animal more susceptible to enzootic pneumonia and any infections of the respiratory tract.

In order to prevent the condition a vaccine may be used or continuous preventive medication may be instituted with a suitable antibiotic, such as Tylan, or a chemotherapeutic drug such as a sulphonamide. It is better to improve the environment and especially ventilation and management than to rely on measures of vaccination and medication.

Diseases largely of fatteners

Salmonellosis

All forms of salmonellosis can occur in pigs. Symptoms vary tremendously depending on the type and virulence of the salmonellae. The worst outbreaks are usually after weaning, in large groups of 'stores'. This often shows itself in the septicaemic form, causing sudden death or acute fever with blue (cyanotic) discoloration of the ears and limbs. Another form of disease shows itself as an acute diarrhoea, also with a fever, but these are not all the signs. Pneumonia may occur with abnormal breathing, and also nervous abnormalities, such as incoordination and paralysis. Also the skin is often affected and parts may even 'slough-off' later in the disease.

Treatment will require the use of antibiotics with known activity against salmonella. Salmonellosis is a notifiable disease under the Zoonoses Order (*see* p. 474).

The only form of salmonella that can be prevented by vaccination is *S. cholerae suis*.

Enzootic pneumonia

One of the most serious scourges and causes of economic loss on the pig farm is the disease of pneumonia caused by a species of mycoplasma *M. hyopneumoniae*. Though this is recognised as the principal cause of the disease, other organisms are common secondary invaders, including other types of mycoplasma and bacteria, such as *Pasteurella multocida* and *Bordetella bronchiseptica*. The lesions of enzootic pneumonia tend to be in the anterior lobe of the lungs which become consolidated (solid) and grey in colour.

Treatment should be instituted in pigs showing serious clinical signs. Broad spectrum antibiotics, such as chlortetracycline and oxytetracycline, may be injected but may not have a permanent beneficial effect as no immunity follows an attack.

Whilst the disease may kill a proportion of pigs its most serious economic effect is to lower the growth rate and food conversion efficiency of nearly all the animals in the building. The longer-term measures to deal with the disease are to improve the housing and environment and, or, to consider re-stocking with enzootic pneumonia-free pigs.

Swine dysentery ('bloody scours')

This is a disease largely of fattening pigs and the infection causes inflammation of the intestines which leads to dysentery (diarrhoea together with some blood effusions), and a general debilitation of the pigs. It is highly contagious, being spread via contact with infected faeces.

The main causal organism involved in swine dysentery is a bacterium, *Treponema hyodysenteriae*, though it is not the only one involved and others such as *Campylobacter*, *E. coli* and salmonellae of various species may always be present as secondaries.

Treatment

There are a number of drugs which may be given via the food or water such as the macralides, in the form of tylosin, erythromycin or spiramycin or lincomycin, which may be used as a feed additive or in combination with spectinomycin may be given in water. Further effective drugs are tiamulin and dimetridiazole.

Diseases of pigs largely without age incidence

Swine erysipelas ('The diamonds')

This disease is caused by a bacterium known as *Erysipelothrix insidiosa* or *E. rhusiopathiae*. Most pigs are under risk of this disease because the organism can live in and around piggeries and 'pig-sick' land for very many years in the sporulated form. In the peracute form there may be hardly any signs at all, and the pig may be found dead after a very brief period of severe illness. In the acute form the pig is also ill, has a high temperature and the skin is inflamed: it will not eat. Then there is a sub-acute form which is the more typical disease, showing characteristic discoloration of the skin with raised purplish areas said to be roughly diamond shaped throughout the back and on the flanks and belly. Finally, there are two 'chronic' forms of the disease. In one the joints are affected which causes the pig severe lameness, whilst in the other the organism causes erosion of the valves of the heart, leading to cauliflower-like growths, so that the pig will often show signs of heart dysfunction or even heart failure and death. A feature of swine erysipelas is that it tends to affect pigs during hot and muggy weather conditions.

The intestinal haemorrhage syndrome ('bloody gut')

This condition, often known as 'bloody gut', affects fatteners in the latter stages of growth and after a short period of depression, with an appearance of paleness and an enlarged abdomen, they die rapidly. Postmortem shows the small intestine full of blood stained fluid and also gas in the large intestine. There may also be a twisting of the gut. The condition is more particularly common in whey fed pigs, but not exclusively so, yet its cause remains a mystery.

Mulberry heart disease

A curious disease of fatteners causing sudden death or the pigs become depressed and weak, collapse and die within about 24 h of the onset of the disease. Few recover. On postmortem examination the signs are of an enlarged and mottled liver and the surface of the heart is streaked with haemorrhages running longitudinally and also occurring in the endocardium. The cause is not really known but there is a suggestion that a deficiency of vitamin E or selenium may be involved. There is little evidence that treatment does much good but the use of multi-vitamin injections and broad spectrum antibiotics may be of some benefit.

Swine fever (hog cholera)

A very serious virus disease now eliminated from the UK. Affected pigs show many symptoms, especially a very high fever, great depression and discoloration of the skin and usually diarrhoea of a particularly foetid type. Pigs also have a depraved appetite and thirst. It is a highly contagious disease. If any cases occurred in the UK the herd would be slaughtered.

Aujeszky's disease

This is a herpes virus which leads to nervous and respiratory symptoms with a fever. Mortality can be high in young pigs. It is a notifiable disease in the UK and affected herds are slaughtered.

Tuberculosis in pigs

Three forms of tuberculosis can infect the pig, the bovine, avian and human forms. Because of the much reduced incidence of tuberculosis in cattle and man, there is now very little in pigs but it does still exist as a clinical entity. Of the cases that occur, most are of the avian type doubtless infected from wild birds since tuberculosis is virtually non-existent in the domestic fowl.

The symptoms are rather non-specific and would rarely be suspected by the pigman. Loss of weight, coughing and discharge from the nose will hardly lead to suspicion so that it is unusual for the disease to be recognised before it is seen in the slaughterhouse by the occurrence of swollen and infected lymph nodes in the throat, chest and abdomen. If samples are taken the disease can then be diagnosed positively.

Treatment is rarely called for but careful note should be given to any lessons there may be in isolating stock from sources of infection and also in improving the hygiene of the premises. The organisms that cause tuberculosis in all species are very persistent and resist destruction by most means; it is therefore very important to take such measures that eliminate them. It is also pertinent to stress the risk to the human population and whilst it is not a highly contagious agent of disease, it could represent a very undesirable challenge to man if the infection became widespread in a pig herd.

A group of important virus infections affecting pigs of various ages

Pig enterovirus infections, including 'SMEDI'

A number of enteroviruses cause stillbirths, mummification, embryonic death and infertility – the initials SMEDI – and hence this name is given to this group of diseases.

Pig enteroviruses may also be associated with nervous symptoms such as incoordination, followed by stiffness, tremors and convulsions. The acute form of this disease, known as Teschen disease, is probably not present in the UK but it is a notifiable disease and would be dealt with by a slaughter policy to attempt to eradicate it. The mild form, known as Talfan disease, is certainly present in the UK but no special control policies are instituted and reliance is made on the pigs developing a natural immunity, as with many other virus diseases of rather mild and uncertain symptoms.

Rotavirus infection

This causes very severe scouring in piglets which can lead to quite a heavy mortality. The symptoms are anorexia, vomiting, diarrhoea, yellow or dark grey in colour, and very profuse. The diarrhoea causes rapid dehydration, which can kill, otherwise the pigs can recover in about 7–10 d.

These clinical signs are very similar to other conditions of young pigs causing profuse diarrhoea, such as TGE and *E. coli* infection and only a laboratory diagnosis which identifies the virus will confirm the cause.

Vomiting and wasting disease

This is a disease of the newborn pig caused by a coronavirus similar to the TGE virus. Affected pigs firstly vomit, huddle together with general illness, are depressed, run a high temperature and show little interest in suckling. Only a proportion of piglets in a litter are affected and only a proportion of all litters may be affected at all.

Epidemic diarrhoea

This is a very contagious disease of pigs caused by a virus which produces a profuse diarrhoea, vomiting, wasting and inappetance. The disease is similar to TGE but affects largely the older pigs; younger pigs are affected very much less or not at all.

Parvovirus infection

Parvovirus infection of pigs is a cause of infertility, stillbirths, small litters and mummification. It is a relatively new condition that has been recognised in a number of large units. It tends to be most commonly seen in young gilts or newly introduced pigs which have no resistance to the 'local' infection. It may also infect boars which then act as spreaders of infection. It appears to be far more of a problem in those housing systems where the sows are kept as individuals (as in tethering or sow stalls) and where the lack of contact fails to produce a passing infection and then resistance. In this type of condition there is no treatment that is of any real use. Licensed vaccines have been produced in the UK. An alternative procedure is to 'infect' a young gilt before service and the most effective way is to 'feed' homogenised placenta from known infected cases. This is, however, a crude method of 'vaccination' and has great dangers of spreading other infections indiscriminately.

General control measures for virus infections when no vaccines are available

The upsurge in the number of virus infections in pig herds in recent years is worrying and provides us with lessons of great importance. They appear to be very similar to those conditions that have affected poultry but the pig industry is much more vulnerable. It tends to be more careless about hygiene, has no general policy of depopulation of sites and moves pigs around during their lifetime much more generally than is the case with poultry.

To minimise the effects, the following preventive measures should be considered:

(1) Introduce a minimum of new stock from outside the unit.
(2) Depopulate the housing of young pigs as frequently as possible.
(3) Keep adult breeders closely but cleanly housed.
(4) Do not use sow stalls or tethers but kennels or yards which group the gilts, sows and boars.
(5) Limit the size of each self-contained unit to a minimum. This will ensure speedy effects of any virulent viruses in the area.
(6) Practise careful isolation of the site in terms of feed deliveries, visitors, collecting lorries for fat pigs and any other potential danger.

External parasites

Pigs can be quite badly infected with lice (*Haematopinus sius*), the mange mite (*Sarcoptes scabiei var. suis*) and the stable fly (*Stomoyx calcitrans*).

Lice

Lice tend to be most common on the folds of the skin of the neck, around the base of the ears, on the insides of the legs and on the flanks. The constant irritation causes the pig to rub and scratch and this reduces growth and food conversion efficiency. Also the lice may be vectors of other disease agents.

Mange

Mange is an infinitely more worrying problem than lice. The parasite which is most common in the pig (*Sarcoptes scabiei var. suis*) burrows in the skin and lives in so-called 'galleries'. The mange parasites are about 0.5 mm long and lay their eggs in the galleries and develop through larval and two nymphal stages to adults within a period of about 15 d. Whilst the parasites can only multiply on the pig, the mites can survive up to two to three weeks in piggeries.

Ascariasis

Ascaris lumbricoides is the common large roundworm that lives in the small intestine. Some can achieve a length of up to $1\frac{1}{2}$ ft (450 mm). Infection can be so bad that the intestines are literally blocked with large numbers of these worms. The life-cycle is direct and after the eggs are laid and ingested by pigs kept under unhygienic conditions, the eggs hatch out and the larvae, in their development to the adult stage, which is always in the intestine, actually pass through the lungs and liver. Coughing and pneumonia result from the migration of the larvae in the lungs and tracking of the larvae through the liver also causes damage and leaves small white areas known as 'milk spots'. Treatment of the pigs with anthelmintics will eliminate the worms and improved hygiene can destroy the eggs.

Other intestinal worms of pigs occur in various parts of the intestine. There are the two stomach worms *Ostertagia* and *Hyostrongylus rubidus* and the worm *Oesophagostomum*, which lives in the caecum and colon.

Finally there is a lungworm of importance known as *Metastrongylus*. The adult worm lives in the bronchioles of the lungs. In this case the life-cycle is indirect, infection being caused by ingestion of the earthworm which is the intermediate host of the larvae of the *Metastrongylus*. The lungworms undoubtedly cause damage to the lungs and pneumonia, but it is chiefly in the younger pigs that the symptoms are serious. Older pigs are not usually adversely affected but do remain as carriers of the infection.

Treatments can be carried out by in feed medication or by injections using products such as fenbendazole (Panacur, Hoechst), tetramisole (Nilverm, Coopers), thiabendazole and piperazine (Thyrazole, Merck, Sharpe & Dohme) and dichlorvos (Atgard, Squibb, Merck, Sharpe & Dohme).

Pigs can be affected with anthrax (p. 483), foot-and-mouth disease (p. 482), a mild condition but rather similar to foot-and-mouth disease known as swine vesicular disease and rabies (p. 483), but all are only very rare occurrences if at all in the UK. All, however, are notifiable diseases.

THE HEALTH OF SHEEP

Health problems at or near lambing

The first point to stress is the essential need for the highest standard of hygiene at all times. Lambing is the period of greatest risk from infection. Ewes must be carefully observed during lambing to reduce losses due to difficulties in the birth of the lamb (dystocia). If no progress is made within 3 h of the start of lambing the competent shepherd should carry out an examination. He may find he can readily correct any malpresentation or may decide to call in a veterinary surgeon according to his judgement of the position.

E. coli infection (colibacillosis)

There are two types – the enteric and septicaemic forms. Symptoms in the enteric form usually manifest themselves at 1–4 d of age and the lamb becomes depressed, shows profuse diarrhoea or dysentery and dies, usually within 24–36 h of the onset of symptoms. Those affected with the septicaemic form are usually two to six weeks old. Affected lambs have a fever and become stiff and incoordinated in their movements; later they lie down, paddle with their legs and become comatose.

To prevent this disease the ewes should be vaccinated during pregnancy. Treatment is by antiserum or broad spectrum antibiotics.

Navel-ill

This occurs in lambs for the same reason as in calves (*see* p. 474), and requires the identical attention.

Mastitis

Mastitis in the ewe is a very damaging disease which often leads to the complete destruction of the parts affected. Immediate treatment with a broad spectrum antibiotic, given both locally into the teat and by injection, may save the udder and the ewe.

The clostridial diseases

The group of bacteria known as clostridia are spore-forming organisms found universally wherever sheep are kept. As soon as the bacteria pass out of the animal's body they form the tough spores (or capsules) which make them extremely resistant to destruction. Their mode of action is to manufacture toxins which make the animal ill or kill it. Clostridia will multiply only in areas where there is no oxygen and they do this in wounds, especially deep ones, in the intestines of animals, or in organs within the body such as the liver. Clostridial diseases tend to affect animals in very good, thriving conditions. The main clostridial diseases are described below.

Lamb dysentery

Lamb dysentery occurs in lambs during the first few weeks of life and is caused by the organism *Clostridium welchii, Type B*.

The principal symptoms are of bloody diarrhoea (dysentery) and may cause many deaths – up to 30% of the flock if it is unchecked.

Immediate treatment or prevention is by the use of antiserum. To protect from future attacks, the ewe should be given two vaccinations, the last one being about one month before lambing thus transferring a strong immunity to the lambs. In succeeding years ewes will only need one additional 'booster' injection.

Pulpy kidney disease

This is caused by *Clostridium welchii, Type D*. The first sign is often the sudden occurrence of a number of deaths in really good lambs at two to three months of age. The name 'pulpy kidney' is given because on postmortem the kidneys show a high degree of destruction.

Antiserum or vaccination procedures will be similar to lamb dysentery. A major predisposing cause is when sheep are placed on rather too good a pasture and much can be done to prevent trouble by watching their condition and regulating their nutrition.

Blackleg

Caused by *Clostridium chauvoei* is similar to blackquarter in cattle which has been described on p. 477.

Braxy

Caused by *Clostridium septique*. Usually the first sign is the sudden death of some young sheep in good condition on frosty autumn mornings. The predisposing cause is the eating of frosty food which damages the wall of the abomasum and allows the invasion of the clostridial organisms. Vaccination is totally effective as a preventive.

Black disease

The symptoms are sudden death, usually in adult sheep. The organism, *Clostridium oedematiens*, invades the liver after damage by the liverfluke. It usually occurs in the autumn and early winter when flukes migrate in their largest numbers from the intestine to the liver.

The same measures of control are used as in other clostridial diseases. It will also help to prevent black disease if the fluke infestation is controlled.

Tetanus (lockjaw)

Caused by the organism *Clostridium tetani* which finds entry through a wound which heals and the organism multiplies within this anaerobic atmosphere producing the toxin which causes nervous symptoms, spasms of muscular contraction, stiffening of the limbs and eventually death.

Control and prevention is on the same lines as other clostridial diseases. Proper treatment of wounds will greatly reduce the likelihood of tetanus.

The use of vaccines and sera with clostridial infections

All the clostridial diseases can be countered by the use of sera and vaccines – sera for immediate prevention or treatment or vaccines to build up an immunity over a period of weeks. Preparations combining several vaccines together, up to at least eight, covering all the common sheep diseases are now available.

Pneumonia

Acute pneumonia infections are especially serious in winter and autumn and are caused by infection with *Pasteurella* bacteria. The predisposing causes of infection can be intensification, harmful weather conditions, transportation under less than ideal conditions and infestation of the lungs with parasites.

Treatment is by appropriate antibiotics, and prevention by vaccination.

Orf (contagious pustular dermatitis)

This is a virus infection of sheep causing vesicles and scabs on the skin and especially over the mouth and legs. It can also cause an unpleasant disease to sheep handlers. After infection older animals develop an immunity.

Deficiency conditions

Swayback (enzootic ataxia)

Swayback affects the nervous system of young lambs causing incoordination and paralysis of the limbs of the body due to a degeneration of the nerve cells in the brain and spinal column; it is associated with a low level of copper in the blood and tissues of the lambs and ewes. The basic cause of the disease is a deficiency of copper, this element being essential in the formation of enzymes for the construction of nervous tissue. The critical level of copper necessary for the proper development of nervous tissue is 5 ppm and in swayback areas levels of copper are generally well below this. To prevent the disease it is necessary to supplement the copper intake of the pregnant ewe. Adult ewes require an average daily intake of 5–10 mg of copper as an adequate allowance. Mineral block mixtures containing 0.5% copper sulphate are satisfactory, or diet supplementation with 0.2 g of copper sulphate fed weekly to each sheep would be adequate.

Pine (pining)

Pine is a disease of sheep which leads to wasting and is caused by a deficiency of cobalamin (vitamin B_{12}) due to a lack of sufficient cobalt in the diet.

A simple way of preventing pine is to add a supplement which has been enriched with cobalt to the normal concentrate feed. Animals may also be dosed individually, the cobalt 'bullet' being the best treatment since it is released gradually over a long period. The pasture may also be treated with 2 kg of cobalt sulphate per acre every five years.

Rickets

Rickets occurs in lambs and is a condition of poor bone construction when the calcification fails to take place efficiently. Proper formation of bones is due to a sufficient quantity and balance of the minerals calcium and phosphorus, together with the vitamin D. In practice usually injections of vitamin D are successful in arresting a problem.

Internal parasites of sheep

The three main types of internal parasites that infest sheep are tapeworms, the flatworms represented by the liverfluke, and the roundworms, which infest both the intestines and the lungs.

Tapeworms

Tapeworms require an intermediate host, as well as sheep, for their existence. The adult tapeworm lives in the sheep's intestines, lays eggs which pass out in the dung on to the pasture and further development occurs in the intermediate host, which is a small pasture mite; lambs become infested when they consume the mites accidentally when grazing.

Sheep may also be infested with a number of bladder worms, infestation being picked up on grazing contaminated with the droppings of dogs and foxes which carry the adult tapeworms in their intestines. The bladder worms occur as thinly-walled, fluid-filled bladders among the intestines, in the lung and liver and in the nervous system.

Liverfluke

Liverfluke can be a serious problem in wet years and has been fully described on p. 475. The snail can exist only in areas where the soil is saturated with moisture for considerable periods though it is not found naturally in running water or ponds. In dry periods or in winter, it tends to burrow into the mud and can survive in the inactive state for a considerable time.

Roundworms

At least ten species of roundworms (nematodes) cause parasitic gastroenteritis. Heavy infestations are disastrous but even light ones can cause great loss of productivity. Principal worms are: *Haemonchus contortus; Ostertagia circumcincta; Trichostrongylus axei; Nematodirus spathiger; Trichuris ovis* and *Oesophagostomum columbianum*.

Control is tackled by treatment of the animal to eliminate the worm burden and by pasture and flock management to minimise infection.

Drugs used to control the infection are thiabendazole, levamisole hydrochloride and organophosphorus preparations. Suitable pasture management is based on rotational grazing allowing up to a week on pasture then closing it for three weeks.

Sheep may also be affected by lungworms and the reader is referred to p. 475 for the description of the similar condition in cattle.

External parasites

These include insects, mites and ticks. The easiest to kill are those which spend their whole time on the sheep – the mites, lice and keds. This group is effectively controlled by a single whole-body treatment with a preparation which is lethal to adults and young stages and persistent enough to kill the larvae as they hatch from resistant eggs, or by repeated use of a material which kills the adult only.

The blowfly, in comparison with the other external parasites, completes only one fairly short stage of its development on the sheep and is thus open to attack by an insecticide for only a short time. Larvae hatch rapidly and feed on the skin and tissues for several days, causing great damage. They fall on to the herbage and form the resting stage, known as the pupa, from which the adult fly emerges several weeks later.

The problem of fly-strike requires additional measures to be effective. All cases of scour should be treated and wounds should be dressed immediately.

In the north of England and in Scotland a serious problem has recently emerged, the '*headfly*'. This fly feeds on the fluid from the eyes, nose and mouth of sheep. As a result of the irritation the sheep rub their heads against fences and walls causing wounds which make them even more attractive to the unwelcome attention of the headfly. The result of this disturbance is a very restless animal and considerable economic loss.

The *sheep tick* attaches itself only for a relatively few days during its three year cycle of development. The preparation that is used must therefore be highly active against all the developmental stages and should protect the sheep over a period of about two months.

Sheep scab, the commonly used name for a mange in sheep caused by the psoroptic mange parasite, is a notifiable disease in the UK. This mange parasite can invade all parts of the body that are coverd with wool, and also the ears. The total effect is extremely damaging, which is the reason for the disease being notifiable. Although at one time it was eliminated from sheep in the UK, it has now returned.

The mite in its progress in the skin causes injuries and it feeds on the serum that oozes out; the wounds develop scabs after a time, hence the name 'sheep scab'. The poison produced by the mite causes intense irritation so the sheep rubs and scratches itself endlessly.

Treatment is instituted by a dipping of the sheep with approved materials. Prevention is also by a regular dipping procedure according to statutory regulations.

Infections of mature sheep

Scrapie

Scrapie is believed to be a virus disease with a very long incubation period which occurs naturally only in sheep and goats. The disease affects the nervous system and the sheep becomes incoordinate, develops a severe itchiness and rubs furiously against posts. No treatment is of any use and all sheep that are infected should be slaughtered.

Foot-rot

This is one of the biggest causes of serious economic loss in sheep everywhere. It is primarily caused by a bacterium *Fusiformis nodosus* and spreads from sheep to sheep via infected soil. The harbourers of infection are in fact the sheep with bad feet in which the organism can survive indefinitely. The conditions that favour the spread of infection are wet weather and soil when the feet are softened and the organisms released to invade other feet through any injured point.

Treatment and control has been given in the section on foul-in-the-foot in cattle (p. 485).

Metabolic diseases of sheep

There are three most important metabolic diseases of sheep: pregnancy toxaemia (or twin lamb disease), milk fever (lambing sickness) and hypomagnesaemia (grass staggers).

Pregnancy toxaemia (twin lamb disease)

This is a disease of ewes which affects them in the last few weeks of pregnancy and nearly always in ewes which are carrying more than one lamb. Affected ewes are incoordinate and then totally recumbent and comatose and will invariably die unless treated.

To some extent the condition is due to excessive demands being made on the ewe's metabolism when associated with insufficient nutrition. Affected ewes may have injections of intravenous glucose solutions together with dosing with glycerine and/or glucose solutions. However, most of the effort should go into prevention. The important and indeed essential requirement is that during the last six weeks of pregnancy the ewe is given a diet which is low in fibre, nutritious and easily digested.

Milk fever and hypomagnesaemia

Milk fever (lambing sickness) and hypomagnesaemia (grass staggers) both occur in sheep and are similar to the condition in cattle (*see* p. 480).

Disease causing abortion

Enzootic abortion

This is caused by a chlamydial organism and tends to be common in certain well-defined areas; in the UK it is especially prevalent in NE England and SE Scotland. The infection is usually introduced into a flock by a 'carrier' which releases a great amount of infected material when it aborts. This infects the ewe lambs which will not abort the first season but will abort in the second. After abortion they are then immune but young sheep and brought-in animals will continue to become infected.

There is no specific treatment for infected animals but it is good policy to isolate the ewes which have aborted for the period when they may be discharging infected material. Thereafter a policy of vaccination may need to be instituted.

Salmonellosis

Salmonella abortion is an acute and contagious condition caused by the bacteria *Salmonella abortus ovis*, and can also be caused by the non-specific salmonellae, such as *S. typhimurium* and *S. dublin*.

Control can be instituted, including antibiotics or chemotherapy, vaccination and hygienic measures.

Vibriosis

Vibrionic abortion, caused by *Vibrio foetus var. intestinalis*, is another bacterial cause of abortion which, like salmonella infections, causes late abortions. There is no specific treatment but vaccines are available to control the condition.

Brucella

The *Brucella* species of bacteria can also cause abortion in sheep, the effects being the same as in cattle. *Brucella abortus*, *Brucella mellitensis* and *Brucella ovis* are all capable of causing the problems. After positive diagnosis, control relies on hygiene and vaccination.

Toxoplasmosis

There is also a parasitic cause of abortion in sheep which is due to a toxoplasm, *Toxoplasma gondii*. The best procedure in an infected flock is to mix infected ewes with non-pregnant ones to stimulate natural immunity before pregnancy.

The notifiable disease of anthrax and foot-and-mouth disease also occur in sheep and the reader is referred to pp. 483 and 482.

Maedi–Visna

This is a new 'slow' virus infection, also known as ovine progressive pneumonia and characterised by an insidious incurable respiratory disease. It is not a notifiable disease but there is an official M–V Accredited Flocks Scheme which enables farms to set up and maintain sheep known to be free after appropriate blood tests which are continued for member flocks.

Part 3

Farm equipment

20

Services

R.P. Heath

DRAINAGE INSTALLATIONS

Empirical rules developed from historical example ensure satisfactory service (Ref: BS 5502: 1978 Section 3.1). Drainage bye-laws used to express these rules must be observed. Modern services accommodate changes in practice of farming, domestic living and the new materials and installation. Where workers occupy a building for 4 h or more daily, lavatory accommodation is required. Human waste must be disposed of separately. A tank of 0.1 m³/person/d is required (*see* CP 301 and BS 5502). Problems such as rodent attack on plastics pipes under concrete aprons provide a new challenge to installers. Economical use of pipe must be matched to the accommodation of hydraulic pressures, the need for human access and ventilation. Increasingly, it is becoming desirable to isolate clean from foul sources.

Building and paved area to 0.5 ha will yield 65 m³ of 'clean' water with a precipitation of 13 mm ($\frac{1}{2}''$). A storm is deemed to be 75 mm/h rain intensity over most of the UK; this is found to occur over a 5-min period once in four years and over a 20-min period once in 50 years. This 'clean' water may be separately stored and used for washing down prior to a more rigorous wash, diluting slurry under controlled conditions or as a buffer source.

For all clean and foul drainage systems a satisfactory standard of materials and workmanship is required. This will include watertight pipes with smooth internal bores; satisfactory junctions which avoid the risk of blockage; and properly aligned joints with no ridges to catch solid matter. All pipes must discharge immediately and thus falls must be continuous.

Rainwater pipes and gutters

Factors other than rain on the roof area will determine the length of gutter to each downspout, which is not normally more than 6 m. The vagaries of fluid flow ensure that full bore flow down a downspout, no matter how full the gutter, never occurs. Gutters should have a slight (1 in 360) fall. In any case the fall should not create a roof finish to gutter space of more than 50 mm. Valley gutters present a particular problem for human access, snow and debris blockage and 'run-off' velocity. CP 308 and Building Research Station Digest Nos. 188 and 189 give size and falls of gutters and associated rainwater goods. Problems of downspouts in stockyards or where vehicle movement occurs can be overcome by building them in. The downspouts should be joint sealed if installed inside a building.

Materials other than asbestos cement or cast iron enable extrusion and fitting continuous lengths to fit. BS 1430 covers aluminium alloy goods.

As a general rule paved, or similar, areas should be drained and should be given a minimum fall of 1 in 60 and preferably away from building walls. Where pressure or steam washing plant is used an open channel of some 0.3 m square section and 3 m long with a shallow fall to a bucket trap, will cause soil and other debris to settle out and be easily removed mechanically rather than block the system.

Snow is not a major problem in the UK. On steep surfaces where roof skin heating and resulting melting and movement of the mass of snow likely, it must be accommodated with gutters and other features built to withstand the snow spill.

Below ground drainage

Discharge from the clean or foul drain to the public sewer is unlikely for most farmsteads. Correct provision to accommodate foul water in a slurry compound is essential from a practical and legal point of view. Drains must be water-tight, not affected by the flows they carry or soil chemicals, able to resist root penetration, able to withstand earth pressures, settlement and earth movement, capable of withstanding thermal and moisture movement and vermin damage. Previously, rigidity of structure was considered important. Now, increasingly, longitudinal and angular flexure is provided on installations with modern materials. Vitrified clay (previously salt glazed) ware, though brittle, is serviceable and now is installed with plastics collars and fittings in many cases. This speeds work and removes the need to caulk and cement hog joints. The relevant standard is BS 65. Pitch

495

fibre provides a durable drain for other than continuous hot or acid flows. Installation needs particular care right through to the latter stages of back filling. BS 2760 applies

Under buildings and through structural walls cast iron may give the structural strength and integrity of flow required. In any case rotational and translational loading from the wall must be avoided. The relevant British Standards are 437,1130 and 1211. For lightness and ease of assembly uPVC drains are popular. Solvent welding of joints may be difficult in some circumstances but the alternative of push-fit rubber seal fittings is available. The supply of plastic manholes and fittings to minimise the substantial cost and inconvenience of masonry manholes is now well recognised.

Drains should be laid in straight lines. Inspection chambers should be at all major angle changes and confluences. Rodding points should be included. These too may be designed to be flush with the surface. Ventilation of the system is required and stack pipes along with at least 50 mm water-seal traps ensure appropriate venting of foul gases. Clean water velocity should not fall below 0.75 m/s. To avoid solid separation a flow of not less than 1.5 m/s should occur in a foul drain.

ENERGY

Compressed air

Compressed air services can be extremely useful in workshops. Safety can be brought more readily to air-operated than to electrically-operated hand tools, though compressed air can be extremely dangerous. No laxity in maintenance can be tolerated. Insurers pay close attention to compressed air services. Condensate can be dealt with adequately by pipe and fixture design and prevent corrosion.

Though a very useful source of power, compressed air can be expensive to run if leakage is allowed. Flow meters may prove useful loss monitors on large systems. Oilers for hand-tools and pressure regulators should be included in supply lines.

BS 1710 depicts a colour code for pipes which can be used to show contents and the use to which those contents may be put.

Electricity

Rotation at 3000 rev/min (50 rev/s) of an independently excited magnetic field inside three sets of windings leads to 50 Hz alternating supply. Voltage pushes current through a resistance (or impedance). Losses occur proportional to the transmission materials, its length and area of section. The basic principle is to keep conductors as short and as large a cross-section area as possible within the cost constraint; with a voltage as high and current as low as possible in order to minimise losses within a safety constraint.

Power for single phase supply,

$$W = VApf$$

Power for three phase supply,

$$W = \sqrt{3}VApf$$

The line amperage (A) a system demands

$$= \frac{kW \times 1000}{\text{line voltage} \times 1.732 \times pf}$$

$$= \frac{\text{brake horse power} \times 746}{\text{line voltage} \times 1.732 \times pf}$$

where W, watt; V, volt; A, ampere (amp); pf, power factor are electrical terms.

Where windings occur in alternating current (AC) supplies, such as in motors, welding sets and transformers, losses occur. Power factor (pf) is the term for this inefficiency. In such cases more current, with its inherent heating effect, is required for the same power output. This current passes through the distribution system designed for normal flow. This adverse situation carries a cost penalty to the consumer because of the dangers. Electricity bills are adjusted according to the pf level below the ideal of 1.0 but generally with a leeway to 0.8. Capacitors or condensers can be used to correct adverse pf. However, they 'consume' electricity and can cost as much as equipment such as a welder.

Assessment of demand prior to installation is essential. Mechanical, chemical or heat damage to wires, cables fixtures and fittings must be avoided. Wiring diagrams and switch labelling associated with the installation are strongly recommended.

New installations must be checked before a supply is connected. Electricity Boards can refuse to make connection where there has not been adherence to Institution of Electrical Engineers (IEE) regulations.

Cables come in sizes characterised by their cross-sectional area of conductor, bearing relevance to their current carrying capacity at the stated voltage and to the 'earth' requirement (*see Table 20.1*).

Dimensions of cables have relevance to voltage drop from source to consumption point. This must not exceed 2.5% of nominal voltage. A 6.0 mm^2 conductor in a conduit has a voltage drop of 7.0 mV/m per ampere, thus 210 mV/m, 2.5% of a 240 V supply is 6 V thus the nominal maximum length at maximum rating is 28.60 m for this case.

Fuses protect installations from electrical damage. Their action is too slow to protect life and an earth leakage circuit breaker (ELCB) or a similar device should be used. Various names for similar devices can be found, e.g. residual current circuit breaker (RCCB). Their action is to sense an imbalance between the feed and return wires which indicates that the current is seeking an alternative route, possibly to earth through a person. Whilst the Electricity Board will test a circuit before connection and provide an earth they will not guarantee it. An ELCB is essential.

Electric motors must be fitted with a suitable starter and isolator within reach of the operator. With three phase motors failure to revolve may be due to one phase failing; the other two phases will still be live. With a capacitor induction motor, failure to rotate may be due to capacitor failure. Again the motor will still be 'live'. Incorrect direction of rotation with a three phase motor can be corrected by reversing any of its two supply wires. Systematic connection of three phase supply where a motor can be plugged into one of several outlets is to be recommended.

A good power output to frame size can be gained from using three phase supply for a comparable one phase current

Table 20.1 Conductor data

Nominal area cross-section (mm²)	Rating (A)	Use
Cable		
1.0	11	Lighting and small heaters
1.5	13	Ring circuits
2.5	18	Heaters to 3 kW
4.0	24	Radial circuits
6.0	31	Cooker circuits
10.0	42	Large heaters
Flexible		
0.05	3	
0.75	6	
1.00	10	
1.50	15	
2.50	20	

Colour codes

Dimension	Old	New	Wire
2 Core flexible	Red	Brown	Live
	Black	Blue	Neutral
3 Core flexible 1 phase	Red	Brown	Live
	Black	Blue	Neutral
	Green	Green and yellow stripe	Earth
3 Core flexible 3 phase	Red	Brown ⎫	Brown ⎫ Live
	Yellow	Brown ⎬ Ends coded	Yellow ⎬ Live
	Blue	Brown ⎭	Blue ⎭ Live

supply. Particularly above 3 kW capacity three phase is to be recommended.

A voltage of 110 V needing greater current for a particular power consumption may seem wrong, as it is current that kills. However, 110 V has less chance of pushing that current through a resistance. The 'building trade' standard for hand tools is, for that reason, 110 V.

Mains supply

Generated electricity is transformed to 400000 V for transmission. At places through the grid-network it is transformed to 275000 V, 132000 V, to 33000 V, to 11000 V as the supply gets closer to consumer. It may be received by the consumer at 11000 V for transforming at the holding. For large consumers of electricity there is a price advantage in doing it this way. The normal process is to receive 410 V or 240 V from a pole type transformer close to the holding. Only at this stage does the three wire system need to accommodate a 'neutral'. The neutral is picked up from the centre tap of the transformer and between any phase and neutral 240 V potential is available. Naturally balanced three phase is carried on the three 'live' wires to point of consumption. Single phase load balancing between phases is desirable.

There are various price structures for electricity used based on peak and overall demand and time of day. Advice can be obtained from the selling agents for the CEGB produced power, the area Electricity Boards.

Private supply

Alternators for private supply can cover for 'mains' failure through temporary fault, industrial action or the like. Equipment costs are high but freedom from supply interruption may make the choice worthwhile.

The starting of electric motors demands some six times as much current as 'run' current even if a starter such as a 'star-delta' unit is used. This must be borne in mind when choosing and using an alternator, especially where motors greater than 3 kW or welding sets are used.

Fuel oil

Oil of 28 second viscosity is normally used for domestic style burners though normal engine fuel oil may be used if the appliance permits. Standard tank sizes are available and are typically self supporting, being set upon piers or stands with suitable proof against corrosion and contaminant separation. Gravity discharge to vehicles is to be recommended to facilitate ease of working and reduce in-vehicle fuel contamination.

For burning oil the tank may be situated conveniently and connected typically by a 8 mm diameter soft copper pipe to BS 1386. Should a small gravity head not be available a header tank and small pump unit can be used.

Liquified petroleum gas (LPG)

A product of the oil industry the price of LPG rises with oil price rises. Convenient as a source of heat energy, being some three times that of natural gas, it may be a plausible possibility for farm use. The gas is supplied in small bottles to 50 kg mass and large above ground storage tanks. The storage system must follow simple rules. The valve must be uppermost to allow gas and not liquid into the system. The store should not be subject to too great an extraneous heat nor should the bottles be insulated. Propane boils at −42°C and butane at −0.5°C. Heating of the 'liquid' occurs as it releases from the bottle pressure. Large demands cause ambient heating of the tank outlet and in cold weather this may cause air moisture to freeze on the outlet and subsequently stop gas flow. Propane may prove the better material to use for high winter demands.

Pipes can be of smaller bore than for natural gas. Drawn copper tube to BS 1386 should be used for preference. Red lead paint should not be used as a pipe jointing compound. Solder joints should be replaced by brazed joints near the heat source. Blockage of the smaller jets may occur and flame failure in draughts is likely, due to the low pressure.

The British Gas Corporation, the supplier of LPG, or approved agents should install any fixtures, furniture or appliances. Refer to the Building Regulations and amendments, the Gas Safety Regulations 1972 (SI 1972: 1178), the Gas Act 1972, CP 331, CP 338, CP 339 and BS 5258.

Natural gas

Installation of gas (natural) supply is the responsibility of the British Gas Corporation or its approved contractors. Compliance with Regulations Parts I to VII is necessary. The specific responsibility and their property ends at the outlet of the meter. The gas safety regulations apply throughout.

Gas is almost pollution free with a calorific value of $37 \, MJ/m^3$ and supply pressure of 1 to 2 kPa. Condensate may occur and pipework is laid to drain back to the meter. Electric and gas supplies must not be laid alongside one another and the output side of the meter must be linked to the electric power earth system.

Flues carry water vapour and carbon dioxide (on correct combustion) outside the building, and should a light failure occur, ventilate gas to atmosphere. Burning is generally noisy because of the air flow for combustion.

Unplasticised PVC is now more commonly used for bulk supply below ground, but copper, typically 15 mm, suffices at consumption points. All piperuns should be well ventilated to prevent build-up pockets from what could be a relatively insignificant seepage of gas from the pipework. Flexible links to the appliance are usually fitted to aid service. Sleeves through walls and support for surface pipes should be provided and alkali attack on copper avoided by screening.

Petrol

The Petroleum (Consolidation) Act 1928 controls the storage and the licensing system. Enactment is through Local Authorities whose advice should be sought.

FIRE

When involved in a construction exercise, reference must be made to BS 5502: 1978 and Building Regulations Part G and L. Insulation, claddings or structural skins should not be of a material that produces toxic fumes, is highly flammable, produces dense smoke or constitutes a major hazard to the building concerned. Consideration should be given to easily taken stock exit routes, allowing escape largely unaided by humans in an emergency. Working areas in buildings should have adequate means of escape and preferably should be illuminated. Smoke vents designed into a structure can only aid livestock and human escape.

Alarm systems may be considered unnecessary or expensive, but their inclusion can only be to advantage. Their use may be linked to a fail-safe device on, for example a controlled environment house.

Building layout should provide fire breaks, ease of access for emergency services and have ample water between 6 and 100 m away. Static tanks of $20 \, m^3$ capacity for suction hose use should be provided if adequate pipe supply is not available. Hose reels should comply with BS 5274 and hydrants to BS 3251.

High fire risk areas such as fuel stores should provide 2 h fire resistance by walls, or have a 12 m separation from any other boundary. Boiler rooms, maintenance workshops, plant material drying compartments and potentially explosive fertiliser stores should provide 1 h fire resistance by walls or 6 m separation. The separation of buildings physically to provide fire breaks can be extended to cover the inside of a particular building. Compartmentalisation is

Table 20.2 Portable fire-lighting appliances

Water	5–9 litre capacity, the contents of which (water) are ejected by puncturing a compressed gas container or by nozzle control on the previously air pressurised unit.
Foam	5–9 litre capacity of foam making liquid ejected by puncturing a compressed gas container or by nozzle control on the previously air pressurised unit. The nozzle design creates foam from the liquid.
Dry powder	2–12 kg of dry powder expelled, completely or partially if necessary, through control of the nozzle in the larger sizes, by compressed gas.
CO_2	1–6 kg of carbon dioxide gas under pressure. Ejection by nozzle control.
BCF	1–5 kg of BCF gas under pressure. Ejection by nozzle control.
Blankets	Asbestos sheets of $1 \, m^2$ or preferably glass fibre sheet.

Old units are coloured red. New units have colours depicting contents. Red — water, Yellow — foam, Red — fire blanket. Blue — dry powder, Black — CO_2 gas, Green — BCF. They should be suitably sighted to give ease of vision, access and use.

Suitability

Carbonaceous materials (e.g. wood, paper) where recombustion may occur due to contained heat	Water
Flammable liquids	Foam, dry powder, BCF
Electrical	CO_2, dry powder, BCF
Special risks	Obtain local fire brigade assistance

advisable, particularly where livestock are concerned, with undivided floor areas exceeding $500 \, m^2$.

Pesticides can be particularly dangerous in a fire. Even if they are not involved in the fire, their proximity can cause problems to fire defence. A separated, lockable fireproof bunker should be provided. Chemical contents list should be kept elsewhere from the store. (*See* Control of Pesticides Regulations 1986.)

Feed grinding and mixing rooms should offer 0.5 h fire resistance or be 3 m from other boundaries. A list of portable fire fighting apparatus is given in *Table 20.2*. Water or foam extinguishers should not be used where an electrocution hazard is present from live electrical services. Dry powder gains its action from smothering the flames. The powder may settle in the appliance with storage, especially if on a vehicle or vibrating structure. BCF should be used with caution inside buildings.

HEATING

Comfortable living conditions, acceptable working environment, or an environment conducive to crop or animal production, are provided by heating systems when ambient temperatures fall below an acceptable level. Typically, heavy work areas require some 13°C, sedentary work 18°C and crop and animal units special conditions.

Humans provide from 100 to 350 W and electric lights their nominal rate of wattage. Due account should be taken of these and other usually discounted sources.

Humidity, ventilation, insulation and fuels supply and storage are integral elements of any system. Failsafe devices are important.

Heating units should be positioned to provide an equitable environment and no great temperature gradients. Open flues, or water heater vents may create draughts, which may be detrimental to animals and plants.

When designing systems macro-considerations, such as resultant condensate, venting of hot water pipework or vibration in air ducts, should be taken into account.

A range of fuels is available (*see Table 20.3*). Many are self stacking. Alternatives to traditional fuels are gaining merit, but generally require courage to install and technical expertise to utilise fully.

Table 20.3 Energy value of various fuels

Fuel	Gross calorific value (MJ)
Anthracite and good coal	35/kg
Biogas (60–70% methane)	22–26/m³
Methane	37/m³
Oil:	
Class C Kerosine	46.4/kg
Class G Heavy	42.5/kg
Peat (14% DM)	19/kg
Sawdust	18/kg
Wheat straw (10% DM)	17.7/kg

Coal-fired independent boilers or room heaters with high output back boilers are popular domestic heaters. Gas heaters with a flue to the outside through a wall or conventional flues can provide hot water or direct space heating. Electricity, gas or hot water can be used to heat air for warm air installations. These tend to be cheaper than indirect hot water systems, but may not be practical because of safety problems.

Oil burning is about 75% efficient and is a useful concentrated form of stored heat, little subject to industrial action or national network breakdown. Overhead infrared or underfloor cable electric heaters are useful for partial heating systems. Wood and other vegetable matter burners can provide useful quantities of heat. Solar gain units for low grade heating can be useful but are expensive to install. Recovery of capital cost is unlikely if new units are installed. Heat may be stored in 'rocks' or used directly as hot water in short-term storage.

Heat pumps offer great opportunity on farms. Recovered heat from milk is a useful source of low grade heat. Pumping heat from other heat sources such as solar panels can be at a coefficient of performance of 3:1. Conventional heat units have a life expentancy of some ten years. Better than double this may reasonably be expected from a heat pump, but a relatively low energy concentration is a major problem.

Biogas

Biological degradation of animal faeces and urine can provide a useful source of heat but extreme caution is advised with regard to design, capital cost, payback period and level and timing of heat output. A constant 30–35°C is held in a 3–15% solid content mix. Temperature maintenance consumes some 35% of the gases produce. This may rise to 100% in winter. Regular throughput is essential with stirring to avoid scum or sediment formation. Pig, cattle and poultry slurries are held for about 8, 17 and 17 d, and produce about 0.3, 0.2 and 0.28 m³/kg waste, respectively. Light fractions from pig excreta can be a particular problem.

The system has the advantage that otherwise waste products of the enterprise can be used in it. The gas of some 65% methane content has a value of 22 to 26 MJ/m³. A 50 cow dairy unit produces some 3.75 kW. Gas storage is a problem and immediate use is inevitable.

Water

Water power can be useful. The capital investment tends to be high and water use is as determined by the area Water Authority. A 60% efficiency for electricity production would be typical from an impulse turbine/alternator set.

LIGHTING

Artificial light can be aided by the innate attribution to humans that natural light provides. Artificial light can give safe, even level to buildings, plant and equipment where usually non-reflective agricultural building surfaces occur. Glare must be avoided and correct levels for detailed work or human access provided, though for many classes of livestock a 'night-light' standard is all that is required. A relationship of light level for task, immediate surroundings and walls and ceilings of 10:4:3 (the Bodmann ratio) has been of great value in eliminating glare and eye strain, whilst accommodating a balance between level and cost of provision for safe egress, work and social needs.

The level of illuminance is measured in lux (lx). Avoiding glare, generally the higher the level of illuminance the better the visibility (*Table 20.4*).

The luminaire provided should be consistent with the level of illumination required, the environment into which it is fitted, the cost and the system flexibility desired (*see Table 20.5* for range of luminaire).

Tungsten filament and tungsten–halogen lamps work

Table 20.4 Illuminance intensity

Area or task	Intensity (lx)
For selection by colour or detailed assembly	500–1000
For 'close' inspection	300
Farm workshop (general level)	100
Milking premises and passages	100
Others	50

After BS 5502: Sect. 3.5 1978.

Table 20.5 Luminaire life and efficacy

Nominal life (h)	Lamp type	Efficacy (lm/W)
7500	High pressure sodium (gold colour light)	70–100
7500	HID mercury vapour (+ halide colour renderer)	62–72
7500	Fluorescent tubes ('white' colour)	54–67
7500	HID mercury vapour (+ fluorescent coat)	35–50
7500	Fluorescent tubes (colour rendered)	33–40
2000	Tungsten-halogen	16–22
1000	Tungsten–filament	10–18

Technical assistance can be gained from the Area Electricity Board. Reference may be made to CP 324.101 and the Electrical Development Association.

direct on line. Fluorescent tubes and high intensity discharge lamps (HID) need additional electrical components to enable the current to 'strike' along the tube. HID lamps take some 20 min to reach their operational lighting intensity and are not suitable for frequent switching. Quartz halogen can be an explosion hazard in dust laden atmospheres.

With a range of fittings and luminaire available advice from Area Electricity Board, reference to CP 324, CP 101 and Electrical Development Association should be made

Switches pertinent to the load are available. Watertight or rotary switches may be necessary to enable safe, or to avoid accidental, switching. Two-way switches can save much time and effort, but in any case careful planning can lead to higher efficiency of operation.

The use of more expensive than average luminaires should not be discounted as their running cost and serviceability will make them more cost effective overall.

LUBRICATING OIL AND GREASE

Supplied in 25 litre (5 gal) or 205 litre (45 gal) containers the fluid can be drawn by gravity through cradle support or by hand pump. Provision for reserve drums, the various classes of lubricant for farm vehicles and the adequate disposal of used oil make careful planning of this facility rewarding in control of mess and ease of access.

WATER SUPPLIES

Public utility supplies are made to some farms but remoteness, demand and cost make it unlikely for agricultural use. Private supply may be a necessity. Underground sources, aquifers, may be utilised. Reference must be made to the relevant Water Authority before use is made of any source and especially if a borehole is intended. Avoidance of pollution is all important. Though stock will drink from many sources considered unsuitable for human consumption, as a general rule the supply of potable water is the principal aim.

The demand by various classes of livestock, crop and agricultural task varies (*Table 20.6*). Generally 13 mm per week for crop irrigation is a useful rule of thumb, but in any case guide averages are only useful so far. The volumes

Table 20.6 Daily water use for farm production and related storage

	Water required (litres)
Cow consumption, 22 litres milk/d	70
Cow consumption, dry	35
Cow allowance, dairy cleaning	20–50
Cow allowance, milk cooling	60–120
Beef animal, drinking	25–45
Calves (up to 6 months), drinking	15–25
Sows drinking, in milk	18–23
Sows drinking, in pig	5–9
Boars drinking	9
Pigs drinking, growing, fattening	2–9
Sheep drinking, growing, fattening	2.5–5
Sheep dipping, all sizes/dip	2.5
Poultry drinking (10 birds), layers	20–30
Poultry drinking (100 adult birds), fattening	13
Turkeys drinking (100 adult birds), fattening	55–75
Storage capacity allowance for sink	90
Storage capacity allowance for WC	90
Storage capacity allowance for hot water system	130
Storage capacity minimum for farm office/canteen	225
Storage capacity minimum for farm workshop	450

involved make purification impractical in the UK. A suitable source of potable water should be found.

Supply of water is typically by alkethene, PVC, copper or ferrous metal and decreasingly lead. The pipe diameter relates to the length of run and the supply demands correlated to the surface nature of the pipe. The essence is that flow in m^3/s will relate to pressure head from source to draw off point. Typically the rate of flow required at appliances is:

12 mm draw off	0.19 litre/s
18 mm draw off	0.30 litre/s
25 mm draw off	0.60 litre/s

Friction tables may be used to design out problems such as too low a flow rate or water hammer: This velocity impact on an elbow or end stop such as closing a tap or valve results from too high a flow rate from a service. Where several draw off points occur on one service the flow will fluctuate.

Water storage and distribution

On completion of carcass work the water service will be brought from the Boards main via a 'stop-cock' (globe-valve). Entering the building 0.76 m below the ground level a drain pipe installed in foundations makes a useful conduit. Attention should be paid to prevent corrosion from soil sulphates or cement based products.

From this stage on, avoidance of frost (CP 99) and animal or mechanical damage is of prime importance, though access and plaster and render deterioration from condensation need accommodating. Where outside the heated volume of the building, or where draught prevention at the site of a

Table 20.7 Copper tube for water, gas and sanitation

Nominal diameter (mm)	(in)	Wall thickness (mm) Table X	Table Z	Maximum working pressure. MN/m² (N/mm²) Table X	Table Z
10	($\frac{3}{8}$)	0.6	0.5	7.7	7.8
15	($\frac{1}{2}$)	0.7	0.5	5.8	5.0
22	($\frac{3}{4}$)	0.89	0.59	5.1	4.1
28	(1)	0.89	0.59	4.0	3.2
35	($1\frac{1}{4}$)	1.185	0.685	4.2	3.0
42	($1\frac{1}{2}$)	1.185	0.785	3.5	2.8
54	(2)	1.185	0.885	2.7	2.5

Table X tube — half hard, light gauge⎱ BS 2871, Part 1, 1971
Table Z tube — hard drawn, thin wall⎰
Table Z tube should not be formed into bends but jointed to change direction.

Table 20.8 Availability of polythene pipe to BS 1972 and BS 3284 (low and high density)

Pipe class and colour code		B, Red		C, Blue		D, Green	
Working pressure head		60 m		90 m		120 m	
Density type — 32 (low) and 50 (high)		32	50	32	50	32	50
Nominal size (in)	Permitted range of outside diameter (mm)						
$\frac{1}{4}$	17.0–17.3	×	×	×	√	√	√
$\frac{1}{2}$	21.2–21.5	×	×	√	√	√	√
$\frac{3}{4}$	26.2–26.9	×	√	√	√	√	√
1	33.4–33.7	√	√	√	√	√	√
$1\frac{1}{4}$	42.1–42.5	√	√	√	√	√	√
$1\frac{1}{2}$	48.1–48.5	√	√	√	√	√	√
2	60.1–60.5	√	√	√	√	√	√

Other sizes used in agriculture 3, 4 and 6 in.

suitable run cannot be accommodated, insulation must be provided.

Pipes must contain the pressure of the system and comply with appropriate legislation. Asbestos cement, though chemically inert suffers from being brittle and is outclassed by plastics and metals. Copper pipe is lightweight and used for clean and dirty water. High strength and ductility with easy installation make its use worth considering, but bear in mind cement-based product attack on copper. BS 659 covers above ground and BS 1386 underground (coil) pipes. CP 310 shows recommended support spacings (see Table 20.7).

Mild steel, 'barrel' pipes to BS 1387 are cheap, strong and generally resist corrosion. Open systems should be galvanised; closed systems, such as heating systems, should be of 'black' ungalvanised pipe. Where water is soft corrosion may be high. Class A, brown band; B, yellow band; and C, green band are available. Authorities will specify the Class, but generally A is used for waste; B for distribution and hot water and C for rising mains. Though structurally sound it needs threading equipment to install. Wrought iron pipe (BS 788) is little used. Malleable iron (BS 2156) should be galvanised. Larger sizes for mains can be made from cast iron to BS 416.

Plastics pipes are now available for hot, cold and waste services (see Table 20.8). Polythene of two densities (type 32 and 50) is used due to its flexible handling, long lengths and ease of compression fixing. Not affected by soil acids the pipe may be moled into place. Softening at 70°C and melting at 115°C its use is really restricted to cold services. In classes B, C, D, E for pressures of 600, 900, 1200 and 1500 kPa the pipes have improved in quality, reduced wall thickness and at the same time, to distinguish from other services, changed to a sky blue colour. Care must be taken to use the correct size of fittings when linking into existing services. BS 1972 and BS 3284 apply and are printed with other data in colours red, blue, green or brown on the pipe to correspond to Class code and pressure of service.

Mechanical support and damage protection are a must. Electrical integrity with existing systems must be maintained.

Semi rigid uPVC in 6 and 9 m lengths can be cold cemented together. Lighter in section and higher in strength for a wall thickness it is found in service for pressure or warm water. It softens at 80°C. Fittings have to be cemented or 'O' ring sealed and with its lower resistance to water expansion

on freezing it is less used than polythene. BS 3505 applies and CP 310 applies for installation.

ABS (acrylonitrile butadiene styrene) melts at a higher temperature (93°C) and has made some inroads. BS 3505 applies as its substitutes for uPVC; however, a separate BS will be available.

Aluminium alloy is used for irrigation but aided by the high price, it is stainless steel that has been used to replace copper for domestic use. Copper coated steel for closed central heating systems is a cheaper alternative especially where sealed, pressurised systems have been installed for space heating.

Water storage on a local basis is more important in some areas than others. Normally one day's supply would be stored locally. In Britain hot water supply installations must be from storage (except instantaneous heaters). Storage in header tanks adds to the static loading, has generally a lower head and thus requires larger diameter distribution pipes and may become contaminated. Advantages come from the mains failure reserve, particularly where stock are concerned; balancing out peak demands from unusually dubious main supplies; reduced pressure of supply, with its own cut in operating noise for stock; and localising hot service safety.

Storage vessels are of galvanised steel (BS 417), polythene and polypropylene (BS 4123) or asbestos cement (BS 2777) with sizes, coded C1 to C21 as BS 417 table 1, and grades A and B. Plastics tanks avoid electrolytic corrosion but suffer the need for support and rodent protection. Though only a base pressure of 9 kPa from 1.5 m depth of water the C21 size holds 3.37 tons of water (actual capacity 4.55 tons).

Isolation of supply from contamination is a must. Ball valves must empty above the free surface of the water and vent to avoid back syphoning.

Taps, cocks and valves are terms loosely used in common building language. Cocks operate by a quarter turn on to off and by such abruptness can lead to water hammer. They are not normal on a main supply, but are seen on such systems as washing machines for isolation after water flow has ceased. Valves control flow in a pipeline. Globe valves, typified by a capstan head (bar lever) are used on water supply pipework where adequate head is available and complete cut-off is essential. With a less angular flow passage the straight through gate valve, typified by a wheel head, offers less resistance to water flow when fully open. They are employed where head is limited ssuch as on header tank systems output. The loose jumper washer is expected to close the supply should backflow occur.

Hot water is expensive to obtain and to deliver. Every effort should be made to minimise pipe losses through draining excess cold before hot water is at the tap and the consequential loss of hot water in the pipe after the tap has been turned off. Pipe of 28 mm diameter would, for example, give a faster flow than a 15 mm, but per meter run has 3.5 times as much water. Generally 22 mm would be used to supply domestic baths or to supply an indirect hot water cylinder from a boiler on a buoyancy rather than pumped system. Generally flows to 4 kW (14000 BTU) should be in a pipe to 15 mm diameter.

Further reading

MITCHELL, G.A. and MITCHELL, A.M. (1988). *Building Construction*. Ed Burberry, P. *Environment and Services*, London: Batsford

21

Farm machinery

P. H. Bomford

The availability of suitable farm machinery gives the farmer the means to carry out essential farming operations, and the capacity to complete these operations within the time available. Successful management of machinery involves the selection of equipment of the correct function and capacity, and the supervision of its efficient and safe operation in order to achieve quality work at an economic cost.

THE AGRICULTURAL TRACTOR

Tractors account for about two-thirds of total farm machinery sales in the UK. On many farms, the tractor is the most costly machine.

The tractor provides power for almost all mobile operations on the farm, as well as for many stationary processes. Power is defined as the rate of doing *work*, and work is done when a *force* acts through a *distance*. One Newton metre (Nm) of work (or 1 Joule) is done when a force of one Newton acts through a distance of one metre. The same units define *energy*, which is the potential to do work. For the measurement of power, a time factor is included, and the unit of measurement is the joule per second, or watt (W). A rate of work of 1000 Newton metres per second is one kilowatt (kW).

Power for the tractor is produced by its engine, and this power is made available in two main forms; as pull at the drawbar, to operate trailed equipment, or as rotary power at the power take-off (PTO) shaft. A small proportion of the engine's power is available through the tractors's hydraulic system.

The 'size' of a tractor is generally described by quoting the power produced by its engine. Since the full power of the engine is not available to do work outside the tractor, power which is available at the PTO is a more useful indication of the work a tractor may be able to carry out.

The engine

The tractor's engine converts the chemical energy contained in diesel fuel into rotary power at the flywheel. Two-thirds or more of the energy value of the fuel is lost as waste heat via the exhaust and cooling systems. Flywheel power, often called brake power because it is measured by applying a braking load to the engine, is the product of the engine *speed* (N rev/min) and the *torque* (T Nm), or twisting effort that the engine can maintain at that speed. The relationship between these factors is:

$$\text{power (kW)} = \frac{2\pi N \text{ (rev/min) } T \text{ (Nm)}}{60\,000}$$

A typical tractor engine produces little torque below 500 rev/min. Maximum torque is reached at about 1400 rev/min, and torque then decreases with increasing speed to 90% or less of maximum at full speed, which is 2000–2800 rev/min. This torque reduction is called 'torque backup'. A large torque backup indicates a flexible engine with good 'lugging ability', and a reduced need for gear changing during work. Brake power, however, increases almost linearly with engine speed, and maximum power is produced only at maximum rated engine speed.

The conventional engine is the best mobile power source at present available in terms of efficiency, weight, availability of fuel and cost. However, it is by no means perfect. It is noisy, it vibrates and it produces exhaust gases which pollute the atmosphere. It is made up of many hundreds of individual components and has many points of wear. The life of a well maintained tractor engine, before it needs a major overhaul to renew worn parts, is between 4000 and 7000 h of work. The life of other mobile engines ranges from 200 h for small air-cooled engines, to 1500–2000 h for a car engine, to a maximum of 12 000 h for heavy duty industrial engines.

The faster an engine rotates, the greater the amount of power (and fuel) that is consumed in just keeping the engine turning at that speed, in relation to the power that is available at the flywheel. If it is not necessary to run an engine at full speed, because maximum power is not needed, operating the engine more slowly will save fuel. Slower operation also increases the engine's reliability and prolongs its working life.

Many diesel engines are fitted with turbochargers, to increase power output. A turbocharger is an exhaust-driven rotary compressor which forces more air into the engine.

This allows more fuel to be burnt, releasing more energy and producing more power. A power increase of 30% or more may be achieved, but this will subject the engine to greater thermal and mechanical stresses. Mechanical components, and cooling and lubricating systems, must be upgraded accordingly. Since the increase in power is achieved with no increase in engine speed or size, frictional losses do not increase in proportion, and the increase in power also results in an increase in engine efficiency.

The transmission

In order to deliver a full range of engine power at a wide range of forward speeds, the engine is connected to the wheels by a transmission offering from 6 to 30 gear ratios. With so many ratios available it is necessary for the driver to change gear often in order to match power and forward speed to changing conditions. Various devices are provided to make this task easier. Synchromesh transmissions synchronise the speeds of rotating parts, to allow quiet gear-changing on the move. Semi-automatic transmissions change gear hydraulically by releasing one clutch and engaging another. The commonest application of a semi-automatic transmission is in a 'high-low' change which inserts an extra ratio between each pair of existing gears; complete transmissions can operate this way, providing up to 15 forward and four reverse ratios. Semi-automatic transmissions deliver less of the engine's power to the wheels than conventional systems because power is lost due to friction and in operating the hydraulic control system of the transmission itself.

Some tractors are fitted with hydrostatic transmissions which provide an infinitely variable range of ratios with a stepless single-lever change even from forward to reverse. This system is ideal for the operation of trailed PTO-driven machines such as balers or forage harvesters. Forward speed can be continuously adjusted to match crop and ground conditions while maintaining a constant engine (and PTO) speed. For the same reason, hydrostatic transmissions are fitted to many combines and self-propelled forage harvesters. Where a major proportion of the tractor's power is to be used in traction, the lower efficiency of the hydrostatic transmission means that less power is available at the wheels from a given engine power.

Drawbar power

A tractor develops drawbar (DB) pull as a result of the gross tractive effort developed between its drive wheels and the soil. After this tractive effort has overcome the rolling resistance of the tractor's own wheels moving the tractor along, any remaining force is available at the drawbar.

DB pull = gross tractive effort − rolling resistance

As will be seen, maximum drawbar pull is achieved both by maximising the gross tractive effort, and by minimising the 'parasitic' effect of rolling resistance.

Drawbar power is the product of drawbar pull and forward speed, as expressed by the equation:

$$\text{DB power (kW)} = \frac{\text{DB pull (kN)} \times \text{speed (km/h)}}{3.6}$$

Gross tractive effort

When the lugs of a drive wheel or track bite into the soil, the rearwards thrust of the lugs tends to push the soil back. As the soil trapped between the lugs is sheared from the underlying soil, a horizontal force is developed. The further the soil is displaced, the greater is this force.

As the tractor moves forward, exerting a pull on some following attachment, the soil is pushed backwards a little. The percentage the soil is moved back, in relation to the distance the tractor would move forward on a rigid surface with no drawbar load, is called the slip. A wheeled tractor develops its maximum drawbar pull at 20–25% slip, but maximum tractive efficiency occurs at about 10–12% slip (Gee-Clough *et al.*, 1982). Any slip at all means that some of the tractor's power is being lost in pushing soil backwards instead of pushing the tractor forwards, but since no pull can be generated without some slip, this must be accepted.

The shear strength of the soil under the tractor's wheel or track depends to a small degree on the rate of shearing; a tractor operating at a higher speed can generate a slightly greater pull because of this property. The major factors affecting soil strength are its coefficient of internal friction and its cohesion.

The more weight that is applied to the soil under the driving wheel, the greater will be its frictional strength and the greater will be the tractive effort generated at a particular level of slip. Coefficients of internal friction range from below 0.2 for a plastic clay, to over 0.8 for a coarse sandy soil. An 'average' figure is 0.6, which means that 60% of the vertical force applied to the soil by the driving wheels would be available as gross tractive effort. Excess water acts as a lubricant between the soil particles, and can reduce the coefficient of friction almost to zero.

Cohesion is the strength with which the soil clings together, even when no weight is applied to it. A typical value for a friable soil is $30 \, \text{kN/m}^2$, with a range from zero for very coarse-textured soils to a maximum of $60 \, \text{kN/m}^2$ in some clay soils. The greater the area of cohesive soil that is put in shear, the greater will be the tractive force generated. Compacting a loose soil, for example by running a wheel over it, will increase its cohesive strength so that a following wheel can generate a greater tractive effort than it could if running on uncompacted soil (Rackham and Blight, 1985). Cohesive strength is high in undisturbed soils.

Frictional tractive effort is increased by increasing the weight on the driving wheels or tracks. The loading may be by means of iron weights, or water ballast in the tyres, or the tractor can be made to carry part of the weight of the implement it is pulling. This last approach has the advantage that when the implement is detached from the tractor, so is the extra weight.

As the tractor pulls a load, the resistance of the load tends to tip the tractor backwards about a point on the ground beneath the rear axle. This has the effect of transferring weight from the front to the rear wheels, and thus increasing the available tractive effort. The height of the hitch point must be kept low enough to eliminate any risk of overturning the tractor. Front-end weights may be fitted to ensure that at least 20% of the tractor's weight remains on the front axle to give steering control. Four-wheel drive tractors, and crawlers, gain no benefit from weight transfer, since all the weight of the machine is carried on the driving members. On four-wheel drive versions of two-wheel drive tractors, weight transfer can remove much of the weight from the front axle under good tractive conditions so that the powered front axle contributes little to traction unless it is ballasted by front-end weights or front-mounted implements.

Cohesive tractive effort is increased by increasing the contact area between drive member and soil. The fitting of larger section rear tyres or dual wheels, or the use of four-wheel drive or crawler tracks all have this effect. Additional soil may be put in shear by the use of grousers, strakes or spade lugs, which can penetrate through a slimy surface layer into stronger underlying soil. However, traction in this case is increased at the expense of reduced tractive efficiency since power is lost in digging into the soil. The use of tyres at no more than the recommended inflation pressure for the load carried will ensure the maximum safe contact area between tyre and soil.

Ballasting

Correct ballasting is essential if a tractor's maximum tractive power output is to be realised on typical frictional-cohesive agricultural soils. A major research programme at the National Institute of Agricultural Engineering (NIAE) (now re-named the AFRC Institute of Engineering Research) has resulted in the production of very clear recommendations (Dwyer and Dawson, 1984). The total weight to be carried on the driving wheels (*w*) is found where

$$w \text{ (kg)} = \frac{650 \times \text{PTO power of tractor (kW)}}{\text{working speed (km/h)}}$$

Much of this weight will be provided by the tractor itself and by any mounted implements that it carries, but additional iron weights and water-ballasting of tyres are normally required. In the case of four-wheel drive tractors, front to rear weight distribution should be in proportion to the tyre-maker's recommended carrying capacity of front and rear tyres at equal inflation pressures. If ballasting is correct with the tractor stationary, the effect of weight transfer will not be large enough to reduce tractive performance.

When the correct ballasting has been established, the tractor must be equipped with drive wheels and tyres large enough to carry the necessary weight at a low inflation pressure (preferably no more than 1.0 bar) (Dwyer, 1983). In work, the tractive load of the tractor should be adjusted so that wheel slip is 10–12%. This is not easy to judge, but it can be measured, and tractor mounted slip indicators and slip control systems are available for this purpose.

Rolling resistance

In order to roll a wheel along a surface, a force must be applied to overcome its rolling resistance. Rolling resistance increases with the load carried by the wheel. In agricultural conditions the force ranges from 5% of the weight carried by the wheel when travelling over a dry field after a cut of silage, to 20% or more when harvesting root crops (Dwyer, 1985).

Rolling resistance is minimised by using the largest available tyre size inflated to the minimum pressure recommended for the load being carried, and by using more wheels to support the load in dual or tandem formation. Where radial tyres are available, a 5% reduction in rolling resistance can be attained by their use (McAllister, 1983).

Tyres

Most tractors rely on rubber tyres to transmit the power of the engine to the soil and to generate tractive effort. So long

as lug height is not less than 20 mm, tread pattern has little effect on overall tractive performance. There is negligible tractive advantage in increasing lug height above this value, as taller lugs will deform under load and lose their bite into the soil (Gee-Clough *et al.*, 1977a).

The familiar chevron tread pattern has the advantage that it is self-cleaning under quite sticky conditions so that the tread bars can continue to bite into the soil when other tread patterns would become completely clogged with mud. Because the tread is more rigidly supported, and the side walls are more flexible, radial tyres have less rolling resistance, give 5–15% better tractive performance, last longer and produce lower peak soil pressures than similarly loaded cross-ply tyres (Gee-Clough *et al.*, 1977b; Plackett 1984).

The size of a tractor tyre is described by two dimensions (usually in inches). The first of these gives the maximum width of the tyre section, and the second indicates the diameter of the wheel rim on which the tyre is mounted. A 50 kW tractor can be fitted with 13.6 × 38, or 18.4 × 30 rear tyres, which are similar in overall diameter. However, the wider 18.4 × 30 tyre has a larger ground contact area and a greater carrying capacity and, when ballasted to take advantage of this, will give a 5–15% increase in tractive performance. It is not possible to use wide section tyres for row-crop work, but for most tillage and haulage operations there is no restriction on tyre width.

Compaction

Soil compaction by heavy machines is a problem not only of rutting the soil surface, but also of compression of the soil itself which reduces pore space, inhibits water movement, and increases the formation of clods. Compaction, under any particular combination of soil conditions, is largely a function of ground pressure; the higher the pressure, the more dense the soil becomes, and the greater the depth to which compaction occurs. Ground pressure is reduced by spreading the weight of the machine over a greater ground contact area, using wider section tyres or dual wheels. Although the degree and depth of compaction is reduced, more soil is compacted by the wider contact surface. Minimum degree, depth and volume of compaction is achieved by a long, narrow contact patch, or by the use of wheels in tandem.

Compaction is also increased under a wheel which is operating at high slip, the maximum effect occurring at a slip range of 15–25% (Raghavan *et al.*, 1977). Since this is above the level at which maximum tractive efficiency occurs, it is advisable, for reasons of efficiency as well as reducing compaction, to operate a tractor at loads which only require a slip of 10–14%. Working rate is maintained by pulling these lighter loads at higher speeds.

The most effective method of controlling compaction is to use the lightest machines that will do the job, keep off the soil when it is wet, and reduce the number of passes over the field to the minimum.

To eliminate compaction of the cropped soil entirely, the tractor can be run in the same wheel-tracks for all operations, while the crop is grown in beds between the wheel-tracks. This system is known as 'controlled traffic' or 'zero compaction'.

Four-wheel drives

The majority of new tractors sold in Europe are fitted with four-wheel drive. Driving all four wheels of a tractor offers a number of benefits:

(1) Greater soil contact area produces a greater cohesive pull, which is particularly advantageous under wet conditions when friction is low. When tractive conditions are good, the advantage is small.
(2) The pull is shared between four drive wheels, reducing each wheel's slip, compaction and sinkage. Ballast is also spread between all wheels, reducing total weight for a given tractor power. Two-wheel drive tractors above 50–60 kW cannot be ballasted sufficiently for maximum performance because of tyre limitations.
(3) The powered front wheels give improved steering control in wet conditions.
(4) The tractor's brakes are effective on all four wheels. Some tractors have front brakes, but normally only rear brakes are fitted, and front-wheel braking is only effective when front-wheel drive is engaged. Systems are available which automatically engage front-wheel drive (if not already engaged) when the brakes are applied.

The best tractive performance is produced by systems where all four wheels are the same size (Dwyer and Pearson, 1976). Smaller front wheels give a tighter steering lock but have smaller contact area and higher rolling resistance than full-sized front wheels. Equal-wheel tractors overcome the turning-circle problem with centre-pivot or four-wheel steering, which also allows front and rear wheels to run in the same tracks during turns.

If front wheels are to contribute fully to tractive performance, they must be correctly ballasted.

The power take-off (PTO)

Test reports show that between 80 and 94% of the power of a tractor's engine is available through the PTO shaft at standard speed, while only 50–70% is available through the wheels under average tractive conditions. The difference is mainly due to wheelslip and rolling resistance losses. There are two internationally standardised shaft sizes and speeds, a six-spline shaft turning at 540 rev/min and a 21-spline shaft turning at 1000 rev/min. Rotation is clockwise when seen from the rear of the tractor.

Since rotary power is the product of torque and rev/min, it can be seen that the faster PTO speed can transmit almost twice the power at any given torque, or through a shaft of a particular size since the power-carrying capacity of a shaft is limited by the torque it can withstand. Many tractors have dual PTO systems to accommodate all types of machines.

A fixed ratio between the engine and the PTO sometimes allows the engine to develop its maximum power at the standard PTO speed. Where this is not so, it is generally possible to over-speed the PTO so that engine power can be maximised.

The PTO is connected to the engine by its own clutch. If this can be operated quite separately from the transmission clutch, the system is called an 'independent' PTO. A two-stage clutch pedal controlling the transmission at half depression and disconnecting the PTO at full depression, gives a 'live' PTO.

Hydraulic systems

According to model and specification, from 15 to more than 30% of a tractor's power is available through the hydraulic system. This is adequate for light work such as the operation of the three-point linkage, tipping trailers, most front-end loaders and a few light duty excavator attachments. Larger hydraulically operated machines such as high-lift loaders or hedge cutters must have their own hydraulic power units, driven by the tractor's PTO.

Hydraulic power (in kilowatts) is a function of fluid flow rate and pressure, and is represented by the expression

$$\frac{\text{flow rate (litre/min)} \times \text{pressure (bar)}}{600} = \text{power (kW)}$$

Tractor hydraulic systems operate at pressures of 140–200 bar, with flow rates of 20–100 litre/min.

The tractor's three-point linkage is of standard dimensions and pin sizes (Category I, II or III or combinations according to tractor size). It is able to carry mounted equipment for transport, and to control the working depth of many types of soil engaging implements. Two alternative control systems are usually available, draught control and position control.

Draught control

This system adjusts the working depth of ploughs or other high-draught implements to maintain a constant draught or tractive load. Draught is sensed by springs in the upper or lower linkage; changes in draught cause the hydraulic system to raise or drop the linkage. A 'response' adjustment controls the rate at which these hydraulic corrections are made; slow response gives the smoothest work, but fast response may be needed if the ground is uneven.

Position control

This system will hold the linkage at a constant height relative to the tractor. This is generally used for transporting equipment in a raised position but may also be used to control the working position of some machines, sometimes in combination with draught control.

Front three-point linkages are available for many tractor models. The use of front-mounted implements can apply ballast to driven front axles, and can allow two light operations to be carried out simultaneously, at front and rear, saving time and labour.

Most tractors have available two or more pairs of external hydraulic couplings, controlled by separate double-acting control valves, and a coupling for the automatic operation of hydraulic trailer brakes. The valves may be used to control a front-end loader, rear fork-lift or other accessory, saving the cost of purchasing separate control valves for each attachment.

External hydraulic hoses are usually connected to the tractor by way of snap-on couplings, which will pull out and seal themselves if the machine becomes detached from the tractor. All couplings are a source of contamination to the tractor's hydraulic system, and care must be taken to avoid the entry of dirt when attaching and storing hydraulic accessories.

Health and safety for the tractor driver

The tractor driver is subject to a number of potential risks due to his occupation. The major risks are suffering hearing loss from long exposure to noise, and injury from tractor overturning accidents.

Tractors are noisy, typically exposing an unprotected driver to a noise level of 95–105 dB(A). In the short term, such noise levels can cause fatigue, and leave a ringing sensation in the ears when the noise has stopped. The long-term effect of exposure to high levels is to produce permanent hearing loss.

The ears can be protected by acoustic ear plugs or ear-phone-type protectors. Both are effective and cheap, but may not be comfortable, especially in hot weather. 'Q' cabs must reduce the tractor's noise to no more than 85 dB(A) inside the cab. Because of the logarithmic scale used, a reduction of 10 dB(A) represents a halving of the noise level to which the tractor driver is exposed.

Since much of the noise from a tractor is airborne, most of the effectiveness of the cab is lost if it is necessary to leave a door or window open, for access to implement controls or for ventilation in hot weather. Remote controls can eliminate the first problem, while the provision of re-frigerated air-conditioning systems can (at a price) do away with the second.

Tractor overturns most commonly occur sideways, when operating on steep slopes or driving too close to gulleys, ditches or steep banks. Trailers, slurry tankers or other heavy, unbraked machines can push the tractor down hills. Backward overturns occur less frequently, usually from attempting to pull from a high hitch point.

Since the introduction of BS approved safety cabs on all new tractors, the number of deaths from tractor overturns has diminished dramatically. There are almost no reorded instances of drivers being killed *inside* safety cabs. In an overturn accident, it is most important to hang on and stay inside the cab until the tractor has come to rest completely.

CULTIVATION MACHINES

In order to convert the surface of a field carrying the remains of a previous crop into an environment tailored to the esta-blishment and growing requirements of the next crop, a sequence of operations from a varying selection of machines must be employed. A detailed understanding of the actions and interactions of the full spectrum of cultivation machines, under a full range of soil situations, is essential if the aim is to achieve the desired result at least cost (Schaffer *et al.*, 1985). Primary cultivations are those involved in breaking up the soil initially, followed by secondary cultivations which refine the soil to produce a final smooth, level, firm seedbed.

Subsoilers

On heavy or poorly-structured soils it is occasionally necessary to loosen the soil to a greater depth than that reached by normal cultivations, in order to improve drainage and root penetration. Subsoilers for this purpose can operate at depths from 300–600 mm, and at spacings as close as 1 m.

The subsoiler consists of one or more heavy vertical tines, with a replaceable point or foot. A knife-edged vertical tine is common, but a flat leading edge in front of a tapering tine requires a lower draught force, as does a tine which is angled forward. Both these latter alternatives have the disadvantage that subsoil will slide up the flat front of the blade and be left on the surface. All tines, but particularly those which are angled forward, lift the soil into a bulge or 'surcharge' ahead

of the tine. It is important that no depth wheel or other component is positioned where it will prevent this action from taking place.

Below a particular depth, called the 'critical depth', at 200–400 mm according to soil conditions, the tine will just cut a slit through the soil, rather than bursting it upwards in a wide V. The width of soil loosened by the subsoiler can be considerably increased with only a small increase in draught force, by the addition to the foot of horizontal wings 300 mm wide. A pair of shallow leading tines, spaced at 0.5 m to either side of the main, winged, tine, loosen the upper soil layers and increase the critical depth for the deeper tine. The result is a large increase in soil disturbance with negligible increase in draft (Spoor and Godwin, 1978).

Ploughs

For centuries, the plough has been the main implement for primary cultivation. As well as loosening and breaking-up the ground, it inverts the top soil to bury weeds and trash. The soil loosening function can be performed by many other machines, but if soil inversion and burying is required, the plough must be used. In many situations, the value of inverting the top 200 mm of soil is being questioned, and some very successful cultivation systems are designed to keep the upper 100 mm, plus organic matter, on the surface by only cultivating to this depth. It was thought that the practice of straw-burning would further reduce the need for ploughing; as growers revert to straw incorporation, ADAS trials have re-emphasised the suitability of the mouldboard plough, without pre-cultivation, for this purpose (Fielder, 1986).

The plough is made up of a number of bodies, each of which turns one furrow, mounted on a rigid frame. Each body covers a width of 300–400 mm; this is normally fixed, but in a few cases furrow width can be adjusted to suit changing field conditions. The total width of all the furrow slices turned by the plough constitutes its effective working width, which can range from 300 mm to more than 4 m. Many different bodies are available to suit a full range of field conditions. A 'general purpose' body has a slow curvature which leaves the furrow slice intact, while a 'digger' body has a very abrupt curvature, to shatter the soil more effectively. 'High-speed' bodies reduce draught and leave neater work at high operating speeds.

Most ploughs can be fitted with pre-loaded release mechanisms which allow the whole body to fold back if it strikes an obstruction. This is particularly necessary where large ploughs are operated at high speeds. A small investment in protective devices can reduce the risk of long and costly delays and expensive repairs at a busy time.

The soil-engaging parts of the plough body are the coulter, a vertical knife or freely rotating disc, which makes a cut to divide the furrow slice from the unploughed ground, the share, which undercuts the furrow slice, and the mouldboard, which lifts and turns the furrow slice to leave it in its final inverted position. A small secondary body, or skim, may be used to shave off an upper corner of the furrow slice to ensure that no surface vegetation remains exposed. All soil-engaging parts are subject to wear, and are individu-ally replaceable.

Since the plough body pushes soil to the side (to the right on conventional ploughs), it follows that the soil pushes the plough equally in the opposite direction. This side force is resisted by a flat plate, the landside, which bears against the vertical edge of the unploughed ground. A correctly adjusted plough exerts no side force on the tractor which pulls it.

While the construction of the plough sets the width of most of the furrow slices, the width of the front furrow slice is governed by the distance between the front body and the inside edge of the tractor rear wheel which is running in the previous furrow. The wheel must be set at the correct width from the tractor's centre line, as specified by the plough manufacturer. A further adjustment can be made to the plough itself, to steer it closer or farther from the tractor wheel until the front furrow slice is at the correct width.

Ploughing depth, which should not exceed two-thirds of the width of the furrow slice if the slices are to turn satisfac- torily, may be set by means of an adjustable depth-wheel, if one is provided. More commonly, the tractor's hydraulic draught control system is used. The use of a depth wheel gives more accurate control, and is not affected by changes in soil conditions, but the use of draught control without a depth wheel puts more weight on the tractor's rear wheels. This improves traction and also eliminates the extra rolling resistance of the trailed wheel. Long ploughs, of four furrows or more, often have a depth wheel at the rear to supplement the draught control system.

In addition to the side-force difficulty mentioned earlier, the one-sided action of the plough poses problems in working a field, since it cannot simply be drawn up and down like most other machines. The field must be marked out in 'lands', which are then worked separately. At the centre of the land a ridge is formed from the soil turned inwards by one or more passes of the plough in each direction, and then the plough works up and down each side of the ridge, 'gathering' the soil, until the land is ploughed. The finished field shows a ridge at the centre of each land, and a double furrow between adjoining lands. The furrows, in particular, can affect subsequent operations carried out in the field at least to the harvesting of the next crop.

The problems of unproductive time spent marking the field out into lands, time wasted travelling along the headland from one land to the next, and the uneven surface produced by the conventional plough have led to the de-velopment and widespread adoption of the reversible plough. In this machine a second set of bodies, which turn the soil to the left, is mounted on the beam in opposition to the right-hand set. The front end of the beam is attached to the tractor's three-point linkage by a headstock which can rotate the plough through almost 180 degrees, when it is raised, and thus transpose the two sets of bodies. By using alternate sets of bodies, the reversible plough can work across a field from one side to the other with no marking-out, no ridges or furrows, and a minimum of idle time on the headland. A higher rate of work can be expected as a result of the better use of time, although the heavier plough will be harder to pull and will cost about twice as much as a conven-tional plough with the same number of bodies. The extra weight gives good penetration into hard soil, but the extra complexity of the rotation mechanism and headstock can sometimes cause trouble.

Rates of work of (0.8–2.0 ha h)/100 kW rated tractor power can be achieved, depending on soil conditions and working depth. Plough and tractor should be matched so that the power of the engine is fully utilised at a speed of 5–8 km/h. Fully loading the tractor at slow speeds causes high losses due to wheelslip, while operating at very high speeds puts great strain on the plough and leads to high

draught force, as draught increases with forward speed (Stafford, 1979).

The chisel plough

This machine is not a plough at all, but was hailed as a replacement for the mouldboard plough when the first units were imported from the USA. A heavy, wheeled, frame carries a number of curved, spring-mounted tines with re-placeable points. The tines, which are spaced out on three or four crossbars of the frame to give good clearance for trash, break the soil at intervals of 300–500 mm across the working width of the machine. Working depth is controlled by adjustable wheels, which also can carry the heavier machines when out of work.

Comparable home-produced machines have rigid tines, often protected by a shear-bolt. Most tines are flat-fronted and some are raked forward, as discussed under 'subsoilers'. Rates of work are double those which can be attained with a mouldboard plough, but at least two passes are necessary to achieve the same amount of soil loosening. There is little inversion, or burying of surface material (Patterson, 1982).

The disc plough

The disc plough resembles a mouldboard plough in layout, but the bodies have been replaced by angled, inclined, free turning concave discs. Large stationary scrapers are fitted to prevent soil build-up on the discs and to increase the turning action on the soil.

The soil is loosened and mixed, rather than inverted; where erosion is a problem, partially-buried crop residues can be very effective in binding and stabilising the soil.

The machine is difficult to adjust, and the work produced does not look like conventional ploughing. Where obstacles such as roots or rocks abound, the discs avoid damage by rolling over the obstructions rather than catching under them.

Although not much used in this country, the disc plough and its derivatives are widely accepted in areas where their special properties can be used advantageously.

The rotary cultivator

The rotary cultivator may be used for primary or secondary cultivation work. The horizontal rotor, up to 3 m wide, carries right- and left-handed L blades and turns at 120–270 rev/min to produce either a coarse or a fine tilth. A rear hood may be raised to allow clods to be thrown out, or lowered to give a further shattering effect as the clods strike the inner surface of the hood. Working depth, to 200 mm, is controlled by a land wheel or a rear crumbler roller. An alternative spiked rotor may be used for seedbed prepara-tion, and a bridge link may be used to operate a drill directly behind the machine to save time and labour (Patterson and Richardson, 1981).

If the ground is hard or stony, some machines can be fitted with front loosening tines which reduce rotor power require-ment and extend blade or spike life. The chopping and mixing action of the rotary cultivator is ideal for the incor-poration of crop residues, fertilisers or chemicals into the soil. Repeated use of the rotary cultivator may break the soil down into fine dust, and there is some risk of polishing the underlying uncultivated soil if conditions are hard. The machine has a high power requirement of 15–30 kW/m of width, but where one pass of this machine can replace several passes with alternative machines this will be acceptable.

The spring-tine cultivator

This is a versatile machine for secondary cultivations. A grid frame 1.5–10 m wide carries a number of S-shaped spring tines distributed over the grid so that one tine passes through each 100–200 mm strip of soil. Replaceable points of various widths are available. The machine is normally tractor-mounted, but is carried in work by adjustable depth control wheels. Machines above 3 m in width are made up of a central section with two hinged wings which fold for transport.

The spring tines vibrate as the machine moves forward, which is very effective in shattering clods. Trash and hard clods are brought up to the surface. The machine leaves the soil furrowed at about 600 mm intervals, corresponding with the spacing of the last row of tines. Crumbler rollers or light harrows, available as extras to fit on the rear of the machine, will produce a smooth surface.

The spring-tine cultivator has a good mixing action at speeds above 7 km/h and may be used for the incorporation of soil chemicals. Two passes are recommended, at 45 degrees to each other. Heavy spring tines (or discs) are a faster-working alternative to the plough when dealing with chopped straw residues on the heaviest land. Work should be to a depth of 100 mm, and the land should be rolled immediately after cultivation to ensure good contact between soil and straw.

The disc harrow

The tandem disc harrow is made up of four 'gangs' of concave discs, each gang clamped to an axle which is free to turn in sealed bearings. The front pair of gangs is angled to turn soil outwards, while the rear gangs draw soil inwards. Seen from above, the four gangs form a wide X.

An alternative layout, known as the offset disc harrow, has only one front and one rear gang of discs, each of which extends across the full width of the machine. The gangs meet at one side and are separated at the other to achieve the same relationship of angles and disc orientation as one-half of the tandem machine.

The front gangs, which must penetrate firm soil, are often fitted with scalloped discs. The rear gangs use plain discs, which last longer and move more soil.

The action of the disc harrow is to cut downwards into the soil, and to turn and mix the material thus loosened. Clods and subsoil are not brought to the surface. Two fast passes of the machine can effectively incorporate straw or chemicals into the soil.

Penetration may be increased by increasing the angle of the discs to the direction of forward travel, but a more effective method of increasing penetration is to add weight to the frame of the machine.

Tractor mounted disc harrows are available but, since much of the penetrating effect depends on weight, heavier trailed machines are better able to deal with tough or hard

soil conditions. A total machine weight of 150 kg per disc is recommended for straw incorporation. Large units can perform primary cultivations, particularly in lighter soils.

These large machines are available in widths up to 4 m, with discs of 760 mm diameter, and weighing 1 t/m of width. A machine 4 m wide, pulled by a tractor of 120 kW, can cultivate at rates exceeding 3 ha/h.

PTO-driven secondary cultivation machines

In addition to the rotary cultivator, described above, other power-driven machines are available for secondary cultivation work. Many use vertical spiked tines rotating in pairs about a vertical axis, or reciprocating across the width of the machine. The speed of the tines is much greater than the forward speed of the tractor and their shattering action is thus more effective than that of rigid tines being pulled through the soil. The action of the tines also levels and compacts the seedbed, without raking up much buried trash or subsoil. Bridge links may be used in conjunction with some of these machines, to combine the operations of seedbed preparation and drilling.

Power requirement is 15–30 kW/m of width. Tine life can be short if soils are abrasive and this can increase the operating cost of the machine. Where one pass of the powered machine can replace several passes with conventional machines, this cost is likely to be justified.

Harrows

A harrow consists of a large number of small tines or spikes carried on rigid or flexible (chain) frames. As they are dragged along, the action of the tines is to sort, level and compact the seedbed, the degree of penetration depending on the weight of the frame and the size of the spikes. Harrows are also used after drilling to ensure seed is covered. Chain harrows may be used on grassland to break up matted swards or to spread dung after grazing.

The power requirement of harrows is very low, so it is uncommon for any tractor to be fully loaded by a set of harrows of normal width. Harrows may be used in combination with other machines – spring-tine cultivators, seed drills – to achieve two operations in one pass.

Rollers

Ridged, or cambridge, rollers, made up of a number of ribbed cast iron wheels on a free turning axle, are often used in seedbed preparation to crush clods, compact the soil and leave a smooth finish. The compacting effect of a roller depends on its weight, decreases with increasing roller diameter and decreases with increasing forward speed. Since cambridge rollers weight 300–400 kg/m of width, and are used at fairly high speeds, their compacting effect is generally small and confined to the top few centimetres of soil only. Light, rigid-tined implements can be just as effective in increasing soil compaction, and do not leave a tight 'skin' at the surface.

Ridged rollers are available in widths to more than 7 m, the larger sizes being made up of a central section and two wings which sometimes fold hydraulically for transport.

Smooth rollers are commonly used for levelling grassland and pressing-in stones in spring to prepare for hay or silage harvesting later in the season. Rollers up to 3.6 m in width are available. By adding ballast to the hollow cylindrical rollers, weights from 0.5–1.3 t/m width can be applied.

Rollers weighing up to 5 t or more can sometimes take control on steep ground. Care must be exercised in matching the roller to a tractor of adequate weight and in operating safely on hillsides.

Some secondary cultivation machines may be fitted with a 'crumbler' roller. This is an open-cage roller, with a surface of spaced straight steel rods held in position by two or more integral wheels. The whole unit is free to turn in bearings at either end. As the roller moves forward, its weight is concentrated successively on each steel rod, which is quite effective in crushing clods at that point. There is also a good levelling effect.

Combination seedbed-preparation machines

Since many seedbed-finishing operations demand little power, a number of manufacturers have developed machines which combine several operations and can thus utilise a tractor's power more completely. Such items as ridged rollers, rigid or spring tines, crumblers or toothed rollers may be combined in such a way as to lift out and crush clods repeatedly, leaving a fine and level seedbed.

Reduced cultivations

Where a deep seedbed is not necessary (as it is for many root crops, for example), there can be savings in energy and increases in work rate if the soil is only cultivated to a depth of 100 mm. This system has been used successfully, even in heavy soils, for cereal production, without loss of yield. In many situations this practice has also led to long-term improvements in the structure of the upper soil layers. Problems with the build-up of annual grass-weed populations can occur unless this is controlled by herbicide applications or by occasional ploughing.

The soil may be worked to 100 mm depth by means of a heavy spring-tined cultivator or a rigid-tined cultivator, with tines spaced at 200 mm overall. Adjustable wheels control working depth. Three or four passes are needed before drilling in heavy soils.

Combination machines, generally using gangs of discs in conjunction with banks of heavy spring tines, can reduce the number of passes needed to produce a satisfactory seedbed. One example, 3 m wide, requires a tractor of 75–105 kW, and can cultivate up to 2.5 ha/h. Two or three passes are generally sufficient to produce a seedbed (Patterson, 1982).

FERTILISER APPLICATION MACHINES

Ninety per cent of the nation's fertiliser is applied as dry granular, prilled, crystalline or powdered materials. In some areas liquid fertiliser solutions are available; they have the advantage of being handled easily by pumping, but the disadvantage is that they are generally less concentrated than solid fertilisers and thus more material must be handled in order to apply a given weight of nutrients to the land. Also,

a storage tank must be installed on the farm to hold at least part of a year's supply of fertiliser.

Liquid fertiliser is applied by spraying machines very similar to those described below under the heading 'Crop sprayers'.

Dry fertiliser may be applied by a combine drill, simultaneously with planting the crop. In this situation, it is not practicable to apply fertiliser at high rates, because the small carrying capacity of most combine drills would necessitate frequent refilling stops.

Broadcasters

Most dry fertiliser is applied by broadcasting machines. A hopper of 250–1000 kg capacity on mounted machines or up to 8 t on trailed machines discharges fertiliser onto a spreading mechanism comprising a single or double spinning disc, or an oscillating spout. If PTO speed is correctly set, the fertiliser can be spread to an effective width of 5–24 m, according to the particular machine and the material used. The material is distributed in a pattern which is heaviest directly behind the machine, gradually reducing to zero at a distance of 4–18 m to each side. As long as this pattern is symmetrical and the reduction of rate is constant with increasing distance from the path of the machine, a return bout at the correct spacing will produce an even application rate across the complete area. Light materials are not spread as widely as heavy ones, so that it is extremely important to follow the manufacturers' bout width recommendations for the type of fertiliser being applied.

Application rates of 20–2500 kg/ha are possible with broadcasting machines, at speeds up to 12 km/h. Application rate is controlled by the rate at which fertiliser is metered onto the spreading mechanism, and by forward speed. Some trailed machines have land-wheel driven metering devices which makes the application rate independent of forward speed.

The machine can be calibrated at a particular setting by operating it, stationary, indoors for a measured time, collecting and weighing the fertiliser delivered. The time needed to cover 1 ha at a known speed and spreading width (*Table 21.1*) can be used to convert the delivery rate/min to that /ha. This test can also show up any difference in the amount of fertiliser thrown to right and to left, although the exact spread pattern cannot be determined.

In addition to the factors of correct PTO speed and forward speed, good condition of spreading mechanism, correct height above targe and machine set level, the accuracy with which the operator maintains his required width of spread is critical to accuracy and evenness of application. Foam marker nozzles on a boom are very satisfactory, and the same attachment can also be used on a sprayer to spread the cost, which only amounts to a few pence/ha. Far more than this amount can be wasted by incorrect fertiliser application.

Fertiliser, particularly in combination with water from the atmosphere, can be very corrosive to mild steel, iron or aluminium components. Manufacturers use corrosion resistant materials such as plastics or stainless steel for some parts of the broadcaster, but it is important also that the machine be easy to dismantle and clean at the end of the season. Vulnerable components can be brushed or washed clean, and coated with oil to protect them in good condition for the following season.

Table 21.1 Time taken to cover 1 ha, according to working width and forward speed (min)

Effective width (m)	*Speed* (km/h)								
	5	6	7	8	9	10	12	14	16
3	40	33.3	28.6	25	22.2	20.0	16.7	14.3	12.5
4	30	25	21.4	18.8	16.7	15.0	12.5	10.7	9.38
5	24	20	17.1	15.0	13.3	12.0	10.0	8.57	7.50
6	20	16.7	14.3	12.6	11.1	10.0	8.33	7.14	6.25
7	17.1	14.3	12.2	10.7	9.52	8.57	7.14	6.12	5.36
8	15	12.5	10.7	9.38	8.33	7.50	6.25	5.36	4.69
10	12	10	8.57	7.5	6.67	6.00	5.00	4.29	3.75
12	10	8.3	7.14	6.25	5.56	5.00	4.17	3.57	3.13

Notes:
(1) Non-productive time for such operations as filling hoppers, adjustment or turning is not included.
(2) Combination of width and speed not shown in the table may be evaluated by means of the formula $t = 600/(w \times s)$ where t is the time in min to cover 1 ha, w is the working width in m and s is the speed, in km/h.

Full-width spreaders

The broadcaster is an inexpensive machine which can be very accurate if calibrated often and operated correctly. However, the variation of application rate with speed, and the difficulty of maintaining the correct bout width have led manufacturers to develop much more sophisticated machines which spread evenly over a known and constant width. Most of these machines take the form of a central hopper (mounted to 1 t, trailed to 5 t) with a metering mechanism, carrying a folding boom up to 30 m wide along which the fertiliser is conveyed by an air blast. The fertiliser emerges from a series of nozzles, spaced as close as 600 mm in some cases, to form an even overlapping pattern across the full width of the boom.

Where the metering is by a wheel-driven force-feed mechanism, the machine may be calibrated by rotating the mechanism by hand for a specified number of turns and collecting the fertiliser in the tray provided. These machines can be extremely accurate, at application rates from 1 kg/ha for certain granules or seeds, up to 2 t or more/ha. However, the machine must be in good condition throughout (MAFF, 1980), PTO speed must be correct, and spacing must be within 0.5 m of the correct width. Inaccurate driving can leave strips without any fertiliser, or apply double rates. Again, the advantages of using a marker device are very clear.

PLANTING MACHINES

The objective of any planting operation is to produce the desired population of vigorous, healthy plants.

Grain drills

These machines can plant a range of seeds from grass or clover to beans. Seed is discharged in rows, in an even trickle, but not as individual seeds. Combine drills also deliver fertiliser, generally to lie close to the seed in the row.

Seed (and fertiliser) hoppers extend across the full width of the machine, which ranges from 2–6 m. Carrying capacity is 100–200 kg/m width on mounted machines, 220–300 kg/m on trailed models. Where a combined grain and fertiliser hopper is provided, it is often possible to reverse a central divider to give ratios of either 1:1 or 1:2 in weight of grain to weight of fertiliser, according to application rate. Removal of the divider allows the full hopper to be used for grain.

Seed is dropped in rows 120–135 mm or 175–180 mm apart, the number of rows per machine being from 15 to 50. Narrower rows put seed further apart in the row (*see Table 21.2*), but make the machine more complicated and expensive. Closely spaced coulters are more prone to blockage, and less weight can be applied to each one if penetration is difficult.

Table 21.2 The number of seeds (or plants)/m² according to row width and spacing within the row

Within-row spacing (mm)	*Row width* (mm)										
	110	*120*	*125*	*130*	*135*	*140*	*165*	*170*	*175*	*180*	*190*
10	909	833	800	769	741	714	606	588	571	555	526
15	606	556	533	513	494	476	404	392	381	370	351
20	455	417	400	385	370	357	303	294	286	278	263
25	364	333	320	308	296	286	242	235	229	222	211
30	303	278	267	256	247	238	202	196	190	185	175
35	260	238	229	220	212	204	173	168	163	159	150
40	227	208	200	192	185	179	152	147	143	139	132
Metres of row/m²	9.09	8.33	8.0	7.69	7.41	7.14	6.06	5.88	5.71	5.55	5.26

The metering mechanism is driven from one of the ground wheels. There is one metering roller per row. As the roller turns in the bottom of the hopper, serrations grip the seeds and carry them out to a position where they are discharged down the seed tube. The seed rate is adjusted by changing the speed of the roller in relation to forward speed, or by changing the proportion of the roller exposed to the seed.

To deal with different sizes of seed, the internal force feed mechanism has alternative openings for small or large seeds. The unwanted opening is closed off by a hinged flap. Some manufacturers can supply alternative sets of metering rollers with small or large serrations for the same purpose.

Individual rows can be shut off by slides in the bottom of the hopper. Thus, a crop such as rape can be planted in wider rows than those used for corn, by shutting off two out of every three rows. Small internal hoppers can concentrate the seed over only those rows that are being planted. Automatic or semi-automatic 'tramlining' attachments can be fitted to close off certain pre-selected rows every two, three or four passes across the field, to leave unseeded strips for further passes of sprayers or fertiliser machines.

Most modern drills are supplied with calibrating trays to collect the seed, and handles to operate the mechanism, for the equivalent of a known area so that the correct setting for a particular batch of seed can be established before drilling begins. Static calibration does not however take into account the effect of forward speed or of wheelslip which can vary from 0.5 to 15% according to soil and tyre conditions. A final calibration where the drill is run at full planting speed over a measured area (*see Table 21.3*) can ensure that the amount of seed delivered is correct under field conditions.

Seed population can be further checked after drilling, and plant population after emergence, with reference to *Table 21.2*.

Seed is released by the metering mechanism up to 600 mm above the ground, and falls downwards through a telescopic or concertina tube. The tube spreads out the flow of seed from small groups which leave some types of metering

Table 21.3 Calibration distances for machines of different widths

Working width (m)	Distance (m) to cover	
	$\frac{1}{10}$ *ha*	$\frac{1}{25}$ *ha*
1	1000	400
1.5	667	267
2	500	200
2.5	400	160
3	333	133
3.5	286	114
4	250	100
5	200	80.0
6	167	66.7
7	143	57.1
8	125	50.0
9	111	44.4
10	100	40.0
11	90.9	36.4
12	83.3	33.3

device, into an even stream. Fertiliser or seed dressing can build up inside these tubes; the action of flexible tubes tends to dislodge these deposits as the coulters move up and down.

The seed is placed in the ground by the coulter. Ideally, each seed will be planted at the same depth, and surrounded by firm, moist, warm soil. Where the soil has been thoroughly prepared, either single-disc or Suffolk coulters are used.

The single curved-disc coulter uses an angled disc to cut a groove in the soil, and the seed drops into this groove. Some soil falls back onto the seed after the coulter has passed. The disc coulter can cut through loose surface trash, but stones can push it up out of the ground, resulting in uneven planting depth.

The Suffolk coulter is a fixed blade, shaped like the front of a boat. It presses a groove in the soil, into which the seed falls. It can work in stony ground, and can give very even planting depth in well-cultivated soil. If loose surface trash is present, the Suffolk coulter will rake it up and frequent blockages will reduce the rate of work.

Both types of coulter are carried on spring-loaded arms; increasing spring tension will increase planting depth. Individual coulters can move vertically to follow uneven ground. In both cases, there is advantage in using a following harrow to improve seed cover; a trailed harrow is ungainly to use, and many manufacturers can supply integral spring-tined harrows for use with their drills. Where additional compaction is required, a very few makes of drill can be fitted with a heavy narrow press wheel behind each coulter.

In order to match up adjacent passes of the machine, most drills are fitted with disc markers which leave a small furrow which is followed by the front wheel of the tractor at the next pass across the field.

Direct drills

Several manufacturers produce machines capable of planting seed into uncultivated land. They are used for planting cereals after the straw of the previous crop has been burnt, and where the soil is suitable. Pasture renovation and stubble-turnip planting can also be carried out.

Most machines use a standard hopper and metering mechanism, but these are mounted on a very heavy chassis, with provision for carrying additional weight to increase penetration of the coulters if ground conditions are particularly hard. Special coulters open the ground and place the seed.

A popular coulter is the triple-disc type. A single vertical disc is followed by a pair of discs in V-formation. The tip of the V is ahead of the lowest point, so that the soil is pushed apart as the coulter moves along. Seed and fertiliser fall from the seed tube into the groove; there is no covering device.

The coulters are carried on independent spring arms; loading can be increased by lowering the frame of the drill onto the springs, or by applying hydraulic pressure to the top of the springs through an equalising linkage. There are many alternative types of coulter, using discs, disc and tine combinations, cultivator tines or narrow rotary cultivators to open up the soil. Spray nozzles may be positioned so that a strip of vegetation can be killed along the line of each row.

Because the direct drill is a heavier machine than the conventional version, rates of work are slower and power requirement is higher. However, since many operations can be eliminated by the practice of direct drilling, the overall requirement for labour and fuel/ha is reduced to less than 20% of that for conventional cultivation and planting systems, and large areas can be covered in a relatively short period (Patterson, 1982).

Precision seeders

Where crops are grown as individual, separate, spaced plants, it is necessary to plant seeds singly, rather than in a stream. This demands a metering mechanism which is capable of picking out individual seeds.

The most common metering device is the cell. Cells of the correct size to match the seed (which should itself be graded or pelleted to improve accuracy) are carried in the outer rim of a wheel, or as perforations in a flexible belt. The cells pass under the seed as it lies in a small hopper, and the seeds drop into the cells by gravity. After removal from the hopper, the seed falls from the cell to the ground by gravity, usually aided by an ejector. The seeds may be thrown rearwards as they leave the seeder, to counteract the machine's forward speed and reduce the tendency of the seeds to roll or bounce along the ground.

The most satisfactory results are achieved at low speeds (3–5 km/h), since this allows a sufficient time for the seeds to find their way into the cells. With increasing speed, more and more cells remain unfilled, and the number of gaps in the row of seeds increases.

An improved design of cell-wheel planter separates the two functions of selecting and feeding the seeds. A slow moving many-celled selector wheel turns in the bottom of the hopper. The seeds are delivered singly into a fast-turning feeder wheel, which ejects the seeds at ground level, with a rearwards velocity equal to the forward speed of the machine. Satisfactory precision planting is possible at speeds of 10–12 km/h.

More positive seed selection is achieved by the vacuum seeder. The vacuum is provided by a PTO driven fan. Single seeds are sucked into a ring of small perforated depressions on one side of a thin circular disc. As the disc rotates, the seeds are retained and lifted out of the hopper. An adjustable finger displaces any doubles. When the seed is directly above the coulter, the vacuum is cut off and the seed drops to the ground, again with a rearwards impetus in many cases. Some vacuum machines can plant seeds accurately at 8–10 km/h. One size of disc can meter a considerable range of seed sizes, so the time and cost normally involved in changing from one set of cells to another between crops is much reduced.

The metering mechanism is driven from a ground wheel on each seeder unit (the cheapest system to buy) or from a master landwheel which drives all units. Changing the drive ratio, by means of stepped pulleys or gears, changes the rotation of the mechanism in relation to the ground covered, and thus the seed spacing. On most machines, cell wheels or belts can be changed for ones with space for more, or less, seeds per turn, which will also change the spacing.

Once the machine has been prepared for work, by matching cells to seed size and by adjusting drive ratios, it is important to check performance in the field, at normal planting speed. A portion of every row should be uncovered so that spacing and evenness may be assessed, and corrected if necessary. *Table 21.4* shows the relationship between spacing, row width and plant population.

Table 21.4 Plant (or seed) population (10^3/ha) according to row width and spacing

Spacing (mm)	*Row width* (mm)					
	400	*500*	*600*	*700*	*800*	*900*
100	250	200	167	143	125	111
125	200	160	133	114	100	88.8
150	167	133	111	95.2	83.3	74.1
175	143	114	95.2	81.6	71.4	63.5
200	125	100	83.3	71.4	62.5	55.6
250	100	80.0	66.7	57.1	50.0	44.4
300	83.3	66.7	55.5	47.6	41.7	37.0
400	62.5	50.0	41.7	35.7	31.3	27.8
km of row/ha	25.0	20.0	16.7	14.3	12.5	11.1

Seed population must not be confused with plant population, which will be lower in proportion to the emergence percentage of the crop involved. The figure for sugar beet is 60–65%; extra seeds must be planted, to allow for this loss.

Precision seeders are made up of a number of single-row units, each with its own seed hopper, attached by flexible links to a tractor-mounted toolbar. By moving the units along the toolbar, row widths down to 200 mm are possible on some machines. Closer spacings are possible by staggering units in two rows. The number of rows of the seeder should be a multiple of the number of rows to be harvested simultaneously, for example a crop to be harvested by a three-row harvester should be planted by a 6, 9, 12, 18 or 24 row planter. This avoids the risk of misalignment which might occur if the harvester had to overlap two passes of the seeder.

The seed is released very close to the ground, so that the spacing is not affected by the presence of a seed tube. This gives the machine very little ground clearance; soil must be smooth, trash free and well cultivated. A typical unit has a front depth wheel, followed by a boat-shaped coulter which presses a groove in the soil. The coulter may be lowered to increase planting depth. Behind the coulter is a covering device, followed by a second depth wheel. As coulters wear, they produce a wider furrow in which seeds can roll or bounce out of position; depth is less constant and seed–soil contact is reduced. Worn coulters must be replaced. Ceramic-tipped coulters last, on average, three times as long as the iron or steel components they replace.

Accessories are available on some machines to push aside loose dry clods, or to level and compact an uneven tilth. Angled discs or blades can draw soil over the seed, and heavy narrow press wheels can improve compaction round the seed. Use of the correct accessories can ensure fast, even germination under a range of soil conditions, but care must be taken not to disturb the seed spacing with any following treatment. Seeder units may also be equipped for the application of liquid or granular chemicals.

Seeder units are very compact and close to the ground, so that it is difficult for the operator to see whether or not they are working satisfactorily. Some machines can be fitted with simple electric monitors to show whether the mechanism of each unit is turning, or whether the seed hoppers are empty. A more sophisticated device causes a light beam to be broken by each seed as it is released. Any interruption in seed flow sets off an alarm, so that the fault can be identified and corrected immediately.

Potato planting machines

The potato planter must deal with a large amount of 'seed' material, typically 40 000 sets or 2.5 t/ha. The material must be handled carefully, especially if chitted seed is used.

Since the crop is grown in a ridge, a deep tilth must be prepared before planting to allow the ridge to be formed. In some cases, ridges are formed before planting but most modern potato planters form ridges from flat ground as part of the planting operation. It is important at all stages of production to avoid any activity which will press the soil into clods, as these are difficult to separate from potatoes at harvest time. If potatoes are to be grown in stony soil, it is desirable to pick up and remove stones from the topsoil, or to gather them into windrows which are buried away from the crop.

Hand-drop planters

These machines, which generally plant two rows a a time into ridged land, are based on the mouldboard type ridger. Seats are provided on the planter for one operator per row, and seed is carried in trays or a hopper. A bell, operated by cams on a ground wheel, sounds at regular intervals for forward travel, and the operators drop a potato into a tube at each ring. The potatotes fall into an open furrow and are then covered with soil as the machine moves slowly forward. A two-row machine can plant 1.25 ha/d.

Semi-automatic planters

The task of the operator is less onerous on this machine as he fills up a series of small trays or cups as they move. These containers subsequently release the tubers at the correct intervals as the machine moves along. One operator per row is still required.

Automatic planters

Operators are not required on automatic planters, as a series of cups or fingers picks up the tubers out of a bulk hopper (375–500 kg capacity for a two-row planter) and releases them close to the ground at the correct spacing. A coulter makes a small furrow to receive the seed, which is then covered by disc or mouldboard ridgers.

The metering mechanism is a single or double row of cups on an endless belt or chain, or a series of spring-loaded fingers on the face of a large disc. Some machines have a make-up device which can add a potato to the metering device if an empty cup is detected. Feeding rates of up to 500 tubers/min are achieved on some cup-feed machines, giving ground speeds to 6.5 km/h, or a rate of work of 1.4 ha/h for a four-row machine. A more typical figure for a four-row machine is 5 ha/d.

Not all automatic planters are gentle enough to deal with chitted seed, but most manufacturers produce special models for this purpose. Machine damage to chits typically causes a yield reduction of 1.8 t/ha (Maunder, 1983). Potatoes with short, green chits are most resistant to damage.

High-speed planting with gentle tuber handling can be achieved on machines where the 'seed' is handled by belts,

and electronic control of spacing is incorporated in one design.

Because of the large weights of seed to be handled, as well as the fertiliser which is often applied by the planter, an efficient transport system is necessary if working rates are to be maximised.

Transplanters

Many vegetable and nursery crops are traditionally planted out as transplants. Bare-rooted transplants are planted by semi-automatic machines. The operator, seated on each single-row unit, selects plants and feeds them roots-up into a gripping device which may consist of a pair of flexible discs or a series of spring-loaded fingers. As this device rotates, it places the plant roots in a furrow which has been opened by a leading coulter, and releases the plants at the correct spacing. A pair of inclined wheels under the operator's seat gathers and firms the soil round the plant roots.

Spacing is adjusted by changing the gearing between the ground-wheel and the planting mechanism, or by changing the number of plant-places on the mechanism. Individual units can be spaced as closely as 600 mm; narrower rows can be planted by two or more banks of units. As with precision seeding, this operation demands a smooth, even soil surface with a good tilth.

Rates of work of up to 1500 plants/h per operator are possible; a five-row machine operated by a team of six, and well serviced with supplies of plants, can plant 1 ha of cabbages in 7 h, or a 1 ha of lettuce in 13 h.

Block transplanters

To avoid the growth check which occurs when quite large plants are uprooted, handled in boxes for a while and then re-planted, many growers are considering transplanting very small plants growing in small blocks of peat compost.

The transplanter presses square studs into the ground, and operators drop individual blocks into the depressions thus produced. Rain or irrigation washes the soil closely round the block, and growth continues without a check. Automatic block transplanting machines are reported to be able to work more than three times as fast as manual machines (Boa, 1984). Because the blocks are handled as well as the plants, materials handling becomes an even larger part of the operation. The use of smaller blocks combats the problem to some extent.

CROP SPRAYERS

Most chemicals are applied as liquid solutions or suspensions, generally water-based. The liquid is applied as fine droplets to give an even cover (*Table 21.5*) and because small particles are more likely to be retained on the leaf than large drops, which can roll or bounce off (Spillman, 1984).

However, very fine drops fall slowly through the air, and are therefore subject to drift (*Table 21.6*).

Spray drift has been measured as far as 6 km from the spraying site (Sharp, 1984). The problem of drift is increased by the fact that water evaporates from droplets in dry weather, making them lighter and more drift prone.

Apart from the problem of drift, it can be seen that reducing droplet size offers two major advantages:

Table 21.5 Effect of droplet size on closeness of spray cover

Diameter of droplet (μm)	Number of droplets/cm² at 45 litre/ha application rate
400	13
250	55
150	254
50	1750

Table 21.6 Effect of droplet size on spray drift (after Spillman, 1984)

Diameter of droplet (μm)	Horizontal drift when released from a height of 0.6 m		
	Wind 0.1 m/s	Wind 1 m/s	Wind 5 m/s
10	19.1 m	191 m	859 m
50	0.84 m	8.4 m	42 m
100	0.24 m	2.4 m	12.3 m
300	0.05 m	0.5 m	2.5 m
1000	0.01 m	0.1 m	0.6 m

(1) more close coverage of plant or soil. A more con-centrated spray solution can be distributed evenly and a tankful of spray will treat a greater area;

(2) better retention on leaves, so that a reduced application of chemical is necessary to give the required protection. There is less risk of the build-up of residues in the soil. Where very fine and uniform droplets can be produced, and with certain approved chemicals only, application rates can be reduced to as little as 0.1 litre/ha (*see* ULV, below). Typical field application rate is 200 litre/ha.

Sprayer nozzles

The orifice of the sprayer nozzle meters the flow of spray solution according to fluid pressure; an increase in pressure gives an increase in flow rate through a particular nozzle. As the nozzle wears with use, its output will increase. If a nozzle is found to be delivering an excess of more than 5%, it should be replaced.

The nozzle breaks up the spray solution into droplets and discharges them in a pattern towards the target. With wear this pattern narrows, and concentrates more spray directly in line with the axis of discharge. Nozzles are classified according to their pattern, output and rated pressure. Patterns include Fan (F), Evenspray Fan (FE), Low Pressure Fan (FLP), Hollow cone (HC) and Deflector (D). The class-ification F/80/1.2/3 denotes a fan spray nozzle, the angle of spread of the fan being 80 degrees, and the nozzle producing an output of 1.2 litres/min at its rated pressure of 3 bar (BCPC, 1986).

The output of a nozzle is also classified as *very fine*, *fine*, *medium*, *coarse*, or *very coarse*. According to pressure, most nozzles can produce more than one grade of spray; manufac-turers produce charts showing the pressures at which one or other grade is produced by a particular nozzle. Chemical manufacturers specify the spray 'quality' that should be used for each product, and this is shown on the product label.

'Fine' spray is used for foliar acting weed control in cereals and for contract acting fungicides and insecticides. Because of the risk of drift, fine sprays must not be used for products labelled 'toxic'. 'Medium' sprays are suitable for most purposes, while 'coarse' are used for soil-incorporated materials. 'Very fine' sprays are used for fogging, and 'very coarse' are used for liquid fertiliser application.

The nozzles are mounted along the boom at intervals of 0.5 m, and must be held at the recommended height above target if application is to be uniform.

The interchangeable nozzle tips may be made of brass, plastics, stainless steel or ceramics; durability of these materials increases in the same order. All nozzles should be replaced at least annually. The cost of a set of nozzles is no more than the cost of 1 litre of some spray chemicals, which is a very small price to pay for accuracy.

Many nozzle holders are equipped with check valves which shut off the flow of spray when pressure drops to 0.2–0.3 bar. This prevents dribble from the nozzles after the machine has been shut off at the end of a run. Corrosion-proof strainers may be fitted to protect the nozzle from blockage. It is important to match the screen size of the strainer to the aperture of the nozzle; if the strainer is too coarse, the nozzle is not protected; too fine, and the screen may clog up with small particles which could otherwise pass freely through the nozzle.

Conventional hydraulic nozzles produce a wide spread of droplet sizes, from those which are too heavy to stick to a leaf, to those which are light enough to be carried away by a breeze. Rotary atomisers have been shown to produce a very narrow spectrum of droplet sizes as the liquid spins off the edge of a rotating disc.

In the controlled droplet application (CDA) sprayer, droplets of 250–300 μm are produced. These resist drift and give good cover using less spray. Application rates of 20–40 litres/ha are recommended, which treats a very large area per tankful.

The ultra low volume (ULV) system uses a droplet size of 70 μm, at very low application rates. These droplets are carried by the wind to give a very thorough cover of the crop with excellent penetration into dense foliage. Herbicides and poisonous materials are not applied, because of the risk of drift, but satisfactory results can be achieved with approved fungicides and insecticides. Special oil-based formulations are used, to reduce the risk of evaporation while the tiny droplets are in the air. Only chemicals approved for this system of application may be used, and manufacturers' re-commendations must be followed.

To propel the spray droplets positively toward the target, thus reducing drift and increasing the effective delivery of spray, a number of systems of electrostatic charging have been developed. A charge of up to 30 kV is applied to the droplets as they leave the nozzle. The voltage gradient from the charged nozzle to the neutral crop or ground propels the droplets toward the target at up to nine times the speed of similar sized non-charged particles (Lake *et al.*, 1980), much reducing the risk of drift (Sharp, 1984). The charged droplets are attracted onto the surface of the crop ensuring that a large proportion of the spray actually ends up on the crop.

Other sprayer components

The tank, generally of a plastic material to resist corrosion, is of 200–1000 litre capacity on mounted machines, and up to 3500 litres on trailed models. The contents can be agitated by a pressure jet or by the more positive mechanical paddle. Agitation is especially important where suspensions are applied. The tank has a filler opening and strainer on top. Since the top of the tank may be quite high, many machines also have a low level hopper from which the chemical can be washed into the main tank by the pump. The tank is fitted with a fine strainer at its outlet to protect other components from blockage.

The pump is of positive displacement, delivering 30–150 litres/min at 540 rev/min, at pressures of 2–4 bar. Roller-vane pumps are cheapest but are subject to wear; diaphragm pumps are used on the majority of machines. Piston pumps are available, and can operate to pressures of 40 bar.

An adjustable pressure regulator maintains the correct spray pressure, and allows excess liquid to return to the tank once a pre-set pressure has been built up. Some pressure regulators contain a forward-speed measuring device which increases or decreases pressure in line with changes in forward speed. This reduces the possibility of errors in application rate because of speed variations.

The boom, which carries the spray nozzles, can be from 6–24 m in width; 12 m is a common size as it fits in with tramline systems. It is generally made in three sections, the outer two of which fold up for transport. Each section of the boom can be shut off independently.

The function of the boom is to carry the nozzles at a set height above the target, usually 0.5 m, with minimal vertical or longitudinal bounce. Boom height can be adjusted to maintain the desired clearance above a growing crop. Vertical bounce is countered by mounting the boom in a damped linkage to isolate it from the rocking of the machine. There may also be considerable longitudinal oscillations at the boom tip. The ground speed of the outer nozzles may vary from backwards, to forwards at 1.8 times tractor speed (Nation, 1982). This effect can also be reduced by appropriately designed flexible mountings (Nation, 1987).

Sprayer calibration

The value of spray chemicals handled by a sprayer in a year is usually far greater than the value of the sprayer itself. Time spent calibrating and checking the machine can yield large dividends in terms of more efficient use of chemicals. The sprayer should be calibrated at least as frequently as every 100 ha of use. The volume of water delivered/min by each nozzle at given gauge pressure can be measured using a special meter, or by collecting the liquid in a graduated 2 litre container for 1 min. This allows the output of the nozzle to be checked against the manufacturer's figure and for uniformity against the performance of the other nozzles on the machine. A minimum of four nozzles should be checked, at least one from each section of the boom. The total output of the machine per minute can also be calculated. From *Table 21.1*, the time taken to spray 1 ha at a given forward speed can be determined, and the output/ha of the sprayer can be calculated and compared with the desired application rate.

Correct forward speed is essential to precision spraying; tractor speedometers, if fitted, must be calibrated to take account of alternative tyre sizes and field conditions. Speed monitors, using independent ground wheels or radar sensors, are available, and some monitors will also display flow rates and application rate/ha.

Servicing sprayers

Although the field sprayer is a wide machine, and high rates of work are possible, a typical work rate for a 12 m machine is only 2.5–3.5 ha/h. A survey (MAFF, 1976a) showed that over 70% of machines studied spent more than half their 'working' time on such tasks as filling the tank, or travelling to the water supply. By good organisation, some machines were able to spend 70% of their time spraying. Fast refilling was achieved by the provision of a water tanker on the headland, often fitted with a 450 litre/min centrifugal pump. This cut down travelling time almost to zero and severely reduced filling-up time in comparison to the time taken to fill the tank via a hose pipe from the mains. Where only a few days are available for the application of a particular chemical, this level of attention to detail can greatly increase the productive capacity of the spraying machine. The survey also pinpointed the importance of driving at the correct bout width, and demonstrated clearly the advantages of a bout-marker system.

Sprayer safety

The use of spraying equipment and spray chemicals is potentially hazardous, both to the operator and to members of the public, livestock, pets, beneficial insects and non-target crops.

The Food and Environment Protection Act 1985 (FEPA) controls all aspects of the importation, advertisement, sale, supply, storage, handling and application of pesticides. An appropriate recognised certificate of competence is required by any person who sells, supplies or uses pesticides approved for agricultural, horticultural, forestry or amenity uses. Anyone applying these pesticides and not holding a certificate must be directly supervised by a certificate holder, unless they were born before 31 December 1964.

Only approved products may be advertised and sold, and users must comply with the conditions of approval relating to use.

The test for the certificate of competence in pesticides application (mandatory from 1 January 1989), is administered by the National Proficiency Tests Council (NPTC). Training is the responsibility of the Agricultural Training Board (ATB); courses are also available at some colleges. The following areas of expertise are included:

(1) Selection of an appropriate approved pesticide, and the use of the correct application method and harvest interval as specified by the manufacturer.
(2) Correct storage of pesticides in a suitable store, with appropriate safety preautions and record keeping.
(3) Correct selection, deployment and use of gloves and other necessary protective clothing and equipment when handling or using pesticides.
(4) Safe handling, transport and mixing of pesticides.
(5) Correct application, machine calibration and operation, including measurement of forward speed, quantity and uniformity of output.
Calculation of treatment rate
(7) Correct procedures for warning neighbours and beekeepers prior to spraying, and for the protection of bees when spraying.
(8) Determination of safe wind conditions for spraying in order to minimise the risk of spray drift.

(9) Exclusion of people and animals from treated areas.
(10) Safe disposal of unused pesticides, surplus mixture, sprayer washings and empty containers.
(11) Decontamination and first aid procedures.

(MAFF, 1986; NPTC, 1987).

HARVESTING FORAGE CROPS

Mowing and swath treatment machines

Whether they are to be conserved as hay or as silage, the majority of forage crops are first cut and then allowed to wilt or dry. The length of this drying period will vary from 24 h to several days, according to the conservation system.

Maximum drying rate of the cut swath depends on maximum use of the drying power of sun and wind. On one hot breezy day, sun and wind can remove more than 3 mm of water, 30 t/ha, from the leaves of a growing crop. If this drying rate could continue after the crop was cut, hay would be made in a single day. As it is, the crop resists water loss, and conditions in the swath do not allow sun and wind to reach all parts of the crop.

The process of 'conditioning' the crop, normally carried out simultaneously with mowing, seeks to deal with both crop and swath factors by

(1) Surface abrasion of the crop to allow water vapour to escape through the outer cuticle layer, once stomata have closed. The surface treatment, the severity of which should be adjustable to match crop requirements, is concentrated more on the thicker butts than on the more easily dried and fragile leaves.
(2) Production of a fluffy, durable swath for maximum wind penetration.
(3) Mixing of the crop to expose thicker butts to wind and sun.

This 'conditioning' must be achieved without the penalty of excessive fragmentation and leaf loss which can result from over-vigorous treatment. Losses increase with the length of time that the crop is drying in the field, so a treatment which gives a rapid drying rate for a 24 h wilted silage crop may well be too severe for a 5 d hay crop.

The ability to leave a very short stubble may seem to be advantageous but in fact a longer stubble, of 50 mm or more, has been shown to give faster drying rates than where the crop is shaved to the ground (Klinner, 1975). The risks of contaminating the crop with soil, and also of damaging the machine, are reduced considerably by leaving a longer stubble.

The cutterbar mower

This machine cuts by shear as the reciprocating knife sections move across the stationary ledger plates. Knives must be sharpened often. Work rate is 0.2–0.6 ha/h for a 1.5 m machine. A leaning or lodged crop reduces the work rate. The machine is very economical to use as it only requires 3–5 kW at the PTO and it is also the cheapest type of mower, being less than half the price of its nearest rival. The cut swath lies quite flat, with the butts covered; it must be tedded or conditioned immediately after cutting if a fast drying rate is to be achieved.

Impact mowers

All other mowers cut by impact. A cutting edge moving at 50–90 m/s smashes its way through the base of the crop. The crop is severed whether or not the machine is sharp, although blunt cutting edges may absorb 50% more power. Power requirement is much greater than for the cutterbar machine; the highest power is consumed when mature, stemmy crops are cut. There is danger both to the machine and to the user if it strikes a stone or other obstacle. Blades can be broken and fragments can be thrown out. Guards and shields must be kept in place.

Impact machines can offer very fast working speeds with little maintenance or downtime, and are little affected by the condition of the crop. Against these advantages must be set the machine's higher initial cost and greater power requirement.

Drum and disc mowers

Two or more rotors turn on vertical spindles. Small flat knives are attached to the periphery of the rotors, which are from 350 mm to 1 m in diameter. As the rotors turn the knives cut a series of horizontal arcs as the machine moves forward. Top speeds of 80–90 m/s ensure a clean cut.

The rotors of a disc machine are driven from below, and the cut material passes over them; the alternative is a drum-shaped rotor assembly driven from above. In the latter case the cut material passes back between adjacent pairs of drums. If the cut material does not pass quickly back off the machine there is a risk of double-cutting, where small fragments are severed from the butt end of the crop and are subsequently lost in the stubble.

Power requirement is 35 kW for a 1.6 m drum machine, with a 15% reduction for disc machines. High forward speeds are possible, and work rates of 0.5–1.5 ha/h/m can be achieved (Raymond *et al.*, 1986). Capital cost per unit width increases with width, with a large increase for trailed machines. For this reason, and as a means of avoiding the reduction in drying rate that is experienced in swaths produced by large machines, the use of twin smaller machines, one front and one rear mounted on the same tractor may be considered (Tuck *et al.*, 1980).

The machines are complex; some have many fast-moving parts which may rotate at speeds in excess of 3000 rev/min. A high level of maintenance is vital if a long working life is to be achieved.

The cut swath is slightly fluffed-up, but the butts are covered. After an initial period of rapid drying, the overall rate is little better than that from a reciprocating mower. Tedding or conditioning is essential.

Mower-conditioners

As mentioned above, the swath produced by cutterbar or drum/disc machines is not 'set up' for rapid drying without further treatment to achieve the conditions described above. The combination of a mower with a 'conditioner' can perform both these operations in a single pass and leave the swath in an ideal state for fast drying. A mower-conditioner requires 35% more power than the equivalent mower and may cost 50% more to purchase. Work rate will be lower than that of the mower alone.

Two types of conditioner may be used. Roller machines have a pair of horizontal rollers which crush and bruise the crop before dropping it into a loose swath. These machines are best suited to stemmy crops such as lucerne, but deal less well with heavy, leafy grass crops.

More recently, flail conditioners have been developed. The cut crop is lifted, mixed, turned and scuffed by a rotor carrying narrow fixed or hinged flails. This machine deals more thoroughly with thick, leafy crops, where the rigidly mounted resilient flails are able to penetrate throughout the material. It is less expensive but more power consuming than the roller machine. The severity of treatment can be easily adjusted by moving a baffle which can delay the passage of the crop so that the flails apply more impacts to it. The 'spoke' conditioner developed at NIAE (Klinner, 1975) has been widely adopted by mower-conditioner manufacturers.

The use of any steel-tined machine carries with it the risk that a broken tine may occasionally be left in the swath and cause serious damage to following machinery. Designs of conditioner which use plastic bristles or toothed plastic sheet eliminate this problem, as well as offering greater improvements in crop drying rates (Klinner and Hale, 1984).

Swath treatment machines

Once cut, and possibly conditioned, further treatment to assist drying requires fluffing-up (tedding) or spreading-out the crop to maintain air circulation, and this must be followed by windrowing the crop to match the next stage of the harvest operation or to reduce the risk of spoilage if rain is expected. The crop should be tedded at least daily in good weather.

Spreading the swath produced by a mower or mower-conditioner allows the cut material to intercept most of the solar energy falling on the field, rather than half or less when gathered into rows. Drying breezes also have better access to all parts of the crop. The disappointing drying rates often experienced where crops are cut by machines over 1.8 m in width can be greatly improved if the crop is spread immediately after cutting (Jones, 1985). A Dutch system involving early morning cutting with a mower-conditioner at a stubble height of 75 mm, spreading immediately, and spreading again after 2 h has been shown to produce wilted material of 35% DM or more, by the end of the day of cutting (Bosma and Verkaik, 1986). Where spreading is used, crop losses are increased both by the scattering action and because a tractor must run over part of the crop in order to gather it back into windrows (Rees, 1982).

Most swath treatment machines use spring-steel tines to move the crop. If these are set too close to the ground the risk of breakage is increased. Farmers often modify their machines by adding small clips or ties so that broken tines are retained rather than being allowed to fall into the windrow. Machine designers have addressed this problem by the use of plastic or rubber crop-handling elements.

Combination machines

Some machines are able to perform all the functions of tedding, turning and windrowing. They cover a 3–5 m width, and generally consist of two PTO-driven tined rotors which rotate on vertical spindles, some with movable swath guides behind. The whole machine is tilted forward to ted or scatter the crop, or levelled to produce a windrow.

Most machines can be made to rotate more slowly if gentle handling is required when the crop becomes dry and brittle or in difficult crops, such as lucerne, where leaf loss is a problem. It is not always easy to produce an even swath, but this is most important if a high harvesting rate is to be achieved by following machines.

'Specialised' machines

An alternative swath treatment system is to use separate machines for the different functions. Spreading machines of 4–7 m width use four or six PTO-driven tined rotors. Wide machines are articulated to follow ground contours.

A strip of spread material 3–5 m in width may be gathered into a windrow by a single-rotor machine on which groups of up to four tines sweep the ground as they pass across the front of the machine, and then fold back to release the crop and leave it in a windrow at the side. A guide, of canvas or of steel rods, assists in the formation of an even swath.

A gentle method of gathering spread material into a windrow, or of combining two or more rows into one, is by means of the finger-wheel side rake. A series of four to eight spring-tined wheels roll along the ground while held at an angle to the direction of travel. The tines sweep the ground and roll the crop from one wheel to the next across the width of the machine. The tines travel slowly in relation to the crop, and the rolling action retains small fragments of the crop that might otherwise be lost in the stubble. This action also tightens the swath and can reduce air circulation and the rate of drying if used at an early stage. The machine is inexpensive to buy and requires little power. Front mounting of the gathering rake can allow two rows to be put into one by the tractor which also powers the baler or the forage harvester.

Barn hay drying

When a hay crop has dried to 35% MC in the field, further drying will only take place if the relative humidity of the air is low enough to extract moisture from the crop against the attraction of the plant material (*Table 21.7*).

This final stage can be speeded up, and exposure time in the field reduced by 1 or 2 d, if the hay is removed from the field at moisture contents of 50–35%, and dried with forced air in a barn or stack. The result will be a better quality product with reduced field losses and weather dependence, although it is produced at a higher cost/t than field cured hay.

Conventional bales are generally dried in the store which they will occupy until used. Batch systems involve a second

Table 21.7 The equilibrium relationship between air relative humidity (RH) and hay moisture content (MC)

Air RH (%)	Hay MC (%) below which moisture will not be removed
95	35
90	30
80	21.5
77	20
70	16
60	12.5

move of the crop which increases cost. The air is blown into the stack by a fan, via a false floor or large duct under the stack, or a vertical duct up the centre of the (square) stack. Mesh floor systems require an airflow of $(0.25 \, m^3 s)/m^2$ of floor area, while the vertical duct system demands $0.005 \, m^3/s$ per bale, at pressures of 60–85 mm water gauge.

Where the air temperature can be raised by 5°C above ambient, drying can be continuous. Without heat, drying can only continue while the ambient air is dry enough to extract water. This requires constant monitoring of the relative humidity, and the drying process will be extended over a longer period.

Forage harvesters

The job of the forage harvester is to gather up crops which are to be conserved as silage or, occasionally, fed directly to livestock. The crop must be delivered into a trailer at an appropriate chop length to suit the conservation system and the feeding method. A short chop length requires more power and more expensive machines. Chopped material becomes more compact in load and store. Fine chopping improves fermentation, particularly of high dry-matter forage.

A $10 \, m^3$ trailer can carry 1500 kg of fresh grass chopped to 200–250 mm, 2300 kg if chopped to 100 mm, and 2500 kg if chopped to 25 mm. For satisfactory compaction and fermentation in a clamp silo, a chop length of 200 mm is recommended for material up to 20% DM, 130 mm for material of 20–25% DM, 80 mm for 25–30% DM and 25 mm for material above 30% DM. For a tower silo a chop length of 25 mm or less is required.

Most forage harvesters discharge directly into trailers and cannot work at all unless a trailer is available. To keep the machine working at full capacity, careful organisation of transport is of paramount importance.

Flail forage harvester

This is a simple and rugged machine which can cut a standing crop or pick up a windrow. A horizontal spindle across the width of the machine carries two or four rows of pivoted flails with sharpened cutting edges. These cut or scoop up the crop as the spindle rotates at 1000–1600 rev/min. The crop is thrown up a chute and directed into the trailer by a rotating spout and a movable deflector flap, controlled by the driver.

Cutting widths range from 1100–1500 mm and the machine requires a tractor of 35–50 kW to produce an overall rate of work of 7–10 t/h. Harvested material is lacerated and broken by the flails; the severity of treatment can be adjusted by moving a baffle closer or further from the flails. Chop length averages 150–300 mm, but particle sizes range from 50 mm to original length.

Side-mounted machines allow the trailer to be pulled directly behind the tractor for good manoeuvrability, stability and ease of hitching. For a one-man operation, the forage harvester is quickly detached, allowing the tractor and trailer to travel to the silo. Trailed machines normally run directly behind the tractor with the trailer behind. This makes a less stable combination and changing trailers takes longer.

Double-chop forage harvester

This machine combines a flail head, using edge-cutting flails, with a flywheel cutter and blower. This gives a more positive chopping action although chop length is not precisely controlled. The machine was originally designed for direct-cutting forage as part of a zero-grazing system, but is also suitable for picking up wilted material.

Typical chop length range is 50–150 mm and the machine requires 40–55 kW to achieve a nett output of 8–12 t/h. This is a trailed offset machine which pulls the trailer directly in line with the tractor.

Precision-chop forage harvester

The precision-chop machine is normally fitted with a tined pickup for harvesting windrowed crops. Alternative headers for direct cutting and for harveting maize are usually available.

'Precision' is attained by controlling the feed of the crop into the cutting mechanism so that the crop moves forward a set distance between each cut of the blades. This distance is known as the 'theoretical chop length', and can normally be adjusted between 5 and 50 mm. The average length of the chopped material will be greater than the set value, because of the way the crop is arranged in the swath. Chop length is adjusted by changing the rate of crop feed, controlled by two or one pair(s) of feed rollers, or by varying the number of knives on the rotary cutter. The shorter the chop length, the greater is the power consumption, or the lower the output of the machine.

Cutting is achieved by the shearing action of a sharp, fast moving, knife against a solid square-edged shear bar. Dull knives or excessive clearance between knife and shear bar waste power and reduce work rate. Machines are fitted with knife sharpeners and these should be used at least daily, followed by re-adjustment of clearances if necessary.

The cutting mechanism is particularly vulnerable to damage from stones or metal objects picked up in the windrow. Damage-resistant designs of cutter have been developed by several makers, and electronic metal detectors can stop the crop intake mechanism instantly if a piece of ferrous metal is detected. The particle of metal has to be picked out before harvesting continues. This device protects the machine from expensive breakdowns, and also prevents small metal fragments from being incorporated into the silage and later fed to livestock. Research workers at NIAE have developed a pickup system which can reject the portion of a windrow containing a stone or a piece of metal, without the need to stop the harvester at all (Klinner and Burgess, 1982).

Most machines use cylinder cutters, which are similar to the mechanism of a traditional lawn mower. Up to 12 knives arranged round the periphery of a cylinder slice the crop against the shear bar. The direction of cut is normally downwards, and the cut material is dragged round under the cutter and then thrown up the chute. Upwards-cutting systems throw the cut crop directly upwards, eliminating drag, and releasing more power for cutting and throwing (Röhrs, 1986). A number of manufacturers use flywheel cutters. Power use is intermediate between the two types of cylinder systems, and the cutters are claimed to work more quietly and to resist damage better.

An experimental design of forage harvester which slices the crop with sharp-edged discs and then conveys it into the trailer with a belt conveyor has been shown to reduce cutting power requirement by 50% and conveying power by 70% and has achieved outputs of 60 t/h (Knight, 1986).

Trailed offset machines come in a range of sizes to match 50–150 kW tractors. Outputs range from 10–40 t/h, according to available power, crop density, chop length and field organisation. Self-propelled harvesters, generally contractors' machines, range in power from 110–220 kW. Outputs of 30–80 t/h are possible, but this depends very much on the availability of large, even windrows and plenty of empty trailers. To reduce overhead costs, the machine's working season may be extended by the harvesting of a succession of crops throughout the summer and autumn.

Self-loading forage (SLF) wagons

An alternative one-man system of loading and transporting crops for clamp silage, the wagon, of 3.5–8 t capacity, loads itself by means of a reel pickup under the drawbar and a short vertical conveyor which packs the crop into the enclosed body. Up to 21 stationary knives may be positioned against the conveyor, to slice up the crop as it passes by. Chop lengths of 80–220 mm are achieved when knives are sharp. A chain and slat conveyor on the floor of the wagon can move the load rearwards as the machine fills and can discharge the contents through a rear gate in 3–10 min at the silo. When the trailer is loaded, the pickup is raised hydraulically and the combination is then driven to the silo.

The SLF wagon has a low requirement for power at the PTO, but a tractor of 35–50 kW is necessary to control the heavy trailer on hilly ground; three or four loads/h can be achieved over a transport distance of 500 m or less (MAFF, 1976b). The chop length is longer than that recommended for wilted silage.

Baled silage

Silage crops may be handled as unit loads if they are baled using large round balers or high-density large rectangular balers. These balers can also be used for packaging straw, but are not suitable for baling hay unless it is drier than 20% MC or treated with a preservative to prevent moulding. The heavy wet crop causes more wear to the baler than a similar amount of dry material.

Bales are handled by tractors with front and rear forks, and may be loaded onto flat trailers for haulage over greater distances. Individual bales may be enclosed in polythene bags of 80–125 μm thickness, or wrapped in four layers of thinner 'cling' material, or a stack may be built and covered with plastic, similar to the conventional clamp but without the cost of the walls. No further compaction is necessary, as the bales are compressed already. A high stack may be built from ground level without any need for a machine to be driven onto the material. Forage should be wilted to 30% DM or more to eliminate effluent problems and minimise settlement when stacked.

Individually wrapped bales offer the advantage of allowing extra forage to be ensiled without the need for additional silo space; batches of silage can be kept separate, and unneeded silage can be sold off the farm as a cash crop. Care must be taken to protect the wrapping membrane from rodent, bird and wind damage.

Silo filling

Handling at the silo must keep pace with harvesting equipment. A tractor with rear push-off buckrake can handle up to 30 t/h. Greater outputs are possible with a front mounted buckrake, and more still from four-wheel drive handling vehicles with fork capacities of 1–2 t. In addition to spreading the crop the machine also compresses it. Final densities of 700–850 kg/m^3, after fermentation, are common.

Silage may also be stored in tower silos, which reduce storage losses and permit every stage of silage handling to be mechanised. Tipping trailers deliver precision-chopped forage into a stationary dump box, which feeds it at up to 30 t/h into a forage blower driven by a tractor of 40–80 kW. The forage is blown up a vertical pipe 225 mm in diameter, over a curved section and against a spreading device in the centre of the roof of the silo. Spreading is essential if the silo is to be filled evenly and its potential storage capacity is to be fully utilised. A minimum filling rate of 2 m depth/d ensures satisfactory fermentation.

During the early stages of fermentation, gases such as carbon dioxide and oxides of nitrogen are evolved. If the silo is partly filled, these gases accumulate above the silage and can kill anyone climbing down onto the silage, perhaps to remove plastic sheeting after a weekend. Great care must be taken to clear any such hazards before entering the silo. *Never* enter a silo or other enclosed space (empty slurry tank, grain silo) without a rope attached, and the supervision of companions capable of giving assistance (without entering the silo themselves) should an emergency arise.

Baling and bale handling

A convenient system for handling hay (and straw) is to package it into bales. These may be moved by hand or grouped into larger unit loads for machine handling. Large bales must be handled by machine. The process adds nothing to the final usefulness or value of the product, indeed this may be reduced. Handling as bales can offer a least-cost solution to the problem of moving bulky materials on the farm.

The conventional pickup baler

This machine produces bales of section 350 × 460 mm, and variable length. Larger sections such as 400 × 450 mm are used on some high-output machines. Bale density, which is adjustable, varies from 160–220 kg/m^3, and bale weights range from 15–40 kg.

The offset spring-tined pickup lifts the crop into the machine. Its height is controlled by skids or small wheels at either side.

The crossfeed moves wads of material from behind the pickup and feeds them laterally into the bale chamber each time the ram draws back.

The heavy ram slices off each wad from any trailing material, and forces it down the bale chamber towards the rear of the machine. As the compressed material moves along, driven by the ram working at 60–95 strikes/min, it carries with it two loops of string which will eventually form the securing bands of the bale.

When sufficient material has passed the bale length sensor, the tying cycle is initiated. This is timed to occur when the ram is forward. The needles bring up the ends of the strings to encircle the bale, place them in the knotters, and withdraw, trailing out a new string ready for the next bale. The knotters tie each bale string by forming a loop and pulling the two ends through. The strings are cut at the knotters to release the bale, which continues rearward past the adjustable restrictor that controls bale density, and is finally discharged via the bale chute.

Many balers have potential work rates of 15 t/h or more; typical field performance is 6–8 t/h. Large, even windrows allow the machine to operate steadily at high rates of throughput.

Handling conventional bales

The conventional bale was originally intended to be a suitable unit for manhandling, and can still be handled that way if labour is available. It remains a convenient unit for feeding to livestock.

Ideally, a bale handling system will deal with bales at all states – field stacking, transport from field to store, loading the store, and unloading from store to final use. It is very unusual for a conventional-bale handling system to fulfil all these requirements; many are only used for loading trailers in the field, which is a very small part of their potential.

Many mechanised bale-handling systems are available. Most make larger units by forming the bales into groups of 8, 10, 20 or multiples of these numbers. A collector sledge may be pulled behind the baler, and the bales released at intervals to form windrows. These windrows are formed into the desired groups by hand. Alternatively, the groups can be formed automatically on an 'accumulator', which releases each group when it is complete. These groups are dotted all over the field unless a further collector is drawn behind.

Groups of bales are picked up by specialised lifting attachments on tractor fore-end loaders or other handling machines, and may be built up into larger stacks or loaded directly onto trailers. Stacks of 40–144 bales may be picked up as units by specialised trailers and moved directly to store.

A survey has indicated the labour requirement for many systems of handling bales from windrow to store (MAFF, 1975, 1976). The figures shown below are averages, and considerable reductions were achieved by the best or 'premium' operators.

Bale/collect; hand-load trailers; hand stack: 83 man min/t
Bale/accumulate 8s; front-end loader to trailer: hand stack: 49 man min/t
Bale/accumulate 10s; field stack 100s: specialised trailer to store 26 man min/t
Bale/self-loading wagon 88s; place stack direct into store: 21 man min/t

It must be remembered that labour requirement is only one of a number of criteria on which a bale-handling system is chosen. Other factors may be overall rate of work, capital costs of the equipment, or the suitability of a system for an existing storage building. The final solution, as with other materials handling problems, must deal with bales at least cost when all forms of cost are taken into account.

Big balers

Handling crops in units weighing from 150 kg to 1 t allows fast rates of work, generally demanding only a single operator. All handling must be by machine, so store layouts have to allow this.

Large round baler

These machines produce a cylindrical bale, either 1.5 m wide × up to 1.8 m diameter or 1.2 m wide × up to 1.5 m diameter. Bale weights range from 150–800 kg, with densities from 90–140 kg/m³. The bale is produced in a cylindrical chamber in the form of a roll. The machine has no crossfeed, so careful swath building and care in aligning windrow and pickup are essential for the production of good-shaped bales.

Fixed-chamber machines accumulate material until a pre-determined tension is attained, while variable chamber systems, generally giving a higher density bale, allow any diameter of bale to be produced up to the full dimension of the bale chamber.

When the bale chamber is sufficiently full, the operator stops forward travel and initiates the wrapping of six or more turns of string round the bale as the baler continues to turn. The string is cut, but not knotted, and the bale is discharged through a rear gate. The bale rolls down a ramp and away from the baler to give room for the rear gate to close. If the windrows are close together it may be advisable to reverse the baler out of the row so that the bale can be dropped in a position where it will not interfere with the next pass of the machine.

The machine can be fitted with equipment for wrapping each bale with about two turns of a self-clinging plastic net. Because of reduced wrapping time, the output of the baler is increased by up to 50%. This advantage can often outweigh the extra cost of the attachment, and of the netting which is more expensive than conventional twine. Continuous baling, with no stops for wrapping, is achieved by the use of a 'feed chamber' where picked-up material can build up while wrapping and ejection are taking place.

Round bales can be carried by a grapple or prong attachment on the front or rear of a tractor, or may be loaded onto flat trailers. The smaller size allows two bales to be loaded across the width of a trailer or lorry, which is not possible with the larger size.

The bales may be self-fed, or can be unrolled on the ground. Mechanical unrollers, shredders and tub grinders may also be used to break up the bale.

Large rectangular high density baler

Packages weighing up to 1 t or more are produced by these very large machines, which require a tractor of at least 75 kW. Bale size varies: one popular machine produces a bale of 1.2 × 1.3 × 2.4 m, at a density of 300 kg/m³ or more. This system is suitable for handling very dry hay, straw or silage crops.

The bales are too heavy to be handled by smaller tractor front-end loaders, but are well within the capabilities of specialised handling vehicles fitted with grapple or squeeze-loader attachments.

GRAIN HARVESTING

The combine harvester

Grain crops (and also pulses, oil seeds and other seed crops) are now universally harvested by the combine harvester. This machine combines four major functions: it gathers the crop, threshes the seed out of the ear, separates the seed from the straw and finally cleans the seed of chaff, weed, seeds and dust. After temporary storage in a hopper on the machine the grain is delivered into a bulk trailer or lorry.

Most machines are self-propelled with cutting widths ranging from 2.1 to 6.7 m: most machines are offered with a choice of two or three alternative cutting widths. Engines of 32–220 kW are fitted. Although the 'size' of a combine is often described by quoting its cutting width, a more useful indication of its harvesting capacity is its power, or the width of the threshing cylinder. Under good working conditions, the factor which normally limits a combine's rate of work is its ability to shake out the grain from the remainder of the 'material other than grain' (MOG) in the separating mechanism. This ability varies from crop to crop according to maturity, moisture content, straw-length and threshing adjustments and so cannot be quoted as a constant figure.

A typical grain harvesting season contains 12–20 dry working days (up to 200 working hours), according to geographical location. While it is common to select a combine size which will provide approximately 1 m of cutting width for every 35 ha of grain (up to 60 ha in a few cases), an alternative approach is to specify a harvesting capacity which can deal with an expected crop within the working time available in the region. As explained earlier, no firm output figures can be given, but the following approximations are satisfactory for planning purposes; machines of 38–60 kW can harvest 4–7 t of grain/h, 60–80 kW, 6–10 t/h; 80–100 kW, 8–12 t/h; more than 100 kW, 11–16 t/h. The largest 'rotary' combines can attain rates in excess of 24 t/h in good conditions.

Inadequate combine capacity leads to an extended harvest period with greater weather risk and rapidly rising levels of crop loss from shedding, both in the field and at the combine intake. A doubling of pre-harvest losses can occur after a delay of only two weeks in some crops (Klinner, 1980).

Most combines operate below full capacity. More than two-thirds of machines are used less than 200 h/year. The availability of excess harvesting capacity is a form of insurance against bad weather at harvest time, which also unfortunately leads to high overhead costs/ha harvested. A further form of insurance against a poor harvest season is the installation of grain drying equipment (*see* p. 306) (*see also* Elrick, 1982).

Gathering

The crop is gathered into the machine at the 'table' or 'header'. It is drawn in by a rotating reel, which can be moved forward, lowered and speeded up to deal with a lodged crop. Unnecessary speed or reach can cause shedding, particularly in over-ripe crops. Crop lifters can be used to slide the crop gently up onto the table; in one trial the use of crop lifters in lodged barley was found to halve grain losses at this point. Header losses can also be reduced by cutting only against the lean of the crop.

The crop is cut by a reciprocating knife, with serrated blades to grip the hard stems. Control of the height of cut, which is adjusted hydraulically, is the most demanding task facing the combine operator. Many larger machines have automatic table-height controllers, and other machines have flotation devices which allow the table to follow ground contours.

The cut crop falls into the table and is drawn towards its centre by a double flight auger. Loss of grain from the table itself is minimised by the length and profile of the deck, which is shaped to retain loose grain should any be threshed out of the ears by the reel or the auger. The crop leaves the header via a conveyor in a band 0.8–1.6 m wide to match the width of subsequent mechanisms in the machine.

To increase the versatility of the machine, most tables can be detached quickly for transport or storage, and alternative gathering equipment is available for such crops as maize, windrowed crops, sunflowers or edible beans. A 'rotary stripper' which picks the ears off the standing straw has been shown to be more effective than a conventional header in gathering-up lodged crops. In addition, the amount of crop material entering the combine is halved, allowing faster working rates (Klinner *et al.*, 1987).

Threshing

The crop is threshed by a rasp-bar cylinder, or 'drum', working against an open grate concave which matches its curvature for almost one-third of its circumference. The impacts of the beaters on the crop at up to 30 m/s, and of the moving crop on the stationary concave, break the seeds loose. Severity of treatment is adjusted to match the type and condition of the crop by altering cylinder speed or varying the gap between concave and cylinder, through which the crop is forced. Excessive threshing severity will break grains, while insufficient treatment will leave grains attached to the ears which will be discharged with the straw and lost.

In addition to the threshing action of the cylinder and concave, 70–90% of the threshed grain is separated from the MOG by passing outwards through the spaces in the concave.

Separation

Separation of the remaining threshed grain from the straw is the function of the straw walkers, which offer a separating area of 1.8–7.9 m^2. To shake out the grain, the threshed mass is agitated as it moves rearwards, finally to be discharged onto the ground. The thicker the mat of straw on the straw walkers, the more difficult is the task of separation, and the situation is even worse if the straw is damp or broken-up. Beyond a certain MOG throughput, different for every machine, satisfactory separation cannot be completed within the length of the straw walkers and grain is carried over the back of the machine in increasing quantities. Electronic loss monitors may be fitted at the rear of the straw walkers; with their aid (if regularly calibrated), the forward speed (and hence throughput) of the machine can be adjusted to maintain an acceptable level of loss at this point.

Most manufacturers have supplemented the action of the straw walkers on their conventional machines with rotary or reciprocating beaters, which shake out the mat of straw and allow trapped grains to escape. At the top end of the performance and price scale, 'rotary' models dispense with straw walkers in favour of one or several rotary separators. The crop is handled fast, in a thin layer, which gives more positive separation, at the cost of a greater power requirement and more broken-up straw behind the machine.

Cleaning

Impurities are removed from the crop by a combination of size and density separation. An adjustable air blast moves less dense particles to the rear of the machine, while two, or three, sieves of total area 2.0–6.5 m^2 prevent particles larger than those required from passing through. Adjustable sieves are convenient, but interchangeable fixed-size sieves are more precise in their dimensions, and also permit greater throughputs of smaller-sized seeds. Grain losses over the sieves can be detected by a grain-loss monitor in the same way as straw walker losses.

Grain losses

Losses of grain from the combine can occur at several points, for some of the following reasons:

(1) gathering losses – reel too aggressive; crop over-ripe; knife too high; crop lodged; inaccurate driving;
(2) threshing losses – cylinder too slow; concave clearance too large;
(3) separating losses – forward speed too high; straw walkers blocked;
(4) cleaning losses – too much wind; sieves insufficiently opened; sieves blocked.

Table 21.8 shows acceptable loss levels. In some cases, a greater straw walker loss may be acceptable, as this will allow harvesting to progress at a faster rate.

Table 21.8 Acceptable grain losses from a combine harvester in good harvesting conditons (Busse, 1977)

Gathering loss	0.2%	14 kg/ha in a 7 t/ha crop
Threshing loss	0.1%	7 kg/ha in a 7 t/ha crop
Straw walker loss	1.0%	70 kg/ha in a 7 t/ha crop
Sieve loss	0.2%	14 kg/ha in a 7 t/ha crop

Grain losses are measured in the field by collecting and quantifying the lost grain from a known area. If the combine is stopped when harvesting an 'average' part of the field, losses may be assessed in the standing crop in front of the combine to give pre-harvest losses. After reversing the combine a short distance, lost grain from beneath the centre of the machine in its initial position is the sum of pre-harvest and gathering losses. Behind the machine lies the straw, with unthreshed grains still attached, and the sum of all types of loss. By subtraction, the extent of each individual type of loss may be determined, and the appropriate corrections can be made.

A convenient unit area is 1 m^2, which should extend across the full width of cut and can be marked out with a string and four pegs. All lost grain within this area is then gathered up. *Table 21.9* shows the relationship between the number of grains found and the loss of grain/ha.

Table 21.9 A method of assessing grain losses

Crop	Seeds/m^2 to equal 1 kg/ha
Barley	1.7
Maize	0.2
Oats	1.6
Rape	20
Wheat	2

Effect of slopes on grain losses

Much grain is grown on sloping land. The combine works best on level ground, and inclination affects the operation of the straw walkers and sieves (Pascal and Hamilton, 1984). In particular, sieve losses increase very rapidly with increasing side slope (PAMI, 1977).

Since the sieves are the combine component most vulnerable to side slopes, sideways-levelling sieves are standard on one range of machines, and another manufacturer offers sieves that can oscillate laterally to counteract the effect of a side slope.

Hillside combines can automatically level the body of the machine on land sloping as steeply as 18 degrees up, 6 degrees down and 22 degrees to the side, a combination also being possible. Side-hill combines can accommodate side slopes only, to 11.5 degrees. Within these limits, the adverse effects of slopes are completely eliminated, and maximum throughput is possible.

Hillside combines cost about 70% more than the equivalent conventional model, while the side-hill type costs 20% more.

Monitors

Many combines are equipped with sealed sound-proof cabs, often air-conditioned. Even without this barrier it is very difficult for the operator to keep in touch with all the processes and components in the very complex machine. In addition to loss monitors, most of the larger machines have as standard, or can be fitted with, a range of 'function monitors', which detect blockages or malfunctions and alert the operator before a major breakdown can occur.

ROOT HARVESTING

Potato harvesters

Lifters

Where labour is available, potatoes, particularly for the valuable early market, may be harvested by hand. The tubers are picked from the ground after the ridge has been broken up by a lifter or 'potato digger'. There are two types, a power driven tined 'spinner', where the ridge is broken up and spread over previously picked ground to expose the tubers, or the elevator-digger on which the ridge is undercut by a flat share and then conveyed up and back by an elevator of round steel crossbars or 'webs'. Loose soil falls between the webs and the tubers are carried over the back to fall on top of the sifted soil.

Complete harvesters

In order to reduce the high labour demand at potato harvest, the elevator-digger has been developed into a complete harvester, which can deliver the potatoes directly into sacks, boxes or bulk trailers. More than 65% of the UK crop area is harvested by 'complete' machines (Statham, 1981). The web elevator has been extended, haulm-removing devices added, and most machines have a divided horizontal conveyor where a gang of four to eight pickers can remove stones and clods, or tubers. The tubers are then directed into the chosen handling system.

Many growers have adopted unmanned machines, typically harvesting two rows at a time, where any clods, stone or trash that cannot be separated by the machine are harvested and dealt with subsequently at the store by hand sorting, or by automatic sorting machines. Rates of work of 0.75–1.25 ha/row daily can be expected.

The mechanisation of the potato harvest has brought with

it the two problems of crop losses and crop damage. Most of the 1.25–2.5 t/ha that are left in or on the ground in a typical potato field are lost at the share of the machine. If the share is too shallow, tubers will be missed or sliced. If soil does not pass freely back over the share, perhaps under loose or wet conditions when working downhill, or when the share is trailing haulm, soil and tubers will be 'bulldozed' ahead of the share and some will be lost by spillage to either side. Improved harvester designs incorporate fine control of share depth, and may include automatic depth control.

Tuber damage, typically as high as 20% severe damage, occurs mainly on the web elevator which is also a frequent source of breakdowns. Forward speed of the machine and the amount of web agitation must be carefully matched to web speed to maintain a cushion of soil over the elevator. Designers are reducing the damage potential of machines by minimising the length of web elevators and eliminating drops and changes of direction of the tubers.

A further site for tuber damage is the drop into the bulk trailer. This should be no more than 200 mm, which requires constant adjustment of discharge conveyor height as the trailer fills. Great care must also be taken when discharging the hopper of tanker type machines.

Tubers can also be damaged by contact with stones during harvesting or transport – in stony ground the removal or windrowing of stones, at planting time, can significantly reduce tuber damage, as well as speeding the harvest operation (McRae, 1977).

Sugar beet harvesters

The entire sugar beet crop is harvested by machine. The harvesting process involves first cutting off the tops, which may be conserved for livestock feeding but are more usually discharged back onto the ground. The roots are then lifted from the soil, most commonly by a pair of inclined, sharp edged, spoked wheels which cut into the soil to either side of the row and lift the root between them while soil can fall away through the spokes. A minority of machines use fixed shares. The beet is then raised by a web elevator which further cleans the crop, before discharging into an integral hopper or into a trailer running alongside.

Harvesters range from one-row trailed or self-propelled models, through three-row trailed machines, popular with larger growers, to self-propelled five- or six-row systems suitable for the largest producers and contractors. Rates of work of 0.125–0.2 ha/h per row can be expected.

Crop losses result from incorrect topping, failing to gather whole roots into the machine, and from leaving the lower parts of roots in the ground. Typical losses are 2.5 t/ha, while over 30% of growers leave as much as 5 t/ha. One-third of these losses are visible as loose beet on the surface, the remainder being below ground. One 0.6 kg root lying on the surface in a 20 m length of row represents a loss of 2 t/ha (Davis, 1977).

The two most common faults in machine operation which lead to crop losses are failing to operate the lifters sufficiently deeply, and failing to position the lifters exactly on the row. The latter fault can be eliminated by the provision of automatic steering on the lifters, or by the use of quite inexpensive guide skids to centre the lifters on the row.

Faulty topping units can damage the beet, and may loosen the roots in the ground so that they are knocked out of the row and are not gathered in by the lifting mechanism. Both the topping knife and the feeler wheels, which adjust the knife to the correct position below the beet crown and steady the root while it is topped, should be sharpened regularly. Lightweight topping units which locate the beet crown by feeler blades or spikes have been introduced. These can respond more rapidly to beet of different heights, and will work accurately at speeds up to 8 km/h. (Traditional toppers are not accurate at speeds beyond 4.5 km/h.) The skew-bar topper is a rotary, power-driven device which scrapes the leaves off the crown of the beet, leaving a domed shape. It is claimed to remove unwanted material thoroughly even at a fast forward speed, but takes off less of the actual root than a knife topper (Billington and Butler, 1984).

Harvested beet are often piled at the roadside for considerable periods before being transported to the factory. Losses are minimised if a concrete area is available for storing and loading the crop. The roots are removed from the pile by front-end loaders; industrial loaders can move up to 90 t/h, rough terrain fork lifts 60 t/h, tractor front-end loaders 20–30 t/h; slatted buckets allow some soil to escape, and this process can be continued by passing the beet through a cleaner-loader on its way to the lorry.

IRRIGATION

The use of irrigation equipment enables the farmer to prevent crop production from being limited by moisture stress at any time during the growing season. Benefits will include good early establishment, even development of the crop, and increased crop value through uniformity and enhanced yield and quality. Additional uses for irrigation equipment include softening the ground to aid seedbed preparation or root harvesting, the application or washing-in of fertilisers and other chemicals and the prevention of wind erosion (Willis and Carr, 1986). Irrigation systems may be installed specifically for the appliation of livestock wastes, or for the protection of sensitive crops from frost.

System capacity

Each irrigation treatment will typically involve the application of the equivalent of 25 mm of rainfall, and the capacity of the system should allow the whole area under irrigation at a particular time to be covered in 6–8 d to allow for a peak depletion rate of over 3 mm/d. With careful management, a modern irrigation system can be kept in operation for 20–22 h/d. The application of 1 mm of water to 1 ha (1 ha mm) represents 10 m^3 of water. Where an irrigation system applies 25 mm to 6 ha in a 22 h day, its water output is 68.2 m^3/h and it has the capacity to cover 48 ha in 8 working days.

With a traditional sprinkler system, a '1 acre setting' applying 25 mm to 0.4 ha in 3 h and covering 4 settings/d, water output will be 33.3 m^3/h and 12.8 ha can be irrigated in 8 working days.

Water supplies

Irrigation is impossible without a reliable supply of water of a quality appropriate to the crops being treated. The local Water Authority must always be consulted before abstracting water (MAFF, 1983a).

Direct abstraction

Because virtually no source works are required, the cheapest water supply is a river or stream from which water can be pumped directly into the irrigation system. Very few new direct abstraction licences are issued by Water Authorities because of the need to maintain flow in rivers during the dry part of the year. In areas where underlying geological structures are suitable, boreholes may offer a source from which water may be pumped directly onto the land.

Water storage

A significant component in the cost of most new irrigation systems is the construction of a reservoir in which winter water can be stored for summer use. Rivers, streams or field drainage systems may be used as sources of supply. Where the supply is irregular, measurements must be taken to ensure that an adequate volume of water will be available to service the planned irrigation system.

Total irrigation need for the season must be calculated, to arrive at the required storage capacity. Likely application quantities, ranging from 25 to 150 mm according to crop, are multiplied by the area of each crop to be irrigated, in order to give the total water requirement. An additional amount, normally equivalent to 500 mm depth in the reservoir, is allowed for seepage and evaporation from the reservoir.

The reservoir may be constructed by enlarging a hollow or closing-off a valley. In flat country, a rectangular bank is built up from the soil excavated from the area enclosed by the bank. Field drainage systems and any porous subsoil layers must be removed. If suitable clay is available, preferably from the excavation itself, this may be used to line the reservoir; if not, a lining of flexible sheet material must be used. The need for a lining of this type can double the cost of a reservoir.

Banks must be protected from the effects of weathering, and from wave action at the waterline. Points of water entry, and spillways to allow for overflow, must be reinforced to resist the erosive effects of concentrated water flow. The advice of a chartered civil engineer should be sought in the design of reservoirs and associated works, especially where reservoir capacity exceeds 20 000 m³, or banks are higher than 5 m.

Ideally, the reservoir will be filled by gravity. A channel taken from far enough upstream can be led directly into the reservoir. Flow is controlled by a sluice at the upper end of the channel. Where this is not feasible, water is raised into the reservoir by a low-lift, high-volume pump positioned as near as possible to the supply water level.

Pumping

Centrifugal pumps

Because of their simplicity and reliability, centrifugal pumps are almost universally used in clean water irrigation systems. One or more vaned rotors spin the water to build up its kinetic energy, and this is converted to pressure energy due to the internal shape of the pump body. This pressure then causes the water to flow through the system. Increasing the rotational speed of the pump increases pressure and flow. Typical rotor speed is 1800–2800 rev/min; this allows direct drive from diesel engines or electric motors, but requires gearing-up if the pump is to be driven by a tractor PTO.

The maximum efficiency of a centrifugal pump is 65–80%, and the peak is at a very specific combination of flow and pressure. The pump for any particular irrigation system should be selected and then operated so that it will work under conditions at which its greatest efficiency will be realised.

In relation to its maximum delivery pressure of 9–15 bar, the inlet suction or lift capacity of a centrifugal pump is small. Most are fitted with a hand-operated priming pump so that they can be filled with water before start-up. The ideal position for a pump is on a floating raft, as close as possible to the water level, whether this is low or high in the reservoir. Where this is not convenient, the pump should be located on the bank, as close as possible to the water.

Pump controls

In addition to manual start and stop controls, irrigation pumps are normally fitted with safety controls which will shut down the system if the motor becomes overloaded (electric drive) or if the engine temperature rises or oil pressure falls beyond a pre-set value (diesel drive). The system will also shut down if water pressure falls, indicating a loss of inlet prime.

To allow adjustments to be made to application equipment without a double visit to the pump, high and low-level pressure switches can stop the motor when hydrants are closed prior to an adjustment, and start it up again once the applicators are re-set and the hydrant is re-opened. Timers can be used to set a given period of operation, allowing the final irrigation setting of the day to finish without supervision.

Pump selection

While operating efficiently, the pump must deliver a sufficient *flow* of water to meet the output requirements of the system, at a *pressure* sufficient to overcome any resistance due to the height the water must be lifted from source to application point, and due to the length and narrowness of the pipes in the distribution system and the restrictions caused by valves, bends and other fittings (MAFF, 1982). This resistance is expressed as a 'head' or pressure; when this is added to the pressure required at the discharge appliance (typically 3–10 bar), the delivery pressure required from the pump may be computed.

Distribution of water

Where appropriate, water is distributed from the pump via a 'skeleton' of permanent underground 'mains', usually of 150 mm diameter, terminating in carefully positioned hydrants. From these points, portable aluminium mains, typically of 100 mm diameter in 9 m lengths, carry the water through the area to be irrigated, with further hydrants at spacings to suit the particular method of irrigation.

Aluminium irrigation pipes are good conductors of electricity, and there have been many serious accidents where pipes, being carried upright, have contacted overhead electricity cables. The greatest care should be taken when irrigating near overhead cables, both to avoid direct contact by pipes, and to avoid contact by solid water jets from large rain guns, which can also conduct electricity.

Application

Nozzles

When water is forced out through a nozzle, it forms a jet which atomises, or breaks up into droplets in a range of sizes. At low pressure, there is little atomisation, and a solid stream of water may result, while at high pressure there is very thorough atomisation, resulting in a wide, feathery spray which does not carry far, and is easily deflected by the wind. Between these extremes, there is an optimum range of pressures at which atomisation and throw are satisfactory, and flow rate is sufficient. It is unusual for too-high pressures to be a problem, but many irrigation systems suffer from lack of pressure, and this can result in poor distribution from nozzles, reduced water application and soil damage from the larger droplets which will be produced.

Larger droplets are produced by larger nozzles, even when operated at the recommended pressure, and the use of such nozzles carries with it the risk of soil damage from impact by droplets larger than 3–5 mm, especially when the soil is not fully protected by a growing crop. For this reason, 22 mm is the largest nozzle diameter recommended for use in beet or potato crops, while 28 or 30 mm nozzles can be used on grassland (Bailey, 1986a).

Rotating nozzles

Most irrigation systems incorporate one or more rotating, upward-facing, nozzles to discharge the water. Sprinklers, incorporating one, or two, nozzles, have a discharge rate of 0.5–5 m³ h and are used at spacings of 12–24 m (*Figure 21.1*). Rainguns, generally with a single large nozzle, have discharge rates of up to 150 m³ h or more and may be spaced up to 140 m apart and irrigate over 1 ha from a single point (*Figure 21.2*).

Rotation is by small increments, caused by reaction with the main water jet. Either a hammer is pushed back by the jet and then swings forward to strike the nozzle round, or an angled plate swings in and out of the jet, and the deflection of the water to one side causes the nozzle to move a small distance in the opposite direction. the resulting pattern of water distribution peaks directly below the nozzle, and decreases with increasing distance from that point. Satisfactory uniformity in still air is achieved by operating nozzles at their recommended spacing.

In windy weather, the circular pattern of a sprinkler or raingun is elongated into an egg-shape in the downwind

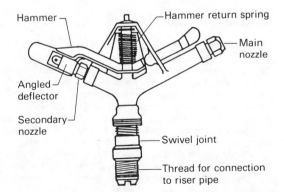

Figure 21.1 A sprinkler (courtesy of Wright Rain Ltd)

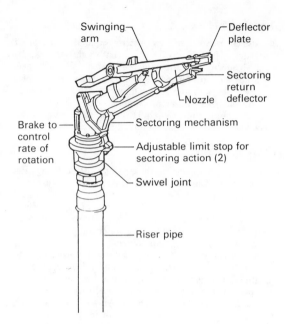

Figure 21.2 A rain gun (courtesy of Wright Rain Ltd)

direction, covering a smaller and narrower area and resulting in a slightly higher application rate within that area. where possible, mobile irrigators should move across the prevailing wind direction; where this is not possible, spacing should be reduced to 70% of that recommended for still air conditions, in order to improve uniformity of application (Bailey, 1986a).

Many rotating nozzles are available with a 'sectoring' attachment, and can be set to cover only a segment of a circle after which the (single) nozzle flicks back to the start point and repeats the segment. This feature is useful when irrigating awkward areas, and is particularly suited to mobile irrigators. The water is sprayed away from the direction of movement of the irrigator so that the trolley carrying the raingun is able to move forward onto relatively dry ground.

Fixed nozzles

Irrigation water can be discharged from many small nozzles spaced along a boom, which itself moves over the cropped area. The small droplets produced by these nozzles are gentle on soil and crop, even though high application rates are often involved. Some of the largest machines use downward-facing nozzles operating at pressures below 2 bar, carried 1–2 m above the crop. the result of this combination of features is to deliver a very uniform water application, little affected by wind, and using less energy for pumping than would be the case with a high-pressure system.

Sprinklers and laterals

A number of sprinklers (often 12), spaced along pipes of 75 mm diameter, is laid out to cover, typically, an area of 0.4 ha. The pipes, or laterals, are available in 6 and 9 m lengths to allow for different spacings, and sprinklers are normally held above crop height by vertical risers every 12–18m (*Figure 21.3*).

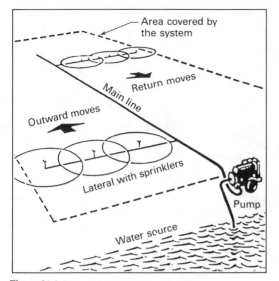

Figure 21.3 Layout of a sprinkler irrigation system (courtesy of Wright Rain Ltd)

Table 21.10 A comparison of irrigation methods (after Curtis, 1980)

Type of system	Approximate relative		
	Cost	Labour	Energy
Hand moved sprinklers	100	100	100
Self-moving sprinklers	200	40	110
Solid set sprinklers	400	20	100
Rotating boom machine	160	60	180
Hosereel mobile	170	20	190
Hose trailing mobile	130	30	150
Centre pivot (low pressure nozzles)	250	15	75

An application of 25 mm of water is delivered in 2–4 h, after which irrigation must stop and two or more people move the lateral to its next position. The work, in wet conditions after irrigation, is not pleasant, and involves the interruption of other activities at what is often a busy time of year.

'Self-moving' lateral systems have been developed, where the entire setting is moved on its own wheels from one position to the next between applications, generally powered by a small petrol engine. Another version comprises a wide rotating boom on a tractor-like self propelled chassis. This irrigates a circular area, and is driven to its next position when the previous area has received sufficient water.

All these systems offer a reduction in labour over that necessary for moving conventional sprinklers, and allow the move to be completed more quickly, but attention is still necessary several times each day.

Solid-set sprinklers

The chore of moving sprinkler laterals three or four times a day can be eliminated where there are sufficient sprinklers and laterals to cover the entire area of a crop. The system is set out at the beginning of the life of the crop, and remains in place until no more irrigation is required.

Semisolid-set systems are also in use. Here, an extended area is laid out with pipes which offer many quick-coupling sockets into which standpipes and sprinklers can be inserted. Sockets not in use are sealed automatically. An 0.4 ha set of sprinklers can be moved from socket to socket across the area as irrigation proceeds, without the need to reposition the pipes. However, as with conventional sprinkler systems, attention is needed three or four times a day.

Mobile irrigators

The popularity of mobile irrigators is due mainly to the great savings in labour they offer over traditional sprinkler systems (*see Table 21.10*). The machine covers a large area per setting, and this allows a long period of operation before attention is required, to move it to a new position. Models are available to cover a great range of areas per setting, from less than 1 to more than 400 ha.

Because mobile irrigators are constantly on the move, they only actually apply water to a fraction of the set area at any one time. This fraction receives the full output of the machine, resulting in an extremely high spot application rate. Typical spot rates for hosereel boom machines are 60–80 mm/h (Bailey, 1986b), while the outer regions of centre-pivot systems can deliver more than 300 mm/h (PAMI, 1984). Such high application rates will cause ponding if the terrain is fairly flat, while runoff and possibly erosion will occur on sloping ground.

Hosereel irrigators

Most new irrigation systems in the UK use hosereel machines. The main component is a large reel, carrying up to 400 m of up to 125 mm diameter hose. On the end of the hose is a wheeled trolley carrying a sectoring raingun capable of irrigating a strip of ground up to 140 m wide (*Figure 21.4*). To prepare the machine for work, the reel is positioned at the headland of the field, and the trolley is pulled across the field by a tractor, unwinding the hose from the reel. The limit of operation is either the length of the hose or the width of the field.

When pressure is applied, the raingun irrigates a sector of ground facing away from the reel, and, after a delay, a water-powered motor begins to turn the reel, drawing-in the trolley and raingun with a force of several tonnes. Application rate depends on nozzle output and rate of travel; it can take over 24 h to draw in the full length of the hose. Most machines have compensators to allow for the fact that the reel increases in diameter as the hose is wound onto it, but it is advisable to check the accuracy and consistency of speed settings from time to time.

When the trolley reaches the reel, a valve stops reel rotation; irrigation continues for a further pre-set period before that also is shut down. The machine must then be repositioned for another period of unsupervised work.

Some machines are fitted with a turntable so that the reel can be swung round to face in any direction without moving the chassis. This allows precise alignment with crop rows where these are not exactly at right angles to the headland, and it also facilitates irrigation to either side of a central laneway where a field is too wide to be irrigated fully from one side headland.

Where delicate crops are to be irrigated, or where poorly structured soils are not fully protected by crop cover, the

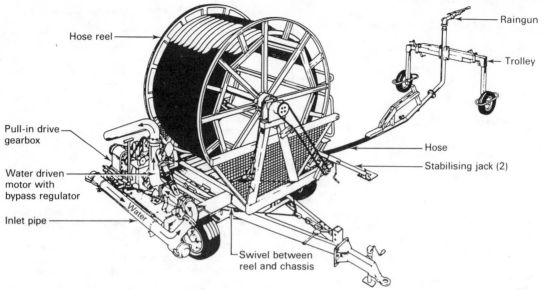

Hose reel

Pull-in drive gearbox

Water driven motor with bypass regulator

Inlet pipe

Water

Swivel between reel and chassis

Raingun

Trolley

Hose

Stabilising jack (2)

Figure 21.4 A hosereel mobile irrigator (courtesy of Wright Rain Ltd)

raingun may be replaced by a boom up to 40 m in width carrying several sprinklers, or many smaller downward-facing nozzles, which will produce fine droplets and minimise impact damage.

An alternative, and cheaper, form of reel irrigator is one in which the reel is carried on the irrigator itself, but instead of winding-in the hose, a small diameter steel cable is used for propulsion and water is supplied by a flexible hose trailed on the ground behind the machine. The 'hose-trailing' machine is less power-consuming than the hosereel type, but requires more labour when moved.

Because of the long runs possible, reel irrigators can be made to operate for almost 24 h/d. Larger machines can work two 11 h runs, while smaller machines can be set for a single 22 h run. Where awkward fields do not allow this, short runs can be worked in the daytime, and long runs saved for night operation (Bailey, 1986a).

Centre pivot and linear-move irrigators

These machines allow a very uniform application of water to be applied automatically to a large area of land. Coefficient of uniformity for centre-pivot machines is 90–95%, compared with 50–60% for hosereel machines, and this allows more precise control of water use (Willis and Carr, 1986). Power requirement is also less, although capital cost is greater than for hosereel machines.

Both types of irrigator consist of a long boom often several hundred metres in length made up of a number of segments, each approximately 30 m in length. Wheeled towers at the outer end of each segment carry the boom above crop height. Sprinklers or downward-facing low pressure nozzles are spaced along the length of the boom.

Centre-pivot machines are anchored to a central mast, through which water and electrical power (for the propulsion system) are fed in. Driven by the wheeled towers, the boom then rotates around the mast, covering a circular area of land which can range from less than 20 to more than 400 ha. Rotational speed of 1–100 h/rev (Kay, 1983) is set by the

travel speed of the outermost drive unit; sensors at each tower ensure that the boom is kept straight. Most makes can be fitted with a hinged outer extension, which will unfold to reach into the corners of a square area, guided by a low-voltage cable buried in the soil below plough depth.

Flat ground is ideal, both for the drive system and because of the heavy application rate, but the drive system can negotiate undulating ground if necessary. Circular tracks are left by the wheels; hedges and banks must be opened up to allow the towers to pass. The area lost to wheel tracks is less than 1% of the whole (PAMI, 1984).

The size and spacing of output nozzles along the boom varies so that all parts of the cropped area receive the same application of water.

The linear-move irrigator follows a straight path, at right angles to the boom itself, and covers a rectangular area. Water enters the boom at one end, either from a trailing or reeled pipe or pumped directly from a supply ditch running down the field. Combination of both principles gives a machine which irrigates a rectangular area, plus a semi-circle at either end.

Irrigation scheduling

During the growing season, the soil is depleted of water at a rate depending on the amount of water taken out by the crop (transpiration) and the amount returned by rainfall. When transpiration exceeds rainfall, a *deficit* develops. This is expressed either as the amount of rainfall or irrigation that would return the soil to field capacity (mm), or the tension with which the remaining water is held in the soil.

As water is depleted, crop growth begins to be affected. The tension or deficit (SMD) at which this takes place is known as the critical SMD, and will depend on the crop itself (*see Table 21.11*), its rooting depth and the available water storage capacity (AWC) of the soil in which it is growing (*see Table 21.12*). The scheduling of the timing and size of

Table 21.11 Soil moisture deficit criteria for the major irrigated crops (MAFF, 1984)

Crops to be irrigated	Recommended water per application (mm)			Critical SMD (mm)			Start SMD[1] (mm)		
	Soil class			Soil class			Soil class		
	A	B	C	A	B	C	A	B	C
Early potatoes and canning: emergence onwards	25	25	25	25	30	30	20	25	25
Second early potatoes from tuber size 6 mm	25	25	25	25	30	30	20	25	25
Main crop potatoes from tuber size 10–20 mm	25	30	30	35	50	50	20	30	30
Sugar beet									
May ⎫ 75% crop cover	25	25	25	25	30	50	20	25	25
June ⎭	25	25	25	35	35	50	20	30	30
July	25	40	50	45	50	100	35	40	90
August	25	50	50	55	75	125	50	60	100
Cereals	25	–	–	50[2]	–	–	35	–	–
Grass-grazing (May onwards)	25	25	25	25	35	35	20	30	30
Grass-conservation (May onwards)	35	50	50	35	50	50	30	40	40

[1] 'Start SMD' is based on assumption that it will take 2 to 4 days to irrigate an area and one has to anticipate the deficit building up.
[2] Up to anthesis

irrigation treatments must be based on an understanding of these variables. A deep-rooting crop on a soil with a high AWC and a high infiltration rate (*see Table 21.13*) can be given heavy, infrequent water applications which are only

Table 21.12 Classification of the available water capacities of common soil textures in approximately increasing order (MAFF, 1982)

A. Low

AWC not more than 12.5% by volume (less than 60 mm/500 mm soil depth)	Coarse sand Loamy coarse sand Coarse sandy loam

B. Medium

AWC more than 12.5% but not more than 20% by volume (60–100 mm/500 mm soil depth)	Sand Loamy sand Fine sand Loamy fine sand Clay Sandy clay Silty clay Clay loam Sandy loam Sandy clay loam Silty clay loam Fine sandy loam Loam

C. High

AWC more than 20% by volume (greater than 200 mm/500 mm soil depth)	Very fine sand Loamy very fine sand Very fine sandy loam Silt loam Silty loam Peaty soils

necessary when a large SMD has developed, while a shallow-rooted crop on a soil of low AWC must be irrigated frequently with smaller applications of water, to maintain a small SMD. Some crops are particularly sensitive to soil water conditions at certain stages of their growth, and this must also be taken into account (*see* MAFF, 1982).

Soil moisture tension may be measured directly by means of tensiometers buried in the ground. This system has the great attraction that no other measurement and no calculation is necessary; some irrigation systems are controlled automatically on the basis of tensiometer measurements. However, several tensiometers must be distributed in each field to give accurate results, and much care is necessary in their use (MAFF, 1983b).

More commonly, a running balance sheet of SMD is kept, either on paper or in a computer. Potential transpiration data are available, based on regional weather measurements, and these figures are combined with local rainfall, crop cover, and irrigation records to arrive at the weekly SMD for each field. Soil information for each field is needed, to indicate the critical SMD and the point at which irrigation should commence. Many consultancy services are available to assist the farmer in scheduling his irrigation applications.

Frost protection by irrigation

Some fruit crops are particularly vulnerable to damage from late frosts. In some cases, the entire crop can be lost. During a frost, developing buds can be prevented from falling below 0°C if they are kept wet by a continuous light application of irrigation water. The latent heat given up by the water as it freezes maintains the temperature at 0°C, and a layer of ice forms over the bush. An application rate of 2.5 mm/h will protect down to −3.3°C; 3.2 mm/h to −5.6°C (Ingram, 1984).

Table 21.13 Infiltration capacities associated with some common soil textures (MAFF, 1982)

Category	Equilibrium infiltration capacity range (mm/h)	Textures
Very high	Greater than 100	Coarse sands, sands, loamy coarse sands, loamy sands
High	20–100	Sandy loams, fine and very fine sandy loams, loamy fine sands and loamy very fine sands
Moderate	5–20	Loams, silt loams, silty loams, clay loams
Low	Less than 5	Clays, silty clays, sandy clays

Quite a large outlay is involved in providing a solid-set sprinkler system and frost controls to protect the whole area of a vulnerable crop, but this could be recovered in a single season if the crop would otherwise have been lost.

Since there is no conflict with conventional irrigation, much of the equipment used for frost protection can also be used for this purpose.

Organic irrigation

Irrigation equipment may be used to transport and spread livestock wastes on the land. This can make manure disposal almost into a 'push button' operation, and also avoids the need to move heavy tankers on the land with attendant risks of soil damage and safety hazards on slopes.

Because of the fibrous nature of manure solids and the probability of hay, silage or straw being present, special pumps, often fitted with chopping attachments, are specified. Also to avoid blockage, rainguns with large nozzles are used, either static or as part of a hosereel irrigator. Only machines recommended for organic irrigation should be used, as liquid manure can affect the water motor and valves of conventional irrigators.

The suitability of liquid manure for handling in this way depends very much on its solids content: generally, slurries containing up to 8–10% DM are suitable for organic irrigation (Schofield, 1984). This corresponds to the dilution of faeces and urine combined with an equal quantity of water, or to the liquid fraction produced by a slurry separator.

FEED PREPARATION MACHINERY

Grinding

Cereals and other ingredients of livestock feeds are ground into fine particles for a number of reasons:

(1) to improve digestion by breaking up the protective outer skin and increasing the surface area of the material;
(2) to increase intake;
(3) to aid mixing of ingredients into homogeneous final diets.

Desirable features of a grinding machine include:

(1) an end product of uniform particle size, with a minimum of dust;
(2) fineness adjustable to suit different classes of livestock;
(3) minimum power requirement in relation to output;
(4) ability to operate reliably without supervision.

Hammer mills

Grain is metered into the mill where it is sheared by sharp-edged beaters rotating at up to 100 m/s. Moving particles from this action are sheared again by the sharp edges of the perforated screen which surrounds the rotor. The process continues until the particles are small enough to pass through the perforations of the screen. A range of screens from 1 to 9.5 mm caters for all requirements. Most mills are equipped with an integral pneumatic conveying system which aids the flow of grain into the mill and also delivers the meal into a hopper or mixer. Less power is needed to convey the meal by auger, and some manufacturers use this alternative system. Throughput of a particular mill decreases with fineness and with increasing moisture content of the grain; 14–16% MC is considered ideal. Edges on rotor and screen become blunt with wear, and must be turned or renewed to restore performance. Typical output would be (45–85 kg/h)/kW of power available; machines range in size from 2.2 to 37 kW, those up to 7.5 kW being the most popular on farms.

The high working speed of the rotor makes the mill vulnerable to damage if metal objects enter it, so most machines are protected by magnets, which will retain ferrous metals. A considerable amount of fine dust is produced by this method of grinding, necessitating the use of cyclones and dust socks.

Hammer mills can operate unattended for several hours; many small machines are run at night, drawing corn from an overhead hopper which has been filled to the required level for making up a batch of feed. When the hopper is empty, a switch in its base is released, and this shuts down the mill.

Plate mills

The plate mill grinds by abrasion and shear. Grain up to 18% MC is broken up between the serrated surfaces of a pair of cast iron discs, one of which rotates as it is pressed against the other. Reduced clearance between the discs results in a finer product and a lower throughput. A more uniform meal is produced, with less dust than the hammer mill. Discharge is by gravity only, so a conveying system must be added unless the mill is positioned directly above a receptacle for its output. The mill produces (60–110 kg of meal h)/kW of power available.

Roller mills

A roller mill crushes grain between a pair of flat surfaced or serrated cylindrical rollers. The rollers are spring-loaded against one another, and spring tension can be adjusted. Dry grain is cracked to produce a coarse meal, while grain of higher moisture content is flattened and crushed into flakes. Moisture content of 18% or more is recommended and the machine lends itself well to a system of high-moisture grain storage. Dry grain can be moistened to improve its rolling characteristics.

Flaked grain is almost dust-free and hence more palatable to livestock than dusty materials. Break-up of the fibrous portion of the grain can be minimised, making a suitable product for feeding to ruminants.

An output of (90–120 kg/h)/kW can be expected, and machines are available in sizes to 7.5 kW with electric drive, as well as larger PTO-driven models. Discharge is by gravity, so that further handling or receiving equipment must be installed. Flaked material is fragile; handling and mixing equipment must be selected and operated with this in mind if the product is to be preserved in this desirable form.

Mixing

When all ingredients of a ration are available in ground form they may be mixed into a homogeneous final product. Mixes are generally prepared in batches of 0.5–1 t. Ingredients are weighed into the mixer in most cases, but some ingredients may be added by volume into a calibrated container. Since different ingredients are of different densities, the container must be calibrated with care if the final diet is to be as required. The volume occupied by 1 t of various ingredients is shown in *Table 21.14 (see also Table 21.15)*.

Larger operators may find it more convenient to produce diets continuously rather than in batches; equipment for this purpose is produced by several manufacturers.

Vertical mixers

The vertical mixer is an upright cylinder, narrowing to a funnel-shaped base. Capacities of 0.5, 1 and 2 t are available.

Table 21.14 The volume of 1 t of various livestock feedstuffs

	m^3
Wheat	1.29
Oats	1.96
Barley	1.43
Beans	1.21
Wheat meal	2.13
Oat meal	2.38
Barley meal	1.96
Bean meal	1.85
Grass meal	2.85–3.90
Dry beet pulp	5.67–3.90
Fish meal	1.83–2.08
Soyabean meal	1.48–1.83
Pelleted ration	1.60–1.69
Crumbed ration	1.83

Table 21.15 Typical annual concentrate consumption of various types of livestock

	Tonnes
Dairy cow	1.25–2.0
Sow or gilt	1.0–1.8
Baconers (from 8 weeks)	0.7–1.0
Porkers (from 8 weeks)	0.5–0.7
1000 laying hens	35–45
1000 broilers	15–20

Ingredients may be added through the top, often from an upper floor or granary, or via a small hopper at the base. A central vertical auger circulates and mixes the ingredients. Adherence to the recommended mixing period of 12–20 min is essential; too short a period results in incomplete mixing, while an excessive time may result in separation of some ingredients. Automatic timers are available.

The mix is discharged from an outlet or sacking spout just above the base of the machine.

Horizontal mixers

The horizontal mixer is a rectangular box with a sloping bottom. A loading hopper is positioned at the low end, while discharge spouts are located under the high end of the machine. The flat top carries the drive motor and can accommodate a grinding machine discharging its output directly into the mixer.

The ingredients are mixed by the action of a chain and slat conveyor which moves up the sloping bottom of the mixer and returns across the top. Mixing time is short (2–3 min) and the gentle action is very suitable for rations containing flaked ingredients.

Blenders

A blender prepares the diet as a continuous process. All ingredients are metered out simultaneously in a stream which is mixed as it is carried away by an auger conveyor.

The metering devices are generally short auger conveyors, the speeds of which are adjusted to give the desired flow rate of each ingredient. Since the metering is in fact volumetric, each auger must be carefully and regularly calibrated to ensure that the proportions, by weight, of each ingredient in the mix remain correct.

Mill mixers

Complete mill mixer units are sold by several manufacturers. They normally consist of a small mill, served by a corn hopper and delivering into a mixer. Controls are pre-wired to the motor, so that the unit can be installed on the farm at minimum cost.

An alternative to fixed equipment, which still permits the farmer to include his own corn in his livestock rations, is the use of a contractor operated mobile mill mixer. This is a large, high-output machine mounted on a lorry which can be brought onto the farm at intervals to prepare batches of feedstuffs to the farmer's requirements.

Cubers

Meal may be made more palatable to livestock if it is pressed into 'cubes', 'crumbs', 'pellets', 'nuts', 'pencils' or 'cobs'. This is achieved by forcing the meal into a funnel-like 'die', of a size appropriate to the desired product. The pressure, plus the heat which is generated by friction, forms a durable cube which may be further strengthened by the use of additives such as molasses, water or beet pulp. Freshly made cubes must be cooled before being stored in bulk, and cubed material must be handled gently.

Cubing is a slow, power-consuming process, and die life is relatively short, making the cuber a rather expensive addition to a farm scale provender plant.

MILKING EQUIPMENT

Milking parlours

Most dairy animals are milked in parlours, a system in which the milking equipment and operator(s) are static, and the cows move from a collecting yard to be milked, and then depart, to be replaced by further batches of cows. This system gives economies in labour over the more traditional cowshed milking, where the cows are tethered in a large building and the operators move the milking equipment from cow to cow as milking proceeds.

The most popular form of parlour is the 'herringbone', where a row of standings is constructed on each side of a sunken operator's pit. The cows face slightly outwards, so that their rear ends are in towards the operator. A feeder is provided for each cow to allow individual rationing. Milking units and milk handling pipelines are centrally installed. In many cases, only one line of cows can be milked at a time, while the other batch is being installed, fed and washed. The provision of a second set of milking units allows milking to begin on all cows as soon as udder washing is complete, regardless of the progress of the previous batch, and can increase the throughput of cows. Herringbone parlours are described by quoting the number of milking units and then the number of standings, for example an 8/16, where there are two rows of eight standings and one set of eight milking units, or a 20/20.

Milking performance in a conventional herringbone parlour is 50 cows/man-h, with up to 75 possible. Further refinements in feeding and milking equipment, and in the shape and layout of the parlour, can reduce the man-time/ cow still more. Outputs of 120 cows/man-h are being achieved, while research workers quote figures of 150 or more as being attainable.

Milking machines

Milk is extracted from the udder by applying a vacuum of 50 kPa to the teats. A higher vacuum will extract the milk faster but the risk of teat damage is also increased. Although a constant vacuum is applied to the interior of the flexible teat-cup liner, this vacuum is only applied to the teat for two-thirds or half of the time. The alternation of vacuum with atmospheric pressure in the space between the liner and the teat-cup, controlled by the pulsator, allows the liner to collapse round the teat and shield it from the vacuum for 0.33 s, at 1 s intervals. This allows blood to circulate more freely, and maintains udder health.

The pulsation ratio, the ratio of the time the teats are exposed to vacuum to the time the liner is collapsed, is typically 2:1 as described above, although some equipment operates at 1:1. Pulsation rate is normally 60 cycles/min. Milking rate can be increased to some extent by widening the pulsation ratio or by increasing the pulsation rate.

The four teat-cups are connected to the claw, which provides pulsation and vacuum connections to each teat-cup. This whole assembly is the cluster. Some claws have a built-in shut-off valve to prevent air entry when the cluster is removed from the cow or if it falls off during milking. Alternatively, a pinch valve on the long milk tube can perform the first function. A small hole in the claw allows air bubbles to enter the milk, making it easier for the vacuum to raise it up the long milk tube.

Most parlours are provided with graduated glass recorder jars of 23 litre capacity, either at operator's head level or below the cows, where low-level pipelines are used. The milk from each cow accumulates in the corresponding jar, so that yield can be recorded or a sample taken.

An individual batch of milk can be rejected, if necessary, without contaminating the entire system. The recorder jar has in some cases been replaced by a milk meter. When installed in the long milk tube, the milk meter indicates milk yield and collects a representative sample of the milk for butterfat analysis. When yields are not being recorded, the meters are not installed and milk passes directly into the pipeline system. A milk flow indicator may be used to let the operator know when a cow has finished milking, or a transparent section of tube or a transparent milk filter may be included in the long milk tube.

A further refinement is to couple the automatic detection of the end of milk flow with the automatic removal of the cluster from the cow (ACR). The cluster is raised by a cord powered by a vacuum operated piston, while the milking vacuum is shut off to release the udder and prevent air entry. If an individual cow has an erratic milk output, ACR can terminate milking before all her milk has been removed.

Milk enters the pipeline either from a transfer valve at the base of the reorder jar, or directly from the milk meter or the long milk tube. An air bleed into the claw or the transfer valve helps the milk to flow towards the source of vacuum until it reaches the releaser vessel (capacity 25 litres) which is normally mounted on the milk room wall. An electric milk pump is actuated by a high level of milk in the vessel to withdraw milk against the vacuum and discharge it through a filter into the bulk tank.

Pipeline milking systems and all their components are designed to be cleaned and sterilised in place. Circulation cleaning is a routine by which the system is first rinsed through with hot water, then a hot solution of detergent and sodium hypochlorite is circulated through the system for 5–10 min followed by a final water or hypochlorite rinse. The various liquids are drawn from a trough or tank into the wash line, and enter each cluster via a set of four jetters which fit over the ends of the teat-cups. From here, the liquid follows the route of the milk, described above, until it reaches the releaser pump which either returns it to the trough for recirculation or discharges it from the system.

A saving in chemicals, but an increased consumption of electricity for water heating is achieved by the process of acidified boiling water (ABW) cleaning. Sterilisation is mainly by heat, as the entire system is raised to 77°C for at least 2 min; 14–18 litres/unit of near-boiling water (96°C) acidified by a small quantity of sulphamic or nitric acid, is drawn once through the system at a rate controlled by an inlet orifice, and discharged by the releaser pump. In addition to a mild sterilising action, the acid helps to keep the glass tubes clear of hard-water salts.

The milking plant is used 365 d of the year; regular maintenance is necessary to ensure that rubber parts are in good condition, pulsators and vacuum regulator are operating correctly and vacuum pump is lubricated. There should be an annual machine test by the Milk Marketing Board or the machine manufacturer's agent. Such care will help to ensure that the installation gives reliable, efficient milking, good milk hygiene and minimum machine-induced mastitis.

Milk cooling and storage

Milk is normally cooled, and stored prior to collection, in a stainless steel bulk tank of 600–4000 litre capacity. When holding 40% of its nominal capacity (from the evening milking) which has been cooled to 4.4°C overnight, the remaining 60% of capacity must be cooled from 35°C to the same level within 30 min of the end of filling, at an ambient temperature of 32.2°C.

The milk is cooled by contact with the chilled inside surface of the tank; an agitator paddle keeps the milk in circulation to assist in this process. The agitator must also be able to homogenise a sample of milk of 4.5% butterfat so that after 2 min operation the milk may be sampled to give an accuracy of ± 0.05 units of butterfat percentage. The tank is insulated so that the chilled milk will not rise in temperature by more than 1.7°C over an 8 h period when the ambient temperature is 32.2°C.

Most bulk tanks are cooled by an ice bank. A medium sized refrigeration unit, the evaporator coils of which are submerged in water in the space between the inner and outer walls of the tank, operates for many hours daily to build up an ice bank. A control will stop the refrigerator when the ice bank is sufficiently large to cope with the expected milk yield and ambient temperature. At milking time, the circulating warm milk inside the tank gives up its heat which is absorbed as latent heat by the melting ice. This system allows a relatively low-powered refrigeration unit to give a rapid rate of milk cooling.

Where very large quantities of milk are handled, it may be more economical to cool the milk in a separate plate-type heat exchanger before it is stored in an insulated tank.

The conventional bulk tank will cool about 45 litres of milk for each kW of energy consumed. This figure may be almost doubled if the milk can be pre-cooled in a heat exchanger before it reaches the bulk tank. If a free source of cold water, such as a spring, is available, or if large quantities of purchased water are to be used elsewhere, this water can be used to carry away as much as half the heat from the milk. A heat pump may also be used to recover some of the energy from the milk and use it to heat washing water.

Bulk tanks are always cold, so that heat sterilisation is impossible; chemical sterilisation must be used. Most tanks are equipped with a programmed cleaning unit which, by means of a spray head, rinses, washes; sterilises and finally rinses the tank. This cycle is set in motion by the tanker driver after he has collected the milk.

Parlour feeding

Feeding of the cows in the parlour occupies them during milking and also allows individual rationing of each cow. In a small parlour, the operator can recognise each cow, remember her ration, and operate a volumetric feeder a certain number of times to deliver the correct ration into a trough in front of her. In some cases the cows also learn how to operate the feeders, and can draw additional rations for themselves by banging the unit.

In order to take up less of the operator's time, parlour feeding systems have been developed in a number of ways:

(1) vacuum operation of the feeders from a central control panel, to reduce the distance walked;
(2) a programmable unit which releases the correct ration at the correct position when the cows' identification numbers are manually dialled or punched in;
(3) machine recognition of individual cows at each standing by means of a coded transponder worn around the neck, followed by the release of the correct quantity of feed. The memory unit is updated regularly, to adjust the ration according to yield and stage of lactation.

Further refinements to the computer systems can include the automatic recording of yield or milk flow rate, and the automatic monitoring of milk conductivity (to detect mastitis) or milk temperature (to detect ill-health or 'heat').

As milking routines speed up, the cow has less and less time to eat her ration. While a typical 6/12 parlour gives her time to eat 5.5 kg, a 12/12 reduces her waiting time, and she only has time to consume 3.5 kg of feed. In order to satisfy the requirements of high-yielding cows, a number of manufacturers offer out-of-parlour feeders, which respond only to cows wearing transponders with a particular coding. A batch of feed is delivered to that individual – and then no further batch can be delivered to that individual – for a pre-set period, in order to control her concentrate intake. Some systems also record the amount of food delivered to each cow in this way.

MANURE HANDLING AND SPREADING

Farmyard manure

The traditional method of dealing with manure is to provide straw or other absorbent material as bedding, and to handle the combined product of faeces, urine and bedding as farmyard manure. This increases the volume of material to be handled by 10–25% in comparison with the figure mentioned in *Table 21.15*.

Where there is sufficient clearance, farmyard manure may be stored in place until it can be spread. Alternatively, a double-handling system is used, involving removal to a temporary storage site followed by spreading onto the land at a convenient time of year.

Using a tractor mounted front-end loader, a rough terrain fork lift with manure bucket, or an industrial loader similarly equipped, farmyard manure can be loaded into trailers or spreaders at a rate of 20–60 t/h.

Spreading on the land is by rear discharge or side discharge, spreader, which have carrying capacities of 2–10 t.

Rear discharge machines resemble trailers. They are equipped with a slatted floor conveyor to move the load rearwards into a beating and shredding mechanism. This distributes a swath of manure behind and to either side of the machine. Some of these spreaders can be adapted to handle other materials.

Side discharge spreaders are cylindrical in shape, with an opening along the top portion for loading and discharge. A shaft through the centre of the cylinder carries numerous chain flails, and a rigid flail at either end. As the shaft rotates, manure is discharged first from the ends of the load, and then progressively towards the centre of the load. Spreading is to one side of the machine only; a wide range of materials from strawy farmyard manure to semi-liquids can be carried and spread.

Liquid manure

In livestock housing systems where little or no bedding is provided, the manure consists only of faeces and urine. This is removed from contact with the animals by allowing it to fall through a slatted floor into a space below, or, as in the case of cubicle housing for dairy cows, dunging areas are scraped daily by a tractor rear-mounted scraper.

It is rarely possible to spread manure daily, so a storage tank must be included in the system to hold, typically, at least three months' production. A longer storage period may permit most of the manure to be applied to stubbles in autumn, or some other seasonal priority.

Size of store

When calculating the size of store necessary for a particular length of storage, it is advisable to add a factor of 40% to the figure given in *Table 21.15*, to allow for spilt drinking water and any washing water which will find its way into the store. If rainwater cannot be excluded, an additional volume *Vl* must be added, where $V \times$ (mm of rainfall over the storage period) $\times$ (m^2 of area draining into the store + m^2 surface area of the store).

Size of tanks

While liquid manure stores can be excavated and lined, or constructed of concrete or masonry to any size, many farmers install prefabricated stores which are bolted together from curved coated steel panels, mounted on a ground level concrete base. Stores are available in diameters of 4.5–25 m, heights of 1.2–6 m, and capacities up to 3000 m^3. When planning a manure handling system, provision must always be made for increasing its capacity, even if no immediate increase in livestock numbers is envisaged. An increase in store height to give greater capacity is only feasible if the original store was designed to accept this increase.

The storage tank may be below the level of the yard, in which case it can be filled directly by the scraper. Alternatively, partially or completely above ground tanks may be used. A small below-ground sump, covered by a heavy grid,

receives the scraped manure from the yard. This is transferred into the main tank at intervals by a pump.

Successful operation of a liquid manure system demands care and good management at all stages, to include the following points.

(1) Exclusion of any fibrous material: hay, silage, straw. This will otherwise block up pumps, and contribute to crust formation.
(2) Pumping of the manure can be facilitated by the addition of water to the dung and urine, although this also increases the volume of material to be handled. Dung and urine together form a slurry at about 12–15% solids content which is barely pumpable; adding water to produce a less viscous consistency at 6−7.5% solids makes the resultant slurry very easy to pump (Schofield, 1984).
(3) If the manure is to be handled by pumps and tanks, the contents of the tank must be thoroughly agitated at least fortnightly to prevent the formation of surface crusts in the case of cattle manure, or sludge deposits in the case of pig manure. The pump which is used for handling the manure can also be used for this purpose. Movable jets can be provided at various levels in the tank, and the output of the pump is concentrated through any jet to break up the surface crust, dislodge sludge deposits, or just mix the contents of the tank. Where regular mixing has been neglected, or mixing power is inadequate, and crusts or deposits have built-up, propeller agitators can be used to homogenise the contents of the tank.

Liquid manure is unloaded from the store directly by a centrifugal or positive-displacement pump, or by vacuum into a tanker of 3500–13500 litre capacity. After transport to the field, a built-in spreading mechanism distributes the manure as the machine moves forward. Considerable odour may be produced, particularly if the liquid manure is thrown high into the air. As an alternative, some tankers can be fitted with heavy tine 'injectors' which can place the manure below the soil surface. This minimises odour and also reduces the loss of nutrients by run-off.

Care must be exercised when operating tankers across slopes, particularly when partly full. The load shifts to the downhill side, moving the centre of gravity nearer to the downhill wheel. Overturns will occur on side slopes where the full tanker, or a trailer, could be used quite safely (Hunter, 1985).

Where the manure is to be spread within 0.5 km of the store, organic irrigation may be considered (*see* p. 530).

If the manure store can be entered by a ramp down from ground level, it can be emptied by a self-loading tanker which is backed down the ramp. Alternatively, a four-wheel drive loader with large capacity bucket may be used to unload the manure at rates of 60–80 m^3/h into tankers or tipping trailers.

Table 21.16 Daily production of faeces and urine by livestock

Calves 3–6 months	7 litres	Pigs 9–12 weeks	3.5 litres
Young stock 6–15 months	14 litres	Pigs 16–20 weeks	7.5 litres
Heifers	21 litres	Sows	11.5 litres
Dairy cows	45 litres	1000 broiler chickens	80 litres
1000 turkey growers	130 litres	1000 laying hens	140 litres

Slurry separators

By passing liquid manure over a series of sieves, rollers or vacuum devices it is possible to remove about half the solids, leaving a liquid that can be stored without problems and pumped easily. The 'solids' can be stacked for later spreading with FYM equipment (Pain *et al.*, 1978).

References

BAILEY, J. (1986a). Choosing and improving an irrigation system. *Br. S. Beet. Rev.* **54**, 3–4

BAILEY, J. (1986b). Report on an irrigation demonstration. *Br. S. Beet. Rev.* **54**, 37–38

BILLINGTON, W. P. and BUTLER, J. (1984). The design and development of a skew-bar topper for sugar beet. *J. agric. Engng. Res.* **29**, 329–335

BOA, W. (1984). The design and performance of an automatic transplanter for field vegetables. *J. agric. engng. Res.* **30**, 123–130

BOSMA, A. H. and VERKAIK, A. P. (1986). Developments in forage conditioning *Ag. Eng.* 86 Conference Papers, 181–2, Wageningen

BRITISH CROP PROTECTION COUNCIL (1986) *Nozzle Selection Handbook*

BUSSE, W. (1977). The design and use of combine harvesters for minimum crop loss. *Agr. Engr.* **32**, 7–9

CURTIS, G. C. (1980) Application equipment — present and future. *Soil & Water* **8**, 17–18

DAVIS, N. B. (1977). The minimisation of crop losses associated with sugar beet harvesting. *Agr. Engr.* **32**, 10–13

DWYER, M. J. (1983). Soil dynamics and the problems of traction and compaction. *Agr. Engr.* **38**, 62–68

DWYER, M. J. (1985). Power requirements for field machines. *Agr. Engr.* **40**, 50–59

DWYER, M. J. and DAWSON, J. (1984). Tyres for driving wheels. Proc. BSRAE Assoc. Members' Day, 4 Oct. 1984

DWYER, M. J. and PEARSON, G. (1976). A field comparison of the tractive performance of two- and four-wheel drive tractors. *J. agric. Engng. Res.* **21**, 77–85

ELRICK, J. D. (1982). How to choose and use combines. The Scottish Agricultural Colleges, publication no. 88

FIELDER, A. (1986) *Straw incorporation – a practical guide. Progress* August, 2–4

GEE-CLOUGH, D., McALLISTER, M. and EVERNDEN, D. W. (1977a). Tractive performance of tractor drive tyres I. The effect of lug height. *J. agric. Engng. Res.* **22**, 373–384

GEE-CLOUGH, D. McALLISTER, M. and EVERNDEN, D. W. (1977b). Tractive performance of tractor drive tyres II. A comparison of radial and cross-ply carcass construction. *J. agric. Engng. Res.* **22**, 385–395

GEE-CLOUGH, D. PEARSON, G. and McALLISTER, M. (1982). Ballasting wheeled tractors to achieve maximum power output in frictional-cohesive soils.*J. agric. Engng. Res.* **27**, 1–19

HUNTER, A. G. M. (1985). Stability analysis of trailed tankers on slopes. *J. agric. Engng. Res.* **32**, 311–320

INGRAM, J. A. (1984). Frost protection of fruit crops. *Irrigation News* No. 6, 17–22

JONES, L. (1985). The effect of ground area cover on the drying rate of mechanically conditioned grass crops.

Grassland Research Institute Final Ann. Rep. 1984–85, 102–3

KLINNER, W. E. (1975). Design and performance characteristics of an experimental crop conditioning system for difficult climates. *J. agric. Engng. Res.* **20**, 149–165

KLINNER, W. E. (1980). Reducing field losses in grain harvesting operations. *Agr. Engr.* **35**, 23–27

KLINNER, W. E. and BURGESS, L. (1982). Separating heavy objects from forage harvester windrows. Proc. BSRAE Assoc. Open Day, April

KLINNER, W. E. and HALE, O. D. (1984). Design evaluation of crop conditioners with plastic elements.*J. agric. Engng. Res.* **30**, 255–263

KLINNER, W. E., NEALE, M. A. and ARNOLD, R. E. (1987). A new stripping header for combine harvesters. *Agr. Engr.* **42**, 9–14

KNIGHT, A. C. (1986). Energy requirement to comminute forages by slicing. *J. agric. Engng. Res.* **33**, 263–271

KAY, M. (1983). *Sprinkler Irrigation. Equipment and Practice*. London: Batsford

LAKE, J. R., FROST, A. R. and WILSON, J. M. (1980). The flight times of spray drops under the influence of gravitational, aerodynamic and electrostatic forces. Proc. BCPC Symposium 'Spraying Systems for the 80s', 119–125

MAUNDER, W. F. (1983). Planting and mechanical handling of seed (potatoes). *Agr. Engr.* **38**, 38–41

McALLISTER, M. (1983). Reduction in the rolling resistance of tyres for trailed agricultural machinery. *J. agric. Engng. Res.* **28**, 127–137

McRAE, D. (1977). The design and operation of potato harvesters for minimum damage and loss. *Agr. Engr.* **32**, 17–19

MAFF (1975). Bale handling methods 1972, 1973 and 1974. Farm Mechanization Studies No. 28.

MAFF (1976a). The utilisation and performance of field crop sprayers 1976. Farm Mechanization Studies No. 29

MAFF (1976b). Self-loading forage trailer performance Pwllpeiran 1975. Mechanisation Dept., ADAS Wales

MAFF (1980). The handling and application of granular fertilizers – 1980. Farm Mechanisation Study No. 35

MAFF (1982). Irrigation. Reference Book 138. London: HMSO

MAFF (1983a). Water for Irrigation. Reference Book 202. London: HMSO

MAFF (1983b). Methods of Short Term Irrigation Planning. Booklet 2118. Alnwick: MAFF Publications

MAFF (1984). Daily Calculation of Irrigation Need. Booklet 2396. Alnwick: MAFF Publications

MAFF (1986). Draft Code of Practice on the Agricultural and Horticultural Use of Pesticides. MAFF July

MAFF (1987). Irrigation of outdoor crops – Enquiry February 1985 – England and Wales. May 1987. Guildford: MAFF Agricultural Census Branch

NATION, H. J. (1982). The dynamic behaviour of field sprayer booms. *J. agric. Engng. Res.* **27**, 61–70

NATION, H. J. (1987). The design and performance of a gimbal-type mounting for sprayer booms. *J. agric. Engng. Res* **36**, 233–260

NPTC (1987). *Pesticides Application*. Stoneleigh: National Proficiency Tests Council

PAIN, B. F., HEPHERD, R. Q. and PITTMAN, R. J. (1978). Factors affecting the performances of four slurry separating machines. *J. agric. Engng. Res.* **23**, 231–242

PAMI (1977). Evaluation Report No. E 0576C – John Deere

Sidehill 6600 Self Propelled Combine. Humboldt: Prairie Agricultural Machinery Institute

PAMI (1984). Evaluation Report No. 388 – Lockwood Model 2265 Centre Pivot Irrigation System with Flexspan Corner System Attachment. Humboldt: Prairie Agricultural Machinery Institute

PASCAL, J. A. and HAMILTON, A. J. (1984). Grain losses associated with combine harvesters operating on sloping land – performance changes over the past fifteen years. *Agr. Engr.* **39**, 124–130

PATTERSON, D. E. (1982). The performance of alternative cultivation systems. *Agr. Engr.* **37**, 8–11

PATTERSON, D. E. and RICHARDSON, C. D. (1981). Bridge links for combined cultivation and drilling. *Agr. Engr.* **36**, 3–6

PLACKETT, C. W. (1984). The ground pressure of some agricultural tyres at low load and with zero sinkage. *J. agric. Engng. Res.* **29**, 159–166

RACKHAM, D. H. and BLIGHT, D. P. (1985). Four-wheel drive tractors – a review. *J. agric. Engng. Res.* **31**, 185–201

RAYMOND, F., REDMAN, P. and WALTHAM, R. (1986). Forage Conservation and Feeding (4th Edn.) Ipswich: Farming Press

RAGHAVAN, G. S. V., McKYES, E. and CHASSÉ, M. (1977). Effect of wheelslip on soil compaction. *J. agric. Engng. Res.* **22**, 79–83

REES, D. V. H. (1982). A discussion of sources of dry matter loss during the process of haymaking. *J. agric. Engng. Res.* **27**, 469–479

RÖHRS, W. (1986). Investigation on energy consumption of cylinder type forage harvesters. Ag. Eng. 86 Conference Papers, 84–5. Wageningen

SCHAFFER, R. L., JOHNSON, C. E., ELKINS, C. B. and HENDRICK, J. G. (1985). Prescription tillage: the concept and examples. *J. agric. Engng. Res.* **32**, 123–129

SCHOFIELD, C. P. (1984). A review of the handling characteristics of agricultural slurries. *J. agric. Engng. Res.* **30**, 101–109

SHARP, R. B. (1984). Comparison of drift from charged and uncharged hydraulic nozzles. Proc. 1984 British Crop Protection Conference, 1027–1031

SPILLMAN, J. J. (1984). Spray impaction, retention and adhesion: an introcution to basic characteristics. *Pesticide Science* **15**, 97–106

SPOOR, G. and GODWIN, R. J. (1978). An experimental investigation into the deep loosening of soil by rigid tines. *J. agric. Engng. Res.* **23**, 243–258

STAFFORD, J. V. (1979). The performance of a rigid tine in relation to soil properties and speed. *J agric. Engng. Res.* **24**, 41–56

STATHAM, O. J. H. (1981). Maincrop potato production in Great Britain. *Agr. Engr.* **36**, 25–28

TUCK, C. R., KLINNER, W. E. and HALE, O. D. (1980). Economic and practical aspects of high-capacity rotary mower and mower-conditioner systems. *Agr. Engr.* **35**, 11–14

WILLIS, S. T. and CARR, M. K. V. (1986). Centre pivot irrigators. *Irrigation News,* Spring 16–36

22

Farm buildings

J. L. Carpenter

DESIGN CRITERIA

Introduction

The great variety of purpose and intensity of use, together with alternative construction methods results in considerable differences of building investment on individual farms. Even when there are similarities of production, it may range from basic weather protection for stock up to elaborate environmental control systems, producing variations from 10 to 90% of a farm's capital. Bearing this potentially heavy, possibly crucial, expenditure in mind it is surprising that agricultural management, unlike that in other industries, often takes on the mantles of specialists such as surveyors and architects when involved in a construction project. The results, commonly prejudiced by the demands of day to day farming activities, are not always satisfactory in an age where techniques are changing rapidly. However, despite this strong amateur element, the great majority of projects are completed reasonably well and work much as intended. No matter how successful they appear to be, this *ad hoc* design style can never ensure that optimal solutions have been found to the problems posed originally. Thus, a strong case can be made in the present economic climate (particularly with regard to protection of the environment) for more care and 'state of the art' advice to be used in the planning and purchase of farm building systems than has been exercised hitherto.

Agricultural building design is currently under considerable pressure from many external factors. Among these are a rapidly awakening environmental consciousness, a need for economic stringency, brought about by general market and legislative forces and the demand for countryside-based (but non-agricultural) industrial and leisure activites. As a result of these forces, farm construction now has a British Standard (BS 5502) to guide (in practice and control) its use of materials and methodology and a series of Welfare Codes for livestock to provide recommendations (and in some cases legislate) for 'good' buildings. Additionally through planning legislation (e.g. Town and Country Planning Acts) pollution control (various 'Water' Acts) and public health regulations (e.g. Health and Safety at Work Act) it must now conform to strict environmental parameters. One of the results of this recent increase in external influence has been a potential increase in cost. This implies that it is more important than ever that the farmer/investor should ensure optimum value for money from any farm building installation. This requires, in turn, that a greater degree of professionalism is used in all phases of the design and construction process. Although, clearly, a hired specialist designer, whether an architect, surveyor or other expert should be in a position to best advise they are relatively rarely employed to do so, partly due to their 'added cost' but also because there are comparatively few of them. However, many organisations exist (*see* later section) to help with building planning on points of general layout, environment and structural method. To be of maximum use they must be consulted as early in the design process as is possible – before mistakes are made rather than after. The next sections outline the principles and procedures to be used and the general order in which they should be adopted in this process.

Planning and design

Despite the great variation implied in the Introduction above, derived from type and scale of farm operations, there are some considerations which are common to the planning processes of all farm buildings. These may be summarised as follows:

(1) suitability for purpose and use,
(2) appropriate cost, relative to permanence and maintenance,
(3) flexibility of design to allow for
    (a) expansion,
    (b) produce changes,
    (c) alternative (or alternate) use,
(4) minimal labour and energy use,
(5) minimal risk
    (a) to health of crop, stock or manpower,
    (b) of fire,
    (c) of polluting the surroundings,

(6) utilisation of common/standard/local materials (with account taken of DIY element),

(7) harmony with surrounding environment,

(8) appropriate use of site topography for aspect/drainage/shelter and gravity movement of materials,

(9) finance available,

(10) impact of legislation and codes of practice,

(11) potential modification or extension of existing buildings allowing for age and technological differences.

Specific design

Whilst the general points above must be borne in mind at all times in any building development, specific use structures will have particular requirements concerned with their productive functions e.g. precise environmental control. The route by which they may be included in the overall design is indicated below in the scheme of work represented by steps A–G and the chart in *Figure 22.1.*

A. Initial preparation

(Confer at all appropriate stages with advisory organisations – the major ones of which are listed at the end of this chapter).

(1) Establish the business motives for the building investment and the level of acceptable cost relative to the return on the capital expended.

(2) Consider alternative systems to achieve similar results and consult other existing users (if any) having experience of them.

(3) Estimate the overall size of the completed structure(s) and ancillaries required and hence assess the *approximate overall* cost *(see Table 22.1)*.

(4) Obtain information on
   (a) grant aid potential,
   (b) tax concessions, etc.,
   (c) need to conform to legislative requirements such as planning/building/safety/pollution/health regulations *(see p. 584)*.

(5) Examine possible sites for the building(s), considering their interactive effects on the farm business and the rural environment both during and after construction. Take particular account of:
   (a) size (area and height – note planning permission restrictions),
   (b) access (for vehicles, animals, operatives, building and other materials handling),
   (c) circulation space (and expansion/change potential),
   (d) drainage (within, from and around building – note anti-pollution requirements),
   (e) aspect (climate and its environmental effects – note also visual effects in areas of particular sensitivity),
   (f) availability of services (provision in remote areas may be very costly),
   (g) utilisation of slopes (gravitational movement of materials, use of storage space and drainage).
   (h) potential for expansion or change (space of surroundings, layout of services).

(6) Confirm the building concept by sketching and listing the requirements of the building. Include provision for all necessary activities within and surrounding it.

(7) Re-estimate the cost in the light of decisions taken about siting and building system/methods, which may modify the original concepts or facilities required.

B. Design criteria establishment

(1) Consider the space, labour, mechanisation and energy inputs required to operate the building system within and around the building. Note and list requirements for activities such as feeding, cleaning, handling, emptying, filling, heating, cooling, weighing, lighting.

(2) Draw up a list of relevant criteria to be met such as living or storage space requirement, insulation standards, slurry capacity, mechanical handling clearances, ventilation rates, cleanability of materials, circulation areas, etc.

C. Concept detailing

(*See* pp. 559–580 for details of specific building types)

(1) Calculate
   (a) *areas* required (of floor, pens, circulation, access, slats, handling equipment, feeders, fixed equipment, etc.) Remember that Net essential area = Gross building area − Structure occupancy.
   (b) *heights* (of ceilings, doorways, walls, mechanical equipment, services, etc.). Remember to allow for essential clearances, e.g. trailer tipping, pig reach.

(2) Arrange the *interior layout* for stock control, materials movement, equipment installation, and environmental effects.

(3) Derive the *external shape* from (1) and (2) taking account of weather resistance (roof shape and wall presentation) and impact on the surrounding rural environment, including any adjacent buildings (odours, contrasts, textures, etc.).

(4) Check and include *safety* elements which are intended to limit hazards (to operatives, stock and goods) from fire, structure failure, poor atmospheric composition, and themselves.

(5) Decide on *preferred* methods, types and manufacturers of materials and structures. Note that individual preference for quality standard, dimensional system, localisation of supply source and workability of various materials will give a very great range of options.

D. Dimensional formulation

Dimensional formulation of plans should now be possible. This entails accurate drafting, preferably to a standard scale (1:10, 20, 50, 100, 200 as appropriate) all the external and internal aspects, in detail, which will enable the builder to construct without further instruction. At this point it may be advisable to employ specialist help with knowledge of drawing interpretation. It is perfectly possible for those without formal training to develop adequate building instructions in the form of sketches. However, it should always be remembered that experts in the building trade are not experts in agriculture and so thorough going instructions should be given wherever possible. (*See Figure 22.66* for an example of the drawings which might be required for a relatively simple case and p. 581 for a fuller explanation of the purpose served by drawings and their accompanying information.)

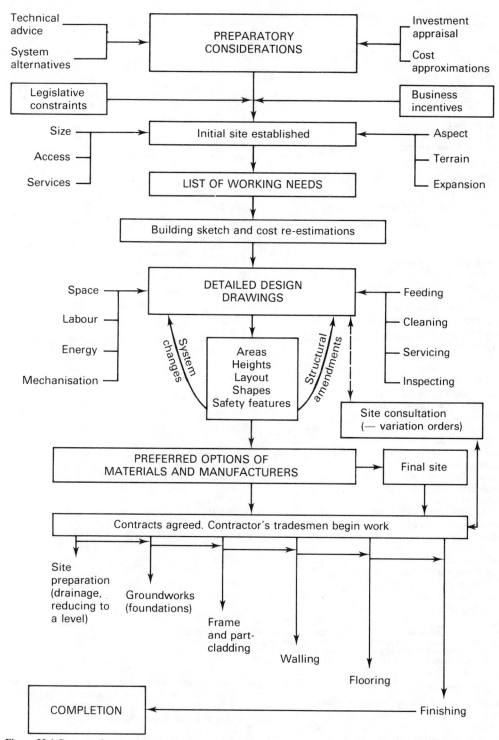

Figure 22.1 Steps to the completion of a farm building

Table 22.1 New building cost ranges, per production unit at 1987 prices

	Gross cost (£)
Cattle	
Cow accommodation (per cow)	
Housing and feed system	250–800
Parlour + collecting area (open)	100–180
Parlour equipment	100–150
Additional automatic parlour equipment	150–200
Bulk milk tank	60–100
Slurry collection and containment	200–300
Possible total variation	800–1700
Beef accommodation, housing only (per beast)	300–600
Calf accommodation, (per calf) up to 3 months	80–170
3–6 months	60–100
Bull accommodation, pen + handling area	7500+
Pigs	
Sows, dry, housing (per sow) in stalls	280–400
Sows, dry, housing in yard with feeders	400–550
Sows, farrowing, housing + equipment (per sow)	650–1600
Boar accommodation, single pen	1800+
Weaner pigs, (per weaner) 3 weeks +	30–60
Fattening/growing pigs, 2–5 months (per pig)	70–100
Fattening/growing pigs 2–5months (per pig) in controlled housing with auto-feeding	120–180
Sheep	
Ewes and lambs, housing only (per ewe)	60–90
Yearling sheep, housing and feedstore (per animal)	40–60
Yearling sheep in 'temporary' (plastic) housing (per animal)	20–40
Sheep handling systems including dip tank (/m^2)	30–50
Crops	
Potatoes, bulk storage (/tonne)	60–100
Potatoes, box storage with cooling (/tonne)	120–140
Grain, bulk storage ('on-the-floor') (/tonne)	40–60
Grain, bin containment + drying (/tonne)	70–110
Silage, clamp silo (/tonne)	20–40
Silage, tower + handling equipment (/tonne)	70–100
Hay, bale store + drying (/tonne)	50–80
General	
'Pole' barn, with roof only (/m^2)	20–30
Portal frame, steel with part-cladding (/m^2)	30–50
'Nissen' type, with end doorways (/m^2)	15–25
Slurry storage, lined lagoon (/m^3)	15–20
Slurry storage, steel 'ring' (/m^3)	15–35

NB These estimates are subject to variations such as scale economies, specification generosity and erratic tender pricing, but the lower end of the ranges indicates simple, cheap systems with minimal sophistication and large elements of DIY, whereas the high end of the ranges presumes a long life expectancy, complex environmental control, low labour input for operation and an intensive level of mechanisation.

E. Final site preparation

This should now be carried out by way of a *survey* (*see* p. 581) using the dimensions of the building established previously. Account should be taken of the area required, the existing slopes and drainage pattern together with the soil characteristics governing the foundation type and sizes. From the results of this survey estimates may be made of soil quantities to be removed and the problems if any, which might arise in the installation of services on the site. At this point the precise position of the building outline can be marked by pegs and string or chalk lines (described in the next section) as a first step in the construction process.

Design amendment

Although the above stages in the design have been taken in logical order, one by one, it is likely that there will be a continual need throughout, to readjust ideas and objectives in attempts to match conflicting (even contradictory) requirements. Compromise between farm business aims and building techniques is often unavoidable, as for instance in the situation where the drainage of slurry may create foundation design problems. Even when construction has started cases arise where the three-dimensional nature of the building work demonstrate that the two-dimensional planning process is inadequate and necessitates changes made in the instructions to the builder. As this will escalate the cost of the work significantly, a contingency sum should be set aside to cover these eventualities, but they should be avoided wherever possible.

G. Contracting for the construction site work

This can proceed as soon as the drawings and building instructions (with, perhaps, a 'bill of quantities') are drawn up. Negotiations can be entered into with potential contractors or building/material suppliers and formal agreement can be reached as to the precise conditions under which the contract will be carried out including, hopefully, a fixed and final cost. However builders and their estimators may prefer to operate on a 'cost plus' (cost of materials and labour plus a reasonable addition for expenses and profit) or a 'price, ex-works, at date of dispatch' basis but it is more usual to negotiate a contractual sum for the provision of completed buildings, including all site works required. This latter element often provides some nasty surprises for the unwary if anything but a straightforward total contract sum is agreed beforehand. (It is also worth noting that there may be extra administrative complications through VAT when building costs for materials, etc. are obtained piecemeal.)

Agricultural construction is very commonly carried out in separate stages by different types of commercial organisation. For instance, frame-members and part-cladding for the structure might be provided first, by a large-scale manufacturer. This would be erected by a sub-contracted site crew, whilst the rest of the work is subsequently carried out by a local building firm, who may, in their turn, sub-contract specialist jobs, such as electrical installations. This system of construction is 'flexible' and allows for elements of DIY or the use of 'labour-only' sub-contracting. It gives potentially lower cost but may lead to lack of contract responsibility and site control. An alternative arrangement (commonly used outside agriculture) in which a 'main-contractor' is responsible for most of the site works and usually for all supervision, generally leads to less confusion and controversy over things like trade's attendance and timeliness. This militates also for higher quality in workmanship. The work expected of all parties on a site should be made clear before contracts are signed and work commences. Above all the temptation to make adjustments or improvements to the design, during construction should be resisted, unless they correct obvious failings which will impair the building system's efficiency later. Amendments to the work in progress are nearly always dones hastily, without sufficient regard to their total impact and are relatively costly, as the resulting 'variation orders' are not constrained by the original pricing for the contract tender.

Further reading

BARNES, M. and MANDER, C. (1986). *Farm Building Construction, the Farmer's Guide*. Ipswich: Farming Press.

NOTON, N.H. (1982). *Farm Buildings*. Reading: College of Estate Management (Distributed by F. Spon).

STAGES IN CONSTRUCTING A BUILDING

Preparation – setting out a simple rectangular building

(1) The turf and surface soil is stripped from the site, along with trees, shrubs, bracken, complete with roots, if possible. The subsoil thus exposed is 'levelled' over an area exceeding that of the building. If the site slopes then the surface must be 'reduced' to a satisfactory gradient as required by the plant (usually close to the horizontal).

(2) Divert existing drains which cross the site (or re-lay in a straight line, surrounded by 150 mm of C20P concrete – *see* later section on concrete – deeper than the proposed works and on suitable falls, or construct a 'barrier' drain trench up-slope from the site – this latter is sometimes known as a 'french' drain).

(3) On the prepared subsoil surface stretch a line between pegs to represent the position of the longest side of the building but much greater in length such that the pegs are well clear of the proposed corners as at (a) in *Figure 22.2*.

(4) Mark one of the corner positions with another peg (b, *Figure 22.2*) and at right angles from the first line stretch another to mark a second building side (c, *Figure 22.2*). An accurate right angle may be made by using a builder's square, a surveyor's site square, or the method of Pythagoras (3:4:5 triangulation giving 6 m along one side, 8 m along a second and 10 m across the measured hypotenuse).

(5) In a similar manner mark the third and fourth sides (d). The dimensions and position of the building are now shown in plan by the lines and corner intersections. It is usual for these to be the outside edges of the building above ground but they may represent centrelines or inside surfaces. It is important to plan according to fixed reference surfaces and levels throughout the construction and to avoid changing them unless it is absolutely necessary.

(6) Check the diagonals (e) of the building outline – they should be the same (for a rectangular building).

(7) Profile boards (*Figure 22.2*, f, and *Figure 22.3*) are now sited beyond each corner, well clear of the building area. These are pieces of timber which have permanent markers (nails or sawcuts) indicating the position of one or more aspects of construction, for example the edges of the foundation trench, the position of the wall to be built on the foundation and internal features such as doorways and walls (h, *Figure 22.2*).

(8) Lines (g) stretched from marker to marker on the profile boards provide an accurate facsimile of the ground plan full size on the surface to be worked on. Before construction begins, they are replaced by trickled silver sand, chalk dust, paint lines or other easy to follow methods of ground marking. The boards, however, are retained until such time as the reference marks on them have no value.

(9) The heights and levels over the site should be set out and checked against a vertical reference stake driven firmly into the ground, clear of all site activity. Marks on the stake should indicate ground level when the stake was

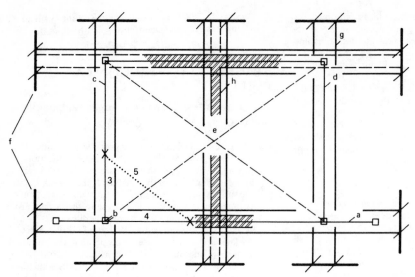

Figure 22.2 Setting out

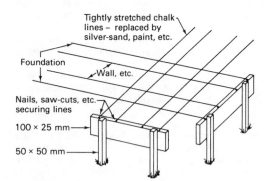

Figure 22.3 Use of profile boards in marking out

driven in, a general reference height, say 1 m, above ground level and other points necessary to the construction such as damp course level. (A clearly visible reference line painted on a nearby permanent structure might be preferable in some cases.)

Construction – groundworks

(*See* sections on Foundations and Drainage)

(1) Trenches and/or pad support areas are dug out to profiled width and correct depth with vertical sides and firm bottom. Any soft areas should be stabilised with stone ballast or concrete and rammed down until firm. Water and water-softened soil should be removed prior to the foundation concrete being poured. In very soft soil conditions it may be necessary to support the trench sides with struts and board temporarily, not least to ensure safety of workers.

(2) Foundations should always be horizontal; on sloping

sites a 'stepped' foundation is required. The depth of each step should, for convenience, be a multiple of whole building units e.g. block(s) or brick(s). Thus:

$$\text{Distance between steps} = \frac{\text{Depth of step} \times 100}{\text{Slope \%}}$$

Each step must project over the slab below substantially more than the step 'depth'.

(3) To ensure horizontal levelling and the correct thickness of materials used, drive pegs into the trench bottom so that the tops represent the finished foundation surface. Use a spirit level and a straight board over short distances but a surveyor's level, either telescopic, 'automatic' or water tube type, (*see* section on Surveying p. 581) should be used when the distances to be pegged exceed 10 m in total.

(4) The concrete mixture used for most farm works will normally be C20P (*see* section on Mass Concrete, p. 545) but may be C7.5P in certain situations where large masses are used, such as in foundation trenches with substantial thicknesses of concrete to be laid down (Reference may be made by contractors and others to the use of mixes described as 1:2:4, 1:2½:4, 1:3:6. These outmoded descriptions relating to volumetric proportions, do not imply satisfactory quality standards and should be avoided.)

(5) The top surface of the foundation concrete must be level, smooth and horizontal to facilitate the laying of brick, block or framework.

(6) Drainage trenches are dug out and pipes laid where appropriate, at the same time as the foundations are excavated. Any work which affects both is carried out first. For instance any pipes to be positioned under or through strip foundations should be placed and fixed by, for instance, encasing in 150 mm of C20P before the mass of concrete is poured around the pipework.

(7) Services are laid onto the site and led to the main distribution points for connection on a temporary basis during construction.

Construction – frame and related cladding (*see Figures 22.4 and 22.5*)

(1) Steel, concrete and timber frames have different shapes and material characteristics which require different assembly and fixing techniques (*see Figure 22.5*). The manufacturer of the frame will give specific guidance on the fixing and foundation requirements of particular structures.

(2) Each foundation pad supports and locates a frame member. The location system, of which there are many, using sockets, bolt assemblies, should be positioned exactly so as to receive the stanchion of the frame without distortion. Some frame manufacturers may wish to cast pad foundations themselves to avoid potential site difficulties. Small frames of 10 m span or less may not require a fixing system but rely on their weight to hold them in position.

(3) The frame members will arrive on site, sectionalised, for assembly prior to lifting into position by mobile crane. The span sections are loosely bolted in place and stabilised spatially by use of longitudinal members, e.g. the purlins, until the whole structure is complete at which time the joints are tightened up and the structure becomes rigid.

(4) Once the frame is secured roof cladding is fixed, starting from one (usually leeward) gable end. Care is needed to align the corrugated sheets squarely with the verge and eaves of the building, otherwise there may be a danger of too little or too much overhang as the laying of sheets progresses down the length of the building.

(5) After both slopes are covered the ridge capping is fixed in place. The overlap on both sheets and capping should be such as to minimise penetration by the prevailing wind

(6) The gable ends and side sheeting are positioned, cut as necessary and secured, followed by the guttering, at a suitable slope, and the 'down' pipes to channel the surface water to ground level. (Note the need for 'surface-water' drains to be laid prior to frame installation.)

(7) Special features like translucent sheeting, eaves fillers, ventilation ducting, insulation or anticondensation materials may be incorporated as the roof is clad.

Construction – masonery cladding and floor laying

(1) With the frame in place blocks or bricks may be laid on strip foundations between stanchion members up to 'damp-course' (DPC) level which should be at least 150 mm above the outside ground surface.

(2) Doorways and other gaps are left in the wall, as appropriate, with timber 'liners' or with fixing blocks left in the masonry for subsequent connections.

(3) The walls will be built according to the design specification based on the original requirements, and may incorporate a 'cavity', support piers and/or other reinforcements, insulation, water proofing or a cleanable surface.

(4) When the walls have reached damp-course level it is often convenient to lay the concrete floor. (For lightweight agricultural buildings the method of construction sometimes used is the 'ring-beam' foundation system, in which the floor is cast at the same time as the frame and wall supports with reinforcement incorporated at the edges.) The ground surface should already be level and free of topsoil so pegs are placed to indicate the finished floor surface. Hardcore is laid onto the soil to the required thickness (75–125 mm) and vibration-rolled into a hard, dense mass. The top of this is then 'blinded' by sand to fill all surface cracks and provide a smooth plane on which to place the damp proof membrane. Polythene sheet, 125 μm thick, or damp proof 'building' paper, is laid, with a minimum overlap of 150 mm at the sheet edges, to cover the floor and is bonded to any adjacent vertical wall and its DPC by bitumen-based paint or mastic. (This also serves as a 'slip' plane for the concrete mass.)

(5) The floor concrete (*see* Mass concrete later) should be poured in convenient shuttered bays, not exceeding 10 m × 3 m in area, for manageability and timeliness. It

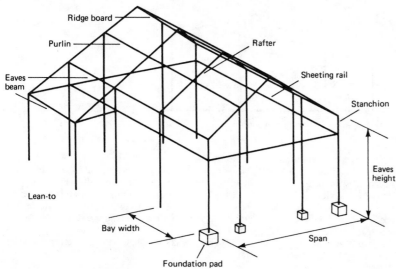

Figure 22.4 Structure elements

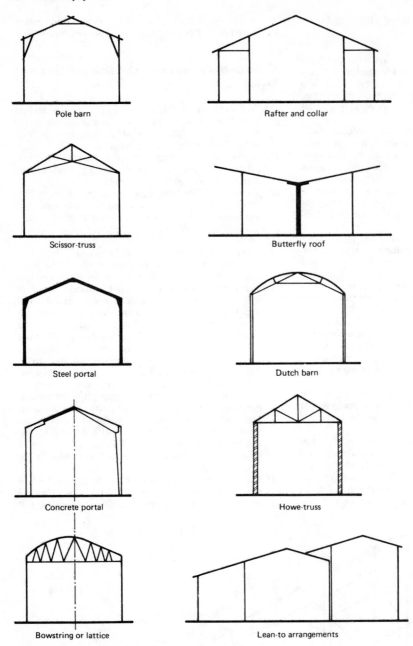

Figure 22.5 Frame systems

should be to C20P or C25P (BS 5328/1976) specification and of a suitable thickness, usually a minimum of 100 mm. It must be tamped and surfaced by floating or other method, preferably with the use of vibration equipment for consolidation. After which it should be left to 'cure' for at least 7 d and preferably 14 d, before the enclosing shuttering is removed or it is required to bear a load. (If the curing period is very warm or dry then a gentle re-wetting of the surface or spraying with resin sealant is recommended to preserve the mix moisture level.)

(6) An additional surface may be called for in some buildings, for instance in a farm dairy where a granolithic 'screed' may be necessary to provide anti-slip characteristics. This is a thin (up to 50 mm) top layer of fine granite aggregate concrete mix, applied to the base mix before curing has finished, and preferably within 72 h of laying. This is a skilled job, usually best left to specialist tradesmen.

Construction – internal finishing and services

(1) The internal fittings such as water supplies, electrical installations, joinery work and masonry finishing (plastering, etc.) are normally the province of professional tradesmen, whose skill can make or marr a building. In this work there is a flexible element of design which is not possible in other areas of construction. 'On-site' decisions are quite usual and in any case craftsmen would be expected to think for themselves in respect of details like pipe and wire runs and fixing methods on all but the largest farm construction. However, there are some pitfalls on which fixing advice needs to be given, for example on door security in piggeries, moisture proofing of electrical equipment in dairies and drinking bowl positions.

(2) The complexity of the work to be done and the need for a careful order of procedure to be followed means that the timeliness of the attendance of the specialists is often important to the smooth transition in the stages of building construction to avoid time wasting and mistakes. The employment of an experienced professional to provide overall supervision may be crucial to the successful completion of a complicated building.

Further reading

See Barnes and Mander (1986), also Noton (1982) in Design criteria above.

BUILDING MATERIALS AND TECHNIQUES

Mass concrete

Concrete is usually a mixture composed of ordinary Portland cement, (as opposed to rapid hardening Portland cement and other types), fine aggregate, coarse aggregate and water. Aggregate may consist of gravel or crushed stone, of a light or heavy nature, and may be round or sharp (angular). The fine aggregate (or sand) fills in the voids between the coarse particles (stone, gravel), the cement coating adhering to both. This should produce an almost solid matrix when set if air is expelled by good mixing and consolidation. All aggregates must be clean and free from vegetable matter, salt and clay. They must be dry for accurate volume or weight measurements to be made prior to mixing. 'All-in' or 'as-dug' aggregates should not be used. Always use clean water, preferably from a drinking source.

Although it is possible to mix concrete by hand in small batches it is not recommended practice. The most reliable way of apportioning mix ingredients is by 'weigh batching', that is, weighing out enough materials to mix with 1 bag (50 kg) of cement, or a multiple number of bags, and to load the quantities into a suitable sized mixing machine in one whole batch.

The time of mixing should be long enough to ensure that all the aggregate particles are coated evenly with cement to give a consistent greyish coloration, for which, normally, a minimum of 2 min is required after the last ingredient has been placed in the drum. Most mixers work best when the coarse aggregate is placed in the drum first, followed by the water, then half the sand, then the cement and finally the second half of the sand. Some 'sticky' aggregates may require that this order be varied to achieve the best possible mixture.

The quantities of cement, aggregate and water will vary according to the job to be done.

Table 22.2 shows the normal proportions by volume, for the work loads expected in three types of circumstance, on farm sites.

The size of aggregate must be taken into account when selecting an appropriate mix. Thin section or reinforced concrete requires stone which is relatively small in size, graded from 6 to 20 mm, whereas large mass concrete can satisfactorily enclose aggregate of up to 40 mm size grade. Fine aggregate is 5 mm or less in size, usually but must not contain too much dust.

Water quantitites are critical – almost as important to the ultimate strength of the concrete as the cement itself. Too much weakens the concrete by leaving cracks when it dries out, too little will give ineffective bonding due to lack of chemical reaction. Workability is also governed by the water content of a mix and as a consequence very often concrete is mixed and laid much too wet in order to increase the facility of placement. The moisture content can be checked by use of the 'slump' test. Special testing cones can be purchsed or hired but a steel bucket may be used as follows instead to provide a rough guide. The bucket is filled with concrete and is then inverted onto a flat surface (steel/wood plate), the bucket is drawn gently away from the contained concrete leaving a cone-shaped heap, the height of which has 'slumped'. A slump of more than 25% of the height of the bucket shows the mix is too wet. Using the correct test equipment 50 mm ± 25 mm should be attained at the correct water content. If such a mix is made it will require vibrating equipment to place and consolidate (tamp) it satisfactorily. This equipment may be hired.

All concrete should be laid as soon as possible after mixing

Table 22.2 Concrete mixes for various purposes proportional by volume

	C7P Large bulk volumes, e.g. deep foundations		C20P Average use, floors, roadways		C25P Heavy use and thin sections	
Cement	2 vol	1 bag	3 vol	1 bag	3 vol	1 bag
Fine aggregate (damp)	5 vol	190 kg	5 vol	115 kg	4 vol	90 kg
Coarse aggregate	8 vol	270 kg	9 vol	195 kg	7 vol	170 kg
Water	1–1.25 vol	20 litres	1.25–1.5 vol	23 litres	1 vol	17 litres
Approx. yield	7–8 vol	200 litres	9–10 vol	170 litres	7–8 vol	140 litres
Previous designation	1:3:6		1:2:4		1:1.5:3	

and always before 1 h has elapsed. It should be carefully positioned from a low height avoiding any movement which might cause desegregation of the constituents, for instance, rough barrowing. There is a brief period of 'setting' (about $\frac{1}{2}$ to $2\frac{1}{2}$ h) followed by 'hardening' during which it gradually achieves greater strength. During this latter 'curing' period which is indefinite, but for practical purposes is said to be 28 d, the concrete should be undisturbed, kept from drying out and in a frost-free environment. Under no circumstances should concrete be mixed when the aggregates are frosted, but in a very cold conditions (less than 4°C) a solution of builder's antifreeze (*not* motor car type) may be used to enable concreting to be carried out.

Pre-cast concrete

Many concrete items used in construction are 'pre-cast', that is, they are made away from the point of use, for example, concrete blocks, drain covers, frame members.

Concrete blocks

These are supplied in various common forms, standard sizes (*see Figure 22.6* for a typical range) and aggregate materials to suit the requirements of the structure. The majority of those used in farm construction work are of 'dense' aggregate, although 'cellular' blocks are used from time to time for insulation purposes. These latter, have a bath sponge appearance, due to the foaming method of manufacture. Although of low heat conductivity they are relatively weak and are porous to water passage. (Note however, that all concrete blocks are poor moisture barriers, compared with bricks.)

When calculating quantities to be purchased it should be noted that the actual sizes of the blocks may be less than the 'nominal' size by 10 mm, the mortar gap allowance.

Concrete block walling

Most walls constructed in farm buildings are of 140 mm or 215 mm wide units laid in 'stretcher' bond pattern (*see Figure 22.7*). For some purposes however, they may require strengthening. This is carried out, traditionally, by either the incorporation of piers or the insertion of steel rods or netting into the bond pattern of the wall. Piers are part of the wall which are thickened by the inclusion of blocks such that a vertical 'ridge' is formed, normally one block wide up the wall. The various forms of steel reinforcement provide strength in tension which the concrete block-work requires but lacks under bending loads (*see Figure 22.8*). The steel must be bonded into the wall such that it all behaves as one composite material and yet is correctly positioned to resist the tension forces. In practice, piering and reinforcement are used together to achieve maximum strength, one method supplementing the other.

See Table 22.3 regarding wall stability and 'rule of thumb' relationships of wall width, strength and use.

It should be noted that many agricultural buildings carry wall loads which are lateral and quite severe. 'Rules of thumb' are not recommended for these and expert advice should be sought for their design.

'Cavity' walls are not common in agriculture, although they are good insulators, as they are complicated to build, are weak to lateral pressures and can be expensive. They consist of two 'skins' (or leaves) of 100 mm concrete blocks 'tied' together by means of steel or plastic connectors (ties) across a 50–60 mm gap left between the skins which may be partially filled by insulation board adjacent to the inner leaf.

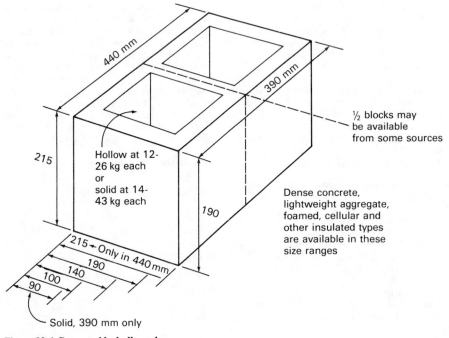

Figure 22.6 Concrete block dimensions

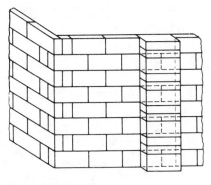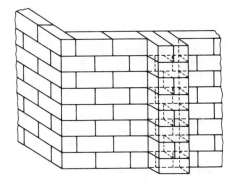

Figure 22.7 Wall bonds

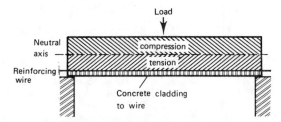

Figure 22.8 Simple reinforcement in a short concrete beam

Table 22.3 Block thickness for unreinforced walls (but which may include piers)

	Block size (mm)	100	140	215
Walls lacking any support at top	Max length	3.6 m	4.5 m	9 m
	Normal max related height	1.8 m	3 m	3.6 m
With continuous top support, e.g. rafters, framing, wall-plates, eaves beams	Max length	4.5 m	6 m	7.5 m
	Normal max related height	4.5 m	6 m	7.5 m

NB Walls designed to withstand wind and stock pressures; not to retain stored grain. For larger stock use the large block sizes.

The ties are placed in the wall, as it is built, about 1 m apart horizontally and 0.5 m vertically. It is important to ensure that the ties are kept scrupulously clean and free from mortar droppings. The internal skin may be of cellular block to increase the thermal insulation level yet further.

Where the length of a wall exceeds its height by $1\frac{1}{2}$–2 times or where there is no natural bond break within 6 m an expansion break must be included to accommodate thermal and moisture movement to which concrete blocks are particularly prone. This may be provided by a slot cut in the outer block surface at least 50 mm deep or, possibly, completely through the wall. In the latter case the stability is retained by the use of steel dowels set in the mortar on one side of the break and greased before laying in the other side. The slot should be filled with mastic at the outer surface to provide a weather-seal.

The mortar used in concrete block laying should be in the volume proportions of 1:2:9 or 1:1:6 mixed from cement, lime and builder's sharp sand. The lime may be replaced with proprietary ingredients known as plasticisers, which make the mortar 'fatty', that is, slip easily into place between the blocks. These mixes are weak, relative to block strength, so that if there is a tendency to crack it will appear less unsightly and be of less significance to wall stability when it is confined to the mortar gaps.

Concrete block walls may, depending on their purpose, require waterproofing. This is primarily done by use of render coatings of mortar-like material, usually in two coats; the first, at 10 mm thick of 1:1:6 mix as described before and the second at 6 mm thick, of 1:1/4:3 mix. Alternatively, the block pores can be filled in and the mortar gaps flush pointed to give a flat surface which can be painted with chlorinated rubber, epoxy-resin or acrylic-resin bonding paints.

A new approach to block laying combines the simplicity of 'dry-laying' (with no mortar), and render-coating for water proofing, by the use, as a wall surface fixative, of a mix of cement, sand, specially prepared glass fibre and bonding chemical laid on to produce a hard, durable coat which stabilises and creates the strength of the wall from the outside surfaces.

Concrete frame members

These are almost always precast, although very simple ones such as door lintels may be cast *in situ*. They contain steel reinforcement at a position determined by the need to carry tensile loads (*see Figure 22.8*) for which concrete is relatively weak having only one-tenth of its capacity to resist compression. The requirement for a high degree of reliability and thus quality control ensures that most are factory made. They may be 'pre-stressed', in which steel wire is pre-tensioned before it has concrete settled round it in the moulds. This technique increases the possible compressive load which can be imposed before the tension borne reaches a limiting value. The net effect is to enable economies to be made in all parts of a frame including its foundations due to reduction in the weight of 'surplus' concrete (that which is not working hard).

Timber

Timber used in construction should be well seasoned, pressure treated with biocides and fire retardants, free from warping, knots, sapwood, bark and shakes (splits). Softwood (*see Figure 22.9* for available size ranges) is used for nearly all work other than decorative or highly specialised functions due to the high cost and difficult working of most hardwoods.

Lengths of timber are generally purchased by the 'metric-foot' – actually 300 mm but for large quantities it may be sold by the cubic metre with additional sawing charges. Pieces in excess of 5.4 m long may exact a premium price. Sawing and planing 'allowances' must be taken into account when sizing timber sections as these may reduce the dimensions by up to 2 mm.

Probably the biggest drawback to the use of timber in construction is biotic decay. In the UK it is affected by three main agents, 'woodworm' (actually the larvae of a beetle, *Anobium punctatum*) 'dry rot' (a fungus, *Merulius lacrymans*) and 'wet rot', which has several possible causal fungi. All these problem pests can be prevented from doing damage by treating the timber initially with a cocktail of potent chemical biocides, but the life-span of timber can, in any case, be prolonged by the avoidance of poor climatic or environmental conditions.

Repair of timber which is already infected is difficult and sometimes, as in the case of dry rot, almost impossible without wholesale replacement.

Recently the use of 'stress-graded' timber has become a requirement in some types of structure. This is timber which has been inspected visually for flaws such as knots, size imperfections, shakes and other defects and is stamped with a grade category which indicates the limitations to its use in structural work. GS (general structural) has been passed as acceptable for general use, for example as floor joisting, SS (special structural) implies a more particular usage (*see* BS 4978 on stress grading of timber).

For information about screw and nail sizes for timber work *see Tables 22.4* and *22.5*. Recently there have been a number of new types of fixing method devised, most of which combine the driving ease of the nail with the firm hold

Table 22.4 Wood screw sizes (slotted countersunk steel)

Gauge	Diameter of unthreaded shank (mm)	Lengths available (mm)
0	1.58	6, 10, 12
1	1.78	5, 6, 8, 10, 12, 16
2	2.08	6, 8, 10, 12, 16, 19, 25
3	2.39	6, 8, 10, 12, 16, 19, 22, 25
4	2.74	6, 8, 10, 12, 16, 19, 22, 25, 32, 38, 45, 50
5	3.10	6, 8, 10, 12, 16, 19, 22, 25, 32, 28, 45, 50
6	3.45	8, 10, 12, 16, 19, 22, 25, 32, 38, 45, 50, 57, 63, 70, 76
7	3.81	10, 12, 16, 19, 22, 25, 32, 38, 45, 50, 57, 63, 76, 89, 101
8	4.17	10, 12, 16, 19, 22, 25, 32, 38, 45, 50, 57, 63, 76
9	4.52	12, 16, 19, 22, 25, 32, 38, 45, 50, 57, 63, 76, 82, 89, 101, 124, 127
10	4.88	12, 16, 19, 22, 25, 32, 38, 45, 50, 57, 63, 76, 82, 89, 101, 124, 127, 152
12	5.59	16, 19, 22, 25, 32, 38, 45, 50, 57, 63, 76, 82, 89, 101, 124, 127, 152
14	6.30	19, 22, 25, 32, 38, 45, 50, 57, 63, 76, 89, 101, 124, 127, 152
16	7.01	25, 32, 38, 45, 50, 57, 63, 70, 76, 89, 101, 127, 152
18	7.72	32, 38, 45, 50, 63, 76, 101, 114, 127
20	8.43	38, 50, 63, 76

From BS 1210: 1963

of the screw. A review of these innovations is recommended before extensive timberworks are carried out.

There are two further general design concerns expressed about structural timber. The first is its apparent susceptibility to fire damage which owing to its characteristic of charring rather than burning is not as threatening as might at first be thought. If proper safety factors are used in design

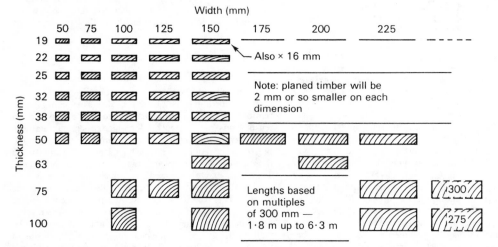

Figure 22.9 Sawn softwood sizes

Table 22.5 Common nail sizes

Length (nominal) (mm)	Round plain head nails, diameter (approx.) (mm)	Approximate number of nails/kg
100	3.7–4.9	22–14
75	2.6–4.1	57–23
50	2.0–3.3	147–58
40	1.8–2.6	236–118
30	1.6–2.3	383–172
25	1.6–2.0	434–294
	Cut floor brads	
75	3.3	20
50	2.6	55
40	2.3	80

From BS 1202: 1966

considerable parts of the cross-section of timber beams must be burnt away before failure occurs.

The second concern relates to its flexibility. Timber lacks 'stiffness' compared with structural concrete and steel so its use is restricted in simple beam or lintel framing to short spans. In larger spans the timber must be used in a lattice girder or similar pattern, to achieve the required strength and rigidity.

Timber boards and composites

The cost of timber has risen rapidly in recent years and will continue to do so in the foreseeable future. As a consequence composite materials have been developed which utilise the lesser quality portion of the timber previously thrown to waste, and at the same time make better use of the high quality material. The original form these took was that of 'plywood' or 'chipboard' but now they may be made partly or even mostly from sources other than wood, for example,

recent building boards which are combinations of plasterboard and glass- or mineral-fibres, also plastic-based insulation combined with wood particle board.

The structural framework used for mounting these boards (usually supplied in 2.44 m × 1.22 m sheets) is generally based on the traditional 'studwork' used for timber planking (*see Figure 22.10a*) but they are sometimes supported on rails in a similar manner to cement-fibre corrugated sheeting (*see Figure 22.10b*). External grade (WBP, weather and boil proof) plywood sheets make excellent cladding for walls and roofs (the latter with a felt covering) though care must be taken when fixing if warps are to be avoided. All edges should be secured with galvanised nails or screws, with intermediate nailing for hardboard (or other types of 'particle' board) at 600 mm centres. Generally, the main studs should be at 1.2 m centres, equal to the sheet widths with noggins at 600 mm vertical centres. Oil-tempered hardboard, an inferior but cheap alternative, should be nailed every 200 mm or screwed every 300 mm to a frame specially provided with studs and noggins to accommodate it. Traditional timber boards are used as an outside wall cladding for poultry houses, barns, etc., either horizontally or vertically (with tongued and grooved joints). The boards may be horizontal and parallel-faced with lapped joints (e.g. 'shiplap' boarding) or, preferably, horizontal and taper-faced with lapped joints (weather boarding). All boards used should be 25 mm thick though 19 mm is not unusual on cheaper work. In the latter case they should be 'clout', galvanised nailed to vertical studs placed at 450 mm centres. For proofing against wind and snow the boards should be lined with building paper trapped to the studwork.

Vertical boards may be used to provide ventilation to cattle yards (space-boarding/Yorkshire boarding) being nailed to horizontal rails, with about 13 mm gaps between. These boards, 125 or 150 × 19 or 25 mm, should be above stock height and should be associated with ridge ventilation to assist air circulation. Yorkshire boarding is also used to roof cattle yards, being an alternative to slotted steel or

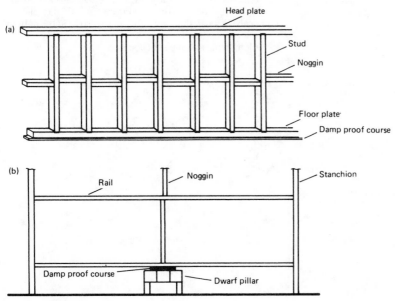

Figure 22.10 Wall framing

steel or aluminium or gapped cement-fibre sheets. Boards used in this manner should be 150 × 25 mm with two longitudinal drainage grooves, one near each topside edge. They should be nailed over a spacer which lifts the boards 6 mm above the purlins. Gaps of 6–12 mm are left between them.

Various materials are available for internal lining boards, including flat cement-fibre sheets (2.4 × 1.2 m) in several qualities and ranging from 3 mm up to 12.5 mm thick. These are suitable for lining timber frames or panelling gates and are often used for controlled-environment housing. Similarly, oil-tempered hardboard in thicknesses of 3–8 mm and sheet sizes up to 1.2 × 3.6 m may be a useful, cheap cladding for this purpose as it resists water absorption. Another common material is manufactured from compressed plastic waste to form a strong, dense but flexible and waterproof board of 12 mm thickness or more.

Insulating fibreboards, usually 12–18 mm thick and up to 1.2 × 3.6 m have low impact resistance and structural strength, absorb moisture and are therefore less useful, though they can be obtained with a foil-lining or bitumen impregnation and are relatively cheap. Additionally, there are many insulation boards formed from single materials such as foamed PVC or combinations of timber, plastic and metals bonded with plastic glues.

Non-timber lightweight cladding

Frames may be clad in simple fashion with uninsulated sheets of cement-fibre, steel, plastic and bitumen-coated hessian, or combinations of these, (aluminium rarely) and fixed with nails or special bolts to wall rails or roof purlins. Corrugated cement-fibre (replacing asbestos-cement), together with plastic (PVC) coated steel are the usual claddings preferred.

The former, however, has low impact strength, becomes brittle with age and requires 'crawling' boards for maintenance procedures. The most common type of this material, so-called Big Six, with standard lengths in increments of 300 mm from 1.5 m to 3 m, have, when laid with a single corrugation lap, an effective width of 1 m. The minimum end lap should be 150 mm and for roofs of under 15 degree pitch should be sealed with a bitumen cord.

Maximum purlin and rail spacing for this sheet type should be 1.35 and 1.8 m, respectively. These sheets are available with slots in the crown of the corrugation which can assist ventilation of roofs for cattle. Other types of sheet are available with smaller 'Standard' and larger (Double Six) dimensions as shown in *Figure 22.11*. It should be noted that different manufacturers do not observe the same profile for similar sheet sizes, thereby making a match difficult for assembly or repair.

Translucent plastics/glass fibre sheets for rooflights or wall panels follow the same profile and sheet size as cement-fibre. Clear plastic ('Perspex') sheets are also available and increase light transmittance, but being much more expensive are rarely used.

Sheets are fixed with hook bolts over steel and concrete rails or purlins or with drive screws into timber, the latter being cheaper. Bolts and screws should be galvanised, protected externally with plastic washers and caps. Holes for fixings should always be drilled in the top of the sheet corrugation (except in the case of steel sheet – see later). It is usual to support each sheet in at least three places. Where four cement-fibre sheets coincide (and overlap), it is necessary to cut two of the sheets on a mitred angle to lay them as flat as possible. There are many specially-formed accessory panels available to cover awkward areas and difficult shapes at for example, roof edges and on curved roofing.

Corrugated steel, available in galvanished, plastic-coated, painted or very rarely, plain, versions, may be supplied in lengths from 2–3 m, gauges of 22, 24 and 26, with a net covering width of 610 mm. It is a lighter and more impact resistant alternative but however well-protected is still subject to rusting processes after ageing. The great majority of these sheets are used coated with coloured PVC to match the requirements of owners or local planners. Exposed sites should be accommodated by use of double corrugation laps. In addition sheet lengths of up to 7.5 m are obtainable to special order, obviating the need for frequent end laps. Fixing methods are similar to those of cement-fibre sheets except that corner mitring is unnecessary and holes may be punched through the sheet – they must also be at the bottom of the corrugation troughs.

Aluminium, although not common owing to cost, is useful

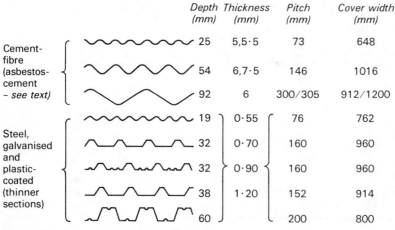

		Depth (mm)	Thickness (mm)	Pitch (mm)	Cover width (mm)
Cement-fibre (asbestos-cement – *see text*)		25	5,5·5	73	648
		54	6,7·5	146	1016
		92	6	300/305	912/1200
Steel, galvanised and plastic-coated (thinner sections)		19	0·55	76	762
		32	0·70	160	960
		32	0·90	160	960
		38	1·20	152	914
		60		200	800

Figure 22.11 Common corrugated sheet profiles

for its lightweight characteristics and solar reflectance.

There has been a recent revivial of interest in slotted and slatted roofing for ventilation in extensive livestock housing. New, cheaper methods have been devised taking the form of corrugated sheeting laid in normal fashion from ridge to eaves but with no side overlap and leaving a small gap (15–20 mm wide) between the sheet 'columns'. The sheets are laid with the edges of the corrugations upwards at the slots. Roofing costs are reduced and the cost of a purpose-built ventilation system let into the roof may also be saved.

Frames and simple roof support structures

The frequently used term 'portal' should specifically be applied to support frames providing clear, unimpeded headroom to the ridge inside the building, but it is often corrupted to mean any factory made, site assembled, large span system (*Figure 22.5*). Common components (*Figure 22.4*) include the upright supports (stanchions), the arms (rafters) running from the eaves to the ridge, purlins, sheeting rails, eaves beams and ridge boards. Many types of connector are used in assembling the structural members, including bolts, clamps, nails, staples, welds and even glue, each relating to an appropriate material and structure system.

The common and 'preferred' size ranges for major frame dimensions to which most manufacturers adhere are shown in *Figure 22.12*. Eaves height is comparatively simple to adjust during manufacture or assembly and so provides

many options which are suited to individual requirements. As a general rule wider buildings have shallower slopes to their roofs to reduce overall height but the framework elements must become stiffer to compensate. Thus it is often more economical to create very wide structures from main central spans with 'lean-to's' at one or both eaves or, alternatively, multi-span arrangements with roof valleys between slopes. A lean-to may extend as much as 11 m from the main structure.

Normally purlins will be of the same material as the frame except that timber may be supplied with steel frames of smaller bay widths and where easy sheet fixing (by nailing) is required. Purlin spacing and size depends more on cladding sheet length than on frame rigidity considerations, as each sheet must be supported at the top, centre and bottom, resulting in a normal arrangement at approximately 1.35 m intervals down a roof.

Occasionally, in a narrow building, it may be more appropriate and economic to use a roof truss (*Figure 22.5*) supported by the walls instead of a frame system. This may be of steel, bolted or welded type, or of timber in a number of patterns, from the simple traditional nailed or bolted variety to the sophisticated design of glued or stapled latticework. The simplest type of frame of all is the 'beam and post' construction of which there are many versions from the rolled steel joist (RSJ) to the box beam in 'glue-lam' plywood. The choice of these is usually left to structural engineers and should not be attempted by the non-professional, but in very simple cases where timber joisting is required for suspended floors a traditional 'rule of thumb' may be used:

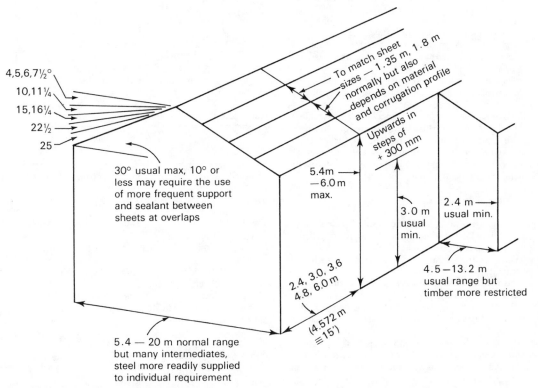

Figure 22.12 Common frame dimensions

$$\text{Depth of timber (mm)} = \frac{125 \times \text{span (m)}}{3} + 50$$

This provides the required depth of a joist system of stress-graded timber, measuring 50 mm thick spaced at 400 mm centres along a strong support wall and strutted by means of noggin pieces halfway across the span. As timber is expensive, particularly in 'heavy' sections, economics will dictate the limitation to the span capacity of this method but timber's characteristic of 'giving' under load will also limit the maximum distance over which it may be unsupported, usually 6 m. (It must be stressed again that any plan using this method should be checked by a qualified structural engineer for safety reasons.)

Foundations

Foundations are intended to sustain building structures against three main loading problems in agriculture:
(Note: Reference to 'soil' implies subsoil, i.e. below the agricultural soil horizon.)

(1) Sinkage – when the building weight (its own structure load and that imposed upon it by contents) exceeds the support capacity of the soil beneath, downward or sideward movement may occur. This can be differential in nature, varying from place to place, creating bending loads on the structure.
(2) Water effects:
   (a) Volume changes due to soil moisture content variation (sand soils best, clay and peat worst) cause structure stress and movement by soil 'heave'.
   (b) Frost action giving volume and structure changes in soil (minimum depth for foundations related to climatic considerations), by expansion/contraction.
   (c) Leaching and erosion undermining the base for support.
(3) Cartwheeling – where lateral (side) loads applied by building contents, e.g. stored grain, turn the structure base through 90 degrees.
(4) Foundations must resist other forces such as vibration and load cycling but fortunately these are comparatively rare in agriculture and can be ignored for practical design purposes. (They must also resist chemical attack from soil acids and sulphates – but these are not directly structural matters, although important.)

Thus foundations must be

(1) correct in area (load imposed ÷ soil bearing capacity × safety factor) which is sometimes, for simplicity, to be a standard recommended size,
(2) at a correct depth (about 600 m to 1.2 m according to soil and climate),
(3) of a correct width to prevent cartwheeling and provide for easy construction practices, e.g. trench and hole sizes are dictated partly by the skill and ease of digging, also masonry units require a margin of measurement accommodation,
(4) of a thickness to be strong enough to resist failure in compression, i.e. to avoid a hole being punched through by the load applied (it may be necessary to reinforce with steel to achieve this),
(5) of sufficient material quality to resist failure by chemical reaction and premature loading during curing.

The actual size and depth is therefore inherently related to particular sites and building circumstances, and there can be no safe 'rules of thumb'.

Foundation design advice should always be obtained from an expert source familiar with the site and the proposed building before commitment to a type of construction is made.

Strip foundations (Figure 22.13)

As agricultural buildings are usually lightweight with few load complications and offer minimal risk to the general public, some restricted soil situations can be provided for in the limited case where simple blockwalls are to be erected. The recommendations in *Table 22.6* apply only where the walls supported are for boundary purposes, carry no imposed load from above and there is no risk of human injury. (Again, to be safe, have the dimensions checked by an expert.)

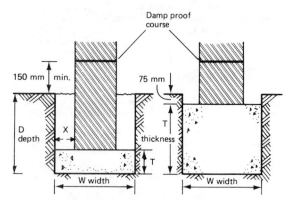

Figure 22.13 Strip and trenchfill foundations

Table 22.6 Foundation dimensions

Strip foundations	Clay	Soft subsoils	Sand, gravel, chalk
Depth (normal climate)	1 m	600 mm	500 mm
Depth (frost prone)	1.5 m	1.1 m	1 m
Minimum width (100 mm wall)	300 m	500 m	300 m
Width with working space (100 mm wall)	400 mm	500 mm	400 mm
Minimum width (270 mm wall)	600 mm	700 mm	450 mm
Width with working space (270 mm wall)	600 mm	700 mm	600 mm
Minimum thickness	200 mm	300 mm	150 mm
(For strip foundations extension X must never exceed thickness T – *Figure 22.13*)			(200 mm in soft conditions)

For 'trench-fill' systems read as strip foundations for depth, always use minimum width. Thickness is depth minus 75 mm. To be safe in uncertain soil conditions use minimum width required in soft soil.
NB This table should only be used where the soil on which the foundation rests is firm, undisturbed, not subject to frost or water action and can support the loads implied by the table conditions.

Pad foundations

These provide support for frame uprights and locate them laterally. Manufacturers of frames will provide detailed drawings of the socket, bolt or dowel fixings with their positions, in pads of recommended size. It is not possible to give guidelines for these as they vary tremendously with shape, dimension and type of frame and the properties of the soil on which they rest.

Drainage

Two categories of drainage must be considered (and treated) separately in practice (*see* Chapter 20 on Services for further information). They are:

(1) Surface water, that is water from rainfall run-off, which should be 'clean' and may be discharged directly into a water course for disposal. In effect this implies water from roofs only.
(2) Foul water (or soiled water) including sewage and slurry flows from all sources and other 'dirty' water such as cow yard washings. This category must in no circumstances be discharged into water courses without purification which will ensure that it conforms with the various anti-pollution requirements before release.

Roadways and external works

Within a farm there are usually three types of road: the access road connecting the farmstead with the highway, the roads which link buildings together and the lanes connecting with the fields. Each serves a different purpose and requires a different investment approach to be economic.

The access road is the main artery of the farm and needs a good layout and surface, justifying the use of concrete which is normally too expensive for other road purposes. The width should be not be less than 3 m to prevent concentrated wear and to allow for the passage of wide vehicles. Passing bays should be provided at not more than 150 m intervals. The 'weight' of traffic on a farm road is light, being related to the frequency of use rather than wheel loading, as the majority of vehicles travelling along it have comparatively low ground contact pressures. However, fast moving traffic will wear surfaces rapidly so it is advisable to limit permissible road speeds for vehicles such as bulk tankers, which have high weight carrying capacities. Particular problems for all farm roads result from silage effluent attack and the combination of slurry droppings and the passage of cloven hoofs, but it is difficult to see how the latter can be minimised without handicapping the farm transport network. Concrete, being the most resistant common surfacing material may be used to combat these effects but only at a cost which may be inappropriate.

Roadways around the buildings are an important factor in the planning process and will vary in width according to the amount of traffic expected to use them. They should be laid well clear of the buildings, except where they serve as direct links. Large vehicles entering a cul-de-sac must be provided with space to turn. All bends and turning circles should be

not less than 16 m diameter and should be strengthened to accept the greater loads induced.

Internal roads usually serve to link fields and buildings together and take mostly tractors, trailers and stock, none of which justify the building of an elaborate road. Thus, to keep down cost, local material should be used wherever possible: chalk, quarry waste, ashes, hardcore from old buildings, flints, burnt shale, limestone chippings, are all materials which make satisfactory internal roads. The finished surface may then be treated with a coat of tar or bitumen plus chippings or a more costly layer of precoated stone (Macadam). These will keep the water out of the subsoil, but may not be very hard wearing.

Keeping the subsoil dry is one of the key factors in successful roadmaking.

It is not usual therefore to excavate into the ground to make a road, though fibrous/humus soils should be removed; it should, instead, be built up from the ground surface. Where a road is made down into the subsoil because of poor support conditions the lower 'in-soil' part is made of introduced stone or hardcore and the surface should still be above the surrounding ground.

On sloping sites, a land drain shoud be provided on the rising gradient side of the road. All work should be carried out in dry weather, preferably during the summer months. Any junction with a highway must be carried out to the satisfaction of the County Council, possibly in pre-coated stone laid on a suitable stone base. It is often feasible to have this work done by the highway authority when a roadmaking gang are working in the area. An existing farm road can be given a wearing surface of pre-coated stone laid between 50 and 75 mm thick and then well rolled. The work should be completed whilst the stone material is warm. Alternatively, uncoated aggregate can be laid first and grouted with hot tar afterwards. All sharp bends should be constructed in concrete as the greatest amount of wear takes place on corners.

Wheel track roads in concrete cost almost as much as a full width road, owing to the labour required in forming the track. They require a lot of maintenance, can lead to tyre-wall damage and are therefore not usually recommended.

Concrete for roadways should normally be from 100 to 150 mm thick on at least 150 mm of compacted hardcore and of C20P, with a mesh reinforcement when bulk tankers use the road. Internal roads around the buildings may be constructed in the cheaper C7.5P concrete, 100 mm thick, laid on at least 100 mm compacted hardcore after all fibrous soil material has been removed, and in C20P if there is a chance either of use before a reasonable period of curing has been allowed or that crawler tractors might run on it.

Reinforcement is only necessary in soil of low bearing capacity such as peat, fenland silt and the like. Expansion joints are usually provided every 5 m length of road when laying continuously.

To reduce expense, passing bays can be constructed of hardcore: they should be 2.5 m wide by 12 m long.

Very light use internal farm roads may be made from 100 to 150 mm of hardcore laid on top of the subsoil surface after spraying with herbicide and rolling well. Larger stone should be used at the edges with smaller stone filling the middle. A drainage furrow may be ploughed alongside the edge which also acts as a support against the surface spreading. Any wet spots should be drained before the stone is laid down and consolidated.

Bridges, culverts and cattle grids

Spans of up to 6 m can be effectively and economically bridged in various simple ways using pre-cast beams or culvert piping of pre-cast concrete covered by thick layers of mass concrete with reinforcement but, as these may result in hazards to the general public, the advice of a structural engineer should always be sought before proceeding with construction.

A cattle grid soon pays for itself in time saved opening and shutting a gate on a busy farm road, but it needs careful thought before siting and construction, owing to the legal complications which can arise from interference with 'rights of way' and perhaps, more importantly, safety considerations. For this latter reason DIY grids are not desirable. Any cattle grid must be stock proof against all types found in the locality, it should be 3 m long and the full road width, with permanent fencing both sides along the full length of the grid. At its side there should be a by-pass gate 1.5 m away measuring at least 3.6 m wide.

Fences

Permanent fencing is specified in BS 1722, Fences, and BS 3954, Farm stock fences. To contain livestock (with the exception of deer), fences should be 1.1–1.25 m high. They can have concrete or timber posts with strained wire or woven wire or with timber rails between. The precise type of fence and its rail spacing, depends largely on the type of stock to be contained. Angle irons and rabbit wire can be used for light work. Deer fencing needs to be at least 1.8 m high (2.2 m where there is a 'deer run'), of about 15 strands of strained wire with mesh attached, and fixed to posts at approximately 15 m intervals.

Reinforced pre-cast concrete post fences

The posts should be bought from a reputable firm manufacturing to BS 1722 as it is impossible to make them properly using farm labour. Corner and straining posts should be set 900 mm in concrete and supported with 100×75 mm struts. Straining posts are spaced at 20–40 m centres and intermediate posts at 2–3 m centres. When metal spacers are used (at less than 3.3 m centres) intermediate posts may be at 50 m centres and straining posts at 200 m centres. Seven galvanised wires are needed to contain most stock starting with the bottom one 100 mm above the ground. The other wires are spaced at 125, 150, 175, 200, 225 and 250 mm centres but six, or even four wires may be sufficient for some types of animal, the top wire may be barbed. (Note – young livestock such as lambs and piglets will not be held by these arrangements.) The wires are usually spaced by means of galvanised, eyeletted strainers or winders.

Timber posts

Timber post and wire fences are constructed in a similar manner to concrete. All softwood timber posts should be pressure treated with preservative at the factory of origin.

Galvanised woven wire for 'stockproof' fences

These can be obtained in a variety of patterns and in many sizes. They are secured to timber posts at 2–3 m centres and a strand of barbed wire is usually run along the top of the posts.

Post and rail fences

Posts should be 125×100 mm at 2.7 m centres; with four 75×32 mm sawn rails, the first 150 mm above the ground and the remainder at 275, 300 and 300 mm centres. An intermediate 75×50 mm post is driven into the ground between the main posts.

Rabbit-proof fencing

This comprises 32 mm mesh fencing nailed to timber posts, the construction of which is similar to the timber post and wire fence, with the exception of the intermediate posts which may be increased in distance to 3.8 m centres. The top of the netting is clipped to a strand of barbed wire set 900 mm above the ground, and 150 mm above that a tripwire is stretched. The bottom 150 mm of the netting should be bent to lay flat on the ground, towards the rabbit side of the fence, on top of which turves should be laid; a further wire should be set about 450 mm above the ground to support the netting in the centre.

Gates

The standard timber five-barred gate is 3.6 m wide by 1.2 m high. Rails and braces should be bolted together with galvanised bolts. The top rail, or back, and second rail from the bottom should be the only two rails morticed through the head.

Steel-framed gates are very variable in size-range and *patterns* but being lighter, if less visually acceptable, tend to provide for the demand in larger sizes.

There are a variety of fasteners and hinges on the market, suitable for all types of gates and hangings. Most gates are hung 18–25 mm off centre to ensure self-closing. Implements such as combine harvesters, balers and drills may need a 5 m wide gate opening. A single gate cannot be economically made in this width, in either timber or steel, but many proprietary tubular or framed mesh gates are available in pairs. They are usually graded for 'heavy stockyard' or 'light duties', with openings of 2.5–5 m.

ENVIRONMENTAL CONTROL

Environmental control in agriculture is generally understood to mean the manipulation of temperatures (and occasionally humidities) by ventilation and heating. Its main use is in the optimisation of livestock housing climates, but in most limited forms it can be used for 'conditioning' in crop stores. Horticulturally, the term has a much wider implication and may infer, among other things, adjustment of gas levels (CO_2, O_2, N_2), supplementation of light and 'natural' control of pests. These activities call for a precision of monitoring and control which is not usually justifiable economically for less 'sensitive' agricultural produce such as potatoes and pig-meat. However, environmental control is still appropriate to these latter, though on a lesser scale, because it is only by the achievement of reasonable energy flow relationships between inputs and outputs that satisfactory profits can be made.

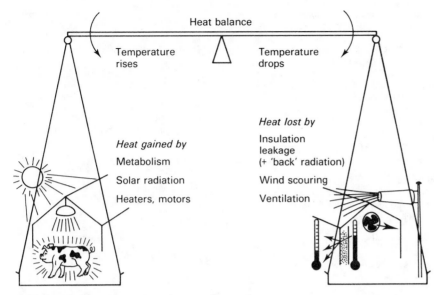

Figure 22.14 Major elements in environmental control

The major elements in this balancing 'act' are represented in *Figure 22.14*. As can be seen some of these factors are the result of management decisions, whilst others are fixed by the nature of a building's design. The resulting 'control' is thus simplified but at the same time evidently more difficult to achieve. It is made simple because only the status of ventilation and heating can be readily altered. On the other hand, the energy flows represented depend to a great extent on the external environmental circumstances, which are constantly changing. This apparent conflict which must be resolved by the building design if a satisfactory environment is to result can, in part, be settled by the use of a 'buffer' between exterior and interior environment – the insulative effect of the building envelope. Thus, the starting point for all environmental control systems should be consideration of the standard of insulation to be used. In addition, it is now being realised that animal welfare is often prejudiced by inadequate insulation, particularly in flooring, this reinforces the need to review this factor first.

Insulating methods

Virtually any structure which encloses an environment is acting as an 'insulator' but the term is generally reserved for materials which have the specific property of resisting, or opposing, heat transfer. This implies that they are capable of preserving a high (relative) temperature difference between external and internal environments, irrespective of which of these is the higher. The practical effect of a high temperature gradient may be seen from the example shown in *Figure 22.15*. Some insulators are clearly better than others (e.g. timber as opposed to dense concrete) but the great majority of specific purpose materials cannot be judged by instinct or on 'common knowledge' grounds. They need, therefore, to be compared by numerical values, of which there are several standard alternatives. The most frequently used of these is the 'U' value, a measure of the rate of heat flow expected through a structure under given temperature conditions;

better insulators are associated with lower values. *Table 22.7* gives some examples and *Figure 22.16* shows how they may be applied to calculate heat losses. Perhaps more logically, the 'R' value – for resistivity – is quoted for some materials. This is the inverse of 'U' and as such gives larger numbers to represent greater insulating effect. Both 'U' and 'R' are affected by their working 'conditions'. For instance, on exposed sites the 'U' will be greater and the 'R' smaller than normal, due to heat being scoured away from the insulating surfaces, creating more rapid transmission effects. In still air, however, the values will be appreciably better. Thus, they reflect the insulation properties for typical situations of material, climate and aspect rather than indicating absolute values which hold good at all times. Care should therefore be exercised in their interpretation for unusual situations.

Two further factors should be accounted for in the

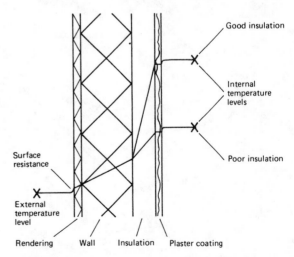

Figure 22.15 Temperature gradients through insulation

Table 22.7 'U' and 'R' (= 1/U) values of common insulators

Wall	'U' W/°C m²	('R')
1. 215 mm dense, hollow, concrete blockwork	2.05	(0.49)
2. As (1) + outside rendering + 12 mm expanded polystyrene slab	1.14	(0.88)
3. 275 mm 'cavity wall' of 100 mm dense concrete blocks + 25 mm expanded polystyrene slab	0.68	(1.47)
4. 215 mm solid wall of foamed blockwork + rendering both sides	0.45	(2.22)
5. 6 mm cement-fibre sandwich on 50 × 50 mm timber frame with expanded polystyrene core	0.45	(2.22)
Roof		
6. Corrugated 'Big Six' cement-fibre sheet	6.53	(0.15)
7. Corrugated sheet steel + fibre 'insulation' board	4.82	(0.21)
8. Corrugated 'Big Six' cement-fibre sheet + 60 mm mineral wool (+ vapour barrier) + 4.5 mm cement-fibre lining board	0.60	(1.67)
9. Corrugated 'Big Six' cement-fibre sheet + 55 mm extruded, foil-faced, polystyrene sheet	0.59	(1.69)
10. Corrugated 'Big Six' cement-fibre sheet + 40 mm foil-faced polyurethane board	0.50	(2.0)

Floor	'R$_{f45}$' (°C m²/W)
11. Dense concrete floor	0.042
12. Concrete slatted panel	0.086
13. 18 mm screed on 150 mm lightweight aggregate	0.17
14. Wooden slats, 58 × 70 mm with 10 mm gaps	0.23
15. 60 mm dry straw on dense concrete	0.66

Note that R$_{f45}$ is not directly equivalent to R(= 1/U) above as it takes into account sideways heat movement from 'point' body contacts.

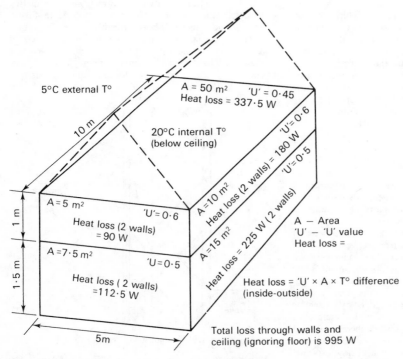

Figure 22.16 Example of heat losses from a warm environment

practical use of insulators. Firstly, the majority of the better ones contain large quantities of air, usually in tiny bubble formations; these cavities should not be allowed to fill with water by condensation or wetting otherwise their insulating properties will be lost. Secondly, radiant heat (positive solar gain and negative 'back' radiation loss to sky) can be an embarrassment which is difficult to counteract. Insulators which inherently have anti-radiation properties, being light in colour or having silvered surfaces, will help to limit heat transfer from this cause.

A good insulator, therefore, usually takes the form of a lightweight, hydrocarbon-plastic mass, filled with minute, discrete, air bubbles and covered with a shiny, preferably silvered, surface which is also a water-vapour proofing coat.

Ventilation methods

Whereas insulation is normally an inseparable part of a building, ventilation systems are often 'add-ons' or at least intrusions (e.g. ridge vents) into the structural envelope. They are, therefore, much more variable in pattern, but can broadly be divided into two types, although there is overlap between them. The simplest of these from the point of view of equipment and installation is 'natural' ventilation, the power source of which is derived from wind force or warm air buoyancy (or more generally a combination of the two). 'Forced' ventilation, the alternative, is more complicated, needs mechanical/electrical power to operate and may involve the use of fans, ducts, diffusers, filters and controllers. Its advantage, in contrast to 'natural' ventilation systems, which may outweigh these complications, is that it provides positive air movement and exchange irrespective of climatic or housing conditions. In livestock housing each of these systems has its appropriate place directly related to economic levels of investment, whereas in crop storage only the more positive powered methods are likely to be successful.

Natural ventilation

Natural ventilation (*Figure 22.17*), created from 'stack' effects (buoyancy) and wind flow, is generally used for housing animals that have low 'critical' temperatures and where stocking rates are sufficient to produce satisfactory flows/exchange rates. Two general problems arise from the use of this system, the first occurs where there are no wind flows of any consequence as for instance in anticyclonic weather conditions. The second is where the building is so wide from eaves to ridge that air flow patterns through the building are sluggish due to inertial effects. Both of these situations result in stagnant air masses which, once establish-

ed, are difficult to move or exchange, and thus create health and structure hazards from condensation and general dampness.

It is acknowledged that natural ventilation is, at best, unpredictable and at worst impossible to control, the situation not being improved by common variables such as building shape and size. Many structures, having the traditional slotted or open ridge, are under-ventilated much of the time. In recognition of this there has been a resurgence of interest in slatted (or slotted) roofing for large livestock buildings. A simple way of creating this effect is to leave gaps of 10–20 mm between each column of sheets when covering the roof (having first turned them upside down so that the edge corrugations conduct rain satisfactorily to the gutters!). This provides good ventilation rates and patterns, balancing adverse wind forces on the roof but nevertheless still lacks the essential element of controllability.

Automatically controlled natural ventilation (ACNV), a fairly recent development based on improved electrical control equipment (electronics), is now commonly used for small volume, high stocking density situations such as weaner housing, where expensive external heat sources are often required. This system uses a combination of wind flow and stack effect to achieve good air flow rates (and control) by mechanically actuated ventilation flaps arranged in optimal structure positions. It may also be used in conjunction with fan power assistance, as a backup, when the wind flow fails. This circumvents some of the drawbacks inherent in a purely 'natural' system.

Forced ventilation

Forced ventilation which relies solely on fan power for air exchange and distribution can be divided into 'suck' (exhaust) and 'blow' (pressurising) methods. Each has advantages for specific types of building but generally those that exhaust are simpler and cheaper but less effective, particularly in achieving good air distribution. In contrast, pressurising methods will exchange air efficiently, enabling better scouring and control. They also offer the possibility of air re-circulation, filtration and silencing but almost invariably cost more. The potential for creating draughts will also be less with pressurised air flow, so long as it is conducted in properly designed ducting at appropriate (low) velocities. *Figure 22.18* demonstrates some of the air-flow pattern variants which are possible with both these systems.

Ventilation rates (see next section)

The need to achieve a reasonable temperature stability in a building, at a set level can result in very different (and variable) air flow needs, possibly requiring alteration from the minimum to remove noxious gases in winter up to a high rate sufficient to reduce heat buildup in still-air summertime conditions. This ratio of demand may be as much as 30 or 40:1, making successful ventilation very difficult to achieve, even without interference from adverse wind flows or the inadequacies of control equipment.

The actual air-flow rates required can be calculated on two bases, either whole building exchange (house volumes/h) or the needs of the particular species concerned (m³/kg liveweight, m³/t, litres/bird). Both these methods have had their merits but recently the latter approach has been more popular.

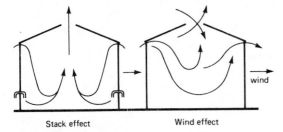

Stack effect   Wind effect

Figure 22.17 Natural ventilation

558

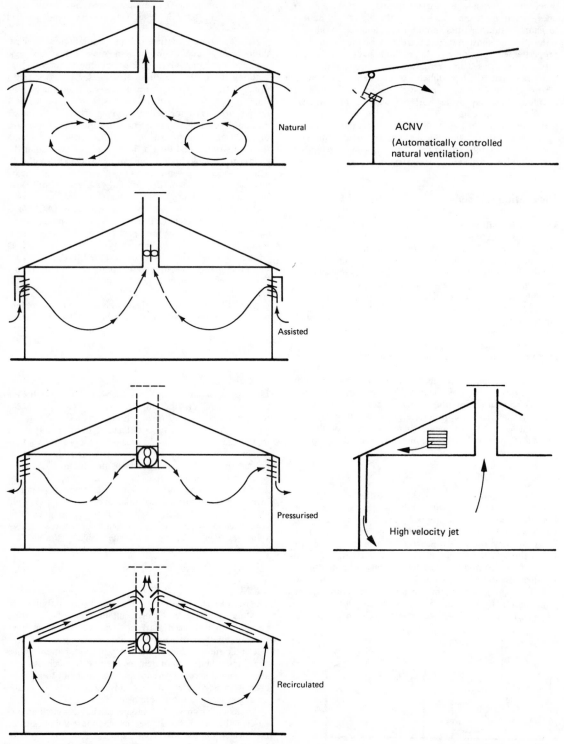

Figure 22.18 Ventilation systems

Humidity control

This has not, hitherto, figured very largely in environmental control design, due mainly to the difficulty of sensing any change in conditions with reasonable precision. Added to this, the reaction of stockmen to humidity tends to be 'anthropomorphic' and so controllers are often over-ridden when the atmosphere is felt by them to be 'uncomfortable'. However, as there is a growing awareness of the effects of 'damp' conditions in disease transmission and general animal welfare more emphasis is bound to be placed on this in future. One particular problem of high humidities lies in the implied risk to structures from condensation effects, both on cold structure surfaces and within insulation under 'dewpoint' conditions. These situations if permitted over long periods will depreciate building materials and reduce their effectiveness, ultimately affecting stock and crop health.

SPECIFIC PURPOSE BUILDINGS

(The reader is referred to Noton, N.H. (1982), *Farm Buildings*, Reading, College Estate Management for general information.)

1 Cattle accommodation

1.1 Dairy cows

Space for loose housing	Between 5.0 m²/small animal, and 8.0 m²/large animal (of which 75% strawed) + 1.0–3.0 m² when housing not completely covered
Space for cattle on slats	3.0–4.0 m². Slats 125 mm wide with 40 mm gap (*see Figure 22.24*)
Cubicle/kennel dimensions (*see Figures 22.19, 22.20, 22.21, 22.22, 22.23*)	1.05–1.2 m wide × 2.0–2.3 m deep (for small and large cows) + passageway (1.5–2.5 m)
Feeding and circulation space	2.0–5.0 m²/cow
Minimum height to eaves	2.2 m (for cubicle systems)–3 m. Depending on bedding depth allowance, access requirements, slatted floor height, to be added to basic minimum.
Temperature and insulation	Not important down to 5°C, no control of environment usually. 2°C is 'lower critical' temperature of small (50 kg) animal. Insulation limited to bedding or, occasionally, flooring.
Ventilation	Natural, by open ridge or space boards in upper wall area (and roof). Usually underestimated, with too little ridge vent provided. Where mechanical 10–50 litres/s per 50 kg liveweight (winter–summer).
Bedding, water and feed consumption	Straw 750–1500 kg/cow per winter. Wood shavings 500–1000 kg/cow per winter. Silage 20–40 kg/d per cow Hay 5–10 kg/d per cow Concentrates 5–7 kg/d per cow (for 15–20 litres of milk) Water 35–70 litres/d per cow (from trough with allowance of 20 cm²/cow or one drinking bowl/10 cows).
Effluent output, drainage and storage	Urine 25–60 litres/d per cow } depends on feed and housing system Dung 10–35 litres/d per cow } Drainage slopes on floors from 1:50 to 1:100 Slats may be part or whole of floor, may have drainage channel or tank under, to contain 3–6 months accumulation, may be raised 2 + m to provide total storage for permanently housed stock. Slats may be of timber, concrete or steel with general dimensions of 125 mm wide with 38 mm gap, 150 mm deep and 25 mm taper. (*see Figure 22.24*).
Feed space	600 mm trough length/cow. 100–150 mm self-feed silage face/cow with unrestricted access 250–300 mm self-feed silage face/cow with restricted access 600–700 mm silage and hay manger/cow.

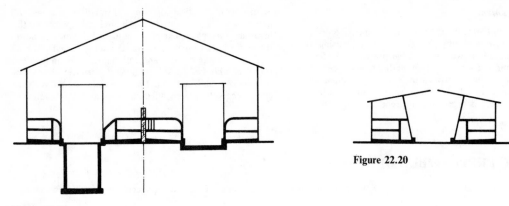

Figure 22.19

Figure 22.20

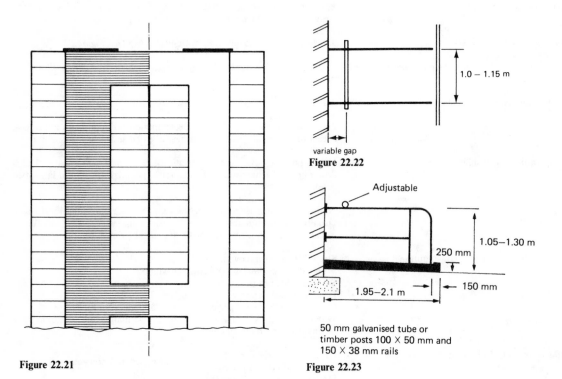

Figure 22.21

variable gap

Figure 22.22

Adjustable

1.0 – 1.15 m

1.05–1.30 m

250 mm

150 mm

1.95–2.1 m

50 mm galvanised tube or
timber posts 100 × 50 mm and
150 × 38 mm rails

Figure 22.23

Figure 22.24

Other consideration	Races and handling equipment should be adjacent to and integrated with the housing
	Non-slip floors and steps required
	Need calving and isolation boxes at 10 m²/animal, separated from rest of buildings
General	Consider cow management and handling, effluent problems, feed store siting, cow comfort, access for mechanical equipment

Further reading

CLARKE, P.O. (1980). *Buildings for milk production*. Slough: Cement and Concrete Association.

HARRISON, R. (1979). *Planning dairy units*. NAC, Kenilworth: Farm Buildings Information Centre.

1.2 Calves

Space for individual pen, controlled environment housing up to 4 weeks old	1.1 m²–1.5 m² in single pens of approximately 0.8 m × 1.5 m (minimum depth) for small calves (*see Figure 22.25*).
Group housing in controlled environment space, up to 3 months	1.5 m² increasing to 2.5 m²
Space in open-fronted housing, 3–6 months	2.5–3.0 m² with three or four per pen (*see Figure 22.26*)
Height recommendations	2.5 m to ceiling, minimum of 6 m³/calf. Barrier height 1.0 m minimum, up to 1.4 m for larger calves to provide isolation between pens.
Temperature	20°C for young calves, 15°C at 1 month old and 12°C at 3 months in controlled environment housing. Note that lower critical temperature varies greatly with bedding type and dryness. Thus, if kept in 'non-controlled' accommodation may be very close to LCT, particularly if draught-prone position
Insulation of small calf environments	Floor damp-proofed and having 'U' value of 0.5 W/m²°C. Roof, double-skinned or with ceiling, of 'U' value 0.6 W/m²°C maximum. Walls, cavity or other, with 'U' value of 0.9 W/m²°C maximum.
Ventilation	Adequate, draught-free ventilation essential for good health. Controlled environment should be capable of 25–30 litres/s per 50 kg liveweight in winter but air-flow should not exceed 0.25 m/s. Natural ventilation in housing for stock over 3 months old, with hopper windows or space boarding and ridge ventilation.
Feed and water consumption	Feed from bucket or machine at pen front, or trough for older calves (350 mm/animal), all near passageway. Milk substitute and small quantity of concentrate (0.7 kg/d) then up to 2.5 kg/d to 3 months of age. 15–25 litres/day per calf from water bowls or bucket near passage.
Effluent output	10 litres/d–30 litres/d depending on food, size and age.

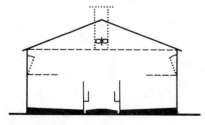

Figure 22.25

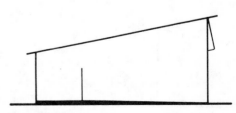

Figure 22.26

Flooring	Hard, cleanable, durable, resistant to disinfectant. Surfaced with wood float in exposed areas if concrete. Drain to 'centre dome' passageway at 1:20, and via step of 100 mm at edge of pen to drain channel at passage-edge. Slats at 3 months and above only. 100 mm wide with 30 mm gap.
Other considerations	Solid or 'see-through' pen divisions? Opinions vary as to value of each because of hygiene and social aspects.
	Care with surfaces as calves chew and lick everything.
	Drainage must never interconnect pen to pen
	Relative humidity should be kept down to 70% or less. (May be more easily achieved with 'open' than 'controlled environment' accommodation.)

Further reading

THICKET, W. *et al.* (1986) *Calf Rearing.* Ipswich: Farming Press.

1.3 Beef animals

Space on straw (*see Figures 22.27* and *22.28*)	2.0–3.5 m^2 per animal at 12 months, 4.5–6.0 m^2 per 2 year-old bedded animals. (4.0 m^2 with self-feed silage systems)
	10–20 in group to one court or pen.
Space for animals on 'sloped' (lightly littered) floors	2.5–3.5 m^2 per animal 12–24 months. (Sloped area at 1:16)
Space for beasts on slats	1.4–1.9 m^2 per animal at 12 months, 100 mm wide with 30 mm gap. 2.0–2.5 m^2 per 2 year old. 125 mm wide with 40 mm gap. 20–25 animals per court or pen (*see Figure 22.24*).
Height to the eaves	3.0 m minimum (allow extra for over-winter dung build up of 1.2 m in strawed yards, or slurry depth under slatted floor)
Roofless units	3.0–5.0 m^2 per animal, strawed; 1.0–2.0 m^2, slatted; 1.7–3.0 m^2, with sloping concrete and litter. (May have cubicles.)
Temperature and insulation	Not critical within range 2–20°C. Insulation not required.

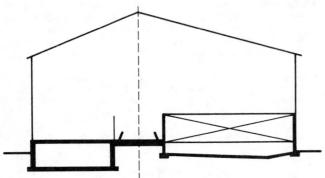

Figure 22.27

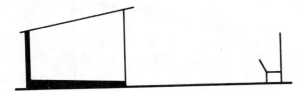

Figure 22.28

Ventilation	Natural, with space-boarded upper walls and ventilated ridge. Avoid draughts. (Slatted or slotted roofing appropriate.)
Bedding and fodder consumption	Straw 750–1500 kg/beast per winter Silage 30–40 kg/beast daily Rolled barley 7–10 kg/beast daily.
Trough space	500 mm length/9–12 month old beast. 600 mm length/18 month old beast. 700 mm length/2 year old beast. 150 mm/animal for *ad libitum* fed young stock.
Water consumption	25–45 litres/d per head Allow 20 cm^2 trough per head or one drinking bowl per ten animals.
Other considerations	Kennels and cubicles may be appropriate (*see* 1.1 Dairy cows) Safety of personnel; (especially with bull beef animals,) requirement for sturdy barriers and doors.

Further reading

HARDY, R. *et al.* (1986). *Indoor Beef Production.* Ipswich: Farming Press Ltd.

MAFF (1985). *Beef Cattle Housing.* Booklet 2512. Alnwick: MAFF Publications.

1.4 Milk production building and bull pens

Note: many aspects controlled by the Milk and Dairies Regulations.

Abreast parlour (six standings) (*see Figure 22.29*)	Overall depth 4.8–5.8 m Overall width 6.9 m Side exit passage 0.3 m Minimum headroom over standings 2.0 m (plus 0.45 m step to standings gives total height minimum 2.45 m)
Tandem parlour (six standings) (*see Figure 22.30*)	Overall width 5.1 m Overall length 9.3 m Passage widths 1.0 m (one end, across, and two exit passages) Minimum headroom over standings 2.0 m Depth of operator pit 0.75 m Width of operator pit 1.8 m
Herringbone parlour (eight standings) (*see Figure 22.31*)	Overall width 4.8 m Overall length 6.9 m (Passageways where required 1.0 m) Depth of operator pit 0.75 m Width of operator pit 1.2 m Minimum headroom over standing 2.0 m Breast rail clearance from wall 0.4 m May be open at entry end
Chute parlour (*Figure 22.30*)	As tandem, less end and exit side passages and with narrower pit. Six standing measures 7.4 m × 3.1 m overall.
Rotaries, tandem (*Figure 22.32*)	Space occupied overall from 5.5 m × 5.5 m for six milking points up to 15.5 m × 15.5 m for 18 points.
abreast	Space occupied overall from 8.1 m × 8.1 m for 12 milking points up to 14.0 m × 14.0 m for 30 points
herringbone	Space occupied overall from 7.0 m × 7.0 m for 12 milking points up to 14.0 m × 14.0 m for 28 points.
Polygon parlours (*see Figure 22.33*)	Four ranks (or three in 'trigon') of standings with up to eight cows per rank, in flattened diamond shape arrangement, 16–40 cows at one time in parlour. 16–18 m wide × 18–33 m long, with 7–9 m at pit centre.

564

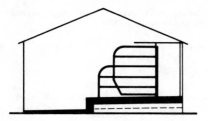

Figure 22.29

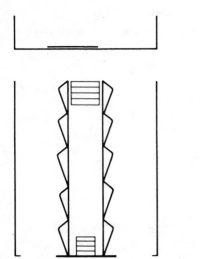

Figure 22.31

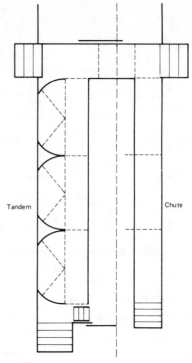

Figure 22.30

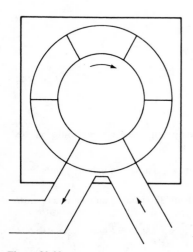

Figure 22.32

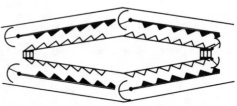

Figure 22.33

Collecting yards and circulation space	Circular or rectangular, open or covered (in which case ventilation should not be through the parlour) allowance of 0.9–1.4 m² per cow.
Drainage	Smooth floated channels or vitrified pipe. Overall falls 1:100 or 150 over lengths of buildings. 1:25 for standings and falls to outlets and gullies. Channels to have 50 mm high minimum rise at edges.
Water supplies	0.5–1.0 litres/cow per d udder wash allowance. 10–20 litres/milking unit per d circulation cleaning allowance. 0.5–1.5 m³/d for pressure hose facility.
Services	Pipes and fittings required for vacuum, milk transfer, cleaning systems; storage facilities for hot water, warm water, milk. All accessible electricity supplies and fittings should be low voltage (110, 24 or 12 V) or leakage protected. Consider heating installations against freezing of pipes and for operator comfort.
Light levels and sources	Natural preferred from above for general illumination. Local high intensity artifical light at udder level. Visual contact to be maintained between parlour, dairy and collecting yard.
Flooring	Hard, durable, resistant to detergents and disinfectants with non-slip surface (granite chippings and carborundum dust) which is cleanable by pressure hose.
Walls	Rendered with waterproof, cleanable, smooth surface to a height of 1.37 m minimum (1.7 m preferable) painted with epoxy resin or chlorinated rubber or other acceptable finish. No insulation provided.
Other considerations	Steps for cows up and down should be non-slip with 150–200 mm 'rise' and 600 mm 'going'. Heelstones on pit edges should be a minimum of 150 mm above standing level. Steps for operatives should not exceed 220 mm rise and should not be less than 250 mm going. Sliding doors or rubber/plastic flap doors required with automatic opening and closing as appropriate.
Dairies	*See Figure 22.34.* Many features controlled by Milk and Dairies Regulation for essential hygiene.
Bull pens	*See Figure 22.35.* All housing elements must be designed to resist aggressive behaviour. Provide escape routes from areas used by operatives.

2 Pig housing

2.1 Pig fattening accommodation

Space within controlled environment housing

	Total area (m²)	Dunging area (m²)	Alternate slatted (m²)
Porkers	0.6–0.7	0.25	0.12
Baconers	0.75–0.9	0.3	0.15
Heavy hogs	0.9–1.0	0.35	0.2

10–20 pigs/pen normal some systems 30–40.

Pen dimensions and arrangement (*see Figures 22.36, 22.37, 22.38, 22.39, 22.40*)	Optimum depth of lying area 1.8 m. Trough width 300–400 mm. Trough length/pig 250, 300, 350 mm porkers, baconers and heavy hogs respectively. *Ad libitum* feeding systems only require 225 mm for all classes of fattening pig). Some systems have no trough allowance – floor feeding, etc. (Trough length usually governs length and shape of pen and thus the house.)
Height of building	As low as possible. Passageways for operatives give minima. 2.15 m normal, 2.7 m minimum with scraping, 3.0 m with catwalk over pen walls for feeding, etc. Eaves height 1.7 m over slatted areas or 1.1 m with strawed/ open yards and 'pop-hole' access.

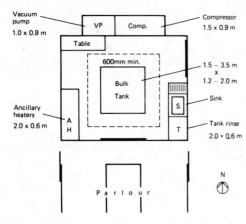

Figure 22.34

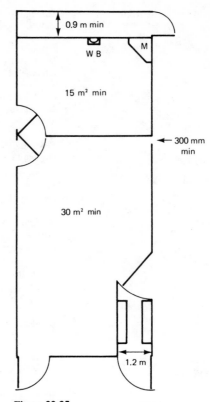

Figure 22.35

Passage widths	Dung passages 1.05–1.35 m slatted outer, 1.15–1.65, slatted central, 1.15–1.5 m concrete solid, outer, 2.1 m minimum for tractor. Feed passages, 1.0–1.2 m.
Temperature and insulation	Minimum of 21°C mean throughout environment. Straw bedding affects lower critical temperature, to reduce thermal risks. 2°C less, generally, for pigs over 4 months old. Roof and walls should have 'U' value of 0.5 and 1.0 W/m²°C maximum respectively. Floor in lying area should be insulated and vapour proofed to a particularly high standard.
Ventilation	2.5 litres/s/50 kg liveweight in winter 12.5 litres/s/50 kg liveweight in summer
Feed and water consumption	2–4 kg/pig per d of meal 4.5–9 litres/pig per d through trough with food, + bowls or drinkers in dunging area or over trough at one per six pigs. (Minimum of two drinkers per pen).
Effluent and drainage	7 litres/d from baconer on dry feed. 14 litres/d from baconer on *ad libitum* wet feed. Floor falls 1:15 in lying area; 1:50 to 1:100 for dung passages or channels. Slatted area – part or whole. Slats–concrete, 75–100 mm top width + 15–20 mm gaps. Steel T bars, 35–50 mm top width + 10–12 mm gaps. Step of 50 mm to slatted area sometimes.
Other considerations	Chew proof doors and partitions with secure latching and fixings.

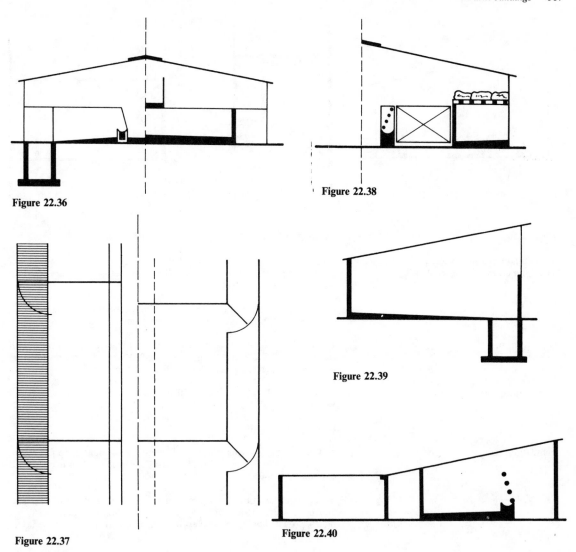

Figure 22.36

Figure 22.38

Figure 22.39

Figure 22.37

Figure 22.40

Further reading

BAXTER, S. (1984). *Intensive Pig Production, Environment Management and Design*. London: Granada.

BRENT, G. (1986). *Housing the Pig*. Ipswich: Farming Press Ltd

2.2 Farrowing accommodation

Space for crate (*see Figure 22.41*)	2.4–2.95 m × 0.6 m wide (+2 × 400 mm side clearance) × 1.4 m high. (Plinth may be provided 200–300 mm high.)
Space for circulation space round crate	Passage at rear minimum of 1.4 m. Feed passage at head end minimum of 0.8 m. Eaves height 2.4 m from passage floor.
Space for creeps surrounding crate	Up to 0.1 m per piglet allowance. Maximum is likely to be 0.4–0.5 m wide × 1.4 m long (length of crate). Covered with lid and heat source above piglets below lid. Bottom rail or pophole access for piglets. Rail height 250–300 mm or three to four popholes along creep base.

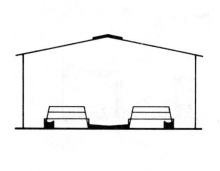

Figure 22.41

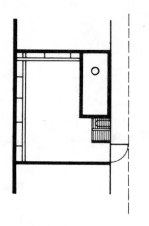

Figure 22.42

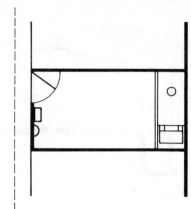

Figure 22.43

Traditional farrowing places/pens with weaning space and creep (*see Figure 22.42*)

2.4 m deep × 3.0 m long (pens either side of feed passage). Eaves height 2.4 m. Creep rails on rear and side wall nearest creep, set 250 mm out from walls and 250 mm above floor.

Creep at pen front, 0.9 m × 0.8 m area with access from pen via popholes and creep rail. Lid over, with heat source below.

Open front pen (*see Figure 22.43*)

1.5 m wide × 4.5–5.5 m deep × 1–1.5 m at rear eaves (mono-pitch with front eaves of 2.1–2.4 m). against rear wall. Removable farrowing crate rails linked to creep, arranged along the pen at centre rear.

Temperature and insulation

General areas 16–20°C. Creep at 30°C initially, reducing to 24°C over 3–4 weeks from birth. 'U' values of 0.5 and 1.0 W/m² °C for roof and walls respectively. Crate and creep areas to be well-insulated concrete, may have in- or under-floor heating

Ventilation

Natural in open systems and in small, pen-type housing (up to five sows). 5 litres/s per sow in winter and 15 litres/s per sow in summer in controlled environment houses.

Feed and water consumption

5–8 kg/d per lactating sow. 0.6 m long trough with drinker over. 250 mm above floor level. 18–25 litres/per lactating sow. Supplementary water for piglets provided by 'tube' drinkers or trough away from creep feeder, over slats if present.

Effluent and drainage

20–25 litres/d per sow
Drainage falls in lying and creep areas to be 1:20, 1:40 in general areas. Steps to drain or dung areas not to exceed 50 mm rise. Slats or mesh area in rear half of crate. Slats consist of slabs of concrete with 10 mm slots and 75–100 mm slat tops. Steel tube or cast iron, 15–50 mm with 10–12 mm gaps. Meshes of steel, plastic, plastic-coated steel of various conformations and aptitudes, care needed for foot comfort of piglets. Crate raised 400–500 mm above surrounding levels at rear to accommodate slurry build up and drainage flows.

Materials of walls and floor

Easy cleaned and resistant to strong disinfectants. Floor flat but not smooth trowelled – no tamping grooves to impede drainage.

Other considerations

Robust fittings, including water bowls required. Strong post and frame fixings. May have crates in an 'open' arrangement with or without low partitioning between each, situated on a plinth or on a well-sloped floor. All crates should be removable for easy cleaning of crate sections and floor.

Further reading

See 2.1 above.

2.3 Dry sows and boar accommodation

Space allowances in open yards	2.7–3.7 m²/sow of which 1.2–1.5 m² is covered and possibly strawed. Group size normally up to 20–40/yard. Automatic feeders require 4–5 m² +
sows in pens	2.1 m × 1.5 m/individual sow.
sow stalls	2.1 × 0.6 m + 1.8 or less slatted area under rear and at back when 'voluntary'
in boar pens (*see Figure 22.44*)	4.5 m × 3.0 m pen of which 4.5 m² is lying area (2.4 m × 1.8 m minimum).
Temperature and insulation	21°C maximum, 14°C minimum. Roof and walls of 0.5 and 1.0 W/°C m² 'U' value respectively.
Ventilation	Natural usually adequate but forced ventilation in some totally enclosed houses at 50 litres/s per pig maximum.
Feed and water consumption	2–3 kg/d per pregnant sow in inidividual feeders. 3–5 kg/d per boar from a trough of 0.6 m length. 9–18 litres/d per sow or boar.
Effluent and drainage	20–25 litres/d per animal. Floor falls laid to 1:40 generally; 1:20 to discharge into open gullies. House end falls 1:50–100. (Slats may be used, *see* 2.2 Farrowing accommodation.)
Other considerations	Boars kept with visual contact of sows in extensive houses. Sow restraint may reduce bullying, can produce sores

Further reading

See 2.1 above.

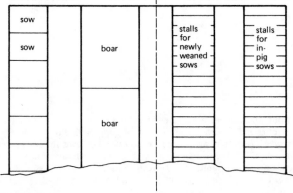

Figure 22.44

2.4 Weaner housing

Space requirements	Tiered cage systems are often used for pigs up to 15 kg. Up to three tiers used 1–1.8 m high. Cages contain up to eight pigs at 0.15 m² each. Good environmental control essential + careful hygiene.
	0.15–0.25 m² per pig in 'flat deck' accommodation. Eaves or ceiling height 2.0 m. Groups of 10, 15, 20/pen are common (*see Figure 22.45*).
	0.5–0.6 m² per pig in strawed yards of which up to 0.3 m² is lying area if open fronted or 'verandah' type (*see Figure 22.46* for slatted version). Eaves height 0.9 m to 2 m depending whether kennel or framed type. Groups of 10–20/pen are usual.
Temperature and insulation	23–27°C inside kennels and in controlled environment housing. Roofing or kennel lid 'U' value should be 0.3–0.5 W/m² °C. Wall 'U' value should be 0.5 W/m² °C. Well insulated floors with in-floor heating using electric cable or hot water piping are strongly recommended. Extra heat usually need in winter – gas heaters common for air flow systems.

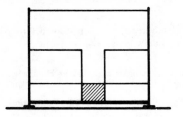

Figure 22.45

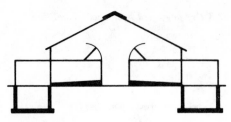

Figure 22.46

Ventilation

2.5 litres/s per 50 kg liveweight in winter
15.0 litres/s per 50 kg liveweight in summer
Care to ventilate avoiding passage of noxious gases from dung through house in total control systems.

Feed and water consumption

Ad libitum feeding from 50 kg capacity hoppers with 50 mm/pig access. 1–2 kg/d per pig. Some rationing systems use troughs at 75–100 mm/pig, should be spill-proof. 1–2 litres/d per pig from nozzle drinkers (2/pen minimum) over well-drained area. Very young pigs might need temporary ramp, etc. to reach drinkers.

Effluent and drainage

1–2 litres/d per pig. Either open dunging area outside kennel or weldmesh/ expanded metal/plastic mesh part or whole floor, e.g. Weldmesh is 76 mm × 13 mm × 8 swg (10 swg for smaller) over slurry pit 1–2 m deep or raised 0.6 m in total control housing. Many alternatives available.

Materials

Prefabricated buildings with easy erect, easy clean and control facilities.

Further reading

See 2.1 above.

3 Sheep accommodation

3.1 Sheep housing

Space requirements (*see Figure 22.47*)

Slatted floor allowance for large ewes 0.9–1.1 m²/ewe; for small ewes, 0.7–1.0 m²/ewe; and for ewe hoggs, 0.5–0.8 m²/hogg. Solid floor allowance for large ewes 1.2–1.4 m²/ewe; for small ewes 1.0–1.3 m²/ewe; for ewe hoggs 0.7–1.0 m²/hogg; for Welsh mountain lamb 0.4 m²/lamb; and for Scotch Blackface lambs 0.5 m²/lamb. 20–30 ewes/pen usual. Feeding passage 0.9–1.1 m wide. Lambing pens 1.5–2 m². Ewe and lambs in pen 2–2.5 m².

Temperature and insulation

Ambient of surroundings, down to 2–5°C. No insulation necessary.

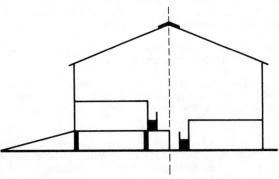

Figure 22.47

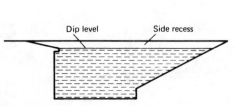

Figure 22.48

Ventilation	Adequate air exchange essential to maintain health of stock. Natural from space boarding walls and ridge ventilation (open 300–600 mm, relative to building span) slatted or slotted roofs possible. Must avoid draughts. Building may be open at one side.
Bedding, feed and water consumption	0.8–1.5 kg/d of straw/ewe in yards. 0.25–1.0 kg/d per ewe of concentrate. 1–1.5 kg/d per ewe of hay. Trough and manger space 250–300 mm/lamb, rising to 400–500 mm/ewe, accessible from outside pen. Up to 5 litres/d per animal from a water trough 0.3 × 0.6 m in each pen or one bowl to 50 ewes. (Running water to overflow desirable, particularly with hill breeds.)
Effluent and drainage	4.5 litres/d per adult animal on average. Slats are wood, 38–60 mm top width × 25–30 mm thickness, with 12–20 mm gaps according to size of breed, etc. Leave 0.65 m clearance underneath for one season's holding capacity, 0.2–1.0 m common. Floor falls, of concrete, to be 1:100 to drainage collection point.
Special areas (Lambing) (Shearing)	Clean area, isolated from general holding area. Adequate services. Pens of 0.75–1.2 m × 1.2–1.8 m. 0.5–0.6 m²/animal holding area. 1.5 m²/shearer.
Other considerations	Polythene or polyester 'tunnels' provide low-cost short-term accommodation. Provide races and satisfactory entry and exits to dipping and handling areas. Ramps to and from house to be 1:10 maximum. Provide long-term fodder storage within building envelope to provision against poor weather periods.

Further reading

BRYSON, T. (1984). *The Sheep Housing Handbook*. Ispwich: Farming Press Ltd

3.2 Sheep handling areas

Space allocations	Gathering and dispersal pens to contain groups of up to 100 sheep at 0.5–0.6 m²/animal.
	Catching pens to contain 10–15 sheep at 0.3–0.4 m²/animal, may be circular with centre pivot gate to crowd into race, dip, etc., or rectangular.
	Central races 3.0–6.0 mm long × 0.45 m wide × 0.8–1.1 m high. Footbath and race 3 m long minimum × 0.25 m wide at bath (0.15 deep with 0.1 m fluid depth) and 0.45 m wide at top of boards, about 1.0 m high.
	Dip tank (*see Figure 22.48*) capacity from 0.75 m³ for 300 ewe flock, to 2.5 m³ for flocks of 1000 + (*see* below for general advice).
	Ancillary holding pens for up to 50 sheep, short term, at 0.4 m²/animal.
Other considerations	Allow 2.5 litres/sheep (minimum 350 litres, then 1000 litres/500 ewes up to 4000 litres) of dip fluid. Circular dip best for large flocks. Floors should be roughened concrete with slope to drain channels of 1:50–100, or may be hard, dry soil in well drained areas, or of introduced stone. Slope from bath should be 30 degrees.
	Drain baths, tanks to an approved holding or disposal point. (Care to avoid pollution.)
	Pen posts for rails should be away from sheep, either on non-sheep side or with shoulder protector boards.

Further reading

See 3.1 above.

4 Poultry housing

4.1 Laying hens

Space requirements (*see Figures 22.49–22.52*)

Deep litter birds, 0.27–0.36 m²/bird, 0.2–0.25 m²/bird when one-third floor area is slatted. Eaves height 1.5–1.8 m

Cages for battery layers measure 430 mm deep × 450 mm high × 300/400/500 mm for single/double/treble occupants per cage. Arrange in tiers of up to three cages high. Passages 1.2 m wide along house (at sides and between rows of cages), 1.5 m across end of house. Eaves height 2.1–2.4 m.

Temperature and insulation

13–16°C in traditional deep litter. 21–24°C in controlled environment housing with 'U' value of 0.5 W/m²°C or less for roof and walls.

Ventilation

Rates are feed-intake related – *see* Chapter 17 on Poultry. 0.3 litres/s per hen in winter, 4.0 litres/s per hen in summer according to weight and housing density. Relative humidity should not exceed 70%, and air delivery speed should not exceed 2 m/s. Danger of ammonia build up with low rates.

Litter, feed and water consumption

250 mm depth of wood shavings, chopped straw or newspaper maintained throughout production cycle. 0.1–0.3 kg/d per bird from trough (60 mm/bird in deep litter or full width of cage in batteries). 20–30 litres/d per 100 birds from constant level troughs or two nipple drinkers per cage in batteries. Drinkers no more than 3 m apart in deep litter houses.

Effluent and drainage

About 0.1 litres/d per bird. Mechanical scraper removal from vertical tiers of cages, dropping pit under staggered tiering. Drain channel for trough overflow let into floor. General floor falls of 1:100. Deep-pits in some slatted floor houses.

Other considerations

Lighting method is crucial as it affects laying performance – refer to specialist manuals for artificial lighting services.

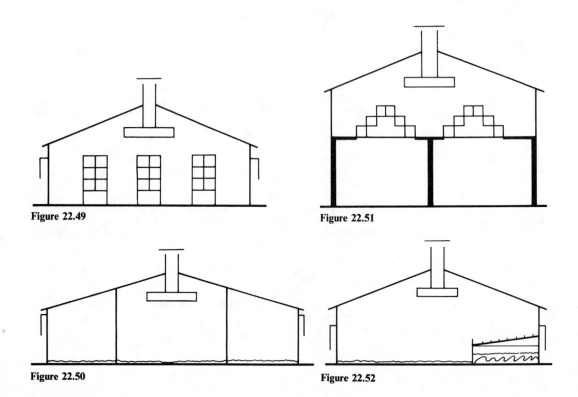

Figure 22.49

Figure 22.51

Figure 22.50

Figure 22.52

Egg room floor allowance at $9.0\,m^2/1000$ birds.

Egg boxes should measure $320\,mm \times 310\,mm \times 310\,mm$.

Further reading

MAFF (1976) *The climate and environment of poultry houses,*
Bulletin 212. London: HMSO.

4.2 Broiler and capon housing

Space requirements (*see Figure 22.50* for general arrangement)	$0.09\,m^2$/bird, up to ten weeks old, kept in batches of 5000–7000. $0.2\,m^2$/bird up to 16 weeks old, kept in batches of up to 2500. Eaves height 1.5 m minimum.
Temperatue and insulation	18–24°C with a maximum 'U' value of $0.5\,W/°C\,m^2$ for roof and walls.
Ventilation	Natural possible for larger birds when controlled by baffling and ducting to avoid draughts. $0.2\,m^2$/100 birds of vent area requirred. 0.2–3.0 litres/s per bird required in controlled environment houses for winter–summer conditions according to weight.
Feed and water consumption	0.2–0.4 kg/d per bird with 150 mm length of trough allowed. 13–20 litres/d per 100 birds from 600 mm minimum of water trough length per 100 birds.
Effluent and drainage	About 0.5 litres/d per 100 birds up to ten weeks old. 1.0 litres/day per 100 birds up to 16 weeks. General floor falls to outlet of 1:100 for washing and disinfecting.
Other considerations	Hard floors preferable under litter 100 mm deep. Platform-tiered so-called 'aviary' system can increase stocking rate. Light intensity and day length is important for maximum growth rate – *see* specialist texts. Unusual noise may create panic among the birds.

Further reading

See 4.1 above.

4.3 Turkey housing

Space requirements	Breeding stock, 0.5–$0.9\,m^2$/bird in groups of 25–30. Growing stock, 0.1–$0.5\,m^2$/bird from three weeks to 16 weeks +. Up to $0.3\,m^2$/bird in controlled environment housing. Eaves height 1.8 m for 16 week olds.
Environment	Natural, usually, with limited control to provide weather protection and shelter. Insulated roof and walls for winter growing systems with a target temperature of 17°C. 'Verandah' building (with partial 'open' sides) usually adequate.
Feed and water consumption	0.4–0.6 kg/d per bird at 16 weeks old. 3.5–7.0 m trough/100 birds or 50 mm/ adult if tubular feeders. 12–70 litres per 100 birds from 3 weeks to 16 weeks old. Spill resistant water troughs required.
Effluent and drainage	0.3–0.6 litres/d per adult bird. Floor falls at 1:100 where concrete. Where meshed/slatted allow 0.6 m space below floor. Area of wire or slats must not exceed one-third of total flooring.
Other considerations	Lighting may be used in 'open' housing systems as a deterrent to foxes. Fox proof netting walls are essential.

Further reading

See 4.1 above.

5 Crop storage

5.1 Grain storage

		Barley	Beans	Herbage seeds	Linseed	Maize	Oats	Peas	Rape	Rye	Wheat
Space requirements (Base on conventional storage MC	(m³/t)	1.45	1.2	3.0–5.0	1.45	1.35	2.0	1.3	1.6	1.45	1.3
and 20–30° repose angle	(kg/m³)	705	833	200–500	705	740	513	785	625	705	785
Capacity of level, 2 m fill, of loose grain, per 4.8 m bay and 6 m span		42	51		42	47	29	44	40	42	44
Capacity of fill when heaped above 2 m, per 4.8 m bay and 6 m span		57				64	39				
Tonnes of product 2 m high, level fill/m² of floor area		1.34	1.67	0.4–0.6	1.34	1.48	1.0	1.54	1.33	1.34	1.54
Capacity, of 4.5 m diameter silo 9 m high		102					76	220		98	110

Note: Further moisture adds weight/m³ at 1–1.5% extra per 1% addition of moisture over standard conventional.

Store types (*see also* Chapter 11 on grain storage)

Circular bins 2–6 m in diameter, up to 6 m high (height not to exceed twice the diameter) – *see Figure 22.53* and *22.54*. Materials may be hessian and wire mesh, oil-tempered hardboard, galvanised corrugated steel, galvanised slotted steel, plyboard, welded steel plate, glass- or enamel-surfaced steel, or concrete stave with steel hoops. Floor may be plain concrete, aerated mesh or sloped steel (45 degrees). Emptied by 'air-sweep', slope, or auger conveyor.

Square or rectangular bins 2.4–4.5 m with heights up to 9 m, (6 m commonly for drying purposes). Materials may be corrugated steel sheet (square profile), galvanised steel pressed panels or reinforced concrete block with substantial piering and frame support. Flooring will be plain, ventilated or sloped as with circular type.

Loose bulk ('on-the-floor') storage uses standard building types and frames; with or without simple retaining walls for containment at the edges or at dividing partitions (*see Figure 22.55*). 2–3 m at edges, 5 m maximum at centre.

Wet grain storage, in sealed silos (or with non-sealed roofing), using circular, glass fused or enamelled steel. (Allow for density of product at 14% MC plus 5–10%.) Danger from CO_2 build up.

Wet grain heaped 'on-the-floor' in clear span framed buildings, preserved by propionic acid application. (Density as above.)

Refrigerated grain stored, at 18–22% MC in bins at a density of 3–5% more than at 14% MC.

Ventilation and drying (*see* Chapter 11 on grain storage)

Either high or low volume flow available for most grain storage systems giving either a drying or a conditioning effect. Require main ducts at centre or sides with laterals and diffusers to distribute air evenly throughout the mass. Air systems may also act as grain conveyors. Fans may be noisy.

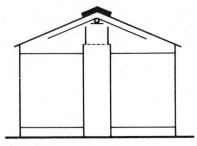

Figure 22.53

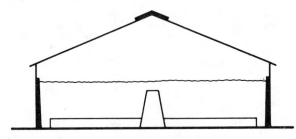

Figure 22.55

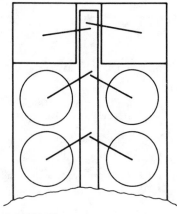

Figure 22.54

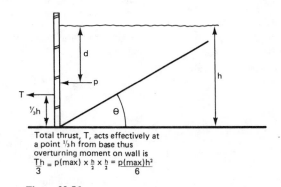

Total thrust, T, acts effectively at
a point ¹⁄₃h from base thus
overturning moment on wall is
$$\frac{Th}{3} = p(max) \times \frac{h}{2} \times \frac{h}{2} = \frac{p(max)h^2}{6}$$

Figure 22.56

Floor

Hard, durable, smooth, dust-free and dry. 150 mm of concrete, reinforced on soft, difficult soils and at edges when supporting bins; steel floated and incorporating a damp proof membrane. May have to accommodate air ducts or conveyors.

Walls

Thrust resistant and air tight. Must be designed with adequate foundations. Building frame may support thrust *if designed to do so*. Thrust calculations used are based on Rankine's formula:

$$p = \frac{wh(1 - \sin\theta)}{(1 + \sin\theta)}$$

where p is pressure at height h from grain, etc., top surface, w is density and θ is the natural angle of repose. When this is 30 degrees, this formula becomes

$$p = \frac{wh(1 - \frac{1}{2})}{(1 + \frac{1}{2})} = \frac{wh}{3}$$

See Figure 22.56 for further explanation on the effect of this on a retaining wall.

Other considerations

High power consumption (e.g. 30–60 kW) likely in some stores

Reception pit needed for bin systems, 2.5 m × 1.5 m × 1.5 m deep, with a sloping delivery floor at 60 degrees to the horizontal

Overhead hoppers to fill lorries rapidly, with high delivery rate conveying systems to match. Large headroom and strong framing required.

Building should be vermin and bird-proof. Doors should provide clearance for tipped trailers (up to 6 m). Provision should be made for heavy vehicles in the flooring in the vicinity of grain stores. Special consideration should be given to fire hazards in grain drying areas. Deep-pit elevator equipment and temporary storage prior to drying may be required. (Service access needed below ground level.)

5.2 Hay, silage, straw

Space required for bulky low density materials (m^3/t) ('Average' figures provided – there will be considerable variation from sample to sample, dependent on moisture content, species)

	m^3/t
Hay, loose	9.0
Hay, medium density baled	6.0
Hay, barn dried	7.5
Barley straw, medium density baled	11.5
Wheat straw, medium density baled	13.0
Grass, wilted	2.2
Grass, silage, consolidated	1.4
Grass, tower silage, high dry matter	1.25
Grass, high temperature dried, 75 mm long	8.5
Grass, high temperature dried, ground	1.5

Volumes and tonnages

A 4.8 m bay of a building will contain:

35 t of baled hay at a height of 4.8 m

69 t of consolidated silage at a height of 2.4 m

20 t of baled barley straw at a height of 5.4 m.

Store types

Standard frame with open or closed sides, may have special walling and floor system (*see Figures 22.57* and *22.58*).

Flooring

Earth or rammed hardcore on well-drained soils, for simple storage of baled hay and straw. Raised floor weldmesh, etc. to provide plenum chamber for drying of hay in batch or storage driers. (*See also* Chapter 21 on Farm Machinery – Barn hay drying.) Silage requires durable, acid-resistant surface of heavy duty concrete, laid to adequate side and end falls and incorporating drainage gullies, leading to holding or disposal systems which must be approved. (Consider using slurry containment and disposal system.)

Walls

Thrust resistant for silage (*see* 5.1) and are a specialist design problem. Should have rails on top for safety of operatives during silage consolidation. Tower silos are similar to wet grain silos (*see* 5.1) except their heights may exceed 20 m.

Hay barn walls may be removable in the form of wooden baulks or steel sheet or they may, in the case of loose material be of wire netting. Some types of walling must be air tight in storage drying. Top of wall near eaves should be slatted for up to 1 m to allow air movement during drying.

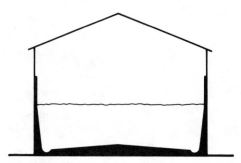

Figure 22.57   **Figure 22.58**

Roofing

No roof required for silage clamps. Sealed tops for some tower silos with air pressure release bag incorporated. Hay barn roofing should be ventilated and of non-condensation-drip forming type.

Other considerations

Fire risk should be minimised. Ease of access to farm roads and livestock buildings. Power requirement may be high in storage driers or tower silo systems (20–100 kW). Drying fans tend to be noisy in operation. Danger from CO_2 build up in silage tower air spaces.

5.3 Concentrated and root crop feed material

Space for 'concentrates' ('Average' values provided – there will be considerable variation from sample to sample, dependent on moisture content, consolidation)

Meal, loose, 2.0 m^3/t or 500 kg/m^3
Pellets, loose, 1.7 m^3/t or 600 kg/m^3
Nuts, loose, 1.4 m^3/t or 700 kg/m^3
(50 kg bags stack 2 high, occupy 2.0 m^3/t or 500 kg/m^3)

Beet pulp 1.5 m^3/t or 650 kg/m^3
Fodder beet 1.7 m^3/t or 600 kg/m^3
Turnips 1.8 m^3/t or 550 kg/m^3
Brewers' grains (dry) 2.0 m^3/t or 500 kg/m^3

Space for equipment used in milling and mixing

Horizontal mixer, 4–10 m^2 × 2–3 m high.
Vertical mixer, 3–5 m^2 × 3–6 m high.
Crushing mill, 2–3 m^2 × 1–1.5 m high.
Hammermill 1–6 m^2 × 1–1.5 m high (5 m with cyclone above).
Allow 1–1.5 m headroom for overhead conveying

Other considerations

Self-emptying bins should have floors at 60 degrees to the horizontal. Noise and dust are potential health hazards. Risk of explosion in a dusty atmosphere. Very high power requirement by most barn machinery (5–40 kW). Mechanical handling requires smooth running, hard flooring. Use of gravity desirable in conveying – overhead hoppers for lorry and trailer filling with 10–20 t capacity for big units. High clearance needed (up to 6 m).

5.4 Vegetable storage

Space requirements and environment		m^3/t	Mean temperature to store long term	Maximum height
	Potatoes, bulk (*see Figures 22.59* and *22.60*)	1.5 to 1.6	Ware 4–5°C	4.0 m (6 m if air flows can reduce temperature gradient)
	Potatoes, in pallet boxes (1.2 m × 1.4 m × 0.8–1.0 m) (*see Figure 22.61*)	2.4	Ware – as above. Crisping and processing, 6–8°C	4–6 boxes (4.0–5 m)
	Onions, dry (*see Figures 22.59* and *22.60*).	1.5 to 2.0	2–3°C	4–5 m
	Onions, green topped	3.0	Ambient temperature + 3–6°C until dry	2.5 m
	Onions, re-frigerated,	2.0	0.5–1.5°C	3.0 m

Space requirements and environment	m^3/t	Mean temperature to store long term	Maximum height
Onions dry, in boxes. (1.1 m × 1.3 m × 1.05 m high)	3.0	Ambient temperature +3 to 6°C.	Boxes stacked to eaves

Store types

Standard, clear span frames.

Floor

As 5.1 plus smooth running heavy duty floor for pallet handling equipment and which may incorporate air-flow and conveying ducts.

Walls

Thrust, resistant (*see* 5.1 and *Figure 22.56* for Rankine's formula – angle of repose for potatoes is 30–40 degrees) In pallet box stores cheaper walling may be used. 'U' value of 1.0 maximum W/m² °C for walls (and roof) is necessary to protect potatoes from frost. Care must be taken to avoid 'cold bridging' across the walls via reinforcing or framing systems. Walls should be smooth and cleanable.

Other considerations

Avoid materials in structure with a high thermal conductivity. Use straw as insulation layer on top of stack, and bales as temporary wall insulators. Should be a concrete apron of 2.5 m minimum width in front of loading doors.

Porch desirable to enable store to be filled to end wall, potatoes rest on removable boards supported on porch side walls which carry stack weight as it is built up.

Chitting houses for potatoes
(*see Figure 22.61*)

Potatoes stacked in trays, with access to illumination from movable lighting columns suspended from ceilings. Trays are 750 × 450 × 150 mm on pallets 1.5 × 0.9 m in area, stacked to depth of 3.5 m maximum. Alleyways 0.5 m minimum between rows. 1 m side alleys. Generous end space in building. Should be frost-proof and force ventilated, with heating equipment available when necessary

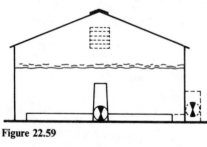

Figure 22.59

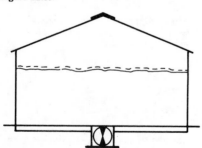

Figure 22.60

Figure 22.61

Further reading

FARM ELECTRIC CENTRE (1983). *Potato storage* NAC, Kenilworth, The Electricity Council

FARM ELECTRIC CENTRE (1985). *Vegetable storage.* NAC Kenilworth, The Electricity Council

6 Servicing and general buildings

6.1 Fertiliser storage

Space required

Loose, bulk, 0.9–1.1 m³/t average for prilled fertiliser.

50 kg bags, stacked 1.0–1.2 m³/t (at 180 mm thickness per bag, with maximum of 10 high).

'Big' bags containing 0.5 t, stack two or three high; are nearly equivalent to loose.

Other considerations

Stored in standard frame building with weather protection and anti-condensation provisions to prevent dripping onto bags. Thrust resistant walls for bulk fertiliser with corrosion resistant surfacing. Ends and top of heap covered with plastic sheeting. Walls and floor clearance of 0.5 m and 0.15 m respectively for bags (on pallets or boards). Provide temporary divisions to isolate different fertiliser types.

6.2 Slurry and farmyard manure storage

Space required

Straw and dung (farmyard manure), 1.0–1.4 m³/t or 700–1000 kg/m³.

Storage systems (NB All systems proposed to hold 5000 m³ of slurry, or water above the original ground level must be designed by a 'certificated' engineer, etc).

40–50% dry matter, handled straight from bedded area or stored in dungstead, (walled enclosure with moisture retaining floor falls). Material may be moved by fore-end loader.

25–35% dry matter, glutinous, non-flowing, in thin layers on flat, hard surfaces, moved by scraping into pit for pumping to enclosure as above with solid matter or into above-ground tank.

10–20% dry matter, plus washing water pumped from sumps, holding tanks and catch pits in drainage and slat systems, into above-ground slurry stores, circular (20–1000 m³) or rectangular (100–2000 m³) or reservoirs or lagoons, excavated into the ground and sealed against leakage.

Up to 10% dry matter (excluding straw) held in lagoon or reservoir for active oxidation or passive storage.

Further reading

GRUNDY, K. (1980). *Tackling Farm Waste*. Ipswich: Farming Press.

ROBERTSON, A.M. (1977). *Farm Waste Handbook*, Aberdeen. Scottish Farm Buildings Investigation Unit.

6.3 Machinery storage

General

Height and width are often more of a problem with storage than area or volume occupied, due to doorway restrictions, eaves heights and turning circles. Site building near good hard road, central to farm operations, adjacent to workshop and cleaning facilities. Provide sufficient light, natural and artificial, for inspection and manoeuvring.

Clearance requirements for tractors and implements

Medium size tractor – 3.0 m high × 1.85 m width (may be extended to 2.5 m wheel track), area 6.75 m², with a turning circle of 7.5 m diameter (exhaust should have 0.5 m minimum height clearance from structure). Medium size tractor plus loader, turning circle, 10 m diameter.

Combine harvester, 3.0–6.0 m wide × 7.5–10 m long × 3.0–4.0 m high.

Trailers 2.0–2.43 m wide × 4.0–5.5 m long × up to 4.5 m high (tipped).

Lorries 2.43 m wide × 4.5–9.1 m long (rigid chassis) or 10.0–12.5 m long (articulated) × up to 4 m high (unusually 5 m high by 16 m long).

Plough, reversible, 3 furrow, 1.5 m wide × 2.5 m long. Disc harrow, 2.5–5 m wide × 1.9 m long. Seed and fertiliser drills, 2.5–4 m wide (trailed) or 2.5–6.5 m wide (mounted) × 2.0–3.0 m long.

Balers, medium density, 2.5–3.0 m wide × 4.5–5.5 m long.

Beet harvesters 2.5–4 m wide × 5.5–7 m long × 3.5 m high.

Other considerations

Hard, well-drained, floor (but not necessarily concrete). High eaves at doorway. Consider upstand in roof at this point for better access). Translucent sheeting in roof to give high level of natural light. Concrete at point of entry to building with generous hard standing outside, adjacent. Limited number of socket outlets for electricity supply (low voltage output desirable).

Further reading

MAFF (1971). *Farm Buildings Pocketbook*. London: HMSO.

6.4 Fuel storage

Diesel store only (petrol storage restricted and licensed – see later, p. 586) required).

Tank sizes 1.0 m × 1.0 m × 1.0 m – 1137 litres (250 gallons)
             1.5 m × 0.9 m × 0.9 m – 1137 litres (250 gallons)
             2.0 m × 1.25 m × 1.25 m – 2728 litres (600 gallons).
Black mild steel to BS 699, sited high enough on support walls (225 mm width minimum) to discharge to highest tank-filler when almost empty. Tank should have fall away from delivery pipe to drain cock. Dip stick reading with ladder access or sight-gauge

6.5 Workshop accommodation

Area required and facilities

One or two bays of standard frame building of 9–15 m span. Pit, 2.0 m × 0.8–1.0 m wide × 1.7 m deep with steps at each end, water-proofed, with drain sump: placed centrally in one of the bays with access from main door to workshop. Clear wall length of 5 m for benching with electric power (110 V) and spot lighting. Concrete area of 50 m² minimum outside main workshop door with good drainage falls to sump or trapped catchpit. Floor inside of 150 mm concrete, reinforced to take heavy equipment. Overhead gantry for crane desirable.

Further reading

See 6.3 above.

6.6 Farm office accommodation

Space required (may be controlled by Offices, Shops and Railway Premises Act, 1973)

10 m² minimum, 15–20 m² preferred, with washroom area and WC of 3 m² adjacent, all served by porch lobby of 3 m² minimum. Ceiling 2.5 m high. Natural light required through one wall of office, preferably with view of main farm access road. Electric power, heating, water and telephone services essential. Large area of clear wall space desirable. Drainage and other building features may be regulated by Building Regulations and inspected during construction by Local Authority Building Control Officer.

6.7 Domestic dwellings

These are controlled by legislation to achieve a high building and environmental standard. Although they are considered beyond the scope of this book an introduction to their planning and structural requirements can be found in Noton (1982). (*See* Design criteria, p. 537)

DRAWINGS AND SITE MEASUREMENT (SURVEYING)

Drawings for construction (*see* Figure 22.62)

(1) Drawings should be prepared to scales of 1:50 or 1:100 showing:
   (a) the layout in plan view indicating position of doors, barriers, feeders, circulation space, drainage, water and electricity supplies in relation to frame members, walls and other structural features at or above ground level. (Some of these might be more clearly shown on separate additional drawings to avoid confusion of lines and shading.)
   (b) the building in elevation to demonstrate shape, volume, cladding materials and external features likely to be obtrusive.
   (c) cross-sections of the structure demonstrating material details and thicknesses, positions of service fixings and the nature of space usage in the building.
(2) A scale map of 1:2500 should be used to show the location and general arrangement of the building relative to its surroundings. (Permission may be granted for a copy to be made for personal use from an Ordnance Survey Map of this scale.)
(3) Further drawings to scales of 1:10, 1:20 and 1:50 may be prepared to show constructional features such as foundations, framing and fixings, drainage schemes, damp and vapour proofing, insulation methods and mechanical services. These are generally the province of professional experts, surveyors and architects or may be provided by commercial building firms as part of a sales service (*see Figure 22.62*).
(4) Planning and grant-aiding authorities may need to see some or all of these drawings to ensure that the law is being complied with in every respect and that the building conforms to the requirements of BS 5502 (*see* next section). (In practice, it is advisable to consult the appropriate bodies at an early stage of design thinking. It is not good, or safe, procedure to request retrospective approval.)

SURVEYING

Ranging out lines

On flat land

To range a line between two points A and B on fairly level land, sight from A, insert ranging poles at C, D and E, working from B towards A (*see Figure 22.63*). The bottom of the ranging poles should be observed.

In hilly areas

To range a line between two points A and B with undulating land between so that B cannot be seen from A (*see Figure 22.64*).

Two assistants take positions C′ and D′, so that C′ can see B and D′ sees A. They then direct each other in turn into line with A and B until points C and D are reached, when no further shift is possible. ACDB is then a straight line.

On steep slopes

To range out a line between two points A and B with a steep slope intervening, as in *Figure 22.65*.

Range out a trial line AB′ in the general direction of B, marking stations C′, D′, E′. Measure AB′, noting the distances from A to C′, D′ and E′. B′B, the ranging error, is also measured. Consider E′E, D′D and C′C drawn parallel to B′B. Triangles AB′B and AE′E are similar – hence E′E/AE′ = B′B/AB′

Therefor E′E = (B′B × AE′)/AB′

D′D and C′C can be calculated in the same way. E′E, D′D and C′C are then moved parallel to B′B to give straight line ACDEB.

When view is obstructed

To continue a chain line obstructed by a building. Range out the chain line to the building. Choose two points A and B on this line, erect perpendiculars AC and BD of equal length to clear the building using accurate site squaring equipment (see later). From C, range a line through D, past the building and choose two points E and F. Drop perpendiculars EG and FH equal in length to AC and DB (*see Figure 22.66*). From G, continue the chain line through H. To give accuracy AJB and GH should be three time the length of the perpendiculars.

Area measurement (note that 10 000 m² = 1 ha)

For small areas a simple chain survey may be made, using the metre chain. This chain is 20 or 30 m in length and subdivided every 200 mm, with tallies every 2 m from the ends.

The area to be surveyed is divided into any convenient number of triangles.

Stations ABCD (*Figure 22.67*) are chosen in the corners of the fields. Lines DB, AB, BC, CD and DA are ranged out, dividing the greater portion of the area into two triangles. Line AC may also be ranged out and measured as a check line. When measuring AB, BC, CA and DC offset measurements are taken to the boundary, wherever it changes direction, as indicated by the short lines.

Taking line AB as an example, measurements in links of 200 mm could be entered in the field book as shown in *Figure 22.68*. From the field notes the area may be drawn to convenient scale, and the area calculated from the plan as shown for *Figure 22.69*.

Curved boundaries are replaced by equalising lines, so that figure EFGH is equal in area for all practical purposes to the original field.

Draw in diagonal HF and erect perpendiculars EJ and KG and (with the aid of sighting squares or Pythagorus, *see below*). Scale off these lines. Thus
Area = (EJ + KG)/2 × HF.

Establishing lines at right angles

This is crucial to accurate area measurement and the laying out of building sites, experimental plots, drainage and mapping grids. The use of Pythagorus (3, 4, 5 or 5, 12, 13 sided triangulations) provides the simplest method of doing this in concept but in practice it is often easier to use purpose-built precision optical equipment, having either

582

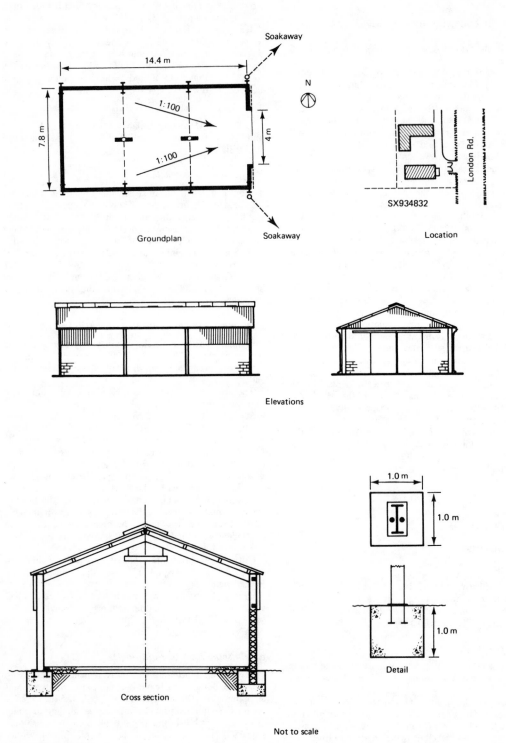

Soakaway

14.4 m

1:100

7.8 m

1:100

4 m

N

Soakaway

Groundplan

Location

SX934832

London Rd.

Elevations

1.0 m

1.0 m

1.0 m

Detail

Cross section

Not to scale

Figure 22.62 Building drawings

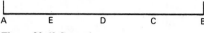

Figure 22.63 Surveying

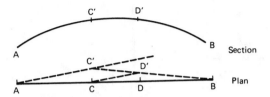

Figure 22.64 Surveying

Figure 22.65 Surveying

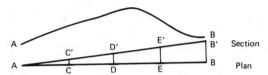

Figure 22.66 Surveying

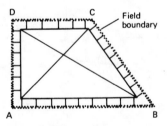

Figure 22.67 Surveying

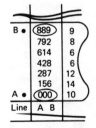

Figure 22.68 Surveying

telescopes or mirrors set in a frame at right angles. These provide 'lines of sight' which can be recorded on the ground or on a map in two planes at 90 degrees to one another, and their point of coincidence through the instrument frame is at the 'point' of the corner.

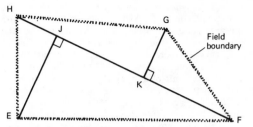

Figure 22.69 Surveying

Ordnance survey

The most important maps to agriculturalists and estate owners are the 1:10 000, (which replaces the 6 in) and the 1:2500.

The 1:10 000 map covers an area 5 km square but some coastal sheets are larger. All boundaries – country, parlimentary, urban, rural and parish are shown. Also all enclosures, buildings and the like are included. In the vicinity of towns, streets are widened to accommodate street names. Instrumentally determined contour lines are drawn in at 10 m, 5 m or as a direct metric equivalent to the older imperial measure. Bench marks are given to two places of decimals of a metre, whilst numerous spot levels are shown.

The map of 1:2500 (25.344 in to the mile) is termed the parish or farmer's map. Each map covers an area 1000 m square, some sheets consisting of two adjacent maps. All features present on the 1:10 000 map are shown except contour lines. Enclosures bear a reference number with the area to three places of decimals of a hectare.

Simple levels used on building sites and small drainage surveys

Three types of instruments are used commonly by builders for levelling on small-area construction sites; these are fluid levels (using water, usually), 'automatic' levels and 'dumpy' levels.

(1) The fluid level is the simplest in principle, being two transparent end tubes, equipped with millimetre scales and connected by a flexible tube containing coloured solution. As the end tubes are raised and lowered the fluid levels remain the same relative to each other and vertical distances can be referenced to the liquid surface positions and tube calibration (*see Figure 22.70*). These can only be used up to 30 m distant, but can measure out of line of sight.

(2) So-called 'automatic' levels require minimal adjustment once they are placed suitably on a tripod. Their optical

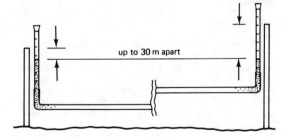

Figure 22.70 Liquid level

system is designed to obviate the need for the precise horizontal alignment which is necessary with other equipment, and to indicate via the viewfinder when the tripod requires repositioning to give the correct field of view. They, also, are short distance instruments (up to 20 m) but can only be used with a clear line of sight. They are, however, simple to interpret. The second operator holding the sight staff reads the measurements on the staff in this system (*see Figure 22.71*).

(3) 'Dumpy' levels are telescopes with the addition of siting crosswires built into the optics, which are used to pin-point readings on a target staff (*see Figures 22.72* and *22.73*). The telescope must be set up precisely to the horizontal. The staff used consists of a telescopic scale of three or more sections, which can extend rigidly to a height of up to 6 m from the ground. The markings on the scale are graduated in centimetres, decimetres and metres, the latter two in figures.

To take satisfactory measurements from any staff the 'plane of collimation' (*Figure 22.72*) must be exactly horizontal at all times, thus particular care is needed when this type of level is set up, to make sure that the spirit bubble on top of the instrument is used in two planes at right angles to position the telescope correctly. Unlike the automatic level the operator viewing the staff through the sight records the measurements and logs their variation. Care should be taken to choose a 'station' for the level which will scan as much of the site as possible and that the base lines used are interrelated if the tripod must be moved, as shown in *Figure 22.19*.

It is usually convenient for building sites to be 'pegged' as levels are taken, such that the bottom of the staff rests on top of each peg and records the same distance to the line of collimation seen through the level as in *Figure 22.71*.

Further reading

IRVINE, W. (1974). *Surveying for Construction*. London: McGraw-Hill.

MANDATORY AND ADVISORY DESIGN REQUIREMENTS (excluding finance)

Principal acts

Town & Country Planning Act (1971) (T and CP)

This Act follows, replaces and/or consolidates those of 1947, 1959, 1962, 1968 and all Orders made under those previous Acts.

General principle

This is one of control of development; requires permission to be sought before any building or 'built environment' engineering operation is carried out or any change of use or purpose of land or building is instituted, or new access is made on to classified road. Agricultural construction works (included in Class VI of the Town and Country Planning general Development Orders 1977–1986) are broadly permitted under the Act through Article 3 of the Town and Country Planning General Development Order of 1973, subject to certain limitations:

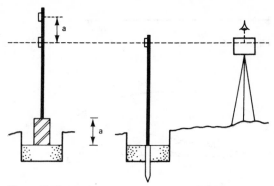

Figure 22.71 Automatic level

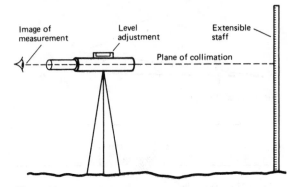

Figure 22.72 Dumpy level

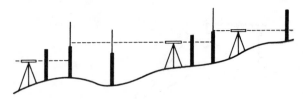

Figure 22.73 Use of levels

(1) the holding is of more than 1 acre,
(2) the sole purpose is agriculture (but intensive livestock buildings, not depending on 'land' for their maintenance are as yet a contentious issue; e.g. horse-riding facilities are not deemed to be agricultural for this purpose),
(3) the development is not in any way connected with a dwelling house,
(4) the area does not exceed 465 m² in aggregate with any other works engineered within 90 m and within two years of the start of construction,
(5) the height does not exceed 12 m (3 m if within 3 km of an airfield),
(6) no part is within 25 m of the metalled portion of a classified road,
(7) the building, etc. is permanent, i.e. not 'mobile' (e.g. caravan),
(8) it is proposed that a 'cordon sanitaire' of 400 m be put around domestic dwellings where farm buildings, (particularly those with effluent 'associations' are to be developed on new sites).

(Article 4 of the General Development Order permits a Local Planning Authority to rescind the general permission in special circumstances e.g. despoilation of attractive areas. In addition the T & CP Landscape Areas Special Development Order, 1986, controls the design and appearance of all farm buildings in National Parks, etc.).

Planning permission is not required for earthworks and excavations so long as they are 25 m at least away from classified roads. Planning permission is not required for milk collection stands, etc. along roadsides except for those abutting a classified road. Planning permission *is required* for roadside selling stands and advertisement hoardings (T and CP (Control of Advertisements) Regulations 1969 and (Amendment) 1972).

Where planning permission is required, certificates, ensuring that owners and tenants affected by the development have been notified of the intentions, must accompany the application. Outline permission may be applied for, in which case no development may start until full permission is granted. Five years is the normal length of time for planning permission to be granted. Re-application is required subsequently. Charges are levied, dependent on the scale of the intended development, the Local Authority (LA) will advise on the appropriate level of fees but there is special provision for these to be set aside for Class VI (agricultural) buildings. Applicants may appeal against refusal of permission within six months of the original decision.

Town & Country Amenities Act 1974

This takes over some of the functions of T & CP Acts in respect of conservation areas; with particular reference to historic buildings and areas of special scenic beauty, e.g. National Parks and Areas of Natural Outstanding Beauty (ANOB).

Lists of buildings which may not be redeveloped without permission are maintained by Local Authorities and the owners are now notified. Planning permission must be sought for any development in an area of 'special scenic beauty', ANOB and Special Scientific Interest (SSSI).

All buildings in a designated conservation area are subject to control and demolition: alteration or improvement requires an application for listed building consent.

Tree preservation orders

These may be made for the preservation of trees at any time and also for the planting of trees during development. Preservation includes action against topping, trimming or wilful destruction. A tree must be of a minimum size and preservation is subject to Ministerial Confirmation. Application can be made to cut down, etc., preserved trees.

The Forestry Act 1967

This is intended to control tree husbandry but has some bearing on development work. May cut down trees not listed up to 23.4 m³, (or 4.25 m³ for sale) in any quarter of a year. Further felling requires Forestry Commission agreement unless it is done under planning permission or to obviate danger.

Building Act 1984

This consolidates much legislation previously found in the Public Health Acts, the Health and Safety at Work Act and the Housing Acts, concerned with the quality and soundness of structure, the health of occupants of buildings and the safety of the built environment. It provides wide powers for the control of building structures to achieve public health and safety, through the regulation of structural strength, drainage and sanitation, area, volume, environment and permanence. Some buildings are partially or totally exempt including Class III (farm buildings): single storey buildings exclusively used for the accommodation or storage of plant, machines, animals, crops, in which only agricultural workers will be employed, which are detached from other buildings and which must be not less than 1.5 times its own height from any building containing human sleeping accommodation and provided with fire exits. Any agricultural building used for retailing, packing or exhibiting is not exempt. Permission to proceed with construction may be sought by providing the Local Authority with a 'Building Notice' instead of 'plans' under this Act. Consent is required before farm effluent is discharged into public sewers. Intensive farm enterprises are also covered by Public Health (Drainage of trade premises) Act 1937.

Control of Pollution Act 1974

Consolidates previous legislation. It is an offence to pollute streams (any ditch or watercourse, in practice, even if dry for a proportion of the year) or to discharge liquid waster underground, except by agreement with the Water Authority, who usually apply a BOD level of 20–25 ppm as a pre-requisite condition of such discharge.

Health and Safety at Work Act 1974 (see Chapter 27)

Sets out safety provisions:

(1) The Agriculture (Ladders) Regulations, 1957:
   (a) Enclosed stairs – one handrail.
   (b) Handrails on each open side of all stairs which are 1 m or more high and 30 degrees or more from vertical.
   (c) Handholds provided at top of any stairs which are less than 30 degrees to vertical.
   All handrails to be strong, smooth and rigid.
(2) The Agriculture (Safe-guarding of workplaces) Regulations, 1959:

   (a) Pits or openings in floors more than 1.53 m deep, should have a rail of between 0.92–1.07 m high or a fence not less than 0.92 m high or a cover (including a grid) so long as it gives no less protection than the rails.
   (b) Floor edges (other than wall openings) to have guard rail or fence.
   (c) Wall openings – guarded by door or fence or guard rail (not openings of less than 1.22 m from top to bottom or those to stairs or those with sills more than 1.61 m from floor). Handhold provided for, when doors etc. are off for the purpose of, e.g. movement of materials.
   (Do not apply during periods of construction or alteration)
(3) The Agriculture (Stationary Machinery) Regulations 1959
   (a) Prime movers to have cut off, clearly labelled with directions.
   (b) No worker contact with belts, shafts, pulleys, etc. when moving.

(c) Adequate natural or artificial light where machine is used.

It should be noted that responsibility for safety lies with employer and employed *equally*.

Food and Drugs Act 1955

The Milk and Dairies (General) Regulations 1959

Part V: Regulates – water supplies and general building construction in food producing area.
Part VI: Suitable structural finishes
    Good drainage with traps, falls, pipes etc.
    Cooling of milk.

Petroleum (Consolidation) Act 1928

Petroleum Spirit (Motor Vehicles) Regulations 1929 Regulating Licensing of petrol storage in excess of 9 litres.

Agriculture (Miscellaneous Provisions) Act 1968

Codes of practice in animal environments to avoid so-called 'factory farming' conditions and in particular fire hazards – flame proofing and escape routes being required. The Animal Welfare Codes (*see also* BS 5502) are becoming more stringent and precise by laying down the limits of acceptability of animal housing, to become mandatory in the near future under the guidance of the Farm Animal Welfare Committee (FAWC) and its 'Regulation Recommendation Subcommittee'.

Other acts and orders bearing on construction in rural areas

Housing Acts – fitness for habitation of dwellings.
Offices, Shops and Railway Premises Acts – staff working conditions.
Caravan Sites (Licences Application) Order 1960 under T & CP Act 1959 consolidated in T & CP Act 1971 – controls through Local Authority licensing the precise management of sites in respect of area, numbers and hygiene. Some sites exempt from control, e.g. workers caravans.
Water Resources Act 1968 – Water Authorities set up and abstraction rigidly controlled.
Land Drainage Act 1930 – controls, bridges and roadworks.
Highways Acts – rights of way.
Highway (Provision of Cattle Grids) Act 1950 – empowers Local Authority to construct grids on classified roads and rights of way.
National Parks Act 1969 (*see also* T & C Amenities Act 1974) – controls pathways, etc. in the countryside.

Additional regulations and constraints

The following regulations and constraints not given similar power of law which may have the same effect however are:
Institution of Electrical Engineers (IEE) Regulations – control of electricity supply systems and fixtures in building structures and elsewhere before supplies are connected. Ensures safe installation procedures and fittings.
BS Codes of Practice in both construction and use of buildings and equipment. Note BS 5502 particularly (*see later*). Advisory in nature but are used in mandatory insistence on quality, e.g. in the Building Act.
Farm Capital Grant Scheme Standards may be very forceful and all-embracing for construction (but are subject to rapid changes caused by variations in economic and political climates).
Private Wayleaves obtained by private Act of Parliament to cross land or boundaries, e.g. electricity pylon systems, or gas pipelines.
BS 5502:1980 is intended to embrace all agricultural structures and impose upon them precise structural and environmental requirements which will result in satisfactory operation for a predicted and assured lifetime whilst providing safe working conditions during that lifetime. All buildings conforming to the BS will carry a plate showing their classification, manufacturer and date. Buildings not conforming to BS will not be acceptable for grant aid directly. Classification relates to occupancy, predicted safe lifespan, type of activity intended for, and the construction material's safe stressing limitations together with its general suitability for the purpose and use (*see below Table 22.8*).

Part 2 of BS 5502 incorporates animal welfare guidance and lays down environmental standards desirable for healthy livestock in humane production systems.

ADVISORY ORGANISATIONS

Asbestos Cement Manufacturer's Association
16 Woodland Avenue
Lymm
Cheshire

British Constructional Steelwork Association
1 Vincent Square
London SW1P 2PJ

British Glasshouse Manufacturer's Association
Simpsons of Spalding Ltd
Branton's Bridge
Bourne Road
Spalding
Lincs PE11 3LP

British Precast Concrete Federation Ltd
60 Charles Street
Leicester

Building Research Establishment
Buckwalls Lane
Garston
Watford
Herts

Cement and Concrete Association
Wexham Springs
Slough
Bucks SL3 6PL

Chipboard Promotion Association
7A Church Street
Esher
Surrey

Table 22.8. Summary of provisions for BS 5502 classification

Class (and potential example)	Expected structure life (minimum) – reflecting strength of components (years)	Human occupancy h/d at population density (persons/m^2)	Situation (m minimum from classified road or non-owned dwelling place)
1. (Vegetable processing shed)	50	Unrestricted	Unrestricted
2. (Milking parlour)	20	6 at 2	10
3. (Pig fattening)	10	2 at 1	20
4. (Plastic envelope sheep house)	Temporary	1 at 1	30

Construction Steel Research and Development
Organisation (CONSTRADO)
12 Addiscombe Road
Croydon
Surrey CR9 3JH

Council for Small Industries in Rural Areas
141 Castle Street
Salisbury
Wilts SP1 3TP

The Country Landowners Association
16 Belgrave Square
London SW1X 8PQ

Electricity Council Marketing Department
Farm Electric Centre
National Agricultural Centre
Stoneleigh
Warwickshire CV8 2LS

The Farm Buildings Association
Roseleigh
Deddington
Oxford

The Farm Buildings Information Centre
NAC Stoneleigh
Kenilworth
Warwickshire

Fibre Building Board Development Organisation Ltd
Stafford House
Norfolk Street
London WC2

The Finnish Plywood Association
21 Panton Street
London SW1Y 4DR

The Fire Protection Association
Aldermary House
Queen Street
Londn EC4

The Institute of Agricultural Engineers
West End Road
Silsoe
Bedford MK45 4DU

The Milk Marketing Board
Thames Ditton
Surrey

Ministry of Agriculture, Fisheries & Food
(ADAS Land and Water Services)
Great Westminster House
Horseferry Road
London WC1
(and local regions)

AFRC Engineering Institute
Wrest Park
Silsoe
Bedfordshire

Plywood Manufacturers Association of British Columbia
81 High Holborn
London WC1

Poultry Research Centre
West Mains Road
Edinburgh 9

The Royal Institute of British Architects
66 Portland Place
London W1N 4ADZ

The Royal Institution of Chartered Surveyors
12 Great George Street
London SW1P 3ADZ

Scottish Farm Buildings Investigation Unit
Craibstone
Bucksburn
Aberdeen AB2 9T2

Steel Cladding Association
European Profiles Ltd
Llandybie
Ammanford
Dyfed

Timber Research & Development Association
Hugendon Valley
High Wycombe
Bucks HP14 4ND

Part 4

Farm management

23

The Common Agricultural Policy of the European Economic Community

Paul Brassley

BACKGROUND, INSTITUTIONS AND THE LEGISLATIVE PROCESS

The need for a Common Agricultural Policy

When the European Economic Community was created it needed a Common Agricultural Policy for two reasons: first, the agricultural industries in each of the member states were subject to government intervention, and second, if this intervention was to continue, it had to be compatible with the other provisions of the EEC.

Left to themselves, the markets for agricultural products in developed economies will produce fluctuating, and, in terms of purchasing power, gradually declining incomes for farmers, as a result of fluctuating and declining real prices of agricultural products. The price of agricultural products, like that of any other product in a free market, depends upon the balance of demand and supply. If demand increases less than supply the price will fall. The demand for agricultural products in a developed country does not usually increase rapidly, unless those products can be sold cheaply on the world market. As a result of technological change, producing increased output per hectare and per unit of labour, supplies tend to increase more quickly than demand. So in the long run prices tend to fall. Demand is also relatively unresponsive to price changes, so short-run supply variations, caused by fluctuations in climate and the incidence of disease, produce short run fluctuations in prices.

Short-term price fluctuations in a free market may result in the demise of businesses which would be viable at average levels of input and output prices, and the possibility of this event creates a disincentive to investment in such farms. Long-run price falls result in lower incomes for those farms which are unable either to reduce costs or increase output, and if such low incomes are unacceptable they may cease trading altogether. It is usually found that farm incomes have to fall to very low levels before farmers leave the agricultural industry, and in any case the land that they no longer farm is often taken over by another farmer. The total output of the industry is therefore maintained, so there is no tendency for prices to rise. While farmers with low incomes may be found in all areas, there may be some regions which

are particularly disadvantaged by physiographical or structural factors, and if agriculture is the major industry in such regions the whole region may be affected by income and outmigration problems.

In a free trading economy farm income problems may be further intensified by the availability of imported agricultural products at prices below the domestic supply price. Imports will not only restrict domestic price rises but will also increase the total import bill, which may be considered a problem in countries with balance of payments difficulties. However, the government of a country which has no difficulty in exporting enough to pay for food imports may still consider it unwise to be totally reliant on imported food supplies: unforeseen emergencies may give rise to difficulties in obtaining supplies, or long-term world price rises due, for example, to population increases, may in time reduce the price advantages of imports. A capacity for rapid expansion of domestic agricultural production may therefore be thought desirable, even if considerable reliance is placed upon imports. In short, free markets for agricultural products in developed free market economies may produce low farm income, balance of payments and potential supply security problems.

By the middle of the 1950s the agricultural industries in all western European countries faced all these problems to some extent, and governments had evolved a number of policies for dealing with them. In the continental European countries agricultural policies usually involved a system of import controls and price support which kept price levels higher than they would have been in a free market and so maintained agricultural incomes and self-sufficiency levels. When the discussions which led to the formation of the EEC were held in 1955 and 1956 it was agreed that the exclusion of agriculture from the general common market was impossible: without free trade in farm products national price levels could differ, and those countries with the lowest food prices would have the lowest industrial costs, thus undermining the common policies which would be introduced for other industries. Although there was a general similarity between the existing agricultural policies, the differences in detail were so numerous that the simple continuation of existing policies would have been

591

impossible. Not only did Community countries require an agricultural policy, they required a common agricultural policy.

The original member states of the Community, and those which have joined more recently, are shown below:

Date of accession	Member states
1958	Belgium, France, West Germany, Italy, Luxemburg, the Netherlands
1973	Denmark, Ireland, the United Kingdom
1981	Greece
1986	Portugal, Spain

The formation of the Common Agricultural Policy

The Treaty which established the EEC was signed in Rome on 25 March 1957. Articles 38–47 of the Treaty apply directly to agriculture. The products covered by the CAP are listed in Annex II of the Treaty of Rome. In practice, some products not listed there (and thus referred to as 'non-Annex II products') are occasionally affected by CAP rules. Article 39.1 lays down the objectives of the Common Agricultural Policy:

(1) to increase agricultural productivity by promoting technical progress and by ensuring the rational development of agricultural production and the optimum utilisation of the factors of production, in particular labour;
(2) thus to ensure a fair standard of living for the agricultural community, in particular by increasing the individual earning of persons engaged in agriculture;
(3) to stabilise markets;
(4) to ensure the availability of supplies;
(5) to ensure that supplies reach consumers at reasonable prices.

Article 40 lays down the broad guidelines for the various policy instruments by which these objectives are to be achieved, and Article 43 indicates the procedure to be followed in reaching agreement on the detailed provisions of the CAP. Article 43 also provides for discussions to be held in order that member states and EEC institutions can decide upon their requirements, after which the Commission is required to submit proposals to the Council of Ministers. The Commission produced these proposals in 1960, and although they have been altered in detail they form the basis of the CAP described below. It is interesting to note that four principles which are often assumed to lie behind the CAP system (a single market, joint financing, Community preference, and a comparable earned income for efficient farms), appear neither in the Commission proposals nor in the Treaty of Rome, but seem to have appeared early in the working life of the CAP. These developments, and the ability of Community policy makers to produce new policies to cope with new problems, mean that the decision-making process in the Community must be examined before the mechanisms of the CAP can be outlined.

The process of decision making in the European Community

The Treaty of Rome lays down the basis of the CAP, but decisions still have to be made about what policy shall be (primary legislation) and how it should work in detail (secondary legislation). There is therefore a large number of formal bodies involved in this process, of which the two most important are the Commission and the Council of Ministers.

The *Commission of the European Communities* is presided over by 17 Commissioners. It is divided into 20 Directorates-General (DGs): DG VI is responsible for agriculture, and is the responsibility of one Commissioner. In some ways the Commission is like a national civil service, but it has extra functions: it has the right to propose primary legislation; it handles the day to day administration of Community laws and policies resulting from this primary legislation, and may enact secondary legislation in order to do so; and it represents the Community in its relations with non-member states.

Proposals for primary legislation produced by the Commission are referred to the *Council of Ministers*. This is the major legislative body of the Community, and consists of a minister from each member state, depending on the subject under discussion: agricultural ministers for agricultural matters, finance ministers for financial matters, and so on. It is normally the final decision making body for all primary legislation, although when problems cannot be resolved they are now often passed on to the European Council, a meeting of the heads of goverment of the member states which has no legal basis in any treaty and which developed from earlier summit meetings. Its decisions have been passed back to the Council of Ministers to be given legal validity.

Since the Community is a partnership of the member states, it makes its decisions by a process of negotiation in a series of committees. The initiation of primary legislation is normally the responsibility of the Commission, although it may also be instigated by several other bodies, from the Council of Ministers to a trade association in a member state. In the Commisssion, the first moves are made by the appropriate department in Directorate General VI, usually in consultation with national civil servants or other experts, and representatives of trade associations and other pressure groups such as the Committee of Professional Agricultural Organisations (COPA), which acts for farmers' unions in the Community. There is then a further process of consultation with other Directorates General (e.g. those concerned with competition or relations with non-member states) before the proposal is presented to all 17 Commissioners. If they agree, it becomes a Commission proposal and is formally presented as such to the Council of Ministers.

The Council immediately passes on this draft legislation, if it is concerned with agriculture, to the *Special Committee – Agriculture* (SCA) for consideration. Non-agricultural legislation is the responsibility of the *Committee of Permanent Representatives* (COREPER), which is a meeting of the ambassadors of the member states to the Community. Like COREPER, the SCA consists of representatives of the member states, together with a representative of the Commission. Detailed discussion of the proposed legislation is carried out by one or more Working Parties, composed of nationally appointed experts, which report their findings to the SCA. The SCA then debates the proposal, so that only those parts of it on which it has not been possible to reach agreement will need to be debated by the Council of Ministers.

For most proposals the Council, before it makes a decision, is required to receive the opinion of the *European Parliament*. The Parliament carries out its work by nominating committees to produce reports on proposed legislation, and it is these reports which are debated in the monthly plenary sessions of the Parliament, before becoming the Opinion which is passed on to the Council. The Parliament also has considerable indirect impact on the formulation of policy through its questioning of members of the Commission, both formally in committees and informally.

The Council may also take note of the opinions of the *Economic and Social Committee*, which is a body set up under the Treaty of Rome to advise the Council and the Commission. It consists of members who are appointed by member states in their personal capacities, and who may broadly be divided into employers, trade unionists and independent members.

Thus, any proposal, before a decision is taken on it by the Council of Ministers, will have been the subject of comment by a wide variety of formal and informal, corporate and individual, Community and national, expert and lay sources. It is therefore hoped that the technical and drafting problems in a proposal will have been sorted out, so that the only issues which remain to be resolved by the Council are the basic political ones. After debating a proposal, a decision must be made. Some agricultural legislation requires the unanimous approval of the Council, but increasingly decisions are adopted by a *qualified majority*, which requires 54 or more votes in favour. In this voting system, France, Italy, the UK and West Germany have ten votes each, Spain has eight votes, Belgium, Greece, the Netherlands and Portugal have five votes each, Denmark and Ireland three each and Luxemburg two votes.

The proposals thus adopted fall into three categories: Regulations, Directives and Decisions. A *Regulation* is directly applicable to all people and governments in all member states. A *Directive* is binding as to its intention on governments, who must then pass legislation to give it effect, so that it is more flexible than a Regulation. A *Decision* is as binding and immediately applicable as a Regulation, but only on the people or governments to whom it is addressed.

The process described up to this point is concerned with the production of primary legislation, but secondary legisla-, tion, establishing the practical rules by which the primary legislation shall be applied, must also be produced. This particularly affects the activities of processors and traders. For example, import levies and export refunds must be fixed, and the standards required for produce bought into intervention must be laid down. This secondary legislation is largely the responsibility of the Commission. An initiative from the appropriate section of the Directorate General for Agriculture is passed to other departments of the Commission for consultation (e.g. on legal, budget or external affairs implications). At the same time the Commission usually takes the advice of *Consultative Committees* (also known as Advisory Committees). These consist of representatives of European producers, processors, traders and consumers of the commodity under discussion. At the time of writing about 150 different groups are recognized as eligible for consultation. The Commission also consults with experts from the administrations of the member states.

After this process of consultation, the proposal is then passed to the relevant *Management Committee*, which is a statutory body made up of officials from the ministries of Agriculture in the member states, under the chairmanship of a senior Commission official. There is a Management Committee for each of the major commodities, and for the more important commodities the Committee may meet once per week. The detailed administration of the monetary arrangement of the Community is also carried out by a Management Committee, and there are others which deal with, for example, structural policy, research, plant health, and so on, although these meet less frequently. Voting in a Management Committee is by qualified majority, using the same weights used for Council voting, but the procedure is reversed, in that 54 votes against are required before a proposal must be referred to the Council of Ministers. After a proposal has passed through the Management Committee it is formally adopted by the Commission, published and becomes, in effect, law.

Community law is thus the primary and secondary legislation produced by the Council and the Commission, and also the law embodied in the founding treaties of the Community, such as the Treaty of Rome. It takes precedence over national law. It is the responsibility of the *Court of Justice* to arbitrate on whether or not Community legislation has been correctly applied, is being applied, or is flawed. Some of the decisions of the Court have had a substantial effect on the way in which the Community is run, from deciding upon the powers of the Parliament to defining what means may legitimately be employed to prevent trade in food products.

CAP PRICE MECHANISMS

General outline

The underlying principle of the CAP price mechanism is that the producer should receive a price for his produce which is determined by market forces. But these market forces are controlled so that the market price fluctuates only within predetermined upper and lower limits. Thus the farmer is protected against excessively low prices, and the consumer against excessively high prices.

The mechanism by which this protection is carried out recognises that farm products may be produced either within the EEC or outside it. The demand for farm products is relatively constant, so if the market was supplied only from within the EEC the major reason for low prices would be an excess supply. The effects of this can be mitigated by artificially increasing demand. Therefore the EEC enters the market to buy up agricultural products for storage when prices are low, a process known as *'intervention'*. If prices subsequently rise this stored produce can be released on to the market again. If they do not it may be sold on the world market with the aid of subsidies known as *'export restitutions'* or 'refunds'.

Many farm products can also be supplied from countries outside the EEC (known as 'third countries'). If this non-EEC supply was available at less than the EEC market price it would reduce the EEC market price, and so the entry of such produce is subject to *import levies* to raise its price. When EEC prices are high, perhaps as a result of supply shortages, the entry of produce from third countries serves to reduce prices by increasing supplies.

The national agency which deals with intervention, import

levies, export restitutions, and all the other payments made under the Guarantee section of the EAGGF (*see below*) in the UK is the *Intervention Board for Agricultural Produce* (IBAP). It does not physically handle any goods itself, but supervises the work of other bodies such as the Home Grown Cereals Authority for cereals and the Meat and Livestock Commission for beef. Butter sold into intervention is stored in commercial cold stores under IBAP contracts.

The financial resources needed for the activities of the IBAP are obtained from the *European Agricultural Guidance and Guarantee Fund* (EAGGF – often referred to by its French acronym FEOGA). The Guarantee section of the Fund, which is the larger part, finances the measures used to support commodity prices within the community. The Guidance section finances structural policy (*see below* p. 599). The Fund is in turn financed from the EEC budget (of which it forms the major part), which has three main sources of revenue: levies on imports of agricultural products from third countries, common Customs Duties, and up to 1.4% of the Value Added Tax collected in member states, calculated on a uniform base.

Essentially, therefore, the price received by the EEC farmer will be at least the price at which intervention begins, and at most the price at which comparable imported goods can be sold, assuming that EEC prices are higher than world prices. Between these two extremes market prices will fluctuate according to the balance of supply and demand. For some commodities the Commission has acquired the power to determine when intervention will operate, and in some years intervention payments have been delayed, so that farmers may not always, in practice, receive the full intervention price, but the basic concept of the system remains in force.

This basic system is used for regulating markets in most of the major products produced by EEC farmers, including cereals, milk products, sugar, beef and veal, pig meat, poultry meat and eggs. Different systems are used for sheepmeat, fruit and vegetables and other crops. There is no EEC price mechanism for potatoes or wool. There are significant differences between the detailed regulations for each commodity, and these are described below, together with the terms applied to the management of each market.

Cereals

The price which the EEC wishes the farmer to receive for his cereals is known as the *target price*. When the market price of cereals falls below the target price to the *intervention price*, intervention agencies may buy cereals for storage, thus preventing further price falls. Cereals sold to intervention stores must be of intervention quality, and in 100 tonne lots, so it is usually merchants and cooperatives rather than farmers which sell into intervention, and the farmer receives the intervention price indirectly. Intervention prices increase in monthly steps throughout the year to encourage orderly marketing. There is a single intervention price for common wheat, barley and maize, another price for durum wheat (used in the manufacture of pasta) and a *reference price* for wheat of breadmaking quality which functions as an intervention price but compensates for the lower yield expected from breadmaking varieties by being higher than the common wheat intervention price. From the mid-1980s, faced with a surplus of cereals and an unwillingness on the part of the Council to reduce intervention prices, the

Commission began to use its power to determine the details of intervention regulations to affect prices. Thus intervention stores may only be opened in some months, and the moisture content specified for grain of intervention quality may be reduced or quality standards changed.

The price at which cereals imported to the EEC from third countries can be sold is also determined from the target price. It used to be assumed that the highest price of cereals will be found at Duisburg in the Ruhr district of West Germany, and that they would be imported at Rotterdam. In order to ensure that the price of imported grain would not be less than the target price a minimum import price, known as the *threshold price*, was calculated by subtracting the transport costs from Rotterdam to Duisburg from the target price. Although the basic scheme remains in use, a separate centre of shortage for each cereal product is now used in the calculation. If imported grain price quotations are less than the threshold price a *variable levy* is imposed, which is equal to the difference between the threshold price and the minimum import offer price, calculated daily. The calculation of this import offer price is made by the Commission, on the basis of information supplied in confidence from a number of traders.

The export of cereals from the Community is also regulated. When the EEC market price is higher than the world market price (as it has been for most of the period for which the EEC has existed) an *export refund*, of the difference between the EEC market price and the world market price, is paid. If the EEC market price is lower than the world market price an *export levy* is charged. Thus the Community producer will always receive the Community price, no more and no less.

These systems are summarised in the diagram below:

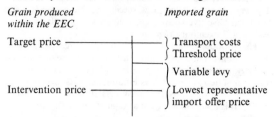

Milk and milk products

Market support for milk and milk products is also based on a system of target and intervention prices for EEC production together with a threshold price for imports. The system is necessarily more complex than support for cereals because a greater range of products is involved, and because a wide variety of additional measures has been introduced in an attempt to overcome the problems of surplus which have existed in the milk market for several years. In theory it is the overall aim that producers should receive the target price for milk delivered to a dairy; in practice they receive rather less, sometimes only 85% of the target price.

The target price is supported by intervention in the market for butter, skimmed milk powder, and three varieties of Italian cheese. The intervention price set for these products reflects the quantity of milk required for their production, so that the intervention price for butter is normally about two and a half times higher than that of skimmed milk powder. Intervention may take place either by buying in to intervention store or by aids to private storage by processors. In periods of surplus changes have been introduced so that intervention is no longer mandatory, but can be suspended, at the discretion of the Commission, once stocks have reached a certain level. However, if market prices fall below 92% of intervention price in any member state, intervention is reinstated in that country. In the UK the price received by producers is controlled by the activities of the Milk Marketing Boards. Some of the activities of the MMBs appeared to contradict the provisions of the Treaty of Rome, but the Community permitted their continued existence because they appeared to maintain the consumption of liquid milk at a higher level than that found in most other European countries. Their constitutions were also changed so that they became, in effect, cooperatives. The returns to UK farmers therefore result from the pool price determined by the MMBs from their sales of milk to the liquid and manufacturing markets. The manufacturing price of milk realised by the MMBs depends upon the intervention price of butter and skimmed milk powder. The retail price of liquid milk is set by the dairy which sells it.

As with cereals, the intra-EEC price of imported milk products is fixed by the threshold price determined by the Community. Threshold prices are determined for 12 'pilot' dairy products (three types of powder, five types of cheese, two types of condensed milk, butter and lactose) and from the lowest representative offer price for these, import levies for imported milk products are determined. Export refunds are also available to enable milk products to be sold on the world market. The importance of dairy farming to community processors, and in particular to those operating small businesses, has meant that the target and intervention prices set by the Council of Ministers have been high enough to maintain production and restrict demand. A number of additional measures have therefore been introduced at various times in an attempt to increase consumption or reduce supplies. Among those designed to increase consumption are: disposal subsidies for skimmed milk powder used in the production of animal feeds or casein; consumer subsidies for butter; sales of butter at reduced prices from intervention stocks to the armed services, non-profit making institutions (e.g. hospitals) and persons receiving social assistance; and sales of liquid milk at reduced prices to schoolchildren.

All previous schemes to reduce supply have been superseded by the system of quotas and superlevies which was introduced initially for a five-year period, in 1984. Each producer is allocated a quota, the size of which depends upon his previous output. The way in which the size of the superlevy is calculated has varied. At the time of writing the MMB determines the extent to which national production has exceeded national quota. If it has not, no producer has to pay superlevy, no matter how much he has exceeded his individual quota. If it has, the MMB calculates a levy threshold of x% of individual quota. Each producer whose production has exceeded $100 + x$% of his individual quota has to pay a superlevy equal to milk target price on each litre in excess of his quota $+ x$%. The value of x depends on the number of producers under quota, the number over, the extent to which each of them is under or over quota and the amount by which national production exceeds national quota. Thus a producer is not penalised for slightly exceeding quota, but penalised heavily for blatantly exceeding quota. An increase in butterfat content is treated as an increase in output. Quotas may sometimes be leased through schemes operated by the MMB, or may, under certain circumstances, be transferred with land that is sold. There are also schemes for compensating producers who

surrender their quota. The details of all these schemes vary from time to time and it is important to consult the MMB to determine the arrangements presently in operation.

Sugar

The sugar regime is basically similar to that for cereals, but production quotas are imposed and special arrangements are made for imports from the sugar producing African, Caribbean and Pacific (ACP) countries which are signatories of the Lomé Convention.

The target price is fixed for white sugar, and from this the intervention price is derived. The minimum price for beet paid by sugar factories for production within the *basic quota* (the A quota, which is approximately equal to domestic consumption) is calculated from the intervention price, taking into account the sugar yield of the beet, together with processing and delivery costs. Production outside the A quota, but within the *maximum quota* (the B quota), is subject to a levy of up to 30% of the intervention price, designed to assist in financing export restitutions. Production in excess of both quotas must be exported to third countries without the aid of export refunds.

The calculation of the threshold price for white sugar imports is different from that of cereals. It is obtained by adding the storage levy and transport costs to the target price, the import levy being the difference between the minimum import offer price and the threshold price. However, the ACP countries are allowed to sell up to 1.2 million tonnes of white sugar (or its equivalent in raw sugar) to Community countries at or above the guaranteed price.

Beef and veal

Market support for beef and veal is based on intervention, but with some significant differences from the systems described above. The average price which should be realised over the year (effectively, the equivalent to the target price) is known as the *guide price*, and intervention may take the form of aids for private storage or buying in, usually of primal cuts rather than live animals. The producer receives the *buying in price*, which is calculated from the intervention price by using coefficients and the killing out percentage. *Reference prices*, or weighted average prices in representative markets, are calculated each week for each member state and for the Community as a whole. Intervention may occur when the Community reference price falls below 91% of the intervention price and the reference price in the member state is less than 87% of the intervention price.

Imports of beef and veal from third countries are subject to Customs Duties (e.g. 16% on live animals and 20% on fresh, chilled or frozen carcasses) in addition to variable levies. The *basic levy* is the difference between the guide price and the fee at frontier offer price (duty paid), but the amount of the basic levy which is payable by importers is dependent upon the relationship between the Community reference price and the guide price, ranging from 114% of the basic levy when the reference price is less than 90% of the guide price to zero when the reference price is more than 106% of the guide price. In practice, therefore, most imported beef enters the Community under quota agreements under the terms of either the General Agreement on Tariffs and Trade or the Lomé Convention.

In the UK only there is a variation on the standard regime, although its continued existence is under almost constant review. A *target price* is set, which must be no more than 85% of the guide price when averaged over the year. If the UK reference price is less than the target price a *variable premium* may be paid to producers selling cattle in that week to make up the difference, subject to a maximum level of variable premium which is decided by the Community. Beef may still be sold into intervention, but in this case the variable premium is deducted from the buying in price.

In addition to these price support schemes a headage payment on suckler cows is paid by the Community in each member state, and individual member states are authorised to increase the size of the payment made from national funds. In some countries (but not the UK) a headage payment is made on calves, and in countries with neither a calf headage payment scheme nor a variable premium scheme there is a headage premium on the first 50 male animals on each farm, paid once in the animal's life or at slaughter.

Sheep meat

The sheep meat regime, which was introduced in 1980, revolves around the *basic price*, the level of which depends upon the market situation in sheep meat and its close substitutes, and costs of production. The market price is supported by aids to private storage which may be paid if the EEC market price falls below 90% of the basic price. If the EEC market price falls below 85% of the basic price, sheep meat may be bought into intervention between mid July and mid December.

Annual compensatory premiums are paid, normally as a ewe headage payment, to make up the difference between the basic price and the representative market price for the marketing year in question. In the UK any expenditure on variable premiums (*see below*) must be deducted from the total annual compensatory premiums before payment. As is customary, these institutional prices are denominated in European Currency Units (ECUs), which are converted into national values at representative rates of exchange, but, since no monetary compensatory amounts are payable on trade between member states, currency fluctuations can produce differences in institutional prices between member states (the reasons for this are explained on p. 597 below). EEC producers are protected from competition from non-member states by voluntary restraint arrangements agreed between the Community and traditional supplying countries, in return for which sheep meat from such countries is subject to a reduced customs duty of 10%.

The regime provides for the use of a different system in the UK. This replaces private storage and intervention by a *variable premium* system. A *guide price* is set at the same level as the EEC intervention price. If the average market price falls below this a variable premium is paid to bring the price received by the producer up to the guide price. However, this only applies to sheep meat bought and consumed within the UK, and an export charge (known as 'clawback') equal to the variable premium, is paid on sheep meat exported to other Community countries. In addition the annual compensatory premium is reduced by an expenditure on variable premiums.

Pig meat

Pig meat production is cyclical, and therefore each year a *basic price* for standard quality carcasses is set at a level which will produce reasonable stability in the market without producing structural surpluses. The market price may be supported by intervention, when the *reference price* (i.e. the weighted average EEC price) falls below 103% of the basic price, in one of two ways: either by the provision of aids to private storage by traders, or by buying in by intervention agencies. In fact, such buying in, at a *buying in price* of between 78 and 92% of the basic price, happens only rarely. The provision of private storage aids is more common, although these are not mandatory, and the Commission may allow reference prices to remain some way below the basic price, particularly when pig numbers in the Community are tending to increase.

Pig meat imports from third countries are controlled by *sluicegate prices*, designed to prevent offers of pig meat to Community traders at prices less than world production costs, and *basic levies*, which take account of the fact that cereal feed costs in third countries may be below those within the EEC. The sluicegate price is therefore an estimate of world production costs, made up of cereal and other feed costs together with overhead costs. The basic levy represents the difference between world and EEC production costs together with an additional 7% of the sluicegate price to protect the domestic industry. Thus the minimum price at which imported pig meat may be sold within the EEC is the sluicegate price plus the basic levy, since if pig meat is offered at less than the sluicegate price a *supplementary levy* is applied.

Poultry meat and eggs

There are no internal support measures for poultry meat or eggs, and all market support is by regulation of the trade with third countries, which itself is not extensive. Sluicegate prices and basic levies are calculated in the same way as those for pig meat, and applied in the same way.

Supplementary levies are also applied if the import offer price is less than the sluicegate price.

Fruit and vegetables

Market support for fruit and vegetables is normally operated through the activities of producers' organisations and by import duties. A *basic price* was arrived at by examining market prices in the Community over the preceding three years but now it is set by the Council of Ministers. From this a *buying in price* is calculated for those countries in which intervention operates. More commonly, a *withdrawal price* is calculated, modifying the buying in price by quality, variety and size coefficients. This determines the compensation received by producer groups for taking produce meeting EEC quality standards off the market. This is the system used in the UK where only cauliflowers, tomatoes, apples and pears are eligible for support. The market for fresh produce is also supported by giving aids to processors buying specified quantities at minimum price. Imports from third countries are subject to *ad valorem* import duties, and for some products there is a system of minimum import prices, called *reference prices* to prevent third country products undercutting domestically produced products.

Other commodities

Although the markets for the major commodities which are produced in the EEC are supported by various systems involving some form of intervention and import control, there are other commodities, less important in terms of total output, which are supported by other means. The crops considered below are those which are grown commercially in the UK, but the CAP also includes regimes for a wide variety of other products, such as rice, tobacco, wine, a number of oilseeds, and olive oil, which do not fall in to this category, although several of them are of major significance in the areas where they are grown.

Oilseeds

Oilseeds, including oilseed rape and sunflower seeds, are supported by a production subsidy which is paid to the crusher. A target price is set each year and the size of the subsidy is calculated from the difference between the target price and the world market price. At the time of writing there is a proposal to introduce a tax on oils for human consumption.

Peas and field beans

These are supported by an aid paid to feedstuffs manufacturers purchasing these products from, and paying a minimum price to, growers under contract. An *activating price* (equivalent to a target price) is fixed each year and a subsidy equal to 45% of the difference between the activating price and the world price of soyabean meal is paid.

Dried fodder

Dried fodder includes both dehydrated potato products which are unfit for human consumption, and artificially dried legume forage crops. A guide price is fixed each year and a variable aid, based on the relationship between the guide price and the world price, is paid to processors. They also receive a production aid based on a flat rate per kilogramme.

Hemp and flax

Hemp and flax grown mainly for fibre also receive a fixed payment per hectare sown and harvested.

Seeds

Seeds, including grass, legume, flax and hemp seeds produced for sowing, receive a flat rate payment per 100 kg of seed.

Hops

Hops are supported by a fixed payment per hectare of registered area. In addition, launching aids are provided to encourage the formation of hop producer groups.

MONETARY ARRANGEMENTS IN THE CAP

Institutional prices in the CAP, such as target or intervention prices, are denominated in European Currency Units (ECUs). The ECU is a currency basket, comprised of fixed amounts of the currencies of member states of the Community (excepting, until the early 1990s, those of Spain and Portugal). For the purposes of everyday transactions, such as sales into intervention, ECU prices need to be converted into their equivalents in national currencies. If exchange rates were fixed for significant periods of time, as they were when the CAP was set up, this would present no problems. It would be sufficient to *multiply* the ECU price by the fixed ECU/national currency exchange rate to determine the national currency price. But exchange rates are not fixed. Some member states (West Germany, the Netherlands, Belgium, Luxemburg, France, Denmark and Eire) belong to the Exchange Rate Mechanism of the European Monetary System (EMS). This means that they declare an exchange rate, known as the 'central rate', between their national currency and the ECU, and then undertake to maintain the cross rates based on these central rates within a margin of $\pm 2.25\%$ until further notice. These are known as narrow band currencies. The Italian lire conforms to the same principle, but within a wider band. The pound sterling and the Greek drachma are not, at the time of writing, members of the Exchange Rate Mechanism, and thus the respective governments give no undertakings about maintaining the value of these currencies. Thus the intervention price in any country would vary from hour to hour in response to changing currency prices on the international money markets. Clearly, this would produce enormous administrative difficulties, so for agricultural purposes a fixed exchange rate is used. This is known as the representative, or green, rate, e.g. the 'green pound'.

The system has now developed to the point where, in most member states, multiple green rates may be used. For example, in the UK there is one green rate for crop products and others for pig meat, milk and poultry, beef and sheep meat. The size of the green rate is determined by the individual member state but changes must be approved by the Council of Ministers. Obviously changes in the green rate will affect the price received by farmers. For example, if the green pound is devalued (so that more pounds are required to buy one ECU), prices will *rise*:

If soft feed wheat intervention price is 170.47 ECU at a green rate of 1 ECU = £0.61855
Intervention price will be £105.46.
After devaluation, green rate is 1 ECU = £0.626994
then intervention price will be £106.88.

It follows that if the market exchange rate of a currency falls faster than the green rate, thus requiring a green rate devaluation to re-align the two, then prices received by farmers will rise, and vice versa. Consequently farmers in weak currency countries often argue that green and market rates should be kept in alignment, while those in strong currency countries are more concerned to maintain the difference.

In the absence of any corrective mechanism, differences between market and green rates would also affect trade between member states. For instance the intervention price of soft feed wheat in West Germany would be:

ECU price multiplied by green rate
(DM per ECU)
i.e. 170.47 × 2.39792 = 408.77 DM

If, at the same time, the market rates of exchange are:

West Germany: 1 ECU = 2.31728 DM
UK   : 1 ECU = £0.776820
then £1 = 2.31738 ÷ 0.776820 = 2.983 DM

So the value of the West German intervention price in pounds sterling would be:

408.77 ÷ 2.983 = £137.03

This compares with the UK intervention price calculated above of £106.88. Under these circumstances it would clearly pay traders to sell into intervention in West Germany rather than the UK. Similarly, export restitutions would be maximised by exporting through West German ports, while import levies would be minimised by importing through UK ports, and trade flows between member states would reflect this. Moreover, such a pattern would minimise the income of FEOGA and maximise its costs.

In order to overcome these problems a system of Monetary Compensatory Amounts (MCAs) was introduced. The effect of the MCA system is that in no matter which member state a product is sold into intervention, the seller in the country of origin receives the price that he would have received from an intervention store in that country of origin. In other words, the MCA reflects the difference between market rates and green rates. The calculation of fixed MCAs for narrow band currencies is relatively straightforward.

Real Monetary Gap = Market Rate – Green Rate
i.e. $(1 - (MR/GR)) \times 100\%$

Since green rates may differ from one commodity to another, so may the Real Monetary Gap.

MCA percentage = Real Monetary Gap
                 – Franchise (1 or 1.5%)

The franchise was introduced in 1979 to produce a small competitive margin as a result of currency depreciation for countries with negative MCAs (*see* Further Reading). If the Real Monetary Gap is negative the franchise is 1.5 percentage points, and if positive, it is 1.0 percentage points but there have been suggestions that it might be increased.

The MCA (denominated in the national currency) is then calculated by the formula:

MCA = Intervention price × Green Rate
                              × MCA percentage

If the MCA is positive (because the green rate is undervalued relative to the market rate) it is applied as a tax on imports and a subsidy on exports, and if negative as a charge on exports and a subsidy on imports. The way in which this works may be seen by taking the example of a trader who exports a tonne of soft feed wheat from the UK to an intervention store in West Germany using the prices and exchange rates mentioned above. For West Germany, the Real Monetary Gap would be:

$(1 - 2.31728/2.39792) \times 100 = 3.363$

So the MCA would be:

170.47 × 2.39792 × (3.363 − 1.0/100) = 9.66 DM

Intervention price in W. Germany is 408.77 DM
− W. German MCA   9.66 DM
= price received by the trader
  on leaving W. Germany   399.11 DM

Converting this to sterling at the exchange rate calculated above, price on arriving in

UK = 399.11 ÷ 2.983 = £133.79

For the UK, if the market rate was

1 ECU = £0.776820, the Real Monetary Gap would be
$(1 - 0.776820/0.626994) \times 100 = -23.896$

So the MCA would be

170.47 × 0.626994 × (23.896 − 1.5/100) = £23.94

This is the tax imposed on the money received from the West German intervention store, so the trader receives (£133.79 − 23.94) = £109.85 in the UK. This compares with the UK intervention price, calculated above, of £106.88. The difference in price between the West German intervention price modified by MCA subsidies and the UK intervention price is simply a result of the franchise. If the calculation above is repeated, omitting the franchise, the difference between the two figures disappears.

In the example above the market rate of exchange between the pound sterling and the ECU has been assumed, but in practice it needs to be calculated. Although exchange rates fluctuate from day to day, for administrative simplicity the calculation is made weekly. Then, since the ECU is not a widely traded currency, a UK Real Monetary Gap is calculated, not against the ECU, but against each narrow-band currency, and the unweighted average of these is used as the UK Real Monetary Gap for the week. From this the MCA can be calculated. Thus a non-narrowband currency will have variable MCAs.

In practice, several modifications have been applied to this basic system. One is the creation of a 'green ECU', which was introduced in 1984 in order to prevent any increase in positive MCAs. When a country with a positive MCA revalues its currency the central rate, for CAP purposes only, is revalued by the amount necessary to maintain the MCA unchanged. This is done by applying a new correcting factor, which is calculated by multiplying the old correcting factor by the percentage revaluation of the strongest currency against the ECU. Secondly, MCA percentages of less than 0.5 are treated as zero, while those between 0.5 and 1.0 are treated as 1.0. Thirdly, MCAs are only changed if the Real Monetary Gap has changed by more than one percentage point since the last time MCAs were set, and when green rates vary between sectors, so that there are several Real Monetary Gaps in a single member state, it is the smallest of these gaps that must change by more than one percentage point if the MCA is to change. Finally, it should be noted that MCAs are not applied at all in trade in sheep meat (in an attempt, which seemed wise at the time, to keep the sheep meat regime as simple as possible), oilseeds, fruit and vegetables, but that they are, in effect, also applied to import levies and export refunds on all other products.

In most circumstances, collection and payment of MCAs in the UK is carried out by the Intervention Board for Agricultural Products at Reading.

STRUCTURAL POLICY

In addition to the common price and market measures, financed through the guarantee section of the EAGGF, the CAP also includes a number of structural measures, financed through the guidance section of the EAGGF. In this context the term 'structural' refers to aspects of agricultural policy which are concerned with farm size, vocational training, cooperation, processing and marketing.

When, in 1960, the Commission first made proposals outlining the form that the CAP should take, they accepted that there were several structural problems within the Community. In 1964, when the detailed rules concerning the operation of the EAGGF were formulated, it was decided that the operation of the guidance section should not exceed one-third of the expenditure on the guarantee section. (In fact, as guarantee expenditure rapidly expanded, this limit was never reached, and in 1966 it was replaced by a restriction to a fixed sum each year.) Of the assistance available under the terms of the first guidance section regulations (R17/64), about half was used for projects concerned with the structure of production, such as land improvements or irrigation and drainage. Most of the rest was used for improving market structures.

It was originally expected that assistance from the guidance section would only be available for projects which formed part of a wider Community project, the framework of which was to have been laid down by the Commission. However, the Commission proposals were never finally adopted, and grants were given for individual projects. By the late 1960s it became apparent that much of the available grant aid was being allocated to the wealthier farming regions, and that a structural problem remained which was not being solved. The major structural problem in the Community arises from the existence of many small farm businesses. This does not automatically imply that the larger business is necessarily more efficient than the small, but simply that the larger business can make lower profits per unit of output and survive because it produces more units of output to cover fixed costs. Other conclusions follow from this: small farm businesses may have difficulty in acquiring capital, and they are forced to concentrate on the more intensive crop and livestock enterprises. The price and market policies of the CAP are not designed to solve these problems. Indeed, insofar as they support farm incomes, and prevent the disappearance of small business from the farming industry, they perpetuate them. As has been pointed out above (p. 591) there are several reasons for doing this, such as the desire to maintain national food self sufficiency and prevent rural depopulation. But by the late 1960s it was felt that the balance might have tilted too far, and that measures designed to decrease the number of small businesses and increase the viability of those remaining might be required.

Accordingly, the Commission Memorandum on the Reform of EEC Agriculture (usually known as the Mansholt Plan), published in 1968, proposed widespread changes designed to reduce the number of a small farmers. Although these proposals were not fully accepted they led, after much discussion, to the formulation of a number of directives designed to ameliorate the structural problems of the Community. These included directives on the modernisation of farms, payments to outgoers and the provision of socio-economic guidance, which were all replaced in 1985 by Regulation 797/85 *on improving the efficiency of agricultural structures.*

In general, the purpose of this regulation is similar to those directives which it replaces. However, under the new regulation there are several changes: it is easier for part-time farmers to become eligible for grants; more grant aid is available for the assistance of tourism and craft enterprises on farms, and for landscape and nature conservation; and higher rates of grant are payable to young farmers. Thus the regulation attempts to deal with allegations that previous schemes had detrimental environmental effects but did little to help new entrants and those with small farm businesses. Only Title 1 of Regulation 797/85, providing for a system of aid for investments in agricultural holdings is mandatory; the other provisions may be introduced or not according to the wishes of individual member states.

In the UK the regulation is put into effect by the Agricultural Improvement Scheme (AIS). The Scheme provides for capital grants at various rates to be given to tenants, owner-occupiers, partners or companies running farms or horticultural businesses. The basic rate of grant depends upon the work to be done, but is generally either 15 or 30%. To be eligible for the scheme a producer must submit an improvement plan which demonstrates that the assisted investment will generate a 'lasting and substantial improvement' in farm income. In addition the farm's income per labour unit (one labour unit = 2200 hours of work) at the start of the plan must be less than the Reference Income (which is based on the average income of full-time workers outside agriculture in Great Britain). At the end of the plan, income per labour unit must be less than 120% of Reference Income. The person who runs the farm from day to day must spend at least 1100 hours per year working on the holding, earn at least half his/her annual income from it, and either have been in farming at least five years or hold a suitable training certificate. There are special rules affecting (mainly reducing) the eligibility of holdings with dairy cows, pigs and poultry. Farmers under 40 years of age who hold a suitable training certificate and who can comply with certain other criteria may have their grant increased by one-quarter.

Under the provisions of the AIS, grants are available for investments other than those which are strictly agricultural. Such investments need not necessarily be part of an Improvement Plan. They fall into three groups. Conservation grants may be paid towards the provision, replacement or improvement of hedges, walls built of materials traditional in the area, trees or shelter belts. They are also available for bracken control and heather regeneration. Second, there are waste storage and treatment grants. Third, grants are available for a variety of energy saving measures such as thermal insulation, heat pumps, straw burning boilers and wind or water powered pumps or generators. There are also measures to encourage farmers in Environmentally Sensitive Areas to maintain certain traditional farming methods.

A subsequent package of structural measures was introduced in 1987. It permitted the provision of aids to farmers who undertake to reduce their output of beef and cereals by using more extensive farming methods, further aids for farmers in Environmentally Sensitive Areas and an extension of the principle of compensatory allowances in Less Favoured Areas (*see below*) to some crop producers who could demonstrate that their yields were low.

Given the size of the European Community it is not surprising that there are areas in which the productivity of the land is low for reasons of climate, altitude, slope, infertility or other natural difficulties. If common prices are set at levels which are appropriate to land of average productivity, then farms in these disadvantaged areas would have to be very large in order to produce reasonable incomes. This could result in a large proportion of those currently employed in agriculture leaving the land, and in some of the land no longer being used for agriculture. But this could give rise to further problems: those leaving the land might also have to leave such areas in order to find work, and the appearance of the countryside might also change significantly. In short, depopulation and conservation problems would arise, and these would affect not only the remaining residents but also the population of the Community as a whole, since these areas are often those which are heavily used by tourists. In order to overcome these problems the Community produced Directive 72/268 *on mountain and hill farming and farming in certain less favoured areas*.

In the UK the areas designated as less favoured under this directive are the hill areas in danger of depopulation in Scotland, Wales, Northern Ireland, northern England and south west England. Several types of assistance are available. Farmers with at least 3 ha of land who undertake to remain in agriculture for at least five years may receive an annual compensatory allowance based on the number of breeding beef cows and sheep kept, up to a stocking rate of nearly one and a half cows/ha. Rates of grant payable through the AIS (including environment and energy saving grants) are usually doubled. Farmers in Less Favoured Areas (LFAs) who are investing in their farms under an Improvement Plan are also eligible for grants of 25% on investments in tourism and crafts, such as altering farm buildings into holiday accommodation or workshops.

When introducing the first guidance section regulations the Community recognised that agricultural prosperity is not only dependent upon price support and efficient farm structure, but also that both producers and consumers benefit from efficient marketing. The CAP therefore provides for grants of up to 25% for projects to improve the processing and marketing of agricultural products. Grants are also available to assist the formation and operation of producer groups concerned with the marketing of fruit, vegetables and hops.

Further details of all these schemes may be obtained from Divisional Offices of the Ministry of Agriculture, Fisheries and Food.

THE PROBLEMS OF THE CAP

The European Community began as a group of six states, all of which had roughly similar agricultural histories, all of which had roughly similar agricultural structures, all of which had been used to a policy of agricultural self sufficiency over nearly a century, and all of which had roughly similar agricultural policies which did not have to be changed very much to conform to the new Common Agricultural Policy. By the middle of the 1980s the Community had expanded to 12 states, some of which were traditional exporters of farm products, some traditional importers, some of which were technically advanced, and some of which were extremely backward. It would have been surprising if the CAP had not had problems.

There are five major problems which affect the CAP. It has produced surpluses of agricultural products which are expensive to store. In the spring of 1987 intervention stocks in the Community included 11 840 000 tonnes of cereals, of which 7 309 000 were wheat, 533 000 tonnes of beef, 1 188 000 tonnes of butter and 765 000 tonnes of skimmed milk powder (figures for March and April 1987 from the IBAP, Reading). By themselves these figures mean little, but they may be set in context in various ways. Towards the end of 1986 butter stocks represented nine months of consumption, or 830% of annual commercial exports, and the cost of supporting milk products in 1985 was approaching 6 billion ECU. Cereal stocks towards the end of 1986 represented 2.6 months of consumption, or 80% of annual commercial exports, and the cost of cereals support in 1985 was approaching 2.3 billion ECU. Thus spending on agriculture as a whole has normally taken up about two-thirds of the Community budget, and the Commission predicted, early in 1987, that the resources provided within the 1.4% VAT limit would be insufficient from 1988.

The second problem is that it has created tensions between the Community and its trading partners both in the Third World and developed traditional exporters. For example, in early 1987, as a result of the accession of Spain and consequent retrictions on maize imports, there was a serious disagreement between the Community and the USA. Thirdly, the principal beneficiaries of agricultural expenditure have not been the poorer farmers, regions, or, in some cases, countries of the Community, but the richer ones.

Fourthly, it has created problems for the rural environment and for rural society, insofar as a declining farm labour force constitutes a problem. Whereas in 1960 there were about 19 million people employed in agriculture in the Community (excluding Spain and Portugal) by 1983 the corresponding figure was about 8 million. (Source: Eurostat). And finally, it can no longer be termed 'common', since as a result of the monetary arrangements of the community the prices received by farmers are now dependent upon the value of national green rates as much as upon the ECU prices decided upon by the Council of Ministers (*see* the section on Monetary Arrangements, *above*).

There are four main reasons why all of these problems exist. The first is that incomes of agricultural producers in the Community are supported by the use of the price mechanism, rather than by any direct means of income support. The second is that the measures employed by the Community to support prices involve intervention and storage for domestically produced products, with export levies and restitutions applied on markets outside the Community. The third is concerned with the way in which the Community makes decisions about the level of agricultural prices, and the fourth is the Green money system. All of the major problems of the CAP may be accounted for by one of, or a combination of, these reasons. The justifications for this apparently sweeping statement are set out below.

It should be noted that much of the following discussion compares the results of the existing CAP with the likely outcome in the absence of any farm support. This is simply a matter of convenience for purposes of comparison, in the absence of any general agreement on alternative methods of farm income support. It is not intended to suggest anything about the desirability of having some system of farm support or another. Most developed countries do support their agricultural industries by a variety of methods, and in most there are arguments about the justifiability of such a policy.

In theory, the existence of a single market price prevailing throughout the Community should have produced a pattern of production which reflected the physical, structural and economic strengths and weaknesses of each of the farming regions. Thus, to take an obvious example, a region such as southern Italy might be expected to have low production costs for tomatoes, but relatively high production costs for dairy products. Over time, therefore, it should have been possible to observe a gradual increase in the extent to which Italian farmers specialised in the production of tomatoes and the extent to which consumers in southern Italy bought milk products which had been produced in other regions of the Community. In general, it would have been expected that regions such as East Anglia and the Paris Basin would have specialised in cereals, while Mediterranean regions would have specialised in fruit and vegetable crops and the cooler, wetter north and west of the Community would have specialised in grass-based products. In practice, there has been some specialisation, mainly in products without strong market regulation, and some increase in intra-Community trade, but not very much. The maintenance of high levels of support prices throughout the Community has meant that those producers with relatively high production costs have not been forced to cease production, while those with low costs have been able to make high profits. In addition, the ideal of a Community-wide market has not, in practice, been realised. A number of factors have permitted the persistence of national markets. These include the sysem of monetary compensatory amounts, the continuation of national policy measures (from time to time it has been alleged that the total amount spent by member states on national measures was as much as total Community spending on agriculture), and non-tariff barriers to trade between member states, such as health and veterinary regulations, some of which are more easily justified than others on purely technical grounds. As a result of all of these factors low cost producers have expanded their output, high-cost producers have not been forced out of business, and production has exceeded consumption in many products.

In addition to producing surpluses, the system of income support through the price mechanism has other side effects. Since prices are kept at higher levels than they would attain in a free market and since, therefore, the more that farmers produce the more support they receive, consequently agricultural production is encouraged on land which would otherwise be left uncultivated, or only extensively cultivated. So hedges are removed, wetlands drained, moors and heaths reclaimed, and so on. Moreover, the returns to applications of fertilisers and pesticides are increased, so provoking their greater use. The value of the land affected by these developments as a wildlife habitat is thus reduced. While this might be quite acceptable when food is in short supply, it is perceived as a misuse of resources when there are surpluses of farm products and increasing numbers of people interested in landscape and wildlife as a recreational resource.

Even if the CAP might be criticised as a producer of surpluses, the resultant waste of resources might in some cases be justified if it brought about a more desirable distribution of income. This might apply to the distribution of income between individuals, regions, or even member states. It is usually assumed that to produce a welfare gain, the poor in any of these categories should gain at the expense of the rich. If the rich gain at the expense of the poor in the same or any other category, the redistribution is regressive and reduces the total welfare of society. Unfortunately, the effect of the CAP is to produce welfare losses in all three categories. Individuals lose because they pay higher prices for their food than they would if they were free to buy from any producer in the world market who was willing to sell. But the producers who gain most are those who produce most, since, in effect, support is given to each unit of output. Thus small farmers, who might be expected to have the greatest income problems, receive less support than farmers

with large businesses. Since the richer regions of the Community are those in the north, and farm products produced there happen to have higher levels of support than products produced in Mediterranean regions, the richer regions benefit more from the CAP than poorer regions. Finally, there are the gains and losses between member states. These are notoriously difficult to calculate, and have given rise to considerable political controversy. However, it is generally agreed that gains and losses do not accurately reflect differences in national incomes. This is hardly surprising, since countries which do not collect much in the form of import levies but receive significant sums in export restitutions will gain most from the CAP, and this will happen regardless of the level of national income.

The system of import levies and export restitutions also affects agricultural markets outside the Community. The effect of import levies is to discourage the sale within the Community of food produced outside it. The effect of export restitutions is to encourage the sale outside the Community of surpluses which would otherwise have to be stored within it. Thus, the demand for food from the world market is reduced while the supply of food to it is increased. Consequently world market prices are likely to be depressed. This has the effect of reducing the returns received by producers in countries which traditionally sold agricultural products to the member states of the Community.

In response to all of these shortcomings there have been numerous proposals to change the CAP from a wide variety of sources, including the Commission itself. Some have involved changing the details while retaining the basic framework of intervention and import controls. Among these are proposals for the gradual reduction of intervention prices, or to extend the system of quotas from dairy products to cereals, or to extend the A, B and C quota system used for sugar to other products. There have also been proposals to remove land from production in return for a compensatory payment, and to reduce the potential output of land by imposing fertiliser taxes or quotas. All of these measures, if adopted, would reduce the supply of farm products. An alternative approach is to increase average farm size, so that producers can maintain their total revenue from a lower level of prices, and this is the purpose of proposals to increase the impact of structural policy. Among the more radical suggestions which have been made are: the replacement of the present system with the kind of deficiency payments scheme which once operated in the UK; the replacement of intervention and import controls by direct support of farmers' incomes; and the cessation of any attempt to retain a common policy for agriculture, allowing each member state to support its own farming industry as it wishes.

Clearly, it is likely that the CAP will be changed in the future; precisely how much it will be changed, and in which direction, is much more difficult to predict. What should be remembered is that not all member states have been convinced, in the past, that change is desirable. Moreover, the CAP has, hitherto, changed only slowly.

Further reading

FENNELL, R. (1979). *The Common Agricultural Policy of the European Community*. London: Collins

HARRIS, S., SWINBANK, A and WILKINSON, G. (1983). *The Food and Farm Policies of the European Community*. Chichester: Wiley

AGRA EUROPE (weekly)

AGRA EUROPE, The CAP Monitor (continuously updated)

24

Farm business management

Martyn Warren

WHAT IS BUSINESS MANAGEMENT?

Business management is difficult to define precisely, and is best described in terms of the activities performed by the manager of a business – whether employee or owner.

A manager's job is to make decisions – not just the day-to-day ones that everyone has to make, but decisions concerning the next month, the next year, or even the next ten years. The manager is deciding how best to use the resources at his or her disposal in trying to achieve the objectives of the business.

A manager's job is also concerned with ensuring that his decisions are having the desired effect. The outcome of each decision must be measured and checked to see that it matches up to expectations. If actual progress does not compare well to the expected performance, the manager should take the appropriate action to bring the two into line.

Viewed in this way, the manager's job can be seen as *planning* (making decisions which affect the future operation of the business) and *control* (monitoring progress of the decision and taking necessary corrective action), in order to achieve the *objectives* of the business.

In a one-person business, where the owner is also the manager and the work-force, the business has only to satisfy the owner's objectives. In most farm businesses, however, there are other people involved – family, partners, employees, maybe even a paid manager. These people have their own objectives, and this adds another dimension to the manager's job – making sure that everyone is working towards the same end. A major objective of many farm workers, for instance, is to do a job to be proud of. The task of the manager is to use that fact to help the business achieve its goals rather than allow it to cause a nuisance.

This dimension of the manager's job – marshalling resources, coordinating and harmonising objectives and efforts, can be summed up as *organisation*. Management thus becomes a process of planning, organising and controlling (*Figure 24.1*).

This chapter is concerned largely with the planning and control functions: the following chapter includes some of the skills needed in organisation.

SETTING PRIMARY OBJECTIVES

Setting objectives is the key to the management process. Unless the manager is sure what he or she is trying to achieve, the business can have no clear direction.

Objectives can be roughly grouped into primary and secondary objectives. Primary objectives are the overall aims of the business – the long-term goals for which the business is striving. Secondary objectives are the shorter-term goals that the business needs to meet in order to achieve the primary objectives.

A primary objective might be, for instance, to increase the wealth of the business by 10% in the next five years: one of the associated secondary objectives might be to achieve an average wheat yield of 10 t/ha over this period.

Primary objectives are reflections of the long-term expectations of the manager. These expectations will range far and wide, but for convenience can be classified into personal, physical, responsibility and financial.

(1) *Personal expectations* are those which are concerned with the lifestyle and personal preferences of the manager. They often include time for family, social and leisure activity, and acquisition of status or influence through, for instance, being a magistrate or county chairman of the NFU.

Figure 24.1 The management process

(2) *Physical expectations* relate to the productivity of the farm, to innovation, and to general technical prowess.

(3) *Responsibility expectations* involve satisfaction of the responsibility the manager feels towards other groups in society. These can include employees, customers, neighbours and the local community.

(3) *Financial expectations* relate to the generation of income and wealth. These are the crucial expectations, not because they are particularly important in themselves (although some people are highly motivated by the acquisition of money for its own sake), but rather because they govern the ability of the business to satisfy the personal, physical and responsibility expectations.

It is because of the importance of financial expectations that the following sections relate particularly to financial management.

A further reason is that, to plan and control effectively, a manager needs a widely accepted and flexible measure by which to judge performance. Money provides that measure.

Financial measures

Where the finances of a farm are concerned, three factors are crucial: survival, efficiency and stability. These will dictate the financial objectives of the business, and the way in which the manager plans and monitors progress.

(1) Survival, in financial terms, is a matter of having sufficient cash available at any time to pay for essential inputs, such as labour, fertilisers, diesel, finance charges and so on. Cash availability is measured through net *cash flow* – the difference between cash coming into the business and cash going out.

(2) Operational efficiency is also reflected in the money coming in and going out of the business. There are other factors to take into account, however, such as changes in the levels of stocks and of bills outstanding. The financial measure that attempts to measure efficiency by taking these and other factors into account is known as net *profit*.

(3) Financial stability is imparted by wealth. Wealth implies ownership of assets – money and things of value – but it must also take account of outstanding debts, or liabilities. A business which has £1 000 000 assets and £999 000 debts is likely to be less stable than one with £10 000 assets and £50 debts. The appropriate measure of financial stability is thus net *capital* – the difference between assets and liabilities.

PLANNING TO ACHIEVE PRIMARY OBJECTIVES

Once the primary objectives have been established, the planning process can get under way. The first, and usually most protracted, stage is to formulate an overall or long-term plan. The time-period covered by such a plan will be one or more years, and it will indicate the decisions and actions needed for the business to have the best possible chance of achieving the primary objectives.

This process relies much on subjective judgement and can never be made foolproof. A structured approach can, however, minimise the risks of making grossly erratic judgements. One such approach is as follows:

(1) Assess the farm and its resources.

Take a hard look at the *strengths* and *weaknesses* of the business. Assess the quantity and quality of the resources available, such as the land, buildings, expertise of the employees, ease of access to markets, and so on. Take account of factors that could limit freedom of action; not only physical and financial factors such as the slope of the ground and the amount of capital available, but also legal and administrative constraints such as town and country planning laws, restrictive clauses in a lease, or the likelihood of obtaining a quota or contract.

(2) Appraise the present and future environment of the business.

Identify the *opportunities* that are open to the business, and the *threats* to its welfare. There is more to this than keeping abreast of technological change. It involves being aware of the economic and political climate affecting the business, and making judgements about how that climate is likely to change during the period planned for (*see* Chapter 23). It also requires an awareness of changes in the habits and attitudes of society in general (e.g. food consumption habits and attitudes towards farming). The likely actions of competitors must be anticipated. The search for opportunities should not be confined to farming, but should include all possibilities for using the business resources effectively.

(3) Formulate alternative plans

List all the possible ways of using the resources of the business to achieve its primary objectives. 'Possible', in this context, must be judged by reference to the strengths and weaknesses, opportunities and threats listed earlier. It is essential to look further than the 'obvious' solution, and consider as wide a range of alternatives as possible. This is, in many ways, the most difficult part of the planning process: it is hard to be objective and imaginative about a business which one knows intimately. One way of overcoming this problem is to enlist the help of family and/or friends in a 'talking shop' aimed at stimulating new ideas.

(4) Appraise the alternative plans and select the one most likely to succeed.

The likely outcomes of the alternative plans must be measured and compared. Given the importance of financial objectives, and the fact that money provides a common unit of measurement, it is critical at this stage to prepare budgets. A budget is a calculation of the financial effects of a course of action, and given that the financial objectives of a business can be measured in terms of cash flow, profit and capital, budgets should be expressed in these terms.

The plan whose budgets indicate the nearest result to the financial objectives will then be chosen, unless another plan more nearly achieves the non-financial objectives specified, and the manager is content to forgo the extra financial benefit for the sake of achieving non-financial expectations. He or she may happily put up with a low-key, low-profit system for the sake of minimising mental and physical strain.

The resulting overall plan can then be used as a focus for short-term planning. The manager will need to:

(1) set secondary objectives

These are the targets which must be achieved for the overall plan to work. They may be expressed purely in financial terms, as the margin which a particular enterprise should achieve in a given year. They may also include key physical targets, such as the yield produced by a major enterprise.

(2) formulate short-term plans

This is a continual process aimed at enabling these targets to be achieved. It ranges from formal budgeting of enterprise margins for a season, or cash flow for a month, down to day-to-day decisions about organisation of the workforce.

Once plans have been formulated and acted upon, the control process can start. On a week-by-week, month-by-month basis, progress can be monitored by comparing key results with the short-term targets (most valuable in dairy and intensive livestock enterprises).

On an annual basis, control will involve monitoring the progress of overall plans. Given the importance of financial objectives at this level, this implies comparing financial expectations with what has actually happened. This *budgetary control* can only take place if the manager has available two key sets of information – the *budgets* which show the detail of financial expectations, and the *accounts* which record the actual performance. Both budgets and accounts use the same measures – cash flow, profit and capital.

INFORMATION FOR FINANCIAL MANAGEMENT

Budgets

Budgeting for profit

A profit budget in conventional form

A budget shows the expected financial performance of a business over a future time period. This period is normally a year and, for convenience, is usually the financial year used by the farm's accountant. Many farm financial years start from Michaelmas (end of September) or Lady Day (end of March): other popular dates are the beginning of January and the beginning of April.

The easiest way of beginning the budgeting process is to start from the practical basics of the farming system – how many hectares of crops are to be grown during the year in question, how many head of livestock are to be kept?

On this framework can be built a detailed picture, by estimating the quantities of inputs which are likely to be consumed and their prices, and the quantities of products which are likely to be produced and their prices.

To reflect fully the business performance of the farm, though, some adjustments need to be made to these figures. The first is that allowance should be made for any goods or livestock likely to be on the farm at the end of year. Although these 'stocks' will not yet have been sold, they will have been produced or purchased by the business during the year, and thus ought to be included. Similarly, any bills yet to be paid at the end of the year (creditors) or income due from goods which have been sold and not yet paid for (debtors) must be included as 'belonging' to the budget year. Conversely, any stocks, creditors or debtors likely to be outstanding at the *beginning* of the budget year are taken out

of the calculation – they 'belong' to last year's production.

The second type of adjustment is to remove any payment or receipt concerned with personal, tax or capital matters. The amount the owner of a business takes for his or her own use is not determined by the efficiency of the business, but by lifestyle: such the same applies to tax payments, which also vary according to how good the accountant is and the policy of the prevailing government. Capital items relate to lump sums, such as the receipt and repayment of loans, and purchase of land, buildings and machinery. Since these do not arise out of the normal trading of the business, they are left out of the calculation, with the exception of 'wasting assets' such as buildings and machinery. The cost of these is spread over their useful life in the business: the resulting annual cost is known as 'depreciation' (see 'estimating outputs and costs' below).

Finally, any farm products consumed without payment by the farmer, family, and employees should be added in. These 'benefits in kind' or 'household consumption' are just as much part of the farm production as those sold. Similarly, goods and services paid for on the farm account but consumed without payment should be credited back to the farm.

The result (shown in *Figure 24.2*) is a budget showing the expected net profit for the year concerned. (*Please note that all the tables in this chapter are based on the same 'model farm' and can be directly related to each other.*)

Enterprise budgeting

A profit budget of the type just illustrated is useful in a limited way. One can tell whether the operation of the farm is likely to make a positive or negative contribution to the owner's financial welfare (i.e. profit or loss). It is also possible to see whether the result represents an improvement on previous years. It is difficult, however, to see how particular costs and returns relate to each other, and to see how the financial quantities reflect physical performance. The *value* of milk produced is easily apparent, for instance, but not the *amount* of milk produced in total and, more important, per cow.

To overcome these problems, it is possible to rearrange the information in the budget on an enterprise basis. Conceptually, the easiest way of doing this is by calculating a net profit for each enterprise (an enterprise being an output-producing part of the business, such as a particular crop or livestock type, or a service such as holiday accommodation or contracting).

The form of enterprise accounting most commonly found in UK farming is the gross margin system. This involves identifying for each enterprise the value produced (the *output*) and the *variable costs*. The latter are costs which both are easily attributable to the enterprise concerned, and also vary in direct proportion to small changes in the scale of the enterprise. Thus for a livestock enterprise, livestock purchase, feed, veterinary and medical costs, transport, and so on are regarded as variable costs. Each one is readily identifiable with the enterprise, and a 5% increase in the size of the flock or herd, for example, would tend to increase each of these by a similar proportion. Typical variable costs for a crop enterprise include seed, fertiliser and spray costs.

Subtracting the variable costs from the output gives, for each enterprise, a *gross margin*. When the gross margins from the various enterprises are added together, the farm overhead or *fixed costs* can be deducted to give the net profit

OPENING VALUATION OF STOCKS	£	CLOSING VALUATION OF STOCKS	£
Dairy cows	50000	Dairy cows	50800
Calves	280	Calves	280
Crops in store	42700	Crops in store	41950
Crops in ground	3000	Crops in ground	3000
Miscellaneous stocks	9000	Miscellaneous stocks	9000
TOTAL VALUATION (a)	104980	**TOTAL VALUATION (b)**	105030
EXPENDITURE		**REVENUE**	
Feed	7200	Milk	98159
Livestock	14400	Calves	7700
Vet and med, AI, etc.	4800	Cull cows	8750
Fertilisers	14740	Grain	30000
Seeds	3590	Straw	2500
Sprays	4120	Miscellaneous	1500
Regular labour	20000		
Machinery running costs	15000	Benefits in kind	1151
Rent and rates	14000		
Miscellaneous costs	10650		
Bank interest	10633		
Depreciation	9000		
TOTAL EXPENDITURE (c)	128133	**TOTAL REVENUE (d)**	149760
BALANCE - NET PROFIT	21677		
	149810		149760

Figure 24.2 Example of profit/loss statement

for the business (*Figure 24.3*). This net profit should be the same as that shown by the budget in conventional form described above – it is only the layout of the information that has changed, not the information itself.

The main advantage of this method is that it avoids the allocation of the fixed costs between enterprises. Such allocation can involve difficult and arbitrary decisions, particularly when general farm costs are concerned – such as the cost of the farm Landrover, office expenses, finance charges, etc. Gross margins are also useful when considering small changes to the farm system. The financial effects of switching one field from spring barley to winter wheat can be seen by substituting the gross margin of one crop for that of the other in the budget (*Figure 24.4*).

There are situations when the gross margin method is insufficient. If it is important to know exactly how much a particular enterprise contributes to the business (when pricing a contract, for instance), it will be necessary to calculate a *net margin*. In its extreme form, this involves allocating *all* the costs of the business (including the general overheads mentioned earlier) between the enterprises. A useful compromise is to allocate to enterprises all those fixed costs which can easily be attributed, leaving only such costs as general overheads and finance charges to be deducted on a whole-business basis (Giles, 1986).

Where the farm is large enough to support its own secretarial services and/or its enterprises are fairly self-contained, the benefits of using this system can outweigh the effort involved in allocating labour and machinery costs (including analysis of timesheets when recording). For most general-purpose management accounting, though, the gross margin system remains a valuable compromise, provided that fixed cost implications are not overlooked.

Preparing an enterprise gross margin budget

Estimating an enterprise gross margin involves two tasks – identifying the information needed, and organising the information so that it can easily be used.

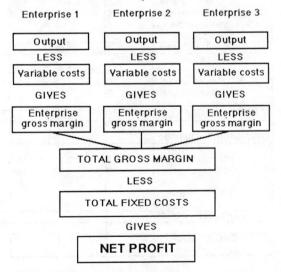

Figure 24.3 The gross margin system

	£
Gross margin gained: 10 ha wheat at £467.00/ha	4670
Gross margin lost: 10 ha barley at £327.50/ha	3275
NET GAIN	1395

Figure 24.4 Net gain from substituting 10 ha winter wheat for spring barley

Calculating the likely output of an enterprise needs information about any value produced by that enterprise. As well as sales (adjusted for changes in debtors), therefore, the calculation must include transfers of products between enterprises within the business. Thus home-grown barley fed to livestock enterprises must be valued at market value at time of transfer, and credited to the barley enterprise as part of its output. In addition, any produce consumed as benefits in kind (milk consumed in the farmhouse, for instance) should be counted as output. Finally, allowance should be made for any increase or decrease in the valuation of stocks of output items (crops in store or in ground, livestock and livestock products).

Similar conditions apply to the variable costs. Allowance must be made for transfers of home-grown inputs from other enterprises, for household consumption, and for changes in stock valuation of each input.

The manner of organising this information is partly a matter of taste, but the important criteria are that the result should be easy to read, and show readily the type of information that is important to the manager. One way of laying out a crop gross margin is shown in *Figure 24.5*. Note that the five columns give information not just on financial amounts in total and per unit of production, but also physical in-

formation such as yields and usage per head or hectare. This degree of detail is necessary for the budget to be used effectively in budgetary control, as it enables the manager not only to identify discrepancies between actual and planned results, but also to explain how they have arisen.

Figure 24.6 shows a similar calculation for a breeding livestock enterprise. Note that the cost of buying and transferring in livestock, though strictly a variable cost, is deducted in calculating the enterprise output. There is no particular reason why you should follow this practice, but it has become an established convention in agricultural management, and most published gross margin information is expressed in this way.

When budgeted gross margins have been prepared for each of the planned enterprises, the net profit can be calculated by totalling the gross margins and deducting the fixed costs. The major fixed costs are likely to be those of regular labour, machinery running costs and depreciation, rent and/or property depreciation and maintenance costs, and finance costs such as interest and leasing charges. Overdraft interest charges, being dependent on fluctuating overdraft levels, can be left on one side for now. This results in a budgeted net profit before overdraft interest: the final net profit can be calculated after using the cash flow budget to estimate interest costs.

Estimating outputs and costs

The value of a budget depends on the quality of the information used in its compilation. Estimates of outputs and variable costs can be built up from forecasts of prices and quantities. The best guide for yields and quantities of inputs is usually the performance of previous years, taking bad years with the good. A good farm management handbook will also help, particularly where the manager has little previous experience of a particular enterprise (*see*, for instance textbooks mentioned under Farm management

Ha 50	Total Quantity	Tonnes/ Hectare	Price/ Tonne	Total £	£/Ha
Grain					
Sales	300	6.0	100	30000	600.00
Transfers out	0				
Valuation change	0				
Private consumption	0				
Total grain	300	6.0	100	30000	600.00
Straw					
Sales	125	2.5	20	2500	50.00
Transfers out	0				
Valuation change	0				
Private consumption	0				
Total straw	125	2.5		2500	50.00
ENTERPRISE OUTPUT				32500	650.00
Fertiliser (nitrogen)	13	0.3	120	1500	30.00
Fertiliser (compound)	23	0.5	100	2250	45.00
Seed	11	0.2	220	2420	48.40
Sprays				2750	55.00
Other costs				230	4.60
VARIABLE COSTS				9150	183.00
GROSS MARGIN				23350	467.00

Figure 24.5 Example of gross margin calculation for winter wheat

Average numbers 120	Total Quantity	Quantity/ Head	Price/ Unit	Total £	£/Head
Milk					
Sales	648000	5400	0.151	98159	817.99
Private consumption	1000	8.3	0.151	151	1.26
Total milk	649000	5408	0.151	98311	819.25
Calves					
Sales	110	0.9	70	7700	64.17
Transfers out	0				
Valuation change	0				
Total calves	110	0.9	70	7700	64.17
Cull sales					
Sales	25	0.2	350	8750	72.92
Valuation change	2			800	6.67
Less l/stock purchase	-32	-0.3	450	-14400	-120.00
Net replacement cost	-5			-4850	-40.42
ENTERPRISE OUTPUT				101161	843.01
Concentrate feed: homegrown	100	0.8	90	9001	75.01
Concentrate feed: bought	60	0.5	120	7200	60.00
Total concentrates	160	1.3	101	16201	135.01
Vet and med, AI				4800	40.00
Bedding: home-grown	40	0.3	30	1200	10.00
Other				2160	18.00
Forage costs				10890	90.75
VARIABLE COSTS				35251	293.76
GROSS MARGIN				65910	549.25
Stocking rate (head/ha)		2.2	Gross margin/ha		1208.35

Figure 24.6 Example of gross margin calculation for a dairy enterprise

data, p. 623. The biggest danger in estimating is that of being too optimistic – the most realistic estimate is usually produced by incorporating a healthy pessimism into budget figures.

Prices are affected by a great many factors, including the state of world trade, the political climate, changes in consumption habits, and quality of the product. The individual manager is unlikely to have the time or the expertise to weigh up all these factors for every commodity. The best procedure is to make use of the opinions of those who *do* have the time and expertise. Economic researchers are employed by a number of organisations: their conclusions are frequently reported in the farming and other media. The job of the manager then becomes one of locating the various price forecasts, reconciling any conflicts between them, and relating them to the business in question.

In estimating fixed costs, previous years can again be a useful guide, allowing for intervening wage awards and increases in prices. The handbooks mentioned above contain guidance where previous years' figures are not available or not relevant (where the system has been changed, for instance). Tables show the likely number of man- and tractor-hours needed for particular enterprises, from which can be calculated likely running costs.

Two methods are in common use for estimating depreciation. The straight-line method takes the original cost of the asset (machine, building or property improvement) divided by the likely life of the asset. Thus, an asset costing £10 000 and expected to last 20 years would have an annual deprecia-

tion of £500 per year. This method is useful for budgeting several years in advance, the constant annual payments making estimation quick and easy. It does not, however, result in a realistic estimate of the depreciation of machines, which tend to lose value more quickly in the first years of their lives than in the later. To allow for the latter effect, the diminishing balance method can be used. Here a constant annual percentage depreciation rate is applied to the depreciated value from the end of the previous year. *Figure 24.7* shows the depreciation pattern of an asset costing £10 000, depreciated at rate of 20% per year.

Ideally, the depreciation rate should be assessed individually for each machine, taking account of its likely life and resale value. Some farm management handbooks contain tables of appropriate rates. In practice, this is too complicated a procedure for most management purposes, and in both budgets and accounts, machines and buildings tend to be grouped or 'pooled' with other similar assets. A depreciation rate is chosen for the type of assets in the 'pool' (e.g. tractors), and is applied to the written-down value from the last year, adjusted for sales and additions during the current year. *Figure 24.8* illustrates the process.

Assessing the profit budget

Before moving on to the cash flow budget, it is important to make an initial assessment of the implications of the profit budget. The first obvious check is whether the budget shows a net profit or a loss, and whether that figure matches up to

Year	Opening Value	Depreciation	Closing Value
1	10000	2000	8000
2	8000	1600	6400
3	6400	1280	5120
4	5120	1024	4096
5	4096	819	3277
6	3277	655	2621
7	2621	524	2097
8	2097	419	1678
9	1678	336	1342
10	1342	268	1074
11	1074		

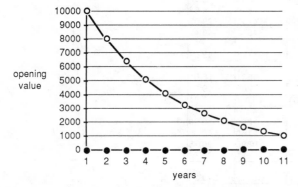

Figure 24.7 Example of diminishing balance depreciation

	£
Value of machinery at beginning of year	30000
plus purchases of machinery	20000
less sales of machinery	-5000
Value before depreciation	45000
Depreciation at 20%	9000
Value of machinery at end of year	36000

Figure 24.8 Example of 'pool' calculation

the profit objective. Another useful check is whether the profit is sufficient to cover basic family drawings and known tax commitments. A comparison between the result of the budget and known results for previous years will help to give a 'feel' as to whether the budget is realistic (it is very easy to be over-optimistic in budgeting).

It is likely that such an assessment of the budget will indicate deficiencies, and encourage adjustments in the projected gross margins as a result. Indeed, several such adjustment stages may be necessary before the manager is sufficiently satisfied (or convinced that no more profitable plan is possible) and moves on to test the effects on cash flow.

Budgeting for cash flow

Once a profit budget in gross margin form has been compiled, preparation of a cash flow budget is relatively easy. The aim is to present a picture of the flow of cash in and out of the business over the budget year.

An outline cash flow budget is illustrated in *Figure 24.9*. For each month, the likely payments and receipts of cash are estimated. By taking the total cash payments from the total cash receipts for the month, an estimate is formed of the likely net cash flow for that month. When the net cash flow is added to the bank balance at the beginning of the month, the estimated bank balance at the end of the month is achieved. If this is done for each month of the budget year, the result is a picture of the way in which the bank balance of the business is likely to vary during that year.

The first stage in compiling a cash flow budget is to make a list of cash receipts and cash payments likely to affect the business bank balance during the budget year. A profit budget, especially one in gross margin form, makes an excellent basis for this list. It is important to remember,

	TOTAL	Oct	Nov	Dec	Jan	Feb	Mar	Apr	May	June	July	Aug	Sep
Income													
Milk	97803	8910	8894	8823	8751	8680	8609	8446	7705	6530	6417	7430	8608
Calves	7700	2100	1400	1050	700	350			350	700		700	1400
Cull cows	8750		350	350		350		700	350	700	3850	2450	
Wheat	30000				30000								
Wheat straw	2500	2500											
Sundries	1500		300			200		500	200	200		300	
VAT charged on outputs	0												
VAT refunds	0												
Capital - grants	0												
machinery sales	5000				5000								
Personal receipts	0												
SUB-TOTAL	153253	13510	10594	10223	44451	9580	8609	9646	8055	7430	10267	10880	10008
Payments													
Feed: cows	7200	3000		2500		1000		700					
Vet and Med	4800	500	700	600	600	400	300	200	200	200	200	400	500
Livestock purchases: cows	14400										6750	3600	4050
Misc. dairy costs	2160	180	180	180	180	180	180	180	180	180	180	180	180
Seed	3590	2420					1170						
Fertiliser	14590		14590										
Spray	4120		4120										
Misc. crop variable costs	490								490				
Wages - permanent	20000	2000	1500	1500	1500	1500	2000	1500	1500	1500	1500	2000	2000
Power and machinery	15000	2000	1000	1000	1000	1000	1000	2000	1000	1000	1000	1000	2000
Rent and rates	14000						7000						7000
Misc. fixed costs	8000	650	750	650	650	650	650	750	650	650	650	650	650
Overdraft interest	6357			1626			1630			1498			1604
Loan interest	4276		2138							2138			
Capital - buildings	0												
machinery	20000				20000								
capital repayments	7000			3500						3500			
Personal - drawings	15000	1250	1250	1250	1250	1250	1250	1250	1250	1250	1250	1250	1250
tax	5000				2500						2500		
SUB-TOTAL	165983	12000	26228	12806	27680	5980	15180	6580	5270	11916	14030	9080	19234
NET CASH FLOW	-12730	1510	-15634	-2583	16771	3600	-6571	3066	2785	-4486	-3763	1800	-9226
OPENING BANK BALANCE	-50000	-50000	-48490	-64124	-66707	-49936	-46336	-52907	-49841	-47056	-51542	-55305	-53505
CLOSING BANK BALANCE	-62730	-48490	-64124	-66707	-49936	-46336	-52907	-49841	-47056	-51542	-55305	-53505	-62730
Overdraft interest @ 12%	500	485	641	667	499	463	529	498	471	515	553	535	627

Figure 24.9 Example of a cash flow budget

however, that one is now concerned only with cash items. Changes in debtors, creditors and stock valuations must be excluded, as must depreciation and benefits in kind. All personal, tax or capital receipts or payments must be included.

Once the list is complete, the pattern of timing of receipts and payments can be worked out. Each type of receipt in turn is examined, and a judgement made concerning the way in which that receipt will be split between the months of the year. With many items, this will require careful estimation of the pattern of physical production – the way in which milk receipts will occur during the year, for instance, will be governed by the combined lactation curves of the cows in the herd.

It is a normal trade practice to allow between two and four weeks' credit on sales, so many cash receipts will be paid into the bank the month after the actual sale has taken place. This must be allowed for in deciding in which month the cash is likely to enter the bank account. Payments by opening debtors should be assumed to be received in the first month of the year.

The process is repeated with each payment item, again ensuring that allowance is made for credit taken. It is sometimes possible to negotiate extended credit (three months, for instance) with suppliers of items such as fertiliser and machinery. This must be allowed for in deciding timings.

The budget can now be completed by entering receipts and payments in the body of the table according to the timings decided, and calculating monthly bank balances. Beginning with the column for the first month of the budget year, total payments for the month are subtracted from total receipts to give the month's net cash flow. When added to the bank balance expected for the beginning of the year, this gives the expected bank balance for the end of the first month. A positive figure denotes money in the bank: a negative figure indicates an overdraft.

This closing bank balance for the first month becomes the opening balance for the second month, and the above calculation is repeated for this and each of the remaining ten months of the year. If the closing balance of any month is negative (i.e. shows an overdraft), the interest due for that month can be estimated by multiplying the balance by one-twelfth of the likely overdraft rate. Interest is generally charged quarterly in March, June, September and December. Thus the 'overdraft interest' entry for December

will show the estimated interest for the months September to November.

A vital part of the budget is the 13th column, headed '*total*', used to summarise the cash flow for the whole year. This is compiled by adding the contents of each line across the sheet, and then finding the annual net cash flow and closing balance as for a monthly column (the opening balance being that used in the first month's column. As well as providing a useful summary of the year's expected cash flow, and a means of cross-checking with the contents of the profit budget, it provides an important internal check. If the closing balance calculated in this manner matches that shown in the final month's column, one can be confident that the arithmetic in the budget is correct. If it does not, the calculations must be checked and rechecked until the error is located.

A valuable aid to the interpretation of the cash flow budget is a graph of the monthly closing bank balances, as shown in *Figure 24.10*.

Assessing the cash flow budget

Before moving on to the budget for capital, it is important to weigh up the implications of the cash flow budget. As before, the main test is whether the budget indicates that the cash flow objective (probably expressed as the maximum tolerable overdraft) will be achieved. A parallel test is whether the overdraft at any time is likely to exceed the existing limit agreed with the bank manager. The next concern should be to check for significant variations in bank balance between months, and for any unusual movements in key items.

If these checks indicate potential problems, there are four possible strategies open to the manager. Ways can be sought of *increasing cash flow*, and thus lifting the overdraft away from problems. Possibilities include cutting or postponing private drawings and capital expenditure, and a general reduction of overhead costs. Another possibility is that enterprise gross margins could be examined for possibilities of increasing output and/or reducing variable costs, although most avenues should have been explored when assessing and adjusting the profit budget.

The second opportunity is that of *delaying payments* to reduce overdraft peaks. Buying fertiliser as it is needed, for instance, or negotiating longer trade credit from a machinery

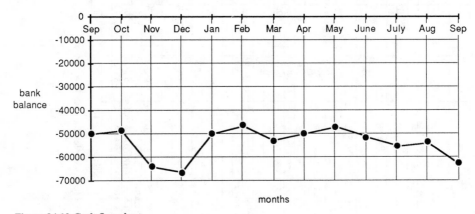

Figure 24.10 Cash flow chart

supplier, could avoid the overdraft limit being exceeded and help to cut interest charges.

Another way of reducing overdraft peaks is by *advancing receipts*, such as selling the wheat off the combine rather than storing, or pressing debtors for earlier payment.

These possibilities should be tested by reworking the cash flow budget and seeing the effect on the cash flow. Each change will have a cost – lost discounts on fertiliser, lower prices on corn sold early – and the costs and benefits of the change should be examined (a partial budget is an ideal tool for this – see 'Planning for Change' later in this chapter). Some changes antagonistic to the manager's objectives (for instance pressing debtors for prompt payment, or cutting personal expenditure) may be totally unacceptable.

The final possibility, when all other options have been considered and the overdraft is still over the bank's existing limit, is to seek to negotiate a new overdraft limit. The existence of a carefully worked and reworked cash flow budget will be of great help in this process

If this fails, and/or the new limit goes against personal objectives with respect to the maximum borrowing, a complete rethink of the business objectives and plan is necessary.

Otherwise, the next stage is to prepare a budgeted balance sheet.

Budgeting for capital

The budgets for profit and cash flow provide the raw material for a projected balance sheet. This is a statement of the wealth of the business at a particular time. Wealth denotes ownership of assets – money and other things of value, such as stocks, machinery and land. Against these assets must be set liabilities, however – debts such as outstanding trade creditors, overdraft and loans.

A typical farm balance sheet is shown in *Figure 24.11*. Assets are listed on the right, liabilities on the left. The difference, or balance, is shown beneath the liabilities as *net capital* (also known as 'net worth' or 'owner equity'). This figure represents the net value of the business to its owner(s). In a company business net capital will be expressed in a

slightly different form, as a combination of share capital and reserves.

The figure used for bank balance (asset side) or overdraft (liabilities side) is the closing balance from the last column of the cash flow budget. The debtors and stock valuations are those used previously in compiling the profit budget. The values of machinery and buildings are those after depreciation (used in the profit budget). The value of land can be taken as that at the beginning of the year, plus or minus any sales or additions during the year (which will be shown in the cash flow budget).

Apart from the overdraft, the liabilities side includes creditors, which have already been determined in compiling the profit budget. The value of loans can be taken as the amounts outstanding at the beginning of the year, plus or minus any additions or repayments during the year (shown in the cash flow budget).

Figure 24.11 also shows the previous year's balance sheet figures (the 'opening' figures) in separate columns to the left of the closing assets and liabilities, thus enabling an easy assessment of likely changes over the year.

A further refinement is the inclusion of a capital account. This traces the growth or otherwise of net capital over the year, by reference to profit and injections of money from outside the business (both adding to net capital), and private drawings, tax payments and benefits in kind (all reducing net capital). As well as showing the relationship between profit and drawings, the capital account acts as a useful arithmetical check. The result of the calculation should be identical with the closing net capital shown by the balance sheet: if it is not, the budget calculations should be checked and rechecked until the error is found.

Assessing the projected balance sheet

Much of the message of the projected balance sheet can be seen at a glance. A positive net capital (i.e. assets more than liabilities) confirms that the business is still likely to be solvent (in business) at the end of the year. Comparison of the opening and closing balance sheets shows how the financial position of the business is likely to change over the

OPENING	LIABILITIES	£	£	£	OPENING	ASSETS	£	£
	PROJECTED BALANCE SHEET AT 30/9/198Y							
	Long-term Liabilities					**Fixed Assets**		
32000	Long-term loans			25000	70000	Bungalow	70000	
					30000	Machinery and plant	36000	
	Current Liabilities				50000	Breeding l/st: Dairy cows	50800	
50000	Overdraft		62730		150000	TOTAL FIXED ASSETS		156800
1200	Trade creditors		1349			**Current Assets**		
51200	TOTAL CURRENT LIABILITIES			64079				
					280	Trading livestock (calves):	280	
83200	TOTAL LIABILITIES			89079	39000	Crops in store: grain	38100	
					3700	straw	3850	
	Capital Account				3000	Crops in ground:	3000	
	Opening net capital	130690			9000	Miscellaneous stores	9000	
	Net profit (loss)	21677						
	Capital injections		152367		8910	Trade debtors	9265	
					0	Cash in bank	0	
	Private drawings	15000			0	Cash in hand	0	
	Income & capital tax payts	5000						
	Benefits in kind	1151	21151		63890	TOTAL CURRENT ASSETS		63495
130690	**CLOSING NET CAPITAL**		131216	131216				
213890				220295	213890	**TOTAL ASSETS**		220295

Figure 24.11 Example of a balance sheet (tenant)

budget year – whether net capital is likely to increase, how much total assets and total liabilities are likely to change, and to what degree major changes in individual items are responsible.

Additional clues can be derived from the application of various ratios. Space precludes mention of all such, and indeed it is doubtful whether it is useful to know more than a few easily-remembered calculations (but *see* Hutchinson and Dyer (1979) and Warren (1986) for more detailed discussion). To check on overall stability of the finances, the *percentage owned* can be calculated (net capital as a percentage of total assets). This both shows the owner's stake in the business, and also gives an impression of the 'safety margin' enjoyed by the business. A business with 40% owned has rather less buffer against changes in financial fortunes than one with 80% (typical of many long-established owner-occupied farms). The degree to which this percentage is likely to change during the course of the budget year is a vital indication of the financial health of the business. The example farm (*Figure 24.11*) has a low and declining percentage owned of 60%.

An additional overall check is provided by the *gearing ratio*. This expresses long-term borrowing (loans and any other finance with a fixed repayment pattern) as a percentage of the net capital. Such borrowing imposes demands for both interest and repayment, irrespective of the general cash flow of the business. Thus the higher the gearing, the higher the risk faced by the business, and the greater the need for a high return on capital (*see below*) to cope with these demands. In an occupation such as farming, which commonly gives low returns on capital, a gearing above about 25% would give rise to concern – but this is only a very general guide. The example farm has a gearing ratio of 19% – close to the worrying state.

Short-term stability can be measured by the ratio between the liquid assets of the business (very short-term assets such as cash and trade debtors) and current liabilities (those at short call such as overdraft and trade creditors). This is the *liquidity ratio*. A ratio of 1:1 indicates that even if the current liabilities were suddenly called in, the business would have enough short-term reserves to cope. The lower this ratio, the more susceptible is the business to running out of cash. As before, changes over the year are more important than static measures – a declining liquidity ratio is a warning signal. The example farm has a liquidity of only 0.14, and this is down on the opening figure of 0.17.

Finally, budgeted net profit and net capital may be linked by calculating a return on capital, such as *return on owner equity*. This is net profit expressed as a percentage of net capital, and indicates how efficiently the owner's capital is used within the business. The farmer who finds that his return on net capital will be less than 3% while he could get an assured 10% by investing in government securities (say), needs to ask himself some searching questions – such as whether satisfying personal objectives (for instance pleasure in running a farm) is worth the financial cost involved. Profits can be highly variable between years, and before making a decision based on return on capital, outline budgets should be used to check likely trends over the following two or three years. The example farm has a return on net capital of 16.5%, which will go some way to compensating for the low stability measures.

As with the other budgets, if the picture indicated by the assessment of the projected balance sheet does not meet the objectives of the owner, alternative plans must be investigated and budgeted. First, though, the valuations of the assets in the balance sheet should be checked. If these are based on past balance sheets, asset values may be out of date. This should be checked, assets conservatively revalued, and ratios recalculated, before any decision to rebudget is made.

Accounts

While budgets estimate the likely future effect of decisions, accounts provide the historic information to show the actual effects of those decisions.

Recording cash flow

The primary source of historic information about a business over a year is the record of the business transactions made during that year. Various methods exist of making this record, ranging from merely collecting invoices in a cardboard box, to using a computer accounting system. That which is most commonly found on farms is the *cash analysis* system of book-keeping.

As its name suggests, this is a method of recording net cash flow. It is generally based on monthly intervals, recording expenditure and receipts as the payment is made or received. The main features of a cash analysis book are shown in the example of a monthly payments and receipts analysis shown in *Figure 24.12*. The broad pages accommodate a large number of columns. Those at the left are used to record the essential information about each transaction: date, details, cheque number (paying-in slip number for receipts) and the amount of money involved.

The remaining columns allow the amount paid or received to be categorised, ideally by reference to eventual use in gross margin accounts. Thus payment analysis columns could be divided into variable and fixed cost groups, with the former including columns for feed, livestock purchase, veterinary and medical, seeds, sprays, fertilisers, contract hire and so on, and the fixed cost columns including those for regular labour, machinery running costs, rent and rates, finance charges, and miscellaneous overheads. In addition to these 'trading' items, some of the transactions will relate to personal, capital or tax items (VAT, income and capital taxes): hence the columns at the right of the sheet.

The receipts sheet is laid out in a similar manner, with the analysis columns categorising receipts by type of output. The extent of the division is limited partly by the number of columns on the sheet, and partly by the need for clarity and simplicity.

At the end of each month, a line should be drawn beneath the last entry on each page, and the total found for each column. Taking the total of the payments page 'bank' column from the total of that on the 'receipts' page gives the bank account at the end of the month – a negative balance denoting an overdrawn account. The accuracy of this balance (and thus the recording for the whole month) can be checked by comparing the balance with that shown on the bank statement (which should be requested to show the cleared transactions up to the end of each month). The two figures will rarely be identical, due to delays between cheques being written and their reaching the bank, and to non-cheque items such as standing orders and direct debits and credits, which may not have been recorded in the cash analysis book. If, though, after making adjustment for these items, the balances in cash analysis book and bank

613

PAYMENT ANALYSIS

Date	Details	stub no.	Bank	Dairy feed	Vet and med	L/stock purchase	Misc. dairy	Crop costs	Wages	Power	Rent, rates, misc o/hds & charges	Interest & charges	Capital	Personal and tax	VAT paid
1/10	Opening bank balance b/fwd (o/d)		48723.24												
5/10	Wages	120	438.40						438.40						
5/10	A. Dealer: tractor repair	121	563.07							489.63					73.44
5/10	Farmco Ltd: 25t dairy feed @ £114/t	123	2850.00	2850.00											
9/10	Adams: vet bill	124	396.75		345.00										51.75
13/10	Newgro Ltd.: corn & grass seeds	125	2560.32					2560.32							
13/10	Fastfuel Ltd.:tractor diesel	126	532.58							532.58					
13/10	telephone bill	127	159.38								138.59				20.79
13/10	wages	128	484.56						484.56						
20/10	wages	129	523.87						523.87						
20/10	High St. Garage: petrol & car service	130	294.26							255.88					38.38
20/10	Browns: dairy supplies	131	539.11				468.79								70.32
28/10	wages	131	427.35						427.35						
28/10	fencing wire	132	437.55								380.48				57.07
29/10	A Dealer: tractor	133	21275.00										18500.00		2775.00
29/10	Transfer to personal a/c	134	1250.00											1250.00	
	TOTAL PAYMENTS		81455.44	2850.00	345.00	0.00	468.79	2560.32	1874.18	1278.09	519.07		18500.00	1250.00	3086.75

INCOME ANALYSIS

Date	Details	stub no.	Bank	Milk	Calves & culls	Crop sales	Sundry income	Capital	Personal	VAT charged	VAT refunds
5/10	J. Smith: 20 calves	927	1568.00		1568.00						
20/10	M.M.B. : milk sales (direct credit)		9234.32	9234.32							
20/10	F. Bloggs: contract ploughing	928	320.85				279.00			41.85	
29/10	A. Dealer: sale of tractor	929	7475.00					6500.00		975.00	
	TOTAL INCOME		18598.17	9234.32	1568.00	0.00	279.00	6500.00	0.00	1016.85	0.00
	BALANCE CARRIED FORWARD (O/D)		62857.27								
			81455.44								

Figure 24.12 Example of cash analysis for one month

statement still disagree, a mistake is indicated (either at the bank or in the book-keeping) and must be located before proceeding.

Monitoring cash flow

The totals of the analysis columns can now be transferred to the 'actual' column for that month in the cash flow budget (*see* 'Budgeting for cash flow' *above*). Ideally the headings used in both cash analysis book and cash flow budget will be the same, making the transfer of information between the two as easy as possible. Once this is done, and the actual closing balance entered at the bottom of the column, the two sets of information can be compared to detect discrepancies. The aim in this comparison should be to establish the main causes of discrepancy, to anticipate the likely effect over the remainder of the year, and to identify appropriate remedial action (*see Figure 24.13*).

The cash flow budget is an essential tool for the latter two tasks. If there is little that can be done to bring actual performance into line with the budget, it may be time to revise the budget for the rest of the year. Should this revision imply a higher overdraft facility than has been agreed with the bank, it can be used to good effect in renegotiating that facility well before it is actually needed.

For a more complete description of the mechanics of cash recording and control, a specialist text is recommended such as Atkinson, Hastings and Warren (1987a), Hosken (1982), or Warren (1986).

Recording profit and capital

Actual profits or losses, and actual net capital, are measured by exactly the same methods as used in the budgets. Thus a profit and loss account, in conventional or gross margin form, is used to measure net profit or loss, and a balance sheet to measure the net capital.

It was suggested above that the selection of headings in the cash analysis book should be influenced by the needs of gross margin accounts, with payments divided into fixed and variable categories. The result will be that, at the end of a year, the monthly totals of each of the 'trading' analysis columns (i.e. excluding capital, personal and tax columns) can be accumulated to give the raw material for the gross margin enterprise accounts. Given the limitations on the number of analysis columns in the cash analysis book, it may be necessary to use a simple coding system during the year to distinguish between items within the same column. A letter 'P' could be written alongside appropriate items in the 'feed' payments column, for instance, to indicate pig feed as opposed to dairy or sheep feed.

From this point, it is a relatively simple matter to compile gross margin accounts for the year, using the procedure described for the preparation of budgeted gross margins. Fixed costs are likewise derived from the relevant analysis column totals, and deducted from total gross margin to arrive at an actual net profit.

Analysis and interpretation of profit and loss accounts

Interpretation of a profit statement depends on comparison: with previous years' figures, with other businesses, and with the relevant budget. All of these methods can be used with a profit and loss account in conventional form, but provide much more useful management information when used with

| | OCTOBER | |
	Budget	Actual
Income		
Milk	8910	9234
Calves	2100	1568
Cull cows		
Wheat		
Wheat straw	2500	
Sundries		279
VAT charged on outputs		1017
VAT refunds		
Capital - grants		
machinery sales		6500
Personal receipts		
SUB-TOTAL	13510	18598
Payments		
Feed: cows	3000	2850
Vet and Med	500	345
Livestock purchases: cows		
misc. dairy costs	180	469
Seed	2420	2560
Fertiliser		0
Spray		0
misc. crop variable costs		
Wages	2000	1874
Power and machinery	2000	1278
Rent and rates		
Misc. fixed costs	650	520
Overdraft interest		
Loan interest		0
VAT payments		3087
Capital - buildings		
machinery		18500
capital repayments		
Personal - drawings	1250	1250
tax		
SUB-TOTAL	12000	32733
NET CASH FLOW	1510	-14135
OPENING BANK BALANCE	-50000	-48723
CLOSING BANK BALANCE	-48490	-62858
Overdraft interest @ 13%	-485	-681

Figure 24.13 Example of cash flow monitoring

a profit statement in enterprise account form, such as the gross margin accounts discussed above.

Comparison with previous years gives a valuable picture of trends within the business, and management 'alarm bells' will be rung by major differences between the performance of the immediate past year and those preceding it. As with any comparison, the comparison should start with the 'bottom line' – the net profit – and work back through total revenues, expenses and stock valuations to build up a picture of cause and effect concerning the variations. Given the physical and financial detail shown in each enterprise gross margin account, such comparisons can be used in building up a list of clues to help management explain how discrepancies arose, and how to correct them in the future (or capitalise on them, where performance is better than before).

Previous-year comparisons are limited by variations in the general environment (e.g. weather, prices) between years, and are also insular, relating as they do to just one business. To avoid the latter problem in particular, comparisons can be made with other businesses. Various comparison media are available in UK agriculture. Some are based on surveys, such as the annual Farm Management Survey (FMS) conducted by Regional Agricultural Economics Centres. The results of this survey are published in regional reports, most of which present their results in enterprise account as well as whole-farm form. Comparing farm gross margin results to these survey results thus gives rise to a number of new clues as to where farm profitability could be improved.

The value of survey results is often limited by the time taken to collect and process the information, so that the latest FMS results available in a region are likely to be at least one year out of date at publication. Survey regions are large, and the number of farms of a particular type in each sample usually relatively small. A further drawback is that such surveys are often not designed to collect detailed physical enterprise information: this inevitably limits the diagnostic possibilities of comparisons.

An alternative source of comparison information is a costing service. Such services are offered to farmers by a number of organisations, including quasi-government agencies (such as the MLC Pigplan service), commercial companies (such as ICI's Dairymaid service) and other agricultural institutions (such as the NFU poultry costing services). These services, particularly those relating to dairy and other intensive livestock, usually depend on monthly updating of information from the farmer, and use computers for the collation and analysis of that information.

As a result they can process information rapidly and provide reasonably up-to-date comparisons. Moreover, being designed specifically as management aids for the farmer, they usually contain a thorough breakdown of the most important physical information concerning the enterprises concerned. Thus a farmer using MLC's Pigplan can trace a low gross margin back to such details as low gradings, high feed used, low liveweight gain, poor market prices, and so on. In these respects, the use of comparisons via a costing scheme has advantages over the use of survey data. On the other hand, it must be remembered that the results of a costing scheme relate to a self-selected sample of farmers, and that the accuracy of the information fed into the costing scheme by farmers is not subject to the same rigorous control over accuracy as would be the data collected for survey purposes. As long as the user remembers that *any* form of comparison is merely a process of looking for clues rather than answers, this should not pose too great a problem.

The third, and potentially most useful, method of comparison is that with budgets. This is part of the budgetary control described earlier as an integral part of the management process, and is analagous to the budget/actual comparisons explained in the section on monitoring cash flow. The process of comparison is as before, but this time using the budgets prepared at the beginning of the year in question, rather than previous years' figures, or those derived from other farms. This form of comparison has the advantage of being specific to the farm in question, and relating to the most recent year for which results are available. Most importantly, it measures the farm's performance against the objectives set at the beginning of the year, as part of the planning process.

Similar structures apply to this as to the other forms of comparison – that it is concerned with collecting clues rather than answers, and that to follow up the clues needs the application of down-to-earth farming knowledge, such as that provided by the other chapters in this volume.

Recording and analysis of net capital

Once cash analysis book and profit statement have been completed for a particular year, the compilation of a closing balance sheet is a simple matter, using information contained in those accounts and the methods described in the earlier section on budgets.

Similarly, analysis of the balance sheet is exactly as described for the projected balance sheet, with the sole additional facility of comparison between budgeted and actual balance sheets.

In summary, effective control of a business depends on *budgets* anticipating the outcome of decisions in terms of cash, profit and capital. It also requires *accounts* to show actual performance in the same terms, allowing the manager to monitor progress and take corrective action.

Physical information

Calculation of reliable gross margin accounts, particularly in the detail suggested above, depends on good physical records. Examples include records of output, such as milk yields, pig liveweights, and grain yields; and records of inputs such as amounts of feed consumed and fertiliser applied.

Physical records are important for other reasons. Breeding records and charts, for instance, are vital in maintaining and improving the genetic potential of a breeding livestock enterprise. Short-term control often depends on the use of physical records. While the cost of water may not be sufficiently large to justify recording solely for enterprise accounts, for example, water records can give valuable advance warning of changes in the health of an intensive poultry flock or pig herd. Other physical records, such as those of labour use and management time, may be useful in making the use of time more efficient.

There is no 'right' way of keeping physical records, but certain guidelines should be borne in mind:

(1) Never record for the sake of recording – be sure that the benefit gained from using each record will justify the trouble taken. It is easy to find oneself keeping records that are rarely, if ever, used, and which could be safely discontinued.
(2) Make it as easy as possible for the records to be entered. This implies careful design of recording sheets, but also careful positioning of the sheets, near to the place where measurement is made, well-lit and with a pencil handy.
(3) Make it as easy as possible for the records to be interpreted. This is particularly important where the records are used in day-to-day control: key information should be highlighted and use made of graphs so that problems are clearly and rapidly shown up. It also applies where records are used as the basis for financial accounts. A little thought at the design stage about the way in which the information is to be used could save a great deal of trouble later.

For more information on the keeping of physical records, Hosken (1982) is recommended.

Planning for change

There are times when to use a full-blown set of budgets would be inappropriate. An example is where a number of partial changes to an established system are being considered. To avoid unnecessary work and potential confusion, a device is needed which will allow consideration of only the *net* financial effects of the change to be calculated. Such a device is the partial budget.

Partial budget principles can be applied to any of the three financial measures – profit, cash flow and capital – but are most commonly used to test the effects on profitability of a given change. These effects are grouped under four headings:

    revenue gained
    costs saved
    revenue lost
    extra costs

The result is a calculation such as that shown in *Figure 24.14*. This is a budget for a 'normal' year – in other words, when the change has been fully established. It is possible to compile a partial budget for a specific year, such as the first year the change is implemented, if profitability is likely to vary significantly between years. The normal year budget is the most important, though – if this does not show a benefit, there is no point in looking more closely at particular years.

Note that all costs and revenues are expressed on an annual basis. Thus expenditure made (or saved) on buildings and machinery is represented by depreciation (initial cost divided by the useful life in years), and the cost of building up (or reducing) a breeding herd or flock of livestock is

shown by the extra annual costs of replacement incurred by those animals.

Note also that the budget is kept clear and uncluttered by including only those items that are likely to change. If a change is likely to reduce workloads, but it is unlikely that a worker can be laid off or overtime reduced, wages costs can safely be omitted.

The best use of a partial budget of this sort is as an 'initial screening' device, allowing a quick and simple check of the relative merits of a number of alternative changes to the farm system. Those changes which partial budgets show to be likely to lose money, or make insufficient profit to justify the effort and inconvenience caused, can be scrapped without further ado, and the remainder built in to revised whole-farm budgets to test the effects on cash flow and balance sheet, as well as profit.

Where the change requires investment, it is important to relate the extra profit to the capital used. There are, broadly, two ways of doing this – by including an interest charge in the partial budget, and by calculating a marginal return on capital.

The ideal way of achieving the former is to compile a cash flow budget. For this, however, the cash flow budget would have to relate to a particular year, rather than a 'normal' year, and its complexity of calculation militates against its convenient use in initial screening. The alternative is to make an estimate of the amount borrowed, and apply to this an interest rate.

Rather than estimate the amount of borrowing outstanding year by year (which again would lose the simplicity that makes the partial budget so effective), a common practice is to take the initial amount borrowed, and assume it will be paid off in equal instalments over the life of the project. The

COSTS OF CHANGE	£	BENEFITS OF CHANGE	£
Revenue Lost		***Extra Revenue***	
cows: milk: 10 @ 5400l		milk: 50 ewes @400l @60p/l	12000
@ £0.15148/l	8193	lambs: £105/ewe	5250
culls: 2.5/year @ £350	875	cull ewes:15/year @ £30	450
calves: 0.95/cow @ £70	665	cull rams: 1/year @ £40	40
		wool: 3kg/ewe @ £1.05/kg	158
		ewe premium: £6/ewe	300
Sub-total	9733	Sub-total	18198
Extra Costs		***Costs Saved***	
ewe purch: 15/year @ £200	3000	cows: concs. (1.33t/cow)	1350
ram purch: 1/year @ 700	700	vet & med & AI etc.	400
concs.:200kg/ewe @ £145/t	1450	other	180
vet & med	300	forage: £198/ha x 5 ha	990
other	250		
forage: £120/ha x 5 ha	600		
extra labour:	1100		
capital: 5000 over 5 years	1000		
Sub-total	8400	Sub-total	2920
TOTAL COSTS	18133	TOTAL BENEFITS	21118
BALANCE -EXTRA PROFIT	**2985**		

Figure 24.14 Partial budget for reducing the dairy herd by ten cows, and releasing 5 ha for 50 milking ewes and four rams

average borrowing outstanding in a 'normal' year is then taken as the initial borrowing divided by two. At worst it is possible to assume that all the net initial capital is to be borrowed, but only if no firm information is available concerning the amount that is to be borrowed.

Net initial capital is the amount of money to be invested in the first instance. It includes both fixed capital (items such as property, machines, breeding stock, etc.) and working capital (the cash needed to finance the running costs of the change until it begins to bring cash in to defray those costs). It also takes account of any fixed and working capital set free by the introduction of the change. An example, related to the earlier partial budget, is shown in *Figure 24.15.*

```
FIXED CAPITAL
  Building and plant            5000
  Breeding stock - ewes        10000
                rams            2800
LESS sale of cows: 10 @ £400   -4000
                                           13800

WORKING CAPITAL
  half running costs of sheep   1850
LESS half running costs of cow -1460
                                             390

TOTAL NET INITIAL CAPITAL                  14190
```

Figure 24.15 Calculation of net initial capital

Using this example, and assuming all the capital needs to be borrowed, the average borrowing would be £14 190 ÷ 2 = £7095. At an interest rate of, say, 15%, the interest charge in a 'normal' year would be 7095 × 15/100 = £1064, and the extra profit after interest would be 2827 − 1064 = £1763.

There are various problems with this type of calculation, not least because of the many assumptions and generalisations which have to be made in the process. Moreover, it only measures the *cash* cost of borrowing, and puts no value on any non-borrowed capital used.

An alternative is to relate profit to capital through calculating return on capital. The appropriate measure here is *marginal return on capital*, calculated by dividing the extra profit before interest (from the partial budget) by the net initial capital required, and multiplying by 100. In the example above, for instance, the rate of return would be 2827 ÷ 14 190 × 100 = 20%.

The resulting percentage rate can firstly be compared with interest rates payable on borrowed money. If the rate of return is significantly more than the interest rate, it looks financially worthwhile: if less, the change is a non-starter from a financial point of view – the finance costs will outweigh the return. If the rate of return is only marginally more than interest rates, the farmer must decide whether it is worth the risk that profit might be less than forecast (or capital needs greater).

If the change passes this test, its marginal return on capital can be compared with those of other possible changes as a basis for selection between projects.

Among the drawbacks of this use of marginal return on capital are that it rests on large assumptions, takes no account of the pattern of costs and returns either within or between years and can give pessimistic results. The latter can be countered by using average rather than initial capital in

calculating the return, but this tends to give over-optimistic results – it is perhaps better to tolerate a pessimistic result than to risk making decisions based on a result which may never be achieved.

More sophistication is available in the shape of 'discounted cash flow' techniques. These are essentially partial budgets which take account of differences in costs and returns between years of a project, and allow flows of money in the near future to be weighted more heavily than those in the far future. Tax payments can also be incorporated. The result is a 'net present value' showing the consolidated margin of future cash flows over the amount invested, after allowing for the time factor. This can in turn be used in calculating an 'internal rate of return', which can be used in the same way as the marginal return on capital described above.

The very sophistication of the techniques gives rise to difficulties in application, calculation and interpretation. Without training, and preferably a computer, it is likely to be better to use the crude but simple devices described earlier in this section. If these are treated purely as initial screening measures, to be followed by full budgets where these appear to be warranted, the crudity is acceptable. Further reading is suggested at the end of the chapter.

One additional device can be useful in planning for change. This is *payback* – the number of years needed for a change to repay the net initial capital.

For this it is necessary to estimate the net cash inflow for each year of the project's life. One way of doing this is merely to take the extra profit from the partial budget, exclude depreciation (the main non-cash expense) and include interest. It is possible to make more precise estimates using simple cash flow budgets. The net cash flows are then progressively totalled, starting from year one (allowing for negative flows), until they accumulate to the net initial capital invested. The payback of the project is the number of years it takes to get to this point. A quicker and cruder method is shown in *Figure 24.16*, where a 'normal year' cash flow is calculated and divided by the net initial capital.

```
Extra profit from change in a typical year       2985
  add back depreciation                          1000
                      Annual cash surplus         3985

Net initial capital                             14190

          Payback ( 14190 /3985) :    3.6 years
```

Figure 24.16 Example of simple payback calculation

Payback is of use in its own right, in giving a 'feel' of the worthwhileness of the project. Compared with that of other possible projects, it gives a measure, though imprecise, of the relative riskiness of the projects. If size, marginal returns and personal preferences were equal between three projects with paybacks of three, six and ten years, the rational decision-maker would go for that with the lowest payback.

Whatever partial budgeting devices are used for the testing of the effects of a change from an established system or a whole-farm budget, it will be necessary, for all but the smallest of changes, to eventually test the effects of the change using whole-farm budgets for profit, cash flow and capital.

Sources of capital

To put a business plan into operation (and to maintain it) usually needs investment of lump sums of money – capital. That capital may be wholly or partly on hand, in the form of savings, retained profits or the proceeds of the sale of assets. The farmer may be lucky enough to be able to claim grants to defray some of the costs of investments, or even to have relatives or other well-wishers willing and able to make outright gifts of money. Though such capital may appear to be free of charge, there are costs attached to the use of any capital sum. If nothing else, there is the opportunity cost arising from the lost benefits from using the money in some other way. Other, non-financial costs arise from legal obligations, in the case of a grant, or 'moral' obligations in the case of a gift, for instance. Nevertheless, the costs are usually relatively low, and these sources should be exploited to the full before turning to raising money from outside the business through borrowing or shared ownership.

Borrowing

The most common alternative is to borrow money. Borrowed money incurs opportunity costs and obligation costs as described above. In addition it incurs a direct cost – that of interest (the exception being some private loans from relatives). This is a fee charged by the lender of the money, expressed as an annual percentage of the amount borrowed. The rate of interest depends on both external factors such as the state of the economy, and internal factors such as the security offered and the apparent creditworthiness of the business.

Interest rates are often expressed in relation to the clearing banks' base rate, which reflects the level of interest in the economy as a whole and is published in the financial sections of quality newspapers. An interest rate may be variable (i.e. fluctuates over time with changes in the base rate) or fixed (i.e. stays the same over the length of the borrowing term).

Rates can be applied in various ways. For instance they can be related to the daily balance outstanding (as with an overdraft), to the amount outstanding at the beginning of a particular year, or applied to the initial amount borrowed ('flat rate'). To allow the cost of different forms of borrowing to be compared, each rate should be expressed as an Annual Percentage Rate (APR). The APR on an overdraft is normally close to its stated rate, while that of a flat rate can be nearly double the expressed rate.

A lender will usually require some form of security against default on payments by the borrower. This security can be in the form of title to specific fixed assets, enabling the bank to sell those assets to recover the debt if necessary. A 'floating charge', commonly used for overdrafts, is a generalised version including more movable assets of the business. Alternatively, security can be provided in the form of a guarantee from an individual or an institution (such as the Agricultural Credit Corporation) that if the borrower defaults, the guarantor will repay the money outstanding.

Sources of borrowed capital

It is difficult to avoid generalising when referring to the characteristics of various sources of borrowing. The following notes should be judged in this context.

The 'high street' banks are the main source of borrowed capital for farming, and the largest part of this is in the form of *overdrafts*. An overdraft is created when more money is drawn out of an ordinary bank current account than was deposited there in the first instance. Interest rates are low (typically 2–3% above base rate), and are applied to the daily balance outstanding. Unauthorised overdrafts are charged a penal rate which can be more than 15% above the base rate, so it is important to agree an overdraft facility with the bank manager in advance of needing it. When an overdraft facility is negotiated or reviewed, an arrangement fee may be charged. This fee can be as much as 1% of the overdraft facility.

Overdrafts are intended for short-term finance, and the bank will normally expect the account to return to credit at some stage during the year. It is normally secured by a 'floating charge' on all the assets of the business, and is technically repayable on demand.

Bank loans, on the other hand, are subject to a contract binding each party. When a loan is arranged, the borrower is given the amount of the loan as a lump sum. This amount is repaid over time in regular instalments, with interest payments (usually at a fixed rate) reflecting the decline in the loan outstanding. As long as these instalments are paid, and the other terms of the contract are met by the farmer, a loan cannot be called in before it is due (or the business becomes insolvent). To protect itself in the event of insolvency, the bank will normally expect the loan to be 'secured' on a specific asset – in other words, if the business defaults on the loan, the bank has a legally valid claim to the title of the asset, and can sell it in order to recover the money owed.

A loan is normally intended for longer-term finance than an overdraft. Interest rates are higher (e.g. 3–6% above base rate) and arrangement fees are again likely to be payable. Against the higher costs must be set the greater security of the loan, and the regular and predictable interest and capital payments, making budgeting easier.

The high street banks are not the only institutions offering loans to farmers. Loans secured on land and property ('mortgages') are available from a variety of sources, including building societies (for house purchase and building) and the Agricultural Mortgage Corporation (for land purchase, buildings, and other longer-term needs).

Trade credit is a popular form of short-term finance, being quick and easy – it is obtained by not paying bills immediately. Most suppliers will allow a period of grace for payment of invoices, usually two to four weeks, though sometimes as short as seven days. Using this period to the full can help keep down the business overdraft. Extending credit beyond this limit will, however, incur penalties – at the least, loss of goodwill from the supplier, and at the worst a credit charge or 'loss of cash discount'. These charges are invariably much more expensive than overdraft finance.

Two forms of borrowing are particularly relevant to machinery purchase. *Hire purchase* or lease purchase uses the asset purchased as the security for the borrowing – repayments are arranged so that the recoverable value of the asset is always greater than the debt outstanding. It is convenient, as purchase and finance can be arranged in one operation, but it is usually considerably more expensive than bank finance.

Finance leasing involves a machine (or other asset) being bought by a finance company (often a subsidiary of a major bank) and rented to the farmer. After the initial rental period of two to five years (depending on the life of the asset) the farmer may be able to continue the lease into a 'secondary' rental at a nominal rent, or even to buy it. If the machine is sold on behalf of the finance company, he is usually able to retain between 80 and 100% of the proceeds. Maintenance, servicing, insurance are all the farmer's responsibility. Thus the effect is of purchase, even though the arrangement is technically a lease. Rentals are quoted in terms of interest rates, and the finance companies' ability to exploit tax concessions helps to keep rates low even compared with overdraft finance. An attractive characteristic is that spreading the cost over several years enables machines to be acquired without immediate effects on the overdraft. This can be dangerous if done to deceive the bank manager as to the true state of affairs – at the time of writing it is not obligatory for sole traders and partnerships to show leased assets on their balance sheets.

Calculating the cost of borrowing

It is crucial, if considering borrowing capital, to be able to estimate the likely costs to the business. Apart from administration costs (such as arrangement fees), the annual costs are composed of two elements – the repayment of the amount borrowed (needed for the cash flow calculations), and interest charges (needed for both the profit and the cash flow budgets).

In calculating the financial commitments arising from an overdraft, there is no substitute for a cash flow budget, as described earlier in this chapter.

Where interest on loans or other forms of finance is quoted at a 'flat' rate, the annual interest charge is given by the initial amount borrowed multiplied by the flat interest rate. The annual capital repayment is given by the initial amount borrowed divided by the number of years of the term. Thus £10 000 borrowed over a term of five years at a flat rate of 8% will give rise to interest payments of £800 per year (10 000 × 8%) and capital repayments of £2000 per year (10 000 ÷ 5 years).

A very common method of repayment is that known as the 'annuity' or 'normal' method. This adjusts the proportions of interest and repayment throughout the term of borrowing, to keep the combined annual payment constant while allowing interest payments to reflect the gradual repayment of the capital. Tables are provided in most farm management handbooks and financial textbooks to enable estimation of interest and repayment charges for a particular year in the life of a loan. Part of such a table is shown in *Figure 24.17*. From this it can be calculated that in year 3 of a 5-year loan of £10 000 at 12% a business will have to pay £800 interest (80 × 10) and £1970 capital repayment (197 × 10) (*see Figure 24.17*).

Shared ownership

An alternative to borrowing is to share all or part of the business with others, in return for an injection of capital. The simplest form of sharing is a *partnership*, where two or more people agree to run the business in common, sharing profits and control in return for investment of capital.

Each of the partners is jointly and severally liable for debts incurred by the other partners as well as by himself. In other words, if the business fails, the partners may have to sell all their personal assets if that is the only way that the creditors can be repaid – even if the cause of the failure was the action of only one of the partners. This is the motivation behind the formation of many private *limited companies*. A company can be formed with as few as two people, and has the advantage of liability being limited to the capital invested in the company – at the expense of higher formation and administration costs, and a certain lack of privacy.

Formation of a *cooperative* can allow shared investment in just part of a business, ranging from a single machine to a complete marketing and distribution service for produce. Liability is limited, and formation costs are similar to those of a company.

Any form of shared ownership raises complex and difficult questions, particularly where taxation is concerned. Generalisation is impossible, and a reputable solicitor and accountant should be consulted before entering into any such arrangement (*see* Centre for Management in Agriculture (1986), Furlong (1987) and Midland Bank (1979)).

Coping with risk and uncertainty

Implicit in all the above discussion is the difficulty of planning and budgeting for the future. This arises from the fact that nothing about the future is certain.

The sources of this uncertainty, and the consequent risk to the business owner on making decisions about future events, are many and varied. In farming, physical factors are prominent, such as the possibility of disease, pest damage, drought, flood, fire, health of farmer and employees, and so on. Economic fluctuations also have effects – the state of world trade, the strength of the pound, and the growth of incomes of the population as a whole are some of the influences on product and input prices at the farm gate.

Just as important can be political factors, affecting such things as tax allowances, agricultural support prices and quotas on production. Social and personal aspects also have an effect – such as changes in general food consumption

	Interest rate					
	10%		12%		14%	
Year	Interest	Repayment	Interest	Repayment	Interest	Repayment
1	100	164	120	157	140	151
2	84	180	101	176	119	172
3	66	198	80	197	95	197
4	46	218	56	222	67	224
5	24	240	30	248	36	256
£ PAYMENT PER £1000 BORROWED						

Figure 24.17 Extract from a loan repayment and interest table

habits, hardening attitudes of non-farming neighbours to noisy or smelly activities, marriage of the farmer's son or daughter, or the divorce of the farmer.

There are ways of organising the business to minimise some of the effects of uncertainty. Some of the factors mentioned can be insured against (fire, for instance). Prices of some products and inputs can be protected by selling and buying forward, or by using the futures markets. The farm system can be designed with a wide spread of different enterprises, so that if one enterprise performs badly, the others can compensate. The farmer can avoid trying any new products or production methods to avoid the risk of their failure.

Not all risks can be avoided by such methods, and reducing risk usually has a cost. A farmer may prefer to take the risk than incur that cost. Insurance incurs the cost of the premium. Forward buying and futures trading incur the cost of losing potential revenue if spot market prices improve and the farm is tied to a contract price. Diversification and conservatism involve lost opportunities to increase profits – either by losing the chance of specialising in products that the farm and farmer are good at producing, or by forgoing the uptake of new, potentially profitable enterprises and production methods.

If risk cannot be limited beyond a certain point, it can at least be allowed for in the planning and control process. Reference has earlier been made to the importance of erring on the pessimistic side when budgeting, and to the need for adequate research to ensure that the forecasts on which the budgets are based are as accurate as possible. Most important of all is the process of budgetary control, whereby the effects of a deviation from the expected performance can be readily identified and appropriate action taken.

None of these methods gives the farmer much advance warning of *how much* risk he or she faces – and this can be crucial in making a decision. The key question concerns the 'downside risk': 'how much do I stand to lose if things go wrong?'.

The simplest solution in concept is to prepare two sets of budgets, one based on the inputs and outputs that the farmer thinks are most likely, and the other based on the worst possible performance. The result is effective, with the farmer being able to see what he or she stands to lose if things go wrong, and to decide whether or not to take this risk. Preparing two sets of budgets is time-consuming, though, unless one has available a suitably programmed computer. Even with a computer, interpreting the results can be difficult: so many variables may have been changed that it is impossible to pick out the factors most responsible for the variations.

The latter problem may be overcome by the calculation of *breakeven points*. Those factors which seem most likely to affect the outcome of the profit budget are identified. Two such factors on a dairy/arable farm might be the amount of concentrate fed to the cows, and the likely price for winter wheat. The budgetary margin (net profit or loss in a whole-farm budget, extra profit or loss in a partial budget) is then expressed in terms of that factor. For instance, in the partial budget example used earlier, the extra profit estimated was £1763 after interest. At a price of 60 p/litre of ewe's milk, this represents 1763 ÷ £0.60 = 2938 litres (59 litres/ewe). This is the amount by which the milk yield would have to fall before the extra profit started become a loss of profit. The breakeven milk yield is thus 400 − 59 = 341 litres/ewe.

Working out breakeven points for various components of

a budget does two things. By looking at each breakeven point, the farmer can make subjective judgements as to which of the breakeven points is most likely to be reached: this gives an indication of those items which most contribute to risk, and which thus should be monitored most closely. It also helps the farmer to judge whether or not the risk is worth taking. In the example above, for instance, the farmer can ask himself, 'how likely is it that I will achieve only 341 litres/ewe, and am I prepared to take that risk?'.

The strength of the breakeven point technique – that it allows the effect of particular factors to be singled out – is also its weakness. It is not possible to identify the combined effect of more than one factor varying at the same time. While this may not be a huge problem in a simple partial budget, it is a severe limitation in budgets with many component variables.

An alternative is a form of *sensitivity analysis*. This uses tables to show the way in which the budgetary margin is likely to vary with changes in the value of one component of the budget. The example in *Figure 24.18* is based on the partial budget from earlier in the chapter. It shows how the extra profit is likely to change from the value in the budget (£1763) if ewe milk yield were to vary from the 'most likely' assumption of 400 litres used in that budget. The lower figure is, in the farmer's view, the worst likely outcome, and the higher figure is the best that he or she feels is possible.

MILK YIELD PER EWE		
300 litres	400 litres	450 litres
-1237	**1763**	**3263**
Extra profit from change, after interest		

Figure 24.18 Sensitivity analysis – 1-way table

This gives the farmer a feel for the 'sensitivity' of the budgetary margin to variations of a particular component of the budget, and thus the degree to which the successful outcome of the plan depends on the farmer's ability to achieve this level of performance. Better still, this table can be combined with others to show the effect of interaction between components. In *Figure 24.19*, for instance, the milk yield table has been extended to incorporate variations in milk price. Here the decision-maker has the opportunity to see what the consequences would be of poor performance on both fronts – and to decide whether or not to take the risk. At the cost of simplicity and clarity, such a table can be extended to show the combined effects of variation in three or even four budget components.

MILK PRICE	MILK YIELD PER EWE		
	300 litres	400 litres	450 litres
50p/litre	-2737	-237	1013
60p/litre	**-1237**	**1763**	**3263**
65p/litre	-487	2763	4388
	Extra profit from change, after interest		

Figure 24.19 Sensitivity analysis – 2-way table

The methods described merely present the farmer with information about the likely consequences of deviation from the forecast performance assumed in the budgets. They leave the farmer to use subjective judgement as to how likely these deviations are to occur. Other techniques exist which allow probabilities of different outcomes to be incorporated in the calculation, and these are described in texts such as Hull (1980) and Barnard and Nix (1979). For most practical purposes in farm management the simpler devices described above will be sufficient, particularly if combined with a rigorous system of budgetary control.

Marketing

The welfare of a business depends on it being able to sell the goods and services produced at a profit. Crucial to this is the process of marketing, which can be conveniently, if predictably, divided into four functions:

(1) determining the *products* to be produced;
(2) deciding the *price* at which they should be sold;
(3) deciding the *place* at which they would best be sold;
(4) and determining the most appropriate forms of *promotion* to use in persuading potential customers to buy.

For these functions to be performed effectively, they must be considered from the viewpoint of the consumer of the product. The business needs a product wanted by consumers at the price and location wanted by consumers, and promotion which will appeal to the consumer. A 'market-orientated' business such as this has a far better chance of maximising profitability than one whose management is totally concerned with getting production techniques right (the 'production-orientated' business).

For many farm products, the farmer has traditionally been a 'price-taker'. One farm's products are very similar to another's, and any individual business is such a small part of a very large market that it can have little or no influence over the price. In such circumstances marketing decisions at farm level are limited to deciding the products to be produced – these are then sold automatically to an institution such as the MMB or a merchant, which then makes decisions about place, price and promotion. Even here, though, it is important that the business is market-orientated, responding to the information relayed back from the consumer by the wholesaling institution. A classic example is the way in which some sheep farmers have been able to boost profits of their flocks by producing the leaner meat requested by supermarkets, who are themselves responding to the wishes of their customers.

Increasingly, farmers are having to look beyond these traditional markets in a quest for profitability (*see* Chapter 29). This brings them to consider products which are particular to that business – for instance contracting, special food products, tourist facilities, farmed deer and trout, and so on. Here the market is small enough to be influenced by the farmer, who then has to consider all aspects of the 'marketing mix'.

If a business in this position is to supply what consumers want, it must first find out some of the characteristics of those consumers. This implies some form of market research. The aim is not only to identify possible new products, but also to establish facts such as who is likely to buy a particular type of product, the characteristics they would like that product to have, and the price they would be prepared to pay. Once these facts have been ascertained,

decisions can be made about the exact form of the product, and how best to promote it.

The term 'market research' implies large and complicated surveys, but can be applied to any ordered and critical examination of the likely market for a product. It should start at the simplest level, with the questions, 'why should anyone want to buy my product rather than those of my competitors?' and, if there are not apparent competitors, 'why is no-one supplying this product already'. Honest, objective answers to such questions may avoid the danger of the emotional decision; the obsessive attachment to a particular idea coupled with a refusal to give full consideration to its limitations.

The next stage is to consider the factors which are likely to affect the demand and supply of the product concerned, since these factors govern the prices which can be charged.

From the *supply* point of view, the most important factor is likely to be the number and type of competitors. It is essential to consider all possible competitors, rather than just those who are providing direct substitutes. For instance, a contractor needs to take account of the presence in the area of an active machinery leasing firm; a pick-your-own enterprise needs to consider greengrocers and supermarkets in the towns he or she hopes to service, and a day-visitor enterprise needs to bear in mind the proximity of National Parks and other areas which can be visited without charge.

It is vital to anticipate the way in which the identified competitors are likely to react to the new challenge. A competitor with substantial financial reserves may adopt aggressive pricing policies in order or safeguard his or her markets; thus many producer-retailers of milk have been forced out of operation by large dairies which can afford to give loss-making discounts for long periods. A long-established firm with a large share of the market may be more complacent about competition than a relatively new business which has built up its trade by hard graft and has no intention of giving any of it up.

Among the factors which may affect the *demand* for a product are changes in the number of potential customers within the area of sale, changes in the level and distribution of their incomes, and changes in their tastes and lifestyles. Other influences will be the presence of businesses producing complementary products, the promotion of the product, and the level at which prices are set in comparison to those of competitors.

Clues to many of these factors can be found in local libraries: residential population and incomes information is published by both central and local government; seasonal population by many of the regional tourist boards; changes in consumption (e.g. the types of food, or cars, or freezers) in the National Incomes and Expenditure Survey (Her Majesty's Stationery Office). Other sources of information may come to light when pressure is put on organisations and individuals with particular interests. It is possible, for instance, that the county council has recently measured the flow of traffic along nearby roads, and is prepared to release the results. For a visitor enterprise, this could be crucial information.

Published information can be supplemented by personal investigation, given common sense and a little imagination. Traffic levels can be assessed roughly by first-hand observation. The Yellow Pages in the telephone directory, together with the advertisement pages of the local newspaper, can be used to identify potential competitors, whose prices can be determined by making telephone enquiries as if from a potential customer. The same sources may be used to identify complementary, rather than competitive businesses. The golf club down the road could encourage the development of a sport fishery. One contractor in a locality might be geared up for operations for which his neighbour is not, and vice versa. The presence of a neighbouring caravan site could provide a ready-made market for a pick-your-own enterprise. At the least such information should influence the way in which the enterprise is promoted; at the best it could be an avenue to profitable cooperation.

This basic research may give a rough idea of the numbers of people who come into the target population of the enterprise, but will not give a reliable estimate of the proportion who might be tempted to buy the produce on offer. The market must be tested in some way in order to see whether a reasonable number of people are interested. One way of doing this is to start the enterprise in a small way, with little commitment of capital. This is easy in some enterprises, such as bed and breakfast in a half-empty farmhouse, or using underemployed men and machines for contracting. In other cases, however, the effort and expense of establishing even a small enterprise are such that full commitment is needed from the start; where planning permission and/or licences are necessary, for instance, or where major capital works are called for. In these circumstances, a market survey of some form is needed. At its simplest, this can take the form of a series of informal discussions with potential customers, or with people who are likely to be 'in the know', such as tourist information officers or officials of the local chamber of commerce (as long as they are not likely to be competitors themselves). If the risks are great and the financial commitment large, it may be worth employing a market research bureau to conduct a formal survey among potential customers.

Once the enterprise has begun in operation, a check can be kept on the market by various types of feedback from customers, ranging from a verbal enquiry 'was everything all right?', through visitors' books and suggestion boxes, to analysis of sales records in detecting significant variations from target.

Promotion

Promotion can be defined as any tactic which is used in attempting to stimulate demand for a given product, and can include devices such as free gifts and special offers as well as advertising. Most forms of promotion cost money and effort, and so discrimination is needed with respect to the amount and type of promotion that is employed. Generally speaking, the effectiveness of promotion depends on the degree of product differentiation that exists. If a business is offering an identical product to those of several competitors in the same area, its promotion may benefit its competitors as much as, or even more than, itself. If, on the other hand, the business can convince the customers that it is offering something unique, the chances of promotion being successful are much higher. This uniqueness can be brought about in a variety of ways, including developing an 'own brand', growing exotic or out-of-season produce, offering a high quality of product and/or service, or providing a particularly attractive environment.

The cheapest and possibly most effective form of advertising is that of word of mouth. Since this takes some time to build up, however, the new enterprise has to use other media. These can include press, TV and radio advertisements and

editorial, direct mailing and hand-posting of leaflets, personal calls, membership of professional associations, directories and guide-books, printed tee-shirts and sponsorship of sports and cultural events.

An important rule in advertising is to match the medium to the market. Market research of even the simplest form will indicate the target population of the enterprise, and it remains to decide which form of advertising would be most effective in reaching that population. A pick-your-own enterprise relies on local trade, so media such as local newspapers and leaflets are appropriate, while national newspapers are not. The reverse is true of a holiday accommodation enterprise. A farm contractor would probably get better results from use of a circular letter followed by a telephone call, than from an advertisement tucked away in the classified advertisement page of the local paper.

Next it is vital to plan the advertising carefully, leaving nothing to chance. The strengths of the product should be assessed so that they may be brought out in the advertisement. The wording should be pared down to leave a clear message. This message should be hard-hitting, even brash. Good taste does not always win customers. It is worth taking pains over design of advertisements, making several mock-ups before deciding on the final version.

Some of the most effective advertising is free, or at least comparatively cheap. This applies particularly to editorial in newspapers and the broadcast media: a news-hungry editor can often be persuaded to publish articles about local businesses. The key lies in developing an interesting story – a grand opening, for instance, or a free afternoon for the 'Over-Sixties Club'.

Planning alone is not enough to get the best out of advertising – its effectiveness must be continually monitored. Where orders or bookings are made by post, this is easily effected by means of a code inserted in the address quoted in the advertisement. Thus respondents to two separate advertisements might be asked to write to M. *A*. Warren and M. *B*. Warren respectively, where the *A* represents an advertisement in the *Barset Chronicle* and the *B* that in the *Barset Advertiser*.

Pricing

Two landmarks govern the process of pricing a product – the cost of production of the goods or service, and the prices charged by competitors for similar products. These landmarks must be determined, which implies careful research and budgeting. In particular it is important to be able to distinguish between fixed and variable costs, since in the short term it may be appropriate to set prices to cover variable costs only.

How the prices are set in relation to these landmarks will depend on one's own estimation of how the market will react, and one's own promotional objectives. The market reaction is dependent on the responses of both potential customers and competitors. If the competition is weak, and the customers are likely to be relatively insensitive to price, 'skimming pricing' can be used, setting an initially high price which can be progressively dropped as new competitors come on the scene (by which time one's own enterprise should be well established). If, on the other hand, competition is strong and customers are likely to be responsive to price levels, the new enterprise can carve out a market by 'penetration pricing'. This is charging low prices at first in order to attract custom away from the established competition. Use of this tactic requires reserves, in the form of cash or the support of other business enterprises, to enable survival in the initial stages of low or negative profits.

Consistently low prices may be used as a long-term strategy, if the objective is to generate high-volume sales which can withstand low margins per unit sold. On the other hand, it may be appropriate to set prices high to establish a high-quality image.

The owner of a well-established enterprise might be tempted to rely on a mark-up method of pricing, applying a constant percentage margin on top of production costs in order to keep the pricing decision simple. The rigidity imposed by this sort of formula reduces the ability of the enterprise to adapt to changing market conditions, however, and could lead to a dangerous decline in competitiveness.

Conclusion

'Management' was described at the beginning of this chapter as a process of making decisions, and monitoring their effect in order to make better decisions in the future. Viewed from some standpoints, this could be considered to need merely the application of common sense. It is to be hoped, however, that this chapter has shown, on the one hand, the potential complexity of the management task, and on the other the way in which various techniques and ways of thinking can clarify issues and aid decision-making.

Sometimes common sense is not enough.

References

ATKINSON, J. M., HASTINGS, M. R., WARREN, M. F. (1987). *The Family Farm Series:* a. *Keeping the Books*, b. *Cash Flow Budgeting*, c. *Using Your Accounts*, d. *Improving Your Profits*. Stoneleigh, Warwickshire: Agricultural Training Board/Seale-Hayne College/Open Tech Unit

BARNARD, C. S. and NIX, J. S. (1979). *Farm Planning and Control*, 2nd edition. Cambridge University Press

CENTRE FOR MANAGEMENT IN AGRICULTURE (1986). *Practical Share Farming*, London: British Institute of Management

FURLONG, L. A. C. (1987). *Farming as a Limited Company, Farm Management* **6**, No 5

GILES, A. K. (1986). *Net Margins and All That*, Study No. 9, University of Reading: The Farm Management Unit.

GILES, A. K. and STANSFIELD, J. M. (1980). *The Farmer as Manager*, London: George Allen and Unwin

HOSKEN, M. (1982). *The Farm Office*, 3rd edition, Ipswich: Farming Press

HULL, J. C. (1980). *The Evaluation of Risk in Business Investment*, London: Pergamon

HUTCHINSON, H. H. and DYER, L. S. (1979). *Interpretation of Balance Sheets*, 5th edition, London: Institute of Bankers

MERRETT, A. J. and SYKES, A. (1973). *The Finance and Analysis of Capital Projects*, 2nd edition, London: Longman

MIDLAND BANK (1979). *Farm Partnerships*, London: Midland Bank

NIX, J., HILL, P. and WILLIAMS, N. (1987). *Farm and Estate Management*, Chichester: Packard Publishing, Ltd

WARREN, M. F. (1986). *Financial Management for Farmers*, 2nd edition, London: Hutchinson.

Farm management data

AGRICULTURAL BUSINESS CONSULTANTS, LTD. *The Agricultural Budgeting and Costing Book*. Warwickshire: Atherstone, (published twice-yearly)

NIX, J. S. *The Farm Management Pocketbook*, Wye College, Kent: Farm Business Unit. (published annually)

THE SCOTTISH AGRICULTURAL COLLEGES, *Farm Management Handbook* (published annually)

EACH REGIONAL AGRICULTURAL ECONOMICS CENTRE (usually based at a prominent university or college) also publishes annual reports and handbooks

25

Farm staff management

E. J. Sobey

THE NEED

The farmer, as a business manager, needs to apply professional management techniques to the small, but vital labour force. The abilities of each employee should be known and utilised in the most effective and economic way. Thus the employer is involved periodically in exercising skill and knowledge of staff management. The application of staff skills may be related to the seasonal requirements of the farm through the use of such techniques as labour profiles and gang-work schedules. The skills may be related to day to day control and motivation of the labour force, to the administration of personal needs – pay, training, safety and equipment or recruitment to meet vacancy or expansion in accordance with a workforce-plan.

Relevant techniques will be discussed in the following order:

(1) farm requirement: labour profiles, gang-work schedules:
(2) workforce planning: the labour plan, job descriptions, personal specification:
(3a) obtaining the right labour: recruitment and selection, induction training, continuation training;
(3b) reduction of labour;
(4) motivation: motivation theory, pay policy, including job evaluation, incentives:
(5) working conditions: need for procedures, facilities.

FARM REQUIREMENTS

Most farms operate with regular labour acquired over a period of years, but additionally have periods when extra labour may be required, e.g. silage making, fruit or potato picking. If the farm needs are not carefully calculated timeliness of operations is likely to suffer or the manager may pay for extra labour or, worse, create extra enterprise complications to 'use up' spare labour capacity.

Labour profiles

Seasonal labour profiles can be calculated so that exact requirements are known and may be plotted as a bar chart (*Figure 25.1*). These charts are also extremely helpful when considering a change in the size of an enterprise or the introduction of a new one. A labour profile can also show casual labour requirements.

Planning for skilled labour and machines

Gang-work day (GWD) charts (*Figure 25.2* and *Table 25.1*) aim to ensure that enough labour and machines are available to carry out operations at the correct time and within the time available for the operation.

Construction of gang-work day charts

It is seldom necessary to chart the whole year, only those periods which are peak periods usually need be charted.

Stage 1

Collect data:

(1) cropping and stocking,
(2) regular labour employed,
(3) machinery and implements available,
(4) gang size and machines required for each operation,
(5) rate of work of each operation,
(6) earliest start and latest finish date for each operation,
(7) number of days on which field work is possible in each month.

Stage 2

Tabulate the information for each operation, under the headings – area, time period, working days available, work rate and gang size. Calculate the gang-work days required for each operation.

Stage 3

Draw chart. Operations using large gangs are the least flexible and should be put first.
Initially assume operations can start at the earliest start date. Charting is a matter of trial and error, like a jigsaw. A chart

Man hours/month

Crop	ha	Jan	Feb	Mar	Apr	May	Jun	Jly	Aug	Sept	Oct	Nov	Dec
Winter wheat	15			15	15				68	119	93	38	3
Spring barley	48.5		48	195	122	35			195	342	68		
Maincrop potatoes	11			82	168	22	43	43	10	296	540		
Sugar beet	6	7		43	26	75	81	9		26	94	109	16
Grassland	38.5			25	25		225	125	25	25			
Kale	10.5						105	26	5				
Total		7	48	360	356	132	454	203	303	808	795	147	19

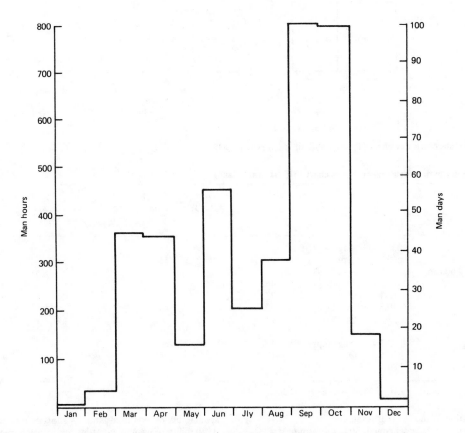

Figure 25.1 Day worker requirements (hours per month measured in the example: average and premium performances may be obtained from the *Farm Management Pocket Book* by John Nix)

with no spaces has no flexibility for poorer than average seasons or other delays.

Casual labour may be written in or 'ballooned' on top.

WORKFORCE PLANNING

One problem for farmers, often not recognised in time, is that they and their workforce grow old together. The position then facing the heir or, equally, an appointed manager is that all the staff may have to be replaced fairly quickly with the consequent loss of knowledge.

Age is, of course, not the only factor; changes in farm size, type of enterprise, technology, or national agricultural policy can all affect farm staffing.

As has been demonstrated already by the use of labour demand charts the day to day requirement should be known. Planning the work force for the future requires different techniques:

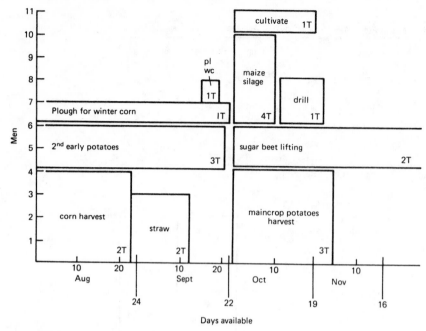

Figure 25.2 Gang-work day chart (T = number of tractors required)

Table 25.1 Information for gang-work day chart (315 ha arable land)

Activity	(ha)	Period	(d)	Work rate (ha/GWD)	GWD reqd	Gang size Regular	Casual	Tractors
Corn harvest	221	Aug–Sept 15	34	10.1	22	4	–	2
Straw cart	60.7	Sept	22	4	15	3	1	2
Second early potato harvest	18.2	Aug–Sept	46	0.4	45	2	4	3
Main potato harvest	12.1	Oct–Nov 7	23	0.5	24	4	6	3
Sugar beet	30.5	Oct–Dec	50	0.7	44	2	–	2
Maize silage	33	Oct 1–20	14	3.3	10	4	–	4
Winter corn								
Ploughing	80.1	Aug–Sept	46	1.6	50	1	–	1
Cultivate	80.1	Oct	19	4	20	1	–	1
Drill	80.1	Oct	19	8.1	10	2	–	1

Performance figures obtained by work study measurement (after Niemeyer).

(1) examination of the external resources;
(2) examination of internal constraints and resources.

External resources

People

National and local statistical returns of school leavers.

National and local information on (apprentice, college, university) trained leavers.

Information on local (and national) availability of skilled, semi-skilled and general labour (Department of Employment and trade journals).

Neighbours' and competitors' pay rates and employment conditions – pensions, perquisites, etc.

Availability of 'contract' labour – short or medium term.

Availability of casual, part-time and pensioner staff.

Vacation period labour, students of all disciplines and nationalities.

Technology

Development work currently taking place in AFRC, institutes and commercial companies which may reduce or increase labour requirement, remove the need for old 'skills' or introduce new skills.

Major changes in farming fashions and markets.

Government policy – national and international, in particular legislation affecting conditions under which the work is done, e.g. safety legislation.

Internal constraints and resources

People

Age distribution of current labour force.

Present skills of labour force.

Health and mobility of present labour force.

Ability of management to train, or to obtain services of appropriate commercial, training board or college training for staff.

Willingness of present staff to undertake new, expanded or diminished work loads.

Potential, i.e. internal promotional possibilities.

By careful examination of all factors, the family farm or the large estate can formulate a plan for the future so that retirement, resignation and sickness are handled with the minimum of 'disturbance' to the farm business. Training for present and future needs can then be given at the appropriate time and if necessary recruitment, or at the other extreme, redundancy, can be given the maximum time and attention. The plan should demonstrate:

(1) the required number of staff,
(2) the required skills for every major task,
(3) that staff effort is concentrated on the business objectives,
(4) that staff are deployed in appropriate tasks.

Internal succession

Formalised succession charts are a management tool in larger farm businesses and allied industries. A simple example (*Figure 25.3*) based on an arable farm employing one foreman, one head tractor driver and six other tractor drivers (TD) illustrates the principles.

In this situation the foreman is due to retire within three years, there appears to be a ready-made successor in the head tractor driver, who is gaining experience of management with responsibility for two other tractor drivers. Should the head tractor driver decide to leave, TDs 1 and 2 are probably too old to promote, TD 6 too young, TDs 3, 4 and 5, should be given careful consideration; if their supervisory potential is considered unsuitable it would reduce risk to replace TD 2, who retires within the year, with a person of foreman potential. The replacement of TD 2 on retirement is a key appointment.

Job analysis – job description

This is the basis of several management techniques and is a statement of:

(1) all the parts of the work to be done,
(2) the nature and extent of responsibilities for people, materials, machines and/or events,
(3) the physical effect and endurance required, and
(4) the working conditions under which the job is performed.

The statement can then be used in three primary ways:

General description of job and title

Key tasks   Date

1

2

3

4

5   (preferably not more than 8 - 10)

6

7

8

Accountable to . . .

Responsibilities . . .

Supervisory:

Tools:

Materials:

Reports:

Records:

Working conditions

Hazards

Safety:

Health:

Figure 25.4 Job analysis form

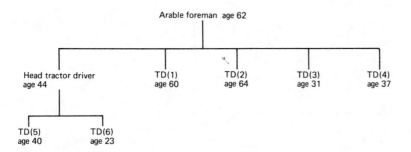

Figure 25.3 Succession plan

---

1. Personal data

   Preferred age:
   Home circumstances:

---

2. Physique

   A. Health and stamina
   B. Disabilities (not acceptable)
   C. Speech
   D. Manner
   E. Appearance

---

3. Basic education

   Essential:
   Preferred:

---

4. Technical training

   Essential:
   Preferred:

---

5. Work experience

   Essential:
   Preferred:

Job requirement	Example from job content	Means of checking
6. Sensory discrimination A. Vision B. Hearing C. Touch D. Taste and smell		
7. Mental requirements A. General aptitudes 1. Comprehension of terms 2. Reasoning and learning 3. Memory for names, faces and detail 4. Initiative & planning 5. Care, attention and accuracy. B. Special aptitudes 1. Verbal fluency 2. Verbal expression 3. Numerical aptitude 4. Spatial aptitude 5. Mechanical comprehension		

Figure 25.5 Personal specification

Job requirement	Example from job content	Means of checking
8. Motor requirements		
9. Character traits required A. Stability B. Industry C. Perseverence D. Loyalty E. Self-reliance F. Get along with Others G. Leadership		
10. Other requirements 1. Military commitments 2. Union membership 3. Licences		

Figure 25.5 cont.

(1) *Job evaluation* for the purpose of relating pay to job level. There is an intermediary stage in this, which is *job assessment*. At this point there is some determination of the factors in skills, responsibility, physical effort, mental effort and working conditions which are involved in meeting the demands of the *job description*.
(2) In the development of training programmes for groups of workers or individuals by testing actual performance levels against optimum standards and applying planned training to cover deficiencies and improve the speed of learning.
(3) In the basic analysis of a job prior to recruitment under any set of circumstances, vacancy or expansion to form a platform upon which to build the selection process (*Figure 25.4*).

Personal specification

This stage is specific to the recruitment process in determing what kind of person the employer wants to fill the job (*Figure 2.5.5*).

It will be noted that sections 3, 4 and 5 have two levels: essential = the minimum necessary standard to perform the job; preferred = the ideal level of education, training or experience to be able to meet all the requirements of the job at present and to be able to expand if required beyond the present work situation.

Sections 6–10 have two vertical divisions to assist the interviewer to consider the importance of any particular factor, e.g. 6A vision – if employing a hill shepherd it is essential that the shepherd should have good eyesight. Means of checking – take the applicant to a viewpoint and ask for identification of a distant object.

Section 8 – motor requirements refers to the coordination of body, hand and feet.

OBTAINING THE RIGHT LABOUR

Recruitment and selection

Objectives of selection

To select a new employee who is likely to be able to perform the tasks to the employer's satisfaction after training: in other words – that the person can do the job, or can learn the job, and that there is an intention to do the job. Therefore the selector's task is to predict each applicant's probable ability in the job.

Preparation and process

The process of recruitment and selection thus needs care, particularly if this is a relatively infrequent operation, and the opportunities for practice are limited. Proceed as follows:

(1) prepare a job description – on paper;
(2) prepare a personal specification – on paper.

The combination of these two tasks is *job analysis* and should take into consideration the main parts of the work to be done, the difficulty of learning or teaching any part, the areas of high risk (damage to stock, crops) and then the special attributes that a person will require in order to carry out the work.

(3) (a) Draft advertisement,
    (b) put out contact enquiries,
    (c) look at situations wanted;
(4) decide where and when to place advertisement;
(5) decide where and when to hold interviews.

Timing is of great importance at this planning stage in order to give possible applicants the opportunity to arrange time off for interview, etc.

(6) Examine and *answer* all replies;
(7) decide who you want to interview and when;
(8) prepare to hold interviews, including tests if applicable.

It should not be necessary to remind a potential employer to answer all letters and telephone enquiries.

(9) Hold interviews (and tests);
(10) *decide*
    (a) *yes*, there is a suitable candidate, *or*
    (b) *no*, re-examine job/personal specification and advertisement;
(11) make offer – take up references – reply to interviewees;
(12) prepare for arrival of new employee and induct on arrival.

Practice

The interview is generally used as the main selection technique.

Whilst it is unimportant as to which particular form of interview is adopted, the selector has certain factors to establish in order to make a prediction on suitability to perform the task.

(1) Check facts stated on letter or application form.
(2) Ensure that there are no unexplained career 'gaps'.
(3) If the selector is knowledgeable in the technical areas relevant to the job – assess the candidate's level of acquired skills.
(4) Assess the consistency of the candidate and realism in terms of the goals set, i.e. motivation.
(5) Decide what his/her likely impact would be in the work situation.

The question of impact is often given too much prominence in selection; however if you are unable to decide which of two candidates, who are technically and otherwise acceptable, to choose try asking the following questions of yourself:

(1) Which employee would you prefer to invite into your own home?
(2) Which family would be most acceptable in your community?
(3) How do you think each person would react if he/she did not get his/her own way?

Preparing for interview

Administrative arrangements should include:

(1) timetable of interviews;
(2) notification to candidates, and to others taking part in the selection;
(3) adequate waiting and toilet facilities;
(4) a light cool room in which the conversation cannot be overheard;
(5) complete freedom from any form of interruption.

The interviewer should then prepare by re-reading the job description, personal specification, and the personal history sheet filled in by the applicant. The interviewer can then decide the topics to be covered during the interview, and any areas which need special clarification or confirmation. Prepare a check list of these points.

Conducting an interview

Observing a set of rules will not automatically bring about a good result. A knowledge of some basic principles and good preparation will however help the interviewer.

Every interview should have a beginning – setting the candidate at ease and showing an interest in the candidate, a middle which is a thorough and logical exploration of experience, background, interests, medical history, attitudes and plans. The conclusion should also be planned to enable the candidate to raise points which may not have been covered in the middle stage. At this point the candidate should be told what is the next stage of the process and how he/she will be informed, also travel or other claims should be dealt with and the applicant looked after until leaving the farm. The social skills of the interviewer lie in encouraging the candidate to talk freely so that open-ended questions, which give the candidate the opportunity to expand are preferred to closed questions which require only yes/no answers or leading questions which suggest to the candidate the expected answer.

Induction training

Introduction

Recruitment and selection is not complete, when an offer of employment, having been made, is accepted. The new employee now has to be introduced to the business, an event which may be only days after the interview or could be months in the case of management or senior technical staff. Unless the introduction is thoroughly prepared and controlled the whole exercise of recruitment could be wasted, that is, the newly-appointed employee may leave the business thoroughly disheartened by attitudes formed in the first few days or weeks.

Industrial psychologists have examined this situation and have produced the following simple analysis. The new employee is tied by fear of an unknown situation so is afraid to ask questions when necessary. This, in turn, limits the ability of the employee to think clearly, take in information and may result in presenting a wrong impression to fellow workers and supervisors.

If this is kept in mind by the manager, supervisor or trainer the planning of what is called Induction Training becomes straightforward. The new employee has the opportunity and guidance to learn about the people with whom and for whom he/she works, the rules and working conditions, the products/crops of the business and the unfamiliar parts of the business. To this end each manager/supervisor responsible for new staff should have a written plan for each type of employee, which must be kept up to date. The detail will be different in each business and in different parts of each business but should always have a simple objective. The sooner a new member of staff can begin to earn wages the better for both employee and company.

Aims and objectives

The general aim of an induction programme is therefore to assist the newcomer to become familiar with work place conditions and be effective in his/her job as soon as possible.

The detailed objectives and content varies greatly

Table 25.2

On completion of induction the trainee will be acquainted with:
(1)   The general objectives of the business.
(2)   The main products of the business.
(3)   The general lay-out of the holding and the identity of the various units.
(4)   Other employees and their authority.
(5)   Conditions of employment.
(6)   The main duties of his/her job.
(7)   Future training to be received.
(8)   General rules and conditions required by the employer and matters concerning trainees' welfare, e.g. housing.
(9)   Safety regulations and their application to the holding.

depending on the individual and the job. Some examples are shown in *Table 25.2*.

Organisation

Induction training must be planned. The objectives above indicate that a great deal of information is required and various people will be involved during this period of learning. A programme should therefore be drawn up to ensure that this learning takes place at the right time, in the right place and at an acceptable rate. The main factors to consider are shown in *Table 25.3*.

Table 25.3

(1)   Who will organise and conduct induction training?
(2)   When will it start?
(3)   How long will the programme need to be?
(4)   Where will the various parts of the programme take place?
(5)   If equipment for demonstration is required, is it available?
(6)   Is transport available if required?
(7)   Are the various people involved in the induction programme free when required?
(8)   Will all new employees have the same programme?
(9)   Can visual aids be used in presenting information, i.e. farm maps, recording systems?

Summary

Induction training should be carefully tailored to suit the individual concerned. Objectives should have been considered carefully and a programme of activity set out to achieve these objectives. It is essential that the programme is carried out smoothly to ensure effective learning and to gain the goodwill of the new employee. The programme must therefore be carefully organised by a person of authority.

Continuation training

The establishment of training needs is derived from three main sources:

(1)   a continuous review of employee performance throughout the year;
(2)   an assessment of future supervisory requirement;
(3)   an assessment of future skills requirement.

The first of these is not an opportunity to apportion blame nor an assessment of personality but an ongoing and participative understanding of the task performance and involvment. The second is normally evident from a workforce plan, particularly in a farm expansion phase. The last is a constant requirement to keep up with technological change or a major change of farm policy, e.g. the introduction of a new enterprise.

Improving present performance

The best place for most training in relation to present performance is on the farm. There are many agents for the farmer to use as well as his/her own skill. Manufacturers, particularly manufacturers of agricultural machinery, find it is in their own interest, as well as that of their customer farmers, to ensure that the farm employees who use the equipment can use it properly, safely and effectively. Similarly it is possible, with the help of the Agricultural Training Board, to obtain the services of a trained instructor for most other farm tasks.

Alternatively, there are many 'off the farm' short and day release courses available. Local Education Authority courses in County Agricultural Colleges or in the Local Technical College, ATB courses locally or at Stoneleigh and commercially sponsored short courses, together with demonstrations and discussion groups, etc. organised by ADAS, NIAB, research stations and so on.

For these and all other training courses a word of caution. It is necessary to examine carefully the stated aims and objectives of any course and the level of trainee for whom it is designed. Much money and time is wasted sending people on unsuitable courses, often to their own frustration and especially damaging if at too high a level for the employee away from the home environment possibly for the first time for many years.

Training for future needs or potential

The dangers in training for future needs lie mainly in the timing of courses, too far in advance leads to frustration of the employees and possible loss; too late, to anxiety that they will be unprepared for the next step when it occurs. It involves an examination not only of present skills but also of attitudes and almost certainly will involve the employee in training away from the home farm. Certainly such courses as 'Effective Supervision' mounted by the ATB have as one objective a number of changes of attitude in potential and newly appointed supervisors regarding increased responsibility as well as knowledge and skill in labour management techniques.

Training programmes

After consideration of the factors above an annual training programme for all staff can be drawn up as a programme to fit farm need and course availability as well as individual leave arrangements. It would normally include:

(1) the individual training targets,
(2) the training to take place on the farm,
(3) the training to take place off the farm, e.g. courses,
(4) long-term training, e.g. full time or block release courses at college of agriculture, university, etc. where appropriate.

A comprehensive and job-related training programme is a helpful motivating force if used correctly.

Validation and evaluation of training

Training itself is of little use without the opportunity to follow initial training with practice and experience. Validation of training therefore is an exercise which only asks the question 'Did the training given achieve its stated aim?'. Evaluation asks the farmer to consider the benefits which can be listed as:

New staff becoming more effective more quickly.
A more effective and reliable work force.
Improved staff relations.
Better work organisation.
Greater safety.
Lower operating costs.
Better supervision.
Increased job satisfaction and pride in work.
Greater confidence to undertake present and new work.
Overall more effective farming.

REDUCTION OF LABOUR

A farmer or manager may be faced with the problem of reducing the labour force. Such a reduction may be due to a variety of causes. For example the termination of a major enterprise, going out of dairying completely but retaining and expanding other less intensive enterprises thus no longer needing the dairy herd staff. Procedures in the case of redundancy are properly set out in employer's and employees' guides available from the offices of the Department of Employment or from HMSO.

Greater difficulty is presented by the need to reduce labour because of low profitability related to high labour costs or a policy decision to switch some tasks from regular employees to contractors. Such changes must be made in ways that cause minimum hardship. There are procedures laid down for terminating a contract of employment.

(1) Notification of the relevant trade union when the number of employees is ten or more.
(2) Full consultation with all employees affected.
(3) A provision for retraining where this is appropriate.
(4) A call for volunteers – although management must retain the right to decide who is dismissed.
(5) If it is not possible to resolve the problem voluntarily then the criteria for dismissal must be clearly stated, i.e.
    (a) first in – last out,
    (b) over normal retiring age,
    (c) competence in relevant skills,
    (d) some other definite basis.

There are organisations and consultants from whom advice may be sought and statutory conditions with which an employer must comply.

MOTIVATION THEORY

In the early twentieth century the developers of management science looked upon the worker as a flexible but inefficient machine which would work reasonably well if the tasks were planned, orders given and work was supervised. Later it was discovered that pay for work done, even when financial incentives were applied, did not increase the quality or quantity of work performed for more than a short period of time and, with the development of social security in the western world – work or starve is not a single option. More recent management theorists have examined the traditional approach to labour management and suggested some practical alternatives.

Abraham Maslow

He evolved the theory of the 'hierarchy of needs'. This suggests that as a lower order of need is satisfied a new higher need is revealed. Thus a satisfied need ceases to motivate and is often illustrated in the form of a pyramid (*Figure 25.6*).

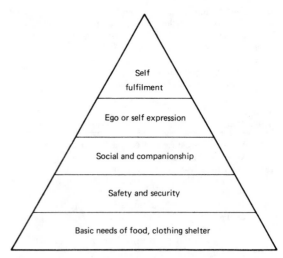

Figure 25.6

Douglas McGregor

He developed a theory to explain the traditional approach to direction and control of staff which he called Theory X. He compared this with the idea that people at work prefer involvement as *Theory Y*.

Theory X

(1) The average human being has an inherent dislike of work and will avoid it if he/she can,
(2) Because of this characteristic, most people must be coerced, controlled, directed, threatened with punishment to get them to put forth adequate effort toward the achievement of organisational objectives.
(3) The average human being prefers to be directed, wishes to avoid responsiblity, has relatively little ambition and wants security above all.

(Often known as the 'carrot and big stick' method of management as if people were donkeys.)

Theory Y

(1) The expenditure of physical and mental effort in work is as natural as play or rest.
(2) External contrόl and the threat of punishment are not the only means for bringing about effort toward organisational objectives. Human beings will exercise self-direction and self-control in the service of objectives to which they are committed.
(3) Commitment to objectives is related to the rewards associated with their achievement.
(4) The average human being learns, under proper conditions, not only to accept but to seek responsibility.
(5) The capacity to exercise a relatively high degree of imagination, ingenuity and creativity in the solution of organisational objectives is widely, not narrowly, distributed in the population.
(6) Under the conditions of modern industrial life, the intellectual potentialities of the average human being are only partially utilised.

Frederick Herzberg

He was the originator of the Motivation/Hygiene theory. He discovered that well motivated employees derived satisfaction through five factors all related to the job itself:

(1) 'doing the job';
(2) 'liking the job';
(3) 'achieving success in doing the job';
(4) 'receiving recognition for doing the job'; and
(5) 'thinking that doing the job will lead to better prospects'.

The fact that they might be paid more for continued success at the job came sixth in this table to motivation. These factors are called 'satisfiers'.

'Hygiene factors'

At the other end of the motivational scale there are 'dissatisfiers'. These all relate to the context in which the job is done. Bad policy and administration, poor supervision, waste, duplication of effort, unfair personnel policy and constant criticism are examples that create dissatisfaction. Remuneration again appears in the scale – insufficient salary or wage is a 'dissatisfier'.

Herzberg calls the dissatisfiers the 'hygiene factors' likening them to the role of public health in the control of disease. Good hygiene, eliminating disease factors does not cure diseases but it is important in preventing their occurrence. Similarly, the absence of dissatisfiers in a work situation may prevent an employee being unhappy in his/her employment, but their absence will not motivate an employee. It is important to realise that the causes of satisfaction are not the same factors as the causes of dissatisfaction. If a factor giving satisfaction is absent, the result is not dissatisfaction but just no satisfaction and similarly if a cause of dissatisfaction is absent the result is not satisfaction but no dissatisfaction.

Equity theory (ascribed to several authors)

It is often said that Herzberg's theory, so relevant to the farm situation does not take sufficient account of money as a motivator. This is not the case as has been shown above but he does disregard money as a comparison. Equity theory suggests that people at work can be demotivated by comparing their earnings with the earnings of others and the work performed by each.

My pay for *my efforts* should be comparable to *his pay* for *his efforts*. This type of motivator equation takes us into the world of pay differentials and the human rationalisation of '*I* would not do that job: *He* (the comparison) deserves how (little/much) he gets'. The difficulty arises when the manager is unable or unwilling to adjust the pay and the operator reduces effort to make the equation more equal again.

Management needs to search for the factors within the farm business which will motivate their work force. Removing demotivating factors usually involves direct expenditure of money (improvements in accommodation, etc.) and supervisory training. Motivating the staff needs identification of goals, stimulating the interest in achieving personal and organisational targets and making sure that these goals, once identified are realisable in some measure.

Payment policy and incentives

General

The payment of a wage or salary for work performed is a factor in motivating an employee. It is at the same time a major tool to use in achieving the personal objectives of a business. Therefore, the objectives of any payment policy may be clearly expressed as;

(1) to attract and retain sufficient staff of sufficient quality to meet the objectives of the business;
(2) to provide, with other forms of encouragement, sufficient incentive to staff to give more than minimum effort;
(3) to achieve the first two objectives at minimum long-term cost.

In practice this means that the policy must provide competitive wages with other businesses, considered to be 'fair' internally between different grades and skills, not out of proportion to the overall cost/returns of the business and yet offering at the same time the necessary incentive mentioned before.

Internal constraints

The policy should be logical, tidy and internally consistent. This requires careful planning and at times careful adjustment to bring into line those accidents of payment, such as the high wage or salary a single specialist can expect until there are more equally skilled specialists available. Currently in the UK there is a shortage of mathematics and physics teachers in education. Educational employers are therefore offering incentives of both salary above basic and accelerated promotion to attract specialists. If the employers are successful, it will take a number of years before the anomalies of payment and apparent injustice to other skill areas works its way out of the system.

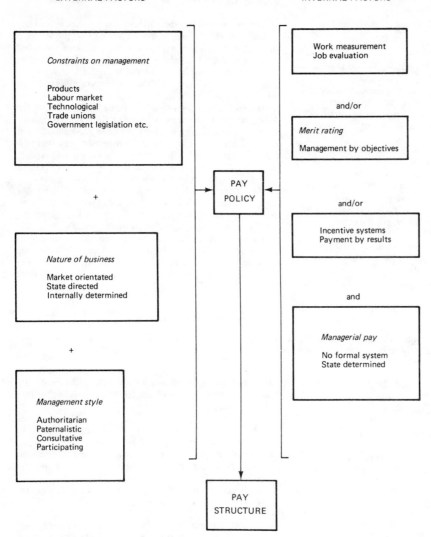

EXTERNAL FACTORS

INTERNAL FACTORS

Constraints on management

Products
Labour market
Technological
Trade unions
Government legislation etc.

Work measurement
Job evaluation

and/or

Merit rating

Management by objectives

PAY
POLICY

and/or

Nature of business

Market orientated
State directed
Internally determined

Incentive systems
Payment by results

and

Management style

Authoritarian
Paternalistic
Consultative
Participating

Managerial pay

No formal system
State determined

PAY
STRUCTURE

Figure 25.7

Similarly computer programmers were in great demand a few years ago throughout most of western Europe. As programmers become more available their wages tended not to be increased at the same rate as other technologists and managers, so that today the wage is generally in line with their contribution to the business and other staff.

The exception to this is still found in those companies where there was no policy, or it was not adequately controlled. The policy should be easily understood and operated; should have flexibility to deal with the planning and cost control systems on the business.

Payment strategy

There are two groups of factors, both of which change with time, which have to be examined before a policy can be established. These are 'external factors' and 'internal

factors'. Examples of these are given in *Figure 25.7*. It is possible to see almost any combination of these factors leading to a payment policy and adding the question 'does the business want to be a market leader or follower on pay'? to arrive at a pay structure which meets all the requirements of the three clauses of the general statement.

Pay

Agricultrual wage minima are determined by national negotiations. The Agricultural Wages Board defines categories of workers and lays down the minimum rate which must be paid to full-time and part-time workers with differential rates, according to age to 20, against agreed hours. It also sets minimum rates for classes of workers, for overtime and night work and the calculation of appropriate stoppages for board and lodging, even a rate for the shepherd's first and subsequent dogs. These rates are normally negotiated annually and are due for payment on publication of the Wages Board order or from the set date specified therein.

The wages board in no way prevents an employer from paying a higher rate if wished or the job market compels.

Wage payment

Wages, as opposed to salaries, are normally calculated on an hourly basis and paid in cash weekly. Employers may ask hourly paid staff to accept payment by bank or giro transfer to reduce the risk in having large sums of money in the farm safe, but employees have the right to insist on cash payment. Employers may not ask any employee to accept payment in kind.

Incentives

Types of incentives

Incentives can be classified under three broad headings:

(1) non-financial;
(2) semi-financial;
(2) financial.

Non-financial

Security of employment
    Working conditions, hours and holidays.
    Opportunities for promotion.
    Opportunities for training and education.
    Job satisfaction.
    Good management.
    Good communications.

These are difficult to measure but are more vital to the success of a payment structure than any specially constructed incentive schemes. The best incentive, in all circumstances, is the evidence of good management continuously in action.

Semi-financial

Pension, insurance and sick pay.
    Subsidised meals and social facilities.
    Farm perquisites – free milk, fuel, etc.
    Free or low-rent housing.
    Company car.
    Prize award schemes.

This is the area of fringe benefits. As distinct from the benefits in the section above these rewards can be given a firm financial value.

Prize award schemes

This type of scheme is the odd one out in the list of fringe benefits because it depends upon the attainment of a specific target.

Financial

Wages and salaries – job evaluation
Ad hoc   – annual bonuses
                  – profit sharing
                  – suggestion schemes

Performance based – merit rating
                       – share of production
                       – rate fixing
                       – related to specific targets
Work study based – directly proportional to output
                       – geared schemes
                       – high stable earnings

Job evaluation   Job evaluation is a formal process used to compare one job with another in terms of skill, responsibility, physical requirements in order to arrive at a consistent and equitable pay structure in an organisation. A number of grading systems are in use. The most widely used is the 'points rating system' where each job is analysed under a number of headings and a points value put upon each factor. Certain key jobs are analysed in depth to provide benchmarks as a reference point for the remainder.

Merit rating   Merit rating or performance appraisal aims at a periodic assessment of an individual's performance based on such factors as:

(1) quality of work,
(2) consistency of output,
(3) reliability,
(4) willingness to cooperate.

Individuals assessed in this way are paid weekly fixed sums until re-examination takes place.

Share of production   These are usually company wide with incentive schemes based on added value of reduction in costs. A standard labour cost is formulated and the actual labour cost in any one accounting period is related to this to determine the bonus distribution.

Rate fixing   Prices or times for jobs are estimated from experience, comparison with other jobs, or from rough and ready timings. They are usually agreed after some bargaining between operatives and management.

In industry this method is often used in heavy engineering or maintenance work and in farming most piece-work rates have no firmer base than this.

Related to specified targets   This describes those financial incentives related to output or a performance factor and not time or the work content of a job, e.g. quantity of milk.

Implementation

It is essential to ensure that an incentive scheme is both necessary and will make a contribution to profitability either directly by increased output or lower costs or indirectly by improved quality or lower waste. In general a bonus scheme should meet the following conditions:

(1) The results against which the incentives are to be paid must be influenced by the participant's skill, judgement or effort.
(2) The rewards offered should be related to the effectiveness of the effort applied.
(3) The scheme should be easily understood and the terms and conditions set out and agreed.
(4) The bonus earnings should be simple to calculate and to verify.

(5) Payments should be made as frequently and soon as practicable after the performance.

(6) The bonus earning opportunities should offer a significant addition to the basic wage. (A target of 25% of the basic wage is suggested as a normal base.)

(7) Quality standards must be rigorously maintained.

Other desirable factors are:

(1) On the whole individual incentive schemes are more effective than group schemes.

(2) Bonuses should be the same for all taking part in a particular scheme.

(3) There should be little or no mixing of day work and incentive work.

WORKING CONDITIONS

Need for procedures

Unfortunately, in any employment situation, there may be a need to dismiss an employee. It may be that the job has simply ceased to exist as discussed earlier, or possibly there may be severe misconduct. Much legislation since the Contract of Employment Act 1963 has provided justifiable protection for employees and has laid down guides for employers. Most particularly minimum standards are set for the form and period of notice to be given under differing circumstances. As this legislation is liable to be changed in detail by a party in government and to meet changing political and economic conditions, current minimum requirements are not stated here. The details and supporting employers' and employees' guides are fully available from the offices of the Department of Employment or from HMSO.

It must be made quite clear that Government lays down minimum standards, there is nothing to prevent an employer offering better terms either in length of notice or for severance money. There is one definite onus on the employer to record in detail every disciplinary action and have clear procedures, steps to follow, as should the employees to deal with grievances or appeals on a disciplinary decision. Again guidelines are available from the Department of Employment, from the NFU and the T and GWU.

Facilities

The farmer is one of the small number of employers who may have an interest not only in employees at work but also after work; the worker's home frequently being on or part of the estate.

As has been mentioned under motivation a sound and comfortable house will not necessarily make an employee happy at work, but bad housing will certainly make him unhappy. Similarly provision of washing and changing facilities may be essential under safety provisions; it requires more imagination then money to make facilities acceptable and pleasant for staff.

Retirement

Approaching retirement age, staff should have the subject discussed with them by the employer. It may be that the retiring employee has everything planned or he/she may be resentful. It is a subject requiring much tact on the employer's part but it can be greatly eased by having a check list.

Has the employee a house to live in after retirement? (The farm-provided house may be required for a successor.)

Will the employee want part-time or occasional employment for a few more years and can the farm offer the opportunity?

Will assistance be required in moving?

Has the farm in the past given rights (e.g. rabbit shooting) which may prove an embarrassment if continued or terminated?

Has the employee sufficient pension (this should be discussed as far before retirement as possible if the farm does not have a pension scheme for employees)?

Are any of the perquisites to be continued and for how long if offered?

Whilst the labour force on a farm may be small in numbers by comparison with industry, and relationships between employer and employee less distant, there is no less requirement for the employer to have a sound policy for the most flexible resource – people.

References

HERZBERG, F. (1966). *Work and the Nature of Man.*- Cleveland: World Publishing Co.

MCGREGOR, D. (1960). *The Human Side of Enterprise.* London: McGraw Hill

MASLOW, A. (1954). *Motivation and Personality.* New York: Harper Bros

26

Agricultural law

G. Spring

INTRODUCTION

This section is designed to provide an introduction to law as it affects the agricultural industry. It should, however, be recognised at the outset that most people concerned in it will need to take professional advice on legal matters sooner or later and it is important for those involved in management to establish relations of mutual confidence with a firm of solicitors well versed in rural business. Solicitors may be regarded as the general practitioners of the legal profession but through them it is possible to obtain the services and advice of specialists in advocacy and in specific areas of law; such specialists practise as barristers. Nevertheless, some knowledge of the relevant law will help agriculturalists to assess when professional advice is necessary and also to alert them to the heavy duties laid upon them by the law.

These duties are such that the farmer will need not only the assistance of lawyers but of insurers and it is essential that the skills of a competent insurance broker are utilised in covering the risks involved. In this connection also the work of the industry's own organisations should be recognised, for example, the NFU, the CLA and the TGWU (Ag Section). Not only do they advise and support their members in handling their own legal problems, but they also act as pressure groups that can lobby for alterations in the law and monitor changes proposed by government and by institutions of the European Community. Law is not static; if a sufficient number of people want it changed, it can and will be changed.

This chapter contains an outline of the law as it relates to ownership, possession, development and use of agricultural land and to the liabilities of those who live by it as employers, employees and as self-employed persons. It does not seek to provide a comprehensive guide to all the rules and regulations that affect the industry but to show the general pattern of law as it affects agriculture and to point to the sources of fuller information and new decisions and enactments.

Whereas some ten years ago most lawyers, when speaking of agricultural law, thought in terms of agricultural tenancies and little else, it has now been recognised that agriculture has a special place in environmental law. Its role as creator and

victim of damage caused by pollution, a matter inextricably linked to the changes in production seen during recent years, the problems posed by public demand for greater access to the countryside and the role of agriculture in relation to the protection of natural flora and fauna and overall landscape values have prompted new developments in legislation. These have made lawyers and agriculturalists alike recognise the greatly expanded scope of agricultural law. This is reflected in the pages that follow.

THE ENGLISH LEGAL SYSTEM

English law shares with our agriculture a long history. Many of its features retain traces of medieval origin, but in recent times large accretions and alterations have shown it to be as lively and innovative as modern farming. Our law is very different in substance and procedure from continental European practice – although this is not to say that ideas of justice differ – and it should be noted that in the rulings of the Court of Justice of the European Communities there now exists a unifying factor of particular importance in relation to agricultural law. But agricultural law is not a separate part of English law; the same general principles apply in agricultural as in other areas. Thus an appreciation of the setting in which it occurs is necessary for its understanding.

Law consists of those rules of conduct that the courts will enforce and in this connection we include in the term 'court' all those bodies recognised by the judges as having an obligation to act judicially. So in addition to the civil and criminal courts we recognise various tribunals and other bodies set up by act of Parliament. Civil courts decide disputes between fellow citizens where one, the plaintiff, alleges that he has suffered injury or loss by the unlawful act or omission of another, the defendant. If the defendant is adjudged legally responsible for the injury he will be required to compensate the plaintiff, normally by a money payment called damages. When a state agency causes such an injury an action brought against it will also be a matter of civil law. The criminal courts decide cases where the state is involved as prosecutor against a citizen accused of committing a criminal offence. Accused persons, if found guilty, are punished by fine or imprisonment or both. The two sets of courts are kept separate; Magistrates Courts and the Crown Court hold criminal trials while County Courts and the High Court hear ordinary civil cases. Certain matters are however reserved for specialised tribunals of which the Agricultural Land Tribunal, the Lands Tribunal and Industrial Tribunals are particularly relevant to the agricultural industry. At the top of the hierarchy of courts the Court of Appeal and the House of Lords, sitting as a court, provide an appeal structure with the possibility of reference to the European Court on questions of European Community law.

Most hearings in court are concerned with matters of fact, in criminal trials the prosecution will have to prove 'beyond reasonable doubt' that the defendant has committed the crime of which he is accused; in civil trials the court will decide 'on the balance of probabilities' whether the events concerned took place as alleged by the plaintiff or not. But in some cases the decision depends on a question of law – for example the Dairy Produce Quotas Regulations 1984 refer to 'land used for milk production'. In Puncknowle Farms Ltd v Kane (1985) 3 All ER 790, the High Court had to decide whether this phrase included only the farm area used for current milk production or whether it also included land used for dry cows and for young heifers, i.e. land used to support the dairy herd as a whole. The court decided that, in the context of apportionment of quota on partial transfer of a holding, the phrase included land used for the support of the dairy herd as a whole, and thus included the land used by followers, as well as by cows in-milk.

It is to be hoped that when an agriculturalist wishes to know the law on a certain topic a source will already exist in clear terms and be readily available without the need for litigation. There are two principal sources of law – decided cases and statute. Case law has been built up into a system termed the common law because, when a superior court makes a decision turning on a point of law, that decision becomes a precedent binding on judges dealing with subsequent cases involving the principle. Statute consists of Acts of Parliament and Statutory Instruments, i.e. orders and regulations made under the authority of acts of Parliament by duly authorised Ministers. EEC Treaty provisions are incorporated into English law by virtue of the European Communities Act 1972 and that act also gives statutory force to Regulations passed by the EEC Council. EEC Directives normally require legislative action by the UK Parliament before implementation but they are capable in certain circumstances of having direct effect. Decisions of the European Court also create precedents which UK courts must follow. For example, in Von Menges (Klaus) v Land Nordrhein-Westfalen No. 109/84 1986 CMLR 309, the plaintiff had kept a herd of dairy cows until 1980 when he obtained a premium for going out of milk production. He undertook, as required by Regulation 1078/77 EEC, not to market milk or milk products for five years. In 1981 he let his farm to another farmer who planned to use it for milk-sheep. He asked in the German courts for a declaration that the marketing of sheep's milk would not be contrary to the undertaking given. The local administrative court referred to the European Court under Art. 177 EEC the question of the meaning of 'milk and milk production' in the Regulation. The court held that the words concerned ewe's milk and ewe's milk products as well as cow's milk and cow's milk products, since otherwise the premium payments would only encourage the replacement of dairy cows by milk-sheep leading to new surpluses. This decision is binding upon

courts throughout the European Community which of course includes UK courts, who will in future follow this interpretation of 'milk and milk products' in this context.

A statutory provision overrules any common law precedents that directly conflict with it. Nowadays the definition of criminal offences and the powers of courts to penalise offenders depend almost exclusively upon statute and a vast range of legislation covers fiscal and commercial matters together with the social activities of government, e.g. housing, employment, public health, and social security. Even in the field of property law, contract and tort (*see below*), formerly the domain of common law, parliament has codified or amended some of the rules developed in the courts.

Thus, to find the law relevant to a particular topic it is necessary to know if statutory rules apply – for instance in respect of security of tenure for full-time farm workers occupying service cottages the position is governed by the Rent (Agriculture) Act 1976, and questions concerning the rights of the farmer and worker when employment comes to an end can only be resolved by reference to that act. When the topic is one that is covered by common law the decisions of relevant cases can be found in the Law Reports where decisions of significance are recorded. In practice, a legal practitioner will depend upon books of reference and texts dealing with specific areas of the law to enable him or her to find the statutes and cases that are relevant. The agriculturalist must know about changes in the law relating to his or her business. Details of many new statutory rules are publicised by government agencies, e.g. on employment law by the Department of Employment, on safety regulations by the Health and Safety Commission, but to keep up to date a farmer should read the professional journals. Criminal offenders are liable to punishment, those committing civil wrongs to pay damages or to have their activities stopped by an order termed an injunction made by a court. These are the means by which the law is enforced. By the standards of other legal systems enforcement of civil judgements in England is reasonably effective. Debtors may have their property sold up, bankrupt persons are subject to severe business disabilities. Flouters of injunctions may be imprisoned for contempt. Although legal procedures can be protracted and nothing is to be gained by suing an impecunious defendant the farmer should be prepared, when it is businesslike to do so, to assert legal rights just as much as he or she should be careful to fulfil legal duties.

LEGAL ASPECTS OF THE OWNERSHIP, POSSESSION AND OCCUPATION OF AGRICULTURAL LAND

Most agricultural enterprises are run by owner-occupiers or tenants of agricultural holdings. In many cases these will be single individuals or partnerships and the persons involved will be legally responsible for any obligations that arise. Where the enterprise is run by a limited company the company is recognised as a legal person with rights and duties separate from those of its members. The directors of an enterprise organised in this way must however recognise their obligations under the Companies Acts which contain provisions designed to prevent them from using the advantages of corporate personality to defraud creditors, members and employees of the company.

English law recognises only two ways by which land may be held: freehold and leasehold. If it is desired to tie up freehold land in the ownership of succeeding generations within a family this can only be done by the creation of a trust set up in accordance with certain rules – known as Equity – developed in the Court of Chancery, now a part of the High Court. This definitely requires professional expertise.

A freeholder has the largest possible freedom to decide how to use his land that the law recognises. It has never been possible for a freeholder to do whatever he pleases to the detriment of neighbours but in recent times to the limits already imposed by the law of nuisance have been added the constraints of town and country planning and the compulsory acquisition of land. Public and private rights of way may also diminish a freeholder's privacy and land can be lost by adverse possession. A freeholder's rights may be reduced by restrictive covenants and his obligations increased by a mortgage. Although ownership of land normally includes rights over what lies in and beneath the soil, mineral rights may be held separately from the freehold; coal and oil deposits belong to the state. Water may be taken from surface streams by riparian owners for normal agricultural purposes such as watering stock (not for aerial spraying) but it is generally the case that for other purposes it may only be taken under a licence from the Water Authority and the same applies to most water taken from underground sources.

The freedom of landlords and tenants to negotiate the terms on which property is let has been greatly circumscribed by statute in recent years – in no area more so than agriculture – and this is true not only as between landowner and tenant farmer but between farmer and farm employee. An account of the present position is given below but it should be emphasised that important regulations are made from time to time adjusting the details.

Both the owner-occupier and the tenant farmer have duties as occupiers towards neighbours, visitors and trespassers. These duties are in general the result of principles developed from case law but they also arise as a result of Public Health Acts and from Health and Safety legislation. Recent legislation has imposed new constraints on agricultural activity and the use of land in the interests of production limitation or conservation or public access. This new feature of agricultural law is described briefly below.

Farm tenancies

The law relating to the letting of agricultural property differs from the general law on leaseholds. It is therefore appropriate to consider it in more detail. In the first half of the twentieth century the normal agricultural tenancy developed under the common law was a tenancy from year to year terminable at six months' notice by landlord or tenant. Statutory provisions have now changed this law radically.

Security of tenure

Security for tenant farmers was first introduced as a permanent measure in the Agricultural Holdings Act 1948 by placing restrictions on the landlord's right to give notice to quit to his tenant. This Act together with all others dealing with agricultural tenancies has now been consolidated in the Agricultural Holdings Act 1986. If the landlord serves a 12

months' notice to quit and the tenant does nothing, the notice will be legally effective, but the tenant may, within one month, serve a counter-notice if he does not wish to go. In this case the landlord's notice will not have effect unless the Agricultural Land Tribunal consents to its operation.

The tenant will be deprived of his security of tenure only on a limited number of grounds, i.e. that the purpose for which the landlord requires possession is:

(1) in the interests of good husbandry;
(2) in the interests of sound estate management;
(3) in the interests of agricultural research or education, or for the provision of small-holdings and allotments; or
(4) that greater hardship will be caused by the withholding than by the granting of consent.

If the Tribunal finds one or more of these grounds proved, it has nevertheless to refuse consent 'if in all the circumstances it appears to them that a fair and reasonable landlord would not insist on possession.'

The tenant's right to serve a counter-notice is excluded by a strictly limited list of cases covering consent by the Tribunal for the reasons given above, failure to remedy a breach of the contract of tenancy or bankruptcy of the tenant, or bad husbandry, or failure to pay rent due by the tenant, also the intended use of the land for non-agricultural purposes for which planning permission has been granted or is not needed. Under the Act there are special rules regarding notices to remedy breaches of the tenant's obligations to keep fixed equipment including hedges, ditches, roads and ponds in good order and the tenant may go to arbitration if he disputes the landlord's claim. An arbitrator can be chosen by agreement of the parties or one of them can ask the President of the Royal Institution of Chartered Surveyors to appoint one from an approved panel.

Until the Agriculture (Miscellaneous Provisions) Act 1976 the death of a tenant enabled the landlord to serve an incontestable notice to quit. This was then changed to enable security of tenure to be claimed for up to three generations including the first occupier, i.e. there can be two succession tenancies. In 1984 the law was changed to enable new tenancies to be created without such security and the rules relating to succession are now included in the Agricultural Holdings Act 1986. This means that security of tenure for tenants' successors now applies in effect only to tenancies that were in existence during the period 1976 to 1984, but most tenancies existing today are in fact succession tenancies. Under these rules eligible persons can apply to the Tribunal for a tenancy of a holding whose tenant has died. To be eligible a person must be the widow, widower, brother, sister or child, natural or adopted, of the decreased, have derived his or her livelihood from agricultural work on the holding for five years out of the previous seven (up to three years spent at college or university will count) and not be the occupier of another viable commercial unit. An eligible applicant must also be found by the Tribunal to be suitable to take over the holding; suitability is judged by the agricultural experience, age, health and financial standing of the applicant.

If a landlord serves a notice to quit within three months of the death of a tenant and if no application is made by a suitable successor this will end the tenancy. If an application is made the landlord can dispute the application on grounds of unsuitability or ineligibility of the applicant or on grounds of good estate management or hardship but the 'fair and reasonable landlord test' still applies. Since 1984 succession is also possible subject to the above conditions on the retirement of a tenant aged 65 or over either by agreement between landlord and tenant or on the tenant's application to the Agricultural Land Tribunal.

The law on security of tenure and succession is complicated and there have been many rulings given by the courts on matters of detail. Moreover the issuing of the relevant notices must be made according to strict time limits and in prescribed form. Both landlord and tenant should insist upon skilled advice from qualified lawyers or surveyors.

Obligations of landlord and tenant

The landlord and tenant of an agricultural holding will be bound by the terms of their agreement but there has been considerable intervention by statute so as to modify and extend their contractual obligations. Either party can insist on a written agreement and the terms may be fixed by arbitration. The parties can fix the rent at the start of the agreement and change it at any time by agreement. The 1986 Act provides for the rent reviews at three year intervals and either party can demand an arbitration to fix it. The rules are to be found in Schedule 2; the rent has to be fixed by reference to a number of factors including the productive capacity and related earning capacity of the holding. Scarcity of lettings has to be disregarded when taking the rents of similar holdings into consideration as comparables. Although the parties may agree their respective maintenance and insurance obligations 'model repair clauses' are set out in the Agriculture (Maintenance, Repair and Insurance of Fixed Equipment) Regulations. Improvements undertaken by the landlord with the tenant's agreement may lead to an increase in rent. If the tenant carries out improvements he will be entitled under certain circumstances to compensation at the end of the tenancy in according with the Act and relevant regulations. The Act also deals with the tenant's right to remove fixtures and the landlord's right to purchase them. When a tenancy comes to an end there will normally be a settlement of claims as between landlord and tenant, by arbitration if necessary, for disturbance and delapidations. When milk quotas were introduced in 1984 it became evident that the value of a landlord's property could be greatly diminished if a tenant gave up a tenanted farm's quota, in general however, for a tenant to do so now would be a breach of the contract of tenancy. Tenants meanwhile felt that, where the milk output of a farm had been increased during their term of occupation of a holding, the consequent size of the quota allocated should be reflected in compensation payable to them by the landlord when a tenancy was terminated. In the Agriculture Act 1986, S. 13 and Schedule 1 provision is made for such compensation for existing tenancies.

In general it may be said that a tenant farmer has freedom to crop a holding as he sees fit despite contrary indications in the tenancy agreement. There are, however, limits on a tenant's freedom in the last year of a tenancy when items of manurial value may not be sold or removed from the holding and the tenant may not establish a cropping scheme at any time that is not in accordance with the practice of good husbandry.

Occupation of agricultural land without security of tenure

If prior approval is given by the Minister of Agriculture, which will only be done for special reasons, an agreement for a letting for a specified period will be terminable when the period expires. By a quirk of the Agricultural Holdings Act a tenancy for more than one year but less than two does not have statutory protection, nor does a short grazing agreement. Thus an agreement for grazing or mowing provided it is for a period of less than a year gives no protection to the taker but it must not permit ploughing or re-seeding or use of buildings unless they are simply shelters for grazing stock. Permission given to an employee to use agricultural land by a landlord employer during the period of employment does not create an agricultural lease. Partnerships between farmers and owners of agricultural land for agricultural business purposes do not create tenancies in favour of farmers. It must, however, be remembered that a partner is liable for his fellow partner's debts while a lease creates no such liability as between landlord and tenant. Share farming agreements are also possible but such contractual agreements require expert drafting.

Service occupation agreements

When houses are provided for employees by an owner-occupier or a tenant farmer the arrangement between them will normally be a service agreement acceptable to the parties. In accordance with the Protection from Eviction Act 1977 an employee other than a full-time farm worker will have six months' security of tenure when his contract of employment ends, subject to recourse to the County Court by the employer if the latter has urgent need of the accommodation. A full-time farm worker however may have rent deducted from his wages only as laid down by the current Agricultural Wages Board Order and has security of tenure of his house under the Rent (Agriculture) Act 1976. Provision is made in the Act, which otherwise gives the farm worker the same protection as that enjoyed by statutory tenants under the Rent Act 1977, for the County Court to grant an order for the employer to recover possession where suitable alternative accommodation is provided by another person, e.g. a housing authority. Local housing authorities are in effect obliged under the 1976 Act to rehouse the outgoing worker when the employer needs the house for another farm worker on grounds of agricultural efficiency and on this question the employer, the employee or the local authority can obtain the advice of the Agricultural Dwelling House Advisory Committee.

Occupier's liability

Any person recognised by the law as the occupier of land owes duties towards persons who enter upon it and towards those who are in its vicinity. These duties are imposed by the law of tort which covers wrongs caused by a person's failure to carry out duties imposed by law as contrasted with duties imposed by a contract. The common law has evolved a number of such duties which have been recognised in judicial decisions as falling within the categories of trespass, nuisance, negligence and strict liability to which further rules have been added by statute.

A farmer is more likely to suffer from trespassers against

whom an action for damages or an injunction may be brought, than to be a trespasser himself. However, if his animals trespass onto neighbouring land he will be liable for damage they cause unless it was the fault of his neighbour or another person and could not have been reasonably anticipated. In this connection it should be remembered that a farmer is responsible for keeping his own stock in; he cannot complain if his animals escape through his neighbour's fence. Much of the law on liability for animals is covered by the Animals Act 1971 which also changed the rules about animals straying onto or off the highway. A person who negligently allows this to happen is now liable for damage done. But there is no duty to fence in animals grazed on common land by those who have the right to do so if it is customarily unfenced. The Act also permits a farmer to shoot dogs worrying livestock. However, he must notify the police within 48 hours in order to have a defence if he is sued by the dog's owner.

Persons entering land lawfully, known technically as 'visitors', as guests or for payment or because they have a statutory right of entry (e.g. Health and Safety Inspectors) are owed a duty by the occupier defined in the Occupiers Liability Act 1957 as 'a duty to take such care as in all the circumstances of the case is reasonable to see that the visitor shall be reasonably safe in using the premises for the purposes for which he is invited or permitted by the occupier to be there.' The Act permits the occupier to 'restrict, modify or exclude' the common duty of care by contract or adequate notice but since the passing of the Unfair Contract Terms Act 1977 it is normally impossible for an occupier to exclude liability for personal injury caused by his own negligence.

Trespassers were regarded by the common law as entering upon other people's land at their peril until in the 1970s (see British Rail Board v Herrington (1972) AC 877), the courts began to make rulings that implied that a farmer would be expected to take care to prevent child trespassers from encountering hazards. This idea has been confirmed and widened by the Occupiers Liability Act 1984 which imposes a duty of care on occupiers towards 'persons other than visitors' such as trespassers. The duty is to take reasonable care to give such persons protection against known dangers, for example by putting up warning notices, although this would probably not be sufficient where child trespassers are known to be at risk. Farmers allowing persons to enter their land for recreational or educational purposes are allowed to exclude their duty of care towards such persons so long as they are not allowed entry as part of the farmer's business enterprise.

'Mass trespass' has caused considerable loss to farmers in certain areas since the formation of 'hippy convoys'. Trespass is in itself not a crime but under the Public Order Act 1986, S. 39 police have powers to order trespassers to leave land where two or more have entered as such and done damage or used threatening behaviour or brought 12 vehicles onto the land. Failure to leave in response to such an order is an offence.

If a farmer's methods create a nuisance such as excessive smells, noise or pollution then a neighbouring owner or occupier can bring an action for damage, and for an injunction to prevent repetition. Country dwellers are expected by the common law to put up with a reasonable degree of inconvenience arising from agriculture but intensive livestock practices have resulted in several successful cases against farmers in recent years. When a nuisance interferes with the rights or convenience of the public or a section of it then it is a public nuisance. An individual who suffers more than most can sue but the perpetrator is open to prosecution and local authorities have powers under the Public Health Acts 1936 and 1961 and the Control of Pollution Act 1974 to order abatement through the Magistrates' Courts. Decayed buildings, contaminated water sources, cesspools, rubbish dumps and waste land all fall under the rules in the Public Health Acts and there are specific provisions in the Refuse Disposal (Amenity) Act 1978 that give local authorities power to deal with abandoned vehicles and other rubbish. There is no liability at common law for the normal spread of weeds or wild pests, but the Ministry of Agriculture can take proceedings under the Weeds Act 1959 and the Pests Act 1954 against the occupier of land harbouring them.

As a result of the leading case of Rylands v Fletcher (1868) LR 3 HL 330 an occupier is strictly liable for the escape of any potentially dangerous thing kept on his land if it escapes and injures neighbouring property or people. Defences to such a claim are very limited and the occupier is liable even if the escape was caused by a contractor. The liability is, however, limited to non-natural used of land and this probably means that things commonly brought onto agricultural land for agricultural purposes give rise to liability only if released negligently. An example of negligent use of pesticides leading to liability in damages is Tutton v Walter (1985) 3 WLR 797. In this case a farmer sprayed oilseed rape in flower without warning neighbouring bee-keepers. The farmer who was aware of the danger to bees had failed to follow the recommendations on the insecticide containers and as a result wiped out his neighbour's bees.

A farmer will cover his liabilities in tort as occupier by insurance but it should be remembered that insurance covers only for legally enforceable claims. Agriculture as an industry has a bad record so far as accidents are concerned and the utmost vigilance is required for the law does not attempt to compensate for the incompensatable such as the death of a child drowned in a slurry pit and damages cannot make good injury caused to an active adult crushed under a runaway machine.

Legal constraints on the development and use of agricultural land

In recent years there has been a change in the attitude of the public towards the agricultural industry as technology has facilitated higher production and greater intensification. This is perceived as a danger to wildlife and also to human health and amenity because of an increase in pollution, particularly of water supplies. There is also greater public pressure for access to the countryside, but this aspect is beyond the scope of this chapter.

Planning law, which is based on the Town & Country Planning Act 1971, controls the development and use of land. Use of land for agriculture is permitted because this activity does not constitute development under the Act. Other activities, i.e. building, mining, quarrying or the carrying out of engineering works are development and so, according to this Act, is a material change of use, e.g. conversion of farm buildings to dwellings or for industry. For such development, planning permission must be obtained from the planning authority – normally a District Council.

Minor additions to buildings do not usually require permission but the extent of such additions is defined in legislation and may be limited in conservation areas. Agriculture enjoys a special position in that planning permission is deemed to have been granted for agricultural buildings not exceeding 465 m² in area, subject to conditions as to height and proximity to roads. Changes in planning law have been projected recently affecting agriculture in various ways, giving more freedom for changes in the use of redundant buildings but reducing the freedom to build close to land used for housing. The law on this topic is expected to show some alteration in the later 1980s.

The Wild Life and Countryside Acts 1981 and 1985 greatly extend the protection of birds and wild animals. Apart from the destruction of common pest birds it is an offence to kill wild birds, destroy their nests or disturb them near their nests. The Secretary of State for the Environment, working with the Nature Conservancy Council, has powers to make orders to protect Sites of Special Scientific Interest, and either the Minister or a local authority may enter into a management agreement with owners and occupiers of land to preserve and enhance the natural beauty of the countryside. The land involved will be subject to restrictions as to use for which compensation is payable, that will be binding not only upon the owner or occupier who makes the management agreement, but also upon successors in title.

The law relating to pesticides has been changed by the Food and Environment Protection Act 1985 which replaced the voluntary system of control by a statutory one. The supply, storage and use of pesticides, a term which includes herbicides and fungicides, has been regulated under the act by the Control of Pesticides Regulations. These regulations also prescribe the training necessary for users of such products and contain special rules for aerial spraying. Provisions of the Control of Pollution Act 1974 which give powers to Water Authorities to prosecute persons who allow polluting substances to enter water courses were brought into force in 1985. This has had serious consequences for farmers who have allowed slurry, silage effluent and dairy wastes to enter rivers and streams. It is however normally a defence for a farmer to show that what has been done has been in accordance with a Code of Good Agricultural Practice published under statutory authority.

Limitation of production is intimately related to the Common Agricultural Policy. The introduction of Milk Quotas by EEC Council Regulations in 1984 has introduced what is likely to become an increasingly important area of agricultural law as production limitation becomes more general. As has been noted above, it has brought a complicating factor into agricultural tenancies. It has also affected land transfers and mechanisms have been invented to effect the transfer of quota without land. The legal nature of milk quotas is still the subject of debate among legal practitioners. Another approach to production limitation coupled with conservation is to be found in the establishment of Environmentally Sensitive Areas under the Agriculture Act 1986. Payments are made to farmers in such areas for farming in accordance with traditional methods thus maintaining established landscape patterns.

EMPLOYER'S LIABILITIES

A farm enterprise has the same legal obligations towards its employees as any other business. The law on contracts of employment, job security and safety has become increasingly technical over the past 20 years, and although its main features are indicated below, farm management will require more detailed information – which must be kept up to date – than is here outlined.

The obligations which a farmer undertakes as an employer – whether towards an employee or other people – will arise only when there is a contract of employment between him and the person employed. When work is done by an independent contractor the employer will not normally be responsible for wrongs done by the contractor unless on express instructions from the employer. If a third party, such as a road user run down by a negligent tractor driver, suffers injury because of the act of an employee done in the course of employment, then the employer will be vicariously liable for his employee's wrongful act. The employee will be liable personally but this will usually not help the employer greatly as the injured person will almost certainly proceed against him as he, the employer, is likely to be insured against claims.

In fact, while it is not compulsory for an employer to be insured against claims by non-employees, he is obliged by the Employers Liability (Compulsory Insurance) Act 1969, to insure against liability for injury or disease sustained by his employees in the course of employment in Great Britain. Such injury will, more often than not, come about because of the negligence of another employee.

A contract of employment, known as a contract of service, is often hard to distinguish from a contract for services such as those rendered by a contractor but most of the modern employment Acts only apply to the former. A farmer has a duty to take reasonable care in choosing a contractor of repute but there it ends. With 'labour only' contracts, on a relief milking scheme for example, the distinction between employee and contractor may be a fine one; the law however looks at the reality of the situation and not at the words used by the parties to describe themselves. This means that a farmer would be likely to be responsible for the tortious acts of, say, a 'self-employed' herdsman that affected third parties and arose out of that person's work.

There are statutory restrictions on the recruitment of employees and contract workers, notably the Sex Discrimination Act 1975 which makes it unlawful for employers and managers to discriminate against women or men or married persons when advertising for, or engaging employees, and the Race Relations Act 1976 which makes it unlawful to discriminate on grounds of colour, race, ethnic or national origins. The first Act does not apply to firms where five or fewer persons are employed but the second applies universally. This legislation has to be taken into account also when promotion or redundancy is considered.

Under the Protection of Employment (Consolidation) Act 1978 an employee is entitled to be given a written statement containing specified particulars of the terms and conditions of employment. It must contain details of pay, hours of work, holidays, pensions, sick pay, job description and grievance procedures. Any changes made must be notified. Personal correspondence is not necessary, for example the current Agricultural Wages Orders can be displayed to inform staff of changes in rates of pay. Other statutory obligations of an employer include the giving of time off for certain purposes, e.g. pregnancy, trade union activities and public duties and, so far as agricultural workers are concerned, the payment of wages at or above the rates laid down under the Agricultural Workers Act 1948 by Agricultural Wages Orders.

When a contract of employment ends, in the absence of prior arrangements, the common law provides only for summary dismissal of an employee for misconduct or the giving of reasonable notice. Because of the onesidedness of this position legislation has been passed to give full time employees the right to minimum periods of notice, redundancy payments and to compensation or reinstatement if unfairly dismissed. First introduced by the Contracts of Employment Act 1963, Redundancy Payments Act 1965 and Industrial Relations Act 1971 the rules have been constantly revised so that the current legislation is to be found in the Employment Protection (Consolidation) Act 1978, as implemented by subsequent Orders and amended by a number of Employment Acts. In effect an employee under retirement age is entitled to notice according to length of service, to redundancy payments in accordance with age and length of service and to remedies for unfair dismissal when the employer dismisses for an inadequate reason or in an unfair manner, e.g. gives the employee little warning or opportunity to justify his actions. When a farmer takes on with a farm its previous owner's or tenant's workers he will also take on responsibility for their previous service on that farm so far as notice and redundancy payments are concerned. A farmer who has given recognition to a trade union must consult in advance about redundancies. The unfair dismissal provisions normally only apply to employees of over two years' standing. All disputes arising under the Employment Protection (Consolidation) Act 1978 are heard by Industrial Tribunals and, despite some folklore to the contrary, tribunals take account of the practicalities of agricultural employment and employers who take care to act with circumspection and follow sensible disciplinary rules are seldom penalised.

All employers and employees owe each other mutual duties during the course of employment; the employee to give faithful service and the employer to take reasonable care for the employees' safety. This common law duty has been reinforced by statute, notably the Health and Safety at Work Act 1974 which imposes general obligations on employers to ensure as far as is reasonably practicable that persons in their employment are not exposed to risks to their health and safety. The Act imposes a similar duty on employees and the self-employed with regard to themselves, fellow workers and other persons including children. Under this and earlier acts many Regulations have been made covering inter alia Stationary and Field Machinery, Workplaces, Tractor-cabs, Power Take-offs, Circular Saws and Poisonous Substances. The Act empowers Health and Safety Inspectors to enter farms and issue Improvement or Prohibition Orders in respect of dangerous equipment. Contravention of the Act may lead to prosecution, fine and imprisonment of both owners, and managers of defaulting firms; moreover an employee injured on account of an employer's failure to provide safe working conditions or equipment will be able to sue for damages. Under the Employers Liability (Defective Equipment) Act 1969 it will be no defence for an employer to plead that he purchased faulty equipment from a reputable manufacturer if the manufacturer was negligent. Every business is required to have a Safety Policy and in general the law not only expects employers to provide employees with healthy and safe conditions of work but to take reasonable steps to see that safety equipment such as protective clothing is used and safe procedures followed.

CONCLUSION

The Agriculture Act 1986 includes a section (S.17) that lays upon Ministers a duty to implement policies relating to agriculture, conservation and the rural economy that will achieve a reasonable balance between them. It seems likely that this concept, if translated into enforceable rules, is bound to alter agricultural law as we know it in the relatively near future.

Many more aspects of law than those mentioned so far have effects on agriculture and agricultural businesses. Farmers, as business men, are closely affected by the law of contract, sale of goods, consumer protection and fair trading and, in their particular profession, by rules concerning agricultural cooperatives, animal health and poisonous wastes, to name some of the more obvious. In addition, there are the regulations on competition, monopolies, free movement of goods and the Common Agricultural Policy under the Treaty of Rome.

It is difficult to point to any area of agricultural law today and declare that it is likely to stand still. Certainly as economic, ethical and recreational pressures grow the significance of law in its impact on agriculture will increase and ignorance of it will be no excuse.

Further reading

As a general introduction to the English legal system and the general principles of English law:
Introduction to English Law (1985). 11th edition, P.S. James, London: Butterworths.

For an up-to-date coverage of commercial law:
Mercantile Law (1984) 14th edition, C.M. Schmitthoff and D.A.G. Sarre, London: Stevens.

On labour law:
Selwyn's Law of Employment (1988). 6th edition, N.M. Selwyn, London: Butterworths.

On most aspects of farm tenancies, land ownership and occupation:
Essential Law for Landowners and Farmers (1987), 2nd edition, M. Gregory and M. Parrish, London: Collins Professional Books.

For wider coverage and an updating see:
The Business of Farming; Law & Finance (1984). Professional Publishing.

For basic knowledge of EEC law:
Law and Institutions of the European Communities (1987), 4th edition, D. Lasok and J.W. Bridge, London: Butterworths.
and for further detail:
Law of the Common Agricultural Policy (1985), 1st edition, F.F.G. Snyder. London: Sweet and Maxwell.

On farm tenancies:
Agricultural Tenancies, Law and Practice (1985). C.P. Rodgers, London: Butterworths.

On environmental aspects, e.g. planning, water supplies and pollution:
Environmental Law (1986). David Hughes. London: Butterworths.

27

Health and safety in agriculture

C. Boswell

INTRODUCTION

The earliest legislation affecting the health and safety of workers in agriculture was enacted towards the end of the last century. The Threshing Machines Act 1878 and the Chaff Cutting Machines (Accidents) Act 1897 were extremely limited in scope and only applied in England. They were not repealed however until 1961.

Attempts to include provisions for agricultural health and safety during the preparation of the Factories Act 1937 were not successful and it was not until the 1950s that further legislation was introduced in the form of the Agriculture (Poisonous Substances) Act 1952 and the Agriculture (Safety, Health and Welfare Provisions) Act 1956. The former enabled Regulations to be made to protect workers from the risk of poisoning when using certain specified substances. The latter, as the title suggests, had a wider application and some of the Regulations made under it are still in force.

In 1972 the Committee on Safety and Health at Work, under the Chairmanship of Lord Robens, published a report containing far reaching recommendations for improving the health and safety of persons at work. Many of those recommendations were subsequently embodied in the Health and Safety at Work, etc. Act 1974.

In the meantime the agricultural industry had been making a remarkable transition from a labour intensive, low productivity industry, to one which deploys far fewer people and yet achieves considerably greater output. Increased mechanisation, electrification and the extensive use of chemicals have undoubtedly been important factors in achieving high productivity but the growth in their use has also had a considerable effect on the nature and scale of workplace hazards.

An analysis of fatal accidents which occurred in the period 1981–85 provides an indication of the main source of these hazards.

Table 27.1 is compiled from fatalities reported to HM Agricultural Inspectorate and the totals cannot be directly compared with those for other industries because of the method of collection.

Table 27.2 provides a general comparison between Agri-

Table 27.1 Fatal accidents in agriculture 1981–85 (England, Scotland and Wales)

Accident category	Total	% total
Self-propelled machines	114	30
Other field machines	49	13
Falls	40	11
Falling objects	45	12
Animals	29	8
Drowning and asphyxiation	27	7
Electrical equipment	20	5
Stationary machines	12	3
Powered hand tools	10	3
Diseases	17	4
Not elsewhere classified	15	4
Total	378	100

Table 27.2 Fatal occupational injuries to employees, self-employed and non-employed persons by industry, 1981–85 (England, Scotland and Wales)

Agriculture – order I published in stats reports	335	
– reported to AI and included diseases	378	
Mining and quarrying – order II	281	
– of which:		
mining		163
quarrying		65
offshore, etc.		53
Construction – order XX	676	

culture, Mining and Quarrying and the Construction Industry. The orders in the table refer to the 1968 Standard Industrial Classification (SIC). This is a system of classification of establishments according to industry. It provides a means of securing uniformity and comparability in the statistics published by Government Departments.

There is therefore little doubt that the Agricultural Industry is extremely hazardous. Farmers and workers are increasingly engaged in complex tasks requiring more detailed knowledge and training and more careful planning than was ever the case in the past. Moreover the reduction in the size of the labour force creates conditions in which workers are expected to be competent in a wide range of tasks and are often required to work alone and without supervision. In these circumstances knowledge and understanding of health and safety matters and the development of a positive attitude to them are prerequisites to the successful prevention of accidents and disease.

The nature of the industry and the fact that many families actually live at the place of work has led to some unique accident patterns. People of 80 years of age or more feature in accident statistics as do childen who work and play on the farms. The law governing the employment of children is generally more permissive for agricultural work than for work in other industries and can lead to very young children working for their parents or guardians, on light agricultural and horticultural duties sometimes with tragic results.

The responsibility for ensuring healthy and safe working conditions in the industry rests primarily on those who create the risks and legislation has been framed accordingly.

The Health and Safety at Work, etc. Act 1974 provides a legislative framework to promote, stimulate and encourage high standards of health and safety at work.

THE HEALTH AND SAFETY AT WORK, ETC. ACT 1974 (HSW ACT)

The HSW Act is in four parts:

Part 1 – dealing with health, safety and welfare in relation to work.
Part 2 – relating to the Employment Medical Advisory Service.
Part 3 – contains matters relating to Building Regulations.
Part 4 – contains a number of miscellaneous and general provisions

Part 1 is the most relevant for the purposes of this chapter. The objectives of Part 1 of the Act are:

(1) securing the health, safety and welfare of people at work;
(2) protecting people other than those at work against risks to their health and safety arising out of work activities;
(3) controlling the keeping and use of explosive or highly flammable or otherwise dangerous substances, and generally preventing people from unlawfully having and using substances;
(4) controlling the release into the atmosphere of noxious or offensive substances from premises to be prescribed by Regulations.

The HSW Act applies to employment generally. Thus duties are placed on all people at work, that is employers, employees and the self-employed; manufacturers, suppliers, designers and importers of materials used at work; and people in control of premises. The Act does not distinguish between industries.

The HSW Act is superimposed on earlier related Acts such as the Agriculture (Poisonous Substances) Act 1952 and the Agricultural (Safety, Health and Welfare Provisions) Act 1956. For the time being these earlier Acts and many of the Regulations made under them remain in force and become 'relevant statutory provisions' and thus, for example, a farmer must ensure he complies both with the general duties contained in the HSW Act and the more specific duties laid down in Regulations.

An objective of the HSW Act is eventually to replace the requirements of the earlier Acts and Regulations by Regulations and Approved Codes of Practice made under the HSW Act. Parts of the two Agricultural Acts and the Regulations made under them have already been repealed.

The HSW Act established two bodies, the Health and Safety Commission and the Health and Safety Executive. The Commission consists of a full-time Chairman and up to nine part-time members, all of whom are appointed by the Secretary of State for Employment after consulting employers organisations about three members, employees organisations about three other members and Local Authority and other organisations about the rest. It is important to note that the responsibility for developing policies in the agricultural health and safety field rests with the Commission and not with the Ministry of Agriculture, Fisheries and Food or the Department of Agriculture for Scotland.

The Commission's duties include promoting the objectives of the Act, carrying out and encouraging research and training, providing an Information and Advisory Service and putting forward to Government proposals for Regulations under the Act. The Commission has developed arrangements whereby there is wide consultation on any proposals to either change existing legislation or introduce new Regulations.

The Commission has also established a number of Advisory Committees one of which is the Agriculture Industry Advisory Committee (AIAC). The Chairman of the AIAC is HM Chief Agricultural Inspector and its 12 members are drawn from both sides of the Agricultural Industry. Other people with a wider knowledge and experience of the industry or particular expertise are co-opted as necessary to working parties and MAFF and DAFS and a representative of Northern Ireland are appointed as assessors.

The terms of reference of the Committee are to consider and advise the Commission on:

(1) the protection of people at work from hazards to health and safety arising from their occupation within the agricultural industry and the protection of the public from related hazards arising from such activities; and
(2) other associated matters referred to them by the Commission or the Health and Safety Executive.

The Health and Safety Executive consists of three full-time members who are appointed by the Health and Safety Commission with the approval of the Secretary of State.

The Executive's duties include making arrangements for enforcement of the legislation and carrying out any of the Commission's functions which the Commission asks the Executive to take on. In practice the Executive and its staff are the operating arm of the Commission.

The Executive's staff can conveniently be divided into three parts: (i) Policy Divisions; (ii) Technical, Scientific and Medical Group and (iii) the Inspectorates.

The Policy Divisions advise the Commission, through the Executive, on all matters which concern the future direction of its affairs. In particular they keep under review the state of safety and health and as necessary initiate changes in the Commission's and Executive's response. They maintain contact with other bodies, national and international and are responsible for evaluating the effect of existing and new policies. One Division in particular is charged with planning and monitoring the resources made available to the Executive and Commission.

The Technical, Scientific and Medical Group are principally responsible for promoting excellence in the scientific and technical advice available to other parts of the Executive and to government and industry on matters of industrial safety and health. The Group is staffed with national and international experts on health and safety matters and is particularly concerned with the assessment of the extent and nature of risk, the development of technical standards, research to promote knowledge both within and outside the Executive and the development of practical measures and techniques for the solution of problems. The Medical Group has specific responsibility both for policy on occupational health and for the management of the Employment Medical Advisory Service (EMAS). Amongst other things, the service provides information and advice on health in relation to employment and training for employment.

The Inspectorates interpret and implement new control packages and existing legislation and guidance. They are the main enforcement arm of the Executive and secure compliance with legal requirements and accepted standards, interpreting and applying the latter so far as is possible on a national basis. They do this through inspection, advice, the investigation of accidents and, if necessary, enforcement.

An important aspect of their work is contributing, through practical experience and direct knowledge of the facts of industrial life, both to policy and to the development of standards. They thus advise the Policy Divisions on enforceability and practical acceptability of proposed measures and the need for new approaches.

HM Factory Inspectorate and HM Agricultural Inspectorate are grouped together in one Division and share a common senior management chain in the field.

HM Agricultural Inspectorate is headed by the Chief Agricultural Inspector and covers England, Wales and Scotland. There are approximately 160 Inspectors working full-time on health and safety of which a small number are at Headquarters and the remainder work in the field. The field force is divided into 30 inspection groups whose leaders report to HM Chief Agricultural Inspector through the HSE Area Directors.

An innovation in 1987 was the introduction of three groups which were assigned national responsibilities and work with the Inspectorate's Headquarters in the development of enforcement policies, advice on such matters as machinery, forestry, chemicals, the use of publicity and the Inspectorate's training programmes. They thus provide a focal point in the field for contact and liaison with the industry. These groups are located in Edinburgh, Nottingham and at the National Agricultural Centre at Stoneleigh where a permanent Exhibition and Advice Centre is maintained.

SOME OF THE STATUTORY REQUIREMENTS OF THE HSW ACT

The HSW Act imposes duties on everyone concerned with work activities, employers, the self-employed and employees. The duties are imposed on individuals, such as farmers and workers and on companies, partnerships, etc. The scope includes manufacturers, designers, suppliers and importers of machinery and materials for use at work.

The duties in the Act are expressed in general terms with more specific requirements covered by Regulations. Thus there are special Regulations for industries such as agriculture and also Regulations having a more general application.

Some of the duties are qualified by the term '*so far as is reasonably practicable*'. This is an important qualification which is not defined in the Act but has been interpreted by the Courts. It implies an assessment of the risk in comparison with the physical difficulties, time, trouble and expense involved in taking steps to avoid the risk. Thus, if the risks to health and safety are very low and the cost or technical difficulties of taking certain steps to avoid the risks are very high, then it might not be reasonably practicable to take them. However if the risks are very high, then less weight can be given to the cost of measures needed to avoid them. The comparison does not include the financial standing of those who have the duty of compliance.

A duty which is '*so far as practicable*' without the world '*reasonably*' is stricter and means that the most effective means must be used to comply with the duty, taking into account the conditions and circumstances, the current state of technical knowledge and the financial implications.

In a prosecution alleging failure to comply with those duties, it is up to the accused to show that it was not reasonably practicable or practicable (as appropriate) for him to do more than he had in fact done to comply with the duties.

Some of the duties imposed by the HSW Act are outlined below.

Employers have a general duty to ensure, so far as is reasonably practicable, the health, safety and welfare at work of employees and five of the most important aspects are specified:

(1) maintaining safe systems of work;
(2) ensuring the safe use, handling, storage and transport of articles and substances;
(3) providing adequate instruction, training and supervision;
(4) maintaining the safe premises and other places of work;
(5) providing a safe working environment and adequate welfare facilities.

For example, a safe system of work involving a tractor might include the choice of an appropriate machine with relevant safety features; maintaining it in a satisfactory condition; considering the terrain and operating conditions; the actual operation of the tractor for specific activities and any special precautions which might need to be taken.

The concept of thinking carefully about the nature of the hazard and the workplace and making arrangements to deal with them is also embodied in a requirement for employers with five or more employees to prepare a written safety policy for the undertaking and to bring it to the notice of employees.

Self-employed persons (as well as employers) are required to conduct their undertakings in such a manner as to ensure, so far as is reasonably practicable, that they do not expose people who are not their employees to risks to their health and safety. This applies both on and off the premises and includes for example, the public and children. The self-employed are also required to conduct their businesses in such ways as to ensure so far as is reasonably practicable, that they do not risk their own health and safety.

Employees are required to take reasonable care of their own health and safety and that of others who may be affected by what they do.

They must also cooperate with their employers and others in meeting statutory requirements and must not interfere with or misuse anything provided to protect their health and safety in compliance with the HSW Act.

Persons who have control to any extent of non-domestic premises are required to take such steps as are reasonably practicable to ensure that there are no risks to health and safety when they are used by persons who are not their employees. The duty is not imposed on employees who might have been put in control at any particular time by the employer.

Designers, manufacturers, importers and suppliers of articles and substances for use at work must ensure, so far as is reasonably practicable that they are safe and without risk to health in prescribed circumstances. They are required to carry out such tests as may be necessary for the purpose of their duties and to make information available about the uses for which the product and substances have been designed and tested. (NB: HSW Act, Section 6 duties were amended by the Consumer Protection Act 1987.)

People in general (i.e. the public) have duties not to interfere intentionally with or misuse anything provided in the interests of safety, health and welfare. This might apply, for example, to fences, warning signs, machinery guards, etc.

Enforcement

Inspectors are appointed under the HSW Act and derive their powers from the Act. They are issued with a warrant which specifies those powers and essentially these enable the Inspector to carry into effect any of the statutory provisions for which he or she is appointed. Thus, in general terms, there are powers of entry; powers to make examinations and investigations and to direct that premises or parts remain undisturbed for those purposes; to take samples; to require persons to answer questions.

Inspectors have the power to institute proceedings in England and Wales (different arrangements apply in Scotland) and to issue *Improvement and Prohibition Notices*. The circumstances in which Notices may be issued are circumscribed by detailed requirements in the HSW Act (Sections 21–23) but in summary:

(1) *An improvement notice* may be issued if there is a legal contravention of any of the relevant statutory provisions and requires that the matter be remedied within a specified time.
(2) *A prohibition notice* may have an immediate or deferred effect but requires that an activity shall cease or not be carried out until the matter specified in the notice has been remedied.

Appeals against notices are dealt with by Industrial Tribunals and details of how to appeal are stated on the notice.

Prosecutions for some offences can only be tried summarily but others can be tried summarily or on indictment. The maximum fine for a summary conviction is £2000. There is no limit to the fine for a conviction on indictment and for a limited number of offences a penal sentence of up to two years can be imposed. Failure to comply with a prohibition notice is one of the offences which could result in a penal sentence.

REGULATIONS APPLYING TO AGRICULTURE ACTIVITIES

It is convenient to consider Regulations in two groups, those specifically intended to apply to agriculture and those which are multi-industry or wider in concept.

The distinction is interesting because whilst the former may owe their origin to legislation such as the Agriculture (Safety, Health and Welfare) Act 1956 and contain basic but detailed requirements to suit specific machines, activities, etc., the latter may have been made under the HSW Act and reflect the thrust of developments since the introduction of the Act. They may be fairly general in the regulatory requirements but underpinned by Approved Codes of Practice or guidance documents.

An *Approved Code of Practice* is approved by the Health and Safety Commission and has a particular standing in law. If the code appears to the Courts to be relevant to a case then it is admissible in evidence. If the guidance in the Approved Code has not been followed, it is up to the defendant to show that he has satisfactorily complied with the requirement in some other way.

Regulations which are specific to agriculture include the following:

The Agriculture (Power Take-off) Regulations 1957

These prescribe the obligations of employers of workers for the guarding and safe use of the power take-off and the power take-off shaft on agricultural tractors and machines.

The Agriculture (Avoidance of Accidents to Children) Regulations 1958

These Regulations make it illegal to allow children under the age of 13 to drive or ride on agricultural tractors and self-propelled machines and to ride on machines and implements.

The Agriculture (Circular Saws) Regulations 1959

These set out the requirements for the guarding, maintenance and operation of circular saws. Workers under the age of 16 may not operate or assist at a circular saw and workers between 16 and 18 may do so only when supervised by an experienced person over 18 years of age.

The Agriculture (Safeguarding of Workplaces) Regulations 1959

These prescribe safety requirements in places where agricultural workers are employed and cover the construction and maintenance of floors and stairways, the provision of handrails and the guarding of apertures in floors, walls and the edges of floors.

The Agriculture (Stationary Machinery) Regulations 1959

These lay down guarding and safety requirements for stationary machinery particularly in respect of guarding requirements.

The Agriculture (Lifting of Heavy Weights) Regulations 1959

These prescribe that the maximum weight of a sack or bag and its contents which a worker may lift or carry unaided is 180 lb. (The equivalent weight is 81.6 kg but the Regulation has not been metricated.) The Regulations do not affect the provision of the 1956 Act that a young person under the age of 18 years may not be allowed to lift, carry or move a load so heavy as likely to cause an injury.

The Agriculture (Threshers and Balers) Regulations 1960

These require the guarding of stationary threshers, hullers, balers and trussers. These do not apply to pea viners nor to combine harvesters unless they have been permanently converted for stationary use only.

The Agriculture (Field Machinery) Regulations 1962

These Regulations deal with the guarding of machines which operate while travelling over the ground, trailers and power driven hand tools. A new field machine may not be sold or hired for use in agriculture unless the requirements for guarding are met.

The Agriculture (Tractor Cabs) Regulations 1974 and the Agriculture (Tractor Cabs) (Amendments) Regulations 1984

These Regulations require agricultural tractors to be fitted with approved safety cabs or frames and with noise requirements. Tractors cannot be sold or let on hire for use in agriculture unless properly fitted with a safety cab approved for that tractor and marked with the approval and supplementary marks. All wheeled tractors driven by workers have to be fitted with an approved safety cab.

The Poisonous Substances in Agriculture Regulations 1984

These Regulations prescribe the precautions to be taken (including the protective clothing to be worn) when substances specified in the schedules to the Regulations are used in agriculture. They apply to the self-employed as well as to employers and employees.

Examples of more general regulations are as follows:

The Health and Safety (First-Aid) Regulations 1981

These place general duties on employers and on self-employed persons so that first-aid provisions are made for employees who are injured or become ill at work and so that the self-employed can render first-aid to themselves if they are injured at work.

The Safety Signs Regulations 1980

These Regulations are based on an EEC Directive and provide that safety signs for persons at work and colours in strips identifying places where there is danger to their health or safety shall comply with BS 5378: Part 1 1980.

The Diving Operations at Work Regulations 1981

These cover all diving operations where the HSW Act applies. This therefore includes fish farming establishments or reservoirs where diving operations may take place.

The Reporting of Injuries, Diseases and Dangerous Occurrence Regulations 1985 (RIDDOR)

These Regulations place responsibilities on employers, the self-employed and certain other responsible persons. The Regulations set out reporting arrangements for fatalities and specified major injuries; certain diseases; specified dangerous occurrences and gas incidents. In the case of death or any of the specified serious injuries, the requirement is to notify the responsible authority (HSE in respect of agricultural activities) by the quickest practicable means and forward a report within seven days. Where persons at work are incapacitated for work for more than three consecutive days then the report has to be forwarded within seven days of the accident. The form of the report is specified in the Regulations. The Regulations also specify matters which are to be kept in records.

FOOD AND ENVIRONMENT PROTECTION ACT 1985

The Food and Environment Protection Act 1985 (FEPA) is an enabling Act in 3 parts. Parts 1 and 2 deal with the contamination of food and with deposits in the sea. Part 3 deals with pesticides and takes effect by means of Regulations. The detailed requirements have been introduced in *The Control of Pesticides Regulations 1986*.

These Regulations apply to pesticides (which include products such as herbicides and fungicides) used in agriculture, horticulture and forestry as well as to other uses less directly concerned with the Agricultural Industry. They apply to animal husbandry but not to pesticides administered directly to farm livestock for example sheep dips or warble fly sprays which are already controlled by the Medicines Act 1968.

The Regulations prohibit the advertisement, sale, importation, supply, storage and use of a pesticide unless it has been approved. All manufacturers, importers and suppliers of pesticides must obtain approval for each pesticide product and the control regime included for full or provisional approval. A further category is that for which an experimental permit is granted during development but this cannot be sold or advertised.

Anyone who advertises, sells, supplies, stores or uses a pesticide is affected by the Regulations which came into force 6 October 1986 with some parts taking effect in stages up to 1 January 1989.

Those provisions which came into effect in October 1986 placed obligations on all sellers, suppliers, storers and users to take all reasonable precautions to protect the health of human beings, creatures, plants and the environment. At the same time it became illegal to supply, store or use a pesticide which had not been approved and conditions of approval had to be complied with. The Regulations also dealt with competency. Very detailed·rules were imposed on aerial applications.

Advertising was controlled from January 1987 and Certificates of Competence were required for anyone who *stores* approved pesticides for the purpose of sale and from January 1988 Certificates of Competence are required from anyone selling or supplying approved pesticides. At the same time the use of pesticides is more specifically controlled, e.g. conditions of approval relating to the use of protective clothing, rates of application, etc. must be complied with.

The last tranche of Regulations take effect from 1 January 1989 when contractors must have Certificates of Competence and when the use of adjuvants and tank mixers are controlled. *Enforcement* of FEPA and the Regulations on agricultural premises will generally be carried out by HM Agricultural Inspectors and elsewhere by Factories Inspectors or Local Authority Inspectors.

CONCLUSIONS

The legal framework within which people have to work and which requires them to accept responsibilities for health and safety is now well developed. The Health and Safety Commission, its Advisory Committees and the Health and Safety Executive are deeply committed to the provision of help and guidance to industry and an extensive system of consultation has been developed so that proposals for change or improvement to the law take account of industry's perception of need.

In some circumstances, the law has catered for particularly dangerous situations through the imposition of detailed requirements which effectively prevent someone from making an inappropriate judgement. The compulsory introduction of safety cabs on tractors is an example which has greatly reduced accidents and the risk of noise-induced deafness while at the same time promoting a better and more productive working environment. Similarly, the detailed requirements for pesticide approval and use should do a great deal to reduce risks to the health of people while at the same time safeguarding the environment.

However, in general, the law does not of itself produce safe and healthy conditions. It is people who create the risks, whether they be manufacturers or users of equipment or substances or engaged in the diverse activities of the industry. They can also create the conditions in which agriculture becomes a safer and healthier industry for those who work in it or are affected by it.

Advances in technology and farming practices have radically changed the industry and introduced new hazards. In dealing with these hazards the modern farmer and farm worker need to combine knowledge and competency with a commitment to healthier and safer working conditions and practices as an integral part of the efficient and profitable running of the business.

Progress has certainly been made and a great deal of help and guidance is readily available from the Health and Safety Executive and from organisations representing employers and the Trade Unions. Farmers and farm workers are better informed about health and safety issues than they were in the past but there is a continuing need for training to recognise and deal with the risks which are inherent in an industry which is quick to respond to changing conditions.

This brief description of health and safety in agriculture in Great Britain is by way of guidance only. It is not intended to be a comprehensive or authoritative interpretation of the Acts and Regulations, copies of which can be obtained through HMSO or booksellers.

Details of a wide range of priced and free publications on agricultural health and safety can be obtained by contacting any HSE office listed in the Telephone Directory.

References

The Health and Safety at Work, etc. Act 1974 Chapter 37

The Food and Environment Protection Act 1985 Chapter 48

The Agriculture (Safety Health and Welfare Provisions) Act 1956 Chapter 49

The Agriculture (Ladders) Regulations 1957 SI 1957 No 1385

The Agriculture (Power Take-Off) Regulations 1957 SI 1957 No 1386

The Agriculture (Avoidance of Accidents to Children) Regulations 1958 SI 1958 No 366

The Agriculture (Circular Saws) Regulations 1959 SI 1959 No 427

The Agriculture (Safeguarding of Workplaces) Regulations 1959 SI 1959 No 428

The Agriculture (Stationary Machinery) Regulations 1959 SI 1959 No 1216

The Agriculture (Lifting of Heavy Weights) Regulations 1959 SI 1959 2120

The Agriculture (Threshers and Balers) Regulations 1960 SI 1960 No 1199

The Agriculture (Field Machinery) Regulations 1962 SI 1962 No 1472

The Agriculture (Tractor Cabs) Regulations 1974 SI 1974 No 2034

The Agriculture (Tractor Cabs) (Amendments) Regulations 1980 SI 1980 No 1036

The Agriculture (Tractor Cabs) (Amendments) Regulations 1984 SI 1984 No 605

The Poisonous Substances in Agriculture Regulations 1984 SI 1984 No 1114

The Control of Pesticides Regulations 1986 SI 1986 No 1510

The Reporting of Injuries, Diseases and Dangerous Occurrences Regulations 1985 SI 1985 No 2023

The Health and Safety (First-Aid) Regulations 1981 SI 1981 No 917

The Diving Operations at Work Regulations 1981 SI 1981 No 399

28

Computers in agriculture

A. T. Vranch and J. C. Eddison

The development of digital computers is one of the major technological achievements of the twentieth century. The rapid expansion in the use of computers in all fields of human activity, including agriculture, has made an enormous impact on society, especially during the last 15 years. There is every reason to expect that new computer developments will have an even bigger impact on society in the future.

This chapter describes the role of computers in agriculture and is divided into four sections, each of which attempts to answer the question posed at the beginning of the section, i.e.

(1) What is a computer and in what applications can it be used?
(2) How are computers used in agriculture?
(3) What are the implications of operating an on-farm computer system and what are the criteria for its selection?
(4) What developments might be seen in the future for computer applications in agriculture?

WHAT IS A COMPUTER AND IN WHAT APPLICATIONS CAN IT BE USED?

Introduction

A computer is an electronic device which will perform, implicitly, every instruction given to it by its operator. A series of such instructions is called a computer *program*. A general term for programs is *software*, while the physical parts that make up the computer are generally referred to as the *hardware*.

Mainframes, superminis, minicomputers and *microcomputers* are all categories of computer. The difference between these categories is becoming difficult to define as new computers exhibit more power in terms of their speed of operation and their capacity for working with larger amounts of information, while at the same time, they are becoming less expensive to buy and easier to use.

A very expensive mainframe computer can support many individual users simultaneously (i.e. *multi-user*), each

running his or her own *terminal*, comprising a keyboard and visual display unit (*VDU*). The mainframe computer can also run many software programs simultaneously (i.e. *multitasking*), each user unaware that anyone else is sharing the same computer. In contrast, the cheapest home microcomputer may be suitable only for a single user running a single program at a time. After mainframes, the smallest computers were initially minicomputers, each handling a limited number of users. As computer developments progressed, the computing power of the minicomputer became available in single-user microcomputers. At the same time, however, the multi-user minicomputers developed into the more powerful superminis.

There is an increasing trend to link supermini computers into a *cluster*. This approach has been exemplified by Digital Equipment Corporation (DEC) with their VAX range of computers. A cluster of superminis provides the facility to handle many users in a manner similar to that of a mainframe, but with the built-in flexibility of having several computers, thereby allowing piecemeal replacement of parts of the system rather than periodic total replacement at a high cost.

Advances in computer technology have made it possible to incorporate into the latest microcomputers the multi-tasking and multi-user facilities previously found only in mainframes or minicomputers. At the same time, microcomputers are now also being used as *intelligent terminals*, which means that, in addition to operating as a personal computer (or *PC*), the same hardware can act as a terminal to another computer. This computer link may occur over long distances via telephone lines, fibreoptic cables or satellite communications if the microcomputer is connected to a telephone socket via a *modem* (or *mod*ulator–*dem*odulator). This means that, via the international telephone system, the user of a relatively cheap personal computer has the potential to communicate with other personal computer users with modems users with modems or even with the very largest of mainframe computers throughout the world.

Most computers are operated as *non-dedicated* systems, that is they are not dedicated to one particular application or program and the same hardware can be used with a range of different programs or applications. There is another category (of *dedicated* systems) in which the computer is used for one dedicated purpose only. A example of a dedicated computer system is a microprocessor-based control unit which senses temperatures, performs calculations and controls the operation of heaters or fans. Often, this type of dedicated computer has no keyboard or VDU and may be powered by batteries rather than by mains supply.

General features of a microcomputer system

A part of the 'brain' of a computer is called the *CPU* (central processor unit) – or microprocessor in the case of a microcomputer – and this is the part of the hardware of the computer which controls operations, performs calculations and directs logical steps. Microprocessors are usually classified in terms of whether they are 8-bit, 16-bit or 32-bit. A *bit* (or *binary digit*) is a fundamental unit of computer information which takes the value 0 or 1, equivalent to 'off' or 'on' of a switch. Microprocessors are under constant development and are becoming faster and able to cope with a larger capacity of processing work with each new type which is invented. As a general rule, the bigger the number of bits, the more advanced is the processor, although this is a simplification.

In order for programs to be used on a microcomputer, it must have an adequate capacity of working memory (i.e. *Random Access Memory* or *RAM*) into which any programs will be loaded and run (simultaneously for a multi-user system). The amount of available RAM in a microcomputer is measured in units of memory called *bytes*, where a byte is a group of eight bits. More commonly, microcomputer RAM is measured in *kilobytes*, where a kilobyte is equal to 1024 bytes (i.e. 2^{10}). The first generation of mass-produced microcomputers had 8-bit microprocessors and up to 48 kbytes of RAM. The latest personal microcomputers have, typically, 32-bit microprocessors and 512 kbytes of RAM (which can be expanded even further if required).

Microprocessors and RAM are both made from tiny wafers of silicon, mounted on a small piece of plastic, with rows of pins to connect the microcircuitry on the silicon wafer to other components on computer circuit boards. These microelectronic devices are usually known as '*chips*'.

The microprocessor, RAM chips and circuit boards are usually located in the main body of the computer system, which is often a plastic moulded box. There may be an integral keyboard with a standard (i.e. *QUERTY*) typewriter layout, although it is more common on business microcomputers to have a separate keyboard for operator comfort. The VDU is usually placed on top of the main system box or may form part of it. The VDU may be colour or monochrome and, especially with home microcomputers, a TV set is commonly used as the VDU. The resolution of VDU screens (i.e. the total number of dots or *pixels* which make up the screen) varies, depending on the required application. For high-quality graphics work, a high resolution (perhaps colour) screen is required. For standard text characters, a low resolution screen is adequate. A TV set normally produces a low resolution screen when used as a VDU.

In addition to the requirements for working memory (RAM), which is lost if the microcomputer power supply is switched off, there must also be adequate capacity for storing programs and other information on a permanent basis for future reference. *Disks* or *tapes* are commonly used for

permanent storage and are used in *disk drives* or *tape drives* which are part of the computer system. Cassette tapes, which are very slow, are widely used with home microcomputers, whereas for bigger business systems, tape cartridges or continuous tapes (i.e. *tape streamers*) are preferred because of their improved capacity and speed of operation.

Hard disks (or *fixed disks* or *Winchester disks*) can store large amounts of information or programs and are usually hermetically-sealed units which are often not removable from the computer. As a guide, a page of text on A4 size paper will need about 4 kilobytes of disc storage. A typical microcomputer hard disk has a capacity of up to 20 megabytes (where 1 megabyte = 1024 kilobytes). On a smaller scale, *floppy disks* can be used to store smaller amounts of information or programs (for example, 1.2 megabytes or 360 kilobytes) and have the advantage that they are removable. The high capacity hard disks are also often preferred for their faster speed of access compared with floppy disks and this can be a particular advantage for some program applications.

Some programs are stored permanently on chips called *Read Only Memory* or *ROM* chips. The contents of ROM chips are not lost when the computer is switched off. These programs will run very quickly indeed and ROM chips are used to advantage where the speed of operation is important (as in for example, screen graphics) or where RAM is limited. ROM chips are located on circuit boards in the main box of the microcomputer system.

A very important piece of computer hardware is the *printer*. Printer quality varies from model to model and is usually a compromise with speed of printing. On many *dot matrix* printers the user has a choice between slow speed/ near letter quality output or fast speed/draft quality output. Dot matrix printers are usually noisy, as are *daisy wheel* printers, which give letter quality output at relatively slow speed. *Laser printers* are very fast, silent and produce outstandingly high quality output. Printers can operate with *friction feed* for single sheets of paper or *tractor feed* for continuous stationery with perforated edges to match the lugs of the tractor feed drive wheels. A *sheet feeder* may be used to input single sheets (of, say, headed notepaper) into the printer automatically. Printers can usually handle stationery of varying width, which means that address labels, cheques or wage slips can be printed on the same hardware used for letters or reports.

Colour printers are available for many microcomputers where colour hard copy output is required. *Graph plotters* (sometimes referred to as *X-Y plotters*) can provide a very flexible microcomputer hardware option, especially where multi-colour graphics and text output is required.

An alternative to the VDU screen or paper as a form output is provided by *speech synthesis* units, which are an integral part of some microcomputers, or can be added as an optional extra for others. Speech synthesis techniques have improved during the last few years to the extent that good quality output (in terms of tone and inflexion) can now be achieved at low cost. This form of voice output already has particular relevance to proof reading applications, as an aid to the blind or for giving warnings to the driver of a vehicle fitted with a dedicated computer system. It is likely that speech synthesis units will become more widely available in the future.

An innovation which has been made popular by the introduction of the Apple Macintosh microcomputer is the *mouse*. A mouse is a device which is linked to the computer by a wire and consists of a small plastic moulded box with a button on the top and a freely-rolling ball underneath. The user moves the mouse around the desk, causing the ball to roll accordingly. The movement of the ball is monitored by the microcomputer via the wire and this movement causes an arrow to move on the screen. Hence, the mouse can be used as an alternative to the keyboard. For example, an option displayed on the screen can be selected by pointing the arrow at it using the mouse and pressing the button, instead of typing into the keyboard the appropriate number or letter corresponding to that option. The mouse is particularly effective in drawing graphics applications onto the screen.

In addition to the mouse, there are other alternative devices to the keyboard for inputting information or instructions into a microcomputer, all of which can speed up and simplify this inputting process. *Joysticks*, like a mouse, are used to move a cursor around the computer screen and usually include a button as well. The difference between a joystick and a mouse is that the joystick remains stationary on the desk and it is the angle if the stick or handle which determines the cursor motion, whereas the mouse is moved around the desk to produce the cursor motion. Joysticks have become particularly popular with home computers used for games. *Light pens* are connected to the microcomputer by a wire and enable the user to point the light pen at a part of the screen to select an option or draw a shape as required. *Touch-sensitive screens* enable the user to point at a part of the screen with their finger to select an option. *Graphics tablets* consist of a special flat plate, connected by wire to the microcomputer and linked to a pen. The pen is used to point to sections marked on the graphics tablet plate to input selections required by the user. Graphics tablets are particularly effective for tracing work with maps or for design work. Perhaps the ultimate method of inputting information into a microcomputer is to speak to it. Although it is still in its early stages of development at present, there is no doubt that *speech recognition* will become an effective and widely-used input medium in the future.

Programs, data and files

A *program* is a sequence of instructions which, when executed (or *run*) will perform the required task. Programs vary from the very simple to the very sophisticated in terms of the task which they perform or in terms of the way in which they go about solving a particular problem. Some Programs have many instructions and so require a large amount of RAM in which to run. Other programs may not need a large RAM but, because they have to repeat the same instructions many times to get to the required result, will run very slowly unless the computer has a very fast microprocessor. Nevertheless, a common feature of all programs is that they are all written in a *programming language* of one form or another.

Programming languages available for most microcomputers include versions of BASIC, FORTRAN, COBOL, C, FORTH, PASCAL, LISP, PROLOG and many others. Some programming languages have been developed from earlier languages used many years ago on mainframes, while others have been devised recently for microcomputers. The choice of programming language is often made to suit the

type of application. For example, some languages are more suited to business applications, while others are designed for use in applications involving monitoring of sensors or control of robot arms.

High level languages are written in a form in which the words used for the programming instructions in the program are similar to words used in normal language. In *low level* languages, which work at a level much closer to the binary arithmetic of the microprocessor operation, words used in the programming language bear little resemblance to words used in normal language. However, for the microprocessor to understand the meaning of the instructions in a high level program, the program must be translated into a low level form, either as a once-only translation (using a *compiler*) or as a line-by-line translation (using an *interpreter*) each time the program is run. Compiled programs generally run faster than those which need an interpreter, since the compiled version can be understood by the microprocessor without the need for repeated translation.

Recent advances in programming techniques have led to the development of *program generators*. These are programming systems that allow the user to design and develop applications, stating the requirements of the software application (in terms of screen layouts, input and output options, etc.) in more or less natural language. The program generator then 'writes' a program in a high level language, for example BASIC. The more sophisticated program generators will even put into the final program features like comment statements to explain to the user how the program has been constructed. Additional advanced features may include full logical checking of the program design and traps for possible logical errors.

Factual details used with a program, either in the form of text or numerical values, are generally referred to as *data*. *Data processing* is a term used to describe calculations, sorting and other manipulations performed on the data by a program.

Programs and data are stored in files, usually on disk or tape. A single disk may have many files stored on it, each file being identified by a *filename*. Filenames for microcomputers have a general form, typically made up of three parts for the full file specification:

Drive letter:Filename.Extension
e.g. B:MYFILE.TXT

The *drive letter* (A or B or C) specifies the disc drive in which the disk containing the file resides. For example, A and B may represent two floppy disk drives on a microcomputer system which also has a single hard disk, identified by letter C.

The *filename* is separated from the drive letter by a colon and can be up to, say 8 characters long but must not contain any invalid characters (e.g. symbols like : or > or = .).

The *filename extension* (up to three characters long and separated from the filename by a full stop) identifies the type of file, for example, a text file (TXT) or a data file (DAT) or a program file (COM). Standard filename extensions are automatically generated by default from some programs. Alternatively, extensions can be chosen by the user, if required. The importance of the filename extension is that it provides a convenient means of identifying types of file on a disk. This is very useful for listing, copying or deleting groups of files.

Operating systems

A special set of programs called the *operating system* (which may be stored on floppy disk, hard disk or ROM chips) must be present for any computer to function at all. The operating system controls the working of the computer hardware e.g. keyboard, screen, disc drive, or printer and also acts as a 'bridge' between the hardware and the software being used. On multi-user or multi-tasking systems the operating system manages the multiple user and tasking operations. Operating systems also include *utilities*, which are programs to perform duties such as copying files from one disk to another. An important utility is the *format* or *initialising* program. This is used to mark, magnetically, the surface of a blank disk into sectors and tracks to which the computer can refer for read and write operations. A new, blank disk can be formatted in a number of ways, depending on the formatting procedure, which also depends on the type of computer being used. A common cause of problems arises from mis-matched formatting of disk and disk drive. This is especially true when transferring a disk from one type of computer into a different type of computer or when using disks in a computer which can support several optional disk formats.

Operating systems are usually closely related to specific microprocessors and as a result, some microcomputers will run only with specific operating systems. *Table 28.1* shows examples of operating systems currently in use, with corresponding microprocessor and computer.

Table 28.1 Operating systems and microcomputers with corresponding microprocessors

Operating system	Microprocessor	Microcomputer
CP/M	Z80A	Superbrain
AppleDOS	6502	Apple IIe
MS DOS	8086	ACT Apricot
CP/M or MS DOS	Z80 & 8088	DEC Rainbow 100
CP/M or MS DOS	8085 & 8086	Zenith
PC DOS	8088	IBM PC
PC (MS) DOS	80286	IBM PC AT
MacDOS	68000	Apple Macintosh

The 8-bit business microcomputers of the early 1980s, which were dominated by the Z80 range of microprocessor and *CP/M* (Control Program and Monitor) operating system, have been superseded by 16-bit machines using the 8086 range of microprocessor with *MS DOS* (MicroSoft Disk Operating System) or the 68000 range. Some microcomputers, including the DEC Rainbow 100 and Zenith, have dual microprocessors to enable either CP/M or MS DOS software to be used. MS DOS is very similar to the *PC DOS* operating system used in the IBM PC and this combination of IBM PC compatible hardware and software has become an important standard for business use. Several manufacturers have recently produced very cheap microcomputers that are compatible with the IBM PC and these are often referred to as IBM PC *clones*. the Amstrad PC1512 is an example of an IBM PC clone that has generated much interest in computing for agriculture in the UK.

Some operating systems, notably *BOS* (Business Operating System), can run on many different microprocessors. This feature offers an advantage to the user who may wish to upgrade the computer hardware without the expense and problems involved in upgrading the operating system and software as well.

Software packages

The microcomputer user can develop original programs to solve particular problems or to perform certain tasks. Depending on the nature of the program and the experience of the user, this type of exercise can be very time-consuming and may result in a finished product containing errors in programming (or *bugs*) which could be disastrous. For many common applications, commercially-developed software *packages* are available. Packages have the advantage that many man-hours of programming can be input into their development as this investment can be financed by volume sales of the finished product. At the same time, if many users run a particular package, it is more likely that any bugs in the programs will be identified and corrected. Similarly, other improvements to the package will result from the suggestions and feedback from a large user base. This feedback can be made formal by the formation of user-groups comprising users of the software package and the representatives of the company producing that package.

A good software package is one which performs the task which the user requires it to do, without restrictions. Apart from doing the job which it is required to do, a good software package should also be *user-friendly*, i.e. easy to use, flexible and with clear instructions for the user. Good packages often have an in-built *help facility*, which allows the user to ask for a screen display to explain the options available in the package, error messages or other details which would otherwise only be found in a reference manual.

In the same way that some microprocessors are restricted to specific operating systems, software packages are restricted to certain operating systems. In some cases the producer of a package may offer a range of versions of the package, each of which will run on a different operating system. However, each version must be supported for development and after-sales advisory service, which may not be a worthwhile business strategy for the producer of the package. At the same time, computer manufacturers must ensure that a wide range of good software packages is available to the users who will purchase their hardware. The net result is that there is a comprehensive range of good, general-purpose software packages available for several modern microcomputer operating systems.

Although different operating systems and software packages work in different ways in terms of their prompts and error messages, the approach adopted for the display as seen by the user is now more standard and more user-friendly as menu-driven software has become more common. These packages are based on a series of *menus*, each of which requires a single character input from the keyboard to direct the program to the selected operation. This move towards simple, standardised, user-friendly software has taken a big leap forward with the introduction of the *WIMPS* (or *w*indows, *i*cons, *m*ouse and *p*ull-down menus) environment, made popular by the Apple Macintosh. The WIMPS approach is now available in one form or another on all modern microcomputers. The user moves the *mouse* around the desk and points to graphical symbols (*icons*) representing an instruction of file operation then clicks the button. Options are stored off the screen in *pull-down menus*, which can be displayed by manipulating the mouse and selected by clicking the button. Multiple documents, tables or graphs can be displayed simultaneously by dividing the screen into smaller regions or *windows*, which can be rearranged on the screen by moving the mouse. Some software packages are extremely versatile, which makes them attractive to a wide range of users for a wide range of applications. *Word-*

processors, spreadsheets, databases and graphics packages fall into this category and, for this reason, these packages are commonly found on all modern microcomputers, often working within a WIMPS environment. Indeed, one or more of these four packages is usually supplied free with the microcomputer.

A *wordprocessor* is probably the most widely used general-purpose software package. A wordprocessor enables the user to create documents (for example letters or reports) on the computer VDU screen. The documents can be printed on the microcomputer printer or saved on disk for future reference. A saved copy of the document can be recalled and any alterations or corrections can be made on the VDU before the final version of the document is saved on disk or before a hard copy is produced on the printer. Using a word-processor makes it very easy to rearrange sentences, paragraphs or whole pages on the VDU screen before printing a hard copy on paper. Other features may be present, all of which enhance the capabilities of the word-processor, for example: spelling checkers, word search facilities, variable typeface founts or the ability to print, automatically, mailshots of standard letters.

In a *spreadsheet* package, the VDU screen is divided into columns and rows to produce individual cells. Into each cell the user can enter numerical values, text or formulae. In this way, a wide range of different applications can be developed, including physical or financial records and calculations. Columns of financial data can be entered into cells and processed using formulae or other in-built functions available in the spreadsheet (e.g. to perform sums, averages, depreciation, projected values). Spreadsheets are particularly useful in 'what if?' exercises, where, for example, the effect of changing one value (such as VAT rate) can be determined very rapidly using the recalculation facilities of the spreadsheet. The total spreadsheet is often too big to be viewed at once on the screen, although the screen can act as a window that can be moved around to view any part of the whole spreadsheet. Spreadsheet applications can be saved on disk for future reference and hard copies can be made using the printer. For the user who prefers not to develop original spreadsheet applications, templates can be purchased for some common types of application. These templates include all the appropriate formulae and explanatory text in a suitable layout so that users need only enter their own data values into the spreadsheet, which then produces the required results.

A *database* is a very versatile package which enables files of information to be set up in a format selected by the user. The database program provides facilities for selecting and sorting these files in whatever way the user requires. For example, it may be necessary to keep records of names and addresses of clients in a way that lists of sub-groups of the total list can be generated. By using a database, a set of files can be created with a common layout (or *field*) for each of the categories under which the client details are to be stored. The details about one client (i.e. several fields) comprise one *record*. Once the fields have been set up and the client details entered into the fields, selection or sorting of the database files can proceed. For example, lists of clients in Devon can be generated by a simple sort of the files on the field corresponding to the 'county' part of the address. This is an extremely simple example of what can be done using a database. A very powerful feature of some database packages is the facility for generation of menus, which can direct the user to various selected options as required. Effectively, this means that there is a powerful programming facility within the database, which can be used to create flexible applications very quickly and in a user-friendly form.

Business *graphics* packages can represent data in various, selected, graphical forms, including pie charts, line plots, scatter plots and histograms. These graphs can be displayed on the VDU screen, saved in disk files or reproduced on paper using a printer or X-Y plotter. A graph can often show trends or significant points better than a table of figures and it is for this reason that graphical representation of data is sometimes essential.

Other general-purpose packages include *accounts, payroll, nominal ledger, purchase ledger, project planning* (and critical path analysis), technical *drawing* and *design*, computer-based *training, communications* and *terminal emulation* (for intelligent terminal operation). Some of these general-purpose packages can be adapted to applications in agriculture, while other packages have also been developed specifically for agriculture.

An important development, made more widespread by the introduction of microcomputers with adequate RAM, disk space and processing power, is the *integrated package*. An integrated package combines two or more individual packages, with the advantage that data can be transferred directly from one part of the package to another. For example, an integrated wordprocessing/spreadsheet/database/graphics/communications package gives the user the facility to take data from the spreadsheet add data from the database, then produce pie charts. The wordprocessor can be used to incorporate these charts, along with tables of data from the spreadsheet, into a report which can then be sent to another computer using the communications package and a modem.

The development of integrated packages incorporating word processors, graphics, and communications has led to a recent revolution in publishing. The Apple Macintosh has led the way in the new growth area of computerised *desktop publishing*. This development allows authors and journalists to type their books or articles on a microcomputer and to typeset it thereby creating pages as they would appear in the book, journal, or newspaper. By using advanced digitisers, photographs can be computerised very easily and incorporated into the text by the author. This procedure could take place in several, widely-dispersed offices and, with the aid of the communications facilities, all contributions to a book or newspaper can be brought together centrally for printing. Alternatively, the entire document, once it is compiled, could be sent electronically to regional sites for printing thereby increasing the speed and efficiency of distribution, which is of particular importance in modern newspaper publishing.

Computer communications and interfacing

Communications between computers or between parts of the same computer system is a rapidly growing technology and a suitable *interface* must be present for communications to take place. An interface is usually a combination of hardware (electronic components and chips) and software (stored on ROM chips or on disk) which controls the interface hardware.

A printer may be connected to a computer system by a *serial* (e.g. *RS232*) or *parallel* (e.g. *Centronics*) interface. The difference lies in the way that bits of information are transferred. In serial communications, the data transfer occurs along single wires sequentially, whereas in parallel communications, a larger number of wires take the computer data simultaneously in parallel. Both systems have advantages. Serial links, although slower, are better for long distances, since fewer wires are needed, whereas parallel links provide fast data transfer over short distances. Limitations in speed of the mechanical moving parts of the printer may, however, override any speed limitations in the serial or parallel rate of data transfer.

Serial and parallel links do not mix – i.e. a serial output port will not connect directly to a parallel printer and vice versa. The plugs and connectors for serial communications are not compatible with those for parallel communications. However, it is possible to connect serial and parallel ports if a device called a *protocol converter* is used. A protocol converter acts as a translator from serial data transfer to parallel data transfer and vice versa. For example, a protocol converter is particularly useful for linking a microcomputer parallel printer port to the serial input port of a modem for data communications.

Linking microcomputers (or minis or mainframes) together – so that file transfers can take place between computers or so that printers and disk drives can be shared by several computers – can be achieved very effectively if a *network* is used. On a small scale, the *local area network* (or *LAN*) may only extend to the offices in one building. On a larger scale, the network may link computers throughout the world, perhaps even using satellite communications techniques.

An interface widely used to link microcomputers to sensors for monitoring work is the *analogue-digital converter* (or *ADC*). The ADC converts an analogue voltage (e.g. from a temperature sensor) into a digital value that the (digital) microcomputer can understand. The ADC may have limitations in terms of the speed and accuracy of conversion from analogue to digital representation, which may be important considerations in a particular application. An 8-bit ADC has a resolution of 256 (or 2^8) steps, which means that a voltage range of 0 to 10 V can be represented digitally by an integer value from 0 to 255. Hence, 1 step represents a voltage of 0.039 V, or 0.39% of the voltage range. An 8-bit ADC cannot therefore detect a change in voltage less than 0.39% of the full voltage range of conversion. To improve the resolution of the analogue to digital coversion, a 12-bit ADC can be used. This gives 4096 (or 2^{12}) steps, corresponding to a resolution of 0.025% of the full voltage range.

Any sensor which produces an analogue voltage output can be interfaced to a microcomputer using a suitable ADC. In addition to temperature, this includes sensors for humidity, pH, conductivity, load measurements sound level, position or orientation.

A *digital-analogue converter* (or *DAC*) interface is used to convert digital signals from a microcomputer into analogue voltages. This facility is very useful in applications where a variable control function is needed on a valve, or motor or robot arm. Limitations of accuracy and speed of conversion from digital to analogue signal described for ADC interfaces apply also to DAC interfaces.

In addition to analogue signals for computer monitoring or control, digital signals are also used in the form of on/off switching states or pulsed signals. The simple on/off state of a switch can be used to detect whether a feed storage bin is full or not. Pulses can be monitored to provide information about the speed of rotation of a shaft.

In the same way that new microchip devices are being developed for microcomputers, there is an expanding industry developing new sensors and interface devices for use in monitoring and control applications involving microcomputers. Many of these devices are being incorporated into new equipment and machinery for agriculture.

HOW ARE COMPUTERS USED IN AGRICULTURE?

Agriculture is very diverse, which is reflected in the wide range of uses to which computers are applied in this industry. The industry is considered here in terms of: agricultural production; ancillary industries; advisory and information services; and education, training and research.

Agricultural production

Dedicated and non-dedicated computers are used extensively and in a wide variety of applications in agricultural production, from simple record keeping to sophisticated environmental monitoring and control in glasshouses.

One of the most frequent applications of computers on the farm is in the recording and organisation of *financial* accounts. Simple packages for a farmer or market gardener will include records of payments and receipts, VAT records and will produce monthly or annual reports for the whole farm. Such packages are purely historic, whereas the more sophisticated packages include many more features (e.g., payroll programs) and put more emphasis on facilities for budgeting or planning, in addition to more detailed analysis of individual enterprises within a farm.

There are other types of package available that are concerned with *physical records* of individual enterprises on a farm, especially dairy or pigs and, to a lesser extent, arable. A farmer or stockman may make detailed records but, because of the sheer volume of data, does not have the time to sort, collate and extract important information to the degree that he might need for efficient management. Such analysis could, for example, enhance the investigation of specific problems in a dairy herd by examining the effects of bloodline, feeding performance and time of year. An exercise of this sort would take a long time to complete by hand, whereas computer-based recording systems offer rapid access to the information and the opportunity for analysis in a variety of ways.

For more sophisticated recording systems, the analysis also incorporates financial details, thus enabling the farmer to plan future projections more effectively, by basing them on historical records and trends.

In their simplest form *dairy programs* record:

(1) yield records for individual cows in the herd (for current and previous lactations);
(2) concentrate allocation and cumulative concentrate consumption during the current lactation;
(3) fertility events (e.g. date of last service, pregnancy diagnosis, projected calving date, number of lactations);

(4) health records; and
(5) whole herd results and action/exception lists (e.g. cows due for service).

More sophisticated dairy programs record milk quality from individual cows and combine physical records with financial programs to produce detailed reports. These programs can calculate the margin over concentrates for individual cows and, with the arrival of milk quotas, knowledge of an individual cow's performance is of increasing importance in managing a dairy herd.

Decision-making in *pig production* can, perhaps, benefit from computerisation even more than dairy farming, since the pig breeding cycle is shorter, herds are larger (thus making management of accommodation more critical) and, as a consequence, the time available to sort through records and use the information is shorter.

Simple programs concerned with pig breeding include the recording of feeding, fertility and health of individual sows plus action/exception lists. Additional, integrated features that are provided in more advanced programs include: financial calculations (e.g. margins); bloodline records (e.g. boar performance); and simple least-cost ration programs.

Some programs are designed to record details of pig finishing operations, for example, health, pig weights and, within a pen, feed consumed and average pig weight. There may also be the facility to link piglet growth information with sow records. As with pig breeding programs, financial calculations are also included.

Simple *arable software* packages record current crop, previous crops, spray usage, and financial performance for individual fields. Additional features provided by other arable programs include stock control of, for example, sprays and fertilisers. Forecasting procedures can also be used to create or amend cropping plans, or to produce financial projections which provide a comparison between the performance of the current year's crops with those predicted in the original budget or with those of previous years.

Standard spreadsheet, database or other packages can be used as an alternative to specific packages for agriculture, especially in the more straightforward applications. It is possible to use a spreadsheet or database to record receipts, payments and VAT, or to perform budgeting and cashflow exercises, or to conduct simple simulations (for example, to examine the effects of changes in the price of feed or rate of VAT). However, applications which go beyond this simple level will usually require the use of agricultural packages developed specifically for the purpose.

As software increases in sophistication, there is a corresponding increase in cost. Many farmers, while appreciating the need for such programs, cannot afford to purchase the computer hardware and software to run them, or pay staff to operate them. *Bureau services* offer an alternative to having an on-farm computer. These services are provided by companies who accept farm data sent to them and then undertake to process these records on a central computer. The results of the data processing operation are then sent back to the farmer for his consideration and appropriate action.

There are several advantages to using a bureau. For example, a farmer (or one of his employees) does not have to spend the time entering data into a computer system, although access is provided to a level of computer power much greater than that which he would be likely to purchase. There are disadvantages as postal delays produce a time lag

between sending off records to the bureau and receiving the analysis, thus reducing the frequency with which action lists are available.

In addition to on-farm micros and bureau services, *viewdata systems* provide another major, non-dedicated computing option that is readily available to farmers. Viewdata systems are computer-based, interactive information services that can be accessed using a suitable dedicated terminal (or microcomputer acting as an intelligent terminal) via a telephone socket.

There are several viewdata systems available for agriculture in the UK, e.g. *Prestel-Farmlink*, now incorporating ICI's *Agviser* and the NFU's *Viewdata* group. These viewdata systems are based typically on superminis like the DEC VAX, often linked in a network from various locations in the country.

The information available through viewdata systems is organised in pages in the same way as the various *teletext* services (e.g. Ceefax, Oracle, where pages are transmitted simultaneously with TV signals and these pages can be decoded and displayed on a suitable TV set). Viewdata systems use telephone communications and are interactive, in addition to providing many more pages than teletext. For example, not only are there pages of information (such as up-to-date prices or weather reports), but it is also possible to order goods (e.g. fertilisers or feed) or send messages to other users in the form of 'electronic mail'.

Viewdata systems are relatively cheap information services, with users paying an annual subscription and the cost of a local telephone call when they access the service. Typically, most pages of information are free, although access to some pages incurs a small charge. There are, in addition to the general service, *Closed-User Groups* which provide specialised information or services. A separate annual licence fee is usually payable to join these special interest groups.

Through these viewdata systems it is possible to access up-to-date market prices (up-dated several times per week) or even access sophisticated programs that can be used to analyse data from one's own farm (e.g. least-cost ration formulation). Closed-user groups allow a farmer to compare the performance of his own farm with national or regional figures. Such comparisons really put the performance of a farm into a true perspective and provide further information that can be used in planning and improving farm management. Viewdata systems are likely to play an increasing role as agricultural information services.

Following the advances made in microprocessor technology, another major area of computer use in agricultural production is in the monitoring and control of processes, housing or machinery (e.g. automatic out-of-parlour feeders, or environmental control within glasshouses, or monitoring and display of tractor/implement performance).

Horticultural use of microprocessor monitoring and control is well established. Humidity and temperature are of great importance in the production of glasshouse crops. Microprocessors are used to monitor these important environmental variables and automatically activate water sprays or ventilation when required. In areas where intense sunlight is a problem, blinds can also be activated when necessary.

Local meteorological variables can be monitored and recorded at low cost using computer-based automatic weather stations. More importantly for the farmer perhaps, these devices are able to process the data and report when weather conditions favour particular crop diseases or advise on irrigation schedules.

Computerised feeding systems have been developed particularly for cows and pigs. These systems enable a farmer to program feed dispensers to provide individual animals with a specific concentrate ration in a particular day. The rate of feeding of an individual animal can be regulated so that its daily ration cannot be eaten in one feeding bout, thus forcing more even feeding throughout the day. The animal receives its food in a cubicle (*feeding station*) which protects it from the aggressive behaviour of dominant animals within the herd, thereby circumventing the problem of particular animals not getting their fair share of food. Each cow or pig wears a collar to which is attached an electronic device that identifies the individual animal. There are two types of such device: *responders* and *transponders*. Responders are passive devices that are activated by a magnetic field generated around the feed hopper; an aerial on the feed station receives a signal from the responder that identifies the animal. Transponders, on the other hand, are active devices in that they have their own battery power supply that enables them to generate a signal identifying the animal. The microcomputer controlling the system recognises the animal and, if that individual has not exceeded its pre-programmed ration, dispenses a portion of food.

In addition to these dry feed systems, computerised wet feed systems are available for feeding pigs. These systems allow the farmer to feed the pigs according to a required growth rate that he programs into the computer.

Computerised feeding systems for cattle may also include automatic monitoring of milk yield and temperature recording for detection of heat and general health.

The integration of data from automatic feeding equipment with data from other computer-based recording systems on the farm can provide the farm manager with more detailed, reliable data upon which to base decisions. Communications links between the dedicated microcomputers of feeding systems and non-dedicated microcomputers is currently being carried out at a number of sites involved in research and development in agriculture. Such links will make possible more detailed research into the welfare and performance of pigs and cattle maintained in these husbandry systems and will also lead to the development of advanced management systems (e.g. the automatic integration of information about feed consumption of each cow in a dairy herd directly into financial management packages).

On-farm food processing (e.g. cheese, yoghurt) is becoming more common as is witnessed by the increasing number of farm shops and this is likely to increase further as the effect of quotas is felt and as the wholefood market develops. The food processing industry has used microprocessor control for some time and, although the scale of operation on farms is smaller than in industrial plants, similar control equipment is now being installed on farms. Monitoring and recording of temperatures, automatic weighing, packaging and labelling all fall into this category. As on-farm processing becomes more widespread, so the potential for integration of computerised monitoring/control functions with farm management systems becomes more attractive.

Several tractor manufacturers, following the lead of Renault, are now incorporating dedicated computers into some of their product range. The aim of these in-cab microcomputers is to increase fuel efficiency or work rate. By monitoring such variables as the engine speed, the temperature of the exhaust gases, and the theoretical forward speed of the tractor, the computer can provide the tractor driver with advice for changing gear or changing engine speed in order to improve tractor performance. Some of these computer systems are reputed to give fuel savings of up to 20%. As computer technology advances, developments in this application of computers in agriculture will progress dramatically, perhaps leading to automatic modification of the engine settings of the tractor without the intervention of the driver (with the facility for manual override retained for safety reasons).

Ancillary industries, advisory and information services

The ancillary industries associated with agriculture are many and varied. They include manufacturers and suppliers of equipment or materials for farms (e.g. machinery, fertiliser and spray manufacturers, and also feed and seed suppliers), services (e.g. accountants, veterinarians) and those businesses that buy farm produce for processing or distribution (e.g. creameries, bakeries, MMB, and feedmills).

In all of these industries, computers are used in general business administration (e.g. word processing and accounting), in addition to more specialised applications.

Computer-aided design (*CAD*) and computer-aided manufacturing (*CAM*) systems are used in machinery manufacture, the latter process perhaps involving the use of *robotics* on the production line. In other areas of the manufacturing industry serving agriculture, computers are used to monitor and control production processes. For example, the mixing of the various feed ingredients in feed mills can be regulated by a computer which also generates and prints the labels for the feed bags, listing the precise details of each of the ingredients by weight. The computer controlling the feed mixing operation will also provide a precise stock control system for the feed mill. Similar computer systems will also be found in the manufacture of chemical sprays and fertilisers. Computers are also found to a great extent in quality control in both the manufacturing and processing industries, and also in the weighing and labelling of end-products (e.g. feed and fertilisers). The laboratory support for these industries also involves computers to a large extent in the areas of research and development, laboratory analysis and quality control.

In addition to manufacturing, supplying or processing agricultural products, the ancillary industries and services provide advice to farmers supplementary to that provided by the main advisory agencies (e.g. ADAS, and the Scottish Colleges extension services).

Many veterinary practices now have computerised recording systems. By recording herd performance and deviations from target, effective preventative health measures can be put into operation. Computerised records also enable vets to advise more farms with greater efficiency. Some vets have even increased their computer services to the extent that they offer extensive bureau services to dairy farmers. Often, these bureaux operate at a local level and therefore can offer a rapid turnaround time between data arriving at the bureau and the analysis being returned to the farm.

As agricultural advice becomes increasingly computerised, the way in which advice is provided is changing. At one time, advice would be given by either a personal visit (but without any direct access to centralised information), or by telephone or letter (with no visual contact). With the development of portable microcomputers, an adviser can have the advantage of actually seeing a problem while, at the same time, having access to analysis programs on his or her portable microcomputer. For example, it is now quite common for an adviser or sales representative to run a rationing program on a customer's farm.

The development of computer communications has led to an increase in local access to central computing facilities, thereby allowing access to programs on a central mainframe computer that are far too large to be run on a portable micro. The use of viewdata systems by the advisory agencies has increased because of these developments and will continue to do so. One application of this advisory facility is local crop disease prediction and appropriate spraying advice.

Viewdata systems (e.g. Prestel) provide an important medium through which to disseminate advice, particularly as a result of developments in *Expert Systems*. These are computer programs based on *Artificial Intelligence (AI)* that guide a user through a complex logical problem using the knowledge of an expert and a set of rules instructing the computer how to reason and draw conclusions. On the basis of these rules, the computer requests information until it has enough upon which to base a conclusion. Expert systems also provide details of their paths of reasoning. They have been developed on mainframe and microcomputers and now some are being implemented as a series of viewdata pages for wider access.

Some expert systems are already in use (e.g. medical diagnosis and geological survey interpretation) and there are agricultural expert systems available (e.g. ICI's *Counsellor*) that are concerned with selecting appropriate plant protection strategies that maximise yields, list potential side effects of the treatments, and conduct cost-benefit analyses. The potential for this sort of application in advisory work is enormous.

Education, training and research for agriculture

Microcomputers play an important role in the areas of education, training and research for agriculture. There are several distinct ways in which microcomputers are applied in these areas.

Microcomputers can be used to train personnel in the use of specific software packages. Training in wordprocessing for a farm secretary, for example, will be geared towards meeting the needs of the secretary to gain the most from the wordprocessor package. This will involve intensive coverage of all the appropriate detailed facilities available in this package. The training programme may include computer-assisted-learning (*CAL*) techniques by using special training software related to the package and which guides the trainee through the various features of that package. On a more simple level, the same wordprocessor package may be used to provide the complete novice with basic keyboard skills and overcome the hurdle of using a keyboard and VDU screen for the first time.

This type of training applies to other software packages in which it is important that the user is fully conversant with all the available features in order to get most benefit from using

the package. Examples include databases, spreadsheets, financial packages and farm records packages.

Another area of microcomputer application is the use of particular packages as educational tools for the user in the subject area of that package. For example, commercially-used least costs feed formulation software can provide a very flexible environment for teaching the principles of feed formulation and rationing to the student. The approach in very *interactive*, with the student entering different strategies via the keyboard and seeing the results of those decisions displayed on the screen, with only a short wait for calculations to be completed. Not all commercially-used packages are suitable for use in this technique of *simulation* of a process for education. Many simulation packages are available for education and research and these are usually based on mathematical *models* of the agricultural system or process being studied. These include: farm buildings design and layout, pig growth models, project planning exercises, linear programming techniques, cash flow analysis, depreciation, crop husbandry simulations, irrigation scheduling and models of forage harvesting systems.

Open learning methods using microcomputers provide a further area of application for education and training. Open learning packages enable the trainee to learn about a topic individually, and at his/her own pace. Computer-based open learning packages are generated by the trainer using an *author language*, which enables the trainer to create pages of text or graphics for display onto the VDU screen. The student or trainee works through these pages interactively. Hence, on the basis of the response of the user to questions appearing on the current page, the package displays the next appropriate page. This could be a repeat of a previous page or a page which explains the same point in a different way or a page relating to the next part of the package. Open learning packages are ideal for revision work where the individual user may have specific areas of interest for concentration of effort. *Interactive video* is a further development of open learning incorporating the use of video disks to store very large amounts of video pictures in addition to text and graphics pages.

Students and research workers may also be involved in developing new software. These developments may involve the use of a programming language, for example BASIC, FORTRAN or PASCAL. Alternatively, spreadsheets or databases may be used, either as stand-alone packages or in an integrated form, to create original software applications. This integration may involve using spreadsheet or database applications in conjunction with original programs so that data transfer can take place within the whole application.

Computer integration is not restricted to software integration and neither is it restricted to a single computer. Development of integrated systems of software and hardware is becoming more common in research. For example, in the work mentioned above, involving links between dedicated feeding computers and other microcomputers, this type of integration will enable research workers to discover more about the feeding and welfare of pigs and cattle. By gaining a greater understanding of the feeding patterns of the animals it should be possible to continue to improve the welfare and performance of pigs and cattle reared in computerised husbandry systems.

Analysis of experimental results is an important part of student project work and research work in general. Comprehesive statistics packages are available for microcomputers and also for minis or mainframes. These packages contain the range of statistical functions or methods of analysis required for research work in agriculture and may also include some form of high level programming facility for developing new statistical methods. Experimental measurements are now commonly made using microcomputers connected to analogue-digital interfaces. With suitable software, this can provide an automatic data recording system, suitable for a wide range of different sensors, and which may also include facilities for direct data transfer to other computers for further analysis.

Microcomputer applications provide an important common link between education, training and research in agriculture and the industry itself. Research methods using microcomputers to take measurements and analyse results give rise to equations, which can be included in a mathematical model of the process. This model can then be developed into a microcomputer simulation of the process for use in education and training. Students and trainees then pass on these ideas and techniques to the industry when they graduate.

In the UK the links between organisations involved in education, training and research are coordinated by the National Computer User's Group in Agricultural Education, which includes representatives of colleges, polytechnics, universities, research institutes and training organisations.

WHAT ARE THE IMPLICATIONS OF INSTALLING AN ON-FARM COMPUTER SYSTEM AND WHAT ARE THE CRITERIA FOR ITS SELECTION?

It is very important that anyone who is considering the possibility of installing an on-farm computer system should think carefully about all the implications which could result from such a decision. Apart from the likely benefits in terms of improved efficiency and higher profitability, consideration must also be given to potential disadvantages or problems which could arise from such a development.

In order to operate a microcomputer system successfully, care must be taken to ensure that the hardware and software are both running efficiently. The supplier of the system often provides support in the form of service contracts. A hardware service contract usually covers the costs of all parts and labour in the event of faults occurring in the system hardware. Software support takes the form of automatic supply of updates to the software and new manuals, in addition to after-sales advisory or training. The extreme case of hardware/software replacement due to a major malfunction can, however, be made less likely if the system user follows a few simple procedures.

Microcomputers store information magnetically and so it is imperative to protect the magnetic medium (e.g. disks) from damage. Disks must be treated carefully and not subjected to static charges, magnetic fields, damp or condensation, excessively high or low temperatures, dust, smoke or mechanical damage by bending. it is always essential policy to make regular copies (i.e. *backups*) of software and data onto spare disks or tapes, and to keep these backups in a safe place. If possible, backups should be kept in a fireproof safe or they should be stored in a different building from the microcomputer system in case of fire.

It is good policy to make regular copies to disk when entering large amounts of data. In this way, the amount of data which may need to be re-entered (after the contents of RAM are lost due to a power cut, for example) will only be that entered since the last disk copy was made. Some software makes this type of copy automatically but this is certainly not true of all software products.

User attitude to the adoption of new technology may be the key factor to the successful operation of a new microcomputer system. User comfort is important and consideration must be made of lighting conditions, VDU height/angle adjustment and adjustable seating. There may be a need to change previous practices to ensure that the user will not be required to spend very long periods in front of the VDU screen without a break.

The installation of a microcomputer system may create the need to change the way in which other parts of the business operate and this may also involve consideration of employee attitudes. For example, there may be a need for a new approach to data collection to provide the detail required for a computer-based recording system.

If personal records are to be stored on the system, it may be a legal requirement to be registered under the Data Protection Act and to comply with the requirements of that Act.

The selection of a suitable computer system for agricultural production

Whatever type of farm, whether it be large or small, arable, animal or mixed, the most important consideration is:

Why is a computer needed for the farm?

The answer to this question is crucial. The reasons for purchasing a computer system must be defined quite clearly, always bearing in mind that bad business practices, for example, a chaotic farm recording system, will not be cured by merely installing an on-farm microcomputer.

Having decided on the use of which the computer is to be put on the farm, the next stage of the decision process is to choose the software/hardware combination which will perform the required function. The purchase of the programs and the computer upon which they are to run cannot really be separated. Programs are generally written in the context of a specific operating system which is, in turn, associated with particular microcomputers. Therefore, a particular software package is very often associated with only one or a very small number of microcomputers. Software dealers will usually also be agents for microcomputer hardware and should be able to supply a complete microcomputer system.

Many commercially-written software packages are available for agricultural applications. However, it is essential to know before purchasing a package that it will actually do precisely what is required. The enthusiasm of a salesman should not divert attention away from the farmer's precise requirements. One should always ask for a demonstration of both the computer and programs before purchase; reputable dealers will be happy to provide this. During a demonstration, one should pay attention to as much detail as possible and it is helpful to have a checklist prepared beforehand and always include the person who will be using the computer in the demonstration.

(1) The full range of reports generated by the package should be examined. Do they provide all the information that is required? Is the information easy to understand?

What inputs were necessary to generate those reports and how much time is needed to input that information?

(2) It is a good idea to try to work through parts of the package using the owner's manual provided to establish if the manual is easy to follow. Are jargon terms defined as they occur in the text, or is there a glossary? Is there advice about the programs (i.e. *software support*) available via the telephone? Are training courses included in the cost of the software?

(3) What is the quality of the printer and is it good enough to produce invoices or letters? Will the same printer be able to produce wage slips, cheques, or labels using the package under consideration?

(4) What is the cost of the microcomputer? Does that cost include the VDU, printer and disk drives? Is it possible to add a hard disk? How much would this cost? What is the cost of consumables such as disks, paper and printer ribbons?

(5) What is the cost of the software? Is it just a once-and-for-all payment or is there an annual licence fee?

(6) What is the cost of a maintenance contract (i.e. *hardware support*), if one exists, and how quickly will the engineer respond to a call? What is the duration of the warranty? and does it cover parts and labour? Does the contract apply to all the components in the same way?

(7) Is it possible to expand the system hardware and software if future needs emerge? Will the package allow the user to incorporate any files created by another package of programs running on the same computer?

(8) Can this microcomputer communicate with other computers via a modem and a telephone? How easily is this done? Will the microcomputer act as a Prestel receiver?

(9) Will the purchase of a computer system actually be a good business investment? Will it save *me* money or make *my* business more efficient? This is perhaps the most important consideration of all.

When purchasing any type of equipment or machinery, nothing can better personal experience. Therefore, before purchasing a microcomputer system, it is an advantage to discuss the matter with other farmers who have installed on-farm microcomputers. They will be able to provide first-hand experience of teething problems that they have encountered and, hopefully, overcome. Ask other farmers that have computers if they feel that such a purchase has been a worthwhile investment.

For independent advice on selection and installation of on-farm microcomputers, it may be helpful to contact the local ADAS office. If the local officer cannot provide this advice personally, he will certainly be able to direct any queries to one of the ADAS specialist advisers in this area.

WHAT DEVELOPMENTS MIGHT BE SEEN IN THE FUTURE FOR COMPUTER APPLICATIONS IN AGRICULTURE?

The current trend, towards more powerful (in terms of processing speed and storage capacity) and cheaper micro-computers which are also easier to use, is likely to continue in the future. This will lead to wider use of the more advanced software packages which need this additional computing power (e.g. expert systems).

Perhaps the introduction of the new IBM System 2 computers running under OS/2, the multi-tasking operating system, will set a new standard in computing for the future in the same way that the IBM PC and MS DOS did in the past. At the same time, the emergence of IBM PC clones which offer very good value-for-money in terms of hardware cost has led to a new breed of inexpensive software becoming available to meet the demand from this low-cost end of the market and this trend seems likely to continue. In the end it is the level of computing power and sophistication required by the user's particular application that will dictate whether an 'old-fashioned' IBM PC clone can cope or whether a more advanced computer system is required.

The biggest growth in computer applications in agriculture is likely to be in two areas. First is the area of dedicated (control and monitoring) systems which will benefit especially from new developments in Artificial Intelligence related to new sensors and sensing techniques. The second area is communications, which on a local scale may link dedicated systems directly to the on-farm microcomputer. This in turn may be linked on a wider scale with larger mainframes or superminis for access to advisory and information services.

Perhaps the key to the future expansion in use of computer-based technology in agriculture lies not with the technology itself, but rather with the computer-literate generation of young people now entering the industry. Their expectations will be high and this will create a demand for better quality and more sophistication in future computer applications for agriculture.

29

Alternative enterprises

R. W. Slee

One major component in contemporary discussions about the future of farming in the UK and many other developed countries is the role of alternative enterprises in adjustment strategies. Few now doubt that major changes will take place in the agricultural sector in Europe and similar pressures for change have been operating in North America and Australasia. But, at a time of policy flux and financial stress, attitudes to alternative enterprises are highly variable. Some look to alternatives as a lifeline; others regard their development as an act of desperation, eccentricity or foolishness.

This chapter provides a brief introduction to alternative enterprises. Alternatives must first be defined. The policy context must be understood for it will strongly influence the viability of conventional and alternative enterprises. The main groups of alternatives will be explored and brief comments on their likely role in future agricultural systems. Finally, the particular challenges of managing alternative enterprises must be recognised, for there are some significant differences from mainstream farm management approaches. (For a fuller discussion of all of these *see* Slee, 1987.)

DEFINING ALTERNATIVES

There is no universally accepted notion of what constitutes an alternative enterprise. Any definition begs the question: 'Alternative to what?' The most reasonable response is to suggest that there is an identifiable group of mainstream agricultural products. Normally the farm enterprises associated with these products do not go beyond the primary production of food and fibre. Normally in the EEC these mainstream products are supported by the policy mechanisms of the Common Agricultural Policy. Alternatives are alternative to these products but any attempt to define alternatives raises a number of problems.

Firstly, there are examples of policy supported products (e.g. lupins) which are clearly thought of as alternatives. Futhermore, in the light of new policy initiatives it is likely that a wider range of alternatives will be embraced by Common Agricultural Policy support measures. Over time it

is likely that alternatives will be much more fully incorporated into what are regarded as conventional farming systems. Oilseed rape would be an example of an alternative that has already been fully incorporated.

Secondly, at least some of the discussion about alternatives has focused upon alternative uses of farmland once it has passed out of agricultural ownership. Thus, urbanisation is an alternative use but it is rarely one carried on within an agricultural proprietal unit. Some alternative enterprises such as golf courses or forestry may take place under agricultural proprietorship but are more frequently managed independently.

Thirdly, there are problems in separating out farm-based alternatives from other gainful activities which exist quite independently of the farm. Thus viticulture and on-farm wine production are farm-based alternatives. A farmer who imports French wine as a business activity may have another source of income but it cannot be regarded as a farm-based alternative. (*See* Gasson (1986) for a detailed examination of the extent of these other gainful activities.)

It is important to realise that the conventional image of the farmer as independent yeoman is largely outdated. There has been a substantial influx of urban money into the countryside often for leisure motives, and this can have a significant impact on agricultural systems practised on and on the development of on-farm diversification. Off-farm incomes might provide the funds necessary to establish capital hungry alternatives like deer farming. Although it is outside the remit of this chapter to consider off-farm income sources, the consequences of them cannot be so readily ignored.

It is misleading to see alternative enterprises as something necessarily novel. The past rural economy was much more diverse as were the component farming systems. As late as the Second World War there was widespread evidence of alternative enterprises. These included many value-adding enterprises, particularly of milk and meat products as well as farm tourism. Alternatives were subsequently pushed into the background by the development of agricultural policies that concentrated on a streamlined product range. Their rehabilitation and the reform of agricultural policy go hand in hand.

THE POLICY CLIMATE

There are many reasons why alternatives are receiving greater attention than at any other time in the last 40 years. Post-war agricultural policy, initially in a domestic context, later in a European one, has both transformed the agricultural industry and created a new set of policy problems. Agriculturalists have become very sensitive to the barrage of criticism, but it is important that they should understand the forces that have created this criticism. The same forces have had, and will continue to have, a significant impact on policy.

Economic influences

The contemporary imbalance between supply and demand of major food products in the European Community is well known. The maintenance of prices above those that would normally prevail in a free market has led to increased production. The failure of the market to absorb this increased production has created the lakes and mountains of surplus produce. This is costly to store and is then frequently 'dumped' on Third World countries at less than the costs of production. At present the storage and dumping of surpluses consumes a major slice of the CAP budget, perhaps as much as 50% in 1987.

It has long been recognised that in a mature economy the agricultural sector is likely to decline in importance. Put simply, the capacity of people to eat is less than their ability and desire to acquire additional consumer goods. If incomes are rising, expenditure will be directed towards products associated with affluence. Historically, this has led to a growing demand for animal products including meat and dairy products. However, changes in taste have tended to dampen this process. Such growth as has occurred has been mainly in luxury foods, many of which are imported into the UK.

Until very recently the expansion of domestic food production was still seen as economically desirable. The high cost of pursuing such a policy is evidenced in the figures found in the Annual Review of Agriculture (*see Table 29.1*). The results of that policy are now being borne by farmers.

Table 29.1 Direct costs of CAP in UK farming

	Income (£ million)	Direct support (£ million)
1982/83	1807	1433
1983/84	1440	1729
1984/85	2114	1714
1985/86	1162	2160
1986/87	1410	1577

Source: HMSO, Annual Review of Agriculture

Political influences

The politics of domestic and European agricultural policy making must be recognised as a major influence on the current state of the industry. In the UK the strength of the agricultural lobby has been depleted by the collapse of support from its historic power base in parliament and by the growth of green issues on the political agenda in the country at large. This collapse is scarcely suprising when the majority of the residents of the British countryside no longer have any direct links with the farming industry.

The domestic situation differs markedly from that in the rest of Europe. On the continent farmers remain a powerful pressure group via the ballot box, as they constitute a greater proportion of the population. In addition, political parties in a number of countries woo the agricultural vote and maintain their position on account of it. Consequently, actions which damage the farming communities, however necessary from an economic viewpoint, will rarely be endorsed on mainland Europe, lest votes are lost. Expediency and horse trading take the place of rationally determined policy. In the UK the current political climate favours free markets over protectionism and agriculture is viewed no differently from such industrial dinosaurs as the coal and steel industries.

The consequences of the decline in the political power of the farming industry are far reaching. Whilst EC indecision compounds the problems of European policy, hard nosed, free market economics occupies a significant place in the UK policy agenda. Tough policy on prices and cosmetic environmentalism can be anticipated. At times this will be reinforced by other measures.

Environmental influences

Public interest in environmental issues has grown dramatically in the last decade. Marion Shoard's polemic, *The Theft of the Countryside* (1979) , angered farmers and raised public awareness of rural environmental change. The debate about environmental issues and the countryside received widespread media coverage and three main areas of discussion emerged. Firstly, it was felt by at least part of the public, that landscape had visually deteriorated under the impact of modern agricultural practices. Hedgerow removal, moorland improvement and land drainage emerged as causes célèbres. These and other side effects of intensification caused increased offence to a growing number of people, especially as the costs of supporting agriculture were visibly growing. Secondly, access is seen as threatened, particularly by field enlargement and the extension of arable systems. This debate looks likely to rumble on in the light of a major Countryside Commission study and a new book from Shoard (1987). Thirdly, habitats have been modified by agricultural practices that have reduced their ecological interest at the same time as public interest in wildlife conservation was intensifying.

Wartime food shortages have ceased to occupy an important place in the public's mind. Instead the dominant image is one of surplus, extravagantly produced at considerable environmental cost. The greening of politics will ensure that environmental issues continue to receive widespread attention and remain high on the policy agenda as the development and expansion of Environmentally Sensitive Areas (ESAs) illustrates.

Food and health influences

Attitudes to food have been transformed in the last decade and these attitude shifts have been reflected in dietary habits.

Medical evidence has linked diet with certain major causes of disease and mortality. A number of major investigations have advocated changes in dietary behaviour including reducing saturated fat (and total fat) intake, reducing sugar and salt consumption and increasing fibre consumption. It is immaterial whether the detailed proof of relationships is available. The significant point is that the diet of almost the whole population is being influenced. The demand for what were seen as growth sectors in the food market (red meat, dairy products) has been less than was anticipated.

Coupled with the concern about diet has been an interest in additives in food. Although many of these additives are put into food in manufacture rather than 'in the field', there is also concern about hormones, pesticide residues and certain farm-based 'additives'.

The combined result of these various policy pressures is to challenge the agricultural status quo and to suggest that policy redirection will be needed. The ALURE package (Alternative Land Use and the Rural Economy) launched in 1987 is the first major manifestation of this change in the policy climate. The succeeding years are likely to see new initiatives many of which will tend to support diversification into alternatives but these initiatives will be introduced in the context of declining support costs and a government commitment to a shift towards a more unfettered market.

TYPES OF ALTERNATIVE ENTERPRISE

Some reviews of alternative enterprises have attempted nothing more than an A to Z of alternatives. There is a case for attempting to classify alternatives in order to identify the characteristics that must be considered in their management. Four main groups of alternatives have been identified (Slee, 1986) and examples are given in *Table 29.2*.

At times agriculturalists appear to be seeking a set of recipes for production systems which, coupled with healthy gross margin data, will direct them towards a panacea for their income problems. This is wishful thinking. In the turbulent waters of unprotected markets, with unproven enterprises founded on flimsy production data, returns are

likely to vary dramatically from location to location. It would be misleading and foolish to advocate simplistic solutions based on diversification. It would be equally foolish, however, to ignore such data as are available in trying to elucidate the qualities and the prospects of the main groups of alternatives.

Tourism and recreation

There is a long tradition of the use of farms and farmland for tourism and recreation. Furthermore, there is evidence of relatively favourable demand trends coupled with a commitment of government to support the development of the leisure industry.

Tourism

Many farm households are involved in farm tourism to a greater or lesser degree. Estimates have been made suggesting that one in ten farmers nationally (DART, 1974) and one in five farmers in Less Favoured Areas (Davies, 1983) are involved with tourist or recreational enterprises. Whilst some of these enterprises create little more than pin money, others provide the major source of income for the business. An examination of the prospects for farm tourism must recognise both demand and supply influences.

The demand for rural tourism is influenced by many factors including incomes, exchange rates, the cost of competing destinations, including travel costs, climate and changing tastes. It is extremely difficult to do more than speculate about tourist futures whilst learning lessons from past experience. The 1970s were a boom decade for tourism with increased activity at home and abroad. Numbers of tourists to the countryside grew. Underlying the boom, a restructuring of the tourist sector was taking place, with a growing preference for self-catering rather than serviced accomodation. Tourists were also more likely to tour rather than stay in one place. In spite of the growth of the tourist sector the old resorts were struggling to adapt and the rural sector appeared to be well situated to benefit from these

Table 29.2 A classification of alternatives (with examples by way of illustration)

Tourism and recreation	*Alternative livestock and crops*	*Value-added*	*Ancillary resources*
Tourism:	Livestock:	By marketing:	Buildings:
Self-catering	Sheep for milk	Pick your own	for craft units
Serviced accommodation	Goats	Home delivered	for homes
Activity holidays	Snails	products	for tourist
		Farm gate sales	accommodation
Recreation:	Crops:	By processing:	Woodlands:
Farm Visitor Centre	Borage	Meat products –	for timber
Farm Museum	Evening primrose	patés, etc.	for game
Restaurant/tea room	Organic crops	Horticultural products	for tax avoidance
		to jam	
		Farmhouse cider	
		Farmhouse cheese	
			Wetlands:
			for lakes
			for game

Source: Slee (1986)

changes. However, the early 1980s have proved rather more volatile. Upward projections of tourist numbers in the late 1970s have now been adjusted downwards. More distant locations have experienced falling visitor numbers. Bad summer weather has compounded the problem. Optimism about tourist potential is now more qualified.

Rural tourism in general and farm tourism in particular may benefit from a number of favourable trends. Firstly, the general concern and interest of the public in the countryside may induce them to take rural holidays. This may allow the expansion of farm tourism in relatively new tourist areas such as Herefordshire or Staffordshire. Secondly, there is a growing interest in activity holidays, particularly for young people. Farm settings can provide suitable bases for activities ranging from art to archery or traditional rural pursuits like riding and shooting. Thirdly, the shorter duration second or third holiday may take advantage of a different experience to the main holiday. The final trend of significance is the shift towards self-catering. Rural areas do not suffer from the disadvantage of the decaying remnants of the declining serviced sector.

The supply side of farm tourism has responded to these changes. There are far more operators than there were a decade ago and many more people are exploiting the saleability of countryside images in their tourist marketing. Farmers have to compete with other providers of country holidays and a dramatic increase in supply of bed spaces could easily lead to spare capacity. There is also evidence that the tourism product is becoming more sophisticated and that the farm holiday must have qualities other than cheapness to attract the visitor.

Estimates of the profitability of farm tourist operations show a very wide range of returns (Davies, 1983). However, no explanations of the highly variable returns are offered. It is likely that at least some of the factors influencing profitability, such as occupancy rates, are partially under the influence of the operator's marketing strategies.

Recreation

The demand for recreation in the countryside can be seen to have experienced similar changes to those in the tourist sector. A decade of buoyant growth in the 1970s was followed by a period of instability and appears to have been replaced by a period of rather more selective growth.

However, the demand for rural recreation poses problems for the provider. The landscape is a principal resource and is subjected to largely uncontrolled access on the footpath system and, what might seem to some, totally uncontrolled access off the footpath system. The demand for much rural recreation is not backed up by a transfer of resources. Those who drive through or walk in pleasant countryside pay no price for their pleasure other than transport costs. Consequently, it is in those areas backed up by a willingness to pay that demand must be examined more closely, if farmers are to benefit financially.

The demand for traditional countryside sports has been taken up by 'green-wellied yuppies' and showed signs of growth in the early 1980s even when demand for passive recreation in the countryside had slackened. Demand for contemporary activity pursuits from rock climbing to hang-gliding also appears to be buoyant. But the total demand for these activity pursuits remains relatively small when contrasted to the demand for passive recreation.

The supply side has responded. There are many more visitor attractions than a decade ago including rare breeds farms, visitor centres, museums and working farms. Private, public and voluntary sectors have all been major providers and a growing professionalism in servicing visitor needs will be needed to sustain viability in the future, as the effects of competition begin to bite.

It is impossible to speculate on the profitability of these enterprises. Few entrepreneurs will willingly give evidence of their profitability. Some enterprises can produce a very high return on capital in a very short period of time. Others, managed self-indulgently by operators, are little more than disguised hobbies. That there will be increased competition for visitors cannot be doubted. That there are still opportunities for well located and well managed visitor attractions on farmland is equally true. Amateurism must be replaced by a new professionalism particularly in visitor servicing.

Alternative crops and livestock

The range of alternative crops and livestock is huge as is evidenced in the Centre for Agricultural Strategy Report (Carruthers, 1986). In these pages it is not possible to review the totality of new possibilities. Neither is it possible to explore the alternative uses of existing crops. Instead the main groups of alternative crops and livestock will be identified and brief comments offered on the qualities that should be sought in alternatives. Four main groups are identifiable: ethnic products; health products; luxury products; craft products. In some cases one alternative may fall into more than one category.

Ethnic products

Different ethnic groups exhibit different patterns of demand for food and other products. These ethnic groups may be residents of the UK or of other countries which might or do import UK products. Typical examples of food products include goat meat and milk, feta cheese, Chinese mushrooms or carp.

There is a tendency to think only in terms of domestic ethnic markets but there may be export opportunities in specialist foods which merit exploration. It is also important to recognise the impact of tourism and travel on domestic eating habits and to realise that there may be increasing interest in these exotic foods from the domestic population.

Luxury products

The demand for luxury food is a growth area in the food market. There is evidence, often of a circumstantial nature, to indicate a growing demand for luxury foods, from snails to quails, venison to specialist cheeses. Farm enterprises can either provide the inputs into these luxury foods or, by adding value on the farm, produce the final product.

The principal problem with many luxury products is being able to determine the market with resonable accuracy. Market research is essential, yet prior to a product launch it can only be indicative and in a free market new entrants could dramatically affect the supply of the product. The main unknown is whether demand will increase if and when, for example, supermarket chains start stocking a product regularly. If this is the case there may be certain thresholds of supply level which must be crossed in order to achieve the expansion of demand.

Health products

Few doubt the impact of health conscious eating on the demand for food. But, instead of bemoaning the loss of markets for traditional products, it is possible to develop new products that offer healthy fare. Low fat meats such as venison may have advantage over high fat meats like lamb. Goats' milk may have a growing market with allergy sufferers. Borage and evening primrose both contain gammalinoleic acid which has a wide range of reputed health benefits.

Organic products fit within this category. They are sought by consumers who want an additive-free diet and there has been a significant increase in the demand for organic foods. At present significant amounts of domestically consumed organic foods are produced in other countries. There may be scope for an increase of UK producers' market share.

Craft products

Craft products encompass a range of products including wood for turning, thatching materials, withies, teasels and specialist animal fibres. In many cases there may be opportunities to add value to these products. Craft products may be sought by working craftsmen such as thatchers or by individuals pursuing craft hobbies. Early retirement and the expansion of leisure time are likely to increase the demand for craft materials.

Value added enterprises

Value can be added to conventional farm products by processing or by marketing them in different ways. Both avenues merit exploration but both require careful scrutiny. It is important to realise that the objective must be to increase *net* value added. Adding value entails additional equipment and resources and the capital and running costs must be carefully assessed.

Adding value by marketing

There are a number of conventional marketing channels used by farmers. It is possible to add value by marketing direct to the consumer or by missing certain links in the marketing chain. The simplest examples include pick-your-own (PYO) or farmgate sales. Other examples include sales of eggs or horticultural produce direct to shopkeepers.

In those situations where the customer is expected to come to the farm, location is of prime importance. Where the farmer conducts sales to the public from the back of a van, the precise location of the farm is much less important although proximity to a large market is clearly beneficial.

It should be realised that in many cases the visit to the farm shop or PYO establishment is a recreational visit in addition to being a food shopping trip. Many operators recognise this and cater for the public in a variety of ways. It may also be important to offer particular facilities in order for the farmer to differentiate his establishment from others. There are clearly limits to the potential for PYO which have probably been reach in certain parts of the country.

Adding value by processing

Value can be added by processing milk into cheese or wheat into flour. Historically, most rural households engaged in various forms of food processing and manufacture. After a period of decline there appears to be a reversal of a trend towards mass produced foods and a growing demand for farm produced foods. Following a decline in the quantity of on farm Cheddar production, the market is now relatively buoyant and in a whole range of products the demand for specialist foods has shown considerable growth (Hill, 1985). It is important that on farm processors should understand the changing structure of food retailing and in particular, the pre-eminent position of large supermarket multiples. Adding value in this way entails the acquisition of new technical and processing skills as well as business and marketing skills. It cannot be entered upon lightly but it may offer significant opportunities to an agricultural industry adapting to changing consumer demand.

Ancillary resources

Many farms contain resources which are not used productively for agricultural purposes but which could nonetheless contribute to farm incomes if developed as an alternative enterprise. Woodland, wetland or redundant buildings are all examples of resources that if managed differently, could generate income.

Few farmers are aware of the value of timber crops or of the management needs to enhance product value. As it is likely that government will fund additional woodland on farms it is important that farmers should acquire both understanding and skills (*see* Chapter 7).

'Wetlands' is a loose term covering everything from water bodies to wet pasture. Wetlands have alternative value as game areas, for fishing or for conservation and other forms of recreation.

Many farm buildings are no longer useful for agricultural enterprises and could be developed for alternative uses. Not infrequently these buildings are converted to tourist use but there are possibilities of developing permanent residences or industrial buildings. Such changes of use of buildings require planning permission which may not readily be obtained as local planning authorities have yet to come to terms with the rush of applications and the central government circulars advising that wherever possible permission should be given.

In all the groups of alternatives the farmer must be looking to explore the markets and the production requirements for his particular alternative. There are a number of questions that must be addressed.

(1) Is the enterprise likely to yield a better return on capital invested than conventional farming?
(2) Is demand expanding/likely to expand?
(3) Are there barriers to stop competitors getting involved?
(4) Are there any health/welfare doubts?
(5) Is the enterprise environmentally acceptable/acceptable to planners?
(6) Is there any prospect of adverse government intervention?
(7) Are the business risks acceptable?

In each case the alternative should not be developed until these questions have been answered satisfactorily.

MANAGEMENT OF ALTERNATIVE ENTERPRISES

The management of alternative enterprises has certain similarities with any farm management problem but there are also differences that must be understood. Management problems can conveniently be separated into those of establishment and those of monitoring and control.

Marketing

Marketing skills have not been historically a major part of farm management training, nor have they been in evidence in the farmer's normal repertoire of business skills. In a supported market and with frequently undifferentiated products this is understandable. Farmers have tended to think about quantities of output and have had a production orientation to their business activity. With alternatives a different business philosophy should prevail. Markets are unsupported and products often differentiated and a marketing orientation provides the key to the successful establishment of alternatives. It is not sufficient to think of marketing as an afterthought, born of unsold stocks. Marketing skills and principles should be embodied in a marketing plan which details the business/enterprise strategy with regard to alternatives.

Whilst some alternatives may be founded successfully after an entrepreneurial vision a safer bet is likely to be market research. Demand should be established from published sources or from market research commissioned or carried out personally. Information should be collected on competitors and it may be possible to visit establishments (preferably in another part of the country) offering a similar product. It is important that the market area in which the product will be sold is considered. In the case of farmgate sales it is very local. With a specialist ethnic food, it may be international. The marketing plan should be built on this market research. Marketing experts suggest an examination of the four Ps: product, place, price and promotion. Each should be considered in turn (*see* Chapter 24).

The resource base

Alternative enterprises are founded not just because markets exist but because farmers have resources which can be combined to provide them with a profit (hopefully) and the consumer with a good or service. A thorough appraisal of the resource base of the farm is an essential stage in reviewing options for alternatives.

There is an inevitable tendency for farmers to think about resources in terms of their agricultural value. Land is considered by its Agricultural Land Classification (ALC) grade, not for example, by its silvicultural potential. Yet quality of land must relate to each of the alternative uses. Areas of high tourist and recreational potential may have very low agricultural value. The location of land in the UK, with the exception of outlying islands, has ceased to have a major impact on most types of agricultural production. With respect to alternatives, location is an attribute of land that is of crucial importance in many cases.

The skills of the farm household as a whole should be taken into account when considering alternatives, as should those of any hired workers. Retraining may be necessary but there may be skills relating to dealing with people or livestock husbandry skills which can be deployed effectively in alternative enterprises. When looking at the labour skills, the seasonal demands that some farm systems create should be considered for if there is a clash with peak demands for alternatives the new enterprise cannot be established.

The capital resources consist of the land itself, the buildings associated with the land and any capital resources in the owner's possession that could be committed to the business and any capital that can be raised from normal sources of borrowing and grant aid. The capital value of land or buildings may be much enhanced by planning permission and neither should be sold to raise capital without first ascertaining development possibilities. Capital can also be locked up in woodlands.

Establishing alternatives

The establishment of alternatives should be guided by a marketing orientation and preceded by a thorough resource audit. It is desirable that the fullest use is made of sources of advice. There is unlikely to be a reliable textbook figure to reveal the likely profitability although attempts have been made (Parker, 1986). Any figures should be used with caution. Consequently, advice from public, private or voluntary agencies should be sought. The pre-eminence of ADAS in conventional agriculture is not matched by a pre-eminence with regard to alternatives. Thus, with a prospective tourist enterprise, the farmer could go to ADAS, the regional Tourist Board or the Rural Development Commission or any number of private sector consultants. Searching out the various agencies can take time and it is likely that some agencies, such as NFU Marketing, will develop a signposting role.

As well as being uncertain of whom to ask for advice, it may be equally difficult to ascertain where to seek capital or grant aid. Here again signposting agencies are likely to be important. MAFF is embarking on a broadening of the range of grants for diversification and farmers will be able to experience the dubious pleasure of shopping around for the best level of grant. There has been much talk of streamlining and coordinating the agencies involved but this is unlikely to materialise in the foreseeable future. The most significant new grant that is likely to emerge is the feasibility study grant from MAFF which will enable up to 50% of the costs of such a study to be recouped. Marketing grants are also planned for alternative enterprises. Ironically, however, these grants have not been made available until after a significant number of diversification grants have already been paid out. It is vitally important that diversifying farmers maintain an awareness of grant changes, and this awareness should extend to set aside payments and farm woodland grants amongst others.

It is not possible to recommend financial evaluation techniques that will be appropriate for any alternative. Much will depend on the level of capital invested, the impact on other farm enterprises and the level of riskiness of the investment. For a variety of reasons, alternatives are likely to be more risky. As a consequence, it is desirable to conduct sensitivity analyses which assess the outcomes of changes in the key variables determining profitability. Once this sensitivity analysis has been conducted, it should be possible to make a more rational decision (for fuller details *see* Slee, 1987).

Running alternatives

Once establishment problems are over the management challenge of alternatives does not cease. Indeed many recently established alternatives may have arisen out of desperation rather than detailed planning in the light of the current economic climate confronting the industry. As a result the monitoring and control problems of both new entrants and established practitioners are likely to grow.

The balance between supply and demand for many alternatives has yet to be struck. It is unlikely that it will be struck easily for there is a shortage of accurate supply and demand data. There are no census data for most alternatives to inform us of supply statistics and market research is often rudimentary. Where the total demand for a product is small compared to conventional agricultural products, oversupply could arise. Where markets are unprotected, competition can be harsh, as new entrants into alternatives have found out to their cost.

Operators can come to terms with competition in a number of ways. Farm holiday groups producing collective brochures may reduce advertising costs and offer the potential tourist a package of possibilities. Food organisations like Devon Fare can help individuals to market their products. Individuals can respond by trying to differentiate their product and offer something different or exclusive to the consuming public.

Monitoring and control are important functions of management. They are especially important where there are difficulties of measuring business performance. Conventional enterprises can be contrasted with farm management handbook standards. Many alternatives cannot. Thus, operators must set their own standards, measure performance accurately and explain deviations between actual and anticipated results. Good budgetary control is fundamental to good management. It should provide the information required to enable appropriate adjustments to be made.

One important consideration with alternatives is the determination of the optimal size of enterprise. The 'bigger is better' philosophy does not necessarily apply, especially where the initial development of the alternative is based on exploiting a slack resource. Empty bedrooms may create bed and breakfast space and cost little to put in order. Extending accommodation by extensions to farmhouses can dramatically increase capital requirements. Direct sales can yield retail price premia. Sales to a wholesaler generate lower prices. As an enterprise grows, so it may have to shift from the retail market to the wholesale one, thus reducing margins. There are, of course, cases where economies of size exist, such as in deer farming or woodland management but they cannot be taken for granted.

The control and motivation of the workforce can be especially important where enterprises require contact with the public. This arises with tourist and recreational enterprises or in the case of PYO enterprises and farm shops. Social skills are often more important than technical skills. Technical skills cannot, however, be ignored, especially with new crops like borage where seed bed preparation and careful harvesting are vital.

In many alternative enterprises the raison d'être is a personal interest of the farmer or a member of the family. This interest may lead to the acquisition of basic knowledge and skills. The same interest may misdirect business behaviour towards self-indulgence. The separation of work and play may be a necessary prerequisite for a healthy balance sheet.

There are certain bureaucratic differences between alternatives and conventional farming. If the enterprise is not a farm enterprise, it is likely to be rated. If a change of use is to take place, planning permission may be required. Opening a farm shop entails satisfying a host of public bodies from planners to weights and measures inspectors to public health inspectors. Recent attempts to 'lift the burden' must be balanced against the desire of the consumer for some kind of protection.

SUMMARY

Alternative enterprises are likely to become more important to the farming industry of the 1990s. Pressures of budgetary crisis linked to overproduction in conventional products are so overwhelming that the industry is searching hard for solutions to the problems and alternatives have a part to play.

The interest and the response on the ground make it very difficult to make generalisations about prospects for individual alternatives. The volatility of the free market for most alternatives exacerbates the difficulties of making long-term appraisals of possibilities.

It is all too easy, though, to resort to hackneyed arguments in dismissing even the thought of diversification into alternatives. Demand may be uncertain; risks may be considerable; markets may not be large. At least some of the dismissals of alternatives may result from the ability to take up existing slack in enterprise management. This is not especially difficult in the first instance but subsequent stages may be more painful and difficult.

At the earliest opportunity a thorough resource audit guided by a shift towards a marketing orientation should be carried out. This is not a one-off process but something that should be ongoing as the prices for different products vary in the future. The full range of opportunities embracing both alternatives and conventional farm enterprises should be considered.

Alternatives will not provide salvation for all but they may offer lifelines for some. Those who can adapt their thinking and develop their business skills, may well use alternatives as a stepping stone into an uncertain future.

References

CARRUTHERS, S. P. (Ed.) (1986). *Alternative Enterprises for UK Agriculture*. Reading: Centre for Agricultural Strategy

DART Rural Planning Services (1974) *Farm Recreation and Tourism in England and Wales*. Cheltenham: Countryside Commission

DAVIES, E. T. (1983). *The Role of Farm Tourism in the Less Favoured Areas of England and Wales*. Exeter: University of Exeter

GASSON, R. (1986). *Farm Families With Other Gainful Activities*. London: Wye College

HILL, S. (1985). *Specialist Foods*. Watford: Institute of Grocery Distribution and National Farmers Union

PARKER, R. R. (1986). *Land: New Ways to Profit*. London: CLA

SHOARD, M. (1979). *The Theft of the Countryside*. London: Temple Smith

SHOARD, M. (1987). *This Land is Our Land*. London: Paladin

SLEE, R. W. (1986). *Alternative Enterprises: Their Role in the Adjustment of UK Agriculture*. Paper presented to Rural Economy and Society Discussion Group, Annual Conference, Oxford

SLEE, R. W. (1987). *Alternative Farm Enterprises*. Ipswich: Farming Press

Glossary of units

METRIC UNITS AND 'IMPERIAL' CONVERSION FACTORS

The established metric unitary system used in the UK is the 'Systeme Internationale d'Unites (SI)' which replaces 'Imperial Measure'.

The system consists of basic units from which all measurements are specified by use of multiplying factors. The most important of these factors in common usage are:

Multiplying factor		Prefix	Symbol
0.000 000 000 001	(10^{-12})	pico	p
0.000 000 001	(10^{-9})	nano	n
0.000 001	(10^{-6})	micro	μ
0.001	(10^{-3})	milli	m
0.01	(10^{-2})	centi	c
1000	(10^{3})	kilo	k
1 000 000	(10^{6})	mega	M
1 000 000 000	(10^{9})	giga	G

Examples of the use of these factors are
millilitre (ml), i.e. 0.001 × 1 litre and kilogram (kg), i.e.
1000 × 1 gram. (The latter is often shortened to the 'kilo' in common usage).

In general, measurements should be specified in terms of units such that the 'whole number' part of the measurement is between 1 and 1000, e.g. 100 kg not 100 000 g, and 20 mm not 0.020 m.

The seven primary units used in SI are:

	Unit	Symbol	Conversion factor SI to Imperial	
Length	metre	m	to ft	3.281
Weight	⎰kilogram	kg	to lb	2.204
	⎱tonne	t	to ton	0.984
Time	second	s		1.000
Temperature	degree kelvin or Celsius	K or °C	to °F	1.8 (+32)
Luminous intensity	candela per square metre	cd/m^2	to cd/ft^2	0.0929
Electric current	ampere	A		1.000
Radiation	becquerel	Bq ($\equiv$ 1 spontaneous nuclear transition per second)		

The above are the basis of many derived practical units which include:

	Unit	*Symbol*	*Conversion factor SI to Imperial*	
Volume	litre	$l (\equiv 1000\,cm^3)$	to gallons	0.218
			to pints	1.744
Force	newton	$N \equiv kg\,m/s^2$	to lbf	0.2248
Work energy	joule	$J \equiv Nm$	to ft lbf	0.7376
			to Btu	0.9479×10^{-3}
			to Therms	9.479×10^{-9}
			to Wh	2.7×10^{-3}
Power	watt	$W \equiv Nm/s$	to HP	1.341×10^{-3}
Pressure	pascal	Pa or N/m^2	to in H_2O	4.015×10^{-3}
			to in Hg	2.953×10^{-4}
Pressure is also expressed in the			to lbf/in^2	1.45×10^{-4}
'non-preferred' unit of the millibar			to $tonf/ft^2$	9.324×10^{-6}
(1 mbar is equivalent to $100\,N/m^2$)			to $tonf/in^2$	6.475×10^8

(1 atmosphere $\equiv$ 1 bar $\equiv 100\,kN/m^2 \equiv 100\,kPa \equiv 10.197\,mH_2O \equiv 33.025\,ft\,H_2O \equiv 760\,mmHg \equiv 14.7\,lbf/in^2$)

Other units commonly used in agriculture are:

	Unit	*Symbol and conversion factor*	
Calorific value	kilojoules per cubic metre	kJ/m^3 to Btu/ft^3	2.684×10^{-2}
U-value	watts per square metre per degree Celsius	$W/m^2\,°C$ to $Btu/ft^2h°F$	0.176
Specific heat capacity	kilojoules per kilogram per degree Celsius	$kJ/kg\,°C$ to $Btu/lb°F$	0.239
Moisture capacity	grams per cubic metre	g/m^3 to grains/100 ft^3	43.6
Illumination level	lux	lx to 1 m/ft^2	0.0929
Land area	hectare	ha to acres	2.47
Fuel consumption	kilometres per litre	km/litre to miles/gal	2.825
Rate of usage	grams per square metre	g/m^2 to oz/yd^2	0.0295
	kilograms per hectare	kg/ha to lb/acre	0.892
	litres per hectare	litre/ha to gal/acre	0.089
Crop yield	tonnes per hectare	t/ha to ton/acre	0.398
		to cwt/acre	7.97
Density	tonnes per cubic metre	t/m^3 to ton/yd^3	0.753
	kilograms per cubic metre	kg/m^3 to lb/ft^3	0.0624
Speed	kilometres per hour	km/h to miles/h	0.621
	metres per second	m/s to ft/min	196.9
Flow rate	cubic metres per second (cumecs)	m^3/s to ft^3/min	2.118×10^3
	litres per second	litre/s to gal/min	13.2

With considerable amounts of imperial standard equipment still in use it will be necessary to convert from metric to imperial measure for many years to come. Accurate conversion will always be the best solution, though quick approximation may be of value to the farmer.

The following near equivalences are useful to remember and do not incur great error:

Length

$25 \, mm = 1 \, in$
$300 \, mm = 1 \, ft$
$1 \, m = 1.1 \, yd$
$8 \, km = 5 \, mile$

Rate of use and yield

$4.5 \, kg/ha = 4 \, lb/acre$
$125 \, kg/ha = 1 \, cwt/acre$
$5 \, t/ha = 2 \, ton/acre$
$11 \, litre/ha = 1 \, gal/acre$
$25 \, mm \text{ over } 1 \, ha = 250 \, m^3 =$
$1 \, acre \, in = 22\,600 \, gal$

Mass

$30 \, g = 1 \, oz$
$1 \, kg = 2 \, lb$
$50 \, kg = 1 \, cwt$
$1 \, t \text{ or } 1000 \, kg = 1 \, ton$

Area

$1000 \, mm^2 = 1.15 \, in^2$
$1 \, m^2 = 10 \, ft^2$
$4 \, ha = 10 \, acre$

Rate of flow

$1 \, m^3/s = 2000 \, ft^3/min$
$1 \, litre/s = 13 \, gal/min$

Force and pressure

$4.5 \, N = 1 \, lbf$
$10 \, kN = 1 \, tonf$
$7 \, kN/m^2 = 1 \, lbf/in^2$

Volume

$100 \, cm^3 = 6 \, in^3$
$3 \, m^3 = 100 \, ft^3$
$3 \, m^3 = 4 \, yd^3$
$4 \, m^3 = 100 \, bushel$
$9 \, litres = 2 \, gal$
$1 \, m^3 = 220 \, gal$

Common metric equivalences

$1 \, litre = 1000 \, cm^3$
$1 \, ha = 10\,000 \, m^2$

$1 \, m^3 = 1000 \, litres$
$1 \, km^2 = 100 \, ha$
$1 \, MN/m^2 = 1 \, N/mm^2 (= 1 \, MPa)$

Heat energy

$1 \, kJ = 1 \, Btu$
$3 \, W \text{ or } J/s = 10 \, Btu/h$

$100 \, MJ = 1 \, therm$

Index

AI, *see* Artificial insemination
ATP/ADP, cycle, 329
Abortion in sheep, 417
 diseases of, 491–492
Abscission zone and leaf shedding,
 62–63
Accidents, fatal, statistics, 644
 see also Reporting of Injuries,
 Diseases and Dangerous
 Occurrences Regulations 1985
Accounts, 612–615
 cash flow recording/monitoring,
 612–614
 profit/capital recording, 614–615
 net capital recording/analysis, 615
 profit/loss analysis/interpretation,
 614–615
 see also Budgeting
Acetic acid metabolism, 31
Acetonaemia,
 in cattle, 480
 in sheep, 339
Acid milk replacers for calves, 393
Acidity, *see* pH of soil
Acidosis in ruminants, 340
 and feed fermentability, 327
Actinobacillus ligneresi, 478
Actinomyces necrophorus, see
 Foul-in-the-foot
Actinomycetes bovis, 478
Actinomycetes in soil, 19
Acyrthosiphum pisum, 150, 303
Additives, for poultry production, 454
Adipocyte development, 360–361
Adrenaline,
 and glycogen oxidation after
 slaughter, 331
 and milk ejection, 359
β-Agonists, for pig production, 449
Agricultural Holdings Acts, and
 security of tenure, 639
Agricultural Improvement Scheme
 (AIS), 599

Agricultural Land Tribunal, 639
Agriculture Act 1986, 643
Agriculture (Miscellaneous Provisions)
 Act 1968,
 on animal housing, 586
 and security of tenure on death, 639
Agriculture (Poisonous Substances) Act
 1952, 644
Agriculture (Safety, Health and
 Welfare Provisions) Act 1956,
 644
Agriotes spp., *see* Wireworms
Agrotis segetum, 294
Agrostis spp. in grassland, 178
Aldosterone in potassium/sodium
 balance, 335
Algae in soil, 20
Alsike, 185
Alternaria dauci, 284
Alternaria spp., *see* Leaf spot
Alternative enterprises, 661–668
Alvinox for bloat, 479
Amino acids,
 metabolism, 332–333
 in nutrition, 321
Ammonia,
 danger from litter, 469
 as fertiliser, 79
 toxicity in ruminants, 340
Ammonium fertilisers, 78, 79, 80
Ammonium in soil, 21
Analysis of soil, 30–32
 'available' elements, 31
 laboratory methods, 31
 nutrient index, 31
 'total' elements, 31
Animals Act 1971, 641
Anion adsorption by soils, 14
Anoestrus, lactational, 353
Anthracnose, on peas/beans, 145, 283
Anthrax,
 in cattle, 483–484

Anthrax (*cont.*)
 in pigs, 489
Antibiotics, growth promotors versus
 therapeutic, 471
Aphids,
 on beans, 143, 145, 303
 on cabbage, 300
 on carrots, 158
 on cereals, 291–292
 and insecticide resistance, 297
 on oilseed rape, 300
 on peas, 150, 302
 on potato, 293
 on sugar beet, 297
 on sweet corn, 135
 in virus yellow spread in sugar beet,
 174
Aphis fabae, 143, 303
Apodemus sylvaticus as sugar beet pest,
 295
Appetite, as disease sign, 472
Arable crops, 90–176
 catch crops, 91
 choice of, 90–91
 double crops, 91
 establishment, 92–93
 horticultural, 'cool chain' for,
 98–99
 maincrops, 91
 production, computer packages for,
 656
 protection, plastic covers, 97–98
 seed, 91–92
 for silage, 202
 transplant raising, 93–96
 bedding systems, 93–94
 compost requirements, 94
 operating tray system, 95–96
 seeds/sowing, 94–95
 see also Rotation of crops
 see also by particular crop
Arion hortensis on potatoes, 294
Armellaria media, 239

Artificial insemination,
    for cattle, 349–350
    for pigs, 437
Ascariasis, on pigs, 488
Ascochyta, in peas/beans, 283
Atomaria linearies on sugar beet, 296
Aujeszky's disease of pigs, 487
Azochyta spp., 283
Azotobacter in nitrogen fixation, 18

BST, *see* Bovine somatotrophin
Bacillus anthracis, in cattle, 483–484
Bacteria, 469
    in mastitis, 482
    in silage making, 199
    in soil, 18–19
        classification, 18
        and nitrogen fixation, 18–19
Baked beans, 151
Baling, 520
    of hay, 204, 521
    of silage, 520–521
        big-bales, 201
Ballasting, and tractor power, 505
Bank loans, 618
Barley, 128–129
    in animal feed, 377
    diseases of, 272–274
        ear, 274
        leaf/stem, 272–273
        root/stem base, 274
    harvesting/storage, 129
    husbandry, 128–129
    quality and use, 128
    variety, 128
    winter, vernalisation requirement,
        119
    yield, 129
    virus, yellow dwarf, 273
Barley beef, 401–402
Barley yellow dwarf virus
    on barley, 273
    on wheat, 270
Basidiomycetes in soil, 19
Basin peat formation, 4
Beans,
    broad, 143–144
    diseases of, 283–284
    dwarf, 144–145
    field, 142–143
    navy, 151
    pests of, 301–303
    pricing, Common Agricultural
        Policy, 597
    runner, 145–146
    soya, 151
    value in animal feeds, 380
Beech bark disease, 239
Beef, 399–404
    carcass classification, 399–400
    and dairy herd decline, 399
    feeding, 398
    growth/development, 399
    housing, 562–563
    pricing, Common Agricultural
        Policy, 595–596

Beef (*cont.*)
    production/consumption, EEC,
        389–390
    slaughtering, selection for, 400–401
    systems, 401–404
        choice, factors in, 399
Bees,
    in field bean pollination, 143
    and mustard, 118
    on oilseed rape, insecticide risk, 116
Beet, weed beet control, 251, 257
    see also Beetroot; Mangels/fodder
        beet; Sugar beet
Beet Cyst Nematode Order 1977, 298
    see also Cyst Nematodes
Beetles,
    black pine, 239
    cabbage stem, 200, 299, 301
    click-beetles, 290
    pollen, on oilseed rape, 300
    see also Flea beetles
Beetroot, 153–155
Beijerinckia in nitrogen fixation, 18
Big-bales for silage, 201
Biogas, 499
Bird cherry aphid, 292
Bird damage to maize, 132
Bird pests,
    on maize, 132
    pigeon, on oilseed rape, 299–300
    skylark, on sugar beet, 297
Birth, 352–353
Biscuit making, wheat for, 125
Black bean aphid, 303
    on sugar beet, 297
Black disease, in sheep, 489
Black leg, of potato, 277
Black medick, 185
Black rot of carrot, 284
Black scurf of potato, 279
Blackleg in lambs, 489
Blackleg, powder, in beet/mangold, 282
Blackquarter, of cattle, 477
Blanket bog formation, 4
Blast furnace slags, 89
Blight,
    of cereals, 268
    halo, on beans, 145, 283
    potato, 276
        prevention, 170
    seedling, in oats, 275
Bloat in cattle, 479
    housed, 401
Block grazing system, 197
Blood,
    antibody test, 392
    chemistry and nutritional status, 340
    dried, as fertiliser, 75
    glucose regulation in, 331
'Bloody gut', in pigs, 487
Blowfly, on sheep, 491
Boar taint, 449
Bone fertilisers, 75
Bone growth, 360
Borage, 140–141
Bordetella bronchiseptica in pigs, 486

Boron,
    in brassicas and overliming, 26
    in crop nutrition, 72–73
    deficiency, in sugar beet, 173
    in soil, 29
Botrytis cineria on carrot, 285
Botrytis of peas/beans, 145, 283, 284
Bovine somatotrophin (BST), 358–359
Bovine virus diarrhoea, 477
Bowel oedema of growing pigs,
    485–486
Bradford Count in wool grading, 421
Brain physiology, 317–318
Brassica flea beetle, on oilseed rape,
    116, 299
Brassica pod midge, on oilseed rape,
    301
Brassicas, 99–118
    club root and liming, 87
    diseases of, 280–282
    for fodder, 110–115
        kale, 110–111
        mustard, white, 112
        radish, 110
        rapes, 111–112
        swedes, 112, 113–115
        turnips, 112–113
    as forage crops, 376–377
    for seed processors, 115–118
        mustard, condiment, 117–118
        oilseed rape, 115–117
    as vegetables, 100–110
        Brussels sprouts, 100–102
        cabbages, 102–106
        calabrese, 108–109
        cauliflowers, 106–108
        kohl rabi, 109–110
Braxy in sheep, 489
Bread wheat, 125–126
Breeding,
    of cattle, 410
    of goats, 428
    of pigs, 432
    of poultry, 453
    of sheep, 414–416
Brevicoryne brassicae, 300
Brewing by-products for animal feed,
    378
Bridge construction, 553
British Organics Standards Committee
    and fertilisers, 74
British Standards,
    on building construction/use, 586,
        587
    on compressed air, 496
    on drainage, farm service, 495–496
    on fencing, 554
    on fire systems, 498
    on fuel pipes, copper, 497
    timber stress grading, 548
    on water supply/storage, 501
British Sugar plc, sugar beet
    agreement, 174–175
British Wool Marketing Board, 421
Broilers,
    breeding industry, 453, 455–456

Broilers (*cont.*)
  buildings for, 573
Brown foot rot, in wheat, 271
Brown rust, in wheat, 269
Brucella, abortion in sheep, 492
Brucite, 7
Brussels sprouts, 100–102
Budgeting, 604–612
  for capital, 611–612
  for cash flow, 608–611
    assessing, 610–611
    compiling, 608–610
  partial, for change, 616–617
  for profit, 604–608
    assessing, 607–608
    conventional form, 604
    enterprise gross margins, 604–606
    output/cost estimating, 606–607
Buffer grazing system, 198
Buildings, 537–588
  advisory organisations, 586–588
  constructions, 541–545
    finishing/services, 545
    frame/roof-cladding, 543
    groundworks, 542
    masonry/floor, 543–544
    setting out, 541–542
  design, 537–541
  drainage, 553
  environmental control, 554–559, 555
    humidity, 559
    insulation, 555–557
    ventilation, 557–558
  external works, 554
  foundations for, 552–553
  legislation/regulations for, 584–586
  materials/techniques, 545–552
    cladding, non-timber, 550–551
    concrete, 545–547
    frames/roof supports, 551–552
    timber, 548–500
  for potato storage, 171
  specifications,
    for animals, *see by particular animal*
    for crops, *see by particular crop*
    for services/general, 579–580
  surveying, 581–584
    drawings, 581, 582
    levels, 583–584
    Ordnance Survey maps, 583
    ranging out, 581, 583
    see also Roadways
Bulb fly on wheat, 288–289
Bulk milk tanks, 533
Bullets for mineral supplementation, 333
  cobalt, 335
  magnesium, 340
Bulls,
  beef,
    maize silage system, 401
    storage beef system, 402
    liveweight gain, versus steers, 401
    pens for, 564, 565

Bunt in wheat, 268, 271
Burning of straw,
  and direct drilling, 40
  and reduced cultivation, 122
  see also Fire
Burnt lime, 89
Business management, 602–623
  accounts, 612–615
  budgeting, 604–612
    for capital, 611–612
    for cash flow, 608–611
    for change, 616–617
    for profit, 604–608
  capital, sources of, 617–619
  definition, 602
  marketing, 621–623
  objectives, primary, 602–603
    planning for, 603–604
  records, 615
  and risk/uncertainty, 619–620
Butterflies, and pesticide reduction, 461

CAP, *see* Common Agricultural Policy
CEC, *see* Cation exchange resin
Cabbage aphid, on oilseed rape, 300
Cabbage ringspot, in brassicas, 281
Cabbage seed weevil, on oilseed rape, 300–301
Cabbage stem beetles,
  on oilseed rape, 200, 299
  on kale, 301
Cabbages, 102–106
  cattle, 105–106
  marketing from field, 105
  storage, 105
  summer/autumn, 104
  winter/savoys, 104–105
Calabrese, 108–109
Calcareous sand for liming, 89
Calcitonin, in calcium/phosphorus balance, 334
Calcium,
  in animal metabolism, 325–335
    and milk fever, 339–340
  in crop nutrition, 72
  fertilisers, 79
  in soil, 25–26
  see also Hypocalcaemia
Calves, 391–394
  colostrum, importance of, 392, 474
  housing, 561–562
  moving, 392–393
  replacement heifers, 394–396
    of autumn calver, 395
    and breeding management, 394–395
    of spring calver, 395–396
    two-versus three-year-old calving, 396
  suckling systems, 394
  tetany, 481
  transport, 392–393
  veal,
    pricing, Common Agricultural Policy, 595–596

Calves (*cont.*)
  veal (*cont.*)
    production, 394
    weaning, early, 393
Canker,
  of brassicas, 280
  oilseed rape, 115
  of parsnips, 164, 165
  stem, of potatoes, 279
Canopy light capture, 65–67
Capital, sources of, 617–619
  borrowing, 618–619
  shared ownership, 619
Capons, buildings for, 573
Carbohydrates, 321–323
  in animals, 323
    digestion, 326–327
  in plants, 321–323
    composition/categorisation, 321–322
    storage, 322
    structural, 322–323
Carbon cycle in soil, 12–13
Carbon/nitrogen ratio and nitrogen availability, 21
Carcasses,
  protein fat and β-agonists, 363
  quality/classification,
    beef, 392–400
    for lamb, 419
    for pigs, 446–447
  weight, 360
  see also Slaughter
Carotenoid metabolism, 337
Carrot fly, 156
  on parsnips, 165
Carrots, 155–158
  cultivation requirements, 156
    seedbed/sowing, 156–158
  diseases of, 284–285
  harvesting/post-harvesting, 158
  pests, 158
  types/varieties, 155–156
  yield, 158
Cash flow,
  budgeting, 608–611
  recording/monitoring, 612–614
Cashmere production, 427
Cassava in compound feed, 376
Castration, 347
  and growth, 362
Catch crops, 91
Caterpillar moth, on potato, 294
Cation adsorption by soils, 13–14
Cation exchange capacity,
  and hydrous oxides, 9
  and organic matter, 10
Cattle, 388–410
  ageing, 391
  breeding, 410
    embryo transfer, 354
  breeds, immigrant, 388
  buildings for,
    beef, 562–563
    bull pen, 564, 565

Cattle (*cont.*)
  buildings for (*cont.*)
    calves, 561–562
    dairy cows, 559–561
    milking parlours, 563–564
  definitions, 390–391
  diseases/pests,
    anthrax, 483–484
    blackquarter, 477
    bovine virus diarrhoea, 477
    of calves, 474
    coccidiosis, 477
    foot-and-mouth, 482–483
    infectious bovine rhinotracheitis, 477–478
    infertility, 482
    metabolic disorders, 339–340, 479–482
    parasites, external, 476–477, 484–485
    parasites, internal, 474–476
    rabies, 483
    redwater, 478–479
    respiratory disease, 478
    wooden tongue/lumpy jaw, 478
  feeding, 396–399
    breeding stock, 396–398
    dairy cattle, 404–406
    growing/finishing stock, 398
    performance prediction, 398
    to target gains, 398–399
  see also Beef; Calves; Dairying; Milk; Milking Equipment
Cattle cabbage, 105–106
Cattle grids, 554
Cauliflowers, 106–108
  manuring, 106
  pests/diseases
    mosaic virus, 281
    'whiptail' molybdenum deficiency, 74
  summer/autumn, 107–108
  winter, 106–107
Cereal beef system, 401–402
Cereals, 118–125
  cultivations, 122–123
  direct drilling, 123
  flowering, determinate, 62
  in feed, 377–378
    by-products, 378–379
  for finishing of lambs, 420
  grazing winter cereals, 119
  growth stages, 120, 121
  lodging, 66–67, 119
  manuring, phosphorus/potassium, 124–125
  overwintering and internode growth, 62
  pests of, 288–292
    aphids, 291–292
    cyst nematode, 291
    field slug, 291
    frit fly, 289–290
    leatherjackets, 290–291
    wheat bulb fly, 288–289

Cereals (*cont.*)
  pests of (*cont.*)
    wireworms, 290
    yellow cereal fly, 289
  pricing, Common Agricultural Policy, 594
  production intensity versus profit, 120
  rotations, 120, 122
  seed, 118
    production, 136
    rate and population/yield, 123–124
    seed treatment, 118
  silage, 202
  straw utilisation/disposal, 122
    burning, 40
  'tramlines', 124
  variety, choice of, 118–119
  weed control in, 247–250
    not-undersown, 247–250
    undersown, 250
  winter,
    and organic matter in soil, 11
    vernalisation need, 64, 119
  see also Grain
  see also by individual cereals
Ceutorhynchus, see Weevils
Chaff Cutting Machines Act 1897, 644
Chalk for liming, 88–89
Chisel plough, 508
Chitting of potatoes, 168
Chloride,
  in animal metabolism, 335
  in crop nutrition, 74
Chlormequat for lodging, 119
Chocolate spot of beans, 143
Chorioptes bovis, 476–477
Chylomicrons, 332
Cimeterol and protein/fat in carcasses, 363
Citrus pulp in animal feeds, 379
Clamping,
  of carrots, 156
  of sugar beet, 174
Classification of soil, 41–43
  for cereal direct drilling, 123
Clays,
  and drainage need, 45
  flocculation/deflocculation, 15, 16
  minerals, 5, 7–10
    hydrous oxides on, 9
    and shrinkage of soil, 9–10
    as soil particles, 4, 9
    structure, 7–9
  texture and lime, 27
  tiles versus plastic pipe, 51
Clenbutarol and protein/fat in carcasses, 363
Click-beetles, 290
Clostridium spp. and silage making, 199
Clostridial infections,
  in cattle, 477
  of piglets, 486
  of sheep, 489–490

Clovers, 184
  nitrogen supply to grassland, 188–190
  rots, of legumes, 287
  seed production,
    establishment for, 137
    sward management for, 139–140
  sward management and nitrogen fixation, 19
  see also Herbage seed production
Club root,
  in brassicas, 281
  and liming, 87
  oilseed rape, 115
  persistence in soil, 268
Cooperatives, 619
Cobalt,
  in animal metabolism, 335
  deficiency in sheep, 490
  in crop nutrition, 74
Coccidiosis in cattle, 477
Cocksfoot, 184
  seed production,
    establishment for, 137
    sward management for, 139
Colletotrichum, see Anthracnose on peas/beans
Colloid chemistry of soils, 13–15
  anions, 14
  cation adsorption, 13–14
  flocculation/deflocculation, 15, 16
  Quantity/Intensity relationships, 14–15
  water adsorption, 14
Colostrum,
  for calves, 392, 474
  for piglets, 444
Combine harvesters, 521–523
Common Agricultural Policy, 591–601
  decision making process, 592–593
  formation, 592
  monetary arrangements, 597–598
    Exchange Rate Mechanism, 597
    'green pound', 597–598
    Monetary Compensatory Arrangements, 598
  the need, 591–592
  price mechanisms for produce, 593–597
  principles, 593–594
  problems of, 600–601
  structural policy, 599–600
Common scab of potatoes, 278
Compaction by tractors, 505
Compost,
  as manure, 77
  for transplant raising, 94
Compressed air services, 496
Computers, 650–660
  in advisory services, 657–658
  in ancillary industries, 657
  applications, 655–657
  communications/interfacing, 654–655
  definitions, 650
  expert systems, 658

Computers (*cont.*)
  future developments, 660
  implications, 659
  operating systems, 653
  programs/data/files, 652–653
  in research, 658–659
  selection criteria, 659–660
  software packages, 653–654
  system components, 651–652
  in training/education, 658
Concrete,
  for fenceposts, 554
  frame members, 547
  mass, 545–546
  pre-cast, 546–547
  for roadways, 553
Conservation,
  of game, 461–465
  of grass, and nutritive value, 373
  headlands, for game habitat, 462
  of wildlife, and drainage, 56
Construction, *see* Buildings
Contarinia pisi, 302, 303
Contracts of Employment Act 1963, 643
Control of Pollution Act 1974, 585
  and nitrate leaching, 23
  and nuisance, 641
  and pollution of watercourses, 642
Cooperia oncophora, in cattle, 475
Copper,
  in animal metabolism, 335–336
    deficiency in sheep, 490
  in crop nutrition, 73
  in soil, 29
Copper chrome arsenate timber preservation, 238–239
Coppice woodland management, 224–225
Coronavirus infection of piglets, 488
Countryside Commission woodland grants, 240
Covered smut in barley, 274
Covers, *see* Plastic covers for crops
Cow equivalent and stocking rate, 193
Craneflies, 290
Credit, trade, 618
Crispbread rye, 131
Crop storage, buildings for, 574–578
  grain, 574–576
  hay/silage/straw, 576–577
  root crop concentrate, 577
  vegetables, 577–578
Crops,
  diseases of, 265–287
  nutrition of, 70–89
  pests of, 287–302
  physiology of, 58–69
Crown rot,
  of ryegrass, 286
  of swedes/turnips, 73
Crown rust in oats, 274–275
Cryptorchidism, 347
Cultivation machines, 507–510
  combined seed-bed, 510

Cultivation machines (*cont.*)
  harrows, 509
  ploughs, 507–508
  reduced cultivation, 510
  rollers, 509–510
  rotary cultivator, 508–509
  spring-tine, 509
  subsoilers, 507
Cultivations, 40–41
  aeration, 35–36
  drainage benefits, 46
  and organic matter level, 11
  for seedbed, 40
  subsoiling, 40–41
Culvert design, 50
Cutworms,
  on potato, 294
  on sugar beet, 296
Cydia nigricana, 150, 301
Cylindrosporium concentrium, *see* Leaf spot
Cyst nematode,
  beet, 174, 298
    on oilseed rape, 115
  cereal, 120, 291
    on oats, 129
    on sweet corn, 135
  potato, 293
  pea, 302

D-value of feeds, 365
Dairying, 404–410
  feeding, 405–407
    complete diets, 406
    during lactation, 406
    flat-rate, 406
    silage, self-feed, 406
    steaming-up, 405–406
  lactation curves, 405
  milking process, 409–410
  oestrus detection, 408–409
  see also Milk; Milking equipment
Damping-off,
  of carrots, 284
  of maize, 276
  of peas/beans, 283
Dasyneura brassicae, 301
Daylength responses of plants, 64–65
Deamination, 333
Deficiency symptoms of crops,
  major nutrients, 70–72
  minor/micro-nutrients, 72–74
Delia coarctata, 288–289
Demodex bovis, 476–477
Denitrification, 22
Depreciation estimation, 607
Deroceras reticulatum, *see* Slugs, field
Design of buildings, 537–541
  contracting, 541
  criteria establishment 538
  dimensions/plans, 538
  planning, 537–538
  preparation, 538
    costs per unit, 540
  site survey, 540–541

Dexamethasone for lambing induction, 425
Diarrhoea,
  bovine virus/mucosal disease, 477
  viral, epidemic, in pigs, 488
Diathesia, exudative, in chicks, and selenium, 336
Dictyocaulus viviparus infection, of cattle, 475
Diesel, *see* Fuel oil storage
Digestibility of feed, evaluation, 365–366
Digestible energy measurement, 366
Digestion, physiology of, 325–329
  anatomy,
    monogastric tract, 325
    pre-ruminant, 326
    ruminant tract, 325–326
  carbohydrate,
    monogastric, 326–327
    ruminant, 327
  lipid, 329
  protein,
    monogastric, 327–328
    ruminant, 328–329
Diphtheria, calf, 474
Diquat in dessication of peas, 150
Direct drilling, 40
  drills, 512–513
Disc,
  harrow, 509
  plough, 508
Diseases of animals,
  legislation on, 473–474
  signs of, 472–473
  see also by particular animal
Diseases of Animals Act, 473–474
Diseases of crops, 265–287
  air-borne, 265–267
    chemical control, 267
    cultural control, 266
    integrated control, 266
    fungicides, resistance to, 267–268
    seed-borne/inflorescence, 268
    soil-borne, 268
    vector-borne, 268–269
  see also by particular crop
Dismissal procedures, 636
Distillery by-products for animal feed, 378
Ditch drainage, 49–50
Docking disorder of sugar beet, 298
Dogs, shooting of, 641
Dormancy, 64
Dorset wedge filling of silage clamp, 200
Downy mildew, *see under* Mildew
Drainage, field, 45–47
  benefits, 46–47
  of clay soil, and temperature, 41
  and conservation, 56
  design, 55, 56
  economics of, 56–57
  field capacity duration, national variation, 45–46

Drainage, field (*cont.*)
  historical perspective, 46
  and lime loss, 26–27
  maintenance, 56
  movement of water, 48
  and need, 45
  pumped systems, 56
  recommendations, 495
  site investigation, 55
  and soil water retention, 47
  systems, 48–55
    ditches, open, 49–50
    moling, 53–54
    pipe, 50–53
    subsoiling, 54–55
  for woodland establishment, 228
    ploughing, 229
Drainage, service, 495–496
  and building construction, 553
  rainwater pipes/gutters, 495
Draught control on tractors, 506
  in ploughing, 508
Driers for grain, 306–309
  high temperature, 306–308
  low-temperature/storage driers,
      308–309
Drills,
  direct, 512–513
  grain, 511–512
Drinking water,
  EEC Directive 1985, 22–23
  nitrate pollution, 84
    and grassland farming, 206
  quality standards/reports, 22–23
Drought, and canopy productivity, 67
Ducks, 458–459
Dumpy level, 584
Durum wheat, 127–128
Dutch elm disease, 239
Dysentery,
  lamb, 489
  swine, 487

Earthworms, 20
Eelworms, stem, on leeks, 161
  see also Cyst nematodes; Nematodes,
      free living
Effluent from silage, 200–201, 206
Egg production, 452–453, 454, 455
Electricity, 496–497
  installation regulations, 586
  recommendations, 496
  supply mains/private, 497
Embryo transfer, 354
Employers Liability Act 1969
    (Compulsory Insurance), 642
    (Defective Equipment), 643
Employment Protection
    (Consolidation) Act 1978, 643
Endocrine system, 319
Energostasis in animals, 340–341
Energy in feed, measurement of,
      366–367
Energy requirement of livestock,
  cattle for fetal growth, 397
  for growth, 386
  for lactation, 383–384

Energy requirement of livestock (*cont.*)
  for maintenance, 382
  for reproduction, 384–385
  ewes, 422–423
  pigs,
    growing, 448
    pregnant sows, 440–441
Enterovirus, pig infection, 487–488
Environmentally Sensitive Areas, 642
Enzootic abortion in sheep, 491
Equipment, *see* Buildings; Machinery;
    Services
Ergosterol, vitamin D precursor,
      337–338
Erwinia carotovora, 277
Erysipelas, swine, 487
Erysiphe cruciferarum, see Mildew,
    powdery
Escherichia coli infection,
  in pigs,
    weaners, 485–486
    piglets, 485
  in sheep, 489
European Council, 592
European Economic Community,
  beef production/consumption,
      389–390
  Court of Justice, 637
    precedent creation, 638
  decision making in, 592–593
  milk quotas, 388–389
    legal nature of, 642
    tenant's compensation, 640
  pig carcass classification scheme, 447
  pigmeat policy, 431
  see also Common Agricultural Policy
Evening primrose, 141
Eyespot in cereals,
  barley, 274
  fungicide resistance, 267–268
  oats, 129
  rotation control, 120
  wheat, 270
  see also Sharp eyespot in wheat

FMD, *see* Foot-and-mouth disease
Factories Act 1937, 644
Farm Woodland Scheme, and pheasant
    habitat, 463
Farrowing,
  crate developments, 433
  fever in sows, 485
Fat,
  adipocyte development, 360–361
  in carcasses, β-agonist modification,
      363
  degradability protection with
      protein, 329
  for energy in animal feeds, 379
  metabolism, 331–332
    circulation, 332
  synthesis in milk, 356, 357
  see also Lipids
Fatty acids, 323, 324
  volatile,
    metabolism, 331
    production in rumen, 327

Fatty liver syndrome in ketosis, 339
Fecundin for ewe productivity, 425
Feeds for livestock, 365–387
  nutrient evaluation, 365–371
    chemical analysis, 369–371
    definitions, 371–372
    digestibility, 365–366
    energy, 366–367
    micronutrients, 369
    protein, 367–369
  nutrient requirements, 381–386
    for growth, 385–386
    for lactation, 383–384
    for maintenance, 382–383
    micronutrient, 386–387
    for reproduction, 384–385
  nutritive value of raw materials,
      372–381
    cereal by-products, 378–379
    cereal grains, 377–378
    citrus pulp, 379
    fats/oils, 379
    forage crops, 376–377
    grass, conserved, 373
    grazed herbage, 372–373
    legume seeds, 380
    molasses, 379
    oilseed residues, 379–380
    roots/tubers, 376
    straw, 373, 376
    supplements, 380–381
    tabulated, 374–375
  preparation machinery, 530–532
Fen peat, drainage and oxidation, 4
Fences/fencing, 554
  for woodland, 228
Fenugreek, 151
Fertilisers, inorganic, 78–87
  application machines, 510–511
    broadcasters, 510
    full-width spreaders, 511
  calcium, 79
  composition and the law, 78
  efficiency of use, and drainage, 46
  environmental considerations,
      83–84
  financial return, 84–85
  forms of, 81–82
  for grass,
    nitrogen, 189–190
    for sheep grazing, 424
  compound, 82–83
  cost comparison, 83
  and lime loss, 27
  for moor grazing, 177–178
  optimum rates, 84–85
  versus organic, 74
  physical response, 84
  placement, 83
  potassium, 75, 81
  recommendations, 85–87
    calculation of use from, 86–87
  storage, buildings for, 579
  sulphur for grass, 25
  for trees, 231
  see also Liming; Manures; Manuring
  see also under Nitrogen; Phosphorus

Fertility,
male, factors in, 350
problems, in cattle, 482
Fescues, 178, 184
seed production, 137,139
Fibre production from goats, 427
Field capacity duration, national
variation, 45–46
Finn Dorset ewes for prolifacacy, 425
Fire,
recommendations in buildings, 498
protection of woodland, 239
risk from constructional timber, 548
Firewood from coppicing, 225
Fish products,
as feed supplement, 381
for pigs, and fishing decline, 431
as fertilisers, 75
Flax, 141
pricing, Common Agricultural
Policy, 597
Flea beetle,
in beet, 296
in mustard, 118
in oilseed rape, 116, 299
Flehman response in male mating
behaviour, 350
Flocculation/deflocculation of clays, 15,
16
Flooding, salt water, deflocculation
prevention, 15
Floors,
laying, 543–544
for livestock units, 468–469
Flowering, 62
Fodder, dried, pricing, Common
Agricultural Policy, 597
Fodder beet, diseases of, 282–283
Follicle development, 343
Fomes annosus, infection of trees, 239
Food and Drugs Act 1955, Milk and
Dairies (General) Regulations
1959, 586
Food and Environment Protection Act
1985, Control of Pesticides
Regulations, 246, 642, 648–649
Food intake,
energostasis, 340–341
manipulation, 341–342
prediction, 341
voluntary, 340–342
Foods Act 1984 on goats' milk, 427
Foot of peas/beans, 283
Foot-and-mouth disease, 482–483
in pigs, 489
Foot-rot of sheep, 491
Forage harvesting machinery, 517–521
harvesters, 519–520
hay baling/bale handling, 520–521
big-baling, 521
conventional, 520–521
mowing/swath treatment, 517
self-loading wagons, 520
silage handling/baling, 520
Forests, 225–226
plantation forest, 225

Forests (*cont.*)
selection system, 225–226
weed control in, selective herbicides
for, 260–262
see also Trees on farms
Forestry Grant Scheme, 241
Foul-in-the-foot in cattle, 485
and calf diphtheria, 474
Foundations for buildings, 552–553
Fox control, 464
Freemartin heifer development, 343
Frit fly on cereals, 289–290
sweet corn, 135
Frost,
protection of fruit by irrigation,
529–530
tolerance of crops, 64
Fruit,
calcium deficiency, 72
frost protection by irrigation,
529–530
price, Common Agricultural Policy,
596–597
Fuel oil storage, 497, 580
see also Petrol storage/licensing
Fungi in soil, 19–20
Fungicides,
foliar treatment, 267
resistance to, 267–268
Fungi and plant diseases,
grey mould, on peas/beans, 145, 283,
284
of peas, 150
of trees, 239–240
Fusarium, in peas/beans, 268, 283
Fusarium spp.,
on sweet corn, 135
in wheat, 271
Fusiformis nodosus, foot-rot in sheep,
491

Gaeumannomyces graminis, *see* Take-all
Game conservation, 461–465
economic benefits, 465
habitat requirements, 462
partridge, grey, 462
pheasant, 463–464
predator control, 464
rearing/release, 464
shooting options, 465
shooting in rural economy, 461
Gapon equation in cation exchange, 14
potassium exchange, 24
Gas danger in livestock buildings,
469–470
Gastroenteritis,
parasitic, in cattle, 476
transmissible, in sow/piglets, 485
see also Diarrhoea; Dysentery
Gates, 554
Geese, 459–460
Genetics in sex determination, 342–343
Germination,
physiology of, 60–61
testing of seed, 91

Gibbsite, 7
Globodera, *see* Cyst nematodes
Glucose metabolism, 330–331
gluconeogenesis, 331
in lactose synthesis in milk, 358
Glume blotch in wheat, 270
Glycogen oxidation, and slaughter, 331
Goats, 426–428
feeding, 427–428
production from, fibre/meat/milk,
426–427
reproduction, 428
statistics, 426
Goitre, 336
Gonadotrophin release and season, 344
Gouy–Chapman model of cation
exchange, 13–14
Grain,
aphid, 291–292
drills, 511–512
harvesting, *see* Combine harvesters
storage/handling, 304–313
aeration temperature control,
310–311
cleaners, 310
conveyors, 311–312
drying, 306–309
fans for stores, 309–310
loading lorries, 312
moisture content measurement,
305–306
pest control, 304–305, 313
sealed storage, 312–313
temperature monitoring, 310
throwers, 312
time factor in, 305
weighers, 312
see also Feed preparation
machinery
Graminaea, *see* Cereals and grasses
Grants,
under Agricultural Improvement
Scheme, 599
for 'less favoured areas', 600
for woodland, 241
Grass,
conservation,
green crop drying, 205
hay, 203–205
silage, 199–203
degradation inhibition, 198–199
growth,
and defoliation, 180–181
inflorescence development, 180
vegetative, 179–180
see also Herbage seed production
Grasses, 183–184
cocksfoot, 184
diseases of, 286–287
fescues, 184
ryegrasses, 183–184
timothy, 184
Grassland, 177–208
composition, factors in, 179
distribution, 178–179
drainage benefits, 47

Grassland (*cont.*)
  environment considerations
    pollution of water, 205
    sward composition, 205–206
  glossary, 206–207
  grazing output, 194–196
    and sward height, 195–196
    utilisation, 195
  improvement, 181
  legumes, 184–185
  nutrients for, 188–192
    calcium/magnesium, 191
    nitrogen, 188–190
    phosphate/potash, 190–191
    sulphur, 191
  output, expression of, 193–194
    digestibility, 194
    efficiency of use, 194
    stocking rate, 193
    utilised metabolisable energy,
      193–194
  permanent, 178
    and organic matter in soil, 11–12
  production patterns, 191–193
  rotational, 178
  rough grazing, 177–178
  seed mixtures, 185–188
  soil pH recommendations, 87
  sowing methods, 181–182
  'teart' and molybdenum, 74
  use of, 179
  weed control in, 257
Grazing,
  cereals, winter, 119
  for heifers, 395
  management, for sheep, 424
  systems, 196–198
Greasy pig disease, 486
'Green pound', 597
  in intervention pricing, 597–598
  and Monetary Compensatory
    Arrangements, 598
Grey mould of peas/beans, 145, 202,
    284
Grey speck and manganese deficiency
    in oats, 73
Grinding machinery for feed, 530–531
Growth of animals,
  energy requirement, 386
  physiology of, 359–363
    genetic effects, 361
    and hormones, 362–363
    the nature of, 359–360
    and nutrition, 361–362
    tissue growth, 360–361
  promotors, 471–472
  protein requirement, 386
Growth promoters/regulators,
  for animals, 471–472
  for plants, 69
    lodging, 119
Growth/development of plants,
  analysis, 60
    Harvest Index, 60
  and breeding, 69

Growth/development of plants (*cont.*)
  and competition, 67–69
    interspecific, 68–69
    intraspecific, 67–68
  development cycle, 60–63
    germination, 60–61
    reproductive, 62
    seeds, 62
    senescence, 62–63
    vegetative, 61–62
  environmental influences,
    light, 64–67
    temperature, 63–64
  regulators, 69
    for lodging, 119
  physiology of, 60
  and temperature, 63–64
    dormancy, 64
    heat units, 63–64
    vernalisation, 64

HRG, *see* Ryegrass, hybrid
Hagberg Falling Number for
    breadmaking wheat, 125, 127
Halo blight of beans, 145, 283
Hammer mill, 530
Harrowing, 40
Harrows, 509
Harvest Index, 60
Harvesters,
  for forage, 519–520
  for grain, 521–523
  for roots, 523–524
Hay,
  nutritive value, versus silage, 373
  storage, buildings for, 576–577
Haymaking, 203–205
Headfly on sheep, 490
Health of animals, 466–492
  antibiotic, growth promotion,
    471–472
  disease types, 467
  and disinfection, 472
  gas danger, 469–470
    from gas heaters, 470
    from manure, 469
  and housing design, 467
    depopulation, 468
    floors, 468–469
    group size, 468
    and health status, 468
    isolation facilities, 469
  immunity development, 470
    vaccines, 471
  and intensification, 466
  legislation on disease, 473
  manure disposal, 469
  medicines, 471
  metabolic diseases, 466–467
  pathogenic organisms, 470
  signs of disease, 472–473
  and welfare, 473
  *see also diseases under particular
    animal*

Health and safety of personnel,
    644–649
  historical perspective, 644
  statistics, 644, 645
  in tractor operation, 506–507
  see also Food and Environment
    Protection Act 1985, Control of
    Pesticides Regulations; Health
    and Safety at Work etc. Act
    1974
Health and Safety at Work etc. Act
    1974, 643, 645–647
  bodies under, 645–646
    Inspectorate, 646, 647
  on building design, 585–586
  objectives, 645
  origins, 644
  statutory requirements, 646–647
Hearing loss from tractor operation,
    507
Heat units and crop development,
    63–64
Heating systems, 498–499
Heavy metals from sewage, 28
Helicolbasidium purpureum, see Violet
    root rot
Helminthosporium solani, 279
Herbage seed production, 136–140
  drying/conditioning/storage, 140
  establishment, 136–138
    general recommendations, 136
    indicator plants, 137–138
    manuring, 138
    special requirements, 137
  harvesting, 139, 140
  sward management,
    clovers, 139–140
    grasses, 138–139
Herbicides, 243–246
  for cereals, not-undersown, 248–250
  choice, factors in, 244–245
  cost, 245
  foliar, 243
  paraquat/diquat adsorption in soil,
    17
  purchase/use, 246
  selective treatments, 243
  soil applications, 245–246
  spray application, 245–246
  for stubble cleaning, 247
  see also Weed control
Hereford disease, 480–481
Herpes infection of pigs, 487
Herzberg, Frederick, motivation
    theory, 633
Heteroder, spp., *see* Cyst nematodes
Hill farms as 'less favoured areas', 600
  see also Mountain grazing
Hire purchase, 618
Honey from oilseed rape, 116
Honey fungus, 239
Hoof-and-horn fertilisers, 75
Hoose in cattle, 475
Hops pricing, Common Agricultural
    Policy, 597

Hormones,
  in animals, physiology of, 318, 319
  in ewe productivity improvement,
    425
  in growth regulation, 362–363
  in mammary gland development, 353
  in milk synthesis, 358–359
  in oestrus control, 345–346
  oxytocin, in milk ejection, 359
  parathyroid/calcitonin, 334
  in plant growth/development, 69
  in pregnancy diagnosis, 352, 354
  progesterone, in pregnancy
    maintenance, 351–352
  in sow parturation, 441
  and superovulation, 353
  see also Growth
    promoters/regulators
Horticultural crops, 'cool chain' for,
    98–99
Housing, *see* Buildings
Hungry gap kale, 111–112
Husk in cattle, 474
Hydraulic conductivity, and soil
    texture, 48, 49
Hydroxyapatite, 334
Hyostrongylus rubidus infection of pigs,
    489
Hypocalcaemia in cattle, 479
Hypochlorite in milking equipment
    cleaning, 408
Hypoderma spp. in cattle, 484
Hypomagnesaemia, 72, 191, 334
  in cattle, 480–481
  and potassium from farmyard
    manure, 76
  in sheep, 491
Hypothalamus function, 318
Hypothermia in lambs and mortality,
    417–418

IBP, *see* Infectious bovine
    rhinotracheitis
IRG, *see* Ryegrass, Italian
Illite structure, 9
Immunisation against ovarian steroids,
    353
Industrial Relations Act 1971, 643
Infectious bovine rhinotracheitis, 477
Influenza, swine, 486
Infrared reflectance analysis of animal
    feeds, 371
Insect control in grain storage, 304–305
Insecticides,
  for aphids on sugar beet, resistance,
    174
  for oilseed rape, and bees, 116
  regulations/legislation, 303
Institution of Electrical Engineers
    Regulations, 586
Insulation of buildings, 555–557
  U/R values, 555, 556
Insurance,
  employer's liability for, 642
  and herbicide use, 245

'Intervention', *see* Common
    Agricultural Policy, price
    mechanisms
Intervention Board for Agricultural
    Produce, 594
Interviewing, 629–630
Intestinal haemorrhage syndrome in
    pigs, 487
Iodine in animal metabolism, 336
Iron,
  in animal metabolism, 335
  in crop nutrition, 73
  in soil as micronutrient, 30
Irrigation,
  of beans, 145, 146
  of beetroot, 153
  of calabrese, 109
  of carrots, 156
  of leeks, 160–161
  of parsnips, 165
  of sugar beet, 174
  of sweet corn, 134
  systems, 524–530
    application, 526–528
    capacity, 524
    distribution, 525
    for frost protection, 529–530
    organic irrigation, 530
    pumping, 525
    scheduling and soil water deficit,
      528–529
    water supplies, 524–525
  water balance,
    calculation for, 38–39
    and crop requirements, 39–40
Isobutylidene di-urea, 80
Itersonilia spp., *see* Canker
Ixodes ricinus infection of cattle,
    478–479

Joint-ill of calves, 474

K, *see* Potassium
Kales, 110–111
  pests of, 301
Kaolinite structure, 9
Ketosis, *see* Acetonaemia
Kidney disease, clostridial, in lambs,
    489
Kohl rabi, 109–110

Laboratory analysis,
  of animal feeds, 369–371
  of organic matter in soil, 10–11
Lactation, 354–359
  anoestrus, 353
  curves, 405
  energy requirement for, 383–384
  mammary gland, 354–356
  protein requirement for, 384
Lactose synthesis,
  from glucose, 331
  in milk, 358
Lambs,
  black leg in, 489

Lambs (*cont.*)
  carcass quality, 419
  mortality, and environmental
    conditions, 364
  production systems, 419–420
  swayback in, 326, 490
Land classification, 43
  and grass production, 192–193
Law and agriculture, 637–643
  advice, need for, 637
  and development/use of land,
    641–642
  employer's liabilities, 642–643
  English system, 637–638
    Case versus Statute law, 638
  freeholder's rights, 639
  occupier's liability, 640–641
  tenancies, 639–640
    obligations, landlord/tenant, 640
    occupation without security, 640
    security, 639–640
    service occupation, 640
Leaf Area Index/Duration, 65
Leaf blight of carrots, 284
Leaf blotch,
  of barley, 272–273
  of ryegrass, 286
Leaf diseases, and canopy efficiency,
    67
Leaf roll virus, potato, 277
Leaf spot,
  of beans, 143
  of beet/mangold, 283
  of brassicas, 280
  of oats, 275
  of wheat, 270
Leaf stripe in barley, 273
Leasing, 618
Leatherjackets, 290–291
  on sugar beet, 174, 296
Lecithins, 324
Leeks, 156–161
  cultivation/husbandry, 159–161
  diseases of, 285–286
  harvesting/post-harvest, 161
  pests/diseases, 161
  types/varieties, 158–159
  yield, 161
Legislation,
  on animal disease, 473–474
  on buildings, design/construction,
    584–586
  see also Law and agriculture
  see also by specific statute
Legumes, 142
  herbage, 184–185
    diseases of, 286–287
  and nitrogen fixation, 18–19
  in rotations, 142
  seedrate calculation, 142
  seeds, in animal feeds, 380
  for silage, 202
  see also by individual crop
Leptosphaeria spp., *see* Canker; Glume
    blotch

'Less Favoured Areas', EEC directive on, 600
Let-down, 407
Leys, and soil organic matter, 11
Lice,
  on cattle, 484–485
  on pigs, 488
Lichen in soil formation, 4
Ligand exchange in soil, 14
Light, and growth/development, 64–67
  and canopy capture, 65
  photoperiod, 64–65
Lighting recommendations, 499–500
  intensity, 456–457
  for poultry, 456
Liquified petroleum gas, 497
Lime in soil, 26–27
  requirement, 26, 27
Liming, 26–27, 87–89
  loss of lime, 26–27
  materials, 88–89
  neutralising value, 88
  overliming, 89
  and pH, 26–27, 87
  and soil texture, 27
Linen, 141
γ-Linolenic acid
  in borage, 140
  in evening primrose, 141
Linseed, 141–142
Lipids, 323–324
Liquid fertilisers,
  ammonia, 79
  ammonium nitrate in, 78
  for transplant raising, 95
  urea in, 79
Liscombe star system for silage additive requirement, 200
Liverfluke infection,
  of cattle, 475–476
  of sheep, 490
Lockjaw in sheep, 489
Lodging of cereals, 66–67, 119
  and seedrate, 66
  sweet corn, 135
Longidorus, 298
Loose smut,
  in barley, 274
  in wheat, 271
Lubricating oil/grease on-farm installation, 500
Lucerne, 151–152
  verticillium wilt, 287
Lumpy jaw of cattle, 478
Lungworm infection,
  of cattle, 475
  of pigs, 489
  of sheep, 490–491
Lupins, 150–151
  seed, value in animal feeds, 380

McGregor, Douglas, motivation theory, 632–633
Machinery, 503–536
  for cultivation, 507–510

Machinery (*cont.*)
  for cultivation (*cont.*)
    combined seed-bed, 510
    harrows, 509
    ploughs, 507–508
    reduced cultivation, 510
    rollers, 509–510
    rotary cultivators, 508–509
    spring-tine, 509
    subsoilers, 507
  for drain laying, 52, 53
  for feed preparation, 530–532
  for fertiliser application, 510–511
  for harvesting,
    forage, 517–521
    grain, 521–523
    roots, 523–524
  for irrigation, 524–530
    application, 526–528
    capacity, 524
    distribution of water, 525
    for frost protection, 529–530
    organic irrigation, 530
    pumping, 525
    scheduling and soil water deficit, 528–529
    water supplies, 524–525
  manufacture, computers in, 657
  for manure handling/spreading, 533–535
    farmyard manure, 533–534
    liquid manure, 534–535
  for milking, 532–533
  for planting, 511–514
    direct drills, 512–513
    grain drills, 511–512
    potatoes, 514
    precision drills, 513–514
    transplanters, 514
  sprayers, 514–517
    calibration, 516
    components, 515–516
    droplet size and drift, 514–515
    nozzles, 515
    productivity, 516
    safety, 516–517
  storage, buildings for, 579–580
  tractors, 503–507
    drawbar power, 504–506
    engine, 503–504
    hydraulic systems, 506
    operator health and safety, 506–507
    power take-off, 506
    transmission, 504
Maedi-Visna in sheep, 492
Magnesium,
  in animal metabolism, 334
    and staggers, 340
  in crop nutrition, 72
    deficiency in grassland, 191
    deficiency in sugar beet, 173
  in soil, 25
  see also Hypomagnesaemia
Maize, 131–133

Maize (*cont.*)
  animal feeding value, 377
  diseases of, 276
  forage, 131–133
  milling by-product for feed, 378–379
  photosynthesis, tropical adapted, 59
  production restriction and heat units, 64
  silage, 202, 204
    for beef, 401
  see also Sweet corn
Maleic hydrazide as onion sprout suppressant, 163
Malting, barley quality for, 128
Mammary gland, 354–356
Management, *see* Business management; Staff management
Manganese,
  in animal metabolism, 336
  in crop nutrition, 73
    deficiency and marsh spot of peas, 150
    deficiency in sugar beet, 173
  in soil as micronutrient, 29
    deficiency and overliming, 26
Mange,
  of cattle, 476–477
  of sheep, 491
Mangels/fodder beet, 175–176
Mangolds, diseases of, 282–283
Manioc in compound feed, 376
Mansholt Plan for EEC agriculture, 599
Manures, 75–78
  disposal, 469
    by irrigation, 530
  handling/spreading equipment, 533–555
    farmyard manure, 533–554
    liquid manure, 534–555
  sources,
    livestock, 75–77
    other sources, 77–78
  storage, buildings for, 579
Manuring,
  of barley, 128–129
  of beans,
    broad, 144
    dwarf, 145
    field, 143
  of beetroot, 153
  borage, 140
  Brussels sprouts, 102
  cabbage, winter, 105
  calabrese, 109
  of carrots, 156
  cauliflowers, 106, 107
  of cereals, 130
    phosphorus/potassium, 124–125
  of durum wheat, 127
  for herbage seed production, 137, 138
  of leeks, 159
  of lucerne, 152
  of maize, 132
  of mustard, condiment, 117

Manuring (*cont.*)
  of oilseed rape, 116–117
  of onions, 162
  of parsnips, 165
  of peas, 149
  of potatoes, 170–171
  of rye, 131
  of sainfoin, 152–153
  of sweet corn, 134
  of wheat,
    nitrogen, 125
    winter, 126–127
Maple peas, 147–148
Marketing, 621–623
Marrowfat peas, 147–148
Marsh spot of peas/beans, 150
  and manganese deficiency, 73
Maslow, Abraham, motivation theory, 632
Mastitis,
  bacteria in, 48
  of cattle, 481
  of sheep, 489
Meat and Livestock Commission, in pig breeding, 436
Meat meal, as feed supplement, 380–381
Melatonin and breeding season control, 353–354
  and gonadotrophin secretion, 344
Meligethes spp. on oilseed rape, 300
Metabolic profile testing of dairy herds, 340
Metabolism, physiology of, 329–340
  catabolism/anabolism, 329
  disorders, 339–340
  fat, 331–332
  fatty acids, volatile, 331
  glucose, 330–331
  mineral nutrition and, 331–337
    major, 334–335
    trace, 335–337
  vitamins in, 337–339
Metabolisable energy of feed, 366–367
Metapolophium dirhodum, 291
Metastrongylus, *see* Lungworm infection
Methaemoglobinaemia and nitrate, 84
Metritis/mastitis/agalactia syndrome in sows, 485
Mice as sugar beet pest, 174, 295–296
Micro-organisms in soil, 17–20
  actimomycetes, 19
  algae, 20
  bacteria, 18–19
  classification, 18
  cyclic process, 17
    carbon cycle, 13
  fungi, 19–20
  manipulation of, 20
  nutrition, 18
  and pesticide degradation, 17
Micronutrients, 322–343
  analysis, 33
  livestock requirements, 386–387
  in soil, 28

Midge,
  brassica pod, on oilseed rape, 301
  pea, 303–304
Mildew,
  downy,
    on beet/mangold, 282
    on brassicas, 281
    on onions/leeks, 285
    on peas/beans, 283
  powdery,
    on barley, 272
    on beet/mangold, 282
    on brassicas, 280
    fungicide resistance, 267–268
    on oats, 274
    on ryegrass, 286
    on sugar beet, 174
    on wheat, 269
Milk,
  by-products as feed supplement, 381
  ejection mechanism, 359
  energy value, 383
  fever, 339–340, 479
    in sheep, 491
  from goats, 426–427
  pricing system, 407
    Common Agricultural Policy, 594–595
  production, computerisation in, 655–656
  quality variations, 406–407
  quotas, 388–389
    legal nature of, 642
    tenant's compensation, 640
  regulatory control, 408
  replacers for calves, 393
  scours, in piglets, 485
  from sheep, 420–421
  synthesis, 356–359
    fat, 356, 357
    lactose, 358
    and nutrition, 358–359
    protein, 356, 357
    site, 356
  yields, recent trends, 388–389
  see also Lactation
Milk and Dairies (General) Regulations 1959, 409, 586
Milking equipment, 532–533
  cleaning, 409–410
  cooling/storage, 533
  machines, 532–533
  parlour feeding, 533
  parlours, 532
    construction, 563–564
    misuse, and mastitis, 481
Milling machinery for feed, 530–531
Millipedes as sugar beet pests, 295
Minerals, 333–337
  availability interactions, 333
  calcium/phosphorus/magnesium, 334–335
  functions, 333
  sodium/potassium/chloride, 335
  supplements, 331, 381
  trace minerals, 335–337

Mites,
  control in grain stores, 304
  mange, on cattle, 476–477
  on sheep, 491
Mixers for feed preparation, 531
Mohair production, 427
Molasses for energy in animal feeds, 379
Mole drainage, 41, 53–54
Molinia spp., 177–178
Molybdenum,
  in animal metabolism, 336
  in crop nutrition, 74
  in soil as micronutrient, 30
Montmorillonite structure, 9
Moor grazing, 177–178
Mop top virus in potato, 278
Mortality,
  embryonic/fetal, 352
  and sheep performance,
    embryo, 415–417
    perinatal, 417
Mosaic virus,
  of cauliflower, 281
  of oats, 275
  of potato, 277
  of ryegrass, 287
Moth,
  caterpillars, on potato, 294
  pea, 150, 302
Motivation theories, 632–633
Motley dwarf virus of carrot, 284
Moulds,
  black/sooty,
    in barley, 274
    in wheat, 272
  grey, on peas/beans, 145
  control in grain stores, 304–305
  see also Mildew
Mountain grazing, 177
Mowers, 517
  mower-conditioners, 517–518
Mowing of hay and drying, 204
Muck spreaders, 533–534
Mulberry heart disease,
  in pigs, 487
  and selenium, 335
Mulches, plastic film, 97–98
Muriate of potash, 81
Muscle growth, 360
Muscular dystrophy, nutrition, and selenium, 336
Must, beet, 282
Mustard for condiment, 117–118
Mycobacterium tuberculosis in cattle, 483
Mycoplasma, 469
Mycoplasma hyopneumoniae in pigs, 486–487
Mycorrhizae, 19–20
Mycosphaerella spp., *see* Cabbage ringspot; Leaf spot
Myzus perpicae, *see* Peach potato aphid

N, *see* Nitrogen

Nardus spp., 177–178
National Farmers Union, and British Sugar plc agreement, 174–175
National Soil Inventory, trace elements, 28
Natural gas installations, 498
Navel-ill,
  of calves, 474
  of sheep, 489
Navy beans, 151
Nematode-borne viruses, 268–269
Nematodes, free-living,
  on potato, 294
  on sugar beet, 298
  see also Cyst nematodes
Nervous system of animals, 317–318
  central, 317–319
  peripheral, 317
  signal propagation, 317–318
Net blotch in barley, 273
Neurons, 317
Nitrate in soil, 21
  leaching, pollution hazard, 22–23, 84
Nitrogen,
  in crop nutrition, 70
    and sulphur, 72
  fertilisers, 75, 78–80
    and canopy development, 66
    forms of, 78–80
    and lime loss, 27
    organic, 75
    recommendations, and nitrate pollution, 23
  fixation, 22
    bacteria in, 18–19
  for grassland, sources, 188–190
    clover, 188–189
    fertiliser, 189–190
    recirculated, 188
    soil, 188
  in soil, 21–23
    cycle, 22
    index, 33
    inorganic, 21
    nitrate and water quality, 22–23
    organic, 21
    trees, uptake by, and mycorrhizae, 19–20
Nocardia in soil, 19
Notifiable diseases, 473–474
Nuisance, 641
Nutrient index system, 33
Nutrient requirements of livestock, 381–386
  for growth, 385–386
  for lactation, 383–384
  for maintenance, 382–383
  micronutrient, 386–387
  for production, 384–385
  units, 382
Nutrients from soil,
  major, 70–72
  minor/micro, 72–74
  uptake by plants, and water, 58

Oats, 129–130
  diseases of, 274–275
  in feed, 377
Occupiers Liability Acts, 641
Oesophagostomum infection,
  in cattle, 476
  in pigs, 489
Oestrone sulphate pregnancy assay, 352
Oestrus cycle, 344–346
  detection in cows, 408–409
  hormonal control, 345–346
  and season, 344
  in sheep, 414
  stages of, 344–345
  synchronization, 346
Office building specifications, 580
Oil, *see also* Lubricating oil/grease, on-farm installation
Oilseed rape, 115–117
  overwintering and internode growth, 62
  pests/diseases, 115, 116, 298–301
Oilseeds,
  pricing, Common Agricultural Policy, 597
  residues in feeds, 379–380
Onion fly on leeks, 161
Onion thrip on leeks, 161
Onion white rot, 268
Onions,
  diseases of, 285–286
  dry bulb, 161–164
  storage, buildings for, 577–578
Ontario Units, 63–64
Onychiurus on sugar beet, 295
Oogenesis, 343
Ophiobolus graminae, *see* Take-all
Opomyza florum, 299
Orf in sheep, 490
Organic matter in soil, 3, 10–13
  analysis, chemical, 10
  and the carbon cycle, 12–13
  level,
    determination, 10–11
    factors affecting, 11–12
  pesticide adsorption, 17
  phosphate in, 23–24
  properties/functions, 10
  sulphur in, 25
Oscinella frit, *see* Frit fly
Ostertagia infection,
  of cattle, 476
  of pigs, 489
Overdrafts, 618
Oxygen in soil, 35
Oxytocin, 409
  in milk ejection, 359

P, *see* Phosphorus
PRG, *see* Ryegrass, perennial
psd, *see* Soils, particle size distribution
PTO, *see* Power take-off
Parakeratosis in pigs, and zinc, 336
Paraquat adsorption by soil, 17

Parasites,
  external,
    of cattle, 476–477
    of pigs, 488–489
    of sheep, 491
  internal,
    of cattle, 474–476
    of sheep, 490–491
Parathyroid hormone in calcium/phosphorus balance, 334
Paratrichodorus spp., *see* Nematodes, free-living
Parsnips, 164–165
Partridge, grey, habitat requirements, 462
Parturient paresis, *see* Milk, fever
Parturition, 352–353
Parvovirus infection of pigs, 488
Pasta, *see* Semolina wheat
Pasture, *see* Grassland
Payment policy,
  internal constraints, 633–634
  objectives, 633
  strategy, 634
Pea midge, 302, 303
Pea wilt, 268, 283
Peach potato aphid, 277, 293
  and insecticide resistance, 297
  on sugar beet, 297
Peas, 147–150
  in animal feeds, 380
  harvesting, 149–150
  manuring, 149
  pests/diseases, 150, 283–284, 301–303
  pricing, Common Agricultural Policy, 597
  sowing, 148–149
  types/varieties, 147–148
  yield, 150
Peat formation, 4
Peat blocks for transplant raising, 93, 94
Pellicularia filamentosa on wheat, 271
Peptidases, 327
Peronospora, *see* Mildew
Pesticides,
  adsorption/degradation in soil, 17
  control, and game survival, 462
  and fire, 498
  legislation, 246, 642, 648, 649
  negligent use, 641
Pests Act 1954, 641
Pests of crops, *see by particular crops*
Petrol storage/licensing, 498
Petroleum (Consolidation) Act 1928, 498, 586
pH of soil,
  and liming, 26
    crop tolerances, 87
    recommendations, 87
  measurement, 15–17
    $CaCl_2$ method, 16–17
    difficulties, 15–16

Pheasant, habitat requirement, 463
Phoma betae, 283
Phoma exigua, 278
Phoma lingam, see Canker
Phosphorus, ruminant requirement, 328
Phosphate fixing capacity and hydrous
　　oxides, 9
Phosphorous uptake by trees and
　　mycorrhizae, 19–20
Phosphorus,
　in animal metabolism, 325–335
　in crop nutrition, 70–71
　fertilisers,
　　from organic sources, 75
　　placement and fixation, 23–24
　　for trees, 231
　　water-insoluble, 80–81
　　water-soluble, 80
　in soil, 23–24
　　fixation and fertiliser placement,
　　　23–24
　　in organic matter, 24
　　reserves, 23
Phospholipids, 324
Photoperiod responses of plants, 64–65
Photosynthesis, 58–60
　and the carbon cycle, 12–13
Phyllotreta spp., 116, 298
Physiology of animals, 317–364
　composition of animals,
　　carbohydrates, 321–323
　　endocrine system, 319
　　lipids, 323–324
　　micronutrients, 324–325
　　proteins, 319–321
　　water, 325
　digestion, 325–329
　　anatomy, 325–326
　　carbohydrates, 326–327
　　lipid, 329
　　protein, 327–329
　environmental effects, 363–364
　food intake, voluntary, 340–342
　　energostasis, 340–341
　　manipulation, 341–342
　　prediction, 341
　growth, 359–363
　lactation, 354–359
　　mammary gland
　　　anatomy/development, 354–356
　　milk synthesis, 356–359
　metabolism, 329–340
　　amino acids, 332–333
　　catabolism/anabolism, 329
　　disorders of, 339–340
　　fat, 331–332
　　fatty acids, volatile, 331
　　glucose, 330–331
　　mineral nutrition and, 333–337
　　vitamins in, 337–339
　nervous system, 317–319
　　central, 318–319
　　peripheral, 318
　　signal propagation, 317–318
　reproduction, 342–354

Physiology of animals (*cont.*)
　reproduction (*cont.*)
　　anoestrus, lactational, 353
　　female system, 342–346
　　male system, 347–350
　　mating/fertilisation, 350
　　parturition, 352–353
　　pregnancy, 350–352
　　technology advances, 353–354
Physiology of crops,
　nutrient uptake, 58
　photosynthesis, 58–60
　water uptake, 58
　see also Growth/development
Phytochrome in photoperiod response,
　　64–65
Phytophthora infestans, see Blight
Piccinia allii, 161, 286
Pig Industry Development Authority,
　　436
Pigs, 430–449
　breeding, 434–437
　　artificial insemination, 437
　　breeds, 434–435
　　crossbreeding, 435
　　hybrid companies, 435–436
　　national involvement, 436–437
　buildings for, 432–434, 565–570
　　breeding stock, 432–433
　　dry sows/boars, 569
　　farrowing, 567–568
　　fattening, 565–567
　　growing pigs, 433–434
　　stocking density, 432
　　weaners, 569–570
　diseases,
　　erysipelas, 487
　　of fatteners, 486–487
　　parasites, external, 488–489
　　of sow/piglets, peri-natal, 485
　　swine fever, 487
　　tuberculosis, 487
　　virus infections, 487–488
　　of young/growing pigs, 485–486
　expansion of production, 430
　finishing pigs, 445–449
　　carcass quality/classification,
　　　446–447
　　management, 447–449
　　markets, 445–446
　　nutrition, 447, 448
　intake manipulation, 341–342
　intake prediction, 341
　iron deficiency risk in piglets, 335
　manure, 77
　　and copper/zinc toxicity, 76
　meat, pricing, Common Agricultural
　　Policy, 596
　production, computerisation in, 656
　protein requirements, maintenance,
　　383
　reproduction, sow/gilt, 437–443
　　feed for gilts, 439
　　feeding in lactation, 442–443
　　lactation/weaning, 441–442

Pigs (*cont.*)
　reproduction, sow/gilt (*cont.*)
　　oestrus cycles, 438
　　oestrus and mating, 438–439
　　parturition, 441
　　post-weaning, 443
　　pregnancy, 439–441
　　puberty, stimulus to, 438
　　temperature susceptibility, 364
　　UK industry, 430–432
　　weaner piglets, 443–445
　　immunological development and
　　　colostrum, 443–445
　　post weaning scours, 445
　　world view, 430
Pine of sheep, 490
Pink rot of potatoes, 279
Pipe drainage, 50–53
　installation, 52–53
　materials, 50–51
Piroplasmosis in cattle, 478–479
Pituitary gland, 319
Planning permission, 641–642
Planting machines, 511–514
　direct drills, 512–513
　grain drills, 511–512
　for potatoes, 514
　precision seeders, 513–514
　transplanters, 514
Plasmodiophora brassicae, see Club
　　root
Plastic covers for crops, 97–98
　for beetroot, 153
　for carrots, 156
　for potatoes, early, 169–170
Plate mill, 530–531
Pleospora bjoerlingii, 282
Ploughing, 40
　in woodland site preparation, 229
Ploughs, 507–508
Pneumonia,
　enzootic, in pigs, 486–487
　in sheep, 490
Pod midge, brassica, 301
Pollards, 224
Pollen beetles on oilseed rape, 300
Pollution,
　atmospheric, as sulphur source, 25
　from manure/slurry effluent, 77
　nitrate in drinking water, 84
　see also Control of Pollution Act
　　1974
Polyscythalum pustulans, 279
Potassium,
　in animal metabolism, 335
　in crop nutrition, 71–72
　fertilisers, 79, 81
　　natural sources, 75
　　for trees, 231
　in soil, 24–25
　　exchangeable, 24
　　fixation, 24–25
Potato gangrene, 278
Potato wart, 268
Potatoes, 165–172

Potatoes (*cont.*)
  cultivations, 170
  pests/diseases of, 276–279, 292–294
    blight, prevention, 170
    leaf/stem, 276–278
    scab, common, and liming, 87
    tuber/storage, 278–279
    wart, 268
  harvesting, 171
    machinery, 523–524
  manuring, 170–171
  planting,
    date, 169
    machines, 514
    seedrate, 167–168
  quality for cooking, 166
  seed,
    certification/grading, 166–167
    production, 166, 167, 268
    size, 167
    sources, 167
    treatment, 168
  storage, 171–172
    buildings for, 577–578
  types, 165–166
  yield, 172
Poultry, 450–460
  broiler breeding industry, 453–455
  buildings for,
    broilers/capons, 573
    layers, 572
    turkeys, 573
  duck, 458–459
  egg production, 452–455
  environmental considerations,
      456–457
    lighting, 456–457
    ventilation, 455–456
  geese, 459–460
  manure, 77
  meat production, 450–452
    additives, 454
    broilers, 451–452
    statistics, 450–451
  pricing, Common Agricultural
      Policy, 596
  turkeys, 457–458
  waterfowl, 458
Powder scab of potatoes, 278–279
Powdery mildew, *see under* Mildew
Power take-off, 506
Predator control for game, 464
Pregnancy,
  development, embryonic/fetal, 350
  diagnosis, 352
  maintenance, progesterone in,
      351–352
  mortality, embryonic/fetal, 352
  physiology of, 350–352
  see also Parturition
Preservatives for timber, 238–239
Pricing of products, 622–623
  Common Agricultural Policy on,
      593–597

Progesterone,
  in oestrus control, 345
    administration, 346
    pregnancy assay, 352, 354
Progesterone-releasing Intravaginal
    Device (PRID), 346
Propionic acid,
  as grain preservative, 313
  as hay additive, 205
Prostaglandin,
  in oestrus control, 346
  synthetic, in sow induction, 441
Protection of Employment
    (Consolidation) Act 1978, 642
Protection from Eviction Act 1977, 640
Proteins, 319–321
  amino acids,
    in metabolism, 332–333
    in nutrition, 321
  and bread wheat, 125
  composition/structure, 319–321
  crude protein, 368
  degradability treatment, 329
  digestion,
    monogastric, 327–328
    ruminant, 328–329
  feed supplement,
    animal origin, 380–381
    evaluation, 367–369
  functions, 319
  livestock requirements,
    breeding cattle, 396
    goats, 427–428
    for lactation, 384
    for maintenance, 382–383
    for reproduction, 385
    pregnant sows, 441
    for growth, 385
    growing pigs, 448
    in milk, 356, 357
  quality, 368–369
Protozoa in soil, 20
Proximate system in feed analysis, 369
Pseudocercosporella herpotrichoides, *see*
    Eyespot in cereals
Pseudomonas phaseolicola, 145, 283
Psila rosea, *see* Carrot fly
Psoroptes communis infestation of
    cattle, 476–477
Psylliodes chrysocephala, *see* Cabbage
    stem beetle
Public Health Acts, and nuisance, 641
Public Order Act 1986, and trespass,
    641
Puccinia spp., *see* Crown rot; Rusts
Pulpy kidney disease in lambs, 489
Pumps for irrigation, 525
Pygmy beetle on sugar beet, 296
Pyrenopeziza brassicae, *see* Leaf spot
Pyrenophora avenae, *see* Leaf spot;
    Seedling blight in oats
Pyrenophora graminae, 273
Pyrenophora teres, 273
Pythium in peas/beans, 283

Quotas for milk, 388–389
  legal nature of, 642
  tenant's compensation, 640

Rabbit-proof fencing, 554
Rabies, 483
  in pigs, 489
Race Relations Act 1976, 642
Radish for fodder, 110
Ram performance, 416
Ranularia betocola, *see* Leaf spot
Rapes, 111–112
  see also Oilseed rape
Rapeseed meal in feed, 379–380
Recruitment/selection, 629–630
Red cabbage, 104–105
Redundancy,
  legislation on, 643
  management, 632
  see also Dismissal procedures
Redundancy Payments Act 1965, 643
Redwater in cattle, 478–479
Refuse Disposal (Amenity) Act 1978,
    641
Rent (Agriculture) Act 1976, 638
  and security of tenure, 640
Rent review, 640
Reporting of Injuries, Diseases and
    Dangerous Occurrences
    Regulations 1985, 648
Reproduction, animal,
  anoestrus, lactational, 353
  energy requirement for, 384–385
  female system, 342–346
    anatomy, 342
    development, 343
    oestrus cycle, 344–346
    oestrus synchronization, 346
    puberty, 343–344
    sex determination, 342–343
  male system, 347–350
    anatomy, 347
    artificial insemination, 349–350
    development/puberty, 347
    and fertility, 350
    sperm production, 347–348
    spermatogenesis, 348–349
  parturition, 352–353
  physiology of, 342–354
    mating/fertilisation, 350
  pregnancy, 350–352
    development, embryonic/fetal, 350
    diagnosis, 352
    maintenance, progesterone in,
        351–352
    mortality, embryonic/fetal, 352
    protein requirement for, 385
  technology advances, 353–354
Reservoirs for irrigation, 525
Respiration of plants and
    photosynthesis, 59–60
Respiratory disease in cattle, 478
Retirement, 636
Rhenania phosphate, 80–81

Rhinitis, atrophic, in pigs, 486
Rhinotracheitis infections, bovine, 477–478
Rhizobium melototi inoculation of lucerne, 151
Rhizobium and nitrogen fixation, 19
Rhopalosiphum padi, 292
Rhynchosporium spp., *see* Leaf blotch
Rice bran for animal feed, 378
Rickets in lambs, 490
Rickettsia, 470
Ringspot, cabbage, 281
Ringworm on cattle, 476
Roadways, 553
Rock phosphates, 80–81
Rocks,
    chemical weathering, 4
    soil forming, 3
Roller mill, 531
Rollers, 509–510
Roof cladding of frame building, 543
Root crop and concentrate storage, 577
Root harvesting machinery, 523–524
    for potatoes, 523–524
    for sugar beet, 524
Rose grain aphid, 292
Rosemaund Beef System, 402
Rots,
    of brassicas, 282
    of carrots, 284, 285
    of legumes, 287
    of peas/beans, 283
    of potatoes, 278, 279
    of ryegrass, 286
    of swedes/turnips, 73
    of wheat, 271
Rotary cultivator, 508–509
Rotation of crops, 90
    carrots in, 156
    cereals, 120, 122
    grassland in, 178
    leeks in, 159
    legumes in, 142
    oats in, 129
    and onions, 162
    and soil micro-organisms, 20
    and weed control, 246
Rotavirus infection of pigs, 488
Roundworm,
    in cattle, 476
    in pigs, 488
    in sheep, 490
Ruminants,
    degradable protein evaluation, 368–369
    digestion,
        anatomy, 325–326
        carbohydrate, 327
        lipid, 329
        protein, 328–329
    intake,
        manipulation, 342
        prediction, 341
Rusts,
    on barley, 272

Rusts (*cont.*)
    on beans, 284
    on leeks, 161, 286
    on oats, 274–275
    on wheat, 269
Rye, 130–131
    diseases of, 275
    forage, 130–131
    grain, 131
    see also Triticale
Ryegrass,
    diseases of,
        crown rot/leaf blotch, 286
        mosaic virus, 287
    hybrid, 184
    Italian, 183–184
    perennial, 183
    seed production,
        establishment for, 137
        harvesting, 139
        sward management, 138–139
    Westwolds, 184

S, *see* Sulphur
Safety and sprayer use, 516
    see also Health and safety of personnel
Sainfoin, 152–153
Salmonella infection from farmyard manure, 76
Salmonellosis,
    abortion in sheep, 491–492
    in calves, 474
    in pigs, 486
Salt as fertiliser, 25
    potassium replacement, 81
Sampling of soil, 28, 31
Sand,
    calcareous, for liming, 89
    as soil component, 4
Sarcoptes scabiei, *see* Mange
Savoy cabbages, 104
Scab,
    on beetroot, 153
    of potatoes, 278–279
    sheep, 491
Scabies, *see* Mange
Scarecrows for pigeons, 300
Sclerotinia, *see* Rots
Scrapie of sheep, 491
Screefing in woodland site preparation, 228
Scurf of potatoes, 279
Scutigerella immaculata on sugar beet, 295
Seasoning of timber, 238
Seaweed,
    calcified, for liming, 89
    as manure, 77–78
Seeders, precision, 513–514
Seedling blight, in oats, 275
Seeds,
    cereals, choosing, 118
    for arable crops, 91–92

Seeds (*cont.*)
    development, 62
    dormancy, 64
    grassland mixtures, 185–188
    pricing, Common Agricultural Policy, 597
    processors,
        mustard, condiment, 117–118
        oilseed rape, 115–117
    production, 135–140
        cereals, 136
        certification requirements, 135–136
        herbage, 136–140
            drying/conditioning/storage, 140
            establishment, 136–138
            harvesting, 139, 140
            sward management, 138
        of vetches, 153
    treatment, chemical, 267
Selenium in animal nutrition, 336–337
Semolina wheat, 127–128
Senescence, 62–63
Services, 495–502
    drainage, 495–496
        rainwater pipes/gutters, 495
        recommendations, 495
        underground, 495–496
    energy sources, 496–498
    and fire, 498
    heating, 498–499
    lighting, 499–500
    lubricating oil/grease, 500
    water, 500–502
Sewage sludge,
    types of, 77
    zinc excess in, 74
Sex determination, 342–343, 354
Sex Discrimination Act 1975, 642
Sharp eyespot in wheat, 271
Shearing of sheep, 421–422
Sheep, 411–426
    abortion, 417
    body condition scoring, 428–429
    breeding season control, and melatonin, 353–354
    buildings for,
        handling, 571
        housing, 570–571
    diseases,
        abortion causing, 491–492
        clostridial, 489–490
        deficiency conditions, 490
        of mature sheep, 491
        metabolic diseases, 491
        orf, 490
        parasites, external, 491
        parasites, internal, 490–491
        perinatal, 489
        pneumonia, 490
    ewe feeding, 422–424
        practice, 423–424
        theory, 422–423
    grazing management, 424
    liver damage from copper, 76

Sheep (*cont.*)
  meat pricing, Common Agricultural
      Policy, 596
  meat production, 418–420
    growth and nutrition, 418–419
    systems, 419–420
  metabolic disorders,
    ketosis, 339
    staggers, 340
  milk production, 420–421
  mortality, lamb, and environmental
      conditions, 364
  performance, factors in, 412,
      414–418
    age, 415–416
    hypothermia, 417–418
    mortality, embryo, 416–417
    mortality, perinatal, 417
    nutrition, 415
    oestrus, 414
    ovulation/fertilisation, 415
    season, 416
  productivity improvement per ewe,
      424–426
    hormonal manipulation, 425
    lambing frequency, 425–426, 429
    prolific ewes, use of, 425
    weaning, early, hill lambs, 424–425
  replacements, culling decisions, 422
  statistics,
      UK, 411–412, 413
      world, 411
  wool production, 421–422
Shelterbelts, 237
  see also Windbreaks
Shoddy, 75
Shooting in rural economy, 461, 465
Silage, 199–204
  additives, 200
    and the Liscombe Star system, 202
  bacteria in, 199
  big-bales, 201
  in clamps,
      feeding, 201–202
      making, 200
  crops for, 202, 204
  density, 204
  effluent, 200–201
  losses, 199
  nutritive value, versus hay, 373
  self-feeding system for dairy cows,
      405–406
  storage, buildings for, 576–577
  in towers, 201
  yields, 202, 203
Silt as soil component, 4
Silver scab of potatoes, 279
Sitobion abenae, 291–292
Sitonia linneatus, see Weevils
Skin spot of potatoes, 279
Slaughter,
  beef, selection for, 390–401
  and glycogen oxidation and meat
      pH, 331

Slugs,
  field,
    on cereals, 291
    on oilseed rape, 298–299
    on sugar beet, 174
  garden, on potatoes, 294
Slurry,
  disposal,
    by irrigation, 530
    problem from pig units, 434
  health risk, 469
  separators, 535
  storage, buildings for, 579
Smectite in soil shrinkage, 9–10
SMEDI diseases in pigs, 487–488
Smuts,
  on cereals, 268
    barley, 274
    maize, 276
    oats, 275
    wheat, 271
  on onions/leeks, 285
Sodium,
  in animal metabolism, 335
  in crop nutrition, 25, 72
  for sugar beet, 173
Sodium nitrate fertiliser, 79
Soils, 3–44
  acidity, 15–17
  analysis, 30–32
  atmosphere, 34–35
    aeration and structure, 34–35
  calcium/lime, 25–27
  classifications,
    land, 42–43
    soils, 40–42
  clay minerals, 5, 7–10
    hydrous oxides, 9
    shrinkage of soil, 9–10
    as soil particles, 9
    structure, 7–9
  colloid chemistry, 13–15
    anion adsorption/exchange, 14
    cation adsorption, 13–14
    flocculation/deflocculation, 15, 16
    Quantity/Intensity relationships,
        14–15
  composition, 3, 4
  erosion, and forest management, 224
  fauna, 17
    earthworms, 20
    protosa, 20
  formation, 3, 4
  magnesium in, 25
  micro-organisms in, 17–20
    actinomycetes, 19
    algae, 20
    bacteria, 18–19
    classification, 18
    fungi, 19–20
    manipulation of, 20
    nutrition, 18
  nitrogen in, 21–23
    nitrate and water quality, 22–23

Soils (*cont.*)
  nitrogen in (*cont.*)
    nitrogen cycle, 22
    organic/inorganic, 21
  organic matter in, 10–13
    analysis, chemical, 10
    and the carbon cycle, 12–13
    level in soil, 10–12
    properties/functions, 10
  particle size distribution, 5
    classifications systems, 6
  pesticide adsorption/degradation, 17
  phosphorus in, 23–24
  potassium in, 24–25
  the profile, 4
  sampling, 28, 30
  sodium in, 25
  strength, and tractor power, 504
  structure, 32–35
    profile examination, 33–34
    subsoils, 33
  sulphur in, 25
  survey, 40–42
  temperature, 40
  texture, 4–5
    triangular diagram, 6
  trace elements in, 27–28
    micronutrients, 28, 29
  water in, 35–40
    balance calculations, 38–40
    potential, 35–37
    tension, and irrigation scheduling,
        528–529
    uptake, 37–38
Somatomedins, and growth hormone
    response, 360
Somatotrophin in milk synthesis,
    358–359
Sorghum,
  in feed, 377–378
  photosynthesis, tropical adapted,
      59
Soya beans, 151
Sperm production, 347–348
  spermatogenesis, 348–349
  see also Artificial insemination
Spongospora subterranea, 278–279
Spraing, of potato, 277–278
Sprayers, 514–517
  calibration, 516
  components, 515–516
  droplet size and drift, 514–515
  nozzles, 515
  productivity, 516
  safety, 516–517
Spring-tine cultivator, 509
Springtails as sugar beet pests, 295
Staff management, 624–636
  disciplinary procedures, 636
  farm requirements, 624–625
    gang-work day charts, 624–625,
        626, 630
    seasonal labour profiles, 624, 625
  job analysis, 627

Staff management (*cont.*)
  motivation,
    incentives, 635–636
      payment policy, 633–635
      theories, 632–633
    the need, 624
  personal specifications, 628–629
  recruitment/selection, 629–630
  redundancy, 632
  retirement, 636
  staff facilities, 636
  staff potential, 626–627
    internal succession, 627
  training,
    continuation, 631–632
    induction, 630–631
  workforce planning,
    external labour sources, 626
    and technology/legislation, 626
Staggers, grass,
  in cattle, 340
  in sheep, 341, 491
Stalk rot,
  of maize, 276
  of sweet corn, 135
Starch, 322
Starvation, and deamination, 333
Steaming-up, 395, 405–406
Stemphylium radicinum, *see* Carrot
    black rot
Steroids, 324
  hormones and growth, 362
  ovarian, immunisation against, 353
Stockmanship and pig reproduction,
    440
Stomach worms of pigs, 489
Stomata opening and carbon dioxide,
    59
Storage rots,
  of carrots, 285
  of onions/leeks, 285–286
  of potatoes, 228
Straw,
  burning,
    and direct drilling, 40
    and reduced cultivation, 122
  decomposition and nitrogen
    immobilization, 21, 77
  as manure, 77
  nutritive value, 373, 376
  storage, buildings for, 576–577
  utilisation/disposal, 122
Streptococcal infections, of piglets, 486
Streptomyces in soil, 19
Streptomyces scabies, 278
Stress in pigs and reproduction, 440
Strontium in crop nutrition, 72
Stubble cleaning weed control, 246
Subsoil structure, 33, 34
Subsoilers, 507
Subsoiling, 40, 54–55
Suckler calf production, 394, 403–404
Sugar beet, 172–175
  by-products, 175
    factory sludge for liming, 89
    pulp, feeding value, 376

Sugar beet (*cont.*)
  cultivations/sowing, 172–173
  crop treatments, 173–174
  harvesting, 174
    machinery, 524
  payment for, 174–175
  pests/diseases, 174, 282–283, 295
  pricing, Common Agricultural
    Policy, 595
  in rotation, 172
  varieties, 172
  winter, vernalisation need, 64
  yield, 175
Sugar cane, photosynthesis, tropical
    adapted, 59
Sulphate of potash, 81
Sulphur,
  in animal metabolism, 335
    ruminant requirement, 328
  in crop nutrition, 72
    grassland requirement and
      nitrogen use, 191
  sources, 25
Sunflowers, 142
Superovulation, 353
Superphosphate fertilisers, 80
Survey, soil, 41–42
Surveying of buildings, 581–584
  drawings, 581, 582
  levels, 583–584
  Ordnance Maps, 583
  ranging out, 581, 583
Swayback in lambs, 334, 490
Swedelike kales, 112–113
Swedes, 112, 113–115
  crown rot, 73
Sweet corn, 133–135
  climate/soil, 133–134
  harvesting/marketing, 134–135
  husbandry, 134
  pests/diseases, 135, 276
  varieties/pollination, 133
Swine dysentery, 487
Swine fever, 487
Swine vesicular disease, 489
Symphylids as sugar beet pests, 295
Synchytrium endobioticum, 268, 279

Tacking of sheep, 422
Take-all, 268
  in barley, 274
  control, rotation, 120
  in maize, 276
  in oats, 120, 275
  in wheat, 270–271
Tankage, 75
Tapeworms in sheep, 490
Tares, 153
Temperature
  body, and disease, 473
  and plant growth/development,
    63–64
  of soil, 41
    and drainage benefits, 46
Tenancy, 639–640
  obligations, landlord/tenant, 640

Tenancy (*cont.*)
  security, 639–640
  service occupation, 640
Testosterone and freemartin heifer
    development, 341
Tetanus in sheep, 490
Tetany, grass/lactation, 480–481
Thatch, rye for, 131
Thermoactinomyces in compost heaps,
    19
Thermoregulation, environmental
    factors in, 382
Threshing Machines Act 1878, 644
Thyroxine, 336
Tick, sheep, 491
Tilletia caries, 268, 271
Timber,
  in construction, 548–550
  for fenceposts, 554
  for joists, 551–552
Timothy grass, 184
  seed production
    establishment for, 137
    sward management for, 139
Tine cultivators, 509
Tipula spp., *see* Leatherjackets
Tobacco rattle virus in potato, 277–278
Tocopherols, 338
Town and Country Amenities Act
    1974, 585
Town and Country Planning Act 1971,
    584–585, 641
Toxaemia, pregnancy, *see* Twin lamb
    disease
Trace elements in soil, 27–28
  deficiencies and overliming, 26
  micronutrients, 28
Tractors, 503–507
  computer monitoring, in-cab, 657
  drawbar power, 504–506
  engine, 503–504
  hydraulic systems, 506
  operator health and safety, 506–507
  power take-off, 506
  transmission, 504
Training,
  computers in, 658
  continuation, 631–632
  induction, 630–631
Transpiration,
  and soil water potential, 37, 38
  and water movement, 58
Transplant raising, 93–96
  bedding systems, 93–94
  compost requirements, 94
  machines, 514
  operating tray system, 95–96
  seeds/sowing, 94–95
Transplanting systems, 92–93
Trees on farms, 209–242
  assistance, grants/agencies, 241
    Farm Woodland Scheme, 463
  diseases/pests, 240–241
  establishment of broadleaves,
    228–229
    mixtures, 228

Trees on farms (*cont.*)
  establishment of broadleaves (*cont.*)
    shelters, individual, 228–229
  fertiliser, 232
  harvesting, 236
  management systems, 225–227
    coppicing, 225–226
    high forest, 226–227
    improvement/regeneration, 227
  marketing, 236–237
  measurement of timber, 209, 211
    growth/productivity, 211
    volume, 211, 224
  planting, 230–232
  preservation of timber, 239–240
  protection, fire/wind, 240
  quality of timber and use, 225
  rotation length, 224
  shelterbelts/windbreaks, 237–239
  site assessment, 209
    indicator species, 210–211
  site preparation, 229–230
  species,
    non-timber, 219–220
    selection, 209
    timber, 212–218
  thinning, 233–236
    broadleaves, 235–236
    conifers, 233–235
  weeding/cleaning, 232–233
Trees and mycorrhizae, 19–20
Trefoil, 185
Trenbolone acetate,
  EEC ban, 362–363
  growth promotion, 362
Trespass,
  and duty of care, 641
  liability for damage, 640–641
Trichodorus, *see* Nematode, free living
Trichophyton spp. on cattle, 476
Trichostrongylus spp., *see* Roundworm
Triglycerides, 324
Triticale, 131
  diseases of, 275
Tryponema chyodysenteriae, 487
Tuberculosis,
  in cattle, 483
  in pigs, 487
Tunnels, plastic, for crop
  establishment, 98
Turkeys, 457–458
  buildings for, 573
Turnip moth, on potato, 294
Turnips, 112–113, 114–115
  crown rot, 73
Twin lamb disease, 339, 491
  and ewe nutrition, 417
Tyres for tractors, 505

USDA soil classification, 41–42
Udder,
  washing, 408
  weight/support, 355
  see also Mammary gland

Ultrasonics in pregnancy diagnosis, 352
  in sheep, 417
  in sows, 441
Urea,
  fertiliser, 78–79
  in livestock feed, 381
  ruminant digestion, 328
Urea-formaldehyde slow-release
    fertiliser, 79–80
Urocystis cepulae, 285
Uromyces betae, 282
Uromyces fabae, 284
Ustilago spp., *see* Smuts

VFA, *see* Voluntary fatty acids
Vaccination of piglets against, 446
Vaccines, 471
Van Soest scheme of feed analysis, 368
Variable premiums,
  in beef pricing, 596
  in sheep meat pricing, 596
Vasectomy, 347
Veal,
  pricing, Common Agricultural
      Policy, 595–596
  production, 394
Vegetables,
  pests, 302
  pricing, Common Agricultural
      Policy, 596–597
  storage, buildings for, 577–578
  see also by particular crop
Ventilation of buildings, 557–558
  for poultry, 454–456
Vermiculite structure, 9
Vernalisation, 64
  of cereals, 119
Verticillium wilt of lucerne, 287
Vetches, 153
Veterinary practices, computers in, 657
Vibriosis abortion in sheep, 492
Violet root rot,
  on beetroot, 153
  of carrots, 284
Virginiamycin for poultry growth
    promotion, 452
Virus infections, 469
  of animals, 469
    cattle, 477–478, 483
    pig infection, 487–488
    spread in livestock units, 467
  of crops,
    barley, 273
    beet/mangold, 282
    carrot, 284
    cauliflower, 281
    grasses/cereals, 287
    oats, 275
    potato, 277–278
    prevention, 268–269
    sugar beet, 174, 297
    wheat, 270
Vitamins, 337–339
Voluntary fatty acid metabolism, 331

Wages, *see* Payment policy
Walls, concrete block, 546–547
Warble fly, in cattle, 484
Wart disease, of potatoes, 268, 279
Wart Disease of Potato (Great Britain)
    Order 1973, 268
Water,
  as animal component, 325
  and crop growth, 36
  for power, 499
  in soil, 36–40
    adsorption, 14
    balance calculations, 38–40
    potential, 36–37
    uptake, 37–38
  supply, 500–502
    demand in farm production, 500
    storage/distribution, 500–502
  see also Drainage; Drinking water;
      Irrigation
Waterfowl, 458
Waterlogging,
  and drainage benefits, 46
  pedological symptoms, 35
Watery wound rot of potatoes, 279
Weathering,
  and potassium fixation, 24
  in soil formation, 3–4
  trace element release, 28
Wedge-filling of silage clamp, 200
Weed control, 243–264
  arable crops/seedling leys, 247–248
  in broadleaved crops, 252–256
    broadleaved weeds, 254–256
    grassy weeds, 253
    principles, 252
    volunteers/groundkeepers, 256
    weed beet, 252, 253
  in cereals,
    not-undersown, 248–251
    rotation, 120, 121
    undersown, 251
  in forest/woodland, 258–263
    herbicide equipment/timing, 260
    herbicides, selective, 261–262
    individual tree treatment, 260
    methods, 258–260
    weed types, 258
  general control, 243–244
  in grassland, 258
  herbicides, 244–245
  by straw burning, 122
  unselective in non-crop situations,
      263–264
  weed types, 243
Weeds,
  and competition, 68
  dormancy, 64
Weeds Act 1959, 641
Weende system in feed analysis, 369
Weevils,
  cabbage seed on oilseed rape,
      300–301
  cabbage stem on oilseed rape, 200
  pea, 150

Weevils (*cont.*)
  pea/bean, 301–302
  pine, 239
  rape winter stem, 299
Welfare of animals,
  and behavioural physiological needs, 364
  and health, 473
  pigs, 430
Wheat,
  in animal feed, 377
  canopy development early sowing, 65
  common, 125–126
  diseases of, 269–272
    ear, 271–272
    leaf/stem, 269–270
    root/stem-base, 270–271
  durum, 127–128
  fertiliser recommendations, 86
  for grain, 125–128
  milling by-products for feed, 378
  rotation, 120
  spring, 127
  winter,
    nitrogen manuring, 126–127

Wheat (*cont.*)
  vernalisation need, 64, 119
  see also Triticale
Wheat bulb fly, 288–289
Whiptail of cauliflower, 74
White rot of onions/leeks, 285
White tip disease of leeks, 161
Whiting sand, 89
Wild Life and Countryside Acts, 642
Wind damage to woodland, avoidance of, 239
Windbreaks,
  for crops, 236–238
  for livestock, 238
  wind types, 236
Wireworms,
  on cereals, 290
  on potatoes, 294
  on sugar beet, 174, 296
Wooden tongue of cattle, 478
Woodland Grant Scheme, 241
Woodland management for pheasant, 463
  see also Trees on farms
Woodmouse as sugar beet pest, 295

Woodworm in constructional timber, 548
Wool production, 421–422
  see also Fibre production from goats
Workshop building, 580

Yellow cereal fly, 289
Yellow dwarf virus
  on barley, 273
  on wheat, 270
Yellow rust in wheat, 269

ZST, *see* Zinc sulphate turbidity test, 392
Zadok's code for cereal growth stages, 121
Zeranol growth promoter, 362
Zinc,
  in animal metabolism, 336
  in crop nutrition, 74
  in soil as micronutrient, 29
Zinc sulphate turbidity test (ZST) for blood antibodies, 392